Table 1.2 Unit Conversion Table (*Continued*)

Unit	Conversion
Pressure (stress)	1 atm = 14.696 lb/in.^{2} or psi = 760 torr = 101,325 Pa 1 atm = 30.0 inHg = 407.2 inH$_2$O 1 ksi = 1000 psi 1 mmHg = 0.01934 psi = 1 torr 1 Pa = 1 N/m^2 1 inHg = 3376.8 Pa
Energy per unit mass	1 kJ/kg = 0.4299 Btu/lbm
Specific heat	1 kJ/(kg · °C) = 0.23884 Btu/(lbm · °F)
Temperature	1 K = 1.8°R K = 273.15 + °C °R = 459.69 + °F
Temperature change	1°C = 1.8°F
Specific thrust	1 lbf/(lbm/s) = 9.8067 N/(kg/s)
Specific power	1 hp/(lbm/s) = 1.644 kW/(kg/s)
Thrust specific fuel consumption (TSFC)	1 lbm/(lbf · h) = 28.325 g/(kN · s)
Power specific fuel consumption	1 lbm/(hp · h) = 168.97 g/(MW · s)
Strength/weight ratio (σ/ρ)	1 ksi/(slug/ft^3) = 144 ft^2/s^2 = 13.38 m^2/s^2

$R = 287.05 \, J/kg$
$= 1717 \, ft \cdot lb \cdots$

Elements of Propulsion:
Gas Turbines and Rockets

Second Edition

Elements of Propulsion: Gas Turbines and Rockets

Second Edition

Jack D. Mattingly and Keith M. Boyer
Department of Aeronautics,
US Air Force Academy,
US Air Force (retired)

AIAA EDUCATION SERIES

Joseph A. Schetz, Editor-in-Chief
Virginia Polytechnic Institute and State University
Blacksburg, Virginia

American Institute of Aeronautics and Astronautics, Inc.

American Institute of Aeronautics and Astronautics, Inc.
12700 Sunrise Valley Drive, Suite 200, Reston, VA 20191-5807

1 2 3 4 5

Library of Congress Cataloging-in-Publication Data
Record on file.
ISBN: 978-1-62410-371-1

Jack has been blessed to share his life with Sheila, his best friend and wife. She has been his inspiration and helper, and the one who sacrificed the most to make this work possible. He dedicates this book and the accompanying software to Sheila.

Keith likewise dedicates this book to his wife and best friend Joyce. It is truly a blessing that a blind date and 3-week engagement would lead to 33 years together at the time of this book's printing. She continues to inspire him with her love and support.

We would both like to share with all the following passage Jack received from a very close friend about 40 years ago. This passage provides guidance and focus to our lives. We hope it can be as much help to you.

Fabric of Life

I want to say something to all of you
Who have become a part
Of the fabric of my life
The color and texture
Which you have brought into
My being
Have become a song
And I want to sing it forever.
There is an energy in us
Which makes things happen
When the paths of other persons
Touch ours
And we have to be there
And let it happen.
When the time of our particular sunset comes
Our thing, our accomplishment
Won't really matter a great deal.
But the clarity and care
With which we have loved others
Will speak with vitality
Of the great gift of life
We have been for each other.

Gregory Norbert, O.S.B.

CONTENTS

FOREWORD TO THE SECOND EDITION

Dr. William H. Heiser

In 1995 Dr. Hans von Ohain wrote in his very informative and interesting foreword to the first edition (included in this textbook after this foreword): "The evolutions of both aerovehicle and aeropropulsion systems have in no way reached a technological level that is close to the ultimate potential! The evolution will go on for many decades toward capabilities far beyond current feasibility and, perhaps, imagination." He was writing, of course, about the turbine engine that he helped bring into general use in the 1930s, but the same is true of other devices today. This foreword will focus on propulsion developments since then and some to come. Dr. von Ohain is justly remembered for his many contributions to the development of turbine propulsion, his sense of humor, his excellent memory, his amazing charm, and his outstanding manners, as well as his dedication to the fundamental principles.

Commercial Turbine Engines

Turbine engines have been developed along two different lines, evolutionary (following the dictates of known physical laws) and revolutionary (employing novel configurations and devices). Dr. Hans von Ohain concentrated on the evolutionary line, and he noted that the overall efficiency of the turbine engine was likely to be increased by increasing both the overall pressure ratio or the ratio of the highest pressure in the engine to the ambient pressure (usually referred to as the OPR) [i.e., increasing the thermal efficiency of the cycle, see Eq. (1.13)] and the bypass ratio [i.e., increasing the propulsive efficiency of the cycle, see Eq. (1.14)]. The last equation also reveals that the propulsive efficiency is always increased when the available energy is spread out among greater amounts of fluid (usually by increasing the engine bypass ratio). The overall efficiency is the product of the two [see Eq. (1.17)]. As usual, Dr. von Ohain was correct. Furthermore, these changes have reduced the environmental impact of turbofans by reducing their fuel consumption and jet noise (i.e., jet velocity). Present-day turbofans may have a static OPR as high as about 44:1 and a static bypass ratio as large as about 11 (see Appendix B), vs his original engine, which had a compression ratio of about 3:1 and a bypass ratio of 0.

The static OPR is now reaching its practical upper limit because the aerodynamic end losses of the short airfoils of the latter stages of the high pressure compressor are becoming excessive, the temperatures leaving the compressor are becoming very high (contributing to thermal differential stress failures of the last stage), temperatures entering the turbine are becoming very high (requiring better materials, improved turbine cooling methods, or more turbine cooling air), and the thermodynamic benefits of further increasing the OPR have diminishing returns. The effective bypass ratio of propellers is quite large (about 50 or more), and Buckingham showed more than 90 years ago [1] (correctly) that the turbine engine cannot compete with the propeller when the aircraft speed is low (say about half of the current subsonic cruise speed or about 350 mph). Because civilian travelers have already surrendered free baggage handling, in-flight meals, leg and seat room, and travel time (larger parking lots and security delays), it is possible that they will also choose to surrender speed and return to propeller flight in order to decrease the amount of fuel used and therefore the cost of flying, especially on shorter routes. The omnipresence of the drone and the propeller-driven unpiloted air vehicle (UAV) may also encourage such a choice, and the "excessive" noise and vibration of the propeller may also be decreased to "acceptable" levels by improved materials and design methods.

Combustion temperatures have also increased with time in commercial and military engines, primarily because of higher-temperature-capability turbine materials and more effective cooling techniques. Ceramic turbines that can operate at very high temperatures remain the direction of materials research, but ceramic materials remain brittle (i.e., intolerant to rapidly changing temperatures). Cooling the cooling air has also been seriously considered but not yet implemented for the first airfoils of the high-pressure turbine. In fact, the need to cool the increasing amount of onboard electronics, especially in military aircraft, makes even greater demands on the available cooling air.

Dr. von Ohain also noted correctly that integrating the engine into the aircraft would become important. Modern commercial engines have nacelles designed to improve integration by: (1) providing uniform flow to the entrance fan blades at all flight conditions (in order to prevent fan stall), (2) reducing the drag at the cruise condition (in order to reduce fuel consumption), and (3) creating favorable interactions between the engine airflow and the aircraft. The design and manufacture of nacelles by engine companies has become a major selling point for modern turbofans.

More recently, the emphasis has shifted in the direction of life cycle cost (including acquisition, fuel, maintenance, and repair) of operating the engines. This trend is due to such advances as the qualification testing of engines for their expected usage, improvements in rotating machinery computational fluid dynamics (CFD), improved knowledge of material properties, structural design, and the greatly increased amount of data sent directly from operating engines. Commercial engines tend to operate at high temperatures for long periods of time, which means that most failures are due to creep, or long exposure to high stress and temperature. The need for lightness (i.e., high thrust/weight) remains important, and this property is strongly influenced by structural design tools, improved material properties, and imaginative system design. Accurate prediction of combustion behavior remains elusive, primarily because mixing (including the micromixing at the molecular level that is necessary for the final chemical reaction) is a complex, viscous phenomenon.

Commercial turbofans have advanced along both the evolutionary line and the revolutionary line. The following examples illustrate both.

Figure 1 contains a photograph of the General Electric (GE) GE36 unducted fan (UDF) or counterrotating open rotor (CROR) affixed to the Boeing 727 test aircraft (circa 1986). The GE UDF is based on the GE F404 turbojet, which has a static OPR of about 30:1 (see Table B.2 in Appendix B); the UDF has an effective bypass ratio of about 10 at sea-level static conditions. A special advantage of this configuration is that the downstream open rotor removes any tangential motion created by the upstream open rotor so that the exit velocity is purely axial and no energy is lost in downstream tangential motion.

Figure 2 contains a cutaway drawing of the revolutionary Pratt & Whitney (P&W) 1000G geared turbofan engine (GTF), which has a static overall compression ratio of about 30:1 and a bypass ratio of about 12 at sea-level static conditions. A special advantage of this configuration is that the gearing allows the fan to turn at a relatively low rotational speed, which reduces the stress at the base of the fan blade and decreases the high Mach

Fig. 1   GE UDF affixed to the Boeing 727 test aircraft.

Fig. 2   Cutaway drawing of the Pratt & Whitney 1000G GTF. Note the gearbox between the low-pressure shaft and the high-bypass-ratio fan.

number losses at the fan blade tip, while allowing the low-pressure (fan-drive) turbine to turn at a relatively high rotational speed, thus increasing the work per turbine stage and reducing the number of heavy low-pressure turbine stages (and the initial and maintenance costs, number of parts, and weight of the low-pressure turbine) required to provide the power needed by the fan. This configuration is certified and flying in two versions, one in the 24,000-lb thrust class and one in the 33,000-lb thrust class at sea-level static conditions.

Figure 3 contains a cutaway drawing of the future Rolls-Royce (RR) Ultra-Fan engine, which will use composite materials in order to lighten the components and employ geared turbofan technology in order to improve the overall efficiency of the high-bypass-ratio fan and reduce low-pressure

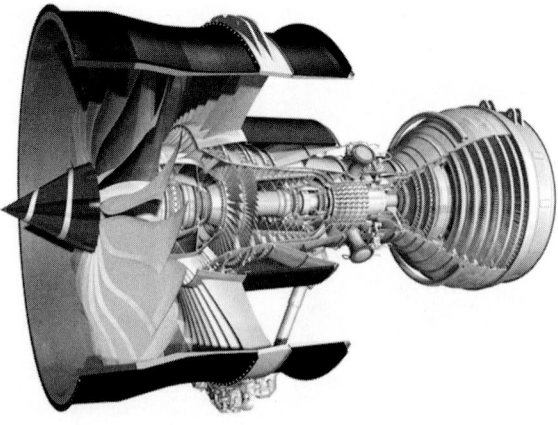

Fig. 3   Cutaway drawing of the advanced Rolls-Royce UltraFan engine.

turbine weight, thus reducing fuel consumption. The static OPR of the Ultra-Fan engine is expected to be about 30:1.

As Dr. von Ohain surmised, commercial turbofans have become very large (i.e., high thrust), especially as increased reliability has allowed the number of engines per aircraft and their corresponding life cycle cost to be reduced while providing safe travel. Today, many large aircraft that fly great distances (such as the Airbus 330WXB and the Boeing 787) may have only two engines, and engine thrust may approach 100,000 lb (rather than the 1000 lb or so of the first von Ohain engines). The two engines installed on the Boeing 777 produce almost as much thrust as the four engines installed on the original Boeing 747.

Around 1960 I experienced two piston engines out of four failing and shutting down on a flight from Los Angeles to St. Louis. Today, the turbine engine companies monitor a steady stream of data from instrumentation on every operating engine in order to detect the signs of impending problems, and the airline company maintenance groups review the situation of every engine regularly in order to replace aging parts or components in an orderly manner. The latter has become known as *removal for cause*. As a result, the odds against two engines shutting down even during a long flight are astronomical; some commercial engines have remained "on the wing" for more than 20,000 h. I have never experienced a single engine shutdown during my thousands of hours of turbine engine aircraft flight.

Stealth has not yet become a major issue for commercial engines, although laser beams, bird strikes, and the interference of civilian drone activity are increasingly causing problems.

Military Turbine Engines

Military turbine engines almost always provide a balance of reasonable fuel consumption and the opportunity for high thrust by combining a moderate fan bypass ratio (say about 1) and high compression ratio with afterburning. Operating the afterburner increases both the total thrust and the fuel consumption greatly, so the afterburner is used only when absolutely required (e.g., during takeoff, transonic acceleration, and combat). It should be noted that the P&W F119 is the first aircraft engine deliberately designed for supercruise (supersonic flight without afterburning), although it has an afterburner to be used when needed. The P&W F119 also has counterrotating turbines, which increases the overall efficiency of the engine.

Military engines have grown in thrust, too [2] (see Table B.2 in Appendix B). The thrust progression has been from the P&W F100 (about 29,000 lb of thrust at takeoff in afterburning circa 1970), to the GE F101 (about 31,000 lb of thrust at takeoff in afterburning circa 1980), to the P&W F119 (about 35,000 lb of thrust at takeoff in afterburning circa 2000), to the P&W F135 (about 40,000 lb of thrust at takeoff in afterburning circa 2010). This progression has allowed the U.S. Air Force F-35A and U.S.

Navy F-35C variants of the Lockheed F-35 Joint Strike Fighter to use a single engine in order to reduce life cycle costs. The U.S. Marine F-35B version also uses an RR lift fan and deflected P&W F135 thrust from a single engine in order to take off in a short distance and land vertically when returning from a mission. Military engine thrust requirements also grow as the weight of the aircraft grows with time (about 1 lb per day). The increasing requirements are usually met by increasing the airflow into the aircraft inlet and the corresponding engine air flow rate and thrust, but they can also be met by altering the engine cycle (e.g., decreasing the bypass ratio) or increasing the combustion temperature.

The U.S. Air Force Lockheed C-5 Galaxy replaced moderate-thrust GE TF-39 engines with more powerful and proven reliable commercial GE CF6-80 engines to create the C-5M Super Galaxy (circa 2006), reducing the maintenance costs considerably and restoring the excess thrust power and productivity intended for the original aircraft.

In the past, most commercial engines were primarily "downrated" (i.e., lower turbine temperature and engine thrust) "derivatives" of military engines. In more recent times, their goals have diverged (mostly because of business and environmental pressures on commercial flight), and modern commercial engines are now significantly different from military engines. Nevertheless, they share advances in such underlying technologies as computer modeling, materials, and manufacturing.

More recently, the military turbine engine emphasis has shifted in the direction of life cycle cost (including acquisition, fuel, maintenance, and repair) of operating the engines, for reasons similar to those of commercial turbine engines. For example, the U.S. Air Force executed life extension programs on the most troublesome parts of the GE F110 and P&W F100 engines, with amazingly good results. The in-flight shutdown rates (IFSD rates) of both engines dropped to much lower levels than had ever been experienced. This was especially beneficial because the airplanes powered by these engines were to remain in service much longer than originally envisioned. Military engines tend to experience many throttle movements, which means that most failures are due to low cycle fatigue (LCF). This also means that training flights can be much more damaging to the engine than combat missions and increases the need to design the engine for its actual usage.

The military maintenance depots have adopted a strategy for engines that anticipates (to the extent possible) the end of useful life for most parts or components. The life is estimated by analysis, from prior field experience, and from cycle usage data recorded and collected during flights. This is a departure from the *static universe* approach of the past that required spare parts to be manufactured and purchased based primarily on past replacement or consumption rates. The goal now is to provide adequate time on wing before the engine must again be removed and returned to the depot for maintenance. This program is known as *on-condition maintenance*, and it is supported by the donation of parts or components with partial useful life remaining from any one group to the entire support system. More recently, modular maintenance (each component could be considered to be a separate module) has replaced whole engine maintenance in order to avoid the cost of shipping an entire engine back to the depot for the repair of a single component. Modular maintenance requires excellent tracking capabilities and could end the concept of the whole engine.

The wear of engine parts, especially compressor and turbine airfoils, is a continuing problem for military engines, especially when they are exposed to a great deal of desert sand and/or salt air. Compressor airfoils not only become less efficient as they corrode or erode, but also can cause the compressor to stall when they become sufficiently degraded. This can make matters worse than they seem. Deterioration of compressor parts can cause the temperatures in the turbine to increase, where the creep life can be halved by a metal temperature increase of only 20°F. Also, small nicks on the trailing edges of compressor airfoils can initiate cracks that can eventually grow to cause release or failure. Finally, the cooling passages of turbine airfoils may no longer function properly because they become clogged with contaminants. Repairing these airfoils in the field can cause other problems, including shaft imbalance. Commercial aircraft always avoid volcanic plumes. One result of this situation is the use of materials or coatings/surface treatments for these airfoils that can better withstand the harsh environment.

Survivability of military aircraft has become and will remain a major issue. Survivability can be increased by decreasing detectability or decreasing vulnerability [3]. Stealth (reduced detectability) has become and will remain a requirement for military engines. The entrance and exhaust regions of the engine usually contain the means to shield the nearby components, and all surfaces are coated to absorb incoming radiation in order to reduce the probability of detection and/or identification. Afterburning is avoided in order to reduce the visible engine exhaust radiation signature. These needs have changed the designs of modern military engines substantially and prevented the radiation "black holes" of the inlet and nozzle from becoming the source of identification and location beacons for the enemy. The Lockheed F-117 Stealth Fighter is a poster child for this type of treatment, and it is said that about 20% of the uninstalled thrust available from the two non-afterburning GE F404 engines is consumed by measures that are necessary in order to provide the necessary stealth. Because the F-117 also has a lot of drag, it is a tribute to modern nonafterburning turbine engines that they can provide enough thrust to propel this aircraft at combat speeds.

The value of speed and altitude to the survivability of military aircraft is highly dependent on the total scenario, including the capability of the enemy. There is no simple way to estimate the importance of speed and altitude.

Military aircraft designers also seek to improve the integration of the aircraft and engine by reducing the *installation drag* or *installation penalty* on the external aircraft surfaces as much as possible. Recently, the U.S. Air Force initiated an engine research program for variable cycle engines that could reduce the installation drag by changing the bypass ratio to tailor the amount of airflow into the aircraft inlet to the amount required by the engine [4].

The acquisition cost of large aircraft engines, both commercial and military, has passed $10 million each, partly because of monetary inflation and engine size growth and partly because material and manufacturing costs have also risen. This seems like a large number, but the fractional cost of engines/aircraft has not increased. The great advantage enjoyed by both commercial and military airbreathing turbine engines is that they accumulate hundreds of thousands of hours of operation in actual service. This allows the unintended consequences of variations of manufacturing, materials properties, the environment, usage, assembly, control systems, and the like to be uncovered and fixed quickly. Even though qualification testing is deliberately designed to reproduce the most demanding parts of the intended flight spectrum, the testing can be carried out on at most only a few engines, and actual service may differ from that originally intended. As a result, many unanticipated problems can be found and fixed before they become important.

Small Turbine Engines

Small turbine engines often introduce revolutionary concepts to the field partly because of their size and cost. For example, the tortuous flowpath (the flow in the combustor is in the direction of flight and the radial compressor and turbine turn the flow away from axial) of the geared turbofan Garrett TFE731 (see Figure 4) allowed a relatively small frontal area with little loss of performance and some cycle variability. The Garrett TFE731 was originally designed about 50 years ago and now has at least nine variants with a static OPR of about 18 that have found both commercial and military applications (see Tables B.2 and B.3 in Appendix B). Also, the Flo Research lobed

Fig. 4   Cutaway drawing of the geared turbofan Garrett TFE731-2.

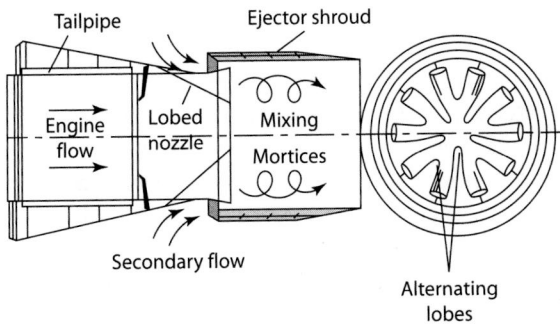

Fig. 5 Schematic diagram of Flo Research lobed mixer/ejector for the RR Spey 511.

mixer/ejector was installed in the two RR Spey 511 engines that propel the Gulfstream II aircraft in order to reduce takeoff noise without reducing engine performance (see Figure 5) [5].

Small turbine engines are frequently found in business jet aircraft, such as the Honeywell 7250G turbofan in the Gulfstream 280 and the P&W JT15D turbofan in the Citation V. They are also used to provide power for military helicopters, such as the GE T700 in the U.S. Army Sikorsky UH-60 Black-hawk and the RR 1107 in the U.S. Marine Boeing V-22 Osprey.

Williams International (WI) is the only new company to enter the turbine engine world in the last 40 years. The U.S. Air Force AGM-86 air-launched cruise missile (ALCM) and the U.S. Navy BGM-109 Tomahawk cruise missile are long-range subsonic missiles propelled by a WI F107 turbofan, and the WI FJ44 turbofan was developed for a variety of small business jets including the Cessna CitationJet.

Rocket Engines

All existing rockets carry onboard both the fuel and oxidizer required for propulsion for their mission. The primary advantage of this approach is that the rocket is independent of the medium through which it travels, meaning that travel outside the sensible atmosphere of the earth (e.g., into space) is the same as travel within the atmosphere (other than the additional atmospheric drag). The primary downside of this approach is that rockets do not take advantage of the oxidizer or mass flow available from the surrounding atmosphere, so their performance at sea level (say, in terms of the specific impulse, or lb of thrust per lb of propellant weight per s—including oxidizer—and therefore in units of s) is less than about 400 s (see Table 1.6 in Chapter 1). Turbine engine specific impulse at sea level is about an order of magnitude larger because the oxidizer is provided by the atmosphere. Consequently, rockets are essential for travel outside the atmosphere but have a relatively small payload fraction. The original concept of the National Aero-Space Plane (NASP), circa 1990, was to trade the reduced fuel plus oxidizer requirement while flying within the atmosphere to provide vehicle operability and reusability, and allow the mission to space to have an increased payload fraction and to be completed with onboard rockets. Unfortunately, the tradeoff was unsuccessful with the technology available at the time, and the NASP program was cancelled.

Today, commercial rocket engines are primarily used for travel to relatively nearby orbits because the payload is a small fraction of the takeoff weight (due to the relatively low specific impulse), so that the takeoff weight would be enormous for a large payload. This will not be true for nuclear rockets, but they have other problems. The search continues for fuel plus oxidizer combinations that are environmentally friendly and could produce a higher specific impulse.

Meanwhile, military rockets (primarily missiles) usually use solid fuel, which reduces specific impulse further (see Table 1.6 in Chapter 1). Thus, they are best for high-speed, short-range weapons designed for specific purposes (e.g., air-to-air, ground-to-air, and ground-to-ground missiles). A strong emphasis in solid rocket research today is to find propellants that are either safer for the user or kinder to the launch environment (aka green propellants).

Novel Concepts

Occasionally, a novel concept is introduced with the promise of superior performance for future applications. The examples that follow are a sample of novel concepts of the past.

One example of this for airbreathing engines was the pulse detonation engine (PDE), which replaces the low-speed, constant-pressure combustion of the turbojet Brayton cycle with unsteady detonation wave combustion. Many analyses have shown that the PDE could have superior performance to the Brayton cycle, especially when the flight speed is subsonic [6]. The biggest hurdles to be overcome by the PDE are the unsteadiness (which reduces structural life, can reduce nozzle performance, and generates noise) and the fact that the PDE cycle has only a small advantage over the Brayton cycle at high flight Mach numbers.

Another example is the airbreathing scramjet (or supersonic combustion ramjet), which employs supersonic combustion above a Mach number of about 5–6 to reduce the static (or thermodynamic) temperature of the flow within the combustor in order to avoid excessive dissociation losses [7]. The scramjet has yet to be proven a practical device for high-speed flight because of the complexities of *unstart* (where the scramjet stops running because a shock wave is disgorged from the inlet separating the boundary layer ahead of the inlet and the flow into the inlet and the thrust are greatly reduced), some of which are prevented by providing a constant area *isolator* between the inlet and the combustor, or by simply reducing the thrust. In contrast, the ramjet-powered Marquardt low-altitude short range missile (LASRM) reached speeds in excess of Mach 5 at an altitude of 40,000 ft in flight tests more than 40 years ago [8]. This would mean that the ramjet could be used for flight Mach numbers up to at least 5.

On the rocket side, Brian Cantwell of Stanford University has developed a class of paraffin-based fuels that burn at surface regression rates that are several times those of conventional polymeric fuels. These new fuels form a thin, hydrodynamically unstable liquid layer on the melting surface of the fuel grain. Entrainment of droplets from the liquid–gas interface can substantially increase the rate of fuel mass transfer, leading to much higher surface regression rates than can be achieved with conventional polymeric fuels. This permits the design of a high-volumetric-loading, single-port hybrid rocket system with a density impulse comparable to a conventional hydrocarbon-fueled liquid rocket propulsion system.

Finally, as noted previously, the propeller is a promising device, primarily for airships and military applications (drones) that require greater range and endurance and much less noise (lower detectability) [9]. The advent of better materials and more accurate CFD has made advances in noise generation possible, especially for variable-pitch propellers, but they have yet to be translated into practice.

Stay tuned, because other novel concepts are certain to be discovered in the future. My strongest recommendation is that their advertised promise first be explored for performance using the fundamental laws and basic principles of propulsion.

Elements of Propulsion: Gas Turbines and Rockets

Dr. Jack Mattingly has packed a lifetime of experience and teaching into this textbook devoted to developing and illustrating the importance of the fundamental laws and basic principles of propulsion. He has also taken the algebraic solution of ideal and real airbreathing engine behavior as far as possible. His methods point the way for the future solution of cycle behavior.

You may be interested to find that *Elements of Propulsion* was the first AIAA Education Series textbook intended to be suitable for an undergraduate audience, and it has more than proven the viability of this concept. The original versions concentrated only on gas turbine airbreathing propulsion, but the latest version also contains material on rocket propulsion (see Chapter 10). Also, the current version contains a new Chapter 3 on compressible flow, written by Dr. Keith Boyer of Practical Aeronautics, with whom we frequently teach propulsions courses.

I hope and trust that you find *Elements of Propulsion* to be as enjoyable, rewarding, and useful as I have.

Dr. William H. Heiser
AIAA Honorary Fellow
June 2016

References

[1]   Buckingham, E., "Jet Propulsion for Airplanes," NACA Report 159, 1924.

[2]   Mattingly, J. D., *Elements of Propulsion: Gas Turbines and Rockets*, 2nd ed., AIAA Education Series, AIAA, Reston, VA, 2016, App. B.

[3]   Ball, R. E., *Fundamentals of Aircraft Combat Survivability Analysis and Design*, AIAA Education Series, AIAA, Washington, DC, 1985.

[4]   "The ADVENT of a Better Jet Engine?," *Defense Industry Daily*, Jan. 2015.

[5]   Presz, W. Jr., Reynolds, G., and Hunter, C., "Thrust Augmentation with Mixer/ Ejector Systems," AIAA Paper 2002-0230, 40th AIAA Aerospace Sciences Meeting and Exhibit, Reno, NV, 2002.

[6]   Zel'dovich, Y. B., "To the Question of Energy Use of Detonation Combustion," *AIAA Journal of Propulsion and Power*, Vol. 22, No. 3, 2006, pp. 588–592.

[7]   Heiser, W. H., and Pratt, D. T., *Hypersonic Airbreathing Propulsion*, AIAA Education Series, AIAA, Washington, DC, 1994.

[8]   Parsch, A., "Martin Marietta ASALM," *Directory of U.S. Military Rockets and Missiles*, 2003, App. 4.

[9]   Morgado, J., Silvestre, M. A. R., and Pascoa, J. C., "Validation of New Formulations for Propeller Analysis," *AIAA Journal of Propulsion and Power*, Vol. 31, No. 1, 2015, pp. 467–477.

FOREWORD TO THE FIRST EDITION

Hans von Ohain
German Inventor of the Jet Engine

Background

The first flight of the Wright brothers in December 1903 marked the beginning of the magnificent evolution of *human-controlled, powered flight*. The driving forces of this evolution are the ever-growing demands for improvements in

1. Flight performance (i.e., greater flight speed, altitude, and range and better maneuverability)
2. Cost (i.e., better fuel economy, lower cost of production and maintenance, increased lifetime)
3. Adverse environmental effects (i.e., noise and harmful exhaust gas effects)
4. Safety, reliability, and endurance
5. Controls and navigation.

These strong demands continuously furthered the efforts of advancing the aircraft system.

The tight interdependency between the performance characteristics of aerovehicle and aeropropulsion systems plays a very important role in this evolution. Therefore, to gain better insight into the evolution of the aeropropulsion system, one has to be aware of the challenges and advancements of aerovehicle technology.

The Aerovehicle

A brief review of the evolution of the aerovehicle will be given first. One can observe a continuous trend toward stronger and lighter airframe designs, structures, and materials—from wood and fabric to all-metal structures; to lighter, stronger, and more heat-resistant materials; and finally to a growing use of strong and light composite materials. At the same time, the aerodynamic quality of the aerovehicle is being continuously improved. To see this development in proper historical perspective, let us keep in mind the following information.

In the early years of the 20th century, the science of aerodynamics was in its infancy. Specifically, the aerodynamic lift was not scientifically well understood. Joukowski and Kutta's model of lift by circulation around the wing and

Prandtl's boundary-layer and turbulence theories were in their incipient stages. Therefore, the early pioneers could not benefit from existent scientific knowledge in aerodynamics and had to conduct their own fundamental investigations.

The most desirable major aerodynamic characteristics of the aerovehicle are a low *drag coefficient* as well as a high *lift/drag ratio L/D* for cruise conditions, and a high *maximum lift coefficient* for landing. In Fig. 1, one can see that the world's first successful glider vehicle by Lilienthal, in the early 1890s, had an *L/D* of about 5. In comparison, birds have an *L/D* ranging from about 5 to 20. The Wright brothers' first human-controlled, powered aircraft had an *L/D* of about 7.5. As the *L/D* values increased over the years, sailplanes advanced most rapidly and now are attaining the enormously high values of about 50 and greater. This was achieved by employing ultrahigh wing aspect ratios and aerodynamic profiles especially tailored for the low operational Reynolds and Mach numbers. In the late 1940s, subsonic transport aircraft advanced to *L/D* values of about 20 by continuously improving the aerodynamic shapes, employing advanced profiles, achieving extremely smooth and accurate surfaces, and incorporating inventions such as the engine cowl and the retractable landing gear.

The continuous increase in flight speed required a corresponding reduction of the *landing speed/cruise speed* ratio. This was accomplished by innovative wing structures incorporating wing slots and wing flaps that, during the landing process, enlarged the wing area and increased significantly the lift coefficient. Today, the arrowhead-shaped wing contributes to a

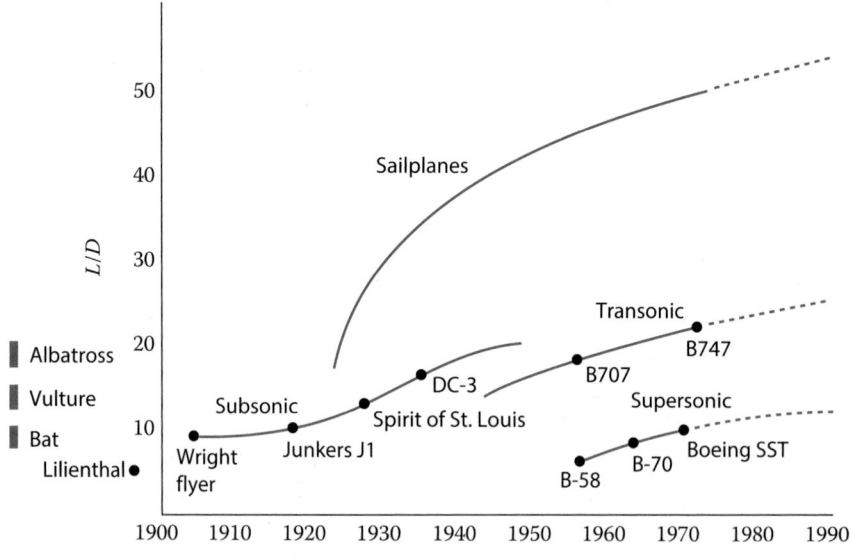

Fig. 1   Progress in lift/drag ratio *L/D*.

high lift for landing (vortex lift). Also, in the 1940s, work began to extend the high L/D value from the subsonic to the transonic flight speed regime by employing the swept-back wing and later, in 1952, the area rule of Whitcomb to reduce *transonic drag rise.* Dr. Theodore von Karman describes in his memoirs, *The Wind and Beyond* [1], how the *swept-back wing* or simply *swept wing* for transonic and supersonic flight came into existence:

> The fifth Volta Congress in Rome, 1935, was the first serious international scientific congress devoted to the possibilities of supersonic flight. I was one of those who had received a formal invitation to give a paper at the conference from Italy's great Guglielmo Marconi, inventor of the wireless telegraph. All of the world's leading aerodynamicists were invited.
>
> This meeting was historic because it marked the beginning of the supersonic age. It was the beginning in the sense that the conference opened the door to supersonics as a meaningful study in connection with supersonic flight, and, secondly, because most developments in supersonics occurred rapidly from then on, culminating in 1946—a mere 11 years later—in Captain Charles Yeager's piercing the sound barrier with the X-1 plane in level flight. In terms of future aircraft development, the most significant paper at the conference proved to be one given by a young man, Dr. Adolf Busemann of Germany, by first publicly suggesting the swept-back wing and showing how its properties might solve many aerodynamic problems at speeds just below and above the speed of sound.

Through these investigations, the myth that sonic speed is the fundamental limit of aircraft flight velocity, the *sound barrier* was overcome.

In the late 1960s, the Boeing 747 with swept-back wings had, in transonic cruise speed, an L/D value of nearly 20. In the supersonic flight speed regime, L/D values improved from 5 in the mid-1950s (such as L/D values of the B-58 Hustler and later of the Concorde) to a possible L/D value of 10 and greater in the 1990s. This great improvement possibility in the aerodynamics of supersonic aircraft can be attributed to applications of artificial stability, to the area rule, and to advanced wing profile shapes that extend laminar flow over a larger wing portion.

The hypersonic speed regime is not fully explored. First, emphasis was placed on winged reentry vehicles and lifting bodies where a high L/D value was not of greatest importance. Later investigations have shown that the L/D values can be greatly improved. For example, the maximum L/D for a "wave rider" is about 6 [2]. Such investigations are of importance for hypersonic programs.

The Aeropropulsion System

At the beginning of this century, steam and internal combustion engines were in existence but were far too heavy for flight application. The Wright brothers recognized the great future potential of the internal combustion engine and developed both a relatively lightweight engine suitable for flight application and an efficient propeller. Figure 2 shows the progress of the propulsion systems over the years. The Wright brothers' first aeropropulsion

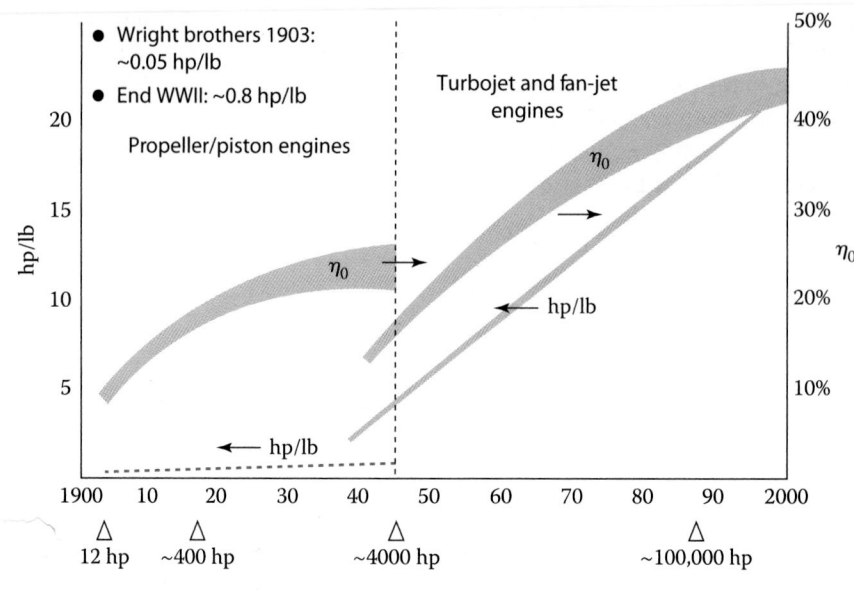

Fig. 2 Trends of power per weight (hp/lb) and overall efficiency (η_0) of aeropropulsion systems from 1900 to 2000.

system had a shaft power of 12 hp, and its power/weight ratio (ratio of power output to total propulsion system weight, including propeller and transmission) was about 0.05 hp/lb. Through the subsequent four decades of evolution, the overall efficiency and the power/weight ratio improved substantially, the latter by more than one order of magnitude to about 0.8 hp/lb. This great improvement was achieved by engine design structures and materials, advanced fuel injection, advanced aerodynamic shapes of the propeller blades, variable-pitch propellers, and engine superchargers. The overall efficiency (engine and propeller) reached about 28%. The power output of the largest engine amounted to about 5000 hp.

In the late 1930s and early 1940s, the turbojet engine came into existence. This new propulsion system was immediately superior to the reciprocating engine with respect to the power/weight ratio (by about a factor of 3); however, its overall efficiency was initially much lower than that of the reciprocating engine. As can be seen from Fig. 2, progress was rapid. In less than four decades, the power/weight ratio increased more than 10-fold, and the overall efficiency exceeded that of a diesel propulsion system. The power output of today's largest gas turbine engines reaches nearly 100,000 equivalent hp.

Impact on the Total Aircraft Performance

The previously described truly gigantic advancements of stronger and lighter structures and greater aerodynamic quality in aerovehicles and greatly advanced overall efficiency and enormously increased power output/ weight ratios in aeropropulsion systems had a tremendous impact on flight performance, such as on flight range, economy, maneuverability, flight speed, and altitude. The increase in flight speed over the years is shown in Fig. 3. The Wright brothers began with the first human-controlled, powered flight in 1903; they continued to improve their aircraft system and, in 1906, conducted longer flights with safe takeoff, landing, and curved flight maneuvers. While the flight speed was only about 35 mph, the consequences of these first flights were enormous:

1. Worldwide interest in powered flight was stimulated.
2. The science of aerodynamics received a strong motivation.
3. The U.S. government became interested in power flight for potential defense applications, specifically reconnaissance missions.

   In 1909, the Wright brothers built the first military aircraft under government contract. During World War I, aircraft technology progressed rapidly. The flight speed reached about 150 mph, and the engine power attained 400 hp. After World War I, military interest in aircraft systems dropped, but aircraft technology had reached such a degree of maturity that two non-military application fields could emerge, namely:

1. Commercial aviation, mail and passenger transport (first all-metal monoplane for passenger and mail transport, the Junkers F13, in 1919, sold worldwide)

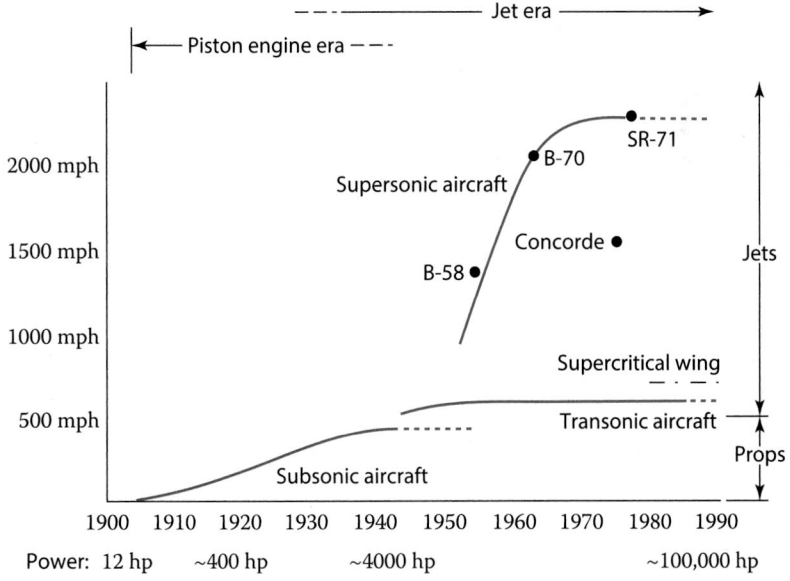

Fig. 3   Aircraft speed trends.

2. Stunt flying leading to general aviation (sport and private transportation)

In the period from 1920 to 1940, the speed increased from about 150 to 350 mph through evolutionary improvements in vehicle aerodynamics and engine technology, as discussed previously. At the end of World War II, the flight speed of propeller aircraft reached about 400–450 mph, and the power output of the largest reciprocating engines was about 5000 hp. This constituted almost the performance limit of the *propeller/reciprocating engine propulsion system*. Today, the propeller/reciprocating engine survives only in smaller, lower-speed aircraft used in general aviation.

In the late 1930s, jet propulsion emerged that promised far greater flight speeds than attainable with the propeller or piston engine. The first jet-propelled experimental aircraft flew in the summer of 1939 (the He-178), and in early 1941, the first prototype jet fighter began flight tests (He-280). In 1944, mass-produced jet fighters reached a speed of about 550 mph (Me-262).

In the early 1950s, jet aircraft transgressed the sonic speed. In the mid-1950s, the first supersonic jet bomber (B-58 Hustler) appeared, and later the XB-70 reached about Mach 3. Also during the 1950s, after more than 15 years of military development, gas turbine technology had reached such a maturity that the following commercial applications became attractive: 1) commercial aircraft, e.g., Comet, Caravelle, and Boeing 707; 2) surface transportation (land, sea); and 3) stationary gas turbines.

In the 1960s, the high-bypass-ratio engine appeared, which revolutionized military transportation (the C5A transport aircraft). At the end of the 1960s, based on the military experience with high-bypass-ratio engines, the second generation of commercial jet aircraft came into existence, the *widebody* aircraft. An example is the Boeing 747 with a large passenger capacity of nearly 400. Somewhat later came the Lockheed L-1011 and Douglas DC10. By that time, the entire commercial airline fleet used turbine engines exclusively. Advantages for the airlines were as follows:

1. Very high overall efficiency and, consequently, a long flight range with economical operation
2. Overhaul at about 5 million miles
3. Short turnaround time
4. Passenger enjoyment of the very quiet and vibration-free flight, short travel time, and comfort of smooth stratospheric flight
5. Community enjoyment of quiet, pollution-free aircraft

By the end of the 1960s, the entire business of passenger transportation was essentially diverted from ships and railroads to aircraft. In the 1970s, the supersonic Concorde with a flight speed of 1500 mph (the third generation of commercial transport) appeared with an equivalent output of about 100,000 hp.

Summary

In hindsight, the evolution of aerovehicle and aeropropulsion systems looks like the result of a master plan. The evolution began with the piston engine and propeller, which constituted the best propulsion system for the initially low flight speeds and had an outstanding growth potential up to about 450 mph. In the early 1940s, when flight technology reached the ability to enter into the transonic flight speed regime, the jet engine had just demonstrated its suitability for this speed regime. A vigorous jet engine development program was launched. Soon the jet engine proved to be not only an excellent transonic but also a supersonic propulsion system. This resulted in the truly exploding growth in flight speed, as shown in Fig. 3.

It is interesting to note that military development preceded commercial applications by 15–20 years for both the propeller engine and the gas turbine engine. The reason was that costly, high-risk, long-term developments conducted by the military sector were necessary before a useful commercial application could be envisioned. After about 75 years of powered flight, the aircraft has outranked all other modes of passenger transportation and has become a very important export article of the United States.

The evolutions of both aerovehicle and aeropropulsion systems have in no way reached a technological level that is close to the ultimate potential! The evolution will go on for many decades toward capabilities far beyond current feasibility and, perhaps, imagination.

How Jet Propulsion Came into Existence

The idea of airbreathing jet propulsion originated at the beginning of the 20th century. Several patents regarding airbreathing jet engines had been applied for by various inventors of different nationalities who worked independently of each other.

From a technical standpoint, *airbreathing jet propulsion* can be defined as a special type of internal combustion engine that produces its net output power as the rate of change in the kinetic energy of the engine's working fluid. The working fluid enters as environmental air that is ducted through an inlet diffuser into the engine; the engine exhaust gas consists partly of combustion gas and partly of air. The exhaust gas is expanded through a thrust nozzle or nozzles to ambient pressure. A few examples of early airbreathing jet propulsion patents are as follows:

1. In 1908, Lorin patented a jet engine that was based on piston machinery (Fig. 4a).
2. In 1913, Lorin patented a jet engine based on ram compression in supersonic flight (Fig. 4b), the *ramjet*.
3. In 1921, M. Guillaume patented a jet engine based on turbomachinery; the intake air was compressed by an axial-flow compressor followed by a combustor and an axial-flow turbine driving the compressor (Fig. 4c).

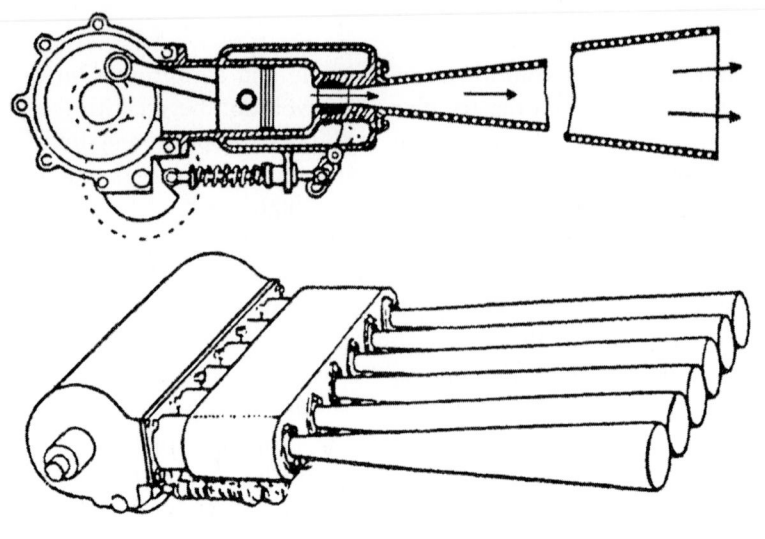

Fig. 4a   Lorin's 1908 patent.

These patents clearly described the airbreathing jet principle but were not executed in practice. The reason lies mainly in the previously mentioned strong interdependency between aerovehicle and aeropropulsion systems. The jet engine has, in comparison with the propeller engine, a high exhaust speed (for example, 600 mph and more). In the early 1920s, the aerovehicle had a flight speed capability that could not exceed about 200 mph. Hence, at that time, the so-called propulsive efficiency of the jet engine was very low (about 30–40%) in comparison to the propeller, which could reach more than 80%. Thus, in the early 1920s, the jet engine was not compatible with the too-slow aerovehicle. Also, in the early 1920s, an excellent theoretical study about the possibilities of enjoying jet propulsion had been conducted by Buckingham of the Bureau of Standards under contract with NACA. The result of this study was clear—the jet engine could not be efficiently employed if the aerovehicle could not greatly exceed the flight speed of 200 mph; a flight speed beyond 400 mph seemed to be necessary. The consequences of the results of this study were that the aircraft engine

Fig. 4b   Lorin's 1913 patent.

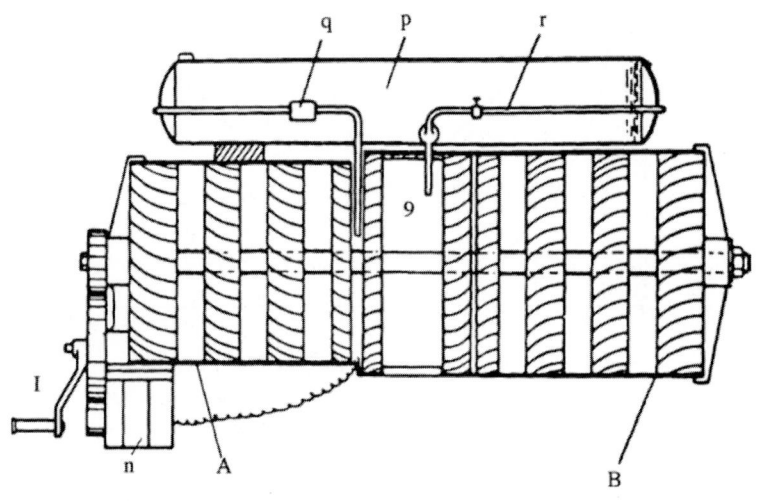

Fig. 4c   Guillaume's 1921 patent.

industry and the scientific and engineering community had no interest in the various jet engine inventions. Thus the early jet engine concepts were forgotten for a long time. They were unknown to Sir Frank Whittle, to me, and to the British and German patent offices. In 1939, however, the retired patent examiner Gohlke found out about the early jet patents and published them in a synoptic review.

The first patent of a turbojet engine, which was later developed and produced, was that of Frank Whittle, now Sir Frank (see Fig. 5). His patent was applied for in January 1930. This patent shows a multistage, axial-flow compressor followed by a radial compressor stage, a combustor, an axial-flow turbine driving the compressor, and an exhaust nozzle. Such configurations are still used today for small- and medium-power output engines, specifically for remote-controlled vehicles.

Turbojet Development of Sir Frank Whittle

Frank Whittle [3] was a cadet of the Royal Air Force. In 1928, when he was 21 years old, he became interested in the possibilities of rocket propulsion and propeller gas turbines for aircraft, and he treated these subjects in his thesis. He graduated and became a pilot officer, continuously thinking about airbreathing jet propulsion. In 1929, he investigated the possibilities of a ducted fan driven by a reciprocating engine and employing a kind of afterburner prior to expansion of the fan gas. He finally rejected this idea on the basis of his performance investigations. The same idea was conceived later in Italy and built by Caproni Campini. The vehicle flew on August 28, 1940, but had a low performance, as predicted by Sir Frank in 1929.

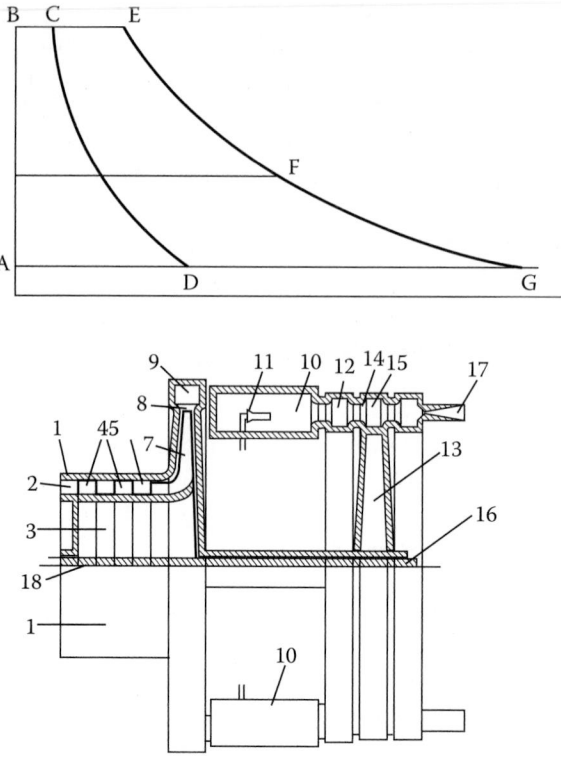

Fig. 5   Whittle's turbojet patent drawing. (Courtesy of National Air and Space Museum.)

Suddenly, in December 1929, Frank Whittle was struck by the idea of increasing the fan pressure ratio and substituting a turbine for the reciprocating engine. This clearly constituted a compact, lightweight turbojet engine. He applied for a patent for the turbojet (Fig. 5) in January 1930.

Frank Whittle discussed this idea with fellow officers and his superior officer. They were very impressed, and a meeting was arranged between him and officials of the British Air Ministry, Department of Engine Development. This department, in turn, sought advice from Dr. A. A. Griffith, who was interested in the development of a propeller gas turbine. Dr. Griffith expressed doubts about the feasibility of Whittle's turbojet concept from a standpoint of too-high fuel consumption. Actually, *in high-speed flight,* the turbojet has great advantages over a propeller gas turbine due to the fact that the turbojet is much lighter than the propeller gas turbine and can fly faster because of the absence of the propeller. Whittle rightfully considered the turbojet as a fortunate synthesis or hybrid of the "propeller gas turbine" and "rocket" principles. As Sir Frank recalls, the department wrote a letter that, in essence, stated that any form of a gas turbine would be impractical in view of the long history of failure and the lack of turbine materials capable of withstanding the high stresses at high temperatures. Whittle's outstanding and very important views were that the flying gas turbine had great advantages over a stationary gas turbine power plant due to the efficient ram pressure recovery, low environmental temperature in high altitude, and high efficiency of the jet nozzle. Unfortunately, these views were ignored by the department.

Frank Whittle (Fig. 6) tried to interest the turbine industry in his concept of jet propulsion, but he did not succeed. Lacking financial support, Whittle allowed his patent to lapse. A long, dormant period was ahead for Frank Whittle's jet propulsion ideas.

After five years, in mid-1935, two former Royal Air Force (RAF) officers tried to revive Whittle's turbojet concept. Whittle was enthused and wrote, "the jet engine had, like the Phoenix, risen from its ashes [3]." At that time, Whittle was under enormous pressure. He was preparing for the examination in mechanical sciences (Tripos); his goal was to graduate with "First-Class Honors." Now, in addition, he had to design his first experimental jet engine in late 1935. In March 1936, a small company, Power Jets Ltd., was formed to build and test Whittle's engine, the W. U. (Whittle Unit). In spite of all the additional work, Whittle passed his exam in June 1936 with First-Class Honors.

In April 1937, Whittle had his bench-test jet engine ready for the first test run. It ran excellently; however, it ran out of control because liquid fuel had collected inside the engine and started to vaporize as the engine became hot, thereby adding uncontrolled fuel quantities to the combustion process. The problem was easily overcome. This first test run was the world's first run of a bench-test jet engine operating with liquid fuel (Fig. 7). In June 1939, the testing and development had progressed to a point that the Air Ministry's Director of Scientific Research (D.S.R.) promised Frank Whittle a contract for building a flight engine and an experimental aircraft, the Gloster E28/29 (Gloster/Whittle). On May 15, 1941, the first flight of the Gloster/Whittle took place (Fig. 8).

Senior ministry officials initially showed little interest, and a request for filming was ignored; however, during further flight demonstrations, interest in jet propulsion increased. Of particular interest was a performance demonstration given to Sir Winston Churchill. At that occasion, the Gloster/Whittle accelerated away from the three escorting fighters, one Tempest and two Spitfires.

Several British aircraft engine corporations adapted the work of Frank Whittle. Specifically, Rolls-Royce, due

Fig. 6 Frank Whittle using slide rule to perform calculations. (Bettman.)

Fig. 7   Whittle's first experimental engine after second reconstruction in 1938.
(Courtesy of National Air and Space Museum.)

to the efforts of Sir Stanley G. Hooker [4], developed the first operational and the first production engine for the two-engine Gloster Meteor, Britain's first jet fighter. In March 1943, the Gloster Meteor prototype made its first flight, powered by two de Haviland (H-1) radial jet engines. In July 1944, the Meteor I, powered with two Rolls-Royce Welland engines, became operational. Its only combat action (in World War II) was in August 1944 in a successful attack against the German V1 flying bomb; it was the only fighter with sufficient level speed for the purpose. Mass production began with the Meteor III powered by two Rolls-Royce Dervents in 1945. The Meteor remained the RAF's first-line jet fighter until 1955.

From the beginning of his jet propulsion activities, Frank Whittle had been seeking means for improving the propulsive efficiency of turbojet engines [4]. He conceived novel ideas for which he filed a patent

Fig. 8   Gloster E28/29. (Courtesy of National Air and Space Museum.)

application in 1936, which can be called a *bypass engine* or *turbofan*. To avoid a completely new design, Whittle sought an interim solution that could be merely "tacked on" to a jet engine. This configuration was later known as the *aft fan*. Whittle's work on *fan jets* or bypass engines and aft fans was way ahead of his time. It was of greatest importance for the future of turbopropulsion.

Whittle's Impact on U.S. Jet Development

In the summer of 1941, U.S. Army Air Corp General Henry H. Arnold was invited to observe flight demonstrations of the Gloster/Whittle. He was very impressed and decided this technology should be brought over to the United States. In September 1941, an agreement was signed between U.S. Secretary of War Stimson and Sir Henry Self of the British Air Commission. The United States could have the engine W1X and a set of drawings of the Whittle W2B jet engine, provided that close secrecy was maintained and the number of people involved were held to a minimum. Under these conditions, open bids for the jet engine development were not possible. General Arnold chose General Electric for jet engine development because of the great experience this company had in the development of aircraft engine turbosuperchargers. The W2B engine was built and tested on March 18, 1942, under the name *GE 1-A*. This engine had a static thrust of 1250 lb and weighed 1000 lb [3]. In the meantime, the Bell Aircomet (XP-59A) was being designed and built. On October 3, 1942, the Aircomet with two GE 1-A engines flew up to 10,000 ft. This aircraft, which in the first tests seemed to have good performance characteristics, had an incurable "snaking" instability and so provided a poor gun platform for a fighter pilot. Also, another serious shortcoming was that the top speed was not sufficiently above that of an advanced propeller fighter. For these reasons, the Bell XP-59A with two GE 1-A engines (W2B) did not become a production fighter. From these experiences, it appeared that an engine of more than 4000-lb thrust was required for a single-engine fighter that would be capable of more than 500 mph operational speed. Lockheed was chosen to design a new jet fighter because when the project was discussed with the engineering staff, Lockheed's Kelly Johnson assured them a single-engine jet fighter in the "5000-plus" mph class could be built on the basis of a 4000-lb thrust engine.

General Electric developed the 4000-lb thrust engine, the 140 (an advanced version of Whittle's W2B engine), and Lockheed built the P80A Shooting Star, which flew on June 11, 1944. Although it did not enter combat during World War II, the Shooting Star became the United States' front-line fighter and outranked the Gloster Meteor with an international speed record (above 620 mph near the ground).

By about 1945, Frank Whittle had successfully completed, with greatest tenacity under the most adverse conditions, the enormous task of leading Great Britain and the United States into the jet age.

Other Early Turbojet Developments in the United States

Independent of European influence, several turbojet and propeller gas turbine projects had been initiated in the United States in 1939 and 1940. Although these projects had been terminated or prematurely canceled, they had contributed significantly to the know-how and technology of aircraft gas turbines, specifically their combustor and turbomachinery components.

One of these projects was the 2500-hp Northrop propeller gas turbine (Turbodyne) and a high-pressure-ratio turbojet under the excellent project leadership of Vladimir Pavelecka. Although the development goal of the large aircraft gas turbine engine was essentially met in late 1940, the project was canceled because the Air Force had lost interest in propeller gas turbines in view of the enormous advancement of the competitive jet engines.

Westinghouse had developed outstanding axial turbojet engines. The first very successful test runs of the Westinghouse X19A took place in March 1943. In the beginning of the 1950s, the Navy canceled the development contract, and top management of Westinghouse decided to discontinue work on turbojet engines.

The Lockheed Corporation began to work on a very advanced turbojet conceived by an outstanding engineer, Nathan C. Price. This engine was so far ahead of its time that it would have needed a far longer development time than that provided by the contract. The development contract was canceled in 1941.

Pratt & Whitney had started to work on its own jet propulsion ideas in the early 1940s but could not pursue these concepts because of the too-stringent obligations during wartime for the development and production of advanced aircraft piston engines. After World War II, Pratt & Whitney decided to go completely into turbojet development using axial-flow turbo machinery. The company began with the construction of a gigantic test and research facility. The government gave Pratt & Whitney a contract to build a large number of 5000-lb–thrust Rolls-Royce Nene engines with a radial compressor of the basic Whittle design. Subsequently, Pratt & Whitney developed its own large axial-flow, dual-rotor turbojet and later a fan-jet with a small bypass ratio for the advanced B52.

Turbojet Development of Hans von Ohain in Germany

My interest in aircraft propulsion began in the fall of 1933 while I was a student at the Georgia Augusta University of Gottingen in physics under Prof. R. Pohl with a minor in applied mechanics under Prof. Ludwig Prandtl. I was 21 years old and beginning my PhD thesis in physics, which was not related to jet propulsion.

The strong vibrations and noise of the propeller piston engine triggered my interest in aircraft propulsion. I felt the natural smoothness and elegance of flying was greatly spoiled by the reciprocating engine with propellers. It appeared to me that a steady, thermodynamic flow process was needed. Such a process would not produce vibrations. Also, an engine based on such a process could probably be lighter and more powerful than a reciprocating engine with a propeller because the steady flow conditions would allow a much greater mass flow of working medium per cross section. These characteristics appeared to me to be most important for achieving higher flight speeds. I made performance estimates for several steady flow engine types and finally chose a special gas turbine configuration that appeared to me as a lightweight, simple propulsion system with low development risks. The rotor consisted of a straight-vane radial outflow compressor back-to-back with a straight vane radial inflow turbine. Both compressor and turbine rotors had nearly equal outer diameters, which corresponds to a good match between them.

In early 1935, I worked out a patent for the various features of a gas turbine consisting of a radial outflow compressor rotor, a combustor, a radial inflow turbine, and a central exhaust thrust nozzle. With the help of my patent attorney, Dr. E. Wiegand, a thorough patent search was made. A number of interesting aeropropulsion systems without a propeller were found, but we did not come across the earlier patents of Lorin, Guillaume, and Frank Whittle. (I learned for the first time about one of Frank Whittle's patents in early 1937 when the German Patent Office held one of his patents, and one patent of the Swedish corporation Milo, against some of my patent claims.)

My main problem was finding support for my turbojet ideas. A good approach, it seemed to me, was to first build a model. This model should be able to demonstrate the aerodynamic functions at very low-performance runs. The tip speed of this model was a little over 500 ft/s. Of course, I never considered high-power demonstration runs for two reasons. The cost for building such an apparatus could easily be a factor of 10 or 20 times greater than that for building a low-speed model. Also, a test facility would be required for high-performance test or demonstration runs. I knew a head machinist in an automobile repair shop, Max Hahn, to whom I showed the sketches of my model. He made many changes to simplify the construction, which greatly reduced the cost. The model was built at my expense by Hahn in 1935 (Fig. 9).

In mid-1935, I had completed my doctoral thesis and oral examination and I received my diploma in November 1935. I continued working in Prof. Pohl's institute and discussed with him my project "aircraft propulsion." He was interested in my theoretical write-up. Although my project did not fit Pohl's institute, he was extremely helpful to me. He let me test the model engine in the backyard of his institute and gave me instrumentation and an electric starting motor. Because the combustors did not work, the model did not run without power from the starting motor. Long, yellow flames leaked out of the turbine. It looked more like a flame thrower than an aircraft gas turbine.

I asked Prof. Pohl to write me a letter of introduction to Ernst Heinkel, the famous pioneer of high-speed aircraft and sole owner of his company. Professor Pohl actually wrote a very nice letter of recommendation. I had chosen Heinkel because he had the reputation of being an unconventional thinker obsessed with the idea of high-speed aircraft. Intuitively, I also felt that an aircraft engine company would not accept my turbine project. I learned later that my intuition was absolutely right. Today, I am convinced no one except Heinkel (see Fig. 10) would have supported my jet ideas at that time. Heinkel invited me to his home on the evening of March 17, 1936, to explain the jet principle to him. He, in turn, gave me a view of his plan. He wanted the jet development to be apart from the airplane factory. For this purpose, he intended to construct a small, temporary building near the Warnow River. I was very enthusiastic about this idea since it gave me a feeling of freedom and independence from the other part of the company and an assurance of Heinkel's confidence in me. Also, he strongly emphasized that he himself wanted to finance the entire jet development without involvement of the German Air Ministry. Finally, he explained to me that he had arranged a meeting between me and his top engineers for the next morning.

On March 18, 1936, I met with a group of 8 to 10 Heinkel engineers and explained my jet propulsion thoughts. Although they saw many problems, specifically with the combustion, they were not completely negative. Heinkel called me to a conference at the end of March. He pointed out

Fig. 9 Max Hahn with model engine. (Courtesy of National Air and Space Museum.)

Fig. 10   Ernst Heinkel (left) and Hans von Ohain (right).
(Courtesy of National Air and Space Museum.)

that several uncertainties, specifically the combustion problems, should be solved before the gas turbine development could be started. He wanted me to work on this problem and to report to him all the difficulties I might encounter. He offered me a kind of consulting contract that stated that the preliminary work (combustor development) could probably be completed in about two months. If successful, the turbojet development would then be started, and I would receive a regular employment contract. I signed this contract on April 3, 1936, and would start working in the Heinkel Company on April 15.

The first experiments with the model in early 1936 had convinced me that the volume of the combustion chambers was far too small for achieving a stable combustion. This was later substantiated in a discussion with combustion engineers at an industrial exhibit. I found a simple way to correct this condition. Cycle analysis of my model clearly showed that for high turbine inlet temperatures, such as 700°C and higher, a centrifugal compressor with a radial inflow turbine was most suitable as a basis for combustor development. My greatest problem was how to develop a functioning combustor in a few months. In my judgment, such a development would need at least six months, more likely one year, while Heinkel's estimate was two months. I had grave doubts whether Heinkel would endure such a long development time without seeing any visible progress, such as an experimental jet engine in operation. However, to avoid any combustor difficulties, I was considering a hydrogen combustor system with a nearly uniform turbine inlet temperature distribution. This hydrogen combustor system should be designed so that it could be built without any risk or need for preliminary testing.

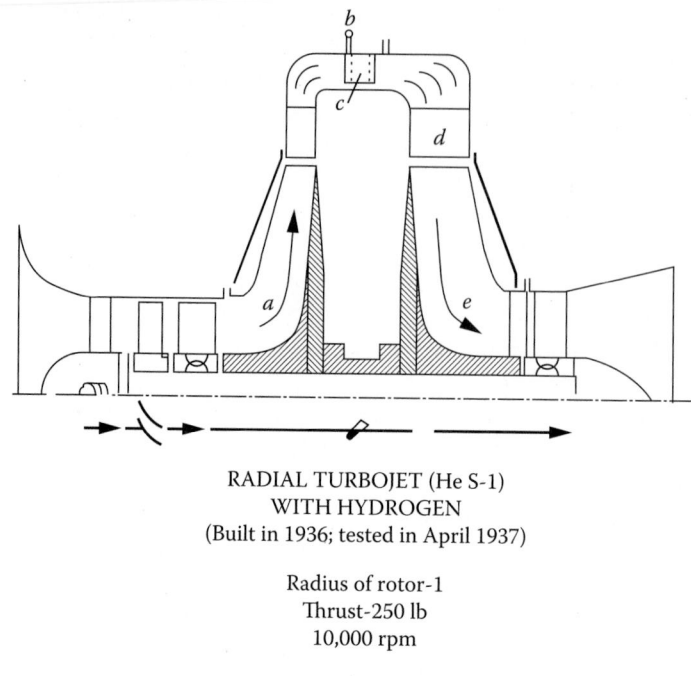

RADIAL TURBOJET (He S-1)
WITH HYDROGEN
(Built in 1936; tested in April 1937)

Radius of rotor-1
Thrust-250 lb
10,000 rpm

Fig. 11 Von Ohain's hydrogen demonstrator engine.

My idea was to separate the compressor and turbine on the rotor by a shaft and to employ an annular connecting duct from the exit of the compressor diffuser to the inlet of the turbine. Within this annular duct, I wanted to place a row of hollow vanes (about 60). These hollow vanes would have blunt trailing edges with many small holes through which hydrogen gas jets would be discharged into the air wakes behind the blunt trailing edges. In this way, the hydrogen combustion would be anchored at the blunt trailing edges of the hollow vanes. I was absolutely certain this combustor system would successfully function without any development or preliminary testing. I was also certain that no pretesting or development was necessary for the simple radial-flow turbomachinery. Testing of the hydrogen demonstrator engine (Fig. 11) showed that my judgment was correct on both points.

By mid-May 1936 I had nearly completed the layout of the hydrogen demonstrator engine. To build this engine was, for me, most important, not only for quick achievement of an impressive demonstration of the jet principle, but also for very significant technical reasons:

1. One reason was to obtain a solid basis for the design of the flight engine and the development of the liquid-fuel combustor, which should be started as a parallel development as soon as possible.

2. To achieve this solid basis, the hydrogen engine was the surest and quickest way when one does not have compressor and turbine test stands.
3. The anticipated step-by-step development approach: First testing the compressor-turbine unit with the "no-risk" hydrogen combustor and then using the tested turbomachine for exploring its interaction with the liquid-fuel combustor system seemed to be good protection against time-consuming setbacks.

Now came the greatest difficulty for me: How could I convince Heinkel that first building a turbojet with hydrogen gas as fuel would be a far better approach than trying to develop a liquid-fuel combustor under an enormous time pressure? According to my contract, of course, I should have worked on the liquid-fuel combustor with the (impossible) goal of having this development completed by June 1936. I briefly explained to Heinkel my reasons for the hydrogen engine and emphasized this engine would be a full success in a short time. I was well prepared to prove my point in case Heinkel wanted me to discuss this matter in a conference with his engineers. Surprisingly, Heinkel asked only when the hydrogen demonstrator could run. My shortest time estimate was half a year. Heinkel was not satisfied and wanted a shorter time. I told him that I had just heard that Wilhelm Gundermann and Max Hahn would work with me, and I would like to discuss the engine and its time schedule with them. So, Heinkel had agreed with my reasons to build the hydrogen jet demonstrator first.

About a week after my discussion with Heinkel, I joined Gundermann and Hahn in their large office. I showed them the layout of the hydrogen engine. Gundermann told me he had attended my presentation to the group of Heinkel's leading engineers in March 1936. He was surprised that I departed from the liquid-fueled turbojet program. I explained my reasons and also told about Heinkel's strong desire to have the hydrogen engine built in less than half a year. After studying my layout, both men came to the conclusion that it would not be possible to build this engine in less than six months, perhaps even longer. Gundermann, Hahn, and I began to work as an excellent team.

The engine was completed at the end of February 1937, and the start of our demonstration program was in the first half of March, according to Gundermann's and my recollections. The first run is clearly engraved in my memory: Hahn had just attached the last connections between engine and test stand; it was after midnight, and we asked ourselves if we should make a short run. We decided to do it! The engine had a 2-hp electric starting motor. Hahn wanted to throw off the belt-connecting starter motor and hydrogen engine if self-supporting operation was indicated. Gundermann observed the exhaust side to detect possible hot spots—none were visible. I was in the test room. The motor brought the engine to somewhat above 2000 rpm. The ignition was on, and I opened the hydrogen valve carefully.

The ignition of the engine sounded very similar to the ignition of a home gas heating system. I gave more gas, Hahn waved, the belt was off, and the engine now ran self-supporting and accelerated very well. The reason for the good acceleration probably was twofold: the relatively low moment of inertia of the rotor and the enormously wide operational range of the hydrogen combustion system. We all experienced a great joy that is difficult to describe. Hahn called Heinkel, and he came to our test stand about 20 minutes later, shortly before 1:00 a.m. We made a second demonstration run. Heinkel was enthused—he congratulated us and emphasized that we should now begin to build the liquid-fuel engine for flying.

The next day and until the end of March, Heinkel began to show further demonstration runs to some of his leading engineers and important friends. The next day following our "night show," Heinkel visited us with Walter and Siegfried Guenther (his two top aerodynamic designers) for a demonstration run. They were very impressed and asked me about the equivalent horsepower per square meter. I replied, "A little less than 1000," but hastened to add that the flight engine would have more than 2500 hp/m^2 because of the much greater tip speed and greater relative flow cross sections. During April, we conducted a systematic testing program.

After the first run of the hydrogen engine, Heinkel ordered his patent office to apply for patents of the hydrogen engine. Because of earlier patents, the only patentable item was my hydrogen combustion system.

I became employed as division chief, reporting directly to Heinkel, and received an independent royalty contract, as I had desired. An enormous amount of pressure was now exerted by Heinkel to build the flight engine.

During the last months of 1937, Walter and Siegfried Guenther began with predesign studies of the first jet-propelled aircraft (He-178) and specified a static thrust of 1100 lb for the flight engine (He.S3). The aircraft was essentially an experimental aircraft with some provisions for armament.

In late 1937, while I was working on different layouts of the flight engine, Max Hahn showed me his idea of arranging the combustor in the large unused space in front of the radial-flow compressor. He pointed out that this would greatly reduce the rotor length and total weight. Hahn's suggestion was incorporated into the layout of the flight engine (see Fig. 12). In early 1938, we had a well-functioning annular combustor for gasoline. The design of the flight engine was frozen in the summer of 1938 to complete construction and testing by early 1939.

In spring 1939, aircraft and engine were completed, but the engine performance was too low: about 800-lb thrust, while a thrust of 1000–1100 lb was desirable to start the aircraft from Heinkel's relatively short company airfield. We made several improvements, mostly optimizing the easily exchangeable radial cascades of the compressor-diffuser and turbine stator. In early August, we had reached 1000 lb of thrust. We made only several

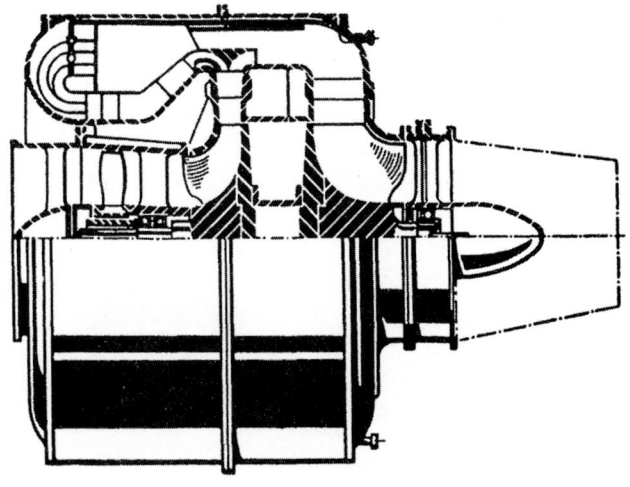

Fig. 12 1937 design of the He.S3 turbojet engine.

one-hour test runs with the flight engine. However, upon suggestion of the Air Ministry, we completed a continuous 10-hour test run with a rotor that was not used for flight tests.

On August 27, 1939, the first flight of the He-178 with jet engine He.S3B was made with Erich Warsitz as pilot (Fig. 13). This was the first flight of a turbojet aircraft in the world. It demonstrated not only the feasibility of jet propulsion, but also several characteristics that had been doubted by many opponents of turbojet propulsion:

1. The flying engine had a very favorable ratio of net power output to engine weight—about 2 to 3 times better than the best propeller/piston engines of equal thrust power.
2. The combustion chambers could be made small enough to fit in the engine envelope and could have a wide operational range from start to high altitude and from low to high flight speed.

The advantages of developing a flight demonstration turbojet in Heinkel's aircraft company were unique. Among the advantages were complete technical freedom, lack of importance attached to financial aspects, no government requirements, and no time delays; the aircraft was, so to speak, waiting for the engine. These great advantages were true only for the initial phases of jet engine development up to the first flight demonstrator. For making a production engine, however, enormous disadvantages included complete lack of experts in fabrication (turbomachinery, etc.), materials, research (turbines), accessory drives, control systems, no machine tools or component test stands, etc. Heinkel was very aware of this situation. His plan was to

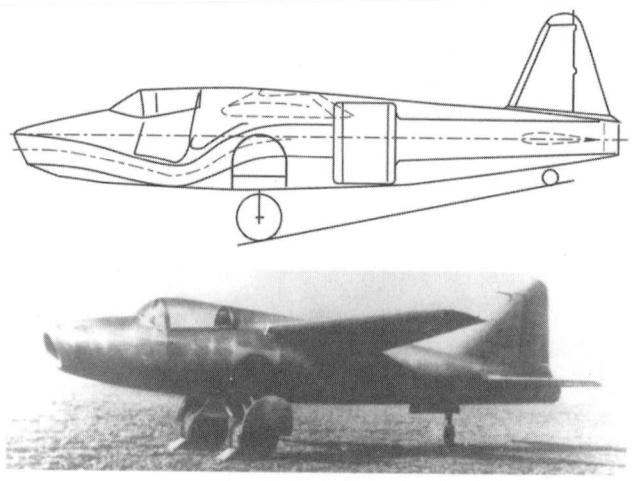

Fig. 13   The world's first jet-powered aircraft, the Heinkel He-178, was powered by the von Ohain–designed He.S3B turbojet engine. (Courtesy of National Air and Space Museum.)

hire engineers from the aircraft engine field and to purchase an aircraft engine company.

Other Early Turbojet Developments in Germany

The following events developed at the same time, which was of great importance for the early phases of the turbojet evolution:

1. Professor Herbert Wagner privately started an aircraft gas turbine development project.
2. The Air Ministry became aware of Heinkel's turbojet project in 1938 and exerted a strong influence on the engine industry to start turbo development projects.
3. Heinkel purchased an aircraft engine corporation and received a contract for development and production of a high-performance turbojet engine.

In 1934, Wagner conceived the idea of an axial-flow propeller gas turbine while he was a professor of aeronautics in Berlin and formed a corporation to pursue these ideas. (I heard about Wagner's project for the first time in spring of 1939.) By introducing a design parameter that was the ratio of propeller power input to total net power output, he had conceived a gas turbine engine that was a cross between a turbojet and a propeller gas turbine.

Wagner first explored what would happen if the propeller power input was 50% of the total power output. This condition was favorable for long-range transport. Then in 1936, he investigated the "limiting case" of zero propeller power input, which constituted a turbojet. This engine was of great interest for high-speed aircraft because of its light weight.

The unique feature of Wagner's design was the utilization of 50% reaction turbomachinery (or *symmetric blading*). A compressor with 50% reaction blading has the greatest pressure ratio and efficiency for a given blade approach Mach number, but the design is difficult because of the inherently strong three-dimensional flow phenomena. This problem was solved by one of Wagner's coworkers, Rudolf Friedrich.

At the time Wagner was working on the turbojet engine, in about 1936, he became technical director of the Junkers Airframe Corporation in Dessau. The jet engine work was conducted in the Junkers machine factory, which was located in Magdeburg. The head of his turbojet development was Max A. Mueller, his former "first assistant."

In late fall 1938, Wagner had decided to leave the Junkers Corporation, but he wanted to obtain funds from the Air Ministry for the continuation of his turbojet development work. The Air Ministry agreed to Wagner's request under the condition that the jet development be continued at the Junkers Aircraft Engine Company in Dessau. This seemed to be acceptable to Herbert Wagner. However, his team of about 12 very outstanding scientists and engineers (among them the team leader, Max A. Mueller, and the highly regarded Dr. R. Friedrich) refused to join the Junkers Aircraft Engine Company under the proposed working conditions. Heinkel made them very attractive work offers that convinced Wagner's former team to join the Heinkel Company. Heinkel added Wagner's axial turbojet to his development efforts (designated as the He.S30). Thus, in early 1939, Heinkel had achieved one goal—to attract excellent engineers for his turbojet development.

In early 1938, the Air Ministry had become aware of Heinkel's private jet propulsion development. The Engine Development Division of the Air Ministry had a small section for special propulsion systems that did not use propellers and piston engines, but rather used special rockets for short-time performance boost or takeoff assistance. Head of this section was Hans Mauch. He asked Heinkel to see his turbojet development in early summer 1938, more than one year before the first flight of the He-178. After he saw Heinkel's hydrogen turbojet demonstrator in operation and the plans for the flight engine, he was very impressed. He thanked Heinkel for the demonstration and pointed out that turbojet propulsion was, for him, a completely *unknown* and new concept. He soon became convinced that the turbojet was the key to high-speed flight. He came, however, to the conclusion that Heinkel, as an airframe company, would never be capable of developing a production engine because the company lacked engine test and manufacturing facilities and, most of all, it lacked engineers experienced in engine development and testing techniques. He wanted the Heinkel team to join an aircraft engine company (Daimler-Benz) and serve as a nucleus for turbojet propulsion development. Furthermore, he stated that Ernst Heinkel should receive full reimbursement and recognition for his great pioneering achievements. Heinkel refused.

In the summer of 1938, Mauch met with Helmut Schelp, who was in charge of jet propulsion in the Research Division of the Air Ministry. Mauch invited Schelp to join him in the Engine Development Division. Schelp accepted the transfer because he saw far greater opportunities for action than in his Research Division. In contrast to Mauch, Schelp was very well aware of turbojet propulsion and was convinced about its feasibility. He was well versed in axial and radial turbomachinery and with the aerothermodynamic performance calculation methods of turbojet, ramjet, and pulse jet. Like Mauch, he was convinced of the necessity that the aircraft engine companies should work on the development of turbojet engines. However, Schelp did not see a necessity for Heinkel to discontinue his jet engine development. He saw in Heinkel's progress a most helpful contribution for convincing the engine industry to also engage in the development of turbojets, and for proving to the higher echelons of the Air Ministry the necessity of launching a turbojet development program throughout the aircraft engine companies.

Schelp worked out the plans and programs for jet propulsion systems, decided on their most suitable missions, and selected associated aircraft types. Schelp's goal was to establish a complete jet propulsion program for the German aircraft engine industry. He also talked with Hans Antz of the Airframe Development Division of the Air Ministry to launch a turbojet fighter aircraft development as soon as possible. This became the Me-262. To implement the program, Mauch and Schelp decided to visit aircraft engine manufacturers—Junkers Motoren (Jumo), Daimler-Benz, BMW Flugmotorenbau, and Bradenburgische Motorenweke (Bramo). Mauch and Schelp offered each company a research contract to determine the best type of jet engine and its most suitable mission. After each study was completed and evaluated, a major engine development contract might be awarded.

The industry's response to these proposals has been summed up by Schlaifer in *Development of Aircraft Engines, and Fuels:* "The reaction of the engine companies to Mauch's proposals was far from enthusiastic, but it was not completely hostile" [4].

Anselm Franz and Hermann Oestrich were clearly in favor of developing a gas turbine engine. Otto Mader, head of engine development at Jumo, made two counter arguments against taking on turbojet propulsion developments. He said, first, that the highest priority of Jumo was to upgrade the performance of its current and future piston engines, and that this effort was already underpowered; and, second, Jumo did not have workers with the necessary expertise in turbomachine engine development! After several meetings between Mader and Schelp, however, Mader accepted the jet engine development contract and put Franz in charge of the turbojet project. At that time, Dr. Anselm Franz was head of the supercharger group. Daimler-Benz completely rejected any work on gas turbine engines at that time. Meanwhile, BMW and Bramo began a merger, and after it was finalized, Hermann Oestrich became the head of the gas turbine project for BMW.

These developments show that the aircraft engine industries in Germany did not begin to develop jet engines on their own initiative, but rather on the initiative and leadership of Mauch, and specifically of Helmut Schelp of the technical section of the German Air Ministry. Without their actions, the engine companies in Germany would not have begun development work on turbojet propulsion. The net result of Schelp's planning efforts was that two important turbojet engine developments were undertaken by the German aircraft engine industry, the Junkers Engine Division and BMW.

The Jumo 004 (shown in Fig. 14), developed under the leadership of Anselm Franz, was perhaps one of the truly unique achievements in the history of early jet propulsion development leading to mass production, for the following reasons:

1. It employed axial-flow turbomachinery and straight throughflow combustors.
2. It overcame the nonavailability of nickel by air-cooled hollow turbine blades made out of sheet metal.
3. The manufacturing cost of the engine amounted to about one-fifth that of a propeller/piston engine having the equivalent power output.
4. The total time from the start of development to the beginning of large-scale production was a little over four years (see Table 1).
5. It incorporated a variable-area nozzle governed by the control system of the engine, and model 004E incorporated afterburning.

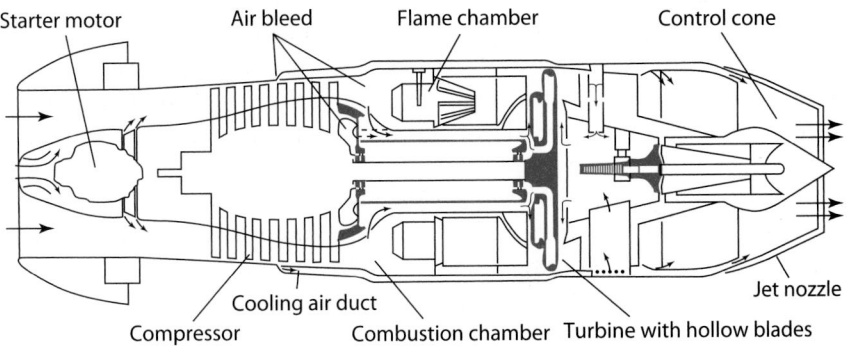

Fig. 14   Drawing of Jumo 004B turbojet engine showing air cooling system [thrust = 2000 lb, airflow = 46.6 lb/s, pressure ratio = 3.14, turbine inlet temperature = 1427°F, fuel consumption = 1.4 (lb/h)/h, engine weight = 1650 lb, diameter = 30 in., length = 152 in., efficiencies: 78% compressor, 95% combustor, 79.5% turbine].

Table 1 Jumo 004 Development and Production Schedule

Start of development	Fall 1939
First test run	11 Oct. 1940
First flight in Me-262	18 July 1942
Preproduction	1943
Beginning of production	Early 1944
Introduction of hollow blades	Late 1944
About 6000 engines delivered	May 1945

The preceding points reflect the design philosophy of Dr. A. Franz for the Jumo 004, which was lowest possible development risk, shortest development time, dealing with a complete lack of heat-resistant materials, and minimizing manufacturing cost. From this design philosophy, it is understandable that the Jumo 004 engine, while fully meeting the requirements, did not have the highest overall performance compared to some contemporary experimental axial-flow engines, such as the He.S30 and others. If it had been possible for the Jumo 004 to employ heat-resistant materials, then the engine thrust, the thrust/weight ratio, and the efficiency would have been increased substantially. Also the engine life could have been drastically increased from about 25 h to well over 100 h. However, because the combat life of a German fighter was well below 25 h, the economical optimum could tolerate a short engine life and the avoidance of nickel. Furthermore, to avoid any development risk or time delay, the compressor type chosen for the Jumo 004 was one where essentially all the static pressure increase occurs in the rotor and none in the stator (a free-vortex type of compressor having constant axial velocity over the blade span). Although such a compressor type does not have the best performance, at that time it was best understood. The previously described points show that the Jumo 004 represented an outstanding compromise among engine performance, the existent design constraints due to materials shortage, the need for short development time, and earliest possible production.

The BMW 003 turbojet engine, which was developed under the leadership of Hermann Oestrich, was also a resounding success. Because its thrust was smaller than that of the Jumo 004, it was ideally suited for the He-162. After World War II, Oestrich and a group of prominent scientists and engineers from Germany went to France and helped lay the foundation for France's turbojet industry.

Now I would like to go back to the end of 1939, when Heinkel began to make plans for buying an engine company. After the first flight of the He-178 on 27 August 1939, Heinkel invited high officials of the Air Ministry to see a flight demonstration of the He-178. This demonstration took place on 1

November 1939. At that occasion, Heinkel offered the development of a jet fighter, the He-280, which had two outboard engines under the wing. Heinkel received a contract for this aircraft in early 1940. In addition, I believe Udet and Heinkel had made an agreement that Heinkel would get official permission to buy the Hirth Engine Company if the first flight of the He-280 could be demonstrated by April 1941.

The He-280 had severe restrictions with respect to distance of the engine nacelle from the ground. It actually was designed for the axial engine He.S30 (the Wagner turbojet engine). It appeared, however, unlikely that the He.S30 would be ready in time. On the other hand, it was impossible to use an engine of the He.S3B type, which had powered the He-178, because the diameter of this engine type would have been far too large. Under these conditions, I could see only one possible solution for succeeding in time, and this solution had extreme high risk. I employed a radial rotor similar to that of the He.S3B and combined it with an axial (adjustable) vane diffuser and a straight throughflow annular combustor.

The company designation of this engine was He.S8A. We had only about 14 months for this development, but we were lucky—it worked surprisingly well, and Heinkel could demonstrate the first flight of the He-280 on 2 April 1941. The government pilot was Engineer Bade. Earlier flights in late March were done by Heinkel's test pilot, Fritz Schaefer.

A few days after this demonstration, Heinkel obtained permission to buy the Hirth Motoren Company in Stuttgart, which was known for its excellent small aircraft engines. This company had outstanding engineers, scientists, machinists, precision machine tools, and test stands. Heinkel relocated the development of the He.S30 to his new Heinkel-Hirth engine company to make use of the excellent test and manufacturing facilities. In the summer of 1942, the He.S30 was ready for testing. It performed outstandingly well. The continuous thrust was about 1650 lb. From a technical standpoint, this engine had by far the best ratio of thrust to weight in comparison to all other contemporary engines. The superiority of the He.S30 was, in large part, the result of its advanced 50% reaction degree axial-flow compressor, designed by Dr. R. Friedrich. However, the success of the He.S30 came too late. The He-280 had been thoroughly tested. While it was clearly superior to the best contemporary propeller/piston fighter aircraft, the He-280 had considerably lower flight performance than the Me-262 with respect to speed, altitude, and range. Also, the armament of the He-280 was not as strong as that of the Me-262. For these reasons, the He.S8 and the He.S30 were canceled in the fall of 1942; the He-280 was officially canceled in early 1943. The Me-262 (Fig. 15) went on to become the first operational jet fighter powered by two Jumo 004B turbojet engines. The Air Ministry did give full recognition to the excellence of the 50% reaction degree compressor type as most suitable for future turbojet developments.

Thus, in the fall of 1942, Heinkel had lost his initial leadership in jet aircraft and turbojet engines. Ironically, this happened when he had just reached

Fig. 15   Messerschmitt Me-262 jet fighter.
(Courtesy of National Air and Space Museum.)

his goal of owning an aircraft engine company with an outstanding team of scientists and engineers. It was the combined team of the original Hirth team, the Heinkel team, and Wagner's team. These conditions made Heinkel fully competitive with the existent aircraft engine industry. The Air Ministry had recognized the excellence of Heinkel's new team and facilities. Helmut Schelp, who in the meantime had become the successor of Hans Mauch, was in favor of the Heinkel-Hirth Company's receiving a new turbojet engine development contract. He clearly foresaw the need for a strong engine for the advanced Me-262, the Arado-234, the Junkers-287, and others. This new engine was supposed to have a thrust of nearly 3000 lb with a growth potential to 4000 lb as well as a high pressure ratio for improving the fuel economy. We began working on the He.S011 engine in the fall of 1942. The He.S011 was an axial-flow design with 50% reaction degree compressor blading, similar to that designed by Dr. Friedrich, and a two-stage, axial-flow, air-cooled turbine. Note that Helmut Schelp not only had established the performance specifications, but also had contributed to the overall design with excellent technical input and suggestions pertaining to the advanced diagonal inducer stage, the two-stage air-cooled turbine, and the variable exhaust nozzle system.

I was in charge of the He.S011 development, while the local director of the Heinkel-Hirth plant, Curt Schif, was in charge of He.S011 production. The top engineer of the Hirth Corporation was Dr. Max Bentele. He was well known for his outstanding knowledge in dealing with blade vibration problems. Upon special request of the Air Ministry, he had solved the serious turbine blade vibration problems of the Jumo 004 in the summer of 1943.

Dr. Bentele was responsible for the component development of the He.S011. After considerable initial difficulties, he achieved excellent performance characteristics of compressor and turbine that made it possible, by the end of 1944, for the performance requirements of the 011 to be met or surpassed. He also contributed to the preparation for production of the 011, which was planned to start in June 1945. The end of World War II, of course, terminated the production plan before it had started. Only a few He.S011 engines are in existence today, and they are exhibited in several museums in the United States and Great Britain.

In other countries, such as Russia and Japan, interesting developments in the field of propeller gas turbines and turbojets had also been undertaken. It is, however, not sufficiently known to what extent these developments were interrelated with, or influenced by, the previously described jet developments and what their development schedules had been.

In summary, at the end of the 1930s and in the first half of the 1940s, turbojet propulsion had come into existence in Europe and the United States. It had been demonstrated that the turbojet is, for high-speed flight, uniquely superior to the propeller/piston engines because of the following two major characteristics:

1. The ratio of power output to weight of the early turbojets was at least 2 or 3 times greater than that of the best propeller/piston engines. This is one necessary condition of propulsion systems for high-speed flight and good maneuverability.
2. Because, in the turbojet, the propeller is replaced by ducted turbomachinery, the turbojet is inherently capable as a propulsion system of high subsonic and supersonic flight speeds.

In the following, it will be discussed how the turbojet engine progressed to the performance capabilities of today.

Evolution of Airbreathing Turbopropulsion Systems to the Technology Level of Today

The early turbojets were used as propulsion systems for high-speed fighter and reconnaissance aircraft. For these applications, the early turbojets (because of their superior power/weight ratios) were far more suitable propulsion systems than the traditional propeller/piston engines. However, the early turbojets were not suitable for those application areas where greatest fuel economy, highest reliability, and a very long endurance and service life were required. For making airbreathing turbojet propulsion systems applicable for all types of aircraft, ranging from helicopters to high-speed, long-range transports, the following development goals were, and still are, being pursued:

1. Higher overall efficiency (i.e., the product of thermodynamic and propulsion efficiencies)
2. Larger-power-output engines
3. Larger ratios of power output to engine weight, volume, and frontal area
4. Greater service life, endurance, and reliability
5. Strong reduction of adverse environmental exhaust gases
6. Reduced noise

To achieve these goals, parallel research and development efforts were undertaken in areas such as

1. Fundamental research in combustion processes, development, and technology efforts for increasing specific mass flow through combustors and reducing the total pressure drop, and for achieving nearly 100% combustion efficiency with a more uniform temperature profile at combustor exit
2. Minimizing the excitation of vibrations (including aeroelastic effects) and associated fatigue phenomena
3. Continuous improvement of the structural design and structural materials, such as composite materials, heat- and oxidation-resistant alloys, and ceramics
4. Increasing the turbine temperature capability by improving air cooling effectiveness; also increasing the polytropic turbine efficiency
5. Improvement of the compressor with respect to greater specific mass flow, greater stage pressure ratio, greater overall pressure ratio, and greater polytropic efficiency
6. Advanced controllable-thrust nozzles and their interactions with the aircraft
7. Advanced control systems to improve operation of existing and new engines

All of these research and development areas were, and still are, of great importance for the progress in turbopropulsion systems; however, the compressor can perhaps be singled out as the key component because its advancement was a major determining factor of the rate of progress in turbo engine development. This will become apparent from the following brief description of the evolution of the turbopropulsion systems.

Development of High-Pressure-Ratio Turbojets

The first step to improve the early turbojets was to increase their overall efficiency. To do this, it was necessary to increase the thermodynamic cycle efficiency by increasing the compressor pressure ratio. The trend of compressor pressure ratio over the calendar years is shown in Table 2.

In the early 1940s, it was well understood that a high-pressure-ratio (above 6:1), single-spool, fixed-geometry compressor can operate with good efficiency only at the design point, or very close to it. The reason is that at the design point, all compressor stages are matched on the basis of the compressibility effects. Consequently, under off-design conditions where the compressibility effects are changed, the stages of a high-pressure-ratio compressor are severely mismatched, resulting in a very low off-design efficiency. For example, at an operational compressor rpm (which would be substantially below the design rpm), the compressibility

Table 2 Trend in Compressor Pressure Ratios

Calendar Years	Compressor Pressure Ratio
Late 1930 to mid-1940	3:1 to about 5:1
Second half of 1940s	5:1 and 6:1
Early 1950s	About 10:1
Middle to late 1960s	20:1 to about 25:1
End of century (2000)	30:1 to about 40:1

effects are small. Under such off-design conditions, the front stages tend to operate under *stalled* conditions, while the last stages tend to work under *turbining* conditions. Such compressor characteristics are unacceptable for the following reasons:

1. There are enormous difficulties in starting such an engine.
2. Very poor overall efficiency (or poor fuel economy) exists under part-load operation.
3. Very low overall efficiency exists at high supersonic flight speed (because the corrected rpm is very low as a result of the high stagnation temperature of the air at the compressor inlet).

By the end of the 1940s and early 1950s, excellent approaches emerged for the elimination of these shortcomings of simple high-pressure-ratio compressors. Pratt & Whitney, under the leadership of Perry Pratt, designed a high-pressure-ratio jet engine (the J57) with a *dual-rotor* configuration. (In later years, triple-rotor configurations were also employed.) The ratio of rpm values of the low- and high-pressure spools varies with the overall pressure ratio and, in this way, alleviates the mismatching effects caused by the changes in compressibility. At approximately the same time, Gerhard Neumann of General Electric conceived of a high-pressure-ratio, single-spool compressor having automatically controlled *variable stator blades.* This compressor configuration was also capable of producing very high pressure ratios because it alleviated the mismatching phenomena under off-design operation by stator blade adjustment. In addition, the controlled variable-stator-compressor offers the possibility of a quick reaction against compressor stall. The variable-stator concept became the basis for a new, highly successful turbojet engine, the J79, which was selected as the powerplant for many important supersonic Air Force and Navy aircraft. A third possibility to minimize mismatching phenomena was *variable front-stage bleeding,* which was employed in several engines.

Before I continue with the evolution of the turbo engines toward higher overall efficiencies, a brief discussion about the research and development efforts on compressor bladings and individual stages is in order. Such efforts existed to a small degree even before the advent of the turbojet (in

Switzerland, Germany, England, and other countries). However, in the mid-1940s, those research efforts began to greatly intensify after the turbojet appeared. It was recognized that basic research and technology efforts were needed to continuously increase 1) stage pressure ratio and efficiency and 2) mass throughflow capability.

Very significant contributions had been made by universities, research institutes, and government laboratories (NACA, later NASA; Aero Propulsion Laboratory; and others), and industry laboratories. Many outstanding research results were obtained from universities in areas of rotating stall, three-dimensional and nonsteady flow phenomena and transonic flow effects, novel flow visualization techniques for diagnostics (which identified flow regions having improvement possibilities), understanding of noise origination, and many others.

Two examples where government laboratories had achieved very crucial advancements that were later adapted by industry were the following: In the early 1950s, the NACA Lewis Research Center, under the leadership of Abe Silverstein, advanced compressor aerodynamics to transonic and supersonic flow, critical contributions for increasing the compressor-stage pressure ratio. He initiated a large transonic and supersonic compressor research program. Many in-depth studies furnished information for utilizing this new compressor concept. Later, the Air Force Propulsion Laboratory, under the leadership of Arthur Wennerstrom, conducted advanced compressor research and introduced a very important supersonic compressor concept, which was particularly applicable for front stages because it solved the most difficult combined requirements of high mass flow ratio, high pressure ratio, high efficiency, and broad characteristics.

The continuous flow of research and technological contributions extended over the last four decades and is still going on. The total cost may have been several hundred million dollars. Some of the results can be summarized as follows: The polytropic compressor efficiency, which was slightly below 80% in 1943, is now about 92%; the average stage pressure ratio, which was about 1.15:1 in 1943, is now about 1.4:1 and greater; the corrected mass flow rate per unit area capability grew over the same time period by more than 50%.

The improvement in compressor efficiency had an enormous impact on engine performance, specifically on the overall engine efficiency. The substantially increased stage pressure ratio and the increased corrected mass flow rate per unit area capability resulted in a substantial reduction of engine length, frontal area, and weight-per-power output.

Let us now return to the 1950s. As previously stated, the new turbojets employing a dual-rotor configuration or controlled variable-stator blades were capable of substantially higher pressure ratios than the fixed-geometry, single-spool engines of the 1940s. Engine cycle analysis showed that the high power-per-unit mass flow rate needed for the advanced engines required higher turbine inlet temperatures. This requirement led to continuous major research efforts to achieve high-efficiency combustion with low pollution, to increase the effectiveness of turbine blade-cooling methods, and to improve the temperature capabilities of materials. Thus the turbojets of the 1950s had made substantial progress in thermodynamic efficiency and propulsive thrust per pound of inlet air per second. The latter characteristic means that the velocity of the exhaust gas jet had increased.

Development of High-Bypass-Ratio Turbofans

For supersonic flight speeds, the overall efficiency of these turbojets was outstanding. However, for high subsonic and transonic flight speeds (around 500–600 mph), the velocity of the exhaust gas jet was too high to obtain a good propulsive efficiency. Under these conditions, the *bypass engine* (also called *turbofan* or *fan-jet*) became a very attractive approach for improving the propulsive efficiency. The first fan-jets had a relatively small bypass ratio of about 2:1. (*Bypass ratio* is the ratio of mass flow bypassing the turbine to mass flow passing through the turbine.) In the early 1960s, the U.S. Air Force had established requirements for military transports capable of an extremely long range at high subsonic cruise speeds. Such requirements could be met only by employing propulsion systems of the highest possible thermodynamic and propulsive efficiencies, which led to the following engine characteristics:

1. Very high compressor pressure ratios between 20:1 and 30:1, which could be achieved by combining the concepts of variable stator and dual rotor
2. Very high turbine inlet temperatures
3. Very high bypass ratios, around 8:1

The first engine of this type was the TF39 (Fig. 16), a military transport engine developed by General Electric under the leadership of Gerhard Neumann. Four of the TF39 turbofan engines powered the Lockheed C5A.

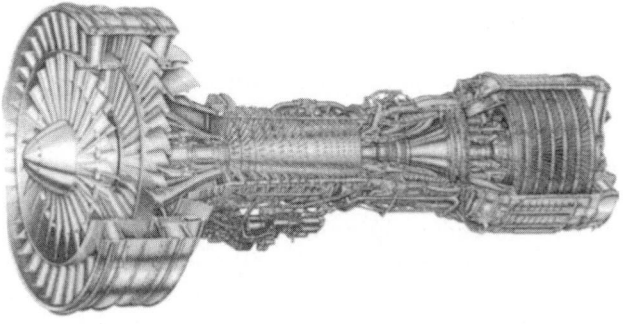

Fig. 16   The TF39 high-bypass-ratio turbofan engine used on the Lockheed C5 transport. (Courtesy of General Electric Aircraft Engines.)

The Air Force Propulsion Laboratory, under the leadership of Cliff Simpson, had played a key role in the establishment of these requirements. Simpson also succeeded in convincing the highest Air Force and Department of Defense echelons of the significance of this new type of airbreathing propulsion system, which he considered a technological breakthrough. (At that time, it was difficult to generate interest in advanced airbreathing propulsion system concepts, since the nation's attention was focused on rocket propulsion for space exploration.)

The advantages of the high-bypass-ratio turbofan engines can be summarized as follows:

1. High overall efficiency, resulting in a long flight range
2. Strong increase in propulsive thrust at low flight speeds, which is important for takeoff, climbing, and efficient part-load operation
3. Lower jet velocity, which leads to great noise reduction
4. Low fuel consumption, which reduces chemical emissions

These characteristics had been demonstrated in the late 1960s and early 1970s. They also became of great interest for commercial aircraft and led to the development of the widebody passenger aircraft.

During the 1980s, the continuous flow of advanced technology of turbo engine components had brought a high degree of maturity to the various aircraft gas turbine engine types, such as the following: 1) the previously discussed high-bypass-ratio engines, 2) the low-bypass-ratio engines with afterburning for supersonic flight, and 3) the pure turbojet for high supersonic flight Mach numbers of approximately 3.

The small and medium aircraft shaft-power gas turbines also benefited enormously from the continuous improvement of turbomachinery technology. These small gas turbines are used in helicopters and subsonic propeller aircraft. The high power/weight ratio and the high thermodynamic efficiency of the advanced shaft-power gas turbine engines played a key role in the advancement of helicopters.

Future Potential of Airbreathing Jet Propulsion Systems

In the future, major advancement of airbreathing jet propulsion systems can be expected from

1. Evolutionary improvements of the established large-bypass-ratio turbofan engines for transonic flight speeds and the low-bypass-ratio turbofan, or pure turbojet engines, for supersonic flight speeds
2. Improvements and new approaches to engine–airplane integration
3. New approaches to airbreathing propulsion systems for high supersonic and hypersonic flight speeds

The evolutionary improvements of established engine types will result in greater fuel economy and better performance characteristics. By the end of the century, one can expect polytropic efficiencies of turbine and compressor of nearly 95%. Furthermore, one will see considerably increased single-stage pressure ratios; significantly higher turbine inlet temperatures resulting from better heat- and oxidation-resistant materials, and more effective blade-cooling methods; and much lighter structural designs and materials (composite materials). This technological progress may result in an overall engine efficiency increase of about 20% and in a weight reduction for given horse-power output by probably a factor of 2 and higher.

For the evolution of high-bypass-ratio engines at cruise speeds between 500 and 600 mph, the following trend is important: the greater the turbine inlet temperature and the higher the polytropic efficiencies of the compressor and the turbine, the higher the optimum pressure ratio of the gas turbine engine and the bypass ratio of the fan. In the future, this trend will lead to larger bypass ratios; hence, the fan shroud will become relatively large in diameter and will contribute substantially to the weight and external drag of the propulsion system. Several solutions are conceivable to alleviate this problem, but at this time it is not possible to predict the most promising approach.

One way is to eliminate the fan shroud by using an unshrouded fan (having a multiplicity of swept-back fan blades, see Fig. 17), also called a *prop-fan.* This configuration is currently in an experimental state and may become very important in the future for improving fuel economy. Another possibility may lie in the development of a transonic airframe configuration in which the large fan shrouds have a dual function: They contribute, in part, to the stability of the aircraft (horizontal and vertical stabilizer surfaces)

Fig. 17 PW-Allison 578DX propulsion system.
(Courtesy of Allison Gas Turbine Division.)

while serving at the same time as a shroud for the fan. Finally, it may be conceivable that, in the future, a practical airframe and wing configuration can be developed that will be capable of extending a high lift/drag ratio to a flight speed regime that is very close to the speed of sound (a kind of well-known "supercritical" airframe configuration proposed by Whitcomb of NASA). If such an airframe configuration could be developed, the bypass ratio at these relatively high flight speeds would be substantially lower than that for flight speeds around 500 mph and, therefore, a shrouded fan would be most applicable.

Perhaps a most interesting question is: Can one expect a future supersonic passenger transport that is economically feasible in view of the progress of future transonic passenger transports? The general trend of the airplane lift/drag ratio is decreasing with increasing flight Mach number, while the overall efficiency of aeropropulsion systems increases with increasing flight Mach number. Currently, the lift/drag ratio of the Boeing 747 compared to that of the supersonic Concorde is about 3:1. In the future, the corresponding ratio may decrease to about 2:1, and the structural weight of the supersonic aircraft will greatly be improved. Such improvements may be the result of advancements in the aerodynamic shape, the structure, the structural airframe materials, and the best use of artificial stability. The overall efficiency of the supersonic flight engine can also be improved, including its capability of cruising subsonically with only one engine. This is important in case one of the two engines fails. Under this condition, the mission must be completed at subsonic speed without requiring an additional fuel reserve. It appears that advanced supersonic-cruise aircraft systems have the potential to achieve these conditions and, thus, become economically acceptable in the future.

New Approaches to Engine–Airplane Integration

In general, the investigations of engine–airplane integration have the goal to avoid or minimize losses due to adverse interface phenomena between the engine inlet stream or exhaust jet and the air vehicle. However, many favorable effects can be achieved by properly integrating functions of the propulsion system with functions of the airplane. Historically, Prof. Ackeret (University of Zurich, Switzerland) had first suggested, in 1921, reaccelerating the boundary-layer air near the trailing edge of the wing. He showed that this method of producing the propulsive thrust can result in substantial gains in overall efficiency of the propulsion system. In the 1970s, various investigations had shown that similar results can be obtained by using momentum exchange of the high-pressure bypass air with the boundary-layer air.

Other uses of bypass air are to energize the boundary layer at proper locations of the wing to prevent boundary-layer separation and to increase the circulation around the wing (often called *supercirculation).* This method may become important for advanced short-takeoff-and-landing (STOL) applications. For future V/STOL applications, new methods of thrust vectoring and thrust augmentation may have very attractive possibilities. Also, bypass air may be used for boundary-layer suction and ejection in advanced laminar systems.

Airbreathing Propulsion Systems for High Supersonic and Hypersonic Speeds

In airbreathing propulsion systems, the combined compression by ram and turbo compressor is of great benefit to the thermodynamic propulsion process up to flight Mach numbers approaching 3. When the flight Mach number increases further, the benefits of the turbo compressor begin to decrease and the engine begins to operate essentially as a ramjet. When the flight Mach number exceeds about 3.5, any additional compression by a turbo compressor would be a disadvantage. Thus, if the engine operates best as a pure subsonic combustion ramjet, it fits in a flight Mach number regime from about 3.5 to 5. Beyond flight Mach numbers of about 6, the pressure and temperature ratios would be unfavorably high if the engine continued to operate as a subsonic combustion ramjet. The reasons are as follows:

1. High degree of dissociation of the combustor exhaust flow, reducing the energy available for exhaust velocity
2. Pressures far too high for Brayton cycle operations or for the structure to withstand

For these reasons, the cycle will be changed from a subsonic to a supersonic combustion ramjet, and hydrogen will be used as fuel because hydrogen has the greatest: 1) combustion heat and fuel/air concentration range, 2) diffusion speed and reaction speed, and 3) heat-sink capabilities.

This supersonic combustion ramjet cycle is characterized by a reduction of the undisturbed hypersonic flight Mach number to a somewhat lower hypersonic Mach number with an increase in entropy, which should be as low as possible. The deceleration process must be chosen in such a manner that the increases in static pressure and entropy correspond to a high-performance Brayton cycle. The internal thrust generated by the exhaust gas must be larger than the external drag forces acting on the vehicle.

To minimize the parasitic drag of the ramjet vehicle systems, various external ramjet vehicle configurations have been suggested. However, theoretical and experimental investigations will be necessary to explore their associated aero-thermochemical problems.

For the experimental research, one may consider investigations using free-flight models or hypersonic wind tunnels with true temperature simulation. Historically, it may be of interest that many suggestions and investigations had been made as to how to achieve clean hypersonic airflows with true stagnation temperatures. I remember much work and many discussions with Dr. R. Mills, E. Johnson, Dr. Frank Wattendorf, and Dr. Toni Ferri about "air accelerator" concepts, which aimed to avoid flow stagnation and the generation of ultrahigh static temperatures and chemical dissociation.

Since the Wright brothers, enormous achievements have been made in both low-speed and high-speed airbreathing propulsion systems. The coming of the jet age opened up the new frontiers of transonic and supersonic flight (see Fig. 3). While substantial accomplishments have been made during the past several decades in the field of high supersonic and hypersonic flight (see Ref. 2), it appears that even greater challenges lie ahead.

For me, being a part of the growth in airbreathing propulsion over the past 60 years has been both an exciting adventure and a privilege. This foreword has given you a view of its history and future challenges. The following book presents an excellent foundation in airbreathing propulsion and can prepare you for these challenges. My wish is that you will have as much fun in propulsion as I have.

Hans von Ohain
German Inventor of the Jet Engine
14 December 1911–13 March 1998

References

[1] Von Kármán, T., *The Wind and Beyond*, Little, Brown, Boston, 1967.

[2] Heiser, W. H., and Pratt, D. J., *Hypersonic Airbreathing Propulsion*, AIAA Education Series, AIAA, Washington, DC, 1994.

[3] Boyne, W. J., and Lopez, D. S., *The Jet Age, Forty Years of Jet Aviation*, Smithsonian Inst. Press, Washington, DC, 1979.

[4] Schlaifer, R., and Heron, S. D., *Development of Aircraft Engines and Fuels*, Pergamon Press, New York, 1970; reprint of 1950 ed.

PREFACE

This edition of *Elements of Propulsion: Gas Turbines and Rockets* was prompted by much feedback from current and prospective users of the text as well as the authors' desire to provide improved treatments and descriptions of the foundational principles applied to gas turbine engine and rocket propulsion systems. This undergraduate text provides an introduction to the fundamentals of gas turbine engines and jet propulsion. These basic elements determine the behavior, design, and operation of the jet engines and chemical rocket motors used for propulsion. The text contains sufficient material for two sequential courses in propulsion: an introductory course in jet propulsion and a gas turbine engine components course. It is based on one- and two-course sequences taught at four universities over the past 35 years. The authors have also used this text for a course on turbomachinery.

The outstanding historical foreword by Hans von Ohain (the German inventor of the jet engine) has been retained because it gives a unique perspective on the first 50 years of jet propulsion. His account of past development work is highlighted by his early experiences. He concludes with predictions of future developments.

An equally outstanding foreword for this text edition is provided by Bill Heiser. Dr. Heiser uses his 60 years of exemplary experience in industry, government, and academia to focus on developments from 1995 and beyond in commercial, military, and small turbine engines as well as rocket engines and novel concepts. Highlighted in his comments is emphasis on understanding fundamental principles to which the authors remain dedicated.

The text gives examples of existing designs and typical values of design parameters. Many example problems are included in this text to help the student see the application of a concept after it is introduced. Problems are included at the end of each chapter that emphasize those particular principles. Two extensive design problems for the preliminary selection and design of a gas turbine engine cycle are included. Several turbomachinery design problems are also included.

The text is divided into five parts:

1. Introduction to aircraft and rocket propulsion (Chapter 1)
2. Review of fundamentals (Chapters 2 and 3)
3. Analysis and performance of air-breathing propulsion systems (Chapters 4–8)

4. Analysis and design of gas turbine engine components (Chapters 9 and 11)
5. Rocket propulsion (Chapter 10)

Chapter 1 introduces the types of air-breathing and rocket propulsion systems and their basic performance parameters with fresh, expanded explanations. The *system perspective* remains an important focus throughout the text. Consequently, uninstalled versus installed engine performance is included early in the chapter. Also included is an introduction to aircraft and rocket systems that reveals the influence that propulsion system performance has on the overall system. This material facilitates incorporation of a basic propulsion design problem into a course, such as new engines for an existing aircraft.

Chapters 2 and 3 provide a fairly comprehensive treatment of fundamental principles intended to enhance foundational understanding for students previously exposed to the material and provide sufficient breadth and depth for those that have not. Thermodynamic fundamentals and the development of the governing equations from the fully three-dimensional, unsteady control volume formulation are presented in Chapter 2. The chapter concludes with the perfect gas relations used throughout the text and fundamentals of thermochemistry. Chapter 3 is all about compressible flow, including isentropic flows, shock waves, expansion fans, Rayleigh flows, and Fanno flows. Application of these aero-thermodynamic principles to propulsion systems are illustrated through many new examples and end-of-chapter problems.

The analysis of gas turbine engines begins in Chapter 4 with the definitions of uninstalled thrust, installed thrust, and installation losses. Uninstalled thrust and additive drag are presented using a new "generalized thrust" development. This chapter also discusses the primary flow path components and reviews the ideal Brayton cycle (and variations), which limits gas turbine engine performance.

Two types of analysis are developed and applied to gas turbine engines in Chapters 5–8: parametric cycle analysis (thermodynamic design point) and performance analysis. New figures and expanded explanations of performance trends are included. The text uses the cycle analysis methods introduced by Frank E. Marble of the California Institute of Technology and further developed by Gordon C. Oates of the University of Washington and Jack Kerrebrock of the Massachusetts Institute of Technology. The steps of parametric cycle analysis are identified in Chapter 5 and then used to model engine cycles from the simple turbojet to the complex, mixed-flow, afterburning turbofan engine. Families of engine designs are analyzed in the *parametric analysis* of Chapters 5 and 7 for ideal engines and engines with losses, respectively. Chapter 6 develops the overall relationships for engine components with losses. The *performance analysis* of Chapter 8 models the actual behavior of an engine and shows why its performance changes with flight conditions and throttle settings. The results of the engine performance analysis can be used to establish component performance requirements. To keep the size of the new edition down, the resulting summary of equations for analysis and the detailed analysis of some engine cycles have been moved to the electronic Supporting Materials available with this book.

Chapter 9 covers both axial-flow and centrifugal-flow turbomachinery. Included are basic theory and mean-line design of axial-flow compressors and turbines, quick design tools (e.g., repeating-row, repeating-stage design of axial-flow compressors), example multistage compressor designs, flow path and blade shapes, turbomachinery stresses, and turbine cooling. Example output from the COMPR and TURBN programs is included in several example problems.

The material on inlets, exhaust nozzles, and combustion systems has been moved to the electronic Supporting Materials to keep the size of the new edition down. The special operation and performance characteristics of supersonic inlets are examined, and an example of an external compression inlet is designed. The principles of physics that control the operation and design of main burners and afterburners are also covered.

Chapter 10 presents the fundamentals of rocket propulsion. Emphasis in this chapter is on chemical propulsion (liquid and solid) with coverage of rocket motor performance. The chapter also covers requirements and capabilities of rocket propulsion.

The appendices contain tables with properties of standard atmosphere and properties of combustion products for air and $(CH_2)_n$. New appendices on compressible flow functions (including Prandtl-Meyer flow), shocks, Rayleigh line flows, and Fanno line flows have been added. There is also material on turbomachinery stresses as well as useful data on existing gas turbine engines and liquid-propellant rocket engines.

Eight computer programs are provided for use with this textbook:

1. AFPROP, properties of combustion products for air and $(CH_2)_n$
2. ATMOS, properties of the atmosphere
3. COMPR, axial-flow compressor mean-line design analysis
4. EQL, chemical equilibrium analysis for reactive mixtures of perfect gases
5. GASTAB, gas dynamic tables
6. PARA, parametric engine cycle analysis of gas turbine engines
7. PERF, engine performance analysis of gas turbine engines
8. TURBN, axial-flow turbine mean-line design analysis

The AFPROP program can be used to help solve problems of perfect gases with variable specific heats. The ATMOS program calculates the properties of the atmosphere for standard, hot, cold, and tropical days. The EQL program calculates equilibrium properties and process end states for reactive mixtures of perfect gases, for different problems involving hydrocarbon fuels.

GASTAB is equivalent to traditional compressible flow appendices for the simple flows of calorically perfect gases. It includes isentropic flow; adiabatic, constant area frictional flow (Fanno flow); frictionless, constant area heating and cooling (Rayleigh flow); normal shock waves; oblique shock waves; multiple oblique shock waves; and Prandtl-Meyer flow. The PARA and PERF programs support the material in Chapters 5–8. PARA is very useful in determining variations in engine performance with design parameters and in limiting the useful range of design values. PERF can predict the variation of an engine's performance with flight condition and throttle (Chapter 8). Both are very useful in evaluating alternative engine designs and can be used in design problems that require selection of an engine for an existing airframe and specified mission. The COMPR and TURBN programs permit preliminary design of axial-flow turbomachinery based on mean-line design.

As already suggested, the authors have found the following two courses and material coverage very useful for planning.

Introductory Course

Chapter 1, all
Chapters 2 and 3, as needed
Chapter 4, all
Chapter 5, Sections 5.1–5.8
Chapter 6, Sections 6.1–6.8
Chapter 7, Sections 7.1–7.4
Chapter 8, Sections 8.1–8.5
Chapter 10, all (if course includes an introduction to rocket propulsion)

Gas Turbine Engine Components Course

Chapters 2 and 3, as needed
Chapter 8, Sections 8.1–8.3
Chapter 9 and Appendix I, all
Chapter 11, all

Chapter 11 on inlets, exhaust nozzles, and combustion systems is provided online.

The material in Chapters 2, 3, and 9 along with Appendix I of this text have also been used to teach the major portion of an undergraduate turbomachinery course.

Jack D. Mattingly and Keith M. Boyer
April 2016

ACKNOWLEDGMENTS

The authors are deeply indebted to Professor Gordon C. Oates (deceased and author of three AIAA propulsion textbooks) and Dr. William H. Heiser (author of the new foreword for this edition). Professor Oates taught Jack the basics of engine cycle analysis during his doctoral studies at the University of Washington and later until his untimely death in 1986. While Dr. Heiser was a Distinguished Visiting Professor at the U.S. Air Force Academy from 1983 to 1985, he and Jack developed an outstanding engine design course and wrote the first edition of the textbook *Aircraft Engine Design*. Both authors have continued to work and teach with Dr. Heiser, and he remains a friend and mentor to both. Keith is deeply indebted to Professor Walter O'Brien, a friend and mentor since 1987 and major advisor during his doctoral studies at Virginia Tech. Professor O'Brien's approach of treating his students as accountable peers continues to be a positive influence in Keith's life. We are both indebted to Dr. Hans von Ohain for his friendship and guidance. We are grateful that he wrote the first historical foreword for this textbook in 1996 before his death in 1998.

Special thanks to Brigadier General Daniel H. Daley (USAF, retired, and former Head, Department of Aeronautics, U.S. Air Force Academy 1965–1984), who sponsored Jack's studies under Professor Oates and guided him during the compilation of the teaching notes "Elements of Propulsion." These notes were used for over a decade in the academy's propulsion courses before the publication of *Elements of Gas Turbine Propulsion* in 1996 and are the basis for that textbook and this new edition. Special thanks also go out to Brigadier Generals (USAF, Retired) Mike Smith and Neal Barlow (former Heads, Department of Aeronautics, U.S. Air Force Academy 1985–2000 and 2000–2015, respectively), both of whom supported and sponsored Keith's studies under Professor O'Brien.

We also thank all of the students who have been in our propulsion courses and our coworkers at the Department of Aeronautics, U.S. Air Force Academy, CO; the Air Force Institute of Technology (AFIT), Wright-Patterson Air Force Base, OH; Practical Aeronautics, Inc., Aurora, CO; the Aerospace Systems Directorate, Air Force Research Laboratory, Wright-Patterson Air Force Base, OH; Tinker Air Force Base, Oklahoma City, OK; Wright-Patterson AFB, OH; Arnold Engineering Development Complex, Arnold Air Force Base, TN; Naval Air Station Patuxent River, Patuxent River, MD; Marine Corps Air Station, Cherry Point, NC; NASA

Glenn Research Center, Cleveland, OH; Seattle University, Seattle, WA; and the University of Washington, Seattle, WA. Many of the students and co-workers from these organizations provided insight and guidance in developing and improving this material.

We also thank users of the previous editions of this textbook and those who responded to the AIAA survey and personal correspondence on this new edition for their many helpful comments and suggestions. Special thanks to Professor Brenda Haven, Embry-Riddle Aeronautical University, Prescott, AZ; Professor Aaron Byerley, U.S. Air Force Academy, CO; Professor Paul King, AFIT, Wright-Patterson AFB, OH; and Professor David T. Pratt, University of Washington, Seattle, WA.

We are indebted to Toni Z. Ackley, our project manager and editor, for her special care on making this the best book—thanks so very much.

NOMENCLATURE

A	= area; constant
a	= speed of sound; constant
b	= constant
C	= effective exhaust velocity [Eq. (1.53), Eq. (11.4)]; circumference; work output coefficient
C_A	= angularity coefficient
C_C	= work output coefficient of core
C_D	= coefficient of drag; discharge coefficient
C_F	= thrust coefficient
C_{fg}	= gross thrust coefficient
C_L	= coefficient of lift
C_p	= pressure coefficient
C_{prop}	= work output coefficient of propeller
C_{tot}	= total work output coefficient
C_V	= velocity coefficient
C^*	= characteristic velocity [Eq. (11.38)]
c	= chord
c_f	= friction coefficient
c_p	= specific heat at constant pressure
c_v	= specific heat at constant volume
c_x	= axial chord
D	= drag
d	= diameter
E	= energy; modulus of elasticity
e	= polytropic efficiency; exponential, 2.7183
F	= force; uninstalled thrust; thrust
F_g	= gross thrust
FR	= thrust ratio [Eq. (5.38)]
f	= fuel/air ratio; function; friction coefficient
g	= acceleration of gravity
g_c	= Newton's constant
g_0	= acceleration of gravity at sea level
HP	= horsepower
h	= enthalpy per unit mass; height
h_{PR}	= low heating value of fuel
$\overline{h}_f^o$	= enthalpy of formation

I	= impulse; impulse function, $PA(1 + \gamma M^2)$
I_{sp}	= specific impulse [Eq. (1.55), Eq. (11.6)]
J	= advance ratio
K	= constant
K_P	= equilibrium constant [Eq. (2.77)]
L	= length
M	= Mach number; momentum
$\dot{M}$	= momentum flux, $\dot{m}\,V$
m	= mass
$\dot{m}$	= mass flow rate
$\mathcal{M}$	= molecular weight
N	= number of moles; revolutions per minute
n	= load factor; burning rate exponent
n_b	= number of blades
P	= pressure
P_c	= electrical output power
P_f	= profile factor
P_s	= weight specific excess power
P_t	= total pressure
Q	= heat interaction
$\dot{Q}$	= rate of heat interaction
q	= heat interaction per unit mass; dynamic pressure, $\rho V^2/(2g_c)$; electric charge
$\tilde{q}$	= dimensionless heat release [Eq. (5.85)]
R	= gas constant; extensive property; radius; additional drag
$\mathcal{R}$	= universal gas constant
r	= radius; burning rate
$°R$	= degree of reaction
S	= uninstalled thrust specific fuel consumption; entropy
$\dot{S}$	= time rate of change of entropy
S_w	= wing planform area
s	= entropy per unit mass; blade spacing
T	= temperature; installed thrust
T_t	= total temperature
t	= time; airfoil thickness
U	= blade tangential or rotor velocity
u	= internal energy; axial velocity
V	= absolute velocity; volume
v	= volume per unit mass; tangential velocity
W	= weight; width
$\dot{W}$	= power

w	= work interaction per unit mass; radial velocity
$\dot{w}$	= weight flow rate
x, y, z	= coordinate system
z_e	= energy height [Eq. (1.25)]
Z	= Zweifel tangential force coefficient [Eq. (9.97)]
α	= bypass ratio; angle; coefficient of linear thermal expansion
α_c	= specific mass [Eq. (10.33)]
β	= angle
Γ	$= \sqrt{\gamma\left(\dfrac{2}{\gamma+1}\right)^{(\gamma+1)/(\gamma-1)}}$; constant
γ	= ratio of specific heats; angle
Δ	= change
δ	= change; dimensionless pressure, P/P_{ref}; dead weight mass ratio; deviation
∂	= partial differential
ε	= nozzle area ratio; rotor turning angle; slip factor
ϕ	= installation loss coefficient; fuel equivalence ratio; function; total pressure loss coefficient
Φ	= function; cooling effectiveness; flow coefficient
η	= efficiency
λ	= payload mass ratio
μ	= Mach angle
ν	= stoichiometric coefficient; Prandtl-Meyer function
θ	= angle; dimensionless temperature, T/T_{ref}
Π	= product
π	= pressure ratio defined by Eq. (5.3)
ρ	= density, $1/\nu$
Σ	= sum
σ	= control volume boundary; dimensionless density, ρ/ρ_{ref}; tensile stress
τ	= temperature ratio defined by Eq. (5.4); shear stress; torque
τ_λ	= enthalpy ratio defined by Eq. (5.7)
ω	= angular speed
ψ	= thermal compression ratio, T_3/T_0

Subscripts

A	= air mass
a	= air; atmosphere
AB	= afterburner

add	= additive
B	= bypass stream
b	= burner or combustor; boattail or afterbody; blade; burning
bo	= burnout
C	= core stream
c	= compressor; corrected; centrifugal; chamber
DB	= duct burner
d	= diffuser or inlet; disk
dr	= disk/rim interface
dry	= afterburner not operating
e	= exit; exhaust; Earth; blade element
ext	= external
F	= fan stream
f	= fan; fuel; final
fn	= fan nozzle
g	= gearing; gas
H	= high-pressure
HP	= horsepower
h	= hub
i	= initial; inside; ideal
int	= internal
j	= jet
L	= low-pressure
M	= mixer
m	= mechanical; mean; middle
max	= corresponding to maximum
N	= new
n	= nozzle
nac	= nacelle
O	= overall; output
o	= overall; outer
opt	= optimum
P	= propulsive; products; power
p	= propellant; propeller
pl	= payload
prop	= propeller
R	= reference; relative; reactants
r	= ram; reduced; rim; rotor
ref	= reference condition
s	= stage; separation; solid; stator
SL	= sea-level
SLS	= sea-level static

T	= thrust
t	= total; turbine; throat; tip
Th	= thermal
vac	= vacuum
w	= forebody; wing
wet	= afterburner operating
x, y, z	= directional component
σ	= control volume
$0, 1, 2, \ldots, 19$	= different locations in space

Superscripts

*	= state corresponding to $M = 1$; corresponding to optimum state
—	= average

Chapter 1 | Introduction

> Even considering the improvements possible . . . the gas turbine [engine] could hardly be considered a feasible application to airplanes mainly because of the difficulty with the stringent weight requirements.
>
> *Gas Turbine Committee, U.S. National Academy of Sciences, 1940, which included Theodore von Kármán*

1.1 Propulsion

The *Random House College Dictionary* defines *propulsion* as "the act of propelling, the state of being propelled, a propelling force or impulse" and defines the verb *propel* as "to drive, or cause to move, forward or onward" [1]. From these definitions, we can conclude that the study of propulsion includes the study of the propelling force, the motion caused, and the bodies involved. Propulsion involves an object to be propelled plus one or more additional bodies, called *propellants*.

The study of propulsion is concerned with vehicles such as automobiles, trains, ships, aircraft, and spacecraft. The focus of this textbook is on the propulsion of aircraft and spacecraft. Methods devised to produce a thrust force for the propulsion of a vehicle in flight are based on the principle of jet propulsion (the momentum change of a fluid by the propulsion system). The fluid may be the gas used by the engine itself (e.g., turbojet), it may be a fluid available in the surrounding environment (e.g., air used by a propeller), or it may be stored in the vehicle and carried by it during the flight (e.g., rocket).

Jet propulsion systems can be subdivided into two broad categories: airbreathing and non-airbreathing. Airbreathing propulsion systems include the reciprocating, turbojet, turbofan, ramjet, turboprop, and turboshaft engines. Non-airbreathing engines include rocket motors, nuclear propulsion systems, and electric propulsion systems. We focus on gas turbine propulsion systems (turbojet, turbofan, turboprop, and turboshaft engines) in this textbook.

The material in this textbook is divided into four parts:

1. Basic concepts and one-dimensional (1-D) gas dynamics
2. Analysis and performance of airbreathing propulsion systems
3. Analysis of gas turbine engine components
4. Rocket propulsion fundamentals

This chapter introduces the types of airbreathing and rocket propulsion systems and the basic propulsion performance parameters. Also included is an introduction to aircraft and rocket performance. The material on aircraft performance shows the influence of the gas turbine engine on the performance of the aircraft system. This material also permits incorporation of a gas turbine engine design problem such as new engines for an existing aircraft.

Numerous examples are included throughout this book to help students see the application of a concept after it is introduced. For some students, the material on basic concepts and gas dynamics will be a review of material covered in other courses they have already taken. For other students, this may be their first exposure to this material, and it may require more effort to understand.

1.2 Units and Dimensions

The engineering world uses both the metric International System of Units (SI) and English unit system, so both will be used in this textbook. One singular distinction exists between the English system and SI—the unit of force is defined in the former but derived in the latter. Newton's second law of motion relates force to mass, length, and time. It states that the sum of the forces is proportional to the time rate of change of the momentum $(M = mV)$. The constant of proportionality is $1/g_c$:

$$\sum F = \frac{1}{g_c}\frac{d(mV)}{dt} = \frac{1}{g_c}\frac{dM}{dt} \tag{1.1}$$

The units for each term in the preceding equation are listed in Table 1.1 for both SI and English units. In any unit system, only four of the five items in the table can be specified, and the latter is derived from Eq. (1.1).

As a result of selecting $g_c = 1$ and defining the units of mass, length, and time in SI units, the unit of force is derived from Eq. (1.1) as kilogram-meters per square second $(kg \cdot m/s^2)$, which is called the *newton* (N). In English units, the value of g_c is derived from Eq. (1.1) as

$$g_c = 32.174\,\text{ft} \cdot \text{lbm}/(\text{lbf} \cdot \text{s}^2)$$

Rather than adopt the convention used in many recent textbooks of developing material for use with *only* SI metric units $(g_c = 1)$, we will maintain g_c in

Table 1.1 Units and Dimensions

Unit System	Force	g_c	Mass	Length	Time
SI	Derived	1	Kilogram, kg	Meter, m	Second, s
English	Pound-force, lbf	Derived	Pound-mass, lbm	Foot, ft	Second, s

all our equations. Thus g_c will also show up in the equations for *potential energy* (PE) and *kinetic energy* (KE):

$$PE = \frac{mgz}{g_c}$$

$$KE = \frac{mV^2}{2g_c}$$

The total energy per unit mass e is the sum of the specific internal energy u, specific kinetic energy ke, and specific potential energy pe:

$$e \equiv u + ke + pe = u + \frac{V^2}{2g_c} + \frac{gz}{g_c}$$

There are a multitude of engineering units for the quantities of interest in propulsion. For example, energy can be expressed in the SI unit of *joule* ($1\,J = 1\,N \cdot m$), in British thermal units (Btu), or in foot-pound force (ft · lbf). One must be able to use the available data in the units provided and convert the units when required. Table 1.2 is a unit conversion table provided to help you in your endeavors.

1.3 Operational Envelopes and Standard Atmosphere

Each engine type will operate only within a certain range of altitudes and Mach numbers (velocities). Similar limitations in velocity and altitude exist for airframes. It is necessary, therefore, to match airframe and propulsion system capabilities. Figure 1.1 shows the approximate velocity and altitude limits, or *corridor of flight*, within which airlift vehicles can operate. The corridor is bounded by a *lift limit*, a *temperature limit*, and an *aerodynamic force limit*. The lift limit is determined by the maximum level-flight altitude at a given velocity. The temperature limit is set by the structural thermal limits of the material used in construction of the aircraft. At any given altitude, the maximum velocity attained is temperature-limited by aerodynamic heating effects. At lower altitudes, velocity is limited by aerodynamic force loads on the vehicle rather than by temperature.

The operating regions of all aircraft lie within the flight corridor. The operating region of a particular aircraft within the corridor is determined by aircraft design, but it is a very small portion of the overall corridor. Superimposed on the flight corridor in Fig. 1.1 are the operational envelopes of various powered aircraft. The operational limits of each propulsion system are determined by limitations of the components of the propulsion system and are shown in Fig. 1.2.

The analyses presented in this text use the properties of the atmosphere to determine both engine and airframe performance. Because these

Table 1.2   Unit Conversion Table

Unit	Conversion
Length	1 m = 3.2808 ft = 39.37 in.
	1 km = 0.621 mile
	1 mile = 5280 ft = 1.609 km
	1 n mile = 6080 ft = 1.853 km
Area	$1 \text{ m}^2 = 10.764 \text{ ft}^2$
	$1 \text{ cm}^2 = 0.155 \text{ in.}^2$
Volume	$1 \text{ gal} = 0.13368 \text{ ft}^3 = 3.785 \text{ l}$
	$1 \text{ l} = 10^{-3} \text{ m}^3 = 61.02 \text{ in.}^3$
Time	1 h = 3600 s = 60 min
Mass	$1 \text{ kg} = 1000 \text{ g} = 2.2046 \text{ lbm} = 6.8521 \times 10^{-2} \text{ slug}$
	$1 \text{ slug} = 1 \text{ lbf} \cdot \text{s}^2/\text{ft} = 32.174 \text{ lbm}$
Density	$1 \text{ slug}/\text{ft}^3 = 515.36 \text{ kg}/\text{m}^3$
	$1 \text{ lbm}/\text{ft}^3 = 16.018 \text{ kg}/\text{m}^3$
Force	$1 \text{ N} = 1 \text{ kg} \cdot \text{m}/\text{s}^2$
	1 lbf = 4.448 N
Energy	$1 \text{ J} = 1 \text{ N} \cdot \text{m} = 1 \text{ kg} \cdot \text{m}^2/\text{s}^2$
	$1 \text{ Btu} = 778.16 \text{ ft} \cdot \text{lbf} = 252 \text{ cal} = 1055 \text{ J}$
	1 cal = 4.186 J
	1 kJ = 0.947813 Btu = 0.23884 kcal
Power	$1 \text{ W} = 1 \text{ J}/\text{s} = 1 \text{ kg} \cdot \text{m}^2/\text{s}^3$
	$1 \text{ hp} = 550 \text{ ft} \cdot \text{lbf}/\text{s} = 2545 \text{ Btu}/\text{h} = 745.7 \text{ W}$
	1 kW = 3412 Btu/h = 1.341 hp
Pressure (stress)	$1 \text{ atm} = 14.696 \text{ lb}/\text{in.}^2$ or psi = 760 torr
	= 101,325 Pa
	$1 \text{ atm} = 30.0 \text{ inHg} = 407.2 \text{ inH}_2\text{O}$
	1 ksi = 1000 psi
	1 mmHg = 0.01934 psi = 1 torr
	$1 \text{ Pa} = 1 \text{ N}/\text{m}^2$
	1 inHg = 3376.8 Pa
Energy per unit mass	1 kJ/kg = 0.4299 Btu/lbm
Specific heat	$1 \text{ kJ}/(\text{kg} \cdot {}^\circ\text{C}) = 0.23884 \text{ Btu}/(\text{lbm} \cdot {}^\circ\text{F})$
Temperature	$1 \text{ K} = 1.8{}^\circ\text{R}$
	$\text{K} = 273.15 + {}^\circ\text{C}$
	${}^\circ\text{R} = 459.69 + {}^\circ\text{F}$
Temperature change	$1{}^\circ\text{C} = 1.8{}^\circ\text{F}$
Specific thrust	1 lbf/(lbm/s) = 9.8067 N/(kg/s)
Specific power	1 hp/(lbm/s) = 1.644 kW/(kg/s)

(Continued)

Table 1.2 Unit Conversion Table *(Continued)*

Unit	Conversion
Thrust specific fuel consumption (TSFC)	1 lbm/(lbf · h) = 28.325 g/(kN · s)
Power specific fuel consumption	1 lbm/(hp · h) = 168.97 g/(MW · s)
Strength/weight ratio (σ/ρ)	1 ksi/(slug/ft^3) = 144 ft^2/s^2 = 13.38 m^2/s^2

properties vary with location, season, time of day, and the like, we will use the International Standard Atmosphere (ISA) [2] to give a known foundation for our analyses. Appendix A gives the properties of the ISA in both English and SI units. Values of the pressure P, temperature T, density ρ, and speed of sound a are given in dimensionless ratios of the property at altitude to its value at sea level (SL), the reference value. The dimensionless ratios of pressure, temperature, and density are given the symbols δ, θ, and σ, respectively. These ratios are defined as follows:

$$\delta \equiv \frac{P}{P_{\text{ref}}} \tag{1.2}$$

$$\theta \equiv \frac{T}{T_{\text{ref}}} \tag{1.3}$$

$$\sigma \equiv \frac{\rho}{\rho_{\text{ref}}} \tag{1.4}$$

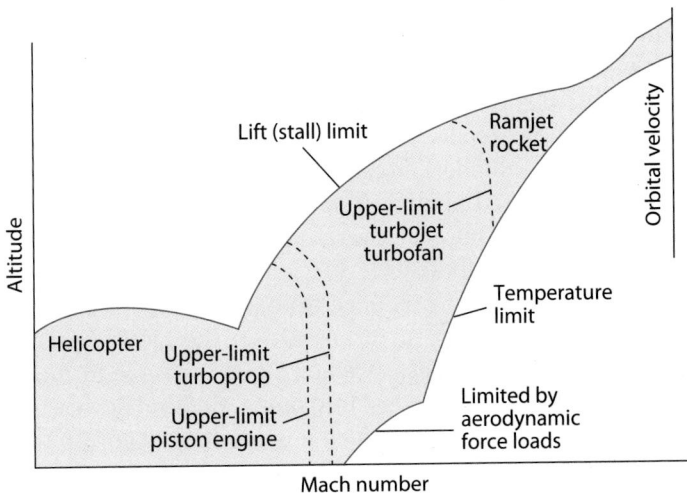

Fig. 1.1 Flight limits.

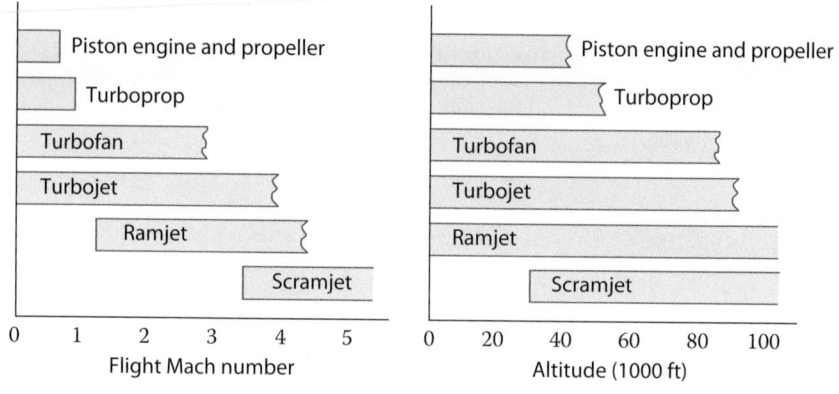

Fig. 1.2 Engine operational limits.

The reference values of pressure, temperature, and density are given for each unit system at the end of its property table.

For nonstandard conditions such as a hot day, the normal procedure is to use the standard pressure and correct the density, using the perfect gas relationship $\sigma = \delta/\theta$. As an example, we consider a 100°F day at 4-kft altitude. From Appendix A, we have $\delta = 0.8637$ for the 4-kft altitude. We calculate θ using the 100°F temperature: $\theta = T/T_{\text{ref}} = (100 + 459.7)/518.7 = 1.079$. Note that absolute temperatures must be used in calculating θ. Then the density ratio is calculated using $\sigma = \delta/\theta = 0.8637/1.079 = 0.8005$.

1.4 Airbreathing Engines

The turbojet, turbofan, turboprop, turboshaft, and ramjet/scramjet engine systems used to power aircraft are discussed in this section of Chapter 1. The listed engines are not all the engine types that are used in providing propulsive thrust to aircraft, nor are they used exclusively on aircraft. Reciprocating (piston-cylinder) engines, rocket engines, and various combination types are also used in aircraft applications. A large file of images is available with the Supporting Material that can be downloaded with the textbook software (see Appendix J).

Fundamentally, the thrust produced from any of the gas turbine engines (turbojet, turbofan, turboprop, or turboshaft) or the ramjet/scramjet is the result of accelerating an air mass as efficiently as possible to meet the aircraft-engine system mission needs. How much air (the air mass flow rate) and how much and the manner in which it is accelerated govern the key distinguishing characteristics among the different engine types we will discuss.

Before beginning this section, the reader is highly encouraged to read the two historical forewords to this book. The first foreword was written by Hans von Ohain (the German co-inventor of the aircraft jet engine), and he gives a detailed history (from its invention to about 2000) from the viewpoint of a noted pioneer. The second foreword is written by William Heiser (a world-renowned expert), who gives a historical update to the first foreword. It is so important in the study of technology to know the history, and we are very fortunate to have these forewords in this textbook.

1.4.1 Gas Generator (Core)

The "heart" of a gas turbine type of engine is the gas generator (also called the core). A schematic diagram of a gas generator is shown in Fig. 1.3a. Figure 1.3b shows the core of a J85 turbojet cutaway. The major components are clearly visible as is the shaft connecting the two-stage turbine to the eight-stage compressor. (More about stages below and in Chapters 4 and 9.) The compressor, combustor (main burner or burner), and turbine are the major components of the gas generator that is common to the turbojet, turbofan, turboprop, and turboshaft engines. The purpose of a gas generator is to supply high-temperature and high-pressure gas.

The compressor increases the pressure and temperature of the incoming air for efficient combustion. Energy is released in the combustor manifested by a large increase in gas temperature, typically through a chemical reaction between oxygen in the air and a hydrocarbon fuel. The high-energy gas expands against a turbine, which supplies the necessary shaft power to drive the compressor. What is done with the "leftover" energy remaining in the gas at the exit of the core turbine dictates the engine cycle—turbojet, turbofan, turboprop, and turboshaft.

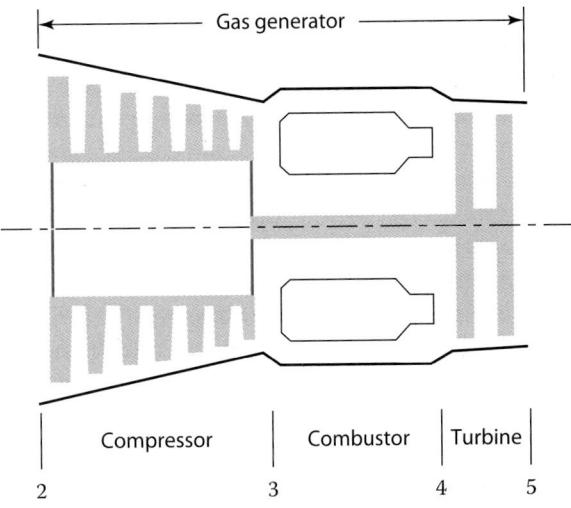

Fig. 1.3a   Schematic diagram of gas generator.

Fig. 1.3b   J85 Turbojet engine. (Courtesy of General Electric Aircraft Engines.)

All engine companies make use of a "common core" concept; that is, a trusted core configuration is used for many different engine types for varied applications. For example, the B-1B, B-2, F-16, and KC-135R engines all use essentially the same core. The common core architecture is used largely because of the tremendous expense associated with the design, manufacture, and sustainment of aircraft gas turbine engines.

1.4.2 Turbojet

By adding an inlet and a nozzle to the gas generator, a turbojet engine can be constructed. (See Jumo 004B in Fig. 14 of the Foreword by von Ohain.) In a turbojet, the leftover gas energy at the core turbine exit is converted to kinetic energy, resulting in very high exhaust gas exit velocities. By adding an after-burner between the core turbine and nozzle, the exhaust gases can be reheated and expanded through the nozzle at even greater velocities (in excess of 2500 mph). Because of these high jet exhaust gas velocities, the term "jet engine" is often used to refer to turbojets (and ramjets), although all four gas turbine engine variations can be considered as jet engines. Of the four gas turbine engine types, the turbojet has the highest specific thrust (thrust per airflow), but at the expense of relatively poor fuel efficiency.

As presented in the Foreword by von Ohain, turbojets were the first gas turbine engines used as a means of aircraft propulsion, first by von Ohain (first flight 27 August 1939) and then by Whittle (first flight 15 May 1941). As development proceeded, the turbojet engine quickly surpassed the power-to-weight of piston engines (which generally plateaued at/around 1.0 hp/lbf) and became more efficient primarily due to higher pressure

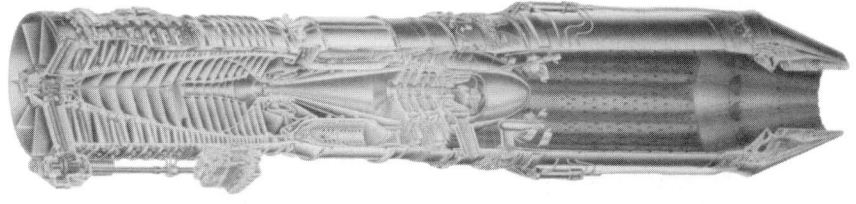

Fig. 1.4a Cross-section of General Electric J79 turbojet with afterburner.
(Courtesy of General Electric Aircraft Engines.)

ratios, often replacing the reciprocating piston-cylinder engine for use on aircraft.

A photograph of the J79 turbojet with afterburner used in the F-104 Starfighter, F-4 Phantom II, B-58 Hustler, and others is shown in Fig. 1.4a. The F-104 and its J79 engine earned the Collier Trophy as the greatest achievement in aeronautics and astronautics in 1958.

The J79 was the first engine to use variable stator vanes, which are highlighted on the compressor case in Fig. 1.4b. Stators are a "row" (circular ring) of stationary vanes that combine with a rotating rotor row of blades to form a stage. In the front of the compressor, the J79 used seven rows of stators whose pitch angle was varied hydro-mechanically as a function of engine throttle position and flight conditions. The pitch angle (the angle between the axial direction and blade chord line) is varied in a way that significantly improves compressor efficiency at off-design operation and aids in engine starting. The vast majority of modern aircraft gas turbine engines use variable stators in the front or early compressor stages, as shown in Fig. 1.4c. (See a further discussion of this topic in Section 9.3.9.)

Variable stator vane
positioning arm and ring

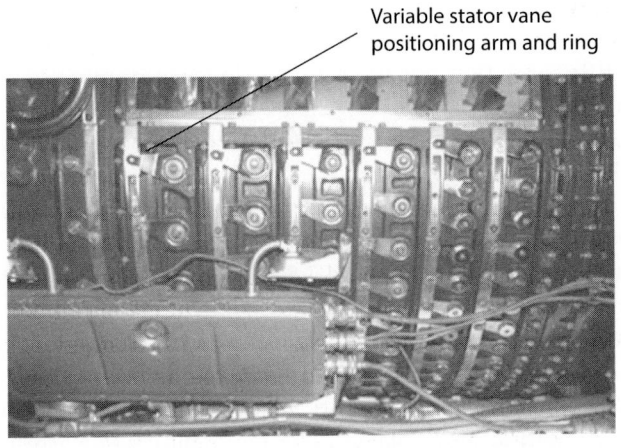

Fig. 1.4b J79 front compressor case with variable stator vane hardware.

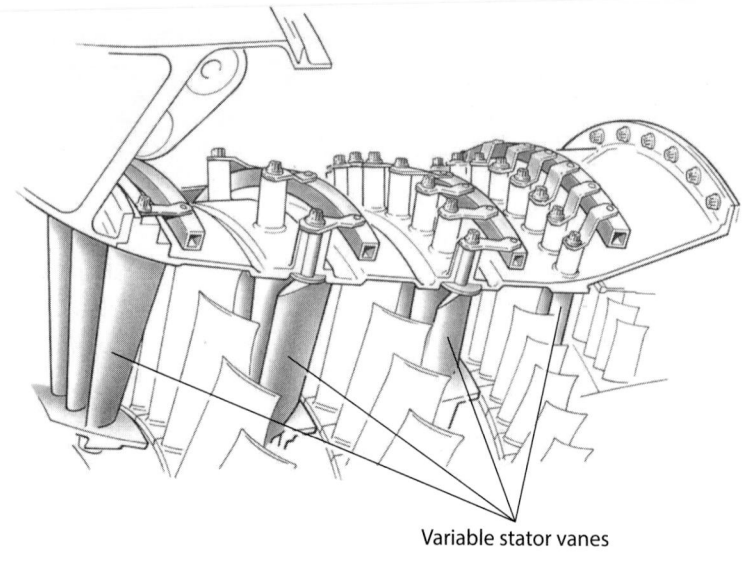

Variable stator vanes

Fig. 1.4c   Sketch of variable stator vanes of compressor in gas flow path.

These early turbojet engines were single-spool engines, like the J79 in Fig. 1.4a. (A spool is a compressor-shaft-turbine assembly.) Later engines had multiple spools—typically two, though Rolls-Royce often uses three—to solve the problems that a single spool placed on design and operation. The multiple spools allow the smaller blades to rotate at a larger rotational speed than the larger blades, significantly improving engine efficiency and operability (stable operation). They also facilitate use of a common core in various engines (see Figs. 1.5a and 1.5b). Also shown in Figs. 1.5a and 1.5b are the station or section numbers used for each primary flow path component in accordance with the industry standard of Aerospace Recommended Practice (ARP) 755A (Ref. 3). We use this standard numbering throughout the text, so the reader would do well to become familiar with it. In Figs. 1.5a and 1.5b, note that the gas generator or core spool is typically referred to as the "high spool" (HPC-shaft-HPT) because it is rotating faster than the "low spool" (LPC-shaft-LPT) and is subjected to the highest gas pressures and temperatures. Also note the commonly used shaft within a shaft configuration indicated in Figs. 1.5a and 1.5b. These two shafts (or three in the case of Rolls-Royce) are mechanically independent of one another.

The pressure, temperature, and velocity variations through a J79 engine are shown in Fig. 1.6. In the compressor section, the pressure and temperature increase as a result of work being done on the air through the action of the rotating rotor blades. The temperature of the gas is further increased by burning fuel in the combustor. In the turbine section, the temperature and pressure decrease as energy is removed from the gas

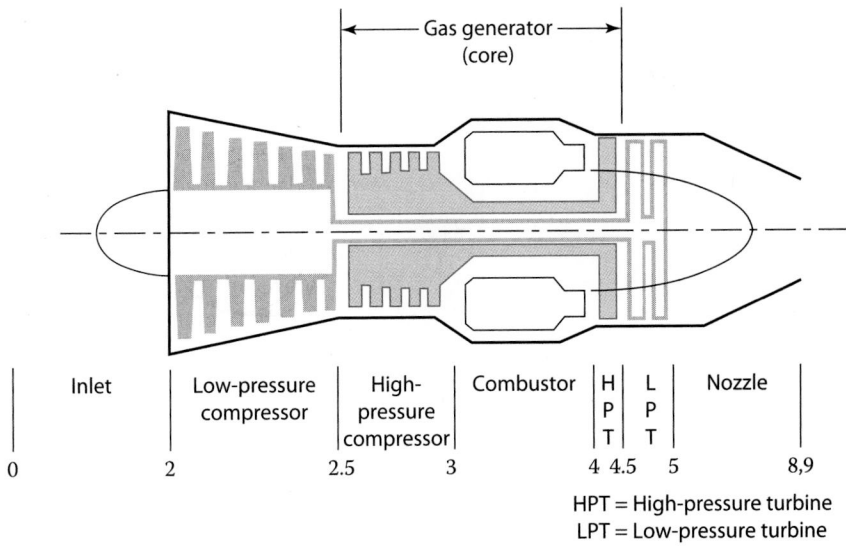

Fig. 1.5a Schematic diagram of a turbojet (dual-spool axial compressor and turbine).

stream as it expands against the rotating turbine blades and is converted to shaft power to turn the compressor. In the exhaust nozzle, the gas stream is further expanded to produce a high exit kinetic energy. The effect of afterburner operation is clearly visible in Fig. 1.6 as exhaust gas temperature and velocity are further increased. All the sections of the

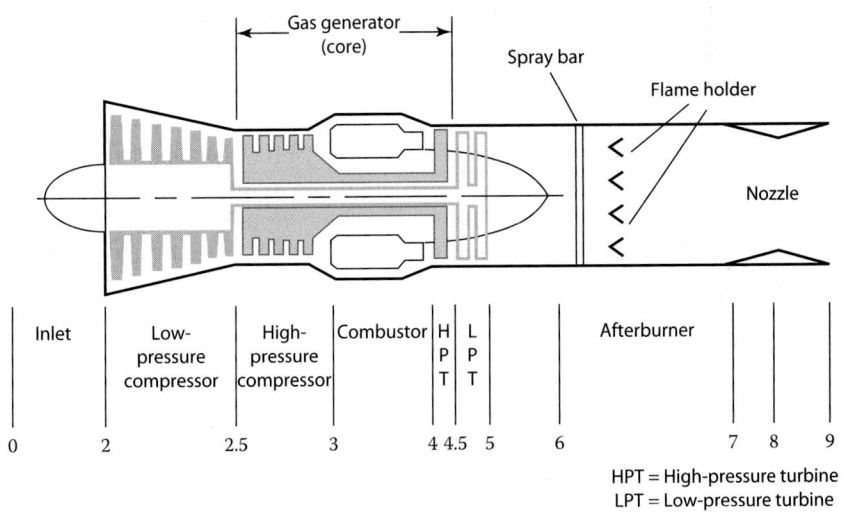

Fig. 1.5b Schematic diagram of a dual-spool turbojet with afterburner.

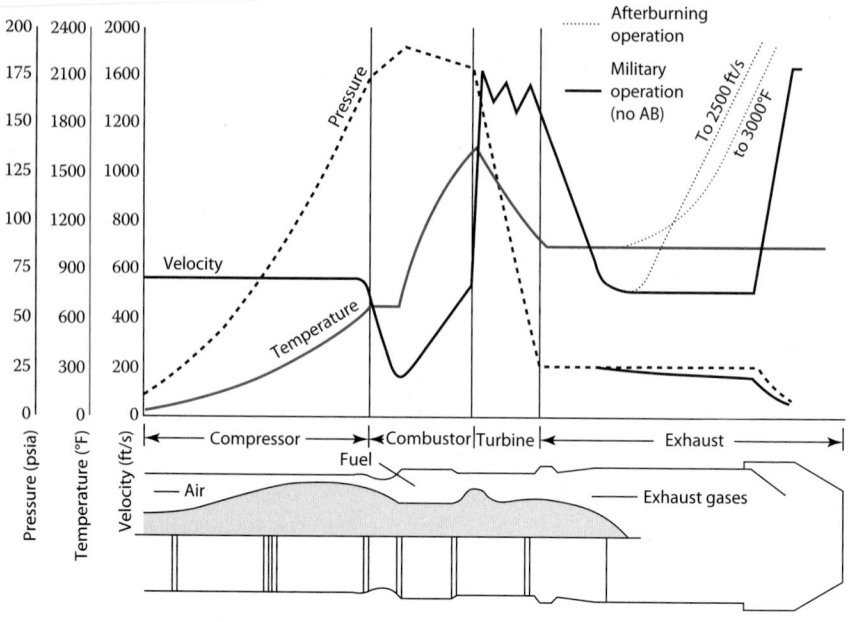

Fig. 1.6 Property variation through the General Electric J79 afterburning turbojet engine. (Courtesy of General Electric Aircraft Engines.)

engine must operate in such a way as to efficiently produce the required amount of thrust for a minimum of weight for a given aircraft-engine system mission.

The adaptations of the turbojet in the form of turbofan, turboprop, and turboshaft engines came with the need for greater thrust generated more efficiently at lower speeds. Some characteristics of different turbojet, turbofan, turboprop, and turboshaft engines are included in Appendix B.

1.4.3 Turbofan

A turbofan engine takes some of the leftover gas energy exiting the core turbine (HPT) and further expands it against a low-pressure turbine to drive a fan and often a low-pressure compressor (LPC) or booster. A schematic diagram of a high-bypass turbofan is shown in Fig. 1.7. Booster stages are often used because the common core practice often necessitates the use of additional compressor stages to achieve the desired overall cycle pressure ratio. These additional stages are placed in the core stream and added on the low spool. Rolls-Royce places them on their own independent spool called the "intermediate" because its rotational speed is between that of the core and the fan.

In a turbofan, all of the air required by the engine passes through the fan, and a portion of that air bypasses the engine core. The choice of bypass ratio (bypass airflow rate/core airflow rate) is a key design choice in dictating the engine cycle characteristics. A low-bypass turbofan (generally bypass ratios less than 2 and often less than 1) takes on performance characteristics more similar to those of a turbojet. A high-bypass turbofan (generally bypass ratios greater than 5) takes on characteristics closer to those of a turboprop; that is, high static thrust that falls off rapidly with flight speed, but very good fuel efficiency at cruise flight speeds higher than those accommodated by a turboprop. Most often, high-bypass turbofans use separate exhaust nozzles in the core and bypass streams ("hot" and "cold" streams), as shown in Fig. 1.7. All of the low-bypass turbofans used in afterburning military applications combine the hot and cold streams downstream of the LPT, and the mixed gases are expanded through a single exhaust nozzle.

Generally the turbofan engine is more economical and efficient than the turbojet engine in subsonic flight. The thrust specific fuel consumption (TSFC, or fuel mass flow rate per unit thrust) is lower for turbofans and indicates a more economical operation. The turbofan also accelerates a larger mass of air to a lower velocity than a turbojet for a higher *propulsive efficiency* (see Sec. 1.4.7). The frontal area of a turbofan is quite large compared to that of a turbojet, and for this reason more drag and more weight result.

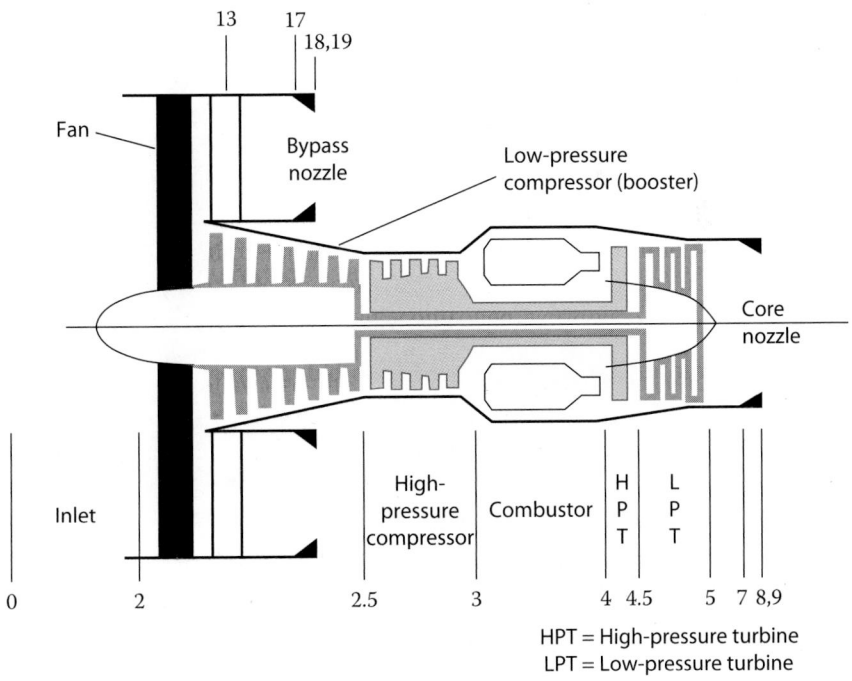

Fig. 1.7 Schematic diagram of a high-bypass-ratio turbofan.

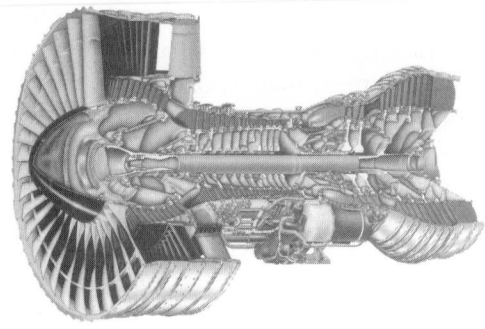

Fig. 1.8a   Pratt & Whitney PW4000 turbofan. (Courtesy of Pratt & Whitney.)

Fig. 1.8b   General Electric GE90 turbofan. (Courtesy of General Electric Aircraft Engines.)

Fig. 1.8c   Rolls-Royce Trent 1000 turbofan. (Courtesy of Rolls-Royce.)

Fig. 1.8d   General Electric GEnx turbofan. (Courtesy of General Electric Aircraft Engines.)

The fan diameter is also limited aerodynamically by compressibility effects resulting from high blade tip velocities (see Section 1.4.4).

Several of the current high-bypass turbofan engines used in subsonic aircraft are shown in Figs. 1.8a–1.8g. All the high-bypass turbofan engines shown here are two-spool engines except the three-spool Rolls-Royce Trent 1000 in Fig. 1.8c.

Figures 1.8a–1.8d show high-bypass turbofan engines with high thrust (see Appendix B) that are used in wide-body (twin-aisle) aircraft like the Boeing 747, 777, and 787, and the largest of them all, the Airbus 380. The PW4000 shown in Fig. 1.8a actually represents a series of engines ranging in thrust ratings from 52,000 lb to 99,000 lb. It is a nice example of an engine that uses the common-core concept with fan diameters ranging from 94 in. to 112 in. The bypass ratio is increasing from generation to generation to increase their efficiency and reduce fuel consumption.

Fig. 1.8e   SNECMA LEAP (Leading Edge Aviation Propulsion) turbofan. (Courtesy of SNECMA.)

Fig. 1.8f Pratt & Whitney PW1100g geared turbofan and its epicyclic gear system with 3:1 gear ratio. (Courtesy of Pratt & Whitney.)

Figures 1.8e and 1.8f are the newest high-bypass turbofan engines in the 25–30 klbf (110–130 kN) thrust class (the most popular). Different variations of the CFM LEAP (Leading Edge Aviation Propulsion) engine (Fig. 1.8e) are being used in both the Airbus 320neo (new engine option) and Boeing 737 MAX. The PW 1100g (Fig. 1.8f) is the newest configuration of this class of engine with a gear system between the fan and low-pressure compressor (booster). This allows higher bypass ratios and also allows the fan to rotate at a lower speed than the booster stages on that spool (Ref. 4). The gearing for this engine (Fig. 1.8f) transfers 22 MW (30,000 hp) of power to the fan and reduces rotational speeds needed for efficient fan operation. The speed reduction gear allows the low-pressure turbine to spin at desired high rotational speeds, thereby reducing the number of both low-pressure compressor (LPC) and low-pressure turbine (LPT) stages.

Rolls-Royce announced plans in 2014 for development of geared turbofans (Ref. 5). Expect other large engine companies to also use gearing to allow much larger bypass ratios. Geared turbofan engines are not something new; small engine manufacturers (e.g., Honeywell TFE731) have used them for many years. As the bypass ratio increases, the diameter of an engine must increase to provide the same thrust, thereby requiring longer landing gear or a new engine installation location.

The turbofan is the most prominent aircraft gas turbine engine in use today. The choice of bypass ratio allows designers to tailor engine performance characteristics in the most desirable manner to meet mission needs—from medium- to long-range transports to high-performance military fighters. It should not surprise you that the trend in turbofan engine design for medium- and long-range transports is higher bypass ratios (10 and beyond) for improved propulsive efficiency and higher overall compression system pressure ratio (as high as 70) for increased *thermal efficiency*

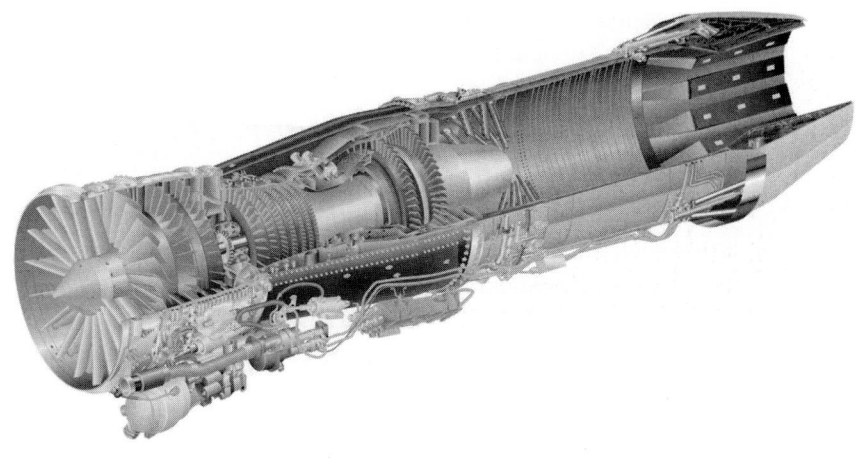

Fig. 1.9a General Electric F414-GE-404 afterburning turbofan.
(Courtesy of General Electric Aircraft Engines.)

(more on efficiencies in Sec. 1.4.7). All of the turbofans shown in Figs. 1.8a through 1.8f as well as their planned future variants are trending in this direction.

Figures 1.9a and 1.9b show the General Electric F414 turbofan and the Pratt & Whitney F135 turbofan, respectively. The F414, used on F/A-18 E/F Super Hornet supersonic fighter aircraft, represents the engine configuration most common on modern-day fighters, the *dual-spool, augmented, mixed-flow, low-bypass turbofan* with a multistage fan and no booster stages. The mixing of the bypass and core streams prior to gas expansion out a single exhaust nozzle facilitates cooling and low observability (stealth) while affording additional oxygen for afterburner operation. Thrust augmentation

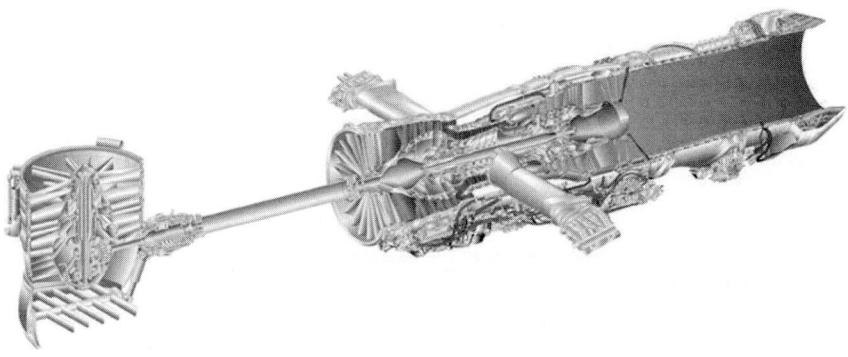

Fig. 1.9b Pratt & Whitney F135-PW-600 turbofan engine with Rolls-Royce lift fan.
(Courtesy of Pratt & Whitney.)

(afterburning) is used for brief periods of time for high g performance, heavy takeoff, and transonic/supersonic flight. Two Pratt & Whitney F119 low-bypass turbofans provide the F-22 Raptor (Fig. 1.22) with enough non-afterburning thrust to cruise supersonically without the use of afterburners, the so-called "supercruise capability."

The F135 turbofan engine (shown in Fig. 1.9b) is used in the F-35B Lightning II, the short takeoff and vertical landing (STOVL) variant of the Joint Strike Fighter. The main engine shown in Fig. 1.9b is also a dual-spool, augmented, mixed-flow, low-bypass turbofan and is used in the F-35A (Air Force variant) and F-35C (Navy variant). The overall engine configuration has some obvious variants to accommodate the STOVL mission, most notably the lift fan, the two-stage, vertically oriented fan shown in Fig. 1.9b, which is powered off the low spool when connected. Also in Fig. 1.9b, note the roll bars which use bleed air from the core compressor (HPC) to aid in roll control during STOVL operation. The 40,000-lb thrust class F135 integrated lift fan propulsion system shown in Fig. 1.9b won the Collier Trophy as the greatest achievement in aeronautics and astronautics in 2001.

1.4.4 Turboprop and Turboshaft

A schematic diagram of a turboprop is shown in Fig. 1.10a. A turboprop uses much of the leftover gas energy exiting the core turbine to drive a propeller. The propeller provides the majority of the propulsive thrust

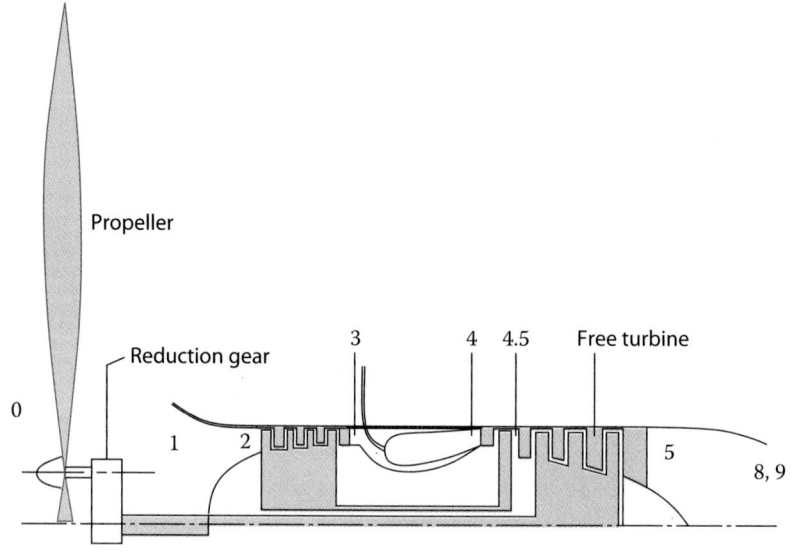

Fig. 1.10a   Schematic diagram of a turboprop.

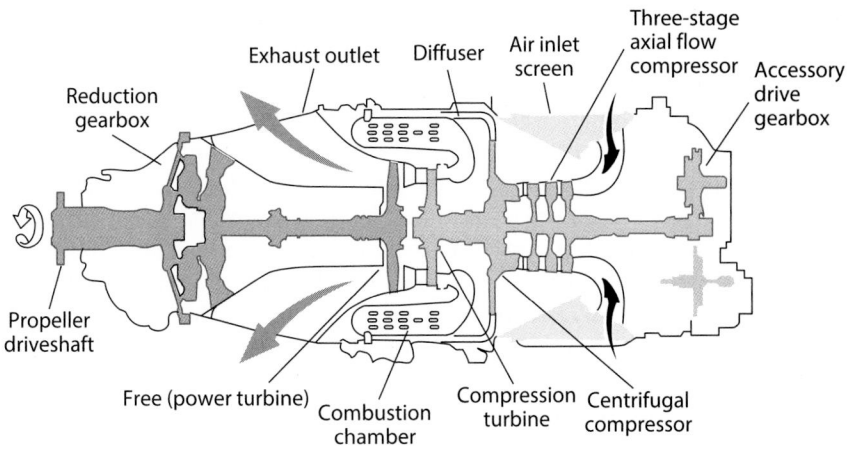

Fig. 1.10b Schematic of Canadian Pratt & Whitney PT6 turboshaft. (Courtesy of Pratt & Whitney of Canada.)

needed by the aircraft. The gas turbine engine is there to provide the necessary shaft power for the propeller, which is provided by expansion of the gas through either an additional turbine often called a "free" or "power" turbine (Figs. 1.10a and 1.10b) or more stages on the HPT (e.g., T56 engine used on the C-130, Fig. 1.10c).

The turboshaft engine is similar to the turboprop except that power is supplied to a shaft rather than a propeller. The turboshaft engine is used quite extensively for supplying power for helicopters. The turboprop engine may find application in vertical takeoff and landing (VTOL) transporters like the tiltrotor V-22 Osprey. The reduced rotational speed required of propeller and helicopter rotor blades for high efficiency and durability

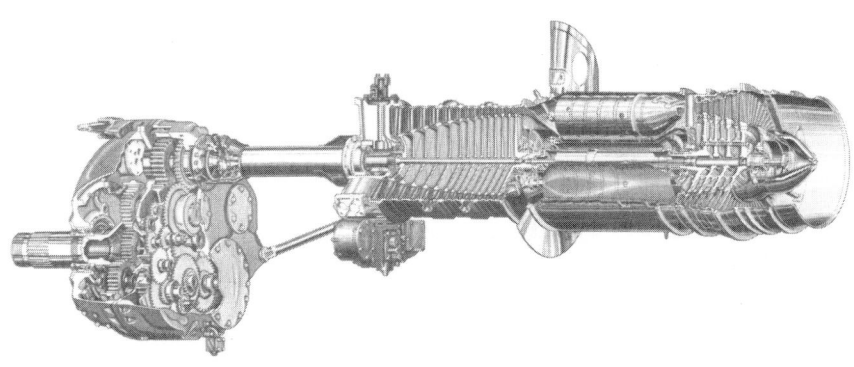

Fig. 1.10c Allison T56 turboshaft. (Courtesy of Rolls-Royce.)

(life) necessitates the use of a speed reduction gearbox between the power turbine shaft and the rotor blades. The gearbox is indicated in all three schematics of Fig. 1.10.

The limitations and advantages of the turboprop relate to the propeller. For low-speed flight and short-field takeoff, the propeller has a performance advantage predominantly due to its ability to produce a lot of thrust efficiently at very low-speed conditions. Disadvantages include high noise levels from the propellers, gearbox sustainment issues, and high propeller losses at flight speeds approaching Mach numbers of 0.6 to 0.7. Because of the rotation of the propeller, the relative air velocity at the tip (i.e., relative to the rotating blade tip) will approach the speed of sound well before the flight speed. Shock waves and other *compressibility effects* will result in rapid reduction of propeller efficiency due to these increased aerodynamic losses. The same is true for helicopter rotor blades, but their efficient operation is limited to flight Mach numbers of 0.3 to 0.4 because of much longer blades.

At high subsonic speeds, the high-bypass turbofan has better aerodynamic performance than the turboprop due to the casing around the fan rotor and closer spacing (increased number) of fan blades. In fact, the turbofan can be thought of as a ducted turboprop. For some perspective, turboprops have "bypass ratios" generally between 30 and 80 if one considers bypass ratio to be propeller airflow rate divided by engine airflow rate. Figure 1.10b shows the Canadian Pratt & Whitney PT6 turboshaft/turboprop engine used in many small commuter aircraft, and Fig. 1.10c the Allison T56 used to power the C-130 Hercules and the P-3 Orion.

1.4.5 Ramjet/Scramjet

The ramjet engine consists of an inlet, a combustion zone, and a nozzle. A schematic diagram of a ramjet is shown in Fig. 1.11. The ramjet does not have the compressor and turbine as the turbojet does. Air enters the

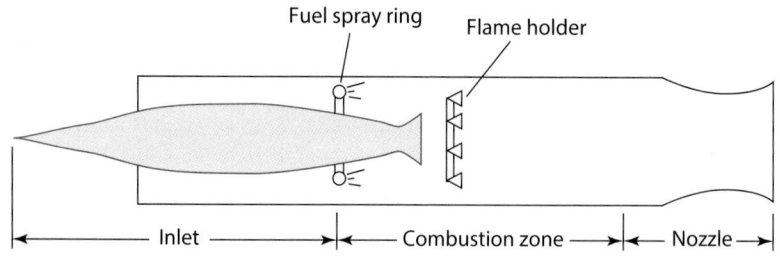

Fuel spray ring   Flame holder

Inlet   Combustion zone   Nozzle

Fig. 1.11 Schematic diagram of a ramjet.

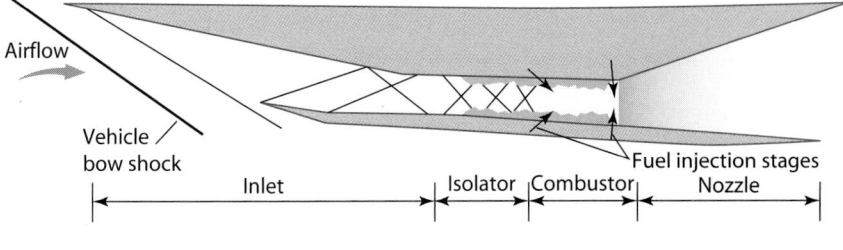

Airflow

Vehicle
bow shock

Inlet

Isolator Combustor

Fuel injection stages
Nozzle

Fig. 1.12a   Schematic diagram of a scramjet.

inlet, where it is compressed, and then enters the combustion zone, where it is mixed with the fuel and burned. The hot gases are then expelled through the nozzle, developing thrust. The operation of the ramjet depends on the inlet to decelerate the incoming air to raise the pressure in the combustion zone. The pressure rise makes it possible for the ramjet to operate. The higher the velocity of the incoming air, the greater the pressure rise. It is for this reason that the ramjet operates best at high supersonic velocities. At subsonic velocities, the ramjet is ineffcient, and to start the ramjet, air at a relatively higher velocity must enter the inlet.

The combustion process in an ordinary ramjet takes place at low subsonic velocities. At high supersonic flight velocities, a very large pressure rise is developed that is more than suffcient to support operation of the ramjet. Also, if the inlet has to decelerate a supersonic high-velocity airstream to a subsonic velocity, large pressure losses can result. The deceleration process also produces a temperature rise, and at some limiting flight speed, the temperature will approach the limit set by the wall materials and cooling methods. Thus when the temperature increase due to deceleration reaches the limit, it may not be possible to burn fuel in the airstream.

In the past several decades, research and development have been done on a ramjet that has the combustion process taking place at supersonic velocities. By using a supersonic combustion process, the temperature rise and pressure loss due to deceleration in the inlet can be reduced. This ramjet with supersonic combustion is known as the *scramjet* (supersonic combustion ramjet). Figure 1.12a shows the schematic of a scramjet engine similar to that used in the NASA supersonic research vehicle, like the X-43 shown in Fig. 1.12b. Further development of the scramjet for applications will continue (e.g., X-51 WaveRider) if research and development produces a scramjet engine with suffcient performance gains. Remember that because it takes a relatively high velocity to start the ramjet or scramjet, another engine system is required to accelerate aircraft like the X-43 to ramjet velocities. (The X-43 was accelerated by a Pegasus rocket launched from a B-52.)

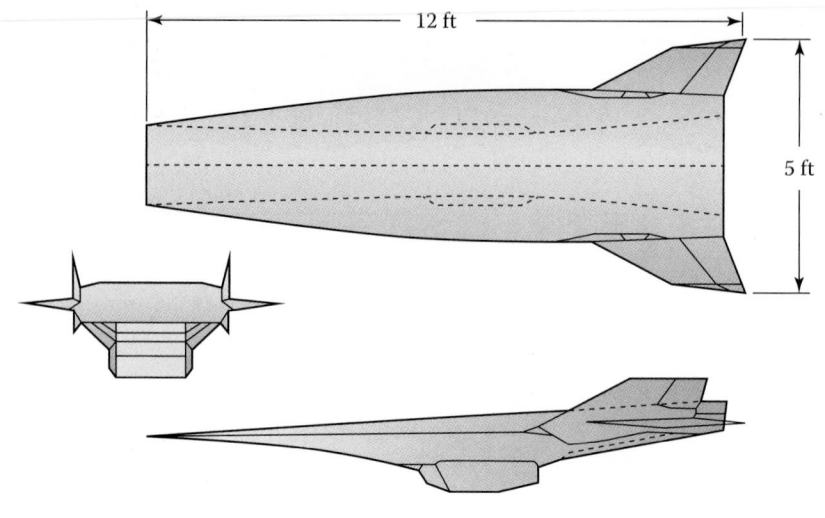

Fig. 1.12b   Conceptual drawing of the X-43. (Courtesy of NASA.)

1.4.6 Variable-Cycle Engines

Two of the Pratt & Whitney J58 turbojet engines (see Fig. 1.12c) were used to power the Lockheed SR-71 Blackbird (see Fig. 1.12d). This was the fastest aircraft (Mach 3+) when it was permanently retired in 1998. The J58 operates as an afterburning turbojet engine until it reaches high Mach level, at which point the six large tubes (Fig. 1.12c) bypass flow to the afterburner. When these tubes are in use, the compressor, burner, and turbine of the turbojet are essentially bypassed, and the engine operates as a ramjet with the afterburner acting as the ramjet's burner.

The advanced needs of future combat aircraft have renewed interest in variable-cycle or adaptive engine technology and development. Research

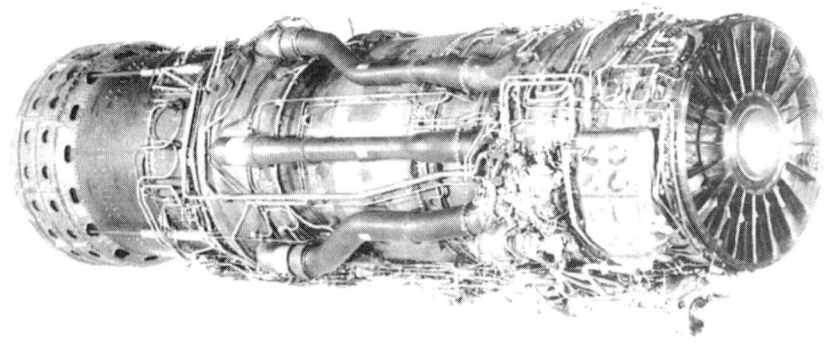

Fig. 1.12c   Pratt & Whitney J58 turbojet. (Courtesy of Pratt & Whitney.)

Fig. 1.12d Lockheed SR-71 Blackbird. (Courtesy of Lockheed.)

and development continues on turbofan engines that can change their bypass ratio to improve performance in subsonic and supersonic operation.

1.4.7 Aircraft Engine Performance Parameters

This section presents several of the airbreathing engine performance parameters that are useful in aircraft propulsion. The first performance parameter is the thrust of the engine that is available for sustained flight (thrust = drag), accelerated flight (thrust > drag), or deceleration (thrust < drag).

We define a *propulsion system* as a unit submerged in a fluid medium about and through which the fluid flows. The propulsion system contains an energy-transfer mechanism that increases the kinetic energy of the fluid passing through the system. This mechanism is called the *engine*. In Fig. 1.13, the engine is shown schematically in a nacelle housing that forms

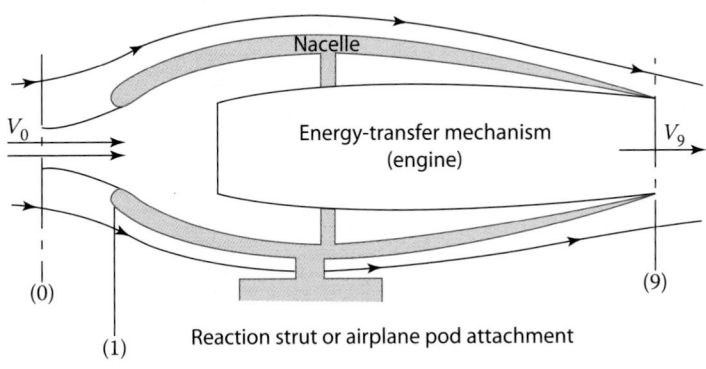

Fig. 1.13 Propulsion system.

the second portion of the propulsion system. Thus the propulsion system contains:

1. An engine (the nozzle is considered to be part of the engine in our terminology)
2. Housing about the engine (nacelle or duct)

Several different aircraft systems may use the same engine submerged in different-shaped nacelles. Thus one propulsion system may use engine X in a pod installation hanging from a wing whereas in another system engine X may be surrounded by a nacelle that is, in fact, the airplane's fuselage— examples are the F-15 vs F-16 propulsion systems that use the F100 turbofan engine. The thrust of a propulsion system will depend on 1) its engine and 2) its nacelle.

As a result, it is conventional to speak of *uninstalled engine thrust and installed engine thrust*. The uninstalled engine thrust should depend on the engine alone and hence must be independent of the nacelle. The installed engine thrust is the thrust produced by both the engine and the nacelle. Installed engine thrust T is defined as the shear force in the reaction strut of Fig. 1.13.

As derived in Chapter 4, the uninstalled thrust F of a jet engine (single inlet and single exhaust) is given by

$$F = \frac{(\dot{m}_0 + \dot{m}_f)V_9 - \dot{m}_0 V_0}{g_c} + (P_9 - P_0)A_9 \qquad (1.5)$$

where

$\dot{m}_0, \dot{m}_f$ = mass flow rates of air and fuel, respectively
V_0, P_0 = freestream (or undisturbed) air velocity and pressure, respectively
V_9, P_9 = exhaust nozzle gas exit velocity and pressure, respectively

For an engine to produce its maximum thrust, it is most desirable to expand the exhaust gas to the ambient pressure, which gives $P_9 = P_0$. In this case, the uninstalled thrust equation becomes

$$F = \frac{(\dot{m}_0 + \dot{m}_f)V_9 - \dot{m}_0 V_0}{g_c} \qquad \text{for } P_9 = P_0 \qquad (1.6)$$

The installed thrust T is equal to the uninstalled thrust F minus the inlet drag D_{inlet} and minus the nozzle drag D_{noz}, or

$$T = F - D_{\text{inlet}} - D_{\text{noz}} \qquad (1.7)$$

Dividing the inlet drag D_{inlet} and nozzle drag D_{noz} by the uninstalled thrust F yields the dimensionless inlet loss coefficient ϕ_{inlet} and nozzle loss coefficient ϕ_{noz}, or

$$\phi_{\text{inlet}} = \frac{D_{\text{inlet}}}{F}$$

$$\phi_{\text{noz}} = \frac{D_{\text{noz}}}{F}$$

(1.8)

Thus the relationship between the installed thrust T and uninstalled thrust F is simply

$$T = F(1 - \phi_{\text{inlet}} - \phi_{\text{noz}})$$

(1.9)

The second performance parameter is the thrust specific fuel consumption (S and TSFC). This is the rate of fuel use by the propulsion system per unit of thrust produced. The uninstalled fuel consumption S and installed fuel consumption TSFC are written in equation form as

$$S = \frac{\dot{m}_f}{F}$$

(1.10)

$$\text{TSFC} = \frac{\dot{m}_f}{T}$$

(1.11)

Note that because the value of g_c derived in the English unit system (Section 1.2) is the same value as g_0 (the acceleration of gravity at sea level), the mass flow rates and weight flow rates have the same magnitude. Consequently, engine manufacturers typically state thrust specific fuel consumptions in units of per hour (per h or h raised to the -1). Throughout this text, we use (lbm/h)/lbf for S and/or TSFC.

The relation between S and TSFC in equation form is given by

$$S = \text{TSFC}(1 - \phi_{\text{inlet}} - \phi_{\text{noz}})$$

(1.12)

Values of uninstalled thrust F and fuel consumption S for various jet engines at sea-level static (SLS), standard day conditions are listed in Appendix B. It is important for you to recognize that unless otherwise stated, engine performance is provided at SLS, standard day conditions, and it is *uninstalled* engine performance. These tables are useful for comparison to results that you obtain in your work.

The predicted variations of uninstalled engine thrust F and uninstalled thrust specific fuel consumption S with Mach number and altitude for an advanced low-bypass-ratio, mixed-flow, augmented turbofan fighter engine [6] are plotted in Figs. 1.14a–1.14b. These figures present the performance for maximum power (afterburner on) and military power (afterburner off). Note that at any given Mach number, the thrust decreases with altitude (due to decreases in air density) and the fuel consumption S also decreases with altitude until 36 kft (the start of the isothermal layer of the atmosphere).

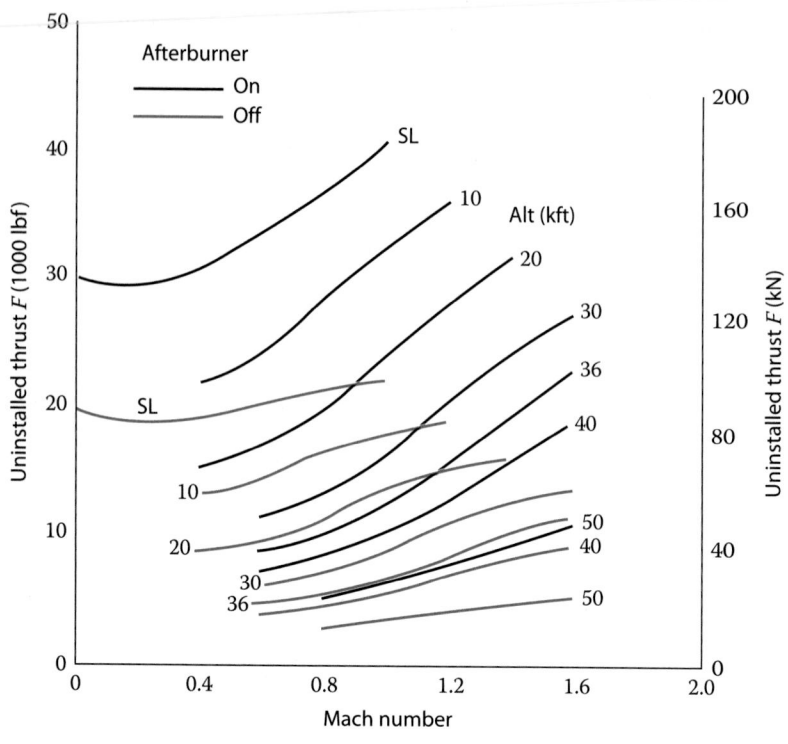

Fig. 1.14a   Uninstalled thrust F of an advanced afterburning fighter engine at maximum (afterburner on) and military (afterburner off) power settings. (Extracted from Ref. 6.)

Also note that the thrust generally increases with Mach number, precisely why the low-bypass-ratio, augmented turbofan is the engine of choice for fighter applications. Most striking from Fig. 1.14b is the large increase in fuel consumption in afterburning operation. In fact, Figs. 1.14a and 1.14b nicely show a good rule-of-thumb for afterburner operation; namely, 50% increase in thrust at the expense of 3–4 times the fuel flow rate. The predicted partial-throttle performance of the advanced fighter engine is shown at three flight conditions in Fig. 1.14c. Afterburner operation is again quite apparent, as indicated by the sharp increase in fuel consumption at the high thrust values.

The takeoff thrust of the JT9D high-bypass-ratio turbofan engine is given in Fig. 1.15a vs Mach number and ambient air temperature for two versions. The JT9D engines were the first high-bypass-ratio turbofans to power a wide-body commercial airliner (Boeing 747) and are still in use today. The PW4000 engines (Fig. 1.8a) were the follow-on engines to the JT9D series. Note the rapid falloff of thrust with rising Mach number that is characteristic of this engine cycle and the constant thrust at a Mach number for temperatures of 86°F and below. (This is often referred to as a *flat rating*.)

The partial-throttle performance of both engine versions is given in Fig. 1.15b for two combinations of altitude and Mach number.

Although the aircraft gas turbine engine is a very complex machine, the basic tools for modeling its performance are developed in the following chapters. These tools are based on the work of Gordon Oates [7]. They permit performance calculations for existing and proposed engines and generate performance curves similar to Figs. 1.14a–1.14c and Figs. 1.15a and 1.15b.

The value of the installation loss coefficient depends on the characteristics of the particular engine/airframe combination, the Mach number, and the engine throttle setting. Typical values are given in Table 1.3 for guidance.

The thermal efficiency η_{Th} of an engine is another very useful engine performance parameter. *Thermal efficiency* is defined as the net rate of organized energy (shaft power or kinetic energy) out of the engine divided by the rate of thermal energy available from the fuel in the engine. The fuel's

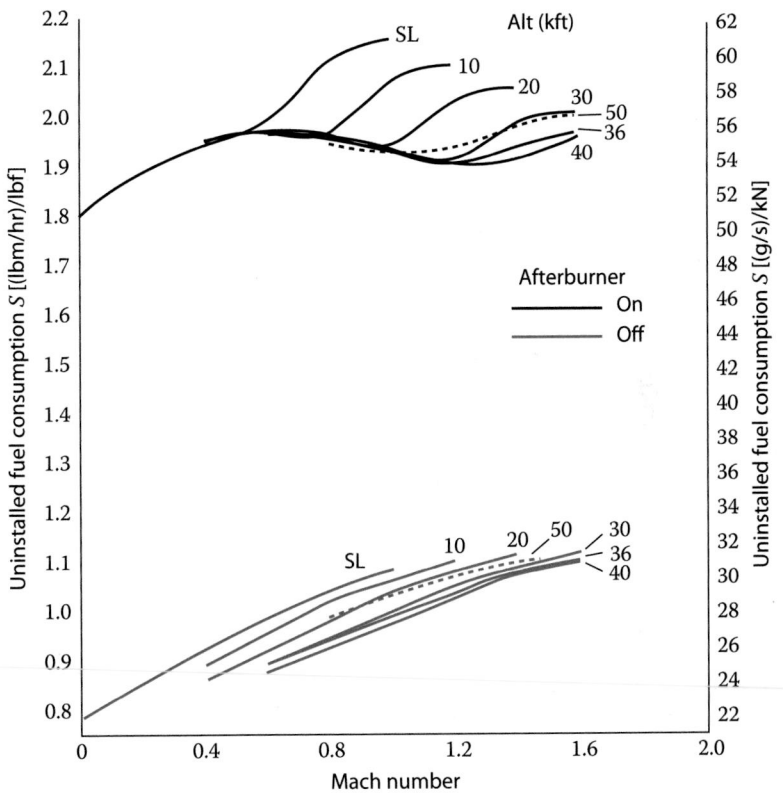

Fig. 1.14b Uninstalled fuel consumption S of an advanced afterburning fighter engine at maximum (afterburner on) and military (afterburner off) power settings. (Extracted from Ref. 6)

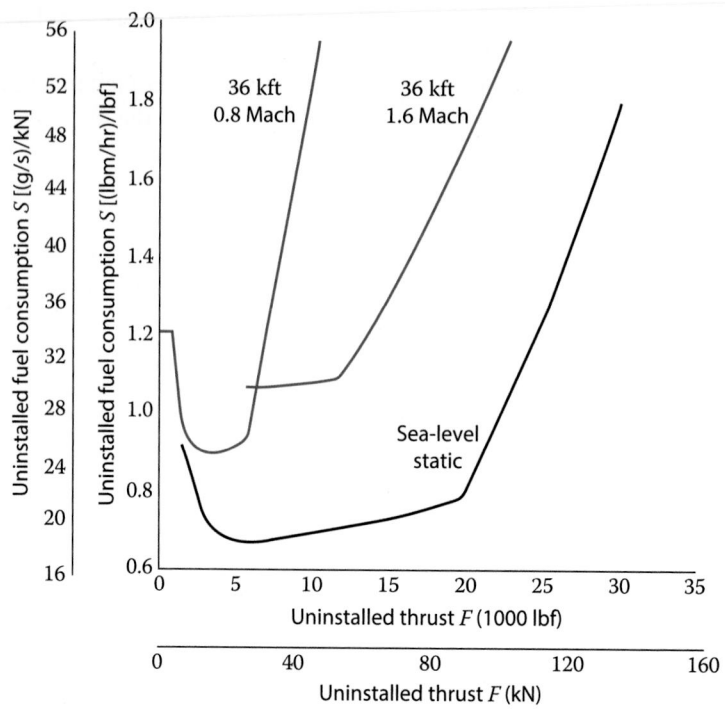

Fig. 1.14c   Partial-throttle performance of an advanced fighter engine. (Extracted from Ref. 6.)

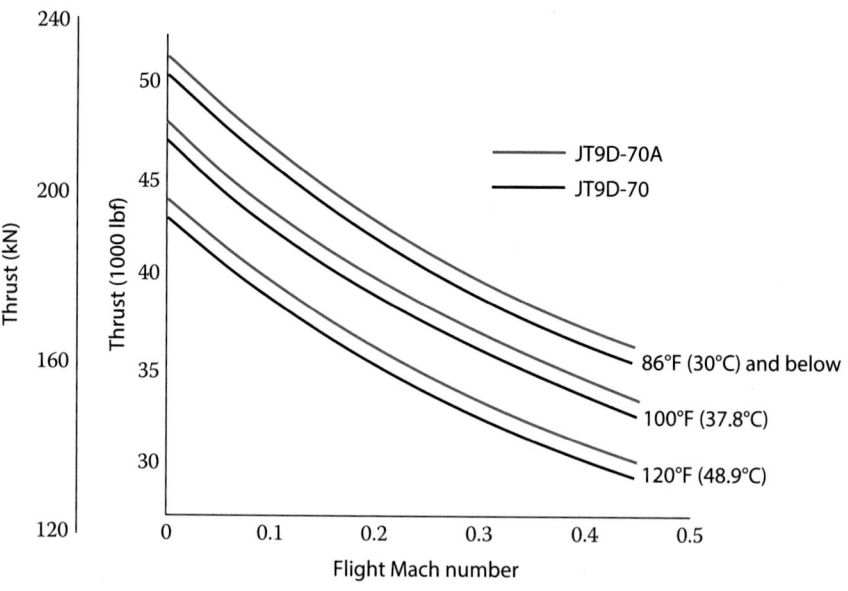

Fig. 1.15a   JT9D-70/-70A turbofan takeoff thrust. (Courtesy of Pratt & Whitney.)

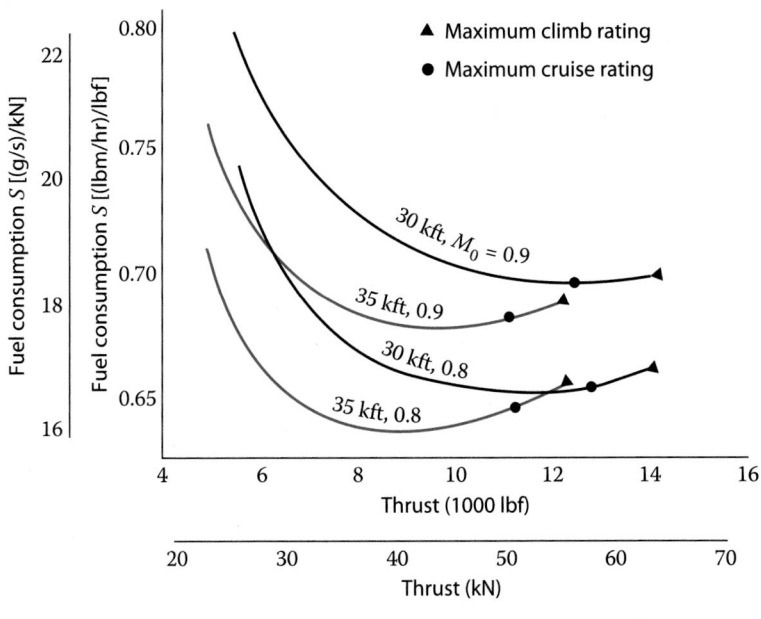

Fig. 1.15b   JT9D-70/-70A turbofan cruise-specific fuel consumption.
(Courtesy of Pratt & Whitney.)

available thermal energy is equal to the mass flow rate of the fuel $\dot{m}_f$ times the fuel lower-heating value h_{PR}. As pointed out in the von Ohain Foreword (in the "Development of High-Pressure-Ratio Turbojets" section) and shown later in Chapter 4 (in Section 4.5), increasing the compressor pressure ratio is the key to increasing the thermal efficiency. Thermal efficiency can be written in equation form as

$$\eta_{Th} = \frac{\dot{W}_{out}}{\dot{Q}_{in}} \tag{1.13}$$

where

η_{Th} = thermal efficiency of engine
$\dot{W}_{out}$ = net power out of engine
$\dot{Q}_{in}$ = rate of thermal energy released ($\dot{m}_f h_{PR}$)

Table 1.3   Typical Aircraft Engine Installation Losses

Flight Condition	M < 1		M > 1	
Aircraft type	ϕ_{inlet}	ϕ_{noz}	ϕ_{inlet}	ϕ_{noz}
Fighter	0.05	0.01	0.05	0.03
Passenger/cargo	0.02	0.01	—	—
Bomber	0.03	0.01	0.04	0.02

Note that for engines with shaft power output, $\dot{W}_{out}$ is equal to this shaft power. For engines with no shaft power output (e.g., turbojet engine), $\dot{W}_{out}$ is equal to the net rate of change of the kinetic energy of the fluid through the engine. The power out of a jet engine with a single inlet and single exhaust (e.g., turbojet engine) is given by

$$\dot{W}_{out} = \frac{1}{2g_c}[(\dot{m}_0 + \dot{m}_f)V_9^2 - \dot{m}_0 V_0^2]$$

The propulsive efficiency η_P of a propulsion system is a measure of how effectively the engine power $\dot{W}_{out}$ is used to power the aircraft. *Propulsive efficiency* is the ratio of the aircraft power (thrust times velocity) to the power out of the engine $\dot{W}_{out}$. In equation form, this is written as

$$\eta_P = \frac{TV_0}{\dot{W}_{out}} \tag{1.14}$$

where

$\eta_P =$ propulsive efficiency of engine
$T =$ installed thrust of propulsion system
$V_0 =$ velocity of aircraft
$\dot{W}_{out} =$ net power out of engine

For a jet engine with a single inlet and single exhaust and an exit pressure equal to the ambient pressure, the propulsive efficiency is given by

$$\eta_P = \frac{2(1 - \phi_{inlet} - \phi_{noz})[(\dot{m}_0 + \dot{m}_f)V_9 - \dot{m}_0 V_0]V_0}{(\dot{m}_0 + \dot{m}_f)V_9^2 - \dot{m}_0 V_0^2} \tag{1.15}$$

For the case when the mass flow rate of the fuel is much less than that of air and the installation losses are very small, Eq. (1.15) simplifies to the following equation for the propulsive efficiency:

$$\eta_P = \frac{2}{V_9/V_0 + 1} \tag{1.16}$$

Equation (1.16) is plotted vs the velocity ratio V_9/V_0 in Fig. 1.16 and shows that high propulsive efficiency requires the exit velocity to be approximately equal to the inlet velocity. Turbojet engines have high values of the velocity ratio V_9/V_0 (especially at subsonic flight speeds) with corresponding low propulsive efficiency, whereas turbofan engines have low values of the velocity ratio V_9/V_0 with corresponding high propulsive efficiency. It is propulsively more efficient to "spread out" a decreased exit velocity over an increased mass flow rate. This is precisely why high-bypass-ratio turbofans dominate the civil aviation market and why engine manufacturers continue to increase bypass ratios.

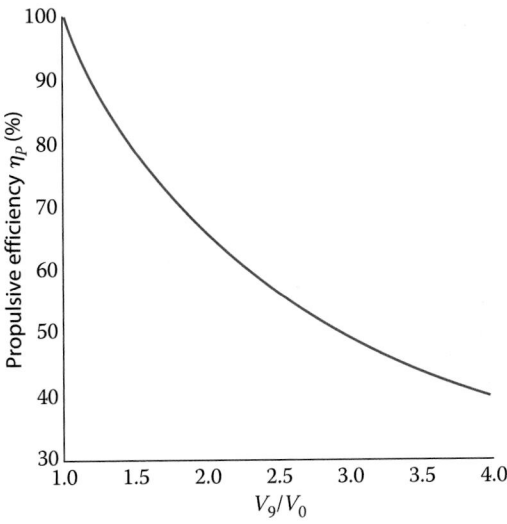

Fig. 1.16   Propulsive efficiency vs velocity ratio (V_9/V_0).

The thermal and propulsive effciencies can be combined to give the *overall efficiency* η_O of a propulsion system. Multiplying propulsive efficiency by thermal efficiency, we get the ratio of the aircraft power to the rate of thermal energy released in the engine (the overall efficiency of the propulsion system):

$$\eta_O = \eta_P \eta_{Th} \tag{1.17}$$

$$\eta_O = \frac{TV_0}{\dot{Q}_{in}} \tag{1.18}$$

Several of the preceding performance parameters are plotted for general types of gas turbine engines in Figs. 1.17a, 1.17b, and 1.17c. These plots can be used to obtain the general uninstalled trends of these performance parameters with flight velocity for each propulsion system.

Because $\dot{Q}_{in} = \dot{m}_f h_{PR}$, Eq. (1.18) can be rewritten as

$$\eta_O = \frac{TV_0}{\dot{m}_f h_{PR}}$$

With the help of Eq. (1.11), this equation can be written in terms of the thrust specific fuel consumption as

$$\eta_O = \frac{V_0}{\text{TSFC} \cdot h_{PR}} \tag{1.19}$$

Using Eqs. (1.17) and (1.19), we can write the following for TSFC:

$$\text{TSFC} = \frac{V_0}{\eta_P \eta_{Th} h_{PR}} \tag{1.20}$$

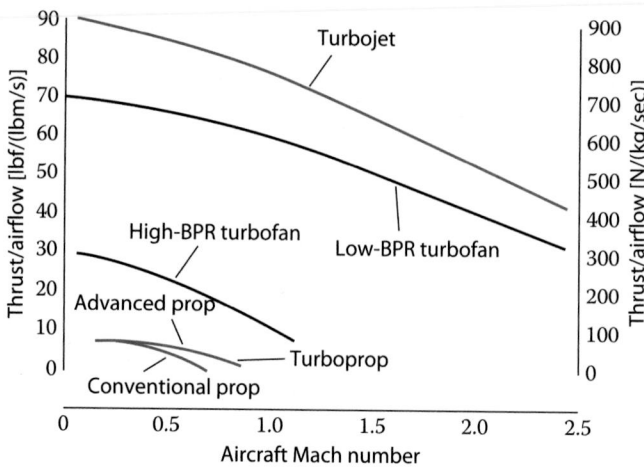

Fig. 1.17a   Uninstalled specific thrust characteristics of typical aircraft engines. (Courtesy of Pratt & Whitney.)

The thermal efficiency of a gas turbine engine is directly related to the compressor pressure ratio—a higher pressure ratio corresponds to higher thermal efficiency. The propulsive efficiency of the gas turbine engine is related to the

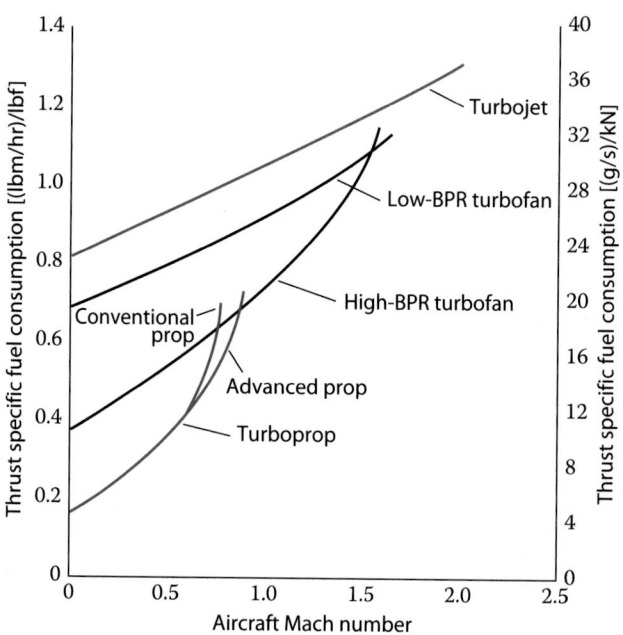

Fig. 1.17b   Uninstalled thrust specific fuel consumption characteristics of typical aircraft engines. (Courtesy of Pratt & Whitney.)

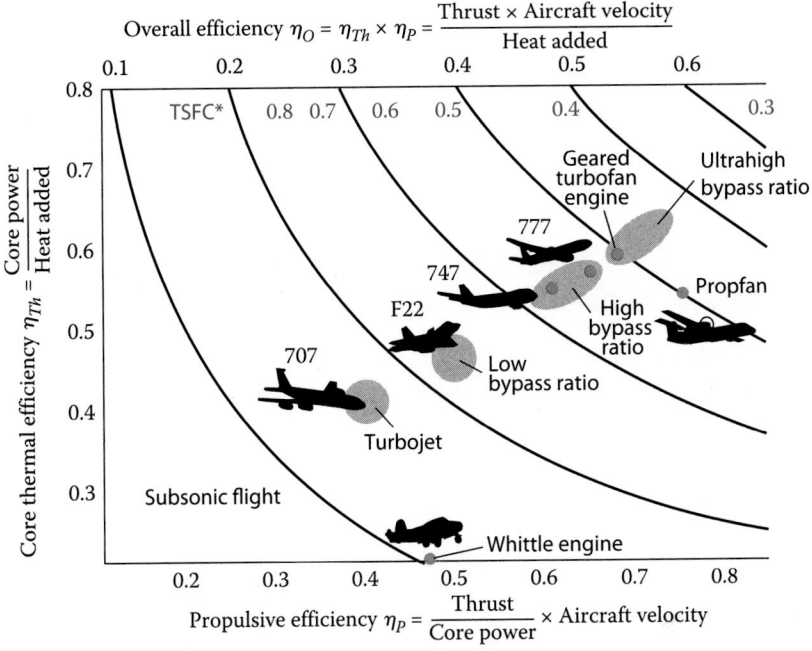

Fig. 1.17c Efficiency characteristics of typical aircraft engines.
(Courtesy of Pratt & Whitney.)

ratio of the exit velocity to the inlet velocity, as shown in Eq. (1.16) and Fig. 1.16. The high-bypass turbofan engine obtains high propulsive efficiency by producing high power in the engine core and transferring this to the bypasss stream of air while keeping its exit velocity low. The amount of

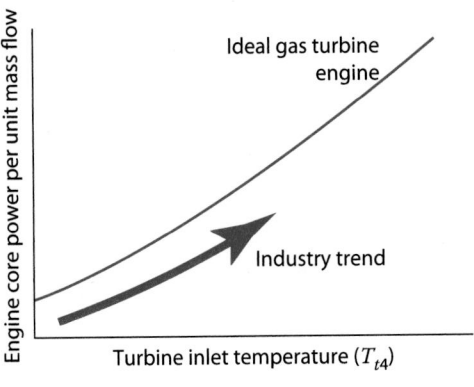

Fig. 1.17d Trend for engine core power.

core power has increased with combustor exit temperature (T_{t4}), which is needed to accommodate higher compressor pressure ratios, as shown in Fig. 1.17d. Table 6.2 shows the increase in maximum allowable T_{t4} in 20-year time intervals starting in 1945 and estimated out to 2045. Continued advances in materials, manufacturing, cooling, and the like make engine operation at these impressive temperatures possible.

Note on Propulsive Efficiency

The kinetic energy of the fluid flowing through an aircraft propulsion system is increased by an energy-transfer mechanism consisting of a series of processes constituting an engine cycle. From the point of view of an observer riding on the propulsion unit (see Fig. 1.18a), the engine cycle output is the increase of kinetic energy received by the air passing through the engine, which is $(V_9 - V_0)/(2g_c)$. From this observer's point of view, the total power output of the engine is the kinetic energy increase imparted to the air per unit time. On the other hand, from the point of view of an observer on the ground (see Fig. 1.18b), one sees the aircraft propulsion system's thrust moving at a velocity V_0 and observes the still air to receive an increase in kinetic energy, after passing through the engine, by an amount $(V_9 - V_0)^2/(2g_c)$. From this point of view, therefore, the total effect of the engine (and its output) is the sum of the propulsive power FV_0 and the kinetic energy per unit time imparted to the air passing through the engine. The sole purpose of the engine is to produce a propulsive power, and this is called the useful power output of the propulsion system. The ratio of the useful power output to the total power output of the propulsion system is called the propulsive efficiency [see Eq. (1.14)].

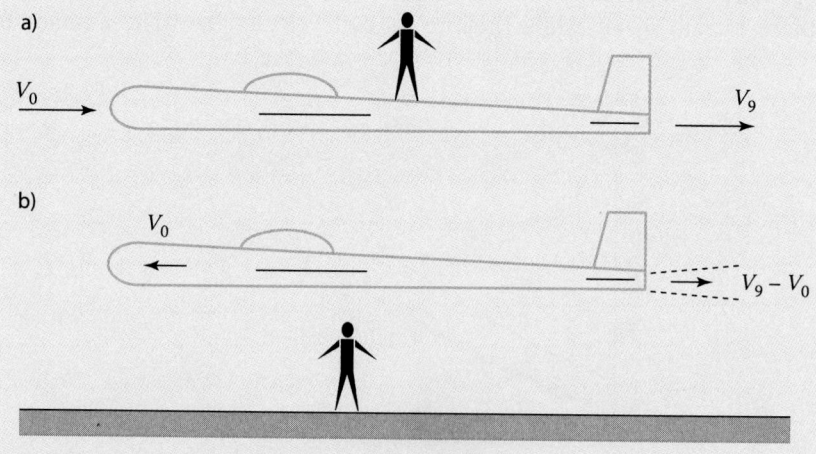

Fig. 1.18 Velocity change by observer a) on aircraft, b) on ground.

Example 1.1

An advanced fighter engine operating at Mach 0.8 and 10-km altitude has the following uninstalled performance data and uses a fuel with $h_{PR} = 42,800\,\text{kJ/kg}$:

$$F = 50\,\text{kN} \quad \dot{m}_0 = 45\,\text{kg/s} \quad \dot{m}_f = 2.65\,\text{kg/s}$$

Determine the specific thrust, thrust specific fuel consumption, exit velocity, thermal efficiency, propulsive efficiency, and overall efficiency (assume exit pressure equal to ambient pressure).

Solution

$$\frac{F}{\dot{m}_0} = \frac{50\,\text{kN}}{45\,\text{kg/s}} = 1.1111\,\text{kN/(kg/s)} = 1111.1\,\text{m/s}$$

$$S = \frac{\dot{m}_f}{F} = \frac{2.65\,\text{kg/s}}{50\,\text{kN}} = 0.053\,(\text{kg/s})/\text{kN} = 53\,\text{mg/N}\cdot\text{s}$$

$$V_0 = M_0 a_0 = M_0\left(\frac{a_0}{a_{\text{ref}}}\right)a_{\text{ref}} = 0.8(0.8802)340.3 = 239.6\,\text{m/s}$$

From Eq. (1.6) we have

$$V_9 = \frac{Fg_c + \dot{m}_0 V_0}{\dot{m}_0 + \dot{m}_f} = \frac{50,000 \times 1 + 45 \times 239.6}{45 + 2.65} = 1275.6\,\text{m/s}$$

$$\eta_{Th} = \frac{\dot{W}_{\text{out}}}{\dot{Q}_{\text{in}}} = \frac{(\dot{m}_0 + \dot{m}_f)V_e^2 - \dot{m}_0 V_0^2}{2g_c \dot{m}_f h_{PR}}$$

$$\dot{W}_{\text{out}} = \frac{(\dot{m}_0 + \dot{m}_f)V_e^2 - \dot{m}_0 V_0^2}{2g_c}$$

$$= \frac{47.65 \times 1275.6^2 - 45 \times 239.6^2}{2 \times 1} = 37.475 \times 10^6\,\text{W}$$

$$\dot{Q}_{\text{in}} = \dot{m}_f h_{PR} = 2.65 \times 42,800 = 113.42 \times 10^6\,\text{W}$$

$$\eta_{Th} = \frac{\dot{W}_{\text{out}}}{\dot{Q}_{\text{in}}} = \frac{37.475 \times 10^6}{113.42 \times 10^6} = 33.04\%$$

$$\eta_P = \frac{FV_0}{\dot{W}_{\text{out}}} = \frac{50,000 \times 239.6}{37.475 \times 10^6} = 31.97\%$$

$$\eta_O = \frac{FV_0}{\dot{Q}_{\text{in}}} = \frac{50,000 \times 239.6}{113.42 \times 10^6} = 10.56\%$$

1.4.8 Uninstalled Specific Thrust vs Fuel Consumption

For a jet engine with a single inlet and single exhaust and exit pressure equal to ambient pressure, when the mass flow rate of the fuel is much less than that of air and the installation losses are very small, the specific thrust $F/\dot{m}_0$ (thrust per unit airflow) can be written as

$$\frac{F}{\dot{m}_0} = \frac{V_9 - V_0}{g_c} \tag{1.21}$$

Then the propulsive efficiency of Eq. (1.16) can be rewritten as

$$\eta_P = \frac{2}{Fg_c/(\dot{m}_0 V_0) + 2} \tag{1.22}$$

Substituting Eq. (1.22) into Eq. (1.20), and replacing TSFC with S because we are interested in uninstalled thrust specific fuel consumption, we obtain the following very enlightening expression:

$$S = \frac{Fg_c/\dot{m}_0 + 2V_0}{2\eta_{Th}h_{PR}} \tag{1.23}$$

Aircraft manufacturers desire engines having low thrust specific fuel consumption S and high specific thrust $F/\dot{m}_0$. Low engine fuel consumption can be directly translated into longer range, increased payload, and/or reduced aircraft size. High specific thrust reduces the cross-sectional area of the engine and has a direct influence on engine weight and installation losses. This desired trend is plotted in Fig. 1.19a. Equation (1.23) is also plotted in Fig. 1.19a and shows that fuel consumption and specific thrust are directly proportional. Thus the aircraft manufacturers have to make a tradeoff.

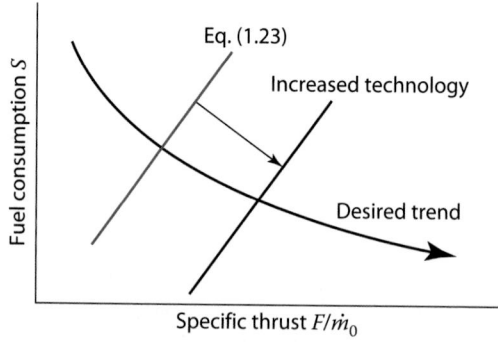

Fig. 1.19a   Relationship between specific thrust and fuel consumption.

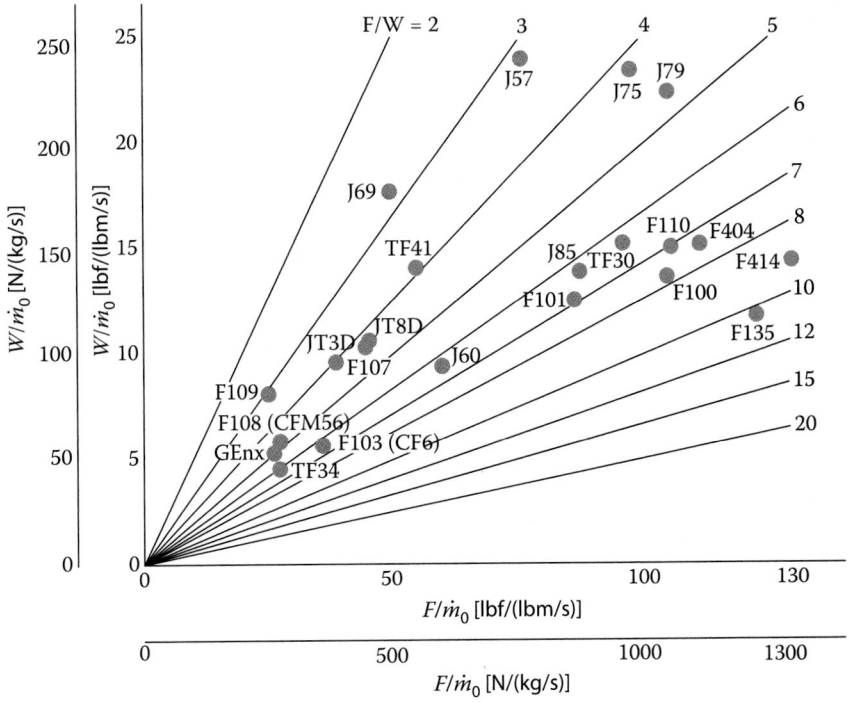

Fig. 1.19b   Engine thrust/weight ratio F/W.

The line of Eq. (1.23) shifts in the desired direction when there is an increase in the level of technology (increased thermal efficiency) or an increase in the fuel heating value.

Another very useful measure of merit for the aircraft gas turbine engine is the thrust/weight ratio F/W. For a given engine thrust F, increasing the thrust/weight ratio reduces the weight of the engine. Aircraft manufacturers can use this reduction in engine weight to increase the capabilities of an aircraft (increased payload, increased fuel, or both) or decrease the size (weight) and cost of a new aircraft under development.

Engine companies expend considerable research and development effort on increasing the thrust/weight ratio of aircraft gas turbine engines. This ratio is equal to the specific thrust $F/\dot{m}_0$ divided by the engine weight per unit of mass flow $W/\dot{m}_0$. For a given engine type, the engine weight per unit mass flow is related to the efficiency of the engine structure, and the specific thrust is related to the engine thermodynamics. The weights per unit mass flow of some existing gas turbine engines are plotted vs specific thrust in Fig. 1.19b. Also plotted are lines of constant engine thrust/weight ratio F/W. Not surprisingly, we find the engines for fighter applications have the highest specific thrusts and engine thrust/weight ratios.

1.5 Aircraft Performance

This section on aircraft performance is included so that the reader may get a better understanding of the propulsion requirements of the aircraft [8]. Also, as discussed in Section 1.4.7, the *installed* performance of an engine is entirely dependent on the application; thus, a foundational understanding of aircraft-engine system performance is important. The coverage is limited to a few significant concepts that directly relate to aircraft engines. It is not intended as a substitute for the many excellent references on this subject (see Refs. 9–12).

1.5.1 Performance Equation

Relationships for the performance of an aircraft can be obtained from energy considerations (see Ref. 13). By treating the aircraft (Fig. 1.20) as a moving mass and assuming that the installed propulsive thrust T, aerodynamic drag D, and other resistive forces R act in the same direction as the velocity V, it follows that

$$
\underbrace{[T - (D+R)]V}_{\substack{\text{rate of} \\ \text{mechanical} \\ \text{energy} \\ \text{input}}} = \underbrace{W\frac{dh}{dt}}_{\substack{\text{storage} \\ \text{rate of} \\ \text{potential} \\ \text{energy}}} + \underbrace{\frac{W}{g}\frac{d}{dt}\left(\frac{V^2}{2}\right)}_{\substack{\text{storage rate} \\ \text{of kinetic} \\ \text{energy}}}
\tag{1.24}
$$

Note that the total resistive force $D+R$ is the sum of the drag of the clean aircraft D and any additional drags R associated with such protuberances as landing gear, external stores, or drag chutes.

By defining the energy height z_e as the sum of the potential and kinetic energy terms

$$
z_e \equiv h + \frac{V^2}{2g}
\tag{1.25}
$$

Eq. (1.24) can now be written simply as

$$
[T - (D+R)]V = W\frac{dz_e}{dt}
\tag{1.26}
$$

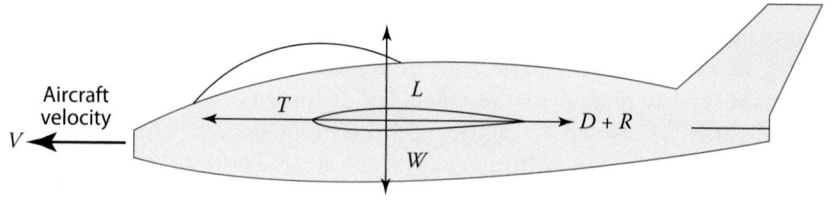

Fig. 1.20   Forces on aircraft.

By defining the *weight specific excess power* P_s as

$$P_s \equiv \frac{dz_e}{dt} \tag{1.27}$$

Equation (1.26) can now be written in its dimensionless form as

$$\frac{T - (D + R)}{W} = \frac{P_s}{V} = \frac{1}{V} \frac{d}{dt} \left(h + \frac{V^2}{2g} \right) \tag{1.28}$$

This is a very powerful equation that gives insight into the dynamics of flight, including both the rate of climb dh/dt and acceleration dV/dt.

1.5.2 Lift and Drag

We use the classical aircraft lift relationship

$$L = nW = C_L q S_w \tag{1.29}$$

where n is the load factor or number of g perpendicular to V ($n = 1$ for straight and level flight), C_L is the coefficient of lift, S_w is the wing planform area, and q is the dynamic pressure. The dynamic pressure can be expressed in terms of the density ρ and velocity V or the pressure P and Mach number M as

$$q = \frac{1}{2} \rho \frac{V^2}{g_c} = \frac{1}{2} \sigma \rho_{\text{ref}} \frac{V^2}{g_c} \tag{1.30a}$$

or

$$q = \frac{\gamma}{2} P M_0^2 = \frac{\gamma}{2} \delta P_{\text{ref}} M_0^2 \tag{1.30b}$$

where δ and σ are the dimensionless pressure and density ratios defined by Eqs. (1.2) and (1.4), respectively, and γ is the ratio of specific heats ($\gamma = 1.4$ for air). The reference density ρ_{ref} and reference pressure P_{ref} of air are their sea-level values on a standard day and are listed in Appendix A. We also use the classical aircraft drag relationship

$$D = C_D q S_w \tag{1.31}$$

Figure 1.21a is a plot of lift coefficient C_L vs drag coefficient C_D, commonly called the lift-drag polar, for a typical subsonic passenger aircraft. The drag coefficient curve can be approximated by a second-order equation in C_L written as

$$C_D = K_1 C_L^2 + K_2 C_L + C_{D0} \tag{1.32}$$

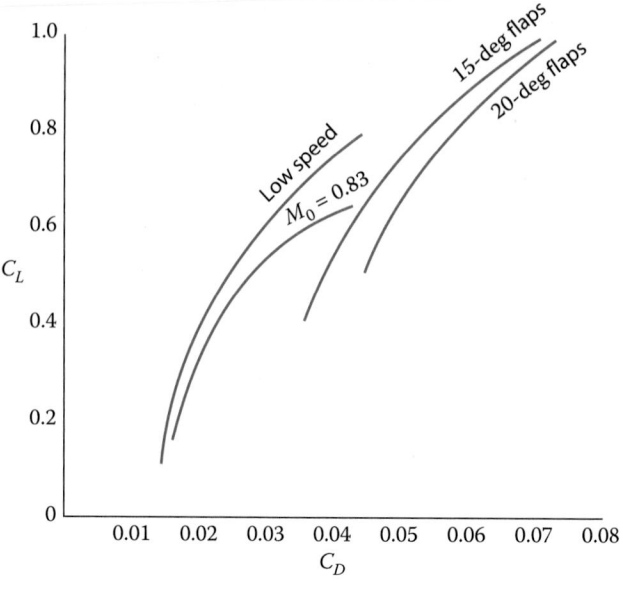

Fig. 1.21a   Typical lift-drag polar for subsonic aircraft.

where the coefficients K_1, K_2, and C_{D0} are typically functions of flight Mach number and wing configuration (flap position, etc.).

The C_{D0} term in Eq. (1.32) is the zero lift drag coefficient that accounts for both frictional and pressure drag in subsonic flight and wave drag in supersonic flight. The K_1 and K_2 terms account for the drag due to lift. Normally K_2 is very small and approximately equal to zero for most fighter aircraft.

Given the installed thrust T performance of an engine and the drag D of an aircraft, the thrust-related performance of the aircraft can be calculated for a weight W and load factor n. Using Eq. (1.28), one can generate the very useful aircraft performance plot showing the resultant P_s. One such plot for a typical fighter aircraft is shown in Fig. 1.21b. Contours of constant P_s are shown in dark blue and lines of constant energy height (z_e) are shown in light blue. This plot shows regions of flight where the aircraft has excess thrust ($P_s > 0$) and regions where it cannot sustain its altitude and speed ($P_s < 0$).

One can see from Fig. 1.21b that the aircraft system (engine/airframe) has a P_s of about zero at a velocity of 1500 ft/s, an altitude of 60 kft, and a P_s of about 250 ft/s at 1500 ft/s and 40 kft. In addition, we can see that the aircraft system has enough P_s (> 0) to fly supersonic ($M_0 > 1$). If this performance is for nonafterburning engine operation, then this figure shows *supercruise* (flying supersonic without afterburner). How long the aircraft can cruise supersonically d epends on fuel consumption. (We will look at this later in Section 1.5.4.)

With P_s known, we can determine the rate of climb (dh/dt with V constant) or horizontal acceleration (dV/dt with h constant) from Eqs. (1.25) and (1.27). Thus a P_s of 250 ft/s can give an aircraft rate of climb of 250 ft/s (dh/dt) or horizontal acceleration of 0.167 g or 5.36 ft/s^2 at a velocity of 1500 ft/s ($dV/dt = P_s\, g/V$). Performance at other flight conditions can easily be determined in a similar manner. Equation (1.27) can be integrated to get the time between different contours of P_s, giving

$$\Delta t = \int_1^2 dt = \int_{z_{e1}}^{z_{e2}} \frac{dz_e}{P_s}$$

This equation shows that the minimum time to climb from $z_{e1}(h_1, V_1)$ to the higher energy level $z_{e2}(h_2, V_2)$ corresponds to the flight path that produces the maximum specific excess power P_s at each z_e. This maximum occurs at the point of tangency between a line of constant energy height and the contour line of the maximum P_s attainable for that energy height. For $z_e = 70$ kft, for example, this tangency occurs at $P_s = 320$ ft/s, $V = 1480$ ft/s, and $h = 36$ kft. Continuing in this vein, one can find the minimum time-to-climb flight path, from a z_e of 13 kft at sea level to a z_e of 96 kft at an altitude of 57 kft, as shown in Fig. 1.21b.

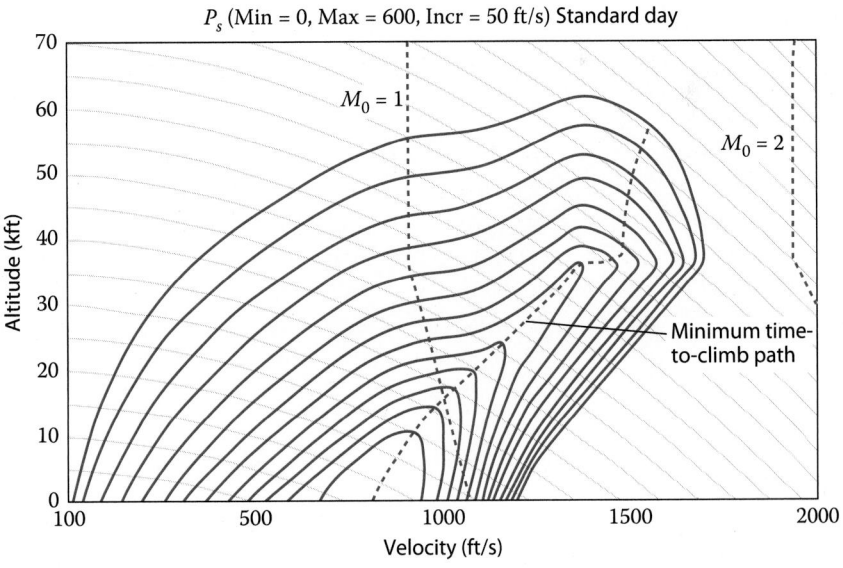

Fig. 1.21b   Typical fighter aircraft performance plot of P_s for $n = 1$.

Example 1.2

For all the examples given in this section on aircraft performance, two types of aircraft will be considered.

a) *Fighter aircraft (HF-1)*. An advanced fighter aircraft is approximately modeled after the F-22 Advanced Tactical Fighter shown in Fig. 1.22. For convenience, we will designate our hypothetical fighter aircraft as the HF-1, having the following characteristics:

> Maximum gross takeoff weight W_{TO} = 40,000 lbf (177,920 N)
> Empty weight = 24,000 lbf (106,752 N)
> Maximum fuel plus payload weight = 16,000 lbf (71,168 N)
> Permanent payload = 1600 lbf (7117 N, crew plus
> return armament)
> Expended payload = 2000 lbf (8896 N, missiles plus ammunition)
> Maximum fuel capacity = 12,400 lbf (55,155 N)
> Wing area S_w = 720 ft^2 (66.9 m^2)
> Engine = Low-bypass-ratio, mixed-flow turbofan with afterburner
> Maximum lift coefficient C_{Lmax} = 1.8
> Drag coefficients given in Table 1.4.

b) *Passenger aircraft (HP-1)*. An advanced 253-passenger commercial aircraft approximately modeled after the Boeing 787 is shown in Fig. 1.23. For convenience, we will designate our hypothetical passenger aircraft as the HP-1, having the following characteristics:

Fig. 1.22 F-22 Advanced Tactical Fighter. (Photo courtesy of Boeing Defense & Space Group, Military Airplanes Division.)

(Continued)

Example 1.2 *(Continued)*

Table 1.4   Drag Coefficients for Hypothetical Fighter Aircraft (HF-1)

M_0	K_1	K_2	C_{D0}
0.0	0.20	0.0	0.0120
0.8	0.20	0.0	0.0120
1.0	0.20	0.0	0.0173
1.4	0.28	0.0	0.0280
2.0	0.40	0.0	0.0270

Maximum gross takeoff weight W_{TO} = 1,645,760 N (370,000 lbf)
Empty weight = 822,880 N (185,500 lbf)
Maximum landing weight = 1,356,640 N (305,000 lbf)
Maximum payload = 420,780 N (94,600 lbf, 253 passengers plus
                   196,000 N of cargo)
Maximum fuel capacity = 716,706 N (161,130 lbf)
Wing area S_w = 282.5 m² (3040 ft²)
Engine = High-bypass-ratio turbofan
Maximum lift coefficient C_{Lmax} = 2.0
Drag coefficients given in Table 1.5.

Fig. 1.23   Boeing 787. (Photo courtesy of Boeing.)

(Continued)

Example 1.2 (Continued)

Table 1.5   Drag Coefficients for Hypothetical Passenger Aircraft (HP-1)

M_0	K_1	K_2	C_{D0}
0.00	0.056	−0.004	0.0140
0.40	0.056	−0.004	0.0140
0.75	0.056	−0.008	0.0140
0.83	0.056	−0.008	0.0150

Example 1.3

Determine the drag polar and drag variation for the HF-1 aircraft at 90% of maximum gross takeoff weight and the HP-1 aircraft at 95% of maximum gross takeoff weight.

a) *Fighter aircraft (HF-1).* The variation in C_{D0} and K_1 with Mach number for the HF-1 are plotted in Fig. 1.24 from the data of Table 1.4.

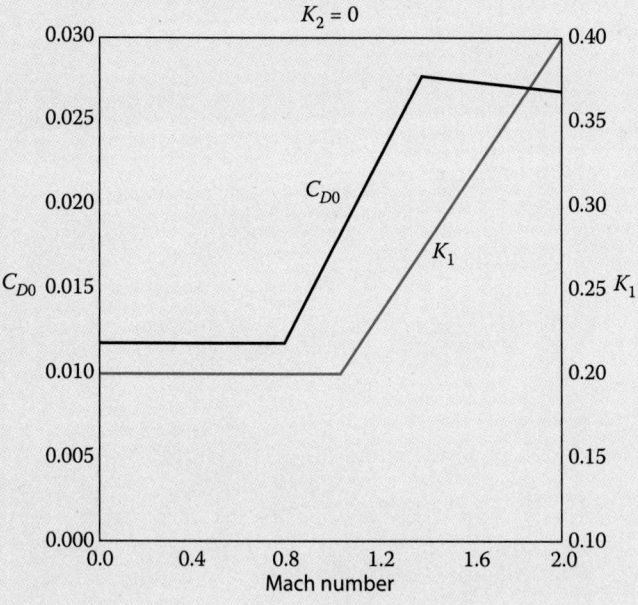

Fig. 1.24   Values of K_1 and C_{D0} for HF-1 aircraft.

(Continued)

Example 1.3 *(Continued)*

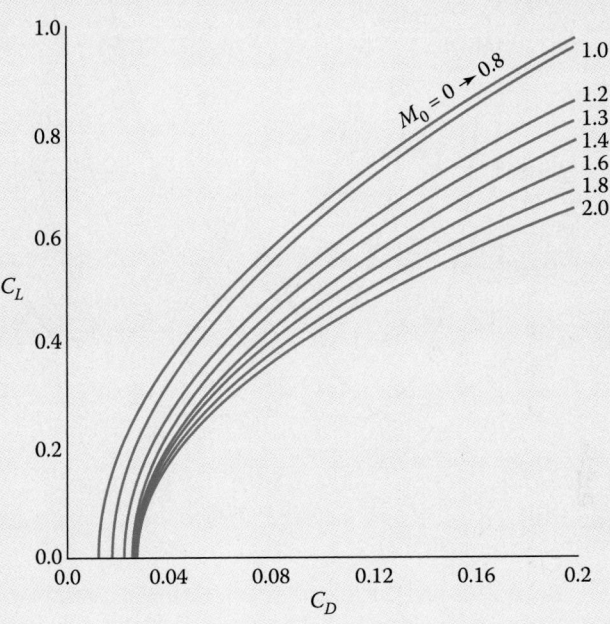

Fig. 1.25   Lift-drag polar for HF-1 aircraft.

Figure 1.25 shows the drag polar at different Mach numbers for the HF-1 aircraft. Using these drag data and the preceding equations gives the variation in aircraft drag with subsonic Mach number and altitude for level flight ($n = 1$), as shown in Fig. 1.26a. Note that the minimum drag is constant for Mach numbers 0 to 0.8 and then increases. This is the same variation as C_{D0}. The variation of drag with load factor n is shown in Fig. 1.26b at two altitudes. The drag increases with increasing load factor, and there is a flight Mach number that gives minimum drag for a given altitude and load factor.

b) *Passenger aircraft (HP-1).* The variation in C_{D0} and K_2 with Mach number for the HP-1 is plotted in Fig. 1.27 from the data of Table 1.5. Figure 1.28 shows the drag polar at different Mach numbers for the HP-1 aircraft. Using these drag data and the preceding equations gives the variation in aircraft drag with subsonic Mach number and altitude for level flight ($n = 1$), as shown in Fig. 1.29. Note that the minimum drag is constant for Mach numbers 0 to 0.75 and then increases. This is the same variation as C_{D0}.

(Continued)

Example 1.3 *(Continued)*

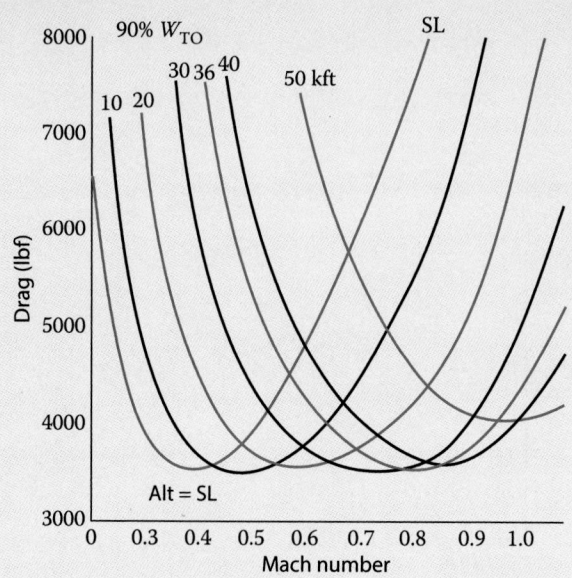

Fig. 1.26a   Drag for level flight ($n = 1$) for HF-1 aircraft.

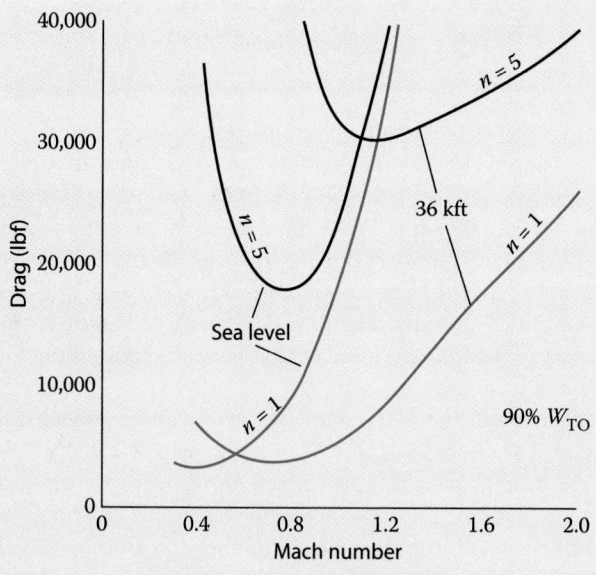

Fig. 1.26b   Drag of HF-1 aircraft at sea level and 36 kft for $n = 1$ and $n = 5$.

(Continued)

Example 1.3 *(Continued)*

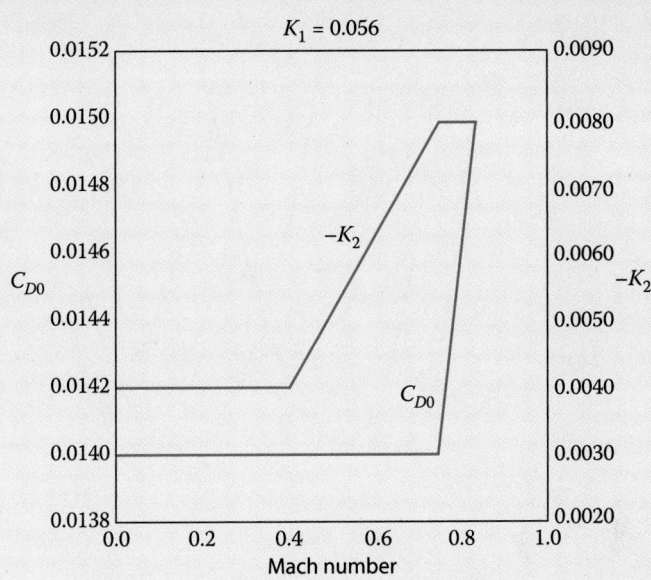

Fig. 1.27 Values of K_2 and C_{D0} for HP-1 aircraft.

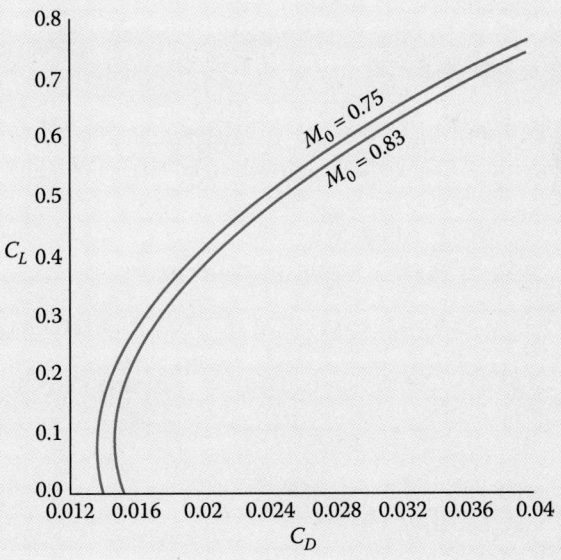

Fig. 1.28 Lift-drag polar for HP-1 aircraft.

(Continued)

Example 1.3 *(Continued)*

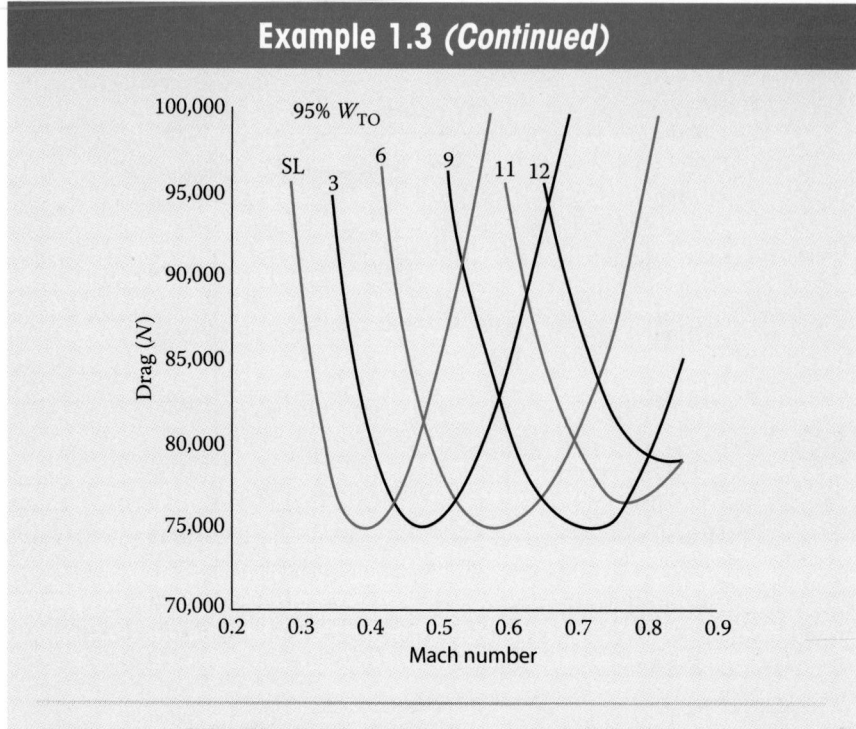

Fig. 1.29 Drag for level flight ($n = 1$) for HP-1 aircraft.

Example 1.4

Calculate the drag at Mach 0.8 and 40-kft altitude of the HF-1 aircraft at 90% of maximum gross takeoff weight with load factors of 1 and 4.

Solution

We begin by calculating the dynamic pressure q:

$$q = \frac{\gamma}{2}\delta P_{ref}M_0^2 = 0.7 \times 0.1858 \times 2116 \times 0.8^2 = 176.1 \, \text{lbf/ft}^2$$

From Fig. 1.24 at $M = 0.8$, $C_{D0} = 0.012$, $K_1 = 0.20$, and $K_2 = 0$.

Case 1: $n = 1$

$$C_L = \frac{nW}{qS_w} = \frac{1 \times 0.9 \times 40,000}{176.1 \times 720} = 0.2839$$

$$C_D = K_1 C_L^2 + K_2 C_L + C_{D0} = 0.2(0.2839^2) + 0.012 = 0.0281$$

$$D = C_D q S_w = 0.0281 \times 176.1 \times 720 = 3563 \, \text{lbf}$$

(Continued)

Example 1.4 (Continued)

Case 2: $n = 4$

$$C_L = \frac{nW}{qS_w} = \frac{4 \times 0.9 \times 40,000}{176.1 \times 720} = 1.136$$

$$C_D = K_1 C_L^2 + K_2 C_L + C_{D0} = 0.2(1.136^2) + 0.012 = 0.2701$$

$$D = C_D q S_w = 0.2701 \times 176.1 \times 720 = 34,247 \text{ lbf}$$

Note that the drag at $n = 4$ is about 10 times that at $n = 1$.

1.5.3 Stall, Takeoff, and Landing Speeds

Stall is the flight condition when an aircraft's wing loses lift. It is an undesirable condition because vehicle control is lost for a time. During level flight (lift = weight), stall will occur when one tries to obtain a lift coefficient greater than the wing's maximum $C_{L\max}$. The *stall speed* is defined as the level flight speed that corresponds to the wing's maximum lift coefficient, easily obtained from Eqs. (1.29) and (1.30a)

$$V_{\text{stall}} = \sqrt{\frac{2g_c}{\rho C_{L\max}} \frac{W}{S_w}} \tag{1.33}$$

To keep away from stall, aircraft are flown at velocities greater than V_{stall}.

Takeoff and landing are two flight conditions in which the aircraft velocity is close to the stall velocity. For safety, the takeoff speed V_{TO} of an aircraft is typically 20% greater than the stall speed, and the landing speed at touchdown V_{TD} is 15% greater:

$$\begin{aligned} V_{\text{TO}} &= 1.20 V_{\text{stall}} \\ V_{\text{TD}} &= 1.15 V_{\text{stall}} \end{aligned} \tag{1.34}$$

Example 1.5

Determine the takeoff speed of the HP-1 at sea level with maximum gross takeoff weight and the landing speed with maximum landing weight.

From Appendix A we have $\rho = 1.255 \text{ kg/m}^3$ for sea level. From Example 1.2b we have $C_{L\max} = 2.0$, $W = 1,645,760$ N, $S_w = 282.5 \text{ m}^2$, and

$$V_{\text{stall}} = \sqrt{\frac{2 \times 1}{1.225 \times 2.0} \frac{1,645,760}{282.5}} = 69.0 \text{ m/s}$$

(Continued)

Example 1.5 (Continued)

Thus

$$V_{TO} = 1.20V_{stall} = 82.8 \, \text{m/s} \, (\approx 185 \, \text{mph})$$

For landing, $W = 1,356,640$ N, and

$$V_{stall} = \sqrt{\frac{2 \times 1}{1.225 \times 2.0} \frac{1,356,640}{282.5}} = 62.6 \, \text{m/s}$$

Thus

$$V_{TD} = 1.15 \, V_{stall} = 72.0 \, \text{m/s} \, (\approx 161 \, \text{mph})$$

1.5.4 Fuel Consumption

The rate of change of the aircraft weight dW/dt is due to the fuel consumed by the engines. The mass rate of fuel consumed is equal to the product of the installed thrust T and the installed thrust specific fuel consumption. For constant acceleration of gravity g_0, we can write

$$\frac{dW}{dt} = -\dot{w}_f = -\dot{m}_f \frac{g_0}{g_c} = -T(\text{TSFC})\left(\frac{g_0}{g_c}\right)$$

This equation can be rewritten in dimensionless form as

$$\frac{dW}{W} = -\frac{T}{W}(\text{TSFC})\left(\frac{g_0}{g_c}\right) dt \tag{1.35}$$

1.5.4.1 Estimate of TSFC

Equation (1.35) requires estimates of installed engine thrust T and installed TSFC to calculate the change in aircraft weight. For many flight conditions, the installed engine thrust T equals the aircraft drag D. The value of TSFC depends on the engine cycle, altitude, and Mach number. For preliminary analysis, the following equations (from Ref. 13) are curve fits of general engine types and can be used to estimate TSFC in units of (lbm/h)/lbf, (θ is the dimensionless temperature ratio T/T_{ref}):

1. High-bypass-ratio turbofan

$$\text{TSFC} = (0.4 + 0.45M_0)\sqrt{\theta} \tag{1.36a}$$

2. Low-bypass-ratio, mixed-flow turbofan
   Military and lower power settings:

$$\text{TSFC} = (0.9 + 0.3M_0)\sqrt{\theta} \tag{1.36b}$$

Maximum power setting:

$$\text{TSFC} = (1.6 + 0.27M_0)\sqrt{\theta} \tag{1.36c}$$

3. Turbojet
   Military and lower power settings:

$$\text{TSFC} = (1.1 + 0.3M_0)\sqrt{\theta} \tag{1.36d}$$

Maximum power setting:

$$\text{TSFC} = (1.5 + 0.23M_0)\sqrt{\theta} \tag{1.36e}$$

4. Turboprop

$$\text{TSFC} = (0.18 + 0.8M_0)\sqrt{\theta} \tag{1.36f}$$

1.5.4.2 Endurance (Loiter)

For straight level unaccelerated flight, thrust equals drag ($T = D$) and lift equals weight ($L = W$). Thus Eq. (1.35) is simply

$$\frac{dW}{W} = -\frac{C_D}{C_L}(\text{TSFC})\left(\frac{g_0}{g_c}\right)dt \tag{1.37}$$

We define the endurance factor (EF) as

$$\text{EF} \equiv \frac{C_L}{C_D(\text{TSFC})}\frac{g_c}{g_0} \tag{1.38}$$

Then Eq. (1.37) becomes

$$\frac{dW}{W} = -\frac{dt}{\text{EF}} \tag{1.39}$$

Note that the minimum fuel consumption for a time t occurs at the flight condition where the endurance factor is maximum.

For the case when the endurance factor is constant or nearly constant, Eq. (1.39) can be integrated from the initial to final conditions and the following expression obtained for the aircraft weight fraction:

$$\frac{W_f}{W_i} = \exp\left(-\frac{t}{\text{EF}}\right) \tag{1.40a}$$

or

$$\frac{W_f}{W_i} = \exp\left[-\frac{C_D}{C_L}(\text{TSFC}\times t)\frac{g_0}{g_c}\right] \tag{1.40b}$$

Note that the weight fraction W_f/W_i is approximately 0.7 when the time t is one-third of the endurance factor EF. For an aircraft with a 30% fuel fraction, all the fuel would be used during an endurance (loiter) flight of a time equal to EF/3. Minimum fuel usage will occur where EF is maximum (which is near the aircraft's maximum L/D, as will be shown in our two example aircraft).

1.5.4.3 Range

For portions of aircraft flight where distance is important, the differential time dt is related to the differential distance ds by

$$ds = V\,dt \tag{1.41}$$

Substituting into Eq. (1.37) gives

$$\frac{dW}{W} = -\frac{C_D}{C_L}\frac{\text{TSFC}}{V}\frac{g_0}{g_c}\,ds \tag{1.42}$$

We define the range factor (RF) as

$$\text{RF} \equiv \frac{C_L}{C_D}\frac{V}{\text{TSFC}}\frac{g_c}{g_0} \tag{1.43}$$

Then Eq. (1.42) can be simply written as

$$\frac{dW}{W} = -\frac{ds}{\text{RF}} \tag{1.44}$$

Note that the minimum fuel consumption for a distance s occurs at the flight condition where the range factor is maximum.

For the flight conditions where the RF is constant or nearly constant, Eq. (1.42) can be integrated from the initial to final conditions and the following expression obtained for the aircraft weight fraction:

$$\frac{W_f}{W_i} = \exp\left(-\frac{s}{\text{RF}}\right) \tag{1.45a}$$

or

$$\frac{W_f}{W_i} = \exp\left(-\frac{C_D}{C_L}\frac{\text{TSFC} \times s}{V}\frac{g_0}{g_c}\right) \tag{1.45b}$$

This is called the *Breguet range equation*. For the range factor to remain constant, C_L/C_D and V/TSFC need to be constant. Above 36-kft altitude, the ambient temperature is constant, and a constant velocity V will correspond to constant Mach and constant TSFC for a fixed throttle setting. If C_L is constant, C_L/C_D will remain constant. Because the aircraft weight W decreases during the flight, the altitude must increase to reduce the density of the ambient air and produce the required lift ($L = W$) while maintaining C_L and velocity constant. This flight profile is called a *cruise climb*. Note that the weight fraction (W_f/W_i) is approximately 0.7 when the distance s is one-third of the range factor RF. For an aircraft with a 30% fuel fraction, all the fuel would be used during a cruise climb flight of a distance equal to RF/3. Minimum fuel usage will occur where RF is maximum, which is near the aircraft's maximum L/D at high subsonic speed.

Example 1.6

Calculate the endurance factor and range factor at Mach 0.8 and 40-kft altitude of hypothetical fighter aircraft HF-1 at 90% of maximum gross takeoff weight and a load factor of 1.

Solution

$$q = \frac{\gamma}{2}\delta P_{\text{ref}}M_0^2 = 0.7 \times 0.1858 \times 2116 \times 0.8^2 = 176.1\,\text{lbf}/\text{ft}^2$$

From Fig. 1.24 at $M = 0.8$, $C_{D0} = 0.012$, $K_1 = 0.20$, and $K_2 = 0$:

$$C_L = \frac{nW}{qS_w} = \frac{1 \times 0.9 \times 40{,}000}{176.1 \times 720} = 0.2839$$

$$C_D = K_1 C_L^2 + K_2 C_L + C_{D0} = 0.2(0.2839^2) + 0.012 = 0.0281$$

Using Eq. (1.36b), we have

$$\text{TSFC} = (0.9 + 0.3M_0)\sqrt{\theta} = (0.9 + 0.3 \times 0.8)\sqrt{0.7519} = 0.9885\,(\text{lbm}/\text{h})/\text{lbf}$$

Thus

$$\text{EF} = \frac{C_L}{C_D(\text{TSFC})}\frac{g_c}{g_0} = \frac{0.2839}{0.0281 \times 0.9885}\frac{32.174}{32.174} = 10.22\,\text{h}$$

$$\text{RF} = \frac{C_L}{C_D}\frac{V}{\text{TSFC}}\frac{g_c}{g_0}$$

$$= \frac{0.2839}{0.0281}\frac{0.8 \times 0.8671 \times 1116\,\text{ft}/\text{s}}{0.9885\,(\text{lbm}/\text{h})/\text{lbf}}\frac{3600\,\text{s}/\text{h}}{6080\,\text{ft}/\text{nm}}\frac{32.174}{32.174}$$

$$= 4685\,\text{nm}$$

Note: The aviation industry uses nautical miles (nm) because 1 nm equates to 1 min of latitude or 6080 ft.

Example 1.7

Determine the variation in endurance factor and range factor for the two hypothetical aircraft HF-1 and HP-1.

a) *Fighter aircraft (HF-1)*. The endurance factor (EF) is plotted vs Mach number and altitude in Fig. 1.30a for our hypothetical fighter aircraft HF-1 at 90% of maximum gross takeoff weight. Note that the best endurance Mach number (minimum fuel consumption) increases with altitude, and the best fuel consumption occurs at altitudes of 30 and 36 kft at a Mach number of 0.7. The range factor (RF) is plotted vs Mach number and altitude in Fig. 1.30b for the HF-1 at 90% of maximum gross takeoff weight. Note that the best cruise Mach number (minimum fuel

(Continued)

Example 1.7 *(Continued)*

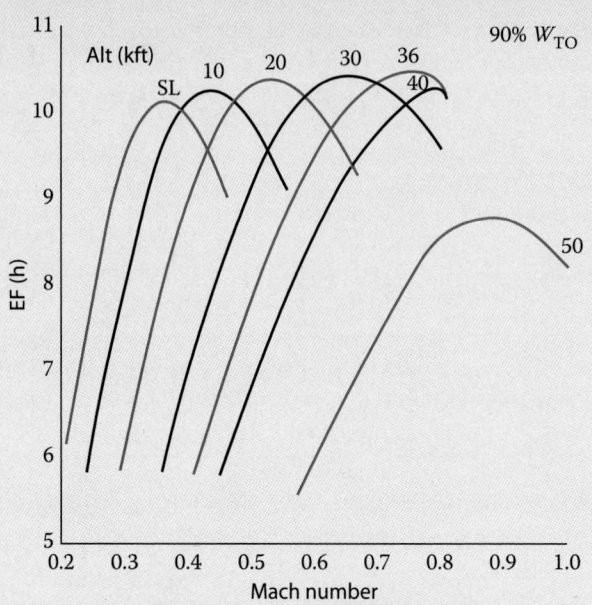

Fig. 1.30a   Endurance factor for HF-1 aircraft.

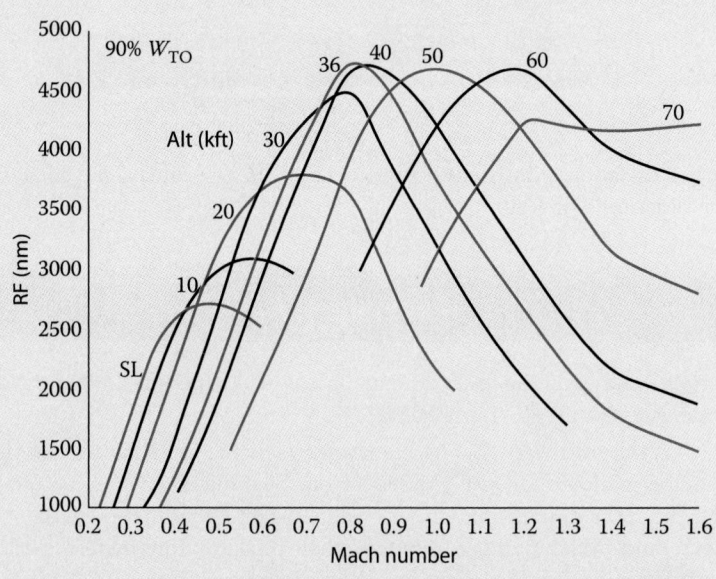

Fig. 1.30b   Range factor for HF-1 aircraft.

(Continued)

Example 1.7 *(Continued)*

consumption) increases with altitude, and the best fuel consumption occurs at an altitude of 36 kft and Mach number of 0.8. Note that the range factor at supersonic speeds is nearly as good as the subsonic values, but the aircraft must operate at high altitude to obtain these high values. The best endurance and cruise flight conditions can be easily seen when contours of EF and RF are plotted on altitude vs Mach number, as is done for this example in Fig. 1.31.

The contours of Fig. 1.31 show that an aircraft can have very good fuel consumption at supersonic cruise. Note that for the HF-1 aircraft, an RF of 4200 nm occurs at Mach 2.0 vs the maximum subsonic RF of 4700 nm. This characteristic of optimum range factor at supersonic cruise can be shown with simple algebra and basic calculus (Ref. 69), which show the *supercruise* characteristics of many types of aircraft. These contour islands of high range factor at high speed drive home the supercruise ranges of the magnitude of the subsonic ranges. Good supercruise fuel consumption can be obtained depending on the engine size and operation at high speed. [Note that higher wing loading (W/S_w) will shift the EF, RF, and max L/D curves down to lower altitudes.]

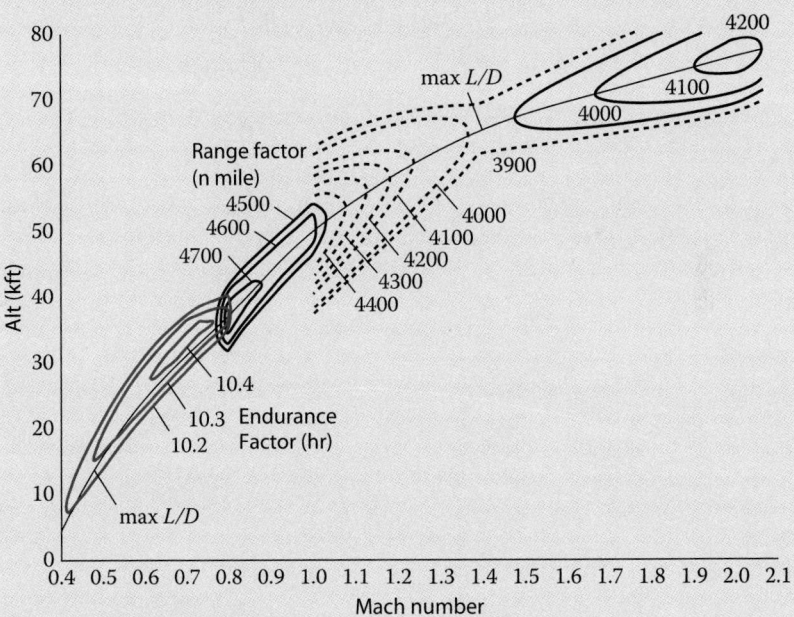

Fig. 1.31 Endurance factor (EF) and range factor (RF) contours for HF-1 aircraft at 90% W_{TO}.

(Continued)

Example 1.7 *(Continued)*

One note of caution: The results of Figs. 1.30 and 1.31 can be misleading because our basic analysis does not consider the thrust available from the engine. The aircraft system needs to have $P_s > 0$ for a viable result (see Fig. 1.21b). The curve of $P_s = 0$ from the performance results like Fig. 1.21b can be superimposed on Fig. 1.31 for a known engine thrust to aircraft weight ratio (T/W), yielding results for that system.

b) *Passenger aircraft (HP-1)*. The endurance factor is plotted vs Mach number and altitude in Fig. 1.32 for our hypothetical passenger aircraft HP-1 at 95% of maximum gross takeoff weight. Note that the best endurance Mach number (minimum fuel consumption) increases with altitude, and the best fuel consumption occurs at sea level [see Eq. (1.38)]. The range factor is plotted vs Mach number and altitude in Fig. 1.33 for the HP-1 at 95% of maximum gross takeoff weight. Note that the best cruise Mach number (minimum fuel consumption) increases with altitude, and the best fuel consumption occurs at an altitude of 11 km and Mach number of about 0.83.

Because the weight of an aircraft like the HP-1 can vary considerably over a flight, the variation in range factor with cruise Mach number was determined for 95% and 70% of maximum gross takeoff weight (MGTOW) vs altitude and Mach number and is plotted in Fig. 1.34. One can see from this figure that the proper cruise altitude can dramatically affect an aircraft's range.

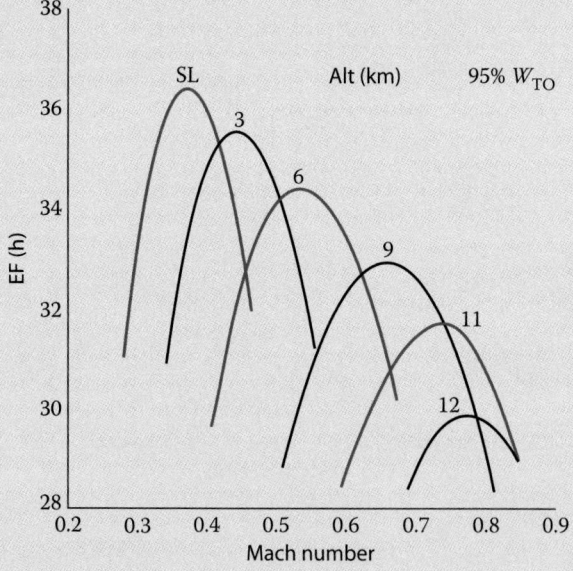

Fig. 1.32 Endurance factor for HP-1 aircraft.

(Continued)

Example 1.7 *(Continued)*

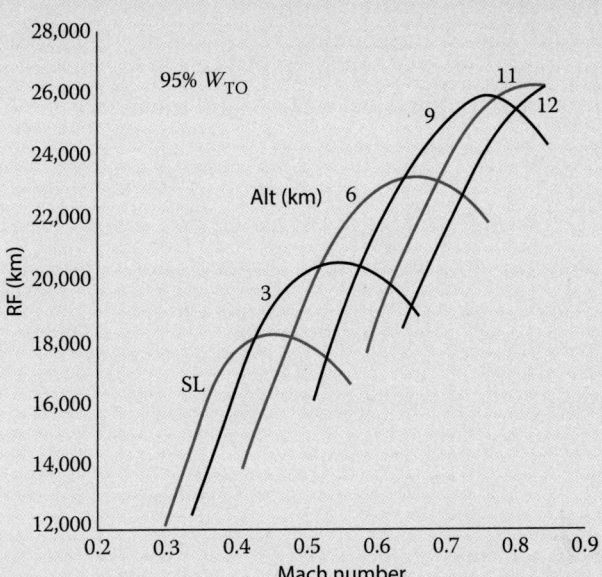

Fig. 1.33 Range factor for HP-1 aircraft for various altitudes.

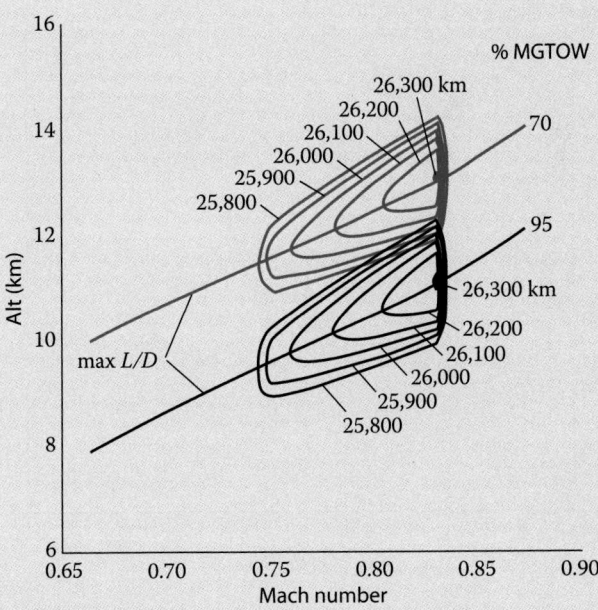

Fig. 1.34 Range factor for HP-1 aircraft at 70% and 95% MGTOW.

1.5.4.4 Maximum L/D or C_L/C_D

For flight conditions requiring minimum fuel consumption, the optimum flight condition can be approximated by that corresponding to maximum C_L/C_D (the equivalent of maximum L/D). From Eq. (1.32), the maximum C_L/C_D (minimum C_D/C_L) can be found by taking the derivative of the following expression, setting it equal to zero, and solving for the C_L that gives minimum C_D/C_L:

$$\frac{C_D}{C_L} = K_1 C_L + K_2 + \frac{C_{D0}}{C_L} \tag{1.46}$$

The lift coefficient that gives maximum C_L/C_D (minimum C_D/C_L) is

$$C_L^* = \sqrt{\frac{C_{D0}}{K_1}} \tag{1.47}$$

and maximum C_L/C_D is given by

$$\left(\frac{C_L}{C_D}\right)^* = \frac{1}{2\sqrt{C_{D0}K_1} + K_2} \tag{1.48}$$

The drag D, range factor, endurance factor, and C_L/C_D vs Mach number at 36,000 ft are plotted in Fig. 1.35 for the HF-1 aircraft and in

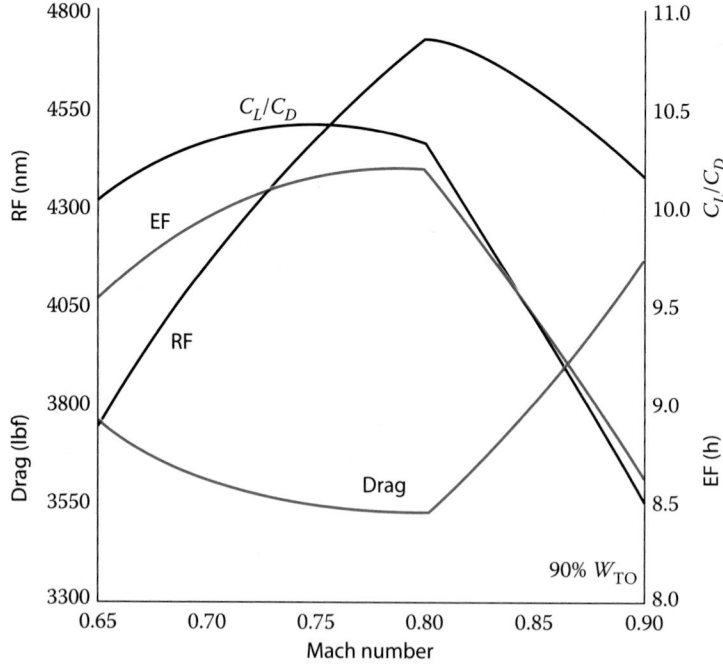

Fig. 1.35 Comparison of drag, C_L/C_D, endurance factor, and range factor for the HF-1 at 36-kft altitude.

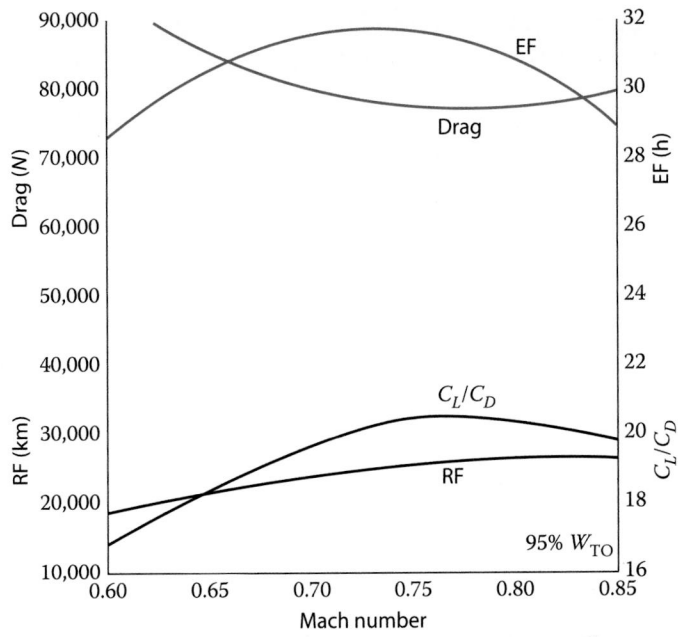

Fig. 1.36 Comparison of drag, C_L/C_D, endurance factor, and range factor for the HP-1 at 11-km altitude.

Fig. 1.36 for the HP-l. Note that the maximum C_L/C_D occurs at Mach 0.8 for the HF-1 and at Mach 0.75 for the HP-1, the same Mach numbers where drags are minimum. The endurance factor is a maximum at a substantially lower Mach number than that corresponding to $(C_L/C_D)^*$ for the HF-1 due to the high TSFC and its increase with Mach number [see Eq. (1.36b)]. The endurance factor for the HP-1 is a maximum very close to the same Mach number that C_L/C_D is maximum due to the lower TSFC of the high-bypass-ratio turbofan engine [see Eq. (1.36a)].

The Mach number for an altitude giving a maximum range factor is called the *best cruise Mach* (BCM). The best cruise Mach normally occurs at a little higher Mach than that corresponding to $(C_L/C_D)^*$. This is because the velocity term in the range factor (RF) [Eq. (1.43)] normally dominates over the increase in TSFC with Mach number. As a first approximation, many use the Mach number corresponding to $(C_L/C_D)^*$ for the best cruise Mach.

1.5.4.5 Accelerated Flight

For flight conditions when thrust T is greater than drag D, an expression for the fuel consumption can be obtained by first noting from Eq. (1.28) that

$$\frac{T}{W} = \frac{P_s}{V[1 - (D + R)/T]}$$

Example 1.8

Calculate the Mach giving maximum C_L/C_D at 20-kft altitude for the HF-1 aircraft at 90% of maximum gross takeoff weight and a load factor of 1.

Solution

From Fig. 1.24 at $M_0 < 0.8$, $C_{D0} = 0.012$, $K_1 = 0.20$, and $K_2 = 0$:

$$C_L^* = \sqrt{\frac{C_{D0}}{K_1}} = \sqrt{\frac{0.012}{0.2}} = 0.2449$$

$$q = \frac{W}{C_L S_w} = \frac{0.9 \times 40{,}000}{0.2449 \times 720} = 204.16 \, \text{lbf/ft}^2$$

$$M_0 = \sqrt{\frac{q}{(\gamma/2)\delta P_{\text{ref}}}} = \sqrt{\frac{204.16}{0.7 \times 0.4599 \times 2116}} = 0.547$$

We define the ratio of drag $D + R$ to thrust T as

$$u \equiv \frac{D+R}{T} \tag{1.49}$$

The preceding equation for thrust to weight becomes

$$\frac{T}{W} = \frac{P_s}{V(1-u)} \tag{1.50}$$

Now Eq. (1.35) can be rewritten as

$$\frac{dW}{W} = -\frac{\text{TSFC}}{V(1-u)}\frac{g_0}{g_c} P_s \, dt$$

Because $P_s \, dt = dz_e$, the preceding equation can be expressed in its most useful forms as

$$\frac{dW}{W} = \frac{\text{TSFC}}{V(1-u)}\frac{g_0}{g_c} dz_e = -\frac{\text{TSFC}}{V(1-u)}\frac{g_0}{g_c} d\left(h + \frac{V^2}{2g}\right) \tag{1.51}$$

The term $1 - u$ represents the fraction of engine power that goes to increasing the aircraft energy z_e, and u represents that fraction that is lost to aircraft drag $D + R$. Note that this equation applies for cases when u is not unity. When u is unity, either Eq. (1.39) or Eq. (1.44) is used.

To obtain the fuel consumption during an acceleration flight condition, Eq. (1.51) can be easily integrated for known flight paths (values of V and z_e) and known variation of $\text{TSFC}/[V(1-u)]$ with z_e.

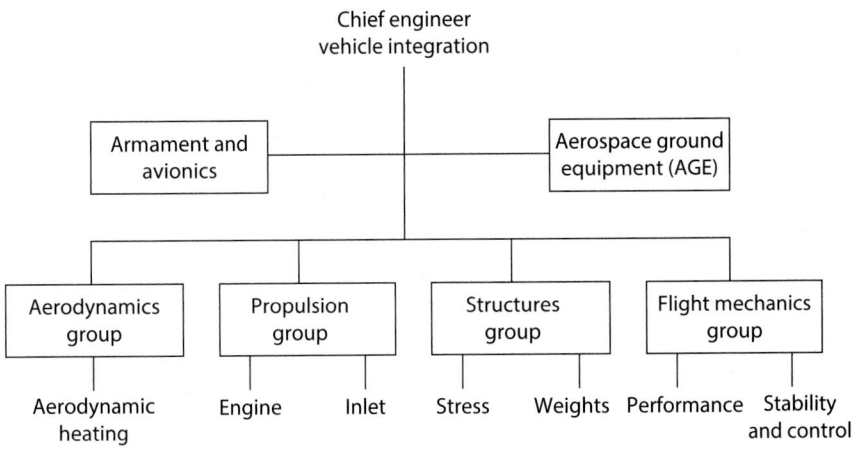

Fig. 1.37   *Organization of a typical vehicle design team.*

1.5.5 Aerospace Vehicle Design: A Team Effort

Aeronautical and mechanical engineers in the aerospace field do many things, but for the most part their efforts all lead to the design and operation of some type of aerospace vehicle. The design team for a new aircraft may be divided into four principal groups: aerodynamics, propulsion, structures, and flight mechanics. The design of a vehicle calls on the extraordinary talents of engineers in each group. Thus the design is a team effort. A typical design team is shown in Fig. 1.37. The chief engineer serves as the referee and integrates the efforts of everyone into the vehicle design. Figure 1.38 [9] shows the kinds of aircraft designs that might result if any one group were able to dominate the others.

1.6 Rocket Engines

Non-airbreathing propulsion systems are characterized by the fact that they carry both fuel and the oxidizer within the aerospace vehicle. Such systems thus may be used anywhere in space as well as in the atmosphere. Figure 1.39 shows the essential features of a liquid-propellant rocket system. Two propellants (an oxidizer and a fuel) are pumped into the combustion chamber where they ignite. The nozzle accelerates the products of combustion to high velocities and exhausts them to the atmosphere or space.

A solid-propellant rocket motor is the simplest of all propulsion systems. Figure 1.40 shows the essential features of this type of system. In this system, the fuel and oxidizer are mixed together and cast into a solid mass called the *grain*. The grain, usually formed with a hole down the middle called the *perforation*, is firmly cemented to the inside of the combustion chamber.

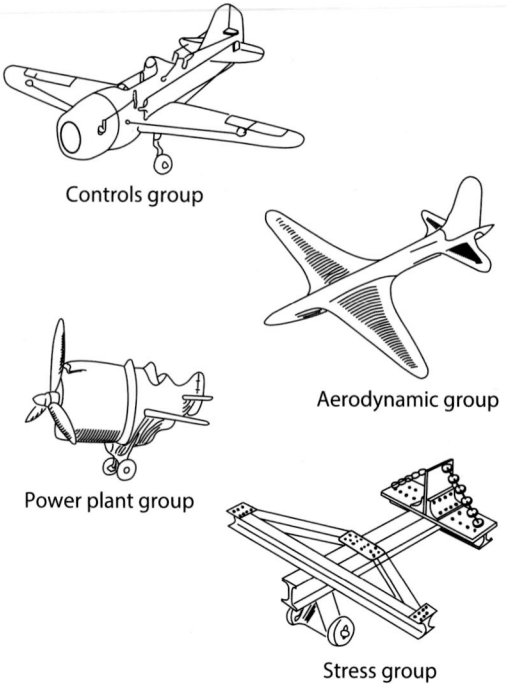

Controls group

Aerodynamic group

Power plant group

Stress group

Fig. 1.38 Aircraft designs.

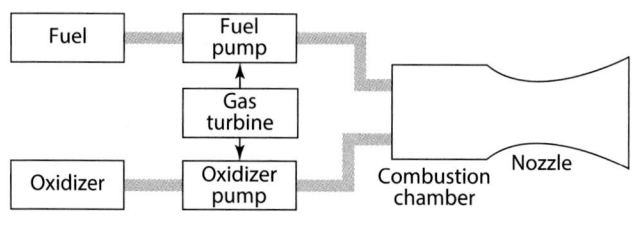

Fig. 1.39 Liquid-propellant rocket motor.

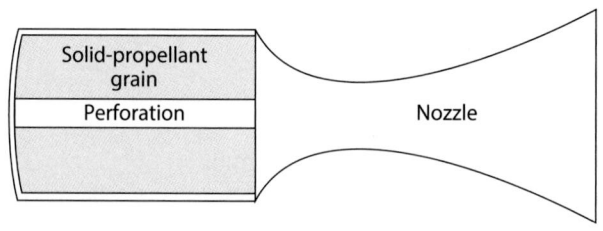

Fig. 1.40 Solid-propellant rocket motor.

After ignition, the grain burns radially outward, and the hot combustion gases pass down the perforation and are exhausted through the nozzle.

The absence of a propellant feed system in the solid-propellant rocket is one of its major advantages. Liquid rockets, on the other hand, may be stopped and later restarted, and their thrust may be varied somewhat by changing the speed of the fuel and oxidizer pumps.

1.6.1 Rocket Engine Thrust

A natural starting point in understanding the performance of a rocket is the examination of the static thrust. Application of the momentum equation developed in Chapter 2 will show that the static thrust is a function of the propellant mass flow rate $\dot{m}_p$, the exhaust velocity V_e and pressure P_e, the exhaust area A_e, and the ambient pressure P_a. Figure 1.41 shows a schematic of a stationary rocket to be considered for analysis. We assume the flow to be steady (invariant with time) one-dimensional (1-D), with a steady exit velocity V_e and propellant flow rate $\dot{m}_p$. Inside this rocket we place a control volume σ whose control surface intersects the exhaust jet perpendicularly through the exit plane of the nozzle. Thrust acts in the direction opposite to the direction of V_e. The reaction to the thrust F necessary to hold the rocket and control volume stationary is shown in Fig. 1.41.

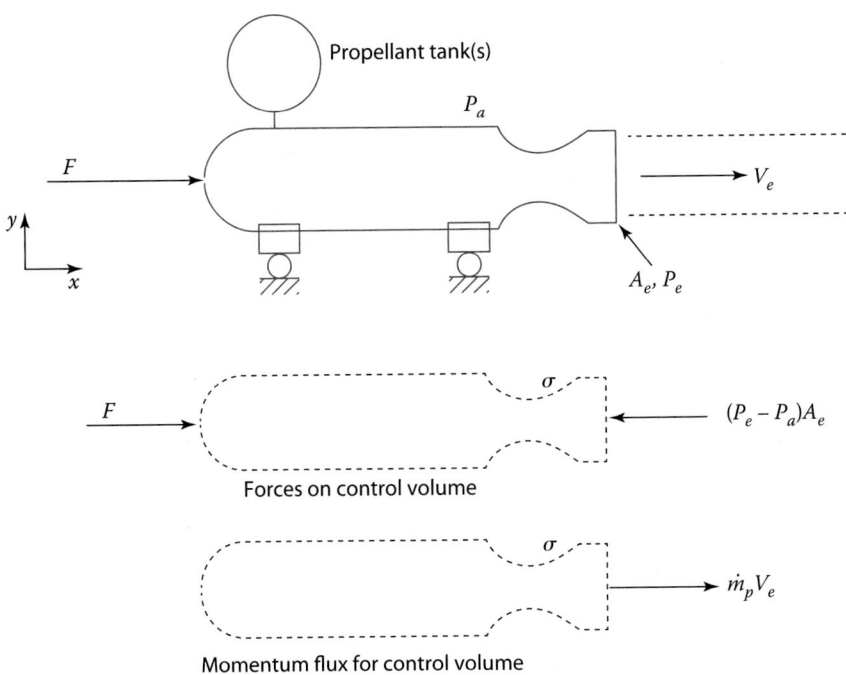

Fig. 1.41 Schematic diagram of static rocket engine.

The momentum equation applied to this system gives the following:

1. Sum of forces acting on the outside surface of the control volume:

$$\sum F_x = F - (P_e - P_a)A_e$$

2. Net rate of change of momentum for the control volume:

$$\Delta(\text{momentum}) = \dot{M}_{\text{out}} = \frac{\dot{m}_p V_e}{g_c}$$

Because the sum of the forces acting on the outside of the control volume is equal to the net rate of change of the momentum for the control volume, we have

$$F - (P_e - P_a)A_e = \frac{\dot{m}_p V_e}{g_c} \qquad (1.52)$$

If the pressure in the exhaust plane P_e is the same as the ambient pressure P_a, the thrust is given by $F = \dot{m}_p V_e/g_c$. The condition $P_e = P_a$ is called *on-design* or *optimum expansion* because it corresponds to maximum thrust for the given chamber conditions. It is convenient to define an *effective exhaust velocity* C such that

$$C \equiv V_e + \frac{(P_e - P_a)A_e g_c}{\dot{m}_p} \qquad (1.53)$$

Thus the static thrust of a rocket can be written as

$$F = \frac{\dot{m}_p C}{g_c} \qquad (1.54)$$

1.6.2 Specific Impulse

The *specific impulse* I_{sp} for a rocket is defined as the thrust per unit of propellant weight flow:

$$I_{sp} \equiv \frac{F}{\dot{w}_p} = \frac{F \, g_c}{\dot{m}_p \, g_0} \qquad (1.55)$$

where g_0 is the acceleration due to gravity at sea level. The unit of I_{sp} is the second. Note that specific impulse is essentially the inverse of thrust specific fuel consumption, as given in Eq. (1.10). From Eqs. (1.54) and (1.55), the specific impulse can also be written as

$$I_{sp} = \frac{C}{g_0} \qquad (1.56)$$

Typical specific impulses for some rocket engines are listed in Table 1.6. Other performance data for rocket engines are contained in Appendix C.

Table 1.6 Specific Impulse I_{sp} for Typical Rocket Engines

Fuel/Oxidizer	I_{sp}, s
Solid propellant	250
Liquid O_2: kerosene (RP)	310
Liquid O_2: H_2	410
Nuclear fuel: H_2 propellant	840

Example 1.9

Find the specific impulse of the space shuttle main engine (SSME) shown in Fig. 1.42a that produces 470,000 lbf in a vacuum with a propellant weight flow of 1030 lbf/s. By using Eq. (1.55), we find that the SSME has a specific impulse I_{sp} of 456 s (=470,000/1030) in vacuum.

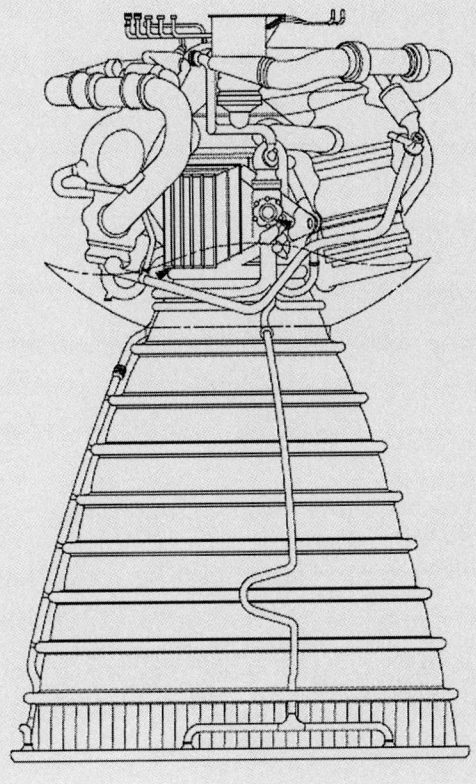

Fig. 1.42a Space shuttle main engine (SSME).

(Continued)

Example 1.9 *(Continued)*

An estimate of the variation in thrust with altitude for the SSME is shown in Fig. 1.42b. Note that the maximum thrust occurs at about 100 kft altitude and the thrust curve changes shape at about 20 kft altitude. (The reason for this is covered in Chapter 10.)

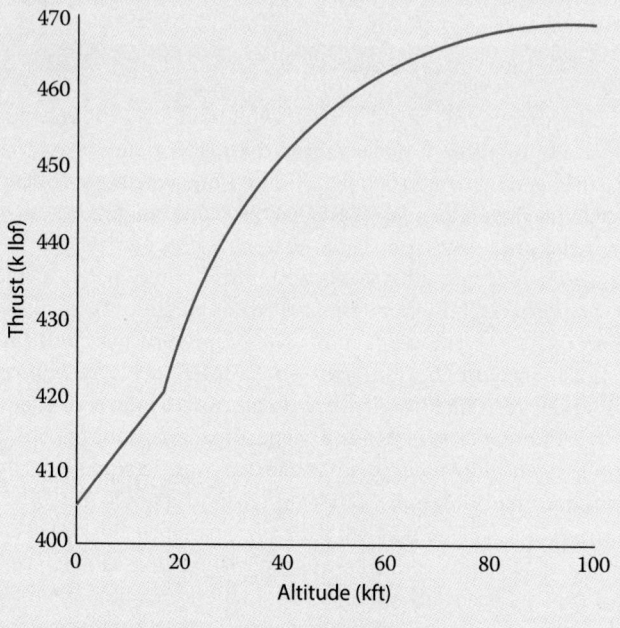

Fig. 1.42b   Rocket thrust variation with altitude.

1.6.3 Rocket Vehicle Acceleration

The mass of a rocket vehicle varies a great deal during flight due to the consumption of the propellant. The velocity that a rocket vehicle attains during powered flight can be determined by considering the vehicle in Fig. 1.43.

The figure shows an accelerating rocket vehicle in a gravity field. At some time, the mass of the rocket is m and its velocity is V. In an infinitesimal time dt, the rocket exhausts an incremental mass dm_p with an exhaust velocity V_e relative to the rocket as the rocket velocity changes to $V + dV$. The net change in momentum of the control volume σ is composed of the momentum out of the rocket at the exhaust plus the change of the momentum of the rocket. The momentum out of the rocket in the V direction is $-V_e dm_p$, and the change in the momentum of the rocket in the V direction is $m dV$. The forces acting on the control volume σ are composed of the

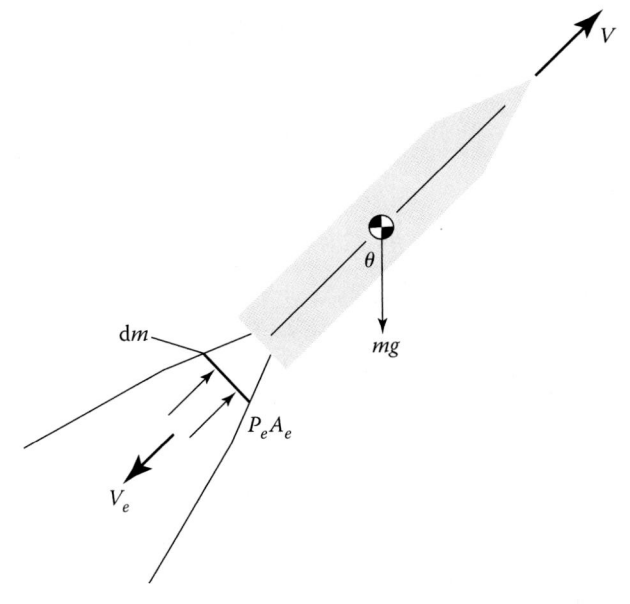

Fig. 1.43   Rocket vehicle in flight.

net pressure force, the drag D, and the gravitational force. The sum of these forces in the V direction is

$$\sum F_V = (P_e - P_a)A_e - D - \frac{mg}{g_c} \cos \theta$$

The resultant impulse on the rocket $(\sum F_V)dt$ must equal the momentum change of the system $\Delta(\text{momentum}) = (-V_e dm_p + m dV)/g_c$. Thus

$$\left[(P_e - P_a)A_e - D - \frac{mg}{g_c} \cos \theta \right] dt = \frac{-V_e\, dm_p + m\, dV}{g_c}$$

From the preceding relationship, the momentum change of the rocket ($m\, dV$) is

$$\frac{m\, dV}{g_c} = \left[(P_e - P_a)A_e - D - \frac{mg}{g_c} \cos \theta \right] dt + \frac{V_e\, dm_p}{g_c} \qquad (1.57)$$

Because $dm_p = \dot{m}_p\, dt = -(dm/dt)dt$, then Eq. (1.57) can be written as

$$\frac{m\, dV}{g_c} = \left[(P_e - P_a)A_e + \frac{\dot{m}_p V_e}{g_c} - D - \frac{mg}{g_c} \cos \theta \right] dt$$

By using Eq. (1.53), this relationship becomes

$$\frac{m\,dV}{g_c} = \left(\frac{\dot{m}_p}{g_c}C - D - \frac{mg}{g_c}\cos\theta\right)dt$$

or

$$dV = -C\frac{dm}{m} - \frac{Dg_c}{m}dt - g\,\cos\theta\,dt \qquad (1.58)$$

The velocity of a rocket along its trajectory can be determined from the preceding equation if C, D, g, and θ are known.

In the absence of drag and gravity, integration of Eq. (1.58) gives the following, assuming constant effective exhaust velocity C:

$$\Delta V = C\,ln\frac{m_t}{m_f} \qquad (1.59)$$

where ΔV is the change in velocity, m_i is the initial mass of the rocket system, and m_f is the final mass. Equation (1.59) can be solved for the mass ratio as

$$\frac{m_i}{m_f} = \exp\frac{\Delta V}{C} \qquad (1.60)$$

Example 1.10

We want to estimate the mass ratio (final to initial) of an H_2–O_2 ($C = 4000$ m/s) rocket for an Earth orbit ($\Delta V = 8000$ m/s), neglecting drag and gravity. Using Eq. (1.60), we obtain $m_f/m_t = e^{-2} = 0.132$, or a single-stage rocket would be about 13% payload and structure and 87% propellant.

Problems

1.1 Calculate the uninstalled thrust for Example 1.1 using Eq. (1.6).

1.2 Develop the following analytical expressions for a turbojet engine:

   a) When the fuel flow rate is very small in comparison with the air mass flow rate, the exit pressure is equal to ambient pressure, and the installation loss coefficients are zero, then the installed thrust T is given by

$$T = \frac{\dot{m}_0}{g_c}(V_e - V_0)$$

b) For the preceding conditions, the thrust specific fuel consumption is given by

$$\text{TSFC} = \frac{Tg_c/\dot{m}_0 + 2V_0}{2\eta_{Th}h_{PR}}$$

c) For $V_0 = 0$ and 500 ft/s, plot the preceding equation for TSFC [in (lbm/h)/lbf] vs specific thrust $T/\dot{m}_0$ [in lbf/(lbm/s)] for values of specific thrust from 0 to 120. Use $\eta_{Th} = 0.4$ and $h_{PR} = 18{,}400$ Btu/lbm.

d) Explain the trends.

1.3   Repeat 1.2c using SI units. For $V_0 = 0$ and 150 m/s, plot TSFC [in (mg/s)/N] vs specific thrust $T/\dot{m}_0$ [in N/(kg/s)] for values of specific thrust from 0 to 1200. Use $\eta_{Th} = 0.4$ and $h_{PR} = 42{,}800$ kJ/kg.

1.4   A J57 turbojet engine is tested at sea-level, static, standard-day conditions ($P_0 = 14.696$ psia, $T_0 = 518.7°$R, and $V_0 = 0$). At one test point, the uninstalled thrust is $10{,}200$ lbf while the airflow is 164 lbm/s and the fuel flow is 8520 lbm/h. Using these data, estimate the exit velocity V_e for the case of exit pressure equal to ambient pressure ($P_0 = P_e$) and calculate the thermal efficiency. (Assume the heating value of jet fuel is about $18{,}400$ Btu/lbm.)

1.5   Using data for the F110-GE-100 engine from Appendix B for both thrusts listed, calculate the fuel flow rate, the exit velocity, and the thermal efficiency in either unit system. (Assume the heating value of jet fuel is about $18{,}400$ Btu/lbm or $42{,}800$ kJ/kg.)

1.6   The uninstalled thrust for a turbofan engine with separate exhaust streams is equal to the sum of the thrust from the engine core F_C and the thrust from the bypass stream F_B. The bypass ratio of the engine α is the ratio of the mass flow through the bypass stream to the core mass flow, or $\alpha \equiv \dot{m}_B/\dot{m}_C$. When the exit pressures are equal to the ambient pressure, the thrusts of the core and bypass stream are given by

$$F_C = \frac{1}{g_c}[(\dot{m}_C + \dot{m}_f)V_{Ce} - \dot{m}_C V_0]$$

$$F_B = \frac{\dot{m}_B}{g_c}(V_{Be} - V_0)$$

where V_{Ce} and V_{Be} are the exit velocities from the core and bypass, respectively, V_0 is the inlet velocity, and $\dot{m}_f$ is the mass flow rate of fuel burned in the core of the engine.

Show that the specific thrust and thrust specific fuel consumption can be expressed as

$$\frac{F}{\dot{m}_0} = \frac{1}{g_c}\left(\frac{1 + \dot{m}_f/\dot{m}_C}{1 + \alpha}V_{Ce} + \frac{\alpha}{1 + \alpha}V_{Be} - V_0\right)$$

$$S = \frac{\dot{m}_f}{F} = \frac{\dot{m}_f/\dot{m}_C}{(F/\dot{m}_0)(1 + \alpha)}$$

where $\dot{m}_0 = \dot{m}_C + \dot{m}_B$.

1.7 The CF6 turbofan engine has a rated thrust of 40,000 lbf at a fuel flow rate of 13,920 lbm/h at sea-level static conditions. If the core airflow rate is 225 lbm/s and the bypass ratio is 6.0, what are the specific thrust [lbf/(lbm/s)] and thrust specific fuel consumption [(lbm/h)/ lbf]? If the exit velocity from the core and bypass are equal, determine the exit velocity at sea-level static conditions. (Use the equations in Problem 1.6.) What is the thermal efficiency? (The heating value of jet fuel is about 18,400 Btu/lbm.)

1.8 Repeat Problem 1.7 using SI units. (The heating value of jet fuel is about 42,800 kJ/kg.)

1.9 The JT9D high-bypass-ratio turbofan engine at maximum static thrust ($V_0 = 0$) on a sea-level, standard day ($P_0 = 14.696$ psia, $T_0 = 518.7°$R) has the following data: the air mass flow rate through the core is 247 lbm/s, the air mass flow rate through the fan bypass duct is 1248 lbm/s, the exit velocity from the core is 1190 ft/s, the exit velocity from the bypass duct is 885 ft/s, and the fuel flow rate into the combustor is 15,750 lbm/h. Estimate the following for the case of exit pressures equal to ambient pressure ($P_0 = P_e$):

a) The thrust of the engine

b) The thermal efficiency of the engine (heating value of jet fuel is about 18,400 Btu/lbm)

c) The propulsive efficiency and thrust specific fuel consumption of the engine

1.10 Repeat Problem 1.9 using SI units. (The heating value of jet fuel is about 42,800 kJ/kg.)

1.11 For the CFM56-5C data of Appendix B, estimate the average exit velocity at takeoff.

1.12 For the GE90-B4 data of Appendix B, estimate the average exit velocity at takeoff.

1.13   One advanced afterburning fighter engine, whose performance is depicted in Figs. 1.14a–1.14c, is installed in the HF-1 fighter aircraft. Using the aircraft drag data of Fig. 1.26b, determine and plot the variation of weight specific excess power (P_s in feet per second) with and without afterburner vs flight Mach number for level flight ($n = 1$) at 36-kft altitude. Assume the installation losses are constant with values of $\phi_{inlet} = 0.05$ and $\phi_{noz} = 0.02$.

1.14   Determine the takeoff speed of the HF-1 aircraft and the P_s with two engines whose performance is given in Fig. 1.14a with afterburner. Assume $\phi_{inlet} = 0.05$ and $\phi_{noz} = 0.02$.

1.15   Determine the takeoff speed of the HP-1 aircraft at 90% of maximum gross takeoff weight.

1.16   Derive Eqs. (1.47) and (1.48) for maximum C_L/C_D. Start by taking the derivative of Eq. (1.46) with respect to C_L and finding the expression for the lift coefficient that gives maximum C_L/C_D.

1.17   Show that for maximum C_L/C_D, the corresponding drag coefficient C_D is given by

$$C_D = 2C_{D0} + K_2\sqrt{\frac{C_{D0}}{K_1}}$$

1.18   An aircraft with a wing area of 800 ft^2 is in level flight ($n = 1$) at maximum C_L/C_D. Given that the drag coefficients for the aircraft are $C_{D0} = 0.02$, $K_2 = 0$, and $K_1 = 0.2$, find

a)   The maximum C_L/C_D and the corresponding values of C_L and C_D

b)   The flight altitude [use Eqs. (1.29) and (1.30b)] and aircraft drag for an aircraft weight of 45,000 lbf at Mach 0.8

c)   The flight altitude and aircraft drag for an aircraft weight of 35,000 lbf at Mach 0.8

d)   The range for an installed engine thrust specific fuel consumption rate of 0.8 (lbm/h)/lbf, if the 10,000-1bf difference in aircraft weight between parts b and c is due only to fuel consumption

1.19   An aircraft weighing 110,000 N with a wing area of 42 m^2 is in level flight ($n = 1$) at the maximum value of C_L/C_D. Given that the drag coefficients for the aircraft are $C_{D0} = 0.03$, $K_2 = 0$, and $K_1 = 0.25$, find the following:

a)   The maximum C_L/C_D and the corresponding values of C_L and C_D

b)   The flight altitude [use Eqs. (1.29) and (1.30b)] and aircraft drag at Mach 0.5

c)   The flight altitude and aircraft drag at Mach 0.75

1.20 The Breguet range equation [Eq. (1.45b)] applies for a cruise climb flight profile with constant RF. Another range equation can be developed for a level cruise flight profile with varying RF. Consider the case where we keep C_L, C_D, and TSFC constant and vary the flight velocity with aircraft weight by the expression

$$V = \sqrt{\frac{2g_c W}{\rho C_L S_w}}$$

Using the subscripts i and f for the initial and final flight conditions, respectively, show the following:

a) Substitution of this expression for flight velocity into Eq. (1.42) gives

$$\frac{dW}{\sqrt{W}} = -\frac{\sqrt{W_i}}{RF_i} ds$$

b) Integration of the preceding between the initial i and final f conditions gives

$$\frac{W_f}{W_i} = \left[1 - \frac{s}{2(RF_i)}\right]^2$$

c) For a given weight fraction W_f/W_i, the maximum range s for this level cruise flight corresponds to starting the flight at the maximum altitude (minimum density) and maximum value of $\sqrt{C_L}/C_D$.

d) For the drag coefficient equation of Eq. (1.32), maximum $\sqrt{C_L}/C_D$ corresponds to $C_L = (1/6K_1)(\sqrt{12K_1 C_{D0} + K_2^2} - K_2)$.

1.21 An aircraft begins a cruise at a wing loading W/S_w of 100 lbf/ft² and Mach 0.8. The drag coefficients are $K_1 = 0.056$, $K_2 = -0.008$, and $C_{D0} = 0.014$, and the fuel consumption TSFC is constant at 0.8 (lbm/h)/lbf. For a weight fraction W_f/W_i of 0.9, determine the range and other parameters for two different types of cruise.

a) For a cruise climb (maximum C_L/C_D) flight path, determine C_L, C_D, initial and final altitudes, and range.

b) For a level cruise (maximum $\sqrt{C_L}/C_D$) flight path, determine C_L, C_D, altitude, initial and final velocities, and range.

1.22 An aircraft weighing 70,000 lbf with a wing area of 1000 ft² is in level flight ($n = 1$) at 30-kft altitude. Using the drag coefficients of Fig. 1.24 and the TSFC model of Eq. (1.36b), find the following:

a) The maximum C_L/C_D and the corresponding values of C_L, C_D, and Mach number (*Note:* Because the drag coefficients are a function of Mach number and it is an unknown, you must first guess a value for the Mach number to obtain the drag coefficients. Try a Mach number of 0.8 for your first guess.)

b) The C_L, C_D, C_L/C_D, range factor, endurance factor, and drag for flight Mach numbers of 0.74, 0.76, 0.78, 0.80, 0.81, and 0.82

c) The best cruise Mach (maximum RF)

d) The best loiter Mach (maximum EF)

1.23   An aircraft weighing 200,000 N with a wing area of 60 m^2 is in level flight ($n = 1$) at 9-km altitude. Using the drag coefficients of Fig. 1.24 and the TSFC model of Eq. (1.36b), find the following:

a) The maximum C_L/C_D and the corresponding values of C_L, C_D, and Mach number (*Note:* Because the drag coefficients are a function of the Mach number and it is an unknown, you must first guess a value for the Mach number to obtain the drag coefficients. Try a Mach number of 0.8 for your first guess.)

b) The C_L, C_D, C_L/C_D, range factor, endurance factor, and drag for flight Mach numbers of 0.74, 0.76, 0.78, 0.80, 0.81, and 0.82

c) The best cruise Mach (maximum RF)

d) The best loiter Mach (maximum EF)

1.24   What is the specific impulse in seconds of the JT9D turbofan engine in Problem 1.9?

1.25   Determine the specific impulse in seconds of the PW2037 at cruise using the data in Appendix B.

1.26   A rocket motor is fired in place on a static test stand. The rocket exhausts 100 lbm/s at an exit velocity of 2000 ft/s and pressure of 50 psia. The exit area of the rocket is 0.2 ft^2. For an ambient pressure of 14.7 psia, determine the effective exhaust velocity, the thrust transmitted to the test stand, and the specific impulse.

1.27   A rocket motor under static testing exhausts 50 kg/s at an exit velocity of 800 m/s and pressure of 350 kPa. The exit area of the rocket is 0.02 m^2. For an ambient pressure of 100 kPa, determine the effective exhaust velocity, the thrust transmitted to the test stand, and the specific impulse.

1.28   The propellant weight of an orbiting space system amounts to 90% of the system gross weight. Given that the system rocket engine has a specific impulse of 300 s, determine:

a) The maximum attainable velocity if all the propellant is burned and the system's initial velocity is 7930 m/s

b) The propellant mass flow rate, given that the rocket engine thrust is 1,670,000 N

1.29   A chemical rocket motor with a specific impulse of 400 s is used in the final stage of a multistage launch vehicle for deep-space exploration. This final stage has a mass ratio (initial to final) of 6, and its single rocket motor is first fired while it orbits the Earth at a velocity of 26,000 ft/s. The final stage must reach a velocity of 36,700 ft/s to escape the Earth's gravitational field. Determine the percentage of fuel that must be used to perform this maneuver (neglect gravity and drag).

Gas Turbine Design Problems

1.D1   *Background (HP-1 aircraft).* You are to determine the thrust and fuel consumption requirements of the two engines for the hypothetical passenger aircraft, the HP-1. The twin-engine aircraft will cruise at 0.83 Mach and be capable of the following requirements:

- Takeoff at maximum gross takeoff weight W_{TO} from an airport at 1.6-km pressure altitude on a hot day (38°C) uses a 3650-m (12-kft) runway. The craft is able to maintain a 2.4% single-engine climb gradient in the event of engine failure at liftoff.

- It transports 253 passengers and luggage (90 kg each) over a still-air distance of 11,120 km (6000 n mile). It has 30 min of fuel in reserve at end (loiter).

- It attains an initial altitude of 11 km at beginning of cruise ($P_s = 1.5$ m/s).

- The single-engine craft cruises at 5-km altitude at 0.45 Mach ($P_s = 1.5$ m/s).

All of the data for the HP-1 contained in Example 1.2 apply. Preliminary mission analysis of the HP-1 using the methods of Ref. 13 for the 11,120-km flight with 253 passengers and luggage (22,770-kg payload) gives the preliminary fuel use shown in Table P1.D1.

Analysis of takeoff indicates that each engine must produce an installed thrust of 214 kN on a hot day (38°C) at 0.1 Mach and 1.6-km pressure altitude. To provide for reasonable-length landing gear, the maximum diameter of the engine inlet is limited to 2.2 m. Based on standard design practice (see Chapter 11, online), the maximum mass flow rate per unit area is given by

$$\frac{\dot{m}}{A} = 231.8 \frac{\delta_0}{\sqrt{\theta_0}} \ (kg/s)/m^2$$

Table P1.D1

Description	Distance, km	Fuel Used, kg
Taxi		200[a]
Takeoff		840[a]
Climb and acceleration	330	5880[a]
Cruise	10,650	50,240
Descent	140	1090[a]
Loiter (30 min at 9-km altitude)	2	2350
Land and taxi		600[a]
TOTAL	11,120	61,200

[a]These fuel consumptions can be considered to be constant.

Thus on a hot day (38°C) at 0.1 Mach and 1.6-km pressure altitude, $\theta = (38 + 273.1)/288.2 = 1.079$, $\theta_0 = 1.079 \times 1.002 = 1.081$, $\delta = 0.8256$, $\delta_0 = 0.8256 \times 1.007 = 0.8314$, and the maximum mass flow through the 2.2-m-diam inlet is 704.6 kg/s.

Calculations (HP-1 Aircraft)

a) If the HP-1 starts out the cruise at 11 km with a weight of 1,577,940 N, find the allowable TSFC for the distance of 10,650 km for the following cases:
   1. Assume the aircraft performs a cruise climb (flies at a constant C_D/C_L). What is its altitude at the end of the cruise climb?
   2. Assume the aircraft cruises at a constant altitude of 11 km. Determine C_D/C_L at the start and end of cruise. Using the average of these two values, calculate the allowable TSFC.

b) Determine the loiter (endurance) Mach numbers for altitudes of 10, 9, 8, 7, and 6 km when the HP-1 aircraft is at 64% of W_{TO}.

c) Determine the aircraft drag at the following points in the HP-1 aircraft's 11,120-km flight based on the fuel consumptions just listed:
   1. Takeoff, $M = 0.23$, sea level
   2. Start of cruise, $M = 0.83$, 11 km
   3. End-of-cruise climb, $M = 0.83$, altitude = ? km
   4. End of 11-km cruise, $M = 0.83$, 11 km
   5. Engine out (88% of W_{TO}), $M = 0.45$, 5 km

1.D2 *Background (HF-1 Aircraft).* You are to determine the thrust and fuel consumption requirements of the two engines for the hypothetical fighter aircraft HF-1. This twin-engine fighter will supercruise at 1.6 Mach and will be capable of the following requirements:
   1. Takeoff at maximum gross takeoff weight W_{TO} from a 1200-ft (366-m) runway at sea level on a standard day

2. Supercruise at 1.6 Mach and 40-kft altitude for 250 n mile (463 km) at 92% of W_{TO}

3. Perform 5-g turns at 1.6 Mach and 30-kft altitude at 88% of W_{TO}

4. Perform 5-g turns at 0.9 Mach and 30-kft altitude at 88% of W_{TO}

5. Perform the maximum mission listed in the following

All of the data for the HF-1 contained in Example 1.2 apply. Preliminary mission analysis of the HF-1 using the methods of Ref. 13 for the maximum mission gives the preliminary fuel use shown in Table P1.D2.

Analysis of takeoff indicates that each engine must produce an installed thrust of 23,500 lbf on a standard day at 0.1 Mach and sea-level altitude. To provide for optimum integration into the airframe, the maximum area of the engine inlet is limited to 5 ft^2. Based on standard design practice (see Chapter 11, online), the maximum mass flow rate per unit area for subsonic flight conditions is given by

$$\frac{\dot{m}}{A} = 47.5 \frac{\delta_0}{\sqrt{\theta_0}} \text{ (lbm/s)/ft}^2$$

Thus at 0.1 Mach and sea-level standard day, $\theta = 1.0$, $\theta_0 = 1.002$, $\delta = 1.0$, $\delta_0 = 1.007$, and the maximum mass flow through the 5-ft^2 inlet is 238.9 lbm/s. For supersonic flight conditions, the maximum mass flow rate per unit area is simply the density of the air ρ times its velocity V.

Table P1.D2

Description	Distance, n mile	Fuel Used, lbm
Warmup, taxi, takeoff		700[a]
Climb and accelerate to 0.9 Mach and 40 kft	35	1800[a]
Accelerate from 0.9 to 1.6 Mach	12	700[a]
Supercruise at 1.6 Mach and 40 kft	203	4400
Deliver payload of 2000 lbf	0	0[a]
Perform one 5-g turn at 1.6 Mach and 30 kft	0	1000[a]
Perform two 5-g turns at 0.9 Mach and 30 kft	0	700[a]
Climb to best cruise altitude and 0.9 Mach	23	400[a]
Cruise climb at 0.9 Mach	227	1600
Loiter (20 min at 30-kft altitude)		1100
Land		0[a]
TOTAL	500	12,400

[a]These fuel consumptions can be considered to be constant.

Calculations (HF-1 Aircraft)

a) If the HF-1 starts the supercruise at 40 kft with a weight of 36,800 lbf, find the allowable TSFC for the distance of 203 n mile for the following cases:
   1. Assume the aircraft performs a cruise climb (flies at a constant C_D/C_L). What is its altitude at the end of the cruise climb?
   2. Assume the aircraft cruises at a constant altitude of 40 kft. Determine C_D/C_L at the start and end of cruise. Using the average of these two values, calculate the allowable TSFC.

b) Find the best cruise altitude for the subsonic return cruise at 0.9 Mach and 70.75% of W_{TO}.

c) Determine the loiter (endurance) Mach numbers for altitudes of 32, 30, 28, 26, and 24 kft when the HF-1 aircraft is at 67% of W_{TO}.

d) Determine the aircraft drag at the following points in the HF-1 aircraft's maximum mission based on the fuel consumptions just listed:
   1. Takeoff, $M = 0.172$, sea level
   2. Start of supercruise, $M = 1.6$, 40 kft
   3. End of supercruise climb, $M = 1.6$, altitude $= ?$ ft
   4. End of 40-kft supercruise, $M = 1.6$, 40 kft
   5. Start of subsonic cruise, $M = 0.9$, altitude $=$ best cruise altitude
   6. Start of loiter, altitude $= 30$ kft

Chapter 2 Review of Fundamentals

> Dr. von Ohain is justly remembered for his many contributions to the development of turbine propulsion, his sense of humor, his excellent memory, his amazing charm and outstanding manners, as well as his dedication to the fundamental principles.
>
> *Bill Heiser in the Foreword to* Elements of Propulsion, *second edition*

2.1 Introductory Comments

2.1.1 Introduction

The operation of gas turbine engines and of rocket motors is governed by the laws of mechanics and thermodynamics. The field of mechanics includes the mechanics of both fluids and solids; however, because the process occurring in most propulsion devices involves a flowing fluid, our emphasis will be on fluid mechanics or, more specifically, *gas dynamics*.

Understanding and predicting the basic performance of gas turbine engines and rocket motors requires a closed set of governing equations (e.g., conservation of mass and energy, Newton's second law, and second law of thermodynamics). For transparency, understanding, and capturing the basic physical phenomena, we model the gas as a perfect gas and the flow as 1-D flow, although the initial presentation will cover the full 3-D unsteady control volume formulation.

This chapter and the next are intended as a fairly extensive overview of primarily thermodynamics and 1-D gas dynamics applied to airflow (*aerothermodynamics*) that are the foundation of this textbook. The intent is to provide enough detail and background for you to gain a foundational understanding of the underlying assumptions and engineering applications of these basic principles to gas turbine engines and rocket motors.

2.1.2 Importance of Engineering Solution Method

Unlike homework problems, rarely is there only one correct solution to an engineering problem. Engineers solve problems by modeling real-world systems either physically (e.g., an experiment) or analytically (mathematical model). Regardless of the type of model, a logical solution methodology is essential because it

* Properly frames the problem
* Guides the solution development
* Bounds the solution (says something about the general applicability of the answer)
* Is expected of the profession

Engineers are expected to think through a problem in an effective and efficient manner, using tools at our disposal to arrive at a solution. Computer analysis tools, be they fairly simple like GASTAB (included with this text) or a sophisticated computational fluid dynamics (CFD) code, are never a substitute for sound engineering analysis. We strongly advise your consistent use of a solid engineering analysis to work homework problems not only from this textbook, but also in all your engineering classes, so you can continually develop this critical skill set. A sound engineering analysis includes the following key elements:

* *System:* A clear definition of the problem with knowns and unknowns. A system sketch often helps here.
* *Governing equations:* These provide the appropriate mathematical formulation of the foundational physical laws that you will use to model the problem, along with applicable assumptions.
* *Assumptions:* To include a thorough understanding of the *implications* of each of the assumptions and their influence on the governing equations.
* *Solving:* Using applicable tools to arrive at a logical solution.
* *Testing:* Critically assessing your solution for reasonableness, a "sanity check" against expected or perhaps known ranges of solutions. Often, a process diagram (e.g., a *T-s* diagram) can be helpful here or elsewhere in the analysis.

These key engineering analysis elements are intended to guide your thinking, not substitute for it. You can think of these as solution steps, but it is better to think of them as necessary elements of any good engineering analysis; as such they need to be given proper consideration, as we demonstrate in the examples that follow in this book.

2.2   Thermodynamics Fundamentals

Thermodynamics is the study of *energy E* and what it can and cannot do. The first law of thermodynamics essentially states that although energy can take on different forms, its total quantity remains the same (energy is conserved). Our focus will be on three forms of energy—kinetic, potential, and internal. Thermodynamics also gives limits of performance such as efficiencies, tells us whether a process can occur naturally, and indicates energy that is in a "nonuseful" form—a form not available to cause the desired effect (often referred to as a loss). *Entropy S*, associated with the second law of thermodynamics, is the thermodynamic property that describes this lost ability to cause change. Like energy, all matter has entropy.

A *property* is a quantifiable characteristic of matter. An intensive property is independent of the size or mass of the system. Extensive (mass-dependent) properties can be made intensive simply by dividing by the mass; for example, $e = E/m$ = specific energy; $u = U/m$ = specific internal energy. Although most are familiar with the mechanical concepts of kinetic and potential energies, internal energy is unique to thermodynamics. Internal energy represents the *macroscopic* collection of all the system's microscopic forms of energy, which are energies related to molecular activity (vibrational, rotational, translational). From the kinetic theory of a perfect gas, the internal energy depends only on the *temperature* of the molecules.

The *state* of a system is described by specifying the values of the properties of the system. A *system* is any collection of matter of fixed identity within a prescribed boundary. A *process* describes how a system changes from one state to another. A *cycle* begins and ends at the same state, typically going through many processes in the interim. Pressure P, temperature T, specific internal energy u, and density ρ or specific volume $v = 1/\rho$ are some basic thermodynamic properties. The specification of any two independent intensive properties will fix the thermodynamic state of a simple (meaning in the absence of motion and force fields) compressible system and, therefore, the values of all other thermodynamic properties of the system. The values of the other properties may be found through equations of state. The most common working fluid addressed in this book is a gas that is modeled using the perfect gas equations of state (refer to Section 2.4).

A function relating one dependent and two independent thermodynamic properties of a simple system of unit mass is called an equation of state. When the three properties are P, v, and T, as in

$$f(P, v, T) = 0 \qquad (2.1)$$

the equation is called the *thermal equation of state*. In general, we cannot write the functional relationship in Eq. (2.1) in the form of an equation in which specified values of the two properties will allow us to determine the value of the third. Although humans may not know what the functional relation in Eq. (2.1) is for a given system, one does exist and nature always knows what it is. When the solution set of Eq. (2.1) cannot be determined from relatively simple equations, tables that list the values of P, v, and T (elements of the solution set) satisfying the function may be prepared. This has been done for water (in all of its phases), air, and most common gases.

The functional relation between the internal energy u of a simple system of unit mass and any two independent properties for the set P, v, T

is called the specific *energy equation of state*. This equation can be written functionally as

$$u = u(T, v) \quad \text{or} \quad u = u(P, v) \quad \text{or} \quad u = u(P, T) \tag{2.2}$$

As with the thermal equation of state, we may not be able to write an analytical expression for any of the functional relations of Eq. (2.2). The important thing is that energy is a property; hence, the functional relations exist.

If the solution sets of the thermal and energy equations of state of a simple system of unit mass are known, all thermodynamic properties of the system can be found when any two of the three properties P, v, T are specified. From the solution set, we can form a tabulation of v and u against specified values of P and T for all states of the system. From these known values of P, T, v, and u, we can determine any other property of the simple system. For example, the value of the specific enthalpy h, a very important property in gas dynamics, is found for any state of the system by combining the tabulated values of P, v, and u for that state by

$$h \equiv u + Pv = u + \frac{P}{\rho} \tag{2.3}$$

Enthalpy is a combination or convenience property defined because any time we analyze mass flow systems, in addition to the internal energy of the mass, flow work Pv, is required to get the mass into and out of the system.

Five other definitions are listed here for use in the later sections of this chapter and the following chapter: specific heat at constant volume c_v, specific heat at constant pressure c_p, ratio of specific heats γ, the speed of sound a, and the Mach number M:

$$c_v \equiv \left(\frac{\partial h}{\partial T} \right)_v \tag{2.4}$$

$$c_p \equiv \left(\frac{\partial h}{\partial T} \right)_p \tag{2.5}$$

$$\gamma \equiv \frac{c_p}{c_v} \tag{2.6}$$

$$a \equiv \sqrt{g_c \left(\frac{\partial P}{\partial \rho} \right)_s} = \sqrt{\gamma g_c \left(\frac{\partial P}{\partial \rho} \right)_T} \tag{2.7}$$

$$M \equiv \frac{V}{a} \tag{2.8}$$

Partial derivatives are used in Eqs. (2.4), (2.5), and (2.7) because, in general, the specific heats and speed of sound are functions of more than one variable.

There are three ways to transfer energy across a system boundary:

1. *Energy flux:* Energy transferred with any mass flow.

2. *Heat transfer Q (or q = Q/m)*: Energy transfer between the system and surroundings due to a temperature difference between the system and surroundings. In addition to the temperature difference, a mechanism for heat transfer must also be available (conduction, convection, radiation). Note that a "well-insulated" or *adiabatic* system implies that heat transfer is negligible even though there may be a significant temperature difference between the system and its surroundings.
3. *Work interaction W (or w = W/m)*: Energy transfer between the system and surroundings associated with a force acting through a distance. More generally, energy transfer not due to energy flux or heat transfer can be classified as work interactions and includes boundary work (like a piston-cylinder), rotating shaft work, and electrical work. In this text, our focus will be on rotating shaft work.

It is important to recognize that heat Q and work W are *not* properties; they represent energy in transit or mechanisms for energy transfer. A quantity is a property if its change in value between two states is independent of the process; properties are path independent. Heat and work interactions are both path dependent; they depend on the process. For example, work is force through a distance, so certainly the amount of work is dependent on the path taken.

2.3 Control Volume Approach to Governing Equations

There are two approaches for formulating the governing equations of motion of a fluid. One approach follows a fixed mass (control mass) of particles as it moves throughout the flow field. The second approach fixes a volume within the flow field (control volume) and relates movements of mass, momentum, and energy across the control volume boundaries to changes that occur within the control volume. Although the control mass approach is intuitively appealing, it can be challenging to implement. From a practical standpoint, "tagging" particles to measure their changes or even determining those changes analytically can often be difficult if not impossible. Consequently, we will use the control volume formulation for our study of gas turbine engines and rocket motors. Note that this approach aligns nicely with most typical measurement devices that are fixed in space (fixed probes or rakes of probes). For example, one might have pressure and temperature measurements upstream and downstream of a compressor, thus providing a logical choice for a control volume.

The fundamental laws of nature are described for a fixed mass, so we must have a relationship between control mass and control volume formulations; that relationship is provided by the Reynolds transport theorem. Derivation of the Reynolds transport theorem is provided in many texts such as [15]. Here we apply the results in the general form of Eq. (2.9) applicable to conservation of mass (continuity), the linear momentum relation (Newton's second law), and the first and second laws of thermodynamics.

$$\underbrace{\frac{\partial}{\partial t} \iiint_\sigma x\rho\,d\forall}_{\text{storage term}} + \underbrace{\iint_{CS} x\rho(\vec{V}\cdot\hat{n})dA}_{\text{flux terms}} = \text{source terms} \qquad (2.9)$$

In Eq. (2.9):

$$x = \text{Property per unit mass or velocity vector}$$
$$\text{source terms} = \text{External influences on control volume}$$

Note: Both x and the source terms take on different values based on the governing equation, as shown in Table 2.1. We will discuss each governing equation in detail in the following sections.

Equation (2.9) is a valuable general relation that represents the fully 3-D, unsteady control volume formulation of the governing fundamental laws for gas dynamics. In the most general sense as accommodated by Eq. (2.9), the flow can be unsteady and vary in space in three dimensions so that any property is a function of time and whatever coordinate system is used (thus, the reason for the partial derivative, $\frac{\partial}{\partial t}$). The *storage term* represents the time-dependent rate of storage of mass, momentum, energy, or entropy within the control volume; the *flux terms* account for those same properties crossing *any and all boundaries of the control surface*.

A sketch of a generic control volume with one inlet and one outlet is shown in Fig. 2.1. Note that the vector $\hat{n}$ is the *unit outward normal vector* perpendicular to the cross-sectional area where mass, momentum, energy, or entropy flux occurs. So the dot product, shown in Eq. (2.10), is accounting for the velocity component perpendicular to dA; it is the mathematical representation of the physical fact that mass or any other property can be transported across a control surface only via the component of velocity normal to that surface.

$$\vec{V}\cdot\hat{n} = |\vec{V}||\hat{n}|\cos(\alpha) \equiv V_\perp \qquad (2.10)$$

Table 2.1   Values of x and Source Terms in Eq. (2.9)

Governing Equation	x	Source Terms
Continuity	1.0	$= 0$
Momentum	$\vec{V}$ (In practice, we use each velocity component.)	$= \sum_{\text{on }\sigma} \vec{F}$
1st law: energy	$e = h + ke + pe$	$= \dot{Q} - \dot{W}$
2nd law: entropy	s	$= \iint_{CS} \frac{\dot{q}dA}{T_{\text{source}}} + \dot{P}_s$

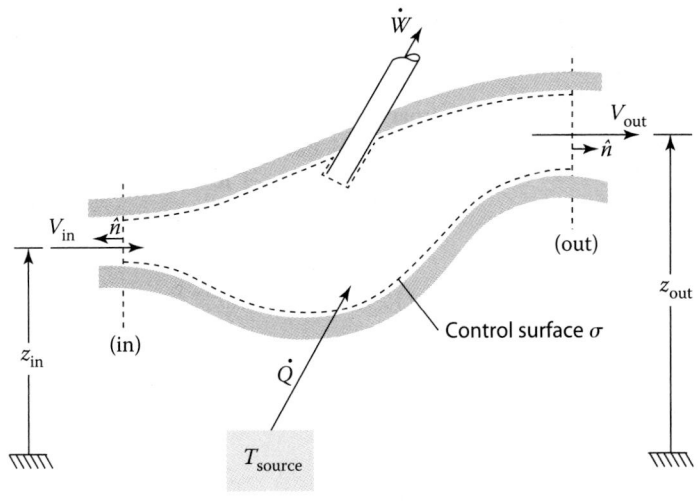

Fig. 2.1   Generic control volume: one inlet and one outlet.

It is important to note also that even if the velocity and unit outward normal were aligned at all inlets and exits, the *signs* would be different because of the definition of the dot product—inflows would be negative (cos 180 deg = −1), and outflows would be positive (cos 0 deg = 1).

To develop a foundational understanding of the basic physical phenomena, our focus will be on steady, 1-D flow—much can be learned even with these simplifying assumptions! Steady flow assumes time dependency of properties is negligible, $\frac{\partial}{\partial t}$ (any property) = 0; consequently, with this assumption the storage triple integral term goes away in Eq. (2.9). One-dimensional flow assumes property changes in one direction, the flow direction, dominate so much that the changes in the other two directions are negligible. *At a control surface cross section, 1-D flow is synonymous with uniform flow*; that is, properties do not vary across the section. From an engineering standpoint, it may help to consider 1-D properties at a cross section as average representative properties at that section. The consequence of this 1-D or uniform flow assumption is that the $x\rho(\vec{V} \cdot \hat{n})$ in Eq. (2.9) can be pulled outside the double integral leaving $\iint_{CS} dA$, which of course integrates out to A, the cross-sectional area. Note that the dot product must still be accounted for in terms of the signs (+ or −) of the flux terms (which we will do in the next sections) and any angle between the velocity vector and unit outward normal vector (see Example 2.1).

2.3.1   Conservation of Mass (Continuity)

Applying the general control volume relation Eq. (2.9) to conservation of mass to a flowing gas, commonly called *continuity*, with $x = 1.0$ and source terms $= 0$ gives us Eq. (2.11).

$$\text{Continuity} \quad \frac{\partial}{\partial t} \iiint_\sigma \rho \, d\forall + \iint_{CS} \rho(\vec{V} \cdot \hat{n}) dA = 0 \qquad (2.11)$$

Equation (2.11) is a mathematical statement that says the time rate of storage of mass in the control volume σ (first term) must equal the net mass flow rate out of the control volume σ (second term). For steady, 1-D flow:

$$\sum_{\text{outflows}} \dot{m} - \sum_{\text{inflows}} \dot{m} = 0 \quad \text{where: } \dot{m} \equiv \rho A V_\perp \qquad (2.12)$$

For steady, 1-D flow with one inlet and outlet, like Fig. 2.1:

$$\dot{m}_{\text{out}} - \dot{m}_{\text{in}} = 0 \quad \text{or} \quad \dot{m}_{\text{out}} = \dot{m}_{\text{in}} \qquad (2.13)$$

P. 118

Example 2.1

Consider 1-D flow through the convergent duct shown in Fig. 2.2. If $A_2 = 0.5A_1$ and V_1 enters section 1 at a 30-deg angle to the cross section (V_2 is aligned with section 2), find an expression for V_2 in terms of V_1. Assume steady flow and negligible density change.

Solution

For steady, 1-D flow: $\dot{m}_2 = \dot{m}_1 \quad$ or $\quad \rho_2 A_2 V_{\perp 2} = \rho_1 A_1 V_{\perp 1}$

$$\text{Because } \rho_2 = \rho_1: A_2 V_{\perp 2} = A_1 V_{\perp 1}$$

$$V_{\perp 2} = V_2 \cos(0) = V_2$$

$$V_{\perp 1} = V_1 \cos(30) = 0.866 V_1$$

$$\text{So: } V_2 = \frac{A_1 V_{\perp 1}}{A_2} = (2.0)(0.866) V_1 = 1.732 V_1$$

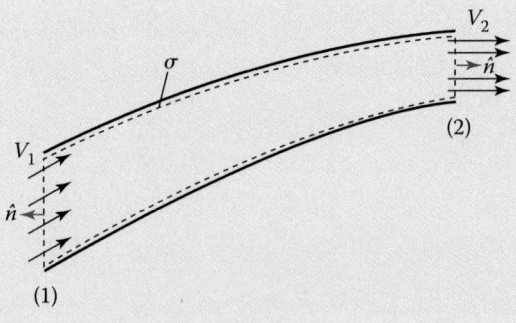

Fig. 2.2 Convergent duct.

Generally, for 1-D flow, we will select a control volume for analysis that results in the velocity being aligned with the unit outward normal vector at all control surfaces; consequently, for the rest of this text, we will drop the $V_\perp$ notation and just use V in our treatment of 1-D flows. As Example 2.1 illustrated, however, there can be instances where the velocity and unit vector do not align even though the problem can be treated as 1-D. The student would do well to remember that it is only the normal component of mass, momentum, energy, or entropy that can physically pass through a control surface.

2.3.2 Steady Flow Linear Momentum

Applying the general control volume relation Eq. (2.9) to the momentum relation (Newton's second law) with $x = \vec{V}$ and source terms $= \sum\limits_{\text{on }\sigma} \vec{F}$ gives us Eq. (2.14).

$$\text{Momentum} \quad \frac{1}{g_c}\frac{\partial}{\partial t}\iiint_\sigma \vec{V}\rho\,d\forall + \frac{1}{g_c}\iint_{CS} \vec{V}\rho(\vec{V}\cdot\hat{n})dA = \sum_{\text{on }\sigma} \vec{F} \quad (2.14)$$

Equation (2.14) is a mathematical statement that says the net force acting *on the gas* of a fixed control volume σ (source terms) is equal to the time rate of increase of momentum within σ (the first term) plus the net flux of momentum from σ (the second term). This very important momentum equation is, in fact, a vector equation, which implies that in the most general sense it *must be applied in three different directions* to solve for unknown quantities.

For steady flow, x-direction:

$$\frac{1}{g_c}\iint_{CS} V_x\rho(\vec{V}\cdot\hat{n})dA = \sum_{\text{on }\sigma} F_x \quad (2.15)$$

Equations similar to that of Eq. (2.15) could be written for the y- and z-directions (or whatever coordinate system is best for the problem). Note that for the dot product, the full velocity vector must be maintained.

For steady, 1-D flow with one inlet and one outlet, like Fig. 2.1:

$$\frac{1}{g_c}(\dot{m}V)_{\text{out}} - \frac{1}{g_c}(\dot{m}V)_{\text{in}} = \sum_{\text{on }\sigma} F_x \quad \text{or} \quad \frac{1}{g_c}\dot{M}_{\text{out}} - \frac{1}{g_c}\dot{M}_{\text{in}} = \sum_{\text{on }\sigma} F_x$$

$$\text{where} \quad \dot{M} \equiv \dot{m}V \quad (2.16)$$

Of course, if there were additional inlets and outlets, the momentum fluxes at those locations would have to be appropriately accounted for.

Applying control volume equations to a steady flow problem gives useful results with only knowledge of conditions at the control surface. Nothing needs to be known about the state of the fluid interior to the control volume. The following examples illustrate the use of the steady, 1-D flow condition and the momentum equation. We suggest that the procedure of *sketching the control volume* and showing the applicable fluxes through each surface as well as applicable forces acting on each surface be followed whenever a control volume equation is used. This situation is similar to the use of free body diagrams for the analysis of forces on solid bodies. We illustrate this procedure in the three examples that follow.

Example 2.2

Water ($\rho = 1000 \text{ kg/m}^3$) is flowing at a steady rate through a convergent duct, as illustrated in Fig. 2.3a. For the data given in the figure, find the force of the fluid $F_{\sigma D}$ acting on the convergent duct D between station 1 and 2.

Solution

We first select the control volume σ such that the force of interest is acting at the control surface. Because we want the force interaction between D and the flowing water, we choose a control surface coincident with the inner wall surface of D bounded by the permeable surfaces 1 and 2, as illustrated in

a) Given data

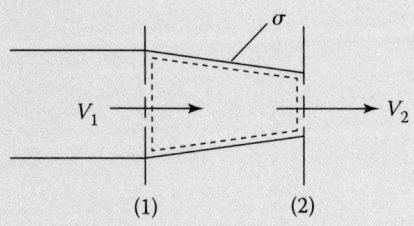

(1)   (2)

$P_1 = 137{,}900 \text{ Pa}$ $P_2 = 101{,}325 \text{ Pa}$
$A_1 = 0.2 \text{ m}^2$ $A_2 = 0.1 \text{ m}^2$
$\rho_1 = \rho_2 = 1000 \text{ kg/m}^3$ $V_2 = 6 \text{ m/s}$

b) Continuity equation

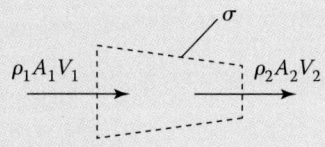

Mass flux in = mass flux out

c) Momentum equation

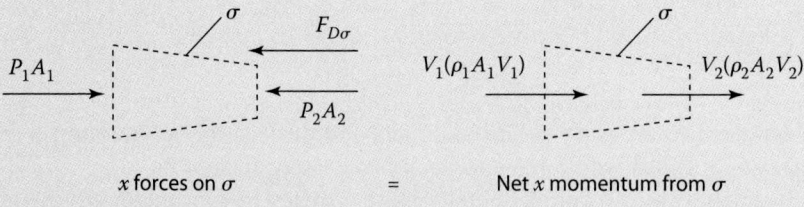

x forces on σ =   Net x momentum from σ

Fig. 2.3   Flow through a convergent duct.

(Continued)

Example 2.2 (Continued)

Fig. 2.3a. By applying the steady 1-D continuity equation [Eq. (2.12)], as depicted in Fig. 2.3b, we find V_1 as follows:

$$\rho_1 A_1 V_1 = \rho_2 A_2 V_2$$

$$V_1 = \frac{A_2}{A_1} V_2 \quad (\rho_1 = \rho_2)$$

$$= 3\,\text{m/s}$$

With V_1 determined, we can apply the momentum Eq. (2.16) to σ and find the force of the duct walls on σ, denoted by $F_{D\sigma}$ ($F_{\sigma D} = -F_{D\sigma}$). By symmetry, $F_{D\sigma}$ is a horizontal force, and so the horizontal x components of forces and momentum fluxes will be considered. The x forces acting on σ are depicted in Fig. 2.3c along with the x momentum fluxes through σ.

From Fig. 2.3c, we have momentum:

$$P_1 A_1 - F_{D\sigma} - P_2 A_2 = \frac{1}{g_c}[(\rho_2 A_2 V_2)V_2 - (\rho_1 A_1 V_1)V_1]$$

And by Fig. 2.3b, continuity:

$$\rho_1 A_1 V_1 = \rho_2 A_2 V_2 = \dot{m}$$

Combining the continuity and momentum equations, we obtain

$$P_1 A_1 - F_{D\sigma} - P_2 A_2 = \frac{\dot{m}}{g_c}(V_2 - V_1)$$

or

$$F_{D\sigma} = P_1 A_1 - P_2 A_2 - \frac{\dot{m}}{g_c}(V_2 - V_1)$$

With $\dot{m} = \rho_1 A_1 V_1 = 1000\,\text{kg/m}^3 \times 0.2\,\text{m}^2 \times 3\,\text{m/s} = 600\,\text{kg/s}$, we have

$$F_{D\sigma} = 137{,}000\,\text{N/m}^2 \times 0.2\,\text{m}^2 - 101{,}325\,\text{N/m}^2 \times 0.1\,\text{m}^2$$
$$- 600\,\text{kg/s} \times (6 - 3)\,\text{m/s}$$

or

$$F_{D\sigma} = 27{,}580\,\text{N} - 10{,}132\,\text{N} - 1800\,\text{N}$$

and

$$= 15{,}648\,\text{N} \quad \text{acts to the left in assumed position}$$

Finally, the force of the water on the duct is

$$F_{\sigma D} = -F_{D\sigma} = -15{,}648\,\text{N} \quad \text{acts to the right (which makes sense)}$$

Example 2.3

Figure 2.4 shows the steady flow conditions at sections 1 and 2 about an airfoil mounted in a wind tunnel where the frictional effects at the wall are negligible. Determine the section drag coefficient C_d of this airfoil. The air may be treated as incompressible for the given conditions.

Solution

Because the fluid is incompressible and the flow is steady, the continuity equation may be used to find the unknown velocity V_B as follows:

$$(\rho A V)_1 = (\rho A V)_2 \quad \text{that is, } \dot{m}_1 = \dot{m}_2$$

or

$$\rho_1 A_1 V_A = \rho_2 \left(\frac{2}{3} V_B + \frac{1}{3} V_C \right) A_2$$

but

$$\rho_1 = \rho_2 \quad \text{and} \quad A_1 = A_2$$

$$V_A = \frac{2}{3} V_B + \frac{1}{3} V_C$$

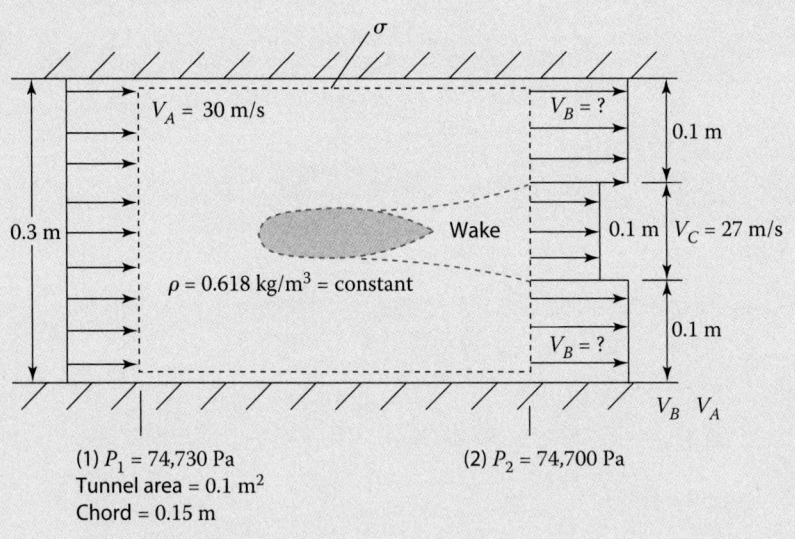

(1) P_1 = 74,730 Pa
Tunnel area = 0.1 m²
Chord = 0.15 m

(2) P_2 = 74,700 Pa

Fig. 2.4 Wind tunnel drag determination for an airfoil section.

(Continued)

Example 2.3 *(Continued)*

thus

$$V_B = \frac{3}{2} V_A - \frac{1}{2} V_C = 31.5 \, \text{m/s}$$

The momentum equation may now be used to find the drag on the airfoil. This drag force will include both the skin friction and pressure drag. We sketch the control volume σ with the terms of the momentum equation as shown in Fig. 2.5.

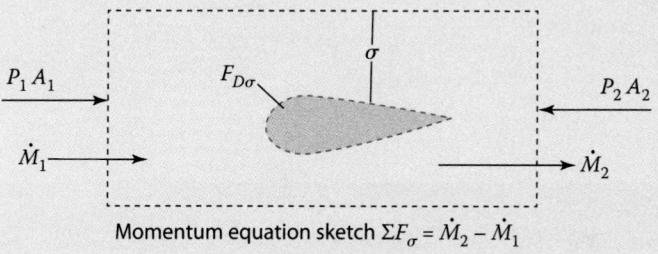

Momentum equation sketch $\Sigma F_\sigma = \dot{M}_2 - \dot{M}_1$

Fig. 2.5   Sketch for momentum equation for airfoil section.

Taking forces to right as positive, we have from the sketch

$$\sum F_\sigma = P_1 A_1 - P_2 A_2 + F_{D\sigma}$$

$$\dot{M}_1 = (\rho_1 A_1 V_A) V_A = \rho_1 A_1 V_A^2$$

$$\dot{M}_2 = \rho_2 \left(\frac{2}{3} A_2 \right) V_B^2 + \rho_2 \left(\frac{1}{3} A_2 \right) V_C^2$$

$$= \rho_2 A_2 \left(\frac{2}{3} V_B^2 + \frac{1}{3} V_C^2 \right)$$

For $\rho = \rho_1 = \rho_2$ and $A = A_1 = A_2$,

$$-F_{D\sigma} = (P_1 - P_2)A + \frac{\rho A}{g_c} \left(V_A^2 - \frac{2}{3} V_B^2 - \frac{1}{3} V_C^2 \right)$$

or

$$= (74{,}730 - 74{,}700)\, 0.1 + 0.618 \times 0.1 \left[30^2 - \frac{2}{3}(31.5^2) - \frac{1}{3}(27^2) \right]$$

$$= 3.0 \, \text{N} - 0.278 \, \text{N}$$

(Continued)

Example 2.3 *(Continued)*

$-F_{D\sigma} = 2.722\,\text{N}$ $\therefore F_{D\sigma}$ acts to left

$F_{\sigma D}$ = drag force for section

$\qquad = -F_{D\sigma}$ and $F_{\sigma D}$ acts to right

$$F_D' = \frac{F_{\sigma D}}{b} = \frac{2.722\,\text{N}}{0.333\,\text{m}} = 8.174\,\text{N/m} \quad (b = \text{tunnel span})$$

$$C_d = \frac{F_D'}{qc} \quad \text{and} \quad q = \frac{\rho V_\infty^2}{2g_c} \quad \text{where } V_\infty = V_A$$

$$C_d = \frac{F_D'}{[(\rho V_\infty^2)/(2g_c)]c} = \frac{8.174}{(0.618 \times 30^2/2)0.15} = 0.196$$

Example 2.4

Figure 2.6 shows a test stand for determining the thrust of a liquid-fuel rocket. The propellants enter at section 1 at a mass flow rate of 15 kg/s, a velocity of 30 m/s, and a pressure of 0.7 MPa. The inlet pipe for the propellants is very flexible, and the force it exerts on the rocket is negligible. At the nozzle exit, section 2, the area is 0.064 m², and the pressure is 110 kPa. The force read by the scales is 2700 N, atmospheric pressure is 82.7 kPa, and the flow is steady. Determine the exhaust velocity at

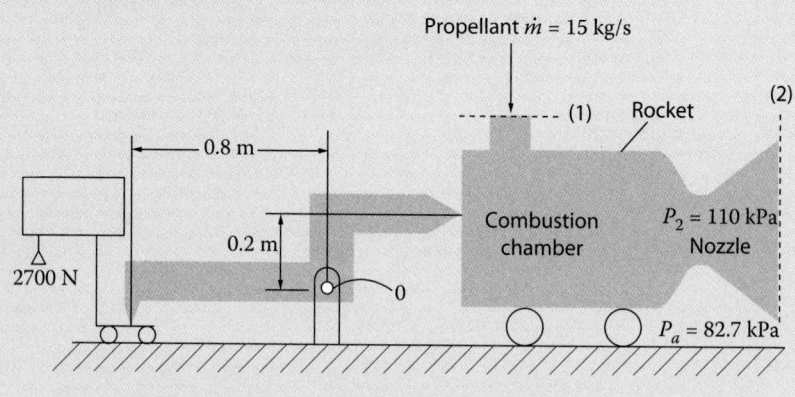

Fig. 2.6 Liquid-fuel rocket test setup.

(Continued)

Example 2.4 (Continued)

section 2, assuming 1-D flow exists. Mechanical frictional effects may be neglected.

Solution

First, determine the force on the lever by the rocket to develop a 2700-N scale reading. This may be done by summing moments about the fulcrum point 0 (see Fig. 2.7a). Sum the horizontal forces on the rocket engine as shown in Fig. 2.7b. Note that the unbalanced pressure force on the exterior of the

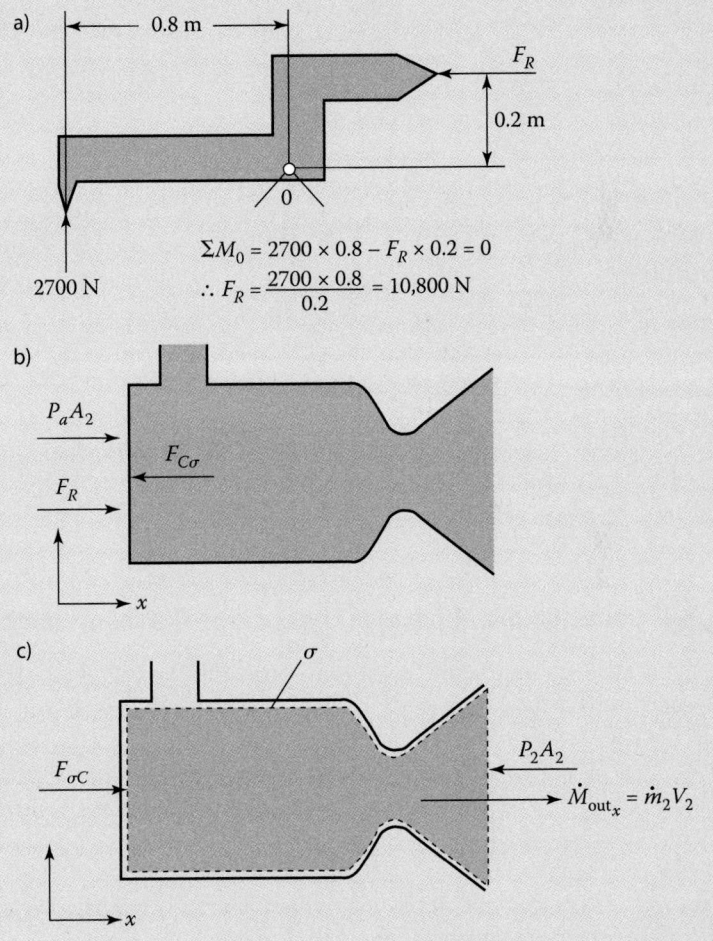

a)

$$\Sigma M_0 = 2700 \times 0.8 - F_R \times 0.2 = 0$$

$$\therefore F_R = \frac{2700 \times 0.8}{0.2} = 10{,}800 \text{ N}$$

Fig. 2.7   Momentum equation sketch $\sum F_\sigma = (\dot{M}_2 - \dot{M}_1)/g_c$.

(Continued)

Example 2.4 *(Continued)*

rocket engine is P_aA_2, and the interior forces (pressure and friction) are contained within the force $F_{C\sigma}$. Next, draw an internal volume σ around the fluid within the rocket engine as shown, and indicate the horizontal forces and momentum flux (see Fig. 2.7c).

Summing the forces on the rocket engine as shown in Fig. 2.7b, we obtain

$$F_R + P_aA_2 = F_{C\sigma}$$

Applying the momentum equation to the control volume σ shown in Fig. 2.7c, we obtain

$$F_{\sigma C} - P_2A_2 = \frac{\dot{m}_2 V_2}{g_c}$$

Combining these two equations to remove $F_{\sigma C}$ gives

$$F_R - (P_2 - P_a)A_2 = \frac{\dot{m}_2 V_2}{g_c}$$

which is the same as Eq. (1.52). Because the flow is steady, the continuity equation yields $\dot{m}_1 = \dot{m}_2 = 15\,\text{kg/s}$. Therefore,

$$
\begin{aligned}
V_2 &= \frac{F_R - (P_2 - P_a)A_2}{\dot{m}_2/g_c} \\[6pt]
&= \frac{10{,}800 - [(110 - 82.7) \times 10^3\,\text{N/m}^2](0.064\,\text{m}^2)}{15\,\text{kg/s}} \\[6pt]
&= 603.5\,\text{m/s}
\end{aligned}
$$

2.3.3 Conservation of Energy (First Law of Thermodynamics)

Applying the general control volume relation Eq. (2.9) to the first law of thermodynamics (conservation of energy) with $x = e$ and source terms $= \dot{Q} - \dot{W}$ gives us Eq. (2.17).

$$\text{Energy} \quad \frac{\partial}{\partial t}\iiint_\sigma e\rho\,d\forall + \iint_{CS} e\rho(\vec{V} \cdot \hat{n})dA = \dot{Q} - \dot{W} \qquad (2.17)$$

Equation (2.17) says that the net rate at which energy is transferred into the control volume σ (source terms) is equal to the time rate of increase of energy within σ (the first term) plus the net flux of energy from σ (the second term). Note the shorthand notation for the source terms in Eq. (2.17); this is common practice in thermodynamics where energy due to *heat transfer in* is considered positive and energy transfer due to *work interaction out* is *considered positive*. So the source terms in Eq. (2.17) should more accurately be written as:

$$\dot{Q} - \dot{W} = \underbrace{\sum \dot{Q}}_{\text{Net in}} - \underbrace{\sum \dot{W}}_{\text{Net out}} = \left(\sum \dot{Q}_{\text{in}} - \sum \dot{Q}_{\text{out}}\right) - \left(\sum \dot{W}_{\text{out}} - \sum \dot{W}_{\text{in}}\right)$$

We will adopt a version of this equation where we drop the summation symbol with the understanding that it is implied. (All energy transfer due to heat and work interactions must be accounted for.)

For steady, 1-D flow with one inlet and one outlet, like that shown in Fig. 2.1:

$$\dot{m}\left(h + \frac{V^2}{2g_c} + \frac{gz}{g_c}\right)_{\text{out}} - \dot{m}\left(h + \frac{V^2}{2g_c} + \frac{gz}{g_c}\right)_{\text{in}} = \dot{Q}_{\text{in}} + \dot{W}_{\text{in}} - \dot{Q}_{\text{out}} - \dot{W}_{\text{out}}$$

(2.18)

You should find the form of Eq. (2.18) intuitively appealing. Note that energy transfer *into* the system due to heat and work interactions will cause the system's energy to increase (hence the positive signs), whereas energy transfer *out* of the system due to heat and work interactions will cause the system's energy to decrease (the negative signs).

Equation (2.18) is the rate form of the steady, 1-D energy equation where the dimensions of $\dot{Q}_{\text{in}}$, $\dot{W}_{\text{in}}$, and the like are those of power or energy per unit time (hp or kW). If the mass flow rate into and out of the system is the same, we can divide by the mass flow rate, which gives

$$\left(h + \frac{V^2}{2g_c} + \frac{gz}{g_c}\right)_{\text{out}} - \left(h + \frac{V^2}{2g_c} + \frac{gz}{g_c}\right)_{\text{in}} = q_{\text{in}} + w_{\text{in}} - q_{\text{out}} - w_{\text{out}} \quad (2.19)$$

where the units of each term are energy per unit mass (BTU/lbm or kJ/kg).

For most of the applications we will encounter, the change in potential energy of the fluid (gravitational effects) as it flows through the control volume (Fig. 2.1) will be negligible compared to the change in specific enthalpy and specific kinetic energy. Also, because h represents the fluid's specific internal energy plus flow work per Eq. (2.3) and $V^2/2g_c$, the fluid's specific kinetic energy, we combine those two terms and define the stagnation or *total specific enthalpy*:

$$h_t \equiv h + \frac{V^2}{2g_c} \qquad (2.20)$$

So, for steady, 1-D flow, no gravity effects:

$$\dot{m}h_{t\,\text{out}} - \dot{m}h_{t\,\text{in}} = \dot{Q}_{\text{in}} + \dot{W}_{\text{in}} - \dot{Q}_{\text{out}} - \dot{W}_{\text{out}} \quad \text{(energy per time)} \quad (2.21)$$

$$h_{t\,\text{out}} - h_{t\,\text{in}} = q_{\text{in}} + w_{\text{in}} - q_{\text{out}} - w_{\text{out}} \quad \text{(energy per mass)} \qquad (2.22)$$

As simple as they look, you will find Eqs. (2.20) to (2.22) extraordinarily useful and insightful as you gain fundamental understanding and experience in analysis of control volume fluid mechanics problems. The following examples will illustrate this. Again, we highlight the importance of an accurate *control volume sketch* showing the applicable energy fluxes through each surface as well as any and all energy transfers due to heat or work interactions. We will hold off on any numerical examples until we introduce the perfect gas relations in Section 2.4.

Example 2.5

Consider a turbojet engine as shown in Fig. 2.8a. We divide the engine into the following control volume regions as shown and apply conservation of energy in the form of Eq. (2.22). We assume the mass flow rate into and out of each control volume remains the same (there are no bleed inflows or outflows).

σ_1: Inlet
σ_2: Compressor
σ_3: Combustor
σ_4: Turbine
σ_5: Exhaust nozzle

a) Adiabatic inlet and exhaust nozzle: σ_1 and σ_5 (Fig. 2.8b)

There are no shaft work interactions at any control surfaces. Heat interactions are negligible as are changes in potential energy.

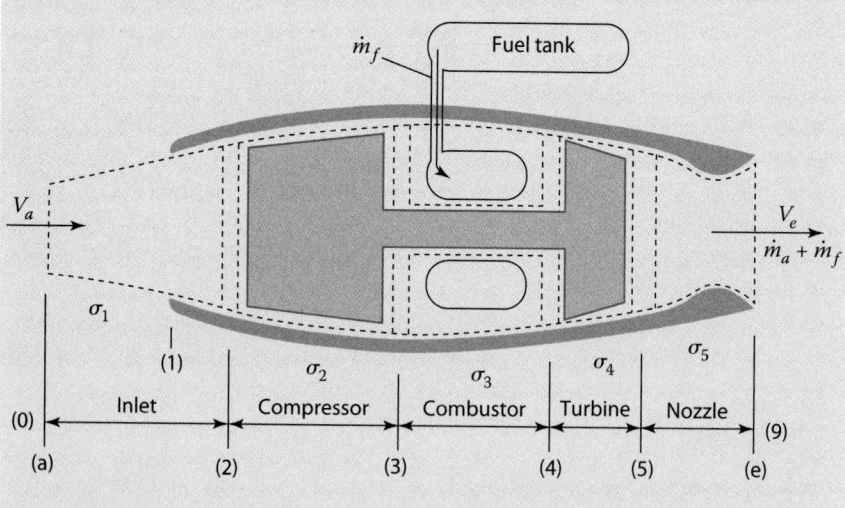

Fig. 2.8a Control volume for analyzing each component of the turbojet engine.

(Continued)

Example 2.5 *(Continued)*

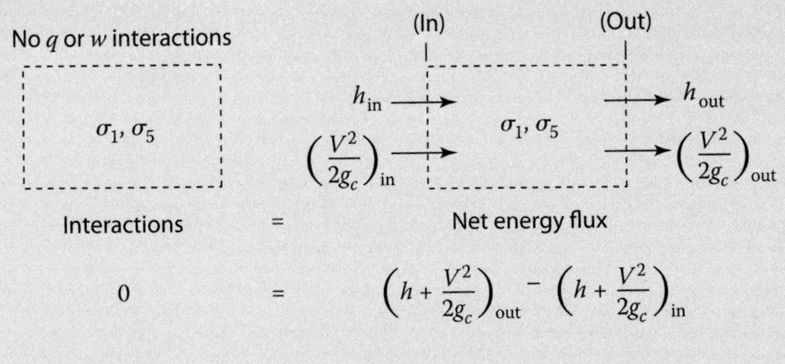

No q or w interactions

(In)   (Out)

$$\sigma_1, \sigma_5$$

$h_{in} \longrightarrow$ $\longrightarrow h_{out}$

$\left(\dfrac{V^2}{2g_c}\right)_{in} \longrightarrow$ σ_1, σ_5 $\longrightarrow \left(\dfrac{V^2}{2g_c}\right)_{out}$

Interactions	=	Net energy flux

$$0 \quad = \quad \left(h + \frac{V^2}{2g_c}\right)_{out} - \left(h + \frac{V^2}{2g_c}\right)_{in}$$

Fig. 2.8b Energy equation applied to control volumes σ_1 and σ_5.

Therefore, the steady, 1-D flow energy equation in Eq. (2.22) gives the result

$$h_{t\,out} - h_{t\,in} = 0 \quad \text{or} \quad h_{t\,out} = h_{t\,in}$$

From the definition of total specific enthalpy, Eq. (2.20)

$$\left(h + \frac{V^2}{2g_c}\right)_{out} = \left(h + \frac{V^2}{2g_c}\right)_{in} \quad \text{or} \quad h_{out} - h_{in} = \frac{1}{2g_c}\left(V_{in}^2 - V_{out}^2\right)$$

So, for an adiabatic inlet and exhaust nozzle, the energy tradeoff is between kinetic energy and specific enthalpy. A properly designed and working inlet will diffuse the flow, resulting in $V_{out} < V_{in}$ and a corresponding rise in specific enthalpy ($h_{out} > h_{in}$). A properly designed and working exhaust nozzle will expand the exhaust gases manifested by an increase in the kinetic energy and a corresponding decrease in specific enthalpy.

b) Adiabatic compressor and turbine: σ_2 and σ_4 (Fig. 2.8c)

Heat interactions are negligible as are changes in potential energy. Shaft work interactions are present because each control surface cuts through a rotating shaft. Therefore, the steady, 1-D flow energy equation in Eq. (2.22) gives the result

$$h_{t\,out} - h_{t\,in} = w_{in} \quad \text{(compressor)}$$

$$h_{t\,out} - h_{t\,in} = -w_{out} \quad \text{or} \quad w_{out} = h_{t\,in} - h_{t\,out} \quad \text{(turbine)}$$

The compressor requires shaft work in, and that energy addition is manifested as an increase in the total enthalpy of the air flowing through the compressor. The turbine provides the shaft work via a shaft connecting the two components. The shaft work out of the turbine is taken from the total enthalpy of the combustion gases flowing through the turbine, resulting in a decrease in that enthalpy.

(Continued)

Example 2.5 *(Continued)*

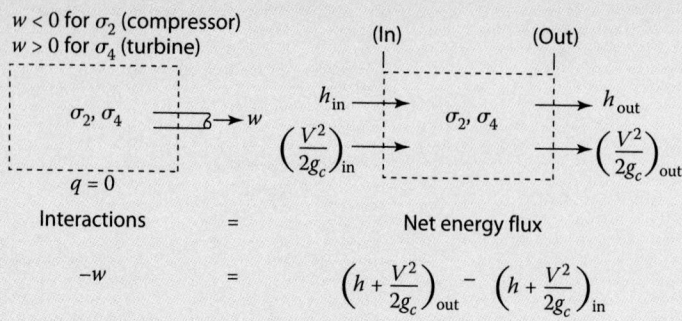

Fig. 2.8c   Energy equation applied to control volumes σ_2 and σ_4.

c) Adiabatic combustor: σ_3 (Fig. 2.8d)

Assume that the fuel and air entering the combustor mix physically in a mixing zone (Fig. 2.8d) to form what we will call reactants (denoted by subscript R). The reactants then enter a combustion zone where combustion occurs, forming products of combustion (subscript P) that leave the

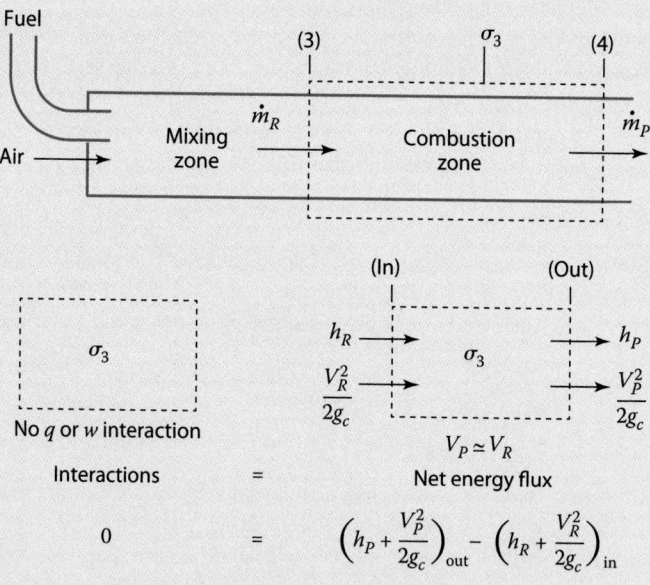

Fig. 2.8d   Energy equation applied to control volume σ_3.

(Continued)

Example 2.5 *(Continued)*

combustor. Because the temperature in the combustion zone is higher than that of the immediate surroundings, there is a heat interaction between σ_3 and the surroundings that, per unit mass flow of reactants, is negligibly small relative to the very high energy release rates due to combustion. There are no shaft work interactions, and changes in potential energy are negligible. Also, the velocities of the products leaving and of the reactants entering the combustion zone are approximately equal. Hence the steady, 1-D flow energy equation in Eq. (2.22) gives the result

$$h_{R3} = h_{P4} \quad (V_{R3} \approx V_{P4}) \tag{2.23}$$

which should be readily verifiable from part a, the adiabatic inlet and exhaust nozzle development.

The seemingly simple Eq. (2.23) is not quite as straightforward as it appears. First, the chemical aggregation of the two states, 3 and 4, are different; the additional subscripts R and P emphasize this point. In the ideal sense, for hydrocarbon–air combustion, air (consisting mostly of nitrogen and oxygen) and hydrocarbon fuel provide the reactants and water and carbon dioxide are the products (with all the nitrogen remaining as N_2). Second, we must measure the enthalpy of each term in the equation relative to the same datum state. To emphasize the second point, we select as our common enthalpy datum a state d having the chemical aggregation of the products at a datum temperature T_d. Then, introducing the datum state enthalpy $(h_P)_d$ into the last equation, we have

$$h_{R3} - h_{Pd} = h_{P4} - h_{Pd} \tag{2.24}$$

Equation (2.24) can be used to determine the temperature of the products of combustion, leaving an adiabatic combustor for given inlet conditions. If the combustor is not adiabatic, simply return to Eq. (2.22) to include the appropriate heat interaction term q. We will continue this example later in this chapter after the perfect gas relations have been introduced.

2.3.4 Production of Entropy (Second Law of Thermodynamics)

We return to our familiar procedure using the general control volume relation Eq. (2.9) and apply it to the second law of thermodynamics with the appropriate source terms to arrive at

$$\text{Entropy} \quad \frac{\partial}{\partial t} \iiint_\sigma s\rho\, d\forall + \iint_{CS} s\rho(\vec{V}\cdot\hat{n})dA = \iint_{CS} \frac{\dot{q}dA}{T_{source}} + \dot{P}_s \tag{2.25}$$

Like the prior three governing equations, the left side of Eq. (2.25) represents the time rate of increase of the property of interest, this time specific entropy, within the control volume σ (the first term) plus the net flux of entropy from

σ (the second term); these terms can be positive, negative, or zero. The first term on the right side of Eq. (2.25) represents the rate of entropy transfer accompanying heat transfer. Just as with the energy Eq. (2.17), the term is positive when heat transfer is into the system and negative when it is out of the system (and of course zero if heat transfer is negligible). There is no entropy transfer associated with work interactions.

The second term on the right side of Eq. (2.25) is the *production rate of entropy due to system irreversibilities*, $\dot{P}_s$. All real systems have irreversibilities. These include friction, heat transfer through a finite temperature difference, spontaneous chemical reaction, spontaneous mixing, and unrestrained expansion of a fluid, just to name a few. Consequently, for all real systems, $\dot{P}_s$ is positive; in the ideal limit with no system irreversibilities, $\dot{P}_s = 0$. A very useful interpretation of the second law of thermodynamics is that *it requires that entropy is produced by irreversibilities and entropy is conserved only in the limit as irreversibilities go to zero*. Read this last statement as many times as needed to let it sink in—it has very powerful implications. The first law of thermodynamics is merely an accounting of energy in its different forms; it says nothing about the natural direction that energy transfer processes follow or even whether or not certain energy transfer processes are possible. The second law of thermodynamics covers these physical aspects as well as others, like placing limits on the maximum possible system performance and accounting for ordered vs disordered states.

Before simplifying Eq. (2.25) into its steady, 1-D form, let us further discuss the first term on the right side, entropy transfer accompanying heat transfer. In practice, this integral is often challenging to evaluate. The term $\dot{q}$ represents the time rate of heat transfer per unit surface area, called the heat flux; in order to evaluate the integral, we must know the variation of the instantaneous temperature with respect to the heat flux at the control volume boundary where energy (and entropy) transfer is occurring due to heat transfer. Here, we assume a thermal energy reservoir at temperature T_{source} (see Fig. 2.1). A thermal energy reservoir (TER) remains at constant temperature even though energy is added or removed due to heat transfer. Although this is an idealization, in practice, the atmosphere, large bodies of water, and a steady combustion process outside the system are all examples that closely approximate the behavior.

For steady, 1-D flow with one inlet and one outlet, like Fig. 2.1:

$$(\dot{m}s)_{\text{out}} - (\dot{m}s)_{\text{in}} = \frac{\dot{Q}_{\text{in}}}{T_H} - \frac{\dot{Q}_{\text{out}}}{T_L} + \dot{P}_s \qquad (2.26)$$

Here we assume a high-temperature TER source T_H and a low-temperature TER sink T_L to allow for entropy transfer out of the system due to heat transfer. Because the production rate of entropy due to system irreversibilities $\dot{P}_s$

must be greater than or equal to zero, we recast Eq. (2.26), assuming the mass flow rate into and out of the system is the same:

$$\dot{P}_s = \dot{m}(s_{out} - s_{in}) + \frac{\dot{Q}_{out}}{T_L} - \frac{\dot{Q}_{in}}{T_H} \geq 0 \qquad (2.27)$$

Equation (2.27) is the steady flow, 1-D equation form of the second law of thermodynamics requiring that *entropy is produced by irreversibilities* ($\dot{P}_s > 0$) *and entropy is conserved only in the limit as irreversibilities go to zero* ($\dot{P}_s = 0$, a reversible process). The first term on the right side of Eq. (2.27) represents the rate of entropy change of the system Δs, and the second term represents the rate of entropy change of the surroundings (outside the system). We see from Eq. (2.27) that an *adiabatic* (negligible heat transfer), *isentropic* ($\Delta s = 0$) process represents an idealized reversible process. We will use adiabatic, isentropic processes frequently to identify maximum theoretical performance.

Example 2.6

As shown in Fig 2.9, a system is undergoing a steady flow, cyclic process. With the numeric quantities provided, is this possible?

Solution
Our control volume system is shown by the dashed line. Thermodynamically, a process is possible if *both* the first and second laws are satisfied. The first law for steady, 1-D flow is shown here, repeated from Eq. (2.18):

$$\dot{m}\left(h + \frac{V^2}{2g_c} + \frac{gz}{g_c}\right)_{out} - \dot{m}\left(h + \frac{V^2}{2g_c} + \frac{gz}{g_c}\right)_{in} = \dot{Q}_{in} + \dot{W}_{in} - \dot{Q}_{out} - \dot{W}_{out}$$

Although there is likely a working fluid flowing through whatever components are being used to accomplish the cycle, those flows are internal to the control volume. No mass is crossing the system boundaries. Consequently, the left side of the equation is zero and the equation reduces to

$$\dot{W}_{out} = \dot{Q}_{in} - \dot{Q}_{out}$$

The second law for steady, 1-D flow is shown here, taken from Eq. (2.27):

$$\dot{P}_s = \dot{m}(s_{out} - s_{in}) + \frac{\dot{Q}_{out}}{T_L} - \frac{\dot{Q}_{in}}{T_H} \geq 0$$

Again, because no mass is crossing the system boundaries, we have

$$\dot{P}_s = \frac{\dot{Q}_{out}}{T_L} - \frac{\dot{Q}_{in}}{T_H} \geq 0$$

(Continued)

Example 2.6 *(Continued)*

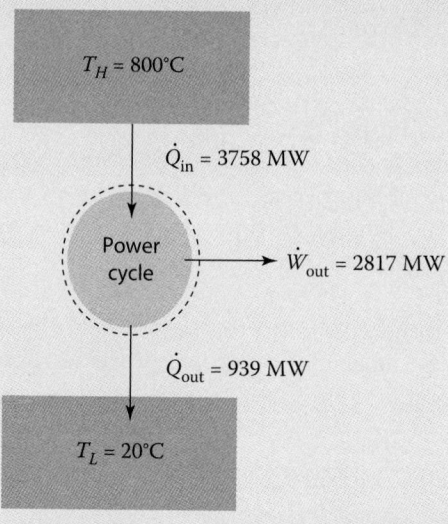

Fig. 2.9   Schematic of a steady flow, cyclic process.

for which we know the second law requirement for all *real* systems holds that the inequality ($>$) be true.

So, from the first law for the quantities given:

$$2817\,\text{MW} = 3756\,\text{MW} - 939\,\text{MW} = 2817\,\text{MW}$$

and indeed, the first law conservation of energy is satisfied.

From the second law for the quantities given:

$$\dot{P}_s = \frac{939\,\text{MW}}{(20+273)\text{K}} - \frac{3756\,\text{MW}}{(800+273)\text{K}} = 3.205\,\frac{\text{MW}}{\text{K}} - 3.500\,\frac{\text{MW}}{\text{K}}$$

$$= -0.295\,\frac{\text{MW}}{\text{K}} \ngtr 0$$

and the second law is not satisfied. Therefore, this process is *not possible* for the quantities given.

2.4   Perfect Gas

Thermodynamics relates energy exchanges (through heat and work interactions and energy flux with mass flow) to property changes. Some properties are measurable, like pressure, temperature, and volume; others are not, such as enthalpy and entropy. In Section 2.2 we discussed a form of the *state postulate*, namely that the state of a simple compressible system is completely specified by two independent intensive properties. This implies that there must be ways to relate properties so that all the other properties can be determined from the two known independent intensive properties. Fundamentally, there are two ways to relate properties—experimentally or analytically (models). In this section, we present the perfect gas relations that will be used extensively throughout this book.

As the name implies, a perfect (or ideal) gas is an approximation, the fundamental assumption being *negligible intermolecular forces*. In essence, a perfect gas is composed of many point particles that interact only when they collide elastically. It happens to be a very good approximation for many gases, including air, over a wide range of conditions. In general, a gas behaves more like a perfect gas under high temperature and low pressure conditions. The interested student is referred to any number of fundamental thermodynamics textbooks, which typically use a quantity called the compressibility factor to quantify the deviation from perfect gas behavior.

2.4.1 General Characteristics

The thermodynamic equations of state for a perfect gas (sometimes called the *ideal gas model*) are

$$P = \rho RT \quad \text{or} \quad Pv = RT \qquad (2.28)$$
$$u = u(T) \qquad (2.29)$$
$$h = h(T) \qquad (2.30)$$

where P is the thermodynamic pressure, ρ is the density, R is the gas constant, T is the thermodynamic temperature, u is the specific internal energy (internal energy per unit mass), and h is the specific internal enthalpy (enthalpy per unit mass), both functions of temperature only. Equation (2.30) arises from the definition of specific enthalpy [Eq. (2.3)], applied to a perfect gas [Eq. (2.28)] to give

$$h = u + RT = h(T) \qquad (2.31)$$

The gas constant R is related to the universal gas constant $\mathcal{R}$ and the molecular weight of the gas $\mathcal{M}$ by

$$R = \frac{\mathcal{R}}{\mathcal{M}} \qquad (2.32)$$

Values of the gas constant and molecular weight for typical gases are presented in Table 2.2 in several unit systems, $\mathcal{R} = 8.31434\,\text{kJ}/(\text{kmol} \cdot \text{K}) = 1.98718\,\text{Btu}/(\text{mol} \cdot {}^\circ\text{R})$.

Differentiating Eq. (2.31) gives

$$dh = du + R\,dT \qquad (2.33)$$

Table 2.2 Properties of Ideal Gases at 298.15 K (536.67°R)

Gas	Molecular Weight	c_p kJ/ (kg·K)	c_p Btu/ (lbm·°R)	R kJ/ (kg·K)	R (ft·lbf)/ (lbm·°R)	γ
Air	28.97	1.004	0.240	0.286	53.34	1.40
Argon	39.94	0.523	0.125	0.208	38.69	1.67
Carbon dioxide	44.01	0.845	0.202	0.189	35.1	1.29
Carbon monoxide	28.01	1.042	0.249	0.297	55.17	1.40
Hydrogen	2.016	14.32	3.42	4.124	766.5	1.40
Nitrogen	28.02	1.038	0.248	0.296	55.15	1.40
Oxygen	32.00	0.917	0.217	0.260	48.29	1.39
Sulfur dioxide	64.07	0.644	0.154	0.130	24.1	1.25
Water vapor	18.016	1.867	0.446	0.461	85.78	1.33

The differentials dh and du in Eq. (2.33) are related to the specific heat at constant pressure and specific heat at constant volume [see definitions in Eqs. (2.4) and (2.5)], respectively, as follows:

$$dh = c_p \, dT$$

$$du = c_v \, dT$$

Note that both specific heats are functions of temperature only, consistent with the perfect gas assumption. These equations can be integrated from state 1 to state 2 to give

$$u_2 - u_1 = \int_{T_1}^{T_2} c_v \, dT \tag{2.34}$$

$$h_2 - h_1 = \int_{T_1}^{T_2} c_p \, dT \tag{2.35}$$

Substitution of the equations for dh and du into Eq. (2.33) gives the relationship between specific heats for a perfect gas

$$c_p = c_v + R \tag{2.36}$$

and γ is the ratio of the specific heat at constant pressure to the specific heat at constant volume, or, as in Eq. (2.6),

$$\gamma \equiv c_p/c_v$$

The following relationships result from using Eqs. (2.36) and (2.6):

$$\frac{R}{c_v} = \gamma - 1 \tag{2.37}$$

$$\frac{R}{c_p} = \frac{\gamma - 1}{\gamma} \tag{2.38}$$

The Gibbs equation relates the entropy per unit mass s to the other thermodynamic properties of a substance. It can be written as

$$ds = \frac{du + P\,d(1/\rho)}{T} = \frac{dh - (1/\rho)\,dP}{T} \tag{2.39}$$

For a perfect gas, the Gibbs equation can be written simply as

$$ds = c_v \frac{dT}{T} + R\frac{d(1/\rho)}{1/\rho} \tag{2.40}$$

$$ds = c_p \frac{dT}{T} + R\frac{dP}{P} \tag{2.41}$$

These equations can be integrated between states 1 and 2 to yield the following expressions for the change in entropy $s_2 - s_1$:

$$s_2 - s_1 = \int_{T_1}^{T_2} c_v \frac{dT}{T} + R\,\ell n\frac{\rho_1}{\rho_2} \tag{2.42}$$

$$s_2 - s_1 = \int_{T_1}^{T_2} c_p \frac{dT}{T} - R\,\ell n\frac{P_2}{P_1} \tag{2.43}$$

If the specific heats are known functions of temperature for a perfect gas, then Eqs. (2.34), (2.35), (2.42), and (2.43) can be integrated from a reference state and tabulated for further use in what are called *gas tables*.

The equation for the speed of sound in a perfect gas is easily obtained by use of Eqs. (2.7) and (2.28) to give

$$a = \sqrt{\gamma R g_c T} \tag{2.44}$$

2.4.2 Calorically Perfect Gas

In addition to the fundamental assumption of negligible intermolecular forces, it is often useful and convenient to assume a *calorically perfect gas* (CPG) to facilitate simple calculations. A CPG is a perfect gas with constant specific heats (c_p and c_v). The implications of the CPG assumption should be readily apparent upon examination of the integral terms in Section 2.4.1.

In this case, the expressions for changes in internal energy u, enthalpy h, and entropy s simplify to the following:

$$u_2 - u_1 = c_v(T_2 - T_1) \tag{2.45}$$

$$h_2 - h_1 = c_p(T_2 - T_1) \tag{2.46}$$

$$s_2 - s_1 = c_v \ln\frac{T_2}{T_1} - R\ln\frac{P_2}{P_1} \tag{2.47}$$

$$s_2 - s_1 = c_p \ln\frac{T_2}{T_1} - R\ln\frac{P_2}{P_1} \tag{2.48}$$

Equations (2.47) and (2.48) can be rearranged to give the following equations for the temperature ratio T_2/T_1:

$$\frac{T_2}{T_1} = \left(\frac{P_2}{P_1}\right)^{R/c_v} \exp\frac{s_2 - s_1}{c_v}$$

$$\frac{T_2}{T_1} = \left(\frac{P_2}{P_1}\right)^{R/c_p} \exp\frac{s_2 - s_1}{c_p}$$

From Eqs. (2.37) and (2.38), these expressions become

$$\frac{T_2}{T_1} = \left(\frac{P_2}{P_1}\right)^{\gamma-1} \exp\frac{s_2 - s_1}{c_v} \tag{2.49}$$

$$\frac{T_2}{T_1} = \left(\frac{P_2}{P_1}\right)^{(\gamma-1)/\gamma} \exp\frac{s_2 - s_1}{c_p} \tag{2.50}$$

2.4.2.1 Specific Heats Recapitulation

Specific heats are properties of a substance. Physically, they represent the amount of energy required to raise a unit mass of a substance by 1 deg under conditions of constant volume c_v and constant pressure c_p. Given this, it should not surprise you that the specific heats for perfect gases are functions of temperature (see, for example, Fig. 6.1b) and that c_p will be greater than c_v. Consequently, when using the calorically perfect gas assumption, *suitable average values for each specific heat should be used*. What are suitable average values? They are representative values based on the temperature range of interest of a given problem. You should find Table 2.3 particularly useful for this purpose. The following examples drive this point home while using the compressor, turbine, and combustor simplified first law development presented in Example 2.5.

Table 2.3   Typical Gas Properties for Air and Products of Combustion

Gas	Temperature Range	R $\dfrac{ft \cdot lbm}{lbm \cdot °R}$	R $\dfrac{kJ}{kg \cdot K}$	c_p $\dfrac{Btu}{lbm \cdot °R}$	c_p $\dfrac{kJ}{kg \cdot K}$	c_v $\dfrac{Btu}{lbm \cdot °R}$	c_v $\dfrac{kJ}{kg \cdot K}$	γ
Air	400–1000°R 220–550 K	53.34	0.286	0.240	1.004	0.171	0.718	1.4
Air	900–1700°R 500–950 K	53.34	0.286	0.258	1.079	0.189	0.794	1.36
Products of combustion*	800–1300°R 440–720 K	53.34	0.286	0.258	1.079	0.189	0.794	1.36
Products of combustion*	1200–1800°R 670–1000 K	53.40	0.287	0.276	1.157	0.207	0.870	1.33
Products of combustion*	1500–2500°R 830–1400 K	54.09	0.290	0.295	1.234	0.226	0.945	1.3

*From air and JP-8 fuel

Example 2.7

a) Adiabatic compressor: Total temperature measurements of the air flowing across a gas turbine engine compressor are shown in Fig. 2.10. The arrows represent the air flow through the compressor. Estimate the energy per unit mass to power this compressor. Be clear with all assumptions and their implications.

Solution

From Example 2.5b, the first law energy conservation equation for steady, 1-D flow through an adiabatic compressor with no mass flow bleeds in or out is

$$h_{t3} - h_{t2} = w_{\text{in}} = \frac{\dot{W}_{\text{in}}}{\dot{m}} \quad \text{(energy per unit mass)}$$

For a CPG:

$$\Delta h_t = c_p \Delta T_t$$

So:

$$c_p(T_{t3} - T_{t2}) = w_{\text{in}} \quad \text{(energy per unit mass)}$$

Note: The total temperature T_t is defined in a similar manner as total specific enthalpy, found in Eq. (2.20). We will formally introduce total temperature (and total pressure) in the next chapter.

What value of c_p should we use? The average temperature of the air is 750°R. From Table 2.3, a suitable value for c_p is 0.24 Btu/(lbm · °R). Therefore,

$$w_{\text{in}} = 0.24 \frac{\text{Btu}}{\text{lbm} \cdot °\text{R}} (1000 - 500)\text{R} = 120 \frac{\text{Btu}}{\text{lbm}}$$

b) Adiabatic turbine: The compressor of Example 2.7a gets its power from a turbine via a connecting shaft (Fig. 2.11). Assuming identical mass flow rate through the turbine and compressor, estimate the turbine inlet total temperature T_{t4} using the compressor–turbine *power balance*. Be clear with all assumptions and their implications.

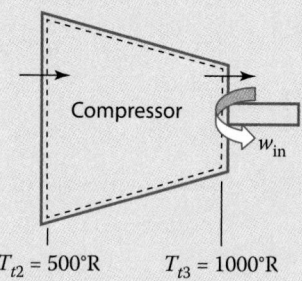

$$T_{t2} = 500°\text{R} \qquad T_{t3} = 1000°\text{R}$$

Fig. 2.10   Schematic of steady flow, adiabatic compressor.

(Continued)

Example 2.7 *(Continued)*

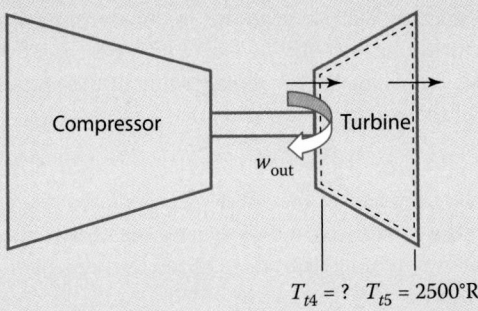

$$T_{t4} = ? \quad T_{t5} = 2500°R$$

Fig. 2.11   Schematic of steady flow, adiabatic turbine powering compressor.

Solution

From Example 2.5b, the first law energy conservation equation for steady, 1-D flow through an adiabatic turbine with no mass flow bleeds in or out is

$$w_{out} = h_{t4} - h_{t5} = \frac{\dot{W}_{out}}{\dot{m}} \quad \text{(energy per unit mass)}$$

For a CPG:

$$\Delta h_t = c_p \Delta T_t$$

So:

$$w_{out} = c_p(T_{t4} - T_{t5}) \quad \text{(energy per unit mass)}$$

From power balance (with equal mass flow rates): $w_{in,compressor} = w_{out,turbine}$
So:

$$w_{in} = 120\,\frac{\text{Btu}}{\text{lbm}} = w_{out} = c_p(T_{t4} - T_{t5}) \quad \text{or} \quad T_{t4} = T_{t5} + \frac{120\,\dfrac{\text{Btu}}{\text{lbm}}}{c_p}$$

As in Example 2.7a, we must determine what value of c_p to use. Although we don't know T_{t4}, we do expect it to be reasonably close to 3000°R because the temperature difference across the compressor is 500° and the specific heat values of the compressor and turbine will be close. So, from Table 2.3, with an average temperature of 2750°R, a suitable value for c_p is 0.295 Btu/(lbm · °R). Therefore,

$$T_{t4} = 2500\,\text{R} + \frac{120\,\dfrac{\text{Btu}}{\text{lbm}}}{0.295\,\dfrac{\text{Btu}}{\text{lbm} \cdot °\text{R}}} = 2500\,\text{R} + 407\,\text{R} = 2907\,\text{R}$$

Note that using the "cold" air c_p of 0.24 Btu/(lbm · °R) would have resulted in a 3% overestimation of the turbine inlet total temperature. (T_{t4} would have been estimated to be 3000°R.)

(Continued)

Example 2.7 *(Continued)*

c) Adiabatic combustor: We return to the combustor model of Example 2.5c with a numerical example (refer to Fig. 2.8d). Because the combustor is modeled as adiabatic, we have that the enthalpy of the reactants at station 3 is equal to that of the products at station 4 or

$$h_{R3} = h_{P4} \qquad (2.51)$$

This can be written in terms of the enthalpy differences to a datum state. We select as our common datum a state d having the chemical aggregate of the products at a datum temperature T_d. Then, introducing the datum state enthalpy $(h_P)_d$ into Eq. (2.51), we have

$$h_{R3} - h_{Pd} = h_{P4} - h_{Pd} \qquad (2.52)$$

For the turbojet engine combustion chamber, 45 lbm of air enters with each 1 lbm of JP-8 (kerosene) fuel. Let us assume these reactants enter an adiabatic combustor at 1200°R. The heating value h_{PR} of JP-8 is 18,400 Btu/lbm of fuel at 298 K. [This is also called the *lower heating value (LHV)* of the fuel.] Thus the heat released $(\Delta H)_{298K}$ by the fuel per 1 lbm of the products is 400 Btu/lbm (18,400/46) at 298 K. The following data are known:

$$c_{pP} = 0.267 \, \text{Btu/(lbm} \cdot \text{°R)} \quad \text{and} \quad c_{pR} = 0.240 \, \text{Btu/(lbm} \cdot \text{°R)}$$

Determine the temperature of the products leaving the combustor.

Solution

A plot of the enthalpy equations of state for the reactants and the products is given in Fig. 2.12. In the plot, the vertical distance $h_R - h_P$ between the curves of h_R and h_P at a given temperature represents the enthalpy of combustion ΔH of the reactants at that temperature. (This is sometimes called the heat of combustion.) In our analysis, we know the enthalpy of combustion at $T_d = 298$ K (536.4°R).

States 3 and 4, depicted in Fig. 2.12, represent the states of the reactants entering and the products leaving the combustion chamber, respectively. The datum state d is arbitrarily selected to be products at temperature T_d. State d' is the reactants' state at the datum temperature T_d.

In terms of Fig. 2.12, the left-hand side of Eq. (2.52) is the vertical distance between states 3 and d, or

$$h_{R_3} - h_{P_d} = h_{R_3} - h_{R_{d'}} + h_{R_{d'}} - h_{P_d}$$

$$\Delta h_R = c_{pR} \Delta T \quad \text{and} \quad h_{R_{d'}} - h_{P_d} = (\Delta H)_{T_d}$$

then

$$h_{R_3} - h_{P_d} = c_{pR}(T_3 - T_d) + (\Delta H)_{T_d} \qquad \text{(i)}$$

(Continued)

Example 2.7 *(Continued)*

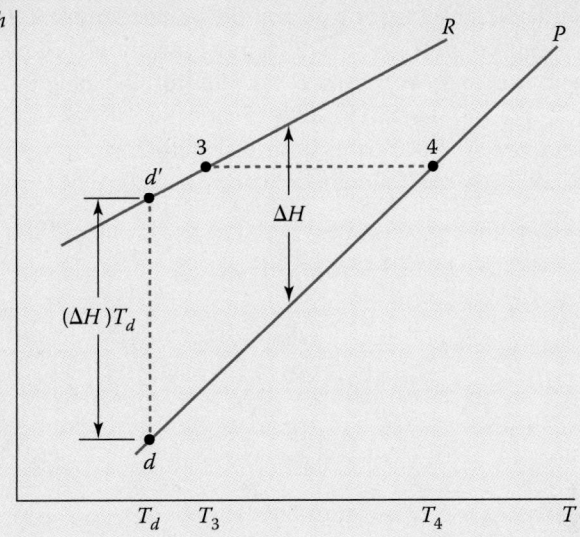

Fig. 2.12   Enthalpy vs temperature for reactants and products treated as perfect gases.

Similarly, the right side of Eq. (2.16) is

$$h_{P_4} - h_{P_d} = c_{pP}(T_4 - T_d) \tag{ii}$$

Substituting Eqs. (i) and (ii) in Eq. (2.16), we get

$$c_{pR}(T_3 - T_d) + (\Delta H)_{T_d} = c_{pP}(T_4 - T_d) \tag{2.53}$$

We can solve this equation for T_4, which is the temperature of the product gases leaving the combustion chamber. Solving Eq. (2.53) for T_4, we get

$$
\begin{aligned}
T_4 &= \frac{c_{pR}(T_3 - T_d) + (\Delta H)_{T_d}}{c_{pP}} + T_d \\
&= \frac{0.240(1200 - 536.4) + 400}{0.267} + 536.4 \\
&= 2631°R \ (2171°F)
\end{aligned}
$$

This is the so-called adiabatic flame temperature of the reactants for a 45:1 mixture ratio of air to fuel weight. For the analysis in portions of this book, we choose to sidestep the complex thermochemistry of the combustion process and model it as a heating process. The theory and application of thermochemistry to combustion in jet engines are covered in many textbooks, such as the classic text by Penner (see Ref. 16).

2.4.3 Isentropic Process

We saw from the second law of thermodynamics (Section 2.3.4) that an adiabatic, isentropic process for a steady-flow device represents a reversible process, an idealized process in which no entropy is produced. We will find isentropic, adiabatic processes very useful for defining component efficiencies.

For an isentropic process from state 1 to state 2, $s_2 = s_1$, and Eqs. (2.49), (2.50), and (2.28) yield the following equations:

$$\frac{T_2}{T_1} = \left(\frac{P_2}{P_1}\right)^{(\gamma-1)/\gamma} \tag{2.54}$$

$$\frac{T_2}{T_1} = \left(\frac{\rho_2}{\rho_2}\right)^{\gamma-1} \tag{2.55}$$

$$\frac{P_2}{P_1} = \left(\frac{\rho_2}{\rho_2}\right)^{\gamma} \tag{2.56}$$

Note that Eqs. (2.54), (2.55), and (2.56) apply only to a calorically perfect gas undergoing an isentropic process. *Be careful not to misuse these equations.*

Example 2.8

Air initially at 20°C and 1 atm is compressed reversibly and adiabatically to a final pressure of 15 atm. Find the final temperature.

Solution

Because the process is isentropic from initial to final state, Eq. (2.54) can be used to solve for the final temperature. The ratio of specific heats for air is 1.4:

$$T_2 = T_1 \left(\frac{P_2}{P_1}\right)^{(\gamma-1)/\gamma} = (20 + 273.15)\left(\frac{15}{1}\right)^{0.4/1.4}$$

$$= 293.15 \times 2.1678 = 635.49\,\text{K } (362.34°\text{C})$$

Example 2.9

Air is expanded isentropically through a nozzle from $T_1 = 3000°$R, $V_1 = 0$, and $P_1 = 10$ atm to $V_2 = 3000$ ft/s. Find the exit temperature and pressure.

(Continued)

Example 2.9 *(Continued)*

Solution

Application of the first law of thermodynamics to the nozzle gives the following for a calorically perfect gas:

$$c_p T_1 + \frac{V_1^2}{2g_c} = c_p T_2 + \frac{V_2^2}{2g_c}$$

This equation can be rearranged to give T_2:

$$T_2 = T_1 - \frac{V_2^2 - V_1^2}{2g_c c_p} = 3000 - \frac{3000^2}{2 \times 32.174 \times 0.240 \times 778.16}$$

$$= 3000 - 748.9 = 2251.1°R$$

Solving Eq. (2.54) for P_2 gives

$$P_2 = P_1 \left(\frac{T_2}{T_1}\right)^{\gamma/(\gamma-1)} = 10\left(\frac{2251.1}{3000}\right)^{3.5} = 3.66\,\text{atm}$$

2.4.4 Mollier Diagram for a Perfect Gas

The Mollier diagram is a thermodynamic state diagram with the coordinates of enthalpy and entropy. Because the enthalpy of a perfect gas depends on temperature alone,

$$dh = c_p \, dT$$

temperature can replace enthalpy as the coordinate of a Mollier diagram for a perfect gas. When temperature T and entropy s are the coordinates of a Mollier diagram, we call it a *T-s diagram*. We can construct lines of constant pressure and density in the *T-s* diagram by using Eqs. (2.42) and (2.43). For a calorically perfect gas, Eqs. (2.47) and (2.48) can be written between any state and the entropy reference state $(s = 0)$ as

$$s = c_v \, \ell n \frac{T}{T_{\text{ref}}} - R \, \ell n \frac{\rho}{\rho_{\text{ref}}}$$

$$s = c_p \, \ell n \frac{T}{T_{\text{ref}}} - R \, \ell n \frac{P}{P_{\text{ref}}}$$

where T_{ref}, P_{ref}, and ρ_{ref} are the values of temperature, pressure, and density, respectively, when $s = 0$. Because the most common working fluid in gas

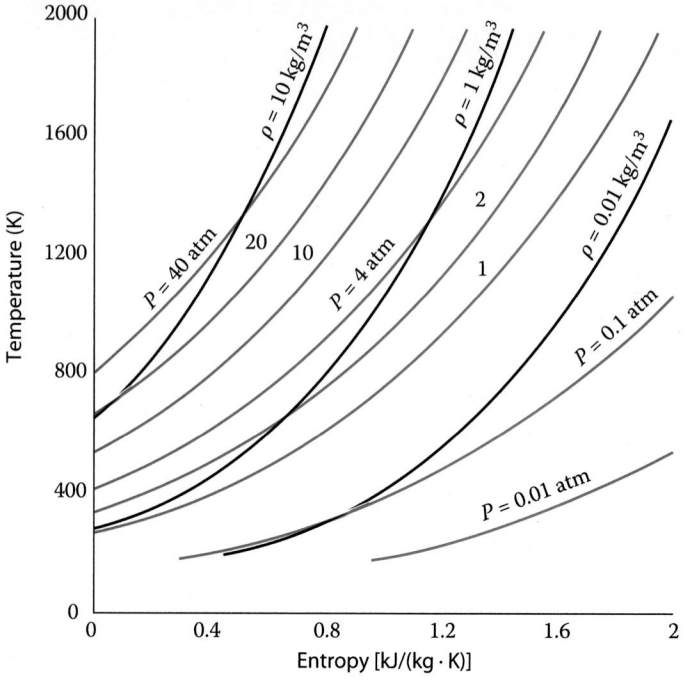

Fig. 2.13 A T-s diagram for air as a calorically perfect gas.

turbine engines is air, Fig. 2.13 has been constructed for air by using the preceding equations with these data:

$$c_p = 1.004 \text{ kJ/(kg} \cdot \text{K)}, \quad T_{ref} = 288.2 \text{ K}, \quad \rho_{ref} = 1.225 \text{ kg/m}^3,$$
$$R = 0.286 \text{ kJ/(kg} \cdot \text{K)}, \quad P_{ref} = 1 \text{ atm} = 101,325 \text{ Pa}$$

2.4.5 Mixtures of Perfect Gases

We consider a mixture of perfect gases, each obeying the perfect gas equation:

$$PV = N\mathcal{R}T$$

where N is the number of moles and $\mathcal{R}$ is the universal gas constant. The mixture is idealized as independent perfect gases, each having the temperature T and occupying the volume V. The partial pressure of gas i is

$$P_i = N_i \mathcal{R} \frac{T}{V}$$

According to the Gibbs–Dalton law, the pressure of the gas mixture of n constituents is the sum of the partial pressures of each constituent:

$$P = \sum_{i=1}^{n} P_i \qquad (2.57)$$

The total number of moles N of the gas is

$$N = \sum_{i=1}^{n} N_i \qquad (2.58)$$

The ratio of the number of moles of constituent i to the total number of moles in the mixture is called the *mole fraction* χ_i. By using the preceding equations, the mole fraction of constituent i can be shown to equal the ratio of the partial pressure of constituent i to the pressure of the mixture:

$$\chi_i = \frac{N_i}{N} = \frac{P_i}{P} \qquad (2.59)$$

The Gibbs–Dalton law also states that the internal energy, enthalpy, and entropy of a mixture are equal, respectively, to the sum of the internal energies, the enthalpies, and the entropies of the constituents when each alone occupies the volume of the mixture at the mixture temperature. Thus we can write the following for a mixture of n constituents:

Energy:

$$E = \sum_{i=1}^{n} E_i = \sum_{i=1}^{n} m_i e_i \qquad (2.60)$$

Enthalpy:

$$H = \sum_{i=1}^{n} H_i = \sum_{i=1}^{n} m_i h_i \qquad (2.61)$$

Entropy:

$$S = \sum_{i=1}^{n} S_i = \sum_{i=1}^{n} m_i s_i \qquad (2.62)$$

where m_i is the mass of constituent i.

The specific heats of the mixture follow directly from the definitions of c_p and c_v and the preceding equations. For a mixture of n constituents, the specific heats are

$$c_p = \frac{\sum_{i=1}^{n} m_i c_{pi}}{m} \quad \text{and} \quad c_v = \frac{\sum_{i=1}^{n} m_i c_{vi}}{m} \qquad (2.63)$$

where m is the total mass of the mixture.

2.4.6 Perfect Gas with Variable Specific Heats

In the case of a perfect gas with variable specific heats, the specific heat at constant pressure c_p is normally modeled by several terms of a power series in temperature T. This expression is used in conjunction with the general equations presented and the new equations that are developed next to generate a gas table for a particular gas (see Ref. 17).

For convenience, we define

$$h \equiv \int_{T_{\text{ref}}}^{T} c_p \, dT \tag{2.64}$$

$$\phi \equiv \int_{T_{\text{ref}}}^{T} c_p \, \frac{dT}{T} \tag{2.65}$$

$$P_r \equiv \exp \frac{\phi - \phi_{\text{ref}}}{R} \tag{2.66}$$

$$v_r \equiv \exp \left(-\frac{1}{R} \int_{T_{\text{ref}}}^{T} c_v \, \frac{dT}{T} \right) \tag{2.67}$$

where P_r and v_r are called the *reduced pressure* and *reduced volume*, respectively. Using the definition of ϕ from Eq. (2.65) in Eq. (2.43) gives

$$s_2 - s_1 = \phi_2 - \phi_1 - R \ln \frac{P_2}{P_1} \tag{2.68}$$

For an isentropic process between states 1 and 2, Eq. (2.68) reduces to

$$\phi_2 - \phi_1 = R \ln \frac{P_2}{P_1}$$

which can be rewritten as

$$\left(\frac{P_2}{P_1} \right)_{s=\text{const}} = \exp \frac{\phi_2 - \phi_1}{R} = \frac{\exp(\phi_2/R)}{\exp(\phi_1/R)}$$

Using Eq. (2.66), we can express this pressure ratio in terms of the reduced pressure P_r as

$$\left(\frac{P_2}{P_1} \right)_{s=\text{const}} = \frac{P_{r2}}{P_{r1}} \tag{2.69}$$

Likewise, it can be shown that

$$\left(\frac{v_2}{v_1}\right)_{s=\text{const}} = \frac{v_{r2}}{v_{r1}} \tag{2.70}$$

For a perfect gas, the properties h, P_r, e, v_r, and ϕ are functions of T, and these can be calculated by starting with a polynomial for c_p. Say we have the seventh-order polynomial

$$c_p = A_0 + A_1 T + A_2 T^2 + A_3 T^3 + A_4 T^4 + A_5 T^5 + A_6 T^6 + A_7 T^7 \tag{2.71}$$

The equations for h and ϕ as functions of temperature follow directly from using Eqs. (2.64) and (2.65):

$$h = h_{\text{ref}} + A_0 T + \frac{A_1}{2} T^2 + \frac{A_2}{3} T^3 + \frac{A_3}{4} T^4 + \frac{A_4}{5} T^5 + \frac{A_5}{6} T^6 + \frac{A_6}{7} T^7 + \frac{A_7}{8} T^8 \tag{2.72}$$

$$\phi = \phi_{\text{ref}} + A_0 \ln T + A_1 T + \frac{A_2}{2} T^2 + \frac{A_3}{3} T^3 + \frac{A_4}{4} T^4 + \frac{A_5}{5} T^5 + \frac{A_6}{6} T^6 + \frac{A_7}{7} T^7 \tag{2.73}$$

After we define reference values, the variations of P_r and v_r follow from Eqs. (2.66) and (2.67), and the preceding equations.

2.4.7 Comment for Appendix L

Typically, air flows through the inlet and compressor of the gas turbine engine whereas products of combustion flow through the engine components downstream of a combustion process. Most gas turbine engines use hydrocarbon fuels of composition $(CH_2)_n$. We can use the preceding equations to estimate the properties of these gases, given the ratio of the mass of fuel burned to the mass of air. For convenience, we use the fuel/air ratio f, defined as

$$f = \frac{\text{mass of fuel}}{\text{mass of air}} \tag{2.74}$$

The maximum value of f is 0.0676 for the hydrocarbon fuels of composition $(CH_2)_n$.

Given the values of c_p, h, and ϕ for air and the values of combustion products, the values of c_p, h, and ϕ for the mixture follow directly from the mixture equations [Eqs. (2.60)–(2.63)] and are given by

$$R = \frac{1.9857117\,\text{Btu/(lbm} \cdot {}^\circ\text{R})}{28.97 - f \times 0.946186} \tag{2.75a}$$

$$c_p = \frac{c_{p\,\text{air}} + f c_{p\,\text{prod}}}{1 + f} \tag{2.75b}$$

Table 2.4   Constants for Air and Combustion Products Used in Appendix L and the AFPROP Program (from Ref. 18)

Air Alone		Combustion Products of Air and $(CH_2)_n$ Fuels	
A_0	2.5020051×10^{-1}	A_0	7.3816638×10^{-2}
A_1	$-5.1536879 \times 10^{-5}$	A_1	1.2258630×10^{-3}
A_2	6.5519486×10^{-8}	A_2	$-1.3771901 \times 10^{-6}$
A_3	$-6.7178376 \times 10^{-12}$	A_3	$9.9686793 \times 10^{-10}$
A_4	$-1.5128259 \times 10^{-14}$	A_4	$-4.2051104 \times 10^{-13}$
A_5	$7.6215767 \times 10^{-18}$	A_5	$1.0212913 \times 10^{-16}$
A_6	$-1.4526770 \times 10^{-21}$	A_6	$-1.3335668 \times 10^{-20}$
A_7	$1.0115540 \times 10^{-25}$	A_7	$7.2678710 \times 10^{-25}$
h_{ref}	-1.7558886 Btu/lbm	h_{ref}	30.58153 Btu/lbm
ϕ_{ref}	0.0454323 Btu/(lbm $\cdot$ °R)	ϕ_{ref}	0.6483398 Btu/(lbm $\cdot$ °R)

$$h = \frac{h_{air} + f\, h_{prod}}{1 + f} \tag{2.75c}$$

$$\phi = \frac{\phi_{air} + f\, \phi_{prod}}{1 + f} \tag{2.75d}$$

Appendix L is a table of the properties h and P_r as functions of the temperature and fuel/air ratio f for air and combustion products [air with hydrocarbon fuels of composition $(CH_2)_n$] at low pressure (perfect gas). These data are based on the preceding equations and the constants given in Table 2.3, which are valid over the temperature range of 300–4000°R. These constants come from the gas turbine engine modeling work of Capt. John S. McKinney (U.S. Air Force) while assigned to the Air Force's Aero Propulsion Laboratory [18], and they continue to be widely used in the industry. Appendix L uses a reference value of 2 for P_r at 600°R and $f = 0$.

2.4.8   Comment for the AFPROP Computer Program

The AFPROP computer program was written by using the preceding constants for air and products of combustion from air with $(CH_2)_n$. The program can calculate the four primary thermodynamic properties at a state (P, T, h, and s) given the fuel/air ratio f and two independent thermodynamic properties (say, P and h).

To show the use of the gas tables, we will re-solve Examples 2.8 and 2.9 using the gas tables of Appendix L. These problems could also be solved by using the AFPROP computer program.

Example 2.10

Air initially at 20°C and 1 atm is compressed reversibly and adiabatically to a final pressure of 15 atm. Find the final temperature.

Solution

Because the process is isentropic from initial to final state, Eq. (2.58) can be used to solve for the final reduced pressure. From Appendix L at 20°C (293.15 K) and $f = 0$, $P_r = 1.2768$ and

$$\frac{P_{r2}}{P_{r1}} = \frac{P_2}{P_1} = 15$$

$$P_{r2} = 15 \times 1.2768 = 19.152$$

From Appendix L for $P_{r2} = 19.152$, the final temperature is 354.42°C (627.57 K). This is 7.9 K lower than the result obtained in Example 2.5 for air as a calorically perfect gas due to the increase in specific heat with temperature.

Example 2.11

Air is expanded isentropically through a nozzle from $T_1 = 3000°R$, $V_1 = 0$, and $P_1 = 10$ atm to $V_2 = 3000$ ft/s. Find the exit temperature and pressure.

Solution

Application of the first law of thermodynamics to the nozzle gives the following for a calorically perfect gas:

$$h_1 + \frac{V_1^2}{2g_c} = h_2 + \frac{V_2^2}{2g_c}$$

From Appendix L at $f = 0$ and $T_1 = 3000°R$, $h_1 = 790.46$ Btu/lbm and $P_{r1} = 938.6$. Solving the preceding equation for h_2 gives

$$h_2 = h_1 - \frac{V_2^2 - V_1^2}{2g_c} = 790.46 - \frac{3000^2}{2 \times 32.174 \times 778.16}$$

$$= 790.46 - 179.74 = 610.72 \, \text{Btu/lbm}$$

For $h = 610.72$ Btu/lbm and $f = 0$, Appendix L gives $T_2 = 2377.7°R$ and $P_{r2} = 352.6$. Using Eq. (2.58), we solve for the exit pressure

$$P_2 = P_1 \frac{P_{r2}}{P_{r1}} = 10 \left(\frac{352.6}{938.6} \right) = 3.757 \, \text{atm}$$

Compare these results to those obtained in Examples 2.8 and 2.9, which assumed calorically perfect gas using suitable values for specific heats. The difference in estimated values is no more than 2.5% using either method. This should increase your confidence in using the CPG model with the appropriate values of specific heats.

2.5 Chemical Reactions

2.5.1 General Characteristics

A chemical reaction from reactants A and B to products C and D is generally represented by

$$N_A A + N_B B \longrightarrow N_C C + N_D D$$

where N_A and N_B are the number of moles of reactants A and B, respectively. Likewise N_C and N_D are the number of moles of products C and D. An example reaction between one mole of methane (CH_4) and two moles of oxygen (O_2) can be written as

$$CH_4 + 2O_2 \longrightarrow CO_2 + 2H_2O$$

and represents *complete combustion* among the fuel, methane, and oxidizer, oxygen. Combustion is complete when all of the carbon in the fuel burns to CO_2 and the hydrogen burns to H_2O. The combustion is incomplete when there is any unburned fuel or compounds such as carbon, hydrogen, CO, or OH. This *theoretical process* of complete combustion is very useful in our analysis because it represents the number of moles of each reactant that are needed for complete combustion to the number of moles of each product. The theoretical or complete combustion process is also referred to as a stoichiometric combustion process and is represented as

$$\nu_{CH_4} CH_4 + \nu_{O_2} O_2 \longrightarrow \nu_{CO_2} CO_2 + \nu_{H_2O} H_2O$$

where ν_{CH_4} and ν_{O_2} are the number of moles of reactants CH_4 and O_2, and ν_{CO_2} and ν_{H_2O} are the number of moles of products CO_2 and H_2O. The ν_i in the preceding equation are called the *stoichiometric coefficients* for the theoretical reaction of component i. Thus the stoichiometric coefficients for the preceding reaction of methane and oxygen are

$$\nu_{CH_4} = 1 \quad \nu_{O_2} = 2 \quad \nu_{CO_2} = 1 \quad \nu_{H_2O} = 2$$

The stoichiometric coefficients give the number of moles of each constituent required for complete combustion. Note that the units of both the mole number N_i and the stoichiometric coefficients ν_i can also be regarded as moles *per unit mass* of the gas mixture.

The *actual process* between reactants normally has different amounts of each reactant than that of the theoretical (stoichiometric) process. For example, consider one mole of methane reacting with three moles of oxygen. The stoichiometric coefficients of the theoretical reaction indicates that only two moles of oxygen are required for each mole of methane. Thus one mole of oxygen will not react and remains on the products side of the chemical equation as shown:

$$CH_4 + 3O_2 \longrightarrow CO_2 + 2H_2O + O_2$$

2.5.2 Chemical Equilibrium

The chemical reactions represented so far were shown as complete reactions. In reality, the reactions go both forward and backward, and the actual process is denoted as

$$N_A A + N_B B \rightleftharpoons N_C C + N_D D$$

Chemical equilibrium is reached when the forward rate of reaction of reactants A and B equals the backward rate of reaction of products C and D. It can be shown that chemical equilibrium corresponds to the minimum value of the Gibbs function $(G = H - TS)$ for the entire gas mixture, as shown in Fig. 2.14. Chemical equilibrium also corresponds to maximum entropy for an adiabatic system.

Chemical equilibrium for a stoichiometric (theoretical) reaction can be represented as

$$\nu_A A + \nu_B B \rightleftharpoons \nu_C C + \nu_D D \tag{2.76}$$

For solution of equilibrium problems, we define the *equilibrium constant* of a perfect gas as

$$K_P \equiv \frac{P_\nu^{\nu_C} P_D^{\nu_D}}{P_A^{\nu_A} P_B^{\nu_B}} \tag{2.77}$$

where P_A, P_B, P_C, and P_D are the partial pressures of components A, B, C, and D; ν_A, ν_B, ν_C, and ν_D are the stoichiometric coefficients shown in Eq. (2.76). For a perfect gas, K_P is only a function of temperature.

From Eq. (2.57), the partial pressure of constituent i can be written as

$$P_i = \frac{N_i}{N_{\text{total}}} P \tag{2.78}$$

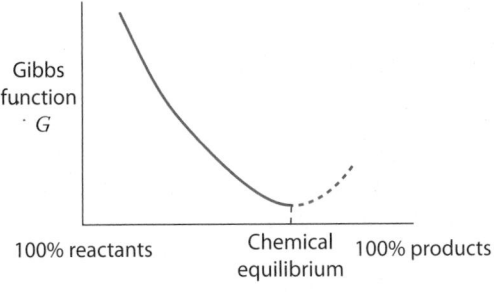

Fig. 2.14   Chemical equilibrium.

where N_{total} is the total number of moles present in the reaction chamber. Using Eq. (2.78), Eq. (2.77) for the equilibrium constant can be rewritten as

$$K_P = \frac{N_C^{v_C} N_D^{v_D}}{N_A^{v_A} N_B^{v_B}} \left(\frac{P}{N_{total}}\right)^{\Delta v} \tag{2.79}$$

where $\Delta v = v_C + v_D - v_A - v_B$. Equation (2.79) can be used to verify equilibrium results from complex computer programs, as will be done later in this section.

Example 2.12

One mole of hydrogen and 0.5 mole of oxygen are in equilibrium at a temperature of 2000 K and a pressure of 10 atm. Determine the number of moles of H_2O, H_2, and O_2 in the products. (Neglect other products such as OH, H, O, etc.)

Solution

We start by writing the chemical reaction equation for both the stoichiometric reaction and the actual reaction being modeled.

Stoichiometric:

$$H_2 + \frac{1}{2}O_2 \rightleftharpoons H_2O \quad (\text{thus } v_{H_2O} = 1, \; v_{H_2} = 1, \; v_{O_2} = 1/2)$$

Actual:

$$H_2 + \frac{1}{2}O_2 \longrightarrow xH_2 + yO_2 + zH_2O$$

Because there are the three unknowns (x, y, and z), three equations are required. Two equations come from the atom balance of each element (H and O) and the third from the chemical equilibrium coefficient (K_p):

Atom balance:

$$H{:}2 = 2x + 2z \quad \text{or} \quad z = 1 - x$$
$$O{:}1 = 2y + z \quad \text{or} \quad y = x/2$$

Now the actual reaction equation can be written with just one unknown:

$$H_2 + \frac{1}{2}O_2 \longrightarrow xH_2 + \frac{x}{2}O_2 + (1 - x)H_2O$$

From the JANNAF tables in Appendix N in the Supporting Material, log $Kp = 3.540$ at 2000 K for the reaction

$$H_2 + \frac{1}{2}O_2 \longrightarrow H_2O$$

(Continued)

Example 2.12 *(Continued)*

Thus

$$K_p = 10^{3.540} = 3467.4 = K_p = \frac{N_C^{\nu_{H_2O}}}{N_A^{\nu_{H_2}} N_B^{\nu_{O_2}}} \left(\frac{P}{N_{total}}\right)^{\Delta \nu}$$

where $N_A = x$, $N_B = x/2$, $N_C = 1 - x$, $\Delta \nu = -1/2$, and $N_{total} = 1 + x/2$.
Thus

$$K_p = \frac{(1-x)}{x^1 \sqrt{x/2}} \left(\frac{10}{1+x/2}\right)^{-1/2} = 3467.4$$

or

$$\frac{(1-x)\sqrt{1+x/2}}{x^{3/2}} = 3467.4\sqrt{10/2} = 7753.3$$

Solving for x gives $x = 0.002550$, and the reaction can be written as

$$H_2 + \frac{1}{2}O_2 \rightarrow 0.002550H_2 + 0.0012750_2 + 0.997450H_2O$$

Thus there are 0.002550 moles of hydrogen, 0.001275 moles of oxygen, and 0.997450 moles of water.

2.5.3 EQL Software

The included EQL software program by David T. Pratt uses the NASA Glenn thermochemical data and the Gordon–McBride equilibrium algorithm (see Ref. 19). This equilibrium calculation is based on finding the mixture composition that minimizes the Gibbs function, which allows rapid solution of reactions resulting in many products.

Example 2.13

The EQL software program gives the following result for 1 mole of hydrogen and 0.5 mole of oxygen at a temperature of 2000 K and a pressure of 10 atm:

$$H_2 + \frac{1}{2}O_2 \longrightarrow 0.002717H_2 + 0.0011180_2 + 0.9952H_2O + 0.000974OH$$

$$+ 2.668 \times 10^{-5}H + 7.016 \times 10^{-6}O + 5.963 \times 10^{-7}HO_2$$

$$+ 1.992 \times 10^{-7}H_2O_2$$

(Continued)

Example 2.13 (Continued)

There are 0.0022 moles less of H_2O produced in this equilibrium mixture than in Example 2.12, and these molecules show up in increases in the number of moles of H_2 and O_2.

The equilibrium constant can be determined from these data and Eq. (2.79):

$$K_P = \frac{N_C^{\nu_{H_2O}}}{N_A^{\nu_{H_2}} N_B^{\nu_{O_2}}} \left(\frac{P}{N_{\text{total}}}\right)^{\Delta\nu} = \frac{0.9952}{0.00271 \times \sqrt{0.001118}} \left(\frac{10}{0.9990}\right)^{-1/2} = 3471$$

or $\log K_p = 3.540$, which agrees with the JANNAF tables in Appendix N.

2.5.4 Enthalpy of Chemical Component, Enthalpy of Formation, Heat of Reaction, and Adiabatic Flame Temperature

For reacting systems, the enthalpy of each component must be written in a form that has the same reference state. The enthalpy of *chemical components* (products or reactants) can be calculated using Eq. (2.50) with $\overline{h}$ for a component written as

$$\overline{h} = \overline{h}_f^\circ + (\overline{h} - \overline{h}^\circ) = \overline{h}_f^\circ + \int_{T_d}^{T} \overline{c}_p \, dT' \qquad (2.80)$$

where $\overline{h}_f^\circ$ is the *enthalpy of formation* (also called the heat of formation) at the reference state (datum) of 25°C (298 K), 1 atm $(\overline{h} - \overline{h}^\circ)$ is the enthalpy change due to the temperature change from the reference state (T_d), and $\overline{c}_p$ is the specific heat at constant pressure per mole. Typical values of $\overline{h}_f^\circ$ are given in Table 2.5 and values $(\overline{h} - \overline{h}^\circ)$ for typical gases are given in the JANNAF tables of the Supporting Material.

The enthalpy of both products and reactants in a reaction can be plotted vs temperature as shown in Fig. 2.14. A reaction typically causes a change in temperature. The *heat of reaction* ΔH is defined as the positive heat transfer to the products that is required to bring them back to the original temperature of the reactants. For ideal (perfect) gases, the heat of reaction ΔH at the standard reference temperature can be calculated using

$$\Delta H = H_P - H_R = \sum_{i=1}^{n_P} N_i \Delta \overline{h}_{f_i}^\circ - \sum_{j=1}^{n_R} N_j \Delta \overline{h}_{f_i}^\circ \qquad (2.81)$$

where N_i and N_j are the number of moles of the products and reactants, respectively, $\Delta h_{f_i}^\circ$ is the heat of formation per mole of species i and j, and

Table 2.5   Enthalpy of Formation $\bar{h}_f^\circ$ for Some Reactants and Product Gases at Datum Temperature 536°R/298 K (Refs. 13, 20)

Gas	Btu/lbmol	kJ/kgmol
Methane, CH_4	−32,192	−74,883
Ethane, C_2H_6	−36,413	−84,701
Hexane, C_6H_{14}	−71,784	−166,978
Octane, C_8H_{18}	−89,600	−208,421
Jet-A, $C_{12}H_{23}$	−152,981	−355,853[a]
Carbon monoxide, CO	−47,520	−110,537
Carbon dioxide, CO_2	−169,181	−393,538
Atomic hydrogen, H	93,717	217,997
Hydrogen, H_2	0	0
Water vapor, H_2O	−103,966	−241,838
Atomic oxygen, O	107,139	249,218
Oxygen, O_2	0	0
Hydroxyl, OH	16,967	39,467
Atomic nitrogen, N	203,200	472,668
Nitrogen, N_2	0	0
Nitrous oxide, N_2O	35,275	82,053
Nitric oxide, NO	38,817	90,293
Nitrogen dioxide, NO_2	14,228	33,096

[a]For heating value $h_{PR} = 18,400$ Btu/lbm = 42,800 kJ/kg.

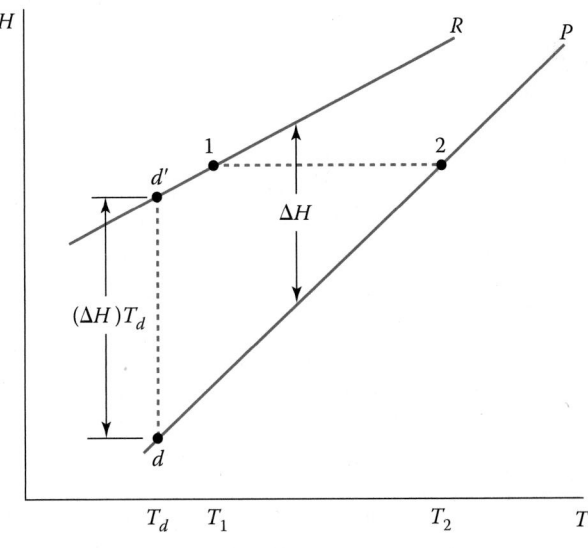

Fig. 2.15   Enthalpy states for a reaction.

ΔH is the vertical difference between the enthalpy line of the reactants and that of the products in Fig. 2.15.

Consider the adiabatic flow through a combustion chamber like that of Fig. 2.8d. The reactants enter at a temperature denoted by 1 in Fig. 2.15 and the products leave at a temperature denoted by 2. Neglecting changes in kinetic energy, the energy balance for this process yields

$$H_1 = H_2 \quad \text{or} \quad H_P = H_R \tag{2.82}$$

The temperature at state 2 is called the *adiabatic flame temperature*. This temperature can be calculated by determining the equilibrium state with the same enthalpy H and pressure P as state 1. To facilitate calculation, we rewrite Eq. (2.82) in terms of the enthalpy difference $(\bar{h} - \bar{h}^\circ)$ and the heat of reaction at the datum (reference) state $[\Delta H]_{T_d}$. First we write

$$H_2 - [H_P]_{T_d} = H_1 - [H_R]_{T_d} + [\Delta H]_{T_d}$$

because

$$H_2 - [H_P]_{T_d} = \sum_{j=1}^{n_P} N_j (\bar{h} - \bar{h}^\circ)_j$$

and

$$H_1 - [H_R]_{T_d} = \sum_{i=1}^{n_R} N_i (\bar{h} - \bar{h}^\circ)_i$$

then

$$H_2 - [H_P]_{T_d} = \sum_{j=1}^{n_P} N_j (\bar{h} - \bar{h}^\circ)_j = \sum_{i=1}^{n_R} N_i (\bar{h} - \bar{h}^\circ)_i + [\Delta H]_{T_d} \tag{2.83}$$

The following steps are used to calculate the adiabatic flame temperature:

1. Assume a final temperature T_2.
2. Calculate the mole fraction of the products for the resulting K_p at T_2.
3. Calculate $\sum_{i=1}^{n_R} N_i (\bar{h} - \bar{h}^\circ)_i + [\Delta H]_{T_d}$.
4. Calculate $\sum_{j=1}^{n_P} N_j (\bar{h} - \bar{h}^\circ)_j$. If its value is greater than that of step 3, reduce the value of T_2 and perform steps 2–4 again.

This calculation by hand is tedious and has been programmed for rapid calculation using computers. The EQL software calculates the adiabatic flame temperature for given reactants and inlet temperature T and pressure P as shown in Example 2.15.

Example 2.14

Calculate the heat of reaction for the reaction of Example 2.12.

Solution

We have

$$H_2 + \frac{1}{2}O_2 \rightarrow 0.002024H_2 + 0.0010120_2 + 0.997976H_2O$$

at 2000 K. From the JANNAF tables in Appendix N in the Supporting Materials, at 2000 K we have $\Delta\bar{h}_{f\,H_2}^{\circ} = 0$, $\Delta\bar{h}_{f\,O_2}^{\circ} = 0$, and $\Delta\bar{h}_{f\,H_2O}^{\circ} = -60.150\,kcal/mole$. Equation (2.78) gives

$$\Delta H = H_P - H_R$$

$$= \sum_{i=1}^{n_P} N_i\Delta\bar{h}_{fi}^{\circ} - \sum_{j=1}^{n_R} N_j\Delta\bar{h}_{fj}^{\circ} = 0.997976 \times -60.150 = -60.028\,kcal$$

Thus 60.028 kcal (251.28 kJ) of energy must be removed during the reaction to keep the temperature at 2000 K.

Example 2.15

Using the EQL software, determine the adiabatic flame temperature for 1 mole of oxygen and 0.5 mole of methane with the reactants at a temperature of 500 K and pressure of 800 kPa.

Solution

We first enter the reactants and inlet pressure and temperature into the opening screen of EQL. Next we select the Equilibrium Processes Tab, which displays the numerous possible combustion processes. We select Adiabatic Flame Temperature and perform the calculations. The resulting adiabatic flame temperature is 3349.85 K, and the resulting principal products are listed here:

Compound	Mole Fraction
H_2O	0.4093
CO	0.1540
CO_2	0.1190
OH	0.0989
O_2	0.0792
H_2	0.0681
H	0.0382
O	0.0330

Problems

2.1 Starting with the fully 3-D unsteady control volume form of conservation of mass (typically called continuity for flow problems), Eq. (2.11), show that it reduces to $\dot{m}_{out} = \dot{m}_{in}$ where $\dot{m} = \rho A V$ for steady, 1-D flow assumptions. Also assume one inlet and one outlet. Be clear with all assumptions and their implications.

2.2 Consider Fig. P2.1. A stream of air with a velocity of 500 ft/s and density of 0.7 lbm/ft^3 strikes a stationary plate and is deflected 90 deg. Use the indicated control volume and determine the force (F_p) necessary to hold the plate stationary. Assume atmospheric pressure surrounds the jet and an initial jet diameter of 1.0 in.

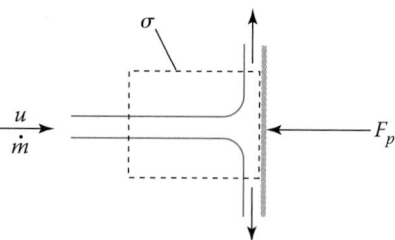

Fig. P2.1 Stream normal to plate.

2.3 An air stream with a density of 1.25 kg/m^3 and velocity of 200 m/s strikes a stationary plate and is deflected 90 deg. Select an appropriate control volume and determine the force (F_p) necessary to hold the plate stationary. Assume atmospheric pressure surrounds the jet and an initial jet diameter of 1.0 cm.

2.4 Consider the flow shown in Fig. P2.2 of an incompressible fluid. The fluid enters (at station 1) a constant-area circular pipe of radius r_o with uniform velocity V_1 and pressure P_1. The fluid leaves (at station 2) with a parabolic velocity profile V_2 given by

$$V_2 = V_{max}\left[1 - \left(\frac{r}{r_o}\right)^2\right]$$

and uniform pressure P_2. Using the conservation of mass and linear momentum equations, show that the force F necessary to hold the pipe in place can be expressed as

$$F = \pi r_o^2\left[P_1 - P_2 + \frac{\rho V_1^2}{3\,g_c}\right]$$

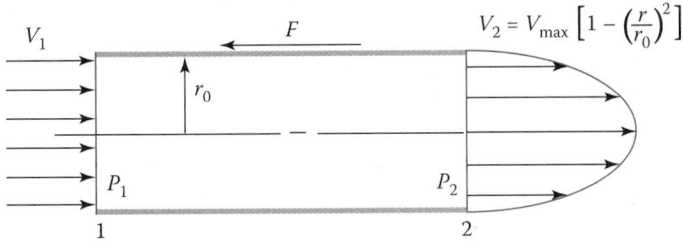

$$V_2 = V_{max}\left[1 - \left(\frac{r}{r_0}\right)^2\right]$$

Fig. P2.2   Entrance flow in circular pipe.

2.5   When a free jet is deflected by a blade surface, a change of momentum occurs and a force is exerted on the blade. If the blade is allowed to move at a velocity, power may be derived from the moving blade. This is the basic principle of the impulse turbine. The jet of Fig. P2.3, which is initially horizontal, is deflected by a fixed blade. Assuming the same pressure surrounds the jet, show that the horizontal force F_x and vertical force F_y by the fluid on the blade are given by

$$F_x = \frac{\dot{m}}{g_c}(u_1 - u_2 \cos\beta) \quad F_y = \frac{\dot{m}u_2 \sin\beta}{g_c}$$

Calculate the force F_y for a mass flow of 100 lbm/sec, $u_2 = 2000$ ft/s, and $\beta = 60$ deg.

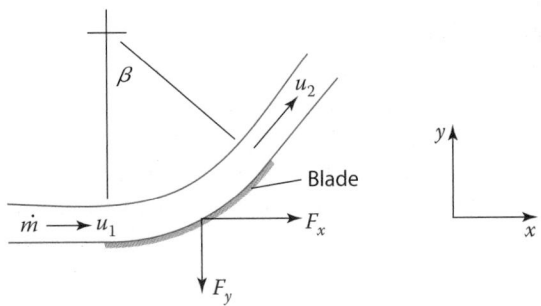

Fig. P2.3   Free jet turbine blade.

2.6   One method of reducing an aircraft's landing distance is through the use of thrust reversers. Consider the turbofan engine of Fig. P2.4 with a thrust reverser of the bypass air stream; 1500 lbm/s of air at 60°F and 14.7 psia enters the engine at a velocity of 450 ft/s and 1250 lbm/s of bypass air leaves the engine at 60 deg to the horizontal, velocity

890 ft/s, and pressure of 14.7 psia. The remaining 250 lbm/s leaves the engine core at a velocity of 1200 ft/s and pressure of 14.7 psia. Determine the force on the strut F_x. Assume the outside of the engine sees a pressure of 14.7 psia.

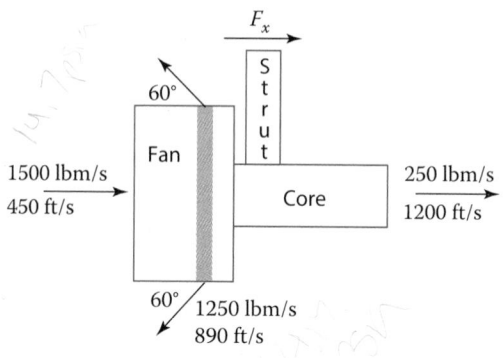

Fig. P2.4   Thrust reverser of high-bypass turbofan engine.

2.7   Air with a density of 0.027 lbm/ft³ enters a diffuser at a velocity of 2470 ft/s and static pressure of 4 psia. The air leaves the diffuser at a velocity of 300 ft/s and static pressure of 66 psia. The entrance area of the diffuser is 1.5 ft² and its exit area is 1.7 ft². Determine the magnitude and direction of the strut force necessary to hold the diffuser stationary when this diffuser is operating in an atmospheric pressure of 4 psia.

2.8   Air equal to 50 kg/s enters a diffuser at a velocity of 750 m/s and static pressure of 20 kPa. The air leaves the diffuser at a velocity of 90 m/s and static pressure of 330 kPa. The entrance area of the diffuser is 0.25 m² and its exit area is 0.28 m². Determine the magnitude and direction of the strut force necessary to hold the diffuser stationary when this diffuser is operated in an atmospheric pressure of 20 kPa.

2.9   Air equal to 100 lbm/s enters a nozzle at a velocity of 600 ft/s and static pressure of 70 psia. The air leaves the nozzle at a velocity of 4000 ft/s and static pressure of 2 psia. The entrance area of the nozzle is 14.5 ft² and its exit area is 30 ft². Determine the magnitude and direction of the strut force necessary to hold the nozzle stationary when this nozzle is operated in an atmospheric pressure of 4 psia.

2.10   Air with a density of 0.98 kg/m³ enters a nozzle at a velocity of 180 m/s and static pressure of 350 kPa. The air leaves the nozzle at a velocity of 1200 m/s and static pressure of 10 kPa. The entrance area of the nozzle is $1.0 \, m^2$ and its exit area is $2.07 \, m^2$. Determine the magnitude and direction of the strut force necessary to hold the nozzle stationary when this nozzle is operated in an atmospheric pressure of 10 kPa.

2.11   Starting with the fully 3-D unsteady control volume form of conservation of energy (first law of thermodynamics), Eq. (2.17), show that it reduces to $\dot{W}_{in} = \dot{m}(h_{t\,out} - h_{t\,in})$ for a compressor with steady, 1-D flow assumptions. Be clear with all assumptions and their implications.

2.12   A gas turbine engine core (compressor-combustor-turbine) is being tested at sea-level static (SLS) conditions. At a particular power setting, the core is receiving energy at a rate of 195,000 hp from the combustion process at average temperature of $3200°R$. Exhaust gases are expelled at a rate of 136,500 hp to ambient air, which is at $70°F$ ($530°R$). Modeling the core as a power cycle (see Example 2.6), with the combustor and ambient air as thermal energy reservoirs,

   a) Estimate its power output and thermal efficiency.

   b) Estimate the rate of entropy production. (Note that entropy production and thermal efficiency are inversely proportional.)

2.13   Air at 1400 K, 8 atm, and 0.3 Mach expands isentropically through a nozzle to 1 atm. Assuming a calorically perfect gas, find the exit temperature and the inlet and exit areas for a mass flow rate of $100 \, kg/s$.

2.14   Air equal to $250 \, lbm/s$ at $2000°F$, 10 atm, and 0.2 Mach expands isentropically through a nozzle to 1 atm. Assuming a calorically perfect gas, find the exit temperature and the inlet and exit areas.

2.15   Air at $518.7°R$ is isentropically compressed from 1 atm to 10 atm. Assuming a calorically perfect gas, find the exit temperature and the compressor's input power for a mass flow rate of $150 \, lbm/s$. Sketch the T-s diagram.

2.16   Air equal to $50 \, kg/s$ at 288.2 K is isentropically compressed from 1 atm to 12 atm. Assuming a calorically perfect gas, find the exit temperature and the compressor's input power. Sketch the T-s diagram.

2.17   Air at $-55°F$, 4 psia, and $M = 2.5$ enters an isentropic diffuser with an inlet area of $1.5 \, ft^2$ and leaves at $M = 0.2$. Assuming a calorically perfect gas, determine:

   a) The mass flow rate of the entering air

b) The pressure and temperature of the leaving air

c) The exit area and magnitude and direction of the force on the diffuser (assume outside of diffuser sees 4 psia)

2.18 Air at 225 K, 28 kPa, and $M = 2.0$ enters an isentropic diffuser with an inlet area of 0.2 m² and leaves at $M = 0.2$. Assuming a calorically perfect gas, determine:

a) The mass flow rate of the entering air

b) The pressure and temperature of the leaving air

c) The exit area and magnitude and direction of the force on the diffuser (assume outside of diffuser sees 28 kPa)

2.19 Air at 1800°F, 40 psia, and $M = 0.4$ enters an isentropic nozzle with an inlet area of 1.45 ft² and leaves at 10 psia. Assuming a calorically perfect gas, determine:

a) The velocity and mass flow rate of the entering air

b) The temperature and Mach number of the leaving air

c) The exit area and magnitude and direction of the force on the diffuser (assume outside of nozzle sees 10 psia)

2.20 Air at 1500 K, 300 kPa, and $M = 0.3$ enters an isentropic nozzle with an inlet area of 0.5 m² and leaves at 75 kPa. Assuming a calorically perfect gas, determine:

a) The velocity and mass flow rate of the entering air

b) The temperature and Mach number of the leaving air

c) The exit area and magnitude and direction of the force on the diffuser (assume outside of diffuser sees 75 kPa)

2.21 It is given that 100 lbm/s of air enters a steady flow compressor at 1 atm and 68°F. It leaves at 20 atm and 800°F. If the process is adiabatic, find the input power, specific volume at exit, and change in specific entropy. Is the process reversible (adiabatic *and* isentropic)? (Assume a calorically perfect gas.)

2.22 It is given that 50 kg/s of air enters a steady flow compressor at 1 atm and 20°C. It leaves at 20 atm and 427°C. If the process is adiabatic, find the input power, specific volume at exit, and change in specific entropy. Is the process reversible (adiabatic *and* isentropic)? (Assume a calorically perfect gas.)

2.23 It is given that 200 lbm/s of air enters a steady flow turbine at 20 atm and 3400°R. It leaves at 10 atm and 2789°R. If the process is adiabatic, determine the output power and change in specific entropy. Is the process reversible? Sketch the T-s diagram. (Assume a calorically perfect gas.)

2.24   It is given that 80 kg/s of air enters a steady flow turbine at 30 atm and 2000 K. It leaves at 15 atm and 1642 K. For an adiabatic turbine, determine the output power and change in specific entropy. Is the process reversible? Sketch the T-s diagram. (Assume a calorically perfect gas.)

2.25   Two kilograms of oxygen at 500 K is mixed in a constant-volume adiabatic container with 8 kg of nitrogen at 800 K. Determine the c_p and γ, and temperature of the mixture. Assume calorically perfect gases.

2.26   Two kilograms of oxygen at 500 K, 1 MPa is mixed in a constant-pressure adiabatic container with 8 kg of nitrogen at 800 K. Determine the c_p and γ, and temperature of the mixture. What is the partial pressure of each gas in the mixture? Assume calorically perfect gases.

2.27   It is given that 10 kg/s of air at 400 K mixes with a 12 kg/s of combustion products at 600 K in a steady flow adiabatic mixer. Determine the c_p and γ, and temperature of the mixture. Assume calorically perfect gases with properties of Table 2.2, and neglect kinetic energy changes.

2.28   Rework Problem 2.14 for variable specific heats using Appendix L or the AFPROP program. Compare your results to Problem 2.14.

2.29   Rework Problem 2.16 for variable specific heats using Appendix L or the AFPROP program. Compare your results to Problem 2.16.

2.30   Rework Problem 2.17 for variable specific heats using Appendix L or the AFPROP program. Compare your results to Problem 2.17.

2.31   Rework Problem 2.20 for variable specific heats using Appendix L or the AFPROP program. Compare your results to Problem 2.20.

2.32   Rework Problem 2.21 for variable specific heats using Appendix L or the AFPROP program. Compare your results to Problem 2.21.

2.33   Rework Problem 2.24 for variable specific heats using Appendix L or the AFPROP program. Compare your results to Problem 2.24.

2.34 Work Example 2.12 for temperatures of 1000 K and 1500 K using equilibrium constant. Comment on your results.

2.35 Work Example 2.13 for temperatures of 1000 K and 1500 K using EQL software. Comment on your results.

2.36 Work Example 2.14 for temperatures of 1000 K and 1500 K. Comment on your results.

2.37 Work Example 2.15 for reactants at 300 K/800 kPa and 500 K/200 kPa. Compare to Example 2.15 and comment on your results.

2.38 Work Example 2.15 for 1 mole of hydrogen and 0.5 mole of oxygen at 300 K/101 kPa. Comment on your results.

2.39 Work Example 2.15 for one mole of $C_{12}H_{26}$ (approximate single molecule model of JP-8 gas turbine engine fuel) and air (3.76 mole of N_2 and one mole of O_2 default software values) at inlet conditions of 900 K and 3 MPa. Comment on your results.

Chapter 3 Compressible Flow

> When I examine myself and my methods of thought, I come to the conclusion that the gift of imagination has meant more to me than any talent for abstract positive thinking.
>
> *Albert Einstein (this quote was inspiration to Hans von Ohain) [74]*

3.1 Introduction

For a simple compressible system, we learned in Chapter 2 that the state of a unit mass of gas is fixed by two independent intensive properties such as pressure and temperature. To fully describe the condition and thus fix the state of this same gas when it is in motion requires the specification of a further property that will fix the speed of the gas. Thus *three independent intensive properties are required to fully specify the state of a gas in motion.*

At any given point in a compressible fluid flow field, the thermodynamic state of the gas is fixed by specifying, at that point, the velocity of the gas and any two independent properties such as pressure and temperature. However, we find that to specify the velocity directly is not always the most useful or the most convenient way to describe 1-D flow. There are other properties of the gas in motion that are dependent on the speed of the gas and that may be used in place of the speed to describe the state of the flowing gas. Some of these properties are the Mach number, total pressure, and total temperature. In this chapter we define these properties, develop the property relationships that you will find very useful, and describe briefly some of the characteristics of compressible flow.

3.2 Compressible Flow Properties

3.2.1 Stagnation or Total Temperature

In Section 2.3.3 we introduced the definition of stagnation or total specific enthalpy:

$$h_t \equiv h + \frac{V^2}{2g_c} \tag{2.20}$$

The stagnation or total temperature of a flowing gas derives directly from Eq. (2.20) and the calorically perfect gas (CPG) assumption introduced in

Section 2.4.2. For a CPG, Eq. (2.20) becomes

$$h_t - h = c_p(T_t - T) = \frac{V^2}{2g_c} \tag{3.1}$$

and then

$$T_t \equiv T + \frac{V^2}{2g_c c_p} \tag{3.2}$$

Thus, the *stagnation or total temperature* T_t is defined by Eq. (3.2). It is the temperature measured if a flow that is modeled as a calorically perfect gas is brought to rest *adiabatically*. The thermodynamic temperature T is sometimes called the *static temperature* to distinguish it from the total temperature T_t. The static temperature T is the temperature we would feel if we were moving along with the fluid at the same velocity. Note that when $V = 0$, the static and total temperatures are identical; all of the kinetic energy of the flowing gas is converted to internal energy (an increase in the average molecular energy).

We saw in Chapter 2 that the first law of thermodynamics is really concerned with changes in energy state. From the definitions in Eqs. (2.20) and (3.2), it follows that for a CPG,

$$\Delta h_t = c_p \Delta T_t$$

and for steady, 1-D flow of a CPG with negligible gravity effects, Eq. (2.22) takes the form

$$h_{t\,\text{out}} - h_{t\,\text{in}} = c_p(T_{t\,\text{out}} - T_{t\,\text{in}}) = q_{\text{in}} + w_{\text{in}} - q_{\text{out}} - w_{\text{out}} \text{ (energy per mass)}$$

If there are negligible heat and work interactions, we see from the above equation that $h_{t\,\text{out}} = h_{t\,\text{in}}$, and for a calorically perfect gas, $T_{t\,\text{out}} = T_{t\,\text{in}}$.

Consider an airplane in flight at a velocity V_0. To an observer riding with the airplane, the airflow about the wing of the plane appears as in Fig. 3.1. We mark out a control volume σ as shown in the figure between a station far upstream from the wing and a station just adjacent to the wing's leading edge stagnation point, where the velocity of the airstream is reduced to a negligibly small magnitude. Applying the steady flow energy equation to the flow through σ of Fig. 3.1, we have

$$q_{\text{in}} - w_{\text{out}} = \left(h + \cancelto{0}{\frac{V^2}{2g_c}}\right)_1 - \left(h + \frac{V^2}{2g_c}\right)_0$$

or

$$0 = c_p T_1 - c_p\left(T + \frac{V^2}{2g_c c_p}\right)_0$$

a) Airflow over a wing

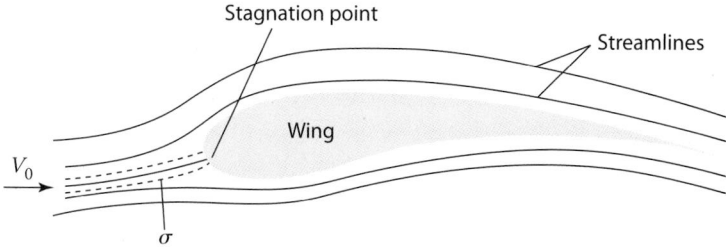

b) Enlarged view of flow in the neighborhood of the stagnation point

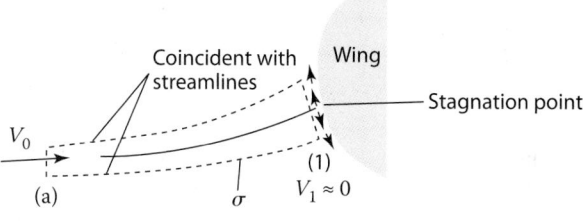

Fig. 3.1   Control volume σ with freestream inlet and stagnation exit conditions (reference system at rest relative to wing).

From this equation, we find that the temperature of the air at the stagnation point of the wing is

$$T_1 = T_0 + \frac{V_0^2}{2g_cc_p} = T_{t0}$$

Thus we see that the temperature, which the leading edge of the wing "feels," is the total temperature T_{t0}.

At high flight speeds, the freestream total temperature T_{t0} is significantly different from the freestream ambient temperature T_0. This is illustrated in Fig. 3.2, where $T_{t0} - T_0$ is plotted against V_0 by using the relation

$$(T_t - T)_0 = \frac{V_0^2}{2g_cc_p} = \frac{V_0^2}{12,000} \approx \left(\frac{V_0}{110}\right)^2 {}^{\circ}\text{R}$$

with V_0 expressed in ft/s. Because the speed of sound at 25,000 ft is 1000 ft/s, a Mach number scale for 25,000 ft is easily obtained by dividing the scale for V_0 in Fig. 3.2 by 1000. (Mach number M equals V_0 divided by the local speed of sound a.) Therefore, Mach number scales are also given on the graphs.

Referring to Fig. 3.2, we find that at a flight speed of 800 ft/s corresponding to a Mach number of 0.8 at 25,000-ft altitude, the stagnation points on an airplane experience a temperature that is about 50°R higher than ambient temperature. At 3300 ft/s ($M = 3.3$ at 25,000 ft), the total temperature is

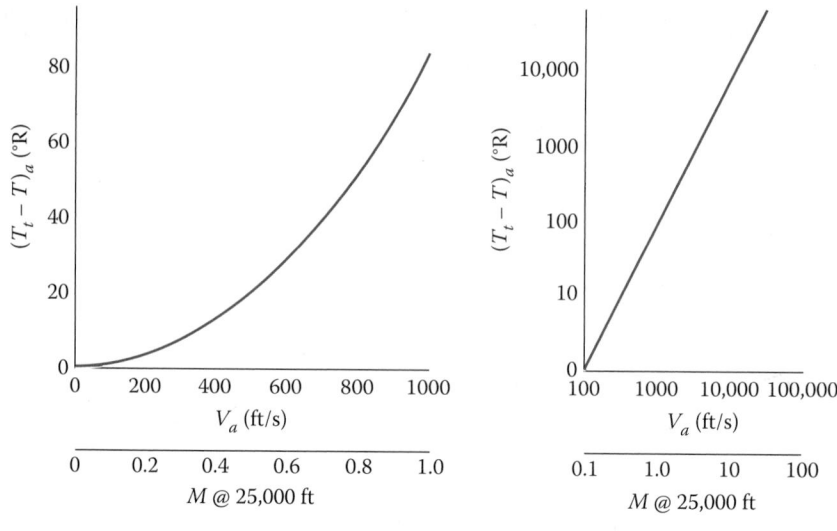

Fig. 3.2 Total temperature minus ambient temperature vs flight speed and vs flight Mach number at 25,000 ft [$g_c c_p$ is assumed constant at 6000 ft^2/(s$^2 \cdot$ °R); therefore, these are approximate curves].

900°R higher than ambient! It should be evident from these numbers that vehicles such as the X-15 airplane and reentry bodies experience high temperatures at their high flight speeds.

These high temperatures are produced as the kinetic energy of the air impinging on the surfaces of a vehicle is reduced and the enthalpy (hence, temperature) of the air is increased a like amount. This follows directly from the steady flow energy equation, which gives, with $q_{in} = w_{out} = 0$,

$$\left(h + \frac{V^2}{2g_c} \right)_{in} = \left(h + \frac{V^2}{2g_c} \right)_{out}$$

or

$$\Delta h = c_p \Delta T = -\Delta \left(\frac{V^2}{2g_c} \right)$$

and

$$\Delta T = -\frac{\Delta V^2}{2g_c c_p}$$

Thus a decrease in the kinetic energy of air produces a rise in the air temperature and a consequent heat interaction between the air and the surfaces of an air vehicle. This heat interaction effect is referred to as *aerodynamic heating*.

Example 3.1

The gas in a rocket combustion chamber is at 120 psia and 1600°R (Fig. 3.3). The gas expands through an adiabatic frictionless (isentropic) nozzle to 15 psia. What are the temperature and velocity of the gas leaving the nozzle? Treat the gas as the calorically perfect gas air with $\gamma = 1.4$ and $g_c c_p = 6000 \text{ ft}^2/(\text{s}^2 \cdot °\text{R})$.

Solution

Locate the state of the combustion chamber gas entering control volume σ on a T-s diagram in the manner depicted in Fig. 3.4. Then, from the diagram, find s_1. Because the process is isentropic, $s_2 = s_1$. The entropy at 2, along with the known value of P_2, fixes the static state of 2. With 2 located in the T-s diagram, we can read T_2 from the temperature scale as 885°R, and we can verify this graphical solution for T_2 by using the isentropic relation Eq. (2.54) with $\gamma = 1.4$. Thus $T_2 = (1600°\text{R})(15/120)^{0.286} = 885°\text{R}$ (checks).

If, in addition to P_2 and T_2, the total temperature T_{t2} of the flowing gas at 2 is known, then the state of the gas at 2 is completely fixed. For with P_2 and T_2 specified, the values of all thermodynamic properties independent of speed (the *static* properties) are fixed, and the speed of the gas is determined by T_2 and T_{t2}.

From the steady flow energy equation [Eq. (2.19)], we find that $T_{t2} = T_{t1} = T_1$ and, hence, V_2 from the relation

$$\frac{V_2^2}{2 g_c c_p} = T_{t2} - T_2$$

We see from this equation that the vertical distance $T_{t2} - T_2$ in the T-s diagram is indicative of the speed of the gas at 2, which is $V_2^2 = 2(6000)(1600 - 885)\text{ft}^2/\text{s}^2$. Thus $V_2 = 2930$ ft/s.

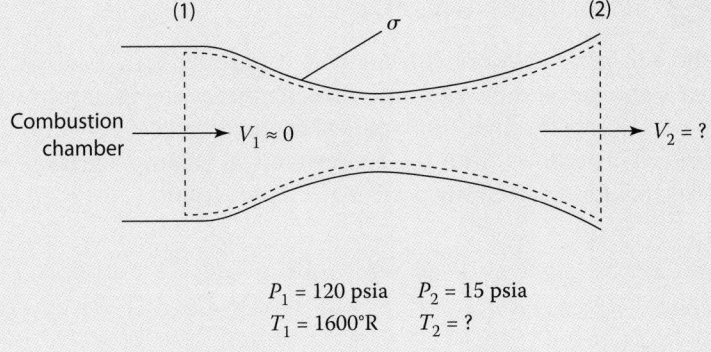

(1) σ (2)

Combustion chamber $V_1 \approx 0$ $V_2 = ?$

$P_1 = 120$ psia $P_2 = 15$ psia
$T_1 = 1600°\text{R}$ $T_2 = ?$

Fig. 3.3   Rocket exhaust nozzle.

(Continued)

Example 3.1 (Continued)

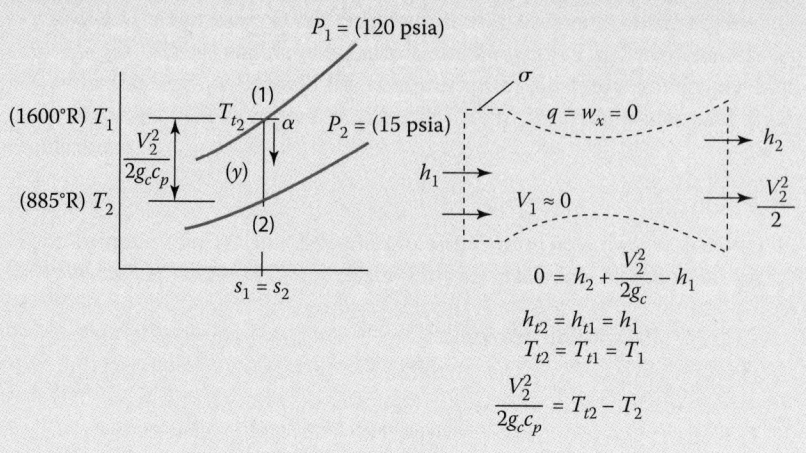

Fig. 3.4 Process plot for example rocket nozzle.

The series of states through which the gas progresses in the nozzle as it flows from the combustion chamber (nozzle inlet) to the nozzle exit is represented by path line α in the $T–s$ diagram. The speed of the gas at any intermediate state y in the nozzle is represented by the vertical distance on the path line from 1 to the state in question. This follows from the relations

$$T_{ty} = T_1 \quad \text{and} \quad \frac{V_y^2}{2g_c c_p} = T_{ty} - T_y = T_1 - T_y$$

3.2.2 Stagnation or Total Pressure

In the adiabatic, no-work slowing of a calorically perfect gas (CPG) to zero speed, the gas attains the same final stagnation temperature whether it is brought to rest through frictional effects (irreversible) or without them (reversible). This follows from the energy control volume equation applied to σ of Fig. 3.5 for a calorically perfect gas. Thus, from

$$q_{\text{in}} - w_{\text{out}} = c_p \left(T_y - T_1 \right) + \frac{V_y^2 - V_1^2}{2g_c}$$

with $q_{\text{in}} = w_{\text{out}} = 0$ and $V_y = 0$, T_y becomes

$$T_y = T_0 = T_1 + \frac{V_1^2}{2g_c c_p}$$

Because the energy control volume equation is valid for frictional or frictionless flow, $T_y = T_0$ is constant and independent of the degree of friction between 1 and y as long as $q_{in} = w_{out} = V_y = 0$.

Although the gas attains the same final temperature T_0 in reversible or irreversible processes, its final pressure will vary with the degree of irreversibility associated with the slowing down process. The second law of thermodynamics [Eq. (2.27)] for an adiabatic process requires that $s_{out} - s_{in} \geq 0$ and for a CPG, we use Eq. (2.48) to determine Δs:

$$c_p \, \ell n \frac{T_y}{T_1} - R \, \ell n \frac{P_y}{P_1} = s_y - s_1 \geq 0 \qquad (3.3)$$

$T_y = T_0 = $ const for this adiabatic process, so the final value of P_y depends on the entropy increase $s_y - s_1$, which in turn is a measure of the degree of irreversibility between 1 and y.

When the slowing down process between 1 and y is reversible, with $s_y - s_1 = 0$, the final pressure is defined as the total pressure P_t. The final state is called the *total state* t_1 of the static state 1. Using this definition of total pressure, we have, from Eq. (2.54),

$$P_t \equiv P \left(\frac{T_t}{T} \right)^{\gamma/(\gamma-1)} \qquad (3.4)$$

These ideas are illustrated in the *T-s* diagram of Fig. 3.5. Let us imagine the flowing gas at station 1 to be brought to rest adiabatically with no shaft work by means of a duct diverging to an extremely large area at station y, where the flow velocity is zero. If the diverging duct is frictionless, then the slowing down process from 1 to y is isentropic with the path line α_r in the

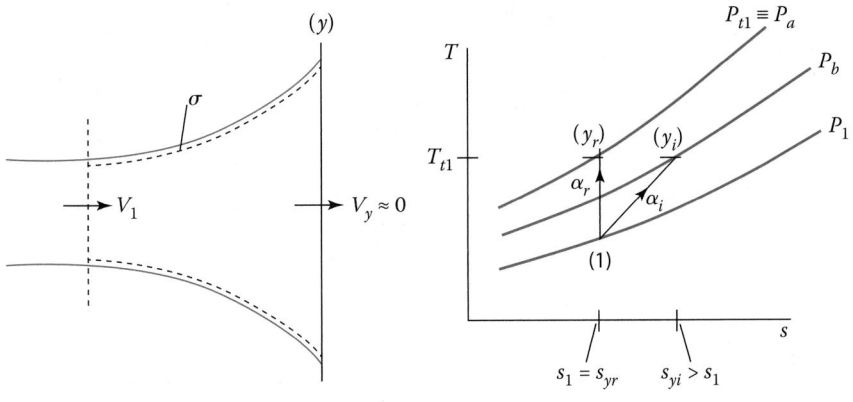

Fig. 3.5 Definition of total pressure.

T-s diagram of the figure. If the diverging duct is frictional, then the slowing down process from 1 to y is irreversible and adiabatic ($s_{yi} > s_1$) to satisfy the entropy control volume equation for adiabatic flow and is shown as the path line α_i in the *T-s* diagram.

The total pressure of a flowing gas is defined as the pressure obtained when the gas is brought to rest *isentropically*. Thus the pressure corresponding to state y_r of the *T-s* diagram is the total pressure of the gas in state 1. The state point y_r is called the *total* or *stagnation state* t_1 of the static point 1. The thermodynamic pressure P is sometimes called the static pressure to distinguish it from the total pressure P_t. The static pressure is the pressure we feel when we are moving along with the fluid at the same velocity.

The concepts of total pressure and total temperature are very useful, because these two properties along with the third property (static pressure) of a flowing gas are readily measured, and they fix the state of the flowing gas. We measure these three properties in flight with pitot-static and total temperature probes on modern high-speed airplanes, and these properties are used to determine speed and Mach number and to provide other data for many aircraft subsystems. We also measure these properties at various locations throughout an engine for engine control and health management.

Consider a gas flowing in a duct in which P and T may change due to heat interaction and friction effects. The flow total state points t_1 and t_2 and the static state points 1 and 2, each of which corresponds to flow stations 1 and 2, are located in the *T-s* diagram of Fig. 3.6. By definition, the entropy of the total state at any given point in a gas flow has the same value as the entropy of the static state properties at that point. Therefore, $s_{t1} = s_1$ and $s_{t2} = s_2$.

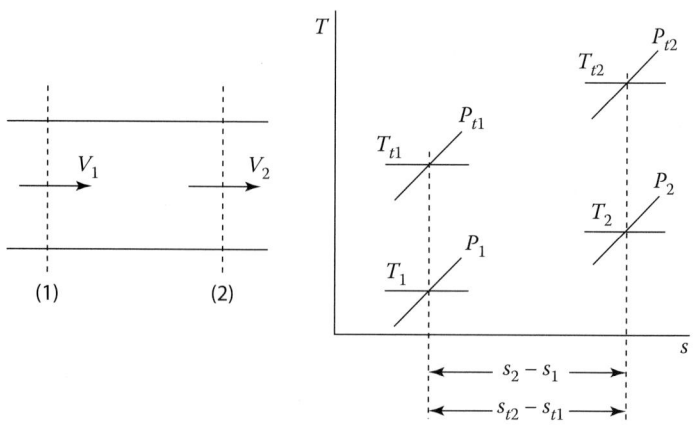

Fig. 3.6   Entropy change in terms of the stagnation properties T_t and P_t.

From the entropy equation of state of a perfect gas, the entropy change between 1 and 2 is

$$s_2 - s_1 = c_p \ell n \frac{T_2}{T_1} - R \ell n \frac{P_2}{P_1}$$

The entropy change between total state points t_1 and t_2 is

$$s_{t2} - s_{t1} = c_p \ell n \frac{T_{t2}}{T_{t1}} - R \ell n \frac{P_{t2}}{P_{t1}} \tag{3.5a}$$

Because $s_{t1} = s_1$ and $s_{t2} = s_2$, we have

$$s_{t2} - s_{t1} = s_2 - s_1$$

Therefore, the change of entropy between two states of a flowing gas can be determined by using total properties in place of static properties.

Equation (3.5a) indicates that in an adiabatic and no-shaft-work constant-T_t flow (such as exists in an airplane engine inlet and nozzle, or flow through a shock wave), we have

$$s_2 - s_1 = -R \ell n \frac{P_{t2}}{P_{t1}} \tag{3.5b}$$

By virtue of this equation and the entropy control volume equation for adiabatic flow, $s_2 - s_1 \geq 0$. Thus, in a constant-T_t flow,

$$\frac{P_{t2}}{P_{t1}} \leq 1 \tag{3.6}$$

The equality of the previous equation holds true ($P_{t2} = P_{t1}$) for the ideal case of isentropic flow of an adiabatic, no-work flowing, calorically perfect gas. Hence the total pressure of air passing through an engine inlet and nozzle or a shock wave cannot increase and must, in fact, decrease because of the irreversible effects of friction or shocks.

3.2.3   Compressible Flow Functions

Using equations introduced previously, many useful relations that give flow property ratios in terms of the flow Mach number and ratio of specific heats can be developed. The speed of sound in a perfect gas is given by

$$a = \sqrt{\gamma g_c R T}$$

Note the use of the static temperature T for determining the speed of sound. From the definition of Mach number provided in Eq. (2.8), $M \equiv V/a$:

$$M^2 = \frac{V^2}{\gamma g_c R T}$$

We use the above relation for Mach number and rearrange Eq. (3.2) to obtain

$$\frac{T_t}{T} = 1 + \frac{V^2}{2g_c c_p T} = 1 + \frac{\gamma g_c R T M^2}{2 g_c c_p T} = 1 + \frac{\gamma R M^2}{2 c_p}$$

But

$$\frac{R}{c_p} = \frac{\gamma - 1}{\gamma}$$

so that

$$\frac{T_t}{T} = 1 + \frac{\gamma - 1}{2} M^2 \quad \text{or} \quad \frac{T}{T_t} = \left(1 + \frac{\gamma - 1}{2} M^2\right)^{-1} \tag{3.7}$$

From Eq. (3.4), we have

$$\frac{P_t}{P} = \left(1 + \frac{\gamma - 1}{2} M^2\right)^{\frac{\gamma}{\gamma - 1}} \quad \text{or} \quad \frac{P}{P_t} = \left(1 + \frac{\gamma - 1}{2} M^2\right)^{\frac{-\gamma}{\gamma - 1}} \tag{3.8}$$

We could obtain a similar expression for ρ/ρ_t using the perfect gas equation of state, Eq. (2.28). Equations (3.7) and (3.8) appear graphically in Fig. 3.7 and form the basis of the compressible flow functions provided in

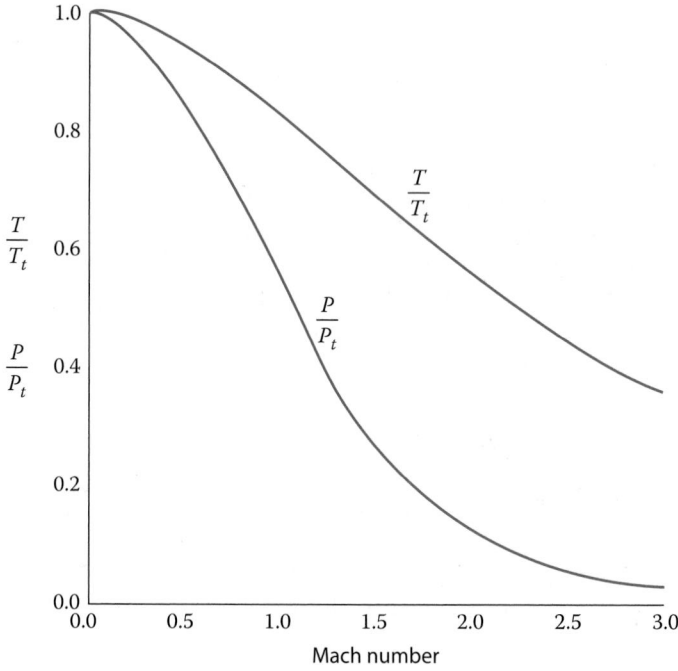

Fig. 3.7 P/P_t and T/T_t vs Mach number ($\gamma = 1.4$).

Appendix D. They are also given in the isentropic flow functions of the GASTAB program for user input γ and M.

Both Fig. 3.7 and the corresponding equations provide Mach numbers and static temperatures for known values of P, P_t, and T_t. For example, suppose we are given the following in-flight measurements:

$$P = 35.4\,\text{kPa} \quad P_t = 60.0\,\text{kPa} \quad T_t = 300\,\text{K}$$

From these data, $P/P_t = 0.590$. From Fig. 3.7 or Appendix D ($\gamma = 1.4$), we find $M = 0.9$ and $T/T_t = 0.86$. Then we obtain the ambient temperature by using $T = (T/T_t)T_t = 258$ K.

Figure 3.7 shows that in a sonic ($M = 1$) stream of gas with $\gamma = 1.4$,

$$\frac{P}{P_t} = 0.528$$

and

$$\frac{T}{T_t} = 0.833$$

and for supersonic flow,

$$\frac{P}{P_t} < 0.528$$

and

$$\frac{T}{T_t} < 0.833$$

Consider the 1-D steady flow of a gas in a duct with T_t and P_t constant at all stations along the duct. This means a total temperature probe will measure the same value of T_t at each duct station, and an isentropic total pressure probe will measure the same value of P_t at each station. The path line of α of the flow is a vertical line in the T-s diagram. The state points on the path line can be categorized as follows.

Subsonic:

$$T > 0.883\,T_t \quad P > 0.528\,P_t$$

Sonic:

$$T = 0.883\,T_t \quad P = 0.528\,P_t$$

Supersonic:

$$T < 0.883\,T_t \quad P < 0.528\,P_t$$

Figure 3.8 delineates the subsonic, sonic, and supersonic portions of path line α.

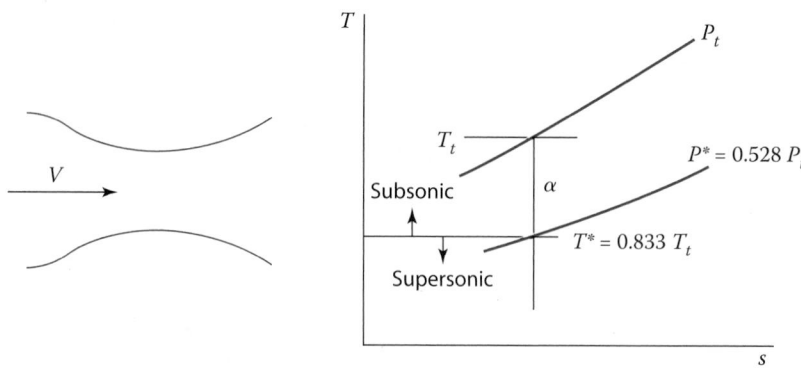

Fig. 3.8   Subsonic and supersonic state points in an isentropic flow.

The thermodynamic properties at the state point where $M = 1$ on α are denoted by P^*, T^*, V^*, and so on (read as P star, etc.) and the state is called the "star state." In addition, the cross-sectional flow area at the $M = 1$ point is indicated by A^*. The magnitude of A^* is determined by the relation

$$A^* = \frac{\dot{m}}{\rho^* V^*} \qquad (3.9)$$

3.2.4   Mass Flow Parameter

The mass flow rate at a cross-sectional area A can be expressed as a function of total pressure and total temperature (both of which are measured quantities), the gas Mach number M, and the ratio of specific heats γ. The 1-D mass flow equation is

$$\dot{m} = \rho A V$$

The mass flow per unit area at duct area A in the flow of a calorically perfect gas is given by the expression

$$\frac{\dot{m}}{A} = \rho V = \frac{PV}{RT} = \frac{V}{\sqrt{\gamma g_c R T}} \frac{P \sqrt{\gamma g_c}}{\sqrt{RT}} = M \sqrt{\frac{\gamma g_c}{R}} \frac{P}{P_t} \sqrt{\frac{T_t}{T}} \frac{P_t}{\sqrt{T_t}}$$

where the 1-D flow equation [Eq. (2.12)], the perfect gas equation of state [Eq. (2.28)], the perfect gas speed of sound equation [Eq. (2.44)], and the definition of Mach number have been used. Replacing the static/total property ratios with Eqs. (3.7) and (3.8) and rearranging, we obtain the grouping defined as the *mass flow parameter (MFP)*:

$$\text{MFP} \equiv \frac{\dot{m} \sqrt{T_t}}{A \ P_t} \qquad (3.10)$$

$$\text{MFP}(M) = \frac{\dot{m}}{A}\frac{\sqrt{T_t}}{P_t} = \sqrt{\frac{\gamma g_c}{R}}M\left(1 + \frac{\gamma - 1}{2}M^2\right)^{-\frac{\gamma+1}{2(\gamma-1)}} \qquad (3.11)$$

Values of the mass flow parameter are plotted in Fig. 3.9 for $\gamma = 1.4$ and $\gamma = 1.3$. One sees readily that the maximum value of the MFP occurs when the Mach number is unity. Thus, for a given total temperature and pressure T_t and P_t, the maximum mass flow rate per area corresponds to a flow $M = 1.0$. Note that for each value of MFP other than that corresponding to maximum mass flow rate, there are two possible solutions, one subsonic and one supersonic. One must use other knowledge associated with the problem, like P/P_t or T/T_t, to determine which solution applies.

3.2.5 Isentropic Area Ratio A/A^*

It is often convenient to reference the properties of a 1-D calorically perfect gas flow at a cross-sectional area A to those at a cross-sectional area where the Mach number is unity A^*. The development is similar to that of the mass flow parameter of the previous section, and again we see

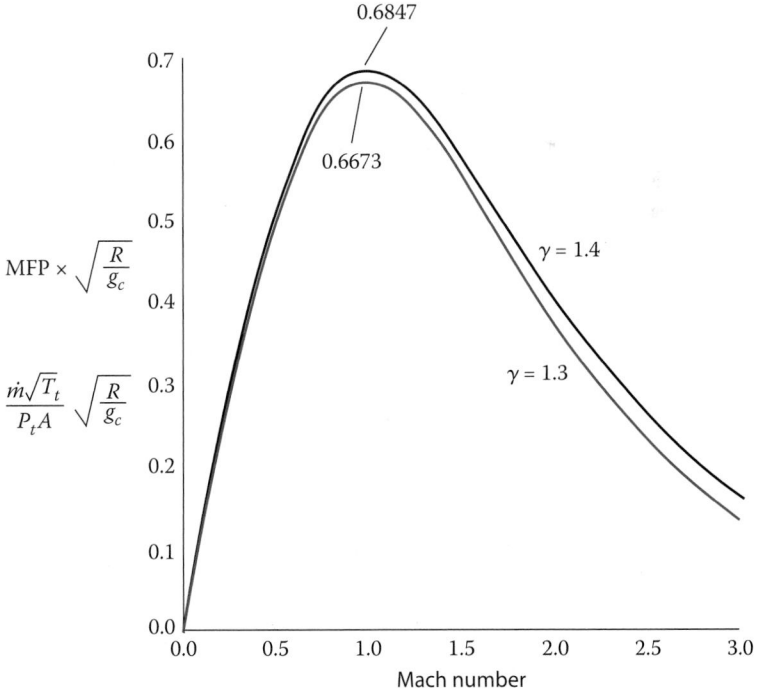

Fig. 3.9 Mass flow parameter vs Mach number ($\gamma = 1.4$ and $\gamma = 1.3$).

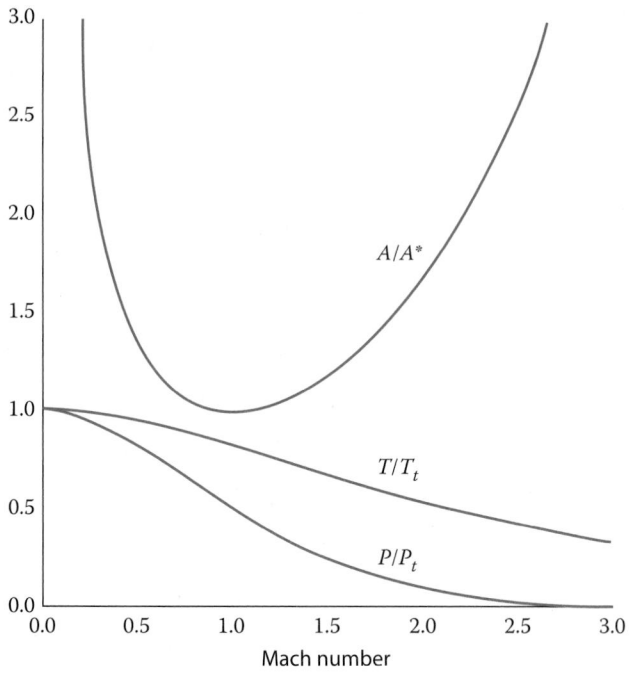

Fig. 3.10 A/A^*, P/P_t, and T/T_t vs Mach number ($\gamma = 1.4$).

that A/A^* becomes a function of γ and M. Consider the isentropic flow of a calorically perfect gas in an isentropic duct from ρ, M, P, T, and A to the sonic state where the properties are ρ^*, $M^* = 1$, P^*, T^*, and A^*. Because both states have the same mass flow, we write

$$\dot{m} = \rho A V = \rho^* A^* V^*$$

Rewriting gives

$$\frac{A}{A^*} = \frac{\rho^* V^*}{\rho V} = \frac{P^*}{T^*} \frac{T}{P} \frac{1}{M} \frac{a^*}{a} = \frac{1}{M} \frac{P^*/P}{T^*/T} \sqrt{T^*/T} = \frac{1}{M} \frac{P^*/P}{\sqrt{T^*/T}} \tag{i}$$

However,

$$\frac{T}{T^*} = \frac{T/T_t}{T^*/T_t} = \left[\frac{2}{\gamma+1}\left(1 + \frac{\gamma-1}{2}M^2\right)\right]^{-1} \tag{ii}$$

and

$$\frac{P}{P^*} = \frac{P/P_t}{P^*/P_t} = \left[\frac{2}{\gamma+1}\left(1 + \frac{\gamma-1}{2}M^2\right)\right]^{-\gamma/(\gamma-1)} \tag{iii}$$

Substitution of Eqs. (ii) and (iii) into (i) gives

$$\frac{A}{A^*} = \frac{1}{M}\left[\frac{2}{\gamma+1}\left(1+\frac{\gamma-1}{2}M^2\right)\right]^{(\gamma+1)/[2(\gamma-1)]} \qquad (3.12)$$

Both the MFP and A/A^* are compressible flow functions included in Appendix D tabulated for γ values of 1.4, 1.33, and 1.3. They are also given in the isentropic flow functions of the GASTAB program for user input γ and M. Figure 3.10 plots A/A^*, T/T_t, and P/P_t vs M for $\gamma = 1.4$.

Example 3.2

Exhaust gases from a gas turbine engine are accelerated through a fixed convergent exhaust duct (nozzle) with properties as shown in Fig. 3.11. (Standard engine station numbering is used.) Find the exit Mach number and static pressure as well as the mass flow rate. The exhaust gases can be modeled as a calorically perfect gas with a ratio of specific heats of 1.33. Assume steady, 1-D, adiabatic, isentropic flow, and use Appendix D.

Solution

For steady, 1-D, adiabatic, no-work flow of a CPG, from the first law:

$$h_{t9} - h_{t5} = c_p(T_{t9} - T_{t5}) = 0 \Rightarrow T_{t9} = T_{t5}$$

For isentropic flow, from the second law:

$$s_{t9} - s_{t5} = 0 = c_p \ln\frac{T_{t9}}{T_{t5}} - R\ln\frac{P_{t9}}{P_{t5}} \Rightarrow P_{t9} = P_{t5} \text{ because } T_{t9} = T_{t5}$$

All we know is the cross-sectional area at the exit plane station 9, so the key to solving this problem is relating properties at stations 5 and 9 to station *,

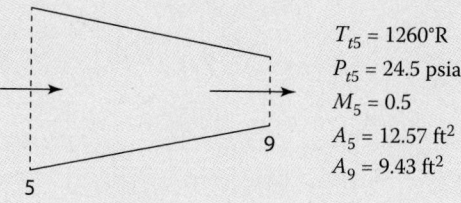

$$T_{t5} = 1260°R$$
$$P_{t5} = 24.5 \text{ psia}$$
$$M_5 = 0.5$$
$$A_5 = 12.57 \text{ ft}^2$$
$$A_9 = 9.43 \text{ ft}^2$$

Fig. 3.11   Nozzle flow and known conditions for Example 3.2.

(Continued)

Example 3.2 *(Continued)*

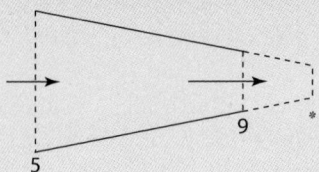

Fig. 3.12 Nozzle of Example 3.2 extended to * reference condition.

the area at which the Mach number would be unity, $M^* = 1.0$ at A^*. First we check to see if $A_9 > A^*$:

For $M_5 = 0.5$, from Appendix D ($\gamma = 1.33$):

$$A_5/A^* = 1.34541 \quad \text{or} \quad A^*/A_5 = 0.7433$$

$$A_9/A_5 = 9.43\,\text{ft}^2/12.57\,\text{ft}^2 = 0.75$$

So, yes, $A_9 > A^*$ and, consequently, $M_9 < 1.0$. We imagine our exhaust nozzle extended to area A^* as sketched in Fig. 3.12.

$$\frac{A_9}{A^*} = \frac{A_9}{A_5}\frac{A_5}{A^*} = (0.75)(1.34541) = 1.0091$$

With $A_9/A^* = 1.0091$, from Appendix D ($\gamma = 1.33$): $M_9 = 0.90$

$$\frac{P_9}{P_{t9}} = 0.6032 \Rightarrow P_9 = 0.6032(24.5\,\text{psia}) = 14.78\,\text{psia}$$

For mass flow rate, use mass flow parameter MFP, again from Appendix D ($\gamma = 1.33$)

$$\text{MFP}\sqrt{R/g_c} = 0.666563 = \frac{\dot{m}\sqrt{T_t}}{P_t A}\sqrt{R/g_c}$$

So, therefore we solve for mass flow rate at station 9, being vigilant with units

$$\dot{m}_9 = \frac{0.666563}{1.28758\,\dfrac{\text{lbf}\cdot\text{s}}{\text{lbm}\cdot\sqrt{\text{R}}}}\frac{\left(24.5\,\dfrac{\text{lbf}}{\text{in}^2}\right)(9.43\,\text{ft}^2)\left(\dfrac{144\,\text{in}^2}{\text{ft}^2}\right)}{\sqrt{1260\,\text{R}}} = 485.2\,\text{lbm/s}$$

Note that because continuity must be satisfied (conservation of mass), we could have solved for the mass flow rate at station 5, or station * for that matter, using the proper values for MFP and the cross-sectional area at each station.

3.2.6 Impulse Function

The impulse function I is a convenient grouping of terms that show up in the linear momentum equation and is defined by

$$I \equiv PA + \frac{\dot{m}V}{g_c} \tag{3.13}$$

Using the 1-D mass flow rate equation and perfect gas equation of state, Eq. (3.13) can be written as

$$I = PA(1 + \gamma M^2) \tag{3.14}$$

For steady flow, from Eq. (2.16), the streamwise axial force exerted on the fluid through a control volume is $I_{\text{out}} - I_{\text{in}}$, while the reaction force exerted by the fluid on the control volume is $I_{\text{in}} - I_{\text{out}}$.

The impulse function makes possible almost unimaginable simplification of the calculation of forces on aircraft engines, rocket engines, and their components. For example, although one could determine the net axial force exerted on the fluid flowing through any device by integrating the axial components of pressure and viscous forces over every infinitesimal element of internal wetted surface area, it is certain that no one ever has. Instead, the integrated result of the forces is obtained with ease and certainty by merely evaluating the change in impulse function across the device. Compressible flow functions tabulate the impulse function as a ratio to that at the star state ($M = 1$) or I/I^*, as is done with A/A^*. We note that

$$\frac{I}{I^*} = \frac{P}{P^*} \frac{A}{A^*} \frac{(1 + \gamma M^2)}{1 + \gamma} \tag{3.15}$$

The GASTAB program provides the value of I/I^* for user input M or vice versa.

3.3 Velocity-Area Variation for Isentropic, Adiabatic Flow

Much can be learned from the study of steady, 1-D, adiabatic, isentropic flow through a varying area duct. Consider the control volume as shown in Fig. 3.13, across which we have differential changes in properties.

We apply our equations of motion from Chapter 2 for steady flow.

For steady flow continuity:

$$\iint_{CS} \rho(\vec{V} \cdot \hat{n})\mathrm{d}A = 0$$

For 1-D flow:

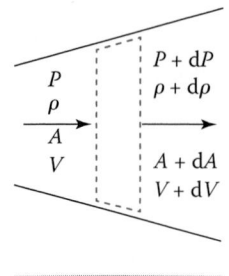

$$(\rho + d\rho)(A + dA)(V + dV) - \rho AV = 0$$

Multiplying out, ignoring products of differential terms, and rearranging, we get

$$\frac{d\rho}{\rho} + \frac{dA}{A} + \frac{dV}{V} = 0 \tag{3.16}$$

Fig. 3.13 Control volume for varying area flow.

For steady flow linear momentum:

$$\frac{1}{g_c}\iint_{CS} V_x \rho (\vec{V} \cdot \hat{n}) dA = \sum_{\text{on } \sigma} F_x$$

For isentropic flow, frictional forces are negligible; the only forces acting on the gas in the control volume are pressure-area forces. We assume that a pressure $(P + dP/2)$ acts on the side surfaces of the control volume, as sketched in Fig. 3.14.

For 1-D, isentropic flow:

$$\frac{1}{g_c}(\rho AV)dV = [PA + \left(P + \frac{dP}{2}\right)dA - (P + dP)(A+dA)]$$

Multiplying out, ignoring products of differential terms, and rearranging, we get

$$dP + \frac{\rho V dV}{g_c} = 0 \tag{3.17}$$

This result is a manifestation of Bernoulli's principle, that *static pressure and velocity are inversely related;* that is, as P increases, V decreases, and vice versa. Note that this principle is generally applicable to high-speed and low-speed flows, whereas application of Bernoulli's equation is limited

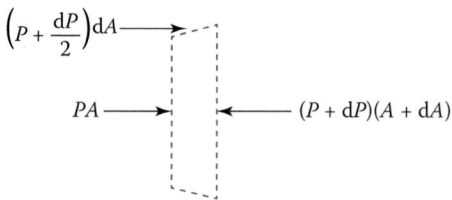

Fig. 3.14 Force balance on differential element.

to incompressible low-speed flows ($M < 0.3$) where density variations are neglected.

Interestingly, application of the first and second laws of thermodynamics to the control volume in Fig. 3.13 under the previous assumptions also yields Eq. (3.17): no new information. This is left to the interested reader, but the relevant governing equations are provided here:

$$\text{Energy} \iint_{CS} \left(h + \frac{V^2}{2g_c}\right) \rho (\vec{V} \cdot \hat{n}) dA = 0 \Rightarrow dh + d\left(\frac{V^2}{2g_c}\right) = 0$$

$$\text{Entropy } Tds = dh - \frac{dP}{\rho} \Rightarrow dh = \frac{dP}{\rho} \text{ for isentropic flow}$$

From linear momentum, we have Eq. (3.17) that can be written as

$$\frac{dP}{\rho} = \frac{-VdV}{g_c}$$

From the chain rule of differential calculus,

$$\left(\frac{dP}{d\rho}\right)_s \left(\frac{d\rho}{\rho}\right) = \frac{-VdV}{g_c} \Rightarrow \frac{a^2}{g_c}\left(\frac{d\rho}{\rho}\right) = \frac{-VdV}{g_c}$$

where a is the speed of sound as in Eq. (2.7).

Now divide by V^2 to introduce the Mach number $M = V/a$

$$\frac{1}{M^2}\left(\frac{d\rho}{\rho}\right) = -\frac{dV}{V}$$

Use the continuity equation for the $d\rho/\rho$ term in Eq. (3.15) to get

$$\frac{dA}{A} + \frac{dV}{V} = \frac{1}{M^2}\frac{dV}{V}$$

or

$$(1 - M^2)\frac{dV}{V} = -\frac{dA}{A} \tag{3.18}$$

Equation (3.18) contains a wealth of information regarding velocity–area variation! For example, it explains the familiar converging-diverging shape of a rocket nozzle. Equation (3.18) tells us that the velocity–area relationship is dependent on the Mach number of the flow. We consider three cases:

Case 1: $M < 1.0$ (subsonic flow): $(1 - M^2)$ term is positive and velocity and area are inversely proportional (as A increases, V decreases; A decreases, V increases).

Case 2: $M = 1.0$ (sonic flow): $(1 - M^2)$ term goes to zero as must the dA/A term (more later).

Case 3: $M > 1.0$ (supersonic flow): $(1 - M^2)$ term is negative and velocity and area are directly proportional (as A increases, V increases; as A decreases, V decreases).

The practical implications of Eq. (3.18) should be clear. Acceleration of a subsonic flow requires a decreasing area in the direction of flow, a converging area duct as in Fig. 3.15a. Acceleration of a supersonic flow requires an increasing area in the direction of flow, a diverging area duct as in Fig. 3.15b.

To accelerate a flow from negligible velocity to supersonic velocity requires a converging-diverging area change, as shown in Fig. 3.16. Such an area change results in a minimum area called the throat where dA/A goes to zero and Mach number is unity (recall Case 2 above). Of course, a converging-diverging area change is not sufficient to cause supersonic velocities; there must also be sufficient pressure difference imposed on the flow (more on this in later sections of this chapter). Note that a converging area can only accelerate a subsonic flow to Mach number unity, regardless of the amount of area change or pressure difference imposed on the flow.

From the development and discussion in this section and the preceding sections, it is clear that the Mach number is a very important parameter in

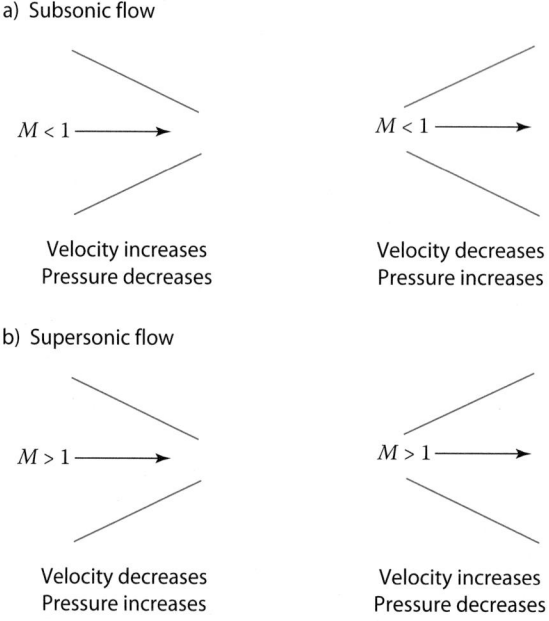

Fig. 3.15 Pressure and velocity variations for subsonic and supersonic flows.

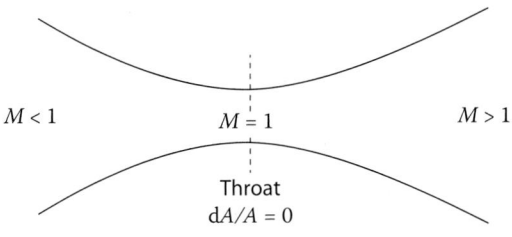

Fig. 3.16   Acceleration from subsonic to supersonic flow.

compressible flows. In fact, the Mach number M is capitalized because it is named after Austrian physicist Ernst Mach, who confirmed the existence of shock waves in supersonic flows in a paper he published in 1877. Mach correctly deduced that for high-speed flows the relationship between the speed of the object and the speed of sound is critical. It is because of the importance of the speed of sound in compressible flows that we next consider the propagation of sound waves.

3.4   Two-Dimensional Small-Amplitude Wave Propagation

Up to this point, we have limited our considerations to 1-D flow. We may obtain further knowledge of the physical significance of the sound velocity, however, by considering 2-D small-amplitude wave propagation. Consider the planar disturbance field produced by an infinitesimally small point source as it sends out weak pressure disturbances via molecular collisions that propagate at acoustic speed. Consider the point source as stationary, moving at subsonic velocity, and moving at supersonic velocity.

The weak pressure waves emanate continuously from the point source and propagate spherically outward at the speed of sound, the physical mechanism being molecular collisions. To facilitate our discussion, we will consider a finite number of waves, as indicated in Fig. 3.17. After any given time interval, the wave loci form the planar patterns shown in Fig. 3.17. The number of each wave indicates its age. Wave 4 is 4 s old, having begun 4 s prior to the time of the picture when the point source was at the center of the circle formed by the wave.

These patterns are easily observed in the form of gravity waves on the surface of a body of water that are distributed periodically by drops of water from the tip of a paddle freshly withdrawn from the water. By moving the paddle at the proper speed relative to the water, any of the patterns shown can be reproduced by the surface wavelets formed as water droplets from the paddle strike the water's surface.

For Fig. 3.17a, the point source is stationary, and concentric circles are formed by the wave loci. When, as in Fig. 3.17b, the source moves at a speed less than that of the wave propagation speed, the wave loci, which

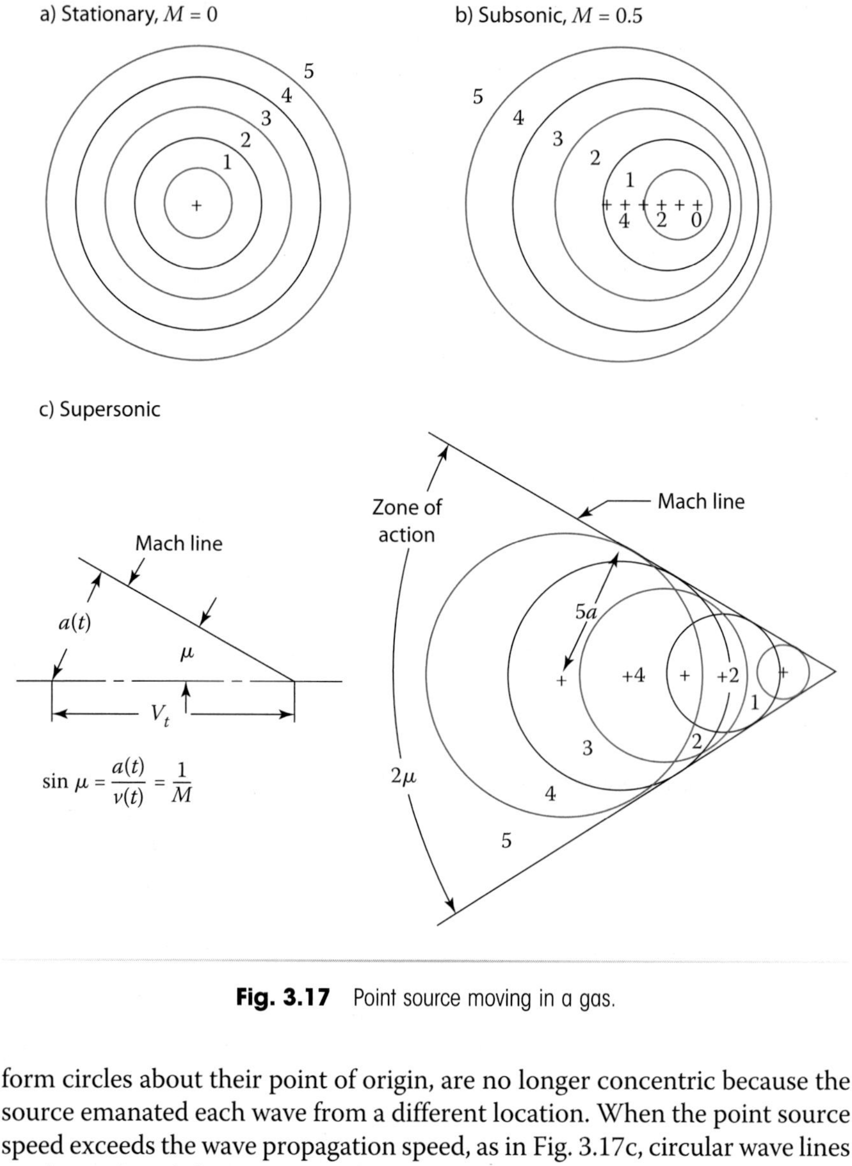

a) Stationary, $M = 0$

b) Subsonic, $M = 0.5$

c) Supersonic

Fig. 3.17 Point source moving in a gas.

form circles about their point of origin, are no longer concentric because the source emanated each wave from a different location. When the point source speed exceeds the wave propagation speed, as in Fig. 3.17c, circular wave lines are formed, and these wave circles are tangent to a line at angle μ, with the direction of the source speed such that

$$\sin \mu = \frac{a(t)}{V(t)} = \frac{1}{M} \tag{3.19}$$

This tangent line is called a Mach line, and the angle μ is called the *Mach angle*. Mach waves are formed by the concentration of infinitesimally small disturbances emanating from a point source moving faster than the speed of sound.

The wave patterns of Fig. 3.17 indicate that for other than supersonic speeds, the disturbance field produced by the point source extends to infinite distances about the source as time progresses (in the absence of viscous effects). At supersonic speeds, however, note that the fluid field is completely undisturbed forward of the Mach cone; the fluid is disturbed only within the cone.

The pattern of Fig. 3.17c illustrates the three rules of supersonic flow given by von Karman in 1947 in the tenth Wright brothers lecture (Ref. 14). These rules are based on the assumption of small disturbances and are applicable at a given instant of time. The rules are:

1. *The rule of forbidden signals:* The effect of pressure changes produced by a point source moving at a speed faster than sound cannot reach points ahead of the point source.
2. *The zone of action and the zone of silence:* All effects produced by a point source moving at a supersonic speed are contained within the zone of action bounded by the Mach cone and extending downstream from the body. The region outside of the cone of action at any instant of time is called the zone of silence.
3. *The rule of concentrated action:* The effects produced by the motion of a point source at supersonic speeds are concentrated along the Mach lines.

Sound pressure waves emanate from all points that make up a finite body. Unlike the earlier point source discussion, the body presents a finite disturbance to the flow; however, the discussion and three rules just presented for infinitesimally small disturbances are qualitatively applicable to finite disturbances.

To a subsonic flow, the pressure waves act as signaling mechanisms, forewarning the molecules of the flowing fluid to the presence of the body; consequently, the molecules are able to adjust smoothly and continuously. For example, the air flow around the 2-D airfoil in Fig. 3.18 has clearly preadjusted to the presence of the airfoil, as indicated by the smoke traces providing a visualization of the streamlines.

For supersonic flow, the sound waves are still present but they cannot make advance headway from the body because the gas molecules moving towards them are traveling at a velocity faster than the propagation speed of the waves, the speed of sound. The sound waves in effect pile into each other, forming a finite compression *shock wave*. The air flow has no mechanism to preadjust to the body and, as indicated in Fig. 3.19, the flow adjustment is sudden and nearly discontinuous across the very thin shock wave [2.5×10^{-5} cm (Ref. 15), thinner than the thickness of a sheet of paper]. Figure 3.19 presents convincing evidence of the qualitative application of rule 3 above, namely, the concentration of effects along a shock wave accompanying a body at supersonic speeds.

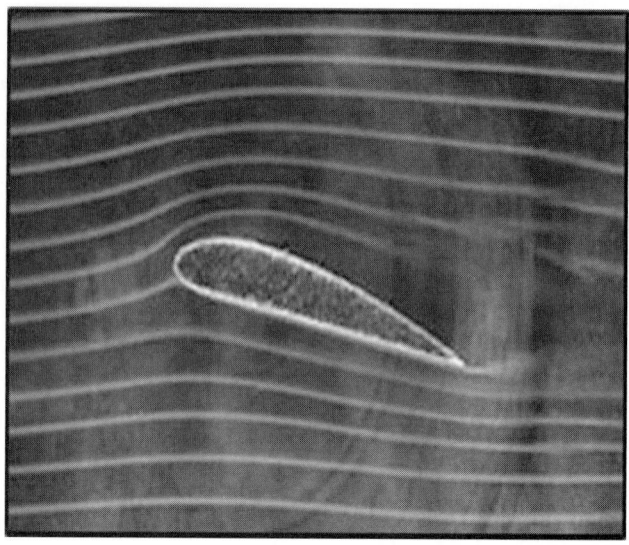

Fig. 3.18   Airfoil in a 2-D smoke tunnel (Courtesy of Department of Aeronautics, U.S. Air Force Academy).

Interestingly, a flow can be termed transonic, characterized by both subsonic and supersonic flow regions in the vicinity of the body. Figure 3.20 presents visual evidence of transonic flow. Although the oncoming flow relative to the projectile is at $M = 0.9$, portions of the flow have clearly accelerated to speeds above sonic velocity, as evidenced by the presence of shock waves on the projectile. Fans in modern turbofan engines are often designed so that the lower span of the rotating fan blade towards the root is subjected to subsonic flow while the outer span towards the

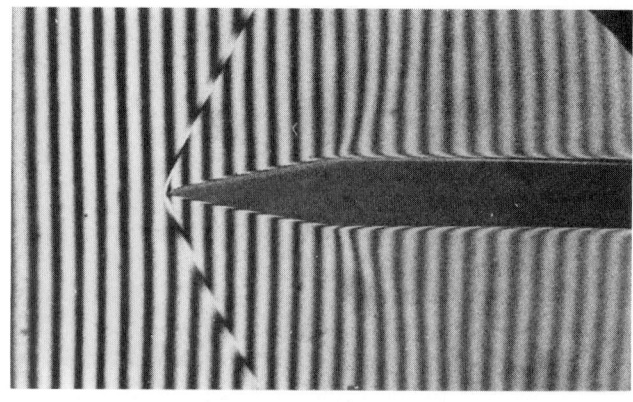

Fig. 3.19   Oblique shock wave (Ref. 30).

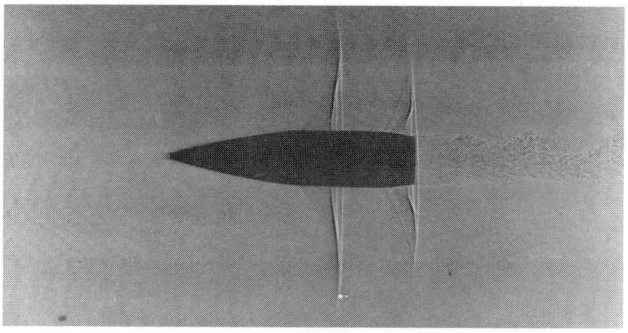

Fig. 3.20 Transonic flow about projectile (Photo by A. C. Charters, Ref. 14).

fan blade tip sees supersonic flow; these designs are termed *transonic fans* (more in Chapter 9).

3.5 Supersonic Compressions and Expansions

Supersonic flow adjusts to the presence of a body by means of compression shock waves (discussed previously) and/or expansion waves (presented in this section). As was pointed out in the previous section, a series of weak compression sound waves emanating from a body can coalesce to form a finite compression shock wave. The adjustment of fluid properties across a shock wave is abrupt, the thickness of the shock wave being of the order of the mean free path of the gas molecules. The shock process is a compression process—the pressure, temperature, and density of the gas will increase across the shock, and the velocity will decrease. A foundational understanding of the shock process and flow property analysis across shocks is essential to a study of compressible flow.

The processes taking place inside a shock wave itself are extremely complex, the study of which is beyond equilibrium thermodynamics and the purpose of this text. Suffice it to say that the *shock process is internally irreversible*; that is, there is a significant increase in entropy across a shock wave. The good news is if we focus our attention on property changes across the shock, the analysis is relatively straightforward with a result that should look familiar–property relations as functions of M_1 and γ. (M_1 is the Mach number just upstream of the shock wave.)

We will examine two types of shock waves: (1) the normal shock that is normal or perpendicular to the flow, and (2) the oblique shock inclined at an angle (not 90 deg) to the oncoming flow. Normal shocks can occur in duct or channel flows, on the surface of wings (similar to Fig. 3.20), or upstream of the leading edge of a body. Oblique shocks can also occur in channel flows, on the surface of sharp corners where the flow is turned towards itself like supersonic inlets, and downstream of a converging-diverging exhaust nozzle. You will encounter many normal and oblique shock applications in the examples and problems that follow.

The analysis of each of these types of shocks is identical, summarized here:

- Select a control volume that encompasses the shock wave and a small amount of fluid upstream and downstream of the wave such that:
  - Area change across the shock wave is negligible.
  - Frictional forces from any applicable surface are negligible (because the selected surface area is so small).
- Apply the governing equations of continuity, linear momentum, and the first and second laws of thermodynamics.
- Apply the perfect gas equation of state, speed of sound in a perfect gas, and definition of Mach number.

The following assumptions apply:

- The flow is steady and 1-D for a normal shock; 2-D for an oblique shock.
- The gas is calorically perfect (constant c_p and c_v).
- The shock process is adiabatic—there is significant heat transfer internal to the shock, but the shock is so thin and gas molecule residence times so small that heat transfer across the shock is negligible.

3.5.1 Normal Shock Waves

We apply the steady, 1-D flow governing equations developed in Chapter 2 to the control volume shown in Fig. 3.21 with the assumptions presented in Section 3.5. A summary of the equations derived from these follow. To gain practice, the reader is encouraged to start with the fully 3-D, unsteady version of the governing equations and apply them to the control volume, being sure to understand the important implications of each applicable assumption to get to the following simplified equations:

$$\text{Conservation of mass (continuity)} \qquad \rho_1 V_1 = \rho_2 V_2 \qquad (3.20)$$

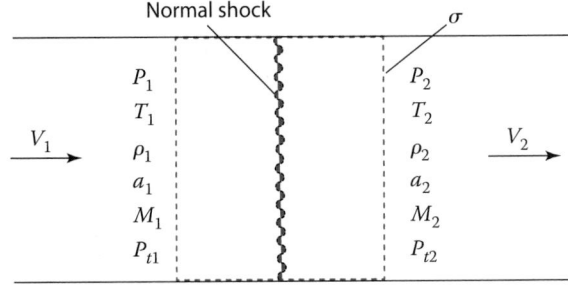

Fig. 3.21   Normal shock wave in σ with steady flow through σ.

Linear momentum
$$P_1 + \frac{\rho_1 V_1^2}{gc} = P_2 + \frac{\rho_2 V_2^2}{gc} \qquad (3.21)$$

Conservation of energy (first law)
$$T_{t1} = T_{t2} \qquad (3.22)$$

Production of entropy (second law) $s_2 - s_1 = -R \ln \dfrac{P_{t2}}{P_{t1}} > 0 \Rightarrow (P_{t2} < P_{t1})$

$$(3.23)$$

The reader would do well to ensure understanding of the results of the first and second laws of thermodynamics per Eqs. (3.22) and (3.23), especially as they relate to the total temperature and total pressure (see Sections 3.2.1 and 3.2.2).

The remaining simplified equations and their origins are shown here:

Equation of state
$$\frac{P_1}{\rho_1 T_1} = \frac{P_2}{\rho_2 T_2} \qquad (3.24)$$

Speed of sound
$$\frac{a_1^2}{T_1} = \frac{a_2^2}{T_2} \qquad (3.25)$$

Mach number
$$\frac{M_1^2 T_1}{V_1^2} = \frac{M_2^2 T_2}{V_2^2} \qquad (3.26)$$

Some algebra is required to solve Eqs. (3.20) through (3.26) and obtain the five functions listed in Eqs. (3.27) through (3.31); this can be found in many texts on compressible flow (see, for example, Ref. 15). The following five functions and the equations $a_2^2 = \gamma R g_c T_2$ and $V_2 = a_2 M_2$ determine the seven exit properties listed in Fig. 3.21 and tabulated in Appendix E for $\gamma = 1.4$.

$$M_2 = \left(\frac{M_1^2 + \dfrac{2}{\gamma - 1}}{\dfrac{2\gamma}{\gamma - 1} M_1^2 - 1} \right)^{1/2} \qquad (3.27)$$

$$\frac{P_2}{P_1} = \frac{2\gamma M_1^2}{\gamma + 1} - \frac{\gamma - 1}{\gamma + 1} \qquad (3.28)$$

$$\frac{\rho_2}{\rho_1} = \frac{(\gamma + 1)M_1^2}{(\gamma - 1)M_1^2 + 2} \qquad (3.29)$$

$$\frac{T_2}{T_1} = \frac{\left(\dfrac{2\gamma}{\gamma - 1} M_1^2 - 1 \right)\left(1 + \dfrac{\gamma - 1}{2} M_1^2 \right)}{\dfrac{(\gamma + 1)^2}{2(\gamma - 1)} M_1^2} \qquad (3.30)$$

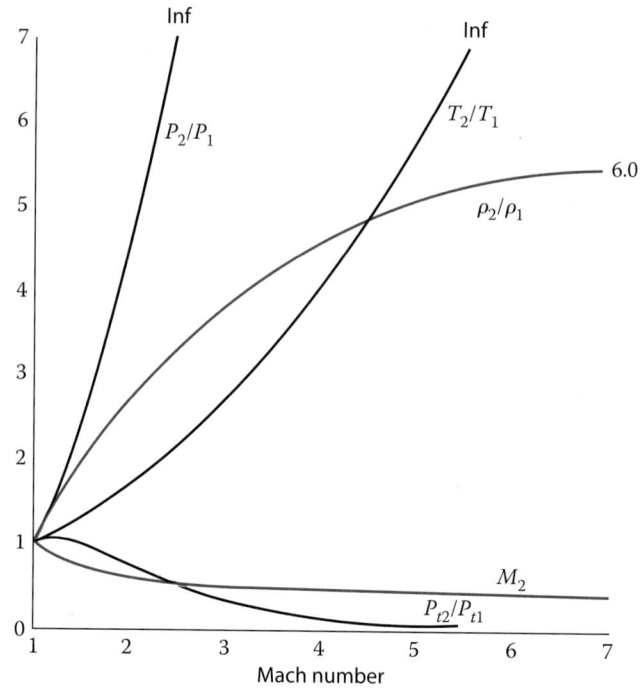

Fig. 3.22   Property variations across a normal shock ($\gamma = 1.4$).

$$\frac{P_{t2}}{P_{t1}} = \left(\frac{\dfrac{\gamma+1}{2}M_1^2}{1+\dfrac{\gamma-1}{2}M_1^2}\right)^{\gamma/(\gamma-1)} \left(\frac{2\gamma}{\gamma+1}M_1^2 - \frac{\gamma-1}{\gamma+1}\right)^{-1/(\gamma-1)} \tag{3.31}$$

Note that the five property relations formulated above are all functions of γ and the Mach number just upstream of the shock wave M_1. Figure 3.22 presents a plot of the five functions versus M ($\gamma = 1.4$).

The static property ratios of the figure may be interpreted as the ratios occurring across a normal shock from the point of view of an observer at rest relative to the wave or at rest relative to the gas into which the wave is propagating. From the latter's point of view, a shock wave advancing through sea-level air at Mach 2.1 from, say, a bomb explosion, produces a static pressure rise of 5:0. The air immediately behind such a shock wave would have a pressure of approximately 73.5 psia (ambient pressure = 14.7 psia, standard day). The curves of M_2 and P_{t2}/P_{t1} of the figure show that the higher the inlet Mach number to a normal shock, the lower the exit Mach number and the lower the total pressure ratio across the shock wave. Contrary to the static property ratio curves, which by definition are independent of an observer's reference velocity, these two curves depend on the observer's reference velocity, in this case, by an observer fixed to the wave.

Some important observations regarding property changes across a normal shock wave:

- The velocity and Mach number just downstream of a normal shock decrease, with the downstream Mach number always *subsonic*.
- The total pressure just downstream of a normal shock is always less than the upstream total pressure. The magnitude of the P_t decrease is indicative of the losses associated with the normal shock and is a manifestation of the second law of thermodynamics.
- The shock is so thin, property changes are assumed to occur discontinuously (think step changes) across the shock—this is a good assumption.

We will see in the next section that all of these observations hold true for the oblique shock wave, but quantitatively, the property changes are not as large; in fact, the velocity just downstream of an oblique shock wave indeed decreases, but typically stays supersonic.

Example 3.3

A steady stream of air passes through a normal shock that stands ahead of the engine inlet of a supersonic airplane flying at Mach 2 at 12 km (see Fig. 3.23). Find the properties of the air at the inlet and exit of the normal shock wave.

Solution
Standard atmosphere tables (Appendix A) give for 12 km

$$\delta = 0.1915 \quad \theta = 0.7519 \quad \frac{a}{a_{\text{ref}}} = 0.8671$$

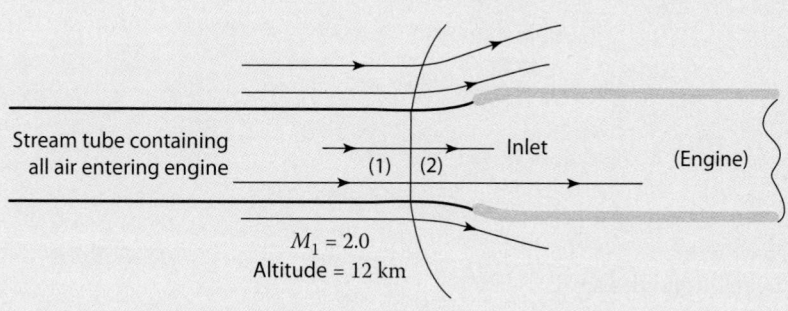

Fig. 3.23 Normal shock in front of engine inlet.

(Continued)

Example 3.3 *(Continued)*

Thus

$$P_1 = \delta P_{\text{ref}} = 0.1915 \times 101{,}303 = 19{,}400\,\text{Pa}$$

$$T_1 = \theta T_{\text{ref}} = 0.7519 \times 288.2 = 216.7\,\text{K}$$

$$a_1 = \frac{a}{a_{\text{ref}}} a_{\text{ref}} = 0.8671 \times 340.3 = 295.1\,\text{m/s}$$

Plots (Fig. 3.10), tabulations (Appendix D), or equations [Eqs. (3.7) and (3.8)] of P/P_t and T/T_t vs Mach number give, for $M_1 = 2.0$ and $\gamma = 1.4$,

$$\left(\frac{P}{P_t}\right)_1 = 0.1278 \qquad \left(\frac{T}{T_t}\right)_1 = 0.5556$$

From these data, we obtain

$$P_{t1} = \frac{P_1}{(P/P_t)_1} = \frac{19{,}400}{0.1278} = 151{,}800\,\text{Pa}$$

$$T_{t1} = \frac{T_1}{(T/T_t)_1} = \frac{216.7}{0.5556} = 390.0\,\text{K}$$

$$V_1 = M_1 a_1 = 2(295.1) = 590.2\,\text{m/s}$$

We have now determined, at the shock inlet,

$$M_1 = 2.0 \qquad V_1 = 590.2\,\text{m/s} \quad T_1 = 216.7\,\text{K}$$
$$T_{t1} = 390.0\,\text{K} \quad P_1 = 19{,}400\,\text{Pa} \quad P_{t1} = 151{,}800\,\text{Pa}$$

From tabulations in Appendix E (for higher accuracy) of Fig. 3.22, at $M_1 = 2.0$, we find the normal shock property ratios (note that our station 1 is table station x and our station 2 is table station y)

$$\frac{P_2}{P_1} = 4.500 \quad \frac{P_{t2}}{P_{t1}} = 0.7209$$

$$\frac{T_2}{T_1} = 1.6875 \quad \frac{V_1}{V_2} = \frac{\rho_2}{\rho_1} = 2.667 \quad (\text{from } \rho V = \text{const})$$

and the exit Mach number

$$M_2 = 0.5774$$

These numbers may be checked, grossly, by Fig. 3.22. The normal shock exit stream properties are

$$P_2 = P_1 \frac{P_2}{P_1} = 19{,}400 \times 4.500 = 87{,}300\,\text{Pa}$$

(Continued)

Example 3.3 *(Continued)*

$$P_{t2} = P_{t1}\frac{P_{t2}}{P_{t1}} = 151{,}800 \times 0.7209 = 109{,}430\,\text{Pa}$$

$$T_3 = T_1\frac{T_2}{T_1} = 216.7 \times 1.6875 = 365.7\,\text{K}$$

$$V_2 = \frac{V_1}{V_1/V_2} = \frac{590.2}{2.667} = 221.3\,\text{m/s}$$

Notice that there is a 28% decrease in total pressure through the normal shock. This is a large decrease. Supersonic inlets are designed to keep this total pressure loss to a minimum.

3.5.2 Oblique Shock Waves

When a wedge-shaped object is placed in a 2-D supersonic flow, a plane-attached shock wave may emanate from the nose of the body, or a detached shock wave may arise. The latter is curved and stands in front of the object (Fig. 3.24) and could very well be treated as a normal shock in the vicinity of the nose. The flow Mach number and the wedge angle θ together determine which of these two types of shocks will occur.

In Fig. 3.25, a flow is deflected through an angle θ from a 2-D wedge (infinite span relative to the flow) as it passes through a straight oblique shock wave that makes an angle β with the upstream flow velocity. A proper

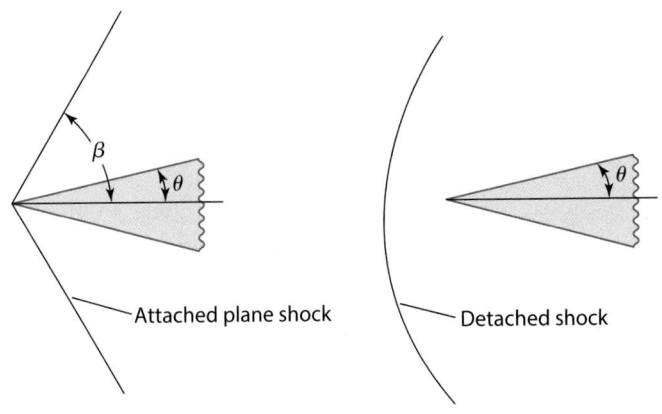

Fig. 3.24 Attached and detached shocks in supersonic flow.

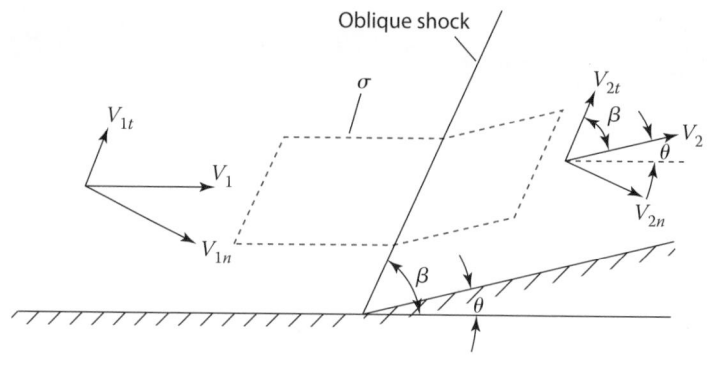

Fig. 3.25 Control volume for oblique shock analysis.

control volume, indicated by the dashed lines, will have its upper and lower sides coincident with the flow streamlines and its ends parallel to the shock front.

The analysis of the property changes that occur across the 2-D shock parallels that of the normal shock in the previous section. Per the control volume shown in Fig. 3.25, we must now consider the components of velocity normal and tangent to the shock wave as dictated by the geometry of the analysis. Application of the governing equations and assumptions previously presented reveal that the oblique shock acts as a normal shock to the component of velocity normal to the wave, while the velocity component tangent to the wave remains unchanged. In other words, it is only the normal velocity component, V_{1n} in Fig. 3.25, that brings about the flow property changes across an oblique shock. Consequently, Eqs. (3.20–3.26) still apply for oblique shock analysis where the normal velocity or normal Mach number components are used in Eqs. (3.20), (3.21), and (3.26). For example,

$$\text{Conservation of mass (continuity)} \quad \rho_1 V_{1n} = \rho_2 V_{2n}$$

Note that although Eq. (3.20) would apply from the application of linear momentum to the normal component of velocity, we would have an additional equation resulting from application of linear momentum in the tangential direction of the shock wave. This would lead to the conclusion that $V_{2t} = V_{1t}$ and is left as a homework exercise. Also, from the geometry of Fig. 3.25

$$\tan(\beta - \theta) = \frac{V_{2n}}{V_{2t}} \tag{3.32}$$

Solution of the governing equations leads to relationships of the functional form $f(M_1 \sin \beta)$, the normal component of M, for a constant γ

for the desired property relations P_2/P_1, T_2/T_1, ρ_2/ρ_1, P_{t2}/P_{t1}, and $M2_n = M_2 \sin(\beta - \theta)$. Additionally,

$$\tan(\beta - \theta) = \frac{2 + (\gamma - 1)M_1^2 \sin^2 \beta}{(\gamma + 1)M_1^2 \sin \beta \cos \beta} \tag{3.33}$$

Here, the functions for the four property ratios and $M2_n$ are identical with those for a normal shock wave where $\beta = 90$ deg and $\theta = 0$ deg. These six property relations are provided in Appendix F.

Because we often know the upstream Mach number and amount of required flow deflection or turning, wedge angle θ, it is convenient and instructive to present Eq. (3.33) graphically as shown in Fig. 3.26. Equation (3.33) is also given in tabular form in Appendix F for $\gamma = 1.4$. When a solution does not exist for a particular combination of Mach number M_1 and wedge angle θ, the maximum value of the wedge angle θ_{max} and the corresponding shock wave angle β are given.

Observe from the graph of Fig. 3.26 that three possible solutions exist for a given wedge or deflection angle θ:

1. Two values of β for a given M_1. For example, $\theta = 20$ deg, $M_1 = 2.0$ gives $\beta = 53.5$ deg or $\beta = 74$ deg. Either value of β may occur depending on the pressure boundary conditions of the flow. Usually, the *weak shock wave* with the smaller shock angle occurs; however, if a physical situation can support a higher pressure differential, then the *strong shock wave* with the larger shock angle can be supported. In external flow problems (flow around objects), we almost always find the weak shock solution. Some internal flow problems (flow bounded by physical boundaries) may be able to support the strong shock solution.
2. One value of β for a given M_1. For example, $\theta = 23$ deg, $M_1 = 2.0$ gives $\beta = 65$ deg. Thus, $\theta = \theta_{max}$ for a straight oblique shock solution with $M_1 = 2.0$.
3. No value of β for a given M_1. For example, in case 2 for $\theta = 23$ deg, any $M_1 > 2.0$ gives no solution for a straight oblique shock wave angle. When this condition exists, a curved detached shock wave results.

Figures 3.27 through 3.30 give the oblique shock stream property ratios and exit Mach numbers for various values of the stream deflection wedge angle θ for $\gamma = 1.4$. The family of curves on each graph is bounded by a normal shock curve and a Mach line curve. Thus these graphs include the curves for the normal shock relations of Fig. 3.22 [excluding T_2/T_1, which may be found from the perfect gas equation of state $T_2/T_1 = (P_2/P_1)/(\rho_2/\rho_1)$]. The normal shock conditions correspond to the limiting strong oblique shock solutions of Figs. 3.27 through 3.30 as the stream deflection angle goes to zero. The Mach line, or Mach wave, conditions are the limiting weak oblique shock solutions of Figs. 3.27 through 3.30 as the stream deflection angle goes to zero. Notice that the stream properties do not change through a Mach line; this is in keeping with the concept of a Mach line being formed from sound waves emanating from an infinitesimally small point disturbance that produce infinitesimally small changes in stream properties (see Section 3.4).

There are numerous references that present the shock relations for the oblique and normal shocks in either graphical or tabular form (see Refs. 7, 31, 32, and 33).

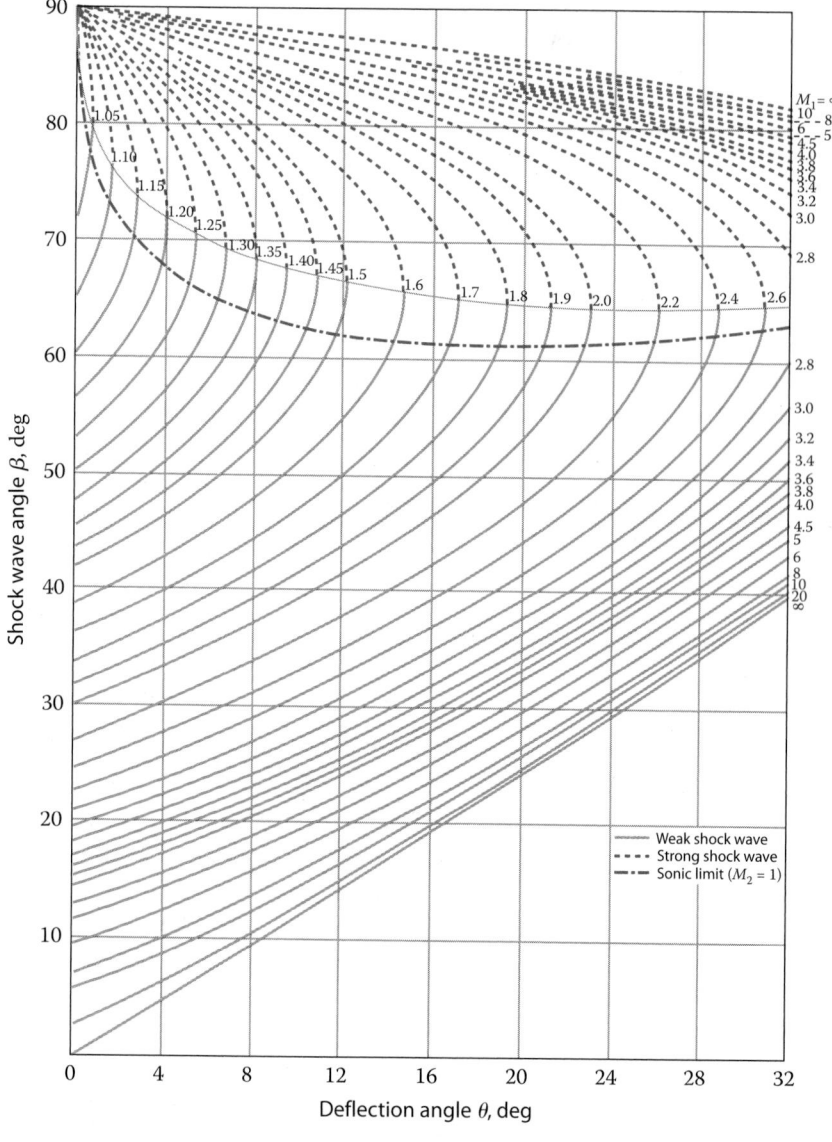

Fig. 3.26   Relationship among M_1, β, and θ for oblique shock ($\gamma = 1.4$).

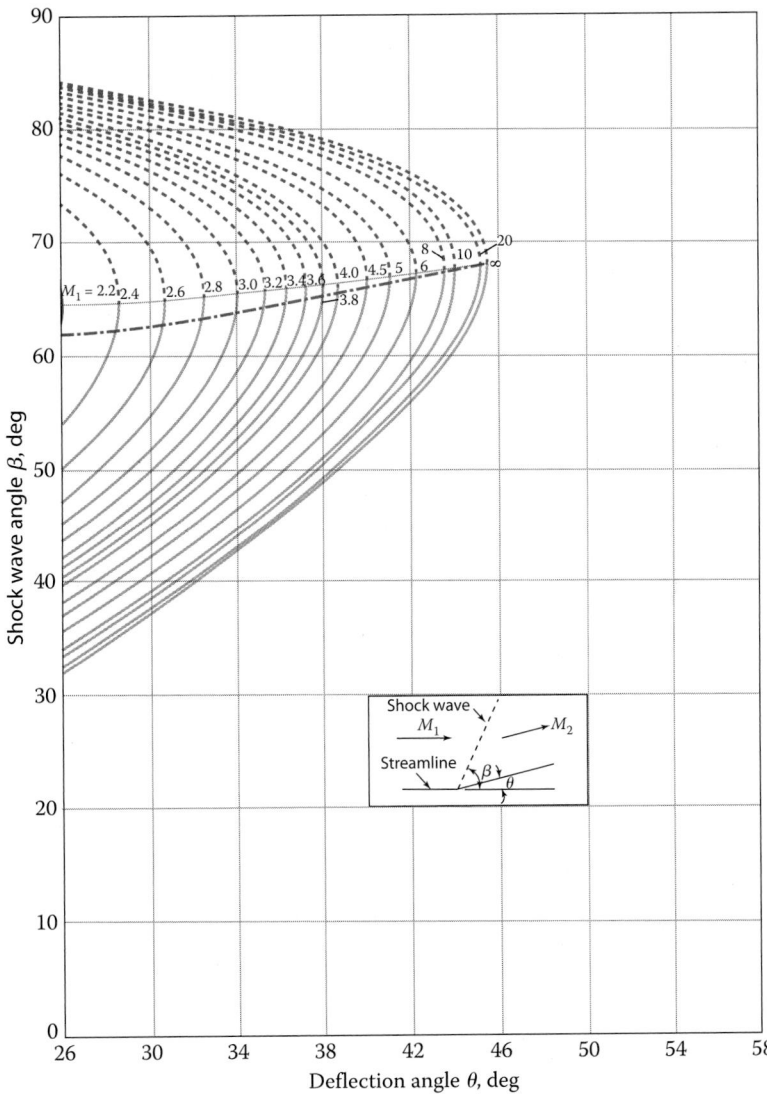

Fig. 3.26 Relationship among M_1, β, and θ for oblique shock ($\gamma = 1.4$) (*Continued*).

Some important observations regarding property changes across an oblique shock wave:

- Generally, the weak shock wave solution occurs because, physically, the large pressure differences that would accompany a strong oblique shock wave (larger shock angle) cannot be sustained when an inlet wedge or an airfoil is subjected to supersonic flow. *Use the weak shock wave solution unless otherwise specified.*

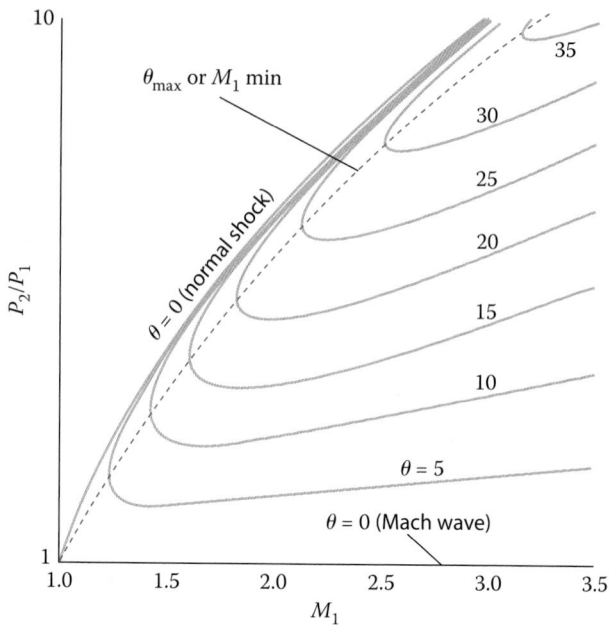

Fig. 3.27 Pressure ratio P_2/P_1 for oblique shock ($\gamma = 1.4$).

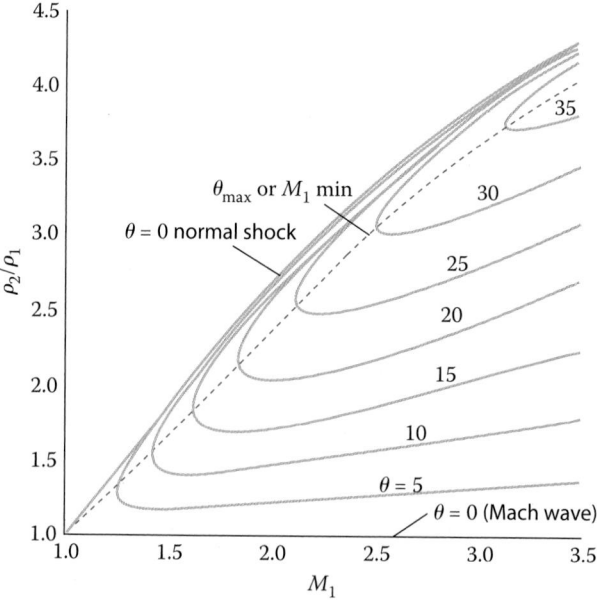

Fig. 3.28 Density ratio ρ_2/ρ_1 for oblique shock ($\gamma = 1.4$).

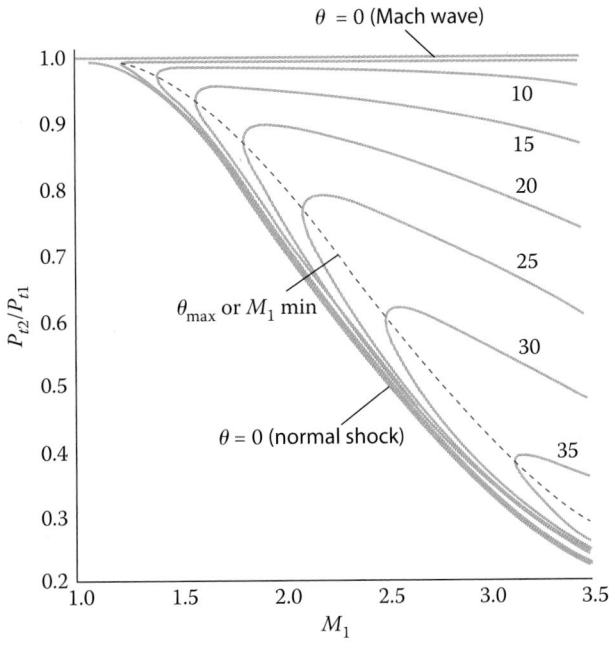

Fig. 3.29   Total pressure ratio P_{t2}/P_{t1} for oblique shock ($\gamma = 1.4$).

- The general characteristics of property changes highlighted at the end of the normal shock Section 3.5.1 hold true for oblique shocks; however, the magnitudes of the changes are reduced. In fact, as seen in Fig. 3.30 for the weak shock solution, the Mach number just downstream of the shock M_2 typically will still be supersonic, reduced from M_1, but still >1.0. Physically, this should make good sense because it is only the normal component of velocity that is reduced across an oblique shock; the tangential velocity component stays unchanged.
- There is a maximum deflection angle that can be tolerated for a given M_1 (or vice versa) for a 2-D oblique shock solution. If this deflection angle is exceeded, a detached curved shock is formed near normal around the leading edge with the shock strength diminishing to that of a Mach wave far from the body (see Fig. 3.24).

Finally, because we have looked at 1-D flow effects through a normal shock and 2-D flow effects around an infinite span wedge, we will include some brief comments regarding 3-D supersonic flow around a right circular cone. Although analysis of conical shock waves is beyond the scope of this text, this more complex flow bears many similarities to wedge flow. For small cone angles, a conical shock will attach to the apex of the cone that can detach if the upstream Mach number M_1 gets too large (or the

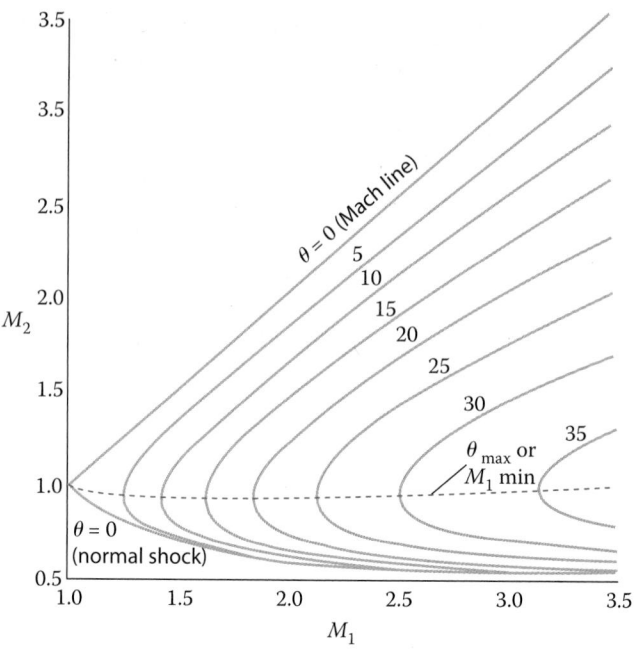

Fig. 3.30 Downstream Mach number M_2 for oblique shock ($\gamma = 1.4$).

cone angle gets too large for a given M_1). For a given deflection angle θ, the conical shock wave angle will be smaller than the oblique shock wave angle from a 2-D wedge; the wedge presents a greater flow disturbance than the cone, as indicated by a stronger wedge shock. Also, like the oblique shock solution, a strong and weak shock solution exist for the conical shock, with the weak shock solution (smaller shock wave angle) being the predominant one observed. For an initial estimate of property changes across a conical shock, the 2-D oblique shock analysis will provide conservative results that will generally be acceptable, especially for preliminary analysis. Conical shocks are included in the GASTAB program.

Example 3.4

A steady supersonic stream of air at an altitude of 12 km and Mach number of 2 (the same conditions as Example 3.3) approaches a wedge with an included angle of 40 deg (Figs. 3.31 and 3.32). Find the stream properties downstream of the weak oblique shock wave attached to the wedge and the angle that the shock makes with the original direction of flow.

(Continued)

Example 3.4 *(Continued)*

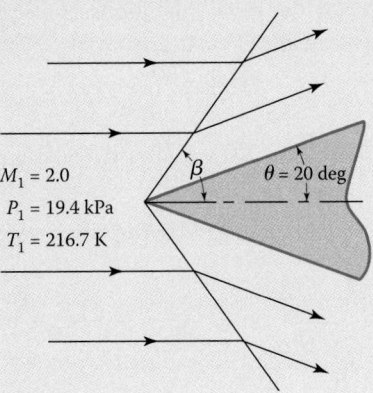

Fig. 3.31 Wedge in supersonic flow.

Solution

From Appendix A, at 12 km we have $P_1 = \delta P_{\text{ref}} = 19{,}400$ Pa and $T_1 = \theta T_{\text{ref}} = 216.7$ K. Figure 3.26 indicates that an attached shock will occur for $\theta = 20$ deg and $M_1 = 2.0$. Thus we may proceed with solution. Of the two

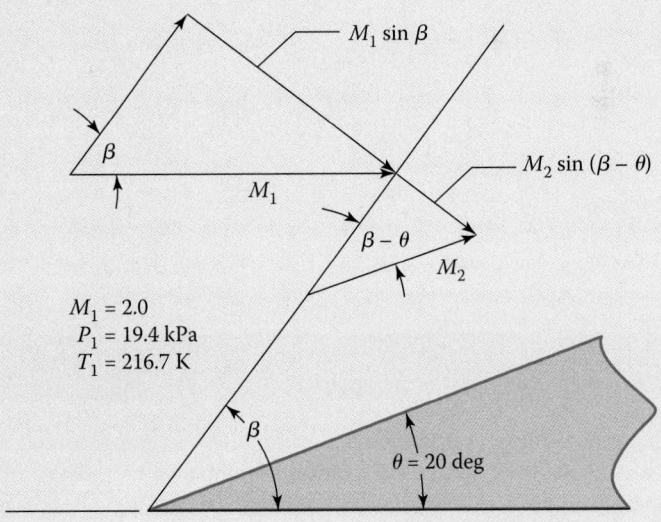

Fig. 3.32 A 20-deg half-angle wedge in supersonic flow.

(Continued)

Example 3.4 *(Continued)*

shock angles possible for the given conditions, the smaller angle generally occurs. Accordingly, we read $\beta = 53$ deg from Fig. 3.26. The following data are obtained from Figs. 3.27 through 3.30:

$$\frac{P_2}{P_1} = 2.8 \quad \frac{\rho_2}{\rho_1} = 2.03 \quad \frac{P_{t2}}{P_{t1}} = 0.89 \quad M_2 = 1.22$$

The approaching total temperature and total pressure were found to be, from Example 3.3,

$$P_{t1} = \frac{P_t}{(P/P_t)_1} = \frac{19,400}{0.1278} = 151,800 \, \text{Pa}$$

$$T_{t1} = \frac{T_t}{(T/T_t)_1} = \frac{216.7}{0.5556} = 390.0 \, \text{K}$$

From these data, we may determine the downstream properties

$$P_2 = P_1 \frac{P_2}{P_1} = 19,400(2.8) = 54,320 \, \text{Pa}$$

$$P_{t2} = P_{t1} \frac{P_{t2}}{P_{t1}} = 151,800(0.89) = 135,100 \, \text{Pa}$$

$$T_2 = T_1 \frac{P_2}{P_1} \frac{\rho_1}{\rho_2} = 216.7 \left(\frac{2.8}{2.03}\right) = 298.9 \, \text{K}$$

To find T_2 alternatively, use $(T/T_t)_2$ for $M_2 = 1.22$ in the relation

$$T_2 = T_{t2} \left(\frac{T}{T_t}\right)_2 = T_{t1} \left(\frac{T}{T_t}\right)_2 = 390(0.7706) = 300.5 \, \text{K}$$

This is a more accurate value of T_2 because T/T_t was read from tables and not from a graph.

Example 3.5

The inlet of the engine of Example 3.3 (normal shocks) incorporates a spike having a cone angle of 40 deg. As a result, the inlet air at 12 km and Mach 2.0 passes through an oblique shock attached to the spike and a normal shock at the inlet's cowl lip (Fig. 3.33). Determine the total pressure and Mach number at the exit of the normal shock. Compare the losses through this inlet with those of Example 3.3.

(Continued)

Example 3.5 *(Continued)*

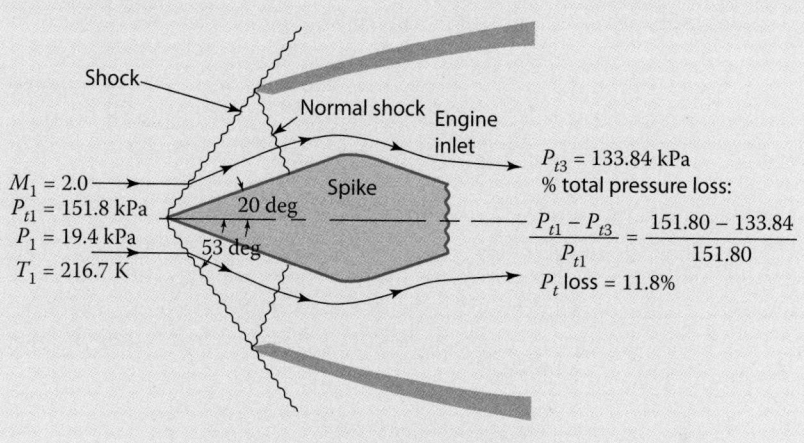

Fig. 3.33   Example 3.5 external compression inlet.

Solution

Although the shock of the spike is a conical shock wave, we may make a sufficiently accurate approximate analysis by using 2-D oblique shock theory. We found the stream properties in zone 2 of Fig. 3.31 (Example 3.4) to be

$$M_2 = 1.22 \quad P_{t2} = 135,100 \, \text{Pa}$$

From Appendix E, the total pressure ratio across a normal shock with an inlet Mach number of 1.22 is $P_{t3}/P_{t2} = 0.9907$, and the Mach number in zone 3 is 0.83. The total pressure in zone 3 is, therefore,

$$P_{t3} = P_{t2} \frac{P_{t3}}{P_{t2}} = 135,100(0.9907) = 133,840 \, \text{Pa}$$

Assuming no total pressure losses except through shock waves, the spike inlet of this example has a 12% total pressure loss. The inlet of Example 3.3 has a 28% total pressure loss. If we define efficiency as the total pressure at the inlet shock pattern exit divided by the initial total pressure of the free stream, the spike inlet is then more efficient. The result of this comparison (inlet of Example 3.3 vs Example 3.5) is not an isolated case. It is generally true that, for a given supersonic flow, the total pressure loss through a series of shocks consisting of oblique shocks terminating in a normal shock is less than the total pressure loss through a single normal shock.

Example 3.6

Normal shock analysis shows that the exit Mach number and the stream property ratios across a normal shock are expressible in terms of the inlet Mach number. Figure 3.22 graphically represents the functional relationships between normal shock properties and the inlet Mach number. Oblique shock analysis indicates that the normal shock functional relationships, and thus Fig. 3.22, are also applicable to the oblique shock wave if, in using the normal shock relations for an oblique shock, the inlet and exit Mach numbers (M_1 and M_2 of Fig. 3.21) are replaced with the normal components of the corresponding Mach numbers for the oblique shock. This example will illustrate how, by using this procedure, one may use tabulations of normal shock relations in oblique shock analysis.

For the oblique shock flow shown in Fig. 3.32, determine the following quantities from normal tabulations of them:

$$\frac{P_2}{P_1} \quad \frac{\rho_2}{\rho_1} \quad \frac{T_2}{T_1} \quad \frac{P_{t2}}{P_{t1}} \quad M_2$$

Compare the results with those of Example 3.4.

Solution

Use Fig. 3.26 to determine the shock angle β. Taking the value of β corresponding to the shock that normally occurs, we get $\beta = 53$ deg. [Note: A more accurate value of β can be obtained by using Eq. (3.33) or Appendix F.] Then the normal component of inlet Mach number $= M_1 \sin \beta = 2 \sin 53$ deg $= 1.6$.

From normal shock tables at a Mach number of 1.6,

$$\frac{P_2}{P_1} = 2.820 \quad \frac{\rho_2}{\rho_1} = 2.032 \quad \frac{T_2}{T_1} = 1.388$$

and the normal component of exit Mach number $= M_2 \sin (\beta - \theta) = 0.6684$, so

$$M_2 = \frac{0.6684}{\sin (53 \text{ deg} - 20 \text{ deg})} = 1.227$$

The total pressure ratio P_{t2}/P_{t1} can be obtained as follows:

$$\frac{P_{t2}}{P_{t1}} = \frac{P_2 \, (P/P_t)_{M_1}}{P_1 \, (P/P_t)_{M_2}} = \frac{P_2 \, (P/P_t)_{M=2.0}}{P_1 \, (P/P_t)_{M=1.227}} = 2.820 \left(\frac{0.1278}{0.3980}\right) = 0.9055$$

These results, based on normal shock tables, agree with those of Example 3.4 that were derived from oblique shock graphs.

3.5.3 Expansion Waves

Normal and oblique shocks represent supersonic compression processes. Unlike the shock process, supersonic expansions occur *isentropically*, that is, with no losses. A detailed analysis of 2-D supersonic expansions, often called *Prandtl–Meyer flow*, is beyond the scope of this text; the interested reader is referred to John & Keith (Ref. 15) and Zucker and Biblarz (Ref. 65). In this section, we will present some highlights in regards to the development of Prandtl–Meyer relations (which once again take the form of functions of γ and Mach number) and focus on the analysis process.

As we have seen, when a supersonic flow has to undergo a finite, sudden change of direction at a concave corner, a compression occurs through an oblique shock (Fig. 3.34a). The flow is turned the amount of the deflection (or wedge) angle, but is also accompanied by losses manifested by a decrease in total pressure across the shock. If the flow is allowed to change direction more gradually, as in Fig. 3.34b, a series of weaker oblique shocks accomplish the turning with less loss; in the limit, these weaker and weaker shocks approach Mach waves (Fig. 3.35), which are defined in Section 3.4 as

$$\sin \mu = \frac{a(t)}{V(t)} = \frac{1}{M}$$

Recall from Section 3.5.2 that the limiting case of the weak oblique shock solution is indeed the Mach wave indicated in Figs. 3.27 through 3.30. There are no losses associated with Mach waves; the total pressure remains constant.

So, if we use a smoothly varying concave corner instead of a sharp one, with the resultant oblique shocks approaching Mach waves, we achieve a continuous compression with vanishingly smaller entropy rise in the vicinity of the surface (Fig. 3.36). Away from the wall, as shown in Fig. 3.36, the very weak compression waves converge into a finite oblique shock wave. Although there is no getting away from the formation of an oblique shock wave, the overall loss associated with a smooth turn as in Fig. 3.36 is less than that associated with a sharp turn (Fig. 3.34a); consequently, smooth contouring is used in practice. Figure 3.37 shows a side view of the F-16 inlet configured with a fixed 6-deg ramp followed by a fixed 6.67-deg "isentropic ramp."

Now consider a supersonic flow subjected to an expansion around a series of infinitesimally small *convex* corners. As shown in Fig. 3.38, a series of Mach

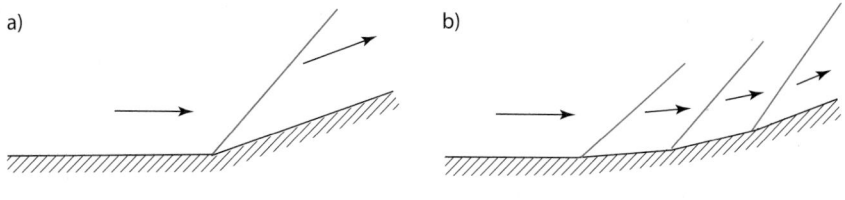

a)   b)

Fig. 3.34   Weak oblique shocks.

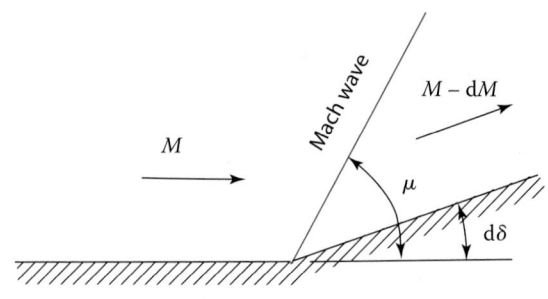

Fig. 3.35   Mach wave.

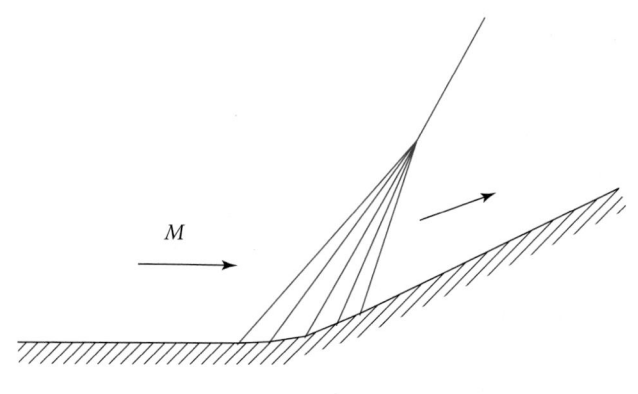

Fig. 3.36   Smooth turn of supersonic flow.

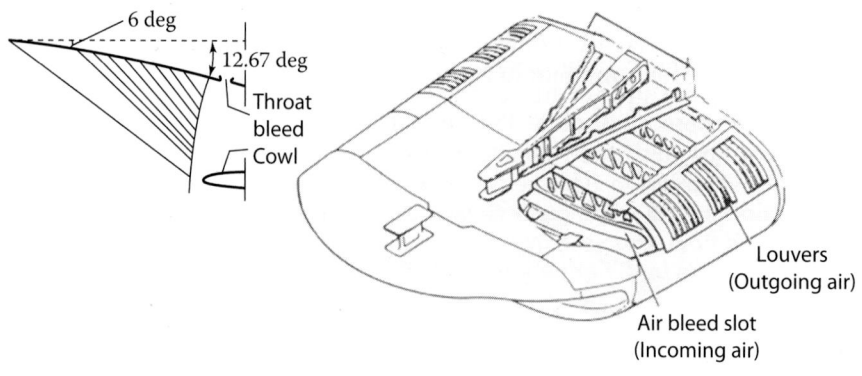

Fig. 3.37   F-16/J79 inlet (Ref. 73).

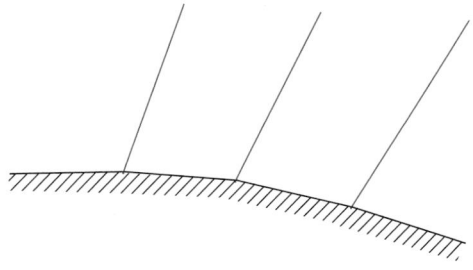

Fig. 3.38   Series of Mach waves.

waves would be generated at each corner with each wave inclined at Mach angle μ to the flow. Unlike the Mach waves emanating from a concave corner, these Mach waves cannot coalesce and reinforce each other; rather, they diverge or fan out. Because flow between each of the waves in Fig. 3.38 is uniform, the length of the wall segments does not matter, and the segments can be taken to a sharp corner without affecting the overall variation of flow properties across the expansion. The resultant series of Mach waves centered at the corner is known as a *Prandtl–Meyer expansion fan* (Fig. 3.39).

To determine the variation in flow properties for a given flow turning angle and upstream Mach number M_1, which are typically known, we determine the 2-D Prandtl–Meyer flow equations in a manner analogous to those of the 2-D oblique shock; velocity components normal and tangent to the initial and final Mach waves must be considered. As indicated in Fig. 3.39, the initial wave is inclined to the upstream flow and the final wave is inclined to the downstream flow at angles

$$\mu_1 = \sin^{-1}\frac{1}{M_1} \quad \text{and} \quad \mu_2 = \sin^{-1}\frac{1}{M_2}$$

respectively. The analysis reveals that the determination of change in Mach number associated with a given flow turning angle necessitates the

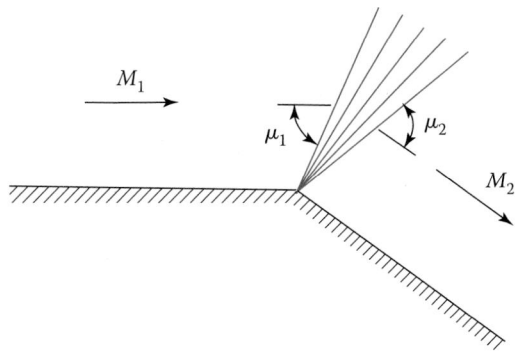

Fig. 3.39   Prandtl–Meyer expansion fan.

definition of a reference state for the purpose of tabulating results. For convenience, the reference state is taken as $v = 0$ at $M = 1.0$, and the desired result is obtained as

$$v = \left(\sqrt{\frac{\gamma + 1}{\gamma - 1}} \tan^{-1} \sqrt{\frac{\gamma - 1}{\gamma + 1}(M^2 - 1)} - \tan^{-1}\sqrt{M^2 - 1} \right) \quad (3.34)$$

The symbol v represents the angle by which a flow initially at Mach 1.0 must be expanded to reach $M > 1.0$. Appendix D presents values of v for Mach numbers from 1.0 to 4.0 for $\gamma = 1.4$, 1.33, and 1.3 Mach wave angles μ are also included. For the likely case that M_1 is not equal to 1.0 for the case under analysis, just subtract the value of v_1 at M_1 from the value of v_2 at M_2, where v_1 and v_2 are found from Appendix D or Eq. (3.34). Figure 3.40 and the example that follows should help facilitate your understanding of the solution process. Prandtl–Meyer flow is also provided in GASTAB.

Because Prandtl–Meyer expansion flow is adiabatic with no change in total temperature, we can find the static temperature change using the compressible flow functions of Section 3.2.3 or tabulated in Appendix D for different values of specific heat ratios γ

$$\frac{T_2}{T_1} = \frac{T_2/T_t}{T_1/T_t} = \frac{1 + \frac{\gamma - 1}{2}M_1^2}{1 + \frac{\gamma - 1}{2}M_2^2}$$

Additionally, because the flow is isentropic with constant total pressure across the expansion fan, we can find the static pressure change using the compressible flow functions

$$\frac{P_2}{P_1} = \frac{P_2/P_t}{P_1/P_t} = \left[\frac{1 + \frac{\gamma - 1}{2}M_1^2}{1 + \frac{\gamma - 1}{2}M_2^2} \right]^{\frac{\gamma}{\gamma - 1}}$$

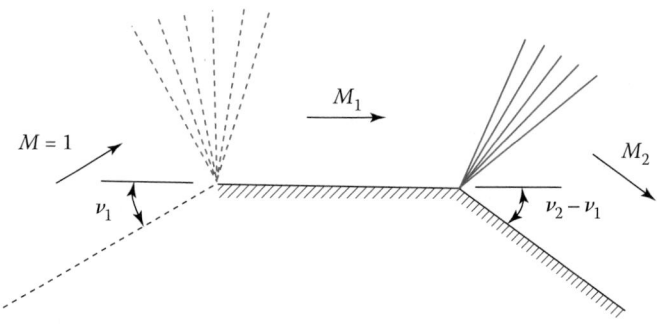

Fig. 3.40 Prandtl–Meyer turning angles.

Some important observations regarding property changes across a Prandtl–Meyer expansion wave:

- Unlike a subsonic flow subjected to a sharp convex corner, a supersonic flow has no problem negotiating such a corner and does so through a Prandtl–Meyer expansion wave process.
- Unlike shock waves, which have irreversible losses associated with them manifested by a decrease in streamwise total pressure, the Prandtl–Meyer expansion process is isentropic; P_t remains constant.
- The velocity and Mach number across a Prandtl–Meyer expansion fan increase; the static pressure, temperature, and density all decrease.

Example 3.7

A uniform supersonic flow with the conditions shown in Fig. 3.41 accelerates around a 15-deg convex corner. Find the flow conditions downstream of the resulting expansion fan to include M_2, P_2, T_2, and the fan angle (shown).

Solution

We assume Prandtl–Meyer (P-M) flow and use the P-M function to find the flow angles v_1 and v_2 and compressible flow functions to find P_2 and T_2. At the static temperature given, $\gamma = 1.4$ is quite appropriate, and we use Appendix D to facilitate the solution.

For $M_1 = 2.0$, from Appendix D:

$$v_1 = 26.38 \text{ deg}$$
$$\mu_1 = 30.0 \text{ deg}$$

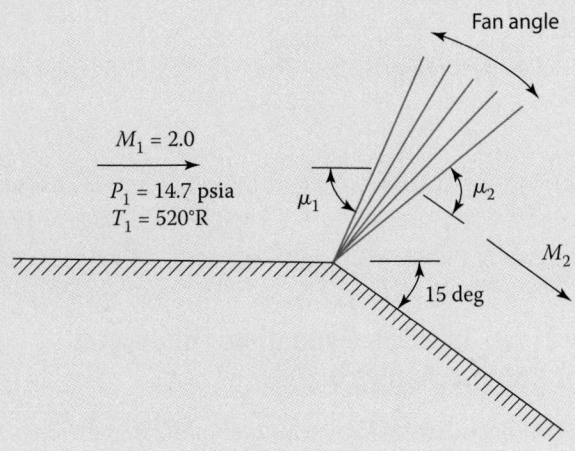

Fig. 3.41 Supersonic expansion.

(Continued)

Example 3.7 *(Continued)*

from Appendix D:

$$T_1/T_{t1} = 0.5556$$
$$P_1/P_{t1} = 0.1278$$

To find M_2 we use the P-M function, recognizing that the angle v is referenced to flow that is initially at $M = 1.0$; consequently,

$$v_2 = v_1 + 15 \deg = (26.38 + 15) \deg = 41.38 \deg$$
$$\text{(total flow turning required of Mach 1.0 flow)}$$

For $v_2 = 41.38$ deg, from Appendix D:

$$M_2 = 2.60$$
$$\mu_2 = 22.62 \deg$$

and for $M_2 = 2.60$, from Appendix D:

$$T_2/T_{t2} = 0.4252$$
$$P_2/P_{t2} = 0.05012$$

So, because the flow is adiabatic (constant T_t) and isentropic (constant P_t):

$$\frac{P_2}{P_1} = \frac{P_2/P_t}{P_1/P_t} = \frac{0.05012}{0.1278} \Rightarrow P_2 = 0.3922 \, P_1 = 0.3922 \, (14.7 \text{ psia}) = 5.76 \text{ psia}$$

$$\frac{T_2}{T_1} = \frac{T_2/T_t}{T_1/T_t} = \frac{0.4252}{0.5556} \Rightarrow T_2 = 0.7653 \, T_1 = 0.7653 \, (520 \text{ R}) = 398.0 \text{ R}$$

Finally, the fan angle is found by knowledge of the Mach angles and required flow turning. A good sketch helps, like Fig. 3.41.

$$\text{Fan angle} = (\mu_1 + 15.0 \deg) - \mu_2 = (30 + 15) \deg - 22.62 \deg$$
$$= 45 \deg - 22.62 \deg = 22.38 \deg$$

3.6   Steady Flow, 1-D Gas Dynamics: Differential Control Volume Analysis

It is useful to formulate the governing equations applied to a differential control volume for a steady flow, 1-D, calorically perfect gas for a general case considering area change, heat interaction, and friction. We considered a differential control volume previously in Section 3.3 for the case of area change of adiabatic, isentropic flow; here, we consider a flow with all three effects. Such a formulation provides general insights into the variation of streamwise properties with each of the three independent parameters, as well as a convenient form of the governing equations that affords property relationships as functions of γ and M between any location in the flow and a convenient reference location, namely, the location where the flow would be choked.

The steady 1-D flow of a chemically inert perfect gas with constant specific heats is conveniently described and governed by the following definitions and physical laws.

3.6.1 Definitions

Perfect gas:

$$P = \rho R T \tag{i}$$

Mach number:

$$M^2 = V^2/(\gamma R g_c T) \tag{ii}$$

Total temperature:

$$T_t = T\left(1 + \frac{\gamma - 1}{2}M^2\right) \tag{iii}$$

Total pressure:

$$P_t = P\left(1 + \frac{\gamma - 1}{2}M^2\right)^{\gamma/(\gamma-1)} \tag{iv}$$

3.6.2 Physical Laws

For 1-D flow through a control volume having a single inlet and exit sections 1 and 2, respectively, we have

One-dimensional mass flow:

$$\rho_1 A_1 V_1 = \rho_2 A_2 V_2 \tag{v}$$

Momentum:

$$F_{\text{frict}} = -\Delta\left(PA + \rho A V^2/g_c\right) \tag{vi}$$

Energy equation (no shaft work):

$$q = c_p(T_{t2} - T_{t1}) \tag{vii}$$

Entropy equation (adiabatic flow):

$$s_2 \geq s_1 \tag{viii}$$

where F_{frict} is the frictional force of a solid control surface boundary on the flowing gas, and A is the flow cross-sectional area normal to the velocity V. The frictional force acting on a length L is traditionally formulated using a frictional coefficient as

$$F_{frict} = c_f \{\rho V^2/(2g_c)\} \text{ wetted perimeter} \times L$$

or

$$F_{frict} = c_f \left(\frac{\rho V^2}{2g_c}\right)\left(\frac{4A}{D}\right)L \qquad (3.35)$$

where

c_f = frictional coefficient

D = hydraulic diameter = $\dfrac{4 \times A}{\text{wetted perimeter}}$

A = flow cross-sectional area

For a circular duct, the hydraulic diameter is the physical diameter of the duct; for a square duct, the hydraulic diameter is the length of one of the sides of the duct. Frictional force and coefficient is developed further in Section 3.7.3.3, or consult [15] or [60].

Now consider the differential element of a duct with length dx, as shown in Fig. 3.42. The independent variables are the area change, total temperature change, and frictional force. The dependent variables are P, T, ρ, V, M^2, and P_t. The application of Eqs. (i–vi) to flow in Fig. 3.42, having the presence of the simultaneous effects of area change, heat

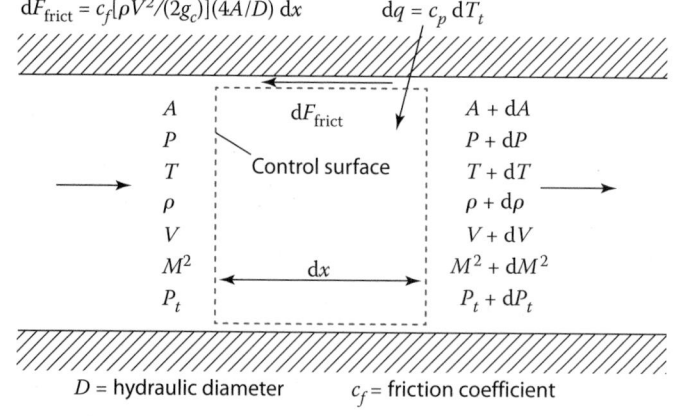

Fig. 3.42 Independent and dependent variables for 1-D flow.

interaction, and friction, results in the following set of equations for the infinitesimal element dx:

Perfect gas:

$$-\frac{dP}{P} + \frac{d\rho}{\rho} + \frac{dT}{T} = 0 \tag{3.36a}$$

Total temperature:

$$\frac{dT}{T} + \frac{[(\gamma-1)/2]M^2}{1 + [(\gamma-1)/2]M^2}\frac{dM^2}{M^2} = \frac{dT_t}{T_t} \tag{3.36b}$$

One-dimensional mass flow:

$$\frac{d\rho}{\rho} + \frac{dA}{A} + \frac{dV}{V} = 0 \tag{3.36c}$$

Total pressure:

$$\frac{dP}{P} + \frac{(\gamma/2)M^2}{1 + [(\gamma-1)/2]M^2}\frac{dM^2}{M^2} = \frac{dP_t}{P_t} \tag{3.36d}$$

Momentum:

$$\frac{dP}{P} + \gamma M^2\frac{dV}{V} + 2\gamma M^2 c_f\frac{dx}{D} = 0 \tag{3.36e}$$

Mach number:

$$2\frac{dV}{V} - \frac{dT}{T} = \frac{dM^2}{M^2} \tag{3.36f}$$

With the exception of the momentum equation, these equations are obtained by taking the derivative of the natural log of Eqs. (i–v).

In these equations, heat interaction effects are measured in terms of the total temperature change according to Eq. (vii). The entropy condition of Eq. (viii) is also applicable if d$T_t = 0$. If dT_t is not zero, then the entropy requirement is ds = dq/T. The six dependent variables M^2, V, P, ρ, T, and P_t in the preceding set of six linear algebraic equations may be expressed in terms of the three independent variables A, T_t, and 4f dx/D. The solution is given in Table 3.1. Note that D is the hydraulic diameter (= 4 × flow cross-sectional area/wetted perimeter).

General conclusions can be made relative to the variation of the stream properties of the flow with each of the independent variables by the relations of Table 3.1. As an example, the relationship given for dV/V at the bottom of the table indicates that, in a constant-area adiabatic flow, friction will increase the stream velocity in subsonic flow and will decrease the velocity in supersonic flow. Similar reasoning may be applied to determine the manner in which any dependent property varies with a single independent variable.

Table 3.1   Influence Coefficients for Steady 1-D Flow

Dependent	Independent		
	$\dfrac{dA}{A}$	$\dfrac{dT_t}{T_t}$	$\dfrac{4\,c_f\,dx}{D}$
$\dfrac{dM^2}{M^2}$	$-\dfrac{2\left(1+\frac{\gamma-1}{2}M^2\right)}{1-M^2}$	$\dfrac{(1+\gamma M^2)\left(1+\frac{\gamma-1}{2}M^2\right)}{1-M^2}$	$\dfrac{\gamma M^2\left(1+\frac{\gamma-1}{2}M^2\right)}{1-M^2}$
$\dfrac{dV}{V}$	$-\dfrac{1}{1-M^2}$	$\dfrac{1+\frac{\gamma-1}{2}M^2}{1-M^2}$	$\dfrac{\gamma M^2}{2(1-M^2)}$
$\dfrac{dP}{P}$	$\dfrac{\gamma M^2}{1-M^2}$	$\dfrac{-\gamma M^2\left(1+\frac{\gamma-1}{2}M^2\right)}{1-M^2}$	$\dfrac{-\gamma M^2[1+(\gamma-1)M^2]}{2(1-M^2)}$
$\dfrac{dp}{p}$	$\dfrac{M^2}{1-M^2}$	$\dfrac{-\left(1+\frac{\gamma-1}{2}M^2\right)}{1-M^2}$	$\dfrac{-\gamma M^2}{2(1-M^2)}$
$\dfrac{dT}{T}$	$\dfrac{(\gamma-1)M^2}{1-M^2}$	$\dfrac{(1-\gamma M^2)\left(1+\frac{\gamma-1}{2}M^2\right)}{1-M^2}$	$\dfrac{-\gamma(\gamma-1)M^4}{2(1-M^2)}$
$\dfrac{dP_t}{P_t}$	0	$\dfrac{-\gamma M^2}{2}$	$\dfrac{-\gamma M^2}{2}$

This table is read:

$$\frac{dV}{V} = \left(-\frac{1}{1-M^2}\right)\frac{dA}{A} + \frac{1+\frac{\gamma-1}{2}M^2}{1-M^2}\frac{dT_t}{T_t} + \frac{\gamma M^2}{2(1-M^2)}\frac{4\,c_f\,dx}{D}$$

Example 3.8

Consider the 1-D flow of a perfect gas in a channel of circular cross section. The flow is adiabatic, and we wish to design the duct so that its area varies with x such that the velocity remains constant.

1. Show that in such a flow, the temperature must be a constant, hence the Mach number must be a constant.
2. Show that the total pressure varies inversely with the cross-sectional area.
3. Show that $2\,dA/C = \gamma M^2 c_f dx$, where C is the circumference.
4. For a constant c_f and a circular channel of diameter D, show that the diameter must vary in accordance with the following equation to keep the velocity constant:

$$D = D_1 + \gamma M^2 c_f x$$

(Continued)

Example 3.8 *(Continued)*

Solution

This problem requires that we apply the relationships of Table 3.1 and other fundamental relationships.

1. Because the flow is adiabatic, the total temperature is constant. From the definition of the total temperature, we have $T_t = T + V^2/(2c_p g_c)$. T_t and V are constant in the duct, so the preceding equation requires that the static temperature remain constant. With the static temperature of the gas constant, the speed of sound will be constant ($a = \sqrt{\gamma R g_c T}$). With constant velocity and speed of sound, the Mach number will be constant.

2. Application of the continuity equation to the constant-velocity flow gives $A_i/A = \rho/\rho_i$. For a perfect gas with constant static temperature, we have $\rho/\rho_i = P/Pi$, and from Eq. (iv) for constant-Mach flow, we get $P_t/P_{ti} = P/P_i$. Thus $P_t/P_{ti} = A_i/A$.

3. To obtain this relationship, we need to get a relationship between two independent properties to keep the dependent property of velocity constant. Table 3.1 gives the basic relationships between dependent and independent properties for 1-D flow. We are interested in the case where velocity is constant, and thus we write the equation listed as an example at the bottom of Table 3.1 for the case where both dV and dT_t are zero:

$$0 = \left(-\frac{1}{1-M^2}\right)\frac{dA}{A} + 0 + \frac{\gamma M^2}{2(1-M^2)}\frac{4c_f\,dx}{D}$$

The area of the duct is equal to the circumference C times one quarter of the diameter D. Thus $A = CD/4$, and the preceding relationship reduces to

$$2\,dA/C = \gamma M^2 c_f\,dx$$

4. For a circular cross section, $A = \pi D^2/4$ and $C = \pi D$. Thus $2dA/C = dD$. Substitution of this relationship into the preceding equation and integration give the desired result (at $x = 0$, $D = D_i$)

$$D = D_i + \gamma M^2\,c_f x$$

3.7   Simple Flows

Gas flows in which a *single independent variable* controls the change in the flow properties are deemed *simple flows*. We have already examined one form of simple flow, that is, isentropic, adiabatic flow of a calorically perfect gas through a duct of changing area, with area treated as the independent variable. In this section, we will explore simple area flow much more thoroughly and introduce two additional simple flows: simple heating flow and simple frictional flow. Note that Table 3.1 can be conveniently used to relate the changes in dependent variables in terms of the *single independent variable* by simply zeroing out the nonapplicable columns for the simple flow under consideration. For example, the relationship for dV/V given at the bottom of Table 3.1 for the case of simple area flow is

$$\frac{dV}{V} = \left(-\frac{1}{1-M^2}\right)\frac{dA}{A}$$

which we first developed in Section 3.3, Eq. (3.18). This functional relationship revealed that for subsonic flow, flow cross-sectional area and velocity are inversely proportional, whereas for supersonic flow, they are directly proportional.

The simple flows to be analyzed are shown in Fig. 3.43. Each of these flows has a direct practical application even though a truly simple flow is seldom encountered in practice. Often, however, in a flow with combined area, friction, or heat effects, one of these greatly outweighs the effect of all others so that the variation in stream properties is determined mainly by a single independent variable. Examples of real flows that may be treated as simple flows are given in the following text for each type of simple flow in Fig. 3.43.

Figure 3.43a could represent exhaust gas acceleration through a nozzle, or by redrawing it so that $A_2 > A_1$, air flow through the inlet (diffuser) of an aircraft gas turbine engine. Both of these flow situations can often be analyzed as simple area flow. In these applications, the effect of the area change on the stream properties typically greatly exceeds the frictional and heating effects; excellent agreement between theory and experiment is obtained by treating the flow as simple area flow.

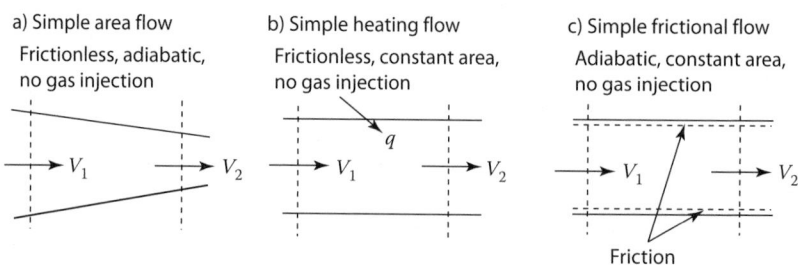

a) Simple area flow
Frictionless, adiabatic, no gas injection

b) Simple heating flow
Frictionless, constant area, no gas injection

c) Simple frictional flow
Adiabatic, constant area, no gas injection

Fig. 3.43   Simple types of flow.

In Fig. 3.43b, the total temperature change occurring in a gas turbine combustor, or in a rocket combustion chamber between the injector plate and the nozzle inlet, can be analyzed as a simple heating flow. By ignoring friction and the change in molecular weight and specific heat of the gases, the adiabatic combustion process can be replaced by an equivalent simple heating process producing the same total temperature rise. This can be an effective model for studying the variation of stream properties with total temperature in these combustion chambers.

In Fig. 3.43c, some rocket propulsion installations employ a relatively long constant-area blast tube between the combustion chamber and nozzle. The gas flow in this blast tube is, for all practical purposes, adiabatic and is an example of simple frictional flow. Another example might be the transport of natural gas through hundreds of miles of pipeline, which will require numerous pumping stations to overcome the dominant frictional effects.

3.7.1 Simple Area Flow: Nozzle Flow

In this section, we concern ourselves with *nozzles*—devices/ducts that employ area change to accelerate a gas flow. Devices/ducts that use area change to slow down or decelerate a flow are called *diffusers*, and are another example where simple area flow analysis could apply. Because area change is the dominant independent variable, the effects of friction, heat interaction, and all others like expansion/contraction of the walls (work interaction), gas injection, and the like are considered negligible. For simple area flow, the applicable functional relations in Table 3.1 are found by zeroing out the dT_t/T_t and $(4c_f dx)/D$ columns.

We previously explored the velocity–area variation for isentropic, adiabatic flow in Section 3.3 using continuity and linear momentum to discover that a converging nozzle could only accelerate a flow to Mach 1.0. A converging–diverging area change is required to accelerate flows beyond Mach 1.0. We pointed out in Section 3.3 that a sufficient pressure difference across the nozzle must be required for such high velocity flows. We are now in a position to define that pressure requirement and, with our knowledge of sound wave propagation (Section 3.4), identify nozzle operating characteristics, which include compression shocks and Prandtl–Meyer expansions.

3.7.1.1 Converging Nozzle

Consider a flow originating in a large storage chamber, like a rocket combustion chamber, which is allowed to expand through a converging nozzle as sketched in Fig. 3.44. The streamwise flow velocity inside the chamber is negligibly small, so the chamber pressure and temperature effectively represent the stagnation or total pressure and temperature as indicated on the *T-s* diagram of Fig. 3.44 ($P_c = P_t$, $T_c = T_t$). For a constant chamber pressure, we wish to determine the mass flow rate through the

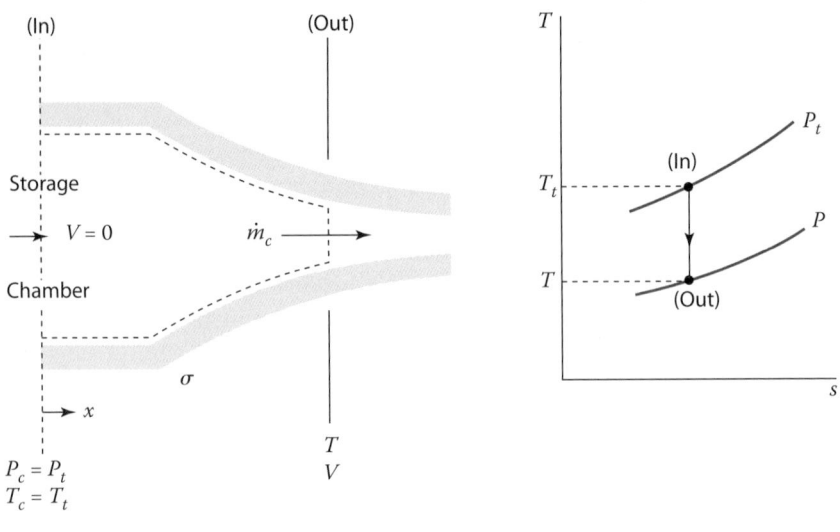

Fig. 3.44 Control volume for simple area flow.

nozzle as a function of exhaust region pressure P_a imposed on the nozzle (often called back pressure). You may find it helpful to think of a rocket accelerating through the atmosphere with decreasing ambient pressure as it gains altitude.

Figure 3.45 shows the pressure distribution and mass flow rate through the nozzle for six different values of back pressure. When the back pressure and chamber pressure are equal (curve 1), there is no mass flow in the nozzle. As the back pressure is reduced (curves 2, 3, and 4), more and more flow is induced through the nozzle with the resulting streamwise static pressure distributions as sketched. The velocity at the nozzle exit plane continues to increase until it reaches the speed of sound at the back pressure P_{a4}. What happens as back pressure is reduced beyond that of P_{a4}? Will more mass flow result? The answer lies in the physical mechanism that allows the gas molecules to sense changes in back pressure; that physical mechanism is, of course, pressure waves, which travel at the speed of sound (Section 3.4). As long as the flow is subsonic through the nozzle, changes in back pressure can be communicated upstream and flow is able to adjust accordingly. Once the flow exits the nozzle at the same velocity as the signaling mechanism (sound waves), further changes in flow *inside the nozzle* cannot be accommodated with further decreases in back pressure. For this reason, the nozzle is said to be *choked*. Choked flow corresponds to maximum mass flow rate and occurs when the exit plane Mach number is one for a convergent nozzle. Note that for back pressures corresponding to curves 5 and 6 (and all back pressures below that of curve 4), the nozzle exit plane static pressure remains unchanged and will *not* equal the back pressure. In fact, the flow will adjust to the back pressure by means of a Prandtl–Meyer expansion *outside the nozzle*. A convergent nozzle can only accelerate a flow to sonic conditions Mach 1.

For a fixed area convergent nozzle with constant chamber conditions, there is one value of exhaust region or back pressure P_a, or better, one nozzle pressure ratio P_c/P_a, that will choke the nozzle with *exit plane static pressure P_e equal to P_a*; we shall denote that pressure ratio as $P_{\bar{n}}$ by industry convention such that

$$P_{\bar{n}} \equiv \left(\frac{P_c}{P_a}\right)_{des}$$

For isentropic, choked ($M = 1.0$) flow, from Eq. (3.8)

$$P_{\bar{n}} = \left(1 + \frac{\gamma - 1}{2}\right)^{\frac{\gamma}{\gamma-1}}$$

So, from the compressible flow equation above or Appendix D: $P_{\bar{n}} = 1/0.52828 = 1.8929$ for $\gamma = 1.4$, and $1/0.54573 = 1.8324$ for $\gamma = 1.3$, like that for a hot gas rocket exhaust stream. The mass flow rate at any condition, including design, can be found using the mass flow parameter MFP [see Eq.

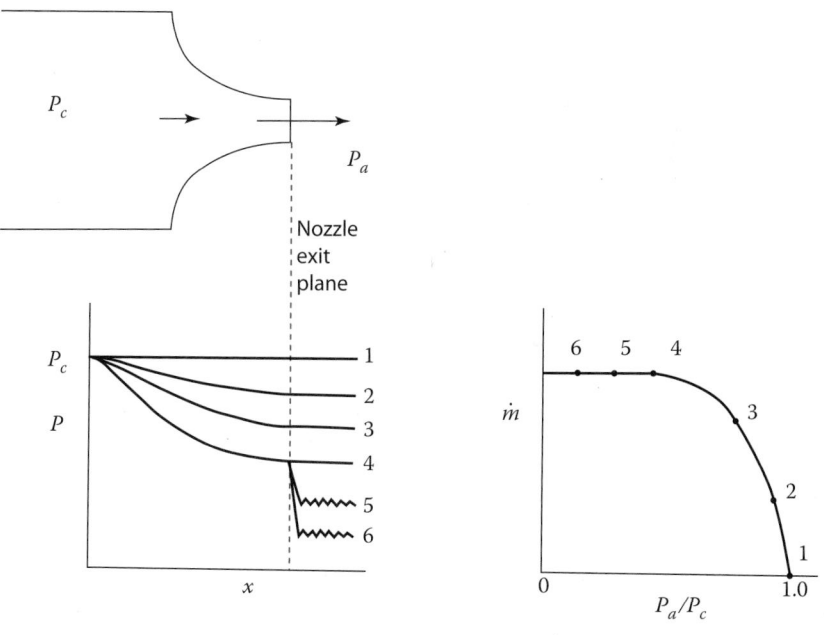

Fig. 3.45 Pressure distribution and mass flow rate for convergent nozzle.

(3.11)] applied at any convenient cross-sectional area because P_t and T_t are constant and equal to the chamber conditions for this simple flow.

3.7.1.2 Converging-Diverging Nozzle

We now examine the much more interesting case of a converging-diverging (C-D) nozzle attached to a storage chamber, as sketched in Fig. 3.46. This could be representative of a rocket nozzle (Fig. 3.47) or exhaust nozzle of a low bypass turbofan engine of a high performance fighter aircraft. Recall that the chamber conditions represent the stagnation or total conditions.

Figure 3.46 shows the pressure distribution and mass flow rate through the nozzle for different values of back pressure or exhaust region pressure P_a. Clearly, a C-D nozzle is designed to produce supersonic exhaust gas velocities, so it is important to understand the nozzle operation as it relates to varied pressure differences imposed on it. In Fig. 3.46, curve 5 represents the nozzle design condition with $P_e = P_{a5}$.

Just as with the convergent nozzle, there is no mass flow in the nozzle and no pressure distribution when the back pressure and chamber pressure are equal (curve 1). As we start to decrease the back pressure

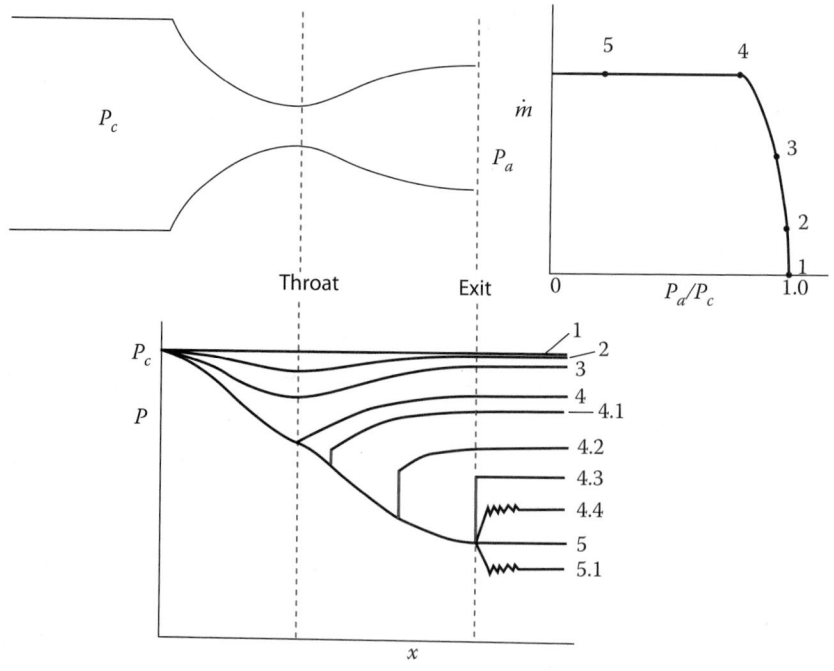

Fig. 3.46 Pressure distribution and mass flow rate for convergent-divergent nozzle.

to that of curves 2 and 3, the flow properties can and do adjust smoothly and continuously with subsonic flow in both the converging and diverging portions of the nozzle, and more and more mass flow induced through the nozzle. For subsonic flow, pressure will decrease in the converging section with a corresponding increase in velocity, and pressure will increase in the diverging section with a velocity decrease. This behavior will continue until the velocity at the *minimum area or throat* of the C-D nozzle equals the speed of sound. When sonic flow occurs at the throat (curve 4), further decreases in back pressure cannot be sensed upstream of the throat, and consequently, the chamber cannot send out any more mass flow than that indicated at condition 4; the C-D nozzle is choked. Note that for all cases represented by curves 1 through 4, the flow in the nozzle is *isentropic* and the nozzle plane exit static pressure $P_e = P_a$.

What happens as back pressure is reduced beyond that associated with curve 4? Further reductions in back pressure beyond that of P_{a4} will result in the formation of a very weak normal shock wave initially positioned just downstream of the throat area when back pressure is just slightly below that of P_{a4}. As back pressure is further reduced, the normal shock will strengthen and continue to move downstream, as denoted by curves 4.1 and 4.2 in Fig. 3.46. At some back pressure denoted as curve 4.3 in Fig. 3.46, the normal shock will be positioned just at the C-D nozzle exit plane. A further reduction in back pressure to that just below curve 4.3 will result in the shock being positioned outside the nozzle exit plane.

The physical mechanism for the formation of the normal shock within the divergent portion of the C-D nozzle is as described in Section 3.4. The weak pressure waves moving at the speed of sound cannot make headway upstream once they encounter flow moving downstream at sonic or super-sonic speeds; the sound waves pile into each other forming a normal shock because the flow is bounded by the walls of the divergent section. It should be intuitively appealing that shock position within the divergent portion of the C-D nozzle is determined by the pressure difference imposed across the nozzle. Note that for all back pressures between those represented by curve 4 and curve 4.3, the flow in the nozzle is *non-isentropic* because of the presence of the normal shock. Further, because the flow is always subsonic following a normal shock, the shock will pos-ition itself to meet the requirement that the nozzle plane exit static pressure $P_e = P_a$.

We will continue our analysis of the C-D nozzle shortly, but for now, let us return to the isentropic area ratio A/A^* function previously described in Section 3.2.5. This function, which had two solutions for Mach number, a subsonic and supersonic solution (Fig. 3.10), is directly applicable to our current analysis of simple area flow through a C-D nozzle. Considering A_e/A^*, where A_e is the nozzle exit plane area and A^* is the throat area *when the nozzle is choked*, what would dictate which Mach number solution would exist at the nozzle exit plane? Clearly, it would have to be the back pressure condition, or better, the P_a/P_c ratio that is imposed across the C-D nozzle. Consequently, P_{a4}/P_c must correspond to the subsonic solution for a given A_e/A^* and P_{a5}/P_c must correspond to the supersonic solution for the same A_e/A^*. We will reinforce our discussion of C-D nozzle operating characteristics with Example 3.9, which has four parts interwoven into the discussion.

Example 3.9a

Given the F-1 C-D rocket nozzle of Fig. 3.47 with $P_c = 965$ psia, $A_e/A^* = 16$, determine the back pressure needed to (1) choke the nozzle, and (2) ensure a supersonic Mach number at the exit plane M_e. Assume the exhaust gas behaves as a calorically perfect gas with $\gamma = 1.3$ and that friction and heat transfer are negligible compared to the kinetic energy increase of the gas. Use the GASTAB program to determine the required property relations.

Fig. 3.47   F-1 rocket engine used on Saturn V rocket (produced 1.5 million pounds of thrust at launch). (Courtesy of NASA.)

(Continued)

Example 3.9a (Continued)

Solution

Because friction and heat transfer are negligible and we have no work interactions, the analysis proceeds as simple area flow. The chamber pressure equals the total pressure, which is constant throughout the nozzle as long as isentropic flow can be assumed in the nozzle.

For $A_e/A^* = 16$, $\gamma = 1.3$, from the Isentropic Flow tab of GASTAB:

Subsonic solution: $M_e = 0.03660$
$$P_e/P_c = 0.99913$$
Supersonic solution: $M_e = 4.00$
$$P_e/P_c = 0.00495$$

Thus, with $P_c = 965$ psia, the back pressure required to *just choke the nozzle* is

$$P_a = 0.99913(965 \text{ psia}) = 964.2 \text{ psia} = P_e$$

The pressure difference required to choke the flow ($M_{\text{throat}} = 1.0$) is remarkably small, 0.8 psi in this case, less than 0.1% of the chamber pressure! This is generally true for rocket exhaust nozzles.

The back pressure required to ensure $M_e > 1.0$ would be slightly less than that required to position a normal shock right at the nozzle exit plane; essentially, P_a just slightly less than $P_{a4.3}$ of Fig. 3.46. So we analyze the case with a normal shock right at the exit plane where the conditions just upstream of the shock are those obtained previously for the supersonic solution because we have uniform isentropic flow all the way to just upstream of the normal shock (see Fig. 3.48).

Because just downstream of the shock the flow would be subsonic, we know that $P_y = P_e = P_a$, so

$$P_e = P_a = \left(\frac{P_x}{P_c}\right)\left(\frac{P_y}{P_x}\right)P_c$$

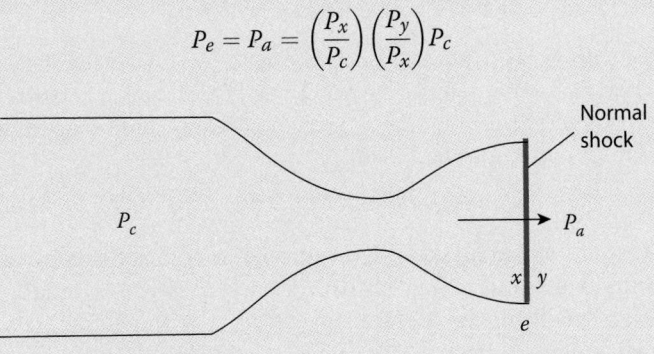

Fig. 3.48   Normal shock at exit plane of C-D nozzle.

(Continued)

Example 3.9a (Continued)

where P_y/P_x is found from the normal shock relations, in this case. For $M_1 = 4.0$, $\gamma = 1.3$, from the Normal Shock tab of GASTAB:

$$P_2/P_1 = 17.957$$
$$M_e = 0.406$$

So

$$P_e = P_a = 0.00495(17.957)(965\,\text{psia}) = 85.78\,\text{psia}$$

Therefore, as long as the back pressure is something slightly less than 85.78 psia, supersonic flow at the nozzle exit plane would occur.

Example 3.9b

What would be the nozzle exit flow conditions for all back pressures below 85.78 psia?

Solution

For all back pressures below 85.78 psia, supersonic, isentropic flow exists throughout the nozzle; consequently, we use the supersonic solution for $A_e/A^* = 16$, $\gamma = 1.3$ from the Isentropic Flow tab of GASTAB.
Supersonic solution:

$$M_e = 4.00$$

$$\frac{P_e}{P_c} = 0.00495 \Rightarrow P_e = 0.00495(965\,\text{psia}) = 4.78\,\text{psia}$$

$$T_e/T_c = 0.29379 \quad \text{(and others as desired)}$$

For all back pressures below 85.78 psia, the conditions at the nozzle exit plane, and for that matter throughout the entire nozzle, will remain invariant. There is no physical mechanism for the properties to change; any required property changes must occur outside the nozzle.

Let's resume our analysis of C-D nozzles with back pressures between those of curves 4.3 and 5 (Fig. 3.46). As back pressure is reduced beyond that associated with curve 4.3, we have seen that the shock gets expelled from the divergent portion of the nozzle, and the nozzle exit conditions remain invariant. What happens to the shock wave? As indicated in Figure 3.49, for $P_{a4.3} > P_a > P_{a5}$, in most cases an oblique shock forms at the exit plane of the nozzle and gets more inclined relative to the flow (weaker) as back pressure gets closer and closer to P_{a5}.

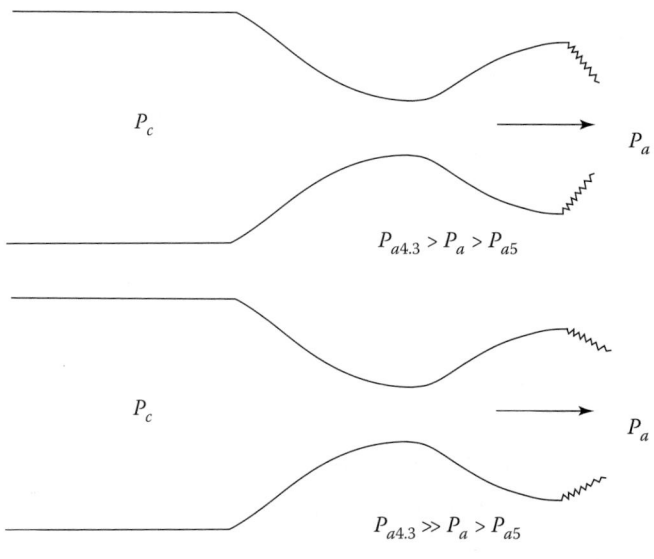

Fig. 3.49 C-D nozzle with oblique shocks at exit.

The reduction in back pressure causes the angle between the shock and the flow to get smaller and smaller, decreasing the shock strength, eventually reaching the fully isentropic case, even outside the nozzle (curve 5). As was previously discussed, this condition represents the nozzle design condition and is called *perfectly expanded flow* or *perfectly expanded nozzle operation*. For this condition, once again, we have $P_e = P_a$.

So, for back pressure conditions between those of curves 4.3 and 5 of Fig. 3.46, what is the relationship between the nozzle exit plane static pressure P_e and the back pressure P_a? Our foundational understanding and Examples 3.9a and 3.9b should provide the needed insights. For back pressures between those of curves 4.3 and 5, $P_e < P_a$. The nozzle is operating *overexpanded*. The exhaust gases have been expanded (accelerated) to a condition where the nozzle exit plane static pressure is below the exhaust region pressure. Matching the exit plane static and back pressures would require a nozzle with a smaller area ratio A_e/A^*.

For back pressures below that of curve 5, the nozzle exit plane static pressure, which remains invariant, is greater than the back pressure $(P_e > P_a)$. The nozzle is operating *underexpanded*. Matching the exit plane static and back pressures would require a nozzle with a larger area ratio A_e/A^*. The condition for $P_a < P_{a5}$ is sketched in Fig. 3.50. Prandtl–Meyer expansion waves form at the nozzle exit, further expanding the flow.

It is important to recognize that with the exception of the perfectly expanded case for curve 5, for all back pressures below that of curve 4.3, the flow adjusts to the back pressure downstream of the nozzle exit plane. For all $P_a < P_{a4.3}$, conditions inside the nozzle remain invariant.

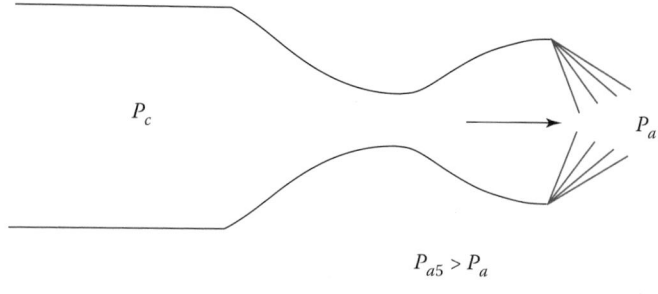

Fig. 3.50 C-D nozzle with Prandtl–Meyer expansion fans at exit.

Example 3.9c

For the C-D nozzle of Example 3.9a, at what altitude would the nozzle be operating perfectly expanded? Assume standard atmosphere and use the program ATMOS.

Solution

For a perfectly expanded nozzle, $P_e = P_a$ at the design Mach number. At the design $M_e = 4.0$, we found $P_e = P_a = 4.78$ psia (Example 3.9b). From ATMOS, standard day, $\delta = 4.78/14.7 = 0.325$: $h = 28{,}000$ ft.

For altitudes below 28,000 ft, the nozzle would be operating overexpanded with $P_e < P_a$. At altitudes above 28,000 ft, the nozzle would be underexpanded with $P_e > P_a$.

The flow adjustment that occurs outside the nozzle for the case of overexpanded or underexpanded nozzle operation is quite interesting, and a first-order analysis can be performed using the foundational concepts presented in this chapter. In each of these cases, the nozzle exit plane static pressure and back pressure are not equal, so the flow will go through a series of expansions and compressions downstream of the nozzle exit plane until the two pressures are equal. The only difference between the overexpanded and underexpanded cases is the phasing of the downstream expansions and compressions. We will examine overexpanded nozzle operation first.

For the case of overexpanded nozzle operation with $P_e < P_a$, we have already seen that in most cases an oblique shock will form at the nozzle exit plane. A shock is a nearly discontinuous compression process, and nature, which seeks equilibrium, desires to increase the static pressure of the flowing gas to that of the exhaust region pressure or back pressure P_a. Figure 3.51 shows a typical flow field downstream of the exit plane of an overexpanded C-D nozzle.

The flow is assumed to exit the nozzle uniformly and parallel to the centerline, which is shown. Consequently, the centerline represents a plane of symmetry. Consistent with our assumptions, no flow can cross the centerline streamline; it can be thought of as a solid boundary for analysis purposes. The oblique shock forms at an angle that is just sufficient to cause the flow static pressure rise to equal the back pressure, as indicated in region 1 of Fig. 3.51. There is a *free boundary* or *jet boundary* in region 1 that, again, can be thought of as a solid boundary for analysis purposes.

Recall that the flow will be turned in region 1, inward as shown, by an amount consistent with the shock angle as determined by oblique shock theory. Because flow cannot cross the centerline streamline, another oblique shock wave must form, turning the flow back parallel to the centerline. So we have ensured that our centerline symmetry boundary condition is not violated, but because the flow has undergone another shock process, the static pressure in region 2 of Fig. 3.51 is now greater than the back pressure, $P_2 > P_a$. The flow is still supersonic, so it now must undergo a supersonic expansion (acceleration) via a Prandtl–Meyer expansion wave to get $P_3 = P_a$ to meet the free boundary condition imposed in region 3 (Fig. 3.51). But, we have the same problem in region 3 as we did in region 1. Although the static pressures are equal, the flow in region 3 is turned, this time outward by an amount consistent with P–M expansion flow. The free boundary condition in region 3 prevents flow from crossing the boundary, so a second P–M expansion wave is formed to turn the flow back to the centerline but with $P_4 < P_a$. You get the picture! This cycle of oblique shocks/Prandtl–Meyer expansions will continue, theoretically to infinity. In actuality, flow mixing at the boundaries and other irreversibilities will cause the wave patterns to die out. In practice, the flow pattern just described often reveals itself as a series of *Mach diamonds* or *shock diamonds* downstream of a rocket or gas turbine engine C-D nozzle exit (see Fig. 3.52).

For the case of underexpanded nozzle operation, because $P_e > P_a$, P–M expansion fans will form first at the exit lip of the nozzle, followed by oblique shocks, and on and on. The same cycle pattern as just described will result; the waves will just be out of phase from the overexpanded case.

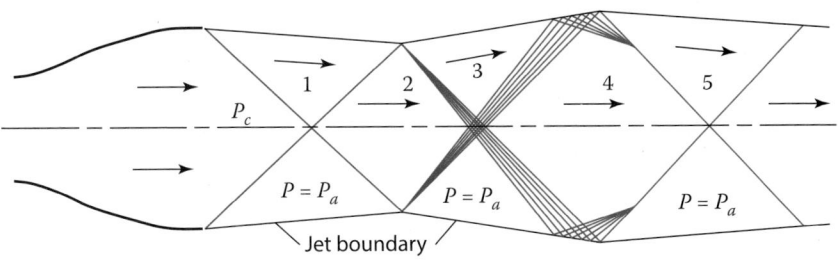

Fig. 3.51   Overexpanded C-D nozzle.

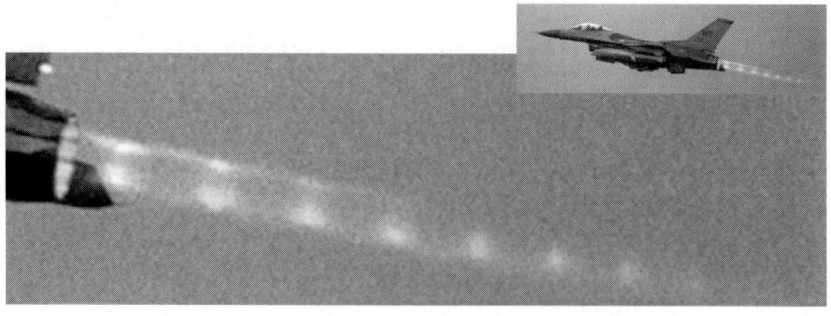

Fig. 3.52   F-16 in afterburner operation shortly after takeoff.

Example 3.9d

Figure 3.51 can be used to represent the flow regions downstream of the rocket nozzle exit plane of Example 3.9a operating at 10,000 ft. Estimate the static pressure, Mach number, flow turning angle, and wave angle of the exhaust gases at the exit plane and in regions 1, 2, and 3. Assume standard day atmospheric conditions, and use GASTAB with $\gamma = 1.3$. The flow leaves the nozzle exit plane uniformly and parallel to the centerline.

Solution

As was shown in Example 3.9c, for standard day altitudes below about 28,000 ft, the nozzle is operating overexpanded ($P_e < P_a$). Indeed, this is the current case:

For 10,000 ft, standard day, from ATMOS: $P_a = 10.1$ psia

We know from Examples 3.9a and 3.9b that as long as the back pressure P_a is below 85.78 psia, the nozzle internal flow conditions are invariant with

$M_e = 4.0$ parallel to centerline; $P_e = 4.78$ psia ($P_e < P_a$)

So Fig. 3.51 is accurate with the exit flow initially decelerating through oblique shocks. The key to working this problem (and those like it) is applying the *pressure equilibrium boundary condition at free or jet boundaries*, and the *flow direction boundary condition at the centerline plane of symmetry*.

The flow in region 1, downstream of an oblique shock, is subjected to a free boundary condition where $P_1 = P_a = 10.1$ psia. Therefore, an oblique shock must be positioned such that

$$\frac{P_1}{P_e} = \frac{10.1\,\text{psia}}{4.78\,\text{psia}} = 2.11$$

(Continued)

Example 3.9d *(Continued)*

and with upstream Mach $M_e = 4.00$, $\gamma = 1.3$, from GASTAB, Oblique Shock (weak):

$$M_1 = 3.494, \beta \text{ (shock angle)} = 20.607 \text{ deg}, \delta \text{ (turning angle)}$$
$$= 8.527 \text{ deg}; P_1 = 10.1 \text{ psia}$$

So in region 1 we have static pressures matched, but the flow has been turned towards the centerline by 8.527 deg. Because no flow can cross the centerline, another oblique shock is formed turning the flow back aligned with the centerline in region 2. Again, we use GASTAB, Oblique Shock, with $\gamma = 1.3$ and with upstream Mach $M_1 = 3.494$, $\delta = 8.527$ deg:

$$M_2 = 3.067 \text{ parallel to centerline}, \beta \text{ (shock angle)} = 22.792 \text{ deg};$$
$$\frac{P_2}{P_1} = 1.941 \Rightarrow P_2 = 1.941(10.1 \text{ psia}) = 19.604 \text{ psia}$$

So in region 2 we have flow back aligned with the centerline, but the flow static pressure is now larger than ambient backpressure. Consequently, the flow expands through a Prandtl–Meyer expansion fan in region 3 such that $P_3 = P_a = 10.1$ psia (a free boundary condition). Therefore, a P–M expansion fan must be positioned such that

$$\frac{P_3}{P_2} = \frac{10.1 \text{ psia}}{19.604 \text{ psia}} = 0.515$$

and with upstream Mach $M_2 = 3.067$, $\gamma = 1.3$, from GASTAB, P–M Flow:

$$M_3 = 3.474, \omega \text{ (turning angle)} = 8.532 \text{ deg}; P_3 = 10.1 \text{ psia},$$
$$\mu_2 \text{ and } \mu_3 \text{ (Mach angles)} = 19.029 \text{ deg and } 16.729 \text{ deg}$$

Clearly, we could continue this process into subsequent downstream regions, but that is unnecessary. Figure 3.53 provides a sketch of the three regions with our results.

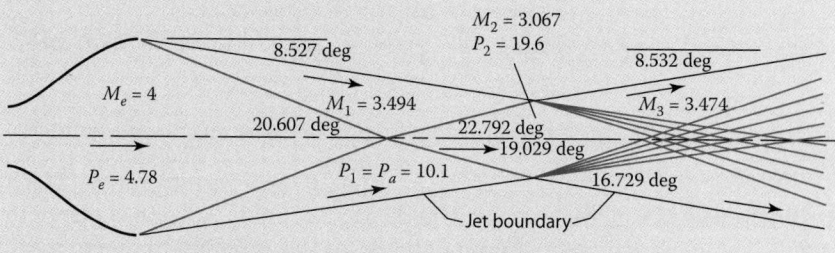

Fig. 3.53 Results showing angles and Mach numbers.

A summary of the different C-D nozzle operating conditions is provided in Fig. 3.54 and Table 3.2, where we use the industry-standard nozzle pressure ratio

$$P_n \equiv \frac{P_c}{P_a}$$

Seven distinct operating conditions are realized, all of which we have discussed in the preceding paragraphs. They are:

1. Subsonic flow throughout the nozzle
2. Sonic limit: sonic at throat, subsonic everywhere else
3. Normal shock inside diverging portion of nozzle
4. Normal shock at exit
5. Overexpanded: $P_e < P_a$
6. Design expansion: perfectly expanded nozzle with $P_e = P_a$, isentropic flow everywhere
7. Underexpanded: $P_e > P_a$

Note that for the overexpanded operation, Fig. 3.54 shows two possible solutions. The transition from P_e to the higher back pressure P_a is produced by either a weak oblique shock system (previously discussed and analyzed in Example 3.9d) indicated in the figure as a "Regular reflection" or a combined oblique-normal shock system indicated as "Mach reflection," which can occur at lower values of P_n in the overexpanded operating region. This nozzle operating condition lies between the design expansion line and normal-shock-at-exit line, as indicated in Fig. 3.54. The strong oblique shock solution cannot be supported in the external exhaust flow field.

Figure 3.55 is a picture of several nozzle exhaust flow patterns for a conical nozzle with an area ratio of $\varepsilon = 1.5$. For $P_n = 1.5$, the normal shock is located well within the nozzle. As P_n is increased to 2.5, the flow is overexpanded in the Mach reflection regime, as evidenced by the clearly defined normal shock internal to the flow. Further increase in P_n causes this normal shock in the Mach reflection pattern to move farther out and diminish in size. It disappears when the regular reflection pattern is obtained for $P_n = 4.5$. The photograph with $P_n = 8.0$ corresponds to an underexpanded operating point with the exhaust gas expanding down to the ambient pressure in the jet plume.

3.7.1.3 Nozzle Characteristics of Some Operational Engines

The operating point of a nozzle is determined by the nozzle pressure ratio P_n and area ratio ε. These ratios are presented in Table 3.3 along with other data for some operational nozzles. The data in Table 3.3 permit us to locate the operating points of the Saturn and Atlas engines at a given ε in the nozzle operating diagram from Fig. 3.54. The Saturn F-1 engine was used in Example 3.9 with $\varepsilon = 16$ and $P_c = 965$ psia. Thus at 46,000 ft, for example,

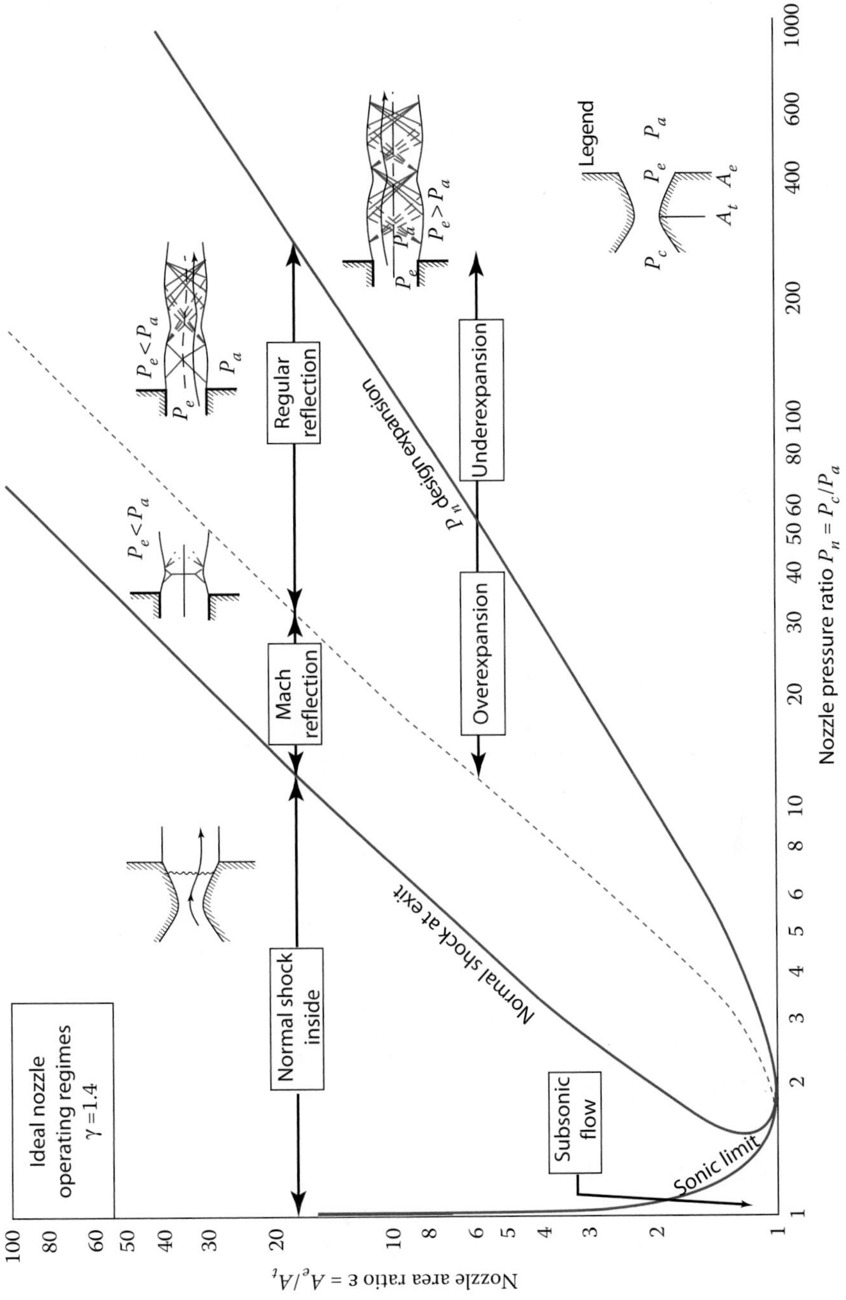

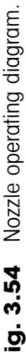

Fig. 3.54   Nozzle operating diagram.

Table 3.2   Nozzle Operating Points[a]

	Operand Point	Exit Section Pressure P_e		Nozzle Pressure Ratio $P_n = P_c/P_a$	Mass Flow Rate
1	Underexpanded	$P_e > P_a$		$P_n > P_{\bar{n}}$	Maximum
2	Design	$P_e = P_a$	$P_e = \dfrac{P_c}{P_{\bar{n}}}$	$P_n = P_{\bar{n}}$	Maximum
3	Overexpanded a) Reg. reflection b) Mach reflection	$P_e < P_a$		$P_n < P_{\bar{n}}$	Maximum
4	Normal shock at exit	$(P_e)_x < (P_e)_y = P_a$		$P_n < P_{\bar{n}}$	Maximum
5	Normal shock in divergent section	$P_e = P_a$		$P_n > P_{\bar{n}}$	Maximum
6	Sonic at throat subsonic elsewhere	$P_e = P_a$		$P_n < P_{\bar{n}}$	Maximum
7	Subsonic everywhere	$P_e = P_a$		$P_n < P_{\bar{n}}$	Less than maximum

[a]$P_{\bar{n}}$ is the design nozzle pressure ratio.

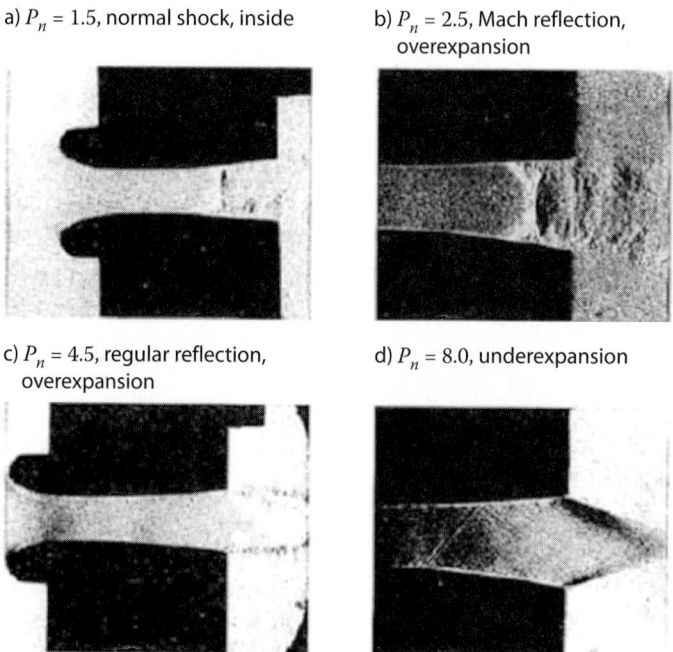

a) $P_n = 1.5$, normal shock, inside

b) $P_n = 2.5$, Mach reflection, overexpansion

c) $P_n = 4.5$, regular reflection, overexpansion

d) $P_n = 8.0$, underexpansion

Fig. 3.55   Spark Schlieren photographs of nozzle exhaust flow patterns for an area ratio of 1.5 (Department of Aero-Mechanical Engineering, Air Force Institute of Technology, Wright-Patterson Air Force Base, Ohio).

Table 3.3 Some Nozzle Characteristics of Rocket and Turbojet Engines

Engine	Chamber Pressure P_c, psia	Area Ratio ε	P_a	$(P_a)_{\text{operate}}$
Saturn				
F-1 without extension	965	10	140	Varies with altitude P_a
F-1 with extension	965	16	275	
J-2	763	27.5	610	
Atlas				
Booster	703	8	100	Varies with altitude P_a
Sustainer	543	25	528	
Subsonic airbreathing turbofan (40,000 ft)	8–14	1.0	1.9	3–5
Supersonic airbreathing turbofan (40,000 ft)	8–56	1–2	1.9–8	3–20

with $P_a = 2$ psia, we have $P_n = 965\ \text{psia}/2\ \text{psia} = 482$. From Fig. 3.54 (assuming $\gamma = 1.4$), we find that the F-1 nozzle is operating underexpanded at 46,000 ft. At sea level for the F-1, P_n is about 65, and the operating point of the engine nozzle (assuming $\gamma = 1.4$) is clearly in the overexpanded regular reflection region.

The turbojet engines of high-performance, air-breathing, subsonic aircraft generally use convergent nozzles ($\varepsilon = 1.0$) and operate with nozzle pressure ratios greater than the design value of 1.9 (assuming $\gamma = 1.4$). These nozzles therefore operate in the underexpanded operating regime. Turbojet/turbofan engines of supersonic aircraft, however, have converging-diverging nozzles, which are typically equipped with hardware to vary the area ratio (within reason) to achieve design expansion or perfectly expanded nozzle operation at different values of altitude and Mach number.

3.7.2 Simple Heating Flow: Rayleigh Line Flow

In this section and the subsequent section on simple frictional flow, we will present only the highlights and a brief discussion primarily aimed at providing a foundational understanding of these simple flows. The interested reader can consult any number of excellent references on compressible flow [see, for example, *Fundamentals of Gas Dynamics*, 2nd ed., Zucker & Biblarz (Ref. 65), and *Gas Dynamics*, 3rd ed., John & Keith (Ref. 15)].

The analysis of these two simple flows should be very familiar to you— select an appropriate control volume; apply the governing physical equations of continuity, momentum, and the first and second laws of thermodynamics with appropriate assumptions; and use the compressible flow functions to get property relations in terms of ratio of specific heats γ and Mach number M. Because simple flows consider the dominating influence of only one independent parameter, simple heating and simple frictional analysis will consider only a constant area duct or channel. Other assumptions applicable to *both flows* include:

- Steady flow: Property changes with respect to time are negligible.
- One-dimensional or uniform flow: Flow property changes in the streamwise direction are dominant.
- Calorically perfect gas (CPG): Constant gas specific heats, or better, suitable average values.
- No work interactions: No shaft work, boundary work, etc.
- Negligible change in potential energy (gravity effects).
- Nonreacting flow: No chemical reactions.

For simple heating flows:

- Negligible frictional effects

Consider the flow of a perfect gas with simple heating effects. This flow alternatively may be called *simple T_t flow* because the total temperature is the independent property controlled through the heating effect, because the assumptions above result in the conservation of energy equation Eq. (2.22)

$$q_{in} \text{ or } - q_{out} = c_p(T_{t\,out} - T_{t\,in})$$

for a CPG. The or appears in the equation above because we will consider either simple flow with heat addition or simple flow with heat rejection from the constant area duct control volume.

The temperature–entropy locus of fluid states that may possibly be attained by a gas in a simple heating flow is called the *simple heating flow line* or, more commonly, the *Rayleigh line*. Let us determine the simple heating flow line and discuss some of its characteristics.

In Fig. 3.56, a calorically perfect gas is shown to flow steadily from a large reservoir through a convergent nozzle and then into a constant-area frictionless duct with heating effects. We designate the duct inlet as station 1.

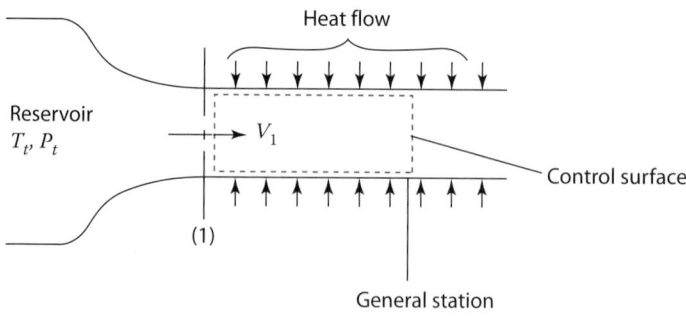

Fig. 3.56   Simple heating flow.

A relation between the stream properties at 1 and the pressure and temperature downstream of 1 at any general station (no subscript) can be obtained by combining the continuity and momentum equations for the flow through the control surface of Fig. 3.56. With the assumptions listed previously, we have

From Eq. (2.13), continuity
$$\rho_1 V_1 = \rho V = \frac{\dot{m}}{A} = \text{constant}$$

From Eq. (2.16), momentum
$$P_1 + \frac{\rho_1 V_1^2}{g_c} = P + \frac{\rho V^2}{g_c}$$

Replacing V in the momentum equation with $\dot{m}/(\rho A)$ from the continuity equation and using the perfect gas equation of state $\rho = P/(RT)$ gives

$$P_1 + \frac{\rho_1 V_1^2}{g_c} = P + \frac{R}{g_c}\left(\frac{\dot{m}}{A}\right)^2 \frac{T}{P} \tag{3.37}$$

This is the equation of a family of lines relating P and T at any general station for parameter values of P_1, V_1, and ρ_1 (hence $\dot{m}/A$). For given inlet conditions, therefore, this equation represents a relation between pressure and temperature that must be satisfied at stations downstream of station 1 in the flow of Fig. 3.56. By assuming different values of P in the flow, the corresponding values of T that must exist in order to satisfy the momentum and continuity equations can be found from Eq. (3.27). These values of P and T can then be substituted into the following equation to determine the corresponding values of entropy:

$$s - s_1 = c_p \ell n \left(\frac{T}{T_1}\right) - R \ell n \left(\frac{P}{P_1}\right) \tag{2.48}$$

A *Rayleigh line* obtained by this procedure is sketched in Fig. 3.57. If the assumed values of P_1, T_1, and V_1 correspond to a subsonic condition, station 1 is located on the upper branch of the Rayleigh line. Supersonic values of state 1 properties for the same $\dot{m}/A$ would place this state point on the lower branch of the Rayleigh line at 1'; higher velocities correspond to lower static temperatures. Thus the Rayleigh line's upper branch corresponds to subsonic state points whereas the lower branch is in the locus of supersonic flow states.

Treating the simultaneous solution of Eqs. (3.37) and (2.48) as a purely mathematical exercise, for the present, we observe the following from Fig. 3.57 starting at point 1: By assuming values of P successively less than P_1, it is found that T and s initially increase until, at P_m, T attains a maximum and, at P_r, s reaches a maximum. For values of P less than P_m the temperature decreases. And for P less than P_r, the entropy is seen to decrease. Finally, we note that by assuming values of P greater than P_1, we get monotonically decreasing values of T and s. Physically, how can the gas flowing through the duct of Fig. 3.56 be made to attain the pressures assumed in

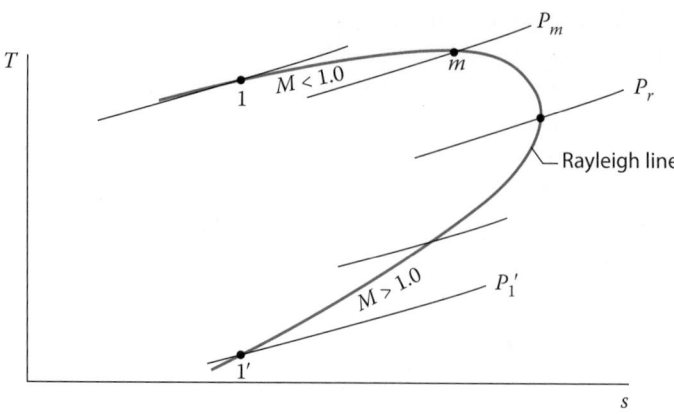

Fig. 3.57   The Rayleigh line.

the mathematical solution of Eqs. (3.37) and (2.48)? That is, beginning at state 1 on the Rayleigh line of Fig. 3.57 and at station 1 in Fig. 3.56, how can the gas be made to proceed through the Rayleigh line state points?

The answer lies in the second law of thermodynamics, which is provided below consistent with applicable assumptions for this simple heating flow:

$$\dot{P}_s = \dot{m}(s_2 - s_1) + \frac{\dot{Q}_{\text{out}}}{T_L} - \frac{\dot{Q}_{\text{in}}}{T_H} \geq 0 \qquad (2.27)$$

Recall what the second law Eq. (2.27) tells us: the rate of entropy production of the system (first term) *and* the surroundings (second two entropy rate due to heat transfer terms) must increase for all real processes; the equality holds true only for idealized reversible processes. Applied to simple heating flow, Eq. (2.27) tells us that for all real processes: (1) energy addition due to a heat interaction *into the gas* $\dot{Q}_{\text{in}}$ must result in an increase in entropy of the gas because there are no other irreversibilities (e.g., friction) present, that is, $s_2 - s_1 > 0$; and (2) energy removal due to a heat interaction *out of the gas* $\dot{Q}_{\text{out}}$ must result in a decrease in entropy of the gas, that is, $s_2 - s_1 < 0$, but with a magnitude less than the entropy increase of the surroundings.

Starting at point 1 on Fig. 3.57:

1. Energy addition due to $\dot{Q}_{\text{in}}$ (heat flow) downstream of station 1 is represented by a progression of the gas along the Rayleigh line to the right to states of higher entropy, lower pressure, and higher temperature up to point *m*. After state *m* is reached, further heat flow into the gas produces a decrease in temperature but will continue to decrease the pressure and, up to point *r*, increase the entropy. It is impossible to reach values of entropy higher than state *r* on the Rayleigh line passing through state 1. It is not possible, therefore, to have heat flow in excess of that

which takes the gas to r and still remain on the same Rayleigh line; this would violate the second law.

2. Heat flow *from* the gas $\dot{Q}_{out}$ downstream of station 1 will cause the gas to proceed to the left of the Rayleigh line to states of lower entropy, higher pressure, and lower temperature.

Let us show that point r is a sonic point on the Rayleigh line. Refer to the analysis in Section 3.6 and Fig. 3.42. Applied to a constant area duct ($dA = 0$):

From continuity:
$$\frac{d\rho}{\rho} + \frac{dV}{V} = 0$$

From momentum:
$$dP + \frac{\rho V dV}{g_c} = 0$$

Combining these two equations:
$$dP = \frac{V^2 d\rho}{g_c} \quad \text{or} \quad \frac{V^2}{g_c} = \frac{dP}{d\rho}$$

But, from Eq. (2.7):
$$\left(\frac{\partial P}{\partial \rho}\right)_s = \frac{a^2}{g_c}$$

So for the special case of $ds = 0$ ($s = $ constant), which corresponds to point r on Fig. 3.57, we have

$$\frac{V^2}{g_c} = \frac{a^2}{g_c} \quad \text{or} \quad V = a \quad \text{at point } r$$

After state point r has been reached by a heat flow to the gas stream of Fig. 3.58 between 1 and r, the gas can theoretically proceed to lower pressure and entropy by a heat flow from the gas downstream of station r, as shown schematically in the figure. (It seems quite probable, however, that one could obtain, experimentally, the transition from subsonic to supersonic flow by this method.) This corresponds to accelerating the gas from subsonic to supersonic flow by reversing the direction of heat flow downstream of point r. This situation is the simple heating counterpart to reversing the area variation from converging to diverging in a simple area flow in order to accelerate the gas from subsonic to supersonic conditions.

Consider a subsonic simple heating flow in which sufficient heat flow is present to produce sonic exit conditions at the duct exit; the flow is *thermally choked*. The process is represented in Fig. 3.59 where 1 represents the duct inlet station and 2 represents the exit. What happens if the heat flow is increased to a value in excess of that required to produce sonic flow at the exit? We have already seen that continuing on the same Rayleigh line would violate the second law. This added heat flow produces a readjustment in the gas flow, which results in a reduced mass flow rate and operation on a different Rayleigh line with a Mach number of 1 still maintained at the exit. The new flow process lies on a Rayleigh line such as the dashed one in the figure with the flow proceeding from $1'$ to $2'$ and with $T_{t2'} > T_{t2}$. This choking phenomenon is analogous to choking in a simple convergent nozzle flow with sonic exit conditions where if the nozzle exit area is

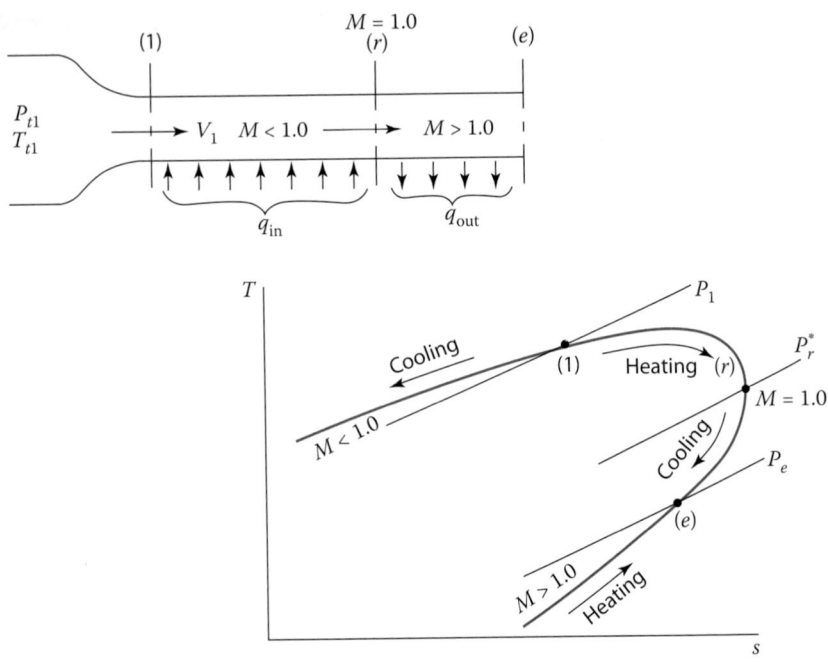

Fig. 3.58   Transition through sonic point in simple heating flow.

reduced, the mass flow is reduced and sonic conditions continue to be maintained at the exit.

Consider a turbojet. The flow is usually choked at the turbine nozzle throat. The Mach number at the exit of the combustion chamber M_{exit} is fixed by the area ratio of the combustor exit to the turbine nozzle throat. If a further heat interaction occurs, M_{exit} cannot increase. Instead, the combustion chamber pressure will increase. Hopefully, the compressor can supply this pressure (with the increased work input from the turbine) without aerodynamic stall occurring in the compressor.

3.7.2.1 Analytical Relations for Simple Heating Flow

For simple heating flow, the applicable functional relations in Table 3.1 are found by zeroing out the dA/A and $(4c_f dx)/D$ columns. General conclusions can be drawn from Table 3.1 concerning the variation of the gas properties in a simple heating flow. For example, we see from the relation

$$\frac{dP_t}{P_t} = -\frac{\gamma M^2}{2}\frac{dT_t}{T_t} \tag{3.38}$$

that increasing the total temperature (heat flow to the gas) causes a decrease in total pressure. This indicates, for example that in a turbojet engine combustion chamber, a total pressure loss occurs due to the rise of total temperature. This loss is over and above the loss in total pressure due to frictional effects.

Each coefficient of dT_t/T_t in Table 3.1, except that of Eq. (3.38), has the term $(1 - M^2)$. As a result, the variation of the dependent properties with total temperature is of opposite sign in subsonic and supersonic flow, the mathematical verification of the property variations already discussed graphically by using the Rayleigh line.

To facilitate the tabulation of Rayleigh flow property relations, it is convenient to let the reference state be the state at which sonic conditions occur (point r in Figs. 3.57 and 3.58), just as was done with simple area flow and A/A^* (and will be done with simple frictional flow). A comparable relation in simple heating flow is T_t/T_t^*, where T_t^* represents the total temperature at the sonic point on a Rayleigh line. The ratio T_t/T_t^* and other similar property relations can be found by integration of the relations in Table 3.1. The first equation in Table 3.1 can be solved for dT_t/T_t and integrated between any general point $M = M$ and $T_t = T_t$ and the sonic point where $M = 1.0$ and $T_t = T_t^*$ to obtain T_t/T_t^* as a function of γ and M. The final result is

$$\frac{T_t}{T_t^*} = \frac{2(\gamma + 1)M^2}{(1 + \gamma M^2)^2}\left(1 + \frac{\gamma - 1}{2}M^2\right) \tag{3.39}$$

In terms of the simple heating flow of Fig. 3.60, Eq. (3.39) tells us that the total temperature at station 1 divided by T_t^* at sonic point r on the Rayleigh line explicitly determines the Mach number at station 1. Similarly, $(T_t/T_t^*)_2$ is related to M_2. It should be made clear that it is not necessary that T_t^*

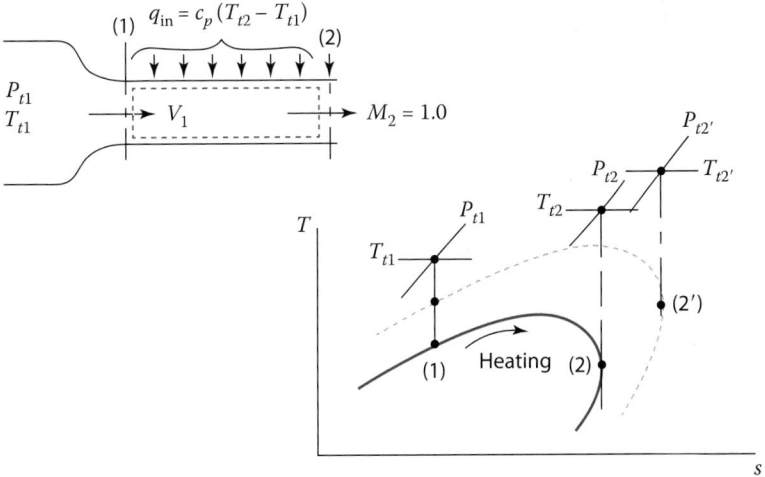

Fig. 3.59 Choked simple heating flow.

physically occur in the flow of any given problem. Just like with A/A^* (Example 3.2), we can imagine sufficient heat flow added to take the flow to sonic conditions along a Rayleigh line.

Through Eq. (3.39), one can evaluate the effect of total temperature changes on the flow Mach number. Suppose, for example, that $M_1 = 0.34$ and $q_{in} = \dot{Q}_{in}/\dot{m}$ is such that $T_{t2}/T_{t1} = 2.0$. From Eq. (3.39) with $M_1 = 0.34$ and $\gamma = 1.4$, we find that $(T_t/T_t^*)_1 = 0.42$. Then

$$\left(\frac{T_t}{T_t^*}\right)_2 = \left(\frac{T_t}{T_t^*}\right)_1 \frac{T_{t2}}{T_{t1}}$$

gives $(T_t/T_t^*)_2 = 0.84$. A tabulation of Eq. (3.39) for $\gamma = 1.4$ provided in Appendix G shows that for this value of $(T_t/T_t^*)_2$, $M_2 = 0.62$. Note that Appendix G provides other property relations referenced to the sonic condition for P_t, T and P acquired in a similar manner as Eq. (3.39). A plot of stream property ratios versus Mach number for simple heating flow is provided in Fig. 3.61 for $\gamma = 1.4$.

Remember to think of T_t as the independent variable (controlled by heat flow) when using the plot of Fig. 3.61 or Appendix G. Either way, we begin with known properties at station 1 and with some known property at station 2, say P_2. From the known M_1, we can find $(T_t/T_t^*)_1$, $(P/P^*)_1$, and so on. The ratio $(P/P^*)_2$ is found in a manner analogous to the example above for $(T_t/T_t^*)_2$, such that

$$\left(\frac{P}{P^*}\right)_2 = \left(\frac{P}{P^*}\right)_1 \frac{P_2}{P_1}$$

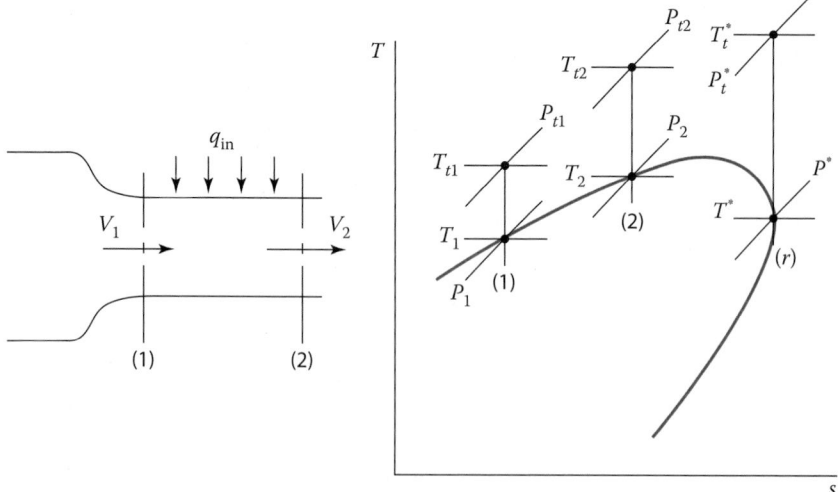

Fig. 3.60 Interpretation of T/T^* in simple heating flow.

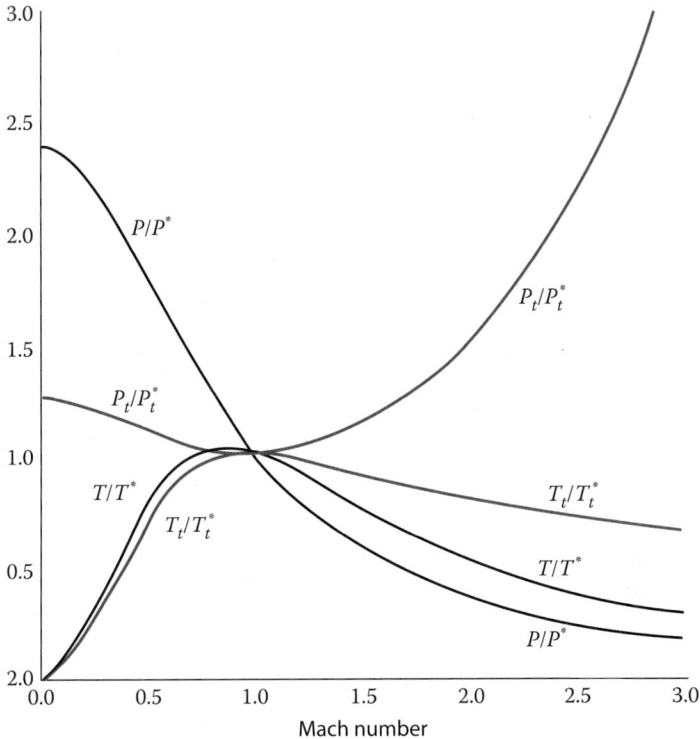

Fig. 3.61 Variation of stream properties with Mach number in simple heating flow ($\gamma = 1.4$).

From this, M_2 can be determined directly and the remaining state 2 properties are found in a similar manner.

For values of γ different from 1.4, of course other tabulations like Appendix G could be constructed, or we could solve Eq. (3.39) for the Mach number (the function $\phi(M^2)$ provided in Appendix G) and then solve for the downstream variables. Rayleigh flow solutions for any γ are included in GASTAB.

Some important observations concerning simple heating flow or Rayleigh line flow:

- A flowing gas can thermally choke (i.e., achieve Mach number of unity due to simple heating)
- In accordance with the first law, heating of the flow causes T_t to increase; cooling of the flow causes T_t to decrease. Heating beyond that required to choke the flow ($M = 1$, $T_t = T_t^*$) causes a change in upstream flow conditions manifested by operation on a different Rayleigh line of lower mass flow rate still with $M = 1$ at the exit plane.
- In accordance with the second law, the flowing gas entropy will increase with heating. The gas entropy will decrease with cooling. In the case of

cooling flow, the entropy rise of the surroundings (outside the system) will increase in greater proportion to the system entropy decrease.

* Some of the interesting, and perhaps counterintuitive, static property changes associated with Rayleigh flow can be explained by our foundational understanding and Fig. 3.61. Given the simplifying assumptions of Rayleigh line flow, there are really only two accountings for where the energy supplied by heating (or removed by cooling) the flow can be manifested. Again, we turn to the first law, which we simplified for this flow as

$$q_{in} = c_p(T_{t\ out} - T_{t\ in})$$

but from the definition of T_t, Eq. (3.2), we see that

$$q_{in} = c_p\left[\left(T_2 + \frac{V_2^2}{2g_c c_p}\right) - \left(T_1 + \frac{V_1^2}{2g_c c_p}\right)\right]$$

or rearranging

$$q_{in} = c_p\left[(T_2 - T_1) + \left(\frac{V_2^2}{2g_c c_p} - \frac{V_1^2}{2g_c c_p}\right)\right]$$

So, the energy supplied by heating goes into raising the macroscopic sum of the molecular energies of the gas manifested by an increase in the static temperature T and/or raising the bulk kinetic energy of the flowing gas manifested by an increase in the streamwise velocity V. Given this, insights into Rayleigh line flow behavior (Fig. 3.57) can be gleaned with the help of Fig. 3.61, which graphically shows rates of changes (slopes) of properties, most notably T/T^* and P/P^* (representative of velocity). For example, starting at point 1 on Fig. 3.57, as heating occurs, T increases and P decreases (indicative of V increasing) until point m, where T begins decreasing as P continues to decrease. From Fig. 3.61, we initially see similar magnitude changes in slopes (but of opposite signs) of T and P until slightly before $M = 1.0$, when T goes through zero slope (this must correspond to point m on Fig. 3.57) and then goes negative slope from that point on. So, up until point m on the Rayleigh line, the energy is shared between the specific internal enthalpy h ($\Delta h = c_p\Delta T$ for a CPG) and specific kinetic energy ke manifested by an increase in both terms in the first law equation. After point m, the continued energy addition due to heating *and* some of the h goes entirely into raising the ke manifested by the drop in static temperature.

To show the similarity of the solution process, example problems for both Rayleigh and Fanno line flow are provided after the introduction of simple frictional (Fanno) flow in the next section.

3.7.3 Simple Frictional Flow: Fanno Line Flow

The analysis of simple frictional flow sketched in Fig. 3.62, or *Fanno line flow*, follows that of simple heating flow just provided. All assumptions provided at the start of Section 3.7.2 apply here with one change. Because we are now concerned with *friction as the dominant independent parameter*, we must consider its influence on flow properties, and so for simple frictional flows, we assume negligible heat transfer.

With this assumption and those previously listed, the first law becomes

From Eq. (2.22), energy

$$T_t = \text{const} = T_1 + \frac{V_1^2}{2c_p g_c} = T + \frac{V^2}{2c_p g_c}$$

The continuity equation provides the same result as it did with Rayleigh line flow:

From Eq. (2.13), continuity

$$\rho_1 V_1 = \rho V = \frac{\dot{m}}{A} = \text{constant}$$

As we did with Rayleigh line flow, we combine these two equations using the perfect gas equation of state for the density to obtain

$$T_t = \text{const} = T_1 + \frac{V_1^2}{2c_p g_c} = T + \frac{1}{2c_p g_c}\left(\frac{\dot{m}}{A}\right)^2 \left(\frac{RT}{P}\right)^2 \qquad (3.40)$$

For given inlet conditions at station 1, this equation represents a relation in terms of pressure and temperature that must be satisfied at any point in the flow. By assuming values of T to exist at points in the flow, it is possible, with Eq. (3.40), to determine the corresponding values of P, thus fixing the state of the gas (P, T, T_t) at these points in the flow. Further, by assuming an arbitrary value of specific entropy s_1 the entropy at any other point of known T and P is

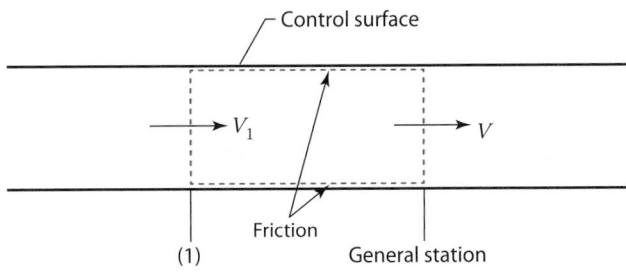

Fig. 3.62 Simple frictional flow.

determined by

$$s - s_1 = c_p \ell n \left(\frac{T}{T_1} \right) - R \ell n \left(\frac{P}{P_1} \right) \qquad (2.48)$$

Fanno lines, the temperature-specific entropy locus of the fluid states satisfying Eqs. (3.40) and (2.48), are sketched in Fig. 3.63. The family of Fanno lines obtained for parametric values of T_t with fixed $\dot{m}/A$ is shown in Fig. 3.63a whereas Fig. 3.63b shows the family of Fanno lines corresponding to

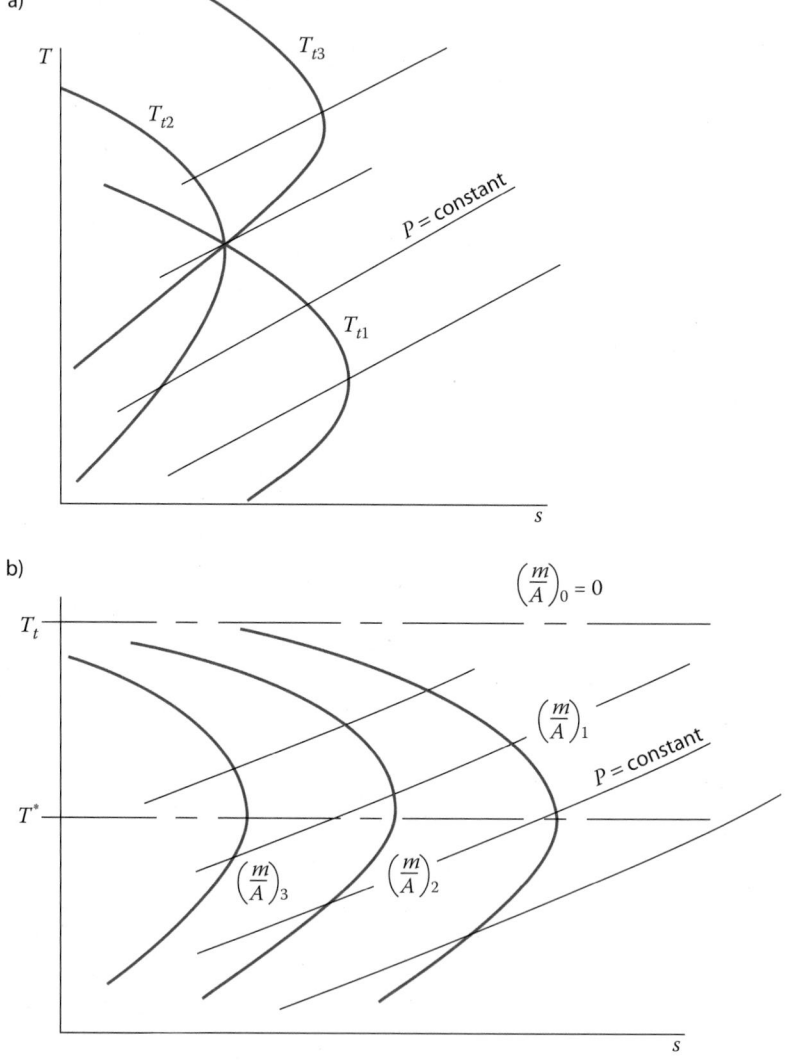

Fig. 3.63   Fanno lines.

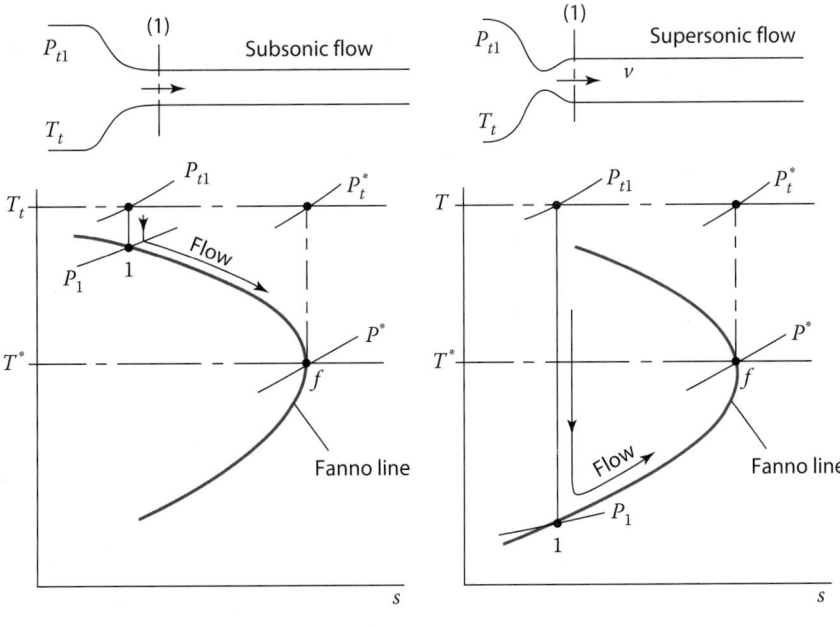

Fig. 3.64 Simple friction flow.

parametric values of $\dot{m}/A$ with T_t held constant. As with Rayleigh line flow, the second law of thermodynamics dictates possible flow excursions on a Fanno line. The second law consistent with applicable assumptions for simple frictional flow is from Eq. (2.27):

$$\dot{P}_s = \dot{m}(s_2 - s_1) > 0 \Rightarrow (s_2 - s_1) > 0$$

We note that the equality sign is dropped in this equation because it only holds true in isentropic, adiabatic flows; simple flow with friction is irreversible. Thus, simple frictional flow can proceed only to the right on a Fanno line.

As an illustration of flows along a Fanno line, consider a perfect gas to be flowing from a large reservoir through a nozzle and then through a simple frictional duct of fixed length. Two cases, one subsonic and one supersonic, are illustrated in Fig. 3.64. In the subsonic case, flow begins in the reservoir at state point (P_t, T_t) on the temperature–entropy diagram and proceeds isentropically to the nozzle exit at (P_1, T_1), then along the Fanno line through 1 to states of increasing entropy, decreasing temperature and pressure, and increasing Mach number, tending toward a limiting sonic point f.

The supersonic flow of Fig. 3.64 is for the same T_t and $\dot{m}/A$ and therefore is on the same Fanno line as the subsonic flow of the figure. The flow entering the frictional duct at station 1 progresses along the supersonic branch of the

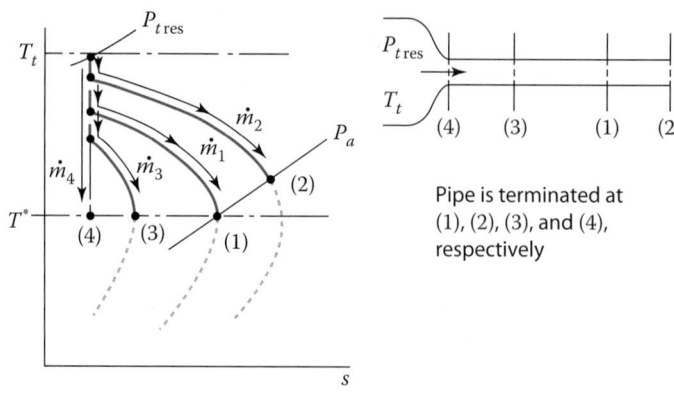

Fig. 3.65 Effect of duct length in subsonic flow.

Fanno line to states of increasing entropy, increasing temperature and pressure, and decreasing Mach number, tending to the same limiting sonic point f. Thus, we see similar flow behavior for Fanno line flow as we do for Rayleigh line heating flow (q_{in}). In all cases of simple frictional flow, P_t decreases and s increases.

For given inlet conditions to a simple frictional duct, there exists a Fanno line representing the *possible* states that the flow may proceed through in the duct. Whether a portion of or all these possible states are attained by the gas as it flows through the duct depends on the *amount of frictional duct length* and pressures imposed on the boundaries (inlet and exit) of the flow system.

3.7.3.1 Effect of Frictional Duct Length (Subsonic Flow)

The flow unit of Fig. 3.65 consists of a reservoir maintained at $P_{t\,res}$ and T_t, an exhaust region of constant ambient pressure P_a, a convergent nozzle, and a constant-area pipe whose length can be adjusted to terminate at station 1, 2, 3, or 4. As a starting point, let the unit be such that at its exit $M = 1$ and the exhaust region is just attained, $P_1 = P_a$, as indicated in Fig. 3.65.

For this case 1, the flow through the unit from ($P_{t\,res}$, T_t) follows isentropically down to the inlet of the simple frictional duct and then along a Fanno line corresponding to $\dot{m}_1$ exiting at $M = 1$ at the duct exit section 1. If now the duct length is increased to 2, everything else remaining the same, we find the flow process to follow along a Fanno line corresponding to a lower mass flow rate of $\dot{m}_2$. For this latter case, the flow is subsonic throughout the length of the frictional duct, including at the exit plane, as indicated in Fig. 3.65 ($T_2 > T^*$). If the duct length is increased further beyond station 2, the mass flow in the unit continues to decrease and, in the limit, tends to zero as the duct length tends to infinity. Now, on the other hand, if the duct length is decreased to station 3, we find the flow process to proceed isentropically down to the duct inlet and then along a Fanno line of mass flow rate $\dot{m}_3 > \dot{m}_1$, still exiting at $M = 1$ but at $P_3 > P_a$. As the duct length goes to zero, the Mach number at the nozzle exit increases to 1, corresponding to a maximum mass rate of flow $\dot{m}_4$ through the nozzle. (As the duct length goes to zero, the unit becomes a simple convergent nozzle.) Beginning at condition 4, we observe that adding duct lengths to the nozzle produces a decrease in mass flow rate, that is, the flow is *choked by friction*.

3.7.3.2 Effect of Frictional Duct Length (Supersonic Flow)

Let the unit of Fig. 3.66 be operating such that the flow leaving the pipe is at $M = 1$ and at the exhaust region pressure with a mass flow rate of $\dot{m}_1$. This condition is indicated on the T-s diagram of Fig. 3.66 where the flow process originates at $(P_{t\,res}, T_t)$, proceeds isentropically to the supersonic branch of the Fanno line corresponding to $\dot{m}_1$, and then follows along this Fanno line to $M = 1$ and $P_1 = P_a$. Now, as the duct length is increased to

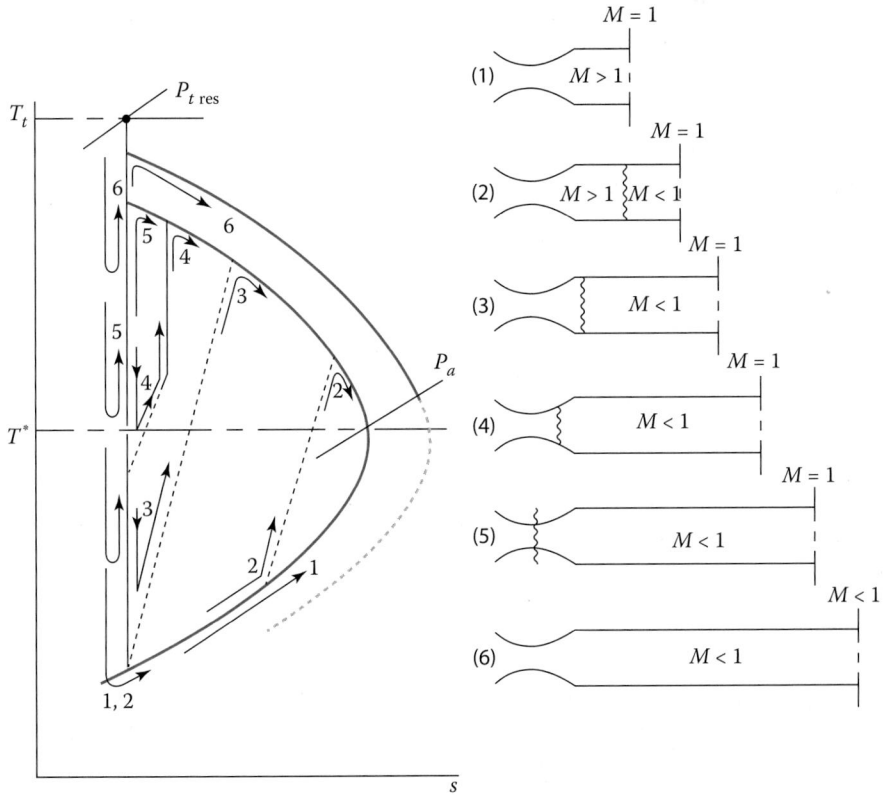

Fig. 3.66 Flow along Fanno line with normal shock.

condition 2, we find that the new boundary condition of increased duct length can be satisfied by assuming a normal shock to occur at a point in the duct such that the combination of duct length preceding and duct length following the flow discontinuity produces a Mach number of 1 at the duct exit. The flow process corresponding to this condition is shown by the arrows numbered 2 (Fig. 3.66). As the duct length is increased further, the normal shock progresses upstream to the duct inlet, then into the nozzle until it reaches the nozzle throat at 5. Further increase in length beyond that corresponding to 5 reduces the mass flow rate through the unit, and the flow progresses subsonically throughout the duct for these cases along Fanno lines of lower mass flow rates exiting at $P_6 = P_a$ but $M_6 < 1$, as indicated by 6 in Fig. 3.66.

Suppose, now, a flow corresponding to condition 1 exists, and let the duct length be reduced. In this case, we find that the stream properties in the remaining portion are unaffected and that as the duct length is reduced to zero, the flow reduces to that through a convergent-divergent nozzle, exhausting to the discharge region through a system of oblique shock waves set up in the exhaust region. These conditions are illustrated schematically in Fig. 3.67. For condition 1, the flow proceeds isentropically to the

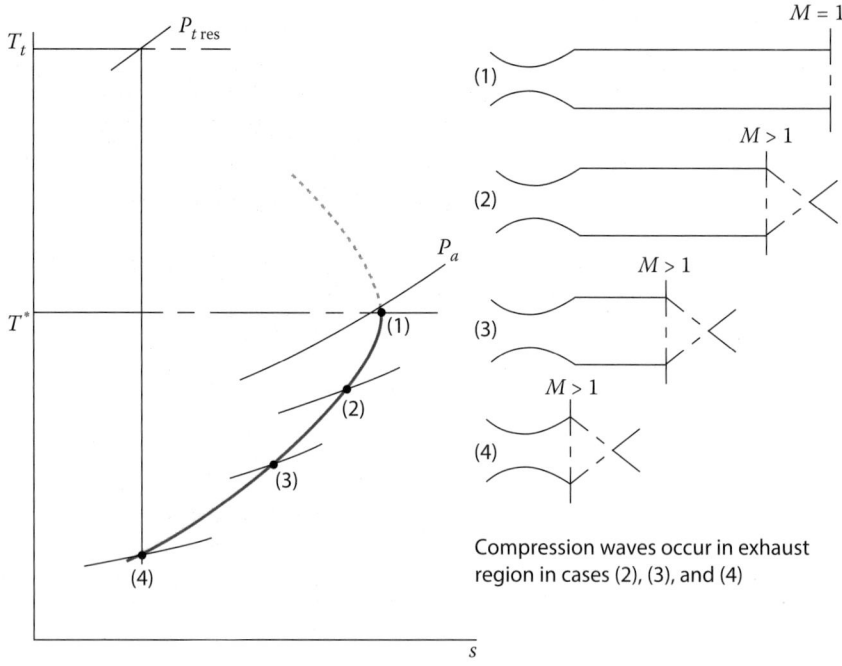

Fig. 3.67 Flow along supersonic branch of Fanno line.

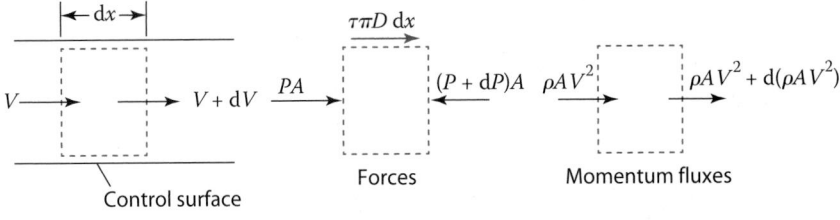

Fig. 3.68   Momentum equation for simple frictional flow.

duct inlet, then along a Fanno line to $M = 1$ at 1. For 2, 3, and 4, the exit conditions from the unit are as indicated on the T-s diagram (Fig. 3.67). Notice that P_2, P_3, and P_4 are each less than P_a. The rise in pressure to P_a in these cases is attained through a series of oblique shock waves set up from the exit of the duct.

3.7.3.3 Analytical Relations for Simple Frictional Flow

The solution of Fanno line flow proceeds in a manner analogous to that of simple area flow and Rayleigh line flow. For simple frictional flow, the applicable functional relations in Table 3.1 are found by zeroing out the dA/A and dT_t columns. The differential form of the governing equations are integrated between any general condition and a convenient reference condition chosen such that $M = 1.0$ at that condition. We have already introduced the idea of frictional choking, so your intuition should tell you that the controlling parameter must have something to do with frictional duct length. To explore this further, we turn to the steady, 1-D form of the linear momentum equation.

We apply the momentum equation to the flow through the control surface in the simple frictional flow of Fig. 3.68 and introduce the duct friction factor c_f and duct length x into the analysis. These terms, initially presented in Section 3.6, arise in the momentum equation by evaluating the force of the duct on the flowing gas. The resulting equation from Eq. (2.16) is

$$-A dP - \tau \pi D dx = \frac{1}{g_c} d(\rho A V^2) \qquad \text{(i)}$$

In Eq. (i), D is the hydraulic diameter introduced in Section 3.6 and τ is the frictional wall shear stress acting over the wetted perimeter area $\pi D dx$. The frictional coefficient c_f is defined as

$$c_f \equiv \frac{\tau}{\rho V^2 / 2 g_c} \Rightarrow \tau = \frac{c_f \rho V^2}{2 g_c} \qquad \text{(ii)}$$

The pipe circumference can be expressed in terms of the pipe area ($\pi D^2/4$) as follows:

$$\pi D = 4 \frac{\pi D^2}{4D} = \frac{4A}{D} \tag{iii}$$

Substituting Eqs. (ii) and (iii) into Eq. (i) and simplifying gives us

$$dP + \frac{\rho V^2}{2g_c} \frac{4c_f}{D} dx + \frac{1}{g_c} d(\rho V^2) = 0$$

To put this in logarithmic differential form, we divide through by P, use $\rho V^2/2g_c = \gamma M^2$, and note that $d(\rho V^2) = 0$ from continuity to obtain

$$\frac{dP}{P} + \gamma M^2 \frac{dV}{V} + \frac{\gamma M^2}{2} \frac{4c_f}{D} dx = 0$$

The quantity $4c_f dx/D$ is selected as the independent variable because it relates to the frictional duct length. From the first equation obtained from the third column of Table 3.1, we solve for $4c_f dx/D$ to obtain

$$\frac{4c_f dx}{D} = \frac{1 - M^2}{\gamma M^2 \left(1 + \frac{\gamma - 1}{2} M^2\right)} \frac{dM^2}{M^2} \tag{3.41}$$

We integrate Eq. (3.41) from any duct length L and Mach number M to the duct length L^* where $M = 1.0$ (Fig. 3.69). To perform the integration, we assume a constant value of c_f so that the left side of Eq. (3.41) becomes

$$\int_L^{L^*} \frac{4c_f dx}{D} = \frac{4c_f}{D}(L^* - L) = \frac{4c_f L_{max}}{D}$$

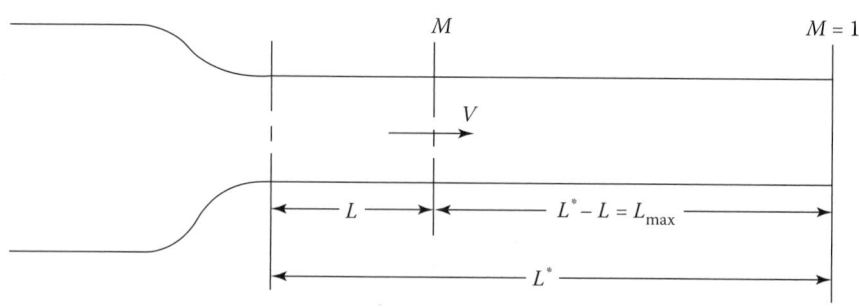

Fig. 3.69 Simple frictional flow with sonic exit conditions.

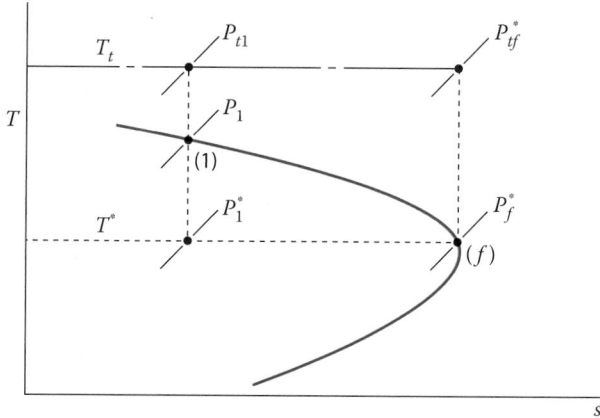

Fig. 3.70 Fanno line interpretation of $4c_f L_{max}/D$, P/P_f^*, and P_t/P_{tf}^*.

Figure 3.69 helps explain L_{max}. It is the maximum pipe length that can be added beyond a given station without affecting the stream properties at the station, regardless of the exhaust region pressure. When this maximum length exists, sonic flow conditions will occur at the pipe exit. Finally, integration of the right side of Eq. (3.41) gives

$$\frac{4c_f L_{max}}{D} = \frac{1 - M^2}{\gamma M^2} + \frac{\gamma + 1}{2\gamma} \ell n \left[\frac{(\gamma + 1)M^2}{2\left(1 + \dfrac{\gamma - 1}{2}M^2\right)} \right] \qquad (3.42)$$

Similarly, the functional relationships for the remaining dependent properties of Table 3.1 can be evaluated for Fanno line flow and presented as property relations relative to the * condition corresponding to the frictional duct length needed to choke the flow on a given Fanno line. These are provided in Appendix H for $\gamma = 1.4$. Fanno flow solutions for any γ are included in GASTAB.

The T-s diagram of Fig. 3.70 helps interpret the relation $4c_f L_{max}/D$. We use a subscript f associated with the * quantities of the Fanno line to distinguish them from the * quantities associated with isentropic flow. As is evident from Fig. 3.70, P_f^* is constant for a given Fanno line whereas the isentropic P^* will vary from state point to state point on a Fanno line. The subscript f will not generally be used with the starred quantities, just as the subscript r will not be used for Rayleigh flow. The way to think about the * reference condition is that it represents the property states where *Mach number of unity has been reached by some particular process*. We have looked at three particular simple processes: area change, heating flow, and frictional flow.

Figure 3.71 shows a plot of several of the property relations for simple frictional flow for $\gamma = 1.4$. Remember to think of $4c_f L_{max}/D$ as the

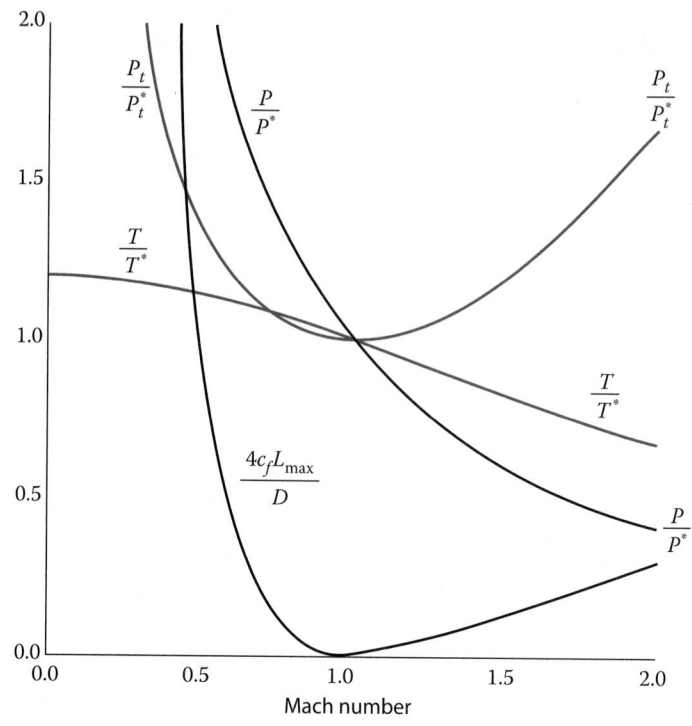

Fig. 3.71 Fanno line relations ($\gamma = 1.4$).

independent variable (controlled by frictional duct length) when using the plot of Fig. 3.71 or Appendix H.

Some important observations concerning simple frictional flow or Fanno line flow:

- A flowing gas can frictionally choke (i.e., achieve Mach number of unity due to simple friction). Physically, this should make sense because boundary layer growth—the region of low momentum flow near the duct walls caused by friction—essentially results in a decreasing duct area in the streamwise direction.

- In accordance with the first law, the total temperature of the flowing gas (the system) stays constant in simple frictional flow. There are no interactions with the surroundings. Frictional duct length added beyond that required to just choke the flow ($M = 1$, $T_t = T_t^*$) causes a change in upstream flow conditions manifested by operation on a different Fanno line of lower mass flow rate.

- In accordance with the second law, there are always irreversible losses that result in an increase in system entropy manifested by a decrease in total pressure of the flowing gas in simple frictional flow.

3.7.4 Examples of Simple Heating Flow and Simple Frictional Flow

Example 3.10

One-dimensional flow of air occurs in a frictionless, constant-area duct while energy is added to the air by a heat interaction. The air enters the duct with a total temperature of 450 K, total pressure of 8 atm, and Mach number of 0.3. Find the heat interaction (kJ/kg), exit total temperature, and total pressure that provides

1. Choked flow at the exit
2. For an exit Mach number of 0.6

Solution

As is typical, we start with a simple system sketch showing pertinent interactions, as shown in Fig. 3.72. We assume Rayleigh line flow and use Appendix G with $\gamma = 1.4$. From the first law applied to Rayleigh line flow:

$$q_{\text{in}} = c_p(T_{t2} - T_{t1})$$

1. Choked flow at the exit implies $M_2 = 1.0$ and represents the $*$ reference condition for Rayleigh line flow. Thus, from Appendix G with $M_1 = 0.3$:

$$\left(\frac{T_t}{T_t^*}\right)_1 = 0.346860 \qquad \left(\frac{P_t}{P_t^*}\right)_1 = 1.198549$$

and with $M_2 = M^* = 1.0$: $T_{t2} = T_{t1}^*$ $P_{t2} = P_{t1}^*$
Thus

$$T_{t2} = \frac{T_{t1}^*}{T_{t1}} = T_{t1} = \frac{1}{0.346860} 450 \text{ K} = 1297.4 \text{ K}$$

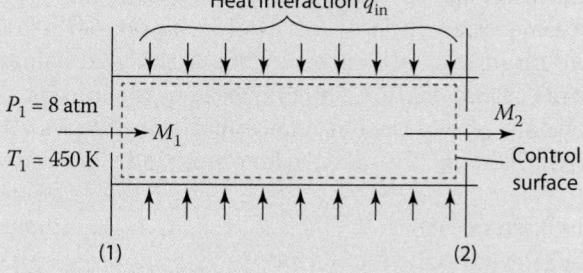

Heat interaction q_{in}

$P_1 = 8$ atm

$T_1 = 450$ K

M_1

M_2

Control surface

(1)   (2)

Fig. 3.72 System sketch.

(Continued)

Example 3.10 *(Continued)*

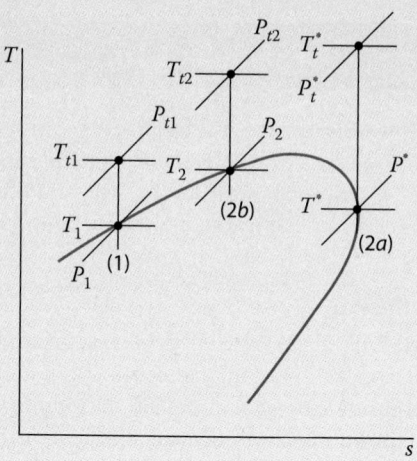

Fig. 3.73 *T-s* diagram sketch showing states 1 and 2a (choked) and 2b.

Likewise

$$P_{t2} = \frac{1}{1.198549} 8 \text{ atm} = 6.675 \text{ atm}$$

and with $c_p = 1.004 \text{ kJ}/(\text{kg} \cdot \text{K})$ from Table 2.1

$$q_{in} = 1.004 \frac{\text{kJ}}{\text{kg} \cdot \text{K}} (1297.4 - 450)\text{K} = 850.8 \frac{\text{kJ}}{\text{kg}}$$

2. For $M_2 = 0.6$, from Appendix G, we have

$$\left(\frac{T_t}{T_t^*}\right)_2 = 0.818923 \qquad \left(\frac{P_t}{P_t^*}\right)_2 = 1.075253$$

A *T-s* diagram showing the three states representative of this example problem helps visualize the Rayleigh line flow under consideration (Fig. 3.73).

Thus

$$T_{t2} = \frac{T_{t2}}{T_{t2}^*} \frac{T_{t1}^*}{T_{t1}} T_{t1} = (0.818923) \frac{1}{0.346860} 450 \text{ K} = 1062.4 \text{ K}$$

(Continued)

Example 3.10 (Continued)

Likewise

$$P_{t2} = (1.075253)\frac{1}{1.198549}8 \text{ atm} = 7.177 \text{ atm}$$

and

$$q_{in} = 1.004\frac{kJ}{kg \cdot K}(1062.4 - 450)K = 614.8\frac{kJ}{kg}$$

All of these part 2 answers make sense because the T_{t2} and q_{in} values should be less than those obtained in part 1. The total pressure obtained in part 2 should be greater than that obtained in part 1 because the additional heat transfer needed to choke the flow would result in increased system irreversibilities (losses) manifested by a lower P_{t2} compared to part 2.

Example 3.11

Figure 3.74 shows the data obtained from an experimental setup to determine the friction factor c_f of the duct material between stations 1 and 2 in an airflow experiment. Estimate the friction factor and the force required to hold the duct in place.

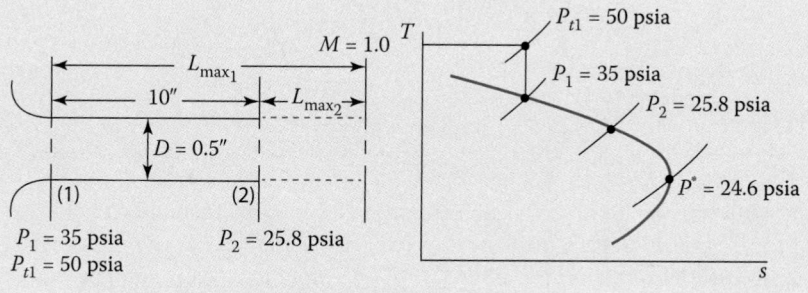

Fig. 3.74   Experimental data for determining the friction factor.

(Continued)

Example 3.11 *(Continued)*

Solution

We assume Fanno line flow, and because we are given no temperature information, we assume $\gamma = 1.4$. This allows us to use Appendix H or GASTAB to obtain our Fanno line property relations.

$$\left(\frac{P}{P_t}\right)_1 = \frac{35 \text{ psia}}{30 \text{ psia}} = 0.7$$

and from Appendix D, $\gamma = 1.4$: $M_1 = 0.73$.

From Appendix H with $M_1 = 0.73$, we get the following property relations:

$$\left(\frac{4c_f L^*}{D}\right)_1 = 0.15605 \quad \left(\frac{P_t}{P_t^*}\right)_1 = 1.0742 \quad \left(\frac{P}{P^*}\right)_1 = 1.4265$$

$$\left(\frac{I}{I^*}\right)_1 = 1.0378$$

Thus

$$\left(\frac{P}{P^*}\right)_2 = \left(\frac{P}{P^*}\right)_1\left(\frac{P_2}{P_1}\right) = 1.4265\left(\frac{25.8 \text{ psia}}{35 \text{ psia}}\right) = 1.05155$$

From GASTAB, with this value of $(P/P^*)_2$ we get

$$M_2 = 0.9576 \quad \left(\frac{4c_f L^*}{D}\right)_2 = 0.00232 \quad \left(\frac{P_t}{P_t^*}\right)_2 = 1.0015 \quad \left(\frac{I}{I^*}\right)_2 = 1.0007$$

Referring to Fig. 3.74, we note that

$$\left(\frac{4c_f L_{\max}}{D}\right)_1 = \frac{4c_f L_{1-2}}{D} + \left(\frac{4c_f L_{\max}}{D}\right)_2$$

$$\Rightarrow \frac{4c_f L_{1-2}}{D} = 0.15605 - 0.00232 = 0.15373$$

and therefore

$$c_f = \frac{0.15373(0.5 \text{ in})}{4(10 \text{ in})} = 0.0019$$

We can use the impulse function to estimate the force required to hold the 10-in segment of duct in place. Referring to Section 3.2.6, the streamwise axial force F exerted on the air is $I_2 - I_1$. We note that for Fanno line flow

$$\frac{I}{I^*} = \frac{P}{P^*}\frac{1 + \gamma M^2}{1 + \gamma}$$

(Continued)

Example 3.11 *(Continued)*

So

$$F_{\text{on air}} = \left[\left(\frac{I}{I^*} \right)_2 - \left(\frac{I}{I^*} \right)_1 \right] I^* \quad \text{where}$$

$$I^* = P^* A (1 + \gamma) \Rightarrow F_{\text{on air}} = (1.0007 - 1.0378)I^* = -0.0371I^*$$

Using the static pressure and area at station 1 to obtain I^*, we obtain

$$F_{\text{on air}} = -0.0371I^* = -0.0371 \frac{35 \text{ psia}}{1.4265} \left(\frac{\pi (0.5)^2 \text{in}^2}{4} \right) 2.4 = -0.43 \text{ lbf}$$

The negative sign tells us the force on the air is to the left of the example figure; the force on the duct by the air is 0.43 lbf to the right. Therefore, a force of 0.43 lbf to the left is required to hold the duct in place.

Notice in simple frictional flow that the basic relationship between station 1 and station 2 properties is governed by the frictional duct length

$$\left(\frac{4 c_f L_{\max}}{D} \right)_2 = \left(\frac{4 c_f L_{\max}}{D} \right)_1 - \frac{4 c_f L_{1-2}}{D}$$

In simple area flow and simple heating flow, the analogous relations are governed by the area and total temperature, respectively, namely,

$$\left(\frac{A}{A^*} \right)_2 = \left(\frac{A}{A^*} \right)_1 \frac{A_2}{A_1} \quad \text{and} \quad \left(\frac{T_t}{T_t^*} \right)_2 = \left(\frac{T_t}{T_t^*} \right)_1 \frac{T_{t2}}{T_{t1}}$$

3.8 Summary of Simple Flows

A simple flow is defined as one in which all but one of the independent variables in Table 3.1 are zero. The three types of simple flows presented in this chapter are summarized in Table 3.4.

In the temperature–entropy diagrams of Table 3.4, the path lines of states corresponding to simple area flow, simple heating flow, and simple frictional flow are shown. These path lines are called the *isentrope, Rayleigh,* and *Fanno lines*, respectively.

To proceed downward along the *isentrope line* from point a of the diagram, the flow area is decreased until the sonic point at b is reached. The area must be increased after point b in order to continue down to point c. By proper adjustments in the flow area and boundary pressures, the flow may be made to proceed through point b in either direction along the isentrope line. Point a represents the isentropic stagnation condition for all points on the isentrope a–c.

Table 3.4 Simple Flows

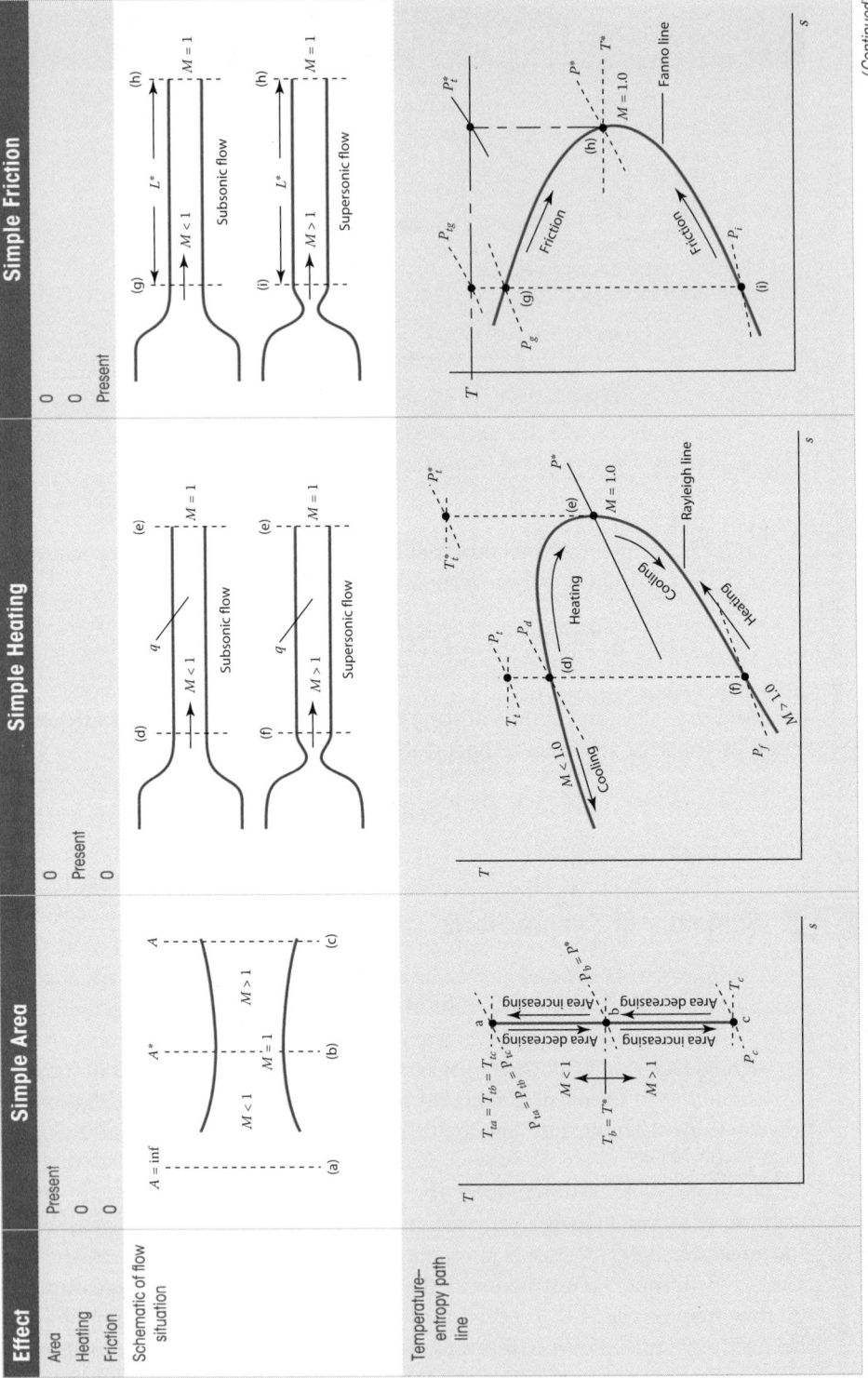

(Continued)

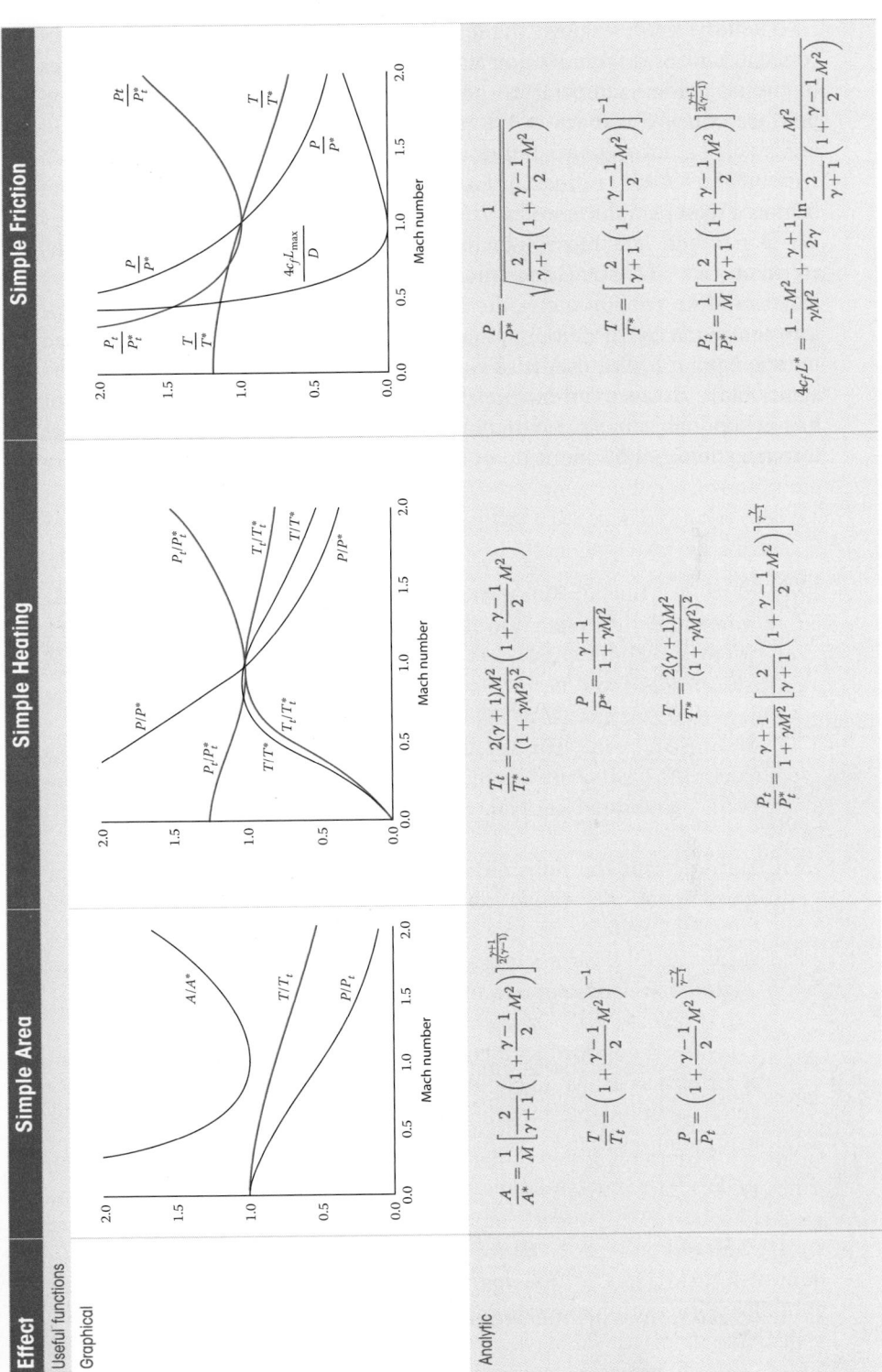

Effect	Simple Area	Simple Heating	Simple Friction
Useful functions			
Graphical	(see figure)	(see figure)	(see figure)

Simple Area — Analytic

$$\frac{A}{A^*} = \frac{1}{M}\left[\frac{2}{\gamma+1}\left(1+\frac{\gamma-1}{2}M^2\right)\right]^{\frac{\gamma+1}{2(\gamma-1)}}$$

$$\frac{T}{T_t} = \left(1+\frac{\gamma-1}{2}M^2\right)^{-1}$$

$$\frac{P}{P_t} = \left(1+\frac{\gamma-1}{2}M^2\right)^{\frac{-\gamma}{\gamma-1}}$$

Simple Heating — Analytic

$$\frac{T_t}{T_t^*} = \frac{2(\gamma+1)M^2}{(1+\gamma M^2)^2}\left(1+\frac{\gamma-1}{2}M^2\right)$$

$$\frac{P}{P^*} = \frac{\gamma+1}{1+\gamma M^2}$$

$$\frac{T}{T^*} = \frac{2(\gamma+1)M^2}{(1+\gamma M^2)^2}$$

$$\frac{P_t}{P_t^*} = \frac{\gamma+1}{1+\gamma M^2}\left[\frac{2}{\gamma+1}\left(1+\frac{\gamma-1}{2}M^2\right)\right]^{\frac{\gamma}{\gamma-1}}$$

Simple Friction — Analytic

$$\frac{P}{P^*} = \sqrt{\frac{1}{\frac{2}{\gamma+1}\left(1+\frac{\gamma-1}{2}M^2\right)}}$$

$$\frac{T}{T^*} = \left[\frac{2}{\gamma+1}\left(1+\frac{\gamma-1}{2}M^2\right)\right]^{-1}$$

$$\frac{P_t}{P_t^*} = \frac{1}{M}\left[\frac{2}{\gamma+1}\left(1+\frac{\gamma-1}{2}M^2\right)\right]^{\frac{\gamma+1}{2(\gamma-1)}}$$

$$4c_fL^* = \frac{1-M^2}{\gamma M^2} + \frac{\gamma+1}{2\gamma}\ln\frac{M^2}{\frac{2}{\gamma+1}\left(1+\frac{\gamma-1}{2}M^2\right)}$$

The *Rayleigh line* shows the series of possible states in a steady, frictionless, constant-area flow. Motion along the Rayleigh line is caused by changes in the stagnation temperature produced by heating effects that, in turn, produce entropy changes in the manner indicated on the line. Heating in an initially subsonic (point d) flow causes the flow Mach number to approach 1 (point e). Neither heating nor cooling can continuously alter the flow from subsonic to supersonic speeds or from supersonic to subsonic speeds.

The *Fanno line* represents the possible series of states in a steady, constant-area, constant-stagnation-temperature flow. Frictional effects alone produce motion along the Fanno line. Consequently, the flow progression along the line must always be one of increasing entropy toward the sonic point h. The flow is subsonic on the Fanno line above h and supersonic below. Because the entropy decreases along the Fanno line from point h, it is impossible, in simple friction flow, to proceed by continuous changes through sonic conditions at point h.

Problems

3.1   Show that the linear momentum equation in Eq. (2.16) for steady, 1-D flow reduces to $I_{out} - I_{in}$, the streamwise force exerted *on the fluid*, where I is the impulse function defined by Eq. (3.13). *Be clear with all assumptions and their implications.*

3.2   Show that the impulse function defined by Eq. (3.13) can be written as Eq. (3.14), a very convenient form for calculations. Hint: Start with the perfect gas equation of state.

3.3   Starting with the fully 3-D unsteady control volume form of linear momentum, Eq. (2.14), show that it reduces to equality of the tangential velocity component $V_{1t} = V_{2t}$ across an oblique shock for steady, 2-D flow assumptions (see Fig. 3.25). *Be clear with all assumptions and their implications.*

3.4   Consider the flow of an incompressible fluid through a 2-D "cascade" of airfoils, as shown in Fig. P3.1. The airfoils are spaced at a distance s and have unit depth into the paper. Application of the conservation of mass requires that $V_i \cos \beta_i = V_e \cos \beta_e$.

a) From the tangential momentum equation, show that

$$F_\theta = \frac{\dot{m}}{g_c}(V_i \sin \beta_i - V_e \sin \beta_e)$$

b) From the axial momentum equation, show that

$$F_z = s(P_e - P_i)$$

c) Show that the axial force can be written as

$$F_z = s \left[\frac{\rho}{2g_c} \left(V_i^2 \sin^2 \beta_i - V_e^2 \sin^2 \beta_e \right) - \left(P_{ti} - P_{te} \right) \right]$$

Note: For an incompressible fluid ($\rho = $ constant), the total pressure can be written as $P_t = P + \dfrac{\rho V^2}{2g_c}$ (see Eq. 3.17).

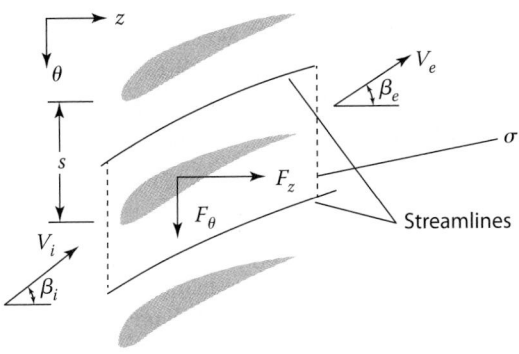

Fig. P3.1   A 2-D cascade of airfoils.

3.5   Aircraft pitot-static measurements indicate the total and static pressures at a particular flight condition to be 5.15 and 3.31 psia, respectively. The total temperature is measured as 443.0°R. Determine the aircraft Mach number and static temperature of the air. Estimate the altitude at which the aircraft is flying.

3.6   Rework Problem 2.13 using total properties and the mass flow parameter (MFP).

3.7   Rework Problem 2.14 using total properties and the MFP.

3.8   Solve Problem 2.17 using total properties and the MFP.

3.9   Solve Problem 2.20 using total properties and the MFP.

3.10   Products of combustion ($\gamma = 1.3$) at a static pressure of 2.0 MPa, static temperature of 2000°K, and Mach number of 0.05 are accelerated in an isentropic nozzle to a Mach number of 1.3, conditions representative of those in a modern combustor-high-pressure turbine. Find the downstream static pressure and static temperature.

If the mass flow rate is 100 kg/s and the gas constant R is 286 J/kg-K, use the mass flow parameter (MFP) and find the flow area for $M = 0.05$ and $M = 1.3$.

3.11   At launch, the space shuttle main engine (SSME) has 1030 lbm/s of gas leaving the combustion chamber at $P_t = 3000$ psia and $T_t = 7350°$R. The exit area of the SSME's nozzle is 77 times the throat area. If the flow through the nozzle is considered to be reversible and adiabatic (isentropic) with $Rg_c = 3800$ ft^2/s$^2 \cdot °$R and $\gamma = 1.25$, find the area of the nozzle throat (in^2) and the exit Mach number. *Hint:* Use the MFP to get the throat area.

3.12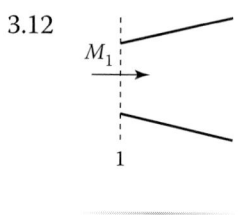

Fig. P3.2 Inlet for Problem 3.12.

The inlet of an aircraft gas turbine engine must not choke; otherwise, the engine would not receive its required air mass flow rate. Suppose an inlet is designed so that the entrance Mach number M_1 is 0.8 at the engine condition requiring maximum mass flow rate (see Fig. P3.2). How much safety margin is provided in terms of the inlet area A_1 relative to the A_1 that would choke the inlet? If the inlet was sized for $M_1 = 0.7$ instead of 0.8, how much larger would the inlet diameter have to be?

3.13   A multistage axial compressor (as shown in Fig. P3.3) is designed to keep the air axial velocity constant at 500 ft/s. The MFP at the compressor inlet (station 2) and exit (station 3) are 0.397 and 0.277 lbm $\cdot \sqrt{R}/(s \cdot$ lbf), respectively. Determine M_2, M_3, and for axial exit flow, the compressor exit static and total temperature T_3 and T_{t3}. If the compressor exit air density is 10 times the inlet air density, how much would the annulus through-flow area have to be reduced to meet these conditions? Give the answer as A_3/A_2.

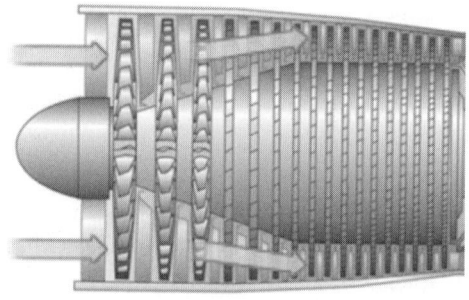

Fig. P3.3 A multistage axial compressor.

3.14 The experimental evaluation of a gas turbine engine's performance requires the accurate measurement of the inlet air mass flow rate into the engine. A bell-mouth engine inlet (shown schematically in Fig. P3.4) can be used for this purpose in the static test of an engine. The freestream velocity V_0 is assumed to be 0, and the flow through the bell mouth is assumed to be adiabatic and reversible. See Fig. P3.4.

Measurements are made of the freestream pressure P_{t0} and static pressure at station 2 P_2, and the exit diameter of the inlet D_2.

a) For the bell-mouth inlet, show that the Mach number at station 2 is given by

$$M_2 = \sqrt{\frac{2}{\gamma-1}\left[\left(\frac{P_{t0}}{P_{t0}-\Delta P}\right)^{(\gamma-1)/\gamma}-1\right]}$$

and the inlet mass flow rate is given by

$$\dot{m} = \frac{P_{t0}}{\sqrt{T_{t0}}}\frac{\pi D_2^2}{4}\sqrt{\frac{2g_c}{R}\frac{\gamma}{\gamma-1}\left[\left(\frac{P_{t0}-\Delta P}{P_{t0}}\right)^{2/\gamma}-\left(\frac{P_{t0}-\Delta P}{P_{t0}}\right)^{(\gamma+1)/\gamma}-1\right]}$$

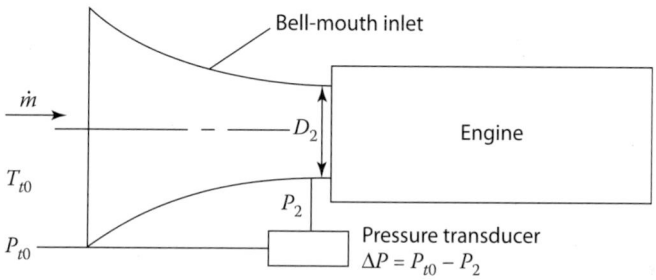

Fig. P3.4 Bell-mouth engine inlet.

b) For the following data, determine the inlet mass flow rate, Mach number M_2, static temperature T_2, and velocity V_2:

$$T_{t0} = 7°C \qquad P_{t0} = 77.80\,\text{kPa}$$
$$\Delta P = 3.66\,\text{kPa} \qquad D_2 = 0.332\,\text{m}$$

3.15 An ideal ramjet engine (see Fig. P3.5) is operated at 50,000-ft altitude with a flight Mach number of 3. The diffuser and nozzle are assumed to be isentropic, and the combustion is modeled as an ideal heat interaction at a constant Mach number with constant total pressure. The cross-sectional area and Mach number for certain engine stations are given in Table P3.1. The total temperature leaving the combustor T_{t4} is 4000°R. Assume ambient pressure surrounding the engine flow passage.

a) Determine the mass flow rate of air through the engine (lbm/s).

b) Complete the table with flow areas, static pressures, static temperatures, and velocities.

c) Find the thrust (magnitude and direction) of the diffuser, combustor, and nozzle.

d) Find the thrust (magnitude and direction) of the ramjet engine.

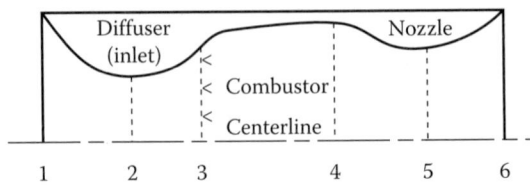

Fig. P3.5   Ideal ramjet engine.

Table P3.1   Ideal Ramjet Design

Station	1	2	3	4	5	6
Area (ft²)	4.235					
Mach	3	1	0.15	0.15	1	3
P (psia)						
T (°R)						
V (ft/s)						

3.16   At Mach 3.2 and an altitude of 76,000 ft, approximately 50% of the SR-71's thrust was produced from the pressure difference imposed across the complex inlet system (see Fig. P3.6). This system included a moveable inlet spike and variable bleed flows to carefully control multiple oblique shock waves followed by a terminal normal shock. Despite this complexity, an estimate of the thrust of the inlet system can be obtained quite readily through the use of the impulse function. Find the thrust given the following data at the entrance (1) and exit (2) of the inlet system: $M_1 = 3.2$, $P_1 = 0.49$ psia, $M_2 = 0.9$, $P_2 = 12.6$ psia, $A_2 = 186$ in². Use the GASTAB program.

3.17   Total property measurements across an experimental aircraft inlet (station 1 to station 2) are as follows:

$$T_{t1} = 550°R \quad P_{t1} = 12.5\,\text{psia} \quad T_{t2} = 550°R \quad P_{t2} = 12.38\,\text{psia}$$

As the lead engineer for this project, you are responsible for explaining why these measurements do or do not make sense. Explain whether

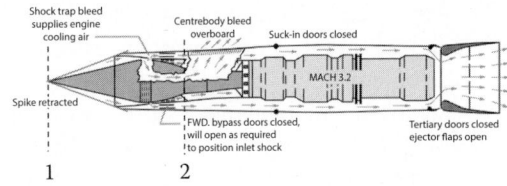

Fig. P3.6   SR-71 aircraft and engine nacelle.

these measurements are reasonable using foundational principles. *Hint:* Think about the first and second laws of thermodynamics.

3.18   The following measurements are obtained across the exhaust nozzle (station 5 to station 9) of an aircraft gas turbine engine:

$$T_{t5} = 1688°R \quad P_{t5} = 51.62 \text{ psia} \quad T_{t9} = 1678°R \quad P_{t9} = 50.07 \text{ psia}$$

Do these measurements make sense? Explain using foundational principles. Show that the energy per unit mass "lost" by the exhaust nozzle due to heat transfer is small when compared to the primary energy transfer of specific enthalpy to specific kinetic energy ke. This is why the *adiabatic assumption* is often used in the analysis of exhaust nozzles (and inlets). The nozzle exit gas velocity V_9 is 2470 ft/s. Compare $q_{out} = \dot{Q}_{out}/\dot{m}$ to ke. Use $c_p = 0.28$ Btu/(lbm · °R).

3.19   Air at 20 kPa, 260 K, and Mach 3 passes through a normal shock. Determine:

a)   Total temperature and pressure upstream of the shock

b)   Total temperature and pressure downstream of the shock

c)   Static temperature and pressure downstream of the shock

3.20   Air upstream of a normal shock has the following properties: $P_t = 100$ psia, $T_t = 100°F$, and $M = 2$. Find the upstream static temperature, static pressure, and velocity (ft/s). Find the downstream total temperature, Mach number, total pressure, static temperature, static pressure, and velocity (ft/s).

3.21   A 15-deg ramp is used on a supersonic inlet at $M = 3.5$ and an altitude of 20 km. Use the GASTAB program.

a)   Determine the angle of the oblique shock and flow properties (Mach number, total and static temperature, total and static pressure) upstream and downstream of the shock.

b)   At what Mach number does the shock detach from the ramp?

3.22 Early supersonic aircraft, like the F-100 (the first U.S. Air Force fighter capable of supersonic speed in level flight; see Fig. P3.12 of Problem 3.38), used "pitot" inlets that resulted in a normal shock upstream of the inlet. More common in today's fighter aircraft are "external compression" inlets with multiple ramps to generate multiple weak oblique shocks followed by a weak terminal normal shock as conditions warrant. For the same inlet Mach number as shown in Fig. P3.7, compare the performance of the pitot inlet to a two-ramp external compression inlet by determining the Mach number M_2 and the inlet total pressure recovery P_{t2}/P_{t1} downstream of the shock system for both cases. The ramp angle for each of the two ramps is 7 deg. What happens with the two-ramp system if the inlet Mach number M_1 is reduced to 1.5? Using vectors, accurately sketch the flow turning and velocity magnitude changes that occur for the two-ramp system for both Mach numbers ($M_1 = 1.6$ and 1.5).

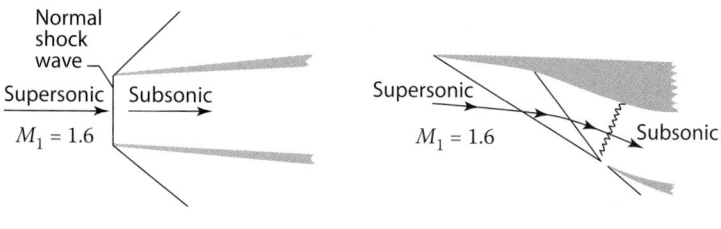

Fig. P3.7 Two different inlets.

3.23 Modern aircraft gas turbine engine fans are often designed to accommodate shock waves at/around the vicinity of the blade tip (more on this in Chapter 9). To minimize the total pressure losses accommodating shock waves, the airfoils in the tip regions are very thin with carefully tailored profiles and sharp leading edges. These features help to ensure a weak oblique shock attached to the blade leading edge, often followed by a weak normal shock within the blade passage at peak efficiency design conditions. Modeling the shock structure as an oblique shock inclined at 55 deg relative to the blade incoming flow at Mach 1.4, followed by a normal shock

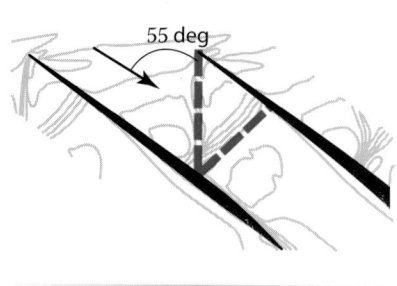

Fig. P3.8 Oblique shock followed by a normal shock (Ref. 67).

(Fig. P3.8), determine the total pressure loss. Compare this to the total pressure loss if only a single normal shock were presented to the oncoming flow. Using vectors, accurately sketch the flow turning and velocity magnitude changes.

3.24   A 2-D flat plate is inclined at a positive angle of attack in a supersonic airstream at Mach 2.0 and static pressure of 20 kPa. Below the plate, an oblique shock wave starts at the leading edge, making an angle of 40 deg with the flow direction. On the upper surface, an expansion occurs at the leading edge. Providing an accurate sketch, determine the following:

a) The angle of attack of the plate

b) The pressure on the lower surface of the plate

c) The pressure on the upper surface of the plate

3.25   A 20-deg wedge (two 10-deg half angles) is to be used in a wind tunnel using air with test conditions of $M = 3.0$, $T_t = 500°R$, and $P_t = 100$ psia. Determine the angle of any oblique shocks and/or expansion waves and the downstream total and static pressure and temperature with the wedge inclined at 10-, 20-, and 25-deg angles of attack to the airstream. Use the GASTAB program. Include accurate sketches.

3.26   Using the influence coefficients of Table 3.1, show that $dP/P = -dA/A$ for Example 3.8.

3.27   A calorically perfect gas undergoes an ideal heat interaction in a duct. The area of the duct is varied to hold the static pressure constant. Using the influence coefficients of Table 3.1, show the following:

a) The area variation required to hold the static pressure constant is given by

$$\frac{dA}{A} = \left(1 + \frac{\gamma - 1}{2}M^2\right)\frac{dT_t}{T_t}$$

b) The relationship between the Mach number and total temperature, for the preceding area variation, is given by

$$\frac{dM^2}{M^2} = -\left(1 + \frac{\gamma - 1}{2}M^2\right)\frac{dT_t}{T_t}$$

c) The relationship for the area of the duct in terms of the Mach number is given by

$$A_2 = A_1 \left(\frac{M_1}{M_2}\right)^2$$

where subscripts refer to states 1 and 2.

d) Show V is constant.

3.28 Air flows from a very large chamber with constant $P_c = 500$ kPa and constant $T_c = 500$ K through a converging nozzle with an exit area of 50 cm². Assume isentropic, steady flow in the nozzle. Determine the mass flow rate through the nozzle and exit plane static pressure for back pressures of 0, 125, 250, and 375 kPa.

3.29 Air at a total pressure of 1.4 MPa, total temperature of 350 K, and Mach number of 0.5 is accelerated isentropically in a nozzle (see Fig P3.9) to a Mach number of 3 (station x), passes through a normal shock (x to y), and then flows isentropically to the exit. Given nozzle throat area of 0.05 m² and the exit area of 0.5 m²:

a) Find the area at the stock.

b) Find the static pressure and static temperature upstream of the shock (station x).

c) Find the Mach number and the total and static pressures and temperatures downstream of the shock (station y).

d) Find the Mach number, static pressure, and static temperature at the exit.

e) Determine the minimum back pressure required to ensure the normal shock is "blown" out of the nozzle (and thus ensure supersonic flow at the nozzle exit plane). *Hint:* See Example 3.9a.

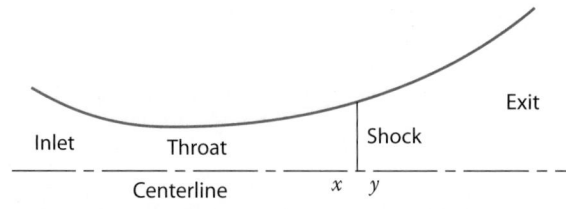

Fig. P3.9 Nozzle with normal shock.

3.30 A variable area convergent-divergent (C-D) exhaust nozzle is often used on fighter aircraft to allow perfectly expanded nozzle operation at various altitudes and Mach numbers. However, other system considerations sometimes preclude perfectly expanded operation. For example, suppose at 40,000 ft (standard day) and $P_{t8} = 48.0$ psia, a particular nozzle is purposefully operated *underexpanded* to keep acoustic pressure levels manageable (afterburner rumble). Flight tests reveal that a nozzle area ratio A_9/A_8 of 2.16 significantly reduces rumble at these conditions. Using $\gamma = 1.3$, do the following:

a) Find the exit plane Mach number and static pressure and verify that the nozzle is operating underexpanded.

b) Find the nozzle area ratio needed for perfectly expanded operation under the given conditions.

c) Sketch the first three regions of the flow downstream of the nozzle exit plane (see Example 3.9d).

3.31 For the space shuttle main engine (SSME) characteristics provided in Prob. 3.11:

a) Determine the altitude at which the nozzle is operating perfectly expanded.

b) Determine the altitudes at which the nozzle is operating overexpanded and underexpanded.

c) At 80,000 ft altitude, estimate the static pressure, Mach number, and flow turning in the first three regions of the flow downstream of the nozzle exit plane. Accurately sketch the flow patterns.

3.32 You are required to design a nozzle to pass a given mass flow rate of air with minimum frictional losses between a storage pressure chamber P_c and an exhaust region P_e with a given variation in pressure between the two regions. The design conditions are

$$\dot{m} = 1000 \, \text{lbm/s} \quad P_c = 3000 \, \text{psia} \quad T_c = 3700°\text{R}$$

a) Using the design conditions, complete Table P3.2.

b) Make a plot of the nozzle contour like Fig. 3.46.

c) Calculate the nozzle design pressure ratio $P_{\bar{n}}$ and the nozzle area ratio ε.

d) Using the altitude table, determine the design altitude for this nozzle ($P_e = P_a$).

e) Determine the thrust of a rocket motor using this nozzle at its design altitude.

Table P3.2 Nozzle Design

Station x, in.	P, psia	P/P_t	M	T/T_t	A/A^*	T, °R	a, ft/s	V, ft/s	A, in.^{2}	D, in.
0	2918									
2	2841									
4	2687									
6	2529									
8	2124									
10 (throat)	1585									
12	727									
16	218									
24	19.8									
41	3.21									
60	1.71									

3.33 A perfect gas enters a constant area heater at a Mach number of 0.3, total pressure of 600 kPa, and total temperature of 500 K. A heat interaction of 500 kJ/kg into the gas occurs. Using the GASTAB software, determine the Mach number, total pressure, and total temperature after the heat interaction for the following gases:

a) $\gamma = 1.4$ and $c_p = 1.004$ kJ/(kg · K)

b) $\gamma = 1.325$ and $c_p = 1.171$ kJ/(kg · K)

3.34 A perfect gas enters a constant-area heater at a Mach number of 0.5, total pressure of 200 psia, and total temperature of 1000°R. A heat interaction of 100 Btu/lbm into the gas occurs, Using the GASTAB software, determine the Mach number, total pressure, and total temperature after the heat interaction for the following gases:

a) $\gamma = 1.4$ and $c_p = 0.24$ Btu/(lbm · °R)

b) $\gamma = 1.325$ and $c_p = 0.28$ Btu/(lbm · °R)

3.35 A convergent-only nozzle is to be used on an afterburning gas turbine engine as shown in Fig. P3.10. Model the afterburner (station 6 to station 7) as a constant-area duct ($A_6 = A_7$) with simple heat interaction q_{in} into the air. The flow through the nozzle (station 7 to station 8) is isentropic. The exit area of the nozzle A_8 is varied with the afterburner setting T_{t7} to keep sonic ($M = 1$) flow at station 8 and the inlet conditions (mass flow rate, P_t, and T_t) constant at station 6.

a) Using the mass flow parameter, show that the area ratio A_8/A_6 is given by

$$\frac{A_8}{A_6} = \frac{P_{t6}}{P_{t7}}\sqrt{\frac{T_{t7}}{T_{t6}}}\frac{\text{MFP}(M_6)}{\text{MFP}(M=1)} = \frac{P_{t6}}{P_{t7}}\sqrt{\frac{T_{t7}}{T_{t6}}}\left(\frac{A^*}{A}\right)_{M_6}$$

b) For $M_6 = 0.4$, determine the area ratio A_8/A_6 for the following values of T_{t7}/T_{t6}: 1.0, 1.1, 1.2, 1.3, 1.4, 1.5, 1.6, and $T_{t7} = T_t^*$.

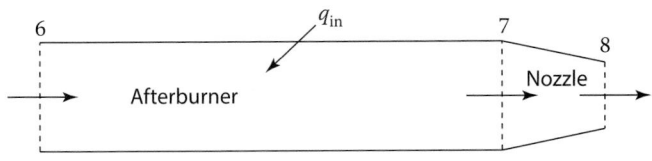

Fig. P3.10 Convergent-only nozzle on an afterburning gas turbine engine.

3.36 Air flows into a constant-area, insulated square duct with a Mach number of 0.6, static pressure of 10.0 psia, and static temperature of 500.0°R. The duct length is 8.0 ft, duct diameter is 1.0 ft, and coefficient of friction $c_f = 0.004$. Determine the Mach number, static pressure, static temperature, and total pressure ratio P_{t2}/P_{t1}. Be sure to include a system sketch.

3.37 A constant-area duct, 25 cm in length by 1.3 cm in diameter, is connected to an air settling chamber through a converging nozzle, as shown in Fig. P3.11. For a constant chamber pressure of 1.0 MPa and constant chamber temperature of 600 K, determine the mass flow rate through the duct for a back pressure of 101 kPa. *Hint:* Assume choked flow, then verify it.

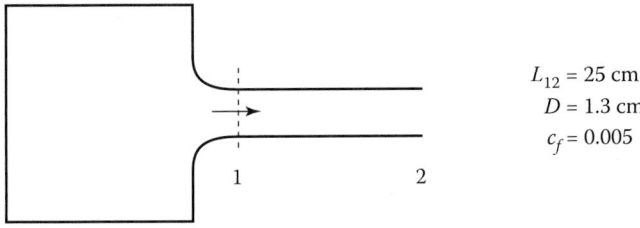

$L_{12} = 25$ cm
$D = 1.3$ cm
$c_f = 0.005$

Fig. P3.11 A constant area duct connected to an air settling chamber through a convergent nozzle.

3.38   The North American F-100 (Fig. P3.12) is 50 ft long and its J57
        afterburning turbojet is 20 ft long. Modeling the approximately
        30-ft-long inlet duct as constant area simple frictional flow, estimate
        the following airflow conditions at the compressor face (station 2)
        for inlet on-design flight conditions of $M_1 = M_0 = 0.8$ and $P_1 = P_0$
        at 36,000 ft. Find M_2, T_{t2}, P_{t2}, and P_2. Use $c_f = 0.0015$ and assume
        a circular area with diameter $D = 39$ in., the diameter of the J57
        engine. Note that this will provide a conservative estimate because
        the actual inlet duct is a subsonic diffuser (proceeds from a smaller
        area to the 39-in. compressor face diameter).

Fig. P3.12   F-100 aircraft.

Chapter 4 | Aircraft Gas Turbine Engine

> ## In thrust we trust.
> *Author unknown*

4.1 Introduction

The introductory fundamentals of aircraft propulsion systems are covered in this chapter. Emphasis is placed on propulsion systems that operate on the so-called Brayton cycle. Such systems include turbojets, turboprops, turbofans, ramjets, and combinations thereof. Before taking up the major flow path components and thermodynamic processes involved in these systems, we consider the forces acting on the propulsive duct and the effect of installation on the net propulsive force.

4.2 Generalized Thrust Equation

4.2.1 Engine Thrust Loads

Companies designing and manufacturing gas turbine engines must make detailed, tedious calculations of component forces in order to properly assess structural loading and the design integrity. These component forces arise from the gas pressures acting on the axial area projection of every part of the engine (for example, the axial area projection of all the rotor blades in a compressor or all the stator vanes in a case assembly) in a manner analogous to gas pressure acting on a piston in a reciprocating engine. Figure 1.6 shows a representative gas pressure distribution through an engine. A breakdown of the approximate axial reaction forces (force of the gas on the structure) on a typical low-bypass, dual-spool, mixed-flow turbofan engine at sea level, standard day, static conditions at "military" power setting (maximum nonafterburning thrust) is shown in Fig. 4.1. Thrust loads are shown on the rotating core or high-pressure spool (HPC-shaft-HPT), rotating fan or low-pressure spool (fan-shaft-LPT), and static engine cases.

Figure 4.1 offers some incredible insights on the magnitude and direction of the component loadings. The engine literally wants to pull itself apart. For example, on the core spool there is 87,900 lbf in the *positive* thrust direction and an 83,700 lbf *negative* thrust on the high-pressure turbine (HPT). The high-pressure spool contributes a net 4200 lbf in the positive thrust direction; the low-pressure spool contributes − 1800 lbf! How can this be? Why would an

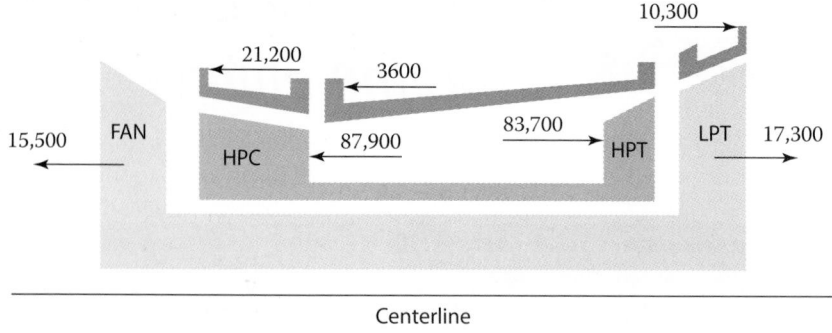

Fig. 4.1   Thrust loads of a typical low-bypass, dual-spool turbofan engine (forces in lbf) (Ref. 66).

engine company use a low-pressure spool in this case? These questions and many others will be answered as you continue your studies with this text.

If we sum all of the component forces shown in Fig. 4.1, we end up with a positive 14,500 lbf on the engine cases that, when added with the forces on the two spools, results in a net force of 16,900 lbf, which is the military uninstalled thrust rating of the engine. All forces on the rotor spools must be transferred through bearings housed in stationary engine frames to the outer cases and engine mount system typically consisting of 2–3 pins. The mount system transfers all of the engine loads—thrust, weight, and torques—to the aircraft.

Due to the extreme complexity of the engine mechanical and aero-dynamic configuration, detailed force calculations of the thrust loads as given in Fig. 4.1 for structural consideration involve a great many assumptions and lack the precision needed for performance estimation. We therefore make use of another characteristic of any propulsion system, namely that all systems involved in aerospace propulsion generate propulsive thrust by acceleration of a mass of fluid (application of Newton's second law). In the typical aircraft application, this mass of fluid is atmospheric air mixed with a little hydrocarbon fuel (typically around 3%). In a rocket engine, the fluid mass is made up of one or more propellants carried by the rocket itself.

4.2.2 Development and Application of Generalized Thrust Equation

In this section, we develop a simple generalized form for determining the thrust of a propulsion device, be it a gas turbine engine or rocket engine, or even just a component, say a compressor or turbine. In doing so, we will stay consistent with our treatment thus far, invoking our familiar steady, 1-D flow assumptions. Engine thrust is, of course, a force, so which of our governing equations is applicable? We apply the linear momentum Eq. (2.14) first to the control volume σ shown in Fig. 4.2 to determine the forces on the gas

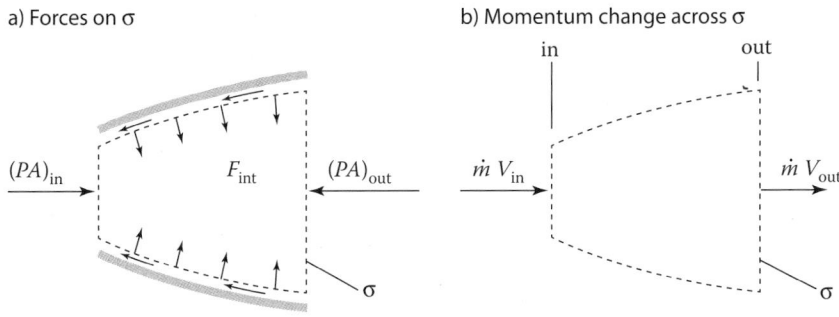

Fig. 4.2   Determination of forces on the gas flowing through a generic propulsion device.

flowing through our propulsion device, then to the free body diagram in Fig. 4.3 to determine the forces on the device.

From Sect. 2.3.2, the general form of the linear momentum equation is

$$\text{Momentum} \quad \frac{1}{g_c} \frac{\partial}{\partial t} \iiint_\sigma \vec{V} \rho \, dV + \frac{1}{g_c} \iint_{CS} \vec{V} \rho (\vec{V} \cdot \hat{n}) dA = \sum_{\text{on } \sigma} \vec{F} \quad (2.14)$$

Recall that the forces on the right side of Eq. (2.14) are those forces acting *on the gas*. The forces on the gas in Fig. 4.2 are composed of internal forces resulting from pressure and shear stresses F_{int}, the details of which we know nothing about (they would have to be obtained using the methods used to get the forces shown in Fig. 4.1), and pressure forces acting on any through flow surfaces (cross sections). With our steady flow assumption, property changes with respect to time are zero; thus, the first term on the left of the equation goes to zero. Our 1-D assumption implies that the only significant property changes occur in the streamwise or x direction at the inlet and exit; thus,

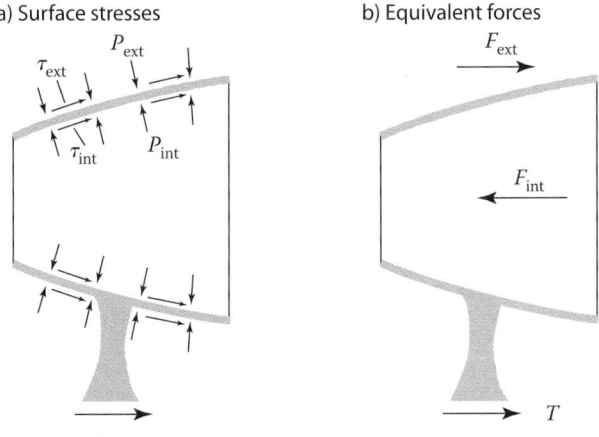

Fig. 4.3   Determination of forces on a generic propulsion device.

properties do not vary at any given cross-sectional area. Applying the steady, 1-D form of Eq. (2.14) to our fixed control volume of Fig. 4.2 in the x direction, we get

$$\frac{\dot{m}V_{out}}{g_c} - \frac{\dot{m}V_{in}}{g_c} = P_{in}A_{in} - P_{out}A_{out} + F_{int} \tag{4.1}$$

From the definition of the impulse function in Sect. 3.2.6, $I \equiv PA + \dot{m}V/g_c$, Eq. (4.1) can be written as

$$F_{int} = I_{out} - I_{in} \tag{4.2}$$

Now, turning our attention to the forces *on the propulsion device* (Fig. 4.3), we have F_{int} on the device (equal to and opposite of F_{int} on the gas) plus external forces F_{ext} including a restraint force T to prevent relative motion between the device and the test stand or vehicle. As indicated in Fig. 4.3, F_{ext} is composed of pressure and shear forces due to friction such that

$$F_{ext} = \int_{in}^{out} P_{ext}dA + \overbrace{\int_{in}^{out} \tau dA}^{\text{Vehicle drag accounting}}$$

Typically, we ignore the external shear force component because it is accounted for as part of the overall air–vehicle drag. We let the external pressure force term be composed of a baseline distribution resulting from ambient pressure P_0 and an "excess" distribution; thus,

$$F_{ext} = \int_{in}^{out} P_{ext}dA = \int_{in}^{out} P_0 dA + \int_{in}^{out} (P_{ext} - P_0)dA$$
$$= P_0(A_{out} - A_{in}) + D_{ext} \tag{4.3}$$

In Eq. (4.3), we have defined

$$D_{ext} \equiv \int_{in}^{out} (P_{ext} - P_0)dA \tag{4.4}$$

The "external drag" D_{ext} is pressure drag from the external air flow over the vehicle surface and is highly dependent on the geometry of the vehicle-propulsion device integration (the housing about the engine or nacelle); for example, the external shape of an inlet lip at station *in*.

Summing the forces acting on the device of Fig. 4.3, we have

$$T = F_{int} - F_{ext}$$

or using Eqs. (4.2) and (4.3)

$$T = I_{out} - I_{in} - P_0(A_{out} - A_{in}) - D_{ext} \tag{4.5}$$

Equation (4.5) represents a generalized thrust equation with T representing the force transmitted to a flight vehicle or the force measured on a test stand. To demonstrate the general ease of use of Eq. (4.5), some common applications follow.

4.2.2.1 Uninstalled Thrust of a Rocket Engine

The uninstalled thrust is defined as the net axial force produced by an engine in an *idealized inviscid fluid*. Consequently, we ignore the D_{ext} term in Eq. (4.5) and distinguish

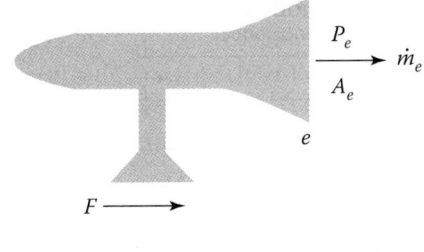

Fig. 4.4   Rocket schematic on a test stand: $D_{ext} = 0$.

the resisting force by F, the *uninstalled thrust*. Note that for a rocket, there is no inlet station *in* (Fig. 4.4) so that I_{in} and A_{in} are both zero; thus with station *out* denoted as e we have from Eq. (4.5)

$$F = I_e - P_0 A_e$$

In most propulsion texts, this equation will be written using the terms that define the impulse function so that we get the equivalent of Eq. (1.52) in Chapter 1

$$F = \frac{\dot{m} V_e}{g_c} + (P_e - P_0) A_e$$

4.2.2.2 Uninstalled Thrust of a Turbojet Engine

We again seek an expression for the *uninstalled thrust F*. Referring to Fig. 4.5, we designate the stations *in* and *out* as 1 and 9, respectively—the industry standard. We note that I_1 of Eq. (4.5) cannot be determined unless we know the flow properties at station 1, which we don't generally have. To get around the problem, it is standard practice to reference entry conditions to station 0, the *freestream capture area*, far enough upstream of the inlet station 1 so that the flow properties equal those of the freestream

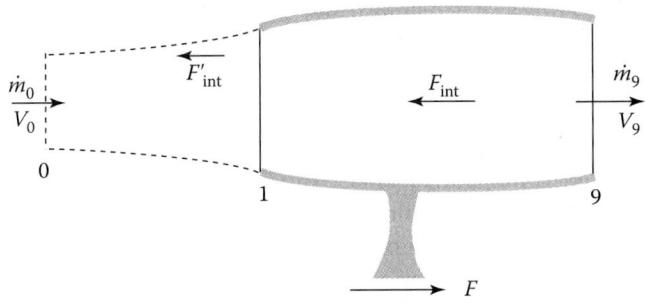

Fig. 4.5   Turbojet engine schematic for determining uninstalled thrust.

(undisturbed) ambient air properties, which we do typically know (∞ is often used in aeronautics texts).

With our new entry control surface station 0, we have to account for any internal forces acting on the stream tube between stations 0 and 1; this is shown in Fig. 4.5 as F'_{int}. We note that F'_{int} is nonzero unless the stream tube from 0 to 1 is a right circular cylinder (i.e., $A_0 = A_1$ and, consequently, $M_0 = M_1$). In general, the air properties change from station 0 to 1 as a result of pressure forces of the surrounding air (see next section). The *uninstalled thrust F* is defined as the sum of the internal engine force from 1 to 9 and the entry stream force from 0 to 1 or

$$F \equiv F_{int} + F'_{int} \tag{4.6}$$

From Eq. (4.5) with $D_{ext} = 0$ applied to Fig. 4.5, we get

$$F = I_9 - I_0 - P_0(A_9 - A_0) \quad \Rightarrow$$

$$F = \frac{\dot{m}_9 V_9}{g_c} + P_9 A_9 - \frac{\dot{m}_0 V_0}{g_c} - P_0 A_0 - P_0 A_9 + P_0 A_0 \quad \Rightarrow$$

$$F = \frac{\dot{m}_9 V_9 - \dot{m}_0 V_0}{g_c} + (P_9 - P_0)A_9 \tag{4.7}$$

Equation (4.7) is identical to Eq. (1.5) with the appropriate subscripts and $\dot{m}_9 = \dot{m}_0 + \dot{m}_f$.

It is important to note that the uninstalled thrust F is a fictitious force, but one that is used as a standard of engine performance. Engine companies commonly provide uninstalled thrust; it is independent of installation effects. The thrust values for different engines provided in Tables B.1 and B.2 of Appendix B are all uninstalled thrusts.

Example 4.1

Figure 4.6 shows the J69 single-spool turbojet engine on a thrust stand at the U.S. Air Force Academy with an aptly named inlet *bellmouth*. The altitude at the U.S. Air Force Academy is approximately 7300 ft above sea level. Bellmouth inlets are commonly used with engines on test stands to obtain an estimation of uninstalled thrust. Their shape aligns nicely with the incoming air stream tube for a static engine test ($M_0 = 0$). The exhaust nozzle of this engine is not choked. The bellmouth measurements provided are at the constant area section of the bellmouth. Use the following measurements to estimate the uninstalled thrust F:

1. Include the measured fuel flow rate of 0.239 lbm/s.
2. Ignore the fuel flow rate.

(Continued)

Example 4.1 *(Continued)*

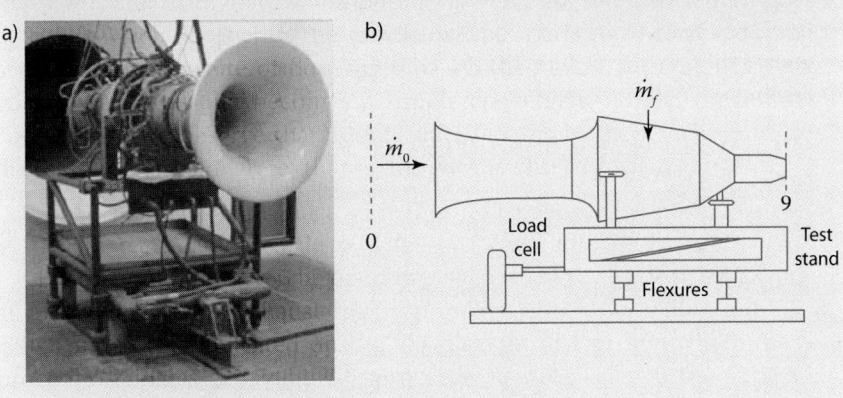

Fig. 4.6   Teledyne J69-T-25 (T-37B) in U.S. Air Force Academy
test cell and through-flow sketch.

3. Compare your answers from 1 and 2 to the measured thrust of 730 lbf.

$$P_0 = 11.44 \, \text{psia}, \quad T_0 = 517°\text{R}, \quad P_{\text{bellmouth}} = 10.96 \, \text{psia},$$

$$D_{\text{bellmouth}} = 13.08 \, \text{in}, \quad T_{t9} = 1558°\text{R}, \quad P_{t9} = 18.594 \, \text{psia}$$

Solution

We apply Eq. (4.7) to the control volume indicated in Fig. 4.6 from station 0 to station 9, making note of the steady, 1-D flow assumptions that led to the equation. We note that for an engine on a thrust stand in an open tunnel, $V_0 = 0$. Also, because we are told that the exhaust nozzle is not choked, we know that $P_9 = P_0$. Applying Eq. (4.7), we have

$$F = \frac{\dot{m}_9 V_9 - \dot{m}_0 \cancelto{0}{V_0}}{g_c} + (\cancelto{0}{P_9 - P_0}) A_9$$

or

$$F = \frac{\dot{m}_9 V_9}{g_c}$$

For the exhaust gas exit mass flow rate

$$\dot{m}_9 = \dot{m}_0 + \dot{m}_f$$

We use the mass flow parameter MFP = MFP (γ, M) [Eq. (3.11)] for determining $\dot{m}_0$, noting that from continuity $\dot{m}_0 = \dot{m}_{\text{bellmouth}}$, convenient because we know the static pressure and area in the bellmouth constant area section. Clearly, $\gamma = 1.4$ is a good choice for the inlet flow. To get MFP, we recognize that because $V_0 = 0$, the static and total pressures and

(Continued)

Example 4.1 *(Continued)*

static and total temperatures are equal. Because there are no heat or work interactions in the bellmouth, the total temperature stays constant. If we assume *negligible total pressure loss due to friction in the bellmouth*, then

$$P_{t\,\text{bellmouth}} = P_{t0} = P_0 \qquad \left(\frac{P}{P_t}\right)_{\text{bellmouth}} = \frac{10.96\,\text{psia}}{11.44\,\text{psia}} = 0.958$$

From Appendix D, with $\gamma = 1.4$ and $P/P_t = 0.958$: $M_{\text{bellmouth}} = 0.25$ and $MFP\sqrt{\dfrac{R}{g_c}} = 0.28498$. So, from the definition of MFP:

$$MFP\sqrt{\frac{R}{g_c}} = \frac{\dot{m}\sqrt{T_t}}{P_t A}\sqrt{\frac{R}{g_c}} = 0.28498 \quad \text{at the bellmouth}$$

Solving for the mass flow rate, noting that area $= (\pi D^2)/4$, and *being careful with units*, we get

$$\dot{m}_{\text{bellmouth}} = 14.96\,\text{lbm/s} = \dot{m}_0$$

For exhaust gas velocity V_9, we use the definition of total specific enthalpy [Eq. (2.20)] applied to a calorically perfect gas (CPG):

$$h_{t9} \equiv h_9 + \frac{V_9^2}{2g_c} \implies V_9 = \sqrt{2g_c(h_{t9} - h_9)}$$

$$= \sqrt{2g_c c_p (T_{t9} - T_9)} \quad \text{for a CPG}$$

At the exhaust gas temperature of $1558°R$ ($\approx 1100°F$), $\gamma = 1.36$ and $c_p = 0.258\,\text{Btu/(lbm} \cdot °R)$ are good approximations (Table 2.3). We estimate T_9 by recognizing we can use the isentropic CPG relation, Eq. (2.54), between the total and static states at exit plane 9. Referring to the *T-s* diagram of Fig. 4.7:

$$T_9 = T_{t9}\left(\frac{P_9}{P_{t9}}\right)^{\frac{\gamma-1}{\gamma}} \implies$$

$$T_9 = 1558°R\left(\frac{11.44}{18.594}\right)^{\frac{0.36}{1.36}} = 1370°R$$

Therefore, using the definition of total specific enthalpy equation above, and *always being careful with units*, we get

$$V_9 = \sqrt{2\left(32.2\,\frac{\text{ft} \cdot \text{lbm}}{\text{lbf} \cdot \text{s}^2}\right)\left(0.258\,\frac{\text{BTU}}{\text{lbm} \cdot \text{R}}\right)\left(778.16\,\frac{\text{ft} \cdot \text{lbf}}{\text{BTU}}\right)(1558 - 1370)\text{R}}$$

$$= 1559\,\text{ft/s}$$

(Continued)

Example 4.1 *(Continued)*

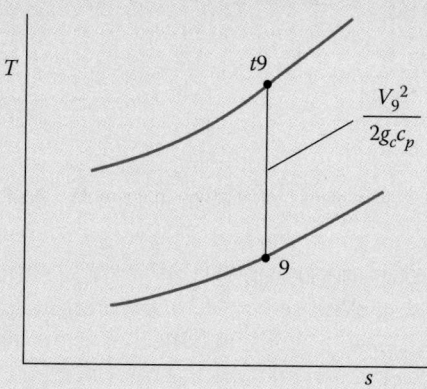

Fig. 4.7   Nozzle T-s diagram.

We note that all of our property estimations make sense: static properties are less than totals at the same location where there is significant gas velocity, and there are reasonable values for air mass flow rate and exhaust gas velocity.

So, from our generalized thrust equation, simplified as applicable to this example, we can answer the questions.

1. F including $\dot{m}_f$:

$$F = \frac{\dot{m}_9 V_9}{g_c} = \frac{(14.96 + 0.239)\dfrac{\text{lbm}}{\text{s}}(1559)\dfrac{\text{ft}}{\text{s}}}{32.2\dfrac{\text{ft} - \text{lbm}}{\text{lbf} - \text{s}^2}} = 736\,\text{lbf}$$

2. F excluding $\dot{m}_f$:

$$F = 724\,\text{lbf}$$

3. Compare 1 and 2 estimations with $F_{\text{measured}} = 730\,\text{lbf}$

The estimated values for uninstalled thrust determined from measured and calculated flow properties differ *less than* 1% from the measured value, whether or not we include the fuel mass flow rate. You should find these results striking! These are real results obtained from real measurements. Despite the complex and challenging flow environment of a gas turbine engine, we are able to get this close in estimating thrust by assuming steady, 1-D flow of a calorically perfect gas.

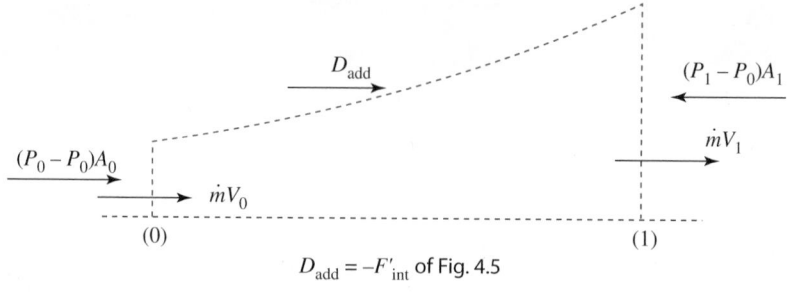

Fig. 4.8 Momentum equation applied to stream tube of engine air from 0 to 1.

4.2.2.3 Additive Drag

Additive drag is the pressure drag on a stream tube bounding the flow between stations 0 and 1. Applying Eq. (4.5) to the stream tube between stations 0 and 1 in Fig. 4.5, we have

$$D_{add} = I_1 - I_0 - P_0(A_1 - A_0) \tag{4.8}$$

Additive drag is the pressure force from the surrounding air that causes property changes of the "captured" stream tube air and dictates the shape of the stream tube. Note that because the stream tube is a "free" boundary (there is no physical boundary), F'_{int} of Fig. 4.5 must be the equal and opposite force of D_{add}, as is noted in Fig. 4.8.

Figure 4.8 shows the momentum and force terms on a schematic of a stream tube of captured engine air. (Due to symmetry, only half of the stream tube is shown.) From Eq. (4.8), using the definition of the impulse function in the form given by Eq. (3.6), we obtain

$$D_{add} = P_1 A_1 (1 + \gamma M_1^2) - P_0 A_0 (1 + \gamma M_0^2) - P_0(A_1 - A_0) \tag{4.9a}$$

which reduces to

$$D_{add} = P_1 A_1 (1 + \gamma M_1^2) - P_0 A_0 \gamma M_0^2 - P_0 A_1 \tag{4.9b}$$

In both Fig. 4.5 and Fig. 4.8, we have arbitrarily chosen a smaller cross-sectional area at station 0 relative to station 1. The actual shape of the stream tube is dependent on flight conditions and engine throttle settings. Additive drag and other forms of pressure drag associated with the engine installation are appropriately referred to as "throttle-dependent drag" and must be accounted for by the engine designers. Additive drag will always be a positive quantity (a drag) and at best can be zero when $A_0 = A_1$, where the stream tube is a right circular cylinder.

4.3 Installed Thrust

As was pointed out in the previous section, the external drag D_{ext} defined by Eq. (4.4) is pressure drag resulting from the external air flow over the vehicle surface and is highly dependent on the engine installation/integration with the vehicle, often accomplished via a nacelle. As was initially presented in Sect. 1.4.7, we must distinguish the *installed* thrust (and thrust specific fuel consumption) from the uninstalled thrust. The installed performance of an engine must depend on each specific application. Thus, this section considers the installation losses (drags) of an installed engine, D_{inlet} and D_{noz} [see Eq. (1.7)].

It is industry standard to split the nacelle drag into two components: the drag associated with the forebody (front part of nacelle from station 1 to m), represented by D_w, and the drag associated with the afterbody (rear part of nacelle from station m to 9, also called the boat-tail), represented by D_{noz}. This is indicated in Fig. 4.9. Point m is located at an arbitrary but convenient midpoint along the nacelle where the external flow is reasonably parallel to the freestream flow and the static pressure nearly equals the freestream static pressure, $P_m = P_0$.

The installed thrust of the engine comes from the sum of forces that gives

$$T - F_{int} + D_{nac} = 0 \quad \text{or} \quad T = F_{int} - D_w - D_{noz} \tag{4.10}$$

Using Eq. (4.6), we get

$$T = F - (D_{add} + D_w + D_{noz}) \tag{4.11}$$

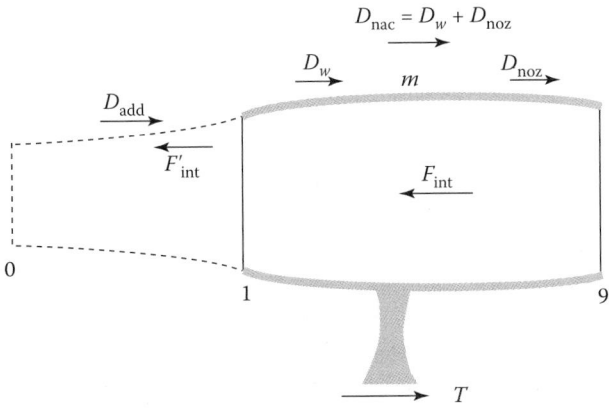

Fig. 4.9   Turbojet engine schematic for determining installed thrust.

Defining the inlet drag D_{inlet} as the sum of the additive and forebody drags

$$D_{\text{inlet}} \equiv D_{\text{add}} + D_w \qquad (4.12)$$

and combining Eqs. (4.11) and (4.12) we get the most useful relationship between *installed thrust T* and *uninstalled thrust F* as

$$T = F - D_{\text{inlet}} - D_{\text{noz}} \qquad (1.7)$$

The additive drag D_{add}, which is greater than or at best equal to zero, can be offset by the forebody portion of the nacelle drag D_w, which is nearly always a negative quantity (i.e., a thrust; see Fig. 11.7). This negative forebody drag is often called "lip suction" and is discussed in more detail in Section 11.3.4. Note that the lip suction force must be very large at times to keep the overall inlet drag small, as demonstrated in Example 4.3. A conservative estimate for D_{inlet} at nearly any flight condition can be obtained by determination of D_{add} from Eq. (4.9) and ignoring any benefit from lip suction D_w.

Figure 4.10 shows streamline paths of air entering an engine inlet at different values of freestream Mach number M_0. Boundary layer separation on the forebody of the nacelle can occur when the inlet must turn the flow through large angles. To reduce the additive drag at low flight Mach

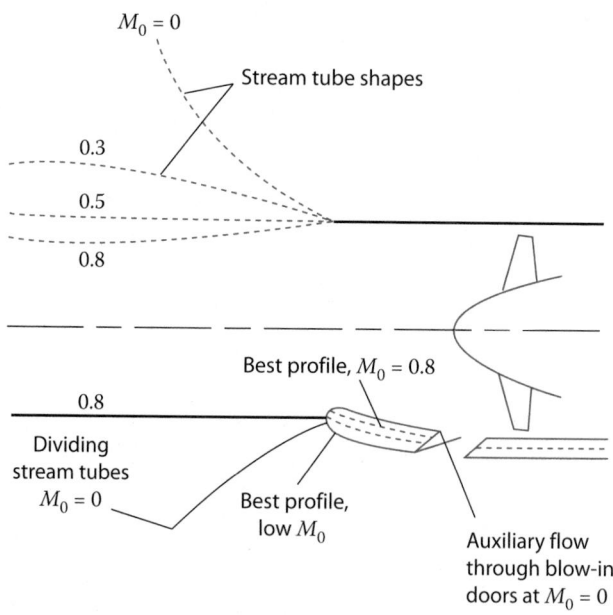

Fig. 4.10   Subsonic inlet at different flight Mach numbers (top) with auxiliary blow-in door (bottom) (from Ref. 21).

numbers, some subsonic inlets and many supersonic inlets have blow-in doors or auxiliary inlets (see Fig. 4.10) that increase the inlet area at the low-flight Mach number/full-throttle flight conditions (i.e., takeoff) and thus reduce the additive drag. Additionally, pilots can reduce the additive drag, hence the required takeoff distance, by rolling into the ground acceleration at partial throttle and progressing to full throttle as ground speed increases rather than starting the ground acceleration at full throttle. Notice how the engine speed is brought up to maximum thrust the next time you are at the end of the runway just prior to accelerating for takeoff on a commercial flight.

Example 4.2

Two nacelles with inlet areas of (1) $A_1 = 0.20 \text{ m}^2$ and (2) $A_1 = 0.26 \text{ m}^2$ are being considered for use with a gas turbine engine X that produces an unin-stalled thrust F of 20,000 N at sea level and $M_0 = 0.5$. Compare the installed engine thrust obtained with nacelle 1 and engine X to the installed engine thrust for nacalle 2 and engine X at $M_0 = 0.5$ and sea level. Engine X has a mass flow of 41.75 kg/s. Nacelles 1 and 2 each have the same nacelle drag of 900 N.

Solution

The installed engine thrust is given by

$$T = F - D = F - D_{\text{nac}} - D_{\text{add}}$$

Because both the uninstalled engine thrust F and nacelle drag D_{nac} are known, only the additive drag D_{add} need be determined for both nacelles. Either Eq. (4.9a) or Eq. (4.9b) can be used to calculate D_{add}, and both require the evalu-ation of P_1, M_1, and A_0. Because the mass flow rate into engine X is the same for both nacelles, A_0 will be determined first. The *mass flow parameter* (MFP) can be used to find A_0 once the total temperature and pressure are known at station 0

$$T_{t0} = T_0\left(1 + \frac{\gamma - 1}{2}M_0^2\right) = 288.2(1 + 0.2 \times 0.5^2) = 302.6 \text{ K}$$

$$P_{t0} = P_0\left(1 + \frac{\gamma - 1}{2}M_0^2\right)^{\gamma/(\gamma-1)} = 101,300(1 + 0.2 \times 0.5^2)^{35} = 120,160 \text{ Pa}$$

$$\text{MFP}(0.5)\sqrt{\frac{R}{g_c}} = \frac{\dot{m}\sqrt{T_{t0}}}{P_{t0}A_0}\sqrt{\frac{R}{g_c}} = 0.511053 \quad \text{from Appendix D}$$

(Continued)

Example 4.2 (Continued)

Thus

$$A_0 = \frac{\dot{m}\sqrt{T_{t0}}\sqrt{R/g_c}}{P_{t0}[\text{MFP}(0.5)\sqrt{R/g_c}]} = \frac{41.75\sqrt{302.6}(16.9115)}{120{,}160(0.511053)}$$

$$= 0.200 \, \text{m}^2$$

With subsonic flow between stations 0 and 1, the flow between these two stations is assumed to be isentropic. Thus the total pressure and total temperature at station 1 are 120,160 Pa and 302.6 K. Because the mass flow rate, area, and total properties are known at station 1 for both nacelles, the Mach number M_1 and static pressure P_1 needed to calculate the additive drag can be determined as follows:

1. Calculate $\text{MFP}\sqrt{R/g_c}$ at station 1, and find M_1 from Appendix D.
2. Calculate P_1, using Eq. (3.8) or P/P_t from Appendix D.

Nacelle 1: Because the inlet area A_1 for nacelle 1 is the same as flow area A_0 and the flow process is isentropic, then $M_1 = 0.5$ and $P_1 = 101{,}300$ Pa. From Eq. (4.11), the additive drag is zero, and the installed thrust is

$$T = F - D_{\text{nac}} - D_{\text{add}} = 20{,}000 - 900 - 0 = 19{,}100 \, \text{N}$$

Nacelle 2: Calculating $\text{MFP}\sqrt{R/g_c}$ at station 1, we find

$$\text{MFP}(M_1) = \frac{\dot{m}\sqrt{T_{t1}}}{P_{t1}A_1} = \frac{41.75\sqrt{302.6}}{120{,}160(0.26)}$$

$$= 0.0232465$$

From Appendix D or GASTAB, $M_1 = 0.3594$ and

$$P_1 = \frac{P_{t1}}{\{1 + [(\gamma-1)/2]M_1^2\}^{\gamma/(\gamma-1)}} = \frac{120{,}160 \, \text{Pa}}{1.09340} = 109.900 \, \text{Pa}$$

Then, using Eq. (4.9a), we have

$$D_{\text{add}} = P_1 A_1(1 + \gamma M_1^2) - P_0 A_0(1 + \gamma M_0^2) - P_0(A_1 - A_0)$$

$$= 109{,}900 \times 0.26(1 + 1.4 \times 0.3594^2) - 101{,}300$$

$$\times 0.2(1 + 1.4 \times 0.5^2) - 101{,}300(0.26 - 0.20)$$

$$= 33{,}741 - 27{,}351 - 6078 = 312 \, \text{N}$$

$$T = F - D_{\text{nac}} - D_{\text{add}} = 20{,}000 - 900 - 312 = 18{,}788 \, \text{N}$$

Conclusion: A comparison of the installed engine thrust T of nacelle 1 to that of nacelle 2 shows that nacelle 1 gives a higher installed engine thrust and is better than nacelle 2 at the conditions calculated.

Example 4.3

An inlet of about $36\,\text{ft}^2$ (about the size of the inlet on one of the C-5M's engines) is designed to have an inlet Mach number of 0.8 at sea level. Determine the variation of the additive drag with flight Mach number from $M_0 = 0$ to 0.9. Assume that M_1 remains constant at 0.8.

Solution

We will be using Eqs. (4.9a) and (4.9b) to solve this problem. The following values are known:

$$P_0 = 14.696\,\text{psia}, \quad M_1 = 0.8, \quad A_1 = 36\,\text{ft}^2, \quad \gamma = 1.4$$

Thus we must find the values of A_0 and P_1 for each flight Mach number M_0. We assume isentropic flow between stations 0 and 1, and therefore P_t and A^* are constant. Using the relations for isentropic flow, we can write

$$P_1 = P_0 \frac{P_{t0}}{P_0} \frac{P_1}{P_{t1}} = P_0 \frac{(P/P_t)_1}{(P/P_t)_0}$$

$$A_0 = A_1 \frac{A_1^*}{A_1} \frac{A_0}{A_0^*} = A_1 \frac{(A/A^*)_0}{(A/A^*)_1}$$

We obtain values of P/P_t and A/A^* from the isentropic table (Appendix D or GASTAB program). At a flight Mach number of 0.9, we have

$$P_1 = P_0 \frac{(P/P_t)_1}{(P/P_t)_0} = 14.696 \left(\frac{0.65602}{0.59126} \right) = 16.306\,\text{psia}$$

$$A_0 = A_1 \frac{(A/A^*)_0}{(A/A^*)_1} = 36 \left(\frac{1.0089}{1.0382} \right) = 34.99\,\text{ft}^2$$

Thus

$$
\begin{aligned}
D_{\text{add}} &= P_1 A_1 (1 + \gamma M_1^2) - P_0 A_0 (1 + \gamma M_0^2) - P_0 (A_1 - A_0) \\
&= 16.306(144)(36)(1.896) - 14.696(144)(34.984)(2.134) \\
&\quad - 14.696(144)(1.016) \\
&= 160{,}269 - 157{,}988 - 2150 \\
&= 131\,\text{lbf}
\end{aligned}
$$

At a flight Mach number of 0, the $A_0 M_0^2$ therm in Eq. (4.9b) is zero and Eq. (4.9b) becomes

$$(D_{\text{add}})_{M_0=0} = P_1 A_1 (1 + \gamma M_1^2) - P_0 A_1$$

We have

$$P_t = (14.696)(0.65602) = 9.641\,\text{psia}$$

(Continued)

Example 4.3 *(Continued)*

Thus

$$D_{add} = 9.641(144)(36)(1.896) - 14.696(144)(36)$$
$$= 94,760 - 76,184$$
$$= 18,576 \, lbf$$

Table 4.1 presents the results of this inlet's additive drag in the range of flight Mach numbers requested. As indicated in this table, the additive drag is largest at low flight Mach numbers for this fixed-area inlet. The CF6-80C2 engine used on the C-5M has a published uninstalled thrust of 59,000 lbf at sea-level/static conditions. Thus the additive drag is 31.4% of the uninstalled thrust, and the inlet was designed with extreme care to recover most of this force on the forebody of the engine nacelle.

Table 4.1 Summary of Additive Calculations for Example 4.3

M_0	$\left(\dfrac{P}{P_t}\right)_0$	P_1, psia	$\left(\dfrac{A}{A^*}\right)_0$	A_0, ft^2	$1 + \gamma M_0^2$	D_{add}, lbf
0.0	1.00000	9.641	—	—	1.000	18,576
0.1	0.99303	9.711	5.8218	201.87	1.014	13,283
0.2	0.97250	9.916	2.9635	102.76	1.056	9101
0.3	0.93947	10.265	2.0351	70.568	1.126	6187
0.4	0.89561	10.768	1.5901	55.137	1.224	3516
0.5	0.84302	11.439	1.3398	46.458	1.350	1838
0.6	0.78400	12.300	1.1882	41.201	1.504	767
0.7	0.72093	13.376	1.0944	37.949	1.686	195
0.8	0.65602	14.696	1.0382	36.000	1.896	0
0.9	0.59126	16.306	1.0089	34.984	2.134	131

4.4 Gas Turbine Engine Components

4.4.1 Inlets

An inlet delivers the required air mass flow rate to the engine *without choking* while *reducing* the entering air velocity to a level (M less than about 0.65) suitable for the compression system—fan and compressor(s), low and high as applicable—with minimal losses. Because the air velocity is reduced with a corresponding increase in static pressure, the inlet is sometimes called a diffuser or inlet/diffuser. The operation and design of the inlet are described in terms of the efficiency of the diffusion process, the external drag of the inlet, and the mass flow into the inlet. The design and operation of the inlet depend on whether the air entering the inlet duct is subsonic or supersonic, but in either case, the inlet generally takes the shape of a divergent duct. As the aircraft approaches the speed of sound, the air becomes more compressed, and at Mach 1, shock waves occur. Shock waves are compression waves, and at higher Mach numbers, these compression waves are stronger. Compression by shock waves is generally inefficient; thus an important consideration in supersonic inlet design must be shock wave control. In subsonic flow there are no shock waves; the air compression typically takes place quite efficiently with friction being the dominant source of losses. Shock waves and the compressibility of air then influence the design of inlets.

4.4.1.1 Subsonic Inlet

The subsonic inlet is typically a divergent duct of fixed geometry, as shown in Fig. 4.11. This duct is satisfactory until the freestream Mach number becomes greater than 1, at which time a normal shock wave occurs at the throat and the process becomes inefficient. The subsonic divergent duct operates best at one freestream velocity (design point); at other velocities the diffusion process is less efficient and the inlet drag is greater. The airflow patterns for the subsonic inlet are shown in Fig. 4.12.

4.4.1.2 Supersonic Inlet

Because shock waves will occur in supersonic flow, the geometry of supersonic inlets is designed to obtain the most efficient compression from the shock process with a minimum of weight. If the velocity is reduced from a supersonic speed to a subsonic speed with one normal shock wave, the compression process is relatively inefficient. If several oblique shock waves are employed to reduce the velocity, the compression process is more efficient. Two typical supersonic inlets are the ramp (2-D wedge) and the centerbody (3-D spike), which are shown in Fig. 4.13. The shock wave positions shown in Fig. 4.13 are for the design condition of shock-on-lip of the inlet. At off-design Mach numbers, the positions of the shock waves change, thus affecting the external drag and the efficiency of compression. A more efficient ramp or centerbody inlet can be designed by using multiple

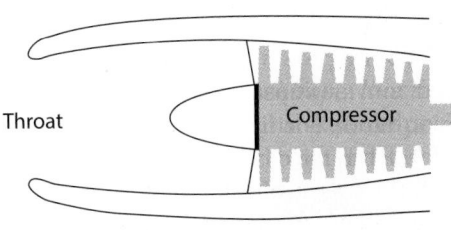

Fig. 4.11   Subsonic inlet.

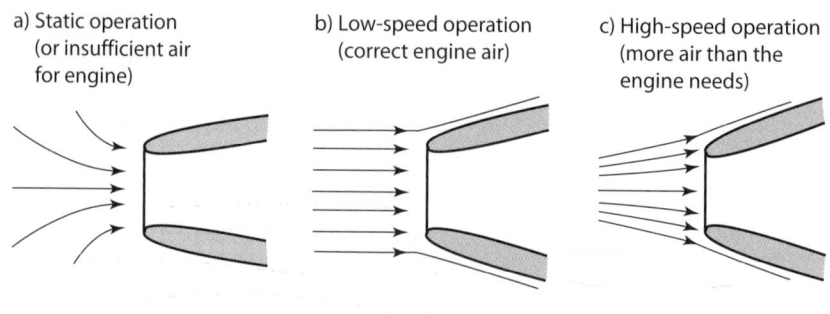

a) Static operation
(or insufficient air
for engine)

b) Low-speed operation
(correct engine air)

c) High-speed operation
(more air than the
engine needs)

Fig. 4.12   Subsonic inlet flow patterns.

shock waves to compress the entering air. Also, if the geometry is designed to be variable, the inlet operates more efficiently over a range of Mach numbers at the expense of added weight and complexity. Note that downstream of the shock system, the inlet area again diverges to provide air to the compression system at a Mach number less than 0.65.

4.4.2 Compressors

The function of the compressor is to increase the pressure of the incoming air with minimal losses so that the combustion process and the power extraction process after combustion can be carried out more efficiently. By increasing the pressure of the air, the specific volume of the air is reduced (density is increased), which means that the combustion of the fuel/air mixture can occur in a smaller volume (think more tightly packed energetic air molecules). The amount of energy required by the compressor is directly proportional to the rotational speed and the amount the flow is turned. This results in an *increase* in total temperature of the air. That energy is provided by the turbine.

4.4.2.1 Centrifugal Compressor

The compressor was the main stumbling block during the early years of turbojet engine development. Great Britain's Sir Frank Whittle solved the

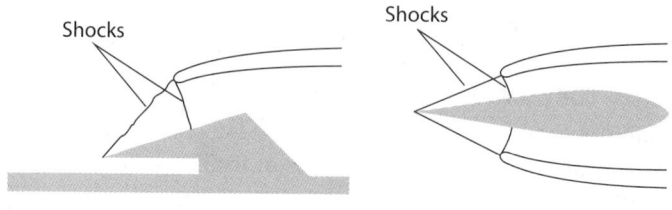

Shocks

Shocks

Fig. 4.13   Supersonic inlets.

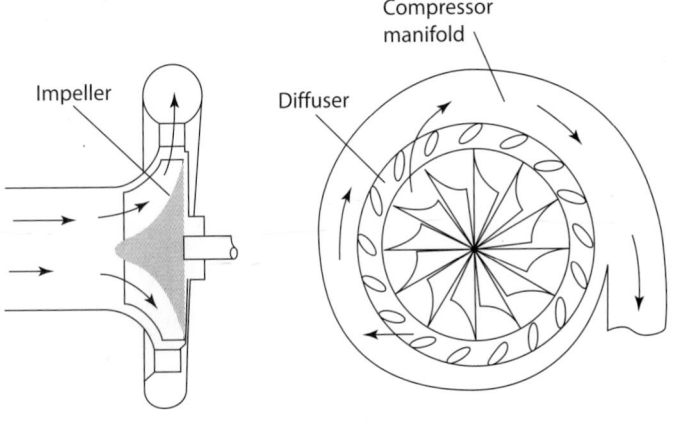

Fig. 4.14 Single-stage centrifugal compressor.

problem by using a centrifugal or radial compressor. This type of compressor is still being used in many of the smaller gas turbine engines like helicopter applications. A typical single-stage centrifugal compressor is shown in Fig. 4.14. The compressor consists of three main parts: an impeller, a diffuser, and a manifold. Air enters the compressor axially near the hub of the impeller and is then compressed by the rotational motion of the impeller exiting radially. The compression occurs by first increasing the velocity of the air (through rotation) and then diffusing the air where the velocity decreases and the static pressure increases. The diffuser also straightens the flow, and the manifold serves as a collector to feed the air into the combustor.

Due largely to the increased turning of the flow (more losses), the single-stage centrifugal compressor has a lower efficiency than a single-stage axial compressor, but is capable of much higher pressure ratios, as high as 7:1. The design of the centrifugal compressor accommodates the air's tendency to centrifuge out radially when subjected to a rotational motion, thus allowing for the higher pressure ratios. In addition to high stage pressure rise, centrifugal compressors are generally more durable, stall resistant, and less sensitive to tip clearance and erosion than axial compressors. However, high losses limit multistaging of centrifugal compressors to generally no more than two stages. Additionally, for air flows of much more than about 100 lbm/s, the size of a centrifugal compressor becomes prohibitive.

4.4.2.2 Axial Compressor

Multistage axial compressors are generally more common in aircraft applications because of their higher air flow per cross-sectional area and higher overall pressure ratio capability (many stages) at increased efficiency over centrifugal compressors. An axial compressor is shown in Fig. 4.15. As the name implies, the air in an axial compressor flows largely in an

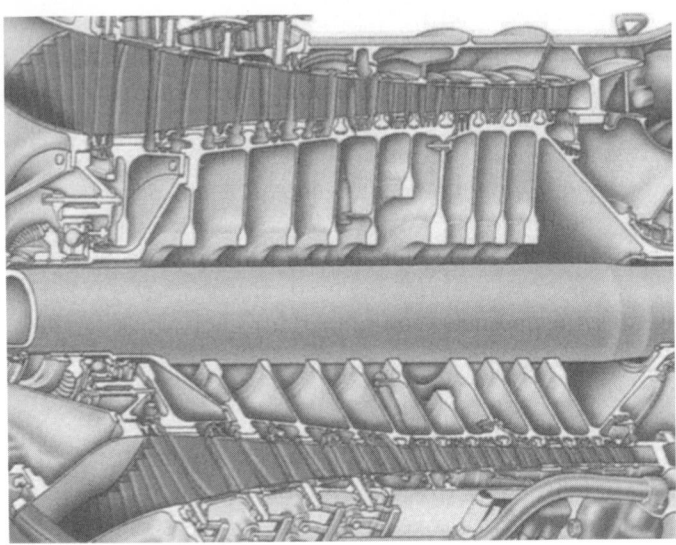

Fig. 4.15   Multistage axial compressor. (Courtesy of Pratt & Whitney.)

axial direction through a series of rotating rotor blades and stationary stator vanes that are concentric with the axis of rotation. Each set of rotor blades and stator vanes is known as a *stage*. The rotor adds energy to the air, turning it away from the axis of rotation and increasing its velocity (observed from a stationary reference frame) while the stator diffuses the air (velocity decreases, static pressure increases), turning it back for proper orientation of the air for the rotor of the next stage. The flow path cross-sectional area in an axial compressor decreases to keep the axial velocity constant. For the best axial compressor efficiency, the compressor operates at a constant axial velocity, as shown in Fig. 1.6. The decrease of area in the direction of flow is in proportion to the increased density of the air as the compression progresses from stage to stage. Due largely to the global adverse pressure gradient (increasing static pressure in the flow direction), each stage of an axial compressor produces a small pressure ratio (generally less than 2.5:1), but at a high efficiency. Therefore, for high pressure ratios, multiple stages are used. For example, the core compressor of the GEnx, which powers the Boeing 787, provides a 23:1 pressure ratio with 10 stages [68].

For a single rotational speed, there is a limit to the proper balance of work distribution between the first and last stages of the compressor. To obtain more flexibility and a more uniform loading of each compressor stage, a dual-shaft compression system with two different rotational speeds is generally used in most modern-day applications requiring high overall pressure rise (40:1); Rolls-Royce uses a three-shaft system (e.g., its Trent series of engines, see Fig. 1.8c). For dual-shaft systems, the core compressor spinning at the higher rotational speed is typically referred to as the high-pressure compressor (HPC), and the other compressor spinning at the lower speed is the low-pressure compressor (LPC); see Fig. 1.7. For three-shaft systems, the spools are referred to as low-intermediate-high, descriptive of the pressures relative to each other. The *fan* of turbofan engines consist of compressor stage(s) on the low-speed shaft. As seen in Fig. 1.7, all of the air mass flows through the fan, but only that portion of the air directed towards the core flows through the LPC and HPC. The LPC compressor is often referred to as a *booster* or *booster stages*.

4.4.3 Combustor or Main Burner

The combustor is designed to efficiently burn a mixture of fuel and air to release energy and deliver the resulting high-energy gases to the turbine at a uniform temperature. The gas temperature must not exceed the allowable structural temperature of the turbine, and is typically limited by desired life concerns (e.g., engine flying hour goal before replacement). A schematic of a combustor is shown in Fig. 4.16. A fraction of the total volume of air entering the main burner mixes with the fuel and burns. The rest of the air—secondary air—is simply heated or may be thought of as cooling the products of combustion and cooling burner and turbine parts. The ratio of total air to fuel varies among the different types of engines from 30 to 60 parts of air to 1 part of fuel by weight. The average ratio in new engine designs is about 40:1, but only about 10 parts are used for burning (because the combustion process demands that the number of parts of air to fuel must be within certain limits at a given pressure for combustion to occur). Most modern combustion chambers are the annular type, but some, especially early designs, may be individual cans or can-annular type. All three types are shown in Fig. 4.17.

For an acceptable burner design, the pressure loss as the gases pass through the burner must be held to a minimum, the combustion efficiency

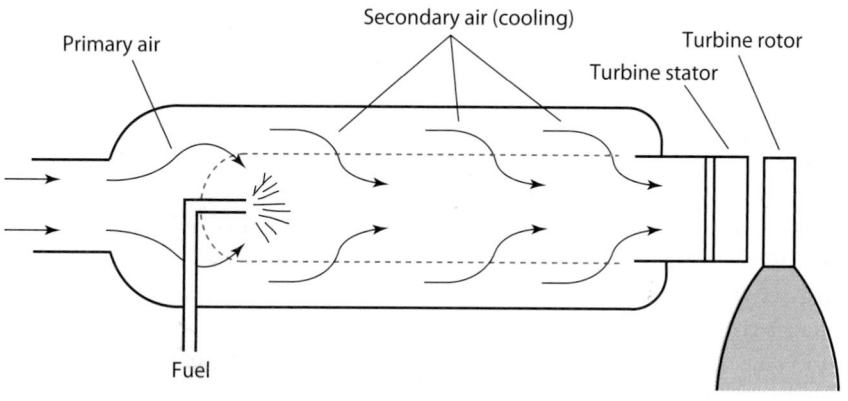

Fig. 4.16   Straight-through flow combustor (main burner).

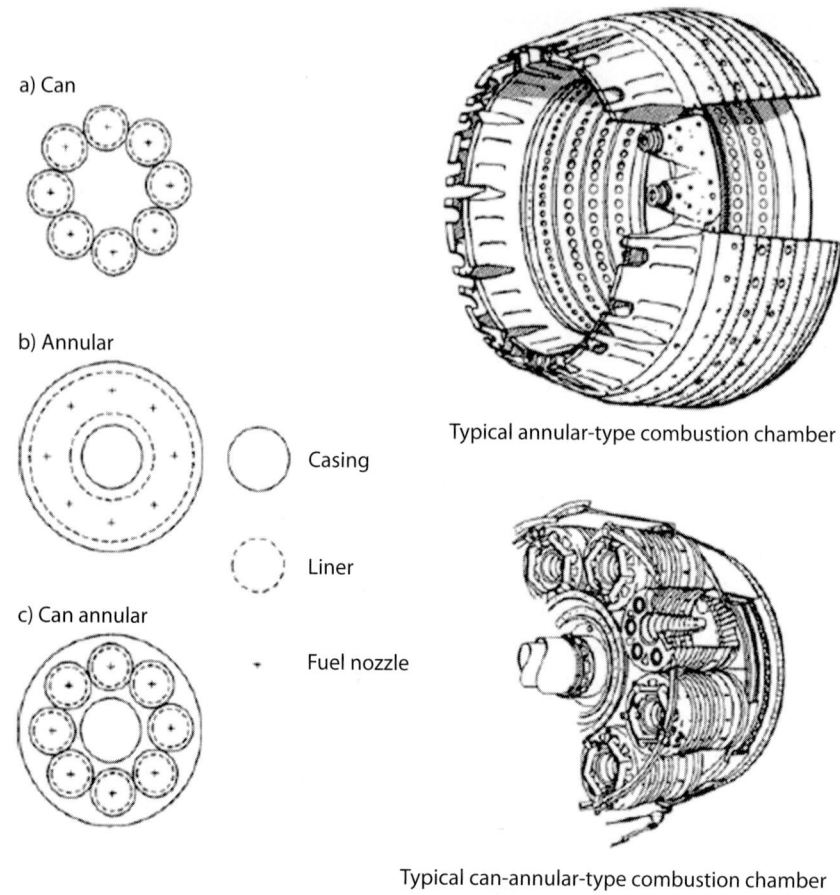

a) Can

b) Annular

Casing

Liner

c) Can annular

Fuel nozzle

Typical annular-type combustion chamber

Typical can-annular-type combustion chamber

Fig. 4.17   Cross sections of combustion chambers. (Courtesy of Pratt & Whitney.)

must be high, and there must be no tendency for the burner to blow out (flameout). Also, combustion generally must take place entirely within the burner, although interturbine burning has been explored.

4.4.4 Turbines

The turbine extracts kinetic energy from the expanding gases that flow from the combustor and converts the kinetic energy into shaft horsepower with minimal losses to drive the compressor and the engine accessories (pumps, generators, etc.). The turbines used in modern jet engines are *axial* type, regardless of compressor type (axial or centrifugal). Nearly three-fourths of all the energy available from the combustion process is typically required to drive the compression system for most gas turbine engines (except turboprop/turboshaft engines).

The axial-flow turbine stage consists of a set of stationary nozzle vanes (stators) followed by the rotating turbine wheel rotor, as shown in Fig. 4.18. (Recall that for the compressor, a stage is rotor-stator.) The nozzle vanes accelerate and direct the hot combustion gases onto the turbine rotor blades at very high velocities. The discharge of the gases onto the rotor allows the kinetic energy of the gases to be transformed to mechanical shaft energy. The amount of energy extracted by the turbine is directly proportional to the rotational speed and the amount the flow is turned. This results in a *decrease* in total temperature of the gases.

Like the axial compressor, the exhaust gas flow through the turbine stays largely in the axial direction, although much greater turning of the flow through the axis can be accommodated by the turbine due to the global favorable pressure gradient (decreasing static pressure in the flow direction). Because the flow turning is proportional to the work extraction, modern core or high-pressure turbines (HPTs) are single-stage or two-stage, providing enough shaft horsepower to drive an HPC of many stages (6 to 10 stages is common).

Of course, if there is an LPC and/or fan on a separate shaft, there must be a low-pressure turbine (LPT) or fan turbine to drive it. As shown in the Chapter 1 engine schematics, the typical dual-shaft arrangement is concentric shaft within a shaft with no mechanical connection between the two.

Fig. 4.18   Axial-flow turbine.

In addition to the low- and high-speed shafts, many Rolls-Royce engines use an intermediate shaft for a shaft within a shaft within a shaft configuration. The LPT of modern high-bypass turbofan engines is often composed of four, five, or even six stages because the fan power requirement can be quite large (on the order of 100,000 hp).

The flow path cross-sectional area in an axial turbine increases to keep the axial velocity approximately constant. The increase of area in the direction of flow is in proportion to the decreased density as the gases expand through the turbine.

4.4.5 Exhaust Nozzles

The purpose of the exhaust nozzle is to collect and straighten exhaust gas flow from the turbine and convert the remaining internal energy (enthalpy) of the gas into kinetic energy (i.e., increase the velocity of the gas) with minimal losses. Overall, the aircraft gas turbine engine converts the internal energy of the fuel to kinetic energy in the exhaust gas stream. As we have seen most recently in Section 4.2, the thrust of the engine is the result of this overall energy exchange. The exhaust nozzle supplies a high exit velocity by accelerating the exhaust gas in an expansion process that requires a decrease in static pressure. Ideally, the exhaust gas static pressure at the nozzle exit plane equals the ambient air pressure. As you learned in some detail in Chapter 3, the pressure ratio across the nozzle controls the expansion process. You will learn that the maximum possible thrust for a given engine is obtained when the exit gas static pressure equals the ambient air pressure. The two basic types of nozzles used in jet engines are the convergent and convergent-divergent nozzles.

4.4.5.1 Convergent Nozzle

The convergent nozzle is a simple convergent duct, as shown in Fig. 4.19. When the nozzle pressure ratio (turbine exit pressure to nozzle exit pressure) is low (less than about 2), the convergent nozzle is used. Because the maximum possible Mach number is 1 for the convergent nozzle, it is generally used in engines for subsonic aircraft.

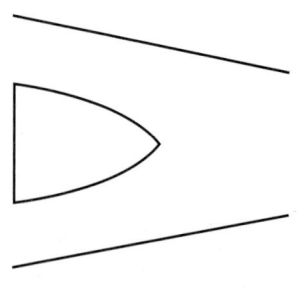

Fig. 4.19 Convergent exhaust nozzle.

4.4.5.2 Convergent-Divergent Nozzle

Acceleration of gases beyond Mach 1 requires a convergent-divergent (C-D) nozzle (Chapter 3). Consequently, aircraft applications like high-performance supersonic fighters that require high thrust per unit air flow require the high exhaust gas velocities only possible with a C-D nozzle. Most convergent-divergent

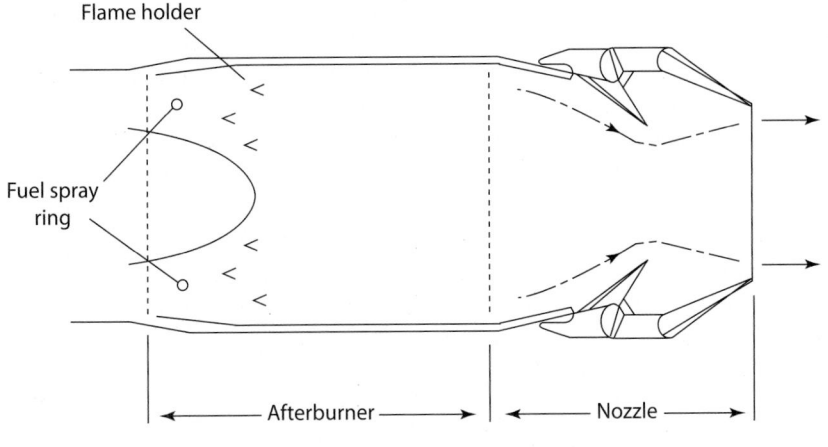

Fig. 4.20   Convergent-divergent exhaust nozzle. (Courtesy of Pratt & Whitney.)

nozzles used in supersonic aircraft are not simple fixed ducts, but incorporate variable geometry and other aerodynamic features, as shown in Fig. 4.20. Only the throat area and exit area of the nozzle in Fig. 4.20 are set mechanically, the nozzle wall positions being determined aerodynamically by the gas flow. The convergent-divergent nozzle is used if thé nozzle pressure ratio is high. If the engine incorporates an afterburner, the nozzle throat and exit area must be varied to match the different flow conditions and to produce the maximum available thrust.

4.4.6   Thrust Augmentation

In this section we present two methods of thrust augmentation: afterburning or reheating and water injection. Afterburning is commonly used in fighter engines whereas water injection has been used in some bomber and tanker aircraft.

4.4.6.1 Afterburning

High-performance military fighter and bomber aircraft often use afterburners as mission needs dictate. As indicated in Fig. 4.20, the afterburner is a section of duct between the turbine and exhaust nozzle. The afterburner consists of fuel injectors, flame holders, and the afterburner duct, which is typically baffled and perforated for noise and cooling concerns. It is possible to have additional burning "after" the main burner because the main combustion products are air-rich. The effect of the afterburning operation is to raise the internal energy (enthalpy) of the exhaust gases manifested by an increase in temperature so that when exhausted through a properly designed C-D nozzle, the gases will reach a higher exit velocity.

A representative pressure/temperature/velocity profile for afterburning operation is shown in Fig. 1.6 for the J79 afterburning turbojet. The J79 has a *maximum* thrust (in full afterburner) of 17,900 lbf and a thrust specific fuel consumption (TSFC) of $1.965\ h^{-1}$; for *military* operation (maximum thrust, no afterburning) it has a thrust of 11,870 lbf and a TSFC of $0.84\ h^{-1}$. This is typical of afterburner operation: a 50% increase in thrust and a tripling of fuel flow rate! Consequently, afterburners are used in short duration for flight segments like heavy takeoff on a hot day, high *g* maneuvers, and transonic/supersonic flight. Also, afterburner operation is typically *staged*, meaning that there are numerous partial throttle afterburner settings controlling various afterburner fuel flow rates so that operation is not just "all or none."

4.4.6.2 Water Injection

Thrust augmentation by water injection (or by water/alcohol mixture) is achieved by injecting water into either the compressor or the combustion chamber. When water is injected into the inlet of the compressor, the mass flow rate increases and a higher combustion chamber pressure results if the turbine can handle the increased mass flow rate. The higher pressure and the increase in mass flow combine to increase the thrust. Injection of water into the combustion chamber produces the same effect, but to a lesser degree and with greater consumption of water. Water injection on a hot day can increase the takeoff thrust by as much as 50% because the original mass of air entering the jet engine is less for a hot day due to reduced air density. The Pegasus turbofan engines of the AV-8B Harrier and the older J57 turbojet engines of the B-52 and KC-135 aircraft used water for thrust augmentation. The Harrier tank carries up to 500 lb of distilled water for a maximum thrust augmentation time of about 90 s.

4.5 Brayton Cycle

A thermodynamic *cycle* is a series of processes that start and end at the same state. The *Brayton cycle* is a model used in thermodynamics for an *ideal gas turbine power cycle*. It is composed of the four following processes, which are also shown on a *T-s* diagram in Fig. 4.21a using standard aircraft engine numbering:

- Isentropic, adiabatic compression (2 to 3)
- Constant-pressure heat addition (3 to 4)
- Isentropic, adiabatic expansion (4 to 9)
- Constant-pressure heat rejection (9 to 2)

Additionally, air is assumed to be the working fluid and the same air mass rate flows through each component. The basic components of the Brayton cycle are shown to the right in Fig. 4.21b. In the ideal cycle, the processes

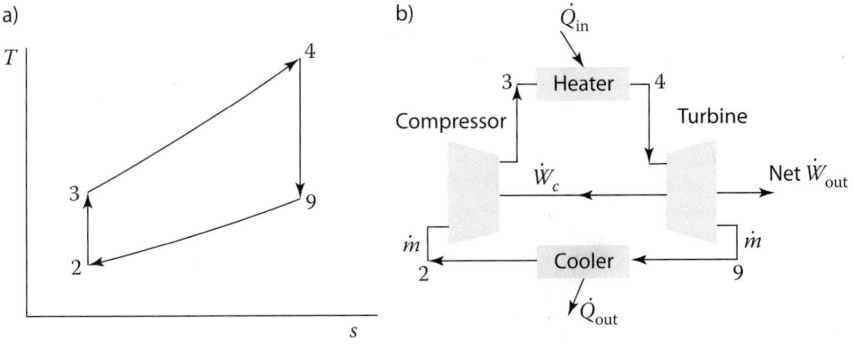

Fig. 4.21   Brayton cycle.

through both the compressor and the turbine are considered to be isentropic and adiabatic (reversible). The processes through the heater and cooler are considered to be constant pressure in the ideal cycle.

As we have seen in Chapters 2 and 3 for a calorically perfect gas (CPG), thermodynamic analysis of the ideal Brayton cycle gives the following equations for the rate of energy transfer of each component:

$$\dot{W}_c = \dot{W}_{in} = \dot{m}c_p(T_3 - T_2) \quad \dot{Q}_{in} = \dot{m}c_p(T_4 - T_3)$$

$$\dot{W}_t = \dot{W}_{out} = \dot{m}c_p(T_4 - T_9) \quad \dot{Q}_{out} = \dot{m}c_p(T_9 - T_2)$$

Here, in keeping with classical thermodynamics, the thermodynamic or static temperatures are shown. In practice, the total temperatures T_t are used because they are more readily measured and account for any kinetic energy changes of the flowing gas (see Example 2.7). The cooler component is the atmospheric air, which dissipates the "wasted" energy of the heated exhaust gases and continuously supplies fresh air to the gas turbine engine.

The overall performance of the Brayton cycle gas turbine engine is quantified by the net power out and the thermal efficiency, both of which were defined in Chapter 1.

$$\dot{W}_{net\ out} = \dot{W}_t - \dot{W}_c = \dot{m}c_p[(T_4 - T_9) - (T_3 - T_2)]$$

$$\eta_{Th} = \frac{\dot{W}_{net\ out}}{\dot{Q}_{in}}$$

Because the compression and turbine expansion processes are adiabatic and isentropic, the isentropic CPG relations apply [see Eq. (2.54)], and one can show that

$$\eta_{Th,\ Brayton} = 1 - \left(\frac{1}{PR}\right)^{\frac{\gamma-1}{\gamma}} \tag{4.13}$$

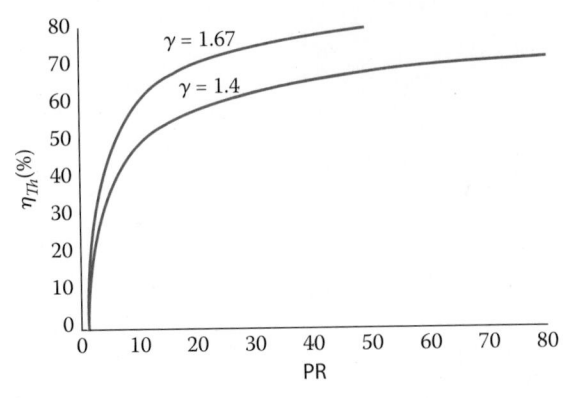

Fig. 4.22   Thermal efficiency of an ideal Brayton cycle.

where PR is the pressure ratio P_3/P_2. Ponder Eq. (4.13) for a moment. We have a four-process *idealized* cycle and the thermal efficiency still is not 100%; it is a function of the cycle pressure ratio! This is a manifestation of one of the very important implications of the second law of thermodynamics; that is, every cycle, even idealized cycles, will have wasted energy due to heat rejection.

The Brayton cycle thermal efficiency is plotted in Fig. 4.22 for two ratios of specific heats. Actual gas turbine engine cycles follow the same trend, but of course with lower magnitudes due to component losses (irreversibilities). Figure 4.22 explains the trend of aircraft gas turbine engine manufacturers designing to higher and higher overall cycle pressure ratios P_3/P_2, as high as 70:1 (Ref. 5). Higher cycle pressure ratio leads to higher thermal efficiency at the expense of additional compressor stages and higher compressor exit temperatures. This has many implications including the use of intercooling to cool the compressor air and increased turbine entry temperatures to ensure high work output per unit mass from the turbines.

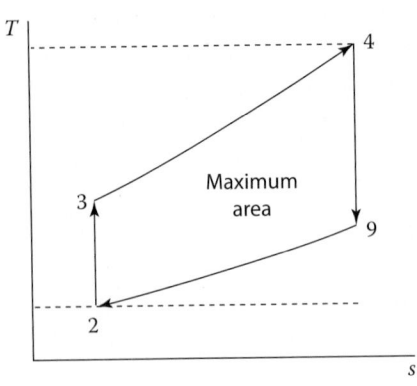

Fig. 4.23   Maximum power output Brayton cycle.

For an ideal Brayton cycle with fixed compressor inlet temperature T_2 and heater exit temperature T_4, basic calculus yields an expression for the pressure ratio P_3/P_2 and associated temperature ratio T_3/T_2 giving the maximum net work output per unit mass $\dot{W}/\dot{m}$. (Take the derivative

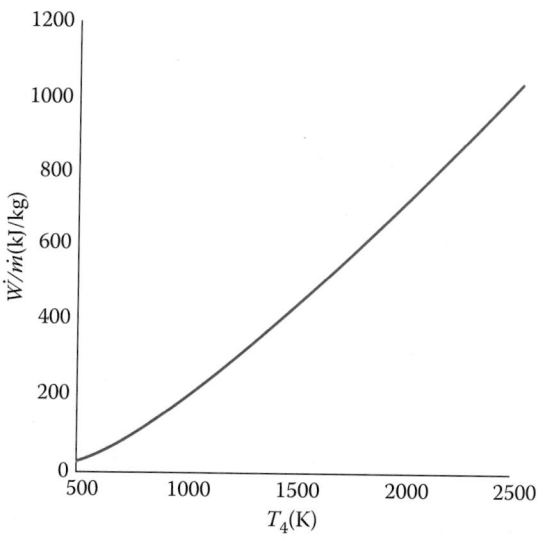

Fig. 4.24 Variation of maximum output power with T_{t4} for ideal Brayton cycle ($T_2 = 288$ K).

of $\dot{W}/\dot{m}$ with respect to P_3/P_2 and set the expression equal to zero.) This optimum compressor pressure, or temperature ratio, corresponds to the maximum area within the cycle on a T-s diagram, as shown in Fig. 4.23. One can show that the optimum compressor temperature ratio is given by

$$\left(\frac{T_3}{T_2}\right)_{\text{max work}} = \sqrt{\frac{T_4}{T_2}} \tag{4.14}$$

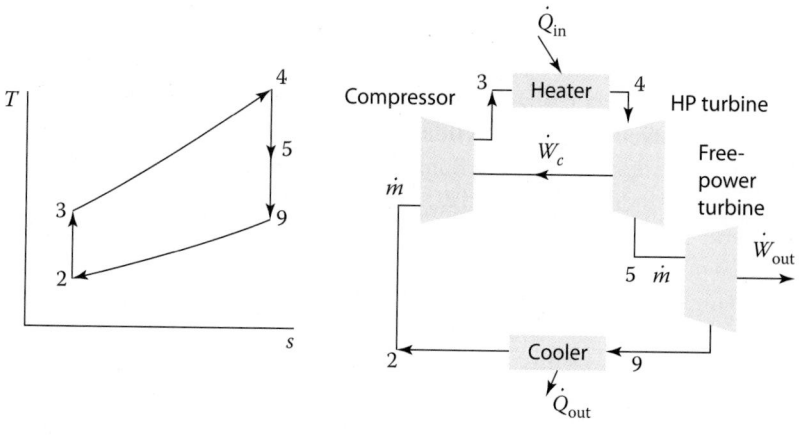

Fig. 4.25 Brayton cycle with free-power turbine.

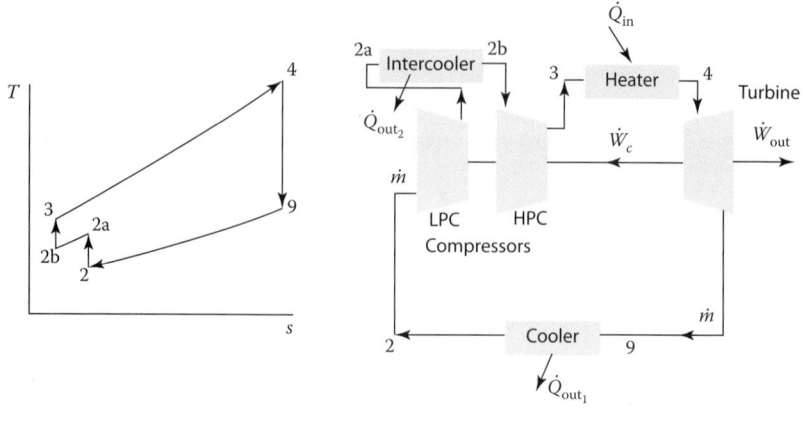

Fig. 4.26   Brayton cycle with intercooling between compressors.

and the corresponding net work output per unit mass is

$$\left(\frac{\dot{W}_{net\,out}}{\dot{m}}\right)_{max} = c_p T_2 \left(\sqrt{\frac{T_4}{T_2}} - 1\right)^2 \tag{4.15}$$

which is plotted in Fig. 4.24 for air with $T_2 = 288$ K (standard day sea level).

Four variations in the basic Brayton cycle are shown in Figs. 4.25–4.28. Figure 4.25 shows the cycle with a high-pressure (HP) turbine driving the compressor and a free-power turbine providing the output power, a common configuration for turboshaft or turboprop engines. This cycle has the same thermal efficiency as the ideal Brayton cycle of Fig. 4.21. Figure 4.26 shows the cycle with intercooling between the LPC and HPC.

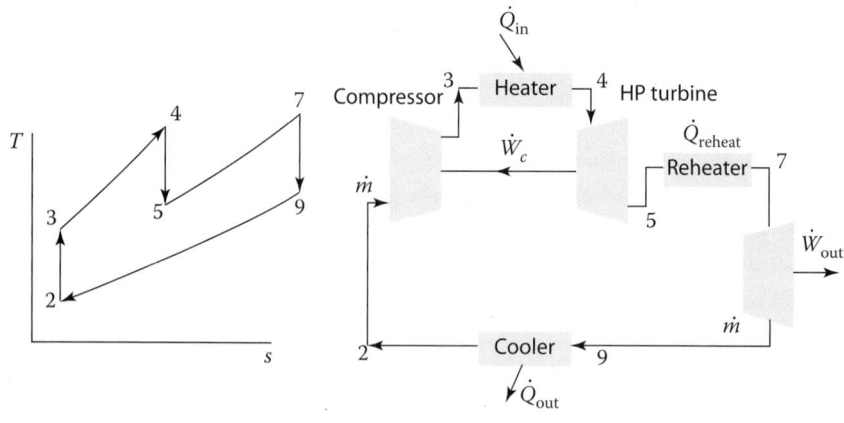

Fig. 4.27   Brayton cycle with reheat.

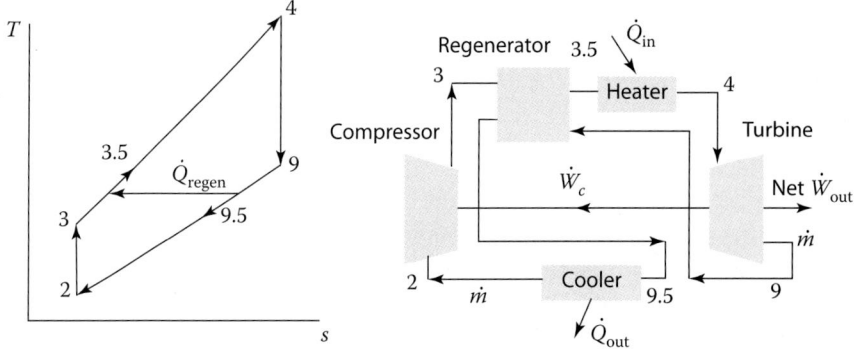

Fig. 4.28   Ideal Brayton cycle with regeneration.

Intercooling reduces the gas temperatures in the HPC and thus the power required by the HPC to reach a desired pressure ratio. Figure 4.27 shows the Brayton cycle with reheat (afterburner). Addition of reheat to the cycle increases the specific power $\dot{W}/\dot{m}$ of the free turbine and reduces the thermal efficiency.

Figure 4.28 shows the ideal Brayton cycle with regeneration, which uses some of the otherwise wasted energy from the exhaust to further

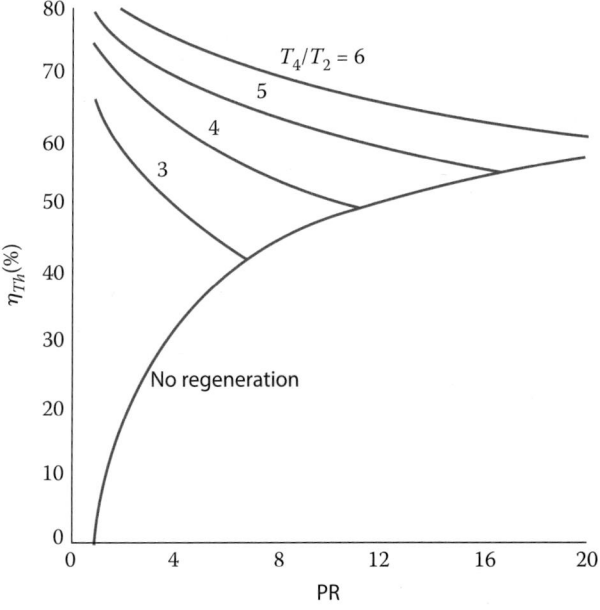

Fig. 4.29   Thermal efficiency of ideal Brayton cycle with regeneration.

preheat the compressor exit air. Regeneration is frequently used in practice in ground-based gas turbine engine applications. When regeneration is added to the basic Brayton cycle, the energy input to the heater is reduced, which increases the cycle's thermal efficiency. For an ideal regenerator, we have

$$T_{3.5} = T_9 \quad T_{9.5} = T_3$$

Note that regeneration is possible only when T_9 is greater than T_3, which requires low cycle pressure ratios. The thermal efficiency of an ideal Brayton cycle with regeneration is shown in Fig. 4.29 for several values of T_4/T_2 along with the thermal efficiency of the cycle without regeneration. The thermal efficiency of the ideal Brayton cycle with regeneration is given by

$$\eta_{\text{Th, Brayton, regen}} = 1 - \frac{(PR)^{\frac{\gamma-1}{\gamma}}}{T_4/T_2} \quad (4.16)$$

4.6 Aircraft Engine Design

This introductory chapter to aircraft propulsion systems has presented a generalized thrust equation and the basic flow path components and ideal cycle model for gas turbine engines. The following chapters present the cycle analysis (thermodynamic design-point study) of ideal and real engines, off-design performance, and an introduction to the aerodynamics of engine components. The design procedure for a typical gas turbine engine, shown in Fig. 4.30, requires that thermodynamic design-point studies (cycle analysis) and off-design performance be in the initial steps of design. The iterative nature of design is indicated in Fig. 4.30 by the feedback loops, although only a few are shown; many more exist. Those items within the dashed lines of Fig. 4.30 are addressed within this textbook.

Problems

4.1 Starting with the generalized thrust equation, Eq. (4.5), find the *uninstalled thrust* (lbf) of the space shuttle main engine (SSME) of Problem 3.11 under the following conditions:

a) At launch

b) At the design altitude of 44,600 ft

c) In a vacuum (ambient pressure $= 0$)

Recall that the uninstalled thrust is the net axial force produced by an engine in an idealized inviscid fluid. The SSME nozzle is choked at all these conditions. *Be clear with all assumptions and their implications.* Refer to Appendix C, Table C.1.

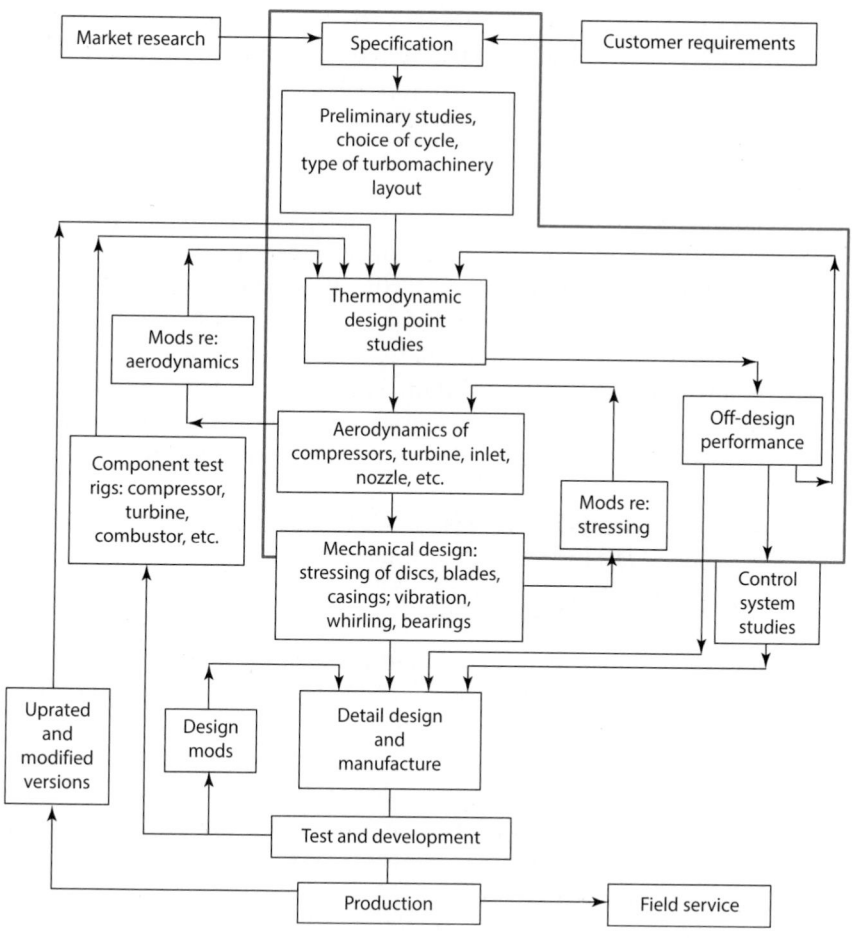

Fig. 4.30 Typical aircraft gas turbine design procedure (from Ref. 22).

4.2   Starting with the generalized thrust equation, Eq. (4.5), find the *uninstalled* thrust (lbf) and *uninstalled* thrust specific fuel consumption (1/hr) of a mixed-flow afterburning turbofan engine on a test stand in

a) Military power (maximum nonafterburning power)

b) Maximum power (maximum afterburner)

The following performance parameters as determined by test measurements are known.

$$\dot{m}_0 = 350\,\text{lbm/s}, \quad \dot{m}_{f,\text{mil}} = 2.715\,\text{lbm/s}, \quad \dot{m}_{f,\text{max}} = 21.033\,\text{lbm/s},$$

$$T_{t9\text{mil}} = 1097°\text{R}, \quad T_{9\text{mil}} = 904°\text{R}, \quad T_{t9\text{max}} = 3400°\text{R}, \quad T_{9\text{max}} = 2889°\text{R}$$

Recall that the uninstalled thrust is the net axial force produced by an engine in an idealized inviscid fluid. Assume the engine exhaust nozzle is operating perfectly expanded in both conditions. Use $c_p = 0.295$ Btu/(lbm · °R). *Be clear with all assumptions and their implications.*

4.3   The inlet for a high-bypass-ratio turbofan engine has an area A_1 of 6.0 m^2 and is designed to have an inlet Mach number M_1 of 0.6. Determine the additive drag at the flight conditions of sea-level static test and Mach number of 0.8 at 12-km altitude.

4.4   An inlet with an area A_1 of 10 ft^2 is designed to have an inlet Mach number M_1 of 0.6. Determine the additive drag at the flight conditions of sea-level static test and Mach number of 0.8 at 40-kft altitude.

4.5   Determine the additive drag for an inlet having an area A_1 of 7000 in.^{2} and a Mach number M_1 of 0.8 while flying at a Mach number M_0 of 0.4 at an altitude of 2000 ft.

4.6   Determine the additive drag for an inlet having an area A_1 of 5.0 m^2 and a Mach number M_1 of 0.7 while flying at a Mach number M_0 of 0.3 at an altitude of 1 km.

4.7   A turbojet engine under static test ($M_0 = 0$) has air with a mass flow of 100 kg/s flowing through an inlet area A_1 of 0.56 m^2 with a total pressure of 1 atm and total temperature of 288.8 K. Determine the additive drag of the inlet.

4.8   In Chapter 1, the loss in thrust due to the inlet is defined by Eq. (1.8) as $\phi_{\text{inlet}} = D_{\text{inlet}}/F$. Determine ϕ_{inlet} for the inlets of Example 4.2.

4.9   Determine the variation of inlet mass flow rate with Mach number M_0 for the inlet of Example 4.3

4.10   In Chapter 1, the loss in thrust due to the inlet is defined by Eq. (1.8) as $\phi_{\text{inlet}} = D_{\text{inlet}}/F$. For subsonic flight conditions, the additive drag D_{add} is a conservative estimate of D_{inlet}.

a) Using Eq. (4.9) and isentropic flow relations, show that ϕ_{inlet} can be written as

$$\phi_{\text{inlet}} = \frac{D_{\text{add}}}{F}$$

$$= \frac{(M_0/M_1)\sqrt{T_1/T_0}\left(1 + \gamma M_1^2\right) - \left(A_1/A_0 + \gamma M_0^2\right)}{(Fg_c/\dot{m}_0)(\gamma M_0/a_0)}$$

b) Calculate and plot the variation of ϕ_{inlet} with flight Mach number M_0 from 0.2 to 0.9 for inlet Mach numbers M_1 of 0.6 and 0.8 with $(Fg_c/\dot{m}_0)(\gamma/a_0) = 4.5$.

4.11   Determine the force on the bell-mouth inlet of Problem 3.14 with the given data for an inlet area that is 8 times the area at station 2. Assume that the inlet wall is a stream tube and that the outside of the bell-mouth inlet sees a static pressure equal to P_0.

4.12   The maximum power out of an ideal Brayton cycle operating between temperatures T_2 and T_4 is given by Eq. (4.15). By taking the derivative of the net work out of an ideal Brayton cycle with respect to the pressure ratio (PR) and setting it equal to zero, show that Eq. (4.14) gives the resulting compressor temperature ratio and Eq. (4.15) gives the net work out.

4.13   For an ideal Brayton cycle with intercooling, determine the pressure ratio for each compressor in terms of the overall pressure ratio (PR) that gives minimum work of compression with $T_{2b} = T_2$.

4.14   Determine the thermal efficiency for an ideal Brayton cycle with reheat for PR $= 20$, $T_2 = 300$ K, $T_4 = 2000$ K, $T_7 = 2200$ K, and $\gamma = 1.4$.

4.15   Show that Eq. (4.16) gives the thermal efficiency for the ideal Brayton cycle with regeneration.

4.16   For the ideal Brayton cycle with regeneration, regeneration is desirable when $T_9 \geq T_3$. Show that the maximum compressor pressure ratio PR_{max} for regeneration $(T_9 = T_3)$ is given by

$$PR_{max} = \left(\frac{T_4}{T_2}\right)^{\gamma[2(\gamma-1)]}$$

Chapter 5 Parametric Cycle Analysis (PCA) of Ideal Engines

> Scientists discover the world that exists; engineers create the world that never was.
>
> *Theodore von Karman*

5.1 Introduction

Cycle analysis studies the thermodynamic changes of the working fluid (air and products of combustion in most cases) as it flows through the engine. It is divided into two types of analysis: *parametric cycle analysis* (also called *design-point* or *on-design*) and *engine performance analysis* (also called *off-design*). Parametric cycle analysis (PCA) determines the performance of engines at different values of design choice (e.g., compressor pressure ratio), component performance (e.g., turbine efficiency), and design limit (e.g., combustor exit temperature) parameters. Engine performance analysis (EPA) determines the performance of a specific engine at all flight conditions and throttle settings.

In both forms of analysis, the components of an engine are characterized by the change in properties they produce. For example, the compressor is described by a total pressure ratio and efficiency. A certain engine's behavior is determined by its geometry, and a compressor will develop a certain total pressure ratio for a given geometry, speed, and airflow. Because the geometry is not included in parametric cycle analysis, the plots of specific thrust $F/\dot{m}_0$ and thrust specific fuel consumption S vs, say, compressor pressure ratio, are not portraying the behavior of a specific engine. *Each point* on such plots represents a different engine. The geometry for each plotted engine will be different, and thus we say that parametric cycle analysis represents a "rubber engine." Parametric cycle analysis is also called design-point analysis or on-design analysis because each plotted engine is operating at its so-called design point [7, 13, 21, 22].

The main objective of parametric cycle analysis is to relate the uninstalled engine performance parameters (primarily specific thrust $F/\dot{m}_0$ and thrust

---

Supporting material for this chapter is available electronically. See the Supporting Materials page at the back of the book for instructions to download.

specific fuel consumption S) to design choices (compressor pressure ratio, fan pressure ratio, bypass ratio, etc.), to design limitations (burner exit temperature, compressor exit pressure, etc.), and to flight environment (Mach number, ambient temperature, etc.). From parametric cycle analysis, we can easily determine which engine type (e.g., turbofan) and component design characteristics (range of design choices) best satisfy a particular need.

The value of parametric cycle analysis depends directly on the specific objectives and realism with which the engine components are characterized. For example, if the primary objective is to narrow the field of design choices to a more manageable range, representing the compressor with a constant efficiency might be appropriate. However, if a compressor is specified by the total pressure ratio and the isentropic efficiency, and if the analysis purports to select the best total pressure ratio for a particular mission, then the choice may depend on the variation of efficiency with pressure ratio. For the conclusions to be useful, a realistic variation of efficiency with total pressure ratio must be included in the analysis.

The parametric cycle analysis of engines will be developed in stages. First the general steps applicable to the parametric cycle analysis of engines will be introduced. Next these steps will be followed to analyze engines where all engine components are taken to be ideal. Trends of these ideal engines will be analyzed, given that only basic conclusions can be deduced. You can often facilitate your understanding of these trends by using the T-s diagram, applicable governing cycle equations, and/or your knowledge of fundamental engine operation. The parametric cycle analysis of ideal engines allows us to look at the characteristics of aircraft engines in the simplest possible ways so that they can be compared. Following this, realistic assumptions as to component losses will be introduced in Chapter 6 and the parametric cycle analysis repeated for the different aircraft engines in Chapter 7. In Chapter 8, engine performance analysis of these engines with losses (real engines) will be performed and performance trends analyzed.

In the last chapter on engine cycle analysis, Chapter 8, models will be developed for the performance characteristics of the engine components. The aerothermodynamic relationships between the engine components will be analyzed for several types of aircraft engines. Then the performance of specific engines at all flight conditions and throttle settings will be predicted.

5.2 Notation

The *total* or *stagnation temperature* is defined as that temperature reached when a steadily flowing fluid is brought to rest (stagnated) adiabatically. If T_t denotes the total temperature, T the static (thermodynamic) temperature, and M the Mach number, we developed in Chapter 3 the following useful equation:

$$T_t = T\left(1 + \frac{\gamma - 1}{2}M^2\right) \tag{5.1}$$

The *total or stagnation pressure* P_t (also developed in Chapter 3) is defined as the pressure reached when a steady flowing stream is brought to rest adiabatically and reversibly (i.e., isentropically). Because $P_t/P = (T_t/T)^{\gamma/(\gamma-1)}$ for the process, then

$$P_t = P\left(1 + \frac{\gamma-1}{2}M^2\right)^{\gamma/(\gamma-1)} \tag{5.2}$$

Ratios of total temperatures and pressures will be used extensively in this text, and a special notation is adopted for them. We denote a *ratio of total pressures* across a component by π, with a subscript indicating the component: d for diffuser (inlet), c for compressor, b for burner, t for turbine, n for nozzle, and f for fan.

$$\pi_a = \frac{\text{total pressure leaving component } a}{\text{total pressure entering component } a} \tag{5.3}$$

Similarly, the *ratio of total temperatures* is denoted by τ, and

$$\tau_a = \frac{\text{total temperature leaving component } a}{\text{total temperature entering component } a} \tag{5.4}$$

There are the following exceptions:

1. We define the total/static temperature and pressure ratios of the freestream (τ_r and π_r) by

$$\tau_r = \frac{T_{t0}}{T_0} = 1 + \frac{\gamma-1}{2}M_0^2 \tag{5.5}$$

$$\pi_r = \frac{P_{t0}}{P_0} = \left(1 + \frac{\gamma-1}{2}M_0^2\right)^{\gamma/(\gamma-1)} \tag{5.6}$$

The pressure ratio of Eq. (5.6) is often referred to as the *ram pressure ratio*. Thus the total temperature and pressure of the freestream can be written as

$$T_{t0} = T_0\tau_r \qquad P_{t0} = P_0\pi_r$$

2. Also, τ_λ is defined as the ratio of the burner exit total enthalpy h_{t4} to the ambient enthalpy h_0:

$$\tau_\lambda = \frac{h_{t4}}{h_0} \tag{5.7}$$

Figure 5.1 shows the cross-section and station numbering of a turbofan engine with both afterburning and duct burning. Note that the freestream (station 0) would be far upstream of the fan and inlet. This station numbering is in accordance with Aerospace Recommended Practice (ARP) 755A (see Ref. 3). Note that the station numbers 13–19 are used for the bypass stream and decimal numbers such as station number 4.5 are used to indicate an intermediate station. If there is a second bypass stream, station numbers 23–29 are used.

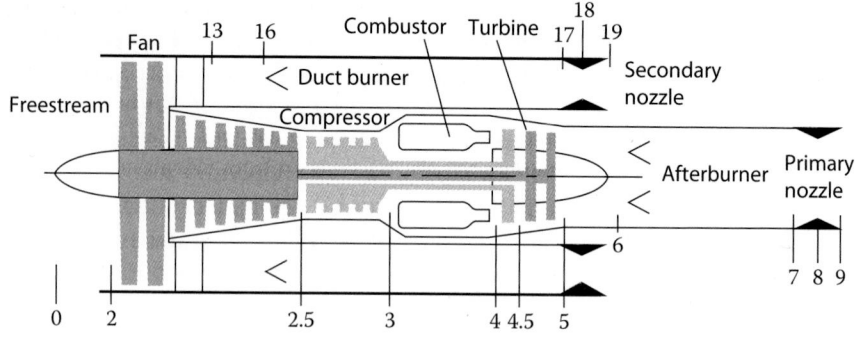

Fig. 5.1 Station numbering for gas turbine engines.

Table 5.1 contains most of the short-form notation total temperature ratios τ and total pressure ratios π that we will use in our analysis. These ratios are shown in terms of the standard station numbering [3] and Fig. 5.1.

Table 5.1 Temperature and Pressure Relationships for all τ and π

Freestream			
$\tau_r = 1 + \dfrac{\gamma-1}{2}M_0^2$		$\pi_r = \left(1 + \dfrac{\gamma-1}{2}M_0^2\right)^{\gamma/(\gamma-1)}$	
Core Stream		**Bypass Stream**	
$\tau_\lambda = \dfrac{h_{t4}}{h_0}$	$\tau_{\lambda AB} = \dfrac{h_{t7}}{h_0}$	$\tau_{\lambda DB} = \dfrac{h_{t17}}{h_0}$	
$\tau_d = \dfrac{T_{t2}}{T_{t0}}$	$\pi_d = \dfrac{P_{t2}}{P_{t0}}$	$\tau_f = \dfrac{T_{t13}}{T_{t2}}$	$\pi_f = \dfrac{P_{t13}}{P_{t2}}$
$\tau_c = \dfrac{T_{t3}}{T_{t2}}$	$\pi_c = \dfrac{P_{t3}}{P_{t2}}$	$\tau_{DB} = \dfrac{T_{t17}}{T_{t13}}$	$\pi_{DB} = \dfrac{P_{t17}}{P_{t13}}$
$\tau_{cL} = \dfrac{T_{t2.5}}{T_{t2}}$	$\pi_{cL} = \dfrac{P_{t2.5}}{P_{t2}}$	$\tau_{fn} = \dfrac{T_{t19}}{T_{t17}}$	$\pi_{fn} = \dfrac{P_{t19}}{P_{t17}}$
$\tau_{cH} = \dfrac{T_{t3}}{T_{t2.5}}$	$\pi_{cH} = \dfrac{P_{t3}}{P_{t2.5}}$		
$\tau_b = \dfrac{T_{t4}}{T_{t3}}$	$\pi_b = \dfrac{P_{t4}}{P_{t3}}$		
$\tau_t = \dfrac{T_{t5}}{T_{t4}}$	$\pi_t = \dfrac{P_{t5}}{P_{t4}}$		
$\tau_{tH} = \dfrac{T_{t4.5}}{T_{t4}}$	$\pi_{tH} = \dfrac{P_{t4.5}}{P_{t4}}$	$\tau_{AB} = \dfrac{T_{t7}}{T_{t5}}$	$\pi_{AB} = \dfrac{P_{t7}}{P_{t5}}$
$\tau_{tL} = \dfrac{T_{t5}}{T_{t4.5}}$	$\pi_{tL} = \dfrac{P_{t5}}{P_{t4.5}}$	$\tau_n = \dfrac{T_{t9}}{T_{t7}}$	$\pi_n = \dfrac{P_{t9}}{P_{t7}}$

Notice in particular the pressure and temperature ratios in multispool engines. We have added the subscript L to denote the low-pressure (LP) spool and H to denote the high-pressure (HP) spool. When calculating the compressor pressure ratio (also called the *overall pressure ratio OPR*), we note that it is the product of the pressure ratio across each spool or $\pi_c = \pi_{cL}\pi_{cH}$.

5.3   Design Inputs

The total temperature ratios, total pressure ratios, and the like, can be classified into one of five categories:

1. Flight conditions: $P_0, T_0, M_0, \tau_r, \pi_r$
2. Design limits: T_{t4}, T_{t7}, etc.
3. Component performance: π_d, π_b, π_n, etc.
4. Design choices: π_c, π_f, etc.
5. Air and fuel properties: c_p, γ, h_{PR}

5.4   Steps of Engine Parametric Cycle Analysis

The steps of engine parametric cycle analysis listed next are based on a jet engine with a *single inlet and single exhaust.* Thus, these steps will use only the station numbers for the core engine flow (from 0 to 9) shown in Fig. 5.1. We will use these steps in this chapter and Chapter 7. When more than one exhaust stream is present (e.g., high-bypass-ratio turbofan engine), the steps will be modified.

Parametric cycle analysis desires to determine how the uninstalled engine performance (specific thrust and fuel consumption) varies with changes in design limits (e.g., main burner exit temperature), component performance (e.g., turbine efficiency), and design choices (e.g., compressor pressure ratio).

1. Starting with an equation for uninstalled engine thrust, we rewrite this equation in terms of the total pressure and total temperature ratios: the ambient pressure P_0, temperature T_0, and speed of sound a_0, and the flight Mach number M_0 as follows:

$$F = \frac{1}{g_c}(\dot{m}_9 V_e - \dot{m}_0 V_0) + A_9(P_9 - P_0)$$

$$\frac{F}{\dot{m}_0} = \frac{a_0}{g_c}\left(\frac{\dot{m}_9}{\dot{m}_0}\frac{V_9}{a_0} - M_0\right) + \frac{A_9 P_9}{\dot{m}_0}\left(1 - \frac{P_0}{P_9}\right)$$

2. Next, express the velocity ratio(s) V_9/a_0 in terms of Mach numbers, temperatures, and gas properties of states 0 and 9:

$$\left(\frac{V_9}{a_0}\right)^2 = \frac{a_9^2 M_9^2}{a_0^2} = \frac{\gamma_9 R_9 g_c T_9}{\gamma_0 R_0 g_c T_0}M_9^2$$

3. Find the exit Mach number M_9. Because

$$P_{t9} = P_9 \left(1 + \frac{\gamma - 1}{2} M_9^2 \right)^{\gamma/(\gamma-1)}$$

then

$$M_9^2 = \frac{2}{\gamma - 1} \left[\left(\frac{P_{t9}}{P_9} \right)^{(\gamma-1)/\gamma} - 1 \right]$$

where

$$\frac{P_{t9}}{P_9} = \frac{P_0}{P_9} \frac{P_{t0}}{P_0} \frac{P_{t2}}{P_{t0}} \frac{P_{t3}}{P_{t2}} \frac{P_{t4}}{P_{t3}} \frac{P_{t5}}{P_{t4}} \frac{P_{t7}}{P_{t5}} \frac{P_{t9}}{P_{t7}}$$

$$= \frac{P_0}{P_9} \pi_r \pi_d \pi_c \pi_b \pi_t \pi_{AB} \pi_n$$

4. Find the temperature ratio T_9/T_0:

$$\frac{T_9}{T_0} = \frac{T_{t9}/T_0}{T_{t9}/T_9} = \frac{T_{t9}/T_0}{(P_{t9}/P_9)^{(\gamma-1)/\gamma}}$$

where

$$\frac{T_{t9}}{T_0} = \frac{T_{t0}}{T_0} \frac{T_{t2}}{T_{t0}} \frac{T_{t3}}{T_{t2}} \frac{T_{t4}}{T_{t3}} \frac{T_{t5}}{T_{t4}} \frac{T_{t7}}{T_{t5}} \frac{T_{t9}}{T_{t7}} = \tau_r \tau_d \tau_c \tau_b \tau_t \tau_{AB} \tau_n$$

5. Apply the first law of thermodynamics to the burner (combustor), and find an expression for the fuel/air ratio f in terms of τ's, etc.:

$$\dot{m}_0 h_{t3} + \dot{m}_f h_{PR} = \dot{m}_0 h_{t4}$$

6. When applicable, find an expression for the total temperature ratio across the turbine τ_t by relating the turbine power output to the compressor, fan, and/or propeller power requirements. This allows us to find τ_t in terms of other variables.

7. Evaluate the specific thrust, using the preceding results.

8. Evaluate the thrust specific fuel consumption S, using the results for specific thrust and fuel/air ratio:

$$S = \frac{f}{F/\dot{m}_0} \tag{5.8}$$

9. Develop expressions for the thermal and propulsive efficiencies.

5.5 Assumptions of Ideal Cycle Analysis

For analysis of ideal cycles, we assume the following:

1. There are isentropic (reversible and adiabatic) compression and expansion processes in the inlet (diffuser), compressor, fan, turbine, and nozzle. Thus we have the following relationships:

$$\tau_d = \tau_n = 1 \quad \pi_d = \pi_n = 1 \quad \tau_c = \pi_c^{(\gamma-1)/\gamma} \quad \tau_t = \pi_t^{(\gamma-1)/\gamma}$$

2. Constant-pressure combustion ($\pi_b = 1$) is idealized as a heat interaction into the combustor. The fuel flow rate is much less than the airflow rate through the combustor, such that

$$\frac{\dot{m}_f}{\dot{m}_c} \ll 1 \quad \text{and} \quad \dot{m}_c + \dot{m}_f \cong \dot{m}_c$$

3. The working fluid is air that behaves as a perfect gas with constant specific heats.
4. The engine exhaust nozzles expand the gas to the ambient pressure ($P_9 = P_0$).
5. There are no bleed or cooling flows.

5.6 Ideal Turbojet

The turbojet engine shown in Figs. 5.2a and 5.2b is the simplest gas turbine engine used for aircraft propulsion. For the ideal turbojet, the thermal efficiency follows directly from the ideal Brayton cycle of Section 4.5. Noting that the pressure ratio (noted as PR in Chapter 4) of the cycle is P_{t3}/P_0, Eq. (4.13) can be rewritten as

$$\eta_{Th} = 1 - \left(\frac{P_0}{P_{t3}}\right)^{\gamma/(\gamma-1)} = 1 - \frac{T_0}{T_{t3}}$$

In terms of the engine temperature ratios (τ's), the thermal efficiency can be written as

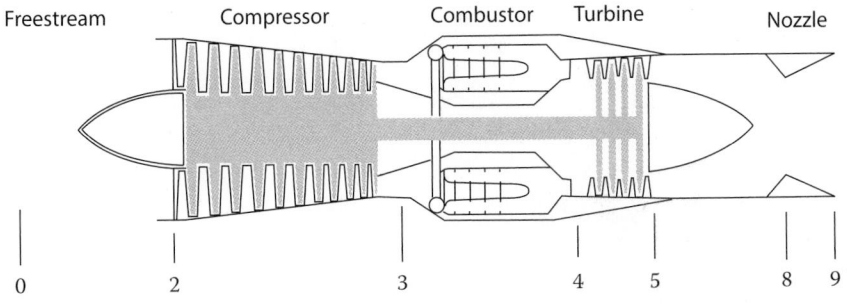

Freestream   Compressor   Combustor   Turbine   Nozzle

0   2   3   4   5   8   9

Fig. 5.2a   Station numbering of ideal turbojet engine.

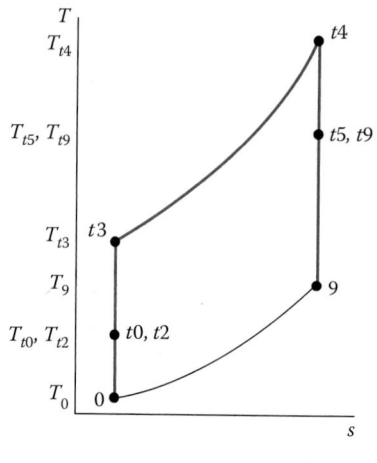

Fig. 5.2b The T-s diagram of an ideal turbojet engine.

$$\eta_{Th} = 1 - \frac{1}{\tau_r \tau_c} \qquad (5.9)$$

An ideal turbojet with a compressor pressure ratio π_c of 10 ($\tau_c = 1.9307$) has a thermal efficiency of 48.2% at a Mach number of 0 ($\tau_r = 1$) and 71.2% at a Mach number of 2.0 ($\tau_r = 1.8$).

For the ideal turbojet,

$$\dot{W}_t = \dot{W}_c \qquad T_{t4} > T_{t3} \qquad P_{t4} = P_{t3}$$

Thus the compressor–burner–turbine combination generates a higher pressure and temperature at its exit and is called, therefore, a *gas generator*. The gas leaving the gas generator may be expanded through a nozzle to form a turbojet, as depicted in Fig. 5.2, or the gases may be expanded through a turbine to drive a fan (turbofan), a propeller (turboprop), a generator (turboshaft), an automobile (gas turbine), or a helicopter rotor (turboshaft).

We will analyze the ideal turbojet cycle and will determine the trends in the variation of thrust, thrust specific fuel consumption, and turbine temperature ratio with compressor pressure ratio and flight Mach number.

5.6.1 Cycle Analysis

Application of the PCA steps to the ideal turbojet engine is presented next in the order listed in Section 5.4 consistent with the assumptions presented in Section 5.5.

Step 1:

$$\frac{F}{\dot{m}_0} = \frac{1}{g_c}(V_9 - V_0) = \frac{a_0}{g_c}\left(\frac{V_9}{a_0} - M_0\right)$$

Step 2:

$$\left(\frac{V_9}{a_0}\right)^2 = \frac{a_9^2 M_9^2}{a_0^2} = \frac{\gamma_9 R_9 g_c T_9 M_9^2}{\gamma_0 R_0 g_c T_0} = \frac{T_9}{T_0} M_9^2$$

Step 3:

$$P_{t9} = P_9\left(1 + \frac{\gamma - 1}{2} M_9^2\right)^{\gamma/(\gamma-1)}$$

and

$$P_{t9} = P_0 \frac{P_{t0}}{P_0} \frac{P_{t2}}{P_{t0}} \frac{P_{t3}}{P_{t2}} \frac{P_{t4}}{P_{t3}} \frac{P_{t5}}{P_{t4}} \frac{P_{t9}}{P_{t5}} = P_0 \pi_r \pi_d \pi_c \pi_b \pi_t \pi_n$$

However, $\pi_d = \pi_b = \pi_n = 1$; thus $P_{t9} = P_0 \pi_r \pi_c \pi_t$, and so

$$M_9^2 = \frac{2}{\gamma - 1} \left[\left(\frac{P_{t9}}{P_9} \right)^{(\gamma-1)/\gamma} - 1 \right]$$

where

$$\frac{P_{t9}}{P_9} = \frac{P_{t9}}{P_0} \frac{P_0}{P_9} = \pi_r \pi_c \pi_t \frac{P_0}{P_9} = \pi_r \pi_c \pi_t$$

Then

$$M_9^2 = \frac{2}{\gamma - 1} [(\pi_r \pi_c \pi_t)^{(\gamma-1)/\gamma} - 1]$$

However, $\pi_r^{(\gamma-1)/\gamma} = \tau_r$ and for an ideal turbojet $\pi_c^{(\gamma-1)/\gamma} = \tau_c$ and $\pi_t^{(\gamma-1)/\gamma} = \tau_t$. Thus

$$M_9^2 = \frac{2}{\gamma - 1} (\tau_r \tau_c \tau_t - 1) \qquad (5.10)$$

Step 4:

$$T_{t9} = T_0 \frac{T_{t0}}{T_0} \frac{T_{t2}}{T_{t0}} \frac{T_{t3}}{T_{t2}} \frac{T_{t4}}{T_{t3}} \frac{T_{t5}}{T_{t4}} \frac{T_{t9}}{T_{t5}} = T_0 \tau_r \tau_d \tau_c \tau_b \tau_t \tau_n = T_0 \tau_r \tau_c \tau_b \tau_t$$

Then

$$\frac{T_9}{T_0} = \frac{T_{t9}/T_0}{T_{t9}/T_9} = \frac{\tau_r \tau_c \tau_b \tau_t}{(P_{t9}/P_9)^{(\gamma-1)/\gamma}} = \frac{\tau_r \tau_c \tau_b \tau_t}{(\pi_r \pi_c \pi_t)^{(\gamma-1)/\gamma}} = \frac{\tau_r \tau_c \tau_b \tau_t}{\tau_r \tau_c \tau_t}$$

Thus

$$\frac{T_9}{T_0} = \tau_b \qquad (5.11)$$

Step 5: Application of the steady flow energy equation to the burner gives

$$\dot{m}_0 h_{t3} + \dot{m}_f h_{PR} = (\dot{m}_0 + \dot{m}_f) h_{t4}$$

For an ideal cycle, $\dot{m}_0 + \dot{m}_f \cong \dot{m}_0$ and $c_{p3} = c_{p4} = c_p$. Thus

$$\dot{m}_0 c_p T_{t3} + \dot{m}_f h_{PR} = \dot{m}_0 c_p T_{t4}$$

$$\dot{m}_f h_{PR} = \dot{m}_0 c_p (T_{t4} - T_{t3}) = \dot{m}_0 c_p T_0 \left(\frac{T_{t4}}{T_0} - \frac{T_{t3}}{T_0} \right)$$

or

$$f = \frac{\dot{m}_f}{\dot{m}_0} = \frac{c_p T_0}{h_{PR}} \left(\frac{T_{t4}}{T_0} - \frac{T_{t3}}{T_0} \right)$$

However,

$$\tau_\lambda = \frac{T_{t4}}{T_0} \quad \text{and} \quad \tau_r \tau_c = \frac{T_{t3}}{T_0}$$

Then

$$f = \frac{\dot{m}_f}{\dot{m}_0} = \frac{c_p T_0}{h_{PR}} (\tau_\lambda - \tau_r \tau_c) \tag{5.12a}$$

or

$$f = \frac{\dot{m}_f}{\dot{m}_0} = \frac{c_p T_0 \tau_r \tau_c}{h_{PR}} (\tau_b - 1) \tag{5.12b}$$

Step 6: The power out of the turbine is

$$\dot{W}_t = (\dot{m}_0 + \dot{m}_f)(h_{t4} - h_{t5}) \cong \dot{m}_0 c_p (T_{t4} - T_{t5})$$

$$= \dot{m}_0 c_p T_{t4} \left(1 - \frac{T_{t5}}{T_{t4}} \right) = \dot{m}_0 c_p T_{t4} (1 - \tau_t)$$

The power required to drive the compressor is

$$\dot{W}_c = \dot{m}_0 (h_{t3} - h_{t2}) = \dot{m}_0 c_p (T_{t3} - T_{t2})$$

$$= \dot{m}_0 c_p T_{t2} \left(\frac{T_{t3}}{T_{t2}} - 1 \right) = \dot{m}_0 c_p T_{t2} (\tau_c - 1)$$

Because $\dot{W}_c = \dot{W}_t$ for the ideal turbojet, then

$$T_{t3} - T_{t2} = T_{t4} - T_{t5} \tag{5.13}$$

or

$$\tau_t = 1 - \frac{T_{t2}}{T_{t4}}(\tau_c - 1)$$

Thus

$$\tau_t = 1 - \frac{\tau_r}{\tau_\lambda}(\tau_c - 1) \tag{5.14}$$

Step 7:

$$\left(\frac{V_9}{a_0}\right)^2 = \frac{T_9}{T_0}M_9^2 = \frac{2}{\gamma - 1}\frac{\tau_\lambda}{\tau_r\tau_c}(\tau_r\tau_c\tau_t - 1) \tag{5.15}$$

However,

$$\frac{F}{\dot{m}_0} = \frac{a_0}{g_c}\left(\frac{V_9}{a_0} - M_0\right)$$

Thus

$$\frac{F}{\dot{m}_0} = \frac{a_0}{g_c}\left[\sqrt{\frac{2}{\gamma - 1}\frac{\tau_\lambda}{\tau_r\tau_c}(\tau_r\tau_c\tau_t - 1)} - M_0\right] \tag{5.16}$$

Step 8: According to Eq. (5.8),

$$S = \frac{f}{F/\dot{m}_0}$$

The thrust specific fuel consumption S can be calculated by first calculating the fuel/air ratio f and the thrust per unit of airflow $F/\dot{m}_0$, using Eqs. (5.12a) and (5.16), respectively, and then substituting these values into the preceding equation. An analytical expression for S can be obtained by substituting Eqs. (5.12a) and (5.16) into Eq. (5.8) to get the following:

$$S = \frac{c_p T_0 g_c(\tau_\lambda - \tau_r\tau_c)}{a_0 h_{PR}\left[\sqrt{\frac{2}{\gamma - 1}\frac{\tau_\lambda}{\tau_r\tau_c}(\tau_r\tau_c\tau_t - 1)} - M_0\right]} \tag{5.17}$$

Step 9: Again the development of these expressions is left to the reader.

Thermal efficiency:

$$\eta_{Th} = 1 - \frac{1}{\tau_r\tau_c} \tag{5.18a}$$

Propulsive efficiency:

$$\eta_P = \frac{2M_0}{V_9/a_0 + M_0} \tag{5.18b}$$

Overall efficiency:

$$\eta_O = \eta_P \eta_{Th} \tag{5.18c}$$

See Section 5.6.SM, Summary of Equations: Ideal Turbojet, in the online supporting material for related information.

Example 5.1

In Figs. 5.3a–5.3c, the performance of ideal turbojets is plotted vs compressor pressure ratio π_c for different values of flight Mach number M_0. Calculations were performed for the following input data:

$$T_0 = 390°R, \quad \gamma = 1.4, \quad c_p = 0.24\,\text{Btu}/(\text{lbm} \cdot °R),$$
$$h_{PR} = 18{,}400\,\text{Btu}/\text{lbm}, \quad T_{t4} = 3000°R$$

Figure 5.3a shows that, for a fixed Mach number, there is a compressor pressure ratio that gives maximum specific thrust. The loci of the compressor pressure ratios that give maximum specific thrust are indicated by the dashed

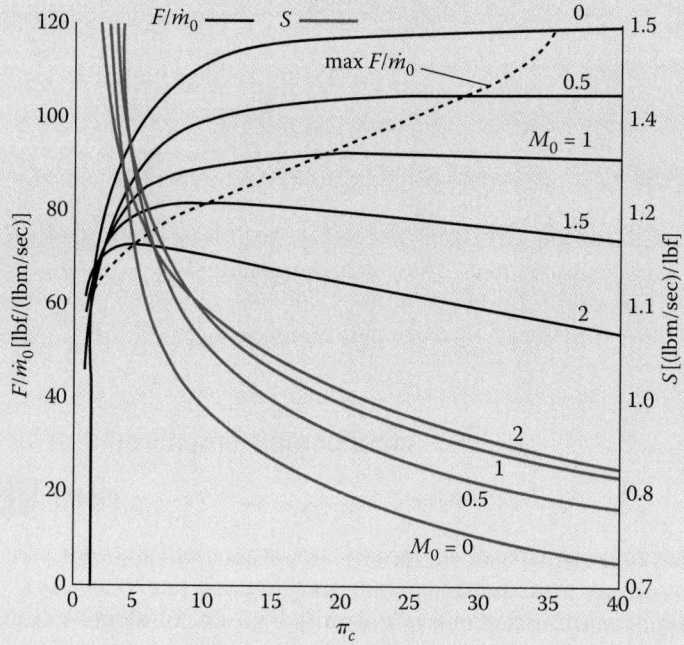

Fig. 5.3a Ideal turbojet performance vs compressor pressure ratio.

(Continued)

Example 5.1 (Continued)

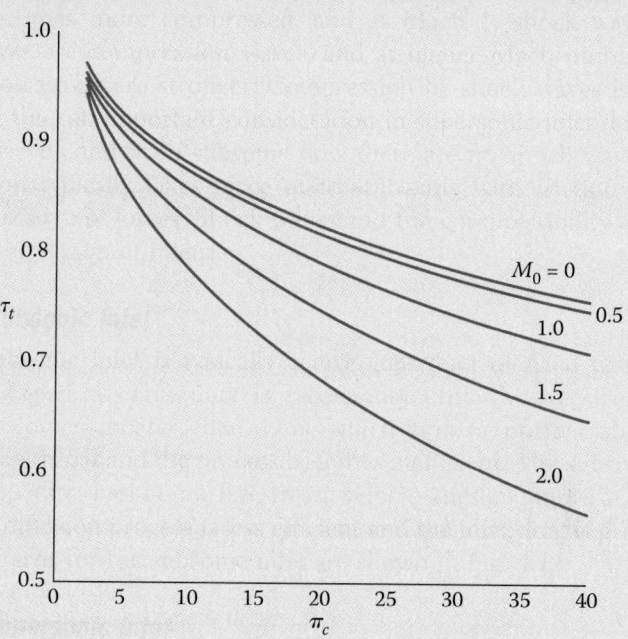

Fig. 5.3b Ideal turbojet performance vs compressor pressure ratio: turbine total temperature ratio.

line in Fig. 5.3a. One can also see from Fig. 5.3a that the maximum specific thrust occurs at lower compressor pressure ratio as Mach number increases, largely because the "ram" total pressure and total temperature, P_{t2} and T_{t2}, respectively, increase with increasing Mach number, as can be inferred on the T-s diagram (Fig. 5.2). At higher supersonic M_0, more compression occurs via the deceleration of the flow through the inlet. Indeed, this trend helps explain why a ramjet engine is preferred at high supersonic Mach numbers. The J58 turbojet on the SR-71 essentially operated as a ramjet at Mach numbers of 3+ (see Fig. 1.12c). Figure 5.3a also shows the trend of improved thrust specific fuel consumption with increasing compressor pressure ratio, as would be expected from Brayton cycle analysis.

Figure 5.3b shows that the turbine total temperature ratio τ_t decreases with increasing compressor pressure ratio π_c and flight Mach number M_0. The total temperature rise through the compressor increases with increasing compressor pressure ratio π_c and flight Mach number M_0 resulting in an increase in total temperature change through the turbine $[T_{t4}(1 - \tau_t)]$

(Continued)

Example 5.1 *(Continued)*

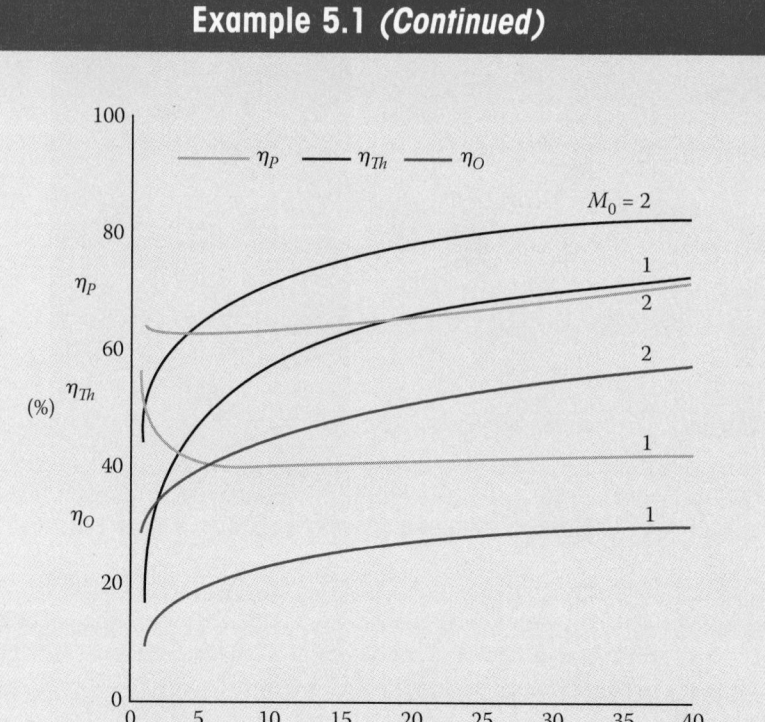

Fig. 5.3c Ideal turbojet performance vs compressor pressure ratio: efficiencies.

needed to power the compressor. Figure 5.3c shows the general increase in propulsive, thermal, and overall efficiencies with compressor pressure ratio and flight Mach number.

5.7 Ideal Turbojet with Afterburner

The thrust of a turbojet can be increased by the addition of a second combuster, called an *afterburner*, aft of the turbine, to reheat the remaining air and combustion exhaust gases prior to expansion through the exhaust nozzle, as shown in Fig. 5.4a. As indicated in the *T-s* diagram of Fig. 5.4b, this energy addition downstream of the turbine affords the opportunity for increased exhaust gas velocity V_9 (vertical distance between state points $t9$ and 9), resulting in increased thrust. The total temperature leaving the afterburner has a higher limiting value than the total temperature leaving the main combustor because the gases leaving the afterburner do not have a turbine to pass through. The station numbering is indicated in Figs. 5.4a and 5.4b, with $9'$ representing the nozzle exit for the case of no afterburning.

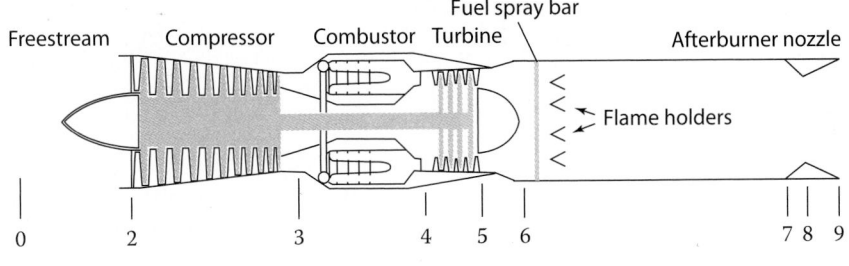

Fig. 5.4a Station numbering of an ideal afterburning turbojet engine.

5.7.1 Cycle Analysis

Rather than go through the complete steps of cycle analysis, we will use the results of the ideal turbojet and modify the equations to include afterburning. The gas velocity at the nozzle exit is given by

$$\frac{V_9^2}{2g_c c_p} = T_{t9} - T_9 = T_{t9}\left[1 - \left(\frac{P_9}{P_{t9}}\right)^{(\gamma-1)/\gamma}\right] \tag{5.19}$$

and for the nonafterburning (subscript dry) and afterburning (subscript AB for afterburning) cases, we have

$$\frac{P_9}{P_{t9}} = \left(\frac{P_{9'}}{P_{t5}}\right)_{\text{dry}} = \left(\frac{P_9}{P_{t9}}\right)_{\text{AB}}$$

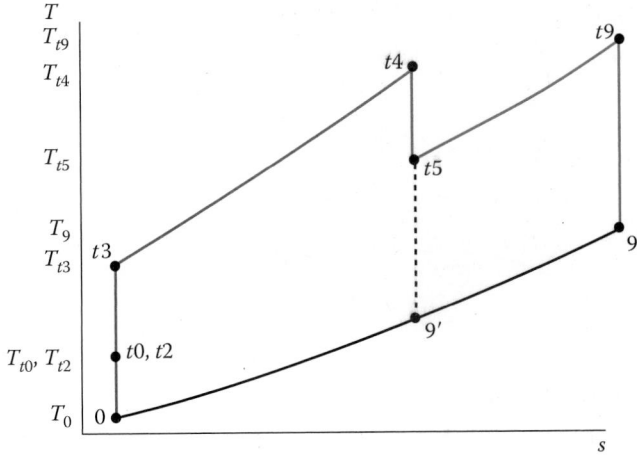

Fig. 5.4b The T-s diagram for an ideal afterburning turbojet engine.

Thus we have

$$\left(\frac{V_9}{V_{9'}}\right)^2 = \frac{T_{t9}}{T_{t5}} \tag{5.20}$$

Consequently, we can find the velocity ratio $(V_9/a_0)^2$ for the afterburning turbojet by multiplying $(V_{9'}/a_0)^2$ from our nonafterburning turbojet analysis [Eq. (5.15)] by the total temperature ratio of the afterburner. That is,

$$\left(\frac{V_9}{a_0}\right)^2 = \frac{T_{t9}}{T_{t5}}\left(\frac{V_{9'}}{a_0}\right)^2 \tag{5.21}$$

We define the temperature ratio $\tau_{\lambda AB}$ as

$$\tau_{\lambda AB} = \frac{T_{t7}}{T_0} \tag{5.22}$$

Thus we can write the total temperature ratio of the afterburner as

$$\frac{T_{t9}}{T_{t5}} = \frac{(T_{t9}/T_{t7})(T_{t7}/T_0)}{(T_{t5}/T_{t4})(T_{t4}/T_0)} = \frac{\tau_{\lambda AB}}{\tau_\lambda \tau_t} \tag{5.23}$$

Using Eqs. (5.14), (5.15), (5.21), and (5.23), we get the following expressions for V_9/a_0:

$$\left(\frac{V_9}{a_0}\right)^2 = \frac{2}{\gamma - 1}\frac{\tau_{\lambda AB}}{\tau_\lambda \tau_t}\frac{\tau_\lambda}{\tau_r \tau_c}(\tau_r \tau_c \tau_t - 1) = \frac{2}{\gamma - 1}\tau_{\lambda AB}\left(1 - \frac{1}{\tau_r \tau_c \tau_t}\right)$$

$$\left(\frac{V_9}{a_0}\right)^2 = \frac{2}{\gamma - 1}\tau_{\lambda AB}\left[1 - \frac{\tau_\lambda/(\tau_r \tau_c)}{\tau_\lambda - \tau_r(\tau_c - 1)}\right] \tag{5.24}$$

We find the total fuel flow rate to the burner and afterburner by an energy balance across the engine from station 0 to station 9, as sketched in Fig. 5.5.

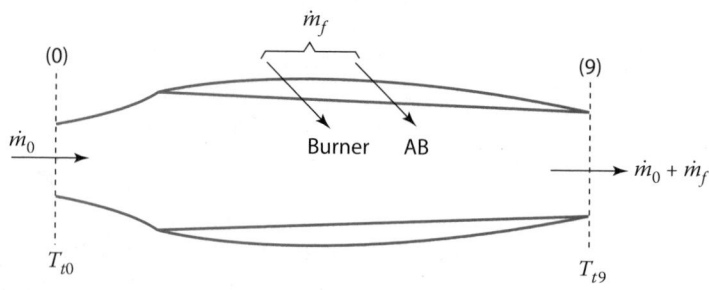

Fig. 5.5 Total fuel flow rate control volume.

The chemical energy of the fuel introduced between stations 0 and 9 is converted to thermal and kinetic energy of the gases, as measured by the total temperature rise $T_{t9} - T_{t0}$. Thus we can write for this ideal engine

$$\dot{m}_{ftot}h_{PR} = \dot{m}_0 c_p (T_{t9} - T_{t0}) = \dot{m}_0 c_p T_0 \left(\frac{T_{t9}}{T_0} - \frac{T_{t0}}{T_0} \right)$$

Thus

$$f_{tot} = \frac{\dot{m}_{ftot}}{\dot{m}_0} = \frac{c_p T_0}{h_{PR}} \tag{5.25}$$

The thermal efficiency is given by

$$\eta_{Th} = \frac{(\gamma - 1)c_p T_0 [(V_9/a_0)^2 - M_0^2]}{2f_{tot}h_{PR}} \tag{5.26}$$

See Section 5.7.SM, Summary of Equations: Ideal Turbojet with Afterburner, in the online supporting material for related information.

Example 5.2

We consider the performance of an ideal turbojet engine with afterburner. For comparison with the nonafterburning (simple) turbojet of Example 5.1, we select the following input data:

$$T_0 = 390°R, \quad \gamma = 1.4, \quad c_p = 0.24 \, \text{Btu}/(\text{lbm} \cdot °R), \quad h_{PR} = 18{,}400 \, \text{Btu/lbm},$$
$$T_{t4} = 3000°R, \quad T_{t7} = 4000°R$$

Figure 5.6 compares the performance of two turbojet engines with a compressor pressure ratio of 10, an afterburning turbojet and a simple turbojet. The graph of specific thrust vs M_0 indicates the thrust increase available by adding an afterburner to a simple turbojet. The afterburner increases the static thrust about 22% for the conditions shown and continues to provide significant thrust as the thrust of the simple turbojet goes to zero at about Mach 3.8. The graph of Fig. 5.6 also shows the cost in fuel consumption of the increased thrust provided by the afterburner. The cost is about a 20–30% increase in the fuel flow rate per unit thrust up to about $M_0 = 2.0$. As will be shown in Chapter 7 for a nonideal turbojet, the same increase in S occurs up to about $M_0 = 2.0$. However, as the thrust of the simple turbojet approaches zero in the real case, its S rises above the afterburning engine's S so that the afterburning engine at the higher M_0 has a lower S and higher specific thrust than the simple turbojet.

(Continued)

Example 5.2 *(Continued)*

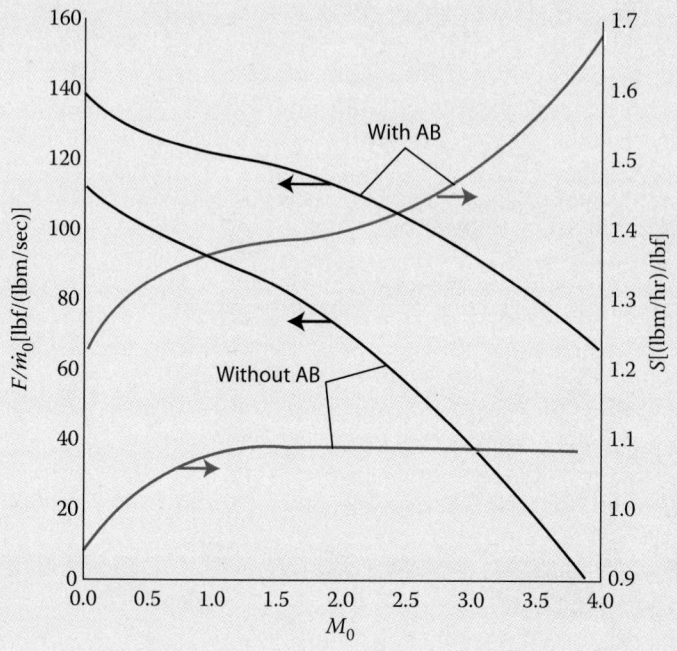

Fig. 5.6 Engine performance vs M_0 for an ideal afterburning turbojet.

Figure 5.7 shows the variation of engine performance with both Mach number and compressor pressure ratio for the afterburning turbojet. This should be compared to its counterparts in Fig. 5.3a. For a fixed flight Mach number, the ideal afterburning turbojet has a compressor pressure ratio giving maximum specific thrust. The locus of these compressor pressure ratios is shown by a dashed line in Fig. 5.7. Comparison of Fig. 5.7 to Fig. 5.3a yields the following general trends:

1. Afterburning increases both the specific thrust and the thrust specific fuel consumption.
2. Afterburning turbojets with moderate to high compressor pressure ratios give very good specific thrust at high flight Mach numbers.
3. The fuel/air ratio of the main burner f is unchanged. The afterburner fuel/air ratio f_{AB} increases with Mach number and compressor pressure ratio. The total fuel/air ratio decreases with Mach number and is not a function of π_c; see Eq. (5.25).

(Continued)

Example 5.2 *(Continued)*

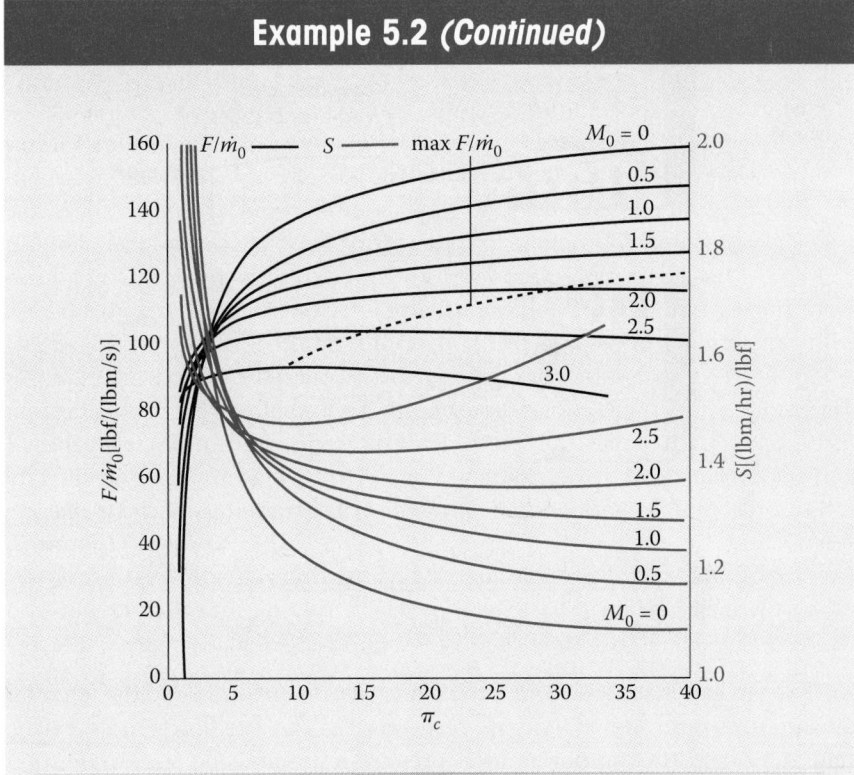

Fig. 5.7   Ideal afterburning turbojet engine performance vs compressor pressure ratio.

5.8 **Ideal Turbofan**

The propulsive efficiency of a simple turbojet engine can be improved by extracting a portion of the energy from the engine's gas generator to drive a ducted propeller, called a *fan*. The fan increases the propellant mass flow rate with an accompanying decrease in the required propellant exit velocity for a given thrust. Because the rate of production of "wasted" kinetic energy in the exit propellant gases varies as the first power with mass flow rate and as the square of the exit velocity, the net effect of increasing the mass flow rate and decreasing the exit velocity is to reduce the wasted kinetic energy production and to improve the propulsive efficiency.

Our station numbering for the turbofan cycle analysis is given in Fig. 5.8, and the *T-s* diagrams for ideal flow through the core engine and the fan are given in Figs. 5.9 and 5.10, respectively. The temperature drop through the turbine $(T_{t4} - T_{t5})$ is now greater than the temperature rise through the compressor $(T_{t3} - T_{t2})$ because the turbine drives the fan in addition to the compressor.

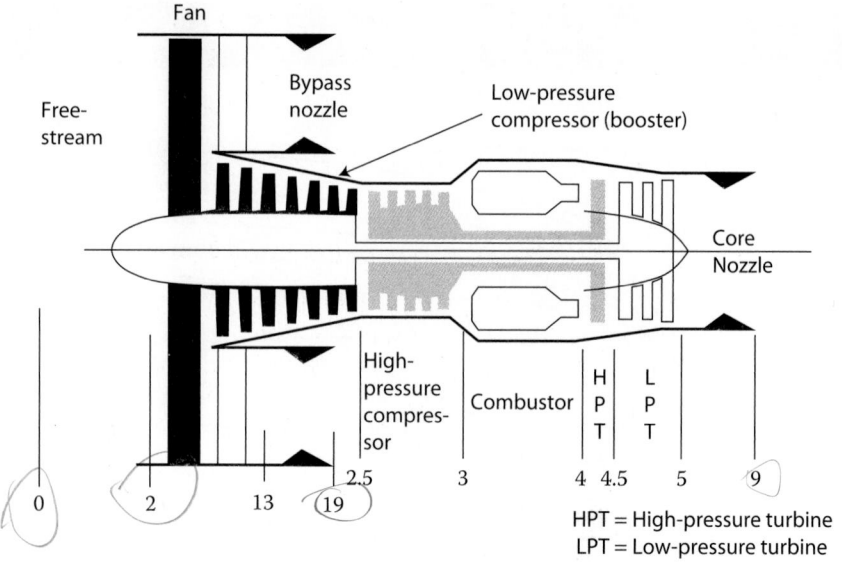

Fig. 5.8 Station numbering of a turbofan engine.

The fan exit is station 13, and the fan total pressure ratio and the fan total temperature ratio are π_f and τ_f, respectively. The fan flow nozzle exit is station 19, and the fan nozzle total pressure ratio and the fan nozzle total temperature ratio are π_{fn} and τ_{fn}, respectively. These four ratios are listed in the following:

$$\pi_f = \frac{P_{t13}}{P_{t2}} \quad \tau_f = \frac{T_{t13}}{T_{t2}} \quad \pi_{fn} = \frac{P_{t19}}{P_{t13}} \quad \tau_{fn} = \frac{T_{t19}}{T_{t13}}$$

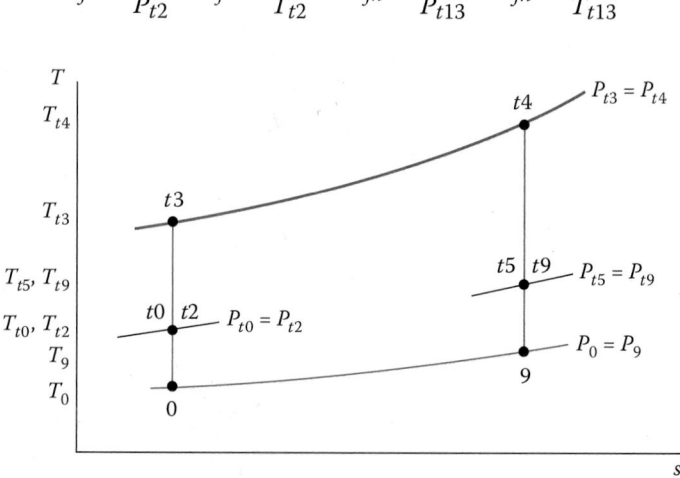

Fig. 5.9 The T-s diagram for core stream of ideal turbofan engine.

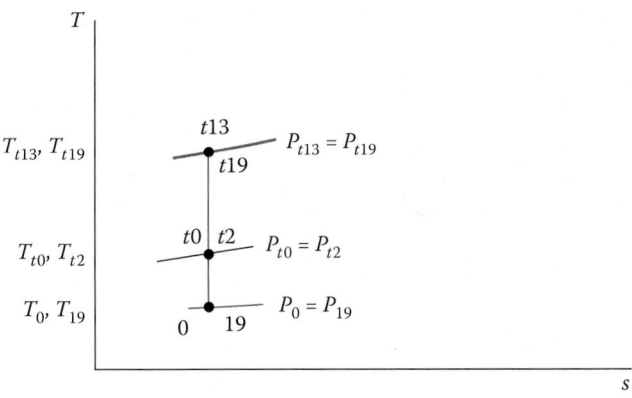

Fig. 5.10   The T-s diagram for fan stream of ideal turbofan engine.

The gas flow through the core engine is $\dot{m}_C$, and the gas flow through the bypass is $\dot{m}_B$. The ratio of the bypass flow to the core flow is defined as the bypass ratio and given the symbol α. Thus

$$\text{Bypass ratio } \alpha = \frac{\dot{m}_B}{\dot{m}_C} \tag{5.27}$$

The total gas flow is $\dot{m}_C + \dot{m}_B$, or $(1+\alpha)\dot{m}_C$. We will also use $\dot{m}_0$ for the total gas flow. Thus

$$\dot{m}_0 = \dot{m}_C + \dot{m}_B = (1+\alpha)\dot{m}_C \tag{5.28}$$

In the analysis of the ideal turbofan engine, the assumptions of Section 5.5 still apply, with both the engine core nozzle and fan nozzle designed so that $P_0 = P_9 = P_{19}$.

5.8.1   Cycle Analysis

Application of the steps of cycle analysis to the ideal turbofan engine of Figs. 5.8–5.10 is presented next in the order listed in Section 5.4.

Step 1: The thrust of the ideal turbofan engine is

$$F = \frac{\dot{m}_C}{g_c}(V_9 - V_0) + \frac{\dot{m}_F}{g_c}(V_{19} - V_0)$$

Thus

$$\frac{F}{\dot{m}_0} = \frac{a_0}{g_c}\frac{1}{1+\alpha}\left[\frac{V_9}{a_0} - M_0 + \alpha\left(\frac{V_{19}}{a_0} - M_0\right)\right] \tag{5.29}$$

Steps 2–4: First, the core stream of the turbofan encounters the same engine components as the ideal turbojet, and we can use its results. We have, from the analysis of the ideal turbojet,

$$\left(\frac{V_9}{a_0}\right)^2 = \frac{T_9}{T_0}M_9^2 \qquad \frac{T_9}{T_0} = \tau_b = \frac{\tau_\lambda}{\tau_r\tau_c} \qquad M_9^2 = \frac{2}{\gamma-1}(\tau_r\tau_c\tau_t - 1)$$

Thus

$$\left(\frac{V_9}{a_0}\right)^2 = \frac{T_9}{T_0}M_9^2 = \frac{2}{\gamma-1}\frac{\tau_\lambda}{\tau_r\tau_c}(\tau_r\tau_c\tau_t - 1) \qquad (5.30)$$

Second, compared to the core stream, the fan stream of the turbofan contains a fan rather than a compressor and does not have either a combustor or a turbine. Thus the preceding equations for the core stream of the ideal turbofan can be adapted for the fan stream as follows:

$$\left(\frac{V_{19}}{a_0}\right)^2 = \frac{T_{19}}{T_0}M_{19}^2 \qquad T_{19} = T_0 \qquad M_{19}^2 = \frac{2}{\gamma-1}(\tau_r\tau_f - 1)$$

$$\left(\frac{V_{19}}{a_0}\right)^2 = M_{19}^2 = \frac{2}{\gamma-1}(\tau_r\tau_f - 1) \qquad (5.31)$$

Step 5: Application of the steady flow energy equation to the burner gives

$$\dot{m}_C c_p T_{t3} + \dot{m}_f h_{PR} = (\dot{m}_C + \dot{m}_f)c_p T_{t4}$$

We define the fuel/air ratio f in terms of the mass flow rate of air through the burner $\dot{m}_C$, and we obtain

$$f = \frac{\dot{m}_f}{\dot{m}_C} = \frac{c_p T_0}{h_{PR}}(\tau_\lambda - \tau_r\tau_c) \qquad (5.32)$$

Step 6: The power out of the turbine(s) is

$$\dot{W}_t = (\dot{m}_C + \dot{m}_f)c_p(T_{t4} - T_{t5}) \cong \dot{m}_C c_p T_{t4}(1 - \tau_t)$$

The power required to drive the compression stages in the core stream is

$$\dot{W}_c = \dot{m}_C c_p(T_{t3} - T_{t2}) = \dot{m}_C c_p T_{t2}(\tau_c - 1)$$

The power required to drive the fan in the bypass stream is

$$\dot{W}_f = \dot{m}_B c_p(T_{t13} - T_{t2}) = \dot{m}_B c_p T_{t2}(\tau_f - 1)$$

Because $\dot{W}_t = \dot{W}_c + \dot{W}_f$ for the ideal turbofan, then

$$T_{t4}(1 - \tau_t) = T_{t2}(\tau_c - 1) + \alpha T_{t2}(\tau_f - 1)$$

$$\tau_t = 1 - \frac{T_{t2}}{T_{t4}}[\tau_c - 1 + \alpha(\tau_f - 1)]$$

$$\tau_t = 1 - \frac{\tau_r}{\tau_\lambda}[\tau_c - 1 + \alpha(\tau_f - 1)] \qquad (5.33)$$

Step 7: Combining Eqs. (5.30) and (5.33), we get

$$\left(\frac{V_9}{a_0}\right)^2 = \frac{2}{\gamma - 1}\frac{\tau_\lambda}{\tau_r \tau_c}\left(\tau_r \tau_c\left\{1 - \frac{\tau_r}{\tau_\lambda}[\tau_c - 1 + \alpha(\tau_f - 1)]\right\} - 1\right)$$

which can be simplified to

$$\left(\frac{V_9}{a_0}\right)^2 = \frac{2}{\gamma - 1}\left\{\tau_\lambda - \tau_r[\tau_c - 1 + \alpha(\tau_f - 1)] - \frac{\tau_\lambda}{\tau_r \tau_c}\right\} \qquad (5.34)$$

Thus the specific thrust of the ideal turbojet is given by Eqs. (5.29), (5.31), and (5.34).

Step 8:

$$S = \frac{\dot{m}_f}{F} = \frac{f}{F/\dot{m}_C} = \frac{f}{(\dot{m}_0/\dot{m}_C)(F/\dot{m}_0)}$$

$$S = \frac{f}{(1 + \alpha)(F/\dot{m}_0)} \qquad (5.35)$$

Step 9: The thermal efficiency of an ideal turbofan engine is the same as that of an ideal turbojet engine; that is [Eq. (5.9)],

$$\eta_{Th} = 1 - \frac{1}{\tau_r \tau_c}$$

This may seem surprising because the net power out of a turbojet is $\dot{m}_0(V_9^2 - V_0^2)/(2g_c)$, whereas for a turbofan it is $\dot{m}_C(V_9^2 - V_0^2)/(2g_c) + \dot{m}_B(V_{19}^2 - V_0^2)/(2g_c)$. The reason the thermal efficiency is the same is that power extracted from the core stream of the turbofan engine is added to the bypass stream without loss in the ideal case. Thus the net power out remains the same.

One can easily show that the propulsive efficiency of the ideal turbofan engine is given by

$$\eta_P = 2\frac{V_9/V_0 - 1 + \alpha(V_{19}/V_0 - 1)}{V_9^2/V_0^2 - 1 + \alpha(V_{19}^2/V_0^2 - 1)} \qquad (5.36)$$

A useful performance parameter for the turbofan engine is the ratio of the specific thrust per unit mass flow of the core stream to that of the fan stream.

We give this *thrust ratio* the symbol *FR* and define

$$FR \equiv \frac{F_C/\dot{m}_C}{F_B/\dot{m}_B} \tag{5.37}$$

For the ideal turbofan engine, the thrust ratio *FR* can be expressed as

$$FR = \frac{V_9/a_0 - M_0}{V_{19}/a_0 - M_0} \tag{5.38}$$

We will discover in the analysis of optimum turbofan engines that we will want a certain thrust ratio for minimum thrust specific fuel consumption.

See Section 5.8.SM, Summary of Equations: Ideal Turbofan, in the online supporting material for related information.

Example 5.3

The turbofan engine has three design variables: 1) compressor pressure ratio π_c, 2) fan pressure ratio π_f, and 3) bypass ratio α. Because this engine has two more design variables than the turbojet, this section contains many more plots of engine performance than were required for the fundamental understanding of the turbojet. We will look at the variation of each of the three design variables for an engine that will operate at a flight Mach number of 0.8, typical subsonic cruise conditions. In all of the calculations for this section, the following values are held constant:

$$T_0 = 216.7 \, \text{K}, \quad \gamma = 1.4, \quad c_p = 1.004 \, \text{kJ/(kg} \cdot \text{K)},$$
$$h_{PR} = 42{,}800 \, \text{kJ/kg}, \quad T_{t4} = 1350 \, \text{K}$$

a) *Figures 5.11a and 5.11b.* Specific thrust and thrust specific fuel consumption are plotted vs the compressor pressure ratio for different values of the bypass ratio in Fig. 5.11a. The fan pressure ratio is held constant at 1.65 in these plots. Figure 5.11a shows that specific thrust remains essentially constant with respect to the compressor pressure ratio over a wide range of values, 10 to 30. Also the specific thrust decreases with increasing bypass ratio, as Eq. (5.29) reveals with $1/(1 + \alpha)$. This we might expect as a smaller and smaller fraction of air is passed through the engine core (higher alpha). Figure 5.11a shows that thrust specific fuel consumption decreases with increasing compressor pressure ratio π_c because of increased thermal efficiency η_{Th} (same trend as the turbojet) and increasing bypass ratio α because of increased propulsive efficiency. Figure 5.11a nicely summarizes the performance tradeoffs associated with increased bypass ratios of turbofan engines—increased fuel efficiency

(Continued)

Example 5.3 *(Continued)*

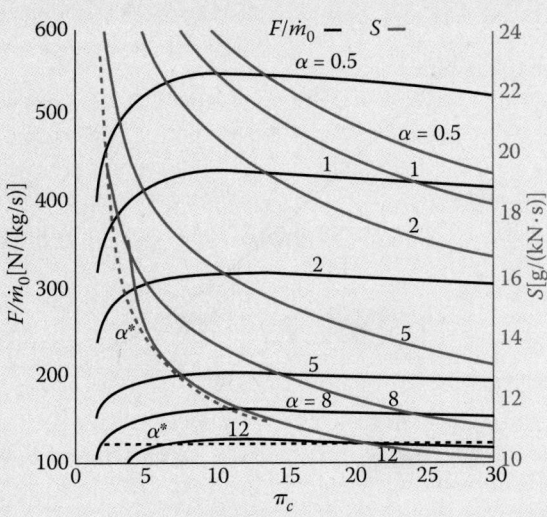

Fig. 5.11a Ideal turbofan performance vs π_c, for $\pi_f = 1.65$ and $M_0 = 0.8$.

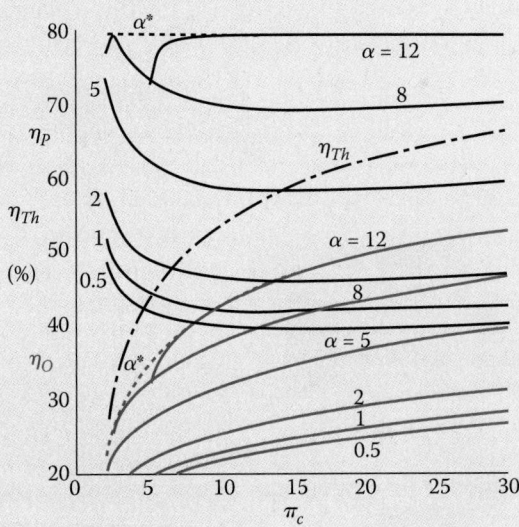

Fig. 5.11b Ideal turbofan performance vs π_c, for $\pi_f = 1.65$ and $M_0 = 0.8$: efficiencies.

(Continued)

Example 5.3 *(Continued)*

(decreased thrust specific fuel consumption) at the expense of a larger diameter engine (decreased specific thrust) for increased air mass flow rate to achieve the required thrust.

Figure 5.11b shows that the thermal efficiency increases with compressor pressure ratio and is independent of the engine bypass ratio. From Fig. 5.11b, we can see that the propulsive efficiency increases with engine bypass ratio and varies very little with compressor pressure ratio. The overall efficiency, shown also in Fig. 5.11b, increases with both compressor pressure ratio and bypass ratio.

b) *Figures 5.12a and 5.12b.* Specific thrust and thrust specific fuel consumption are plotted vs the compressor pressure ratio π_c for different values of the fan pressure ratio π_f. The bypass ratio α is held constant in these plots. Figure 5.12a shows that the specific thrust remains essentially constant with respect to the compressor pressure ratio for values of π_c from 15 to 25 and that the specific thrust has a maximum with respect to the fan pressure ratio π_f. Figure 5.12b shows that thrust specific fuel consumption decreases with increasing compressor pressure ratio π_c and that S has a minimum with respect to fan pressure ratio π_f. We will look at this optimum fan pressure ratio in more detail in Section 5.10.

Propulsive and overall efficiencies are plotted vs compressor pressure ratio and fan pressure ratio in Fig. 5.12b. This figure shows that propulsive efficiency increases with fan pressure ratio until a value of $\pi_f = 3.5$ and then decreases. There is a fan pressure ratio giving maximum propulsive efficiency. Propulsive efficiency is essentially constant for values of the compressor pressure ratio above 15.

c) *Figures 5.13a and 5.13b.* For a fixed compressor pressure ratio, the turbine must produce sufficient energy to power the compressor and fan. The minimum power would correspond to that of a turbojet ($\alpha = 0$). However, the maximum power extracted by the turbine corresponds to the core exit velocity V_9 equal to V_0 (refer to the *T-s* diagram in Fig. 5.9). The range of available turbine temperature ratio τ_t for this example with a compressor pressure ratio π_c of 24 is shown in Fig. 5.13a for a range of fan pressure ratios π_f and bypass ratios α. The region of possible energy extraction for powering the bypass stream is highlighted in blue and corresponds to an exit temperature leaving the turbine T_{t5} from about 990 K to 510 K. The specific region of this plot where the fuel consumption is lower will be shown in Section 5.9.

The variation of both thrust specific fuel consumption and specific thrust can be investigated using the carpet plot calculation feature of the PARA program over ranges of two independent variables. The carpet plot is a very useful analysis tool for helping to determine viable design space for a given reference condition. For fixed flight condition and compressor

(Continued)

Example 5.3 *(Continued)*

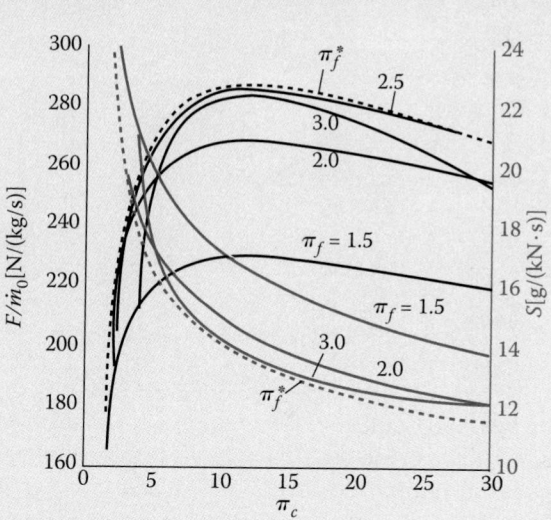

Fig. 5.12a   Ideal turbofan performance vs π_c, for $\alpha = 5$ and $M_0 = 0.8$: specific thrust and thrust specific fuel consumption.

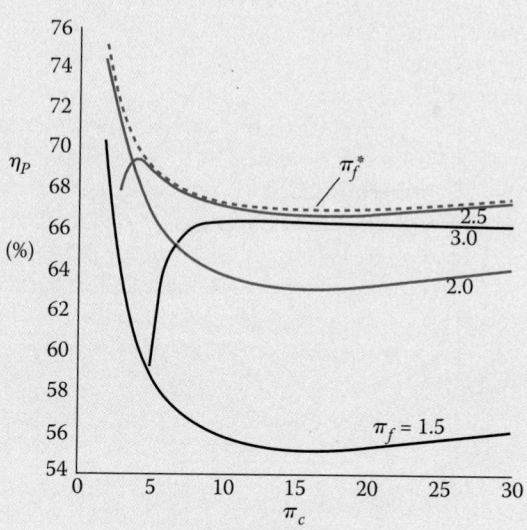

Fig. 5.12b   Ideal turbofan performance vs π_c, for $\alpha = 5$ and $M_0 = 0.8$: propulsive efficiency.

(Continued)

Example 5.3 *(Continued)*

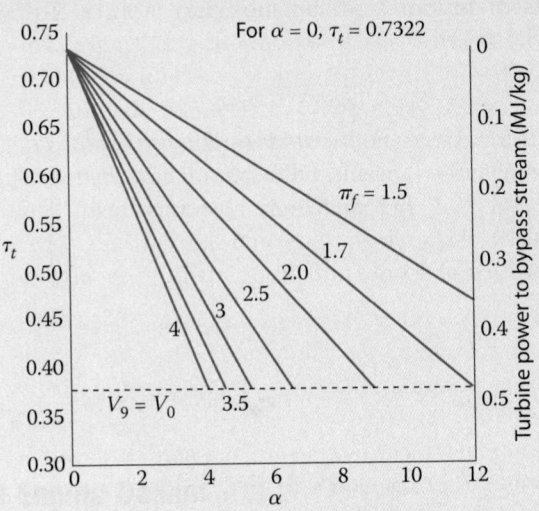

For $\alpha = 0$, $\tau_t = 0.7322$

Fig. 5.13a Available turbine total temperature ratio for $\pi_c = 24$ and $M_0 = 0.8$.

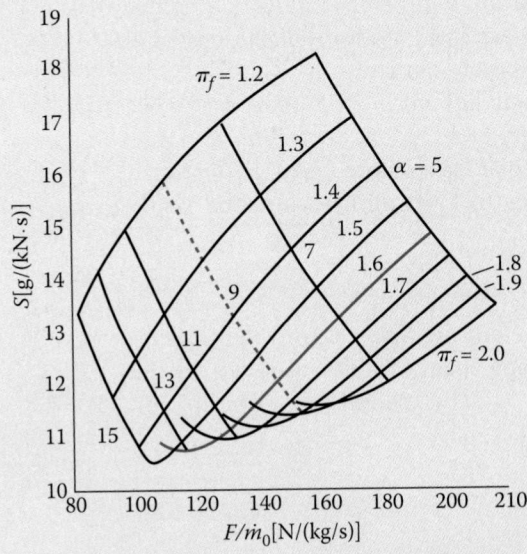

Fig. 5.13b Ideal turbofan engine performance for $\pi_c = 24$ and $M_0 = 0.8$.

(Continued)

Example 5.3 (Continued)

pressure ratio, investigating the range of bypass ratio from 5 to 15 and fan pressure ratio from 1.2 to 2 gives the plot of Fig. 5.13b. This plot shows that there are values of bypass ratio and fan pressure ratio that give low fuel consumption. Notice the solid blue line for a constant fan pressure ratio of 1.6, which shows that the fuel consumption decreases with increasing engine bypass ratio until a bypass ratio of about 13 (its optimum bypass ratio where the fuel consumption is at its minimum). Likewise, following along the dashed blue line in Fig. 5.13b for a constant bypass ratio of 9, you see that the fuel consumption decreases with increasing fan pressure ratio until a fan pressure ratio of 1.8 (its optimum fan pressure ratio where the fuel consumption is at its minimum).

d) *Figures 5.14a and 5.14b.* Specific thrust, thrust specific fuel consumption, propulsive efficiency, and thrust ratio are plotted vs fan pressure ratio for different values of the bypass ratio in Figs. 5.14a and 5.14b, respectively. The compressor pressure ratio is held constant in these plots. Figure 5.14a shows there is an optimum fan pressure ratio for each bypass ratio that will maximize specific thrust while minimizing fuel consumption. Figure 5.14b shows the optimum fan pressure ratio corresponds to maximum propulsive efficiency. We will look at this optimum fan pressure ratio in Section 5.10.

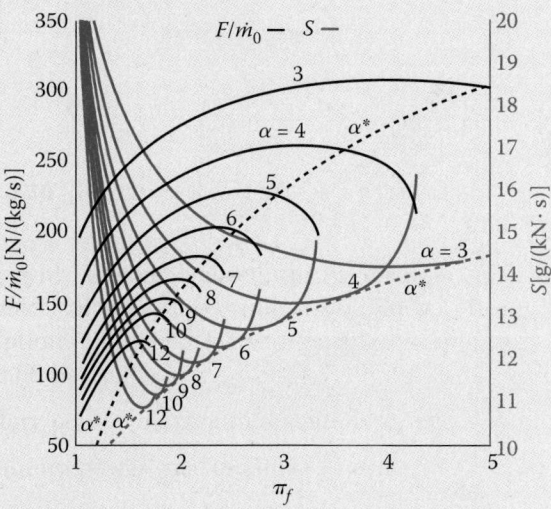

Fig. 5.14a Ideal turbofan performance vs π_f, for $\pi_c = 24$ and $M_0 = 0.8$: specific thrust and thrust specific fuel consumption.

(Continued)

Example 5.3 *(Continued)*

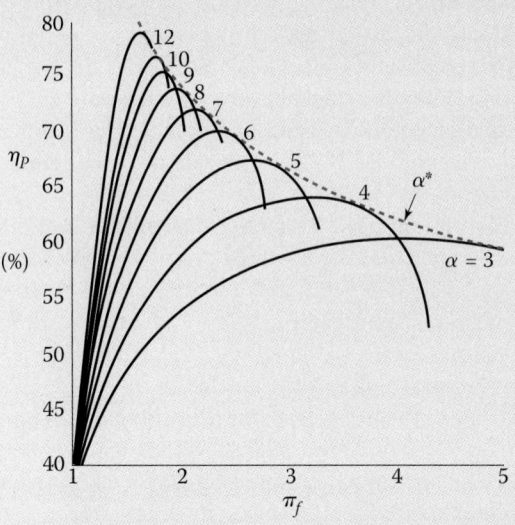

Fig. 5.14b Ideal turbofan performance vs π_f, for $\pi_c = 24$ and $M_0 = 0.8$: propulsive efficiency.

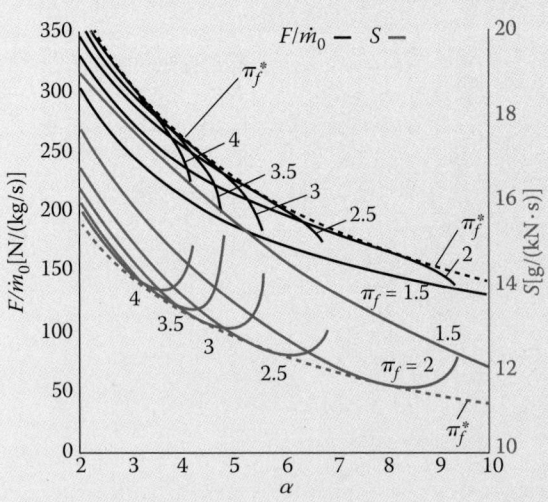

Fig. 5.15a Ideal turbofan performance vs α for $\pi_c = 24$ and $M_0 = 0.8$: specific thrust and thrust specific fuel consumption.

(Continued)

Example 5.3 *(Continued)*

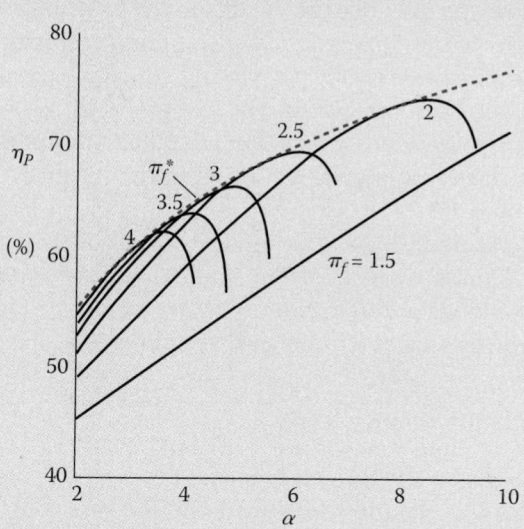

Fig. 5.15b Ideal turbofan performance vs α, for $\pi_c = 24$ and $M_0 = 0.8$: propulsive efficiency.

e) *Figures 5.15a and 5.15b.* Specific thrust and thrust specific fuel consumption are plotted vs the bypass ratio for different values of the fan pressure ratio in Fig. 5.15a. The compressor pressure ratio is held constant in these plots. Figure 5.15a shows the decreasing trend in specific thrust with increasing bypass ratio that is characteristic of turbofan engines.
Figure 5.15a shows that there is an optimum bypass ratio for each fan pressure ratio that will minimize fuel consumption. We will look at this optimum bypass ratio in greater detail in Section 5.9.

Propulsive efficiency is plotted vs the bypass ratio in Fig. 5.15b. This figure shows there is an optimum bypass ratio for each fan pressure ratio that will maximize propulsive efficiency.

5.9 Ideal Turbofan with Optimum Bypass Ratio

In most engine designs there is a choice of design variables that gives optimum performance. Minimum fuel consumption is desired for the high-bypass turbofan used for most subsonic aircraft. We introduce optimum analysis because we can apply calculus to our algebraic engine models. The result is a closed form solution that provides insight into the optimum. In the analysis of Chapter 7, the optimum is not this simple and results in a system of iterative calculations.

For a given set of flight conditions T_0 and M_0 and design limit τ_λ, there are three design variables: π_c, π_f, and α. In Figs. 5.11a–5.11b and 5.12a–5.12b, we can see that by increasing the compressor ratio π_c we can increase the thrust per unit mass flow and decrease the thrust specific fuel consumption. For the data used in these figures, increases in the compressor pressure ratio above a value of 20 do not increase the thrust per unit mass flow but do decrease the thrust specific fuel consumption. In Figs. 5.14a–5.14b, we can see that an optimum fan pressure ratio exists for all other parameters fixed. This optimum fan pressure ratio gives both the maximum thrust per unit mass flow and minimum thrust specific fuel consumption. In Figs. 5.15a–5.15b, we see that an optimum bypass ratio exists for all other parameters fixed. This optimum bypass ratio gives the minimum thrust specific fuel consumption. We will look at this optimum bypass ratio first. The optimum fan pressure ratio will be analyzed in the next section of this chapter.

5.9.1 Optimum Bypass Ratio α^*

When Eq. (5.32) for the fuel/air ratio and Eq. (5.29) for the specific thrust are inserted into the equation for thrust specific fuel consumption S [Eq. (5.35)], an expression for S in terms of bypass ratio α and other prescribed variables results. For a given set of prescribed variables (τ_r, π_c, π_f, τ_λ, V_0), we may locate the minimum S with respect to the bypass ratio α by taking the partial derivative of S with respect to the bypass ratio α and setting the equation equal to zero. The resulting value of the bypass ratio that gives minimum S is denoted as α^*. After some calculus and algebra (included in Chapter 5's supporting material 5.9.SMa, Ideal Turbofan: Optimum Bypass Ratio α^*), we get

$$\alpha^* = \frac{1}{\tau_r(\tau_f - 1)}\left[\tau_\lambda - \tau_r(\tau_c - 1) - \frac{\tau_\lambda}{\tau_r\tau_c} - \frac{1}{4}\left(\sqrt{\tau_r\tau_f - 1} + \sqrt{\tau_r - 1}\right)^2\right] \quad (5.39)$$

At this value of bypass ratio, we get a thrust ratio FR value of $\frac{1}{2}$ (see supporting material).

$$FR = \frac{V_9 - V_0}{V_{19} - V_0} = \frac{1}{2} \quad (5.40)$$

We thus note from Eq. (5.40) that when the bypass ratio is chosen to give the minimum specific fuel consumption, the thrust per unit mass flow of the engine core is one-half that of the fan. Thus the thrust ratio of an optimum-bypass-ratio ideal turbofan is 0.5. Using this fact, we may write the equation for the specific thrust simply as

$$\left(\frac{F}{\dot{m}_0}\right)_{\alpha^*} = \frac{a_0}{g_c}\frac{1 + 2\alpha^*}{2(1 + \alpha^*)}\left[\sqrt{\frac{2}{\gamma - 1}(\tau_r\tau_f - 1)} - M_0\right] \quad (5.41)$$

where α^* is obtained from Eq. (5.39).

One can easily show that the propulsive efficiency at the optimum bypass ratio is given by

$$(\eta_P)_{\min S} = \frac{4(1 + 2\alpha^*)V_0}{(3 + 4\alpha^*)V_0 + (1 + 4\alpha^*)V_{19}} \tag{5.42}$$

See Section 5.9b.SM, Summary of Equations: Optimum-Bypass-Ratio Ideal Turbofan, in the online supporting material for related information.

Example 5.4

The engine we looked at in Section 5.8 is used again in this section for an example. The following values are held constant for all plots:

$$T_0 = 216.7 \text{ K}, \quad \gamma = 1.4, \quad c_p = 1.004 \text{ kJ}/(\text{kg} \cdot \text{K}),$$
$$h_{PR} = 42{,}800 \text{ kJ/kg}, \quad T_{t4} = 1350 \text{ K}$$

The optimum bypass ratio is plotted in Fig. 5.16 vs the compressor pressure ratio for a flight Mach number of 0.8 and fan pressure ratio of

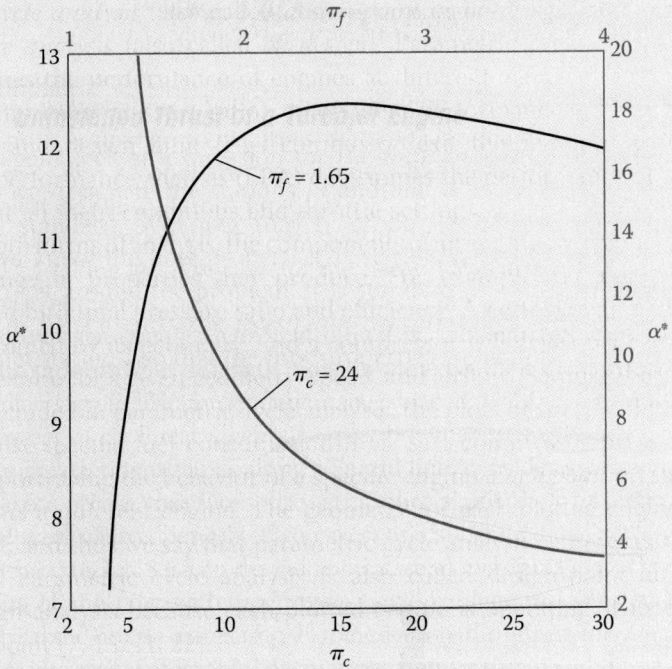

Fig. 5.16 α^* vs π_c, for $\pi_f = 1.65$ and $M_0 = 0.8$.

(Continued)

Example 5.4 *(Continued)*

1.65. From this plot, we can see that the optimum bypass ratio increases with the compressor pressure ratio. The optimum bypass ratio is plotted in Fig. 5.16 vs the fan pressure ratio for a compressor pressure ratio of 24, and this figure shows that the optimum bypass ratio decreases with the fan pressure ratio.

The optimum bypass ratio vs the flight Mach number is plotted in Fig. 5.17 for fan pressure ratios of 2 and 3. From these plots, we can see that the optimum bypass ratio decreases with the flight Mach number. Note that the optimum engine is a turbojet at a Mach number of about 3.0 for a fan pressure ratio of 3.

The plots of specific thrust and thrust specific fuel consumption for an ideal turbofan engine with optimum bypass ratio vs compressor pressure ratio are superimposed on Figs. 5.11a–5.11b by a dashed line marked α^*. The optimum-bypass-ratio ideal turbofan has the minimum thrust specific fuel consumption.

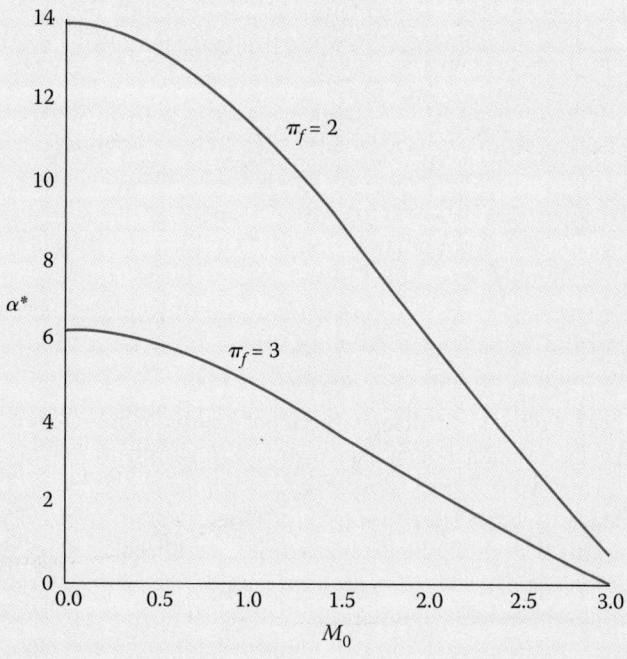

Fig. 5.17   Comparing α^* vs M_0, for $\pi_c = 24$ and $\pi_f = 2$ and 3.

(Continued)

Example 5.4 *(Continued)*

The plots of specific thrust and thrust specific fuel consumption for the optimum-bypass-ratio ideal turbofan vs fan pressure ratio are superimposed on Figs. 5.14a–5.14b by a dashed line marked α^*. As shown in these figures, the plot for α^* is the locus of the minimum value of S for each π_f.

The thermal efficiency of an optimum-bypass-ratio ideal turbofan is the same as that of an ideal turbojet. The bypass ratio of an ideal turbofan affects only the propulsive efficiency. The propulsive efficiency of an optimum-bypass-ratio ideal turbofan is superimposed on those of Figs. 5.11b and 5.14b. As can be seen, the optimum bypass ratio gives the maximum propulsive efficiency. Likewise, the optimum bypass ratio gives the maximum overall efficiency.

5.10 Ideal Turbofan with Optimum Fan Pressure Ratio

Figures 5.15a shows that for given flight conditions T_0 and M_0, design limit τ_λ, compressor pressure ratio π_c, and bypass ratio α, there is an optimum fan pressure ratio π_f that gives the minimum specific fuel consumption and maximum specific thrust. As will be shown, the optimum fan pressure ratio corresponds to the exit velocity V_{19} of the fan stream, being equal to the exit velocity of the core stream V_9. It is left as a reader exercise to show that equal exit velocities ($V_9 = V_{19}$) correspond to maximum propulsive efficiency.

5.10.1 Optimum Fan Pressure Ratio π_f^*

When Eq. (5.32) for the fuel/air ratio and Eq. (5.29) for the specific thrust are inserted into the equation for thrust specific fuel consumption S [Eq. (5.35)], an expression for S in terms of fan pressure ratio π_f and other prescribed variables results. For a given set of prescribed variables (τ_r, π_c, α, τ_λ, V_0), we may locate the minimum S with respect to the fan pressure ratio π_f by taking the partial derivative of S with respect to the fan temperature ratio τ_f and setting the equation equal to zero. The resulting value of fan temperature ratio and pressure ratio that gives minimum S are denoted as π_f^* and π_f^*. After some calculus and algebra (included in the Chapter 5 supporting material online), we get that the exit velocity of the bypass stream is equal to that of the engine core or

$$V_{19} = V_9 \tag{5.43}$$

Also

$$FR = 1 \tag{5.44}$$

The fan temperature ratio that gives the two exit velocities equals

$$\tau_f^* = \frac{\tau_\lambda - \tau_r(\tau_c - 1) - \tau_\lambda/(\tau_r \tau_c) + \alpha \tau_r + 1}{\tau_r(1 + \alpha)} \tag{5.45}$$

An equation for the specific thrust of an optimum-fan-pressure-ratio turbo-fan can be obtained by starting with the simplified expression

$$\left(\frac{F}{\dot{m}_0}\right)_{\tau_f^*} = \frac{a_0}{g_c}\left(\frac{V_{19}}{a_0} - M_0\right)$$

which becomes

$$\left(\frac{F}{\dot{m}_0}\right)_{\tau_f^*} = \frac{a_0}{g_c}\left[\sqrt{\frac{2}{\gamma - 1}(\tau_r \tau_f^* - 1)} - M_0\right] \tag{5.46}$$

The propulsive efficiency for the optimum-fan-pressure-ratio turbofan engine is simply

$$(\eta_P)_{\tau_f^*} = \frac{2}{V_{19}/V_0 + 1} \tag{5.47}$$

See Section 5.10b.SM, Summary of Equations: Optimum-Fan-Pressure-Ratio Ideal Turbofan, in the online supporting material for related information.

5.10.2 Effect of Bypass Ratio on Specific Thrust and Fuel Consumption

Data for several turbofan engines are listed in Table 5.2. The bypass ratio ranges from 0.36 for the engine in the smaller F-15 and F-16 fighter aircraft to over 8 for engines in the commercial transports. The specific thrust for the fighters is about four time the values for the three transport-type airplanes, and thrust specific fuel consumption is about six time the values for the three transport-type airplanes. Let us look at the interrelationship among the bypass ratio, specific thrust, and specific fuel consumption to help explain the trends in the values of these quantities, as exhibited by the table.

Table 5.2 Data for Several Turbofan Engines

Engine	Bypass Ratio α	$F/\dot{m}_0$, [N/(kg/s)] [lbf/ (lbm/s)]	S [(g/s)/kN] [(lbm/h)/ lbf]	π_f	π_c	Aircraft
GE90-85B	8.4	273.5 (27.89)	9.06 (0.32)	1.65	39.3	B777-200
F108-CF-100	6.0	270.3 (27.56)	10.28 (0.363)	1.5	23.7	KC-135R
CF6-80C2	5.15	287.2 (26.06)	9.54 (0.348)	1.7	30.4	C-5M
F100-PW-229	0.36	1146.8 (116.94)	58.07 (2.05)	3.8	32	F-15/F-16

Example 5.5

We can examine how these quantities interact for the ideal turbofan with optimum fan pressure ratio by plotting specific thrust and thrust specific fuel consumption vs bypass ratio. Such plots are given in Fig. 5.18 for

$$M_0 = 0.8, \quad T_0 = 216.7\,\text{K}, \quad T_{t4} = 1350\,\text{K}, \quad \pi_c = 24, \quad h_{PR} = 42{,}800\,\text{kJ/kg},$$
$$\gamma = 1.4, \quad c_p = 1.004\,\text{kJ/(kg}\cdot\text{K)}$$

The optimum fan pressure ratio that gives $V_9 = V_{19}$ is plotted vs the bypass ratio α in Fig. 5.18.

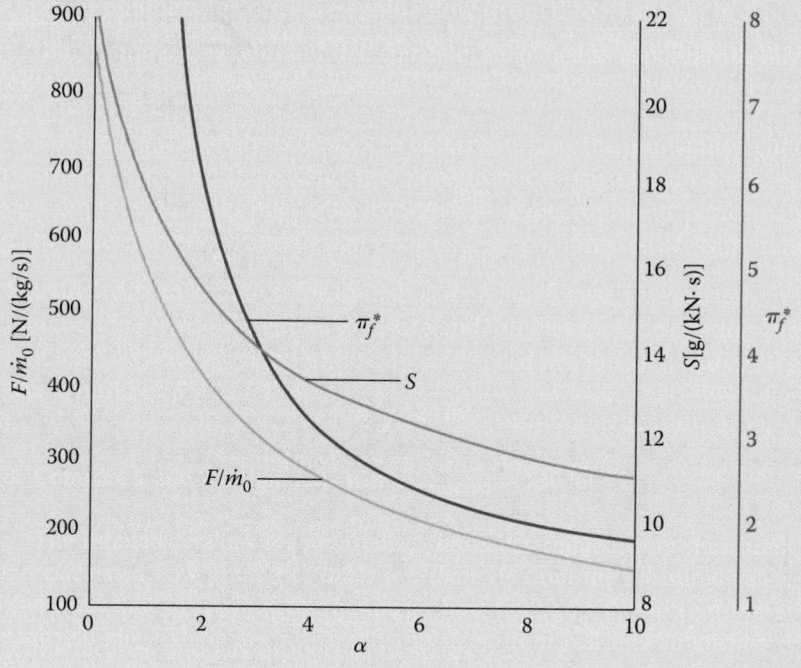

Fig. 5.18 Performance at π_f^* vs α, for $\pi_c = 24$ and $M_0 = 0.8$.

(Continued)

Example 5.5 *(Continued)*

We can see, in Fig. 5.18, a sharp reduction in specific fuel consumption as α increases from zero. An equally marked, but unfavorable, decrease in thrust per unit mass flow occurs. A large fraction of the beneficial decrease in S is obtained by selecting an α of about 6, as was done for the engines in the KC-135R tanker. At a bypass ratio of 8, corresponding to the GE90 engines, a further decrease in specific fuel consumption is realized. Because engine weight is a small fraction of the takeoff gross weight for these airplanes, it is of secondary importance compared to fuel weight and hence specific fuel consumption. Thus we find a relatively large bypass ratio for the engines in the transporters tabulated compared to the 0.36 value of the F-15/F-16 engine.

5.10.3 Comparison of Optimum Ideal Turbofans

A comparison of the two optimum ideal turbofan engines presented in the previous two sections can be made by looking at contour plots of specific thrust $F/\dot{m}_0$ and thrust specific fuel consumption S vs fan pressure ratio π_f and bypass ratio α. For our example problems (Examples 5.3, 5.4, and 5.5), contours of constant specific thrust values and contours of constant thrust specific fuel consumption are plotted in Figs. 5.19a and 5.19b, respectively. Also plotted are the curves for π_f^* and α^* for each optimum engine.

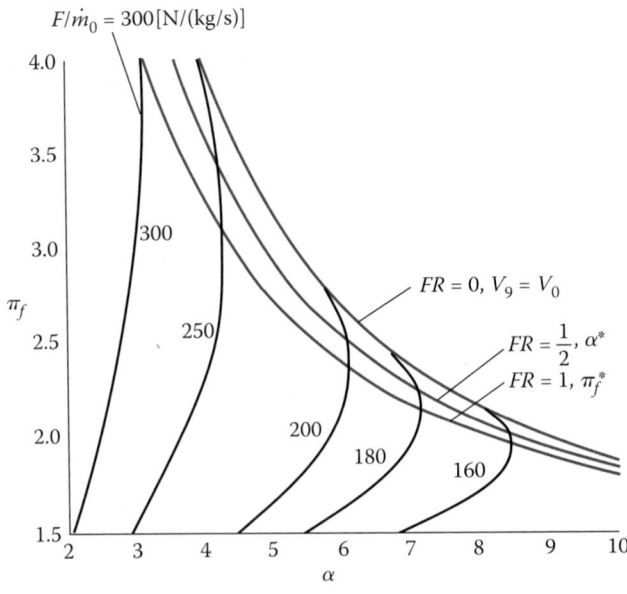

Fig. 5.19a Contours of constant specific thrust and curves for π_f^* and α^*.

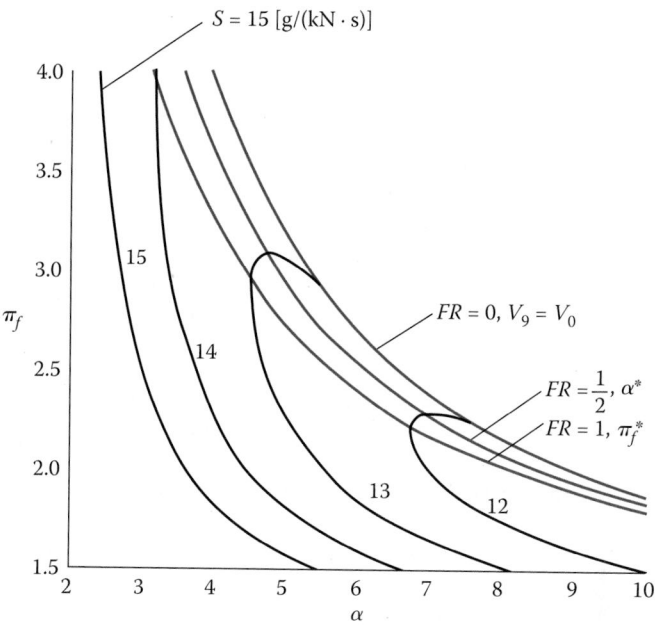

Fig. 5.19b   Contours of constant fuel consumption and curves for π_f^* and α^*.

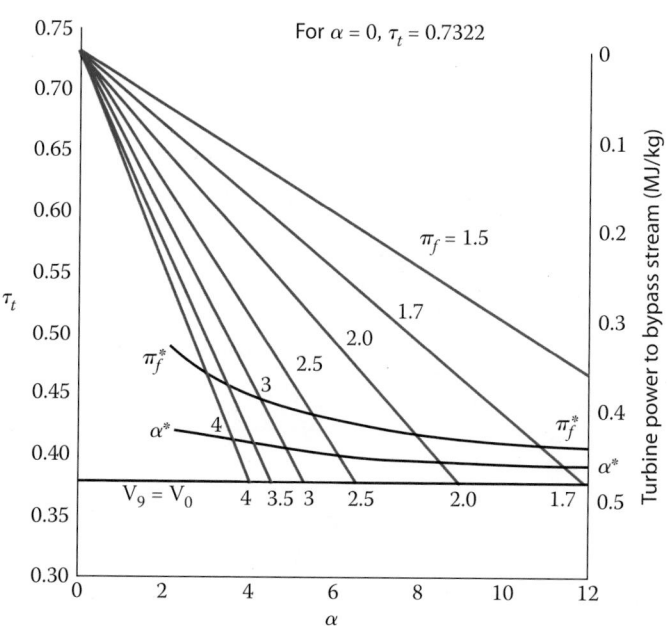

Fig. 5.20   Available turbine total temperature ratio with optimum bypass ratio α^* and optimum fan pressure ratio π_f^*.

Figures 5.19a and 5.19b show that π_f^* is the fan pressure ratio giving maximum specific thrust and minimum thrust specific fuel consumption for a given bypass ratio. Figures 5.19b shows that α^* is the bypass ratio giving minimum thrust specific fuel consumption for a given fan pressure ratio. Also shown in Fig. 5.19b are that the α^* curve is the locus of horizontal tangents to contours of constant S and the π_f^* curve is the locus of vertical tangents to contours of constant S.

The locus of turbine total temperature ratios τ_t's for optimum bypass ratio α^* and optimum fan pressure ratio π_f^* have been superimposed on Fig. 5.13 to give Fig. 5.20. Note in this figure that the optimum fan pressure ratio corresponds to lower power extraction from the engine core than the optimum bypass ratio. One can also see from Fig. 5.20 that the optimum fan pressure ratio is 1.7 for a bypass ratio of 10.8.

5.11 Ideal Mixed-Flow Turbofan with Afterburning

We consider the turbofan engine in which the core flow and bypass flow are mixed together before passing through an afterburner and nozzle. Figure 5.21 shows a view of such a turbofan engine, and Fig. 5.22 shows the ideal engine cycle on a T-s diagram. Modern fighter aircraft use this type of engine because it gives the required high specific thrust with the afterburner on and lower thrust specific fuel consumption than the turbojet engine when the afterburner is off. Because this engine has a single inlet and exhaust, the ideal thrust of this engine is given by

$$\frac{F}{\dot{m}_0} = \frac{a_0}{g_c}\left(\frac{V_9}{a_0} - M_0\right) \tag{5.48}$$

The analysis of this engine requires the definition of the total temperature and total pressure ratios across the mixer. We define

$$\tau_M = \frac{T_{t6A}}{T_{t6}} \qquad \pi_M = \frac{P_{t6A}}{P_{t6}} \tag{5.49}$$

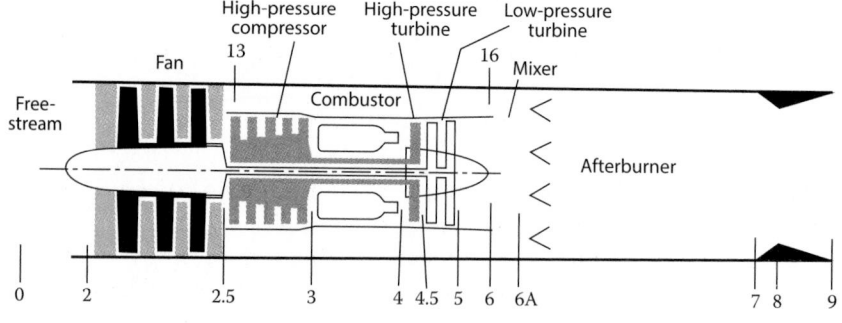

Fig. 5.21   Station numbering for mixed-flow turbofan engine with afterburner.

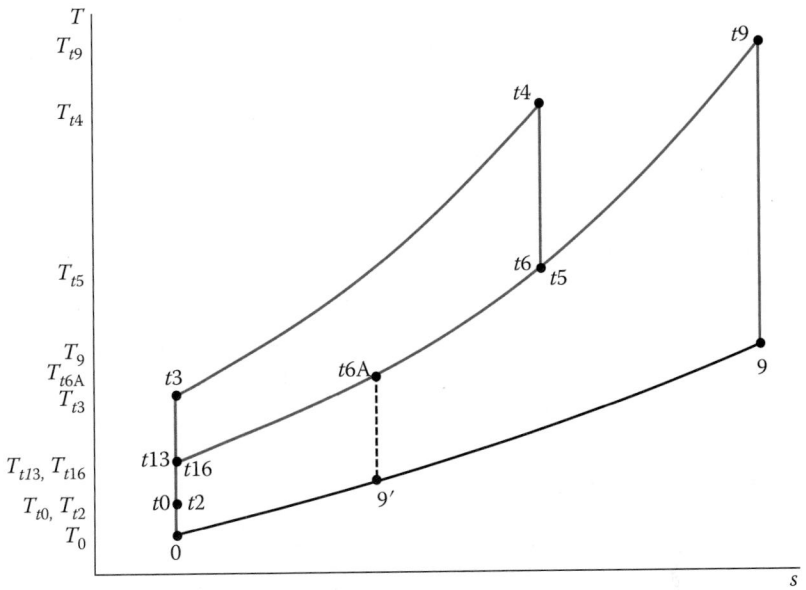

Fig. 5.22   The T-s diagram for mixed-flow turbofan engine.

The flow in the bypass duct from station 13 to 16 is considered to be reversible and adiabatic. The bypass stream enters the mixer at station 16 with the same total properties as the fan discharge. An energy balance of the mixer gives

$$\dot{m}_6 c_p T_{t6} + \dot{m}_{16} c_p T_{t16} = \dot{m}_{6A} c_p T_{t6A}$$

or

$$T_{t6} + \alpha T_{t13} = (1 + \alpha) T_{t6A}$$

Because $T_{t6} = T_{t5}$, then

$$\tau_M = \frac{1 + \alpha T_{t13}/T_{t5}}{1 + \alpha}$$

This equation can be written in terms of the engine τ's and bypass ratio α as

$$\tau_M = \frac{1}{1 + \alpha}\left(1 + \alpha \frac{\tau_r \tau_f}{\tau_\lambda \tau_t}\right) \tag{5.50}$$

Fluid dynamics requires equal static pressures at stations 6 and 16 (Kutta condition) as long as the two streams are subsonic. Normal design of the mixer has the Mach numbers of the two entering streams nearly equal. For this ideal analysis, then, we assume that the total pressures of the two entering streams are equal, or

$$P_{t6} = P_{t16} \tag{5.51}$$

The total pressure ratio of the mixer in this ideal turbofan engine is assumed to be unity (Refs. 13 and 23), or

$$\pi_M = \frac{P_{t6A}}{P_{t6}} = 1 \tag{5.52}$$

Equal total pressures at stations 6 and 16 require that the total pressure ratio of stations 3 to 13 equal that of stations 3 to 6. The total pressure ratio across the burner is unity, which allows us to write

$$\frac{P_{t3}}{P_{t13}} = \frac{P_{t3}/P_{t2}}{T_{t13}/P_{t2}} = \frac{\pi_c}{\pi_f} = \frac{P_{t3}}{P_{t6}} = \frac{P_{t4}}{P_{t5}} = \frac{1}{\pi_t}$$

We note, from Fig. 5.22, that stations 13, 16, 6, 6A, and 9 all have the same total pressure. When the afterburner is *on*, the isentropic process between states $t9$ and 9 represents the flow through the exhaust nozzle. When the after burner is *off*, the flow enters the exhaust nozzle at state $t6A$ and exits at $9'$.

The compression and expansion processes are isentropic, so the preceding expression becomes

$$\frac{1}{\tau_t} = \frac{\tau_c}{\tau_f} \tag{5.53}$$

5.11.1 Cycle Analysis

The power balance among the turbine, fan, and compressor is developed in Section 5.8.1 as Eq. (5.33) and gives the following relationship between the total temperature ratios across these components:

$$\tau_t = 1 - \frac{\tau_r}{\tau_\lambda}[\tau_c - 1 + \alpha(\tau_f - 1)] \tag{5.33}$$

For given values of τ_r, τ_λ, and τ_c there is one value of τ_f for each value of α that satisfies both Eq. (5.53) and Eq. (5.33). This is an important result. Unlike the ideal separate-flow turbofan (Section 5.8), the selection of fan pressure ratio and bypass ratio are *not independent* for the ideal mixed-flow turbofan. From these two equations, we can obtain an expression for the bypass ratio α in terms of the other variables and an expression for the fan total temperature ratio τ_f. From Eqs. (5.53) and (5.33), we have

$$\frac{\tau_f}{\tau_c} = 1 - \frac{\tau_r}{\tau_\lambda}[\tau_c - 1 + \alpha(\tau_f - 1)]$$

or

$$\tau_f = \tau_c - \frac{\tau_r \tau_c}{\tau_\lambda}[\tau_c - 1 + \alpha(\tau_f - 1)] \tag{i}$$

Case 1: Fan pressure ratio π_f is specified. Equation (i), solved for the bypass ratio α, yields

$$\alpha = \frac{\tau_\lambda(\tau_c - \tau_f)}{\tau_r \tau_c(\tau_f - 1)} - \frac{\tau_c - 1}{\tau_f - 1} \tag{5.54}$$

Case 2: Bypass ratio α is specified. Equation (i), solved for τ_f, yields

$$\tau_f = \frac{\tau_\lambda/\tau_r - (\tau_c - 1) + \alpha}{\tau_\lambda/(\tau_r \tau_c) + \alpha} \tag{5.55}$$

The velocity ratio V_9/a_0 is given by

$$\left(\frac{V_9}{a_0}\right)^2 = \frac{T_9}{T_0} M_9^2$$

where

$$M_9^2 = \frac{2}{\gamma - 1}(\tau_r \tau_f - 1) \tag{5.56}$$

The temperature ratio T_9/T_0 depends on the operation of the afterburner. When the afterburner is *on*, then

$$\frac{T_9}{T_0} = \frac{T_{t9}/T_0}{(P_{t9}/P_9)^{(\gamma-1)/\gamma}} = \frac{\tau_{\lambda AB}}{\tau_r \tau_f} \tag{5.57}$$

When the afterburner is *off*, we have

$$\frac{T_9}{T_0} = \frac{T_{t9}/T_0}{(P_{t9}/P_9)^{(\gamma-1)/\gamma}} = \frac{\tau_\lambda \tau_t \tau_M}{\tau_r \tau_f} \tag{5.58}$$

The fuel/air ratio of the burner is given by Eq. (5.12a). The fuel/air ratio of the afterburner is obtained by an energy balance between stations 6A and 7. We write

$$\dot{m}_{fAB}h_{PR} + \dot{m}_0 c_p T_{t6A} = \dot{m}_0 c_p T_{t7}$$
$$f_{AB} = \frac{\dot{m}_{fAB}}{\dot{m}_0} = \frac{c_p T_0}{h_{PR}}(\tau_{\lambda AB} - \tau_\lambda \tau_t \tau_M) \tag{5.59}$$

The overall fuel/air ratio f_O is defined as the total fuel flow rate divided by the total airflow rate, or

$$f_O = \frac{\dot{m}_f + \dot{m}_{fAB}}{\dot{m}_0} \tag{5.60}$$

For this engine, the overall fuel/air ratio can be written in terms of the fuel/air ratios of the burner and afterburner as

$$f_O = \frac{f}{1+\alpha} + f_{AB} \tag{5.61}$$

The thrust specific fuel consumption S is then given by

$$S = \frac{f_O}{F/\dot{m}_0} \tag{5.62}$$

Equation (5.31b) gives the propulsive efficiency η_P for this cycle, and the thermal efficiency η_{Th} is given by

$$\eta_{Th} = \frac{\gamma - 1}{2} \frac{c_p T_0}{f_O h_{PR}} \left[\left(\frac{V_9}{a_0} \right)^2 - M_0^2 \right] \tag{5.63}$$

See Section 5.11.SM, Summary of Equations: Ideal Mixed-Flow Turbofan, in the online supporting material for related information.

Example 5.6

The afterburning and nonafterburning performances of ideal mixed-flow turbofan engines are plotted in Figs. 5.23a and 5.23b vs compressor pressure ratio π_c for different values of the fan pressure ratio π_f at two flight Mach numbers M_0. Figure 5.24 plots the required bypass ratio α vs compressor pressure ratio π_c for the different values of fan pressure ratio π_f at flight Mach numbers of 0.9 and 2, respectively. Calculations were performed for the following input data:

$$T_0 = 390°R, \quad \gamma = 1.4, \quad c_p = 0.24\,\text{Btu/(lbm} \cdot °R), \quad h_{PR} = 18{,}400\,\text{Btu/lbm},$$

$$T_{t4} = 3000°R, \quad T_{t7} = 3600°R, \quad \pi_c = 10 \rightarrow 40, \quad \pi_f = 2 \rightarrow 5, \quad M_0 = 0.9, 2$$

Figures 5.23a and 5.23b show that the performance of the afterburning engine is a function of only the fan pressure ratio and flight Mach number. The fuel/air ratio is proportional to the total temperature rise across the engine $(T_{t9} - T_{t0})$, essentially a constant. The specific thrust increases with the nozzle pressure ratio $(P_{t9}/P_9 = \pi_r \pi_f)$. As a result, increasing the fan pressure ratio increases the specific thrust and decreases the fuel consumption. Figures 5.23a and 5.23b also show the performance trends of the mixed-flow turbofan engine during nonafterburning operation.

Figure 5.24 shows the bypass ratios required for equal total pressures of the two streams entering the mixer. At Mach 2, a fan pressure ratio of 4–5

(Continued)

Example 5.6 *(Continued)*

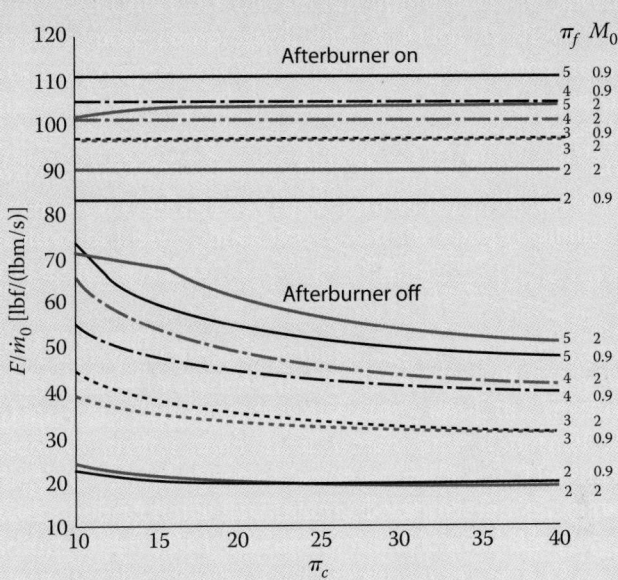

Fig. 5.23a Specific thrust of a mixed-flow turbofan engine.

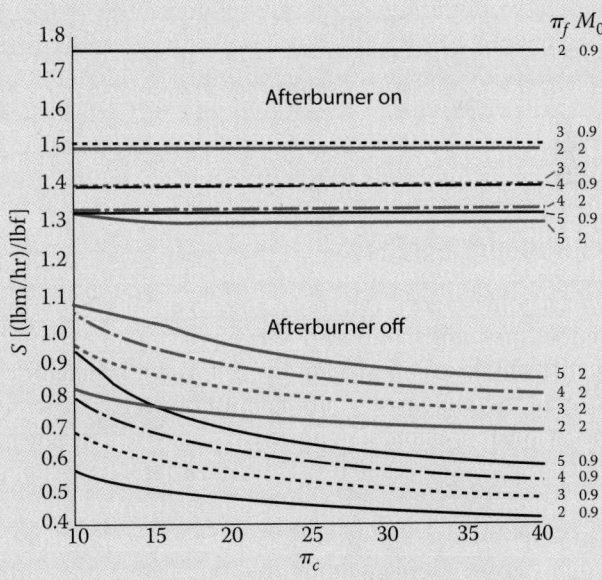

Fig. 5.23b Thrust specific fuel consumption of a mixed-flow turbofan engine.

(Continued)

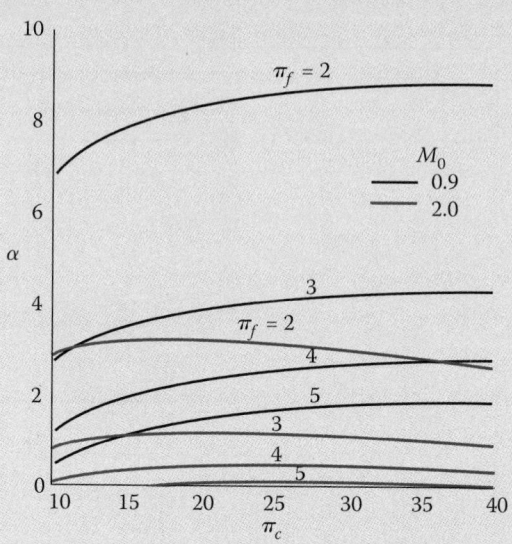

Fig. 5.24   Bypass ratio of a mixed-flow turbofan engine, $M_0 = 0.9$.

requires a bypass ratio less than 0.5 (refer to Table 5.2 and the F100 engine). As can be seen in Eq. (5.54), the bypass ratio can be increased above those shown by increasing the value of T_{t4} above 3000°R.

5.12   Ideal Turboprop Engines

Interest in propeller-driven vehicles has grown considerably recently due to the tremendous increase in the use of unmanned aerial systems or vehicles, the majority of which are powered by relatively low-speed propellers. Although many of these propellers are battery powered, some are powered by reciprocating (piston-cylinder) engines, and some are powered by gas turbine engines (turboprops). Further, continued interest in highly efficient flight transportation has spurred investigation into *very-high-bypass-ratio fans*. Cycle analysis indicates that such bypass ratios (for subsonic flight) could approach those corresponding to the "old" turboprop engines (approximately 100:1). There are several reasons why the turbofan engines became more popular than the turboprop engines, and it is wise to review such reasons so that we can comprehend why similar concepts are again gaining in popularity.

A major reason for the success of the turbofan was its highly efficient operation at high subsonic flight Mach numbers, typically around $M_0 = 0.8$ to 0.85. In a turboprop, the propeller tip Mach numbers become very large, above approximately $M_0 = 0.7$, and the resultant loss in propeller efficiency limits the turboprop use to flight Mach numbers less than 0.7. With a turbofan, the onset of high-Mach-number effects is reduced by the diffusion within the duct. In addition, the individual blade loading can be much reduced by utilizing many blades. A second important benefit of conventional turbofans is that they require no gearbox to reduce the tip speeds of their relatively short blades. (Note, of course, that usually a turbofan engine has multiple spools.) Turboprop gearboxes have, to date, been heavy and subject to reliability problems. Finally, the high tip speed of the turboprops led to high noise levels, both in the airport vicinity and within the aircraft at flight speeds.

Studies of the very-high-bypass-ratio engines, however, have suggested some compromise designs that show great promise. (See Figure 1 in this book's Foreword to Second Edition by Heiser.) Thus if a bypass ratio of, say, 25 is selected, the corresponding cowl could have identified with it both weight penalties and drag penalties that do not compensate for the benefits of the inlet diffusion and of the reduction in tip losses. By considering this "in between" bypass ratio, the required shaft speed reduction will be reduced with the result that a lighter and simpler gearbox may be utilized (see Ref. 70). The effects of tip losses and noise production may be somewhat curtailed by utilizing many (up to eight) of the smaller blades and by sweeping the blades to reduce the relative Mach numbers. An additional benefit is available in that the blades may be made variable-pitch, which will allow high propeller efficiencies to be maintained over a wide operating range. Recall from Eq. (1.16) that the propulsive efficiency of a single-exhaust engine is $\eta_P = 2V_0/(V_j + V_0)$, where $V_0 =$ flight speed and $V_j =$ jet speed.

This expression is appropriate for a propeller also, and it serves to emphasize that we want a large propeller [to reduce V_j for a given thrust $F = \dot{m}_0(V_j - V_0)/g_c$] if the propulsive efficiency is to be high. The *propulsive efficiency* represents the ideal limit of the propeller efficiency, which is defined as

$$\eta_{\text{prop}} = \frac{\text{power of vehicle}}{\text{power to propeller}} = \frac{F_{\text{prop}} V_0}{\dot{W}_{\text{prop}}} \tag{5.64}$$

Thus

$$\eta_{\text{prop}} = \frac{F_{\text{prop}} V_0}{\dot{W}_{\text{prop}}} = \frac{\dot{m}_{\text{prop}}(V_j - V_0)}{1/2\dot{m}_{\text{prop}}(V_j^2 - V_0^2)} \frac{1/2\,\dot{m}_{\text{prop}}(V_j^2 - V_0^2)}{\dot{W}_{\text{prop}}}$$

$$\equiv \eta_P \eta_L \tag{5.65}$$

where $\dot{W}_{prop}$ = propeller power in, η_P is the propulsive efficiency of the propeller, and

$$\eta_L = \frac{1/2\,\dot{m}_{prop}(V_j^2 - V_0^2)}{\dot{W}_{prop}} \quad (5.66)$$

represents the power output of the propeller to the stream divided by the power input to the propeller. Appendix M on propeller theory and design is provided online for those wanting to know more about propellers.

Thus we expect the propeller efficiency to increase with propeller size simply because the ideal propeller efficiency (i.e., the propulsive efficiency) increases. This, of course, relates the propeller efficiency and bypass ratio. The propeller size will, in practice, be limited by the onset of tip Mach number losses and the like, which would be reflected in a reduced η_L.

5.12.1 Cycle Analysis

It is appropriate in analyzing the turboprop class of engine to consider the work supplied to the vehicle, rather than the thrust. To facilitate this, we introduce the dimensionless *work output coefficient C*, defined as

$$C = \frac{\text{power interaction/mass flow of air through engine core}}{h_0} \quad (5.67)$$

where h_0 is the enthalpy of the freestream air.

For the thrust of the core stream F_C, we define its work output coefficient as

$$C_C = \frac{F_C V_0}{\dot{m}_0 c_p T_0} \quad (5.68)$$

For the thrust of the propeller, we define its work output coefficient as

$$C_{prop} = \frac{\eta_{prop}\dot{W}_{prop}}{\dot{m}_0 c_p T_0} \quad (5.69)$$

Thus the work output coefficient for the total turboprop engine is

$$C_{tot} = C_{prop} + C_C \quad (5.70)$$

and the corresponding thrust is

$$F = F_{prop} + F_C = \frac{C_{tot}\dot{m}_0 c_p T_0}{V_0} \quad (5.71)$$

It is usual, with turboprop engines, to have the core stream exit nozzles unchoked, so the pressure imbalance term will not contribute in the expression for the thrust. We consider the numbering stations indicated in

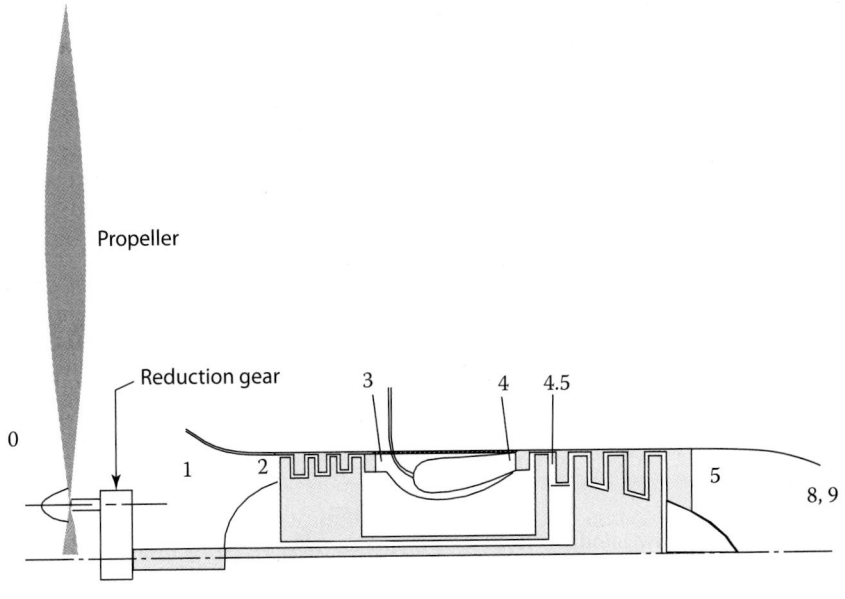

Fig. 5.25a   Station numbering of turboprop engine.

Figs. 5.25a and 5.25b, and we proceed with the analysis in much the same way as with the previously considered engine types.

For a turboprop engine, we have two design variables—the compressor pressure ratio and the low-pressure (free or power) turbine temperature ratio. We want to develop the cycle equations in terms of the pressure and temperature ratios across both the high-pressure turbine (drives the compressor) and the low-pressure turbine. With station number 4.5 between

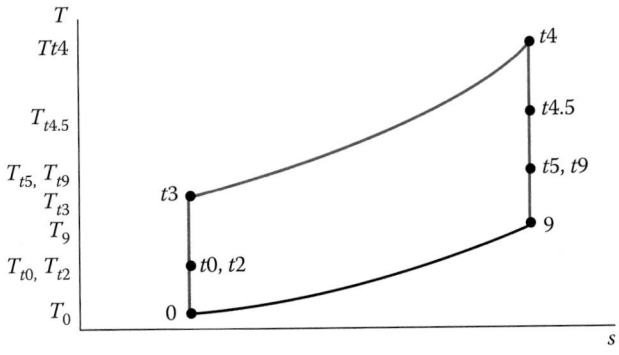

Fig. 5.25b   The T-s diagram of an ideal turboprop engine.

the high- and low-pressure turbines, we define

$$\pi_{tH} = \frac{P_{t4.5}}{P_{t4}} \quad \tau_{tH} = \frac{T_{t4.5}}{T_{t4}} \quad \pi_{tL} = \frac{T_{t5}}{T_{t4.5}} \quad \tau_{tL} = \frac{T_{t5}}{T_{t4.5}} \tag{5.72}$$

Step 1: We have for the engine core

$$\frac{F_C}{\dot{m}_0} = \frac{a_0}{g_c}\left(\frac{V_9}{a_0} - M_0\right)$$

Then

$$C_C = \frac{V_0}{c_p T_0}\frac{a_0}{g_c}\left(\frac{V_9}{a_0} - M_0\right)$$

Thus

$$C_C = (\gamma - 1)M_0\left(\frac{V_9}{a_0} - M_0\right) \tag{5.73}$$

Step 2: Now $V_9/a_0 = \sqrt{T_9/T_0}\, M_9$, where

$$M_9 = \sqrt{\frac{2}{\gamma - 1}\left[\left(\frac{P_{t9}}{P_9}\right)^{(\gamma-1)/\gamma} - 1\right]} \tag{5.74}$$

Step 3:

$$\frac{P_{t9}}{P_9} = \frac{P_0}{P_9}\pi_r \pi_d \pi_c \pi_b \pi_{tH} \pi_{tL} \pi_n = \pi_r \pi_c \pi_{tH} \pi_{tL}$$

$$M_9 = \sqrt{\frac{2}{\gamma - 1}(\tau_r \tau_c \tau_{tH} \tau_{tL} - 1)} \tag{5.75}$$

Step 4:

$$\frac{T_9}{T_0} = \frac{T_{t9}/T_0}{T_{t9}/T_9} = \frac{T_{t9}/T_0}{(P_{t9}/P_9)^{(\gamma-1)/\gamma}}$$

where

$$\frac{T_{t9}}{T_0} = \tau_\lambda \tau_{tH} \tau_{tL} \tau_n = \tau_\lambda \tau_{tH} \tau_{tL}$$

and

$$\left(\frac{P_{t9}}{P_9}\right)^{(\gamma-1)/\gamma} = (\pi_r \pi_c \pi_{tH} \pi_{tL})^{(\gamma-1)/\gamma} = \tau_r \tau_c \tau_H \tau_{tL}$$

Thus

$$\frac{T_9}{T_0} = \frac{\tau_\lambda}{\tau_r \tau_c} \tag{5.76}$$

Step 5: From the energy balance of the ideal turbofan's burner, we have [Eq. (5.32)]

$$f = \frac{c_p T_0}{h_{PR}}(\tau_\lambda - \tau_r \tau_c) \tag{5.32}$$

Step 6: For the high-pressure turbine, we have from the ideal turbojet analysis

$$\tau_{tH} = 1 - \frac{\tau_r}{\tau_\lambda}(\tau_c - 1) \tag{5.77}$$

For the low-pressure turbine (also called *free turbine*), we equate the power out of this turbine to the power into the propeller. Thus

$$\dot{m}_{4.5} c_p (T_{t4.5} - T_{t5}) = \dot{W}_{prop}$$

or

$$C_{prop} = \frac{\eta_{prop} \dot{W}_{prop}}{\dot{m}_0 c_p T_0} = \eta_{prop} \tau_\lambda \tau_{tH}(1 - \tau_{tL}) \tag{5.78}$$

Step 7: For the turboprop engine, we consider both the specific power and the specific thrust:

$$\frac{\dot{W}}{\dot{m}_0} = C_{tot} c_p T_0 \tag{5.79}$$

$$\frac{F}{\dot{m}_0} = \frac{C_{tot} c_p T_0}{V_0} \tag{5.80}$$

Step 8:

$$S = \frac{f}{F/\dot{m}_0}$$

We should note here, also, that for propeller aircraft it is more usual to refer to the power specific fuel consumption in terms of S_P, where S_P is defined by

$$S_P = \frac{\dot{m}_f}{\dot{W}} = \frac{f}{\dot{W}/\dot{m}_0}$$

Thus

$$S_P = \frac{f}{C_{tot} c_p T_0} \tag{5.81}$$

Step 9: The thermal efficiency of the engine cycle is the same as that for the turbojet or turbofan engines. Thus

$$\eta_{Th} = 1 - \frac{1}{\tau_r \tau_c}$$

From Eq. (1.17), the overall efficiency can be simply expressed as

$$\eta_O = \frac{\dot{W}}{\dot{m}_f h_{PR}}$$

Then with Eq. (5.32) for the fuel/air ratio and the definition of C_{tot}, the overall efficiency can be simply expressed as

$$\eta_O = \frac{C_{tot}}{\tau_\lambda - \tau_r \tau_c} \tag{5.82}$$

See Section 5.12a.SM, Summary of Equations: Ideal Turboprop, in the online supporting material for related information.

5.12.2 Selection of Optimal Turbine Temperature Ratio

Turbopropeller or prop-fan engines will be designed primarily to be low-specific-fuel-consumption engines. Thus we select the turbine temperature ratio τ_t to make S a minimum. Equivalently, we locate the maximum of C_{tot},

$$\frac{\partial S}{\partial \tau_t} = 0 \quad \Rightarrow \quad \frac{\partial C_{tot}}{\partial \tau_t} = 0$$

where

$$C_{tot} = C_{prop} + C_C = \eta_{prop} \tau_\lambda (\tau_{tH} - \tau_t) + (\gamma - 1) M_0 \left(\frac{V_9}{a_0} - M_0 \right)$$

Thus

$$\frac{\partial C_{tot}}{\partial \tau_t} = -\eta_{prop} \tau_\lambda + (\gamma - 1) M_0 \frac{\partial}{\partial \tau_t} \left(\frac{V_9}{a_0} \right) = 0 \tag{i}$$

Noting

$$\frac{\partial}{\partial \tau_t} \left(\frac{V_9}{a_0} \right) = \frac{1}{2 V_9 / a_0} \frac{\partial}{\partial \tau_t} \left[\left(\frac{V_9}{a_0} \right)^2 \right] = \frac{1}{2 V_9 / a_0} \frac{2 \tau_\lambda}{\gamma - 1}$$

we see that Eq. (i) becomes

$$\frac{\partial C_{\text{tot}}}{\partial \tau_t} = -\eta_{\text{prop}}\tau_\lambda + \frac{M_0\tau_\lambda}{(V_9/a_0)_{\text{opt}}} = 0$$

Thus

$$\left(\frac{V_9}{a_0}\right)_{\text{opt}} = \frac{M_0}{\eta_{\text{prop}}} \quad \text{or} \quad (V_9)_{\text{opt}} = \frac{V_0}{\eta_{\text{prop}}} \tag{5.83}$$

Hence for a 100-percent-efficient propeller, the optimum exit velocity for the core of the turboprop is equal to the flight velocity, and the propeller produces all the engine thrust.

The optimum turbine temperature ratio is obtained by equating the velocity ratio V_9/a_0 from Eq. (5.15)

$$\frac{V_9}{a_0} = \sqrt{\frac{2}{\gamma-1}\left(\tau_\lambda\tau_t - \frac{\tau_\lambda}{\tau_r\tau_c}\right)}$$

to Eq. (5.83), giving

$$\frac{2}{\gamma-1}\frac{\tau_\lambda}{\tau_r\tau_c}(\tau_r\tau_c\tau_t^* - 1) = \frac{M_0^2}{\eta_{\text{prop}}^2}$$

Thus

$$\tau_t^* = \frac{\tau_\lambda}{\tau_r\tau_c} + \frac{[(\gamma-1)/2]M_0^2}{\tau_\lambda\,\eta_{\text{prop}}^2} \tag{5.84}$$

See Section 5.12b.SM, Summary of Equations: Turboprop (Optimal Work Distribution), in the online supporting material for related information.

Example 5.7

With improved propeller designs, the aircraft industry is considering a turbo-prop business aircraft for flight at Mach 0.8. Consider an engine suitable for use in an 8- to 10-passenger business jet. The assumed altitude, gas properties, propeller efficiency, and the like are given below. The relatively high propeller efficiency has been taken from estimated propeller performance when modern

(Continued)

Example 5.7 *(Continued)*

transonic techniques, such as sweeping the blades, are used in the blade design.

Altitude $= 7.6\,\text{km}\,(25{,}000\,\text{ft})$, $\eta_{\text{prop}} = 0.83$, $T_0 = 240\,\text{K}$, $T_{t4} = 1370\,\text{K}$,

$M_0 = 0.8$, $h_{PR} = 42{,}800\,\text{kJ/kg}$, $\gamma = 1.4$, $c_p = 1.004\,\text{kJ/(kg K)}$

With these parameters, a range of compressor pressure ratios for both specific values and the optimum value of the turbine temperature ratios were considered. The specific thrust, thrust specific fuel consumption, cycle efficiencies, work output coefficients, and optimum turbine temperature ratio are shown in Figs. 5.26a–5.26e. The results corresponding to the optimum turbine temperature ratio τ_t^* are drawn with dashed curves in Figs. 5.26a–5.26e. Note the high value for specific thrust $F/\dot{m}_0$ compared to turbojets or turbofans. This is misleading because, here, $\dot{m}_0$ is mass flow through the core only and not through the propeller area. (Refer to Fig. 1.17a, which more appropriately factors in the mass flow captured by the propeller area.) As shown in Fig. 5.26d, the specific thrust developed by the core increases with π_c for a constant turbine temperature ratio and remains constant for an optimum turbine temperature ratio. On the other hand, the thrust per unit core mass flow developed by the propeller has a shape similar to the total specific thrust curve of Fig. 5.26a. The compressor

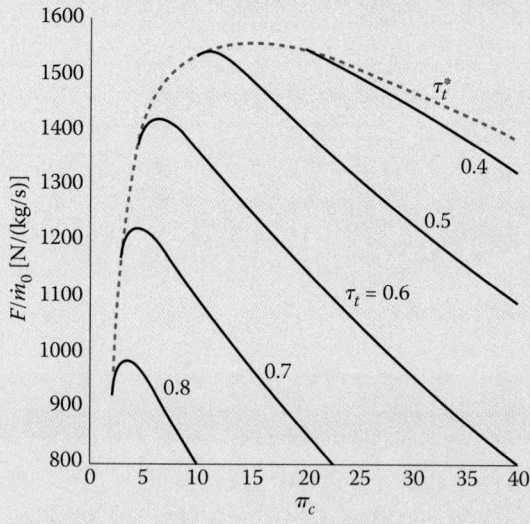

Fig. 5.26a Ideal turboprop performance vs π_c: specific thrust.

(Continued)

Example 5.7 *(Continued)*

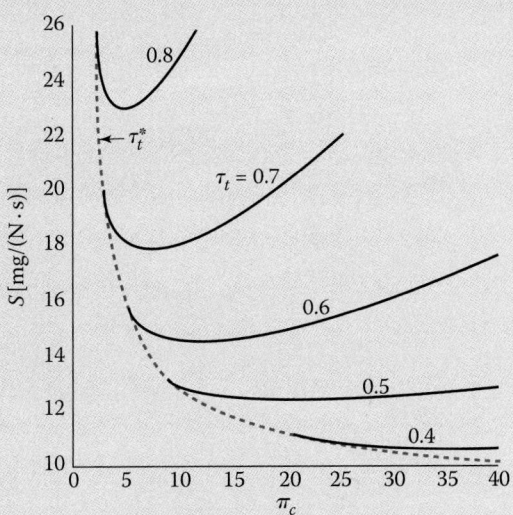

Fig. 5.26b   Ideal turboprop performance vs π_c: thrust specific fuel consumption.

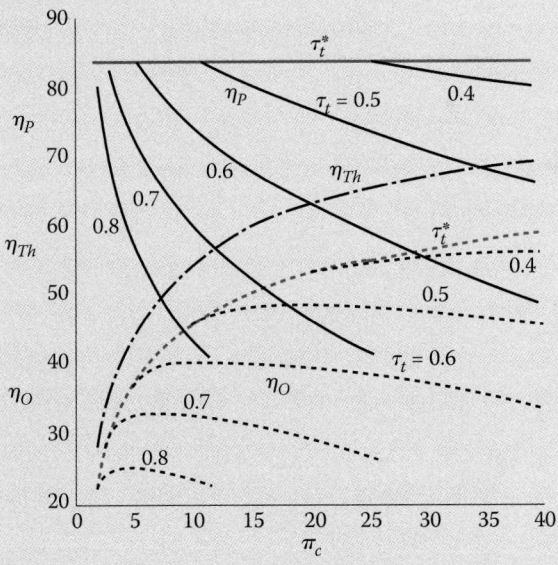

Fig. 5.26c   Ideal turboprop performance vs π_c: efficiencies.

(Continued)

Example 5.7 *(Continued)*

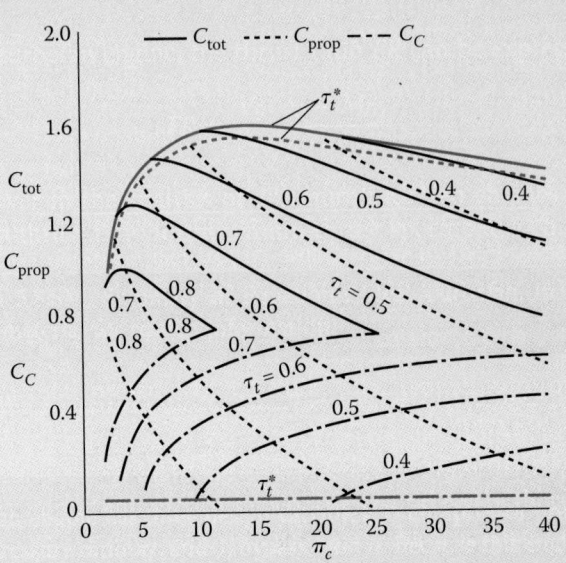

Fig. 5.26d   Ideal turboprop performance vs π_c: work coefficients.

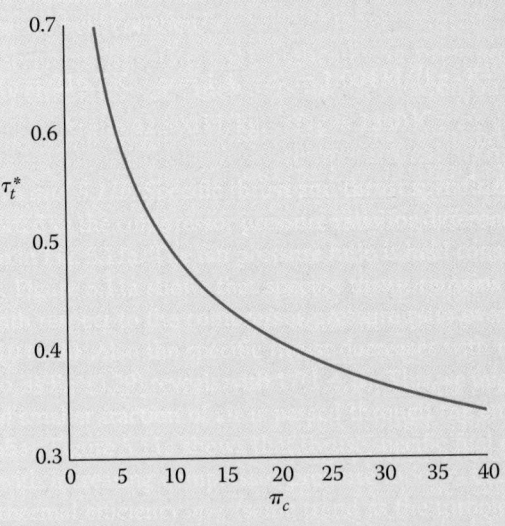

Fig. 5.26e   Ideal turboprop performance vs π_c: optimum τ_r.

(Continued)

Example 5.7 *(Continued)*

pressure ratio for maximum total specific thrust with optimum turbine temperature ratio, therefore, corresponds approximately to 15.

Figure 5.26b shows that the optimum turbine temperature ratio gives the minimum thrust specific fuel consumption. Also, a compressor pressure ratio greater than 15 is desirable to keep fuel consumption low. As you will see in later chapters of the textbook, component losses will increase fuel consumption at compressor pressure ratios higher than 15. As shown in Fig. 5.26c, the optimum turbine temperature ratio gives the maximum propulsive efficiency that is equal to η_{prop}. The optimum turbine temperature ratio τ_t^* is plotted in Fig. 5.26e vs the compressor pressure ratio.

5.13   Ideal Pulse Detonation Engine

The classical, calorically perfect gas, closed thermodynamic cycle analysis should always be considered for the initial evaluation of ideal propulsion devices because it often provides transparent algebraic results that reveal fundamental behavior almost effortlessly. This is particularly true because the *T-s* diagram is not anchored in space or time, but represents the succession of states experienced by every element of the working fluid, and can therefore be applied equally well to steady (e.g., Brayton and Carnot) and unsteady (e.g., Diesel and Otto) cycles.

The pulse detonation engine (PDE) is a contemporary example of a novel propulsion cycle. Because of its inherently unsteady behavior, it has been difficult to conveniently classify and evaluate it relative to its steady-state counterparts by means of unsteady gas dynamic calculations alone. The classical, closed thermodynamic cycle analysis of the ideal PDE is identical to that of the ideal turbojet (see Section 5.6), except that the heat is released by a detonation wave that passes through the working fluid containing fuel that has been brought essentially to rest in the combustor. Please note that the compression process may include both ram compression due to forward flight and mechanical compression driven by the fluid leaving the combustor. The *T-s* diagram for an example PDE is shown in Fig. 5.27, where $\psi = T_3/T_0$ is the cycle thermal compression ratio, $\tilde{q}$ is the dimensionless heat release, and $\gamma = 1.36$. This approach takes advantage of the fact that entropy is a thermodynamic state property, independent of velocity reference frame, as opposed to, for example, the total pressure and total temperature. The complete version of this analysis can be found in Ref. 24.

We therefore direct our attention to the pulse detonation wave process. The reader is advised that the ensuing mathematical analyses will be straightforward, and is therefore encouraged to focus on the physical phenomena involved.

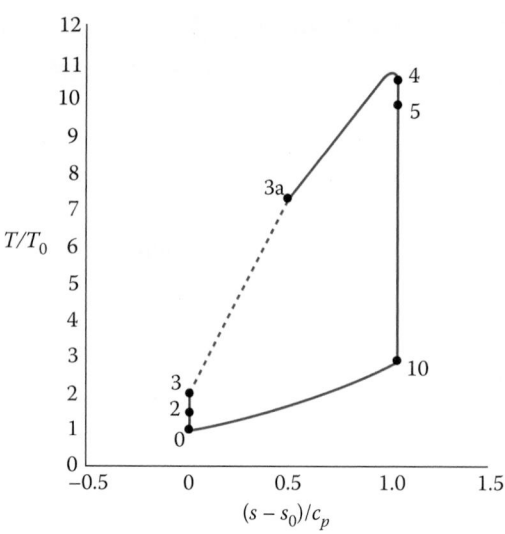

Fig. 5.27 Temperature-entropy diagram for the ideal PDE cycle, for $\psi = 2$, $\tilde{q} = 5$, and $\gamma = 1.36$.

The generally accepted model of a normal detonation wave in PDP devices is that of a Zeldovich/von Neumann/Doering or ZND wave (see Refs. 25 and 26), a compound wave consisting of a normal shock wave progressing into the undisturbed fuel–air mixture, which is nearly at rest at the combustor entry condition (point 3), followed by release of sensible heat in a constant-area region (Rayleigh flow) terminating at point 4. The strength (Mach number, pressure ratio, or temperature ratio) of the leading shock wave, from point 3 to point 3a, is uniquely determined by the initial conditions and the amount of heat added. The entire process is constrained by the Chapman–Jouguet condition, which requires that the local Mach number at the termination of the heat addition region (point 4) be one (sonic or choked flow). The heat addition region is followed by a very complex constant-area region of nonsteady expansion waves, the most important property of which is that they are assumed to be isentropic. This ZND wave structure is stationary in detonation wave coordinates from the undisturbed flow to the end of the heat addition process. To ensure the greatest cycle performance, it is further assumed that 1) the nonsteady expansion (point 4 to point 10) of the detonated mixture is isentropic, 2) every fluid particle experiences the same normal detonation process, and 3) there is no energy penalty to the cycle for whatever spark or ignition torch may be required to initiate the detonation process.

Closed-form, algebraic solutions for the leading normal shock wave (Chapman–Jouguet) Mach number M_{CJ} and the entropy rise in the detonation wave have been derived by Shapiro [27] and others [26, 28] for calorically perfect gases, and are given, respectively, by

$$M_{CJ}^2 = (\gamma + 1)\frac{\tilde{q}}{\psi} + 1 + \sqrt{\left[(\gamma + 1)\frac{\tilde{q}}{\psi} + 1\right]^2 - 1}$$

where

$$\tilde{q} \equiv \frac{q_{supp}}{c_p T_0} = \frac{fh_{PR}}{c_p T_0} \tag{5.85}$$

and

$$\frac{s_4 - s_3}{c_p} = -\ell n \left[M_{CJ}^2 \left(\frac{\gamma + 1}{1 + \gamma M_{CJ}^2}\right)^{\frac{\gamma+1}{\gamma}}\right] \tag{5.86}$$

Consequently, the constant-pressure heat rejected and the cycle thermal efficiency become

$$q_{rej} = h_{10} - h_0 = c_p(T_{10} - T_0) = c_p T_0 \left(e^{\frac{s_{10} - s_0}{c_p}} - 1\right)$$

$$= c_p T_0 \left(e^{\frac{s_4 - s_3}{c_p}} - 1\right) = c_p T_0 \left[\frac{1}{M_{CJ}^2}\left(\frac{1 + \gamma M_{CJ}^2}{\gamma + 1}\right)^{\frac{\gamma+1}{\gamma}} - 1\right] \tag{5.87}$$

and

$$\eta_{Th} = 1 - \frac{\dfrac{1}{M_{CJ}^2}\left(\dfrac{1 + \gamma M_{CJ}^2}{\gamma + 1}\right)^{\frac{\gamma+1}{\gamma}} - 1}{\tilde{q}} \tag{5.88}$$

Because

$$\frac{F}{\dot{m}_0} \equiv \frac{1}{g_c}\left[\sqrt{V_0^2 + 2\eta_{th}q_{supp}} - V_0\right] \tag{5.89}$$

and

$$S \equiv \frac{\dot{m}_f}{F} = \frac{f\dot{m}_0}{F} = \frac{f}{F/\dot{m}_0} \tag{5.90}$$

then generalized ideal PDE performance information, such as that shown in Figs. 5.28 and 5.29, are easily calculated. The ideal Brayton (ramjet/turbojet) is included for reference.

In summary, Figs. 5.28 and 5.29 clearly show that the thermal compression inherent in the PDE normal shock wave, albeit irreversible (see Fig. 5.27), leads to superior propulsion performance for the ideal PDE cycle vs the ideal

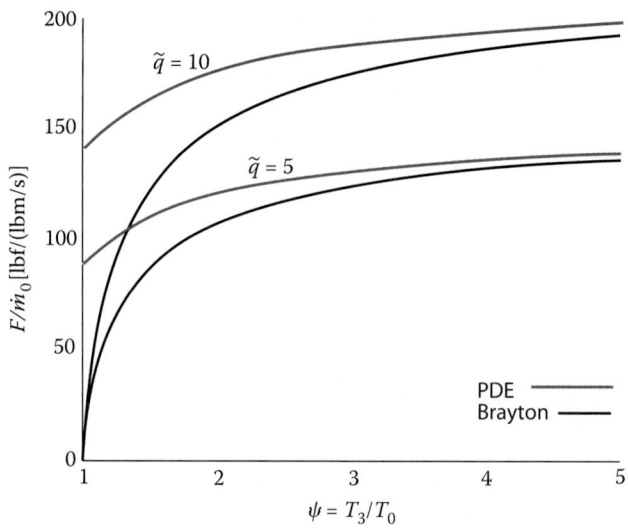

Fig. 5.28   Specific thrust $F/\dot{m}_0$ of ideal PDE and Brayton cycles as functions of ψ, for $\tilde{q} = 5$ and 10, $\gamma = 1.36$, and vehicle speed $V_0 = 0$.

Brayton cycle when $\psi = T_3/T_0$ is small. However, when $\psi \geq$ about 3 (due to combined ram and mechanical compression), the difference between PDE and Brayton performance is significantly reduced. Readers are now empowered to freely investigate comparisons of their own choice.

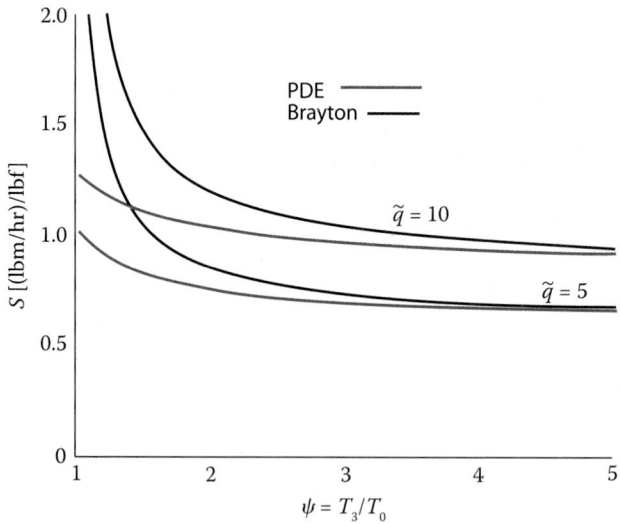

Fig. 5.29   Specific fuel consumption S of ideal PDE and Brayton cycles as functions of ψ, for $\tilde{q} = 5$ and 10, $\gamma = 1.36$, $h_{PR} = 19{,}000$ Btu/lbm, and $V_0 = 0$.

5.14 Summary: PCA of Ideal Engines

The primary goals of this chapter were to introduce you to and apply the general steps of parametric cycle analysis (PCA) to the most common types of aircraft gas turbine engines consisting of ideal engine components. (Refer to Section 5.5 for the specific assumptions.) This ideal PCA afforded us the opportunity to explore basic trends primarily in uninstalled specific thrust and thrust specific fuel consumption in the most fundamental manner possible with the understanding that only very general conclusions can be deduced. In doing so, we gained some practice using T-s diagrams, applicable governing cycle equations, and knowledge of fundamental engine operation to help with our understanding of performance trends. The following general observations emerge from PCA of ideal engines:

- A cycle analysis is a thermodynamic analysis in which the behavior of each component is represented by the change in aero-thermodynamic properties (total and static) it produces.
- The main goal of a PCA ("ideal" or "real") is to narrow down your range of design parameters in terms of design limits at some reference condition or set of reference conditions (altitude, Mach number).
- An ideal PCA helps us determine engine cycle (turbojet, low-bypass turbofan, high-bypass turbofan, etc.) because of the influence of required mission flight conditions (altitude, Mach number) on engine performance.
- The influence of compressor total pressure ratio π_c on both specific thrust and thrust specific fuel consumption S is striking, but not surprising given that this is essentially a Brayton cycle analysis (see Section 4.5).
- The influence of bypass ratio α is equally striking and also not surprising given how strongly it affects the propulsive efficiency. Whereas the thermal efficiency is completely determined by the core (and flight conditions), increasing the bypass ratio increases propulsive efficiency by bringing both the core and bypass exhaust gas velocities closer to the free stream velocity. It does so, of course, at the peril of reduced specific thrust, thereby necessitating an engine of increased diameter to capture higher air mass flow rates to achieve desired mission thrust levels.
- Fan total pressure ratio π_f was the third design parameter we examined. Its influence on specific thrust and thrust specific fuel consumption is also remarkable. We saw that for given values of the other design parameters, there is an optimum choice for π_f in terms of maximizing specific thrust and minimizing S. That optimum value increases with decreased bypass ratio. Given the impact of π_f on both the bypass and core exhaust velocities, this is intuitively appealing.
- For the turboprop cycle, the ideal PCA once again revealed an optimum design choice, this time for the turbine temperature ratio τ_t (indicative of the turbine work per unit mass) in terms of minimizing S. It is often the case with turboprops that the propeller receives its shaft power from a

turbine separate from the core turbine (just like in a turbofan), and in this case, we of course would want to find the optimum choice for the low-pressure turbine temperature ratio τ_{tL}. We also saw the strong interplay between the compressor pressure ratio π_c and turbine total temperature ratio τ_t, which one would expect given that the choices for these two parameters largely dictates the core/propeller work split.

Problems

5.1   The ideal ramjet engine shown in Fig. P5.1 can be modeled as a turbojet with a compressor pressure ratio of 1. For an ideal ramjet engine, do the following:

a) Show that $M_9 = M_0$.

b) Develop an expression for specific thrust in terms of τ_λ, τ_r, a_0, M_0, and g_c.

c) Starting with an energy balance of the combustion, show that the fuel/air ratio is given by

$$f = \frac{c_p T_0}{h_{PR}}(\tau_\lambda - \tau_r)$$

d) Once the inlet flow is supersonic, the exit nozzle of the ideal ramjet is choked and M_8 is 1. Using the mass flow parameter at station 8, calculate the variation in F/A_8 with flight Mach number at the following conditions:

$$T_{t4} = 2000\,\text{K}, \quad \text{Alt} = 15\,\text{km}, \quad \gamma = 1.4,$$
$$c_p = 1.004\,\text{kJ/(kg} \cdot \text{K)}, \quad M_0 = 2\,\text{to}\,6$$

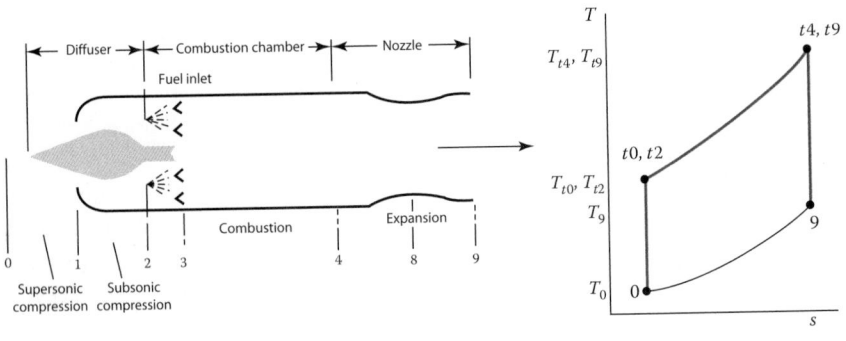

Fig. P5.1   Ideal ramjet.

5.2   Calculate the variation with T_{t4} of exit Mach number, exit velocity, specific thrust, fuel/air ratio, and thrust specific fuel consumption of an ideal turbojet engine for compressor pressure ratios of 10 and 20 at a flight Mach number of 2 and $T_0 = 390°R$. Perform calculations at T_{t4} values of 4400, 4000, 3500, and 3000°R. Use $h_{PR} = 18,400$ Btu/lbm, $c_p = 0.24$ Btu/(lbm · °R), and $\gamma = 1.4$. Compare your results with the output of the PARA computer program.

5.3 Calculate the variation with T_{t4} of exit Mach number, exit velocity, specific thrust, fuel/air ratio, and thrust specific fuel consumption of an ideal turbojet engine for compressor pressure ratios of 10 and 20 at a flight Mach number of 2 and $T_0 = 217$ K. Perform calculations at T_{t4} values of 2400, 2200, 2000, and 1800 K. Use $h_{PR} = 42,800$ kJ/kg, $c_p = 1.004$ kJ/(kg · K), and $\gamma = 1.4$. Compare your results with the output of the PARA computer program.

5.4 Show that the thermal efficiency for an ideal turbojet engine is given by Eq. (5.9).

5.5 An ideal turbojet with a compressor pressure ratio of 24 operates at inlet temperature T_{t2} of 290 K (522°R). Determine the temperature increase across the compressor and the temperature decrease across the turbine.

5.6 Show that the thermal efficiency for an ideal afterburning turbojet is given by Eq. (5.26).

5.7 For the ideal turbojet of Fig. 5.2a, take the partial derivative of the specific thrust with respect to the compressor temperature ratio τ_c and show that the maximum specific thrust is given at the following compressor temperature ratio:

$$(\tau_c)_{\max F/\dot{m}_0} = \sqrt{\tau_\lambda}/\tau_r$$

5.8 A major shortcoming of the ramjet engine is the lack of static thrust. To overcome this, it is proposed to add a compressor driven by an electric motor, as shown in the model engine in Fig. P5.2. For this new ideal engine configuration, show the following:

a) The specific thrust is given by

$$\frac{F}{\dot{m}_0} = \frac{a_0}{g_c}\left[\sqrt{\frac{2}{\gamma-1}\frac{\tau_\lambda}{\tau_r\tau_c}(\tau_r\tau_c - 1)} - M_0\right]$$

b) The compressor power requirement is given by

$$\dot{W}_c = \dot{m}_0 c_p T_0 \tau_r(\tau_c - 1)$$

c) The static thrust of this engine at the following conditions:

$$T_0 = 518.7°R, \quad T_{t4} = 3200°R, \quad \pi_c = 4,$$
$$c_p = 0.24 \, Btu/(lbm \cdot °R), \quad \gamma = 1.4$$

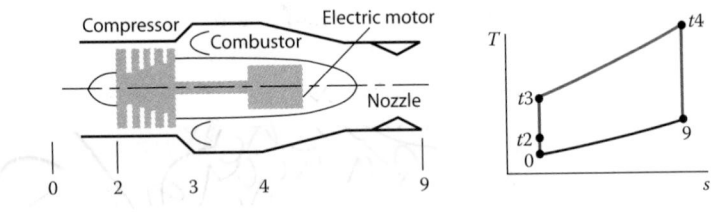

Fig. P5.2 Model engine.

5.9   Determine the optimum compressor pressure ratio (see Problem 5.6), specific thrust, and thrust specific fuel consumption for an ideal turbojet engine giving the maximum specific thrust at the following conditions:

$$M_0 = 2.1, \quad T_0 = 220 \, K, \quad T_{t4} = 1700 \, K, \quad h_{PR} = 42,00 \, kJ/kg,$$

$$c_p = 1.004 \, kJ/(kg \cdot K), \quad \gamma = 1.4$$

5.10   Show that the thermal efficiency for an ideal turbojet engine with optimum compressor pressure ratio is given by $\eta_{Th} = 1 - 1/\sqrt{\tau_\lambda}$. Use results of Problem 5.7.

5.11   For the ideal afterburning turbojet of Fig. 5.4a, take the partial derivative of the specific thrust with respect to the compressor temperature ratio τ_c and show that the maximum specific thrust is given at the following compressor temperature ratio:

$$(\tau_c)_{max \, F/\dot{m}_0 \, AB} = \frac{1}{2}\left(\frac{\tau_\lambda}{\tau_r} + 1\right)$$

5.12   Show that the thermal efficiency for an ideal turbofan engine is given by Eq. (5.9).

5.13   Compare the performance of three ideal turbofan engines with an ideal turbojet engine at two flight conditions by completing Table P5.1. The first flight condition (case 1) is at a flight Mach number of 0.8 at an altitude of 40,000 ft where $T_{t4} = 2520°R$, and the second flight condition (case 2) is at a flight Mach number of 2.5 and an altitude of 50,000 ft where $T_{t4} = 3000°R$. Note that the fuel/air ratio need be calculated only once for each case because it is not a function of α or π_f. The following additional design information is given:

$$\pi_c = 20, \quad \gamma = 1.4, \quad c_p = 0.24\,\text{Btu}/(\text{lbm} \cdot °R),$$
$$h_{PR} = 18{,}400\,\text{Btu/lbm}$$

Table P5.1

Engine	α	π_f	V_9/a_0	V_{19}/a_0	F/m_0	f	S
Turbofan							
(a) Case 1	5	2					
Case 2	1	4					
(b) Case 1, α^*		2					
Case 2, α^*		4					
(c) Case 1, π_f^*	5						
Case 2, π_f^*	1						
Turbojet							
(a) Case 1	0	n/a		n/a			
Case 2	0	n/a		n/a			

5.14   Repeat Problem 5.13 with the first flight condition (case 1) at a flight Mach number of 0.8 and an altitude of 12 km where $T_{t4} = 1350$ K and the second flight condition (case 2) at a flight Mach number of 2.5 and an altitude of 18 km where $T_{t4} = 1670$ K. The following additional design information is given:

$$\pi_c = 20, \quad \gamma = 1.4, \quad c_p = 1.004\,\text{kJ}/(\text{kg} \cdot \text{K}),$$
$$h_{PR} = 42{,}800\,\text{kJ/kg}$$

5.15   Calculate the total temperatures at stations 2, 3, 4, 5, and 13 of Problem 5.13, Turbofan a, Case 1.

5.16   Calculate the total temperatures at stations 2, 3, 4, 5, and 13 of Problem 5.13, Turbofan a, Case 2.

5.17   Show that the propulsive efficiency for an ideal turbofan engine with optimum bypass ratio is given by Eq. (5.42).

5.18   Show that the propulsive efficiency for an ideal turbofan engine with optimum fan pressure ratio is given by Eq. (5.47).

5.19 For an ideal turbofan engine, the maximum value of the bypass ratio corresponds to $V_9 = V_0$.

a) Starting with Eq. (5.34), show that this maximum bypass ratio is given by

$$\alpha_{max} = \frac{\tau_\lambda + 1 - \tau_r \tau_c - \tau_\lambda/(\tau_r \tau_c)}{\tau_r(\tau_f - 1)}$$

b) Show that the propulsive efficiency for this maximum bypass ratio is given by

$$\eta_P = \frac{2}{\sqrt{(\tau_r \tau_f - 1)/(\tau_r - 1)} + 1}$$

5.20 Under certain conditions, it is desirable to obtain power from the freestream. Consider the ideal air-powered turbine shown in Fig. P5.3. This cycle extracts power from the incoming airstream. The incoming air is slowed down in the inlet and then heated in the combustor before going through the turbine. The cycle is designed to produce no thrust ($F = 0$), so that $V_9 = V_0$ and $P_9 = P_0$.

a) Starting with Eq. (5.16), show that

$$\tau_t = \frac{1}{\tau_r} + \frac{1}{\tau_\lambda} \frac{\gamma - 1}{2} M_0^2$$

b) Then show that the turbine output power is given by

$$\dot{W}_i = \dot{m} c_p T_0 \left(\frac{\tau_\lambda}{\tau_r} - 1\right) \frac{\gamma - 1}{2} M_0^2$$

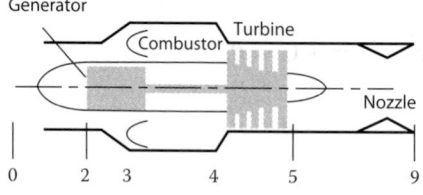
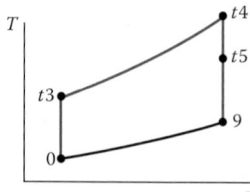

Fig. P5.3 Ideal air-powered turbine.

5.21 In the early development of the turbofan engine, General Electric developed a turbofan engine with an aft fan, as shown in Fig. P5.4. The compressor is driven by the high-pressure turbine, and the aft fan is directly connected to the low-pressure turbine. Consider an ideal turbofan engine with an aft fan.

a) Show that the high-pressure turbine temperature ratio is given by

$$\tau_{tH} = \frac{T_{t4.5}}{T_{t4}} = 1 - \frac{\tau_r}{\tau_\lambda}(\tau_c - 1)$$

b) Show that the low-pressure turbine temperature ratio is given by

$$\tau_{tL} = \frac{T_{t5}}{T_{t4.5}} = 1 - \frac{\alpha \tau_r}{\tau_\lambda \tau_{tH}}(\tau_f - 1)$$

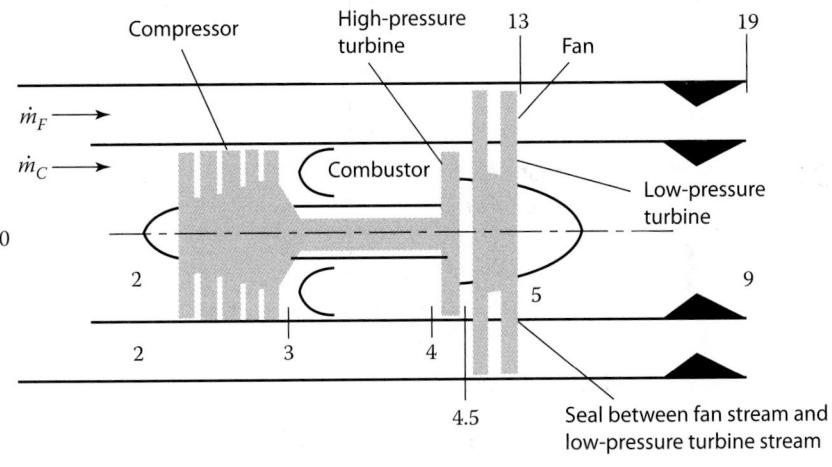

Fig. P5.4 Turbofan engine with aft fan.

5.22 Starting with Eq. (5.29), show that for known values of M_0, T_0, T_{t4}, τ_c, and τ_f, the bypass ratio α giving a specified value of specific thrust $F/\dot{m}_0$ is given by the solution to the quadratic equation

$$A\alpha^2 + B\alpha + C = 0$$

where

$$A = \frac{\gamma - 1}{2}D^2$$

$$B = (\gamma - 1)\left(D^2 + D\frac{V_{19}}{a_0}\right) + \tau_r(\tau_f - 1)$$

$$C = \frac{\gamma - 1}{2}\left(D^2 + 2D\frac{V_{19}}{a_0}\right) - E$$

$$D = \frac{F}{\dot{m}_0}\frac{g_c}{a_0} - \left(\frac{V_{19}}{a_0} - M_0\right)$$

$$E = \tau_\lambda - \tau_r(\tau_c - 1) - \frac{\tau_\lambda}{\tau_r \tau_c} - (\tau_r \tau_f - 1)$$

$$\frac{V_{19}}{a_0} = \sqrt{\frac{2}{\gamma - 1}(\tau_r \tau_f - 1)}$$

5.23 Using the system of equations listed in Problem 5.22, determine the bypass ratio that gives a specific thrust of 400 N/(kg/s) for the following data:

$T_0 = 216.7$ K, $T_{t4} = 1670$ K, $M_0 = 0.8$, $\pi_c = 24$, $\pi_f = 4$,
$\gamma = 1.4$, $c_p = 1.004$ kJ/(kg · K), $h_{PR} = 42{,}800$ kJ/kg

5.24 Using the system of equations listed in Problem 5.22, determine the bypass ratio that gives a specific thrust of 40.0 lbf/(lbm/s) for the following data:

$T_0 = 390°$R, $T_{t4} = 3000°$R, $M_0 = 2$, $\pi_c = 16$, $\pi_f = 5$,
$\gamma = 1.4$, $c_p = 0.24$ Btu/(lbm · °R), $h_{PR} = 18{,}400$ Btu/lbm

5.25 Considerable research and development effort is going into increasing the maximum T_{t4} in gas turbine engines for fighter aircraft. For a mixed-flow turbofan engine with $\pi_c = 16$ at a flight condition of $M_0 = 2.5$ and $T_0 = 216.7$ K, use the PARA computer program to determine and plot the required engine fan pressure, specific thrust, and specific fuel consumption vs T_{t4} over a range of 1600 to 2200 K for bypass ratios of 0.5 and 1. Use $h_{PR} = 42{,}800$ kJ/kg, $c_p = 1.004$ kJ/(kg · K), and $\gamma = 1.4$. Comment on your results in general. Why do the plots of specific fuel consumption vs specific thrust for these engines all fall on one line?

5.26 Considerable research and development effort is going into increasing the maximum T_{t4} in gas turbine engines for fighter aircraft. For a mixed-flow turbofan engine with $\pi_c = 20$ at a flight condition of $M_0 = 2$ and $T_0 = 390°$R, use the PARA computer program to determine and plot the required engine bypass ratio α, specific thrust, and specific fuel consumption vs T_{t4} over a range of 3000 to 4000°R for fan pressure ratios of 2 and 5. Use $h_{PR} = 18{,}400$ Btu/lbm, $c_p = 0.24$ Btu/(lbm · °R), and $\gamma = 1.4$. Comment on your results.

5.27 Calculate the bypass ratio and the total temperature for an ideal mixed flow turbofan engine at engine stations 2, 3, 5, 13, and 6A for $\pi_c = 24$, $\pi_f = 3.5$, $T_{t2} = 288$ K, and $T_{t4} = 2000$ K.

5.28   Calculate the bypass ratio and the total temperature for an ideal mixed flow turbofan engine at engine stations 2, 3, 5, 13, and 6A for $\pi_c = 20$, $\pi_f = 3.2$, $T_{t2} = 520°$R, and $T_{t4} = 3200°$R.

5.29   Show that the thermal efficiency of the afterburning mixed-flow turbofan engine given by Eq. (5.63) can be rewritten as

$$\eta_{Th} = \frac{c_p T_0}{f_0 h_{PR}} \left[\tau_{\lambda AB} \left(1 - \frac{1}{\tau_r \tau_f} \right) - (\tau_r - 1) \right]$$

5.30 `For an afterburning mixed-flow turbofan engine with $\pi_c = 16$ and $T_{t7} = 2200$ K at a flight condition of $M_0 = 2.5$ and $T_0 = 216.7$ K, use the PARA computer program to determine and plot the required engine fan pressure, specific thrust, and specific fuel consumption vs T_{t4} over a range of 1600 to 2200 K for bypass ratios of 0.5 and 1. Use $h_{PR} = 42{,}800$ kJ/kg, $c_p = 1.004$ kJ/(kg · K), and $\gamma = 1.4$. Comment on your results in general. Why do the plots of specific fuel consumption vs specific thrust for these engines all fall on one line? Compare these results to those for Problem 5.25, and comment on the differences.

5.31   For an afterburning mixed-flow turbofan engine with $\pi_c = 20$ and $T_{t7} = 4000°$R at a flight condition of $M_0 = 2$ and $T_0 = 390°$R, use the PARA computer program to determine and plot the required engine bypass ratio α, specific thrust, and specific fuel consumption vs T_{t4} over a range of 3000 to 4000°R for fan pressure ratios of 2 and 5. Use $h_{PR} = 18{,}400$ Btu/lbm, $c_p = 0.24$ Btu/(lbm · °R), and $\gamma = 1.4$. Comment on your results. Compare these results to those for Problem 5.26, and comment on the changes.

5.32   Calculate T_t's for Problem 5.30 with $\alpha = 0.5$ and $T_{t4} = 2000$ K.

5.33   Calculate T_t's for Problem 5.31 with $\pi_f = 2$ and $T_{t4} = 3200°$R.

5.34   Use the PARA computer program to determine and plot the thrust specific fuel consumption vs specific thrust for the turboprop engine of Example 5.7 for $T_{t4} = 1750$ K over the same range of compressor pressure ratios for turbine temperature ratios of 0.8, 0.7, 0.6, 0.5, 0.4, and optimum.

5.35   Use the PARA computer program to determine and plot the thrust specific fuel consumption vs specific thrust for the turboprop engine over the range of compressor pressure ratios from 2 to 40 at $T_0 = 425°$R, $M_0 = 0.65$, $\gamma = 1.4$, $c_p = 0.24$ Btu/(lbm · °R,

$h_{PR} = 18,400$ Btu/lbm, $T_{t4} = 2460°R$ and $\eta_{prop} = 0.8$ for turbine temperature ratios of 0.8, 0.7, 0.6, 0.5, 0.4, and optimum.

5.36 Calculate the total temperatures in an ideal turboprop engine with a compressor pressure ratio of 15, T_{t2} of 290 K, T_{t4} of 1500 K, and optimum τ_t (assume $\eta_{prop} = 0.8$).

5.37 Calculate the total temperatures in an ideal turboprop engine with a compressor pressure ratio of 20, T_{t2} of 520°R, T_{t4} of 3000°R, and optimum τ_t (assume $\eta_{prop} = 0.8$).

Problems for Supporting Material

SM5.1 Use the PARA computer program to determine and plot the power specific fuel consumption vs specific power for the turboshaft engine with regeneration over compressor pressure ratios of 4 to 18 at $T_0 = 290$ K, $M_0 = 0$, $\gamma = 1.4$, $c_p = 1.004$ kJ/(kg · K), $h_{PR} = 42,800$ kJ/kg, and $x = 1.02$ for combustor exit temperatures T_{t4} of 1300, 1400, 1500, and 1600 K.

SM5.2 Use the PARA computer program to determine and plot the power specific fuel consumption vs specific power for the turboshaft engine with regeneration of Example SM5.1 at $x = 1.02$ over the same range of compressor ratios for combustor exit temperatures T_{t4} of 2400, 2600, 2800, and 3000°R or equivilent SI T_{t4}'s.

Gas Turbine Design Problems

5.D1 You are to determine the range of compressor pressure ratios and bypass ratios for ideal turbofan engines that best meet the design requirements for the hypothetical passenger aircraft, the HP-1.

   Hand-Calculate Ideal Performance (HP-1 Aircraft). Using the parametric cycle analysis equations for an ideal turbofan engine with $T_{t4} = 1560$ K, hand-calculate the specific thrust and thrust specific fuel consumption for an ideal turbofan engine with a compressor pressure ratio of 36, fan pressure ratio of 1.6, and bypass ratio of 10 at the 0.83 Mach, 11-km-altitude cruise condition. Assume $\gamma = 1.4$, $c_p = 1.004$ kJ/(kg · K), and $h_{PR} = 42,800$ kJ/kg. Compare your answers to results from the parametric cycle analysis program PARA.

   Computer-Calculate Ideal Performance (HP-1 Aircraft). For the 0.83-Mach, 11-km-altitude cruise condition, determine the performance available from turbofan engines. This part of the analysis is accomplished by using the PARA computer program with

$T_{t4} = 1560$ K. Specifically, you are to vary the compressor pressure ratio from 20 to 40 in increments of 2. Fix the fan pressure ratio at your assigned value of ___. Evaluate bypass ratios of 4, 6, 8, 10, 12, and the optimum value. Assume $\gamma = 1.4$, $c_p = 1.004$ kJ/(kg · K), and $h_{PR} = 42,800$ kJ/kg.

Calculate Minimum Specific Thrust at Cruise (HP-1 Aircraft). You can calculate the minimum uninstalled specific thrust at cruise based on the following information:

1. The thrust of the two engines must be able to offset drag at 0.83-Mach and 11 km altitude and have enough excess thrust for P_s of 1.5 m/s. Determine the required installed thrust to attain the cruise condition using Eq. (1.28). Assuming $\phi_{inlet} + \phi_{noz} = 0.02$, determine the required uninstalled thrust.

2. Determine the maximum mass flow into the 2.2-m-diam inlet for the 0.83-Mach, 11-km-altitude flight condition, using the equation given in the background section for this design problem in Chapter 1.

3. Using the results of steps 1 and 2, calculate the minimum uninstalled specific thrust at cruise.

4. Perform steps 2 and 3 for inlet diameters of 2.5, 2.75, 3.0, 3.25, and 3.5 m.

Select Promising Engine Cycles (HP-1 Aircraft). Plot thrust specific fuel consumption vs specific thrust (thrust per unit mass flow) for the engines analyzed in the preceding. Plot a curve for each bypass ratio, and cross-plot the values of the compressor pressure ratio (see Fig. P5.D1). The result is a *carpet plot* (a multivariable plot) for the cruise condition. Now draw a dashed horizontal line on the carpet plot corresponding to the maximum allowable uninstalled thrust specific fuel consumption S_{max} for the cruise condition (determined in the Chapter 1 portion of this design problem). Draw a dashed vertical line for each minimum uninstalled specific thrust determined in the preceding. Your carpet plots will look similar to the example shown in Fig. P5.D1. What ranges of bypass ratio and compressor pressure ratio look most promising?

5.D2   You are to determine the ranges of compressor pressure ratio and bypass ratio for ideal mixed-flow turbofan engines that best meet the design requirements for the hypothetical fighter aircraft, the HF-1.

Hand-Calculate Ideal Performance (HF-1 Aircraft). Using the parametric cycle analysis equations for an ideal mixed-flow turbofan engine with $T_{t4} = 3250°$R, hand-calculate the specific thrust and thrust specific fuel consumption for an ideal turbofan engine with a compressor pressure ratio of 25 and bypass ratio of 0.5 at the 1.6-Mach, 40-kft-altitude supercruise condition. Assume $\gamma = 1.4$, $c_p = 0.24$ Btu/(lbm · °R), and $h_{PR} = 18{,}400$ Btu/lbm. Compare your answers to the results from the parametric cycle analysis program PARA.

Computer-Calculate Ideal Performance (HF-1 Aircraft). For the 1.6-Mach, 40-kft-altitude supercruise condition, determine the performance available from turbofan engines. This part of the analysis is accomplished by using the PARA computer program with $T_{t4} = 3250$°R. Specifically, you are to vary the bypass ratio from 0.1 to 1.0 in increments of 0.05. Evaluate compressor pressure ratios of 16, 18, 20, 22, 24, and 28. Assume $\gamma = 1.4$, $c_p = 0.24$ Btu/ (lbm · °R), and $h_{PR} = 18{,}400$ Btu/lbm.

Calculate Minimum Specific Thrust at Cruise (HF-1 Aircraft). You can calculate the minimum uninstalled specific thrust at supercruise based on the following information:

1. The thrust of the two engines must be able to offset drag at 1.6 Mach, 40 kft altitude, and 92% of takeoff weight. Assuming $\phi_{\text{inlet}} + \phi_{\text{noz}} = 0.05$, determine the required uninstalled thrust for each engine.

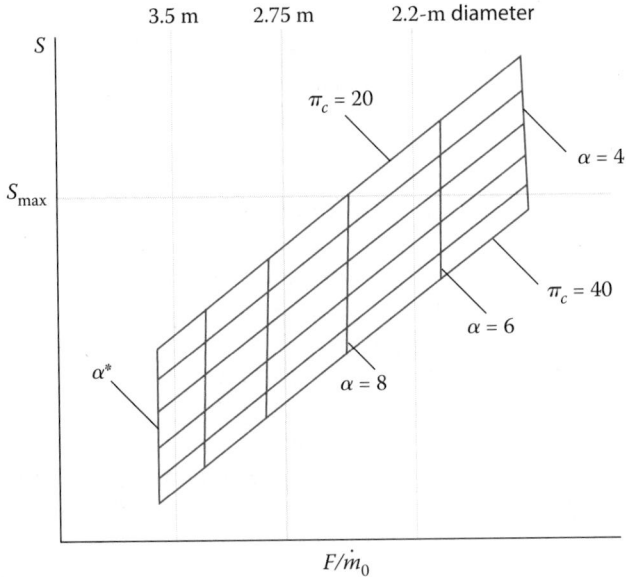

Fig. P5.D1 Example carpet plot for HP-1 aircraft engine.

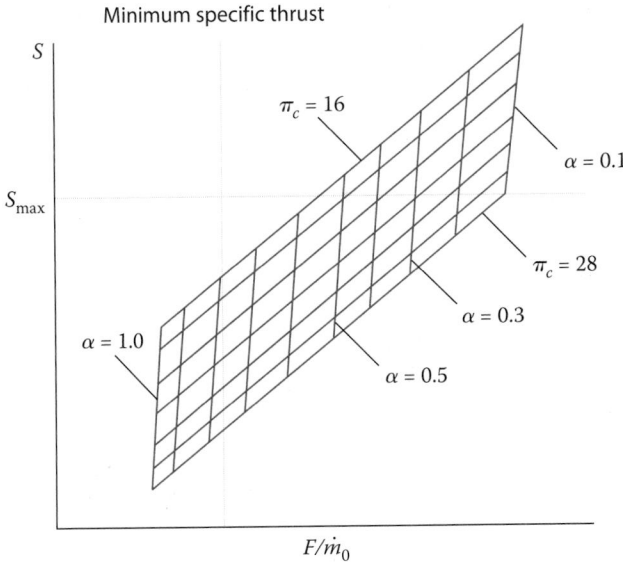

Fig. P5.D2 Example carpet plot for HF-1 aircraft engine.

2. The maximum mass flow into a 5-ft^2 inlet for the 1.6-Mach, 40-kft-altitude flight condition is $\dot{m} = \rho AV = \sigma \rho_{ref} AMa = (0.2471 - 0.07647)(5)(1.6 \times 0.8671 \times 1116) = 146.3$ lbm/s.

3. Using the results of steps 1 and 2, calculate the minimum uninstalled specific thrust at supercruise.

Select Promising Engine Cycles (HF-1 Aircraft). Plot the thrust specific fuel consumption vs specific thrust (thrust per unit mass flow) for the engines analyzed in the preceding. Plot a curve for each bypass ratio, and cross-plot the values of the compressor pressure ratio (see Fig. P5.D2). The result is a carpet plot (a multivariable plot) for the supercruise condition. Now draw a dashed horizontal line on the carpet plot corresponding to the maximum allowable uninstalled thrust specific fuel consumption S_{max} for the cruise condition (determined in the Chapter 1 portion of this design problem). Draw a dashed vertical line for the minimum uninstalled specific thrust determined in the preceding section. Your carpet plots will look similar to the example shown in Fig. P5.D2. What ranges of bypass ratio and compressor pressure ratio look most promising?

Chapter 6 / Component Performance

> The gas generator or core engine (compressor, burner and turbine) is the key to achieving higher performance since it provides the high energy gases used in propulsion systems.
>
> *Bernard Koff, Celebration of the Golden Anniversary of Jet-Powered Flight,*
> *23 August 1989, Dayton, OH*

6.1 Introduction

In Chapter 5, we idealized the engine components and assumed that the working fluid behaved as a perfect gas with constant specific heats. These idealizations and assumptions permitted the basic cycle analysis of several types of engines and the analysis of engine performance trends. In this chapter, we will develop the analytical tools that allow us to use realistic assumptions as to component losses and to include the variation of specific heats.

6.2 Variation in Gas Properties

The enthalpy h and specific heat at constant pressure c_p for air (modeled as a perfect gas) are functions of temperature. Also, the enthalpy h and specific heat at constant pressure c_p for a typical hydrocarbon fuel JP-8 and air combustion products (modeled as a perfect gas) are functions of temperature and the fuel/air ratio f. The variations of properties h and c_p for fuel/air combustion products vs temperature are presented in Figs. 6.1a and 6.1b, respectively. The ratio of specific heats γ for fuel/air combustion products is also a function of temperature and of fuel/air ratio. A plot of γ is shown in Fig. 6.2. These figures are based on Eq. (2.64) and the coefficients of Table 2.4. Note that both h and c_p increase and γ decreases with temperature and the fuel/air ratio. Our models of gas properties in the engines need to include changes in both c_p and γ across components where the changes are significant.

In Chapter 7, we will include the variation in c_p and γ through the engine. To simplify the algebra, we will consider c_p and γ to have constant representative values through all engine components except the main burner (combustor) and afterburner, if applicable. The values of c_p and γ will be

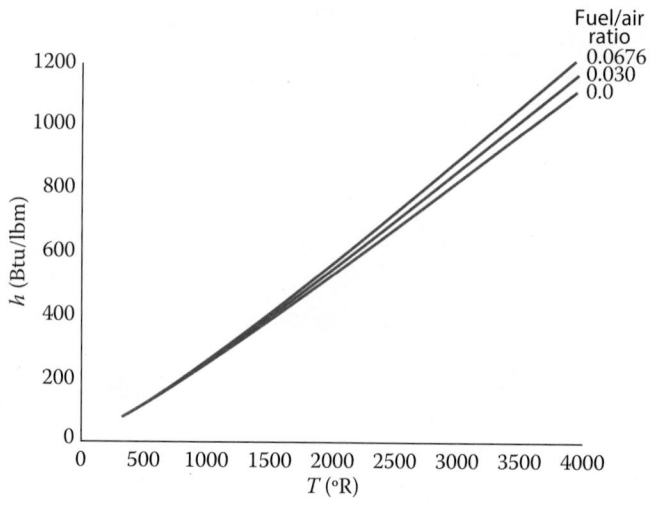

Fig. 6.1a Enthalpy vs temperature for JP-8 and air combustion products.

allowed to change across both of these components. Thus we will approximate c_p as c_{pc} (a constant average value for the air upstream of the burner) and as c_{pt} (a constant average value for the gases downstream of the burner). Likewise, γ will be γ_c upstream of the burner and γ_t downstream of the burner. The release of thermal energy in the combustion process affects the values of c_{pt} and γ_t, but these two are related by

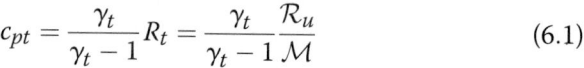

$$c_{pt} = \frac{\gamma_t}{\gamma_t - 1} R_t = \frac{\gamma_t}{\gamma_t - 1} \frac{\mathcal{R}_u}{\mathcal{M}} \qquad (6.1)$$

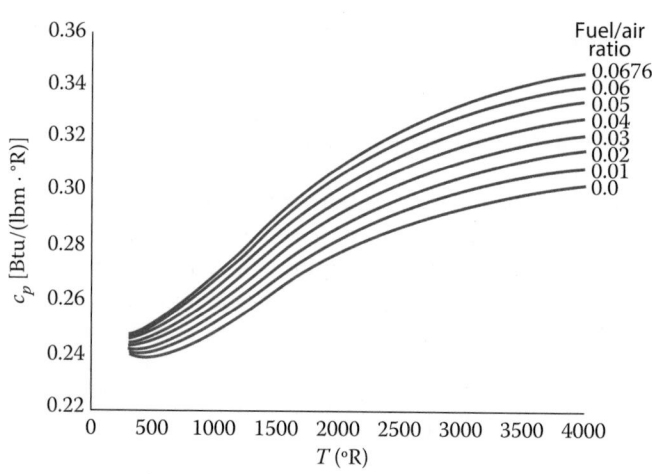

Fig. 6.1b Specific heat c_p vs temperature for JP-8 and air combination products.

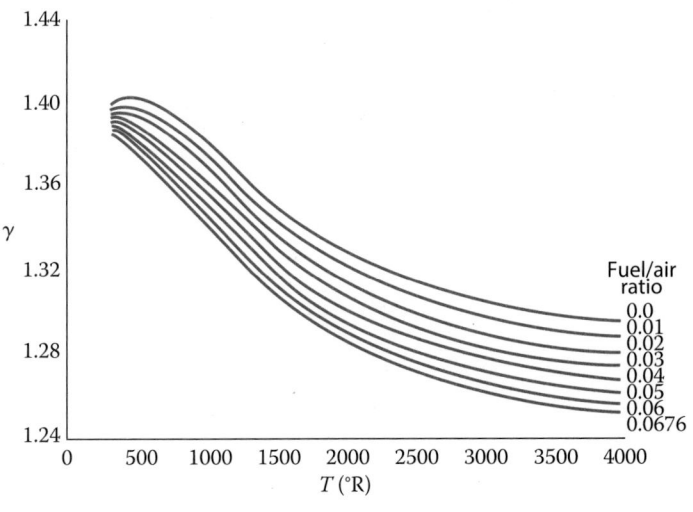

Fig. 6.2   Ratio of specific heats γ vs temperature for JP-8 and air combustion products.

where

$\mathcal{R}_u$ = universal gas constant
$\mathcal{M}$ = molecular weight

Thus if the chemical reaction causes the molecular vibrational modes to be excited but does not cause appreciable dissociation, then the molecular weight $\mathcal{M}$ will be approximately constant. In this case, a reduction in γ is directly related to an increase in c_p by the formula

$$\frac{c_{pt}}{c_{pc}} = \frac{\gamma_t}{\gamma_t - 1} \frac{\gamma_c - 1}{\gamma_c} \tag{6.2}$$

6.3 Component Performance

In this chapter, each of the engine components will be characterized by *figures of merit* that model the component's performance and facilitate cycle analysis of real airbreathing engines. The total temperature ratio τ, the total pressure ratio π, and the interrelationship between τ and π will be used as much as possible in a component's figure of merit.

6.4 Inlet/Diffuser Pressure Recovery

Inlet losses arise because of the presence of wall friction and shock waves (in a supersonic inlet). These irreversibilities result in a reduction in total pressure so that $\pi_d < 1$. Inlets are adiabatic to a very high degree of

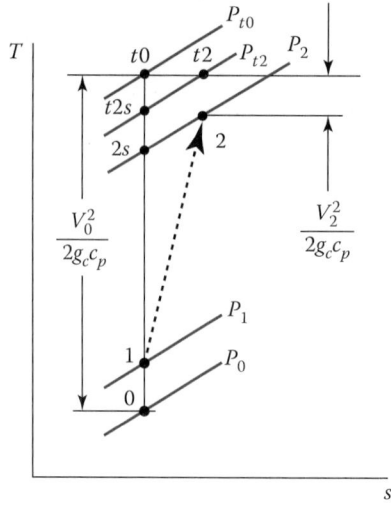

Fig. 6.3   Definition of inlet states.

approximation, and so we have $\tau_d = 1$. The inlet's figure of merit is defined simply as π_d.

The *isentropic efficiency* η_d of the diffuser is defined as (refer to Fig. 6.3)

$$\eta_d = \frac{h_{t2s} - h_0}{h_{t0} - h_0} \simeq \frac{T_{t2s} - T_0}{T_{t0} - T_0} \quad (6.3)$$

This efficiency can be related to τ_r and π_d to give

$$\eta_d = \frac{\tau_r (\pi_d)^{(\gamma-1)/\gamma} - 1}{\tau_r - 1} \quad (6.4)$$

Figure 6.4 gives typical values of π_d for a subsonic inlet. The diffuser efficiency η_d was calculated from π_d by using Eq. (6.4).

In supersonic flight, the flow deceleration in inlets is accompanied by shock waves that can produce a total pressure loss much greater than, and in addition to, the wall friction loss. The inlet's overall total pressure ratio is the product of the shock total pressure ratio and the diffuser total pressure ratio, both of which are less than one.

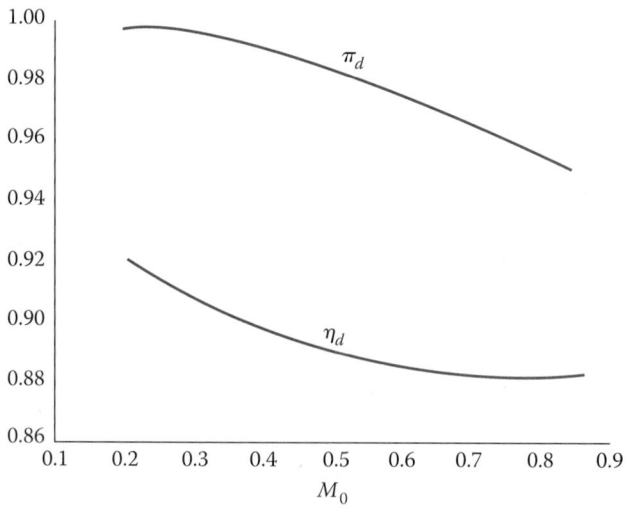

Fig. 6.4   Typical subsonic inlet π_d and η_d.

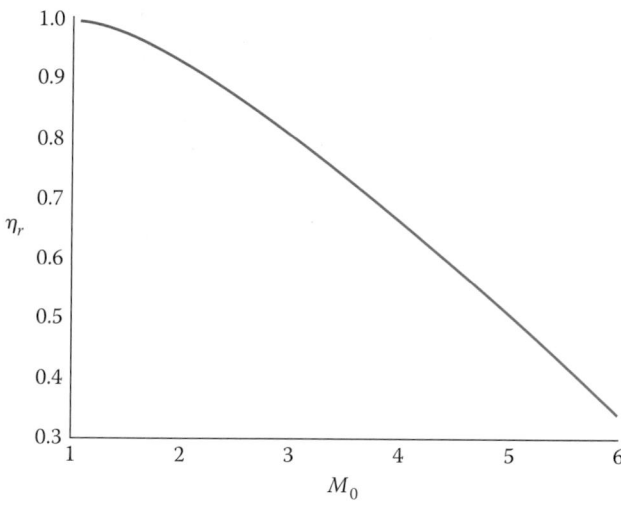

Fig. 6.5   Inlet total pressure recovery η_r of Military Specification 5008B.

We now define $\pi_{d\max}$ as that portion of π_d that is due to wall friction and define η_r as that portion of π_d due to total pressure ratio of the shocks. Thus

$$\pi_d = \pi_{d\max}\,\eta_r \tag{6.5}$$

For subsonic and supersonic flow, a useful reference for the ram recovery η_r is Military Specification 5008B [34] which is expressed as follows:

$$\eta_r = \begin{cases} 1 & M_0 \le 1 \\ 1 - 0.075(M_0 - 1)^{1.35} & 1 < M_0 < 5 \\ \dfrac{800}{M_0^4 + 935} & 5 < M_0 \end{cases} \tag{6.6}$$

Because we often do not yet know the details of the inlet in cycle analysis, it is assumed that Military Specification 5008B applies as a goal for ram recovery. The ram recovery of Military Specification 5008B is plotted in Fig. 6.5 vs M_0.

6.5 Compressor and Turbine Efficiencies

6.5.1 Compressor Isentropic Efficiency

Compressors are, to a high degree of approximation, adiabatic. The overall efficiency used to measure a compressor's performance is the

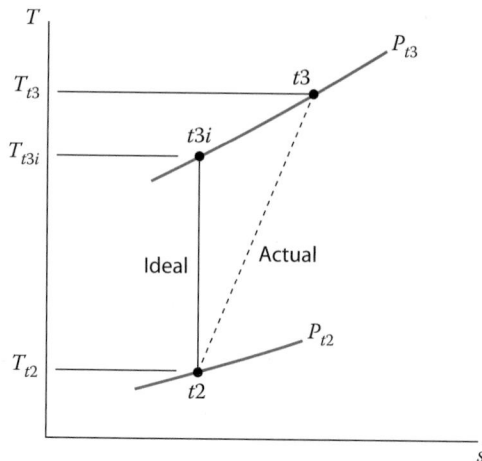

Fig. 6.6   Actual and ideal compressor processes.

isentropic efficiency η_c, defined by

$$\eta_c = \frac{\text{ideal work of compression for given } \pi_c}{\text{actual work of compression for given } \pi_c} \tag{6.7}$$

Figure 6.6 shows both the ideal and actual compression processes for a given π_c on a *T-s* diagram. From the first law of thermodynamics consistent with our assumptions, the actual work per unit mass w_c is $h_{t3} - h_{t2}$ $[=c_p(T_{t3} - T_{t2})]$, and the ideal work per unit mass w_{ci} is $h_{t3i} - h_{r2}$ $[=c_p(T_{t3i} - T_{t2})]$. Here, h_{t3i} is the ideal (isentropic) compressor exit total enthalpy. Writing the isentropic efficiency of the compressor η_c in terms of the thermodynamic properties, we have

$$\eta_c = \frac{w_{ci}}{w_c} = \frac{h_{t3i} - h_{t2}}{h_{t3} - h_{t2}}$$

For a calorically perfect gas, we can write

$$\eta_c = \frac{w_{ci}}{w_c} = \frac{c_p(T_{t3\iota} - T_{t2})}{c_p(T_{t3} - T_{t2})} = \frac{\tau_{ci} - 1}{\tau_c - 1}$$

Here τ_{ci} is the ideal compressor temperature ratio that is related to the compressor pressure ratio π_c by the isentropic relationship for calorically perfect gases (CPGs)

$$\tau_{ci} = \pi_{ci}^{(\gamma-1)/\gamma} = \pi_c^{(\gamma-1)/\gamma} \tag{6.8}$$

Thus we have

$$\eta_c = \frac{\pi_c^{(\gamma-1)/\gamma} - 1}{\tau_c - 1} \tag{6.9}$$

6.5.2 Compressor Stage Efficiency

For a multistage compressor, each stage (set of rotor and stator) will have an isentropic efficiency. Let η_{sj} denote the isentropic efficiency of the jth stage. Likewise, π_{sj} and τ_{sj} represent the pressure ratio and temperature ratio, respectively, for the jth stage. From Eq. (6.9), we can write for the jth stage

$$\eta_{sj} = \frac{\pi_{sj}^{(\gamma-1)/\gamma} - 1}{\tau_{sj} - 1} \tag{6.10}$$

where $\tau_{sj} = T_{tj}/T_{tj-1}$ and $\pi_{sj} = P_{tj}/P_{tj-1}$.

Figure 6.7 shows the process for a multistage compressor. Here η_{sj} can be interpreted as the vertical height from A to B divided by the vertical height from A to C. For counting purposes, subscript 0 outside the parentheses is at the entrance, and subscript N is at the outlet of the compressor. Thus $(P_t)_0 = P_{t2}$, $(T_t)_0 = T_{t2}$, $P_{tN} = P_{tN} = P_{t3}$, and $T_{tN} = T_{t3}$. From Eq. (6.9),

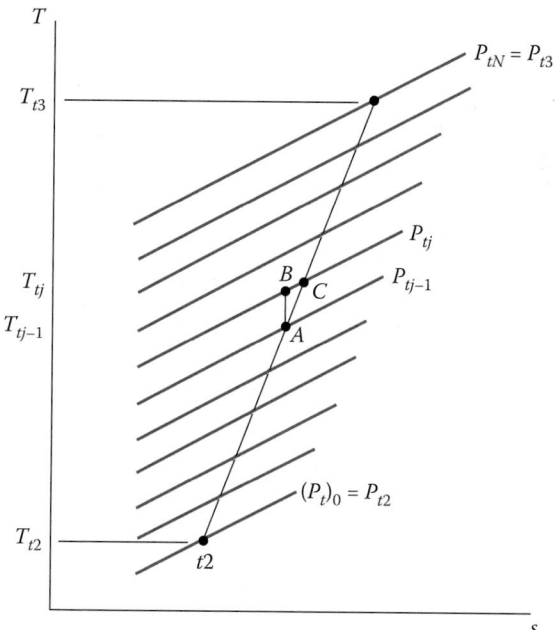

Fig. 6.7 Multistage compressor process and nomenclature.

we have for the overall compressor isentropic efficiency

$$\eta_c = \frac{(P_{t3}/P_{t2})^{(\gamma-1)/\gamma} - 1}{T_{t3}/T_{t2} - 1} = \frac{[P_{tN}/(P_t)_0]^{(\gamma-1)/\gamma} - 1}{T_{tN}/(T_t)_0 - 1} \tag{6.11}$$

From Eq. (6.10), we have

$$\frac{T_{tj}}{T_{tj-1}} = 1 + \frac{1}{\eta_{sj}}\left[\left(\frac{P_{tj}}{P_{tj-1}}\right)^{(\gamma-1)/\gamma} - 1\right]$$

and so

$$\frac{T_{tN}}{(T_t)_0} = \prod_{j=1}^{N}\left\{1 + \frac{1}{\eta_{sj}}\left[\left(\frac{P_{tj}}{P_{tj-1}}\right)^{(\gamma-1)/\gamma} - 1\right]\right\}$$

where $\prod$ means product. We note also the requirement that

$$\frac{P_{tN}}{(P_t)_0} = \prod_{j=1}^{N}\frac{P_{tj}}{P_{tj-1}}$$

where

$$\frac{P_{tN}}{(P_t)_0} \equiv \pi_c$$

Thus Eq. (6.11) becomes

$$\eta_c = \frac{[P_{tN}/(P_t)_0]^{(\gamma-1)/\gamma} - 1}{\prod\limits_{j=1}^{N}\left\{1 + (1/\eta_{sj})\left[(P_{tj}/P_{tj-1})^{(\gamma-1)/\gamma} - 1\right]\right\} - 1} \tag{6.12}$$

We can see from Eq. (6.12) that the isentropic efficiency of a compressor is a function of the compressor pressure ratio, the pressure ratio of each stage, and the isentropic efficiency of each stage. This complex functional form makes the isentropic efficiency of the compressor undesirable for use in cycle analysis. We are looking for a simpler form of the figure of merit that will allow us to vary the compression ratio and still accurately predict the variation of η_c.

Let us consider the special case when each stage pressure ratio and each stage efficiency are the same. In this case

$$\pi_c = \prod_{j=1}^{N} \left(\frac{P_{tj}}{P_{tj-1}} \right) = \pi_s^N$$

Here η_s is the stage efficiency, and π_s is the stage pressure ratio. Also

$$\frac{T_{tj}}{T_{tj-1}} = 1 + \frac{1}{\eta_s} (\pi_s^{(\gamma-1)/\gamma} - 1)$$

and so

$$\frac{T_{tN}}{(T_t)_0} \equiv \tau_c = \left[1 + \frac{1}{\eta_s} (\pi_s^{(\gamma-1)/\gamma} - 1) \right]^N$$

or

$$\tau_c = \left[1 + \frac{1}{\eta_s} (\pi_c^{(\gamma-1)/(\gamma N)} - 1) \right]^N$$

and

$$\begin{aligned} \eta_c &= \frac{\pi_c^{(\gamma-1)/\gamma} - 1}{[1 + (1/\eta_s)(\pi_c^{(\gamma-1)/(\gamma N)} - 1)]^N - 1} \\ &= \frac{\pi_s^{N(\gamma-1)/\gamma} - 1}{[1 + (1/\eta_s)(\pi_s^{(\gamma-1)/\gamma} - 1)]^N - 1} \end{aligned} \qquad (6.13)$$

Note: This is a relationship connecting η_c with η_s for an N-stage compressor with equal stage pressure ratios and equal stage efficiencies.

6.5.3   Compressor Polytropic Efficiency

The *polytropic efficiency* e_c (sometimes referred to as small-stage efficiency) is related to the preceding efficiencies and is defined as

$$e_c = \frac{\text{ideal work of compression for a differential pressure change}}{\text{actual work of compression for a differential pressure change}}$$

Thus

$$e_c = \frac{dw_i}{dw} = \frac{dh_{ti}}{dh_t} = \frac{dT_{ti}}{dT_t}$$

Note that for an ideal compressor, the isentropic relationship gives $T_{ti} = P_{ti}^{(\gamma-1)/\gamma} \times$ constant. Thus

$$\frac{dT_{ti}}{T_t} = \frac{\gamma - 1}{\gamma}\frac{dP_t}{P_t}$$

and

$$e_c = \frac{dT_{ti}}{dT_t} = \frac{dT_{ti}/T_t}{dT_t/T_t} = \frac{\gamma - 1}{\gamma}\frac{dP_t/P_t}{dT_t/T_t}$$

Assuming that the polytropic efficiency e_c is constant, we can obtain a simple relationship between τ_c and π_c as follows:

1. Rewrite the preceding equation as

$$\frac{dT_t}{T_t} = \frac{\gamma - 1}{\gamma e_c}\frac{dP_t}{P_t}$$

2. Integration between states $t2$ and $t3$ gives

$$\ell n\,\frac{T_{t3}}{T_{t2}} = \frac{\gamma - 1}{\gamma e_c}\,\ell n\,\frac{P_{t3}}{P_{t2}}$$

or

$$\tau_c = \pi_c^{(\gamma-1)/(\gamma e_c)} \tag{6.14}$$

For a given level of compressor technology, the polytropic efficiency is essentially constant. Substitution of Eq. (6.14) into Eq. (6.9) gives

$$\eta_c = \frac{\pi_c^{(\gamma-1)/\gamma} - 1}{\tau_c - 1} = \frac{\pi_c^{(\gamma-1)/\gamma} - 1}{\pi_c^{(\gamma-1)/(\gamma e_c)} - 1} \tag{6.15}$$

Equation (6.15) accurately predicts the relationship between the isentropic efficiency of a compressor and the compressor pressure ratio for a given level of technology's polytropic efficiency. This relationship is plotted in Fig. 6.8 for a given value of e_c.

In Fig. 6.8, note how overall compressor isentropic efficiency decreases with increased compressor pressure ratio π_c. The divergence of the constant pressure lines indicated in Fig. 6.6 helps explain this trend. For a given level of technology (given e_c), it will take even more work (ΔT_t) relative to the ideal case to get to a higher pressure ratio because the pressure lines diverge. Increased π_c requires an increased number of stages for a given level of technology. The next section will provide additional insight into the behavior shown in Fig. 6.8.

We will use the polytropic efficiency e_c as the figure of merit for the compressor. Equations (6.14) and (6.15) will be used to obtain the total temperature ratio and isentropic efficiency of the compressor, respectively, in the cycle analysis.

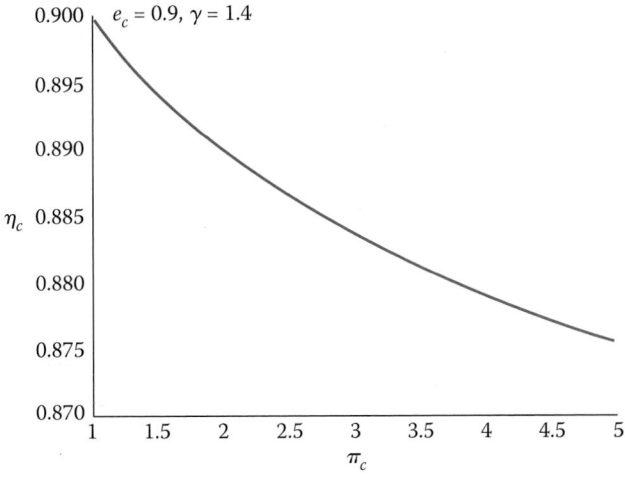

Fig. 6.8   Isentropic efficiency vs compressor pressure ratio for constant polytropic efficiency of 0.9.

6.5.4 Relationship Between Compressor Efficiencies

We have, from Eqs. (6.13) and (6.15), relationships connecting η_c, η_s, and e_c. Oates (Ref. 7) shows that η_s approaches e_c as we let the number of stages get very large and the pressure ratio per stage get very small.

The resulting expression for overall efficiency for a multistage machine is

$$\eta_c = \frac{\pi_c^{(\gamma-1)/\gamma} - 1}{\pi_c^{(\gamma-1)/(\gamma\eta_s)} - 1} \tag{6.16}$$

This expression is identical to Eq. (6.15) with e_c replaced by η_s. Thus for very large N, η_s approaches e_c.

Example 6.1

Say we plan to construct a 16-stage compressor, with each stage pressure ratio the same, given $\pi_c = 25$. Then we have $\pi_s = 25^{1/16} = 1.223$. Say η_s is measured at 0.93. Then, with Eq. (6.14) solved for e_c, we have

$$e_c = \frac{(\gamma-1)/\gamma\,\ell n\,\pi_s}{\ell n[1 + (1/\eta_s)(\pi_s^{(\gamma-1)/\gamma} - 1)]}$$

$$= 0.9320$$

(Continued)

Example 6.1 (Continued)

Then we get two estimates for η_c, one based on constant η_s and another based on constant e_c. From Eq. (6.13), with $\pi_s = 1.223$, we obtain

$$\eta_c = \frac{\pi_c^{(\gamma-1)/\gamma} - 1}{[1 + (1/\eta_s)(\pi_s^{(\gamma-1)/\gamma} - 1)]^N - 1}$$

$$= \frac{25^{1/3.5} - 1}{[1 + (1/0.93)(1.223^{1/3.5} - 1)]^{16} - 1}$$

$$= 0.8965^-$$

From Eq. (6.15) we find, for comparison,

$$\eta_c = \frac{\pi_c^{(\gamma-1)/\gamma} - 1}{\pi_c^{(\gamma-1)/(\gamma e_c)} - 1} = 0.8965^+$$

More simply, if $e_c = \eta_s = 0.93$, then by either Eq. (6.15) or Eq. (6.16), we get $\eta_c = 0.8965$.

The point of all of this is that for a multistage machine, the simplicity and accuracy of using the polytropic efficiency make it a useful concept. Thus from now on we will use Eq. (6.15) to compute the compressor efficiency.

6.5.5 Compressor Stage Pressure Ratio

For a multistage compressor, the energy added is divided somewhat evenly per stage, and each stage increases the total temperature of the flow about the same amount. The total temperature ratio of a stage τ_s that has a total temperature change of ΔT_t can be written as

$$\tau_s = 1 + \frac{\Delta T_t}{T_{ti}}$$

Using the polytropic efficiency e_c to relate the stage pressure ratio π_s to its temperature ratio τ_s, we have

$$\pi_s = (\tau_s)^{(\gamma e_c)/(\gamma-1)} = \left(1 + \frac{\Delta T_t}{T_{ti}}\right)^{(\gamma e_c)/(\gamma-1)} \tag{6.17}$$

Equation (6.17) gives the variation of the compressor stage pressure ratio with stage inlet temperature T_{ti} and total temperature change ΔT_t. This equation shows the decrease in stage pressure ratio with increases in stage inlet temperature for stages with the same total temperature change.

Stage pressure ratio results for $\gamma = 1.4$ and $e_c = 0.9$ are plotted in Fig. 6.9 vs $\Delta T_t/T_{ti}$. By using this figure, a 30 K change with 300 K inlet temperature

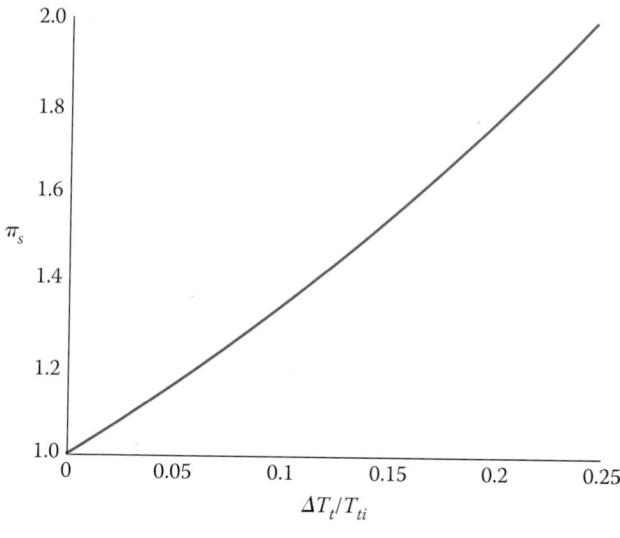

Fig. 6.9 Variation of compressor stage pressure ratio ($\gamma = 1.4$, $e_c = 0.9$).

gives a stage pressure ratio of about 1.35. Likewise, a 60°R change with 1000°R inlet temperature gives a stage pressure ratio of about 1.20. Because stages within a compressor have about the same temperature rise (ΔT_t), Fig. 6.9 helps explain the decrease in stage pressure ratio through a compressor with increasing stage inlet temperature T_{ti}.

Equation (6.17) helps further explain the decrease in compressor isentropic efficiency with higher pressure ratios. Each successive stage in a multi-stage compressor is essentially penalized by the energy addition of the preceding stage. It takes additional work input (additional ΔT_t) to achieve the same pressure rise across each successive compressor stage as revealed by Eq. (6.17). Consequently, η_c decreases with increasing π_c (Fig. 6.8); it costs more (more work) to compress an already compressed gas. To achieve some of the future desired pressure ratios, approaching 30:1 across an HPC and 70:1 overall, will surely require *intercooling* between stages and/or alternative novel concepts. Intercooling involves the use of a heat exchanger to cool the compressed air prior to the next stage or stages of compression. This reduces the needed work input, but does not necessarily increase core thermal efficiency because the temperature of the air entering the combustor would be reduced (good for compressor material considerations though). A lower combustor entrance temperature would require additional energy release in the combustor (additional fuel) to achieve the desired work per unit air mass flow out of the turbine. Thus, an aircraft gas turbine engine with intercooled compressors would likely require the use of regeneration (see Section 6.4) to raise the temperature of the air prior to combustion. This would increase the cycle thermal efficiency, but at the expense of additional weight and complexity associated with the hardware needed for intercooling and regeneration. These tradeoffs would have to be assessed relative to overall aircraft-engine system performance.

6.5.6 Turbine Isentropic Efficiency

Modern turbines are cooled by air taken from the compressors, passed through vanes and rotors, and then remixed with the main flow. From the point of view of the overall flow, the flow is adiabatic; but to be correct, a multiple-stream analysis would have to be applied. Such an analysis is straightforward conceptually, but it is difficult to estimate the various mixing losses and the like that occur. The concept of isentropic efficiency is still utilized in most such analyses (for the mainstream portion of the flow), and in any case, the isentropic efficiency gives a reasonable approximation to the turbine performance when cooling flow rates are small. Hence, in this text, we consider only the adiabatic case.

Analogous to the compressor isentropic efficiency, we define the *isentropic efficiency of the turbine* by

$$\eta_t = \frac{\text{actual turbine work for a given } \pi_t}{\text{ideal turbine work for a given } \pi_t}$$

The actual and ideal expansion processes for a given π_t are shown in Fig. 6.10 on a T-s diagram. From the first law of thermodynamics consistent with our assumptions, the actual turbine work per unit mass is $h_{t4} - h_{t5}$ $[=c_p(T_{t4} - T_{t5})]$, and the ideal turbine work per unit mass is $h_{t4} - h_{t5i}$ $[=c_p(T_{t4} - T_{t5i})]$. Here, T_{t5i} is the total temperature that would exist at the exit of an ideal turbine, as indicated in Fig. 6.10. Writing the isentropic

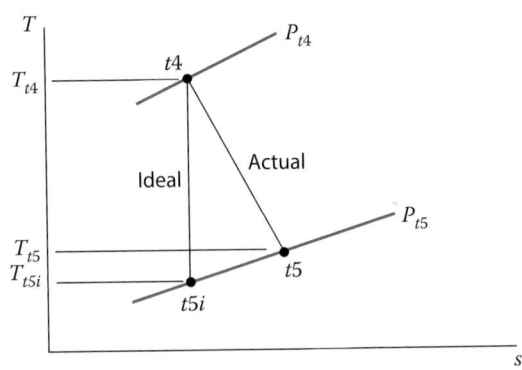

Fig. 6.10 Actual and ideal turbine processes.

efficiency of the turbine in terms of the thermodynamic properties, we have

$$\eta_t = \frac{h_{t4} - h_{t5}}{h_{t4} - h_{t5i}} = \frac{T_{t4} - T_{t5}}{T_{t4} - T_{t5i}}$$

or

$$\eta_t = \frac{1 - \tau_t}{1 - \pi_t^{(\gamma-1)/\gamma}} \tag{6.18}$$

6.5.7 Turbine Stage Efficiency

In a completely similar analysis to that for the compressor, the turbine isentropic efficiency can be written in terms of η_{sj} and π_{sj}:

$$\eta_t = \frac{1 - \prod_{j=1}^{N} [1 - \eta_{sj}(1 - \pi_{sj}^{(\gamma-1)/\gamma})]}{1 - \pi_t^{(\gamma-1)/\gamma}} \tag{6.19}$$

When all stages have the same π_s and η_s, the preceding equation reduces to

$$\eta_t = \frac{1 - [1 - \eta_s(1 - \pi_s^{(\gamma-1)/\gamma})]^N}{1 - \pi_t^{(\gamma-1)/\gamma}} \tag{6.20}$$

6.5.8 Turbine Polytropic Efficiency

The *polytropic turbine efficiency* e_t is defined similarly to the turbine isentropic efficiency as

$$e_t = \frac{\text{actual turbine work for a differential pressure change}}{\text{ideal turbine work for a differential pressure change}}$$

Thus

$$e_t = \frac{\mathrm{d}w}{\mathrm{d}w_i} = \frac{\mathrm{d}h_t}{\mathrm{d}h_{ti}} = \frac{\mathrm{d}T_t}{\mathrm{d}T_{ti}}$$

For the isentropic relationship, we have

$$\frac{\mathrm{d}T_{ti}}{T_t} = \frac{\gamma - 1}{\gamma} \frac{\mathrm{d}P_t}{P_t}$$

Thus

$$e_t = \frac{\mathrm{d}T_t}{\mathrm{d}T_{ti}} = \frac{\mathrm{d}T_t/T_t}{\mathrm{d}T_{ti}/T_t} = \frac{\mathrm{d}T_t/T_t}{[(\gamma - 1)/\gamma]\,\mathrm{d}P_t/P_t}$$

Assuming that the polytropic efficiency e_t is constant over the pressure ratio, we integrate the preceding equation to give

$$\pi_t = \tau_t^{\gamma/[(\gamma-1)e_t]} \tag{6.21}$$

and so

$$\eta_t = \frac{1 - \tau_t}{1 - \tau_t^{1/e_t}} \tag{6.22}$$

or

$$\eta_t = \frac{1 - \pi_t^{(\gamma-1)e_t/\gamma}}{1 - \pi_t^{(\gamma-1)/\gamma}} \tag{6.23}$$

This relationship is plotted in Fig. 6.11 along with Eq. (6.15) for the compressor. Note that the turbine efficiency increases with the turbine expression ratio $1/\pi_t$ for a constant e_t. This should be expected given the discussions provided in the previous sections to explain the e_c behavior. In a multistage turbine, each successive turbine stage benefits from the energy removal (reduced T_t) of preceding stages.

In cycle analysis, τ_t is usually first obtained from the work balance between the turbine and the compressor. Then π_t can be calculated for a known e_t by using Eq. (6.21), and η_t can be calculated by using either Eq. (6.22) or Eq. (6.23). We will use the polytropic efficiency e_t as the figure of merit for the turbine. Just like the compressor polytropic efficiency

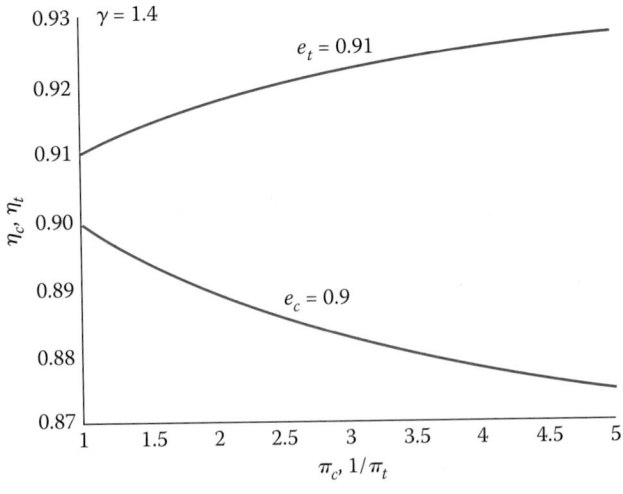

Fig. 6.11 Compressor and turbine efficiencies vs pressure ratio.

e_c, e_t can be thought of as a measure of level of technology for the turbine, with higher polytropic efficiencies representing improved state-of-the-art technology.

6.6 Burner Efficiency and Pressure Loss

In the burner, we are concerned with two efforts: incomplete combustion of the fuel and total pressure loss. *Combustion efficiency* η_b is defined by

$$\eta_b = \frac{(\dot{m} + \dot{m}_f)h_{t4} - \dot{m}h_{t3}}{\dot{m}_f h_{PR}} \tag{6.24}$$

For a known burner efficiency η_b, Eq. (6.24) can be used to calculate the fuel–air ratio f for known temperatures T_{t3} and T_{t4} using Appendix L or the AFPROP program for values of h_{t3}. Because h_{t4} is a function of the fuel–air ratio f, iteration will be required, as shown in Example 6.2.

The total pressure losses arise from two effects: the viscous losses in the combustion chamber and the total pressure loss due to combustion at finite Mach number (see the discussion of Rayleigh line flow in Section 3.7.2). These effects are combined for the purpose of performance analysis in

$$\pi_b = \frac{P_{t4}}{P_{t3}} < 1$$

Note that some total pressure loss across the burner is necessary to facilitate required cooling air flow from the high-pressure compressor (HPC) exit to the high-pressure turbine (HPT) vanes and rotor blades.

We will use both η_b and π_b as the figures of merit for the burner. There are similar combustion efficiencies and total pressure ratios for afterburners (augmenters) and duct burners. Note that one of the performance penalties associated with the use of an afterburner is additional viscous losses associated with friction and pressure drag resulting from the additional hardware (spraybars, flameholders, etc.).

Example 6.2

For a burner efficiency η_b of 0.98, T_{t4} of 3200°R and T_{t3} of 1600°R, determine the fuel–air ratio f for a fuel having $h_{PR} = 18{,}400$ Btu/lbm. Equation (6.24) can be solved for the fuel–air ratio f, giving

$$f = \frac{h_{t4} - h_{t3}}{\eta_b h_{PR} - h_{t4}}$$

Because h_{t4} is a function of the fuel–air ratio, iteration will be required. From Appendix L, $h_{t3} = 395.60$ Btu/lbm. Using Appendix L and an initial guess

(Continued)

Example 6.2 *(Continued)*

of $f = 0.0338$, $h_{t4} = 891.52$ Btu/lbm. Equation (i) gives a new iteration value for the fuel–air ratio of

$$f = \frac{h_{t4} - h_{t3}}{\eta_b h_{PR} - h_{t4}} = \frac{891.52 - 395.60}{0.98 \times 18{,}400 - 891.52} = 0.02893$$

Iteration until successive values do not change more than 0.00001 gives an answer of $f = 0.02855$. The AFPROP program allows for rapid determination of h_{t4} without interpolation of tabulated values.

6.7 Exhaust Nozzle Loss

The primary loss due to the nozzle has to do with the over- or underexpansion of the nozzle. In addition, there will be a loss in total pressure due to viscous effects from the turbine exit to the exhaust nozzle exit plane. Thus we may have

$$P_9 \neq P_0 \quad \text{and} \quad \pi_n = \frac{P_{t9}}{P_{t7}} < 1$$

We still have $\tau_n = 1$, because the nozzle is very nearly adiabatic. We will use π_n as the figure of merit for the nozzle.

6.8 Mechanical Efficiency of a Power Shaft

We define the mechanical efficiency of a shaft to account for the loss or extraction of power on that shaft. The mechanical efficiency η_m is defined as the ratio of the power leaving the shaft that enters the compressor, $\dot{W}_c$, to the power entering the shaft from the turbine, $\dot{W}_t$. This can be written in equation form as

$$\eta_m = \frac{\text{power leaving shaft to compressor}}{\text{power entering shaft from turbine}} = \frac{\dot{W}_c}{\dot{W}_t} \qquad (6.25)$$

The mechanical efficiency η_m is less than one due to losses in power that occur from shaft bearings and power extraction for driving engine acessories like oil and fuel pumps. For multishaft engines, each shaft will have a mechanical efficiency associated with the power transfer on the shaft.

6.9 Mixer

The mixer is a very simple component of a mixed-flow turbofan engine that limits independent choices of the bypass ratio α and fan pressure ratio π_f. In addition, the mixing of the two streams can increase the thrust and thus reduce the fuel consumption. Like the engine inlet, we model the mixer total pressure ratio as the product of that associated with friction ($\pi_{M\,max}$) times that associated with mixing ($\pi_{M\,ideal}$). This section presents the analysis of a constant-area mixer, shown in Fig. 6.12, that would mix the core and bypass flows. This model is used in the analysis of engines with mixers in Chapters 7 and 8.

6.9.1 Ideal Mixer Analytical Model

Consider an ideal (no wall friction) subsonic constant-area mixer (Fig. 6.12) with primary and secondary streams of calorically perfect gases having different c_p and γ values. The flow is assumed to be 1-D at inlet and exit, and the subscripts 6, 16, and 6A are used for the core, bypass, and mixed streams, respectively. We assume that the c_p and γ values of the core and bypass streams are known as well as M_6 and the following ratios:

$$\frac{T_{t16}}{T_{t6}} \quad \frac{P_{t16}}{P_{t6}} \quad \text{and} \quad \alpha' = \frac{\dot{m}_{16}}{\dot{m}_6} \quad \text{or} \quad \frac{A_{16}}{A_6}$$

The Mach number of the secondary stream M_{16} is determined by M_6 and the total pressure ratio P_{t6}/P_{t16}, as shown below. Using the Kutta condition at the splitter plate end ($P_6 = P_{16}$) and, as was shown in Eq. (3.8) in Chapter 3,

$$P_t = P\left(1 + \frac{\gamma - 1}{2}M^2\right)^{\gamma/(\gamma-1)} \tag{3.8}$$

yields

$$\frac{P_{t16}}{P_{t6}} = \frac{\{1 + [(\gamma_{16} - 1)/2]M_{16}^2\}^{\gamma_{16}/(\gamma_{16}-1)}}{\{1 + [(\gamma_6 - 1)/2]M_6^2\}^{\gamma_6/(\gamma_6-1)}} \tag{6.26}$$

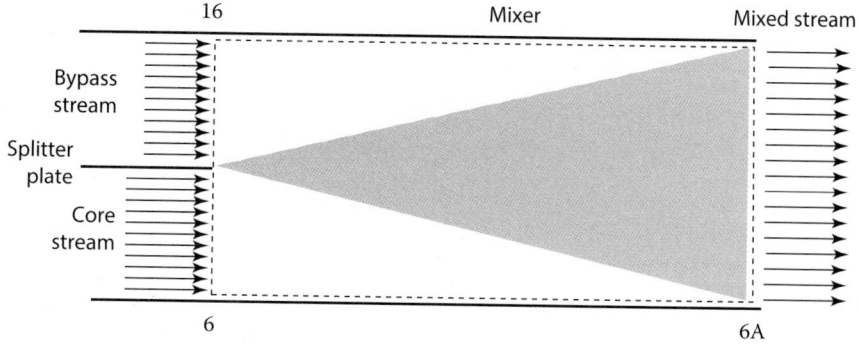

Fig. 6.12 Constant-area mixer.

or

$$
M_{16} = \sqrt{ \frac{2}{\gamma_{16} - 1} \left\{ \left[\frac{P_{t16}}{P_{t6}} \left(1 + \frac{\gamma_6 - 1}{2} M_6^2 \right)^{\gamma_6/(\gamma_6-1)} \right]^{(\gamma_{16}-1)/\gamma_{16}} - 1 \right\} }
$$

$$(6.27)$$

A useful ratio for compressible flows in the *mass flow parameter (MFP)* is defined as Eq. (3.11)

$$
\frac{\dot{m}\sqrt{T_t}}{P_t A} \equiv \mathrm{MFP}(M, \gamma, R) = \frac{M\sqrt{\gamma g_c/R}}{\{1 + [(\gamma - 1)/2]M^2\}^{(\gamma+1)/[2(\gamma-1)]}}
$$

$$(3.11)$$

Combining Eqs. (3.8) and (3.11) gives

$$
\frac{\dot{m}\sqrt{T_t}}{P A} = M\sqrt{1 + \frac{\gamma - 1}{2} M^2}\sqrt{\frac{\gamma g_c}{R}}
$$

$$(6.28)$$

Taking the ratio of Eq. (6.28), written for the bypass stream, to the equation written for the core stream gives the following relationship between the inlet flow properties:

$$
\frac{\alpha'\sqrt{T_{t16}/T_{t6}}}{A_{16}/A_6} = \frac{M_{16}}{M_6}\sqrt{\frac{\gamma_{16}R_6}{\gamma_6 R_{16}} \frac{1 + [(\gamma_{16} - 1)/2]M_{16}^2}{1 + [(\gamma_6 - 1)/2]M_6^2}}
$$

$$(6.29)$$

This equation can be used to solve for the mass flow ratio α' given the area ratio A_{16}/A_6 or vice versa. The inlet flow characteristics of the mixer, defined by Eq. (6.26), is plotted in Fig. 6.13. These figures can give considerable insight into the pumping characteristics of a mixer.

Figure 6.13 shows that the Mach number of the secondary stream M_{16} can be increased by increasing the total pressure ratio P_{t16}/P_{t6} at constant values of M_6. When the total pressure ratio is near unity, the Mach numbers at stations 6 and 16 are nearly equal. For a constant value of the ratio $\alpha'\sqrt{T_{t16}/T_{t6}}/(A_{16}/A_6)$, Eq. (6.29) gives a linear relationship between the Mach numbers at stations 6 and 16.

To obtain the properties of the mixed stream, we start by writing the conservation laws for the ideal constant-area mixer.

Conservation of mass:

$$
\dot{m}_6 + \dot{m}_{16} = \dot{m}_{6A}
$$

$$(i)$$

Conservation of energy:

$$
\dot{m}_6 c_{p6} T_{t6} + \dot{m}_{16} c_{p16} T_{t16} = \dot{m}_{6A} c_{p6A} T_{t6A}
$$

$$(ii)$$

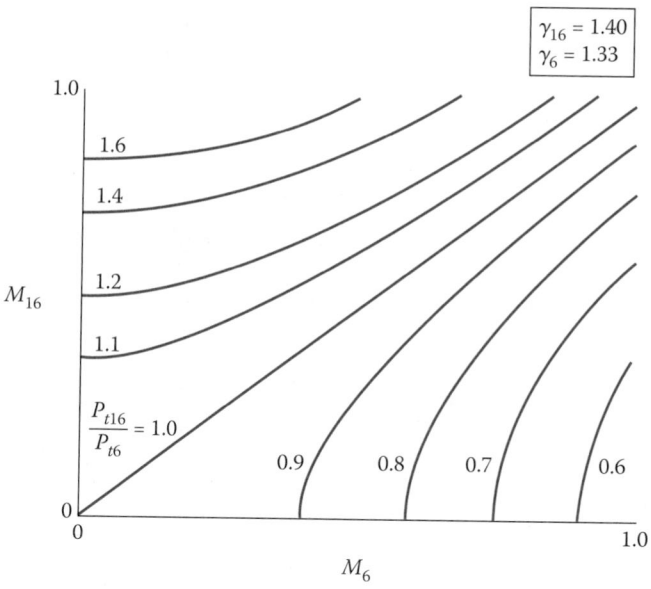

Fig. 6.13   Contours of constant P_{t16}/P_{t6}.

Momentum:

$$P_6 A_6 + \frac{\dot{m}_6 V_6}{g_c} + P_{16} A_{16} + \frac{\dot{m}_{16} V_{16}}{g_c} = P_{6A} A_{6A} + \frac{\dot{m}_{6A} V_{6A}}{g_c} \qquad \text{(iii)}$$

Constant area:

$$A_6 + A_{16} = A_{6A} \qquad \text{(iv)}$$

The mixture law of perfect gases yields, for the mixed stream,

$$c_{p6A} = \frac{c_{p6} + \alpha' c_{p16}}{1 + \alpha'} \qquad \text{(6.30a)}$$

$$R_{6A} = \frac{R_6 + \alpha' R_{16}}{1 + \alpha'} \qquad \text{(6.30b)}$$

$$\gamma_{6A} = \frac{c_{p6A}}{c_{p6A} - R_{6A}} \qquad \text{(6.30c)}$$

From Eq. (ii), the mixer total temperature ratio τ_M is

$$\tau_M \equiv \frac{T_{t6A}}{T_{t6}} = \frac{c_{p6}}{c_{p6A}} \frac{1 + \alpha'(c_{p16}/c_{p6})(T_{t16}/T_{t6})}{1 + \alpha'} \qquad \text{(6.31)}$$

The momentum equation [Eq. (iii)] can be written as

$$I_6 + I_{16} = I_{6A} \qquad \text{(6.32)}$$

where, by Eq. (2.79),

$$I = PA(1 + \gamma M^2) \tag{6.33}$$

Now PA can be replaced in Eq. (6.33) by using Eq. (6.28) to yield

$$I = \dot{m}\sqrt{\frac{RT_t}{\gamma g_c}} \frac{1 + \gamma M^2}{M\sqrt{1 + [(\gamma - 1)/2]M^2}} = \dot{m}\sqrt{\frac{RT_t}{\gamma g_c \phi(M, \gamma)}} \tag{6.34}$$

where

$$\phi(M, \gamma) = \frac{M^2\{1 + [(\gamma - 1)/2]M^2\}}{(1 + \gamma M^2)^2} \tag{6.35}$$

Substituting Eq. (6.34) for each term in Eq. (6.32) and solving for $\phi(M_{6A}, \gamma_{6A})$ yield

$$\phi(M_{6A}, \gamma_{6A}) = \left[\frac{1 + \alpha'}{\dfrac{1}{\sqrt{\phi(M_6, \gamma_6)}} + \alpha'\sqrt{\dfrac{R_{16}\gamma_6}{R_6\gamma_{16}}\dfrac{T_{t16}/T_{t6}}{\phi(M_{16}, \gamma_{16})}}} \right]^2 \frac{R_{6A}\gamma_6}{R_6\gamma_{6A}}\tau_M \tag{6.36}$$

For a given value of M_6 and M_{16}, Eq. (6.36) gives $\Phi \equiv \phi(M_{6A}, \gamma_{6A})$, which, when placed in Eq. (6.35), yields a quadratic equation in M_{6A}^2 with the solution

$$M_{6A} = \sqrt{\frac{2\Phi}{1 - 2\gamma_{6A}\Phi + \sqrt{1 - 2(\gamma_{6A} + 1)\Phi}}} \tag{6.37}$$

An expression for the mixer total pressure ratio π_M in terms of τ_M and the other flow properties at stations 6, 16, and 6A is obtained directly from the mass flow parameter [Eq. (3.11)]. Solving Eq. (2.76) for P_t and forming the ratio P_{t6A}/P_{t6} yields

$$\pi_{M\,\text{ideal}} \equiv \frac{P_{t6A}}{P_{t6}} = \frac{(1 + \alpha')\sqrt{\tau_M}}{1 + A_{16}/A_6}\frac{\text{MFP}(M_6, \gamma_6, R_6)}{\text{MFP}(M_{6A}, \gamma_{6A}, R_{6A})} \tag{6.38}$$

Given the c_p, R, and γ of the core and bypass streams (M_6, T_{t16}/T_{t6}, P_{t16}/P_{t6}, and α'), then Eqs. (6.27), (6.29), (6.30a–c), (6.31), (6.36), (6.37), and (6.38) will give M_{16}, M_{6A}, τ_M, A_{16}/A_6, and $\pi_{M\,\text{ideal}}$. These equations will be used to determine the performance of the mixer in the mixed-flow afterburning turbofan engine.

6.10 Gross Thrust

It is interesting to compare the gross thrust capabilities of mixed and unmixed streams for the case of correctly expanded ideal exhaust nozzles. The gross thrusts are, using subscript e for the exhaust nozzle exit stations,

$$F_{G_{UM}} = (\dot{m}_6 V_{e6} + \dot{m}_{16} V_{e16})/g_c \quad \text{for unmixed streams} \qquad (6.39a)$$

$$F_{G_{MX}} = (\dot{m}_6 + \dot{m}_{16}) V_{e6A}/g_c \qquad \text{for mixed streams} \qquad (6.39b)$$

Noting that for all cases

$$V_e^2 = \frac{2\gamma}{\gamma - 1} R g_c T_t \left\{ 1 - \left(\frac{P_e}{P_t}\right)^{(\gamma-1)/\gamma} \right\}$$

if follows immediately that

$$\frac{F_{G_{MX}}}{F_{G_{UM}}} = \frac{(1+\alpha')\sqrt{\dfrac{\gamma_{6A}}{\gamma_{6A}-1}\dfrac{\gamma_6-1}{\gamma_6}\dfrac{R_{6A}}{R_6}\dfrac{T_{t6A}}{T_{t6}}\left\{1-\left(\dfrac{P_e}{P_{t6A}}\right)^{(\gamma_{6A}-1)/\gamma_{6A}}\right\}}}{\left(\sqrt{1-\left(\dfrac{P_e}{P_{t6}}\right)^{(\gamma_6-1)\gamma_6}}\right.}$$

$$\left. + \alpha'\sqrt{\dfrac{\gamma_{16}}{\gamma_{16}-1}\dfrac{\gamma_6-1}{\gamma_6}\dfrac{R_{16}}{R_6}\dfrac{T_{t16}}{T_{t6}}\left\{1-\left(\dfrac{P_e}{P_{t16}}\right)^{(\gamma_{16}-1)/\gamma_{16}}\right\}}\right)$$

$$(6.40)$$

Example 6.3

Contours for parametric values of $F_{G_{MX}}/F_{G_{UM}}$ are plotted vs T_{t16}/T_{t6} for different mixer mass flow ratios ($\alpha' = \dot{m}_{16}/\dot{m}_6$) in Fig. 6.14 for the following input data to the system of constant area mixer equations developed previously:

$$\frac{P_{t16}}{P_{t6}} = 1.0, \quad M_6 = 0.6, \quad \frac{P_e}{P_{t6}} = 0.5, \quad \gamma_{16} = 1.4,$$

$$c_{p16} = 0.24 \frac{\text{Btu}}{\text{lbm} \cdot {}^\circ\text{R}}, \quad \gamma_6 = 1.33, \quad c_{p6} = 0.276 \frac{\text{Btu}}{\text{lbm} \cdot {}^\circ\text{R}}$$

Although the mixer can actually augment (or increase) the ideal thrust under some conditions, it is most sensitive to T_{t16}/T_{t6}. For high-bypass-ratio turbofan engines, $T_{t16}/T_{t6} \approx 0.4$, the ideal mixer can result in an increase in thrust ($\approx 2\%$) that with frictional losses remains a small increase. Rolls-Royce

(Continued)

Example 6.3 *(Continued)*

high-bypass turbofan engines mix the two exhaust streams for this reason; however, this requires that the outer engine nacelle be longer and thus heavier. For the low-bypass-ratio turbofan engine used in fighter aircraft, $T_{t16}/T_{t6} \approx 0.5$, the ideal mixing gives a small increase in thrust ($\approx 1\%$) that with frictional losses results in a slight reduction in thrust.

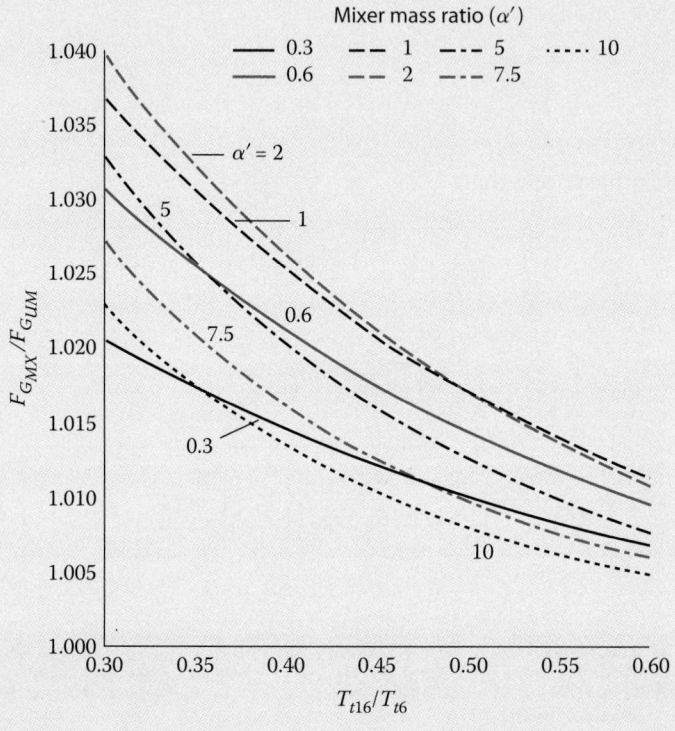

Fig. 6.14 Thrust ratio of mixed-to-unmixed flow.

6.11 ## Summary of Component Figures of Merit

Table 6.1 summarizes the ideal and actual behaviors of gas turbine engine components with calorically perfect gases. Note that for a compressor with constant polytropic efficiency, the isentropic efficiency follows from Eq. (6.15). We will use these figures of merit in the following chapter.

At a particular level of technological development, the polytropic efficiency e_c for the compressor can be considered to be a constant, and thus the compressor pressure ratio π_c determines the compressor efficiency η_c

Table 6.1 Summary of Component Figures of Merit

Component	Ideal Behavior	Actual Behavior	Figure of Merit
Inlet	Adiabatic and reversible (isentropic) $\tau_d = 1, \pi_d = 1$	Adiabatic, not reversible $\tau_d = 1, \pi_d < 1$	π_d
Compressor	Adiabatic and reversible (isentropic) $w_c = c_p T_{t2}(\tau_c - 1)$ $\pi_c = \tau_c^{\gamma/(\gamma-1)}$	Adiabatic, not reversible $w_c = c_{pc} T_{t2}(\tau_c - 1)$ $\pi_c = [1 + \eta_c(\tau_c - 1)]^{\gamma_c/(\gamma_c-1)}$ $\tau_c = 1 + \dfrac{1}{\eta_c}(\pi_c^{(\gamma_c-1)/\gamma_c} - 1)$ $\tau_c = (\pi_c)^{(\gamma_c-1)/(\gamma_c e_c)}$	η_c e_c
Burner	No total pressure loss, 100% combustion $\pi_b = 1$ $\dot{m} c_p T_{t4} - \dot{m} c_p T_{t3} = \dot{m}_f h_{PR}$	Total pressure loss, combustion $<100\%$ $\pi_b < 1$ $(\dot{m} + \dot{m}_f) h_{t4} - \dot{m} h_{t3} = \eta_b \dot{m}_f h_{PR}$	π_b η_b
Turbine	Adiabatic and reversible (isentropic) $w_t = c_p T_{t4}(1 - \tau_t)$ $\pi_t = \tau_t^{\gamma/(\gamma-1)}$	Adiabatic, not reversible $w_t = c_{pt} T_{t4}(1 - \tau_t)$ $\pi_t = \left[1 - \dfrac{1}{\eta_t}(1 - \tau_t)\right]^{\gamma_t/(\gamma_t-1)}$ $\tau_t = 1 - \eta_t(1 - \pi_t^{(\gamma_t-1)/\gamma_t})$ $\pi_t = \tau_t^{\gamma_t/[(\gamma_t-1)e_t]}$	η_t e_t
Nozzle	Adiabatic and reversible $\tau_n = 1, \pi_n = 1$	Adiabatic, not reversible $\tau_n = 1, \pi_n < 1$	π_n

[see Eq. (6.15)]. Similarly, the polytropic efficiency e_t for the turbine can be considered to be a constant, and thus the turbine temperature ratio τ_t determines the turbine efficiency η_t [see Eq. (6.22)]. For the analysis in the following chapter, we will use the polytropic efficiencies and other efficiencies and total pressure loss coefficients as known input data and will calculate the resulting component efficiencies.

The values of these figures of merit have changed as technology has improved over the years. In addition, the values of the figures of merit for the diffuser and nozzle depend on the application. For example, a commercial airliner with engine nacelles and convergent, fixed-area exhaust nozzles will typically have much higher values of π_d and π_n than a supersonic fighter with its engines in the airframe and convergent-divergent, variable-area exhaust nozzles. Table 6.2 lists typical values for the figures of merit that correspond to different periods in the evolution of engine technology (called levels of technology) and the application. This information is based on historical data, experience, and projections. The 20-year increments beginning in 1945 indicated by "Level of Technology" in Table 6.2 are approximate, but about right, dictated by the tremendous expenses and long lead times associated with advances in aircraft gas turbine engine technology.

6.12 Component Performance with Variable c_p

In Section 2.4.6 of Chapter 2, we developed the relationships [Eqs. (2.64), (2.65), and (2.68)] for the thermodynamic properties h, ϕ, and s of a perfect gas with variable specific heat. In addition, the reduced pressure P_r and reduced volume v_r were defined by Eqs. (2.66) and (2.67), respectively. We will use these properties to describe the performance of engine components when the variation of specific heat is to be included. We will use Appendix L or the computer program AFPROP to obtain the thermodynamic properties.

The notation π_a [see Eq. (5.3)] represents a component's total pressure ratio. However, we will use a modified definition for τ_a to represent the ratio of total enthalpies [see original definition given by Eq. (5.4)]. Thus

$$\tau_a = \frac{\text{total enthalpy leaving component } a}{\text{total enthalpy entering component } a} \tag{6.41}$$

6.12.1 Freestream Properties

From the definition of the total enthalpy and total pressure, we can write

$$\tau_r = \frac{h_{t0}}{h_0} = \frac{h_0 + V_0^2/(2g_c)}{h_0} \quad \text{and} \quad \pi_r = \frac{P_{r0}}{P_0} = \frac{P_{rt0}}{P_{r0}} \tag{6.42}$$

Table 6.2 Component Efficiencies, Total Pressure Ratios, and Temperature Limits

Component	Figure of Merit	Type[a]	Level of Technology[b]				
			1945–1965 1	1965–1985 2	1985–2005 3	2005–2025 4	2025–2045 5
Diffuser	$\pi_{d\,max}$	A	0.90	0.95	0.98	0.995	0.998
		B	0.88	0.93	0.96	0.98	0.985
		C	0.85	0.90	0.94	0.96	0.97
Compressor	e_c		0.80	0.84	0.88	0.90	0.91
Fan	e_f		0.78	0.82	0.86	0.89	0.92
Burner	π_b		0.90	0.92	0.94	0.95	0.96
	η_b		0.88	0.94	0.99	0.999	0.999
Turbine	e_t	Uncooled	0.80	0.85	0.89	0.90	0.91
		Cooled		0.83	0.87	0.89	0.90
Mixer	$\pi_{M\,max}$			0.95	0.97	0.98	0.985
Afterburner	π_{AB}		0.90	0.92	0.94	0.95	0.96
	η_{AB}		0.85	0.91	0.96	0.99	0.995

(Continued)

Table 6.2 Component Efficiencies, Total Pressure Ratios, and Temperature Limits (Continued)

Component	Figure of Merit	Type[a]	Level of Technology[b]				
			1945–1965 1	1965–1985 2	1985–2005 3	2005–2025 4	2025–2045 5
Nozzle	π_n	D	0.95	0.97	0.98	0.995	0.997
		E	0.93	0.96	0.97	0.98	0.99
		F	0.90	0.93	0.95	0.97	0.98
Mechanical shaft	η_m	Shaft only	0.95	0.97	0.99	0.995	0.996
		With power takeoff	0.90	0.92	0.95	0.97	0.98
Maximum T_{t4}		(K)	1110	1390	1780	2000	2220
		(R)	2000	2500	3200	3600	4000
Maximum T_{t7}		(K)	1390	1670	2000	2220	2440
		(R)	2500	3000	3600	4000	4400

[a]A = subsonic aircraft with engines in nacelles; B = subsonic aircraft with engine(s) in airframe; C = supersonic aircraft with engine(s) in airframe; D = fixed-area convergent nozzle; E = variable-area convergent nozzle; F = variable-area convergent-divergent nozzle.
[b]Stealth may reduce $\pi_{d\,max}$, π_{AB}, and π_n.

6.12.2 Inlet

Because the inlet is assumed to be adiabatic, then

$$\tau_d = \frac{h_{t2}}{h_{t0}} = 1 \tag{6.43}$$

By using Eq. (2.68), the inlet total pressure ratio can be expressed in terms of the entropy change as follows:

$$\pi_d = \frac{P_{t2}}{P_{r0}} - \exp\left(-\frac{s_2 - s_0}{R}\right) \tag{6.44}$$

6.12.3 Compressor

The variable τ_c represents the total enthalpy ratio of the compressor, and π_c represents its total pressure ratio, or

$$\tau_c = \frac{h_{t3}}{h_{t2}} \quad \text{and} \quad \pi_c = \frac{P_{t3}}{P_{t2}} \tag{6.45}$$

The polytropic efficiency of the compressor e_c can be written as

$$e_c = \frac{dh_{ti}}{dh_t} = \frac{dh_{ti}/T_t}{dh_t/T_t}$$

By using the Gibbs equation [Eq. (2.39)], the numerator in the preceding equation can be expressed as

$$\frac{dh_{ti}}{T_t} = R\frac{dP_t}{P_t}$$

Thus

$$e_c = R\frac{dP_t/P_t}{dh_t/T_t} \tag{6.46}$$

For a constant polytropic efficiency, integration of the preceding equation between states $t2$ and $t3$ gives

$$\phi_{t3} - \phi_{t2} = \frac{R}{e_c}\ln\frac{P_{t3}}{P_{t2}}$$

Thus the compressor pressure ratio π_c can be written as

$$\pi_c = \frac{P_{t3}}{P_{t2}} = \exp\left(e_c\frac{\phi_{t3} - \phi_{t2}}{R}\right)$$

or

$$\pi_c = \left(\frac{P_{rt3}}{P_{rt2}}\right)^{e_c} \tag{6.47}$$

The reduced pressure at state $t2$ can be obtained from Appendix L or the computer program AFPROP, given the temperature at state $t2$. If the values of π_c and e_c are also known, one can get the reduced pressure at state $t3$ by using

$$P_{rt3} = P_{rt2}\pi_c^{1/e_c} \tag{6.48}$$

Given the reduced pressure at state $t3$, the total temperature and total enthalpy can be obtained from Appendix L or the computer program AFPROP.

The isentropic efficiency of a compressor can be expressed as

$$\eta_c = \frac{h_{t3i} - h_{t2}}{h_{t3} - h_{t2}} \tag{6.49}$$

This equation requires that the total enthalpy be known at states $t2$, $t3$, and $t3i$.

Example 6.4

Air at 1 atm and $540°R$ enters a compressor whose polytropic efficiency is 0.9. If the compressor pressure ratio is 15, determine the leaving total properties and compressor isentropic efficiency. From Appendix L for $f = 0$, we have

$$h_{t2} = 129.02 \text{ Btu/lbm} \quad \text{and} \quad P_{rt2} = 1.384$$

The exit total pressure is 15 atm. The reduced pressures at stations $t3$ and $t3i$ are

$$P_{rt3} = P_{rt2}\pi_c^{1/e_c} = 1.384 \times 15^{1/0.9} = 28.048$$
$$P_{rt3i} = P_{rt2}\pi_c = 1.384 \times 15 = 20.76$$

Linear interpolation of Appendix L, with $f = 0$ and using the preceding reduced pressures, gives the following values of total enthalpy and total temperature:

$$h_{t3} = 304.48 \text{ Btu/lbm} \quad \text{and} \quad T_{t3} = 1251.92°R$$
$$h_{t3i} = 297.67 \text{ Btu/lbm} \quad \text{and} \quad T_{t3i} = 1154.58°R$$

The compressor isentropic efficiency is

$$\eta_c = \frac{279.67 - 129.02}{304.48 - 129.02} = 0.8586$$

6.12.4 Burner

The τ_b represents the total enthalpy ratio of the burner, and π_b represents its total pressure ratio, or

$$\tau_b = \frac{h_{t4}}{h_{t3}} \quad \text{and} \quad \pi_b = \frac{P_{t4}}{P_{t3}} \tag{6.50}$$

From the definition of burner efficiency, we have

$$\eta_b = \frac{(\dot{m} + \dot{m}_f)h_{t4} - \dot{m}h_{t3}}{\dot{m}_f h_{PR}}$$

or

$$\eta_b = \frac{(1+f)h_{t4} - h_{t3}}{f h_{PR}} \tag{6.51}$$

where f is the ratio of the fuel flow rate to the airflow rate entering the burner, or $f = \dot{m}_f / \dot{m}$.

In the analysis of gas turbine engines, we determine the fuel/air ratio f and normally specify η_b, h_{PR}, and T_{t4}. The enthalpy at station t_3 (h_{t3}) will be known from analysis of the compressor. Equation (6.51) can be solved for the fuel/air ratio f, giving

$$f = \frac{h_{t4} - h_{t3}}{\eta_b h_{PR} - h_{t4}} \tag{6.52}$$

Note: The value of h_{t4} is a function of the fuel/air ratio f, and thus the solution of Eq. (6.52) is iterative.

6.12.5 Turbine

The τ_t represents the total enthalpy ratio of the turbine, and π_t represents its total pressure ratio or

$$\tau_t = \frac{h_{t5}}{h_{t4}} \quad \text{and} \quad \pi_t = \frac{P_{t5}}{P_{t4}} \tag{6.53}$$

The *polytropic efficiency of a turbine* is defined as

$$e_t = \frac{dh_t}{dh_{ti}}$$

which can be written as

$$e_t = \frac{1}{R} \frac{dh_t / T_t}{dP_t / P_t} \tag{6.54}$$

For constant polytropic efficiency, the following relationships are obtained by integration from state $t4$ to $t5$:

$$\pi_t = \frac{P_{t5}}{P_{t4}} = \exp\frac{\phi_{t5} - \phi_{t4}}{Re_t} \tag{6.55}$$

and

$$\pi_t = \left(\frac{P_{rt5}}{P_{rt4}}\right)^{1/e_t} \tag{6.56}$$

The *isentropic efficiency of the turbine* is defined as

$$\eta_t = \frac{h_{t4} - h_{t5}}{h_{t4} - h_{r5i}} \tag{6.57}$$

where each total enthalpy is a function of the total temperature T_t and fuel/air ratio f.

Example 6.5

Products of combustion ($f = 0.0338$) at 20 atm and 3000°R enter a turbine whose polytropic efficiency is 0.9. If the total enthalpy of the flow through the turbine decreases 100 Btu/lbm, determine the leaving total properties and turbine isentropic efficiency. From Appendix L for $f = 0.0338$, we have

$$h_{t4} = 828.75 \text{ Btu/lbm} \quad \text{and} \quad P_{rt4} = 1299.6$$

The exit total enthalpy is 728.75 Btu/lbm. From linear interpolation of Appendix L, the total temperature and reduced pressure at station $t5$ are

$$T_{t5} = 2677.52°R \quad \text{and} \quad P_{rt5} = 777.39$$

From Eq. (6.40), the turbine pressure ratio is

$$\pi_t = \left(\frac{P_{rt5}}{P_{rt4}}\right)^{1/e_t} = \left(\frac{777.39}{1299.6}\right)^{1/0.9} = 0.5650$$

and the reduced pressure at state $t5i$ is

$$P_{rt5i} = P_{rt4}\pi_t = 1299.6 \times 0.5650 = 734.3$$

Linear interpolation of Appendix L, with $f = 0.0338$ and using the preceding reduced pressure, gives the following values of total enthalpy and total temperature for state $t5i$:

$$h_{t5\iota} = 718.34 \text{ Btu/lbm} \quad \text{and} \quad T_{t5\iota} = 2643.64°R$$

The turbine isentropic efficiency is

$$\eta_t = \frac{828.75 - 728.75}{828.75 - 718.34} = 0.9057$$

6.12.6 Nozzle

Because the nozzle is assumed to be adiabatic, then

$$\tau_n = \frac{h_{t9}}{h_{t8}} = 1 \tag{6.58}$$

By using Eq. (2.41), the inlet total pressure ratio can be expressed in terms of the entropy change as follows:

$$\pi_n = \frac{P_{t9}}{P_{t8}} = \exp\left(-\frac{s_9 - s_8}{R}\right) \tag{6.59}$$

The exit velocity is obtained from the difference between the total and static enthalpies as follows:

$$V_9 = \sqrt{2g_c(h_{t9} - h_9)} \tag{6.60}$$

where h_9 is obtained from the reduced pressure at state 9 (P_{r9}) given by

$$P_{r9} = \frac{P_{rt9}}{P_{t9}/P_9} \tag{6.61}$$

Problems

6.1 Calculate the total pressure recovery η_r, using Eq. (6.6), and the total pressure ratio across a normal shock in air for Mach numbers 1.25, 1.50, 1.75, and 2.0. How do they compare?

6.2 The isentropic efficiency for a diffuser with a calorically perfect gas can be written as

$$\eta_d = \frac{(P_{t2}/P_0)^{(\gamma-1)/\gamma} - 1}{T_{t0}/T_0 - 1}$$

Show that this equation can be rewritten in terms of τ_r and π_d as

$$\eta_d = \frac{\tau_r \pi_d^{(\gamma-1)/\gamma} - 1}{\tau_r - 1}$$

6.3 Determine the isentropic efficiency of a diffuser with $\pi_{d\max} = 0.98$ and η_r of Eq. (6.6) at $M_0 = 2$.

6.4 Determine the isentropic efficiency of a diffuser with $\pi_{d\max} = 0.95$ and η_r of Eq. (6.6) at $M_0 = 3$.

6.5   Starting from Eqs. (6.14) and (6.21), show that the polytropic efficiency for the compressor and turbine can be expressed as

$$e_c = \frac{\gamma_c - 1}{\gamma_c} \frac{\ell n\, \pi_c}{\ell n\, \tau_c} \qquad e_t = \frac{\gamma_t}{\gamma_t - 1} \frac{\ell n\, \tau_t}{\ell n\, \pi_t}$$

6.6   How many stages of compressor are required to produce a pressure ratio of 24 with stages of $\Delta T_t = 50°C$ (90°F) on a sea-level, standard day? Assume $e_c = 0.9$.

6.7   What fan pressure ratio can be obtained with one stage of $\Delta T_t = 50°C$ (90°F) on a sea-level, standard day? Assume $e_f = 0.88$.

6.8   A J57B afterburning turbojet engine at maximum static power on a sea-level, standard day ($P_0 = 14.696$ psia, $T_0 = 518.7°R$, and $V_0 = 0$) has the data listed in Appendix B.

a) Calculate the adiabatic and polytropic efficiencies of the low-pressure and high-pressure compressors. Assume $\gamma = 1.4$ for the low-pressure compressor and $\gamma = 1.39$ for the high-pressure compressor.

b) Calculate the adiabatic and polytropic efficiencies of the turbine. Assume $\gamma = 1.33$.

c) Calculate π_b, τ_b, π_{AB}, and τ_{AB}.

6.9   A J57B afterburning turbojet engine had 167 lbm/s of air at 167 psia and 660°F enter the combustor (station 3) and products of combustion at 158 psia and 1570°F leave the combustor (station 4). If the fuel flow into the combustor was 8520 lbm/h, determine the combustor efficiency η_b, assuming $h_{PR} = 18{,}400$ Btu/lbm.

6.10   A J57B afterburning turbojet engine had 169.4 lbm/s of air at 36 psia and 1013°F enter the afterburner (station 6) and products of combustion at 31.9 psia and 2540°F leave the afterburner (station 7). If the fuel flow into the afterburner was 25,130 lbm/h, determine the afterburner efficiency η_{AB}, assuming $h_{PR} = 18{,}400$ Btu/lbm.

6.11   Engine compressor exit temperatures are limited due to material limits for thermal stresses of transient compressor operation. You will calculate and compare the compressor pressure ratio limit for air at standard-day, static of current technology of about 900 K to future technology of 1300 K Assume $e_c = 0.9$.

6.12   A JT9D high-bypass-ratio turbofan engine at maximum static power on a sea-level, standard day ($P_0 = 14.696$ psia, $T_0 = 518.7°R$, and $V_0 = 0$) has the data listed in Table B.4 of Appendix B.

a) Calculate the adiabatic and polytropic efficiencies of the fan. Assume $\gamma = 1.4$.

b) Calculate the adiabatic and polytropic efficiencies of both the low-pressure and high-pressure compressors. Assume $\gamma = 1.4$ for the low-pressure compressor and $\gamma = 1.39$ for the high-pressure compressor.

c) Calculate the adiabatic polytropic efficiencies of the turbine. Assume $\gamma = 1.35$.

d) Calculate the power (horsepower and kilowatts) into the fan and compressors. Assume $Rg_c = 1716\ \text{ft}^2/(\text{s}^2 \cdot {}^\circ\text{R})$ and the γ of parts b and c.

e) Calculate the power (horsepower and kilowatts) from the turbine for a mass flow rate of 251 lbm/s. Assume $\gamma = 1.31$ and $Rg_c = 1716\ \text{ft}^2/(\text{s}^2 \cdot {}^\circ\text{R})$.

f) How do the results of parts d and e compare?

6.13 A F100-PW-200 mixed-flow turbofan engine has the temperature/pressure data of Table B.4 in Appendix B. Determine the polytropic efficiency of the fan and high-pressure compressor (assume $\gamma = 1.4$) Determine the polytropic efficiency of the high-pressure and low-pressure turbines (assume $\gamma = 1.3$ and uncooled $T_{t4} = 2739^\circ\text{R}$). Calculate the power required by the fan and high-pressure compressor (assume $c_p = 0.24\ \text{Btu/lbm} \cdot {}^\circ\text{F}$).

6.14 The isentropic efficiency for a nozzle η_n is defined as

$$\eta_n = \frac{h_{t7} - h_9}{h_{t7} - h_{9s}}$$

where the states $t7$, 9, and $9s$ are shown in Fig. P6.1. For $P_9 = P_0$, show that the isentropic efficiency for the nozzle in a turbojet engine can be written for a calorically perfect gas as

$$\eta_n = \frac{\text{PIT}^{(\gamma-1)/\gamma} - 1}{\text{PIT}^{(\gamma-1)/\gamma} - (\pi_n)^{(\gamma-1)/\gamma}}$$

where $\text{PIT} = \pi_r \pi_d \pi_c \pi_b \pi_t \pi_n$.

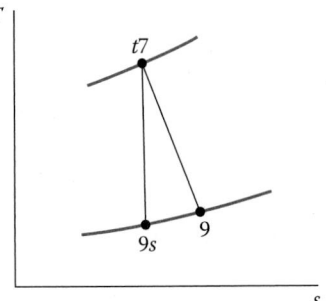

Fig. P6.1 Nozzle *T-s*.

6.15 Gas from the engine core with $\gamma = 1.33$ mixes with bypass air ($\gamma = 1.4$) in a mixer as is shown in

Fig. 6.12. If the total pressure ratio of the two streams is 1, determine M_{16} for M_6 values of 0.3, 0.4, 0.5, 0.6, and 0.7. Comment on the results.

6.16 Gas from the engine core with $\gamma = 1.33$ mixes with bypass air ($\gamma = 1.4$) in a mixer as is shown in Fig. 6.12. If the total pressure ratio of the two streams is 1, the total temperature ratio T_{t16}/T_{t6} is 0.3, the mixer mass flow ratio α' is 0.5, and the M_6 value is 0.5, determine M_{16}, M_{6A}, τ_M, and π_M ideal. Comment on the results.

6.17 Repeat Problem 6.8 with variable gas properties, using Appendix L or the AFPROP program. Compare your results to those of Problem 6.8.

6.18 Repeat Problem 6.11 with variable gas properties, using Appendix L or the AFPROP program. Compare your results to those of Problem 6.11.

6.19 A JT9D high-bypass-ratio turbofan engine at maximum static power on a sea-level, standard day ($P_0 = 14.696$ psia, $T_0 = 518.7°R$, and $V_0 = 0$) has the data listed in Appendix B. Assuming 100% efficient combustion and $h_{PR} = 18{,}400$ Btu/lbm, calculate the fuel/air ratio, using Appendix L or the AFPROP program and starting with an initial guess of 0.03. Now repeat Problem 6.12 with variable gas properties, using Appendix L or AFPROP. Compare your results to those of Problem 6.12.

6.20 Repeat Problem 6.13 with variable gas properties, using Appendix L or the AFPROP program. Assume 100% efficient combustion and $h_{PR} = 18{,}400$ Btu/lbm. Compare your results to those of Problem 6.13.

6.21 The main burner of the turbojet engine was analyzed in Problem 2.5c using different specific heats for the reactants and the products. Based on this analysis and a burner efficiency of η_b,

a) Show that

$$\eta_b = \frac{(1 + f)c_{pP}(T_{t4} - T_d) - c_{pR}(T_{t3} - T_d)}{fh_{PR}}$$

b) Using the data of Problem 6.9, calculate the burner efficiency assuming $c_{pR} = 0.245$ Btu/lbm · R, $c_{pP} = 0.265$ Btu/lbm · R, $h_{PR} = 18{,}400$ Btu/lbm, and $T_d = 536.4°R$. Compare to the results of Problem 6.9.

Chapter 7 Parametric Cycle Analysis (PCA) of Real Engines

> The turbojet engine was a simple concept, but not simple to build. It needed engineers and specialists in materials, manufacturing and casting processes, and controls.
>
> *Hans von Ohain*

7.1 Introductory Comments and Assumptions

7.1.1 Introduction

In Chapter 5, we idealized the engine components and assumed that the working fluid behaved as a perfect gas with constant specific heats. These idealizations and assumptions permitted the basic parametric analysis of several types of engine cycles and the analysis of engine performance trends. In Chapter 6, we looked at the variation of specific heat with temperature and fuel/air ratio and developed component models and performance figures of merit. This allows us to use realistic assumptions as to component losses and to include the variation of specific heats in engine cycle analysis. In this chapter, we develop the cycle analysis equations for many engine cycles, analyze their performance, and determine the effects of real components by comparison with the ideal engines of Chapter 5. We begin our analysis with the turbojet engine cycle and treat the simpler ramjet engine cycle as a special case of the turbojet ($\pi_c = 1$, $\tau_c = 1$, $\pi_t = 1$, $\tau_t = 1$).

As was emphasized in Chapter 5, you must be careful interpreting plot results of a parametric cycle analysis (PCA). Be it a plot of a single iteration (design) variable or a carpet plot with two iteration (design) variables, recognize that each point on the plot represents a different engine operating at the selected condition (i.e., altitude, Mach number). The performance of a selected engine for different flight conditions and throttle settings is presented in Chapter 8. Remember to use T-s diagrams, applicable governing cycle equations, and knowledge of fundamental engine

---

Supporting material for this chapter is available electronically. See the Supporting Materials page at the back of the book for instructions to download.

operation to help with your understanding of performance trends when possible.

7.1.2 Assumptions of Real Cycle Analysis

For analysis of real cycles, we assume the following based on Chapter 6:

1. Perfect gas upstream of main burner with constant properties γ_c, R_c, c_{pc}, and so on.
2. Perfect gas downstream of main burner with constant properties γ_t, R_t, c_{pt}, and so on.
3. All components are adiabatic.
4. The efficiencies of the turbomachinery—compressor(s), turbine(s), fan—are described through the use of constant polytropic efficiencies e_c, e_t, and e_f, respectively.
5. No cooling or bleed flows.

Note that in the development that follows, for all but the turboprop analysis in Section 7.6, the cycle analysis equations are developed for one value of e_c, e_t, and mechanical efficiency η_m. This implies that for two-shaft engines consisting of a high-spool (H) and low-spool (L), $e_{cH} = e_{cL} = e_c$, for example. Different values for the turbomachinery polytropic efficiencies and mechanical efficiencies can be easily accommodated as is done in Section 7.6 (and Section 7.6.SM, the online supporting material) and the cycle analysis program PARA that accompanies this textbook.

7.2 Turbojet

We will now compute the behavior of the turbojet engine including component losses, the mass flow rate of fuel through the components, and

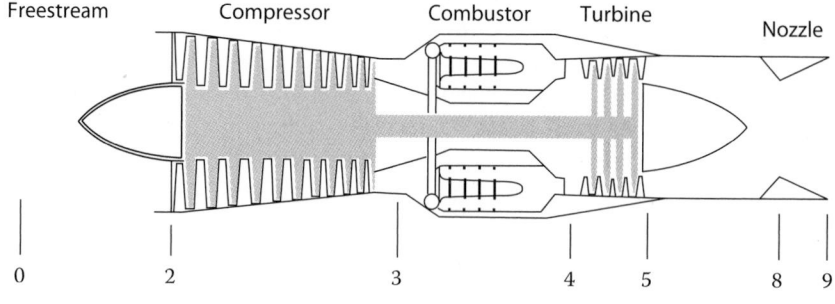

Fig. 7.1   Station numbering of turbojet engine.

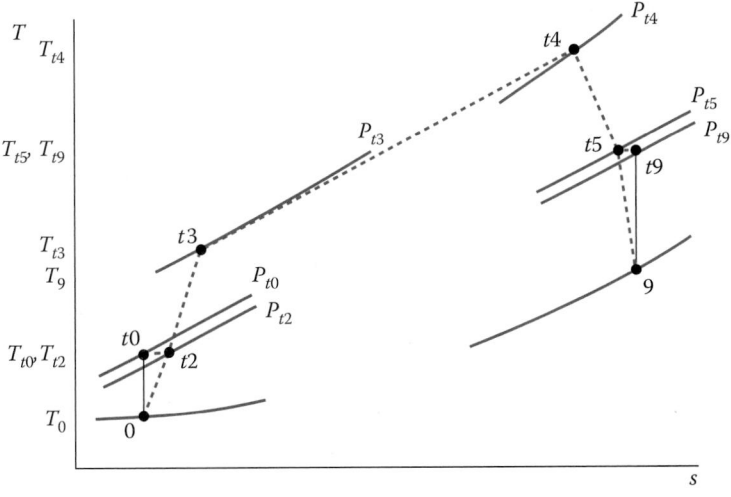

Fig. 7.2   The T-s diagram for turbojet engine.

the variation of specific heats. Our analysis assumes 1-D steady flow at the entrance and exit of each component. The variation of specific heats will be approximated by assuming a perfect gas with constant specific heat c_{pc} upstream of the main burner (combustor) and a perfect gas with different constant specific heat c_{pt} downstream of the main burner.

The turbojet engine with station numbering is shown in Fig. 7.1, and the T-s diagram for this cycle with losses is plotted in Fig. 7.2. Figure 7.2 shows the total states for all engine stations along with the static states for stations 0 and 9.

7.2.1 Cycle Analysis

In this section we develop a system of equations to analyze the turbojet engine cycle exactly as we did in Chapter 5, this time accounting for "real" engine effects as presented in Chapter 6. The steps of cycle analysis are applied to the turbojet engine and presented next in the order listed in Section 5.4.

Step 1: For uninstalled thrust,

$$F = \frac{1}{g_c}(\dot{m}_9 V_9 - \dot{m}_0 V_0) + A_9(P_9 - P_0)$$

$$\frac{F}{\dot{m}_0} = \frac{a_0}{g_c}\left(\frac{\dot{m}_9 V_9}{\dot{m}_0 a_0} - M_0\right) + \frac{A_9 P_9}{\dot{m}_0}\left(1 - \frac{P_0}{P_9}\right) \qquad (7.1)$$

We note that

$$\frac{A_9 P_9}{\dot{m}_0}\left(1 - \frac{P_0}{P_9}\right) = \frac{\dot{m}_9}{\dot{m}_0} \frac{A_9 P_9}{\rho_9 A_9 V_9}\left(1 - \frac{P_0}{P_9}\right)$$

$$= \frac{\dot{m}_9}{\dot{m}_0} \frac{P_9}{[P_9/(R_9 T_9)]V_9}\left(1 - \frac{P_0}{P_9}\right)$$

$$= \frac{\dot{m}_9}{\dot{m}_0} \frac{R_9 T_9}{V_9}\left(1 - \frac{P_0}{P_9}\right)$$

$$= \frac{\dot{m}_9}{\dot{m}_0} \frac{R_9 T_9}{V_9} \frac{\gamma_0 R_0 g_c T_0}{\gamma_0 R_0 g_c T_0}\left(1 - \frac{P_0}{P_9}\right)$$

$$= \frac{\dot{m}_9}{\dot{m}_0} \frac{R_9 T_9}{R_0 T_0} \frac{a_0^2}{\gamma_0 g_c V_9}\left(1 - \frac{P_0}{P_9}\right)$$

$$\frac{A_9 P_9}{\dot{m}_0}\left(1 - \frac{P_0}{P_9}\right) = \frac{a_0}{g_c}\left(\frac{\dot{m}_9}{\dot{m}_0} \frac{R_9}{R_0} \frac{T_9/T_0}{V_9/a_0} \frac{1 - P_0/P_9}{\gamma_0}\right) \qquad (7.2)$$

For the case of the turbojet cycle, the mass ratio can be written in terms of the fuel/air ratio f:

$$\frac{\dot{m}_9}{\dot{m}_0} = 1 + f$$

Using Eq. (7.2) with gas property subscripts c and t for engine stations 0 to 9, respectively, we can write Eq. (7.1) as

$$\frac{F}{\dot{m}_0} = \frac{a_0}{g_c}\left[(1+f)\frac{V_9}{a_0} - M_0 + (1+f)\frac{R_t}{R_c} \frac{T_9/T_0}{V_9/a_0} \frac{1 - P_0/P_9}{\gamma_c}\right] \qquad (7.3)$$

Step 2:

$$\left(\frac{V_9}{a_0}\right)^2 = \frac{a_9^2 M_9^2}{a_0^2} = \frac{\gamma_9 R_9 g_c T_9}{\gamma_0 R_0 g_c T_0} M_9^2$$

For the turbojet cycle, this equation becomes

$$\left(\frac{V_9}{a_0}\right)^2 = \frac{\gamma_t R_t T_9}{\gamma_c R_c T_0} M_9^2 \qquad (7.4)$$

Note: From the definition of τ_λ given in Eq. (5.7), we have

$$\tau_\lambda = \frac{c_{pt} T_{t4}}{c_{pc} T_0} \qquad (7.5)$$

Step 3: We have

$$M_9^2 = \frac{2}{\gamma_t - 1}\left[\left(\frac{P_{t9}}{P_9}\right)^{(\gamma_t - 1)/\gamma_t} - 1\right] \qquad (7.6)$$

where

$$\frac{P_{t9}}{P_9} = \frac{P_0}{P_9} \pi_r \pi_d \pi_c \pi_b \pi_t \pi_n \tag{7.7}$$

Step 4: We have

$$\frac{T_9}{T_0} = \frac{T_{t9}/T_0}{(P_{t9}/P_9)^{(\gamma_t - 1)/\gamma_t}} \tag{7.8}$$

where

$$\frac{T_{t9}}{T_0} = T_{t4} \tau_t \tau_n / T_0$$

Step 5: Application of the first law of thermodynamics to the burner gives

$$\dot{m}_0 h_{t3} + \eta_b \dot{m}_f h_{PR} = \dot{m}_4 h_{t4} \tag{7.9}$$

Dividing the preceding equation by $\dot{m}_0$ gives

$$h_{t3} + f \eta_b h_{PR} = (1 + f) h_{t4}$$

Solving for the fuel/air ratio gives

$$f = \frac{h_{t4} - h_{t3}}{\eta_b h_{PR} - h_{t4}} \tag{7.10}$$

Step 6: The power balance between the turbine and compressor, with a mechanical efficiency η_m of the turbine compressor coupling (shaft and bearing assemblies), gives

$$\text{Power into compressor} = \text{net power from turbine}$$
$$\dot{m}_0 c_{pc}(T_{t3} - T_{t2}) = \eta_m \dot{m}_4 c_{pt}(T_{t4} - T_{t5}) \tag{7.11}$$

Dividing the preceding equation by $\dot{m}_0 c_{pc} T_{t2}$ gives

$$\tau_c - 1 = \eta_m (1 + f) \frac{\tau_\lambda}{\tau_r} (1 - \tau_t)$$

Solving for the turbine temperature ratio gives

$$\tau_t = 1 - \frac{1}{\eta_m (1 + f)} \frac{\tau_r}{\tau_\lambda} (\tau_c - 1) \tag{7.12}$$

This expression enables us to solve for τ_t, from which we then obtain

$$\pi_t = \tau_t^{\gamma_t / [(\gamma_t - 1)e_t]} \tag{7.13}$$

We note that η_t will be given in terms of e_t by [Eq. (6.22)]

$$\eta_t = \frac{1 - \tau_t}{1 - \tau_t^{1/e_t}} \tag{7.14}$$

We also require the calculation of τ_c to allow determination of τ_t. Thus we note from Eq. (6.14)

$$\tau_c = \pi_c^{(\gamma_c - 1)/(\gamma_c e_c)} \tag{7.15}$$

Then also, from Eq. (6.15),

$$\eta_c = \frac{\pi_c^{(\gamma_c - 1)/\gamma_c} - 1}{\tau_c - 1} \tag{7.16}$$

Step 7: Equation (7.3) for specific thrust cannot be simplified for this analysis.
Step 8: The equation for the thrust specific fuel consumption is

$$S = \frac{f}{F/\dot{m}_0} \tag{7.17}$$

Step 9: From the definitions of propulsive and thermal efficiency, one can easily show that for the turbojet engine

$$\eta_P = \frac{2g_c V_0(F/\dot{m}_0)}{a_0^2[(1 + f)(V_9/a_0)^2 - M_0^2]} \tag{7.18}$$

and

$$\eta_{Th} = \frac{a_0^2[(1 + f)(V_9/a_0)^2 - M_0^2]}{2g_c f\, h_{PR}} \tag{7.19}$$

Now we have all of the equations needed for analysis of the turbojet cycle. For convenience, this system of equations is listed (in the order of calculation) in the online Supporting Material.

See Section 7.2.SM, Summary of Equations: Turbojet Engine, in the online supporting material for related information.

7.2.2 Examples: Turbojet Engine

The following examples involve multiple calculations using the PCA equations just developed to investigate performance trends for a real turbojet. Examples 7.2 and 7.3 highlight differences in the performance trends between ideal and real turbojets. Example 7.4 re-introduces (refer to Example 5.3, Fig. 5.13b) you to one of the most useful analysis tools—the "carpet plot"—for assisting in your determination of viable design space for a given reference condition (altitude, freestream Mach number).

Example 7.1

Consider a turbojet engine with losses that are representative of modern engine technology operating at high speed with the following input data.

Inputs:

$$M_0 = 2, \quad T_0 = 216.7\,\text{K}, \quad \gamma_c = 1.4, \quad c_{pc} = 1.004\,\text{kJ}/(\text{kg} \cdot \text{K}), \quad \gamma_t = 1.3,$$

$$c_{pt} = 1.239\,\text{kJ}/(\text{kg} \cdot \text{K}), \quad h_{PR} = 42,800\,\text{kJ/kg}, \quad \pi_{d\,\text{max}} = 0.95, \quad \pi_b = 0.94,$$

$$\pi_n = 0.96, \quad e_c = 0.9, \quad e_t = 0.9, \quad \eta_b = 0.98, \quad \eta_m = 0.99, \quad P_0/P_9 = 0.5,$$

$$T_{t4} = 1800\,\text{K}, \quad \pi_c = 10$$

Equations:

$$R_c = \frac{\gamma_c - 1}{\gamma_c} c_{pc} = \frac{0.4}{1.4}(1.004) = 0.2869\,\text{kJ}/(\text{kg} \cdot \text{K})$$

$$R_t = \frac{\gamma_t - 1}{\gamma_t} c_{pt} = \frac{0.3}{1.3}(1.239) = 0.2859\,\text{kJ}/(\text{kg} \cdot \text{K})$$

$$a_0 = \sqrt{\gamma_c R_c g_c T_0} = \sqrt{1.4 \times 286.9 \times 1 \times 216.7} = 295.0\,\text{m/s}$$

$$V_0 = a_0 M_0 = 295.0 \times 2 = 590\,\text{m/s}$$

$$\tau_r = 1 + \frac{\gamma_c - 1}{2} M_0^2 = 1 + 0.2 \times 2^2 = 1.8$$

$$\pi_r = \tau_r^{\gamma_c/(\gamma_c - 1)} = 1.8^{3.5} = 7.82445$$

$$\eta_r = 1 - 0.075(M_0 - 1)^{1.35} = 1 - 0.075(1^{1.35}) = 0.925$$

$$\pi_d = \pi_{d\,\text{max}}\eta_r = 0.95 \times 0.925 = 0.87875$$

$$\tau_\lambda = \frac{c_{pt}T_{t4}}{c_{pc}T_0} = \frac{1.239 \times 1800}{1.004 \times 216.7} = 10.2506$$

$$\tau_c = \pi_c^{(\gamma_c - 1)/(\gamma_c e_c)} = 10^{1/(3.5 \times 0.9)} = 2.0771$$

$$\eta_c = \frac{\pi_c^{(\gamma_c - 1)/\gamma_c} - 1}{\tau_c - 1} = \frac{10^{1/3.5} - 1}{2.0771 - 1} = 0.8641$$

$$T_{t3} = T_0 \tau_r \tau_c = 216.7 \times 1.8 \times 2.0771 = 810.2\,\text{K}$$

From Appendix L, we have for $T_{t3} = 810.2$ K and the fuel/air ratio $f = 0$ that $h_{t3} = 832.92$ kJ/kg. For $T_{t4} = 1800$ K and an initial estimate for f of 0.0338 (which is 50% of the maximum value of 0.0676), Appendix L gives $h_{t4} = 2103.12$ kJ/kg. A new estimate of f follows

$$f = \frac{h_{t4} - h_{t3}}{\eta_b h_{PR} - h_{t4}} = \frac{2103.12 - 832.92}{0.98 \times 42,800 - 2103.12} = 0.03188$$

(Continued)

Example 7.1 *(Continued)*

After several iterations, we get $f = 0.03173$.

$$\tau_t = 1 - \frac{1}{\eta_m(1+f)} \frac{\tau_r}{\tau_\lambda}(\tau_c - 1)$$

$$= 1 - \frac{1}{0.99 \times 1.0173} \frac{1.8}{10.2506}(2.0771 - 1) = 0.8148$$

$$\pi_t = \tau_t^{\gamma_t/[(\gamma_t-1)e_t]} = 0.8148^{1.3/(0.3\times0.9)} = 0.3731$$

$$\eta_t = \frac{1 - \tau_t}{1 - \tau_t^{1/e_t}} = \frac{1 - 0.8148}{1 - 0.8148^{1/0.9}} = 0.9099$$

$$\frac{P_{t9}}{P_9} = \frac{P_0}{P_9} \pi_r \pi_d \pi_c \pi_b \pi_t \pi_n$$

$$= 0.5 \times 7.824 \times 0.8788 \times 10 \times 0.94 \times 0.3731 \times 0.96 = 11.574$$

$$M_9 = \sqrt{\frac{2}{\gamma_t - 1}\left[\left(\frac{P_{t9}}{P_9}\right)^{(\gamma_t-1)/\gamma_t} - 1\right]}$$

$$= \sqrt{\frac{2}{0.3}(11.574^{0.3/1.3} - 1)} = 2.250$$

$$\frac{T_9}{T_0} = \frac{T_{t4}\tau_t/T_0}{(P_{t9}/P_9)^{(\gamma_t-1)/\gamma_t}} = \frac{1800 \times 0.8148/216.7}{11.574^{0.3/1.3}} = 3.8464$$

$$\frac{V_9}{a_0} = M_9\sqrt{\frac{\gamma_t R_t T_9}{\gamma_c R_c T_c}} = 2.253\sqrt{\frac{1.3 \times 0.2859}{1.4 \times 0.2869}}(3.846) = 4.250$$

$$\frac{F}{\dot{m}_0} = \frac{a_0}{g_c}\left[(1+f)\frac{V_9}{a_0} - M_0 + (1+f)\frac{R_t T_9/T_0}{R_c V_9/a_0}\frac{1 - P_0/P_9}{\gamma_c}\right]$$

$$= \frac{295}{1}\left(1.03173 \times 4.250 - 2 + 1.03173\frac{0.2859}{0.2869}\frac{3.846}{4.250}\frac{0.5}{1.4}\right)$$

$$= 295(2.3849 + 0.3323) = 801.6\,\text{N}/(\text{kg}/\text{s})$$

$$S = \frac{f}{F/\dot{m}_0} = \frac{0.03173}{801.6} \times 10^6 = 39.58\,(\text{mg}/\text{s})/\text{kN}$$

$$\eta_{Th} = \frac{a_0^2[(1+f)(V_9/a_0)^2 - M_0^2]}{2g_c f h_{PR}}$$

$$= \frac{295.0^2[(1.03173)4.250^2) - 2^2]}{2 \times 1 \times 0.03173 \times 42,800 \times 1000} = 46.89\%$$

(Continued)

Example 7.1 *(Continued)*

$$\eta_P = \frac{2g_c V_0 (F/\dot{m}_0)}{a_0^2 [(1+f)(V_9/a_0)^2 - M_0^2]}$$

$$= \frac{2 \times 1 \times 590 \times 801.6}{295^2 [1.03173(4.250^2) - 2^2]} = 74.27\%$$

$$\eta_O = \eta_P \eta_{Th} = 0.4689 \times 0.7427 = 34.83\%$$

Example 7.2

Now we consider the turbojet cycle with losses that are representative of modern turbojet technology over the same range of Mach numbers and compressor pressure ratios as analyzed for the ideal turbojet cycle in Example 5.1 and plotted in Figs. 5.3a–5.3c.

Inputs:

$$M_0 = 0 \rightarrow 3, \quad T_0 = 390°\text{R}, \quad \gamma_c = 1.4, \quad c_{pc} = 0.24 \, \text{Btu}/(\text{lbm} \cdot °\text{R}),$$
$$\gamma_t = 1.33, \quad c_{pt} = 0.276 \, \text{Btu}/(\text{lbm} \cdot °\text{R}), \quad h_{PR} = 18{,}400 \, \text{Btu/lbm},$$
$$\pi_{d\,\text{max}} = 0.98, \quad \pi_b = 0.98, \quad \pi_n = 0.98, \quad e_c = 0.92, \quad e_t = 0.91,$$
$$\eta_b = 0.99, \quad \eta_m = 0.98, \quad P_0/P_9 = 1, \quad T_{t4} = 3000°\text{R}, \quad \pi_c = 2 \rightarrow 30$$

The results of the real turbojet analysis are plotted vs compressor pressure ratio in Figs. 7.3a–7.3c. When compared to the corresponding figures for the ideal turbojet cycle, the following can be concluded for the turbojet cycle with losses.

a) *Specific thrust $F/\dot{m}_0$ and thrust specific fuel consumptions S.* Comparing Fig. 7.3a to Fig. 5.3a, one can see that the variation of specific thrust with compressor pressure ratio for any given Mach number is not appreciably changed and the magnitudes are nearly equal. This is because some of the real effects we have included, like component efficiencies < 1.0, tend to reduce the specific thrust estimation, whereas others, like higher specific heat c_{pt} downstream of the burner, tend to increase the specific thrust estimation. At high Mach numbers, the effect of the losses causes the thrust to go to zero at a lower compressor pressure ratio. This should make sense because the losses (energy in non useful forms) take away from energy available in desired forms. For a given Mach number, the compressor pressure ratio that gives maximum specific thrust is lower than that of the ideal turbojet.

(Continued)

Example 7.2 *(Continued)*

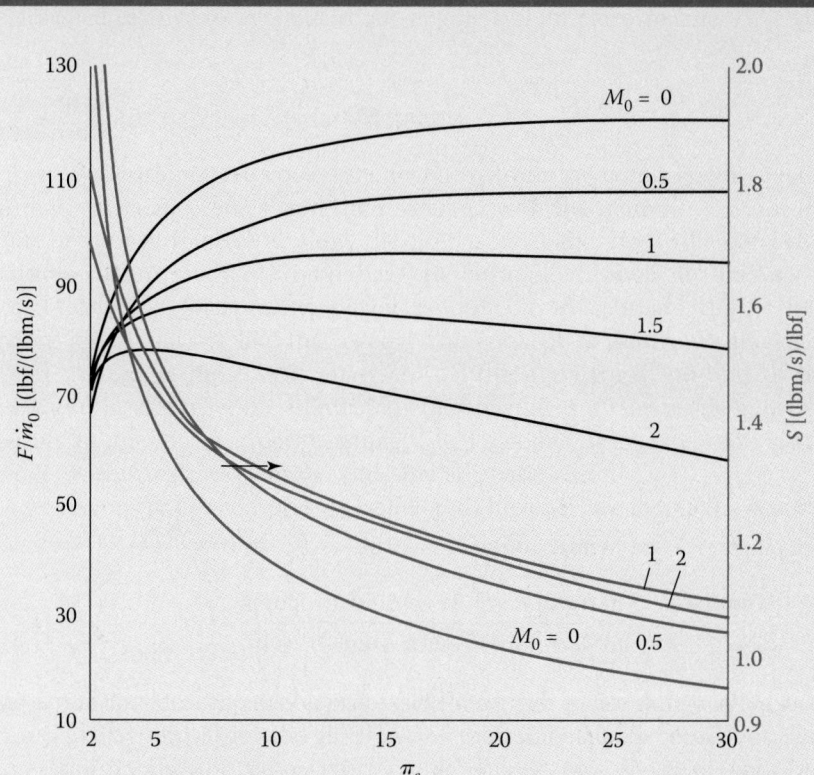

Fig. 7.3a Turbojet performance vs π_c: specific thrust and fuel consumption.

The impact of including real effects is quite striking when one compares the uninstalled thrust specific fuel consumption trends for turbojets. Comparison of Fig. 7.3a to Fig. 5.3a shows that the values of fuel consumption are larger for the engine with losses. The thrust specific fuel consumption no longer continues to decrease with increasing compressor pressure ratio, and there is now a compressor pressure ratio giving minimum fuel consumption for a given Mach number. All of these trends should make sense. To get to the same highest thermodynamic state ($T_{t4} = 3000°R$) in the cycle, more energy must be released in the combustion process of the real turbojet to overcome the component losses.

b) *Turbine temperature ratio τ_t.* Comparing Fig. 7.3b to Fig. 5.3b, we see that the values of the turbine temperature ratio are approximately

(Continued)

Example 7.2 *(Continued)*

the same. The main reason for this is that the increase in specific heat across the main burner accounted for by c_{pt} offsets the inefficiency of the compressor and turbine.

c) *Propulsive efficiency* η_P. Comparison of Fig. 7.3c to Fig. 5.3c shows that the propulsive efficiencies are a little larger for the turbojet with losses. This is due mainly to the decrease in exhaust velocity for the engine with losses, which brings V_9 closer to V_0 [see Eq. (1.16)].

d) *Thermal efficiency* η_{Th}. Comparing Fig. 7.3c to Fig. 5.3c, we can see that the engines with losses have lower thermal efficiency, as would be expected. Also, the thermal efficiency of high-compressor-pressure-ratio engines at high Mach go toward zero because the thrust goes to zero before the fuel flow rate.

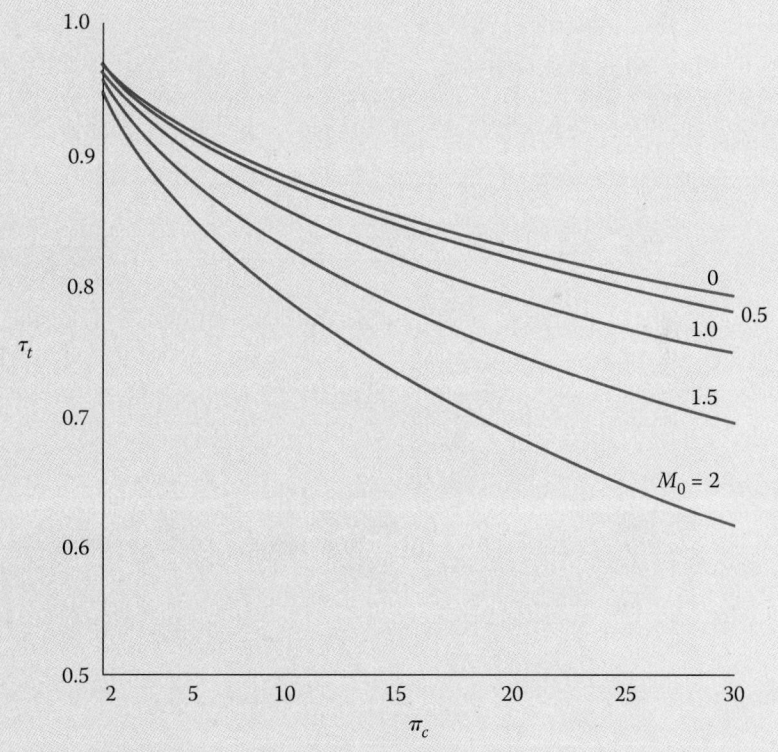

Fig. 7.3b   Turbojet performance vs π_c: turbine temperature ratio.

(Continued)

Example 7.2 *(Continued)*

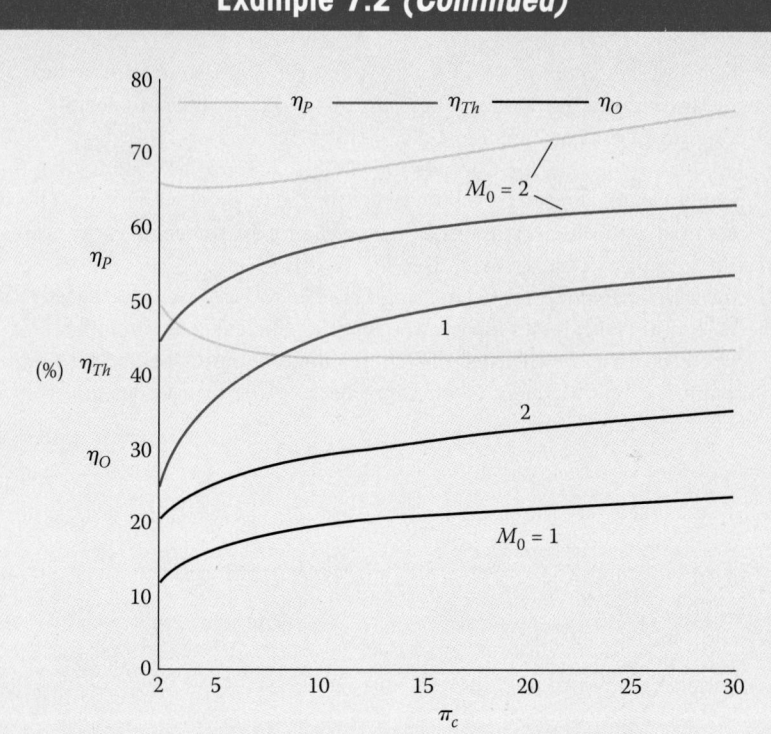

Fig. 7.3c Turbojet performance vs π_c: efficiencies.

e) *Overall efficiency η_O.* One can see that the overall efficiencies are lower for the turbojet engines with losses by comparison of Fig. 7.3c to Fig. 5.3c. This is mainly due to the decrease in thermal efficiency of the engines with losses.

Example 7.3

The effect of compressor efficiency and exhaust nozzle off-design conditions on the performance of a turbojet engine cycle at Mach 2.0 is investigated in this example. The compressor pressure ratio was varied over the range of 2 to 30 for two different compressor polytropic efficiencies to give the results indicated in Fig. 7.4. Also included on this plot are the results of ideal cycle analysis. The input data for these analyses are listed here.

(Continued)

Example 7.3 *(Continued)*

Inputs:

$$M_0 = 2, \quad T_0 = 390°R, \quad \gamma_c = 1.4, \quad c_{pc} = 0.24 \, \text{Btu}/(\text{lbm} \cdot °R),$$
$$\gamma_t = 1.33, \quad c_{pt} = 0.276 \, \text{Btu}/(\text{lbm} \cdot °R), \quad h_{PR} = 18{,}400 \, \text{Btu}/\text{lbm},$$
$$\pi_{d\,max} = 0.98, \quad \pi_b = 0.96, \quad \pi_n = 0.98, \quad e_c = 0.90 \text{ and } 0.87, \quad e_t = 0.90,$$
$$\eta_b = 0.99, \quad \eta_m = 0.99, \quad P_0/P_9 = 1, \quad T_{t4} = 2800°R, \quad \pi_c = 2 \to 30$$

a) *Effect of compressor technology level (polytropic efficiency).* It can be seen from Fig. 7.4 that the ideal turbojet analysis gives the basic trends for the lower values of the compressor pressure ratio. As the compressor pressure ratio increases, the effect of engine losses increases. At low compressor pressure ratios, the ideal analysis predicts a lower value of specific thrust than the engine with losses because the exit momentum of the fuel is neglected in the ideal case.

It can also be seen from Fig. 7.4 that a prospective designer would be immediately confronted with a design choice, because the compressor pressure ratio leading to maximum specific thrust is far from that leading to minimum fuel consumption. Clearly, a short-range interceptor would better suit a low compressor pressure ratio with the resultant high specific

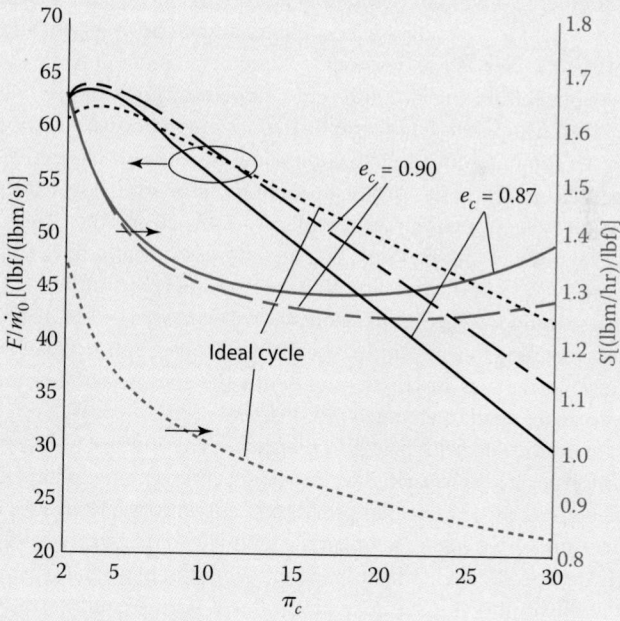

Fig. 7.4 Effect of compressor polytropic efficiency of turbojet cycle.

(Continued)

Example 7.3 *(Continued)*

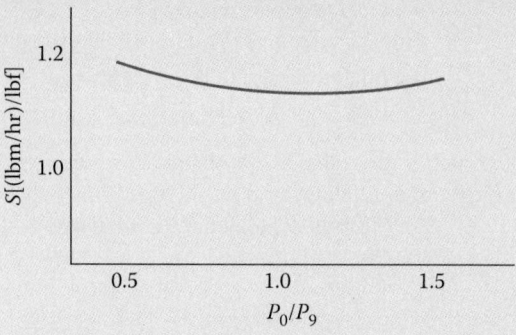

Fig. 7.5   Effect of nozzle off-design conditions on thrust specific fuel consumption.

thrust and lightweight (small compressor) engine. Conversely, the designer of a long-range supersonic transport would favor an engine with high compressor pressure ratio and low specific fuel consumption. Thus we see what should be obvious—before an engine can be correctly designed, the mission (use) for which it is being designed must be precisely understood.

Another aspect of the designer's dilemma becomes apparent when the curves obtained for the two different compressor polytropic efficiencies in Fig. 7.4 are compared. For example, if a designer chooses a compressor pressure ratio of 22 for use in a supersonic transport because the compressor design group has promised a compressor with $e_c = 0.90$, and then the group delivers a compressor with $e_c = 0.87$, clearly the choice $\pi_c = 22$ would be quite inappropriate. Such a compressor would have a higher pressure ratio than that leading to minimum fuel consumption. Thus the designer would have a compressor that was heavier (and more expensive) than that leading to a minimum specific fuel consumption, he or she would also have lower specific thrust, and finally the compressor designer would have an acute need to change employers.

b) *Effect of exhaust nozzle off-design conditions.* The effect of nozzle off-design conditions can be invesigated by considering the engine to have the same parameters as those indicated previously, but with various values of P_0/P_9. As an example, we consider an engine with $\pi_c = 22$ and $e_c = 0.90$ to obtain the thrust specific fuel consumption information plotted in Fig. 7.5.

It is apparent that for small exit nozzle off-design conditions ($0.8 < P_0/P_9 < 1.2$), the thrust and thrust specific fuel consumption vary only slightly with the exit pressure mismatch. This result indicates that the best nozzle design should be determined by considering the external flow behavior

(Continued)

Example 7.4

At a specific flight condition M_0, the most dominant independent variables that impact the performance of the simple turbojet engine are the compressor pressure ratio π_c and the combustor exit temperature T_{t4}. This example investigates these changes for the following input data:

Inputs:

$$M_0 = 2, \quad T_0 = 216.7\,\text{K}, \quad \gamma_c = 1.4, \quad c_{pc} = 1.004\,\text{kJ/(kg}\cdot\text{K)}, \quad \gamma_t = 1.35,$$

$$c_{pt} = 1.096\,\text{kJ/(kg}\cdot\text{K)}, \quad h_{PR} = 42,800\,\text{kJ/kg}, \quad \pi_{d\,\text{max}} = 0.98, \quad \pi_b = 0.96,$$

$$\pi_n = 0.98, \quad e_c = 0.89, \quad e_t = 0.90, \quad \eta_b = 0.99, \quad \eta_m = 0.99, \quad P_0/P_9 = 1,$$

$$T_{t4} = 1600, 1800, 2000, \text{ and } 2200\,\text{K}, \quad \pi_c = 8, 16, \text{ and } 24$$

A plot of the specific thrust and thrust specific fuel consumption vs these two independent variables is given in Fig. 7.6. This plot has a shape of a 3-D surface and is referred to as a "carpet plot." These carpet plots contain a wealth of information. We have already explored the effects of π_c on F/m_0 and S.

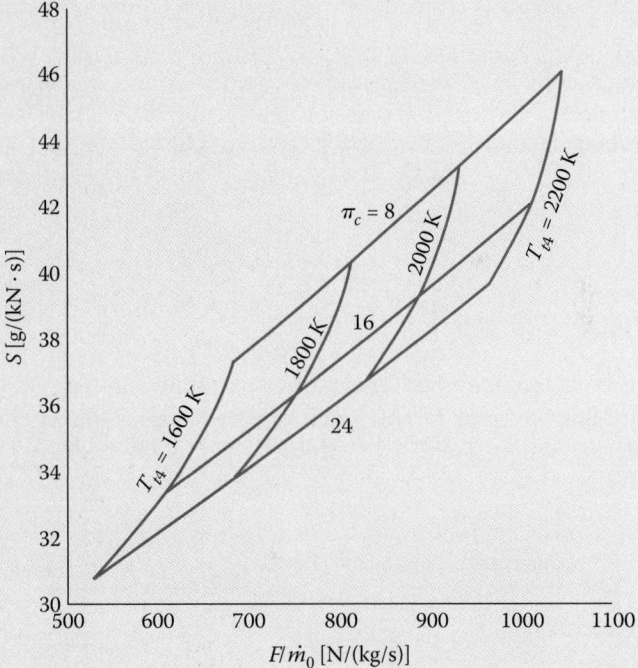

Fig. 7.6   Carpet plot of turbojet performance vs t_{t4} and compressor pressure ratio π_c.

(Continued)

Example 7.4 *(Continued)*

Let's examine T_{t4} effects with the help of the T-s diagram from Fig. 7.2. With everything else held constant, an increased T_{t4} results in an increased energy state at the turbine exit T_{t5} resulting in increased exhaust nozzle exit velocity V_9. Consequently, specific thrust increases, but the additional energy release required by increased fuel addition in the combustor to get to higher T_{t4} results in an increase in S. Looking at Eq. (7.17), the increase in fuel/air ratio f must be proportionately higher than the increase in specific thrust.

Let's examine Fig. 7.6 in some detail.

- In general, one would like to see the performance trends go down and to the right—lower S and higher specific thrust. That, of course, is not the case; as usual, design tradeoffs must be made.
- For a given level of technology (our component performance figures of merit):
  - Increasing T_{t4} increases specific thrust (good), but also increases thrust specific fuel consumption (bad).
  - Increasing π_c increases the cycle thermal efficiency, which decreases thrust specific fuel consumption (good), but also decreases specific thrust (bad).

In practice, one must select design choices (in this case, π_c and T_{t4}) based on the proportional changes in overall engine performance relative to overall mission needs and engine goals. These goals would likely include life-cycle costs (includes development and sustainment), time between overhauls, maintainability, and the like.

7.3 Turbojet with Afterburner

The turbojet engine with afterburner is shown in Fig. 7.7, and the temperature-entropy plot of this engine with losses is shown in Fig. 7.8. The numbering system indicated in these figures is the industry standard.

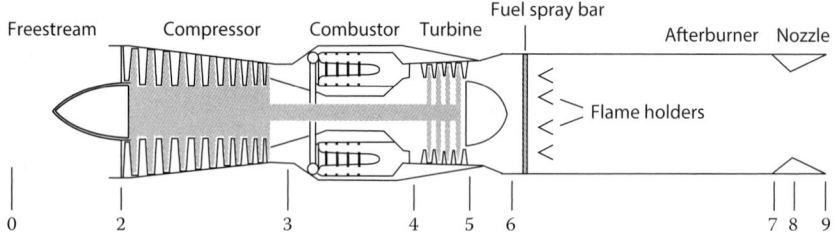

Fig. 7.7 Afterburning turbojet engine with station numbering.

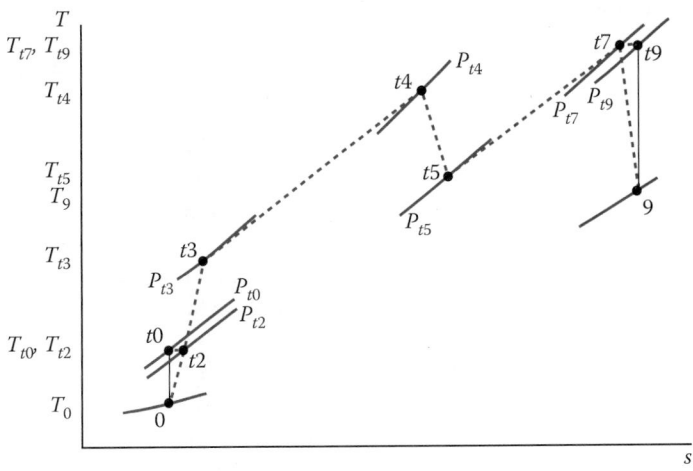

Fig. 7.8 The T-s diagram for afterburning turbojet engine.

For the analysis of this engine, we remind the reader of the following definitions for the afterburner (see Tables 5.1 and 6.1):

$$\pi_{AB} = \frac{P_{t7}}{P_{t6}} \qquad f_{AB} = \frac{\dot{m}_{fAB}}{\dot{m}_0} \qquad (7.20a)$$

$$\tau_{AB} = \frac{T_{t7}}{T_{t6}} \qquad \tau_{\lambda AB} = \frac{h_{t7}}{h_0} \qquad (7.20b)$$

$$\eta_{AB} = \frac{(\dot{m}_0 + \dot{m}_f + \dot{m}_{fAB})h_{t7} - (\dot{m}_0 + \dot{m}_f)h_{t6}\dot{m}_f}{\dot{m}_{fAB}h_{PR}} \qquad (7.21)$$

Note that stations 6 and 7 are used for these afterburner parameters. We assume isentropic flow from station 5 to station 6 in the following analysis. Thus Fig. 7.9 shows the afterburning process going from station 5 to 7.

We also note that

$$\tau_{\lambda AB} = \frac{c_{PAB}}{c_{pt}} \frac{h_{t4}}{h_0} \frac{T_{t5}}{T_{t4}} \frac{T_{t8}}{T_{t5}} = \frac{c_{pAB}}{c_{pc}} \tau_\lambda \tau_t \tau_{AB} \qquad (7.22)$$

and

$$\frac{\dot{m}_9}{\dot{m}_0} = \frac{\dot{m}_0 + \dot{m}_f + \dot{m}_{fAB}}{\dot{m}_0} = 1 + f + f_{AB} \qquad (7.23)$$

7.3.1 Cycle Analysis

The expression for the thrust will be the same as that already obtained for the turbojet without afterburning except that the effects of fuel addition in the afterburner must be included. Application of the steps of cycle analysis (see Section 5.4) is listed next.

Step 1: The specific thrust equation becomes

$$\frac{F}{\dot{m}_0} = \frac{a_0}{g_c} \left[(1 + f + f_{AB}) \frac{V_9}{a_0} - M_0 + (1 + f + f_{AB}) \right.$$
$$\left. \times \frac{R_{AB}}{R_c} \frac{T_9/T_0}{V_9/a_0} \frac{1 - P_0/P_9}{\gamma_c} \right] \qquad (7.24)$$

Step 2: As before,

$$\left(\frac{V_9}{a_0} \right)^2 = \frac{\gamma_{AB} R_{AB} T_9}{\gamma_c R_c T_0} M_9^2 \qquad (7.25)$$

Step 3: We have

$$M_9^2 = \frac{2}{\gamma_{AB} - 1} \left[\left(\frac{P_{t9}}{P_9} \right)^{(\gamma_{AB} - 1)/\gamma_{AB}} - 1 \right] \qquad (7.26)$$

where

$$\frac{P_{t9}}{P_9} = \frac{P_0}{P_0} \pi_r \pi_d \pi_c \pi_b \pi_t \pi_{AB} \pi_n \qquad (7.27)$$

Step 4: We have

$$\frac{T_9}{T_0} = \frac{T_{t9}/T_0}{(P_{t9}/P_9)^{(\gamma_{AB} - 1)/\gamma_{AB}}}$$

where

$$\frac{T_{t9}}{T_0} = \frac{T_{t7}}{T_0} \qquad (7.28)$$

Step 5: Application of the steady flow energy equation to the main combustor gives [Eq. (7.10)]

$$f = \frac{h_{t4} - h_{t3}}{\eta_b h_{PR} - h_{t4}}$$

Application of the steady flow energy equation to the afterburner gives

$$(\dot{m}_0 + \dot{m}_f) h_{t6} + \eta_{AB} \dot{m}_{fAB} h_{PR} = (\dot{m}_0 + \dot{m}_f + \dot{m}_{fAB}) h_{t7}$$

This equation can be solved for the afterburner fuel/air ratio f_{AB}, giving

$$f_{AB} = (1 + f) \frac{h_{t7} - h_{t6}}{\eta_{AB} h_{PR} - h_{t7}} \qquad (7.29)$$

Step 6: The power balance between the turbine and compressor is unaffected by the addition of the afterburner. Thus Eqs. (7.12)–(7.16) apply to this engine cycle.

Step 7: See Eq. (7.24).

Step 8: The thrust specific fuel consumption S is expressed in terms of both the main burner and afterburner fuel/air ratios as

$$S = \frac{f + f_{AB}}{F/\dot{m}_0} \tag{7.30}$$

Step 9: From the definitions of propulsive and thermal efficiency, one can easily show that for the afterburning turbojet engine

$$\eta_P = \frac{2g_c V_0(F/\dot{m}_0)}{a_0^2[(1 + f + f_{AB})(V_9/a_0)^2 - M_0^2]} \tag{7.31}$$

$$\eta_{Th} = \frac{a_0^2[(1 + f + f_{AB})(V_9/a_0)^2 - M_0^2]}{2g_c(f + f_{AB})h_{PR}} \tag{7.32}$$

Now we have all of the equations needed for analysis of the afterburning turbojet cycle. For convenience, this system of equations is listed (in the order of calculation) in the online supporting materials for easier calculation.

See Section 7.3.SM, Summary of Equations: Afterburning Turbojet Engine, in the online supporting material for related information.

Example 7.5

We consider an example similar to that considered for the "dry" turbojet of Example 7.3 so that we can directly compare the effects of afterburning. Thus we have the following input data.

Inputs:

$$M_0 = 2, \quad T_0 = 390°R, \quad \gamma_c = 1.4, \quad c_{pc} = 0.24 \text{ Btu/(lbm} \cdot °R),$$
$$\gamma_t = 1.33, \quad c_{pt} = 0.276 \text{ Btu/(lbm} \cdot °R), \quad h_{PR} = 18{,}400 \text{ Btu/lbm},$$
$$\gamma_{AB} = 1.30, \quad c_{pAB} = 0.295 \text{ Btu/(lbm} \cdot °R), \quad \pi_{d\,max} = 0.98, \quad \pi_b = 0.96,$$
$$\pi_{AB} = 0.96, \quad \pi_n = 0.98, \quad e_c = 0.89, \quad e_t = 0.90, \quad \eta_b = 0.99,$$
$$\eta_{AB} = 0.96, \quad \eta_m = 0.99, \quad P_0/P_9 = 1, \quad T_{t4} = 3000°R,$$
$$T_{t7} = 3500°R, \quad \pi_c = 8 \rightarrow 24$$

In this example, we limit the maximum compressor pressure ratio to 14 because the total temperature leaving the compressor T_{t3} will exceed 1200°F (temperature limit of many compressors without getting too exotic with material alloys) at higher pressure ratios. For the afterburning turbojet, we take

$$\pi_n \pi_{AB} = 0.98 \times 0.96 \qquad \text{afterburner on}$$
$$\pi_n \pi_{AB} = 0.98 \qquad \text{afterburner off}$$

(Continued)

Example 7.5 *(Continued)*

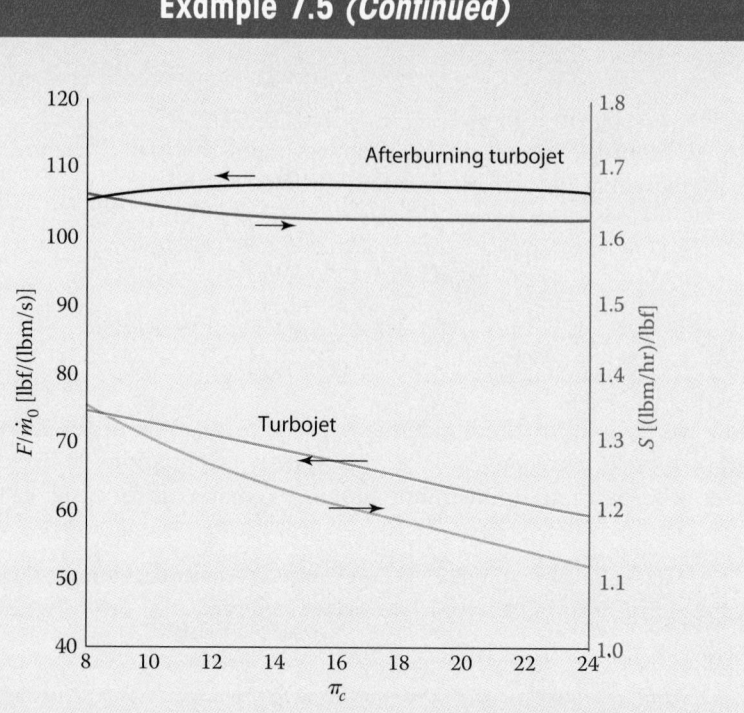

Fig. 7.9 Performance of turbojet engine with and without afterburner.

The results are indicated in Fig. 7.9. Note that operation of the afterburner will increase both the specific thrust and the fuel consumption. The magnitude of the increases depends on the compressor pressure ratio. A compressor pressure ratio of 12 gives good specific thrust and fuel consumption for afterburner operation and reasonable performance without the afterburner.

The design compressor pressure ratio will depend on the aircraft and its mission (use), which requires an in-depth analysis.

7.4 Turbofan: Separate Exhaust Streams

A turbofan engine with industry standard station numbering is shown in Fig. 7.10. A temperature vs entropy plot for the flow through the fan and the engine core is shown in Fig. 7.11. The effect of engine losses can be seen by comparing Fig. 7.11 with Figs. 5.9 and 5.10. The exit velocity of both the fan stream and the engine core stream is reduced by engine losses.

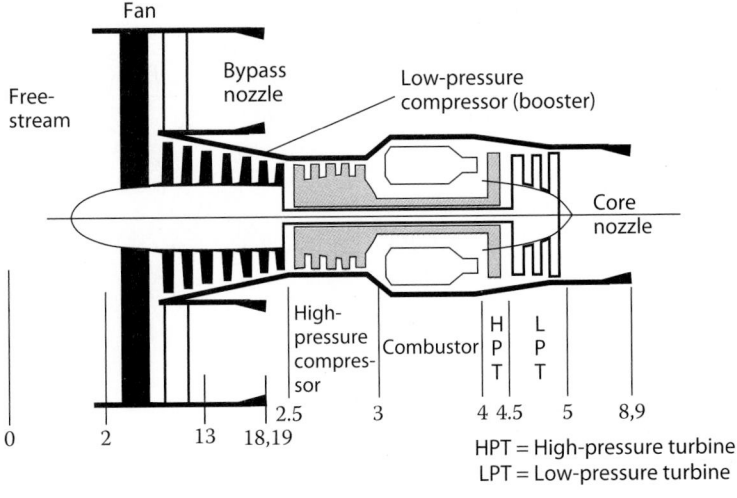

Fig. 7.10   Station numbering of turbofan engine.

7.4.1 Cycle Analysis

The steps of cycle analysis are applied to the turbofan engine and presented next in the order listed in Section 5.4. We will apply the steps of cycle analysis to both the fan stream and the engine core stream.

7.4.1.1 Bypass Stream

Steps 1–4 are as follows.

Step 1: Uninstalled thrust of bypass stream F_B:

$$F_B = \frac{\dot{m}_B}{g_c}(V_{19} - V_0) + A_{19}(P_{19} - P_0)$$

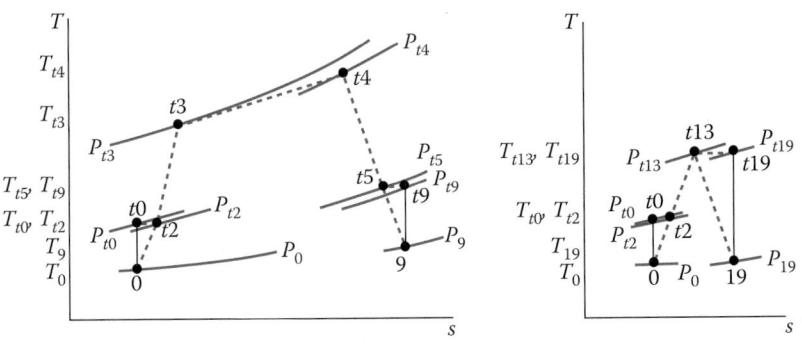

Fig. 7.11   The T-s diagram of turbofan engine with losses (not to scale).

Using Eq. (7.2) for the bypass stream gives

$$\frac{F_B}{\dot{m}_F} = \frac{a_0}{g_c}\left(\frac{V_{19}}{a_0} - M_0 + \frac{T_{19}/T_0}{V_{19}/a_0}\frac{1 - P_0/P_{19}}{\gamma_c}\right) \qquad (7.33)$$

Step 2:

$$\left(\frac{V_{19}}{a_0}\right)^2 = \frac{T_{19}}{T_0}M_{19}^2 \qquad (7.34)$$

Step 3: We have

$$M_{19}^2 = \frac{2}{\gamma_c - 1}\left[\left(\frac{P_{t19}}{P_{19}}\right)^{(\gamma_c - 1)/\gamma_c} - 1\right] \qquad (7.35)$$

where

$$\frac{P_{t19}}{P_{19}} = \frac{P_0}{P_{19}}\pi_r\pi_d\pi_f\pi_{fn} \qquad (7.36)$$

Step 4: We have

$$\frac{T_{19}}{T_0} = \frac{T_{t19}/T_0}{(P_{t19}/P_{19})^{(\gamma_c - 1)/\gamma_c}} \qquad (7.37a)$$

where

$$\frac{T_{t19}}{T_0} = \tau_r\tau_f \qquad (7.37b)$$

7.4.1.2 Engine Core Stream

Steps 1–5 are the same as for the turbojet engine cycle with losses.

Step 1: Uninstalled thrust:

$$F_C = \frac{1}{g_c}(\dot{m}_9 V_9 - \dot{m}_C V_0) + A_9(P_9 - P_0)$$

or

$$\frac{F_C}{\dot{m}_C} = \frac{a_0}{g_c}\left[(1 + f)\frac{V_9}{a_0} - M_0 + (1 + f)\frac{R_t}{R_c}\frac{T_9/T_0}{V_9/a_0}\frac{1 - P_0/P_9}{\gamma c}\right] \qquad (7.38)$$

where the fuel/air ratio for the main burner is defined as

$$f \equiv \frac{\dot{m}_f}{\dot{m}_C} \qquad (7.39)$$

Step 2:

$$\left(\frac{V_9}{a_0}\right)^2 = \frac{\gamma_t R_t T_9}{\gamma_c R_c T_0} M_9^2 \tag{7.40}$$

Step 3: We have

$$M_9^2 = \frac{2}{\gamma_t - 1}\left[\left(\frac{P_{t9}}{P_9}\right)^{(\gamma_t-1)/\gamma_t} - 1\right] \tag{7.41a}$$

where

$$\frac{P_{t9}}{P_9} = \frac{P_0}{P_9}\pi_r \pi_d \pi_c \pi_b \pi_t \pi_n \tag{7.41b}$$

Step 4: We have

$$\frac{T_9}{T_0} = \frac{T_{t9}/T_0}{(P_{t9}/P_9)^{(\gamma_t-1)/\gamma_t}} \tag{7.42a}$$

where

$$\frac{T_{t9}}{T_0} = \tau_r \tau_d \tau_c \tau_b \tau_t \tau_n = \frac{c_{pc}}{c_{pt}}\tau_\lambda \tau_t \tag{7.42b}$$

Step 5: Application of the first law of thermodynamics to the burner gives

$$\dot{m}_C h_{t3} + \eta_b \dot{m}_f h_{PR} = \dot{m}_4 h_{t4}$$

Solving for f, we get

$$f = \frac{h_{t4} - h_{t3}}{\eta_b h_{PR} - h_{t4}} \tag{7.43}$$

Step 6: The power balance among the turbine, compressor, and fan, with a mechanical efficiency η_m of the coupling among the turbine, compressor, and fan, gives

$$\underbrace{\dot{m}_C c_{pc}(T_{t3} - T_{t2})}_{\substack{\text{power into}\\\text{core stream}}} + \underbrace{\dot{m}_B c_{pc}(T_{t13} - T_{t2})}_{\substack{\text{power into}\\\text{bypass stream}}} = \underbrace{\eta_m \dot{m}_4 c_{pt}(T_{t4} - T_{t5})}_{\substack{\text{net power}\\\text{from turbine}}} \tag{7.44}$$

Dividing the preceding equation by $\dot{m}_C c_{pc} T_{t2}$ and using the definitions of temperature ratios, fuel/air ratio, and the bypass ratio [α, Eq. (5.27)], we obtain

$$\tau_c - 1 + \alpha(\tau_f - 1) = \eta_m(1 + f)\frac{\tau_\lambda}{\tau_r}(1 - \tau_t)$$

Solving for the turbine temperature ratio gives

$$\tau_t = 1 - \frac{1}{\eta_m(1+f)} \frac{\tau_r}{\tau_\lambda} [\tau_c - 1 + \alpha(\tau_f - 1)] \tag{7.45}$$

Equations (7.13–7.16) are used to obtain the unknown pressure or temperature ratio and efficiencies of the turbine and compressor. For the fan, the following equations apply:

$$\tau_f = \pi_f^{(\gamma_c-1)/(\gamma_c e_f)} \tag{7.46}$$

$$\eta_f = \frac{\pi_f^{(\gamma_c-1)/\gamma_c} - 1}{\tau_f - 1} \tag{7.47}$$

Step 7: Combining the thrust equations for the fan stream and the engine core stream, we obtain

$$\frac{F}{\dot{m}_0} = \frac{1}{1+\alpha} \frac{a_0}{g_c} \left[(1+f)\frac{V_9}{a_0} - M_0 + (1+f)\frac{R_t T_9/T_0}{R_c V_9/a_0}\frac{1 - P_0/P_9}{\gamma_c} \right]$$

$$+ \frac{\alpha}{1+\alpha} \frac{a_0}{g_c} \left(\frac{V_{19}}{a_0} - M_0 + \frac{T_{19}/T_0}{V_{19}/a_0}\frac{1 - P_0/P_{19}}{\gamma_c} \right) \tag{7.48}$$

Step 8: The thrust specific fuel consumption S is

$$S = \frac{\dot{m}_f}{F} = \frac{\dot{m}_f/\dot{m}_C}{(\dot{m}_0/\dot{m}_C)F/\dot{m}_0}$$

or

$$S = \frac{f}{(1+\alpha)F/\dot{m}_0} \tag{7.49}$$

Step 9: Expressions for the propulsive efficiency η_p and thermal efficiency η_{Th} are listed next for the case of $P_9 = P_{19} = P_0$. Development of these equations is left as an exercise for the reader.

$$\eta_P = \frac{2M_0[(1+f)(V_9/a_0) + \alpha(V_{19}/a_0) - (1+\alpha)M_0]}{(1+f)(V_9/a_0)^2 + \alpha(V_{19}/a_0)^2 - (1+\alpha)M_0^2} \tag{7.50}$$

$$\eta_{Th} = \frac{a_0^2[(1+f)(V_9/a_0)^2 + \alpha(V_{19}/a_0)^2 - (1+\alpha)M_0^2]}{2g_c f h_{PR}} \tag{7.51}$$

7.4.2 Exit Pressure Conditions

Separate-stream turbofan engines are generally used with subsonic aircraft, and the pressure ratio across both core and bypass nozzles is not very large. As a result, often typically convergent-only fixed geometry nozzles are used. In this case, if the nozzles are choked, we have from Eq. (3.8) with $M = 1.0$

$$\frac{P_{t19}}{P_{19}} = \left(\frac{\gamma_c + 1}{2}\right)^{\gamma_c/(\gamma_c - 1)} \quad \text{and} \quad \frac{P_{t9}}{P_9} = \left(\frac{\gamma_t + 1}{2}\right)^{\gamma_t/(\gamma_t - 1)} \tag{7.52}$$

Thus

$$\frac{P_0}{P_{19}} = \frac{P_{t19}/P_{19}}{P_{19}/P_0} = \frac{[(\gamma_c + 1)/2]^{\gamma_c/(\gamma_c - 1)}}{\pi_r \pi_d \pi_f \pi_{fn}} \tag{7.53}$$

and

$$\frac{P_0}{P_9} = \frac{P_{t9}/P_9}{P_9/P_0} = \frac{[(\gamma_t + 1)/2]^{\gamma_t/(\gamma_t - 1)}}{\pi_r \pi_d \pi_c \pi_b \pi_t \pi_n} \tag{7.54}$$

Note that these two expressions are valid only when both P_9 and P_{19} are equal to or greater than P_0, the perfectly expanded or underexpanded cases (see Section 3.7.1). If these expressions predict P_9 and P_{19} less than P_0, the nozzles will not be choked. In this case, we take P_{19}/P_0 and/or P_9/P_0, which represents the physics.

See Section 7.4a.SM, Summary of Equations: Separate Exhaust-Stream Turbofan Engine, in the online supporting material for related information.

Example 7.6

As our first example for the turbofan with losses, we calculate the performance of a turbofan engine cycle with the following input data representative of a typical cruise flight condition of a modern high-bypass turbofan engine. Assume convergent-only exhaust nozzles.

Inputs:

$$M_0 = 0.8, \quad T_0 = 390°R, \quad \gamma_c = 1.4, \quad c_{pc} = 0.240 \text{ Btu/(lbm} \cdot °R),$$
$$\gamma_t = 1.33, \quad c_{pt} = 0.276 \text{ Btu/(lbm} \cdot °R), \quad h_{PR} = 18,400 \text{ Btu/lbm},$$
$$\pi_{d\,max} = 0.99, \quad \pi_b = 0.96, \quad \pi_n = 0.99, \quad \pi_{fn} = 0.99,$$
$$e_c = 0.90, \quad e_f = 0.89, \quad e_t = 0.90, \quad \eta_b = 0.99, \quad \eta_m = 0.99,$$
$$T_{t4} = 2420°R, \quad \pi_c = 36, \quad \pi_f = 1.65, \quad \alpha = 7$$

(Continued)

Example 7.6 *(Continued)*

Equations:

$$R_c = \frac{\gamma_c - 1}{\gamma_c} c_{pc} = \frac{0.4}{1.4}(0.24 \times 778.16) = 53.36 \, \text{ft} \cdot \text{lbf}/(\text{lbm} \cdot {}^\circ \text{R})$$

$$R_t = \frac{\gamma_t - 1}{\gamma_t} c_{pt} = \frac{0.33}{1.33}(0.276 \times 778.16) = 53.29 \, \text{ft} \cdot \text{lbf}/(\text{lbm} \cdot {}^\circ \text{R})$$

$$a_0 = \sqrt{1.4 \times 53.36 \times 32.174 \times 390} = 968.2 \, \text{ft/s}$$

$$V_0 = a_0 M_0 = 968.2 \times 0.8 = 774.6 \, \text{ft/s}$$

$$\tau_r = 1 + \frac{\gamma_c - 1}{2} M_0^2 = 1 + 0.2 \times 0.8^2 = 1.128$$

$$\pi_r = \tau_r^{\gamma_c/(\gamma_c - 1)} = 1.128^{3.5} = 1.5243$$

$$\eta_r = 1 \quad \text{because } M_0 < 1$$

$$\pi_d = \pi_{d\,\max} \eta_r = 0.99$$

$$\tau_c = \pi_c^{(\gamma_c - 1)/(\gamma_c e_c)} = 36^{1/(3.5 \times 0.9)} = 3.119$$

$$\eta_c = \frac{\pi_c^{(\gamma_c - 1)/\gamma_c} - 1}{\tau_c - 1} = \frac{36^{1/3.5} - 1}{3.119 - 1} = \frac{1.784}{2.119} = 84.2\%$$

$$\tau_f = \pi_f^{(\gamma_c - 1)/(\gamma_c e_f)} \, 1.65^{1/(3.5 \times 0.89)} = 1.1744$$

$$\eta_f = \frac{\pi_f^{(\gamma_c - 1)/\gamma_c} - 1}{\tau_f - 1} = \frac{1.65^{1/3.5} - 1}{1.1744 - 1} = \frac{0.1637}{0.1857} = 88.2\%$$

$$\tau_\lambda = \frac{c_{pt} T_{t4}}{c_{pc} T_0} = \frac{0.276 \times 2420}{0.240 \times 390} = 7.136$$

$$\tau_c = \pi_c^{(\gamma_c - 1)/(\gamma_c e_c)} = 36^{1/(3.5 \times 09)} = 3.119$$

$$\eta_c = \frac{\pi_c^{(\gamma_c - 1)/\gamma_c} - 1}{\tau_c - 1} = \frac{36^{1/3.5} - 1}{3.119 - 1} = \frac{1.784}{2.119} = 84.2\%$$

$$\tau_f = \pi_f^{(\gamma_c - 1)/(\gamma_c e_f)} = 1.65^{1/(3.5 \times 0.89)} = 1.1744$$

$$\eta_f = \frac{\pi_f^{(\gamma_c - 1)/\gamma_c} - 1}{\tau_f - 1} = \frac{1.65^{1/3.5} - 1}{1.1744 - 1} = \frac{0.1637}{0.1857} = 88.2\%$$

$$T_{t3} = 1372.3 {}^\circ \text{R}, \, h_{t3} = 335.56$$

$$f = \frac{h_{t4} - h_{t3}}{h_{PR} \eta_b - h_{t4}}$$

(Continued)

Example 7.6 *(Continued)*

After iteration, we have

$$f = \frac{636.86 - 335.56}{18{,}400 \times 0.99 - 636.86} = 0.01714$$

$$\tau_t = 1 - \frac{1}{\eta_m(1+f)} \frac{\tau_r}{\tau_\lambda}[\tau_c - 1 + \alpha(\tau_f - 1)]$$

$$= 1 - \frac{1}{0.99(1.01714)} \frac{1.128}{7.136}[(3.119 - 1) + 7(1.1744 - 1)]$$

$$= 0.47568$$

$$\pi_t = \tau_t^{\gamma_t[(\gamma_t - 1)e_t]} = 0.47568^{1.33/(0.33 \times 0.90)} = 0.03589$$

$$\eta_t = \frac{1 - \tau_t}{1 - \tau_t^{1/e_t}} = \frac{1 - 0.47568}{1 - 0.47568^{1/0.90}} = 92.3\%$$

$$\frac{P_{t9}}{P_0} = \pi_r \pi_d \pi_c \pi_b \pi_t \pi_n = 1.5243 \times 0.99 \times 36 \times 0.96 \times 0.3589 \times 0.99 = 1.853$$

This pressure ratio is just greater than 1.8506 (that required to choke the flow). Thus $M_9 = 1$ and

$$\frac{P_0}{P_9} = 1.8506/1.853 = 0.9989$$

$$\frac{T_9}{T_0} = \frac{T_{t4}\tau_t/T_0}{(P_{t19}/P_0)^{(\gamma_t - 1)/\gamma_t}} = \frac{2420 \times 0.47568/390}{1.853^{0.33/1.33}} = 2.533$$

$$\frac{V_9}{a_0} = M_9 \sqrt{\frac{\gamma_t R_t T_9}{\gamma_c R_c T_0}} = 1.00 \sqrt{\frac{-1.33 \times 53.29}{1.40 \times 53.36}(2.533)} = 1.55$$

$$\frac{P_{t19}}{P_0} = \pi_r \pi_d \pi_f \pi_{fn}$$

$$= 1.5243 \times 0.99 \times 1.65 \times 0.99 = 2.4651$$

This pressure ratio is greater than that required to choke the bypass exhaust nozzle. Thus the exit Mach number of this convergent nozzle will be 1.0 and the exit pressure P_{19} will be greater than P_0.

$$\frac{P_0}{P_{19}} = \frac{P_{t19}/P_{19}}{P_{19}/P_0} = \frac{[(\gamma_c + 1)/2]^{\gamma_c/(\gamma_c - 1)}}{\pi_r \pi_d \pi_f \pi_{fn}} = \frac{1.8929}{2.4651} = 0.7679$$

$$\frac{T_{19}}{T_0} = \frac{\tau_r \tau_f}{(P_{t19}/P_{19})^{(\gamma_c - 1)/\gamma_c}} = \frac{1.128 \times 1.1744}{1.8929^{1/3.5}} = 1.1039$$

(Continued)

Example 7.6 *(Continued)*

$$\frac{V_{19}}{a_0} = M_{19}\sqrt{\frac{T_{19}}{T_0}} = 1.0\sqrt{1.1039} = 1.0569$$

$$\frac{F}{\dot{m}_0} = \frac{1}{1+\alpha}\frac{a_0}{g_c}\left[(1+f)\frac{V_9}{a_0} - M_0 + (1+f)\frac{R_t T_9/T_0}{R_c V_9/a_0}\frac{1 - P_0/P_9}{\gamma_c}\right]$$

$$+ \frac{\alpha}{1+\alpha}\frac{a_0}{g_c}\left(\frac{V_{19}}{a_0} - M_0 + \frac{T_{19}/T_0}{V_{19}/a_0}\frac{1 - P_0/P_{19}}{\gamma_c}\right)$$

$$= \frac{968.2}{1 \times 32.174}\left(1.01714 \times 1.550 - 0.8 + 1.01714\frac{53.29}{53.36}\frac{2.533}{1.55}\frac{0.0011}{1.4}\right)$$

$$+ \frac{7 \times 968.2}{8 \times 32.174}\left(1 - .0569 - 0.8 + \frac{1.10391}{1.0569} - \frac{0.7679}{1.4}\right)$$

$$= 3.7627 \times 0.7766 + 26.33 \times 0.4300 = 14.244\,\text{lbf}/(\text{lbm/s})$$

$$S = \frac{f}{(1+\alpha)F/\dot{m}_0} = \frac{3600 \times 0.01714}{8 \times 14.244} = 0.5415(\text{lbm/h})/\text{lbf}$$

$$\eta_P = \frac{2M_0[(1+f)V_9/a_0 + \alpha(V_{19}/a_0) - (1+\alpha)M_0]}{(1+f)(V_9/a_0)^2 + \alpha(V_{19}/a_0)^2 - (1+\alpha)M_0^2} = 76.34\%$$

$$\eta_{Th} = \frac{a_0^2[(1+f)(V_9/a_0)^2 + \alpha(V_{19}/a_0)^2 - (1+\alpha)M_0^2]}{2g_c f h_{PR}} = 46.69\%$$

$$\eta_O = \eta_{Th}\eta_P = 0.4660 \times 0.7634 = 35.64\%$$

These results are very much representative of the modern high-bypass turbofans of the type shown in Fig. 1.8. Note especially the very high value of propulsive efficiency of 76.3% leading to a low cruise S of 0.5415 lbm/hr/lbf. The thermal efficiency suffers some as a result of the losses in transferring a large amount of core energy to the bypass stream; however, the overall efficiency and hence the S are improved because of the dramatic increase in propulsive efficiency.

The compressor pressure ratio π_c is the principle design variable that affects the engine thermal efficiency η_{Th} and π_c is normally chosen as high as possible for subsonic cruise aircraft. A note of caution: the split of the pressure ratio of the remaining compressor stages [e.g., low-pressure compressor (LPC) π_{cL} and high-pressure compressor (HPC) π_{cH}] are determined when considering the preliminary turbine design. They make little or no difference in the results of the overall engine parametric analysis.

The other two design variables are the fan pressure ratio π_f and the bypass ratio α. These variables influence the propulsive efficiency η_p.

For preliminary design, we generally don't think of T_{t4} as a design choice, but rather a "limit" whose maximum value will be determined often by required takeoff thrust on a hot day and/or the sustained temperature limits for the material of the high-pressure turbine.

When analyzing the performance trends of separate exhaust turbofan engines, one would do well to refer back often to the T-s diagram (Fig. 7.11) and use foundational knowledge. Where does the power to drive a fan and any required booster stages come from? It comes from an additional turbine, the LPT, in the core stream. Thus, we recognize that the two T-s diagrams—one for the core and one for the bypass—cannot be interpreted as independent of each other. They are linked through the thermodynamic cycle. For example, if one desires to explore the overall engine performance effects resulting from an increase in fan pressure ratio, everything else held constant, one must recognize the following (see Fig. 7.11):

- P_{t13} would increase, as would T_{t13} (this is what you are requesting), leading to a change in bypass air exit velocity V_{19}.
- T_{t5} and P_{t5} would change according to your chosen bypass ratio and component efficiencies, leading to a change in core exit velocity V_9.

The examples that follow should help you in discovering and understanding parametric cycle analysis of the most common aircraft gas turbine engines in operation, the separate exhaust turbofan.

Example 7.7

Because the turbofan cycle has three design variables, its performance with losses can be understood by performing a parametric cycle analysis, plotting the results vs values of the design variables, and comparing results to the performance of the ideal turbofan. Figures 7.13–7.16 are plots for turbofan engines with $P_9 = P_{19} = P_0$ and the following input values. Unless shown otherwise, the Mach number, compressor pressure ratio, and fan pressure ratio are the values listed under *Baseline*:

$T_0 = 216.7\,\text{K}$	$\pi_{d\,\text{max}} = 0.98$	$e_c = 0.90$	*Baseline*
$\gamma_c = 1.4$	$\pi_b = 0.98$	$e_t = 0.90$	$M_0 = 0.8$
$c_{pc} = 1.004\,\text{kJ} = (\text{kg}\cdot\text{K})$	$\pi_n = \pi_{fn} = 0.99$	$e_f = 0.89$	$\pi_c = 24$
$\gamma_t = 1.33$	$\eta_b = 0.99$	$h_{PR} = 42{,}800\,\text{kJ/kg}$	$\pi_f = 1.65$
$c_{pt} = 1.156\,\text{kJ}/(\text{kg}\cdot\text{K})$	$\eta_m = 0.99$	$T_{t4} = 1350\,\text{K}$	

Figures 7.12a, 7.12b, and 7.12c show the influence of the compressor pressure ratio and bypass ratio on engine performance. As the bypass ratio

(Continued)

Example 7.7 *(Continued)*

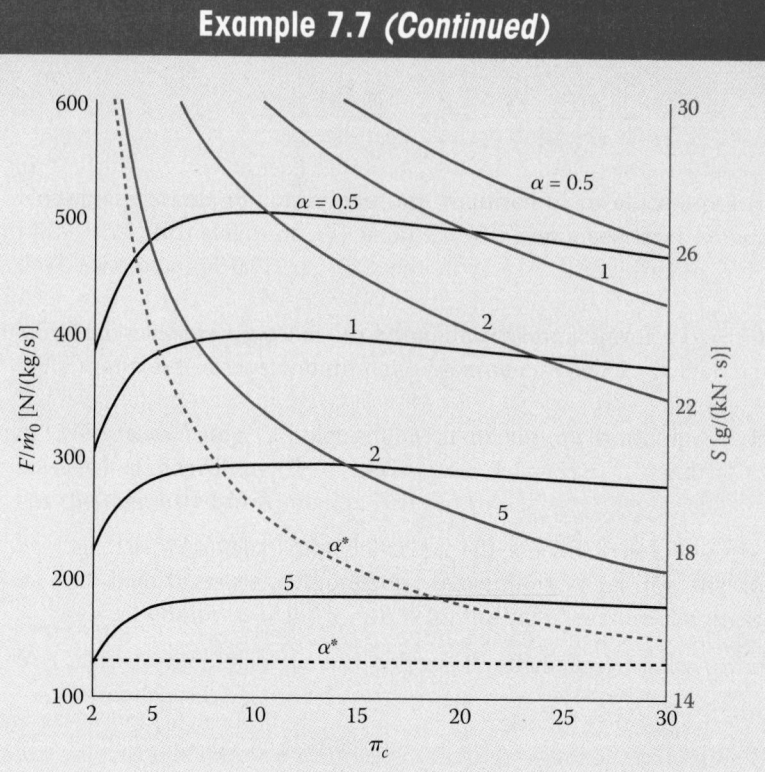

Fig. 7.12a Turbofan engine with losses vs compressor pressure ratio.

increases, the specific thrust decreases while the thrust specific fuel consumption improves (decreases).

For a fixed compressor pressure ratio π_c, the turbine must produce sufficient energy to power the compressor and fan. The minimum power would correspond to that of a turbojet ($\alpha = 0$). However, the maximum power extracted by the turbine corresponds to the core exit velocity V_9 equal to V_0. The range of available turbine temperature ratio τ_t for this example with a compressor pressure ratio π_c of 24 is shown in Fig. 7.13a for a range of fan pressure ratios π_f and bypass ratios α. The region of possible energy extraction for powering the bypass stream is highlighted in blue and corresponds to an exit temperature leaving the turbine T_{t5} from about 985 K to 610 K. The region of this plot where the fuel consumption is lower will be shown and discussed later in this chapter. Comparison to that for the ideal turbofan engine in Fig. 5.13a shows that the lowest value of total temperature ratio is higher in Fig. 7.13a (an exit temperature of 610 K for the real engine versus 510 K for the ideal engine) and the range of power extraction is about equal between the two engine models. Due to the higher specific heat

(Continued)

Example 7.7 *(Continued)*

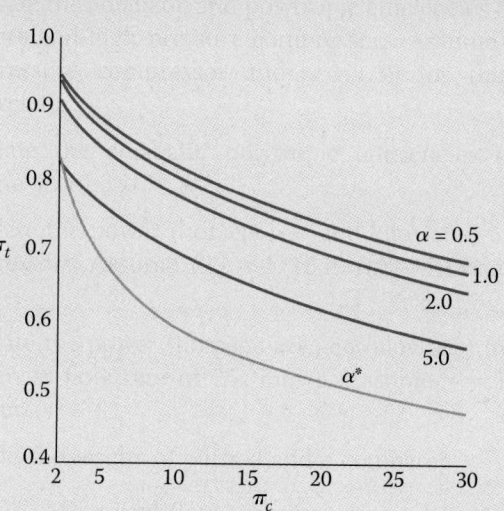

Fig. 7.12b Turbofan engine with losses vs compressor pressure ratio: turbine temperature ratio.

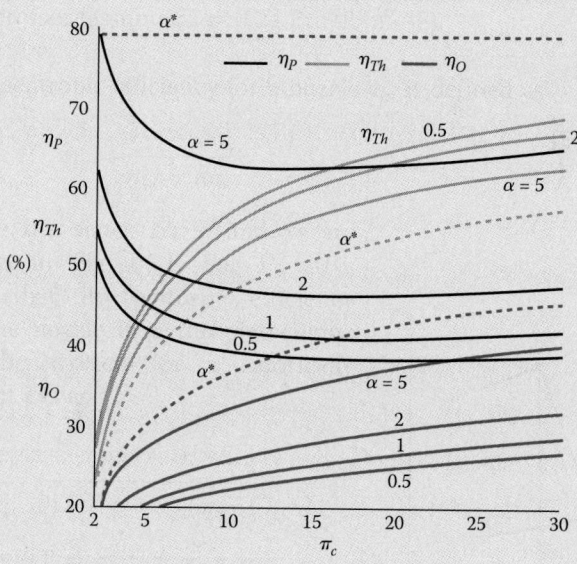

Fig. 7.12c Turbofan engine with losses vs compressor pressure ratio: efficiencies.

(Continued)

Example 7.7 (Continued)

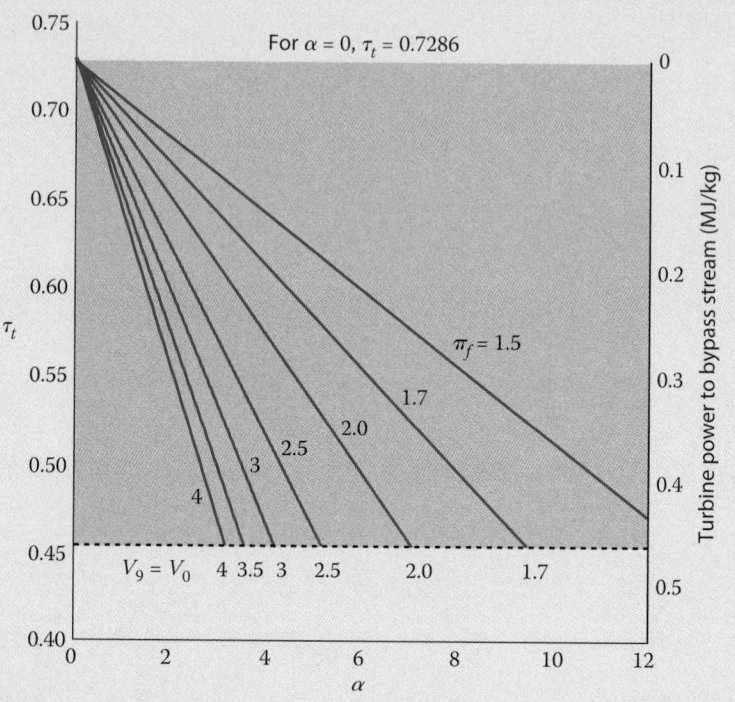

Fig. 7.13a Available turbine total temperature ratio for $\pi_c = 24$ and $M_0 = 0.8$.

of the gas through the turbine, a lower temperature change is needed to extract the same amount of power from the turbine to power the bypass air, which explains the major difference in the required total temperature ratio π_t.

The variation of both thrust specific fuel consumption and specific thrust can be investigated using the carpet plot calculation feature of the PARA program over ranges of two independent variables. For fixed flight condition and compressor pressure ratio, investigating the range of bypass ratios from 5 to 15 and fan pressure ratios from 1.2 to 2 gives the plot shown in Fig. 7.13b. Two clear trends emerge:

1. Unlike the example turbojet carpet plot in Fig. 7.6, we see selection sets of design choices that result in performance trends in the desired direction—lower S and higher specific thrust. Namely, for a given overall compression system pressure ratio (compressor and fan) and bypass ratio, turbofan performance trends move down and to the right, as shown in Fig. 7.13b.

(Continued)

Example 7.7 (Continued)

2. For a given overall compression system pressure ratio and fan pressure ratio, there is an optimum bypass ratio that minimizes S (at the expense of reduced specific thrust). As fan pressure ratio increases, that optimum occurs at lower bypass ratios.

If you look at the solid blue line indicating a constant fan pressure ratio of 1.6, you see that the fuel consumption decreases with increasing engine bypass ratio until a bypass ratio of about 9 (its optimum bypass ratio where the fuel consumption is at its minimum). Likewise, following along the dashed blue line in Fig. 7.13b for the constant bypass ratio of 9, you see that the fuel consumption decreases with increasing fan pressure ratio until a fan pressure ratio of 1.6 (its optimum fan pressure ratio where the fuel consumption is at its minimum). Clearly, if one wishes to minimize fuel consumption, there are optimum design choices. We will explore this in the next section.

Examination of the real turbofan equations for specific thrust and thrust specific fuel consumption, Eqs. (7.48) and (7.49), can give us some insights into the performance trends of Fig. 7.13b.

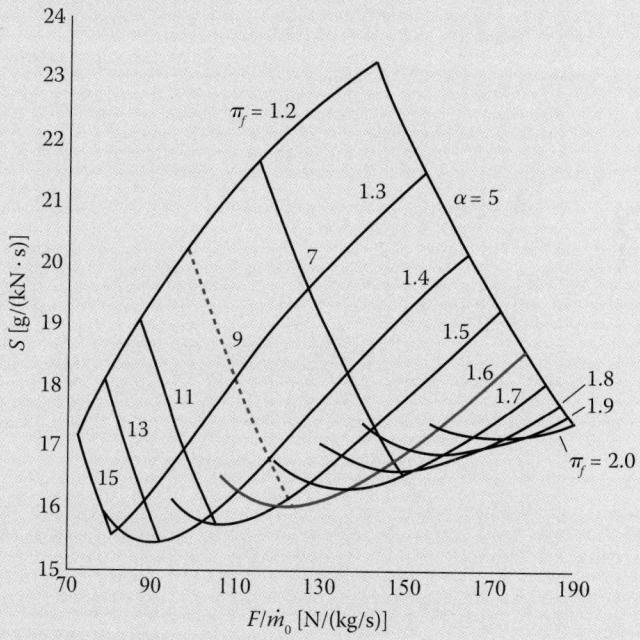

Fig. 7.13b   Performance of turbofan engine with losses for $\pi_c = 24$ and $M_0 = 0.8$.

(Continued)

Example 7.7 *(Continued)*

- As fan pressure ratio increases and all other parameters are held constant, more specific work is pulled from the core, decreasing V_9 and increasing V_{19}. This is because the fan must flow all of the air, the core only a portion of the air (dictated by α). As we increase π_f, we are asking for a larger portion of the air to be increased to the higher pressure. Based on this tradeoff between V_9 and V_{19}, we should not be surprised that there is a limit to how high π_f can go before the trend of increased specific thrust and decreased S reverses, as indicated in Fig. 7.13b.
- As the bypass ratio increases and all other parameters are held constant, the engine specific thrust decreases as the $1/(1+\alpha)$ term in Eq. (7.48) dictating the core specific thrust dominates the $\alpha/(1+\alpha)$ term dictating the bypass specific thrust.

In both cases, the behavior of S, namely that there is an optimum π_f and α for minimizing S, cannot be discerned simply by looking at Eq. (7.49), but again, the fact that there are optimums in each case should make sense. Figures 7.14 and 7.15 display the performance trends just described for π_f and α changes, respectively, all other parameters held constant. Note that these two plots present the same information as the more complex carpet plot, but do so in the more traditional manner of one independent variable changed at a time.

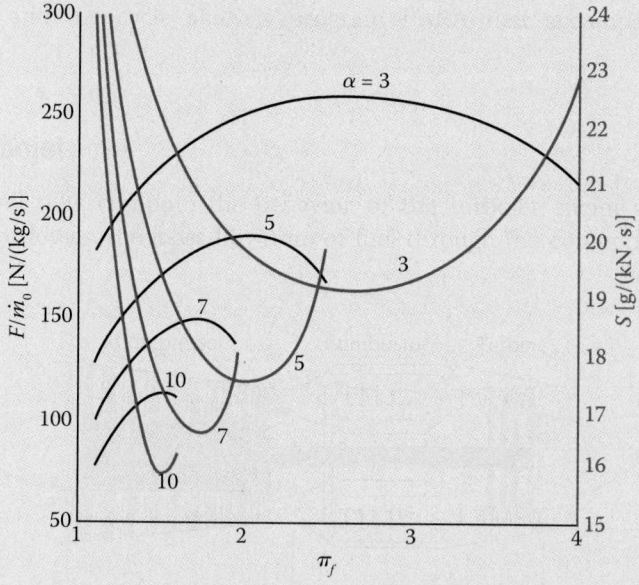

Fig. 7.14 Turbofan engine with losses vs fan pressure ratio.

(Continued)

Example 7.7 *(Continued)*

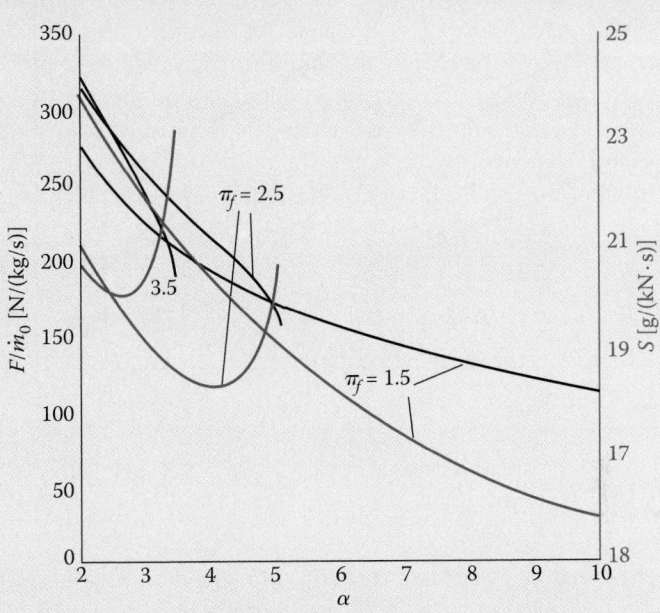

Fig. 7.15 Turbofan engine with losses vs bypass ratio.

One can see the following from a comparison of Fig. 7.13b to the ideal turbofan engine given in Fig. 5.13b:

1. The thrust specific fuel consumption is higher and the specific thrust is lower.
2. For a given bypass ratio, the optimum fan pressure ratio is lower (1.6 vs 1.8 for α of 9).
3. For a given fan pressure ratio, the optimum bypass ratio is lower (9 vs 13 for π_f of 1.6).

Figure 7.14 shows the influence of fan pressure ratio and bypass ratio on engine performance. An optimum fan pressure ratio for maximum specific thrust still exists for the turbofan with losses, and the value of the optimum fan pressure ratio is much lower than that for the ideal turbofan shown in Fig. 5.14.

Figure 7.15 shows the variation in specific thrust and thrust specific fuel consumption with bypass ratio and fan pressure ratio. An optimum bypass ratio for minimum S still exists for the turbofan with losses, and the value of the optimum bypass ratio is much less than that for the ideal turbofan shown in Fig. 5.15.

7.4.3 Optimum Bypass Ratio α^*

As was true for the turbofan with no losses, we may obtain an expression that allows us to determine the bypass ratio α^* that leads to minimum thrust specific fuel consumption. For a given set of such prescribed variables $(\tau_r, \pi_c, \pi_f, \tau_\lambda, V_0)$, we may locate the minimum S by taking the partial derivative of S with respect to the bypass ratio α and setting the resulting equation equal to zero. We consider the case where the exhaust pressures of both the fan stream and the core stream equal the ambient pressure $P_0 = P_9 = P_{19}$. Because the fuel/air ratio is not a function of bypass ratio, we have

$$S = \frac{f}{(1+\alpha)(F/\dot{m}_0)}$$

$$\frac{\partial S}{\partial \alpha} = \frac{\partial}{\partial \alpha}\left[\frac{f}{(1+\alpha)(F/\dot{m}_0)}\right] = 0$$

$$\frac{\partial S}{\partial \alpha} = \frac{-f}{[(1+\alpha)(F/\dot{m}_0)]^2}\frac{\partial}{\partial \alpha}\left[(1+\alpha)\left(\frac{F}{\dot{m}_0}\right)\right] = 0$$

Thus $\partial S/\partial \alpha = 0$ is satisfied by

$$\frac{\partial}{\partial \alpha}\left[\frac{g_c}{V_0}(1+\alpha)\left(\frac{F}{\dot{m}_0}\right)\right] = 0$$

where

$$\frac{g_c}{V_0}(1+\alpha)\left(\frac{F}{\dot{m}_0}\right) = (1+f)\left(\frac{V_9}{V_0}-1\right) + \alpha\left(\frac{V_{19}}{V_0}-1\right)$$

Then the optimum bypass ratio is given by the following expression:

$$\frac{\partial}{\partial \alpha}\left[(1+f)\left(\frac{V_9}{V_0}-1\right) + \alpha\left(\frac{V_{19}}{V_0}-1\right)\right]$$

$$= (1+f)\frac{\partial}{\partial \alpha}\left(\frac{V_9}{V_0}\right) + \frac{V_{19}}{V_0}-1 = 0 \tag{7.55}$$

As shown in the Supporting Material for Chapter 7, solution of Eq. (7.55) gives an expression for the turbine temperature ratio τ_t^* corresponding to the optimum bypass ratio α^*

$$\tau_t^* = \frac{\tau_t^{-(1-e_t)/e_t}}{\Pi} + \frac{1}{\tau_\lambda(\tau_r-1)}\left[\frac{1}{2\eta_m}\frac{\tau_r(\tau_f-1)}{V_{19}/V_0-1}\left(1+\frac{1-e_t}{e_t}\frac{\tau_t^{-1/e_t}}{\Pi}\right)\right]^2 \tag{7.56}$$

where

$$\Pi = (\tau_r\pi_d\pi_c\pi_b\pi_n)^{(\gamma_t-1)/\gamma_t} \tag{7.57}$$

Because Eq. (7.56) is an equation for τ_t^* in terms of itself, in addition to other known values, an iterative solution is required. A starting value of τ_t^*, denoted by τ_{ti}^*, is obtained by solving Eq. (7.57) for the case when $e_t = 1$ (an ideal turbine), which gives

$$\tau_{ti}^* = \frac{1}{\Pi} + \frac{1}{\tau_\lambda(\tau_r - 1)}\left[\frac{1}{2\eta_m}\frac{\tau_r(\tau_f - 1)}{V_{19}/V_0 - 1}\right]^2 \tag{7.58}$$

This starting value can be substituted into the right side of Eq. (7.57), yielding a new value of τ_t^*. This new value of τ_t^* is then substituted into Eq. (7.57), and another new value of τ_t^* is calculated. This process continues until the change in successive calculations of τ_t^* is less than some small number (say, 0.0001). Once the solution for τ_t^* is found, the optimum bypass ratio α^* is calculated by using Eq. (7.45), solved for α:

$$\alpha^* = \frac{\eta_m(1+f)\tau_\lambda(1 - \tau_t^*) - \tau_r(\tau_c - 1)}{\tau_r(\tau_f - 1)} \tag{7.59}$$

When the optimum bypass ratio α^* is desired in calculating the parametric engine cycle performance, Eqs. (7.56), (7.57), (7.58), and (7.59) replace the equation for τ_t contained in the summary of equations, and α^* is an output.

Example 7.8

Because the optimum-bypass-ratio turbofan cycle has two design variables, its performance with losses can be understood by performing a parametric analysis, plotting the results vs values of the design variables, and comparing results to the performance of the optimum-bypass-ratio ideal turbofan. Figures 7.16 and 7.17 are plots for optimum-bypass-ratio turbofan engines with the following input values (the same input used for the parametric analysis of the turbofan engine with losses in Example 7.7). Unless shown otherwise, the Mach number, compressor pressure ratio, and fan pressure ratio are the values listed under *Baseline*:

$T_0 = 216.7\,\mathrm{K}$	$\pi_{d\,max} = 0.99$	$e_c = 0.90$	*Baseline*
$\gamma_c = 1.4$	$\pi_b = 0.96$	$e_t = 0.90$	$M_0 = 0.8$
$c_{pc} = 1.004\,\mathrm{kJ} = (\mathrm{kg}\cdot\mathrm{K})$	$\pi_n = \pi_{fn} = 0.99$	$e_f = 0.89$	$\pi_c = 24$
$\gamma_t = 1.33$	$\eta_b = 0.99$	$h_{PR} = 42{,}800\,\mathrm{kJ/kg}$	$\pi_f = 1.65$
$c_{pt} = 1.156\,\mathrm{kJ} = (\mathrm{kg}\cdot\mathrm{K})$	$\eta_m = 0.99$	$T_{t4} = 1350\,\mathrm{K}$	
$\dfrac{P_0}{P_9} = 1$	$\dfrac{P_0}{P_{19}} = 1$		

Figures 7.16 and 7.17 show the following characteristics of the optimum-bypass-ratio turbofan engine:

1. Increasing the fan pressure ratio increases the specific thrust.
2. Specific fuel consumption increases with increasing π_f.
3. The optimum bypass ratio decreases with increasing π_f.

(Continued)

Example 7.8 *(Continued)*

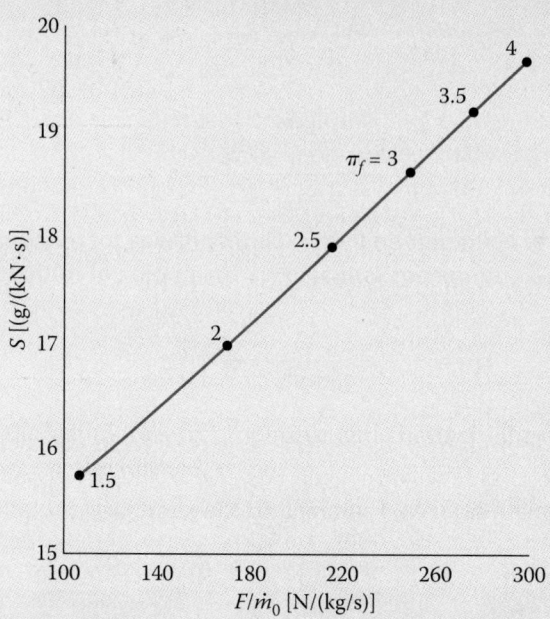

Fig. 7.16   Optimum-bypass-ratio turbofan engine performance vs π_f.

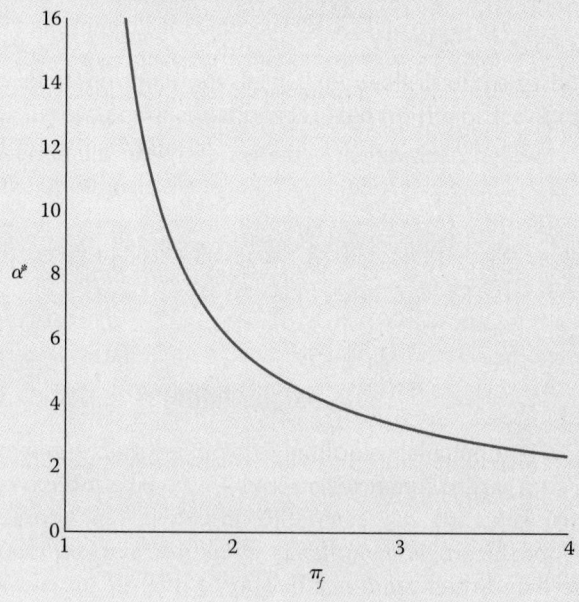

Fig. 7.17   Optimum bypass ratio vs π_f.

7.4.4 Optimum Fan Pressure Ratio

An analytical expression for the optimum fan pressure ratio π_f^* cannot be easily obtained like that of the optimum bypass ratio α^* in the previous section. Therefore, the optimum fan pressure ratio will be obtained by numerous numerical calculations at different fan pressure ratios using the PARA program. Figure 7.13b shows both the optimum fan pressure ratio and the optimum bypass ratio ($\alpha = 9$) as the lines of constant bypass ratio and fan pressure ratio ($\pi_f = 1.6$), respectively, reach their minimum value of thrust specific fuel consumption S.

Note that in this particular case, Fig. 7.13b shows that going to a lower bypass ratio of 7 at the same fan pressure ratio of 1.6 results in an increase in specific thrust of about 18% with a corresponding 3% increase in specific fuel consumption. One may choose to accept slightly worse uninstalled engine fuel burn rate S for an overall smaller diameter engine. Ultimately, such decisions must be made on the basis of mission needs and aircraft-engine system tradeoff studies as briefly highlighted in Section 7.4.6. The more the airframe and engine manufacturers know about each others' systems, the better off the integrated product will be—communication is key.

7.4.5 Comparison of Optimums

Figure 7.18 shows the optimum bypass ratio α^* and optimum fan pressure ratio π_f^* superimposed on Fig. 7.13a, the range of available turbine temperature ratio τ_t for Examples 7.7 and 7.8. This can be compared to that of the ideal turbofan engine in Fig. 5.20. Note that both the optimum bypass ratio α^* and optimum fan pressure ratio π_f^* are lower for the engine with losses. These are a result of the losses that occur in transferring energy from the core stream to the fan stream. The optimum engines occur at higher turbine temperature ratios τ_t.

7.4.6 Aircraft System Optimum

Although our focus in Chapters 5 through 8 is on engine uninstalled performance, it is appropriate to add here a brief discussion relative to optimum analysis and installed engine performance. A much higher bypass ratio is optimum when the engine is considered by itself than when it is included as part of the aircraft system. Figure 7.19 shows that the cruise installed thrust specific fuel consumption (TSFC) decreases with increasing bypass ratio α—the basic characteristic of the turbofan engine. However, both the engine weight and nacelle drag increase with bypass ratio because of the increase in the engine's physical size, resulting in increased aircraft fuel burn. The increase in nacelle drag is likely to be proportional to surface area, hence, engine diameter squared D^2. Note that one study in Cambridge [72] showed the propulsion system weight (engine plus nacelle and pylon) to be proportional to $D^{2.4}$. Both of these would result in increased

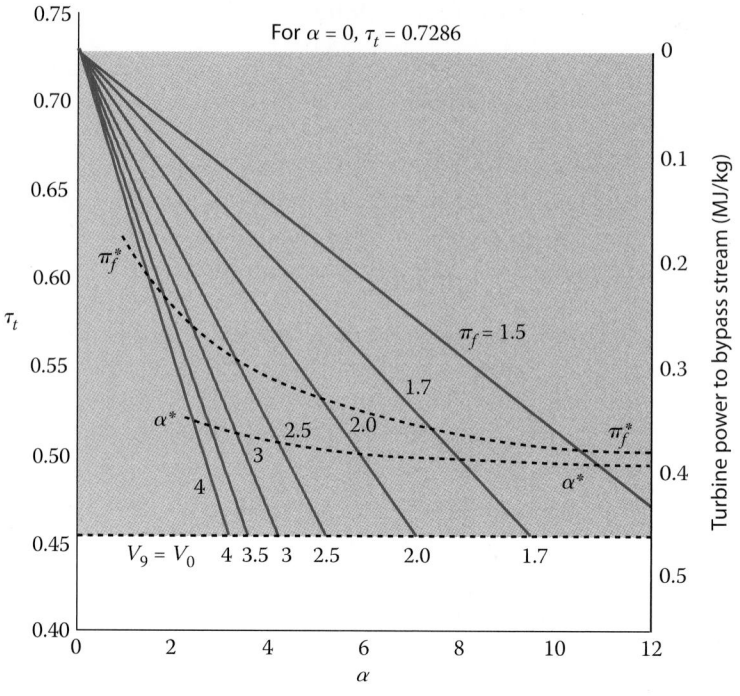

Fig. 7.18   Available turbine total temperature ratio for $\pi_c = 24$ and $M_0 = 0.8$: optimum bypass ratio and fan pressure ratio.

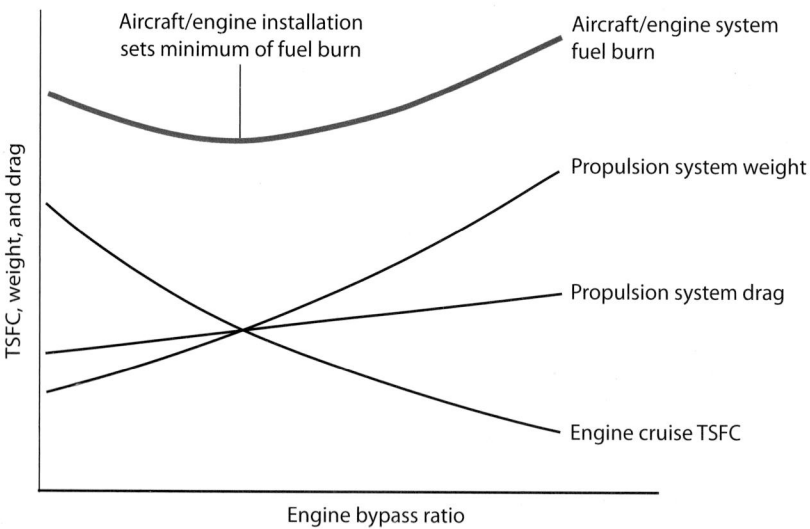

Fig. 7.19   Effect of engine bypass ratio on aircraft fuel burn.

aircraft-engine system drag and, consequently, increased aircraft fuel burn. The overall impact of increasing bypass ratio is a system optimum that is at a much lower bypass ratio than one finds when only considering the engine.

7.5 Turbofan with Afterburning: Mixed Exhaust Stream

The dual-spool, mixed-flow, augmented, low-bypass turbofan is the engine cycle of choice for all modern high-performance fighter aircraft and supersonic bombers like the B-1B. Figures 1.9a and 1.9b show cross-sectional drawings of this engine type in use today. (Refer to Appendix B for examples of the bypass ratios used in these engines.) Most high-speed military aircraft require an afterburner for thrust augmentation at certain flight conditions (e.g., takeoff, high-speed turns, etc.). Figure 7.20a shows the cross section of a mixed-flow afterburning turbofan engine, and Fig. 7.20b shows a T-s plot for this cycle. The mixed-flow turbofan cycle has several advantages over the turbofan with separate exhausts: 1) one variable-area exhaust nozzle, 2) one augmenter (afterburner), and 3) cold bypass air available to cool the afterburner liner and nozzle, and/or for low observables (LO) considerations. Also, mixers can improve engine performance when the afterburner is off or not present. All of these advantages are, of course, at the expense of added complexity in terms of both analysis of the cycle performance and the design and construction of the engine to realize that performance.

7.5.1 Mixer

In our modeling of the ideal cycle for the mixed-flow afterburning turbofan engine (see Chapter 5), we first defined the temperature and pressure ratios of the mixer as

$$\tau_M = \frac{T_{t6A}}{T_{t6}} \qquad \pi_M = \frac{P_{t6A}}{P_{t6}} \qquad (5.49)$$

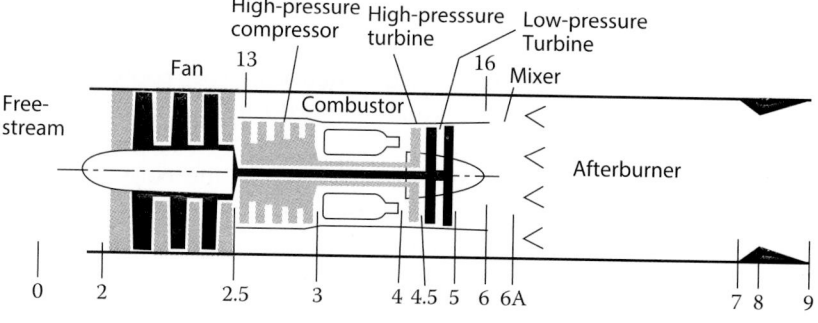

Fig. 7.20a Station numbering for mixed-flow turbofan engine with afterburner.

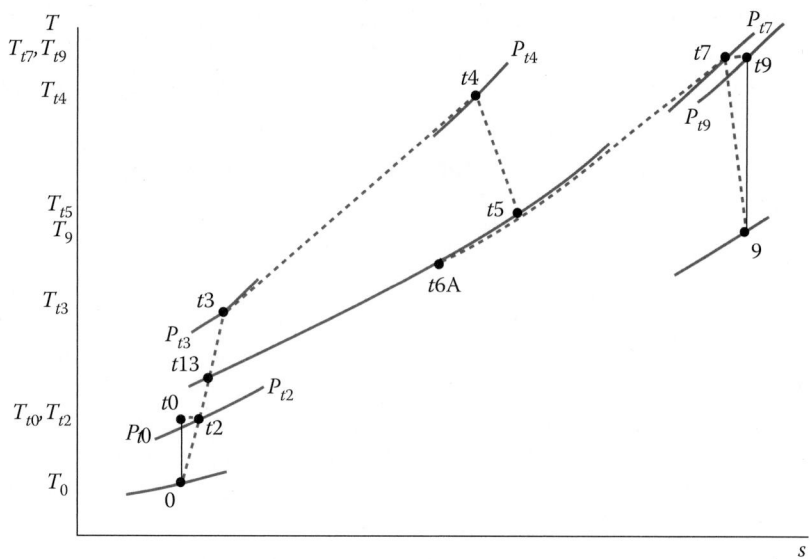

Fig. 7.20b T-s diagram for mixed-flow turbofan engine with afterburner.

We also assumed that the total pressures of the two entering streams are equal ($P_{t6} = P_{t16}$) and that the mixer pressure ratio π_M was unity. The mixer temperature ratio τ_M for the ideal cycle was obtained from an energy balance of the mixer.

The mixer temperature and pressure ratios defined in Eq. (5.49) will be used in the analysis of this engine cycle with losses, using the analysis presented in Section 6.9. Our mixer model is shown in Fig. 7.21, a constant-area mixer. The mixer temperature ratio τ_M was obtained from an energy balance of the mixer. In addition, we assume that the mixer pressure ratio π_M is the product of the pressure ratio of an ideal constant-area mixer ($\pi_{M\,ideal}$) times a total pressure ratio for the frictional losses ($\pi_{M\,max}$)

$$\pi_M = \pi_{M\,max}\,\pi_{M\,ideal} \qquad (7.60)$$

The analysis in Section 6.9 obtained the total pressure ratio of the ideal mixer $\pi_{M\text{ideal}}$ [Eq. (6.38)] and the total temperature ratio of the mixer τ_M [Eq. (6.31)].

Figure 7.21 shows a constant-area mixer where the core and bypass streams are mixed together. At the entrance, the two streams are divided by a splitter plate that suddenly ends. Nature forces the static pressure of the two streams to be equal (Kutta condition)—just like the trailing edge of a wing where the two streams come together. Thus we have

$$P_{16} = P_6 \qquad (7.61)$$

The total pressure of each stream could be independent variables but the *best performance of the mixed-flow turbofan engine occurs when the two total pressures are nearly equal* (Refs. 13 and 23). Thus, for ease of analysis, we will consider that

$$P_{t16} = P_{t6} \tag{7.62}$$

As a result of this selection, the fan pressure ratio π_f and bypass ratio α are no longer independent variables. We get to select one, and the other one is calculated so that Eq. (7.62) is satisfied.

7.5.2 Cycle Analysis

This secton develops the additional equations needed to analyze the mixed-flow afterburning turbofan engine. Because this engine, like the turbojet, has a single inlet and exhaust, Eq. (7.1) can be used for the engine uninstalled thrust, and Eq. (7.2) relates the pressure unbalance term to other flow quantities. We define the overall fuel/air ratio for this engine as the total fuel flow rate (main burner plus afterburner) divided by the inlet airflow rate, or

$$f_O = \frac{\dot{m}_f + \dot{m}_{fAB}}{\dot{m}_0} \tag{7.63}$$

We assume in this analysis that the mass flow rate at station 9 equals the mass flow rate at station 0 plus the fuel added in the main burner and in the afterburner. In equation form, we write

$$\dot{m}_9 = \dot{m}_0 + \dot{m}_f + \dot{m}_{fAB}$$

or

$$\frac{\dot{m}_9}{\dot{m}_0} = 1 + f_O \tag{7.64}$$

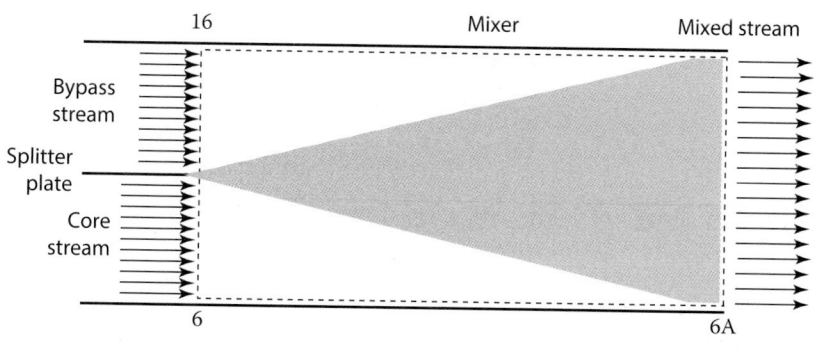

Fig. 7.21 Model of constant-area mixer.

In the following, we use the steps of cycle analysis as listed in Section 5.4.

Step 1: Uninstalled thrust: Using Eqs. (7.1) and (7.2), along with Eq. (7.64), we can write

$$\frac{F}{\dot{m}_0} = \frac{a_0}{g_c} \left[(1 + f_O)\frac{V_9}{a_0} - M_0 + (1 + f_O)\frac{R_9}{R_c}\frac{T_9/T_0}{V_9/a_0}\frac{1 - P_0/P_9}{\gamma_c} \right] \tag{7.65}$$

where the subscript c is used for gas properties at station 0.

Step 2:

$$\left(\frac{V_9}{a_0}\right)^2 = \frac{a_9^2 M_9^2}{a_0^2} = \frac{\gamma_9 R_9 g_c}{\gamma_0 R_0 g_c}\frac{T_9}{T_0} M_9^2$$

Using the subscript c for the gas properties at station 0, we write

$$\left(\frac{V_9}{a_0}\right)^2 = \frac{\gamma_9 R_9}{\gamma_c R_c}\frac{T_9}{T_0} M_9^2 \tag{7.66}$$

Step 3: We have

$$M_9^2 = \frac{2}{\gamma_9 - 1}\left[\left(\frac{P_{t9}}{P_9}\right)^{(\gamma_9 - 1)/\gamma_9} - 1 \right] \tag{7.67a}$$

where

$$\frac{P_{t9}}{P_9} = \frac{P_0}{P_9}\pi_r \pi_d \pi_c \pi_b \pi_t \pi_M \pi_{AB} \pi_n \tag{7.67b}$$

Step 4: We have

$$\frac{T_9}{T_0} = \frac{T_{t9}/T_0}{(P_{t9}/P_9)^{(\gamma_9 - 1)/\gamma_9}}$$

where

$$T_{t9} = T_{t4}\tau_t\tau_M \quad \text{afterburner off} \tag{7.68a}$$

or

$$T_{t9} = T_{t7} \quad \text{afterburner on} \tag{7.68b}$$

Step 5: Application of the first law of thermodynamics to the main burner and solution for the fuel/air ratio f yields Eq. (7.10), where

$$f = \frac{\dot{m}_f}{\dot{m}_C} \tag{7.69}$$

Application of the first law of thermodynamics to the afterburner gives

$$\dot{m}_{6A}h_{t6A} + \eta_{AB}\,\dot{m}_{f\,AB}h_{PR} = \dot{m}_7 h_{t7}$$

Dividing the preceding equation by $\dot{m}_0$ gives

$$\left(1 + \frac{f}{1+\alpha}\right)h_{t6A} + f_{AB}h_{AB}h_{PR} = \left(1 + \frac{f}{1+\alpha} + f_{AB}\right)h_{t7}$$

Solving for the afterburner fuel/air ratio f_{AB} gives

$$f_{AB} = \left(1 + \frac{f}{1+\alpha}\right)\frac{h_{t7} - h_{t6A}}{\eta_{AB}h_{PR} - h_{t7}} \tag{7.70}$$

Iteration is required to solve this equation (like the equation for the fuel/air ratio of the main burner) because h_{t7} is a function of the overall fuel/air ratio of the mixture.

Step 6: The power balance among the turbine, compressor, and fan, with a mechanical efficiency η_m of the turbine shaft, gives Eq. (7.44). This equation, solved for the turbine temperature ratio, gives

$$\tau_t = 1 - \frac{1}{\eta_m(1+f)}\frac{\tau_r}{\tau_\lambda}[\tau_c - 1 + \alpha(\tau_f - 1)] \tag{7.71}$$

This expression allows solution for τ_t, from which we then obtain π_t by using Eq. (7.13) and h_t by using Eq. (7.14). Equations (7.15), (7.16), (7.46), and (7.47) are used to obtain τ_c, η_c, τ_f, and η_f, respectively, from given π_c, π_f, and polytropic efficiencies.

As presented previously in Section 7.5.1, normally, the fan pressure ratio or bypass ratio is selected for this engine cycle so that the total pressures at stations 6 and 16 are equal (Refs. 13 and 23). Assuming isentropic flow in the bypass duct from 13 to 16, we can write

$$\pi_c\pi_b\pi_t = \pi_f$$

or

$$\tau_t = \pi_t^{(\gamma_t-1)e_t/\gamma_t} = \left(\frac{\pi_f}{\pi_c\pi_b}\right)^{(\gamma_t-1)e_t/\gamma_t} \tag{7.72}$$

Equations (7.71) and (7.72) can be solved to obtain the bypass ratio α or the fan temperature ratio τ_f in terms of known quantities. Solution for the bypass ratio gives

$$\alpha = \frac{\eta_m(1+f)(\tau_\lambda/\tau_r)\{1 + [\pi_f/(\pi_c\pi_b)]^{(\gamma_t-1)e_t/\gamma_t}\} - (\tau_c - 1)}{\tau_f - 1} \tag{7.73}$$

Step 7: See Eq. (7.65).

Step 8: The equation for the thrust specific fuel consumption is

$$S = \frac{f_O}{F/\dot{m}_0} \tag{7.74}$$

Step 9: From the definitions of propulsive and thermal efficiency, one can easily show that, for this turbofan engine,

$$\eta_P = \frac{2g_c V_0(F/\dot{m}_0)}{a_0^2[(1+f_O)(V_9/a_0)^2 - M_0^2]} \tag{7.75a}$$

$$\eta_{Th} = \frac{a_0^2[(1+f_O)(V_9/a_0)^2 - M_0^2]}{2g_c f_O h_{PR}} \tag{7.75b}$$

Now we have all the equations needed for analysis of the mixed-flow afterburning turbofan engine. For convenience, this system of equations is listed (in the order of calculation) in the Supporting Material for easier calculation.

See Section 7.5.SM, Summary of Equations: Mixed-Flow Afterburning Turbofan Engine, in the online supporting material for related information.

Example 7.9

The afterburning and nonafterburning performance of the mixed-flow turbofans with losses are plotted in Fig. 7.22 vs fan pressure ratio π_f at two flight Mach numbers. Figure 7.23 plots the required bypass ratio α (to meet our mixer analysis constraint of equal stream total pressures) vs compressor pressure ratio π_c for the different values of fan pressure ratio π_f at flight Mach numbers of 0.9 and 2. Calculations were performed for the following input data:

$$c_{pc} = 0.240 \text{ Btu/(lbm} \cdot {}^\circ\text{R)}, \quad \gamma_c = 1.4, \quad T_0 = 390{}^\circ\text{R}, \quad \pi_{d\,max} = 0.98,$$

$$\pi_{M\,max} = 0.98, \quad e_c = 0.90, \quad e_f = 0.89, \quad e_t = 0.91, \quad \pi_n = 0.98,$$

$$c_{pt} = 0.295 \text{ Btu/(lbm} \cdot {}^\circ\text{R)}, \quad \gamma_t = 1.3, \quad \eta_b = 0.995, \quad \pi_b = 0.96, \quad T_{t4} = 3200{}^\circ\text{R},$$

$$c_{pAB} = 0.295 \text{ Btu/(lbm} \cdot {}^\circ\text{R)}, \quad \gamma_{AB} = 1.3, \quad \eta_{AB} = 0.95, \quad \pi_{AB} = 0.96,$$

$$T_{t7} = 3600{}^\circ\text{R}, \quad h_{PR} = 18{,}400 \text{ Btu/lbm}, \quad M_6 = 0.4, \quad \eta_m = 0.99, \quad \frac{P_0}{P_9} = 1,$$

$$\pi_f = 2 \rightarrow 5, \quad M_0 = 0.9, 2$$

A compressor pressure ratio π_c of 24 was chosen for the engines of Fig. 7.22 as a reasonable choice for π_c in terms of tradeoff between specific thrust and fuel consumption (see Fig. 7.12a). Figure 7.22 is a composite performance plot of fuel consumption vs specific thrust with different fan pressure ratios and Mach numbers—this is another example of a *carpet plot* presented in a different manner than the ones up to this point. This

(Continued)

Example 7.9 *(Continued)*

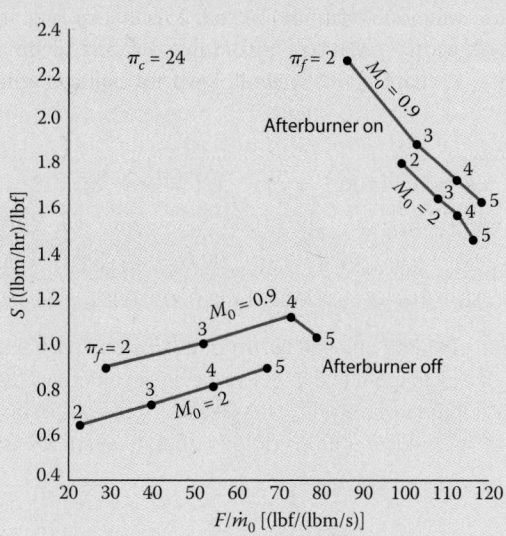

Fig. 7.22   Performance of mixed-flow afterburning turbofan engine.

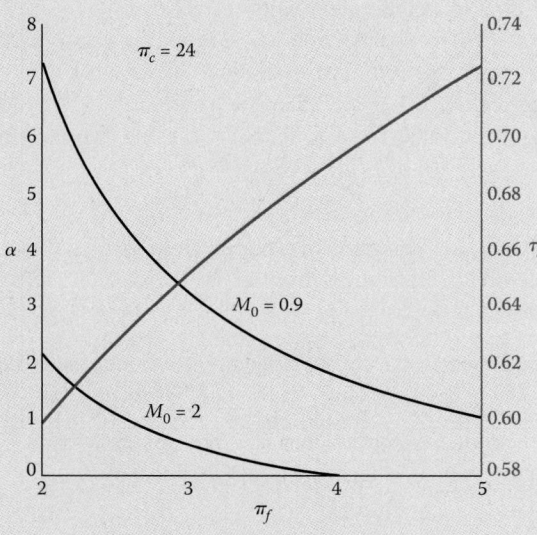

Fig. 7.23   Bypass ratio and turbine temperature ratio of mixed-flow turbofan engine.

(Continued)

Example 7.9 (Continued)

figure shows that increasing the fan pressure ratio reduces fuel consumption and increases specific thrust for the afterburning engine—both desirable.

For nonafterburning subsonic cruise, we see the general trend of S increasing with increased π_f, but it is proportionally much smaller than the increase in specific thrust. Figure 7.22 shows an increase of about 170% in specific thrust for π_f increased from 2 to 5 and only a 22% change in S over that same π_f range for the "afterburner off" condition.

Figure 7.23 shows that the engine bypass ratio corresponding to each fan pressure ratio decreases with increasing π_f. This should make sense. As the fan requirement to do more work on the air increases, less and less bypass air can be accommodated; a greater portion of the air must be directed through the core to provide for the increased power needed by the fan.

Figures 7.22 and 7.23 can be compared to the results for the ideal mixed-flow turbofan engine presented in Figs. 5.23a–5.24. The trends are the same.

Example 7.10

In this example, we look at the effect of increasing T_{t4} on the performance of the mixed-flow turbofan engine without afterburning. The fan pressure ratio π_f was varied from 2 to 5 for values of T_{t4} from 3000 to 4000°R with $M_0 = 0$, $\pi_c = 24$, and the other data of Example 7.9. The results are plotted in Figs. 7.24 and 7.25. The results show the following for increasing T_{t4} from 3000 to 4000°R while keeping the fan pressure ratio at 4:

1. About a 1% decrease in specific thrust
2. Dramatic increases in engine bypass ratio (from about 0.9 to 2.3) resulting in increased fuel consumption [from about 0.68 to 0.70 (lbm/hr)/lbf, a 2.5% increase]

If the engine bypass ratio is held constant at 3.0, increasing T_{t4} from 3000 to 4000°R results in the following:

1. Dramatic increase in fan pressure ratio from about 2.4 to 3.5
2. Increase in fuel consumption from about 0.49 to 0.64 (lbm/hr)/lbf (about a 30% increase)
3. Similar percentage increase (34%) in specific thrust as there is for S

This could translate into an engine that is physically much smaller (less air mass flow rate for required thrust levels).

(Continued)

Example 7.10 *(Continued)*

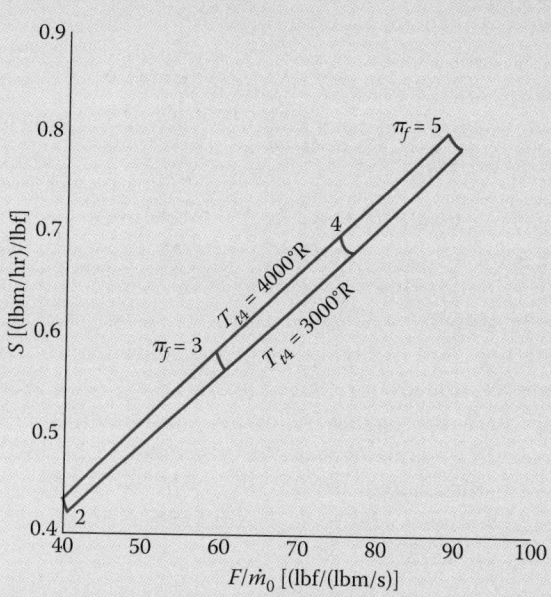

Fig. 7.24 Specific thrust of mixed-flow turbofan engine (no afterburner).

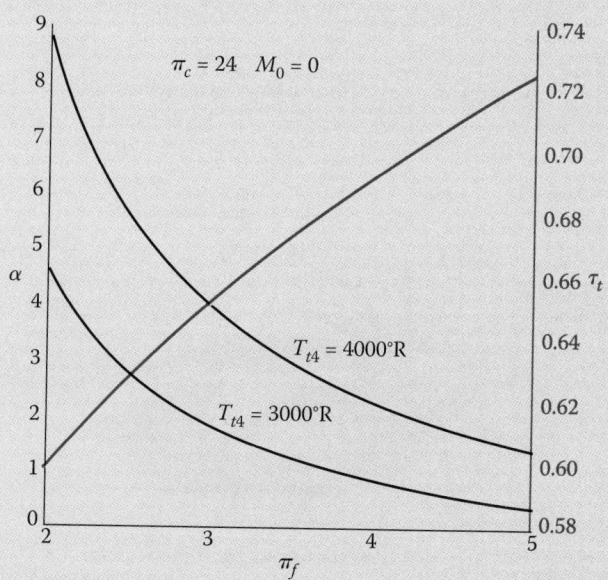

Fig. 7.25 Bypass ratio and turbine temperature ratio of mixed-flow turbofan engine.

7.6 Turboprop Engine

As discussed in Section 5.12, there continues to be interest in turboprop engines due to the ever-expanding uses for unmanned aerial systems and continued interest in high-efficiency, high-subsonic-speed flight. Two particular research and development efforts led to some advancements in turboprop technology that we see today. Figure 7.26 shows a sketch of a turboprop tested under the NASA-sponsored Advanced Turboprop Propulsion System program [71], which included highly swept propeller blades connected by an advanced gearbox to a conventional gas turbine engine. The gearbox allows a better match in rotating speeds between the low-pressure turbine and propeller. Figure 7.27 shows the *unducted fan* (UDF) engine developed by General Electric Aircraft Engines, which has two counterrotating rows of highly swept propeller blades that are directly driven by stages of the low-pressure turbine. Lack of a gearbox reduces engine weight, increases engine reliability, and compromises the speed match of the propeller and low-pressure turbine. Continued interest will likely remain with these "open rotor" or "propfan" configurations because of their potential for very low fuel consumption. Noise and containment issues remain key concerns.

7.6.1 Cycle Analysis

The analysis of the turboprop engine cycle builds on the ideal analysis developed in Section 5.12 We use the dimensionless *work output coefficient* C, defined by

$$C = \frac{\text{power interaction/mass flow of air through engine core}}{h_0} \qquad (5.67)$$

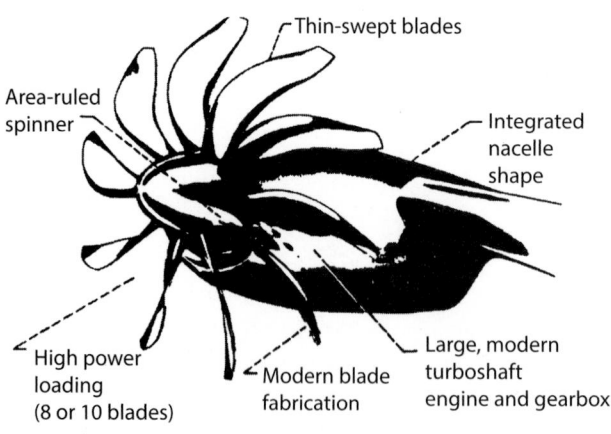

Fig. 7.26   NASA's advanced turboprop propulsion system.

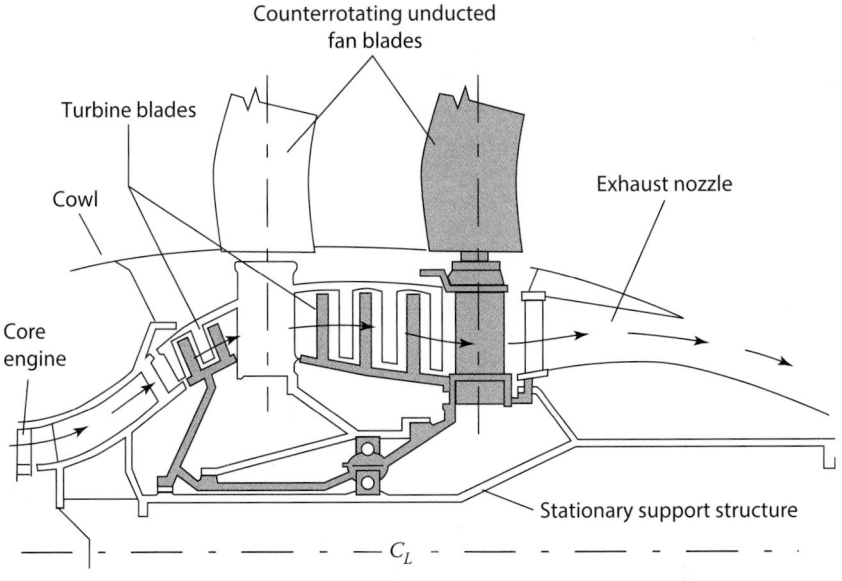

Fig. 7.27   General Electric's unducted fan (UDF) engine. (Courtesy of
General Electric Aircraft Engines.)

For the thrust of the core stream, we have its work output coefficient

$$C_C = \frac{F_C V_0}{\dot{m}_0 c_p T_0} \qquad (5.68)$$

The work output coefficient for the propeller is given by

$$C_{\text{prop}} = \frac{\eta_{\text{prop}} \dot{W}_{\text{prop}}}{\dot{m}_0 c_p T_0} \qquad (5.69)$$

Thus the work output coefficient for the total turboprop engine is

$$C_{\text{tot}} = C_{\text{prop}} + C_C \qquad (5.70)$$

and the corresponding thrust is

$$F = F_{\text{prop}} + F_C = \frac{C_{\text{tot}} \dot{m}_0 c_p T_0}{V_0} \qquad (5.71)$$

With turboprop engines, it is usual to have the core stream exit nozzles
unchoked, so the pressure imbalance term will not contribute in the
expression for the thrust. We consider the numbering stations indicated in
Fig. 5.25a, with station 4.5 being the exit from the high-pressure turbine and
the entrance to the low-pressure or power turbine. For a turboprop engine,
we have two design variables: the compressor pressure ratio and the

low-pressure (free or power) turbine temperature ratio. As with the cycle equation development of ideal turboprops, we desire the cycle equations in terms of the pressure and temperature ratios across both the high-pressure turbine (drives the compressor) and the low-pressure turbine. A sketch of a *T-s* diagram representative of a real turboprop is left as a homework exercise. (Refer to Fig. 5.25b for the *T-s* diagram of an ideal turboprop.)

Step 1: Using Eq. (7.3), we have for the engine core

$$\frac{F_C}{\dot{m}_0} = \frac{a_0}{g_c}\left[(1+f)\frac{V_9}{a_0} - M_0 + (1+f)\frac{R_t}{R_c}\frac{T_9/T_0}{V_9/a_0}\frac{1-P_0/P_9}{\gamma_c}\right]$$

Then

$$C_C = \frac{V_0}{c_p T_0}\frac{a_0}{g_c}\left[(1+f)\frac{V_9}{a_0} - M_0 + (1+f)\frac{R_t}{R_c}\frac{T_9/T_0}{V_9/a_0}\frac{1-P_0/P_9}{\gamma_c}\right]$$

Thus

$$C_C = (\gamma_c - 1)M_0\left[(1+f)\frac{V_9}{a_0} - M_0 + (1+f)\frac{R_t}{R_c}\frac{T_9/T_0}{V_9/a_0}\frac{1-P_0/P_9}{\gamma_c}\right] \quad (7.76)$$

Step 2:

$$\left(\frac{V_9}{a_0}\right)^2 = \frac{\gamma_t R_t T_9}{\gamma_c R_c T_0}M_9^2 \quad (7.4)$$

Step 3: We have

$$M_9^2 = \frac{2}{\gamma_t - 1}\left[\left(\frac{P_{t9}}{P_9}\right)^{(\gamma_t-1)/\gamma_t} - 1\right] \quad (7.6)$$

where

$$\frac{P_{t9}}{P_9} = \frac{P_0}{P_9}\pi_r\pi_d\pi_c\pi_b\pi_{tH}\pi_{tL}\pi_n \quad (7.77)$$

Two flow regimes exist for flow through the convergent exhaust nozzle: unchoked flow and choked flow. For unchoked flow, the exit static pressure P_9 is equal to the ambient pressure P_0, and the exit Mach number is less than or equal to 1. From Eq. (3.8) with $M = 1$, unchoked flow will exist when

$$\frac{P_{t9}}{P_0} \le \left(\frac{\gamma_t + 1}{2}\right)^{\gamma_t/(\gamma_t-1)} \quad (7.78a)$$

and the exit Mach number is given by Eqs. (7.6) and (7.77) with $P_9 = P_0$. On the other hand, choked flow will exist when

$$\frac{P_{t9}}{P_0} > \left(\frac{\gamma_t + 1}{2}\right)^{\gamma_t/(\gamma_t-1)} \quad (7.78b)$$

thus

$$M_9 = 1 \quad \frac{P_{t9}}{P_0} = \left(\frac{\gamma_t + 1}{2}\right)^{\gamma_t/(\gamma_t - 1)} \quad \text{and} \quad \frac{P_0}{P_9} = \frac{P_{t9}/P_9}{P_{t9}/P_9} \tag{7.78c}$$

Step 4: We have

$$\frac{T_9}{T_0} = \frac{T_{t9}/T_0}{(P_{t9}/P_9)^{(\gamma_t - 1)/\gamma_t}} \tag{7.79}$$

where

$$\frac{T_{t9}}{T_0} = \frac{T_{t4}}{T_0} \tau_{tH} \tau_{tL} \tag{7.80}$$

Combining Eqs. (7.4), (7.6), (7.7), and (7.8) gives

$$\frac{V_9}{a_0} = \sqrt{\frac{2\tau_\lambda \tau_{tH} \tau_{tL}}{\gamma_c - 1} \left[1 - \left(\frac{P_{t9}}{P_9}\right)^{-(\gamma_t - 1)/\gamma_t}\right]} \tag{7.81}$$

Step 5: From the energy balance of the real turbojet's burner, we have

$$f = \frac{h_{t4} - h_{t3}}{\eta_b h_{PR} - h_{t4}} \tag{7.10}$$

Step 6: For the high-pressure turbine, we have from Eq. (7.12) of the real turbojet analysis

$$\tau_{tH} = 1 - \frac{1}{\eta_{mH}(1 + f)} \frac{\tau_r}{\tau_\lambda} (\tau_c - 1) \tag{7.82}$$

where η_{mH} is the mechanical efficiency of the high-pressure shaft. For the low-pressure turbine (also called the *free turbine*), we equate the power out of this turbine to the power into the propeller. Thus

$$\eta_{mL} \dot{m}_{4.5} c_{pt} (T_{t4.5} - T_{t5}) = \frac{\dot{W}_{\text{prop}}}{\eta_g}$$

where η_g is the gearbox efficiency. This has to be specified (known) or included in the mechanical efficiency of the free-turbine shaft.

So

$$C_{\text{prop}} = \frac{\eta_{\text{prop}} \dot{W}_{\text{prop}}}{\dot{m}_0 c_p T_0} = \eta_{\text{prop}} \eta_g \eta_{mL}(1 + f)\tau_\lambda \tau_{tH}(1 - \tau_{tL}) \tag{7.83}$$

Step 7: For the turboprop engine, we consider both the specific power and the specific thrust:

$$\frac{\dot{W}}{\dot{m}_0} = C_{\text{tot}} c_{pc} T_0 \tag{7.84a}$$

$$\frac{F}{\dot{m}_0} = \frac{C_{\text{tot}} c_{pc} T_0}{V_0} \tag{7.84b}$$

Step 8:

$$S = \frac{f}{F/\dot{m}_0}$$

We should note here, also, that it is more usual when we refer to propeller aircraft to refer to the power specific fuel consumption in terms of S_p, where S_p is defined by

$$S_p = \frac{\dot{m}_f}{\dot{W}} = \frac{f}{\dot{W}/\dot{m}_0}$$

Thus

$$S_p = \frac{f}{C_{\text{tot}}\, c_{pc}\, T_0} \tag{7.85}$$

Step 9: The thermal efficiency of the turboprop engine cycle is defined as the ratio of the total power produced by the engine to the energy contained in the fuel. Thus

$$\eta_{Th} = \frac{\dot{m}_0\, c_{pc}\, T_0 C_{\text{tot}}}{\dot{m}_f h_{PR}}$$

or

$$\eta_{Th} = \frac{C_{\text{tot}}}{f h_{PR}/(c_{pc} T_0)} \tag{7.86}$$

The propulsive efficiency of the turboprop engine is defined as the ratio of the total power interaction with the vehicle producing propulsive power to the total energy available for producing propulsive power. Thus

$$\eta_P = \frac{\dot{m}_0\, c_{pc}\, T_0 C_{\text{tot}}}{\dot{W}_{\text{prop}}(\dot{m}_9 V_9^2 - \dot{m}_0 V_0^2)/(2g_c)}$$

or

$$\eta_P = \frac{C_{\text{tot}}}{C_{\text{prop}}/\eta_{\text{prop}} + [(\gamma_c - 1)/2][(1+f)(V_9/a_0)^2 - M_0^2]} \tag{7.87}$$

7.6.2   Selection of Optimal Turbine Expansion Ratio τ_{tL}^*

Turboprop or prop-fan engines are designed primarily to be low-specific-fuel-consumption engines. Thus we select τ_{tL} to make S_p a minimum. From Eq. (7.85), this is equivalent to locating the maximum of C_{tot}. We will obtain an expression for the optimum low-pressure turbine temperature ratio τ_{tL}^* that gives maximum C_{tot} with all other variables constant and when $P_9 = P_0$. We have

$$C_{\text{tot}} = C_{\text{prop}} + C_C$$

with

$$C_{\text{prop}} = \eta_{\text{prop}} \eta_g \eta_{mL}(1+f)\tau_\lambda \tau_{tH}(1 - \tau_{tL}) \tag{7.83}$$

and from Eq. (7.76) for the case $P_9 = P_0$

$$C_C = (\gamma_c - 1)M_0 \left[(1+f)\frac{V_9}{a_0} - M_0\right]$$

Thus the total temperature ratio of the low-pressure turbine correspond-ing to minimum fuel consumption is obtained by finding the maximum of C_{tot} with respect to τ_{tL}. Taking the partial derivative of C_{tot} with respect to τ_{tL} (noting that only τ_{tL} is a variable in the equation for C_{prop} and that V_9/a_0 is a function of τ_{tL} in the equation for C_c) and setting the result equal to zero gives

$$\frac{\partial C_{\text{tot}}}{\partial \tau_{tL}} = -\eta_{\text{prop}} \eta_g \eta_{mL}(1+f)\tau_\lambda \tau_{tH} + (\gamma_c - 1)(1+f)M_0 \times \frac{\partial}{\partial \tau_{tL}}\left(\frac{V_9}{a_0}\right) = 0$$

or

$$\frac{\partial}{\partial \tau_{tL}}\left(\frac{V_9}{a_0}\right) = \eta_{\text{prop}}\, \eta_g\, \eta_{mL}\frac{\tau_\lambda \tau_{tH}}{(\gamma_c - 1)M_0} \tag{7.88}$$

The resulting optimum expression, developed in the Supporting Material of Chapter 7 for τ_{tL}, gives

$$\tau_{tL}^* = \frac{\tau_{tH}^{-1/e_{tH}}}{\Pi}\tau_{tL}^{-(1-e_{tL})/e_{tL}} + A\left(1 + \frac{1 - e_{tL}}{e_{tL}}\frac{\tau_{tH}^{-1/e_{tH}}\tau_{tL}^{-1/e_{tL}}}{\Pi}\right)^2 \tag{7.89a}$$

where

$$A = \frac{[(\gamma_c - 1)/2][M_0^2/(\tau_\lambda \tau_{tH})]}{(\eta_{\text{prop}} \eta_g \eta_{mL})^2} \tag{7.89b}$$

and

$$\Pi = (\pi_r \pi_d \pi_c \pi_b \pi_n)^{(\gamma_t - 1)/\gamma_t} \tag{7.57}$$

Because Eq. (7.89a) is an equation for τ_{tL}^* in terms of itself, an iterative sol-ution is required. A starting value of τ_{tL}^*, denoted by τ_{tLi}^*, is obtained by solving Eq. (7.89a) for the case when $e_{tL} = 1$ (ideal low-pressure turbine):

$$\tau_{tLi}^* = \frac{\tau_{tH}^{-1/e_{tH}}}{\Pi} + A \tag{7.90}$$

This starting value can be substituted into Eq. (7.89a) and another new value of τ_{tLi}^* calculated. This process continues until the change in successive calculations of τ_{tLi}^* is less than some small number (say, 0.0001).

See Section 7.6.SM, Summary of Equations: Turboprop Engine, in the online supporting material for related information.

7.6.3 Propellers

The majority of the thrust of the turboprop engine comes from the propeller. It is for this reason that we have included the propeller analysis and design tools developed for Appendix M. If you are interested in propeller theory, operation, and sizing, you are encouraged to read this appendix online.

Example 7.11

For comparison with the ideal turboprop in Example 5.7, we consider the performance of a turboprop engine for use at Mach 0.8. The input data are as follows.

Inputs:

$$M_0 = 0.8, \quad T_0 = 240\ \text{K}, \quad \gamma_c = 1.4, \quad c_{pc} = 1.004\ \text{kJ/(kg} \cdot \text{K)}, \quad \gamma_t = 1.35,$$

$$c_{pt} = 1.108\ \text{kJ/(kg} \cdot \text{K)}, \quad h_{PR} = 42{,}800\ \text{kJ/kg}, \quad e_c = 0.90, \quad e_{tH} = 0.89,$$

$$e_{tL} = 0.91, \quad \pi_d = 0.98, \quad \pi_b = 0.96, \quad \pi_n = 0.99, \quad \eta_b = 0.99, \quad \eta_{mH} = 0.99,$$

$$\eta_{mL} = 0.99, \quad \eta_{\text{prop}} = 0.83, \quad \eta_g = 0.99, \quad T_{t4} = 1370\ \text{K}$$

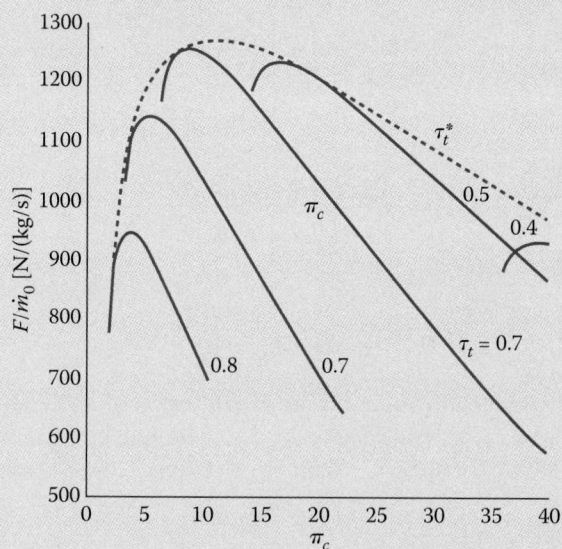

Fig. 7.28a Turboprop performance vs compressor pressure ratio: specific thrust.

(Continued)

Example 7.11 *(Continued)*

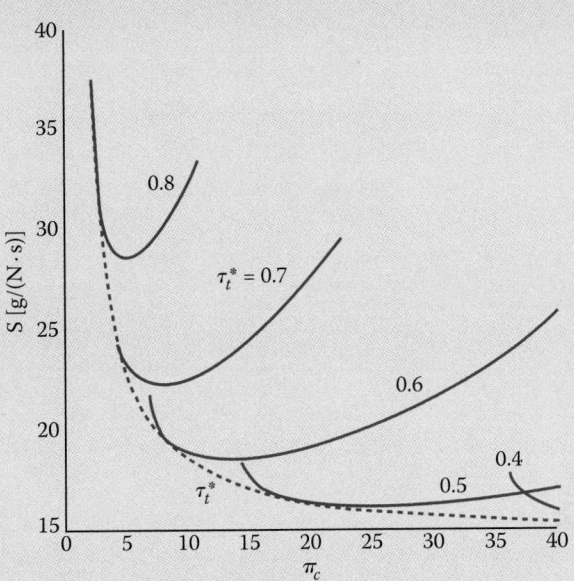

Fig. 7.28b Turboprop performance vs compressor pressure ratio: thrust specific fuel consumption.

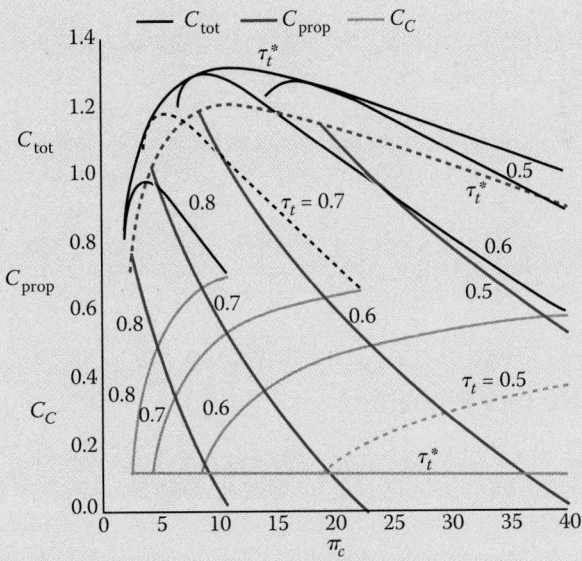

Fig. 7.28c Turboprop performance vs compressor pressure ratio: work coefficients.

Example 7.11 *(Continued)*

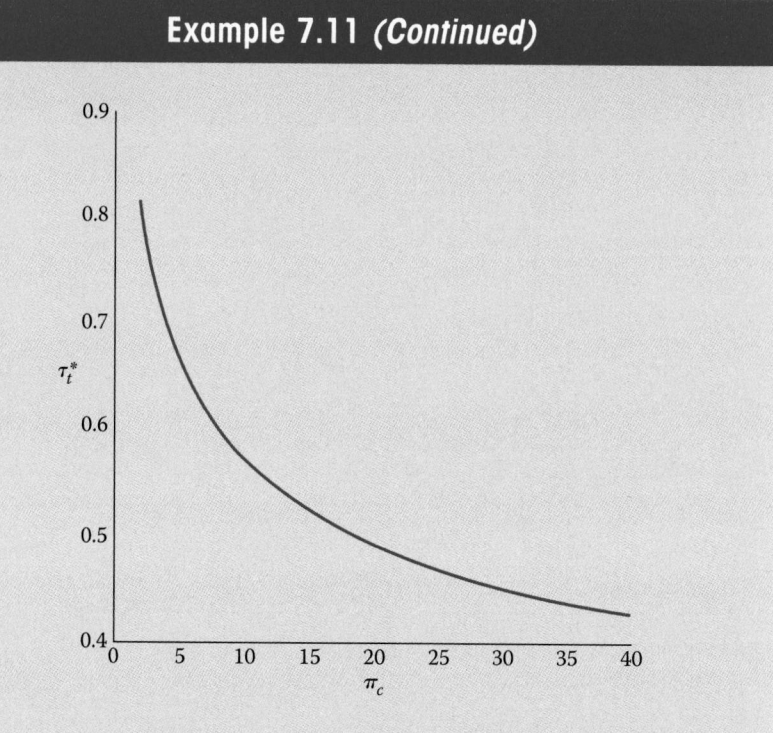

Fig. 7.28d   Turboprop performance vs compressor pressure ratio: optimum τ_t.

Figures 7.28a–7.28d show the variation in performance of a turboprop engine with compressor pressure ratio π_c and turbine temperature ratio τ_t. These figures for the engine cycle with losses can be compared with the performance of ideal turboprop engines (Figs. 5.26a–5.26e) to see the effects of component performance on the overall cycle. The general trends are the same as for the ideal cycle. The engine with losses that gives minimum fuel consumption has a compressor pressure ratio of $\pi_c = 30$ (see Fig. 7.28b) and corresponding optimum turbine temperature ratio of $\tau_t^* = 0.45$.

7.7 Variable Gas Properties

The effect of variable gas properties can be easily included in the computer analysis of gas turbine engine cycles. One first needs a subroutine that can calculate the thermodynamic state of the gas given the fuel/air ratio f and two independent properties. A subroutine for air and products of combustion from air and hydrocarbon fuels of the type $(CH_2)_n$ was developed for use in the AFPROP program first introduced in Chapter 2 and mentioned again in Chapter 6. This subroutine is called *FAIR*.

Table 7.1   Calling Nomenclature for Subroutine FAIR

Symbol	Knowns	Unknowns
FAIR(1, T, h, P_r, ϕ, c_p, R, γ, a, f)	T, f	h, c_p, P_r, ϕ, R, γ, a
FAIR(2, T, h, P_r, ϕ, c_p, R, γ, a, f)	h, f	T, c_p, P_r, ϕ, R, γ, a
FAIR(3, T, h, P_r, ϕ, c_p, R, γ, a, f)	P_r, f	T, h, c_p, ϕ, R, γ, a
FAIR(4, T, h, P_r, ϕ, c_p, R, γ, a, f)	ϕ, f	T, h, c_p, P_r, R, γ, a

The FAIR subroutine contains Eqs. (2.60–2.62), (2.64a–2.64d), and (2.55) and the constants of Table 2.2. This information gives direct calculation of h, c_p, ϕ, and P_r, given the fuel/air ratio f and temperature T. Calculation of the temperature T given the fuel/air ratio f and one of h, P_r, or ϕ is made possible by the addition of simple iteration algorithms. In addition, FAIR determines c_p, the gas constant R, the ratio of specific heats γ, and the speed of sound a.

For convenience, we will use the following nomenclature for FAIR. The primary input of FAIR (T, h, P_r, or ϕ) is indicated by first listing a corresponding number from 1 to 4, followed by a list of the variables. Table 7.1 identifies the four sets of knowns for FAIR. The FAIR subroutine with the first three sets of unknowns is used extensively in the list of equations in Section 7.7.1 for the turbojet engine with afterburner.

An additional property such as pressure P, density ρ, or entropy s is needed to completely define the thermodynamic state of the gas. We will normally use pressure or entropy as the additional property and Eqs. (2.28) and (2.68) to obtain the remaining unknowns:

$$\rho = \frac{P}{RT} \tag{2.28}$$

$$s_2 - s_1 = \phi_2 - \phi_1 - R \quad \ln\frac{P_2}{P_1} \tag{2.68}$$

Section 6.8 develops the basics of component performance with variable gas properties. The relationships among total pressure ratio, polytropic efficiency, and reduced pressure ratio are developed for the compressor and turbine. The relationships between static and total properties are developed for the inlet and nozzle. The energy balance of the combustor gives an expression for the fuel/air ratio f. These basics and others are used in the following section for the variable-gas-properties cycle analysis of a turbojet engine with afterburning.

7.7.1 Cycle Analysis: Turbojet with Afterburning

Figure 7.7 shows a cross-sectional drawing of a turbojet engine with afterburning and its station numbering. The uninstalled specific thrust is given in

Eqs. (7.1) and (7.2) and is written in terms of the engine stations, or

$$\frac{F}{\dot{m}_0} = \frac{a_0}{g_c}\left[(1+f_O)\frac{V_9}{a_0} - M + (1+f_O)\frac{R_9}{R_0}\frac{T_9/T_0}{V_9/a_0}\frac{1-P_0/P_9}{\gamma_0}\right] \tag{7.91}$$

where f_O is the overall fuel/air ratio, defined by Eq. (5.60) as

$$f_O = \frac{\dot{m}_f + \dot{m}_{f\,AB}}{\dot{m}_0} \tag{5.60}$$

The exit velocity V_9 is determined from the total and static enthalpies at station 9 by

$$V_9 = \sqrt{2g_c(h_{t9} - h_9)} \tag{6.60}$$

The total enthalpy at station 9 is obtained from application of the first law of thermodynamics to the engine and from tracking the changes in energy from the engine's inlet to its exit. The static state at station 9 (h_9, T_9, etc.) is obtained by using the following relationship among the reduced pressure at the static state P_{r9}, the reduced pressure at the total state P_{rt9}, and the nozzle pressure ratio P_{t9}/P_9:

$$P_{r9} = \frac{P_{rt9}}{P_{t9}/P_9} \tag{6.61}$$

The nozzle pressure ratio is obtained by using Eq. (7.27) which tracks the ratios of total pressure from engine inlet to exit:

$$\frac{P_{t9}}{P_9} = \frac{P_9}{P_9}\pi_r\pi_d\pi_c\pi_b\pi_t\pi_{AB}\pi_n \tag{7.27}$$

The enthalpy leaving the compressor h_{t3} is obtained by first determining the reduced pressure leaving the compressor P_{rt3}. From a rewrite of Eq. (6.31), we have

$$P_{rt3} = P_{rt2}\pi_c^{1/e_c} \tag{7.92}$$

Application of the first law of thermodynamics to the compressor and turbine gives the required turbine exit enthalpy. Equating the required compressor power to the net output power from the turbine, we have

$$\dot{W}_c = \eta_m \dot{W}_t$$

Rewriting in terms of mass flow rates and total enthalpies gives

$$\dot{m}_0(h_{t3} - h_{t2}) = \eta_m(\dot{m}_0 + \dot{m}_f)(h_{t4} - h_{t5}) = \eta_m\dot{m}_0(1+f)(h_{t4} - h_{t5})$$

Solving for h_{t5} gives

$$h_{t5} = h_{t4} - \frac{h_{t3} - h_{t2}}{(1+f)\eta_m} \tag{7.93}$$

where both h_{t4} and h_{t5} are functions of the fuel/air ratio. The turbine pressure ratio π_t is obtained from Eq. (6.56).

The fuel/air ratio for the main burner is given by Eq. (6.36), where h_{t4} is a function of f:

$$f = \frac{h_{t4} - h_{t3}}{\eta_b h_{PR} - h_{t4}} \tag{6.52}$$

Application of the first law to the afterburner gives the following equation for its fuel/air ratio f_{AB}, where h_{t6} is a function of the entering fuel/air ratio and h_{t7} is a function of the leaving fuel/air ratio ($f_O = f + f_{AB}$):

$$f_{AB} = \frac{h_{t7} - h_{t6}}{\eta_{AB} h_{PR} - h_{t7}} \tag{7.94}$$

Example 7.12

We consider an example with the same input data as those considered for the afterburning turbojet in Example 7.5. Thus we have variable specific heats and the following input data.

Inputs:

$M_0 = 2$, $T_0 = 390°R$, $h_{PR} = 18{,}400$ Btu/lbm, $\pi_{d\,max} = 0.98$, $\pi_b = 0.98$, $\pi_{AB} = 0.98$, $\pi_n = 0.98$, $e_c = 0.89$, $e_t = 0.91$, $\eta_b = 0.99$, $\eta_{AB} = 0.96$, $\eta_m = 0.98$, $P_0/P_9 = 1$, $T_{t4} = 3000°R$, $T_{t7} = 3500°R$, $\pi_c = 2 \rightarrow 14$

The results for the engine with variable specific heats are indicated in Fig. 7.29 with solid lines, along with the data for the engine with constant specific heats (Fig. 7.10), shown in dashed lines. The similarity of the results is remarkable, with no more than 4% difference in specific thrusts or fuel consumptions between the two models. The performance of both the afterburning and nonafterburning engines with constant specific heats is very close to those with variable specific heats because of the selected values for the ratio of specific heats. Note that the ratio of specific heats for the flow through the exhaust nozzle was selected to be 1.3 for the afterburning case and 1.33 for the nonafterburning case. The computations for this engine model with variable heats require about 10 times longer than those for the engine model with constant specific heats. The choice of whether to use the more accurate variable specific heats model should be based primarily on the desired fidelity of the results, not on the computational time. For most preliminary design analysis, use of the constant specific heat model (with judicious choice for specific heats) should be sufficient.

(Continued)

Example 7.12 *(Continued)*

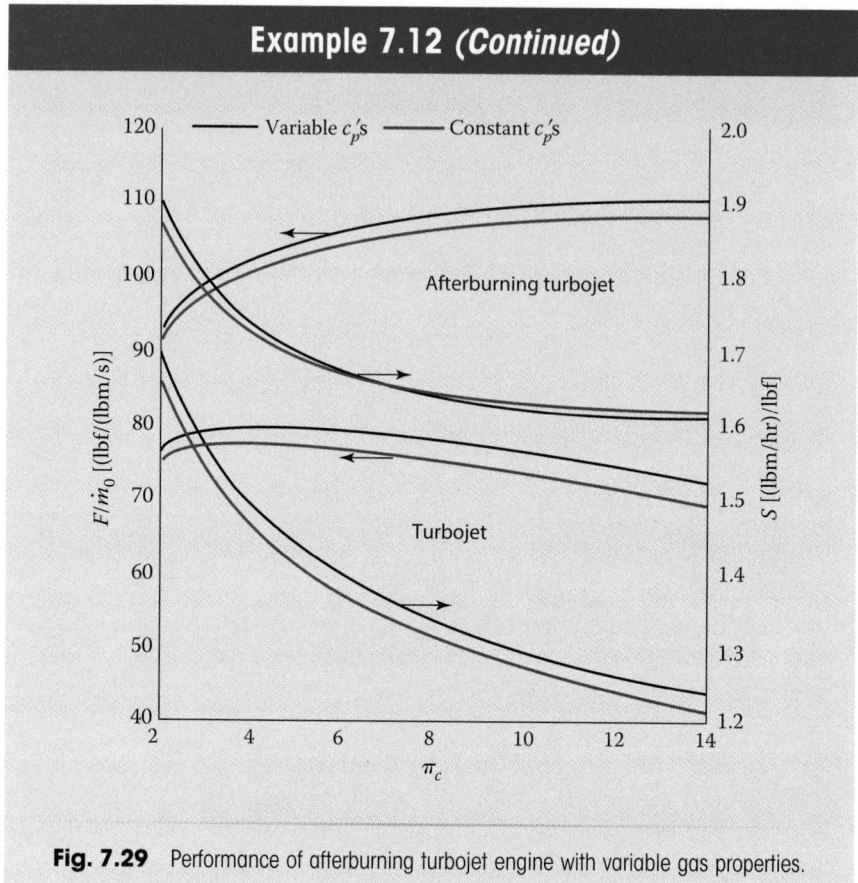

Fig. 7.29 Performance of afterburning turbojet engine with variable gas properties.

See Section 7.7b.SM, Summary of Equations: Afterburning Turbojet with Variable Specific Heats, in the online supporting material for related information.

7.7.2 Equivalent Gas Properties

As shown by Oates [7], one can obtain equivalent constant gas properties and simplify the analysis by selecting the ratio of specific heats that matches the exit velocity (equal specific thrusts) of the variable and constant specific heats models. The resulting equivalent gas properties can be used in the analytical tools developed while using constant gas properties and trends obtained very rapidly.

An expression for the equivalent ratio of specific heats γ_e can be obtained for the case of the turbojet engine by equating the exit velocity of Eq. (6.60) to that given for an engine cycle with constant specific heats. For expansion of a gas with constant specific heats, we can write

$$V_9^2 = 2g_c h_{t9e}\left[1 - \left(\frac{P_9}{P_{t9e}}\right)^{(\gamma_e-1)/\gamma_e}\right] \tag{7.95}$$

where the subscript e is used to represent the equivalent value. Setting this equation equal to Eq. (6.60) gives

$$h_{t9e}\left[1 - \left(\frac{P_9}{P_{t9e}}\right)^{(\gamma_e-1)/\gamma_e}\right] = h_{t9}\left(1 - \frac{h_9}{h_{t9}}\right)$$

or

$$\gamma_e = \left\{1 + \frac{\ell n[1 - (h_{t9}/h_{t9e})(1 - h_9/h_{t9})]}{\ell n(P_9/P_{t9e})}\right\}^{-1} \tag{7.96}$$

For the simple turbojet cycle, we could compare engines with the same values of h_{t4}. In the case of the afterburning turbojet, we would have $h_{t9} = h_{t9e}$, and Eq. (7.95) simplifies. The easiest method to find the equivalent ratio of specific heats γ_e is to first calculate the performance with variable specific heats and then calculate the performance with constant specific heats for several values of γ.

Problems

7.1   Develop a set of equations for parametric analysis of a ramjet engine with losses. Calculate the performance of a ramjet with losses over a Mach number range of 1 to 3 for the following input data:

$$\pi_{d\,max} = 0.95, \quad T_0 = 217\text{ K}, \quad \gamma_c = 1.4, \quad c_{pc} = 1.004\text{ kJ}/(\text{kg}\cdot\text{K}),$$
$$\pi_b = 0.94, \quad \eta_b = 0.96, \quad \gamma_t = 1.3, \quad c_{pt} = 1.235\text{ kJ}/(\text{kg}\cdot\text{K}),$$
$$\pi_n = 0.95, \quad \frac{P_0}{P_9} = 1, \quad T_{t4} = 1800\text{ K}, \quad h_{PR} = 42{,}800\text{ kJ}/\text{kg}$$

Compare your results to those obtained from the PARA computer program.

7.2   Why are the polytropic efficiencies, rather than the isentropic efficiencies, used for the fans, compressors, and turbines in parametric engine cycle analysis?

7.3   Calculate T_t and P_t at each station for Example 7.1.

7.4   Calculate and compare the performance of turbojet engines with the basic data of Example 7.1 for components with technology level 2 values in Table 6.2 (assume cooled turbine and the same diffuser and nozzle values as in Example 7.1). Comment on the changes in engine performance.

7.5   Using the PARA computer program, compare the performance of turbojet engines with the basic data of Example 7.1 for the polytropic efficiencies of component technology levels 1, 2, 3, 4, and 5 in Table 6.2 (assume uncooled turbine). Comment on the improvements in engine performance.

7.6   Using the PARA computer program, compare the performance of turbojet engines with the basic data of Example 7.3 for the polytropic efficiencies of component technology levels 1, 2, 3, and 4 in Table 6.2 (assume uncooled turbine). Comment on the changes in optimum compressor pressure ratio and improvements in engine performance.

7.7   Using the PARA computer program, find the range of compressor pressure ratios that give turbojet engines with specific thrust greater than 88 lbf/(lbm/s) and thrust specific fuel consumption below 1.5 (lbm/h)/lbf at $M_0 = 1.5$, $T_0 = 390°R$, and component performance of technology level 3 in Table 6.2 (assume type C diffuser, cooled turbine, and type F nozzle). Determine the compressor pressure ratio giving maximum specific thrust. Assume $\eta_m = 0.99$, $\gamma_c = 1.4$, $c_{pc} = 0.24$ Btu/(lbm · °R), $\gamma_t = 1.3$, $c_{pt} = 0.296$ Btu/(lbm · °R), $h_{PR} = 18{,}400$ Btu/lbm, and $P_0/P_9 = 1$.

7.8   Using the PARA computer program, find the range of compressor pressure ratios that give turbojet engines with specific thrust greater than 950 N/(kg/s) and thrust specific fuel consumption below 40 (g/s)/kN at $M_0 = 0.9$, $T_0 = 216.7$ K, and component performance of technology level 3 in Table 6.2 (assume type C diffuser, cooled turbine, and type F nozzle). Determine the compressor pressure ratio giving maximum specific thrust. Assume $\eta_m = 0.99$, $\gamma_c = 1.4$, $c_{pc} = 1.004$ kJ/(kg · K), $\gamma_t = 1.3$, $c_{pt} = 1.239$ kJ/(kg · K), $h_{PR} = 42{,}800$ kJ/kg, and $P_0/P_9 = 1$.

7.9   For a single-spool turbojet engine with losses, determine the compressor exit T_t and P_t, the turbine exit T_t and P_t, and the nozzle exit Mach number M_9 for the following input data:

$$M_0 = 0.8, \quad \pi_c = 9, \quad T_{t4} = 1780 \text{ K}, \quad h_{PR} = 42{,}800 \text{ kJ/kg},$$

$$P_0 = 29.92 \text{ kPa}, \quad T_0 = 229 \text{ K}, \quad \gamma_c = 1.4, \quad c_{pc} = 1.004 \text{ kJ/(kg · K)},$$

$$\pi_{d\,max} = 0.95, \quad \pi_b = 0.904, \quad \gamma_t = 1.3, \quad c_{pt} = 1.239 \text{ kJ/(kg · K)},$$

$$e_c = 0.85, \quad e_t = 0.88, \quad \eta_b = 0.99, \quad \eta_m = 0.98, \pi_n = 0.98, \quad \frac{P_0}{P_9} = 0.8$$

Compare your results with those obtained from the PARA computer program.

7.10   Work Problem 7.9 with data in English units.

7.11 Products of combustion enter the afterburner (station 6) at a rate of 230 lbm/s with the following properties: $T_{t6} = 1830°R$, $P_{t6} = 38$ psia, $M_6 = 0.4$, $\gamma = 1.33$, $c_p = 0.276$ Btu/(lbm · °R), and $R = 53.34$ ft · lbf/(lbm · °R). Assume a calorically perfect gas and $\eta_{AB} = 0.95$.

a) Determine the flow area at station 6 in square feet.

b) With the afterburner off, determine the area (ft²) of the exhaust nozzle's choked throat (station 8) for $P_{t8}/P_{t6} = 0.97$.

c) With the afterburner on, determine the afterburner fuel flow rate (lbm/s) and the area (ft²) of the exhaust nozzle's choked throat (station 8) for $P_{t8}/P_{t6} = 0.94$ and $T_{t8} = 3660°R$. Assume that the gas leaving the operating afterburner is a calorically perfect gas with $\gamma = 1.3$, $c_p = 0.297$ Btu/(lbm · °R), and the same gas constant. Also assume the properties at station 6 do not change and $h_{PR} = 18{,}400$ Btu/lbm.

7.12 Work Problem 7.11 with data in English units.

7.13 Calculate and compare the performance of afterburning turbojet engines with the basic data of Example 7.5 but with combustion temperatures of level 4 in Table 6.2 for compressor pressure ratios of 4, 8, and 12. Comment on the improvements in engine performance.

7.14 Using the PARA computer program, find the range of compressor pressure ratios that give afterburning turbojet engines with specific thrust greater than 118 lbf/(lbm/s) and thrust specific fuel consumption below 1.7 (lbm/h)/lbf at $M_0 = 1.5$, $T_0 = 390°R$, and component performance of technology level 3 in Table 6.2 (assume type C diffuser, cooled turbine, and type F nozzle). Determine the compressor pressure ratio giving maximum specific thrust. Assume $\eta_m = 0.99$, $\gamma_c = 1.4$, $c_{pc} = 0.24$ Btu/(lbm · °R), $\gamma_t = 1.3$, $c_{pt} = 0.296$ Btu/(lbm · °R), $\gamma_{AB} = 1.3$, $c_{pAB} = 0.296$ Btu/(lbm · °R), $h_{PR} = 18{,}400$ Btu/lbm, and $P_0/P_9 = 1$.

7.15 Using the PARA computer program, find the range of compressor pressure ratios that give afterburning turbojet engines with specific thrust greater than 1250 N/(kg/s) and thrust specific fuel consumption below 45 (g/s)/kN at $M_0 = 0.9$, $T_0 = 216.7$ K, and component performance of technology level 3 in Table 6.2 (assume type C diffuser, cooled turbine, and type F nozzle). Determine the compressor pressure ratio giving maximum specific thrust. Assume $\eta_m = 0.99$, $\gamma_c = 1.4$, $c_{pc} = 1.004$ kJ/(kg · K), $\gamma_t = 1.3$, $c_{pt} = 1.239$ kJ/(kg · K), $\gamma_{AB} = 1.3$, $c_{pAB} = 0.239$ kJ/(kg · K), $h_{PR} = 42{,}800$ kJ/kg, and $P_0/P_9 = 1$.

7.16   Using the PARA computer program, calculate and compare the performance of afterburning turbojet engines with the basic data of Example 7.5 for the different combustion temperatures and component technologies of levels 2, 3, 4, and 5 in Table 6.2 (assume cooled turbine, type B diffuser, and type F nozzle). Comment on the improvements in engine performance.

7.17   Work Problem 7.16 with input data in SI units.

7.18   Show that the propulsive efficiency and thermal efficiency of a turbofan engine with separate exhausts are given by Eqs. (7.50) and (7.51), respectively.

7.19   Calculate the performance of a turbofan engine with the basic data of Example 7.6 but with a fan pressure ratio of 1.65 and a bypass ratio of 8. Comment on the improvement in engine performance. Compare your results to those of the PARA computer program.

7.20   Work Problem 7.19 with data in SI units.

7.21   Using the PARA computer program, compare the performance of turbofan engines with the basic data of Example 7.6 for the polytropic efficiencies of component technology levels 2, 3, 4, and 5 in Table 6.2 (assume cooled turbine, type A diffuser, and type D nozzle). Comment on the improvement in engine performance.

7.22   Work Problem 7.21 with data in SI units.

7.23   Using the PARA computer program, find the range of compressor pressure ratios and fan pressure ratios that give optimum-bypass-ratio, separate-exhaust turbofan engines with specific thrust greater than 13 lbf/(lbm/s) and thrust specific fuel consumption below 1.0 (lbm/h)/lbf at $M_0 = 0.9$, $T_0 = 390°R$, and component performance of technology level 2 in Table 6.2 (assume type A diffuser, uncooled turbine, and type D nozzle). Assume $\eta_m = 0.99$, $\gamma_c = 1.4$, $c_{pc} = 0.24$ Btu/(lbm · °R), $\gamma_t = 1.3$, $c_{pt} = 0.296$ Btu/(lbm · °R), $h_{PR} = 18,400$ Btu/lbm, and $P_0/P_9 = 1$.

7.24   Using the PARA computer program, find the range of compressor pressure ratios and fan pressure ratios that give optimum-bypass-ratio, separate-exhaust turbofan engines with specific thrust greater than 130 N/(kg/s) and thrust specific fuel consumption below 26 (g/s)/kN at $M_0 = 0.8$, $T_0 = 216.7$ K, and component performance of technology level 2 in Table 6.2 (assume type A diffuser, uncooled turbine, and type D nozzle). Assume $\eta_m = 0.99$, $\gamma_c = 1.4$, $c_{pc} = 1.004$ kJ/(kg · K), $\gamma_t = 1.3$, $c_{pt} = 1.239$ kJ/(kg · K), $h_{PR} = 42,800$ kJ/kg, and $P_0/P_9 = 1$.

7.25 Calculate the performance of an optimum-bypass-ratio turbofan engine with the basic data of Example 7.8 but with a compressor pressure ratio of 30 and fan pressure ratio of 1.7. Compare your results to those of the PARA computer program.

7.26 Using the PARA computer program, compare the performance of optimum-bypass-ratio turbofan engines with the basic data of Example 7.8 for the polytropic efficiencies of component technology levels 2, 3, 4, and 5 in Table 6.2 (assume cooled turbine, type A diffuser, and type D nozzle). Comment on the improvement in engine performance.

7.27 A stationary gas turbine engine with regeneration is shown in Fig. P7.1. The effectiveness of a regenerator η_{rg} is defined by

$$\eta_{rg} = \frac{T_{t3.5} - T_{t3}}{T_{t5} - T_{t3}}$$

The total pressure ratios across the cold and hot gas paths of the regenerator are defined by

$$\pi_{rg\,cold} = \frac{P_{t3.5}}{P_{t3}} \qquad \pi_{rg\,hot} = \frac{P_{t6}}{P_{t5}}$$

Using these definitions and others, develop a set of equations for parametric analysis of this turboshaft engine with regeneration and losses.

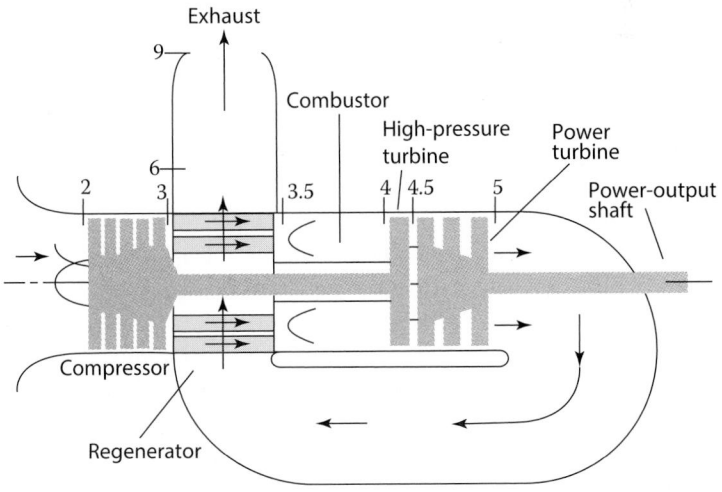

Fig. P7.1 Stationary gas turbine engine with regeneration.

7.28 Calculate the performance of an afterburning mixed-flow turbofan engine with the basic data of Example 7.9 at $M_0 = 0.9$ for a compressor pressure ratio of 30 and a fan pressure ratio of 4. Compare your results to those of the PARA computer program.

7.29 Calculate the performance of an afterburning mixed-flow turbofan engine with the basic data of Example 7.9 at $M_0 = 0.9$ for a compressor pressure ratio of 24 and a fan pressure ratio of 3.5. Compare your results to those of the PARA computer program.

7.30 Using the PARA computer program, find the range of compressor pressure ratios and corresponding fan pressure ratios that give mixed-flow turbofan engines of 0.5 bypass ratio a specific thrust greater than 55 lbf/(lbm/s) and thrust specific fuel consumption below 1.3 (lbm/h)/lbf at $M_0 = 1.8$, $T_0 = 390°R$, and component performance of technology level 3 in Table 6.2 (assume type C diffuser, cooled turbine, and type F nozzle). Assume $\eta_m = 0.99$, $\gamma_c = 1.4$, $c_{pc} = 0.24$ Btu/(lbm · °R), $\gamma_t = 1.3$, $c_{pt} = 0.296$ Btu/(lbm · °R), $M_6 = 0.5$, $\pi_{M\,max} = 0.95$, $h_{PR} = 18,400$ Btu/lbm, and $P_0/P_9 = 1$.

7.31 Using the PARA computer program, find the range of compressor pressure ratios and corresponding fan pressure ratios that give mixed-flow turbofan engines of 0.4 bypass ratio a specific thrust greater than 555 N/(kg/s) and thrust specific fuel consumption below 32.5 (g/s)/kN at $M_0 = 2.0$, $T_0 = 216.7$ K, and component performance of technology level 3 in Table 6.2 (assume type C diffuser, cooled turbine, and type F nozzle). Assume $\eta_m = 0.99$, $\gamma_c = 1.4$, $c_{pc} = 1.004$ kJ/(kg · K), $\gamma_t = 1.3$, $c_{pt} = 1.239$ kJ/(kg · K), $M_6 = 0.5$, $\pi_{M\,max} = 0.95$, $h_{PR} = 42,800$ kJ/kg, and $P_0/P_9 = 1$.

7.32 Using the PARA computer program, find the range of compressor pressure ratios and corresponding fan pressure ratios that give afterburning mixed-flow turbofan engines of 0.5 bypass ratio a specific thrust greater than 105 lbf/(lbm/s) and thrust specific fuel consumption below 1.845 (lbm/h)/lbf at $M_0 = 1.8$, $T_0 = 390°R$, and component performance of technology level 3 in Table 6.2 (assume type C diffuser, cooled turbine, and type F nozzle). Assume $\eta_m = 0.99$, $\gamma_c = 1.4$, $c_{pc} = 0.24$ Btu/(lbm · °R), $\gamma_t = 1.3$, $c_{pt} = 0.296$ Btu/(lbm · °R), $\gamma_{AB} = 1.3$, $c_{pAB} = 0.296$ Btu/(lbm · °R), $M_6 = 0.5$, $\pi_{M\,max} = 0.95$, $h_{PR} = 18,400$ Btu/lbm, and $P_0/P_9 = 1$.

7.33 Using the PARA computer program, find the range of compressor pressure ratios and corresponding fan pressure ratios that give afterburning mixed-flow turbofan engines of 0.4 bypass ratio a specific thrust greater than 1000 N/(kg/s) and thrust specific fuel consumption below 52.25 (g/s)/kN at $M_0 = 2.0$, $T_0 = 216.7$ K, and component performance of technology level 3 in Table 6.2 (assume type C diffuser, cooled turbine, and type F nozzle). Assume $\eta_m = 0.99$, $\gamma_c = 1.4$, $c_{pc} = 1.004$ kJ/(kg · K), $\gamma = 1.3$, $c_{pt} = 1.239$ kJ/(kg · K), $\gamma_{AB} = 1.3$, $c_{pAB} = 1.239$ kJ/(kg · K), $M_6 = 0.5$, $\pi_{M\,max} = 0.95$, $h_{PR} = 42{,}800$ kJ/kg, $M_6 = 0.5$, and $P_0/P_9 = 1$.

7.34   Using the PARA computer program, compare the performance of afterburning mixed-flow turbofan engines with the basic data of Example 7.9 at $M_0 = 0.9$, $\pi_c = 24$, and $\pi_f = 3.5$ for the different combustion temperatures and component technologies of levels 2, 3, 4, and 5 in Table 6.2 (assume cooled turbine, type C diffuser, and type F nozzle). Also assume the same γ, c_p, η_m, and $\pi_{M\,max}$. Comment on the improvement in engine performance.

7.35   For the mixed-flow turbofan engine with the bypass ratio specified, show that the following functional iteration equation for the fan temperature ratio with matched total pressures entering the mixer can be obtained from Eqs. (7.71) and (7.72):

$$(\tau_f)_{i+1} = \frac{\tau_r[\alpha - (\tau_c - 1)] + \eta_m(1+f)\tau_\lambda}{\tau_r\alpha + [\eta_m(1+f)\tau_\lambda/(\pi_c\pi_b)^{(\gamma_t-1)e_t/\gamma_t}](\gamma_t-1)/(\gamma_c-1)(\gamma_c/\gamma_t)e_te_j - 1}$$

with the first value of the fan temperature ratio given by

$$(\tau_f)_1 = \frac{\tau_r[\alpha - (\tau_c - 1)] + \eta_m(1+f)\tau_\lambda}{\tau_r\alpha + \eta_m(1+f)\tau_\lambda/(\pi_c\pi_b)^{(\gamma_t-1)e_t/\gamma_t}}$$

7.36   Sketch a *T-s* diagram representative of a real turboprop engine cycle. Use Figs. 5.25a and 5.25b from the ideal turboprop cycle analysis, and Fig. 7.11, *T-s* diagram for the core of a separate-flow real turbofan, to help you.

7.37   Calculate the performance of a turboprop engine with the basic data of Example 7.11 at a compressor pressure ratio of 20 and turbine temperature ratio of 0.5. Compare your results to those of Example 7.11 and the PARA computer program.

7.38   Work Problem 7.36 at $\pi_c = 12$ and optimum turbine temperature ratio. Compare your results to those of Example 7.11 and the PARA computer program.

7.39   Using the PARA computer program, find the range of compressor pressure ratios that give turboprop engines with optimum turbine temperature ratio τ_t^* a specific thrust greater than 105 lbf/(lbm/s) and thrust specific fuel consumption below 0.8 (lbm/h)/lbf at $M_0 = 0.7$, $T_0 = 447°R$, and component performance of technology level 2 in Table 6.2 (assume type A diffuser, uncooled turbine, and type D nozzle). Assume $\eta_{prop} = 0.83$, $\eta_g = 0.99$, $\eta_{mH} = 0.99$, $\eta_{mL} = 0.99$, $\gamma_c = 1.4$, $c_{pc} = 0.24$ Btu/(lbm · °R), $\gamma_t = 1.35$, $c_{pt} = 0.265$ Btu/(lbm · °R), and $h_{PR} = 18{,}400$ Btu/lbm.

7.40   Using the PARA computer program, find the range of compressor pressure ratios that give turboprop engines with optimum turbine temperature ratio τ_t^* a specific thrust greater than 1300 N/(kg/s) and thrust specific fuel consumption below 18 (g/s)/kN at $M_0 = 0.6$, $T_0 = 250$ K, and component performance of technology level 2 in Table 6.2 (assume type A diffuser, uncooled turbine, and type D nozzle). Assume $\eta_{prop} = 0.83$, $\eta_g = 0.99$, $\eta_{mH} = 0.995$, $\eta_{mL} = 0.995$, $\gamma_e = 1.4$, $c_{pc} = 1.004$ kJ/(kg · K), $\gamma_t = 1.35$, $c_{pt} = 1.108$ kJ/(kg · K), and $h_{PR} = 42{,}800$ kJ/kg.

7.41   Using the PARA computer program, compare the performance of turboprop engines with the basic data of Example 7.11 with component technologies of levels 1, 2, 3, 4, and 5 in Table 6.2. Comment on the improvement in engine performance.

Gas Turbine Design Problems

7.D1   You are to determine the range of compressor pressure ratios and bypass ratios for turbofan engines with losses that best meet the design requirements for the hypothetical passenger aircraft HP-1.

Hand-Calculate Performance with Losses (HP-1 Aircraft). Using the parametric cycle analysis equations for a turbofan engine with losses and component technology level 4 in Table 6.2 (assume cooled turbine, type A diffuser, and type D nozzle) with $T_{t4} = 1560$ K, hand-calculate the specific thrust and thrust specific fuel consumption for a turbofan engine with a compressor pressure ratio of 36, fan pressure ratio of 1.8, and bypass ratio of 10 at the 0.83-Mach and 11-km-altitude cruise condition. Assume $\gamma_c = 1.4$, $c_{pc} = 1.004$ kJ/(kg · K), $\gamma_t = 1.3$, $c_{pt} = 1.235$ kJ/(kg · K), $h_{PR} = 42{,}800$ kJ/kg, and $\eta_m = 0.99$. Compare your answers to results from the parametric cycle analysis program PARA and Design Problem 5.D1.

Computer-Calculate Performance with Losses (HP-1 Aircraft). For the 0.83-Mach and 11-km-altitude cruise condition, determine the performance available from turbofan engines with losses. This part of the analysis is accomplished by using the PARA computer program with component technology level 4 in Table 6.2 (assume cooled turbine, type A diffuser, and type D nozzle) and $T_{t4} = 1560$ K.

Specifically, you are to vary the compressor pressure ratio from 20 to 40 in increments of 2. Fix the fan pressure ratio at your assigned value of ____. Evaluate bypass ratios of 4, 6, 8, 10, 12, and the optimum value. Assume $\gamma_c = 1.4, c_{pc} = 1.004\,\text{kJ}/(\text{kg} \cdot \text{K})$, $\gamma_t = 1.3$, $c_{pt} = 1.235\,\text{kJ}/(\text{kg} \cdot \text{K})$, $h_{PR} = 42{,}800\,\text{kJ}/\text{kg}$, and $\eta_m = 0.99$.

Calculate Minimum Specific Thrust at Cruise (HP-1 Aircraft). You can calculate the minimum uninstalled specific thrust at cruise based on the following information:

1. The thrust of the two engines must be able to offset drag at 0.83 Mach and 11-km altitude and have enough excess thrust for P_s of 1.5 m/s. Determine the required installed thrust to attain the cruise condition, using Eq. (1.28). Assuming $\phi_{inlet} + \phi_{noz} = 0.02$, determine the required uninstalled thrust.

2. Determine the maximum mass flow into the 2.2-m-diam inlet for the 0.83-Mach and 11-km altitude flight condition, using the equation given in the background section for this design problem in Chapter 1.

3. Using the results of steps 1 and 2, calculate the minimum uninstalled specific thrust at cruise.

4. Perform steps 2 and 3 for inlet diameters of 2.5, 2.75, 3.0, 3.25, and 3.5 m.

Select Promising Engine Cycles (HP-1 Aircraft). Plot thrust specific fuel consumption vs specific thrust (thrust per unit mass flow) for the engines analyzed in the preceding. Plot a curve for each bypass ratio and cross-plot the values of the compressor pressure ratio (see Fig. P5.D1). The result is a carpet plot (a multivariable plot) for the cruise condition. Now draw a dashed horizontal line on the carpet plot corresponding to the maximum allowable uninstalled thrust specific consumption (S_{max}) for the cruise condition (determined in the Chapter 1 portion of this design problem). Draw a dashed vertical line for each minimum uninstalled specific thrust determined in the preceding. Your carpet plots will look similar to the example shown in Fig. P5.D1. What ranges of bypass ratio and compressor pressure ratio look most promising? Compare to the results of Design Problem 5.D1.

7.D2   You are to determine the ranges of compressor pressure ratio and bypass ratio for mixed-flow turbofan engines with losses that best meet the design requirements for the hypothetical fighter aircraft HF-1.

Hand-Calculate Performance with Losses (HF-1 Aircraft). Using the parametric cycle analysis equations for a mixed-flow turbofan engine with losses and component technology level 4 in Table 6.2

(assume cooled turbine, type C diffuser, and type F nozzle) with $T_{t4} = 3250°R$, hand-calculate the specific thrust and thrust specific fuel consumption for an ideal turbofan engine with a compressor pressure ratio of 25 and bypass ratio of 0.5 at the 1.6-Mach and 40-kft-altitude supercruise condition. Because the bypass ratio is given, you will need to use the system of equations given in Problem 7.27 to calculate the temperature ratio of the fan. Assume $\gamma_c = 1.4$, $c_{pc} = 0.240$ Btu/(lbm · °R), $\gamma_t = 1.3$, $c_{pt} = 0.296$ Btu/(lbm · °R), $h_{PR} = 18,400$ Btu/lbm, $M_6 = 0.4$, $\pi_{M\,max} = 0.96$, and $\eta_m = 0.99$. Compare your answers to results from the parametric cycle analysis program PARA and Design Problem 5.D2.

Computer-Calculate Performance with Losses (HF-1 Aircraft). For the 1.6-Mach and 40-kft-altitude supercruise condition, determine the performance available from mixed-flow turbofan engines with losses. This part of the analysis is accomplished by using the PARA computer program with component technology level 4 in Table 6.2 (assume cooled turbine, type C diffuser, and type F nozzle) and $T_{t4} = 3250°R$. Specifically, you are to vary the bypass ratio from 0.1 to 1.0 in increments of 0.05. Evaluate compressor pressure ratios of 16, 18, 20, 22, 24, and 28. Assume $\gamma_c = 1.4$, $c_{pc} = 0.240$ Btu/(lbm · °R), $\gamma_t = 1.3$, $c_{pt} = 0.296$ Btu/(lbm · °R), $h_{PR} = 18,400$ Btu/lbm, $M_6 = 0.4$, $\pi_{M\,max} = 0.96$, and $\eta_m = 0.99$.

Calculate Minimum Specific Thrust at Supercruise (HF-1 Aircraft). You can calculate the minimum uninstalled specific thrust at supercruise based on the following information:

1. The thrust of the two engines must be able to offset drag at the 1.6-Mach and 40-kft-altitude condition and 92% of takeoff weight. Assuming $\phi_{inlet} + \phi_{noz} = 0.05$, determine the required uninstalled thrust for each engine.

2. The maximum mass flow into a 5-ft² inlet for the 1.6-Mach and 40-kft-altitude flight condition is $\dot{m} = \rho A V = \sigma \rho_{ref} A M a = (0.2471 \times 0.07647)(5)(1.6 \times 0.8671) \times 1116 = 146.3$ lbm/s.

3. Using the results of steps 1 and 2, calculate the minimum uninstalled specific thrust at supercruise.

Select Promising Engine Cycles (HF-1 Aircraft). Plot thrust specific fuel consumption vs specific thrust (thrust per unit mass flow) for the engines analyzed in the preceding. Plot a curve for each bypass ratio, and cross-plot the values of compressor pressure ratio (see Fig. P5.D2). The result is a carpet plot (a multivariable plot) for the supercruise condition. Now draw a dashed horizontal line on the carpet plot corresponding to the maximum allowable uninstalled thrust specific fuel consumption (S_{max}) for the cruise condition (determined in the Chapter 1 portion of this design problem).

Draw a dashed vertical line for the minimum uninstalled specific thrust determined in the preceding. Your carpet plots will look similar to the example shown in Fig. P5.D2. What ranges of bypass ratio and compressor pressure ratio look most promising? Compare to the results of Design Problem 5.D2.

Chapter 8 Engine Performance Analysis (EPA)

> I want to tell you what thrust means to a fighter pilot—
> how what you do can make the difference between
> living and dying for the people who depend on
> your products.
>
> *John M. Loh, Lieutenant General, USAF, Celebration of the Golden Anniversary*
> *of Jet Powered Flight, 23 August 1989, Dayton, OH*

8.1 Introduction

This chapter is concerned with predicting the performance of a gas turbine engine and obtaining performance data similar to Figs. 1.14a–1.14c, 1.15a, and 1.15b and the data contained in Appendix B. The analysis required to obtain engine performance is related to, but very different from, the parametric cycle analysis (PCA) of Chapters 5 and 7. For example, in parametric cycle analysis of a turbojet engine, we independently selected values of the compressor pressure ratio, main burner exit temperature, flight condition, and the like. The analysis determined the turbine temperature ratio—it is dependent on the choices of compressor pressure ratio, main burner exit temperature, and flight condition, as shown by Eq. (7.12) and Fig. 7.3b. In engine performance analysis (EPA), we consider the performance of an engine that was built (constructed physically or created mathematically) with a selected design or reference compressor pressure ratio and its corresponding turbine temperature ratio. As will be shown in this chapter, the turbine temperature ratio remains essentially constant for a turbojet engine (and many other engine cycles), and its compressor pressure ratio is dependent on the throttle setting (main burner exit temperature T_{t4}) and flight condition (M_0 and T_0). The basic independent and dependent variables of the turbojet engine are listed in Table 8.1 for both PCA and EPA.

In PCA, we looked at the variation of gas turbine engine cycles where the main burner exit temperature and aircraft flight conditions were specified via the design inputs: T_{t4}, M_0, T_0, and P_0, In addition, the engine cycle was

---

Supporting material for this chapter is available electronically. See the Supporting Materials page at the back of the book for instructions to download.

Table 8.1 Comparison of Analysis Variables

Variable	Parametric Cycle	Engine Performance
Flight condition (M_0, T_0, and P_0)	Independent	Independent
Compressor pressure ratio π_c	Independent	Dependent
Main burner exit temperature T_{t4}	Independent	Independent
Turbine temperature ratio τ_t	Dependent	Constant

selected along with the compressor pressure ratio, the polytropic efficiency of turbomachinery components, and so on. For the combination of design input values, the resulting calculations yielded the specific performance of the engine (specific thrust and thrust specific fuel consumption), the required turbine temperature ratio, and the efficiencies of the turbomachinery (fan, compressor, and turbine). The specific combination of design input values is referred to as the engine *design point* or *reference point*. The resulting specific engine thrust and fuel consumption are valid only for the given engine cycle and values of T_{t4}, M_0, T_0, π_c, τ_t, η_c, and so on. When we changed any of these values in PCA, we were studying a "rubber" engine (i.e., one that changes its shape and component design to meet the thermodynamic, fluid dynamic, etc., requirements).

When a gas turbine engine is designed and built, the degree of variability of an engine depends on available technology, the needs of the principal application for the engine, and the desires of the designers. With the exception of the use of variable stator vanes, discussed in Section 1.4.2, and variable area converging-diverging exhaust nozzles (if so equipped), most gas turbine engines have fixed flow path areas. The main variable is the fuel flow rate that directly affects the combustor exit temperature T_{t4}, and sometimes T_{t7} when an afterburner is present. In a simple constant-flow-area turbojet engine, the performance (pressure ratio and mass flow rate) of its compressor depends on the power from the turbine and the inlet conditions to the compressor. As you will see in this chapter, a simple analytical expression can be used to express the relationship between the compressor performance and the independent variables: throttle setting (T_{t4}) and flight condition (M_0, T_0, P_0).

When a gas turbine engine is installed in an aircraft, its performance varies with flight conditions and throttle setting and is limited by the engine control system. In flight, the pilot controls the operation of the engine directly through the throttle and indirectly by changing flight conditions. The thrust and fuel consumption will thereby change. In this chapter, we will look at how specific engine cycles perform at conditions other than their design (or reference) point.

There are several ways to obtain this engine performance. One way is to look at the interaction and performance of the compressor–burner–turbine combination, known as the *pumping characteristics* of the gas generator. In this case, the performance of the components is known because the gas generator exists. However, in a preliminary design, the gas generator has not been built, and the pumping characteristics are not available. In such a case, the gas generator performance can be estimated by using first principles and estimates of the variations in component efficiencies, if those estimates are known and such fidelity is required. In reality, the principal effects of engine performance occur because of the changes in *propulsive efficiency* and *thermal efficiency* (rather than because of changes in component efficiency). Thus a good approximation of an engine's performance can be obtained by simply assuming that the component efficiencies remain constant.

The analysis of engine performance requires a model for the behavior of each engine component over its actual range of operation. The more accurate and complete the model, the more reliable the computed results. Even though the approach (constant efficiency of rotating components and constant total pressure ratio of the other components) used in this textbook gives answers that are perfectly adequate for preliminary design, it is important to know that the usual industrial practice is to use data or correlations having greater accuracy and definition in the form of component "maps." The principal values of the maps are to improve the understanding of component behavior and to slightly increase the accuracy of the results. In Section 8.1.3, we include sample, representative component maps, along with a brief discussion, to illustrate their use and demonstrate that our assumptions are not overly restrictive, especially for preliminary engine design.

The station numbering used for the performance analysis of the turbojet and turbofan shown in Fig. 8.1 is the same as we have used previously. Note that the turbine is divided into a high-pressure turbine (station 4 to 4.5) and a low-pressure turbine (station 4.5 to 5). The high-pressure

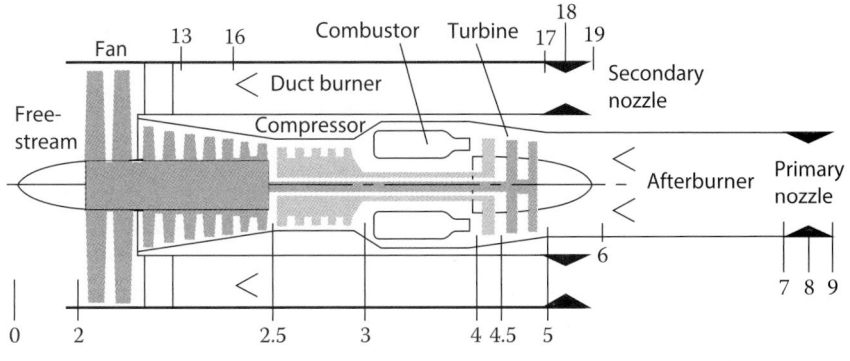

Fig. 8.1   Station numbering for two-spool gas turbine engine.

turbine drives the high-pressure compressor (station 2.5 to 3), and the low-pressure turbine drives the fan (station 2 to 13) and low-pressure compressor (station 2 to 2.5).

8.1.1 Reference Values and Engine Performance Analysis Assumptions

Functional relationships are used to predict the performance of a gas turbine engine at different flight conditions and throttle settings. These relationships are based on the application of mass, energy, momentum, and entropy considerations to the 1-D steady flow of a calorically perfect gas (CPG) at an engine steady-state operating point. Thus, if

$$f(\tau, \pi) = \text{const}$$

represents a relationship between the two engine variables τ and π at a steady-state operating point, then the constant can be evaluated at a reference condition (subscript R) so that

$$f(\tau, \pi) = f(\tau_R, \pi_R) = \text{const}$$

because $f(\tau, \pi)$ applies to the engine at all operating points. *Sea-level static (SLS), standard day, or a flight condition taken from PCA is normally chosen as the reference condition (design point)* for the value of the gas turbine engine variables. This technique for replacing constants with reference conditions is frequently used in the analysis to follow.

For conventional turbojet, turbofan, and turboprop engines, we will consider the simple case where the high-pressure turbine entrance nozzle, low-pressure turbine entrance nozzle, and primary exit nozzle (and bypass duct nozzle for the separate-exhaust turbofan) are choked. In addition, we assume that the throat areas where choking occurs in the high-pressure turbine entrance nozzle and the low-pressure turbine entrance nozzle are constant. This type of turbine is known as a *fixed-area turbine* (FAT) engine. These assumptions are true over a wide operating range for modern gas turbine engines. The following performance analyses also include the case(s) of unchoked engine exit nozzle(s).

The following assumptions will be made in the turbojet and turbofan performance analysis:

1. The flow is choked at the high-pressure turbine entrance nozzle, low-pressure turbine entrance nozzle, and primary exit nozzle. Also, the bypass duct nozzle for the turbofan is choked.
2. The total pressure ratios of the main burner, primary exit nozzle, and bypass stream exit nozzle (π_b, π_n, and π_{fn}) do not change from their reference values.
3. The component efficiencies (η_{cL}, η_{cH}, η_f, η_b, η_{tH}, η_{tL}, η_{mH}, and η_{mL}) do not change from their reference values.

4. Turbine cooling, bleed flow, and leakage effects are neglected.
5. No power is removed from the turbine to drive accessories (or alternately, η_{mH} or η_{mL} includes the power removed but is still constant).
6. Gases will be assumed to be calorically perfect both upstream and downstream of the main burner, and γ_t and c_{pt} do not vary with the power setting (T_{t4}).
7. The term unity plus the fuel/air ratio $(1 + f)$ will be considered as a constant.

Assumptions 4 and 5 are made to simplify the analysis and increase understanding. Reference 3 includes turbine cooling air, compressor bleed air, and power takeoff in the performance analysis. Assumptions 6 and 7 permit easy analysis that results in a set of algebraic expressions for an engine's performance. The performance analysis of an engine with variable gas properties is covered in the Supporting Material Section 8.3.SM.

8.1.2 Dimensionless and Corrected Performance Parameters

Dimensional analysis identifies correlating parameters that allow data taken under one set of conditions to be extended to other conditions. With aircraft gas turbine engines, these correlating parameters result in what is commonly known as *corrected performance* and are derived in a manner that makes sense physically (typically using the Buckingham-Pi theorem in physics). There are several excellent reasons for using corrected engine and component performance parameters:

* It is always impractical to accumulate experimental data for the bewildering number of possible operating conditions (throttle, altitude, Mach number), and it is often impossible to reach many of the conditions in a single, affordable test facility.
* Engine control systems operate with corrected parameters because their use collapses an unmanageable set of operating conditions to a set of performance curves amenable to control logic.
* Corrected parameters allow comparison of engine performance operated at different test facilities with different atmospheric conditions.

The quantities of pressure and temperature are normally made dimensionless by dividing each by its respective standard SLS values. The dimensionless pressure and temperature are represented by δ and θ, respectively, and given in Eqs. (1.2) and (1.3). When total (stagnation) properties are nondimensionalized, a subscript is used to indicate the station number of that property. The only static properties made dimensionless are freestream, the symbols for which carry no subscripts. Thus

$$\delta_i \equiv \frac{P_{ti}}{P_{\text{ref}}} \qquad (8.1a)$$

and

$$\theta_i \equiv \frac{T_{ti}}{T_{\text{ref}}} \tag{8.1b}$$

where $P_{\text{ref}} = 14.696$ psia (101,300 Pa) and $T_{\text{ref}} = 518.69°\text{R}$ (288.2 K).

Dimensionless analysis of engine components yields many useful dimensionless and/or modified component performance parameters. Some examples of these are the compressor pressure ratio, adiabatic efficiency, Mach number at the compressor face, ratio of blade (tip) speed to the speed of sound, and the Reynolds number.

The *corrected mass flow rate* at engine station i used in this analysis is defined as

$$\dot{m}_{ci} \equiv \frac{\dot{m}_i \sqrt{\theta_i}}{\delta_i} \tag{8.2}$$

and is related to the Mach number at station i as shown in the following. From the definition of the mass flow parameter [Eq. (3.11)], we can write the mass flow at station i for a given ratio of specific heats γ as

$$\dot{m}_i = \frac{P_{ti}}{\sqrt{T_{ti}}} A_i \times \text{MFP}(M_i)$$

Then

$$\frac{\dot{m}_{ci}}{A_i} = \frac{\dot{m}_{ci} \sqrt{T_{ti}}}{A_i P_{ti}} \frac{P_{\text{ref}}}{\sqrt{T_{\text{ref}}}} = \frac{P_{\text{ref}}}{\sqrt{T_{\text{ref}}}} \text{MFP}(M_i) \tag{8.3}$$

and the corrected mass flow rate per unit area is a function of the Mach number alone for a CPG. Equation (8.3) is plotted vs Mach number in Fig. 8.2 for three different γ values. Aircraft gas turbine engines need high thrust or power per unit weight that requires high corrected mass flow rates per unit area.

At the entrance to the fan or compressor (station 2), the design Mach number is generally limited to 0.65 to prevent excessive losses due to stronger than desired shock structure in the blade tip region. This M_2 of 0.65 corresponds to approximately 44 lbm/(s · ft²)[214.8 kg/(s · m²)]. A reduction in engine power will lower the corrected mass flow rate and the corresponding Mach number into the fan or compressor.

The flow is normally choked at the entrance to the turbine (station 4) and the throat of the exhaust nozzle (station 8) for most steady-state operating conditions of interest. (The flow is typically unchoked at these stations during engine startup.) When the flow is choked at station 4, the corrected mass flow rate per unit area entering the turbine is constant (it varies only slightly with γ, as shown in Fig. 8.2), which helps define the pumping characteristics of the gas generator. As shown later in this chapter, choked flow at both stations 4 and 8 limits the turbine operation. Even if the flow unchokes

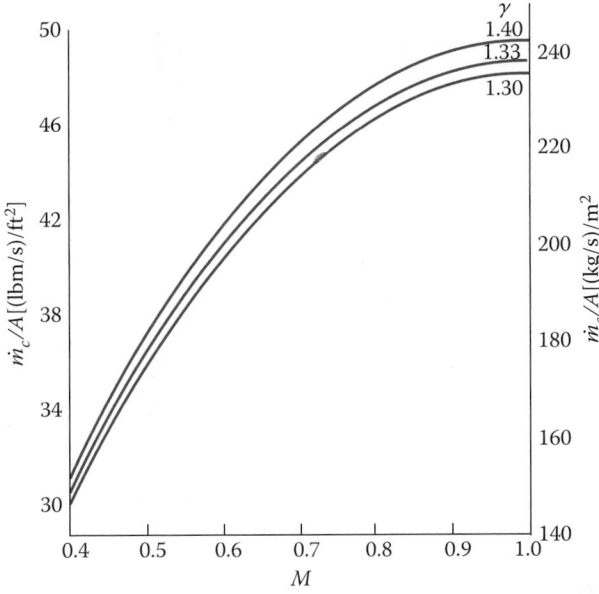

Fig. 8.2   Variation of corrected mass flow per area.

at a station and the Mach number drops from 1.0 to 0.9, the corrected mass flow rate is reduced less than 1%. Thus the corrected mass flow rate is considered constant when the flow is near or at choking conditions.

Choked flow at station 8 is desired in convergent-only exhaust nozzles to obtain high exit velocity and is required in a convergent-divergent exhaust nozzle to reach supersonic exit velocities. When the afterburner is operated on a turbojet or turbofan engine with choked exhaust nozzle, T_{t8} increases—this requires an increase in the nozzle throat area A_8 to maintain the correct mass flow rate/area ratio corresponding to choked conditions [Eq. (8.3)]. If the nozzle throat is not increased, the pressure increases and the mass flow rate decreases, which can adversely impact the upstream engine components.

The *corrected engine speed* at engine station i used in this analysis is defined as

$$N_{ci} \equiv \frac{N}{\sqrt{\theta_i}} \tag{8.4}$$

and is related to the *blade* Mach number, as presented in Section 9.3.4.7.

These four parameters represent a first approximation of the complete set necessary to reproduce nature for the turbomachinery. These extremely useful parameters have become a standard in the gas turbine industry and are summarized in Table 8.2.

Table 8.2 Corrected Parameters

Parameter	Symbol	Corrected Parameter
Total pressure	P_{ti}	$\delta_i = \dfrac{P_{ti}}{P_{ref}}$
Total temperature	T_{ti}	$\theta_i = \dfrac{T_{ti}}{T_{ref}}$
Rotational speed	$N = \text{RPM}$	$N_{ci} = \dfrac{N}{\sqrt{\theta_i}}$
Mass flow rate	$\dot{m}_i$	$\dot{m}_{ci} = \dfrac{\dot{m}_t \sqrt{\theta_i}}{\delta_i}$
Thrust	F	$F_c = \dfrac{F}{\delta_0}$
Thrust specific fuel consumption	S	$S_c = \dfrac{S}{\sqrt{\theta_0}}$
Fuel mass flow rate	$\dot{m}_f$	$\dot{m}_{fc} = \dfrac{\dot{m}_f}{\delta_2 \sqrt{\theta_2}}$

Three additional corrected quantities have found common acceptance for describing the performance of gas turbine engines: corrected thrust F_c, corrected thrust specific fuel consumption S_c, and corrected fuel mass flow rate $\dot{m}_{fc}$.

The *corrected thrust* is defined as

$$F_c \equiv \frac{F}{\delta_0} \tag{8.5}$$

For many gas turbine engines, the corrected thrust collapses the variation in thrust with flight condition into essentially a single curve.

The *corrected thrust specific fuel consumption* is defined as

$$S_c \equiv \frac{S}{\sqrt{\theta_0}} \tag{8.6}$$

and the *corrected fuel mass flow rate is* defined as

$$\dot{m}_{fc} \equiv \frac{\dot{m}_f}{\delta_2 \sqrt{\theta_2}} \tag{8.7}$$

Like the corrected thrust, these two corrected quantities collapse the variation in fuel consumption with flight condition and throttle setting into a single curve or set of curves, making them especially useful for engine control logic.

These three corrected quantities are closely related. By using the equation for thrust specific fuel consumption

$$S = \frac{\dot{m}_f}{F}$$

$\pi_d = P_{t2}/P_{t0}$, and the fact that $\theta_2 = \theta_0$ for an adiabatic inlet, the following relationship results between these corrected quantities:

$$S_c = \pi_d \frac{\dot{m}_{fc}}{F_c} \tag{8.8}$$

These extremely useful corrected engine performance parameters have also become a standard in the gas turbine industry and are included in Table 8.2.

8.1.3 Component Performance Maps

8.1.3.1 Compressor and Fan Performance Maps

The performance of a compressor or fan is normally shown by using the total pressure ratio, corrected mass flow rate, corrected engine speed, and component efficiency. Most often this performance is presented in one map showing the interrelationship of all four parameters, like that depicted in Fig. 8.3. Sometimes, for clarity, two maps are used, with one

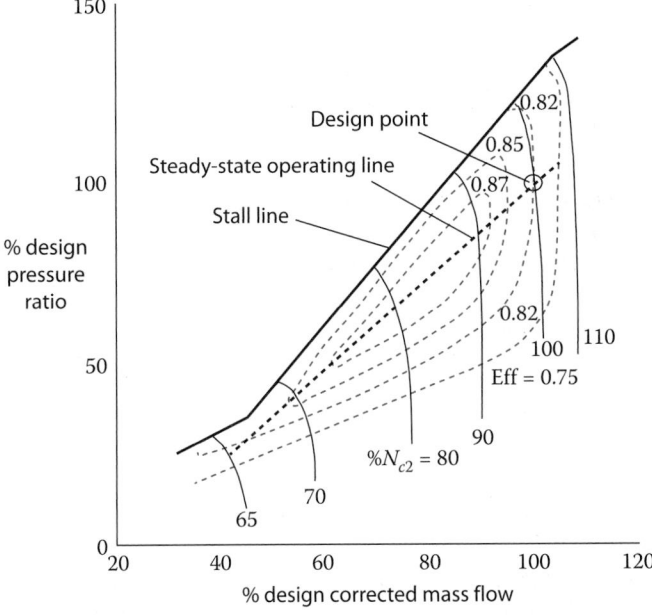

Fig. 8.3 Typical compressor performance map.

showing the pressure ratio vs corrected mass flow rate/corrected speed and the other showing compressor efficiency vs corrected mass flow rate/ corrected speed.

A limitation on fan and compressor performance of special concern is the *stall* or *surge line*. As will become much clearer with the study of velocity diagrams (Chapter 9), anything that causes the axial velocity to decrease relative to a rotating blade or stationary vane can result in aerodynamic stall (i.e., separation of the low momentum boundary layer off the blade/vane leading edge). In an aircraft gas turbine engine, the stall can manifest itself in a couple different ways depending on the energy state (throttle position and flight conditions) and engine volumes. The most common manifestation is surge—a full annulus stall that can be accompanied by reverse flow through the compression system. Surge is generally self-recoverable as long as whatever disturbance that caused the stall has not resulted in permanent damage to the engine. If the surge is not recovered from and/or at some low power settings, a rotating stall can occur, which can be nonrecoverable (often called hung stall or stagnation stall). The rotating stall manifests itself as a partial annulus region or packets of stalled air that can become trapped in the compressor and rotate at a fraction of the rotational speed. Generally, manufacturers try to design the engine so that rotating stall occurs under conditions below that of flight idle (60–70% N_{c2}) to avoid the issue altogether. Regardless of how stall manifests itself in the compressor, steady operation above the line is impossible, and entering the region even momentarily is dangerous to the gas turbine engine.

8.1.3.2 Main Burner Maps

The performance of the main burner is normally presented in terms of its performance parameters that are most important to engine performance: total pressure ratio of the main burner π_b and its combustion efficiency η_b. The total pressure ratio of the main burner is normally plotted vs the corrected mass flow rate through the burner $(\dot{m}_3 \sqrt{\theta_3}/\delta_3)$ for different fuel/air ratios f, as shown in Fig. 8.4. The efficiency of the main burner can be represented as a plot vs the temperature rise in the main burner $T_{t4} - T_{t3}$ or fuel/air ratio f for various values of inlet pressures P_{t3}, as shown in Fig. 8.4.

8.1.3.3 Turbine Maps

The flow through a turbine first passes through stationary airfoils (typically called *nozzles*) that turn and accelerate the fluid, increasing its tangential momentum. The flow then passes through rotating airfoils (called *rotor blades*) that remove energy from the fluid as they change its tangential momentum. Successive pairs of stationary airfoils followed by rotating airfoils remove additional energy from the fluid. To obtain a high output power/ weight ratio from a turbine, the flow entering the first-stage turbine rotor is normally supersonic, which requires the flow to pass through sonic conditions

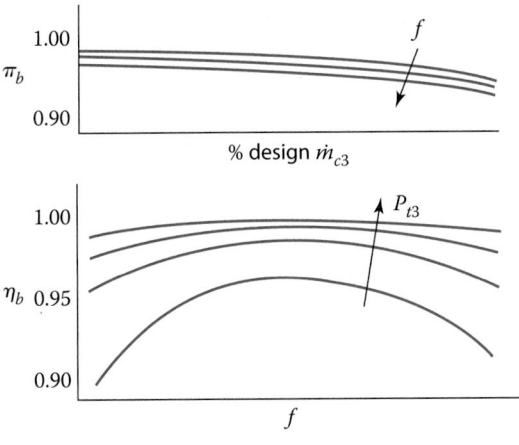

Fig. 8.4 Combustor performance.

at the minimum passage area in the turbine nozzles (see Fig. 8.5a). By using Eq. (8.3), the corrected inlet mass flow rate based on this minimum passage area (throat) will be constant for fixed-inlet-area turbines. This flow characteristic is shown in the typical turbine flow map (Fig. 8.5a) when the expansion ratio across the turbine $[(P_{t4}/P_{t5}) = 1/\pi_t)]$ is greater than about 2 and the flow at the throat is choked [see Appendix E, $M = 1.0$ or Eq. (3.8)].

The performance of a turbine is normally shown by using the total pressure ratio, corrected mass flow rate, corrected turbine speed, and component efficiency. This performance can be presented in two maps or a combined map (similar to that shown for the compressor in Fig. 8.3). When two maps are used, one map shows the interrelationship of the total pressure ratio, corrected mass flow rate, and corrected turbine speed, like that depicted on the left in Fig. 8.5b. The other map shows the interrelationship of turbine efficiency vs corrected mass flow rate/expansion ratio, like that

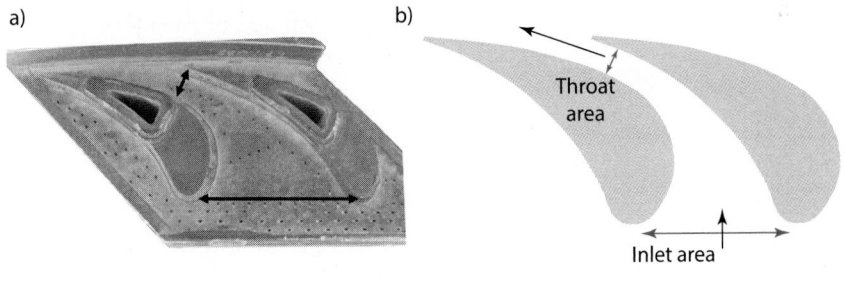

Fig. 8.5a (a) High-pressure turbine (HPT) nozzle segment; (b) inlet area and minimum (choked throat) area.

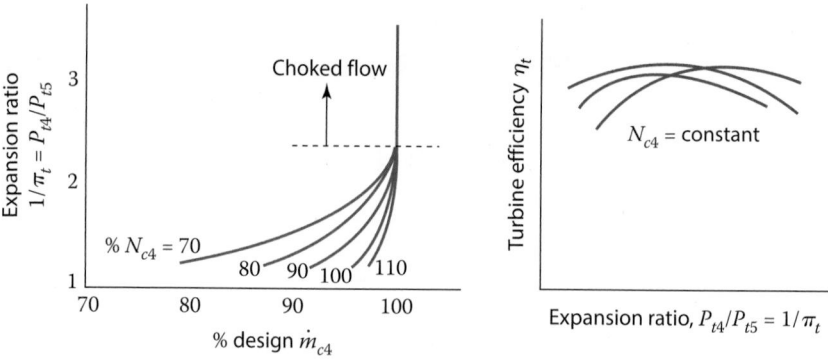

Fig. 8.5b Typical turbine flow map.

shown on the right in Fig. 8.5b. When a combined map is used, the total pressure ratio of the turbine is plotted vs the product of corrected mass flow rate and the corrected speed, as shown in Fig. 8.5c. This spreads out the lines of constant corrected speed from those shown in Fig. 8.5b, and the turbine efficiency can now be shown. If we tried to add these lines of constant turbine efficiency to Fig. 8.5b, many would coincide with the line for choked flow.

For the majority of aircraft gas turbine engine operation, the turbine efficiency varies very little. In the analysis of this chapter, we consider that the turbine efficiency is constant.

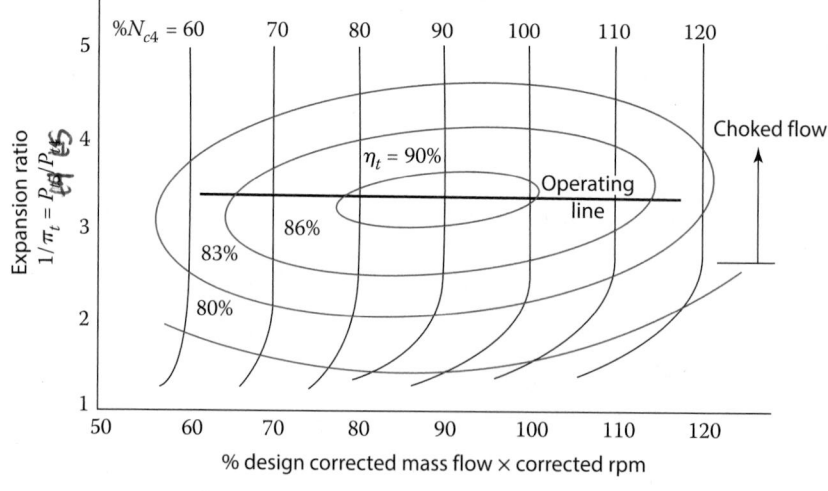

Fig. 8.5c Combined turbine performance map.

8.1.4 Preliminary Engine Sizing

As has been discussed, parametric cycle analysis (PCA) generally seeks to find promising candidate engines in terms of design choices and limits at a design or reference condition. Generally, with PCA, we are dealing with unsized engines, examining engine performance in terms of uninstalled specific thrust $F/\dot{m}$ and uninstalled thrust specific fuel consumption S.

In this section, preliminary engine sizing simply means scaling a candidate engine in terms of inlet air mass flow rate. Of course, this implies that something must be known about the overall aircraft mission requirements. Often, not surprisingly, the sizing condition is takeoff and the engine can be initially sized based on an estimated installed engine thrust loading T/W where W is usually the aircraft maximum gross takeoff weight (GTOW). Multiplying the installed thrust loading by the maximum GTOW and dividing by the number of engines provides the installed thrust per engine T. With a known specific thrust $F/\dot{m}$ for a candidate engine and accounting for installation losses, an estimate for the required inlet air mass flow rate is obtained.

If accurate drag characteristics for the aircraft are available, key mission legs can be evaluated to ensure required engine thrust is obtained at all or at least the most demanding flight conditions. Ultimately, the engine must be sized based on the maximum corrected air mass flow rate for whatever mission segment in which that occurs. For certain missions, that segment might by an acceleration requirement, a "hot day" (hotter than standard day) runway length requirement, or the like.

Once the required maximum air mass flow rate is determined, the mass flow parameter (MFP) becomes a very useful tool for determining required annulus flow path areas at the key engine stations. Recall that for given gas properties (selected γ, c_p), the MFP is a unique function of the Mach number at that station [Eq. (8.3)]. Because there are many engine station locations where we know the Mach number or some maximum not to exceed M (stations 2, 4, 4.5, 8, and 9), the MFP provides a straightforward means to estimate the flow path area. We will see this more in Chapters 9 and 11.

8.2 Gas Generator or Core

The performance of a gas turbine engine depends largely on the operation of its gas generator. Engine manufacturers use a proven core configuration for many different types of engines ("common core"—see Section 1.4.1). In this section, algebraic expressions for the pumping characteristics of a simple (single-spool) gas turbine engine are developed.

8.2.1 Conservation of Mass: Single-Spool Turbojet

We consider the flow through a single-spool turbojet engine with constant inlet area to the turbine ($A_4 = $ constant). Consistent with our assumptions, the mass flow rate into the turbine is equal to the sum of the mass flow rate through the compressor and the fuel flow rate into the main burner. Using the MFP, we can write

$$\dot{m}_2 + \dot{m}_f = (1 + f)\dot{m}_2 = \dot{m}_4 = \frac{P_{t4}A_4}{\sqrt{T_{t4}}}\mathrm{MFP}(M_4)$$

With the help of Eq. (8.3), the preceding equation yields the following expression for the compressor corrected mass flow rate:

$$\dot{m}_{c2} = \sqrt{\frac{T_{t2}}{T_{\mathrm{ref}}}\frac{P_{\mathrm{ref}}}{P_{t2}}}\frac{P_{t4}}{\sqrt{T_{t4}}}\frac{A_4}{1+f}\mathrm{MFP}(M_4)$$

Noting that $P_{t4} = \pi_c\pi_b P_{t2}$, we see that

$$\dot{m}_{c2} = \sqrt{\frac{T_{t2}}{T_{t4}}}\pi_c\pi_b\frac{P_{\mathrm{ref}}}{\sqrt{T_{\mathrm{ref}}}}\frac{A_4}{1+f}\mathrm{MFP}(M_4) \qquad (8.9)$$

Equation (8.9) is a straight line on a compressor map for constant values of T_{t4}/T_{t2}, A_4, f, and M_4. Lines of constant T_{t2}/T_{t4} are plotted on a typical compressor map in Fig. 8.6 for constant values of A_4 and f. Note that these

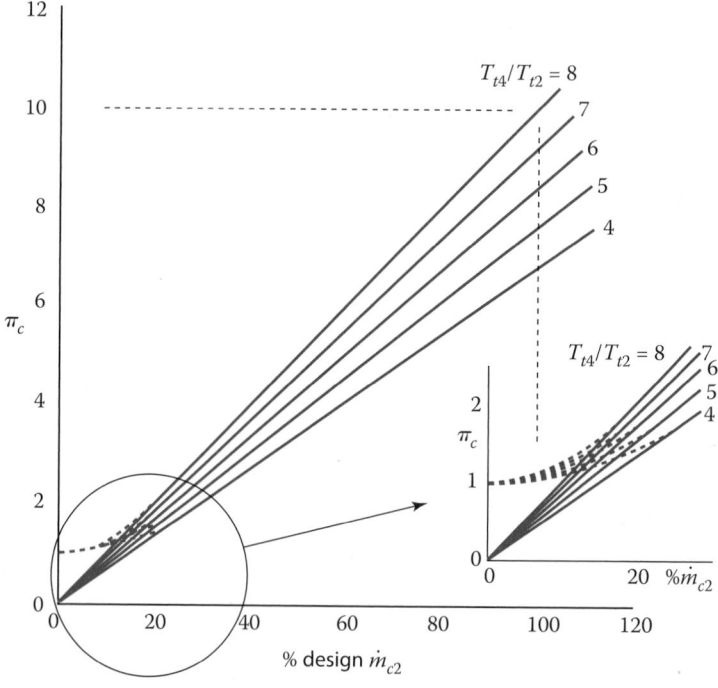

Fig. 8.6   Compressor map with lines of constant T_{t4}/T_{t2}.

lines start at a pressure ratio of 1 and corrected mass flow rate of 0 and are curved for low compressor pressure ratios (see insert in Fig. 8.6) because station 4 is unchoked. Station 4 chokes at a pressure ratio of about 2; the exact value is determined by actual gas properties (see Appendix D). At pressure ratios above 2, these lines are straight and appear to start at the origin (pressure ratio of 0 and mass flow rate of 0). The lines of constant T_{t2}/T_{t4} show the general characteristics required to satisfy conservation of mass and are independent of the turbine. For a given T_{t4}/T_{t2}, any point on that line will satisfy mass conservation for engine stations 2 and 4. The actual operating point of the compressor depends on the turbine and exhaust nozzle.

Equation (8.9) can be written simply for the case when station 4 is choked (the normal situation in gas turbine engines) as

$$\dot{m}_{c2} = C_1 \frac{\pi_c}{\sqrt{T_{t4}/T_{t2}}} \tag{8.10}$$

where

$$C_1 = \frac{\pi_b P_{\text{ref}}}{\sqrt{T_{\text{ref}}}} \frac{A_4}{1+f} \text{MFP}(M_4 = 1)$$

For an engine or gas generator, the specific relationship between the compressor pressure ratio and corrected mass flow rate is called the *engine operating line* and depends on the characteristics of the turbine. The equation for this operating line is developed later in this section.

8.2.2 Turbine Characteristics

Before developing the equations that predict the operating characteristics of the turbine, we write the mass flow parameter at any station i in terms of the mass flow rate, total pressure, total temperature, area, and Mach number. Because

$$\frac{\dot{m}_i \sqrt{T_{ti}}}{P_{ti} A_i} = \text{MFP}(M_i) = \sqrt{\frac{\gamma_i g_c}{R_i}} M_i \left(1 + \frac{\gamma_i - 1}{2} M_i^2\right)^{-(\gamma_i+1)/[2(\gamma_i-1)]}$$

Then, for $M_i = 1$,

$$\frac{\dot{m}_i \sqrt{T_{ti}}}{P_{ti} A_i} = \text{MFP}(M_i = 1) = \sqrt{\frac{\gamma_i g_c}{R_i}} \left(\frac{2}{\gamma_i + 1}\right)^{(\gamma_i+1)/[2(\gamma_i-1)]}$$

$$= \frac{\Gamma_i}{\sqrt{R_i/g_c}} \tag{8.11a}$$

where

$$\Gamma_i \equiv \sqrt{\gamma_i}\left(\frac{2}{\gamma_i+1}\right)^{(\gamma_i+1)/[2(\gamma_i-1)]} \tag{8.11b}$$

For a turbojet engine, the flow is choked ($M = 1$) in the turbine nozzle vanes (station 4) for virtually all engine operating conditions and choked or nearly choked at the throat of the exhaust nozzle (station 8). Thus the corrected mass flow rate per unit area is constant at station 4 and

$$\dot{m}_4 = \frac{P_{t4}A_4}{\sqrt{T_{t4}}}\frac{\Gamma_4}{\sqrt{R_4/g_c}} \qquad \dot{m}_8 = \frac{P_{t8}A_8}{\sqrt{T_{t8}}}\text{MFP}(M_8) \tag{i}$$

For a simple turbojet engine, the mass flow rate through the turbine is equal to that through the exhaust nozzle, or

$$\dot{m}_8 = \dot{m}_4$$

Using Eq. (i) then, we have

$$\frac{\sqrt{T_{t8}/T_{t4}}}{P_{t8}/P_{t4}} = \frac{A_8}{A_4}\frac{\text{MFP}(M_8)}{\Gamma_4/\sqrt{R_4/g_c}}$$

or

$$\frac{\sqrt{\tau_t}}{\pi_t} = \frac{A_8}{A_4}\frac{\text{MFP}(M_8)}{\Gamma_4/\sqrt{R_4/g_c}} \tag{8.12a}$$

where

$$\pi_t = \left(1 - \frac{1-\tau_t}{\eta_t}\right)^{\gamma_t/(\gamma_t-1)} \tag{8.12b}$$

For constant turbine efficiency η_t, constant values of R and Γ, constant areas at stations 4 and 8, and choked flow at station 8, Eqs. (8.12a) and (8.12b) can be satisfied only by constant values of the turbine temperature ratio τ_t and the turbine pressure ratio π_t. Thus we have

$$\tau_t, \pi_t \text{ constant} \qquad \text{for } M_8 = 1 \text{ and constant } A_4 \text{ and } A_8$$

If the exhaust nozzle unchokes and/or its throat area is changed, then both τ_t and π_t will change. Consider a turbine with reference values of $\eta_t = 0.90$ and $\tau_t = 0.80$ when the exhaust nozzle is choked and the gas has $\gamma = 1.33$. From Eqs. (8.12a) and (8.12b), $\pi_t = 0.363174$ and $A_8/A_4 = 2.46281$ at reference conditions (see Fig. 8.7a). Figure 8.7a also shows plots of Eq. (8.12a) for different values of the area ratio A_8/A_4 times the mass flow parameter at station 8 [MFP(M_8)] and Eq. (8.12b). Because of the relative slopes of these equations, the changes of both τ_t and π_t

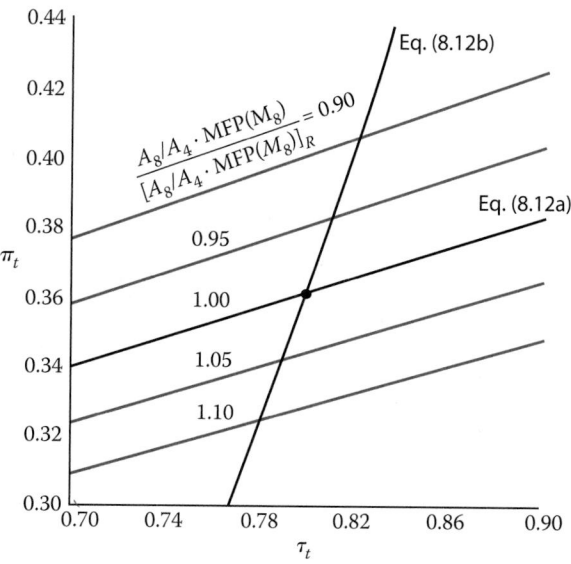

Fig. 8.7a Plot of turbine performance with Eqs. (8.12a) and (8.12b).

with A_8 and M_8 can be found by using the following functional iteration scheme, starting with an initial value of τ_t:

1. Solve for π_t using Eq. (8.12a).
2. Calculate a new τ_t using Eq. (8.12b).
3. Repeat steps 1 and 2 until successive values of τ_t are within a specified range (say, +0.0001).

The results of this iteration, plotted in Figs. 8.7b and 8.7c, show that when the Mach number M_8 is reduced from choked conditions ($M_8 = 1$), both τ_t and π_t increase; when the exhaust nozzle throat area A_8 is increased from its reference value, both τ_t and π_t decrease. A decrease in τ_t, with its corresponding decrease in π_t, will increase the turbine power per unit mass flow and change the pumping characteristics of the gas generator.

8.2.3 Engine Operating Line

From a work or power balance between the compressor and turbine, we write

$$\dot{m}_2 c_{pc}(T_{t3} - T_{t2}) = \eta_m \dot{m}_2(1+f)c_{pt}(T_{t4} - T_{t5})$$

or

$$\tau_c = 1 + \frac{T_{t4}}{T_{t2}} \frac{c_{pt}}{c_{pc}} \eta_m(1+f)(1 - \tau_t) \tag{8.13}$$

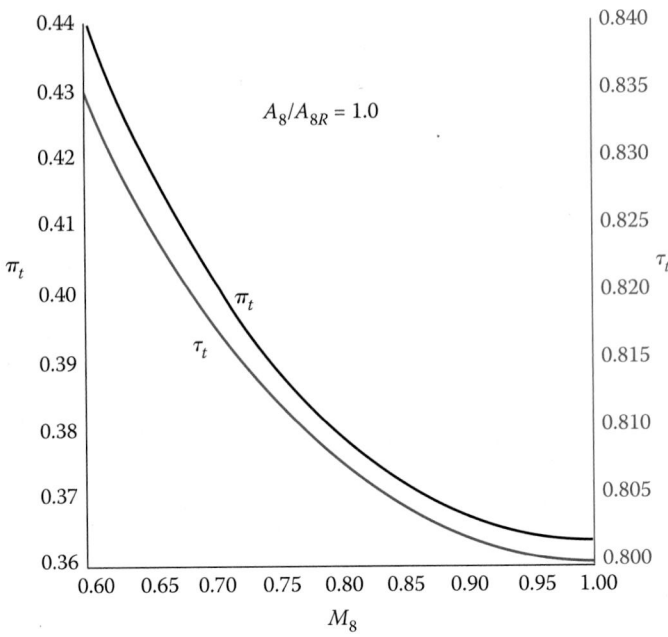

Fig. 8.7b   Variation of turbine performance with exhaust nozzle Mach number.

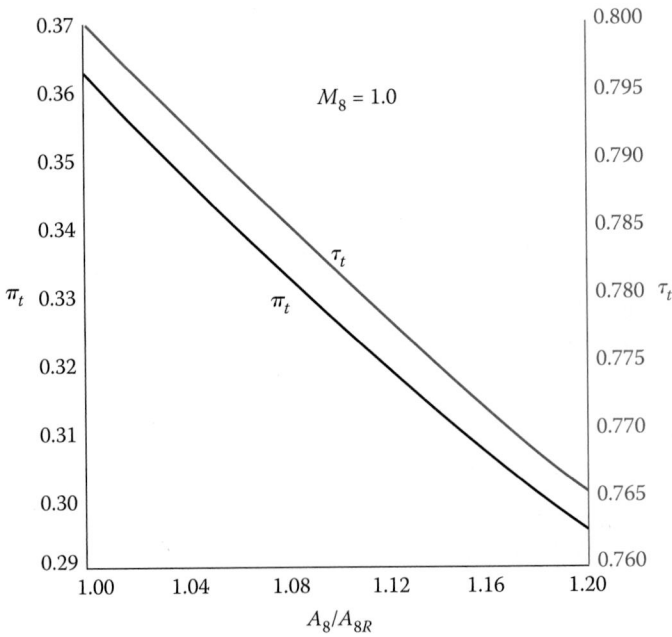

Fig. 8.7c   Variation of turbine performance with exhaust nozzle area.

where

$$\pi_c = [1 + \eta_c(\tau_c - 1)]^{\gamma_c/(\gamma_c-1)} \tag{ii}$$

Combining Eqs. (8.13) and (ii) gives

$$\pi_c = \left\{ 1 + \frac{T_{t4}}{T_{t2}} \left[\frac{c_{pt}}{c_{pc}} \eta_c \eta_m (1+f)(1-\tau_t) \right] \right\}^{\gamma_c/(\gamma_c-1)} \tag{8.14}$$

where the term in square brackets can be considered a constant when τ_t is constant. Solving Eq. (8.14) for the temperature ratio gives

$$\frac{T_{t4}}{T_{t2}} = C_2[(\pi_c)^{(\gamma_c-1)/\gamma_c} - 1]$$

where C_2 represents the reciprocal of the constant term within the square brackets of Eq. (8.14). Combining this equation with Eq. (8.10) gives an equation for the *engine operating line* that can be written as

$$\dot{m}_{c2} = \frac{\pi_c}{\sqrt{\pi_c^{(\gamma_c-1)/\gamma_c} - 1}} \frac{C_1}{\sqrt{C_2}} \quad \text{for constant } \tau_t \tag{8.15}$$

We can plot the engine operating line, using Eq. (8.15), on the compressor map of Fig. 8.6, giving the compressor map with operating line shown in Fig. 8.8. This engine operating line shows that for each value of the

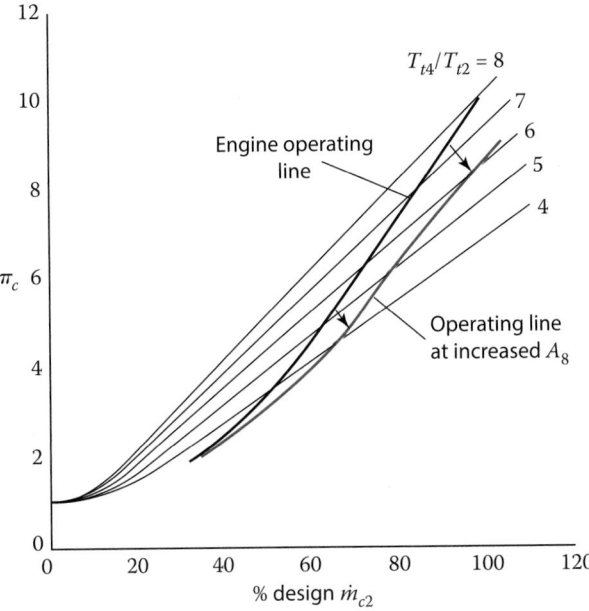

Fig. 8.8 Compressor map with engine operating line.

temperature ratio T_{t4}/T_{t2} there is one value of compressor pressure ratio and corrected mass flow rate. One can also see that for a constant value of T_{t2}, both the compressor pressure ratio and the corrected mass flow rate will increase with increases in throttle setting (increases in T_{t4}). In addition, when at constant T_{t4}, the compressor pressure ratio and corrected mass flow rate will decrease with increases in T_{t2} due to higher speed and/or lower altitude (note: $T_{t2} = T_{t0} = T_0\tau_r$). The curving of the operating line in Fig. 8.8 at pressure ratios below 4 is due to the exhaust nozzle being unchoked ($M_8 < 1$), which increases the value of τ_t (see Fig. 8.7b).

The engine operating line defines the pumping characteristics of the gas generator. As mentioned earlier, changing the throat area of the exhaust nozzle A_8 will change these characteristics. It achieves this change by shifting the engine operating line. Increasing A_8 will decrease τ_t (see Fig. 8.7c). This decrease in τ_t will increase the term within the square brackets of Eq. (8.14) that corresponds to the reciprocal of constant C_2 in Eq. (8.15). Thus an increase in A_8 will decrease the constant C_2 in Eq. (8.15). For a constant T_{t4}/T_{t2}, this shift in the operating line will increase both the corrected mass flow rate and the pressure ratio of the compressor, as shown in Fig. 8.8 for a 20% increase in A_8. For some compressors, an increase in the exhaust nozzle throat area A_8 can keep engine operation away from the stall line (see Fig. 8.3).

8.2.4   Engine Controls

The engine control system will control the gas generator operation to keep the main burner exit temperature T_{t4}, the compressor's pressure ratio π_c, rotational speed N, exit total temperature T_{t3}, and exit total pressure P_{t3} from exceeding specific maximum values. Limiting T_{t4} and π_c has the most dominating effect and is included in the following analysis. An understanding of the influence of the engine control system on compressor performance during changing flight conditions and throttle settings can be gained by recasting Eqs. (8.10) and (8.14) in terms of the dimensionless total temperature at station 0 (θ_0). We note that

$$T_{t0} = T_{\text{ref}} \frac{T_{t0}}{T_{\text{ref}}} = T_{\text{ref}}\,\theta_0$$

and

$$\theta_0 = \frac{T_0}{T_{\text{ref}}}\tau_r = \frac{T_0}{T_{\text{ref}}}\left(1 + \frac{\gamma - 1}{2}M_0^2\right) \tag{8.16}$$

Equation (8.16) and Figs. 8.9 and 8.10 show that θ_0 includes the influence of both the altitude (through the ambient temperature T_0) and the flight Mach number. It represents the total energy of the air captured and delivered to the compressor at station 2 (because $T_{t0} = T_{t2}$) at any flight condition. Although

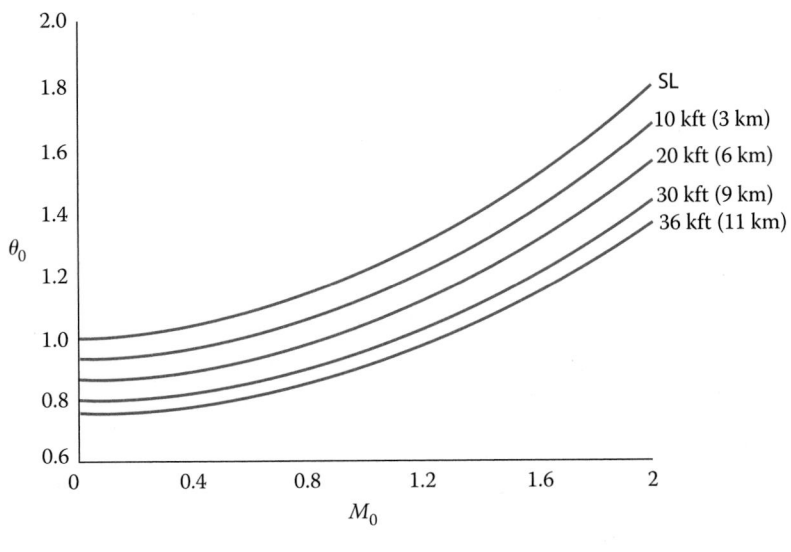

Fig. 8.9 θ_0 vs Mach number at different altitudes (standard day).

Fig. 8.9 shows the direct influence of Mach number and altitude on θ_0, Fig. 8.10 is an easier plot to understand in terms of aircraft flight conditions (Mach number and altitude) and relates directly to the aircraft flight envelope (see Figs. 1.1 and 1.21b).

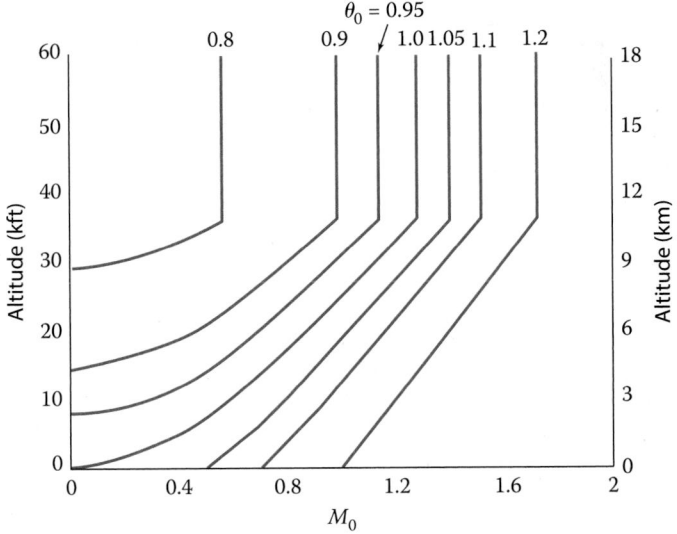

Fig. 8.10 θ_0 vs Mach number and altitude (standard day).

Using Eq. (8.16) and the fact that $T_{t2} = T_{t0}$, we can write Eq. (8.14) as

$$\pi_c = \left(1 + \frac{T_{t4}}{\theta_0} K_1\right)^{\gamma_c/(\gamma_c - 1)} \tag{8.17}$$

where K_1 is a constant. Equation (8.17) is plotted in Fig. 8.11 for the turbojet engine of Example 8.1 later in this chapter. Equation (8.17) is very useful and insightful. It relates the overall compressor total pressure ratio to throttle position (T_{t4}) and altitude and M_0 flight condition (θ_0).

By using Eqs. (8.17) and (8.10), the corrected mass flow rate through the compressor can be expressed as

$$\dot{m}_{c2} = \left(\frac{\theta_0}{T_{t4}}\right)^{1/2} \left(1 + \frac{T_{t4}}{\theta_0} K_1\right)^{\gamma_c/(\gamma_c - 1)} K_2 \tag{8.18}$$

where K_2 is a constant. Equation (8.18) is also plotted in Fig. 8.11 for the turbojet engine of Example 8.1, along with engine control limits as discussed in the next section.

8.2.5 Theta Break and Throttle Ratio

In Fig. 8.11, without limits on π_c and T_{t4}, the performance curves shown would continue past $\pi_c = 15$ on up, and additional curves corresponding to higher T_{t4} values beyond 3200°R would be seen. In practice, of course, all engines have control limits to ensure safe and efficient operation.

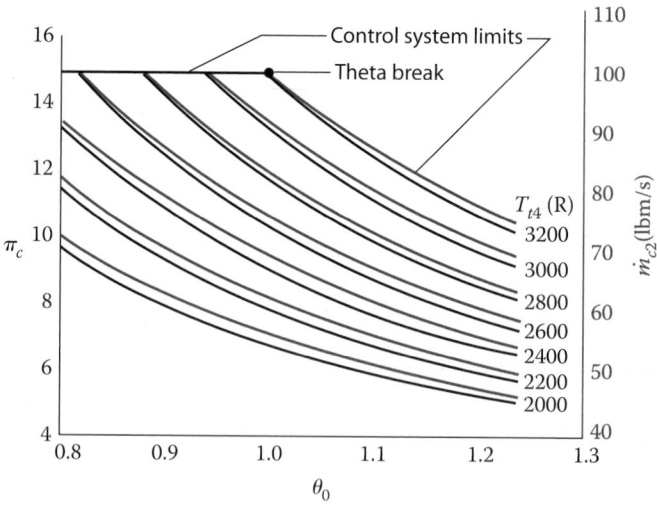

Fig. 8.11 Compressor performance vs θ_0 and T_{t4}.

The unique point, so visually striking in Fig. 8.11, at which the control logic must switch from limiting π_c to limiting T_{t4} is known as the *theta break*, or $\theta_{0\,\text{break}}$. Returning briefly to Fig. 8.11 and focusing on the control limit lines, you will find it very convenient to visualize that at any point in the flight envelope to the left of the theta break $\pi_c = \pi_{c\,\text{max}}$ and $T_{t4} < T_{t4\text{max}}$—the engine controls using the compressor pressure ratio limit. At any point on the control limit line to the right of the theta break, $\pi_c < \pi_{c\text{max}}$ and $T_{t4} = T_{t4\,\text{max}}$—the engine controls using the maximum turbine inlet temperature. The relationship of the instantaneous value of θ_0 to $\theta_{0\,\text{break}}$ has important consequences to engine cycle performance. On the one hand, when $\theta_0 < \theta_{0\,\text{break}}$ and $T_{t4} < T_{t4\,\text{max}}$, the specific thrust of the engine is less than its inherent material capabilities would make possible. On the other hand, when $\theta_0 > \theta_{0\,\text{break}}$ and $\pi_c < \pi_{c\,\text{max}}$, the specific fuel consumption is more than its inherent thermal efficiency would make possible.

The designer would therefore strongly prefer to have the engine always operate at or very near $\theta_0 = \theta_{0\,\text{break}}$, but this is impossible because every aircraft has a flight envelope with a range of θ_0 (see Fig. 8.10). The best available compromise is to choose a $\theta_{0\,\text{break}}$ that provides the best balance of engine performance over the expected range of flight conditions.

It is interesting to note that, because early commercial and military aircraft primarily flew at or near $\theta_0 = 1$, they were successfully designed with $\theta_{0\,\text{break}} = 1$. Consequently, several generations of propulsion engineers took it for granted that aircraft engines always operated at $\pi_{c\,\text{max}}$ and $T_{t4\,\text{max}}$ under standard sea-level static conditions. However, the special requirements of modern commercial aircraft to be "flat-rated" (i.e., constant sea-level takeoff thrust up to a certain ambient temperature) or the newest military aircraft like the F-22 Raptor to supercruise (supersonic cruise without afterburner), where $\theta_0 > 1.2$, have forced designers to select theta breaks different from 1.0. These engines may operate at either $\pi_{c\,\text{max}}$ or $T_{t4\,\text{max}}$ at standard sea-level static conditions, but never both.

The *throttle ratio* (TR) is defined as the ratio of the maximum value of T_{t4} to the value of T_{t4} at sea-level static (SLS) conditions. In equation form, the throttle ratio is

$$\text{TR} \equiv \frac{(T_{t4})_{\text{max}}}{(T_{t4})_{\text{SLS}}} \tag{8.19a}$$

The throttle ratio for the simple turbojet engine and compressor of Fig. 8.11 and many of the following examples has a value of 1.0. Both the compressor performance and engine performance curves change shape at a θ_0 value of 1.0. This change in shape of the performance curves occurs at the simultaneous maximum of π_c and T_{t4}, the *theta break*. The fact that both the throttle ratio and dimensionless total temperature θ_0 have a value of 1.0 at the simultaneous maximum is not a coincidence but is a direct result of compressor–turbine power balance given by Eq. (8.17). At the simultaneous maximum of π_c and T_{t4}, the throttle ratio equals θ_0:

$$\text{TR} = \theta_0 \quad \text{at max } \pi_c \text{ and max } T_{t4}$$

or

$$\text{TR} = \theta_{0\,\text{break}} \tag{8.19b}$$

Example 8.1

We now consider compressor operation at different T_{t4} and θ_0, specifically a compressor that has a pressure ratio of 15 and corrected mass flow rate of 100 lbm/s for T_{t2} of 518.7°R (sea-level standard) and T_{t4} of 3200°R. At these conditions, θ_0 is 1, and constants K_1 and K_2 in Eqs. (8.17) and (8.18) are 3.649×10^{-4} and 377.1, respectively. In addition, we assume that an engine control system limits π_c to 15 and T_{t4} to 3200°R, as indicated in Fig. 8.11. By using Eqs. (8.17) and (8.18), the compressor pressure ratio and corrected mass flow rate are calculated for various values of T_{t4} and θ_0. Figure 8.11 show the resulting variation of compressor pressure ratio and corrected mass flow rate, respectively, with flight condition θ_0 and throttle setting T_{t4}. Note that at θ_0 above 1.0, the compressor pressure ratio and corrected mass flow rate are limited by the maximum combustor exit temperature T_{t4} of 3200°R. The compressor pressure ratio limits performance at θ_0 below 1.0. Thus $\theta_{0\,\text{break}} = 1.0$.

This example represents commercial engines of the past. Modern engines are required to have constant sea-level takeoff thrust up to ambient temperatures of 86°F (20°C)—*flat rating* (see Fig. 1.15a). A flat rating up to 86°F (20°C) corresponds to a $\theta_{0\,\text{break}}$ of 1.053. Flat rating limits the thrust force on the aircraft structure and provides an aircraft takeoff performance that varies only with aircraft weight except on very warm days.

8.2.6 Variation in Engine Speed

As will be shown in Chapter 9, the change in total enthalpy across a fan or compressor is proportional to the rotational speed N squared. For a calorically perfect gas, we can write

$$T_{t3} - T_{t2} = K_1 N^2$$

or

$$\tau_c - 1 = \frac{K_1}{T_{\text{ref}}} N_{c2}^2 \tag{i}$$

where N_{c2} is the compressor corrected speed. The compressor tempera-ture ratio is related to the compressor pressure ratio through the efficiency [Eq. (6.15)], or

$$\tau_c - 1 = (\pi_c^{(\gamma_c-1)/\gamma_c} - 1)/\eta_c$$

Combining this equation with Eq. (i), rewriting the resulting equation in terms of pressure ratio and corrected speed, rearranging into variable and constant terms, and equating the constant to reference values give for con-stant compressor efficiency

$$\frac{\pi_c^{(\gamma_c-1)/\gamma_c} - 1}{N_{c2}^2} = \frac{\eta_c K_1}{T_{\text{ref}}} = \frac{\pi_{cR}^{(\gamma_c-1)/\gamma_c} - 1}{N_{c2R}^2} \tag{ii}$$

Solving Eq. (ii) for the corrected speed ratio N_{c2}/N_{c2R}, we have

$$\frac{N_{c2}}{N_{c2R}} = \sqrt{\frac{\pi_c^{(\gamma_c-1)/\gamma_c} - 1}{\pi_{cR}^{(\gamma_c-1)/\gamma_c} - 1}} \tag{8.20a}$$

Equation (8.20a) can be rewritten using Eq. (8.17) to give the variation of corrected engine speed N_{c2} with the temperature ratio T_{t4}/θ_0 (engine throt-tle and flight condition). The result is

$$\frac{N_{c2}}{N_{c2R}} = \sqrt{\frac{T_{t4}/\theta_0}{(T_{t4}/\theta_0)_R}} \tag{8.20b}$$

From Eq. (8.4), the change in rotational speed N is

$$\frac{N}{N_R} = \sqrt{\frac{T_{t4}}{T_{t4R}}} \tag{8.21a}$$

which can be used to estimate the variation in engine speed N with throttle. Equation (8.21a) is plotted in Fig. 8.12 for Example 8.1. Note that the maximum engine rotor speed N occurs at maximum T_{t4}. We can use Eqs. (8.21a), (8.19a), and (8.19b) to obtain the relationship between full throttle engine speed and engine speed at SLS conditions

$$\frac{N_{\text{max}}}{N_{\text{SLS}}} = \sqrt{\frac{T_{t4\,\text{max}}}{T_{t4\text{SLS}}}} = \sqrt{\text{TR}} = \sqrt{\theta_{0\text{break}}} \tag{8.21b}$$

The ratio $N_{\text{max}}/N_{\text{SLS}}$ is referred to as *overspeed* and often given in per-centage. For example, a throttle ratio (TR) of 1.08 is equivalent to a 4% overspeed.

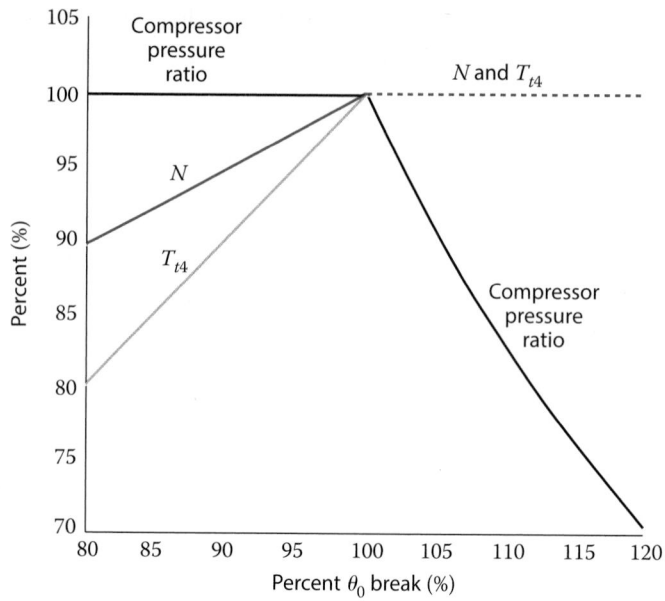

Fig. 8.12   Variation in engine properties with θ_0 at full throttle.

Because the compressor and turbine are connected to the same shaft, they have the same rotational speed N, and we can write the following relationship between their corrected speeds:

$$N_{c2} = \frac{1}{\sqrt{T_{\text{ref}}}} \sqrt{\frac{T_{t4}}{\theta_0}} N_{c4}$$

Comparison of Eq. (8.20b) to the above equation gives the result that the corrected turbine speed is constant, or

$$N_{c4} = \text{const} \tag{8.22}$$

This result may surprise one at first. However, given that the turbine's temperature ratio τ_t, pressure ratio π_t, and efficiency η_t are considered constant in this analysis, the turbine's corrected speed must be constant (see Fig. 8.5b).

8.2.7 Gas Generator Equations

The pumping characteristics of a simple gas generator can be represented by the variation of the gas generator's parameter ratios with corrected compressor speed. The equations for the gas generator's pressure and temperature ratios, corrected air mass flow and fuel flow rates, compressor pressure ratio, and corrected compressor speed can be written in terms of

T_{t4}/T_{t2}, reference values (subscript R), and other variables. The gas generator's pressure and temperature ratios are given simply by

$$\frac{P_{t6}}{P_{t2}} = \pi_c \pi_b \pi_t \tag{8.23}$$

$$\frac{T_{t6}}{T_{t2}} = \frac{T_{t4}}{T_{t2}} \tau_t \tag{8.24}$$

From Eq. (8.10) and referencing, the corrected mass flow rate can be written as

$$\frac{\dot{m}_{c2}}{\dot{m}_{c2R}} = \frac{\pi_c}{\pi_{cR}} \sqrt{\frac{(T_{t4}/T_{t2})_R}{T_{t4}/T_{t2}}} \tag{8.25}$$

where the compressor pressure ratio is given by Eq. (8.17), rewritten in terms of T_{t4}/T_{t2}, or

$$\pi_c = \left[1 + \frac{T_{t4}/T_{t2}}{(T_{t4}/T_{t2})_R} (\pi_{cR}^{(\gamma_c-1)/\gamma_c} - 1) \right]^{\gamma_c/(\gamma_c-1)} \tag{8.26}$$

Equation (8.19b) for the corrected speed can be rewritten in terms of T_{t4}/T_{t2} as

$$\frac{N_{c2}}{N_{c2R}} = \sqrt{\frac{T_{t4}/T_{t2}}{(T_{t4}/T_{t2})_R}} \tag{8.27}$$

An expression for the corrected fuel flow rate results from Eqs. (7.9), (8.2), and (8.7) as follows. Solving Eq. (7.9) for the fuel flow rate gives

$$\dot{m}_f = \dot{m}_0 \frac{h_{t4} - h_{t3}}{\eta_b h_{PR} - h_{t4}}$$

From Eqs. (8.2) and (8.7), this equation becomes

$$\dot{m}_{fc} = \frac{\dot{m}_{c2}}{\theta_2} \frac{h_{t4} - h_{t3}}{\eta_b h_{PR} - h_{t4}} \tag{8.28}$$

By using Eq. (8.13) and referencing, τ_c is given by

$$\tau_c = 1 + (\tau_{cR} - 1) \frac{T_{t4}/T_{t2}}{(T_{t4}/T_{t2})_R} \tag{8.29}$$

Equations (8.23–8.29) constitute a set of equations for the pumping characteristics of a simple gas generator in terms of $T_{t4} = T_{t2}$ and reference values. Only Eq. (8.28) for the corrected fuel flow rate has terms that are not strictly a function of $T_{t4} = T_{t2}$. The first term in the denominator of Eq. (8.28) has a magnitude of about 130, and $T_{t4} = T_{ref}$ has a value of about 6 or smaller. Thus the denominator of Eq. (8.29) does not vary appreciably, and the corrected fuel flow rate is a function of T_{t4}/T_{t2} and reference values. In summary, the pumping characteristics of the gas generator are a function of only the temperature ratio T_{t4}/T_{t2}.

Example 8.2

We want to determine the characteristics of a gas generator with a maximum compressor pressure ratio of 15, a compressor corrected mass flow rate of 100 lbm/s at T_{t2} of 518.7°R (sea-level standard), and a maximum T_{t4} of 3200°R. This is the same compressor we considered in Example 8.1 (see Fig. 8.11). We assume the compressor has an efficiency η_c of 0.8572 ($e_c = 0.9$), and the burner has an efficiency η_b of 0.995 and a pressure ratio π_b of 0.96. In addition, we assume the following gas constants: $\gamma_c = 1.4$, $c_{pc} = 0.24$ Btu/(lbm · °R), $\gamma_t = 1.33$, and $c_{pt} = 0.276$ Btu/(lbm · °R).

By using Eq. (7.10), the reference fuel/air ratio f_R is 0.03435 for $h_{PR} = 18,400$ Btu/lbm, and the corrected fuel flow rate is 12,365 lb/h. From Eq. (7.12), the turbine temperature ratio τ_t is 0.8124 for $\eta_m = 0.995$. Assuming $e_t = 0.9$, Eqs. (7.13) and (7.14) give the turbine pressure ratio π_t as 0.3966 and the turbine efficiency η_t, as 0.910. The reference compressor temperature ratio τ_{cR} is 2.3624.

Calculations were done over a range of T_{t4} with $T_{t2} = 518.7°R$ and using Eqs. (8.23–8.29). Results plotted in Fig. 8.12 show the variation in compressor corrected rotor speed determined by Eq. (8.20b), as well as variation expected in π_c and T_{t4} consistent with Fig. 8.11. The resulting gas generator pumping characteristics are plotted in Fig. 8.13. We can see that the compressor pressure ratio and corrected fuel flow rate decrease more rapidly with decreasing corrected speed than corrected airflow rate. As discussed previously, the gas generator's pumping characteristics are a function of only T_{t4}/T_{t2}, and Fig. 8.13 shows this most important relationship in graphical form.

Because the maximum T_{t4} is 3200°R and the maximum pressure ratio is 15, the operation of the gas generator at different inlet conditions (T_{t2}, P_{t2}) and/or different throttle settings (T_{t4}) can be obtained from Fig. 8.13. For example, consider a 100°F day (T_{t2}) at sea level with maximum power. Here $T_{t2} = 560°R$, $P_{t2} = 14.7$ psia, and $T_{t4} = 3200°R$; thus $T_{t4}/T_{t2} = 5.71$, and Fig. 8.13 gives the following data: $N_c/N_{cR} = 0.96$, $\dot{m}_c/\dot{m}_{cR} = 0.90$, $\pi_c/\pi_{cR} = 0.87$, $\dot{m}_{fc}/\dot{m}_{fcR} = 0.78$, $T_{t6}/T_{t2} = 4.6$, and $P_{t6}/P_{t2} = 4.8$. With these data, the pressures, temperatures, and flow rates can be calculated as follows:

$$\dot{m} = \left(\frac{P_{t2}}{P_{t2R}}\sqrt{\frac{T_{t2R}}{T_{t2}}}\right)\left(\frac{\dot{m}_c}{\dot{m}_{cR}}\right)\dot{m}_{cR} = 1 \times \sqrt{\frac{518.7}{560}}(0.90)(100) = 86.6\,\text{lbm/s}$$

(Continued)

Example 8.2 *(Continued)*

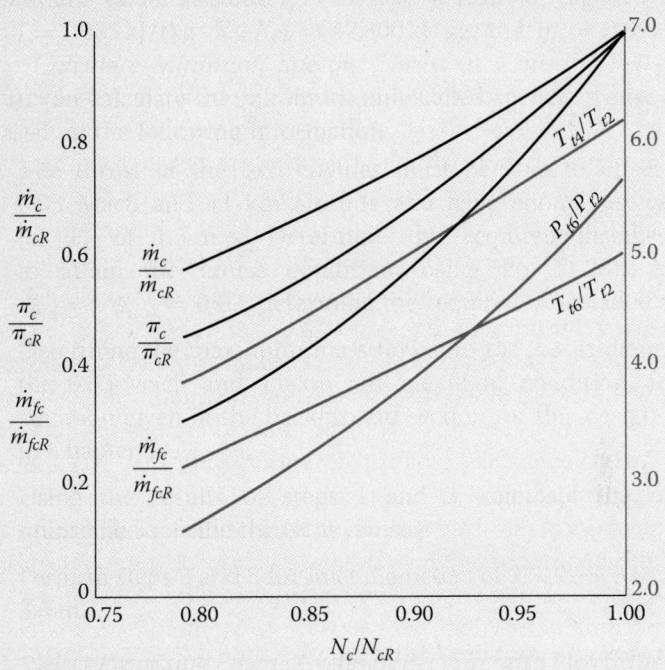

Fig. 8.13   Gas generator pumping characteristics.

$$\dot{m}_f = \frac{P_{t2}}{P_{t2R}} \sqrt{\frac{T_{t2}}{T_{t2R}}} \, \dot{m}_{fcR} = 1 \times \sqrt{\frac{560}{518.7}} (0.78)(12{,}365) = 10{,}020 \, \text{lbm/h}$$

$$T_{t6} = \frac{T_{t6}}{T_{t2}} T_{t2} = 4.6(560) = 2576°\text{R}$$

$$P_{t6} = \frac{P_{t6}}{P_{t2}} P_{t2} = 4.8(14.7) = 70.6 \, \text{psia}$$

$$\pi_c = \frac{\pi_c}{\pi_{cR}} \pi_{cR} = 0.87(15) = 13.0$$

As another example, consider flight at Mach 0.6 and 40 kft ($\theta = 0.7519$, $\delta = 0.1858$) with maximum throttle. Because T_{t2} ($= 418.1°\text{R}$) is less than T_{t2R}, the maximum value for T_{t4} is $2579.4°\text{R}$ ($= 3200 \times 418.1/518.7$), and the compressor has a pressure ratio of 15 and corrected mass flow rate of 100 lbm/s. The air mass flow rate is reduced to 26.3 lbm/s and the fuel mass flow rate is reduced to 2490 lbm/h.

8.2.8 Component Matching

The textbook approach to this subject is best explained by William Heiser in *Aircraft Engine Design*, second edition (Ref. 13). We have made minor modifications to that explanation here so it corresponds to the material in this chapter.

This is the time to clarify what appears to be a shortcoming or internal inconsistency in the off-design calculation procedure of Sections 8.2.1–8.2.3. The heart of the matter is this: the off-design equations are silent with regard to the rotational speeds of the rotating machines, despite the obvious fact that each compressor is mechanically connected via a permanent shaft to a turbine, and thus they must always share the same rotational speed. The rotational speed will, in general, be different from the design point speed and, for this purpose, dimensionless compressor and turbine performance maps (e.g., Figs. 8.3 and 8.5b) also contain data pertaining to rotational speed. It would seem, then, that the off-design equation set lacks some true physical constraints (i.e., $N_c = N_t$) and must therefore produce erroneous results. The following discussion will demonstrate that this appearance is, fortunately, misleading.

The business of ensuring that all of the relationships that join a compressor and turbine are obeyed, including mass flow, power, total pressure, and rotational speed, is known as *component matching*. The off-design calculation procedure of this textbook correctly maintains all known relationships except rotational speed. The simplest and most frequently cited example of component matching found in the open literature is that of developing the "pumping characteristics" for the "gas generator" (i.e., compressor, burner, and turbine) of a nonafterburning, single-spool turbojet (e.g., Refs. 7, 21, and 35). Careful scrutiny of this component matching process reveals that the compressor performance map is used to update the estimate of compressor efficiency and to determine the shaft rotational speed; the turbine performance map and shaft rotational speed are then used only to update the estimate of turbine efficiency. In other words, the main use of enforcing $N_c = N_t$ is to provide accurate values of compressor and turbine efficiency.

If suitable compressor and turbine performance maps were available, they could, of course, be built into the off-design calculation procedure, and the iteration just described would automatically be executed internally. It is extremely difficult to duplicate real compressor and turbine maps in computations. When such performance maps are not available, as is often the case early in a design study, the best approach is to supply input values of η_c and η_t based on experience. This "open-loop" method of using component performance maps can also be used later when such maps become available through component rig tests or advanced computational methods. The principal conclusion is that accurate estimation of η_c and η_t at the off-design conditions has the *same* result as using performance maps and setting $N_c = N_t$. Consequently, the off-design calculation procedure of this textbook is both correct and complete. An important corollary to this conclusion is that the operating line (i.e., π or τ vs $\dot{m}\sqrt{\theta/\delta}$) of every component in the engine is a "free" byproduct of the off-design calculations, even when the engine cycle is arbitrarily complex. We will return to this discussion with a concrete example in Section 9.5.9.

8.2.9 Summary: Engine Performance Analysis of Engine Cores

The material presented thus far in this chapter is so foundational to the understanding of engine performance analysis (EPA), it is worthwhile to summarize the highlights. Consistent with our approach throughout this textbook, we choose assumptions (i.e., constant component efficiencies) to help simplify the math without masking key engine performance behaviors or trends. Up to this point in the chapter, our focus has been on the engine core or gas generator for good reason—all gas turbine engine cycles have a core.

- The main goal of an EPA is to evaluate the "off-design" performance of potential candidate engines to determine if they can successfully accomplish aircraft mission desires and goals. Candidate engines are identified through a parametric cycle analysis (PCA) that narrows down the range of design parameters in terms of design limits at some reference or "design" condition or set of reference conditions (altitude, Mach number).
- For an EPA, the visual one should have is of an engine that has been properly sized and built, and is installed in a test facility that can simulate the required mission flight conditions (altitude, Mach number).
- The primary independent parameters in EPA are the flight conditions (M_0, P_0, T_0) and the main burner exit total temperature T_{t4} set by the throttle position. You do not get to pick a new compressor pressure ratio like in PCA; however, you do get to investigate engine performance at different throttle settings.
- For all but those engine rotational speeds below flight idle (60–70% design speed), the high-pressure or core turbine (HPT) of nearly all engines is choked ($M = 1$) within the initial row of nozzle vanes (see Fig. 8.5a). In many applications, another choked area downstream of the HPT occurs, within the initial row of low-pressure turbine (LPT) nozzle vanes and/or the exhaust nozzle.
- From foundational principles applied to an engine core—namely, conservation of mass—we find that the core turbine pressure ratio and temperature ratio stay constant for conditions of choked flow upstream and downstream of it. This is a remarkable result that one should find intuitively appealing. If flow is choked upstream and downstream of the HPT, changes that occur upstream and downstream cannot be communicated to the turbine (see Section 3.4). The core turbine changes its power output in response to throttle changes, but does so by keeping π_t

and τ_t constant as long as flow remains choked upstream and downstream. If the flow does unchoke upstream or downstream of the core turbine and/or the exhaust nozzle throat area changes, then changes in π_t and τ_t do occur (see Figs. 8.7a–c), accounted for by the governing physics (conservation of mass) applied consistent with our assumptions.

• When we combine another foundational principle—power balance between the turbine and compressor—with conservation of mass, we find that core compressor pressure ratio π_c and corrected mass flow rate through the compressor $\dot{m}_{c2}$ are both functions of T_{t4} (throttle setting) and flight condition (M_0, T_0), another result you should find intuitively appealing. For π_t and τ_t constant, our analysis reveals a single engine operating line, quite representative of aircraft gas turbine engines. As shown in Fig. 8.8, changes to that operating line are accommodated in the analysis.

• Real engines have control limits, and our analysis includes these for π_c and T_{t4}. (Note that the PERF program included with this textbook also allows for control limits on other engine parameters like compressor exit temperature T_{t3}.) The incorporation of control limits in our analysis is quite consistent with industry; namely, through the use of throttle ratio (TR), defined by Eq. (8.19a), or equivalently, theta break $\theta_{0\,\text{break}}$. The total energy entering an engine is captured in a single performance parameter T_{t0} (nondimensionalized by T_{ref} and referred to as theta-zero θ_0) whose value is set by flight conditions [see Eq. (8.16)]. Taking advantage of the fact that $T_{t2} = T_{t0}$ for an engine inlet with negligible heat loss (adiabatic) leads to π_c and $\dot{m}_{c2}$ being specified by T_{t4} and θ_0. From power balance considerations and constant τ_t, both π_c and T_{t4} are at their maximum values simultaneously only when $\theta_0 = \theta_{0\,\text{break}} = \text{TR}$ (see Fig. 8.11). It is important to note that changing the value of TR implies a new core turbine, one that can accommodate a higher T_{t4} than that at SLS, for example, through the use of improved materials. Also, because engine rotor speed N is a function of T_{t4} [see Eq. (8.21a)], engine manufacturers can specify control limits through the use of "overspeed" $(N_{\text{max}}/N_{\text{SLS}})$ as an alternative to TR.

For the remainder of this chapter, we develop the governing cycle equations based on the fundamental principles just summarized, consistent with our assumptions, and analyze off-design performance for the most common types of aircraft gas turbine engines. At the start of each analysis, we identify the principal dependent and independent performance variables as well as the known or constant parameters. The student would be wise to refer back to this section as needed to aid in their understanding.

8.3 Turbojet Engine

In this section, the performance equations of the single-spool turbojet engine, shown in Fig. 8.14, are developed and the results are studied. We

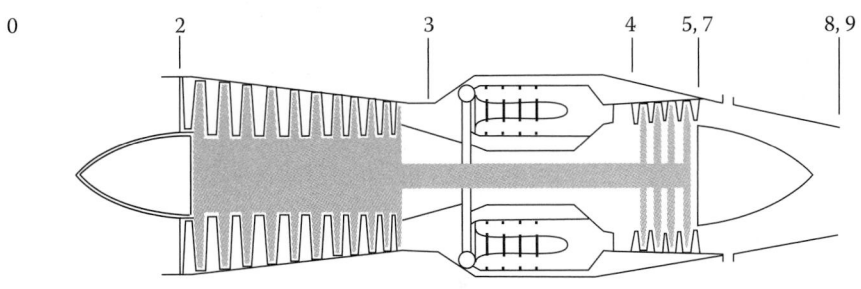

Fig. 8.14 Single-spool turbojet engine.

assume choked flow at stations 4 and 8. In addition, the throttle (T_{t4}), flight conditions (M_0, T_0, and P_0), and ambient pressure/exhaust pressure ratio (P_0/P_9) can be independently varied for this engine. The performance equations for this turbojet can be obtained easily by adding inlet and exhaust nozzle losses to the single-spool gas generator studied in the previous section.

This engine has five independent variables (T_{t4}, M_0, T_0, P_0, and P_0/P_9). The performance analysis develops analytical expressions for component performance in terms of these independent variables. We have six dependent variables for the single-spool turbojet engine: engine mass flow rate, compressor pressure ratio, compressor temperature ratio, burner fuel/air ratio, exit temperature ratio T_9/T_0, and exit Mach number. A summary of the independent variables, dependent variables, and constants or knowns for this engine is given in Table 8.3.

The following seven equations were developed in Section 7.2.1. The specific thrust for this engine is given by

$$\frac{F}{\dot{m}_0} = \frac{a_0}{g_c}\left[(1+f)\frac{V_9}{a_0} - M_0 + (1+f)\frac{R_t}{R_c}\frac{T_9/T_0}{V_9/a_0}\frac{1-P_0/P_9}{\gamma_c}\right] \qquad \text{(i)}$$

Table 8.3 Performance Analysis Variables for Single-Spool Turbojet Engine

| Component | Variables | | |
	Independent	Constant or Known	Dependent
Engine	M_0, T_0, P_0		$\dot{m}_0$
Diffuser		$\pi_d = f(M_0)$	
Compressor		η_c	π_c, τ_c
Burner	T_{t4}	π_b, η_b	f
Turbine		π_t, τ_t	
Nozzle	P_9/P_0	π_n	M_9, T_9/T_0
Total number	5		6

where

$$\frac{T_9}{T_0} = \frac{T_{t4}\tau_t}{(P_{t9}/P_9)^{(\gamma_t-1)/\gamma_t}} \tag{ii}$$

$$\frac{P_{t9}}{P_9} = \frac{P_0}{P_9}\pi_r\pi_d\pi_c\pi_b\pi_t\pi_n \tag{iii}$$

$$M_9 = \sqrt{\frac{2}{\gamma_t-1}\left[\left(\frac{P_{t9}}{P_9}\right)^{(\gamma_t-1)/\gamma_t} - 1\right]} \tag{iv}$$

and

$$\frac{V_9}{a_0} = M_9\sqrt{\frac{\gamma_t R_t T_9}{\gamma_c R_c T_0}} \tag{v}$$

The thrust specific fuel consumption for this engine is given by

$$S = \frac{f}{F/\dot{m}_0} \tag{vi}$$

where

$$f = \frac{h_{t4} - h_{t3}}{h_{PR}\eta_b - h_{t4}} \tag{vii}$$

Equations (i–vii) can be solved for given T_{t4}, M_0, T_0, P_0, P_0/P_9 gas properties with expressions for τ_λ, π_r, τ_r, π_d, π_c, τ_c, and engine mass flow rate in terms of the five independent variables and other dependent variables. In the previous section, we developed Eq. (8.29), repeated here, for the compressor's temperature ratios in terms of T_{t4}/T_{t2} and reference values:

$$\tau_c = 1 + (\tau_{cR} - 1)\frac{T_{t4}/T_{t2}}{(T_{t4}/T_{t2})_R}$$

The compressor pressure ratio is related to its temperature ratio by its efficiency.

An equation for the engine mass flow rate follows from the mass flow parameter (MFP) written for station 4 with choked flow and the definitions of component π values. We write

$$\dot{m}_0 = \frac{P_{t4}}{\sqrt{T_{t4}}}\frac{A_4}{1+f}\text{MFP}(1) = \frac{P_0\pi_r\pi_d\pi_c}{\sqrt{T_{t4}}}\left[\frac{\pi_b A_4}{1+f}\text{MFP}(1)\right]$$

Because the terms within the brackets are considered constant, we move the variable terms to the left side of the equation, and, using referencing, equate the constant to reference values (see Section 8.1.1):

$$\frac{\dot{m}_0\sqrt{T_{t4}}}{P_0\pi_r\pi_d\pi_c} = \frac{\pi_b A_4}{1+f}\text{MFP}(1) = \left(\frac{\dot{m}_0\sqrt{T_{t4}}}{P_0\pi_r\pi_d\pi_c}\right)_R$$

Solving for the engine mass flow rate, we get

$$\dot{m}_0 = \dot{m}_{0R} \frac{P_0 \pi_r \pi_d \pi_c}{(P_0 \pi_r \pi_d \pi_c)_R} \sqrt{\frac{T_{t4R}}{T_{t4}}} \tag{8.30}$$

Relationships for τ_λ, π_r, τ_r, and π_d follow from their basic definitions presented in Section 5.2.

The throat area of the exhaust nozzle is assumed to be constant. With P_0/P_9 an independent variable, the exit area of the exhaust nozzle A_9 must correspond to the nozzle pressure ratio P_{t9}/P_9. An expression for the exhaust nozzle exit area follows from the mass flow parameter and other compressible flow properties. The subscript t is used in the following equations for the gas properties (γ, R, and Γ) at stations 8 and 9. Using Eq. (8.11) for choked flow at station 8 gives

$$\dot{m}_8 = \frac{P_{t9} A_8}{\sqrt{T_{t8}}} \frac{\Gamma_t}{\sqrt{R_t/g_c}} \tag{i}$$

From the equation for the mass flow parameter [Eq. (3.11)], the mass flow rate at station 9 is

$$\dot{m}_9 = \frac{P_{t9} A_9}{\sqrt{T_{t9}}} \frac{\sqrt{\gamma_t}}{\sqrt{R_t/g_c}} M_9 \left(1 + \frac{\gamma_t - 1}{2} M_9^2 \right)^{-(\gamma_t+1)/2(\gamma_t-1)]} \tag{ii}$$

Using the nozzle relationships $T_{t8} = T_{t9}$ and $\pi_n = P_{t9}/P_{t7}$ and equating the mass flow rate at station 8 [Eq. (i)] to that at station 9 [Eq. (ii)] give

$$\frac{A_9}{A_8} = \frac{\Gamma_t}{\sqrt{\gamma_t}} \frac{1}{\pi_n} \frac{1}{M_9} \left(1 + \frac{\gamma_t - 1}{2} M_9^2 \right)^{(\gamma_t+1)/[2(\gamma_t-1)]}$$

Replacing the Mach number at station 9 by using

$$M_9 = \sqrt{\frac{2}{\gamma_t - 1} [(P_{t9}/P_9)^{(\gamma_t-1)/\gamma_t} - 1]}$$

gives

$$\frac{A_9}{A_8} = \Gamma_t \sqrt{\frac{\gamma_t - 1}{2\gamma_t}} \frac{1}{\pi_n} \frac{(P_{t9}/P_9)^{(\gamma_t+1)/(2\gamma_t)}}{\sqrt{(P_{t9}/P_9)^{(\gamma_t-1)/\gamma_t} - 1}} \tag{8.31}$$

Because the throat area A_8 is constant, Eq. (8.31) can be used to obtain the ratio of the exit area A_9 to a reference exit area A_{9R} that can be written as

$$\frac{A_9}{A_{9R}} = \left[\frac{P_{t9}/P_9}{(P_{t9}/P_9)_R} \right]^{(\gamma_t+1)/(2\gamma_t)} \sqrt{\frac{(P_{t9}/P_9)_R^{(\gamma_t-1)/\gamma_t} - 1}{(P_{t9}/P_9)^{(\gamma_t-1)/\gamma_t} - 1}} \tag{8.32}$$

See Section 8.3.SM, Summary of Performance Equations: Single-Spool Turbo-ject Without Afterburner, in the online supporting material for related information.

Example 8.3

We consider the performance of the turbojet engine of Example 7.1 sized for a mass flow rate of 50 kg/s at the reference condition and altitude of 12 km. We are to determine this engine's performance at an altitude of 9 km, Mach number of 1.5, reduced throttle setting ($T_{t4} = 1500°$R), and exit to ambient pressure ratio (P_0/P_9) of 1.0.

Reference:

$$T_0 = 216.7\,\text{K}, \quad \gamma_c = 1.4, \quad c_{pc} = 1.004\,\text{kJ}/(\text{kg}\cdot\text{K}), \quad \gamma_t = 1.3,$$

$$c_{pt} = 1.239\,\text{kJ}/(\text{kg}\cdot\text{K}), \quad T_{t4} = 1800\,\text{K}, \quad M_0 = 2, \quad \pi_c = 10,$$

$$\tau_c = 2.0771, \quad \eta_c = 0.8641, \quad \tau_t = 0.8155, \quad \pi_t = 0.3746,$$

$$\pi_{d\,\text{max}} = 0.95, \quad \pi_d = 0.8788, \quad \pi_b = 0.94, \quad \pi_n = 0.96,$$

$$\eta_b = 0.98, \quad \eta_m = 0.99, \quad P_0/P_9 = 0.5, \quad h_{PR} = 42{,}800\,\text{kJ}/\text{kg},$$

$$f = 0.03183, \quad P_{t9}/P_9 = 11.57, \quad F/\dot{m}_0 = 800.5\,\text{N}/(\text{kg/s}),$$

$$S = 39.64\,(\text{g/s})/\text{kN}, \quad P_0 = 19.40\,\text{kPa}\,(12\,\text{km}), \quad \dot{m}_0 = 50\,\text{kg/s},$$

$$F = \dot{m}_0 \times (F/\dot{m}_0) = 50 \times 800.5 = 40{,}025\,\text{N}$$

Off-design condition:

$$T_0 = 229.8\,\text{K}, \quad P_0 = 30.8\,\text{kPa}\,(9\,\text{km}), \quad M_0 = 1.5,$$

$$P_0/P_9 = 1, \quad T_{t4} = 1500\,\text{K}$$

Equations:

$$R_c = \frac{\gamma_c - 1}{\gamma_c}c_{pc} = \frac{0.4}{1.4}(1.004) = 0.2869\,\text{kJ}/(\text{kg}\cdot\text{K})$$

$$R_t = \frac{\gamma_t - 1}{\gamma_t}c_{pt} = \frac{0.3}{1.3}(1.239) = 0.2859\,\text{kJ}/(\text{kg}\cdot\text{K})$$

$$a_0 = \sqrt{\gamma_c R_c g_c T_0} = \sqrt{1.4 \times 286.9 \times 1 \times 229.8} = 303.8\,\text{m/s}$$

$$V_0 = a_0 M_0 = 303.8 \times 1.5 = 455.7\,\text{m/s}$$

$$\tau_r = 1 + \frac{\gamma_c - 1}{2}M_0^2 = 1 + 0.2 \times 1.5^2 = 1.45$$

$$\pi_r = \tau_r^{\gamma_c/(\gamma_c-1)} = 1.45^{3.5} = 3.671$$

(Continued)

Example 8.3 *(Continued)*

$$\eta_r = 1 - 0.075(M_0 - 1)^{1.35} = 1 - 0.075(0.5)^{1.35} = 0.9706$$

$$\pi_d = \pi_{d\,max}\eta_r = 0.95 \times 0.9706 = 0.9220$$

$$\tau_\lambda = \frac{c_{pt}T_{t4}}{c_{pc}T_0} = \frac{1.2329 \times 1670}{1.004 \times 229.8} = 8.9682$$

$$T_{t2} = T_0\tau_r = 229.8 \times 1.45 = 333.2\,\text{K}$$

$$\tau_{rR} = 1 + \frac{\gamma_c - 1}{2}M_{0R}^2 = 1 + 0.2 \times 2^2 = 1.80$$

$$T_{t2R} = T_{0R}\tau_{rR} = 216.7 \times 1.8 = 390.1\,\text{K}$$

$$\pi_{rR} = \tau_{rR}^{\gamma_c/(\gamma_c-1)} = 1.8^{3.5} = 7.824$$

$$\tau_c = 1 + (\tau_{cR} - 1)\frac{T_{t4}/T_{t2}}{(T_{t4}/T_{t2})_R}$$

$$= 1 + (2.0771 - 1)\frac{1500/333.2}{1800/390.1} = 2.0509$$

$$\pi_c = [1 + \eta_c(\tau_c - 1)]^{\gamma_c/(\gamma_c-1)} = [1 + 0.8641(2.0509 - 1)]^{3.5} = 9.60$$

$$T_{t3} = T_0\tau_r\tau_c = 683.38\,\text{K}, \quad h_{t3} = 695.25\,\text{kJ/kg at}\,f = 0$$

Initial value of h_{t4} obtained for $T_{t4} = 1500\,\text{K}$ and $f = 0$, $h_{t4} = 1635.65\,\text{kJ/kg}$.

$$f = \frac{h_{t4} - h_{t3}}{h_{PR}\eta_b - h_{t4}} = \frac{1635.65 - 695.25}{42,800 \times 0.98 - 1635.65} = 0.02333$$

New value of h_{t4} obtained for $T_{t4} = 1500\,\text{K}$ and $f = 0.02333$, $h_{t4} = 1688.37\,\text{kJ/kg}$.

After several iterations, $f = 0.02493$.

$$\frac{P_{t9}}{P_9} = \frac{P_0}{P_9}\pi_r\pi_d\pi_c\pi_b\pi_t\pi_n$$

$$= 1.0 \times 3.671 \times 0.9220 \times 9.60 \times 0.94 \times 0.3746 \times 0.96 = 10.98$$

$$M_9 = \sqrt{\frac{2}{\gamma_t - 1}[(P_{t9}/P_9)^{(\gamma_t-1)/\gamma_t} - 1]} = \sqrt{\frac{2}{0.3}(10.98^{0.3/1.3} - 1)} = 2.219$$

$$\frac{T_9}{T_0} = \frac{T_{t4}\tau_t/T_0}{(P_{t9}/P_9)^{(\gamma_t-1)/\gamma_t}} = \frac{1500 \times 0.8155/229.8}{10.98^{0.3/1.3}} = 3.062$$

$$\frac{V_9}{a_0} = M_9\sqrt{\frac{\gamma_t R_t T_9}{\gamma_c R_c T_0}} = 2.219\sqrt{\frac{1.3 \times 285.9}{1.4 \times 286.9}(3.062)} = 3.735$$

(Continued)

Example 8.3 (Continued)

$$\frac{F}{\dot{m}_0} = \frac{a_0}{g_c}\left[(1+f)\frac{V_9}{a_0} - M_0 + (1+f)\frac{R_t}{R_c}\frac{T_9/T_0}{V_9/a_0}\frac{1-P_0/P_9}{\gamma_c}\right]$$

$$= 303.8(1.02493 \times 3.7353 - 1.5)$$

$$= 707.3\,\text{N}/(/(\text{kg/s}))$$

$$S = \frac{f}{F/\dot{m}_0} = \frac{0.02493 \times 10^6}{707.3} = 35.25\,(\text{g/s})/\text{kN}$$

$$\dot{m}_0 = \dot{m}_{0R}\frac{P_0\,\pi_r\,\pi_d\,\pi_c}{(P_0\,\pi_r\,\pi_d\,\pi_c)_R}\sqrt{\frac{T_{t4R}}{T_{t4}}}$$

$$\dot{m}_0 = 50\frac{30.8 \times 3.671 \times 0.9220 \times 9.60}{19.4 \times 7.824 \times 0.8788 \times 10}\sqrt{\frac{1800}{1500}} = 41.13\,\text{kg/s}$$

$$F = \dot{m}_0\frac{F}{\dot{m}_0} = 41.13 \times 707.3 = 29{,}090\,\text{N}$$

$$\eta_{Th} = \frac{a_0^2[(1+f)(V_9/a_0)^2 - M_0^2]}{2g_c f h_{PR}}$$

$$= \frac{303.8^2[(1.02493)(3.735)^2 - 1.5^2]}{2 \times 1 \times 0.02493 \times 42{,}800 \times 1000} = 52.10\%$$

$$\eta_P = \frac{2g_c V_0(F/\dot{m}_0)}{a_0^2[(1+f)(V_9/a_0)^2 - M_0^2]}$$

$$= \frac{2 \times 1 \times 455.7 \times 707.3}{303.8^2[(1.02493)(3.735^2) - 1.5^2]} = 57.97\%$$

$$\eta_O = \eta_P\eta_{Th} = 0.5210 \times 0.5797 = 30.20\%$$

$$\frac{N}{N_R} = \sqrt{\frac{T_0\tau_r}{T_{0R}\tau_{rR}}\frac{\pi_c^{(\gamma_c-1)/\gamma_c} - 1}{\pi_{cR}^{(\gamma_c-1)/\gamma_c} - 1}}$$

$$= \sqrt{\frac{229.8 \times 1.45}{216.7 \times 1.8}\frac{9.6^{0.4/1.4} - 1}{10^{0.4/1.4} - 1}} = 0.9130$$

$$\frac{\dot{m}_{c2}}{\dot{m}_{c2R}} = \frac{\pi_c}{\pi_{cR}}\sqrt{\frac{(T_{t4}/T_{t2})_R}{T_{t4}/T_{t2}}} = \frac{9.6}{10}\sqrt{\frac{1800/390.1}{1500/333.2}} = 0.9719$$

$$\frac{A_9}{A_{9R}} = \left[\frac{P_{t9}/P_9}{(P_{t9}/P_9)_R}\right]^{(\gamma_t+1)/(2\gamma_t)}\sqrt{\frac{(P_{t9}/P_9)_R^{(\gamma_t-1)/\gamma_t} - 1}{(P_{t9}/P_9)^{(\gamma_t-1)/\gamma_t} - 1}}$$

$$\frac{A_9}{A_{9R}} = \left(\frac{10.98}{11.62}\right)^{2.3/1.3}\sqrt{\frac{11.62^{0.3/1.3} - 1}{10.98^{0.3/1.3} - 1}} = 0.9185$$

(Continued)

Example 8.3 *(Continued)*

This example is simply meant to show the simplicity of the calculations. With the exception of the iterative solution for the fuel-air ratio f (which we had in PCA as well), all algebraic equations can be solved directly. Note the reduced value of thrust and air mass flow rate at this reduced throttle setting and Mach number from the reference condition, even at a lower altitude where the air density is higher.

Example 8.4

Consider a turbojet engine composed of the gas generator of Example 8.2, an inlet with $\pi_{d\,\mathrm{max}} = 0.99$, and an exhaust nozzle with $\pi_n = 0.99$ and $P_0/P_9 = 1$. The reference engine has the following values.

Reference:

$$T_0 = 518.7°\mathrm{R}, \quad \gamma_c = 1.4, \quad c_{pc} = 0.24\,\mathrm{Btu/(lbm \cdot °R)}, \quad \gamma_t = 1.33,$$

$$c_{pt} = 0.276\,\mathrm{Btu(lbm \cdot °R)}, \quad T_{t4} = 3200°\mathrm{R}, \quad M_0 = 0,$$

$$\pi_c = 15, \quad \eta_c = 0.8572, \quad \tau_t = 0.8124, \quad \pi_t = 0.3943,$$

$$\pi_{d\,\mathrm{max}} = 0.99, \quad \pi_b = 0.96, \quad \pi_n = 0.99, \quad \eta_b = 0.995,$$

$$\eta_m = 0.995, \quad P_0/P_9 = 1, \quad P_0 = 14.696\,\mathrm{psia\ (sea\ level)},$$

$$P_{t9}/P_9 = 5.5981, \quad \dot{m}_0 = 100\,\mathrm{lbm/s}, \quad F/\dot{m}_0 = 113.27\,\mathrm{lbf/(lbm/s)},$$

$$F = \dot{m}_0 \times (F/\dot{m}_0) = 100 \times 113.27 = 11{,}327\,\mathrm{lbf}$$

This engine has a control system that limits the compressor pressure ratio π_c to 15 and the combustor exit total temperature T_{t4} to 3200°R. Calculation of engine performance with full throttle at altitudes of sea level, 20 kft, and 40 kft over a range of flight Mach numbers gives the results shown in Figs. 8.15–8.17. The general trend of increasing thrust with M_0 (at a given altitude) is very evident, a characteristic trend for turbojets. As flight speed V_0 increases, more "ram" air increases the mass flow rate into the engine inlet. This increased mass flow rate tends to dominate the reduced gas acceleration $(V_9 - V_0)$ also resulting from the increased flight speed V_0.

Most striking are the breaks in the plots of thrust, engine mass flow rate, compressor pressure ratio, and station 2 corrected mass flow rate at a Mach/ altitude combination of about 0.9/20 kft and 1.3/40 kft. *These breaks are characteristic of engine control limits and would not be apparent if our analysis had not included control limits.*

To the left of these breaks, the combustor exit temperature T_{t4} is below its maximum of 3200°R, and the compressor pressure ratio π_c is at its maximum of 15. To the right of these breaks, the combustor exit temperature T_{t4} is at its

(Continued)

Example 8.4 *(Continued)*

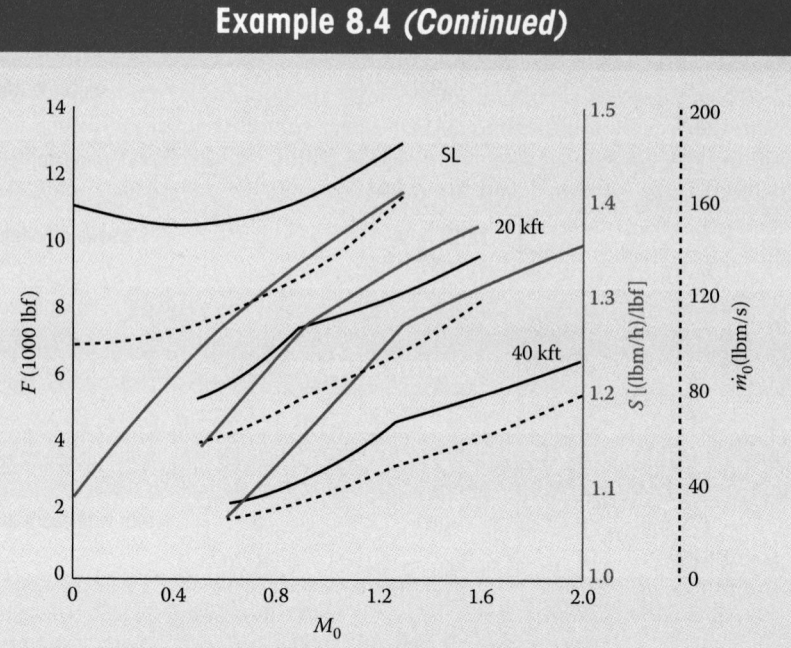

Fig. 8.15   Overall performance of a turbojet vs M_0.

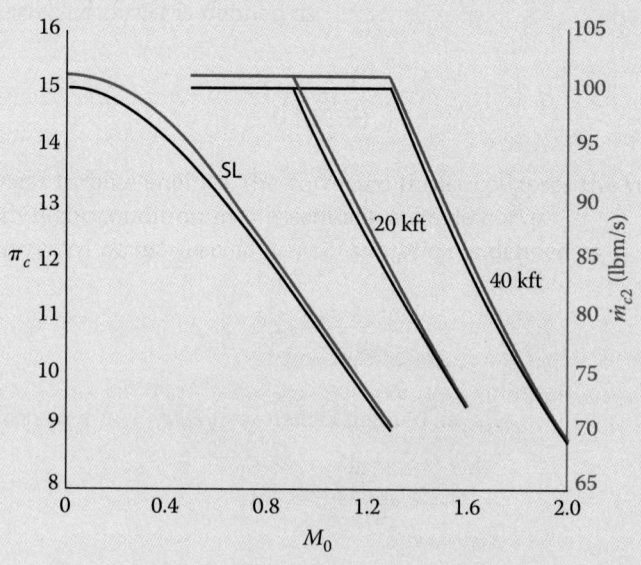

Fig. 8.16   Compressor pressure ratio and corrected mass flow rate of a turbojet vs M_0.

(Continued)

Example 8.4 *(Continued)*

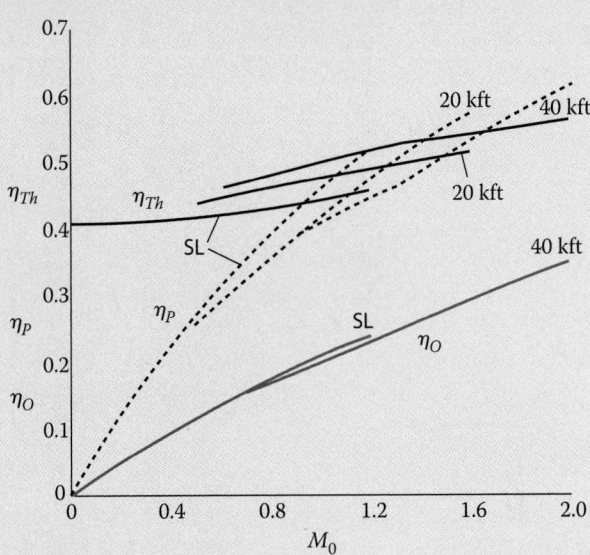

Fig. 8.17   Turbojet efficiencies vs M_0.

maximum of 3200°R, and the compressor pressure ratio π_c is below its maximum of 15. At the break, both the compressor pressure ratio and combustor exit temperature are at their maximum values. This break corresponds to the engine's theta break of 1.0. The interested student can easily verify that $\theta_0 = 1.0$ at SLS and Mach/altitude combinations of about 0.9/20 kft and 1.3/40 kft.

The designer of a gas generator's turbomachinery needs to know the maximum power requirements of the compressor and turbine. Because the turbine drives the compressor, the maximum requirements of both occur at the same conditions. Consider the following power balance between the compressor and turbine:

$$\dot{W}_c = \eta_m \dot{W}_t$$

Rewriting turbine power in terms of its mass flow rate, total temperatures, and the like gives

$$\dot{W}_c = \eta_m \dot{m}_0 (1+f) c_{pt} (T_{t4} - T_{t5})$$

or

$$\dot{W}_c = \dot{m}_0 T_{t4} [\eta_m (1+f) c_{pt} (1 - \tau_t)]$$

Because the terms within the square braces of the preceding equation are considered constant, the maximum compressor power will be at the flight

(Continued)

Example 8.4 *(Continued)*

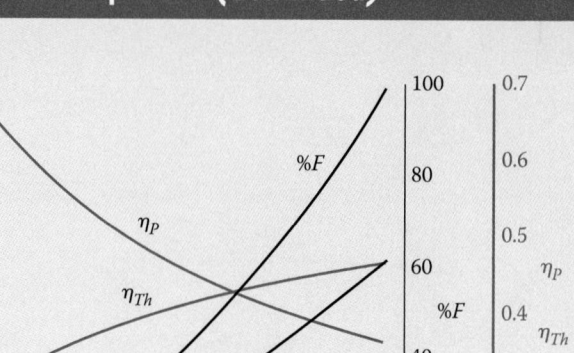

Fig. 8.18 Partial throttle turbojet performance at 20 kft and $M_0 = 0.8$.

condition having maximum engine mass flow rate at maximum T_{t4}. The maximum compressor or turbine power corresponds to the maximum engine mass flow rate at sea level and Mach 1.3.

At an altitude of 20 kft and a Mach number of 0.8, engine performance calculations at reduced throttle T_{t4} were performed, and some of these results are given in Fig. 8.18. The typical variation in thrust specific fuel consumption S with thrust F (or T_{t4} in this case) is shown in this figure. As the throttle is reduced, the thrust specific fuel consumption first reduces before increasing. This plot of thrust specific fuel consumption S vs thrust F is commonly called the *throttle hook* because of its shape.

We stated at the beginning of this chapter that the principal efficiencies that affect engine performance are the thermal efficiency and the propulsive efficiency. Figure 8.18 shows the very large changes in both propulsive and thermal efficiency with engine thrust, and can be used to explain the throttle hook shape. Note that as thrust is reduced from its maximum, the increase in propulsive efficiency more than offsets the decrease in thermal efficiency such that the overall efficiency increases and the thrust specific fuel consumption decreases until about 20% of maximum thrust. Below 20% thrust, the decrease in thermal efficiency dominates the increase in propulsive efficiency and the overall efficiency decreases, and the thrust specific fuel consumption increases with reduced thrust.

8.3.1 Corrected Engine Performance and the Value of Throttle Ratio

The changes in maximum thrust of a simple turbojet engine can be presented in a corrected format that essentially collapses the thrust data. Consider the thrust equation for the turbojet engine as given by

$$F = \frac{\dot{m}_0}{g_c}[(1+f)V_9 - V_0]$$

where

$$V_9 = \sqrt{2g_c c_{pt} T_{t4} \tau_t [1 - (\pi_r \pi_d \pi_c \pi_b \pi_t \pi_n)^{-(\gamma_t - 1)/\gamma_t}]}$$

and

$$V_0 = M_0 a_0 = M_0 \sqrt{\gamma_c R_c T_0}$$

Note that the engine mass flow rate is related to the compressor corrected mass flow rate by

$$\dot{m}_0 = \dot{m}_{c0} = \frac{\delta_0}{\sqrt{\theta_0}} = \dot{m}_{c2} \frac{\delta_2}{\sqrt{\theta_2}} = \dot{m}_{c2} \frac{\pi_d \delta_0}{\sqrt{\theta_0}}$$

The engine thrust can now be written as

$$F = \frac{\dot{m}_{c2} \, \pi_d \delta_0}{g_c \sqrt{\theta_0}}[(1+f)V_9 - V_0]$$

Dividing the thrust by the dimensionless total pressure at station 0 gives

$$\frac{F}{\delta_0} = \frac{\dot{m}_{c2} \pi_d}{g_c}\left[(1+f)\frac{V_9}{\sqrt{\theta_0}} - \frac{V_0}{\sqrt{\theta_0}}\right] \tag{8.33}$$

where

$$\frac{V_9}{\sqrt{\theta_0}} = \sqrt{\frac{T_{t4}}{T_{t2}}}\sqrt{2g_c c_{pt} T_{\text{ref}} \tau_t [1 - \pi_r \pi_d \pi_c \pi_b \pi_t \pi_n)^{-(\gamma_t - 1)/\gamma_t}]} \tag{8.34a}$$

and

$$\frac{V_0}{\sqrt{\theta_0}} = \frac{M_0}{\sqrt{\tau_r}} a_{\text{SL}} \tag{8.34b}$$

The maximum thrust for the turbojet engine of Example 8.4 can be determined by using the preceding equations. Figures 8.15 and 8.16 show the variation of the maximum thrust F, compressor pressure ratio, and corrected mass flow rate from this turbojet engine at full throttle vs flight Mach number M_0. The corrected thrust F/δ_0 of this engine is plotted vs flight condition θ_0 in Fig. 8.19. The variation of T_{t4}/T_{t2}, compressor pressure ratio,

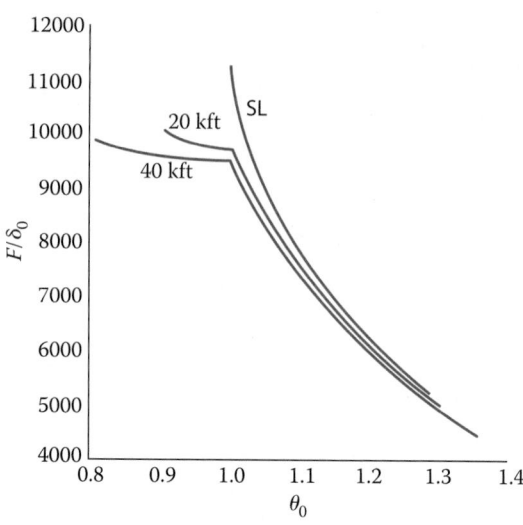

Fig. 8.19 Maximum corrected thrust (F/δ_0) of a turbojet vs θ_0.

corrected mass flow rate, and corrected fuel flow rate are plotted vs θ_0 in Fig. 8.20. Note that in Fig. 8.20 a single line represents the variation of each parameter with θ_0.

The representation of the engine thrust in Fig. 8.19, as corrected thrust vs θ_0, essentially collapses the thrust data into one line for θ_0 greater than 1.0. The discussion that follows helps one see why the plot in Fig. 8.19 behaves as shown. When θ_0 is less than 1.0, we observe the following:

1. The compressor pressure ratio is constant at its maximum value of 15 (see Fig. 8.20).
2. The compressor corrected mass flow rate is constant at its maximum value 100 lbm/s (see Fig. 8.20).
3. The value of T_{t4}/T_{t2} is constant at its maximum value of 6.17 (see Fig. 8.20).
4. The corrected exit velocity given by Eq. (8.34a) is essentially constant.
5. The corrected flight velocity [Eq. (8.34b)] increases in a nearly linear manner with M_0.
6. The corrected thrust [Eq. (8.33)] decreases slightly with increasing θ_0.

When θ_0 is greater than 1.0, we observe the following:

1. The compressor pressure ratio decreases with increasing θ_0.
2. The compressor corrected mass flow rate decreases with increasing θ_0.
3. The value of T_{t4} is constant at its maximum value of 3200°R, explaining the decrease in T_{t4}/T_{t2} shown in Fig. 8.20.
4. The corrected exit velocity given by Eq. (8.34a) decreases with increasing θ_0.

5. The corrected flight velocity [Eq. (8.34b)] increases in a nearly linear manner with M_0.
6. The corrected thrust [Eq. (8.34a)] decreases substantially with increasing θ_0.

As shown in Fig. 8.19, the trend in maximum corrected thrust F/δ_0 of this turbojet dramatically changes at the $\theta_{0\,\text{break}}$ value of 1.0. Both the compressor pressure ratio π_c and combustor exit temperature T_{t4} are at their maximum values when θ_0 is 1.0. The engine control system varies the fuel flow to the combustor to keep π_c and T_{t4} under control. The control system maintains π_c at its maximum for θ_0 values less than 1.0, and T_{t4} at its maximum for θ_0 values greater than 1.0. These same kinds of trends are observed for many other aircraft gas turbine engines.

The thrust specific fuel consumption S of this turbojet at maximum thrust is plotted vs Mach number in Fig. 8.14. If the values of S are divided by the square root of the corrected ambient temperature, then the curves for higher altitudes in Fig. 8.15 are shifted up and we get Fig. 8.21. Note that these curves could be estimated by a straight line. Equations (1.36a–1.36f)

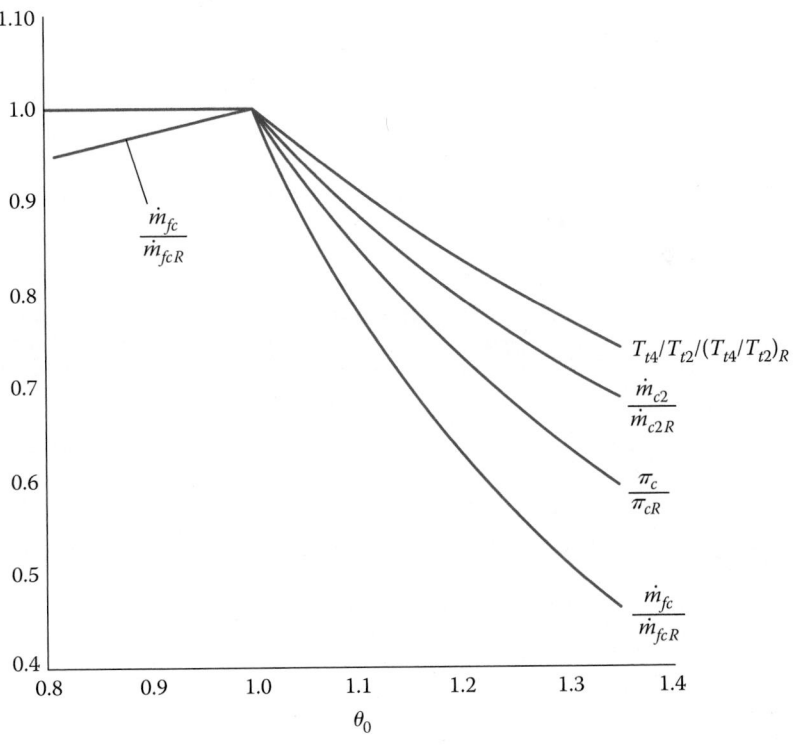

Fig. 8.20 Maximum throttle characteristics of a turbojet vs θ_0.

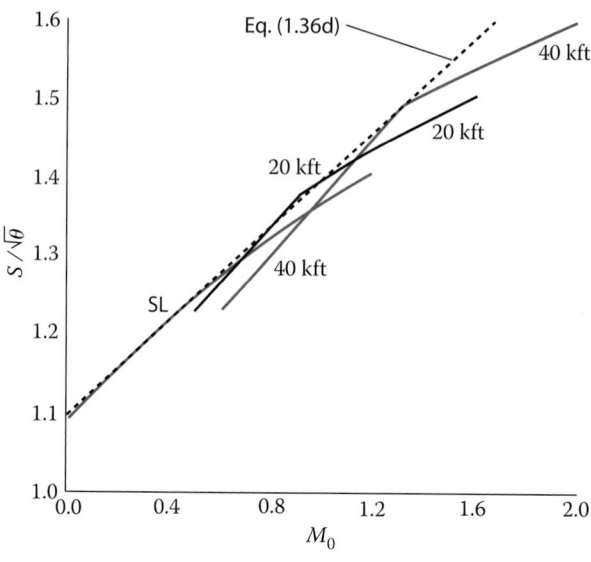

Fig. 8.21   Corrected thrust specific fuel consumption of a turbojet vs M_0.

are based on this nearly linear relationship with flight Mach number M_0. Again, the usefulness of corrected performance parameters is demonstrated.

High-performance fighters want gas turbine engines whose thrust does not drop off as fast with increasing θ_0 as that of Fig. 8.19. The value of θ_0, where the corrected maximum thrust F/δ_0 curves change slope, can be increased by increasing the maximum T_{t4} of the preceding example turbojet engine. We will see the performance of an engine with increased throttle ratio (TR) in the following example.

Example 8.5

Again, we consider the example turbojet engine with a compressor that has a compressor pressure ratio of 15 and corrected mass flow rate of 100 lbm/s for T_{t2} of 518.7°R and T_{t4} of 3200°R. The maximum π_c is maintained at 15, and the maximum T_{t4} is increased from 3200 to 3360°R (TR = 1.05, $\theta_{0\,break} = 1.05$). The variation in thrust, thrust specific fuel consumption, compressor pressure ratio, and corrected mass flow rate of this turbojet engine at full throttle are plotted vs flight Mach number M_0 in Figs. 8.22 and 8.23. Figure 8.24 shows the corrected thrust F/δ_0 plotted vs θ_0. Comparing Figs. 8.15 and 8.22, we note that the thrust of both engines is the same at sea-level static, and the engine with a throttle ratio of 1.05 has higher thrust at high Mach numbers. Figures 8.16 and 8.23 show that the engine with theta break (TR) of 1.05 will be able to operate with a compressor

(Continued)

Example 8.5 *(Continued)*

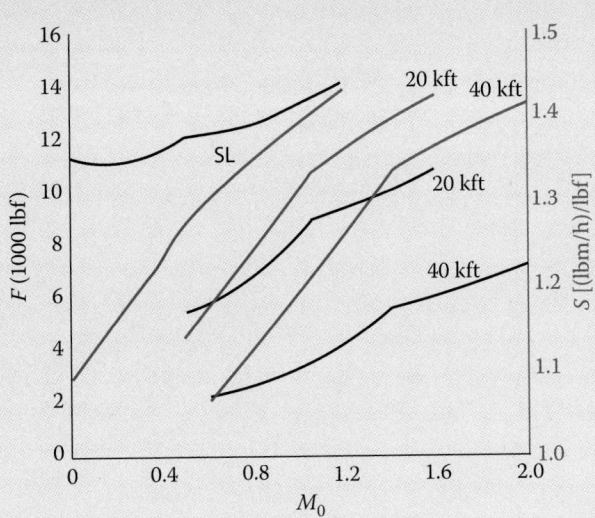

Fig. 8.22 Maximum thrust and thrust specific fuel consumption of improved turbojet (TR = 1.05) vs M_0.

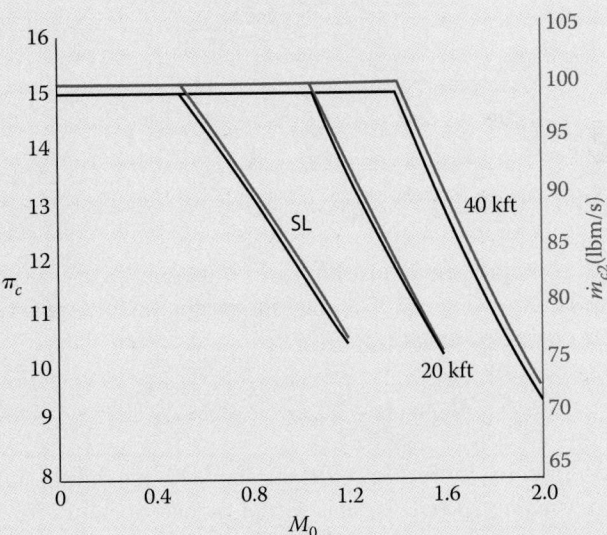

Fig. 8.23 Compressor pressure ratio and corrected mass flow rate of improved turbojet (TR = 1.05) vs M_0.

(Continued)

Example 8.5 *(Continued)*

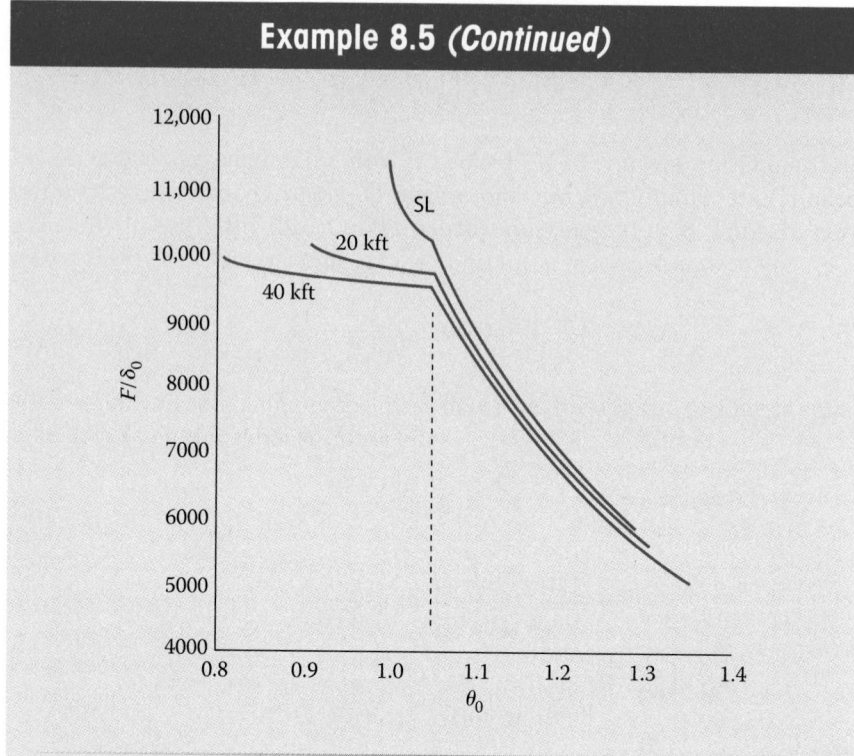

Fig. 8.24 Maximum corrected thrust F/δ_0 of improved turbojet (TR = 1.05) vs θ_0.

pressure ratio of 15 at much higher Mach numbers. Figures 8.19 and 8.24 show that changing the throttle ratio from 1.0 to 1.05 changes the θ_0 value at which the curves change shape ($\theta_{0\,break} = \text{TR}$) and increases the corrected thrust at θ_0 values greater than 1.0.

Because the compressor and turbine are connected to the same shaft, they have the same rotational speed N, and we can write the following relationship between their corrected speeds:

$$N_{c2} = 1/\sqrt{T_{\text{ref}}}\sqrt{T_{t4}/\theta_0}\,N_{c4} \qquad (8.35)$$

Recall that for constant turbine efficiency and choked flow at stations 4 and 8, the correct turbine speed N_{c4} was found to be constant. For maximum thrust engine conditions where θ_0 is less than the throttle ratio, the corrected rotational speed of the compressor N_{c2} and the ratio T_{t4}/θ_0 are constant. Equation (8.35) shows that the corrected speed of the turbine N_{c4} must also be constant at these engine conditions. At $\theta_0 = \text{TR} = \theta_{0\,break}$, T_{t4} is maximum, the corrected rotational speed of the compressor N_{c2} is constant, and the shaft rotational speed N increases by the square root of θ_0. Thus an

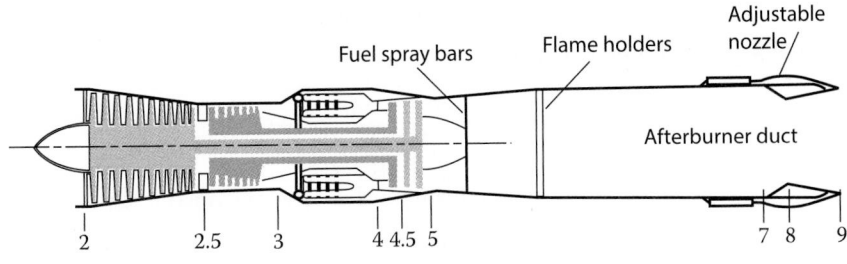

Fig. 8.25 Dual-spool afterburning turbojet engine.

engine with a throttle ratio of 1.05 can have a shaft rotational speed at $\theta_0 = \mathrm{TR} = \theta_{0\,\mathrm{break}}$, T_{t4} that is 1.0247 times the maximum speed at sea-level static conditions. This is commonly referred to as an *overspeed* of 2.47%, an alternative way to specify a theta break or throttle ratio of 1.05.

8.3.2 Turbine Performance Relationships: Dual-Spool Engines

Two-spool engines, like the turbojet engine shown in Fig. 8.25 and the turbofan engine of Fig. 8.1, are designed with choked flow at engine stations 4, 4.5, and 8. Under some operating conditions, the flow may unchoke at station 8. The resulting high-pressure turbine and low-pressure turbine performance relationships are developed in this section for later use.

8.3.2.1 High-Pressure Turbine

Because the mass flow rate at the entrance to the high-pressure turbine equals that entering the low-pressure turbine,

$$\dot{m}_4 = \frac{P_{t4}}{\sqrt{T_{t4}}} A_4 \mathrm{MFP}(M_4) = \dot{m}_{4.5} = \frac{P_{t4.5}}{\sqrt{T_{t4.5}}} A_{4.5} \mathrm{MFP}(M_{4.5})$$

We assume that the areas are constant and the flow is choked at stations 4 and 4.5. Then

$$\frac{P_{t4}/P_{t4.5}}{\sqrt{T_{t4}/T_{t4.5}}} = \frac{\sqrt{\tau_{tH}}}{\pi_{tH}} = \mathrm{const}$$

Thus for constant η_{tH}, we have

$$\text{constant values of } \pi_{tH}, \tau_{tH}, \dot{m}_{c4}, \text{and } \dot{m}_{c4.5} \qquad (8.36)$$

8.3.2.2 Low-Pressure Turbine

Because the mass flow rate at the entrance to the low-pressure turbine equals that at the exit nozzle throat,

$$\dot{m}_{4.5} = \frac{P_{t4.5}}{\sqrt{T_{t4.5}}} A_{4.5} \mathrm{MFP}(M_{4.5}) = \dot{m}_8 = \frac{P_{t8}}{\sqrt{T_{t8}}} A_8 \mathrm{MFP}(M_8)$$

We assume that the areas are constant at stations 4.5 and 8 and the flow is choked at station 4.5, and so

$$\frac{P_{t4.5}/P_{t5}}{\sqrt{T_{t4.5}/T_{t8}}}\frac{1}{\mathrm{MFP}(M_8)} = \frac{\sqrt{\tau_{tL}}/\pi_{tL}}{\mathrm{MFP}(M_8)} = \frac{A_8\pi_{\mathrm{AB}}}{A_{4.5}\mathrm{MFP}(M_{4.5})}$$

Using the reference condition to evaluate the constant on the right side of the preceding equation gives

$$\pi_{tL} = \pi_{tL\,R}\sqrt{\frac{\tau_{tL}}{\tau_{tLR}}\frac{\mathrm{MFP}(M_{8R})}{\mathrm{MFP}(M_8)}} \tag{8.37}$$

where

$$\tau_{tL} = 1 - \eta_{tL}(1 - \pi_{tL}^{(\gamma_t-1)/\gamma_t}) \tag{8.38}$$

If station 8 is choked at the reference condition and at the current operating point, then π_{tL} and τ_{tL} are constant and equal their reference values.

8.4 Dual-Spool Turbojet Engine with Afterburning

The dual-spool afterburning turbojet engine (Fig. 8.25) is normally designed with choked flow at stations 4, 4.5, and 8. For the afterburning turbojet engine, the variable-area exhaust nozzle is controlled by the engine control system so that the upstream turbomachinery is unaffected by the afterburner operation. In other words, the exhaust nozzle throat area A_8 is controlled during afterburner operation such that the turbine exit conditions (P_{t5}, T_{t5}, and M_5) remain constant. Because the exhaust nozzle has choked flow at its throat at all operating conditions of interest, Eqs. (8.37) and

Table 8.4 Performance Analysis Variables for Dual-Spool Afterburning Turbojet Engine

Component	Variables		
	Independent	Constant or Known	Dependent
Engine	M_0, T_0, P_0		$\dot{m}_0$
Diffuser		$\pi_d = f(M_0)$	
Fan		η_{cL}	π_{cL}, τ_{cL}
High-pressure compressor		η_{cH}	π_{cH}, τ_{cH}
Burner	T_{t4}	π_b, η_b	f
High-pressure turbine		π_{tH}, τ_{tH}	
Low-pressure turbine		π_{tL}, τ_{tL}	
Afterburner	T_{t7}	$\pi_{\mathrm{AB}}, \eta_{\mathrm{AB}}$	f_{AB}
Nozzle		π_n	$M_9, T_9/T_0$
	P_9/P_0		
Total Number	6		9

(8.38) for constant efficiency of the low-pressure turbine require that π_{tL} and τ_{tL} be constant:

$$\text{constant values of } \pi_{tL} \text{ and } \tau_{tL} \qquad (8.39)$$

This engine has six independent variables (T_{t4}, T_{t7}, M_0, T_0, P_0, and P_0/P_9) and nine dependent variables. These performance analysis variables are summarized in Table 8.4.

8.4.1 Analysis of Compressors

8.4.1.1 High-Pressure Compressor (τ_{cH}, π_{cH})

The power balance between the high-pressure turbine and the high-pressure compressor (high-pressure spool) gives

$$\eta_{mH} \dot{m}_4 c_{pt}(T_{t4} - T_{t4.5}) = \dot{m}_2 c_{pc}(T_{t3} - T_{t2.5})$$

Rewriting this equation in terms of temperature ratios, rearranging into variable and constant terms, and equating the constant to reference values give

$$\frac{\tau_r \tau_{cL}(\tau_{cH} - 1)}{T_{t4}/T_0} = \eta_{mH}(1 + f)(1 - \tau_{tH}) = \left[\frac{\tau_r \tau_{cL}(\tau_{cH} - 1)}{T_{t4}/T_0}\right]_R$$

Solving for τ_{cH} gives

$$\tau_{cH} = 1 + \frac{T_{t4}/T_0}{(T_{t4}/T_0)_R} \frac{(\tau_r \tau_{cL})_R}{\tau_r \tau_{cL}}(\tau_{cH} - 1)_R \qquad (8.40)$$

From the definition of compressor efficiency, π_{cH} is given by

$$\pi_{cH} = [1 + \eta_{cH}(\tau_{cH} - 1)]^{\gamma_c/(\gamma_c - 1)} \qquad (8.41)$$

8.4.1.2 Low-Pressure Compressor (τ_{cL}, π_{cL})

From a power balance between the low-pressure compressor and low-pressure turbine (low-pressure spool), we get

$$\eta_{mL} \dot{m}_{4.5} c_{pt}(T_{t4.5} - T_{t5}) = \dot{m}_2 c_{pc}(T_{t2.5} - T_{t2})$$

Rewriting this equation in terms of temperature ratios, rearranging into variable and constant terms, and equating the constant to reference values give

$$\frac{\tau_r(\tau_{cL} - 1)}{T_{t4}/T_0} = \eta_{mL}(1 + f)\tau_{tH}(1 - \tau_{tL}) = \left[\frac{\tau_r(\tau_{cL} - 1)}{T_{t4}/T_0}\right]_R$$

Solving for τ_{cL} gives

$$\tau_{cL} = 1 + \frac{T_{t4}/T_0}{(T_{t4}/T_0)_R} \frac{(\tau_r)_R}{\tau_r}(\tau_{cL} - 1)_R \qquad (8.42)$$

where

$$\pi_{cL} = [1 + \eta_{cL}(\tau_{cL} - 1)]^{\gamma_c/(\gamma_c - 1)} \tag{8.43}$$

8.4.2 Mass Flow Rate

Because

$$\dot{m}_4 = \dot{m}_0 + \dot{m}_f = \dot{m}_0(1 + f)$$

and

$$\dot{m}_4 = \frac{P_{t4}}{\sqrt{T_{t4}}} A_4\, \mathrm{MFP}(M_4)$$

thus

$$\dot{m}_0 = \frac{P_{t4}}{\sqrt{T_{t4}}} \frac{A_4\, \mathrm{MFP}(M_4)}{1 + f} = \frac{P_0 \pi_r \pi_d \pi_{cL} \pi_{cH} \pi_b}{\sqrt{T_{t4}}} \frac{A_4\, \mathrm{MFP}(M_4)}{1 + f}$$

For A_4, M_4, $1 + f$, and π_b essentially constant, the preceding expression can be rewritten as

$$\frac{\dot{m}_0 \sqrt{T_{t4}}}{P_0 \pi_r \pi_d \pi_{cL} \pi_{cH}} = \frac{\pi_b A_4\, \mathrm{MFP}(M_4)}{1 + f} = \left(\frac{\dot{m}_0 \sqrt{T_{t4}}}{P_0 \pi_r \pi_d \pi_{cL} \pi_{cH}} \right)_R$$

or

$$\frac{\dot{m}_0}{\dot{m}_{0R}} = \frac{P_0 \pi_r \pi_d \pi_{cL} \pi_{cH}}{(P_0 \pi_r \pi_d \pi_{cl} \pi_{cH})_R} \sqrt{\frac{T_{t4R}}{T_{t4}}} \tag{8.44}$$

See Section 8.4.SM, Summary of Performance Equations: Dual-Spool Turbojet with and without Afterburner, in the online supporting material for related information.

Example 8.6

In this example, we consider the variation in engine performance of the dry turbojet in Example 8.3 with Mach number M_0, altitude, ambient temperature T_0, ambient pressure P_0, and throttle setting T_{t4}. The compressor pressure ratio is limited to 12.3, and the combustor exit temperature T_{t4} is limited to 1800 K, the value at the reference condition. Figures 8.26 and 8.27 show the variations of thrust, thrust specific fuel consumption S, and compressor pressure ratio with Mach number, respectively.

(Continued)

Example 8.6 *(Continued)*

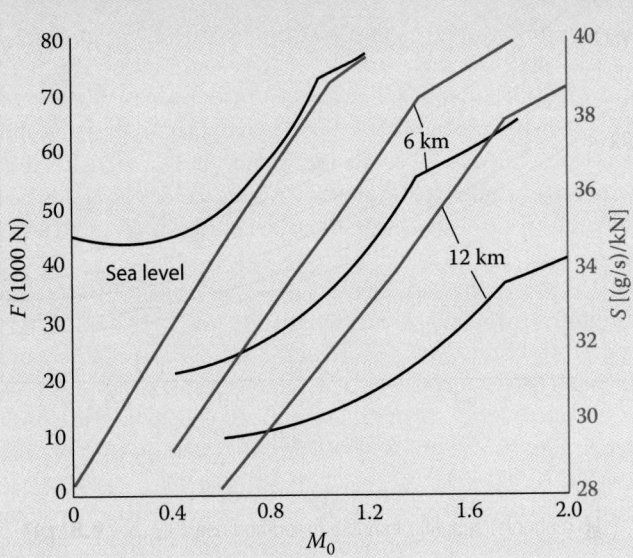

Fig. 8.26   Maximum performance of dry turbojet.

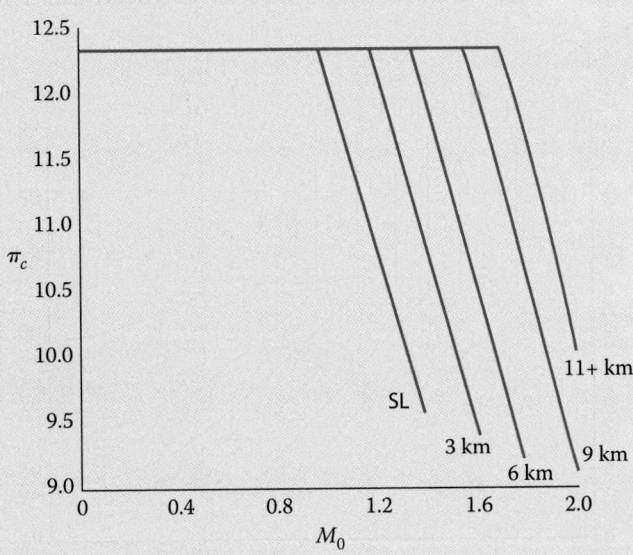

Fig. 8.27   Compressor pressure ratio of dry turbojet at maximum thrust.

(Continued)

Example 8.6 *(Continued)*

At θ_0 values below 1.2, the engine is at the maximum compressor pressure ratio of 12.3. This can most easily be ascertained from Fig. 8.27 at sea level (SL), where the now familiar "break" in π_c occurs at $M = 1.0$. From Eq. (8.16), under these conditions $\theta_0 = 1.2$ must equal $\theta_{0\,\text{break}}$. Thus, at θ_0 values less than 1.2, T_{t4} is below its maximum value of 1800 K. At θ_0 values above 1.2, the engine is at its maximum T_{t4} value of 1800 K, and the compressor pressure ratio is below its maximum value of 12. This engine has a throttle ratio (TR) ($= \theta_{0\,\text{break}}$) of 1.2.

Figure 8.28 shows the variation of engine performance with reduction in engine throttle T_{t4}. Here each engine parameter is compared to its value at the reference condition. As T_{t4} is reduced, S decreases and goes through a minimum value before increasing. The same explanation as was provided in Example 8.4 using the changes in thermal and propulsive efficiencies applies here. The thrust, mass flow rate, and compressor pressure ratio decrease as T_{t4} is reduced.

Figure 8.29 shows the variations of engine performance with changes in ambient temperature T_0. Note that decreases in temperature improve thrust F and thrust specific fuel consumption S. The combined effect of ambient temperature and pressure T_0 and P_0 can be seen in the plot vs altitude shown in Fig. 8.30.

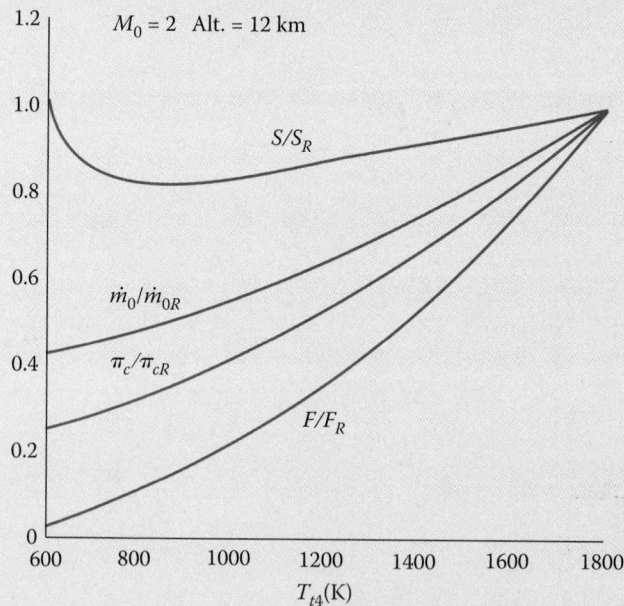

Fig. 8.28   Performance of dry turbojet at partial throttle.

Example 8.6 *(Continued)*

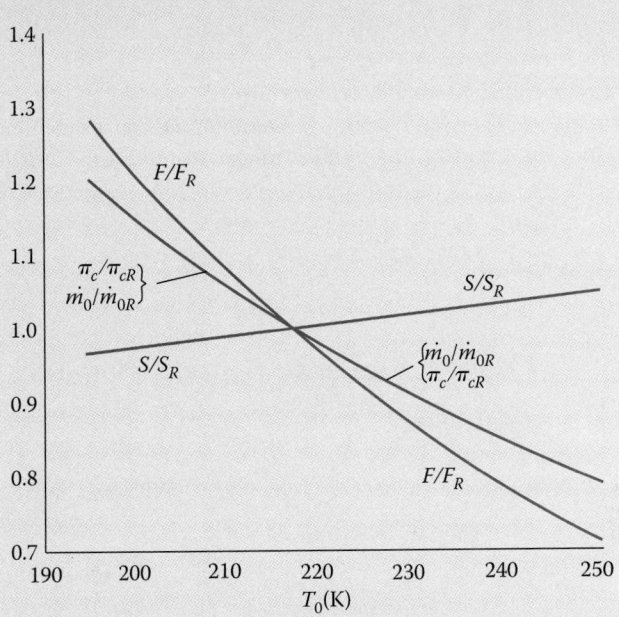

Fig. 8.29   Performance of dry turbojet at maximum thrust vs T_0.

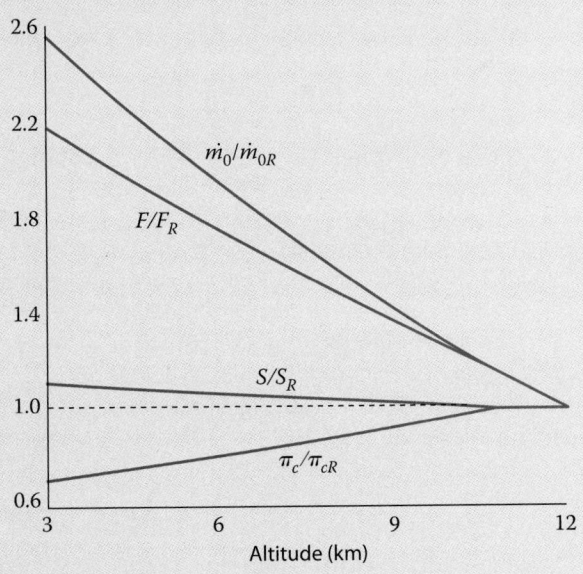

Fig. 8.30   Performance of dry turbojet at maximum thrust vs altitude.

Example 8.7

The performance of an afterburning turbojet with a throttle ratio (TR) or $\theta_{0\,\text{break}}$ of 1 is considered. The engine performance, when the afterburner is operating at its maximum exit temperature, is commonly referred to as *maximum* or *wet*. This afterburning turbojet engine has a maximum thrust of 25,000 lbf. The terms *military* and *dry* refer to the engine's performance when the afterburner is off (not operating) and the engine core is still at maximum operating conditions. The reference conditions and operating limits for the afterburning turbojet engine are as follows.

Reference:

$$\text{Sea-level static } (T_0 = 518.7°\text{R}, \ P_0 = 14.696\,\text{psia})$$
$$\pi_c = 20, \quad \pi_{cL} = 5, \quad \pi_{cH} = 4, \quad T_{t4} = 3200°\text{R},$$
$$\pi_{d\,\text{max}} = 0.98, \quad \pi_b = 0.96, \quad \pi_n = 0.98,$$
$$c_{pc} = 0.24\,\text{Btu}/(\text{lbm} \cdot °\text{R}), \quad \gamma_c = 1.4, \quad c_{pt} = 0.295\,\text{Btu}/(\text{lbm} \cdot °\text{R}),$$
$$\gamma_t = 1.3, \quad \eta_b = 0.995, \quad \eta_{mL} = 0.995, \quad \eta_{mH} = 0.995,$$
$$h_{PR} = 18,400\,\text{Btu}/\text{lbm}, \quad T_{t7} = 3600°\text{R}, \quad c_{pAB} = 0.295\,\text{Btu}/(\text{lbm} \cdot °\text{R}),$$
$$\gamma_{AB} = 1.3, \quad \pi_{AB} = 0.94, \quad \eta_{AB} = 0.95, \quad \eta_{cL} = 0.8755,$$
$$\eta_{cH} = 0.8791, \quad \eta_{tH} = 0.9062, \quad \eta_{tL} = 0.9050, \quad \pi_{tH} = 0.5454,$$
$$\tau_{tH} = 0.8817, \quad \pi_{tL} = 0.6115, \quad \tau_{tL} = 0.9029, \quad M_8 = 1,$$
$$M_9 = 1.824, \quad f = 0.03232, \quad f_{AB} = 0.02070, \quad f_O = 0.05302,$$
$$F = 25,000\,\text{lbf}, \quad S = 1.3905\,(\text{lbm/h})/\text{lbf}, \quad \dot{m}_0 = 181.11\,\text{lbm/s}$$

Operation:

$$\text{Maximum } T_{t4} = 3200°\text{R}, \quad \text{Maximum } T_{t7} = 3600°\text{R},$$
$$\text{Mach number: 0 to 2,} \quad \text{Altitudes (kft): 0, 20, and 40}$$

The wet and dry performances of this afterburning turbojet are compared in Figs. 8.31 and 8.32. Note that the wet thrust is about 20% greater than the dry thrust, and the thrust at 40-kft altitude is about 25% of its sea-level value. The thrust specific fuel consumption for afterburner on at 40-kft altitude is much higher than expected for Mach numbers below about 1.3. This high S is due to the reduction in T_{t4} below maximum for $\theta_0 < \text{TR} \ (\theta_{0\,\text{break}})$, which lowers the temperature of the gas entering the afterburner and increases the temperature rise across the afterburner.

The partial-throttle performance of the afterburning turbojet is shown in Fig. 8.33 at flight conditions of sea-level static and Mach 1.5 at 40 kft. Again, as in Fig. 8.18 of Example 8.4, we see the throttle hook behavior in S vs F. We also see the afterburner's dramatic increase of S indicated by a significant slope

(Continued)

Example 8.7 (Continued)

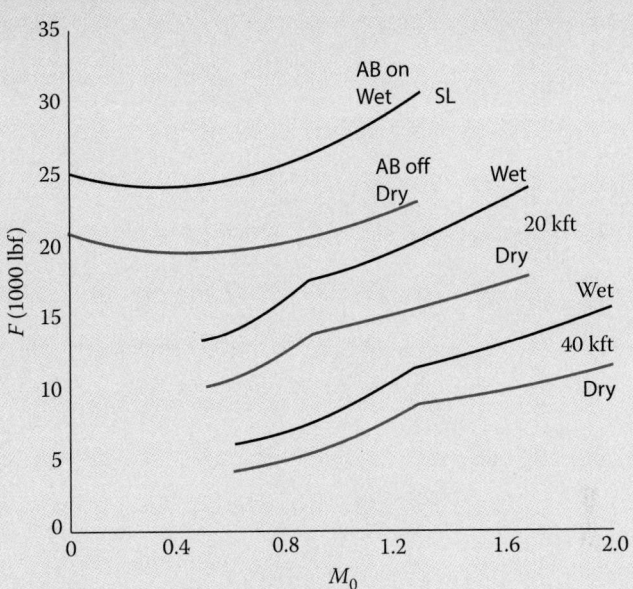

Fig. 8.31 Maximum wet and dry thrust of afterburning turbojet.

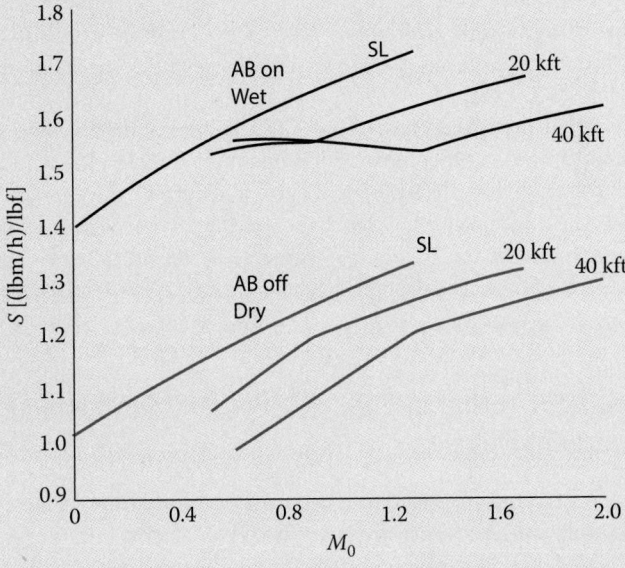

Fig. 8.32 Maximum wet and dry thrust specific fuel consumption of afterburning turbojet.

(Continued)

Example 8.7 (Continued)

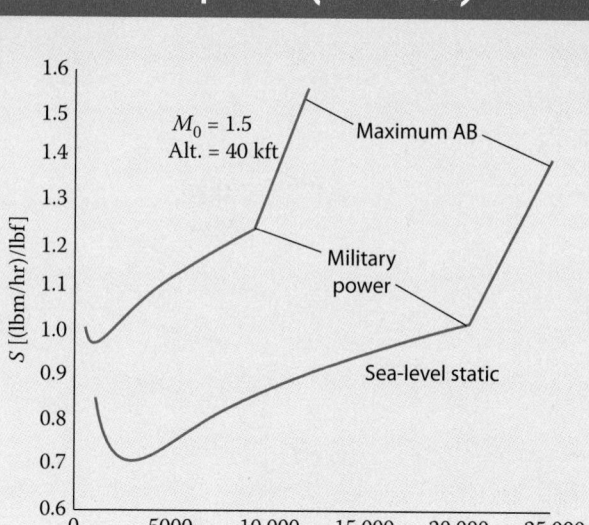

Fig. 8.33 Partial-throttle performance of afterburning turbojet.

increase once the afterburner is engaged. At sea-level static conditions, the minimum thrust specific fuel consumption of about 0.7 (lbm/h)/lbf occurs at a thrust of about 3000 lbf (about 15% of dry thrust). It is worth noting that at partial-power levels this low, the change in component efficiency (refer to Fig. 8.3) can cause the fuel consumption of a real engine to be higher than that predicted here. Because the engine models used to generate these curves are based on constant component efficiencies, the results at significantly reduced throttle settings can be inaccurate. Comparison of this figure with the partial-power performance of the advanced fighter engine of Fig. 1.14c shows that the trends are correct. The advanced turbofan engine of Fig. 1.14c has lower thrust specific fuel consumption mainly because it is a low-bypass-ratio turbofan engine.

8.5 Dual-Spool Turbofan Engine: Separate Exhausts and Convergent Nozzles

The turbofan engines used on commercial subsonic aircraft typically have two spools and separate exhaust nozzles of the convergent type, as shown in Fig. 8.34. For ease of analysis, we will consider a turbofan engine whose fan exit state (13) is the same as the low-pressure compressor exit state (2.5), as indicated in Fig. 8.34. Thus

$$\tau_f = \tau_{cL} \quad \text{and} \quad \pi_f = \pi_{cL}$$

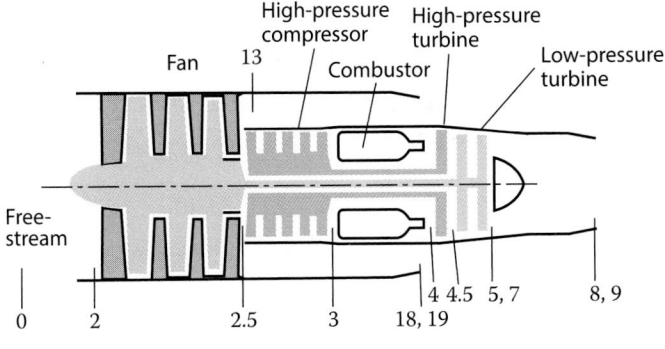

Fig. 8.34   Turbofan engine with separate exhausts.

The exhaust nozzles of these turbofan engines have fixed throat areas that will be choked when the exhaust total pressure/ambient static pressure ratio is equal to or larger than $[(\gamma + 1)/2]^{\gamma/(\gamma-1)}$, according to Eq. (3.8). When an exhaust nozzle is unchoked, the nozzle exit pressure equals the ambient pressure and the exit Mach number is subsonic.

Choked flow at stations 4 and 4.5 of the high-pressure spool during engine operation requires [Eq. (8.36)]

$$\text{constant values of } \pi_{tH}, \tau_{tH}, \dot{m}_{c4}, \text{ and } \dot{m}_{c4.5}$$

Because the exhaust nozzles have fixed areas, this gas turbine engine has 4 independent variables (T_{t4}, M_0, T_0, and P_0). We will consider the case when both exhaust nozzles may be unchoked, resulting in 11 dependent variables. The performance analysis variables and constants are summarized in Table 8.5.

Table 8.5   Performance Analysis Variables for Separate-Exhaust Turbofan Engine

| Component | Variables | | |
	Independent	Constant or Known	Dependent
Engine	M_0, T_0, P_0		$\dot{m}_0, \alpha$
Diffuser		$\pi_d = f(M_0)$	
Fan			π_f, τ_f
High-pressure compressor			π_{cH}, τ_{cH}
Burner	T_{t4}	π_b, η_b	f
High-pressure turbine		π_{tH}, τ_{tH}	
Low-pressure turbine			π_{tL}, τ_{tL}
Core exhaust nozzle		π_n	M_9
Fan exhaust nozzle		π_{fn}	M_{19}
Total Number	4		11

8.5.1 Engine Analysis

8.5.1.1 Low-Pressure Turbine (τ_{tL}, π_{tL})

Equations (8.37) and (8.38) apply for the low-pressure turbine temperature and pressure ratios of this turbofan engine:

$$\pi_{tL} = \pi_{tLR}\sqrt{\frac{\tau_{tL}}{\tau_{tLR}}}\frac{\mathrm{MFP}(M_{9R})}{\mathrm{MFP}(M_9)}$$

where

$$\tau_{tL} = 1 - \eta_{tL}(1 - \pi_{tL}^{(\gamma_t-1)/\gamma_t})$$

If station 9 is choked at the reference condition and at off-design conditions, then π_{tL} and τ_{tL} are constant.

8.5.1.2 Bypass Ratio α

An expression for the engine bypass ratio at any operating condition is obtained by first relating the mass flow rates of the core and bypass streams to their reference values. For the engine core, we have

$$\dot{m}_C = \frac{\dot{m}_4}{1+f} = \frac{P_{t4}A_4}{\sqrt{T_{t4}}}\frac{\mathrm{MFP}(M_4)}{1+f}$$

Thus

$$\frac{\dot{m}_C}{\dot{m}_{CR}} = \frac{P_{t4}}{P_{t4R}}\sqrt{\frac{T_{t4R}}{T_{t4}}} \tag{i}$$

For the bypass stream, we have

$$\dot{m}_B = \frac{P_{t19}A_{19}}{\sqrt{T_{t19}}}\mathrm{MFP}(M_{19})$$

Thus

$$\frac{\dot{m}_B}{\dot{m}_{BR}} = \frac{P_{t19}}{P_{t19R}}\sqrt{\frac{T_{t19R}}{T_{t19}}}\frac{\mathrm{MFP}(M_{19})}{\mathrm{MFP}(M_{19R})} \tag{ii}$$

Combining Eqs. (i) and (ii) to obtain the equation for the bypass ratio α yields

$$\alpha = \alpha_R\frac{\pi_{cHR}}{\pi_{cH}}\sqrt{\frac{\tau_\lambda/(\tau_r\tau_f)}{[\tau_\lambda/(\tau_r\tau_f)]_R}}\frac{\mathrm{MFP}(M_{19})}{\mathrm{MFP}(M_{19R})} \tag{8.45}$$

8.5.1.3 Engine Mass Flow $\dot{m}_0$

The engine mass flow rate can be written simply as

$$\dot{m}_0 = (1 + \alpha)\dot{m}_C$$

Then from Eq. (i), we have

$$\dot{m}_0 = \dot{m}_{0R}\frac{1 + \alpha}{1 + \alpha_R}\frac{P_0\pi_r\pi_d\pi_f\pi_{cH}}{(P_0\pi_r\pi_d\pi_f\pi_{cH})_R}\sqrt{\frac{T_{t4R}}{T_{t4}}} \tag{8.46}$$

8.5.1.4 High-Pressure Compressor (τ_{cH}, π_{cH})

The power balance between the high-pressure turbine and the high-pressure compressor [high-pressure (HP) spool] gives

$$\eta_{mH}\dot{m}_4 c_{pt}(T_{t4} - T_{t4.5}) = \dot{m}_{2.5}c_{pc}(T_{t3} - T_{t2.5})$$

Rewriting this equation in terms of temperature ratios, rearranging into variable and constant terms, and equating the constant to reference values give

$$\frac{\tau_r\tau_f(\tau_{cH} - 1)}{T_{t4}/T_0} = \eta_{mH}(1 + f)(1 - \tau_{tH}) = \left[\frac{\tau_r\tau_f(\tau_{cH} - 1)}{T_{t4}/T_0}\right]_R$$

Solving for τ_{cH} gives

$$\tau_{cH} = 1 + \frac{T_{t4}/T_0}{(T_{t4}/T_0)_R}\frac{(\tau_r\tau_f)_R}{\tau_r\tau_f}(\tau_{cH} - 1)_R \tag{8.47}$$

From the definition of compressor efficiency, π_{cH} is given by

$$\pi_{cH} = [1 + \eta_{cH}(\tau_{cH} - 1)]^{\gamma_c/(\gamma_c-1)} \tag{8.48}$$

8.5.1.5 Fan (τ_f, π_f)

From a power balance between the fan and low-pressure turbine, we get

$$\eta_{mL}\dot{m}_{4.5}c_{pt}(T_{t4.5} - T_{t5}) = (\dot{m}_C + \dot{m}_B)c_{pc}(T_{t13} - T_{t2})$$

Rewriting this equation in terms of temperature ratios, rearranging variable and constant terms, and equating the constant to reference values give

$$(1 + \alpha)\frac{\tau_r(\tau_f - 1)}{T_{t4}/T_0} = \eta_{mL}(1 + f)\tau_{tH}(1 - \tau_{tL}) = \left[(1 + \alpha)\frac{\tau_r(\tau_f - 1)}{T_{t4}/T_0}\right]_R$$

Solving for τ_f gives

$$\tau_f = 1 + \frac{1 - \tau_{tL}}{(1 - \tau_{tL})_R}\frac{\tau_\lambda/\tau_r}{(\tau_\lambda/\tau_r)_R}\frac{1 + \alpha_R}{1 + \alpha}(\tau_{fR} - 1) \tag{8.49}$$

where

$$\pi_f = [1 + (\tau_f - 1)\eta_f]^{\gamma_c/(\gamma_c-1)} \tag{8.50}$$

8.5.2 Solution Scheme

The principal dependent variables for the turbofan engine are π_{tL}, τ_{tL}, α, τ_{cH}, π_{cH}, τ_f, π_f, M_9, and M_{19}. These variables are dependent on each other plus the engine's independent variables—throttle setting and flight condition. The functional interrelationship of the dependent variables can be written as

$$\tau_{cH} = f_1(\tau_f) \qquad\qquad \alpha = f_6(\tau_f, \pi_{cH}, M_{19})$$
$$\pi_{cH} = f_2(\tau_{cH}) \qquad\qquad \tau_f = f_7(\tau_{tL}, \alpha)$$
$$\pi_f = f_3(\tau_f) \qquad\qquad \tau_{tL} = f_8(\pi_{tL})$$
$$M_{19} = f_4(\pi_f) \qquad\qquad \pi_{tL} = f_9(\tau_{tL}, M_9)$$
$$M_9 = f_5(\pi_f, \pi_{cH}, \pi_{tL})$$

This system of nine equations is solved by functional iteration, starting with reference quantities as initial values for π_{tL}, τ_{tL}, and τ_f. The following equations are calculated for the nine dependent variables in the order listed until successive values of τ_{tL} do not change more than a specified amount (say, 0.0001):

$$\tau_{cH} = 1 + \frac{T_{t4}/T_0}{(T_{t4}/T_0)_R}\frac{(\tau_r\tau_f)_R}{\tau_r\tau_f}(\tau_{cH} - 1)_R \tag{i}$$

$$\pi_{cH} = [1 + \eta_{cH}(\tau_{cH} - 1)]^{\gamma_c/(\gamma_c-1)} \tag{ii}$$

$$\pi_f = \left[1 + (\tau_f - 1)\eta_f\right]^{\gamma_c/(\gamma_c-1)} \tag{iii}$$

$$P_{t19}/P_0 = \pi_r\pi_d\pi_f\pi_{fn} \tag{iv}$$

$$\text{If}\quad \frac{P_{t19}}{P_0} < \left(\frac{\gamma_c+1}{2}\right)^{\gamma_c/(\gamma_c-1)} \quad \text{then}\quad \frac{P_{t19}}{P_{19}} = \frac{P_{t19}}{P_0}$$

$$\text{else}\quad \frac{P_{t19}}{P_{19}} = \left(\frac{\gamma_c+1}{2}\right)^{\gamma_c/(\gamma_c-1)} \tag{v}$$

$$M_{19} = \sqrt{\frac{2}{\gamma_c - 1}\left[\left(\frac{P_{t19}}{P_{19}}\right)^{(\gamma_c-1)/\gamma_c} - 1\right]} \tag{vi}$$

$$\frac{P_{t9}}{P_0} = \pi_r\pi_d\pi_f\pi_{cH}\pi_b\pi_{tH}\pi_{tL}\pi_n \tag{vii}$$

$$\text{If}\quad \frac{P_{t9}}{P_0} < \left(\frac{\gamma_t+1}{2}\right)^{\gamma_t/(\gamma_t-1)} \quad \text{then}\quad \frac{P_{t9}}{P_9} = \frac{P_{t9}}{P_0}$$

$$\text{else}\quad \frac{P_{t9}}{P_9} = \left(\frac{\gamma_t+1}{2}\right)^{\gamma_t/(\gamma_t-1)} \tag{viii}$$

$$M_9 = \sqrt{\frac{2}{\gamma_t - 1} \left[\left(\frac{P_{t9}}{P_9} \right)^{(\gamma_t - 1)/\gamma_t} - 1 \right]} \qquad \text{(ix)}$$

$$\alpha = \alpha_R \frac{\pi_{cHR}}{\pi_{cH}} \sqrt{\frac{\tau_\lambda/(\tau_r \tau_f)}{[\tau_\lambda/(\tau_r \tau_f)]_R}} \frac{\text{MFP}(M_{19})}{\text{MFP}(M_{19R})} \qquad \text{(x)}$$

$$\tau_f = 1 + \frac{1 - \tau_{tL}}{(1 - \tau_{tL})_R} \frac{\tau_\lambda/\tau_r}{(\tau_\lambda/\tau_r)_R} \frac{1 + \alpha_R}{1 + \alpha} (\tau_{fR} - 1) \qquad \text{(xi)}$$

$$\tau_{tL} = 1 - \eta_{tL}(1 - \pi_{tL}^{(\gamma_t - 1)/\gamma_t}) \qquad \text{(xii)}$$

$$\pi_{tL} = \pi_{tLR} \sqrt{\frac{\tau_{tL}}{\tau_{tLR}}} \frac{\text{MFP}(M_{9R})}{\text{MFP}(M_9)} \qquad \text{(xiii)}$$

See Section 8.5a.SM, Summary of Performance Equations: Turbofan Engine with Separate Exhausts and Convergent Nozzles, in the online supporting material for related information.

Example 8.8

Given the reference engine (see data) sized for a mass flow rate of 600 lbm/s at 40 kft and Mach 0.8, determine the performance at sea-level static conditions with $T_{t4} = 3200°R$.

Reference:

$$T_0 = 390°R, \quad \gamma_c = 1.4, \quad c_{pc} = 0.24\,\text{Btu/(lbm} \cdot °R), \quad \gamma_t = 1.33,$$
$$c_{pt} = 0.276\,\text{Btu/(lbm} \cdot °R), \quad T_{t4} = 2750°R, \quad M_0 = 0.8,$$
$$\pi_c = 36, \quad \pi_f = 1.7, \quad \alpha = 8, \quad f = 0.02315, \quad \eta_f = 0.8815,$$
$$\eta_{cH} = 0.8512, \quad \tau_{tH} = 0.7341, \quad \pi_{tH} = 0.2505, \quad \tau_{tL} = 0.6895,$$
$$\pi_{tL} = 0.1892, \quad \eta_{tL} = 0.9175, \quad \eta_b = 0.99, \quad \pi_{d\,\text{max}} = 0.99,$$
$$\pi_b = 0.96, \quad \pi_n = 0.99, \quad \pi_{fn} = 0.99, \quad \eta_{mH} = 0.9915,$$
$$\eta_{mL} = 0.997, \quad P_0 = 2.730\,\text{psia (40 kft)}, \quad \dot{m}_0 = 600\,\text{lbm/s},$$
$$F/\dot{m}_0 = 16.20\,\text{lbf/(lbm/s)}, \quad F = 9{,}720\,\text{lbf}, \quad S = 0.588\,\text{lbm/(hr} \cdot \text{lbf)}$$

(Continued)

Example 8.8 *(Continued)*

Performance Conditions:

$$T_0 = 518.7^\circ R, \quad P_0 = 14.696 \, psia \, (sea \, level), \quad T_{t4} = 3200^\circ R, \quad M_0 = 0$$

Detailed calculations are found in the online Supporting Material for this chapter. A summary is given here:

$$\dot{m}_0 = 1906 \, lbm/s, \quad \pi_f = 1.68, \quad \pi_c = 34.9, \quad \alpha = 8.005,$$
$$F = 61,980 \, lbf, \quad S = 0.346 \, lbm/(hr \cdot lbf)$$

The thrust at sea-level static is 6.38 times that at cruise flight conditions. However, we note that the fan pressure ratio, compressor pressure ratio, and bypass ratio do not vary much from their reference values, because these quantities depend on the value of T_{t4}/θ_0, which has increased only 1.32% from its reference value. One could estimate the compressor pressure ratio of this complex engine using Eq. (8.17) for the simple turbojet and get a value of 34.9 at sea-level static with $T_{t4} = 3200^\circ R$. Is this a coincidence? No it is not!

Example 8.9

In this example, we consider the variation in engine performance of a 260,000-N thrust, high-bypass-ratio turbofan engine with Mach number M_0, altitude, ambient temperature T_0, and throttle setting T_{t4}. The engine reference flight condition is sea-level static with $T_0 = 30^\circ C$ (303.15 K) and the following values:

$$\alpha = 8, \quad \pi_f = 1.77, \quad \pi_{cH} = 20.34, \quad T_{t4} = 1890 \, K, \quad \dot{m}_0 = 760 \, kg/s$$

For the performance curves drawn in solid lines, the compressor pressure ratio was limited to 36 and the combustor exit temperature T_{t4} was limited to 1890 K. This engine has a throttle ratio (TR) or $\theta_{0\,break}$ of 1.052. At θ_0 values below 1.052, the engine is at the maximum compressor pressure ratio of 36 and T_{t4} is below its maximum value of 1890 K. For these conditions, the flow in the bypass stream is unchoked at sea level and does not choke at altitude until a Mach number of 0.45.

Figures 8.35–8.39 show the variations of thrust, thrust specific fuel consumption S, engine mass flow, corrected engine mass flow, bypass ratio, fan pressure ratio π_f, and high-pressure (HP) compressor pressure ratio π_{cH} with Mach number and altitude, respectively.

Figure 8.35b shows the variation of the maximum thrust with ambient temperature and freestream Mach number M_0. The large sea-level thrust drop off with M_0 is characteristic of these high-bypass turbofans (as well as

(Continued)

Example 8.9 *(Continued)*

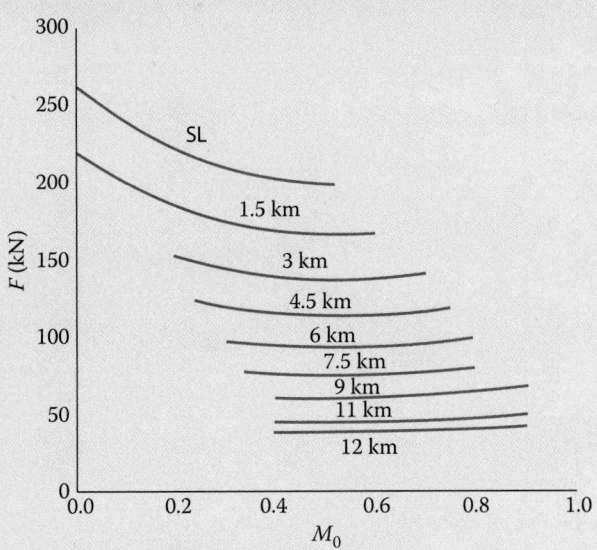

Fig. 8.35a   Maximum thrust of high-bypass-ratio turbofan.

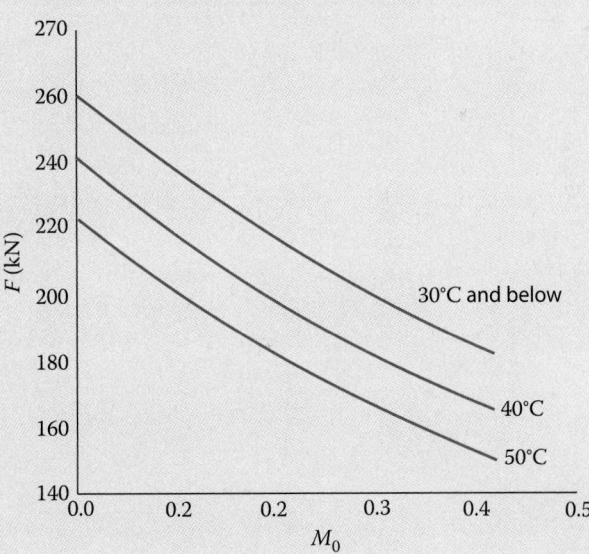

Fig. 8.35b   Maximum thrust and sea-level and elevated temperatures.

(Continued)

Example 8.9 *(Continued)*

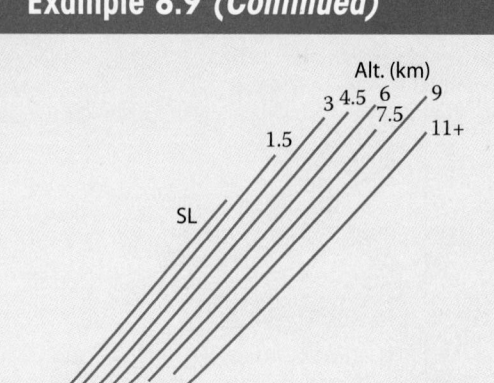

Fig. 8.36 Thrust specific fuel consumption of high-bypass-ratio turbofan at maximum thrust.

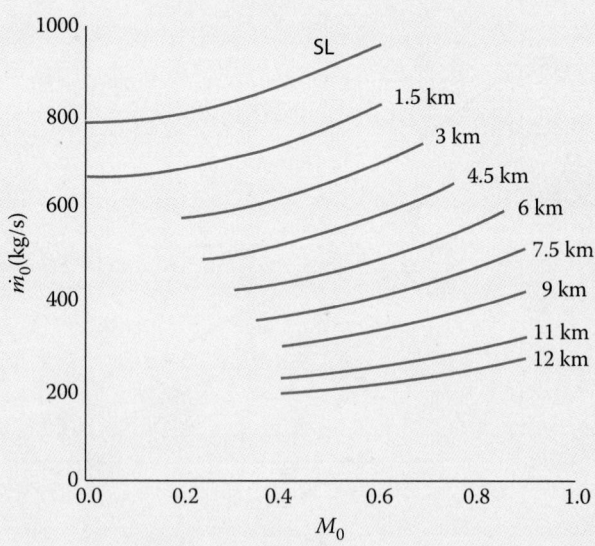

Fig. 8.37 Mass flow rate of high-bypass-ratio turbofan at maximum thrust.

(Continued)

Example 8.9 *(Continued)*

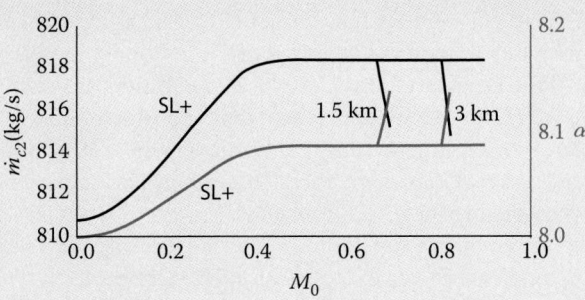

Fig. 8.38   Corrected mass flow rate of high-bypass-ratio turbofan at maximum thrust.

turboprops). Note that the thrust falls off approximately 30% from static to $M_0 = 0.4$ for each of the three curves presented. Clearly, the reduced acceleration of the bypass and core streams as V_0 increases dominates the increased air mass flow rate due to ram effect. At temperatures above 30°C ($\theta_0 > \theta_{0\,break}$), the thrust drops off because the engine is being limited by the combustor exit temperature T_{t4}. At temperatures below 30°C ($\theta_0 < \theta_{0\,break}$), the compressor pressure ratio limits the engine and the thrust varies with Mach number (M_0)—*this engine is flat rated to 30°C*. Compare this figure to Fig. 1.15a and see the same trend.

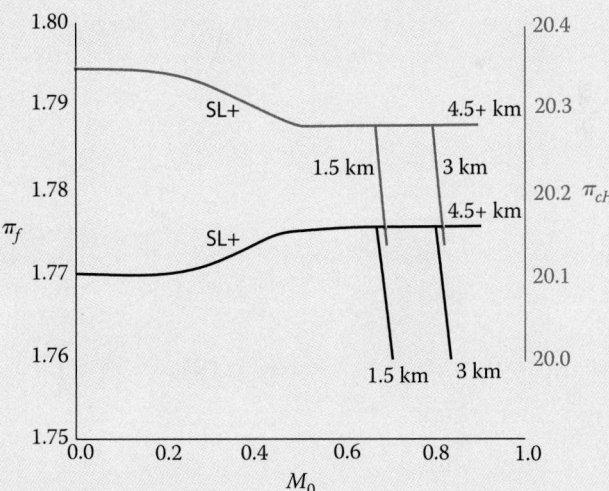

Fig. 8.39   Fan pressure ratio and HP compressor pressure ratio of high-bypass-ratio turbofan at maximum thrust.

(Continued)

Example 8.9 *(Continued)*

The corrected engine mass flow rate of Fig. 8.38 has the same trend with Mach number and altitude as the fan pressure ratio and the high-pressure compressor (HPC) pressure ratio of Fig. 8.39. Both the corrected mass flow rate and the HPC pressure ratio reach their maximum values when the bypass stream chokes (flight Mach of 0.43). Figure 8.38 shows that the engine bypass ratio at maximum thrust has a constant minimum value of about 8.1 when the bypass stream is choked.

The effects of ambient temperature T_0 and altitude on engine performance at maximum thrust are shown in Figs. 8.40 and 8.41, respectively. For ambient temperatures below the reference value of 303.15 K ($\theta_0 < 1.052$), Fig. 8.40 shows that the limit of 36 for the compressor pressure ratio ($\pi_c = \pi_f \pi_{cH}$) holds the engine thrust, bypass ratio, and fan pressure ratio constant. As mentioned earlier, this practice of flat rating engines (holding π_c at its limit by reducing T_{t4} when the ambient temperature is less than the flat-rating temperature) saves on undo wear and tear, thus extending the life of the engine. Engine thrust drops off rapidly with T_0 for $\theta_0 > 1.052$. The decreases in engine thrust, fuel consumption, and air mass flow rate with altitude are shown in Fig. 8.41 for a flight Mach number of 0.5. The decrease in engine thrust with altitude for the high-bypass-ratio turbofan engine is much greater than that of the dry turbojet (see Fig. 8.30). If both a high-bypass-ratio turbofan engine and a dry turbojet engine were sized to produce the same thrust at 9 km and 0.8 Mach, the high-bypass-ratio turbofan engine would have much greater thrust at sea-level static conditions. This helps explain the decrease in takeoff length between the early turbojet-powered passenger aircraft and the modern high-bypass-ratio turbofan-powered passenger aircraft of today.

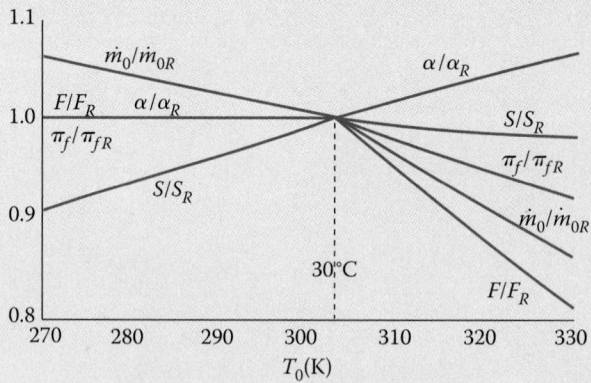

Fig. 8.40 Performance of high-bypass-ratio turbofan vs T_0 at sea-level static, maximum thrust.

(Continued)

Example 8.9 *(Continued)*

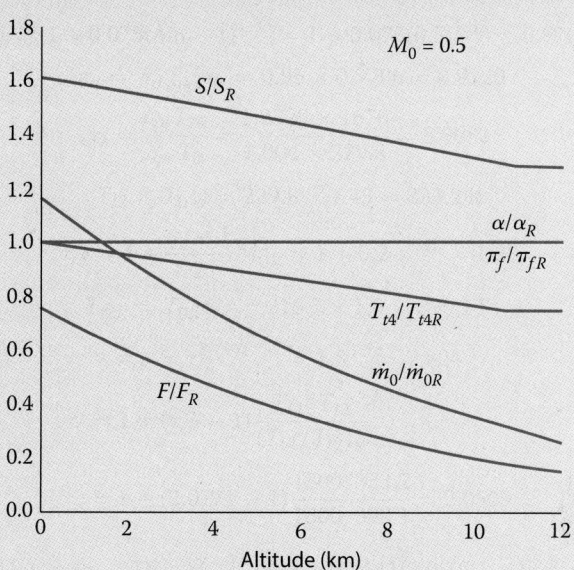

Fig. 8.41 Performance of high-bypass-ratio turbofan at maximum thrust vs altitude.

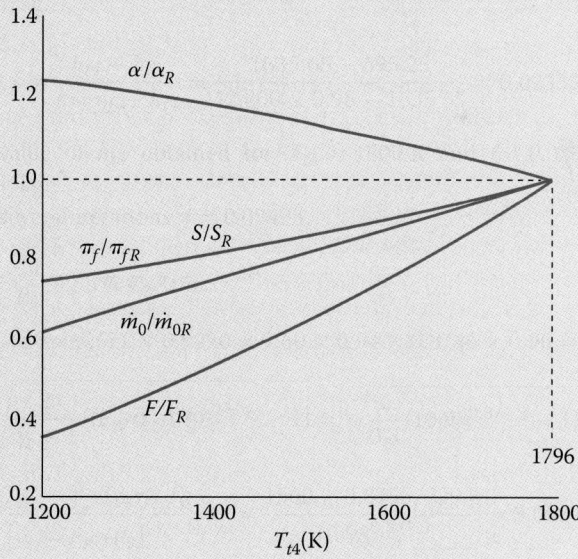

Fig. 8.42 Performance of high-bypass-ratio turbofan vs T_{t4} at sea-level static conditions.

(Continued)

Example 8.9 *(Continued)*

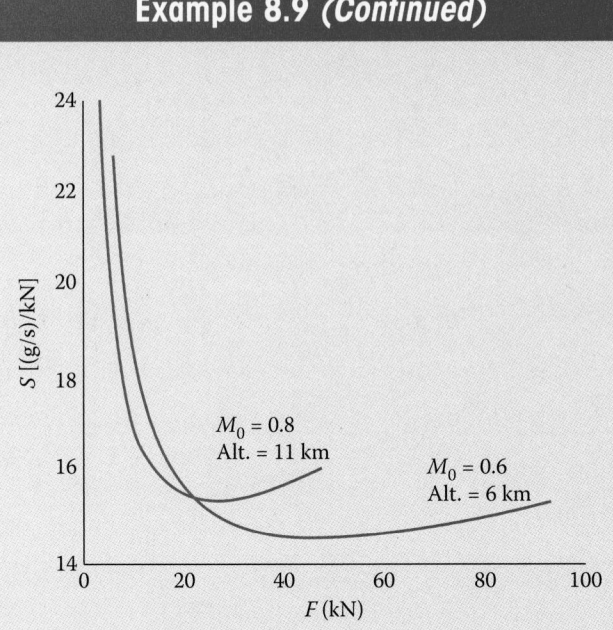

Fig. 8.43   Partial-throttle performance of high-bypass-ratio turbofan.

Figures 8.42 and 8.43 show the effects that changes in combustor exit temperature T_{t4} have on engine performance. As shown in Fig. 8.42, all engine performance parameters except the bypass ratio decrease with reduction in engine throttle T_{t4}. If the throttle were reduced further, the thrust specific fuel consumption S would start to increase. The thrust specific fuel consumption S vs thrust F at partial throttle (partial power) is shown in Fig. 8.43 for two different values of altitude and Mach number. These curves have the classical hook shape that gives them their name of throttle hook. Minimum S occurs at about 50% of maximum thrust. At lower throttle settings, the thrust specific fuel consumption rapidly increases (due to the domination of reduced thermal efficiency over increased propulsive efficiency).

8.5.3   Booster Stages: Compressor Stages on Low-Pressure Spool

Modern high-bypass-ratio turbofan engines and other turbofan engines are constructed with compressor stages on the low-pressure spool, as shown in Fig. 8.44. A quick look at Figs. 1.8a–1.8f verifies this. As was discussed in Section 1.4.3, engine company use of the "common core" often necessitates additional compressor stages in the core stream placed on the low-pressure spool to achieve the desired overall compression system pressure ratio. These stages are often called *booster stages* because they boost the pressure rise a small amount relative to the higher RPM HPC

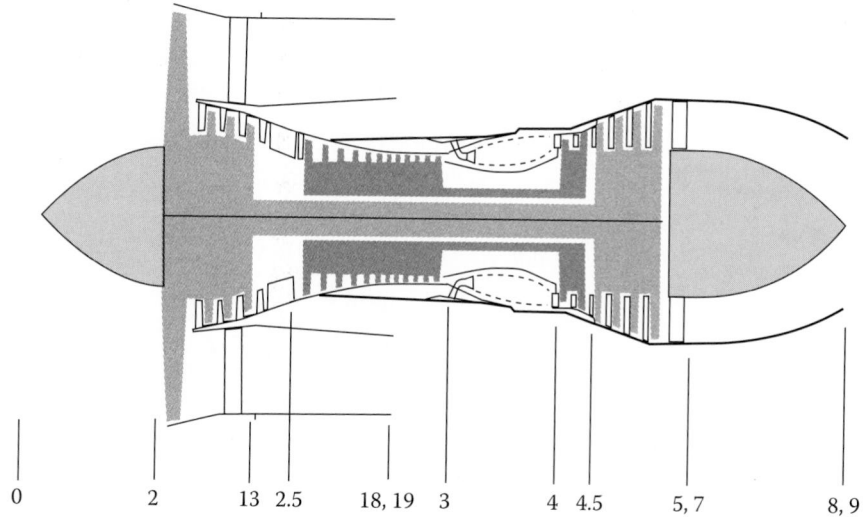

Fig. 8.44 Turbofan engine with compressor stages on low-pressure spool.

stages. The LP spool RPM is limited by the large fan blades. This limitation can be removed by putting the booster stages on their own spool driven by their own turbine—an intermediate spool, as Rolls-Royce continues to do. Alternatively, a speed reduction gearbox placed between the fan and low-pressure shaft (geared turbofan) allows the low-pressure turbine, and hence these booster stages, to spin at a higher RPM than would otherwise be possible, thus eliminating several stages. Both of these alternatives were discussed in Section 1.4.3. This change in engine layout also adds two dependent variables to the nine principle dependent variables we had for the performance analysis of the turbofan engine in the previous section. These two new variables are the low-pressure compressor's total temperature ratio τ_{cL} and total pressure ratio π_{cL}.

Because the low-pressure compressor and the fan are on the same shaft, the enthalpy rise across the low-pressure compressor will be proportional to the enthalpy rise across the fan during normal operation. For a calorically perfect gas, we can write

$$T_{t2.5} - T_{t2} = K(T_{t13} - T_{t2})$$

or

$$\tau_{cL} - 1 = K(\tau_f - 1)$$

Using reference conditions to replace the constant K, we can solve the preceding equation for τ_{cL} as

$$\tau_{cL} = 1 + (\tau_f - 1)\frac{\tau_{cLR} - 1}{\tau_{fR} - 1} \tag{8.51}$$

The pressure ratio for the low-pressure compressor is given by Eq. (8.43):

$$\pi_{cL} = [1 + \eta_{cL}(\tau_{cL} - 1)]^{\gamma_c/(\gamma_c-1)}$$

In a manner like that used to obtain Eq. (8.45), the following equation for the bypass ratio results:

$$\alpha = \alpha_R \frac{\pi_{cLR}\pi_{cHR}/\pi_{fR}}{\pi_{cL}\pi_{cH}/\pi_f} \sqrt{\frac{\tau_\lambda/(\tau_r\tau_f)}{[\tau_\lambda/(\tau_r\tau_f)]_R} \frac{\text{MFP}(M_{19})}{\text{MFP}(M_{19R})}} \qquad (8.52a)$$

By rewriting Eq. (8.46) for this engine configuration, the engine mass flow rate is given by

$$\dot{m}_0 = \dot{m}_{0R} \frac{1+\alpha}{1+\alpha_R} \frac{P_0\pi_r\pi_d\pi_{cL}\pi_{cH}}{(P_0\pi_r\pi_d\pi_{cL}\pi_{cH})_R} \sqrt{\frac{T_{t4R}}{T_{t4}}} \qquad (8.52b)$$

Equation (8.40) applies to the high-pressure compressor of this engine and is

$$\tau_{cH} = 1 + \frac{T_{t4}/T_0}{(T_{t4}/T_0)_R} \frac{(\tau_r\tau_{cL})_R}{\tau_r\tau_{cL}}(\tau_{cH} - 1)_R$$

Equations (8.36), (8.37), and (8.38) apply to the high- and low-pressure turbines.

From a power balance among the fan, low-pressure compressor, and low-pressure turbine, we get

$$\eta_{mL}\dot{m}_{4.5}c_{pt}(T_{t4.5} - T_{t5}) = \dot{m}_B c_{pc}(T_{t13} - T_{t2}) + \dot{m}_C c_{pc}(T_{t2.5} - T_{t2})$$

Rewriting this equation in terms of temperature ratios, rearranging into variable and constant terms, and equating the constant to reference values give

$$\frac{\tau_r[(\tau_{cL} - 1) + \alpha(\tau_f - 1)]}{(T_{t4}/T_0)(1 - \tau_{tL})} = \eta_{mL}(1 + f)\tau_{tH} = \left\{\frac{\tau_r[\tau_{cL} - 1 + \alpha(\tau_f - 1)]}{(T_{t4}/T_0)(1 - \tau_{tL})}\right\}_R$$

Using Eq. (8.53), we substitute for τ_{cL} on the left side of the preceding equation, solve for τ_f, and get

$$\tau_f = 1 + (\tau_{fR} - 1)\left[\frac{1 - \tau_{tL}}{(1 - \tau_{tL})_R} \frac{\tau_\lambda/\tau_r}{(\tau_\lambda/\tau_r)_R} \frac{\tau_{cLR} - 1 + \alpha_R(\tau_{fR} - 1)}{\tau_{cLR} - 1 + \alpha(\tau_{fR} - 1)}\right] \qquad (8.53)$$

8.5.4 Solution Scheme

The principal dependent variables for the turbofan engine are π_{tL}, τ_{tL}, α, τ_{cL}, π_{cL}, τ_{cH}, π_{cH}, τ_f, π_f, M_9, and M_{19}. These variables are dependent on each other plus the engine's independent variables—throttle setting and flight condition. The functional interrelationship of the dependent variables can be written as

$$
\begin{aligned}
\tau_{cH} &= f_1(\tau_{cL}) & M_{19} &= f_7(\pi_f) \\
\pi_{cH} &= f_2(\tau_{cH}) & M_9 &= f_8(\pi_f, \pi_{cH}, \pi_{tL}) \\
\tau_f &= f_3(\tau_{tL}, \alpha) & \pi_{tL} &= f_9(\tau_{tL}, M_9) \\
\pi_f &= f_4(\tau_f) & \tau_{tL} &= f_{10}(\pi_{tL}) \\
\tau_{cL} &= f_5(\tau_f) & \alpha &= f_{11}(\tau_f, \pi_{cH}, M_{19}) \\
\pi_{cL} &= f_6(\tau_{cL})
\end{aligned}
$$

This system of 11 equations is solved by functional iteration, starting with reference quantities as initial values for π_{tL}, τ_{tL}, and τ_f. The following equations are calculated for the 11 dependent variables in the order listed until successive values of τ_{tL} do not change more than a specified amount (say, 0.0001):

$$
\tau_{cH} = 1 + \frac{T_{t4}/T_0}{(T_{t4}/T_0)_R} \frac{(\tau_r \tau_{cL})_R}{\tau_r \tau_{cL}} (\tau_{cH} - 1)_R \tag{8.54a}
$$

$$
\pi_{cH} = [1 + \eta_{cH}(\tau_{cH} - 1)]^{\gamma_c/(\gamma_c-1)} \tag{8.54b}
$$

$$
\pi_f = [1 + (\tau_f - 1)\eta_f]^{\gamma_c/(\gamma_c-1)} \tag{8.54c}
$$

$$
\frac{P_{t19}}{P_0} = \pi_r \pi_d \pi_f \pi_{\mathrm{fn}} \tag{8.54d}
$$

If $\dfrac{P_{t19}}{P_0} < \left(\dfrac{\gamma_c + 1}{2}\right)^{\gamma_c/(\gamma_c-1)}$ then $\dfrac{P_{t19}}{P_{19}} = \dfrac{P_{t19}}{P_0}$

else $\dfrac{P_{t19}}{P_{19}} = \left(\dfrac{\gamma_c + 1}{2}\right)^{\gamma_c/(\gamma_c-1)}$ \hfill (8.54e)

$$
M_{19} = \sqrt{\frac{2}{\gamma_c - 1}\left[\left(\frac{P_{t19}}{P_{19}}\right)^{(\gamma_c-1)/\gamma_c} - 1\right]} \tag{8.54f}
$$

$$
\frac{P_{t9}}{P_0} = \pi_r \pi_d \pi_{cL} \pi_{cH} \pi_b \pi_{tH} \pi_{tL} \pi_n \tag{8.54g}
$$

If $\dfrac{P_{t9}}{P_0} < \left(\dfrac{\gamma_t + 1}{2}\right)^{\gamma_t/(\gamma_t-1)}$ then $\dfrac{P_{t9}}{P_9} = \dfrac{P_{t9}}{P_0}$

else $\dfrac{P_{t9}}{P_9} = \left(\dfrac{\gamma_t + 1}{2}\right)^{\gamma_t/(\gamma_t-1)}$ \hfill (8.54h)

$$M_9 = \sqrt{\frac{2}{\gamma_t - 1}\left[\left(\frac{P_{t9}}{P_9}\right)^{(\gamma_t - 1)/\gamma} - 1\right]} \qquad (8.54\text{i})$$

$$\alpha = \alpha_R \frac{\pi_{cLR}\,\pi_{cHR}/\pi_{fR}}{\pi_{cL}\,\pi_{cH}/\pi_f}\sqrt{\frac{\tau_\lambda/(\tau_r\tau_f)}{[\tau_\lambda/(\tau_r\tau_f)]_R}\frac{\text{MFP}(M_{19})}{\text{MFP}(M_{19R})}} \qquad (8.54\text{j})$$

$$\tau_f = 1 + (\tau_{fR} - 1)\left[\frac{1 - \tau_{tL}}{(1 - \tau_{tL})_R}\frac{\tau_\lambda/\tau_r}{(\tau_\lambda/\tau_r)_R}\frac{\tau_{cLR} - 1 + \alpha_R(\tau_{fR} - 1)}{\tau_{cLR} - 1 + \alpha(\tau_{fR} - 1)}\right] \qquad (8.54\text{k})$$

$$\tau_{cL} = 1 + (\tau_f - 1)\frac{\tau_{cLR} - 1}{\tau_{fR} - 1} \qquad (8.54\text{l})$$

$$\pi_{cL} = [1 + \eta_{cL}(\tau_{cL} - 1)]^{\gamma_c/(\gamma_c - 1)} \qquad (8.54\text{m})$$

$$\tau_{tL} = 1 - \eta_{tL}(1 - \pi_{tL}^{(\gamma_t - 1)/\gamma_t}) \qquad (8.54\text{n})$$

$$\pi_{tL} = \pi_{tLR}\sqrt{\frac{\tau_{tL}}{\tau_{tLR}}\frac{\text{MFP}(M_9R)}{\text{MFP}(M_9)}} \qquad (8.54\text{o})$$

Examples 8.8 and 8.9 of the previous section are representative of the performance of these high-bypass-ratio turbofans with booster stages. Most notably, the engine performance characteristics shown in Example 8.9 are quite representative. Only very subtle changes would result by including any differences in LPC efficiency from that of the fan and/or slight changes in mechanical efficiency of the LP spool.

See Section 8.5c.SM, Summary of Performance Equations: Turbofan Engine with Compressor States on Low-Pressure Spool, in the online supporting material for related information.

8.6   Dual-Spool Turbofan with Afterburning: Mixed Exhaust Stream

Modern fighter aircraft and advanced bombers use the low-bypass-ratio, mixed-flow exhaust turbofan engine with afterburning like that shown in Fig. 8.45. These engines have improved fuel consumption over turbojets when operated dry (afterburner off) with a manageable increase in fan diameter and high specific thrust when the afterburner is turned on. This type of engine is normally designed with choked flow at stations 4, 4.5, and 8. The variable-area exhaust nozzle is controlled by the engine control system so that the upstream turbomachinery is unaffected by the afterburner operation.

The mixing of the core stream and bypass stream in the fixed-area mixer adds six dependent variables to the performance analysis (π_M, τ_M, M_6, M_{16}, M_{6A}, and α'). As first presented in Section 6.9, the static pressures at stations 6 and 16 must be equal for unchoked flow conditions at stations

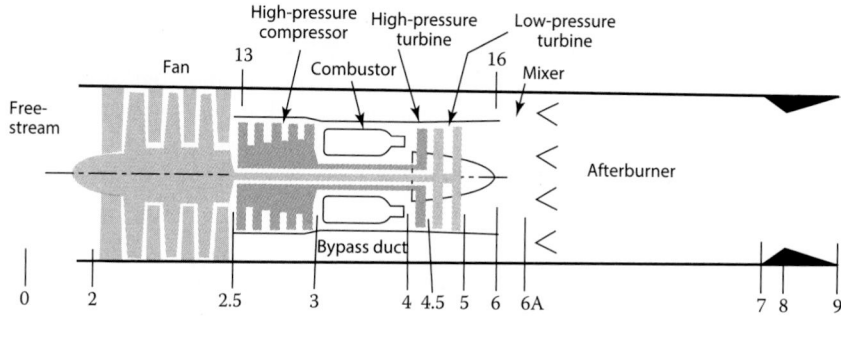

Fig. 8.45 Mixed-flow exhaust turbofan engine with afterburning.

16 and 6 ($P_6 = P_{16}$), which requires the total pressure ratio of the two streams P_{t6}/P_{t16} will be near unity. This dramatically restricts the fan pressure ratio π_f and bypass ratio α of this type of engine.

The independent variables, constant or known values, and dependent variables for the performance analysis of this engine cycle are listed in Table 8.6. Note that for this engine, there are 14 dependent variables and 6 independent variables. As we will show for the choked exhaust nozzle, the temperature ratio T_{t6A}/T_{t2} and pressure ratio P_{t6A}/P_{t2} depend on only the temperature ratio T_{t4}/T_{t2}. For the flow between stations 2 and 6A,

Table 8.6 Performance Analysis Variables for Mixed-Flow Exhaust Turbofan Engine

Component	Variables		
	Independent	Constant or Known	Dependent
Engine	M_0, T_0, P_0		$\dot{m}_0, \alpha$
Diffuser		$\pi_d = f(M_0)$	
Fan			π_f, τ_f
High-pressure compressor			π_{cH}, τ_{cH}
Burner	T_{t4}	π_b	
High-pressure turbine		π_{tH}, τ_{tH}	
Low-pressure turbine			π_{tL}, τ_{tL}
Mixer		π_{Mmax}	$\pi_M, \tau_M, M_6, M_{16}, \alpha',$ M_{6A}
Afterburner	T_{t7}	π_{AB}	
Nozzle	P_0/P_9 or A_9/A_8	π_n	
Total Number	6		14

there are 13 variables (α, π_f, τ_f, π_{cH}, τ_{cH}, π_{tL}, π_M, τ_M, M_6, M_{16}, M_{6A}, and α') that are dependent on 3 independent variables (M_0, T_0, and T_{t4}). To obtain the performance of this engine, we need to write an equation for each dependent variable in terms of the independent variables and other dependent variables. Recall from Section 6.9 that α' is the mixer bypass ratio.

The equations for the performance analysis of the mixed-flow turbofan engine with afterburner are developed by making the assumptions and using the methods for both the turbojet engine and separate-exhaust turbofan engine. The goal is to obtain independent equations relating the dependent variables of Table 8.6. The equations to be developed are represented, in turn, by the following 14 functional relationships among the 14 dependent variables of interest:

$$\tau_{tL} = f_1(\tau_f, M_6, M_{16}, \alpha') \qquad \pi_{tL} = f_2(\tau_{tL})$$
$$\tau_f = f_3(\tau_{tL}, \alpha) \qquad \pi_f = f_4(\tau_f)$$
$$\tau_{cH} = f_5(\tau_f) \qquad \pi_{cH} = f_6(\tau_{cH})$$
$$\alpha = f_7(\alpha') \qquad M_6 = f_8(\tau_{tL}, \pi_{tL})$$
$$M_{16} = f_9(\pi_{cH}, \pi_{tL}, M_6) \qquad \tau_M = f_{10}(\tau_f, \tau_{tL})$$
$$M_{6A} = f_{11}(M_6, M_{16}, \tau_M) \qquad \pi_M = f_{12}(\alpha', \tau_M, M_6, M_{6A})$$
$$\dot{m}_0 = f_{13}(\pi_f, \pi_{cH}, \alpha)$$
$$\alpha' = f_{14}(\pi_{tL}, \pi_M, \tau_{tL}, \tau_M) \qquad \text{\textit{for choked exhaust nozzle}}$$
$$\alpha' = f_{14}(\pi_f, \pi_{cH}, \pi_{tL}, \pi_M, \tau_{tL}, \tau_M) \quad \text{\textit{for unchoked exhaust nozzle}}$$

Please notice that each equation can be solved, in principle, in the order listed for given initial estimates of the two component performance variables τ_{tL} and α'. As the solution progresses, these estimates are compared with the newly computed values and iterated, if necessary, until convergence is obtained. We start the analysis with the turbomachinery.

8.6.1 High-Pressure and Low-Pressure Turbines

The flow is assumed to be choked at stations 4 and 4.5. Thus Eq. (8.36) applies for the high-pressure turbine, or

$$\text{constant values of } \pi_{tH}, \tau_{tH}, \dot{m}_{c4}, \text{ and } \dot{m}_{c4.5} \qquad (8.36)$$

From Eq. (6.28), we can write

$$T_t = \left(\frac{PAM}{\dot{m}}\right)^2 \frac{\gamma g_c}{R}\left(1 + \frac{\gamma - 1}{2}M^2\right)$$

There follows, with $P_6 = P_{16}$ and $\alpha' = \dot{m}_{16}/\dot{m}_6$,

$$\frac{T_{t6}}{T_{t16}} = \left(\frac{A_6}{A_{16}}\frac{M_6}{M_{16}}\alpha'\right)^2 \frac{\gamma_6}{\gamma_{16}}\frac{R_{16}}{R_6}\frac{1 + [(\gamma_6 - 1)/2]M_6^2}{1 + [(\gamma_{16} - 1)/2]M_{16}^2}$$

The ratio T_{t6}/T_{t16} also can be written in terms of τ_{tL} as

$$\frac{T_{t6}}{T_{t16}} = \frac{T_{t4}}{T_{t2}} \frac{\tau_{tH} \tau_{tL}}{\tau_f}$$

Equating the last two equations and solving for τ_{tL} gives

$$\tau_{tL} = \frac{T_{t2}}{T_{t4}} \frac{\tau_f}{\tau_{tH}} \left(\frac{A_6}{A_{16}} \frac{M_6}{M_{16}} \alpha'\right)^2 \frac{\gamma_6}{\gamma_{16}} \frac{R_{16}}{R_6} \frac{1 + [(\gamma_6 - 1)/2]M_6^2}{1 + [(\gamma_{16} - 1)/2]M_{16}^2} \tag{8.55}$$

From the definition of the turbine efficiency, we write

$$\pi_{tL} = \left(1 - \frac{1 - \tau_{tL}}{\eta_{tL}}\right)^{\gamma_t/(\gamma_t - 1)} \tag{8.56}$$

8.6.1.1 Fan Temperature and Pressure Ratio (τ_f, π_f)

Because the low-pressure turbine drives only the fan, Eqs. (8.49) and (8.50) both apply for the mixed-flow turbofan engine:

$$\tau_f = 1 + \frac{1 - \tau_{tL}}{(1 - \tau_{tL})_R} \frac{\tau_\lambda/\tau_r}{(\tau_\lambda/\tau_r)_R} \frac{1 + \alpha_R}{1 + \alpha} (\tau_{fR} - 1) \tag{8.49}$$

where

$$\pi_f = [1 + (\tau_f - 1)\eta_f]^{\gamma_c/(\gamma_c - 1)} \tag{8.50}$$

8.6.1.2 High-Pressure Compressor (τ_{cH}, π_{cH})

The power balance between the high-pressure turbine and the high-pressure compressor (high-pressure spool) for the separate-exhaust turbofan engine also applies to the mixed-flow turbofan engine. Thus

$$\tau_{cH} = 1 + \frac{T_{t4}/T_0}{(T_{t4}/T_0)_R} \frac{(\tau_r \tau_f)_R}{\tau_r \tau_f} (\tau_{cH} - 1)_R \tag{8.47}$$

From the definition of compressor efficiency, π_{cH} is given by

$$\pi_{cH} = [1 + \eta_{cH}(\tau_{cH} - 1)]^{\gamma_c/(\gamma_c - 1)} \tag{8.48}$$

8.6.1.3 Engine Bypass Ratio α

The bypass ratio of the engine α is related to the bypass ratio of the mixer α' by

$$\alpha' = \frac{\alpha}{1 + f} \tag{8.57}$$

8.6.2 Solution for Mixer Performance

8.6.2.1 Mach Number at Station 6 M_6

Because the mass flow rate at the entrance to the low-pressure turbine (station 4.5) equals that at the core entrance to the mixer (station 6),

$$\dot{m}_{4.5} = \frac{P_{t4.5}}{\sqrt{T_{t4.5}}} A_{4.5} \text{MFP}(M_{4.5}) = \dot{m}_6 = \frac{P_{t6}}{\sqrt{T_{t6}}} A_6 \text{MFP}(M_6)$$

We assume that the areas are constant at stations 4.5 and 6, there is isentropic flow from station 5 to 6, and the flow is choked at station 4.5. Then

$$\frac{P_{t4.5}/P_{t6}}{\sqrt{T_{t4.5}/T_{t6}}} \frac{1}{\text{MFP}(M_6)} = \frac{\pi_{tL}/\sqrt{\tau_{tL}}}{\text{MFP}(M_6)} = \frac{A_6}{A_{4.5}\text{MFP}(M_{4.5})}$$

Using the reference condition to evaluate the constant on the right side of the above equation and solving for $\text{MFP}(M_6)$ gives

$$\text{MFP}(M_6) = \text{MFP}(M_{6R})\sqrt{\frac{\tau_{tL}}{\tau_{tLR}}\frac{\pi_{tLR}}{\pi_{tL}}} \tag{8.58}$$

The Mach number at station 6 follows directly from the mass flow parameter MFP.

8.6.2.2 Mach Number at Station 16 M_{16}

The definition of the total pressure yields M_{16} once P_{t16}/P_{16} is evaluated. Because $P_6 = P_{16}$,

$$\frac{P_{t16}}{P_{16}} = \frac{P_{t16}}{P_{t6}}\frac{P_{t6}}{P_6} = \frac{P_{t16}}{P_{t6}}\left(1 + \frac{\gamma_6 - 1}{2}M_6^2\right)^{\gamma_6/(\gamma_6-1)}$$

where, by referencing and assuming both $P_{t13} = P_{t2.5}$ and isentropic flow from station 13 to 16,

$$\frac{P_{t16}}{P_{t6}} = \frac{P_{t13}}{P_{t2.5}\pi_{cH}\pi_b\pi_{tH}\pi_{tL}} = \frac{(\pi_{cH}\pi_{tL})_R}{\pi_{cH}\pi_{tL}}\left(\frac{P_{t16}}{P_{t6}}\right)_R$$

so that

$$\frac{P_{t16}}{P_{16}} = \frac{(\pi_{cH}\pi_{tL})_R}{\pi_{cH}\pi_{tL}}\left(\frac{P_{t16}}{P_{t6}}\right)_R\left(1 + \frac{\gamma_6 - 1}{2}M_6^2\right)^{\gamma_6/(\gamma_6-1)} \tag{8.59a}$$

For the flow to be possible, $P_{t16} > P_{16}$.

With P_{t16}/P_{16} known, then

$$M_{16} = \left\{ \frac{2}{\gamma_{16} - 1} \left[\left(\frac{P_{t16}}{P_{16}} \right)^{(\gamma_{16}-1)/\gamma_{16}} - 1 \right] \right\}^{1/2} \tag{8.59b}$$

where, for the flow to be subsonic, $M_{16} < 1.0$.

8.6.2.3 Mixer Temperature Ratio τ_M

From Eq. (7.74), we have

$$\tau_M \equiv \frac{T_{t6A}}{T_{t6}} = \frac{c_{p6}}{c_{p6A}} \frac{1 + \alpha'(c_{p16}/c_{p6})(T_{t16}/T_{t6})}{1 + \alpha'} \tag{8.60a}$$

where

$$\frac{T_{t16}}{T_{t6}} = \frac{T_{t2}}{T_{t4}} \frac{\tau_f}{\tau_{tH} \tau_{tL}} \tag{8.60b}$$

8.6.2.4 Mach Number at Station 6A M_{6A} and Mixer Pressure Ratio π_M

The Mach number at station 6A is given by

$$M_{6A} = \sqrt{\frac{2\Phi}{1 - 2\gamma_{6A}\Phi + \sqrt{1 - 2(\gamma_{6A} + 1)\Phi}}} \tag{8.61}$$

with Φ and ϕ given by Eqs. (6.35) and (6.36), respectively. Using Eqs. (7.60) and (6.38), we see that the total pressure ratio of the mixer π_M is given by

$$\pi_M = \pi_{Mmax} \frac{(1 + \alpha')\sqrt{\tau_M}}{1 + A_{16}/A_6} \frac{MFP(M_6, \gamma_6, R_6)}{MFP(M_{6A}, \gamma_{6A}, R_{6A})} \tag{8.62}$$

8.6.2.5 Mixer Bypass Ratio α'

The engine exhaust nozzle normally operates in the choked condition but may become unchoked at reduced throttle settings, flight Mach numbers, and altitudes. Each of these operating conditions must be considered separately in determining α'. However, because α' is unaffected by afterburning, it is not necessary to consider the afterburning case separately. This follows from the facts that the afterburner is not used in the unchoked exhaust nozzle condition and that, during choked afterburner operation, A_8 is modulated to keep $A_8 \pi_{AB}/\sqrt{\tau_{AB}}$ constant so that the upstream flow remains uninfluenced by afterburning. The development to follow, therefore, is based on no afterburning, but applies equally well to the choked-exhaust-nozzle afterburning case.

From the conservation of mass and the definition of α'

$$1 + \alpha' = \frac{\dot{m}_8}{\dot{m}_{4.5}} = \frac{\dot{m}_9}{\dot{m}_{4.5}}$$

Case 1: For choked exhaust nozzle and low-pressure turbine inlet nozzles, Eqs. (8.11a) and (8.11b) yields

$$1 + \alpha' = \frac{\dot{m}_8}{\dot{m}_{4.5}} = \frac{P_{t8}A_8}{\sqrt{T_{t8}}} \frac{\sqrt{T_{t4.5}}}{P_{t4.5}A_{4.5}} \frac{\Gamma_8}{\sqrt{R_8}} \frac{\sqrt{R_{4.5}}}{\Gamma_{4.5}}$$

When the afterburner is off, the exhaust nozzle control system holds the throat area of the exhaust nozzle A_8 constant when the nozzle is choked, or

$$\frac{P_{t9}}{P_0} \geq \left(\frac{\gamma_8 + 1}{2}\right)^{\gamma_8/(\gamma_8 - 1)}$$

In addition, the choked area at station 4.5 $(A_{4.5})$ is constant and the gas properties $(R_{4.5}$ and $\Gamma_{4.5})$ are considered constant. Rewriting in terms of τ's and π's and separating out the constant terms give

$$1 + \alpha' = \frac{\dot{m}_8}{\dot{m}_{4.5}} = \frac{\pi_{tL}\pi_M}{\sqrt{\tau_{tL}\tau_M}} \frac{\Gamma_8}{\sqrt{R_8}} \left(\frac{A_8}{A_{4.5}} \frac{\sqrt{R_{4.5}}}{\Gamma_{4.5}}\right)$$

Replacing the constant term within the parentheses with reference conditions and solving for α' give, for the *choked* exhaust nozzle,

$$\alpha' = (1 + \alpha'_R) \frac{\pi_{tL}\pi_M}{(\pi_{tL}\pi_M)_R} \sqrt{\frac{(\tau_{tL}\tau_M)_R}{\tau_{tL}\tau_M}} \frac{\Gamma_8}{(\Gamma_8)_R} - 1 \qquad (8.63\text{a})$$

Case 2: For an unchoked exhaust nozzle and choked low-pressure turbine inlet nozzle at station 4.5, the mass flow parameter at station 8 and Eq. (8.11a) give

$$1 + \alpha' = \frac{\dot{m}_8}{\dot{m}_{4.5}} = \frac{\pi_{tL}\pi_M}{\sqrt{\tau_{tL}\tau_M}} A_8 \text{MFP}(M_8, \gamma_8, R_8) \left(\frac{P_{t8}}{P_{t6}} \frac{1}{A_{4.5}} \frac{\sqrt{R_{4.5}/g_c}}{\Gamma_{4.5}}\right)$$

For engine operation when the exhaust nozzle is unchoked, it is advantageous to adjust A_8 so that $M_8 = M_9$. This requires that $A_8 = A_9\pi_n$. Using $M_8 = M_9$, replacing the constant term within the parentheses above with reference conditions, and solving for α' give, for the *unchoked* exhaust nozzle,

$$\alpha' = (1 + \alpha'_R) \frac{\pi_{tL}\pi_M}{(\pi_{tL}\pi_M)_R} \sqrt{\frac{(\tau_{tL}\tau_M)_R}{\tau_{tL}\tau_M}} \frac{A_s}{(A_8)_R} \frac{\text{MFP}(M_8, \gamma_8, R_8)}{\text{MFP}(M_8, \gamma_8, R_8)_R} - 1 \qquad (8.63\text{b})$$

The Mach number at station 9 in Eq. (8.63b) follows from the definition of the total pressure

$$M_9 = \left\{ \frac{2}{\gamma_9 - 1} \left[\left(\frac{P_{t9}}{P_9}\right)^{(\gamma_9 - 1)/\gamma_9} - 1 \right] \right\}^{1/2}$$

where, because $P_9 = P_0$,

$$\frac{P_{t9}}{P_9} = \pi_r\,\pi_d\,\pi_f\,\pi_{cH}\,\pi_b\,\pi_{tH}\,\pi_{tL}\,\pi_M\,\pi_{AB}\,\pi_n$$

8.6.3 Engine Mass Flow and Exhaust Nozzle Areas

8.6.3.1 Engine Mass Flow $\dot{m}_0$

The engine mass flow rate can be written simply as

$$\dot{m}_0 = (1+\alpha)\dot{m}_C$$

Thus Eq. (8.46) applies:

$$\dot{m}_0 = \dot{m}_{0R}\,\frac{1+\alpha}{1+\alpha_R}\,\frac{P_0\,\pi_r\,\pi_d\,\pi_f\,\pi_{cH}}{(P_0\,\pi_r\,\pi_d\,\pi_f\,\pi_{cH})_R}\sqrt{\frac{T_{t4R}}{T_{t4}}} \tag{8.46}$$

The corrected mass flow at station 2 (fan entrance) can be obtained from the preceding equation and Eq. (8.2), yielding

$$\dot{m}_{c2} = \dot{m}_{c2R}\,\frac{1+\alpha}{1+\alpha_R}\,\frac{\pi_f\,\pi_{cH}}{(\pi_f\,\pi_{cH})_R}\sqrt{\frac{(T_{t4}/T_{t2})_R}{T_{t4}/T_{t2}}} \tag{8.64}$$

8.6.3.2 Exhaust Nozzle Throat Area A_8

Increasing the exhaust nozzle throat area A_8 moves the operating line to the right on a compressor or fan map (see Fig. 8.8), and decreasing the throat area moves this line to the left. For variable-area exhaust nozzles, the throat area is increased during engine start-up and low thrust settings to reduce the turbine backpressure, increase the corrected mass flow rate, and thus keep the compressor from stall or surge (see Fig. 8.3). Once the engine operation has reached high corrected mass flow rates (this normally corresponds to sufficient nozzle pressure to choke its flow), the throat area is reduced to its nominal value, which shifts the engine operating line to the left—a region of improved efficiency (see Fig. 11.53). For analysis of engine operation, we limit ourselves to the steady-state conditions when sufficient nozzle total pressure exists to choke the flow.

The throat area of the choked exhaust nozzle depends on the operation of the afterburner. With the *afterburner off* (dry), we assume that the engine control system maintains a constant throat area, or

$$A_8 = A_{8R} \tag{8.65a}$$

This will keep the operating line on the fan and high-pressure compressor maps constant.

When the afterburner is in operation (wet), the engine control system increases the throat area of the exhaust nozzle to keep the flow properties at station $6A$ constant (P_{t6A}, T_{t6A}, and M_{6A}). In this way, the turbomachinery (fan, high-pressure compressor, high-pressure turbine, and low-pressure turbine) is not affected by afterburner operation—a highly desirable situation.

If the engine control system allows P_{t6A} to vary during afterburning operation, the flow through the turbomachinery is affected, which, in the extreme, could lead to compressor stall or engine overspeed.

An expression for the ratio of exhaust nozzle throat area with the afterburner on to its area when the afterburner is off will now be developed. Using conservation of mass between stations $6A$ and 8 and the mass flow parameter with the afterburner on gives

$$\dot{m}_8 = \dot{m}_{6A} + \dot{m}_{fAB}$$

or

$$\left[\frac{P_{t8}A_8}{\sqrt{T_{t8}}} \text{MFP}(\gamma_8, R_8, M_8 = 1) \right]_{AB}$$
$$= \left(1 + \frac{\dot{m}_{fAB}}{\dot{m}_{6A}} \right) \left[\frac{P_{t6A}A_{6A}}{\sqrt{T_{t6A}}} \text{MFP}(\gamma_{6A}, R_{6A}, M_{6A}) \right]$$

Likewise, conservation of mass with the afterburner off gives

$$\left[\frac{P_{t8}A_8}{\sqrt{T_{t8}}} \text{MFP}(\gamma_8, R_8, M_8 = 1) \right]_{noAB} = \left[\frac{P_{t6A}A_{6A}}{\sqrt{T_{t6A}}} \text{MFP}(\gamma_{6A}, R_{6A}, M_{6A}) \right]$$

Because the flow at station $6A$ is unaffected by the operation of the afterburner, the terms inside square brackets on the right sides of both equations written above are equal. In addition, the γ, R, and T_t at station 8 with the afterburner off are the same as those at station $6A$. Thus

$$\frac{A_{8AB \text{ on}}}{A_{8AB \text{ off}}} = \frac{\pi_{AB \text{ dry}}}{\pi_{AB \text{ wet}}} \sqrt{\frac{T_{t8}}{T_{t6A}}} \left(1 + \frac{\dot{m}_{fAB}}{\dot{m}_{6A}} \right) \frac{\Gamma_{6A}}{\Gamma_{8AB \text{ on}}}$$

Neglecting the afterburner fuel flow rate and differences in Γ, we can write

$$\frac{A_{8AB \text{ on}}}{A_{8AB \text{ off}}} \approx \frac{\pi_{AB \text{ dry}}}{\pi_{AB \text{ wet}}} \sqrt{\frac{T_{t7}}{T_{t6A}}} \qquad (8.65b)$$

8.6.4 Iteration Scheme

The solution of the 13 dependent variables between stations 2 and $6A$ is found by using the iteration scheme described in this section. Initial values are needed for τ_{tL} and α' to start the solution. The initial value of τ_{tL} is selected to be its reference value, τ_{tLR}. When θ_0 is less than the throttle ratio (TR), the initial value of α' is selected to be equal to its reference value, α_R'. However, when θ_0 is greater than the TR, the initial value of α' is calculated by using the following relationship, which accounts for its variation with Mach number and altitude:

$$\alpha' = \alpha_R' \frac{\theta_0}{\theta_{0R}} \qquad (8.66)$$

For repeated calculations where one of the independent variables is changed, the initial values of τ_{tL} and α' for subsequent calculations are taken to be their previously calculated values.

A flowchart of the iteration scheme is shown in Fig. 8.46. Certain features of the programmed iteration methods used in the engine performance computer program PERF deserve highlighting:

1. If either the pressure ratio of the bypass stream entering the mixer P_{t16}/P_{16} or the Mach number M_{16} is out of limits ($P_{t16} < P_{16}$ or M_{16}

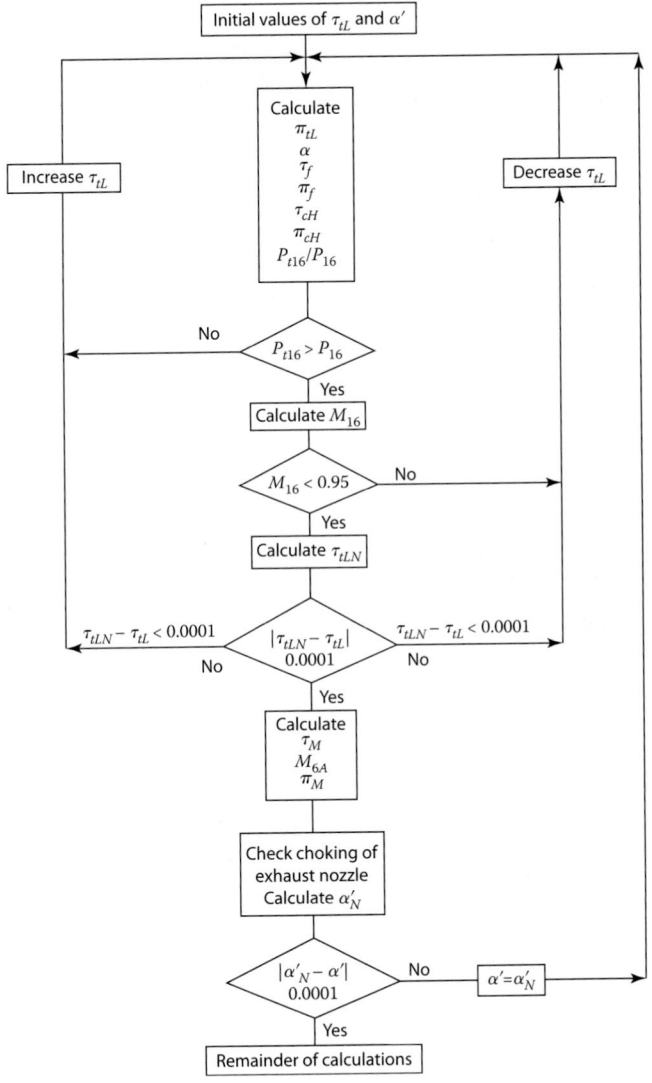

Fig. 8.46 Flowchart of iterative solution scheme.

approaching sonic), the low-pressure turbine temperature ratio τ_{tL} is incremented by 0.0001.

2. If the new value of the low-pressure turbine temperature ratio calculated by Eq. (8.58) is not within 0.0001 of the previous value, Newtonian iteration is used to converge to a solution.

3. Functional iteration is used for the mixer bypass ratio α'.

See Section 8.6.SM, Summary of Performance Equations: Turbofan with Afterburning Mixed-Flow Exhaust, in the online supporting material for related information.

Example 8.10

In this example, we consider the performance of a low-bypass-ratio, mixed-flow, two-spool, afterburning turbofan engine with an uninstalled thrust of 25,000 lbf, π_c of 20, and α of 0.5 at sea-level static (SLS) conditions. An SLS reference is selected with the resulting data listed below. The engine has a T_{t4max} of 3200°R and a T_{t4} at SLS conditions of 2909°R—these give an engine throttle ratio (TR) of 1.10.

The performance of the engine between station 2 and station 6 is considered first. This is followed by the full-throttle performance maps for the engine with and without afterburning. This example concludes with a discussion of several partial-throttle performance curves.

Reference:

$$T_0 = 518.7°R, \quad \gamma_c = 1.4, \quad c_{pc} = 0.24 \text{ Btu}/(\text{lbm °R}), \quad \gamma_t = 1.3,$$

$$c_{pt} = 0.295 \text{ Btu}/(\text{lbm} \cdot °R), \quad M_0 = 0, \quad T_{t4} = 2909°R,$$

$$\gamma_{AB} = 1.3, \quad c_{pAB} = 0.295 \text{ Btu}/(\text{lbm} \cdot °R), \quad T_{t7} = 3600°R, \quad \pi_c = 20,$$

$$\pi_f = 4.16, \quad \alpha = 0.5, \quad M_6 = 0.4, \quad M_{16} = 0.386,$$

$$\eta_f = 0.8663, \quad \eta_{cH} = 0.8761, \quad \tau_{tH} = 0.8548, \quad \pi_{tH} = 0.46591,$$

$$\tau_{tL} = 0.8552, \quad \pi_{tL} = 0.4749, \quad \eta_{tL,} = 0.9169, \quad \eta_b = 0.99,$$

$$\pi_{d\,max} = 0.97, \quad \pi_b = 0.94, \quad \pi_{M\,max} = 0.96, \quad \pi_{AB} = 0.94,$$

$$\pi_n = 0.98, \quad \eta_{AB} = 0.95, \quad \eta_{mH} = 0.99, \quad \eta_{mL} = 0.99,$$

$$P_9/P_0 = 1, \quad P_0 = 14.696 \text{ psia}, \quad h_{PR} = 18{,}400 \text{ Btu}/\text{lbm},$$

$$\dot{m}_0 = 209.9 \text{ lbm}/s, \quad F/\dot{m}_0 = 119.1 \text{ lbf}/(\text{lbm}/s), \quad f = 0.02677,$$

$$f_{AB} = 0.03507, \quad F = 25{,}000 \text{ lbf}, \quad S = 1.599 \text{ (lbm/hr)}/\text{lbf}$$

For flight conditions where $\theta_0 < \text{TR}$, the engine controls off of $\pi_{c\,max}$, and the maximum value of T_{t4} is given by the following equation, which follows from Eqs. (8.34a) and (8.34b):

$$T_{t4} = \frac{\theta_0}{\text{TR}} T_{t4\,max} \qquad (8.67)$$

(Continued)

Example 8.10 *(Continued)*

When the engine operates at the value of T_{t4}, given by Eq. (8.67), the ratio T_{t4}/T_{t2} is constant at its maximum value and the corrected speeds N_c of both the low-pressure spool and the high-pressure spool are constant at their maximum (refer to Section 8.2.6). For flight conditions where $\theta_0 > \mathrm{TR}$ ($= \theta_{0\,\text{break}}$), the maximum value of T_{t4} is $T_{t4\,\text{max}}$, the ratio T_{t4}/T_{t2} is less than its maximum value, and the corrected speeds of both the low-pressure spool and the high-pressure spool are less than their maximum.

The pumping characteristics of the engine are plotted in Fig. 8.47 vs the corrected speed of the low-pressure spool. Note that at reduced corrected speeds of the low-pressure spool, all the quantities plotted, except the bypass ratio α, are less than their maximum values. Also note that the variations of the high-pressure spool's corrected quantities are much less than those of the low-pressure spool. Even for a cycle as complex as the mixed-flow turbofan engine, *there is a one-to-one correspondence between the temperature ratio T_{t4}/T_{t2}, and the engine's pumping characteristics between station 2 and station 6.*

The characteristics of the engine's mixer and low-pressure turbine are plotted in Fig. 8.48 vs Mach numbers at three different altitudes for maximum/military thrust. For flight conditions where $\theta_0 \leq \theta_{0\,\text{break}}$ ($= \mathrm{TR}$),

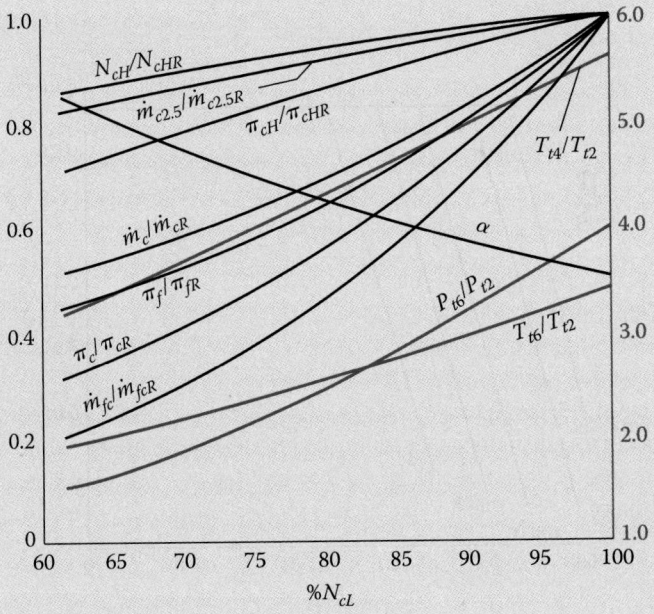

Fig. 8.47   Pumping characteristics of mixed-flow turbofan engine.

(Continued)

Example 8.10 *(Continued)*

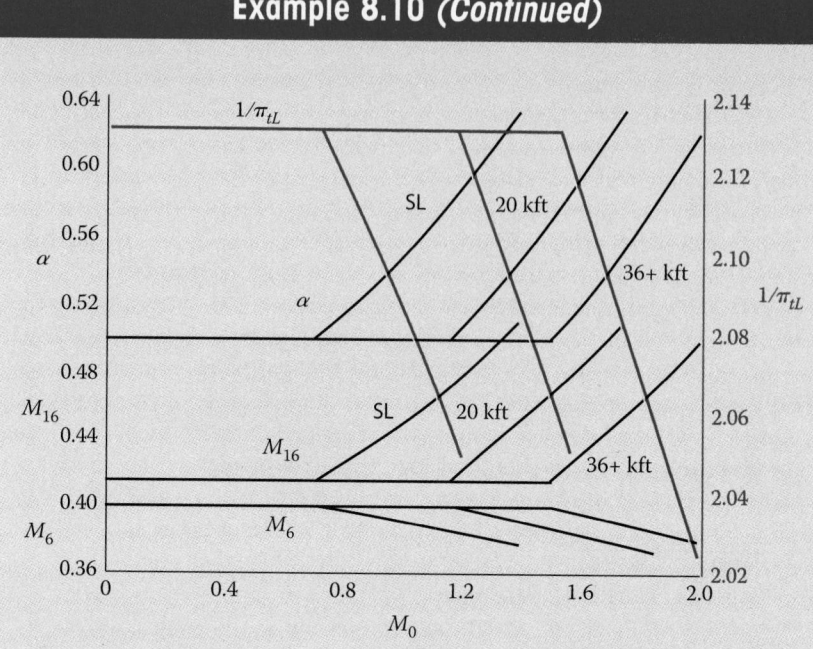

Fig. 8.48 Engine mixer and low-pressure turbine characteristics.

the bypass ratio α, M_6, M_{16}, and low-pressure expansion ratio $(P_{t4.5}/P_{t5} = 1/\pi_{tL})$ are constant. For flight conditions where $\theta_0 > \theta_{0\,\text{break}}$ (=TR), the bypass ratio α, and M_{16} increase and the low-pressure expansion ratio $(P_{t4.5}/P_{t5} = 1/\pi_{tL})$ and M_6 decrease. This change in trend for $\theta_0 > \theta_{0\,\text{break}}$ (=TR) is due to the reduction in fan and high-pressure compressor pressure ratios.

The uninstalled thrust F and uninstalled thrust specific fuel consumption S performance of the mixed-flow turbofan engine at full throttle with afterburner on and off are plotted in Figs. 8.49 and 8.50 vs flight Mach number at three different altitudes. Not surprisingly, the performance of this engine closely resembles that of the turbojet, but with improved (lower) fuel consumption.

One notes from these figures the change in slope of the curves at a combination of Mach number and altitude that corresponds to θ_0 of 1.1 (the throttle ratio of this engine). At flight conditions (Mach and altitude) corresponding to $\theta_0 > \text{TR}$, the engine is operating at $T_{t4\text{max}}$ and reduced corrected speed, whereas when $\theta_0 < \theta_{0\,\text{break}}$ (= TR), the engine is operating at 100% corrected speed and T_{t4} leaving the main burner is given by Eq. (8.67).

Comparison of these performance results based on simple models to those of an engine that includes all the variations can be made by comparing Figs. 8.49–8.50 to Figs. 1.14a–1.14b. The performance curves of Figs. 8.49–8.50 show the correct trends with Mach number and altitude except for the altitude

(Continued)

Example 8.10 *(Continued)*

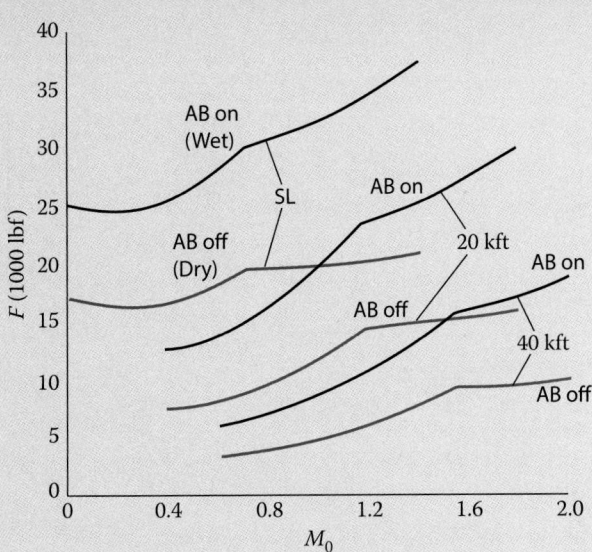

Fig. 8.49 Uninstalled thrust F of mixed-flow turbofan engine at maximum power setting (afterburner on) and military power setting (afterburner off).

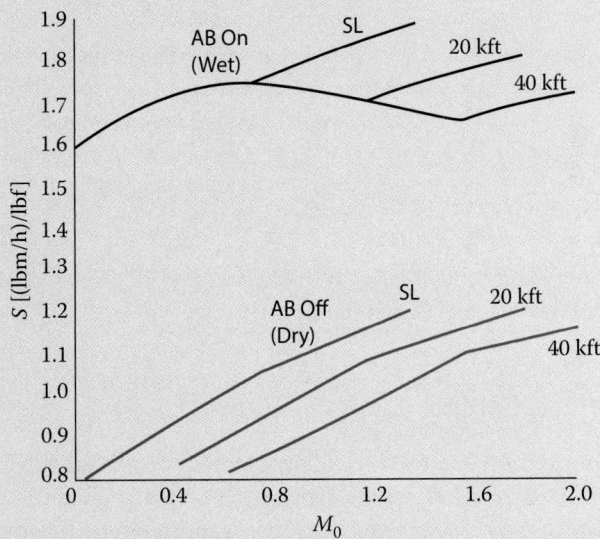

Fig. 8.50 Uninstalled fuel consumption S of mixed-flow turbofan engine at maximum power setting (afterburner on) and military power setting (afterburner off).

(Continued)

Example 8.10 (Continued)

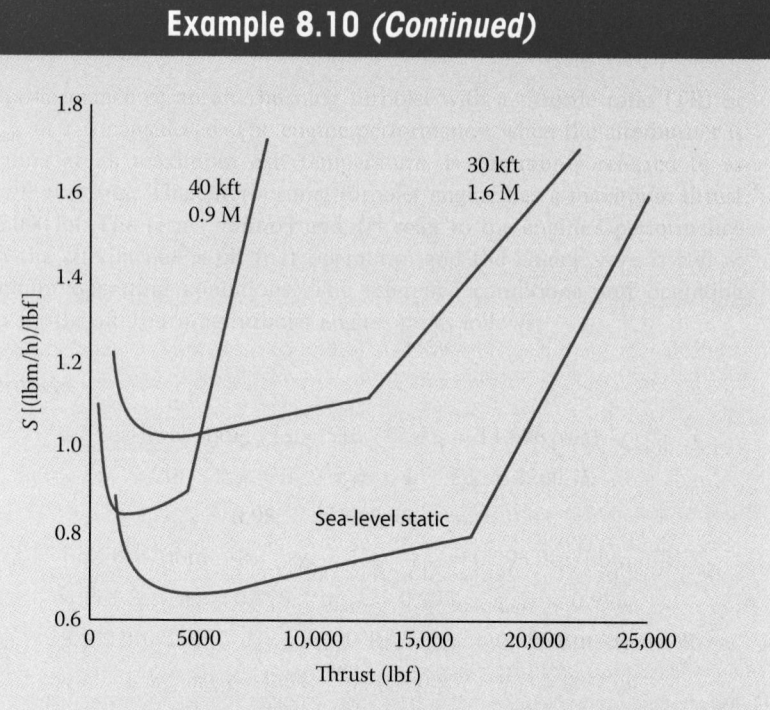

Fig. 8.51 Partial-throttle performance of mixed-flow turbofan engine.

of 50 kft. (The differences at 50 kft are due mainly to the effects of lower Reynolds number at high altitude on the operation of the fan, which are not included in our analysis.) The changes in slope of the performance curves in Figs. 1.14a–1.14b are much smoother than those of our engine model, due mainly to a more complex engine fuel control algorithm.

The partial-throttle performance of the mixed-flow turbofan engine is plotted in Fig. 8.51 at three flight conditions. By comparing Fig. 8.51 to Fig. 1.14c, one can see that the engine model developed in this section predicts the proper trends in partial-throttle performance.

8.7 Turboprop Engine

The turboprop engine used on many small commercial subsonic aircraft is shown in Fig. 8.52. This engine typically has two spools: the core engine spool and the power spool. The pressure ratios across the high-pressure turbine and power turbine are normally high enough to have choked flow in that turbine's inlet nozzle (station 4 and station 4.5, respectively) for most operating conditions of interest.

The convergent exhaust nozzle of these turboprop engines has a fixed throat area that will be choked when the exhaust total pressure/ambient

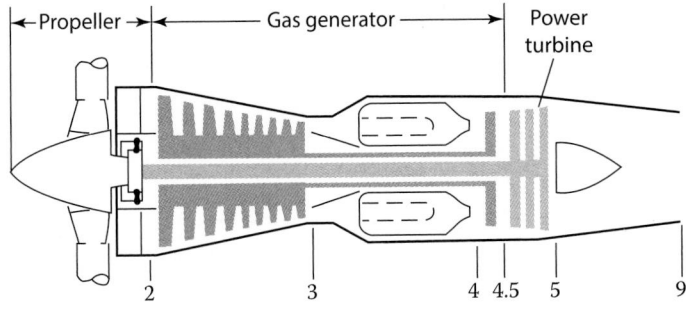

Fig. 8.52   Turboprop engine.

static pressure ratio is equal to or larger than $[(\gamma_t + 1)/2]^{\gamma_t/(\gamma_t-1)}$. When an exhaust nozzle is unchoked, the nozzle exit pressure equals the ambient pressure and the exit Mach number is subsonic.

Choked flow at stations 4 and 4.5 of the high-pressure spool during engine operation and our assumption of constant η_{tH} require

$$\text{Constant values of } \pi_{tH}, \tau_{tH}, \dot{m}_{c4}, \text{ and } \dot{m}_{c4.5} \qquad (8.36)$$

With constant temperature ratio and pressure ratio for the turbine driving the compressor, the analysis of the turboprop's core mass flow rate and compressor pressure ratio follows directly from the single-spool turbojet engine with choked exhaust nozzle.

The independent variables, constant or known values, and dependent variables for the performance analysis of this engine cycle are listed in Table 8.7. Note that for this engine, there are six dependent variables and four independent variables.

Table 8.7   Performance Analysis Variables for Turboprop Engine

Component	Variables		
	Independent	**Constant or Known**	**Dependent**
Engine	M_0, T_0, P_0		$\dot{m}_0$
Diffuser		$\pi_d = f(M_0)$	
Compressor		η_c	π_c, τ_c
Burner	T_{t4}	π_b, η_b	
High-pressure turbine		π_{tH}, τ_{tH}	
Low-pressure (power) turbine		η_{tL}	π_{tL}, τ_{tL}
Nozzle		π_n	P_9/P_0 or M_9
Propeller		$\eta_{prop} = f(M_0)$	
Total Number	4		6

8.7.1 Engine Mass Flow $\dot{m}_0$

Using the mass flow parameter and conservation of mass, we can write for the turboprop engine

$$\dot{m}_2 + \dot{m}_f = \dot{m}_2(1+f) = \dot{m}_4 = \frac{P_{t4}A_4}{\sqrt{T_{t4}}} MFP(M_4)$$

Because the flow is choked ($M_4 = 1$) and the area is constant ($A_4 = $ constant) at station 4, the above equation can be rewritten for the core mass flow rate in terms of component pressure ratios as

$$\dot{m}_0 = \frac{P_0 \pi_r \pi_d \pi_c}{\sqrt{T_{t4}}} \left[\pi_b A_4 \frac{MFP(M_4)}{1+f} \right]$$

where the terms within the square brackets are considered constant. Equating the constants to reference values gives

$$\dot{m}_0 = \dot{m}_{0R} \frac{P_0 \pi_r \pi_d \pi_c}{(P_0 \pi_r \pi_d \pi_c)_R} \sqrt{\frac{T_{t4R}}{T_{t4}}} \tag{8.68}$$

8.7.2 Compressor (τ_c, π_c)

The power balance of the core spool between the high-pressure turbine and the compressor gives

$$\eta_m \dot{m}_4 c_{pt}(T_{t4} - T_{t4.5}) = \dot{m}_2 c_{pc}(T_{t3} - T_{t2})$$

Rewriting this equation in terms of temperature ratios, rearranging into variable and constant terms, and equating the constant to reference values give

$$\frac{\tau_r(\tau_c - 1)}{T_{t4}/T_0} = \eta_m(1+f)(1 - \tau_{tH}) = \left[\frac{\tau_r(\tau_c - 1)}{T_{t4}/T_0} \right]_R$$

Solving for τ_c gives

$$\tau_c = 1 + \frac{T_{t4}/T_0}{(T_{t4}/T_0)_R} \frac{(\tau_r)_R}{\tau_r}(\tau_c - 1)_R \tag{8.69}$$

From the definition of compressor efficiency, π_c is given by

$$\pi_c = [1 + \eta_c(\tau_c - 1)]^{\gamma_c/(\gamma_c - 1)} \tag{8.70}$$

8.7.3 Core Work Coefficient C_C

The expression for the core work coefficient C_C developed in Section 7.6 still applies and is given by Eq. (7.76):

$$C_C = (\gamma_c - 1)M_0 \left[(1+f)\frac{V_9}{a_0} - M_0 + (1+f)\frac{R_t}{R_c}\frac{T_9/T_0}{V_9/a_0}\frac{1 - P_0/P_9}{\gamma_c} \right] \tag{7.76}$$

where V_9/a_0 is given by Eq. (7.4) and T_9/T_0 is given by Eqs. (7.79) and (7.80).

8.7.4 Power (Low-Pressure) Turbine (τ_{tL}, π_{tL})

Equations (8.37) and (8.38) apply for the power (low-pressure) turbine temperature and pressure ratios of this turboprop engine. We write Eq. (8.37) in terms of the Mach number at station 9 as

$$\pi_{tL} = \pi_{tL\,R}\sqrt{\frac{\tau_{tL}}{\tau_{tL\,R}}}\,\frac{\text{MFP}(M_{9R})}{\text{MFP}(M_9)} \tag{8.71}$$

where

$$\pi_{tL} = 1 - \eta_{tL}(1 - \pi_{tL}^{(\gamma_t-1)/\gamma_t}) \tag{8.38}$$

If station 9 is choked at the reference condition and at all performance conditions of interest, then π_{tL} and τ_{tL} are constant.

The power balance of this spool was found in Chapter 7. It still applies, and Eq. (7.83) gives the propeller work coefficient C_{prop}:

$$C_{\text{prop}} = \frac{\eta_{\text{prop}}\dot{W}_{\text{prop}}}{\dot{m}_0 c_{pc} T_0} = \eta_{\text{prop}}\eta_g\eta_{mL}(1+f)\tau_\lambda\tau_{TH}(1-\tau_{tL}) \tag{7.83}$$

8.7.5 Exhaust Nozzle

Two flow regimes exist for flow through the convergent exhaust nozzle (unchoked flow and choked flow). Unchoked flow will exist when

$$\frac{P_{t9}}{P_0} = \pi_r\,\pi_d\,\pi_c\,\pi_b\,\pi_{tH}\,\pi_{tL}\,\pi_n < \left(\frac{\gamma_t+1}{2}\right)^{\gamma_t/(\gamma_t-1)}$$

and choked flow will exist when

$$\frac{P_{t9}}{P_0} = \pi_r\,\pi_d\,\pi_c\,\pi_b\,\pi_{tH}\,\pi_{tL}\,\pi_n \geq \left(\frac{\gamma_t+1}{2}\right)^{\gamma_t/(\gamma_t-1)}$$

For unchoked flow, the exit static pressure P_9 is equal to the ambient pressure P_0, and the subsonic exit Mach number M_9 is given by

$$M_9 = \sqrt{\frac{2}{\gamma_t-1}\left[\left(\frac{P_{t9}}{P_0}\right)^{(\gamma_t-1)/\gamma_t}-1\right]}$$

When the flow is choked, then

$$M_9 = 1, \quad \frac{P_{t9}}{P_9} = \left(\frac{\gamma_t+1}{2}\right)^{\gamma_t/(\gamma_t-1)}, \quad \text{and} \quad \frac{P_0}{P_9} = \frac{P_{t9}/P_9}{P_{t9}/P_0}$$

8.7.6 Iteration Scheme for τ_{tL}, π_{tL}, and M_9

Determination of the conditions downstream of station 4.5 requires an iterative solution. The method used is as follows:

1. Initially assume that π_{tL} equals its reference value π_{tLR}.
2. Calculate τ_{tL}, using Eq. (8.38).
3. Calculate P_{t9}/P_0 and conditions at exit including M_9.
4. Calculate the new π_{tL}, using Eq. (8.69).
5. Compare the new π_{tL} to the previous value. If the difference is greater than 0.0001, then go to step 2.

8.7.7 Propeller Performance

The performance of the propeller can be simply modeled as a function of the flight Mach number (see Ref. 13); one such model, which is used in the PERF computer program, is shown in Fig. 8.53 and expressed algebraically as

$$
\eta_{prop} = \begin{cases} 10M_0 \eta_{prop\,max} & M_0 \leq 0.1 \\ \eta_{prop\,max} & 0.1 < M_0 \leq 0.7 \\ \left(1 - \dfrac{M_0 - 0.7}{3}\right)\eta_{prop\,max} & 0.7 < M_0 < 0.85 \end{cases}
$$

This equation for the Mach range of 0.7–0.85 models the drop in η_{prop} experienced in this flight regime due to transonic flow losses in the tip region of the propeller.

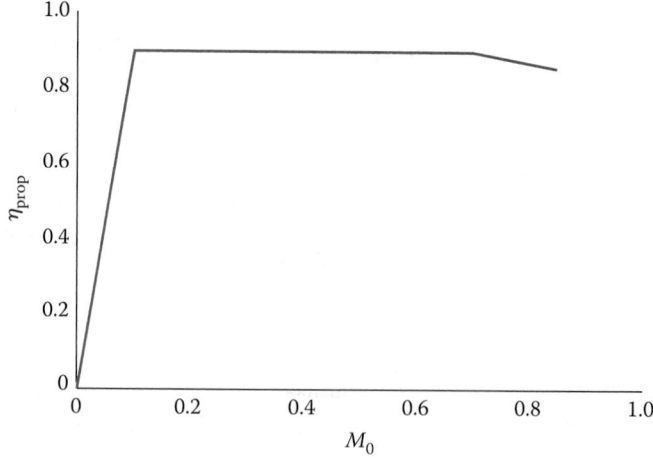

Fig. 8.53   Variation in propeller efficiency with Mach number.

See Section 8.7.SM, Summary of Performance Equations: Turboprop, in the online supporting material for related information.

Example 8.11

In this example, we consider the performance of a turboprop engine with an uninstalled thrust of 140,000 N, π_c of 30, and τ_t of 0.55 at sea level, and $M_0 = 0.1$. Mach 0.1 is selected for reference because the propeller efficiency falls off very rapidly below this Mach number. The resulting reference data are listed below. The engine has a $T_{t4\,max}$ of 1670 K and an engine throttle ratio TR $(=\theta_{0\,break})$ of 1.0.

The full-throttle performance for this turboprop engine is considered first. This is followed by a discussion of several partial-throttle performance curves. This example concludes with a discussion of gas generator performance curves.

Reference:

$$T_0 = 288.2 \text{ K}, \quad \gamma_c = 1.4, \quad c_{pc} = 1.004 \text{ kJ}/(\text{kg} \cdot \text{K}), \quad \gamma_t = 1.3,$$

$$c_{pt} = 1.235 \text{ kJ}/(\text{kg} \cdot \text{K}), \quad M_0 = 0.1, \quad T_{t4} = 1670 \text{ K}, \quad \pi_c = 30,$$

$$\eta_c = 0.8450, \quad \tau_{tH} = 0.7323, \quad \pi_{tH} = 0.2193, \quad \tau_{tL} = 0.7511,$$

$$\pi_{tL} = 0.2521, \quad \eta_{tL} = 0.9137, \quad \eta_b = 0.995, \quad \pi_{d\,max} = 0.98,$$

$$\pi_b = 0.94, \quad f = 0.02558, \quad \pi_n = 0.98, \quad \eta_m = 0.99,$$

$$\eta_{prop} = 0.812, \quad \eta_g = 0.99, \quad P_9/P_0 = 1, \quad P_0 = 101.3 \text{ kPa},$$

$$h_{PR} = 42,800 \text{ kJ/kg}, \quad \dot{m}_0 = 14.6 \text{ kg/s},$$

$$F/\dot{m}_0 = 9590 \text{ N}/(\text{kg/s}), \quad F = 140,000 \text{ N},$$

$$S = 2.666 \text{ (g/s)/kN}, \quad \text{Power} = 4764 \text{ kW}$$

The uninstalled thrust and thrust specific fuel consumption of the turboprop engine at full throttle are shown in Figs. 8.54 and 8.55, respectively. Note that both the thrust and fuel consumption curves are flat between Mach numbers of 0 and 0.1. This is due to the constant engine output power in the Mach range and the linear variation of propeller efficiency shown in Fig. 8.53. The thrust of the turboprop engine falls off rapidly with flight Mach number at low altitudes just like the high-bypass-ratio turbofan.

The partial-throttle performance of this turboprop engine is given at 6-km altitude in Fig. 8.56. Note that the minimum thrust specific fuel consumption for this engine occurs closer to maximum throttle and not at a reduced throttle setting as it did for the turbojet and turbofan engines. This characteristic occurs because the turboprop engine has a high propulsive efficiency, and the reduction in the engine's thermal efficiency at partial throttle is not offset by its increase in propulsive efficiency at partial throttle (refer back to Fig. 8.18 for an example).

(Continued)

Example 8.11 *(Continued)*

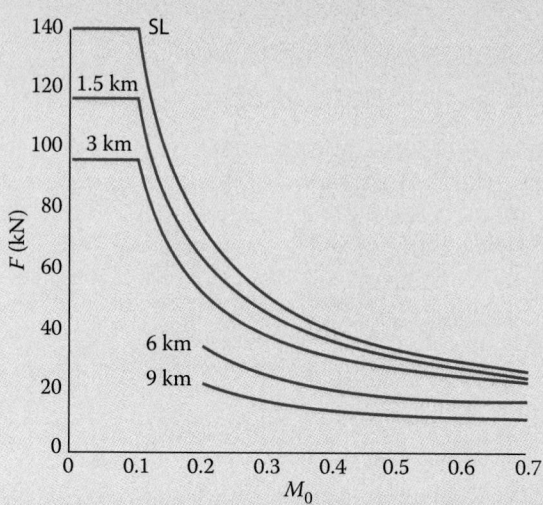

Fig. 8.54 Uninstalled thrust F at maximum power setting.

The variation in the pumping characteristics of the turboprop engine's gas generator is shown in Fig. 8.57. These are the same trends that we observed for the gas generators of the turbojet engine (Fig. 8.14) and mixed-flow turbofan engine (Fig. 8.47). Is this surprising?

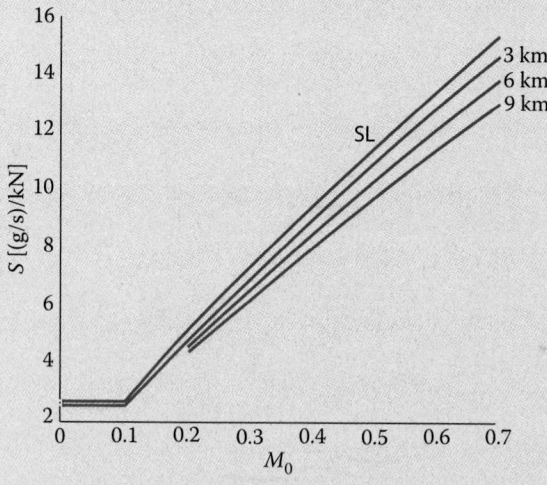

Fig. 8.55 Uninstalled thrust specific fuel consumption S at maximum power setting.

(Continued)

Example 8.11 *(Continued)*

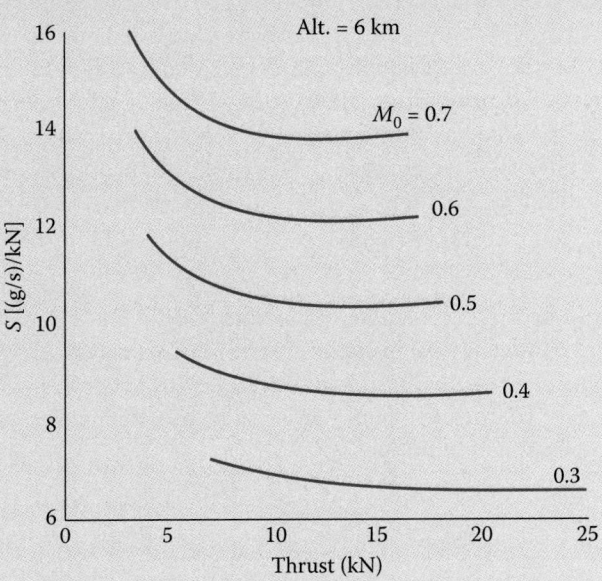

Fig. 8.56   Uninstalled partial power at 6-km altitude.

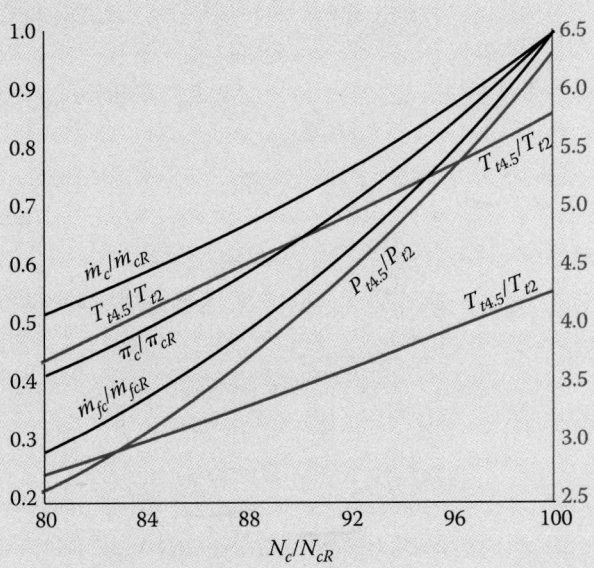

Fig. 8.57   Gas generator pumping characteristics.

8.8 Variable Gas Properties

The effect of variable gas properties can be included in the analysis of gas turbine engine performance. The fundamental thermodynamic relationships are given in Chapter 2 and the basic analysis of components is presented in Section 6.11. Section 7.7 developed parametric analysis of a turbojet engine having variable gas properties. The performance analysis of a turbojet engine having variable gas properties is presented in Section 8.8.SM of the Supporting Material for Chapter 8. This performance analysis builds on the earlier material of Chapters 2, 6, 7, and 8.

Problems

8.1 The flow in a typical single-spool turbojet engine is choked in two locations. This fact is used in performance analysis to predict the variations of the compressor and turbine with changes in T_{t4} and T_{t2}. As a result, the turbine temperature and pressure ratios are constant for all operating conditions.

   a) Identify the two locations where the flow is choked in the engine.

   b) Describe the basic engineering principle that gives the equation for the lines of constant T_{t4}/T_{t2}, sketched in Figs. P8.1 and 8.6.

   c) Sketch the operating line of a typical turbojet on the compressor map of Fig. P8.1.

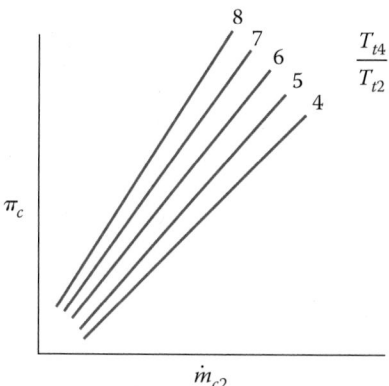

Fig. P8.1 Compressor map.

8.2 If the compressor of a single-spool turbojet has a maximum compressor ratio of 8 with a corrected mass flow rate of 25 kg/s for $T_{t4} = 1800$ K and $\theta_0 = 1$, determine the following (assume $\gamma_c = 1.4$):

   a) Constants K_1 and K_2 of Eqs. (8.17) and (8.18)

b) Compressor pressure ratio and corrected mass flow rate for $T_{t4} = 1200$ K and $\theta_0 = 1.1$

c) The maximum T_{t4} and corresponding compressor pressure ratio and corrected mass flow rate at $\theta_0 = 0.9$ and 1.1 for a throttle ratio of 1.0

d) The maximum T_{t4} and corresponding compressor pressure ratio and corrected mass flow rate at $\theta_0 = 0.9$ and 1.1 for a throttle ratio of 1.05

8.3   The typical operation of the single-spool turbojet engine at maximum throttle for flight conditions where $\theta_0 < \mathrm{TR}$ is such that $T_{t4}/T_{t2} =$ constant. Show that π_c, T_{t5}/T_{t2}, P_{t5}/P_{t2}, and $\dot{m}_{c2}$ are also constant.

8.4   Describe the advantage of a turbojet engine having a throttle ratio greater than 1.

8.5   The gas generator of Example 8.2 has air entering at $-20°$F at sea-level static conditions and the compressor pressure ratio is 12. Use Fig. 8.13 and determine the following in English units: air mass flow rate, T_{t4}, N_c/N_{cR}, fuel mass flow rate, P_{t6}, and T_{t6}.

8.6   The gas generator of Example 8.2 has air entering at $-10°$C at sea-level static conditions and the compressor pressure ratio is 10. Use Fig. 8.13 and determine the following in SI units: air mass flow rate, T_{t4}, N_c/N_{cR}, fuel mass flow rate, P_{t6}, and T_{t6}.

8.7   Perform Problem 8.5 for air entering at $-60°$F at 40 kft/M_0 of 0.8 for $\pi_c = 12$ and 15.

8.8   Perform Problem 8.6 for air entering at $-50°$C at 12 km/M_0 of 0.8 for $\pi_c = 12$ and 15.

8.9   Show that the total temperature ratios for the compressors of the two-spool turbojet engine [Eqs. (8.42) and (8.40)] can be written as

$$\tau_{cL} = 1 + \frac{T_{t4}/\theta_0}{(T_{t4}/\theta_0)_R}(\tau_{cL} - 1)_R$$

$$\tau_{cH} = 1 + \frac{T_{t4}/\theta_0}{(T_{t4}/\theta_0)_R}\frac{\tau_{cLR}}{\tau_{cL}}(\tau_{cH} - 1)_R$$

8.10 In terms of reference conditions (subscript R), show that the following equations apply for the operation of an *ideal* single-spool turbojet engine:

a) The compressor pressure ratio is given by

$$\pi_c = \left\{ 1 + [(\pi_{cR})^{(\gamma-1)/\gamma} - 1]\frac{T_{t4}/T_{t2}}{(T_{t4}/T_{t2})_R} \right\}^{\gamma/(\gamma-1)}$$

b) The engine mass flow rate is given by

$$\dot{m}_0 = \dot{m}_{0R}\frac{P_{t2}}{P_{t2R}}\frac{\pi_c}{\pi_{cR}}\sqrt{\frac{T_{t4R}}{T_{t4}}}$$

c) The corrected mass flow rate at station 2 is given by

$$\dot{m}_{c2} = \dot{m}_{c2R}\frac{\pi_c}{\pi_{cR}}\sqrt{\frac{(T_{t4}/T_{t2})_R}{(T_{t4}/T_{t2})}}$$

d) Given $(T_{t4}/T_{t2})_R = 4$, $\pi_{cR} = 12$, $\dot{m}_{cR} = 200$ lbm/s, and $\gamma = 1.4$, calculate the engine operating line by completing the following table of data; plot the results.

T_{t4}/T_{t2}	4.0	3.6	3.2	2.8	2.4	2.0	1.6
π_c	12						
$\dot{m}_c$ (lbm/s)	200						

8.11 For a ramjet engine with a constant nozzle throat area A_8, develop the performance equations and calculate the performance of the ramjet with the reference data of Problem 7.1.

a) Show that the engine mass flow rate can be written as

$$\dot{m}_0 = \dot{m}_{0R}\frac{P_0 \pi_r \pi_d}{(P_0 \pi_r \pi_d)R}\sqrt{\frac{T_{t4R}}{T_{t4}}}\frac{MFP(M_8)}{MFP(M_{8R})}$$

where the Mach number at station 8 is determined by the value of P_{t8}/P_0.

b) Show that the remainder of the performance equations are given by those developed to answer Problem 7.1.

c) If the ramjet of Problem 7.1 has a reference mass flow rate of 20 kg/s at 15 km and $M_0 = 2$, determine its performance at 20 km, $M_0 = 3$, and $T_{t4} = 1600$ K. Assume $P_9 = P_0$.

8.12 A turbojet engine operated at a flight Mach number of 2.0 and an altitude of 15 km has the following compressor performance with $T_{t4} = 1860$ K: $\pi_c = 8$, $\tau_c = 1.9$, and $\eta_c = 0.9$. Determine τ_c, π_c, and $\dot{m}_2/\dot{m}_{2R}$ for $M_0 = 0.8$, 11-km altitude, and $T_{t4} = 1640$ K. Assume that $\pi_d = 0.88$ at $M_0 = 2$, $\pi_d = 0.98$ at $M_0 = 0.8$, and $\gamma_c = 1.4$.

8.13 Calculate the thrust of the turbojet engine in Example 7.1 at $M_0 = 0.8$, 9-km altitude, reduced throttle ($T_{t4} = 1100$ K), and $P_0/P_9 = 1$ for a reference air mass flow rate of 50 kg/s at a reference altitude of 12 km.

8.14 Calculate the thrust of the turbojet engine in Example 7.1 at $M_0 = 0.8$, 30-kft altitude, reduced throttle ($T_{t4} = 2100°$R), and $P_0/P_9 = 1$ for a reference air mass flow rate of 100 lbm/s at a reference altitude of 40 kft.

8.15 Given a single-spool turbojet engine that has the following reference values:

Reference:

$$\dot{m}_{c2} = 100 \text{ lbm/s}, \quad \pi_c = 6, \quad M_0 = 0, \quad T_{t2} = 518.7°\text{R},$$
$$T_{t4} = 3200°\text{R}, \quad P_0 = 1 \text{ atm}, \quad \pi_d = 0.97, \quad \pi_b = 0.96,$$
$$\pi_n = 0.98, \quad \eta_b = 0.995, \quad \eta_m = 0.99, \quad P_0/P_9 = 1,$$
$$\eta_c = 0.8725, \quad \pi_t = 0.6098, \quad \tau_t = 0.9024, \quad \eta_t = 0.9051,$$
$$f = 0.041895, \quad V_9/a_0 = 2.8876, \quad P_{t9}/P_9 = 3.3390, \quad M_9 = 1.4624,$$
$$F/\dot{m}_0 = 104.41 \text{ lbf (lbm/s)}, \quad \gamma_c = 1.4, \quad c_{pc} = 0.24 \text{ Btu/(lbm} \cdot °\text{R)},$$
$$\gamma_t = 1.3, \quad c_{pt} = 0.296 \text{ Btu/(lbm} \cdot °\text{R)}, \quad h_{PR} = 18{,}400 \text{ Btu/lbm}$$

a) Determine $\dot{m}_0$, P_{t5}/P_{t2}, T_{t5}/T_{t2}, V_9, F, and S at the reference condition.

b) If this engine has a throttle ratio of 1.0 and is operating at maximum T_{t4} at a flight condition where $\theta_0 = 1.2$, determine π_c, $\dot{m}_{c2}$, P_{t5}/P_{t2}, T_{t5}/T_{t2}, and N_c/N_{cR} at this operating point.

c) If this engine has a throttle ratio of 1.1 and is operating at maximum T_{t4} at a flight condition where $\theta_0 = 1.2$, determine π_c, $\dot{m}_{c2}$, P_{t5}/P_{t2}, T_{t5}/T_{t2}, and N_c/N_{cR} at this operating point.

d) Determine the percentage change in performance parameters between parts b and c.

8.16   Given a single-spool turbojet engine that has the following reference values:

Reference:

$$\dot{m}_{c2} = 50 \text{ kg/s}, \quad \pi_c = 8, \quad M_0 = 0, \quad T_{t2} = 288.2 \text{ K},$$

$$T_{t4} = 1780 \text{ K}, \quad P_0 = 1 \text{ atm}, \quad \pi_d = 0.97, \quad \pi_b = 0.96,$$

$$\pi_n = 0.98, \quad \eta_b = 0.995, \quad \eta_m = 0.99, \quad P_0/P_9 = 1,$$

$$\eta_c = 0.8678, \quad \pi_t = 0.5434, \quad \tau_t = 0.8810, \quad \eta_t = 0.9062,$$

$$f = 0.040796, \quad V_9/a_0 = 3.0256, \quad P_{t9}/P_9 = 3.9674, \quad M_9 = 1.5799,$$

$$F/\dot{m}_0 = 1071 \text{ N/(kg/s)}, \quad \gamma_c = 1.4, \quad c_{pc} = 1.004 \text{ kJ/(kg} \cdot \text{K)},$$

$$\gamma_t = 1.3, \quad c_{pt} = 1.24 \text{ kJ/(kg} \cdot \text{K)}, \quad h_{PR} = 42{,}800 \text{ kJ/kg}$$

a) Determine $\dot{m}_0$, P_{t5}/P_{t2}, T_{t5}/T_{t2}, V_9, F, and S at the reference condition.

b) If this engine has a throttle ratio of 1.0 and is operating at maximum T_{t4} at a flight condition where $\theta_0 = 1.4$, determine π_c, $\dot{m}_{c2}$, P_{t5}/P_{t2}, T_{t5}/T_{t2}, and N_c/N_{cR} at this operating point.

c) If this engine has a throttle ratio of 1.15 and is operating at maximum T_{t4}, at a flight condition where $\theta_0 = 1.4$, determine π_c, $\dot{m}_{c2}$, P_{t5}/P_{t2}, T_{t5}/T_{t2}, and N_c/N_{cR} at this operating point.

d) Determine the percentage change in performance parameters between parts b and c.

8.17   Calculate the thrust of the afterburning turbojet engine in Example 8.7 at $M_0 = 2.0$, 40-kft altitude, maximum afterburner ($T_{t7} = 3600°R$), and $P_0/P_9 = 1$ for throttle ratios of 1.0, 1.05, 1.1, 1.15, and 1.2. Comment on the variation in performance with throttle ratio.

8.18   Perform Problem 8.17 using the PERF computer program.

8.19   Calculate the thrust and fuel consumption of the afterburning turbojet engine in Example 8.7 at $M_0 = 0.8$, 40-kft altitude, $P_0/P_9 = 1$, a throttle ratio of unity, and a range of T_{t7} from 3600°R down to T_{t5}. Compare your results to those of the PERF computer program.

8.20   Perform Problem 8.19 using the PERF computer program.

8.21   Using the PERF computer program, find the performance of the afterburning turbojet engine in Example 8.7 for throttle ratios of 1.1 and 1.2 over the same range of Mach numbers and altitudes. Compare these results to those of Example 8.7.

8.22 Calculate the thrust of the high-bypass-ratio turbofan engine in Example 8.8 at $M_0 = 0.8$, 40-kft altitude, and partial throttle ($T_{t4} = 2600°R$). Assume convergent-only exhaust nozzles. Compare your results to those of the PERF computer program.

8.23 Perform Problem 8.22 using the PERF computer program.

8.24 Calculate the thrust of the high-bypass-ratio turbofan engine in Example 8.8 at $M_0 = 0$, sea-level altitude, and increased throttle ($T_{t4} = 3500°R$). Assume convergent-only exhaust nozzles. Compare your results to those of the PERF computer program.

8.25 Perform Problem 8.24 using the PERF computer program.

8.26 Using the PERF computer program, find the performance of the high-bypass-ratio turbofan engine in Example 8.8 for a throttle ratio of 1.0 over the range of Mach numbers from 0 to 0.8 and altitudes of sea level, 20 kft, and 40 kft. Use $3200°R$ for maximum T_{t4}.

8.27 Perform Problem 8.26 for a throttle ratio of 1.05.

8.28 Early jet aircraft for passenger service used turbojet engines (e.g., Boeing 707) and the newer aircraft use high-bypass-ratio turbofan engines (e.g., Boeing 777). The early turbojets required a much longer takeoff distance than the newer turbofan-powered aircraft. This difference is due mainly to the variation in thrust of these different engine types with Mach number and altitude. To get a better understanding, use the PERF computer program to design two engines and determine their variations in thrust with Mach number and altitude. Consider a turbojet with the component performance of technology level 3 in Table 6.2 (assume type A diffuser, uncooled turbine, and type D nozzle), $T_{t4} = 2500°R$, compressor pressure ratio of 12, and sea-level static thrust of 10,000 lbf. Determine the turbojet's performance for TR = 1 at maximum T_{t4}, Mach 0.8, and 30-kft altitude. Now consider a high-bypass-ratio turbofan with the component performance of technology level 3 in Table 6.2 (assume type A diffuser, uncooled turbine, and type D nozzle), $T_{t4} = 3000°R$, compressor pressure ratio of 22, fan pressure ratio of 1.54, bypass ratio of 5, and sea-level static thrust of 56,000 lbf. Determine the turbofan's performance for TR = 1 at maximum T_{t4}, Mach 0.8, and 30-kft altitude, and compare to the turbojet.

8.29 Perform Problem 8.28 for a high-bypass-ratio turbofan with TR = 1.05.

8.30 Develop a set of performance equations for the stationary gas turbine engine with regeneration (see Problem 7.27) depicted in Fig. P8.2 based on the following assumptions:

a) Flow is choked at engine stations 4 and 4.5.

b) Constant component efficiencies (η_c, η_b, η_c, η_{rg}, etc.).

c) Exit pressure equals the ambient pressure ($P_9 = P_0$).

d) Constant specific heat c_{pc} and ratio of specific heats γ_c from stations 0 to 3.5.

e) Constant specific heat c_{pt} and ratio of specific heats γ_t from stations 4 to 9.

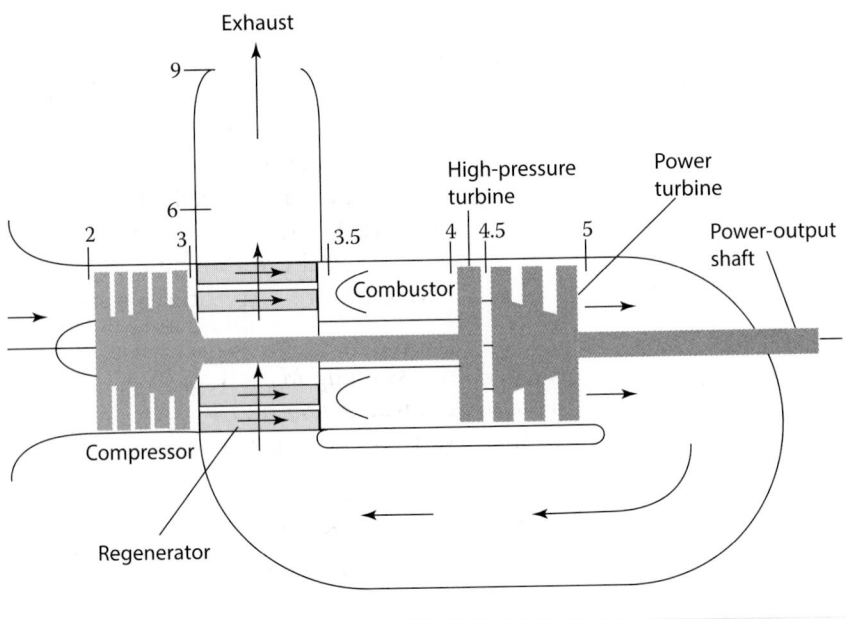

Fig. P8.2 Gas turbine with regeneration.

8.31 Calculate the corrected fuel flow rate [see Eq. (8.7)] and corrected thrust specific fuel consumption [see Eq. (8.6)] for the turbojet engine of Example 8.4 at an altitude of 40 kft and Mach numbers of 0.6, 0.8, 1.0, and 1.2 with maximum T_{t4}. Comment on your results.

8.32 For a single-spool turbojet engine, show that the maximum corrected fuel flow rate is essentially constant for $\theta_0 \leq$ TR. See Eq. (8.28) for a starting point.

8.33 Consider the dual-spool afterburning turbojet engine modeled in Section 8.4. In this development, the area of station 4.5 was modeled as fixed. If this area changes from its reference value, the temperature and pressure ratios of the high- and low-pressure turbines vary from their reference values. As a result, changes occur in the operating points for both the low- and high-pressure compressors. To better understand this, you are asked to

a) Redevelop the turbine and compressor relationships influenced by variation in the area at station 4.5. Consider all other assumptions remain valid.

b) For the reference conditions of Example 8.7, determine the changes in the temperature and pressure ratios of the turbomachinery (compressors and turbines) for a 5% increase in the flow area at station 4.5.

8.34   Variable area turbojet (VAT) engine: The partial throttle (reduced T_{t4}) performance of the typical turbojet engine (see Fig. 8.8) results in reduced compressor pressure ratio π_c and compressor corrected mass flow rate $\dot{m}_{c2}$. The thermal efficiency of the engine is reduced by the lower compressor pressure ratio. Reduced compressor corrected mass flow rate results in increased installation loss from inlet spillage drag (see Chapter 11). We desire to keep compressor mass flow rate $\dot{m}_{c2}$ and compressor total pressure ratio π_c constant at reduced throttle T_{t4} by varying the areas at stations 4 and 8 in a single-spool turbojet engine.

a) Starting from Eq. (8.9) and assuming $M_4 = 1$ and π_b and $(1 + f)$ are constants, show that constant $\dot{m}_{c2}$ and π_c require that the area at station 4 (A_4) be varied in the following way:

$$\frac{A_4}{A_{4R}} = \sqrt{\frac{T_{t4}/T_{t2}}{(T_{t4}/T_{t2})_R}} \qquad \text{(VAT.1)}$$

b) Starting with the work balance between the turbine and compressor of Eq (8.13), show that for τ_c constant the turbine total temperature will vary as

$$\tau_t = 1 - (1 - \tau_{tR})\frac{(T_{t4}/T_{t2})_R}{T_{t4}/T_{t2}} \qquad \text{(VAT.2)}$$

c) Using conservation of mass between stations 4 and 8 (with $M_4=1$ and $M_8 = 1$), show that

$$\frac{\sqrt{\tau_t}}{\pi_t} = \frac{A_8}{A_4}$$

and the area at station 8 must be varied in the following way:

$$\frac{A_8}{A_{8R}} = \frac{A_8}{A_{4R}}\sqrt{\frac{\tau_t \; \pi_{tR}}{\tau_{tR} \; \pi_t}} = \sqrt{\frac{T_{t4}/T_{t2}}{(T_{t4}/T_{t2})_R}}\sqrt{\frac{\tau_t \; \pi_{tR}}{\tau_{tR} \; \pi_t}} \qquad \text{(VAT.3)}$$

where, assuming constant turbine efficiency η_t, the turbine total pressure ratio is given by Eq. (8.12b).

d) Because the corrected mass flow rate into the compressor is constant, show that the engine mass flow rate can be written as

$$\frac{\dot{m}_0}{\dot{m}_{0R}} = \frac{P_0 \pi_r \pi_d}{(P_0 \pi_r \pi_d)_R}\sqrt{\frac{(T_0 \tau_r)_R}{T_0 \tau_r}} \qquad \text{(VAT.4)}$$

8.35   Calculate the maximum thrust of the afterburning mixed-flow exhaust turbofan engine in Example 8.10 at $M_0 = 2$ and 40-kft altitude for throttle ratios of 1.1 and 1.2 using the PERF computer program.

8.36   Calculate the maximum thrust of the afterburning mixed-flow exhaust turbofan engine in Example 8.10 at $M_0 = 1.6$ and 40-kft altitude for throttle ratios of 1.0 and 1.1 using the PERF computer program.

8.37   Using the PERF computer program, find the performance of the afterburning mixed-flow exhaust turbofan engine in Example 8.10 for throttle ratios of 1.0 and 1.2 over the range of Mach numbers from 0 to 2 and altitude of sea level, 10 kft, 20 kft, 30 kft, 36 kft, 40 kft, and 50 kft. Compare your results to those of Example 8.10 for a throttle ratio of 1.1.

8.38   Calculate the thrust and fuel consumption of the turboprop engine in Example 8.11 at $M_0 = 0.5$, a 6-km altitude, and reduced throttle ($T_{t4} = 1400$ K). Assume a convergent-only exhaust nozzle. Compare your results to those of the PERF computer program.

8.39   Using the PERF computer program, find the performance of the turboprop engine in Example 8.11 at partial throttle at $M_0 = 0.2$ and altitudes of sea level, 3 km, and 6 km. Compare these results to those of Example 8.11 (see Fig. 8.56).

Gas Turbine Design Problems

8.D1   Find the best high-bypass-ratio turbofan engine for the HP-1 aircraft from those engines showing promise in Design Problem 7.D1. You are to determine the best engine, sized to meet the required engine thrust at takeoff and/or single engine out (0.45 Mach at 16 kft), and whose fuel consumption is minimum.

Hand-Calculate Engine Performance (Optional). Using the performance analysis equations for a high-bypass-ratio turbofan engine, hand-calculate the performance of the turbofan engine hand-calculated in Design Problem 7.D1 at the flight condition of 0.83 Mach and 11-km altitude at $T_{t4} = 1500$ K.

Required Thrust at Different Flight Conditions (HP-1 Aircraft). Determine both the required installed thrust and the required uninstalled engine thrust for each of the following flight conditions (assume $\phi_{noz} = 0.01$):

1. Uninstalled thrust of 267 kN for each engine at sea-level static while at takeoff power setting ($T_{t4} = 1890$ K). Assume $\phi_{inlet} = 0.05$.

2. Start of Mach 0.83 cruise, 11-km altitude, $P_s = 1.5$ m/s, 95.95% of takeoff weight; $\phi_{inlet} = 0.01$.

3. Mach 0.45, 5-km altitude, engine out, $P_s = 1.5$ m/s, 88% of takeoff weight; $\phi_{inlet} = 0.02$. Assume a drag coefficient increment for engine out $= 0.0035$ (based on wing area).

4. Loiter at 9-km altitude, 64% of takeoff weight; $\phi_{inlet} = 0.03$.

Computer-Calculate Engine Performance (HP-1 Aircraft). For each of the engines showing promise in Design Problem 7.D1, systematically perform the following analysis:

1. Design the reference engine at sea-level static conditions. Size the engine to provide the required uninstalled thrust. (Engine size normally will be determined by either the takeoff flight condition or the engine-out flight condition listed previously.) Check engine operation at takeoff and make sure that $T_{t3} < 920$ K.

2. Determine the uninstalled fuel consumption at the start of Mach 0.83 cruise, 11-km altitude, and $P_s = 0$. You can do this by input of "% thrust" in the PERF computer program and calculate for the required uninstalled thrust. Assuming cruise climb with constant range factor (RF), calculate the weight at the end of the Mach 0.83 cruise, using the Breguet range equation, Eq. (1.45a or b).

3. Determine the loiter Mach number at 9-km altitude for the current aircraft weight at the start of loiter. Find the engine fuel consumption at start of loiter. Assuming a constant endurance factor (EF), calculate the weight at the end of the 9-km loiter. Some engines may not throttle down to the required uninstalled thrust due to the model used in the computer program, and the

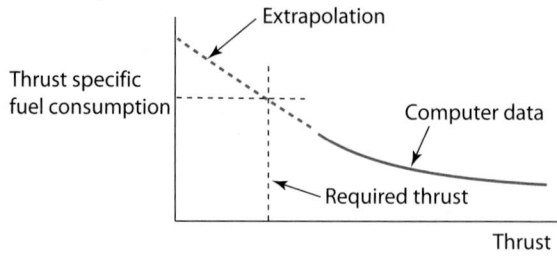

Fig. P8.D1a   Extrapolation method at partial throttle.

uninstalled thrust specific fuel consumption will have to be obtained by *extrapolation*. To extrapolate, use the performance analysis computer program and iterate on T_{t4} in steps of 25 K to the lowest value giving results. Then plot S vs F as shown in Fig. P8.D1a, and draw a tangent line to obtain the extrapolated value of S at the desired uninstalled thrust.

Fuel Consumption (HP-1 Aircraft). During this preliminary analysis, you can assume that the fuel consumption changes only for the Mach 0.83 cruise and 9-km loiter. For every one of the engines you investigate, you must determine whether it can satisfy the fuel consumption requirements by calculating the amount of fuel consumed during the Mach 0.83 cruise climb and 9-km loiter and adding these to that consumed for other parts of the flight. During your analysis of the engines, make a plot of fuel consumed vs the reference bypass ratio like that shown in Fig. P8.D1b. Starting with one compressor pressure ratio π_{c1} and a low-bypass-ratio engine, calculate

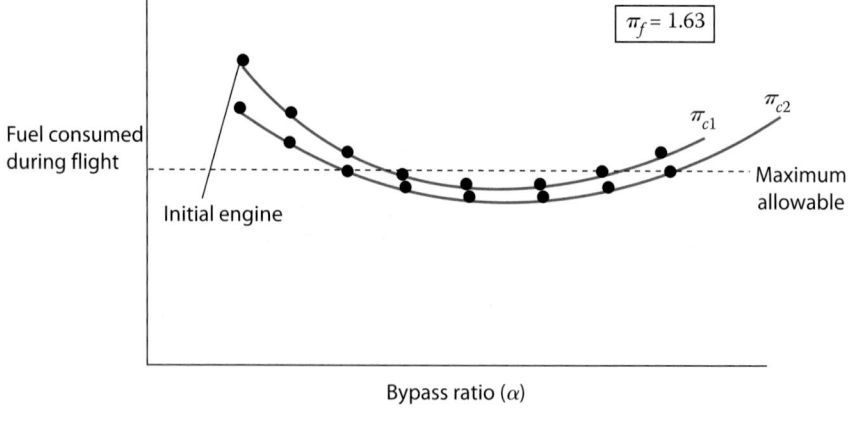

Fig. P8.D1b   Plot for finding the best engine.

the total fuel consumed for the flight. Now increase the bypass ratio, size the engine, and determine this engine's performance.

Engine Selection (HP-1 Aircraft). Select one of your engines that, according to your criteria, best satisfies the mission requirements. Your criteria *must include* at least the following items (other items may be added based on knowledge gained in other courses and any additional technical sources):

> Thrust requirement
> Fuel consumption
> Aircraft performance
> Operating cost (assume 10,000-h engine life and fuel cost of $1.00/lb)
> First cost
> Size and weight
> Complexity

Determine Engine Thrust vs Mach Number and Altitude (HP-1 Aircraft). For the engine you select, determine and plot the uninstalled thrust F at maximum power vs Mach number at sea level, 5-km altitude, and 11-km altitude and at takeoff power at sea level (see Fig. 8.43). Use $T_{t4} = 1780$ K for maximum power and $T_{t4} = 1890$ K for takeoff power.

Summary. Summarize the final choice for the selected engine including a list of design conditions and choices, performance during the mission, and overall mission performance. Include suggestions, if necessary, for overcoming any of the performance shortcomings that may exist in any of the mission legs. In addition, make meaningful comments about the feasibility of building such an engine.

8.D2   Find the best mixed-flow turbofan engine with afterburner for the HF-1 aircraft from those engines showing promise in Design Problem 7.D2. You are to determine the best engine, sized to meet the required engine thrust at takeoff, supercruise, and/or 5-*g* turns, and whose cruise fuel consumption over the maximum mission (see Design Problem 1.D2) is minimum.

Hand-Calculate Engine Performance (Optional). Using the performance analysis equations for a mixed-flow turbofan engine with afterburner, hand-calculate the performance of the mixed-flow turbofan engine hand-calculated in Design Problem 7.D2 at the flight condition of 1.6 Mach and 40-kft altitude at $T_{t4} = 3200°$R with the afterburner turned off.

Required Thrust at Different Flight Conditions (HF-1 Aircraft). Determine both the required installed thrust and the required uninstalled engine thrust for each of the following flight conditions (assume $\phi_{noz} = 0.01$):

1. For takeoff, an installed thrust of 23,500 lbf for each engine at sea level, 0.1 Mach while at maximum power setting (afterburner on with $T_{t7} = 3600°R$). Assume $\phi_{inlet} = 0.10$.

2. Start of Mach 1.6 supercruise, 40-kft altitude, 92% of takeoff weight W_{TO}; $\phi_{inlet} = 0.04$.

3. Start of 5-g turn at Mach 1.6, 30-kft altitude, 88% of W_{TO}; $\phi_{inlet} = 0.04$.

4. Start of 5-g turn at Mach 0.9, 30-kft altitude, 88% of W_{TO}; $\phi_{inlet} = 0.04$.

Computer-Calculate Engine Performance (HF-1 Aircraft). Develop a reference mixed-flow turbofan engine with afterburner based on the data used in Design Problem 7.D2. For the afterburner, use $\gamma_{AB} = 1.3$, $c_{pAB} = 0.296$ Btu/(lbm · °R), $\tau_{AB} = 0.96$, $\eta_{AB} = 0.97$, and $T_{t7} = 3600°R$. For each of the engines showing promise in Design Problem 7.D2, systematically perform the following analysis:

1. Design the reference engine at sea-level static conditions. Size the engine to provide the required uninstalled thrust. (Engine size will normally be determined by the takeoff flight condition, by the start of supercruise flight condition, or by the 5-g turn at Mach 0.9 listed previously.) Check engine operation at takeoff, and make sure that $T_{t3} < 1600°R$ and that the compressor pressure is within a specified limit.

2. Determine uninstalled fuel consumption at the start of the Mach 1.6 supercruise, 40-kft altitude. You can do this by input of "% thrust" in the PERF computer program and calculate for the required uninstalled thrust. Your engine should be able to deliver the required thrust with the afterburner off. Assuming cruise climb with constant range factor (RF), calculate the weight at the end of the Mach 1.6 supercruise, using the Breguet range equation, Eq. (1.45a or b).

3. Calculate the aircraft weight at the start of the Mach 0.9 cruise and the corresponding best cruise altitude (maximum C_L/C_D) and aircraft drag. Calculate the uninstalled required thrust at the start of Mach 0.9 cruise, assuming $\phi_{inlet} = 0.05$ and $\phi_{noz} = 0.01$. Now determine the uninstalled fuel consumption at the start of Mach 0.9 cruise. Assuming cruise climb with constant range factor (RF), calculate the weight at the end of the Mach 0.9 cruise, using the Breguet range equation, Eq. (1.45a or b).

4. Determine the loiter Mach number at 30-kft altitude for the current aircraft weight at the start of loiter. Find the engine fuel consumption at the start of loiter. Assuming constant endurance factor (EF), calculate the weight at the end of the 30-kft loiter. Some engines may not throttle down to the required uninstalled thrust due to the model used in the computer program, and the uninstalled thrust specific fuel consumption will have to be obtained by *extrapolation*. To extrapolate, use the performance analysis computer program and iterate on T_{t4} in steps of 50°R to the lowest value giving results. Then plot S vs F as shown in Fig. P8.D1a, and draw a tangent line to obtain the extrapolated value of S at the desired uninstalled thrust.

Fuel Consumption (HF-1 Aircraft). During this preliminary analysis, you can assume that the fuel consumption changes only for the Mach 1.6 supercruise, Mach 0.9 cruise, and 30-kft loiter. For every one of the engines you investigate, you must determine whether it can satisfy the fuel consumption requirements by calculating the amount of fuel consumed during the Mach 1.6 supercruise, Mach 0.9 cruise, and 30-kft loiter and adding these values to that consumed for other parts of the maximum mission. During your analysis of the engines, make a plot of fuel consumed vs the reference bypass ratio like that shown in Fig. P8.D1b. Starting with one compressor pressure ratio π_{c1} and a low-bypass-ratio engine, calculate the total fuel consumed for the mission. Now increase the bypass ratio, size the engine, and determine this engine's performance.

Engine Selection (HF-1 Aircraft). Select one of your engines that, according to your criteria, best satisfies the mission requirements. Your criteria *must include* at least the following items (other items may be added based on knowledge gained in other courses and any additional technical sources):

Thrust required
Fuel consumption
Aircraft performance
Operating cost (assume 10,000-h engine life and fuel cost of $1.00/lb)
First cost
Size and weight
Complexity

Determine Engine Thrust vs Mach Number and Altitude (HF-1 Aircraft). For the engine you select, determine and plot the uninstalled thrust F and thrust specific fuel consumption at both maximum power (afterburner on) and military power (afterburner off) vs

Mach number at altitudes of sea level, 10 kft, 20 kft, 30 kft, 36 kft, and 40 kft (see Figs. 8.49 and 8.50). Use $T_{t4\,max} = 3250°R$ and the throttle ratio TR for your engine. Also determine and plot the partial-throttle performance (see Fig. 8.51) of your engine at sea-level static, 1.6 Mach and 40 kft, 1.6 Mach and 30 kft, and 0.9 Mach and 30 kft.

Summary. Summarize the final choice for the selected engine including a list of design conditions and choices, performance during the mission, and overall mission performance. Include suggestions, if necessary, for overcoming any of the performance shortcomings that may exist in any of the mission legs. In addition, make meaningful comments about the feasibility of building such an engine.

Chapter 9 Turbomachinery

> The early history illustrates a point connected with the origin of the gas turbine engine that is still valid today: it takes longer to design and develop and bring into production a jet engine than it takes to produce a new airplane system. And it is primarily the advances made in engine work that make it possible to come up with an aircraft to do new and different jobs.
>
> *Gerhard Neumann*

9.1 Introduction

In general, turbomachinery is classified as all those devices in which energy is transferred either to or from a continuously flowing fluid by the dynamic action of one or more moving blade rows [36]. The word *turbo* or *turbinis* is of Latin origin and implies that which spins or whirls around. Essentially, a rotating blade row or rotor changes the total enthalpy of the fluid moving through it by either doing work on the fluid or having work done on it by the fluid, depending on the effect required of the machine. These enthalpy changes are intimately linked with the pressure changes occurring simultaneously in the fluid.

This definition of turbomachinery embraces both open and enclosed turbomachines [36]. Open turbomachinery (such as propellers, wind turbines, and unshrouded fans) influence an often indeterminate quantity of fluid. In enclosed turbomachinery (such as centrifugal compressors, axial-flow turbines, etc.), a finite quantity of fluid passes through a casing in unit time. In this chapter, we will focus on the enclosed turbomachinery used in gas turbine engines: fans, compressors, and turbines. There are many excellent references on enclosed turbomachinery, such as Refs. 3, 7, 21, 22, and 35–46. Open turbomachines are covered in aeronautics textbooks such as Refs. 47, 48, and 49. Fundamentals of propellers are included in Appendix M online. Because the thrust of the turboprop engine is mostly from the propeller, this material helps the reader better understand this engine's thrust.

Supporting material for this chapter is available electronically. See the Supporting Materials page at the back of the book for instructions to download.

Turbomachines are further categorized according to the nature of the flow path through the passages of the rotor. When the path of the through-flow is wholly or mainly parallel to the axis of rotation, the device is termed an axial-flow turbomachine. When the path of the throughflow is wholly or mainly in a plane perpendicular to the rotation axis, the device is termed a radial-flow turbomachine. When the direction of the throughflow at the rotor outlet has both radial and axial velocity components present in significant amounts, the device is termed a mixed-flow turbomachine. As previously discussed in Chapter 4, the majority of turbomachines used for aircraft applications are the axial-flow type, with some use of axial-centrifugal compressors most notably in turboshaft (helicopter) applications. The turbines for aircraft applications are all axial-flow type. Consequently, this chapter will focus on axial compressors and turbines, with some discussion of centrifugal compressors. Information on centrifugal turbines is provided in this chapter's Supporting Material.

9.2 Euler's Turbomachinery Equations

In turbomachinery, power is added to or removed from the fluid by the rotating components. These rotating components exert forces on the fluid that change both the energy and the tangential momentum of the fluid. In this section, we will develop Euler's equations for turbomachinery that relate the change in energy to the change in tangential momentum.

Consider the adiabatic flow of a fluid as shown in Fig. 9.1. In general, the flow will have axial, radial, and tangential velocity components. In Fig. 9.1, only the tangential velocity is shown. The fluid in a stream tube enters a control volume at radius r_i with tangential velocity v_i and exits at r_e with tangential velocity v_e. The tangential velocity components in turbomachines are often called *swirl*. For a compressor or pump with steady

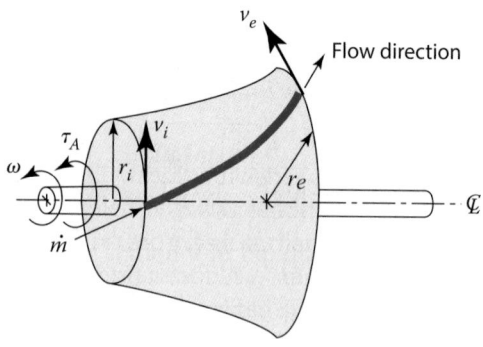

Fig. 9.1 Control volume for a general turbomachine.

flow, the applied torque τ_A is equal to the change in angular momentum of the fluid, or

$$\tau_A = \frac{\dot{m}}{g_c}(r_e v_e - r_i v_i)$$

The input power is $\dot{W}_c = \omega \tau_A$, or

$$\dot{W}_c = \frac{\dot{m}\omega}{g_c}(r_e v_e - r_i v_i) \qquad (9.1)$$

where ω is the angular rotor speed (rad/s).

This equation is often referred to as the *Euler pump equation*. Application of the first law of thermodynamics to the flow through the control volume gives

$$\dot{W}_c = \dot{m}(h_{te} - h_{ti})$$

Combining this expression with Eq. (9.1) gives an expression for the specific work of compression $\dot{W}/\dot{m}$ (Btu/lbm or kJ/kg) related to the change in angular momentum of the gas:

$$h_{te} - h_{ti} = \frac{\omega}{g_c}(r_e v_e - r_i v_i) \qquad (9.2)$$

Likewise, for a steady-flow turbine, the output torque τ_O is equal to the change in angular momentum of the fluid, or

$$\tau_O = \frac{\dot{m}}{g_c}(r_i v_i - r_e v_e)$$

The output power is

$$\dot{W}_t = \omega \tau_O$$

or

$$\dot{W}_t = \frac{\dot{m}\omega}{g_c}(r_i v_i - r_e v_e) \qquad (9.3)$$

This equation is often referred to as the *Euler turbine equation*. Application of the first law of thermodynamics to the flow through the control volume gives

$$\dot{W}_t = \dot{m}(h_{ti} - h_{te})$$

Combining this expression with Eq. (9.3) gives an expression for the specific work pulled out from the turbine $\dot{W}/\dot{m}$ (Btu/lbm or kJ/kg) related to the change in angular momentum of the gas:

$$h_{ti} - h_{te} = \frac{\omega}{g_c}(r_i v_i - r_e v_e) \qquad (9.4)$$

which is clearly the same general format as Eq. (9.2).

In the design of turbomachinery for a compressible gas, the gas is often modeled as having constant specific heats. For this case, we can write Eq. (9.4) as

$$c_p(T_{ti} - T_{te}) = \frac{\omega}{g_c}(r_i v_i - r_e v_e) \qquad (9.5)$$

The Euler pump and turbine equations are also referred to as the *Euler turbine equation*. We will be using this equation throughout the analysis of turbomachinery for a compressible gas. Component efficiencies and the isentropic relationships will be used to relate total temperature changes to total pressure changes.

9.3 Axial-Flow Compressor Analysis

The axial-flow compressor is one of the most common compressor types in use today. Its major application is in large gas turbine engines like those that power today's jet aircraft. A picture of an axial-flow compressor rotor is shown in Fig. 9.2. The compressor is made up of two major assemblies: the *rotating rotor* with its blades and the *stationary stator* with its vanes, each shown in Fig. 9.2, as well as in Figs. 1.4c, 1.8a, and many of the engine pictures included in the online Supporting Materials.

This chapter investigates the relationships of the desired performance parameters to the related blade aerodynamic loading and resultant fluid flow angles. Because the flow is inherently 3-D, the problem of analysis seems almost incomprehensible. *Do not fear!* This most complex flow can be understood by dividing the 3-D flowfield into three 2-D flowfields. The complete flowfield will be the "sum" of these less complex 2-D flows. The 2-D flowfields are normally called the *throughflow field*, the *cascade field*

Fig. 9.2   Axial-flow compressor rotor and stator case.

(or *blade-to-blade field*), and the *secondary flowfield*. Each of these fields is described in more detail next.

The *throughflow field* is concerned with the variation in fluid properties in only the radial r and axial z directions (see Fig. 9.3). No variations in the θ direction occur. A row of blades is modeled as a thin disk that affects the flowfield uniformly in the θ direction. If you imagine that the forces of the blades had been distributed among an infinite number of very thin blades, then you can begin to see the throughflow field model. As a result of throughflow field analysis, one will obtain

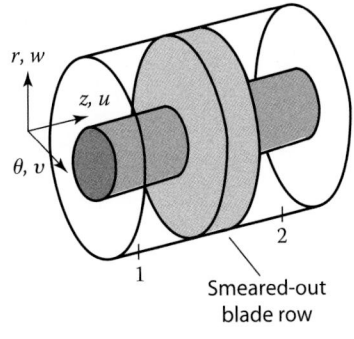

Fig. 9.3 Coordinate system and throughflow representation.

the axial, tangential, and radial velocities as a function of r and z. Typical axial velocity profiles as a result of throughflow analysis are shown in Fig. 9.4a.

When the axial velocity changes, like that shown in Fig. 9.4a, conservation of mass requires that a downward flow of the fluid occur between stations 1 and 2. This downward flow could be shown by the stream surface as drawn in Fig. 9.4b.

The *cascade field* considers the flow behavior along stream surfaces (s coordinate) and tangentially through blade rows. Unwrapping the stream surface (like that shown in Fig. 9.5) gives the meridional projection of blade profiles, as shown in Fig. 9.6—a 2-D flowfield in the θ and the meridional (almost z) coordinates. If the curvature of the stream surfaces in the throughflow field is not too great, the flow at a radial location may be modeled as a meridional projection and a suitable blade profile determined for the flow conditions. By considering a number of such projections, the

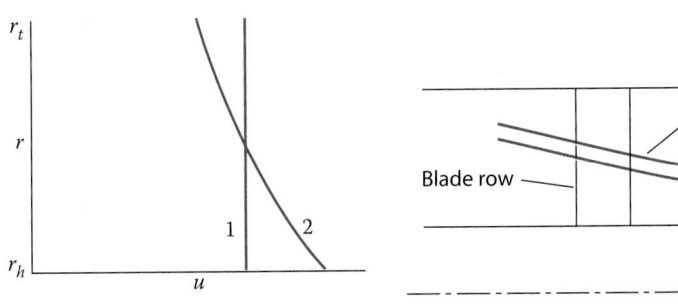

Fig. 9.4a Typical axial velocity profiles.

Fig. 9.4b Typical stream surface.

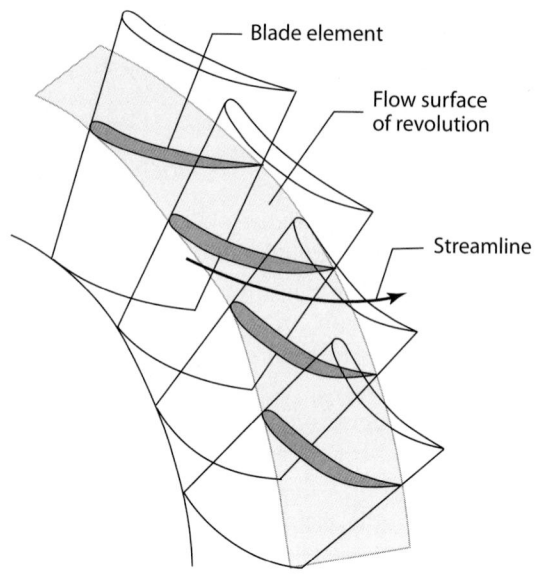

Fig. 9.5   Stream surface and streamlines.

blade profiles for selected radial locations on the blade are determined. The complete blade shape necessary to describe the full 3-D blade can be obtained by blending in the desired blade profiles.

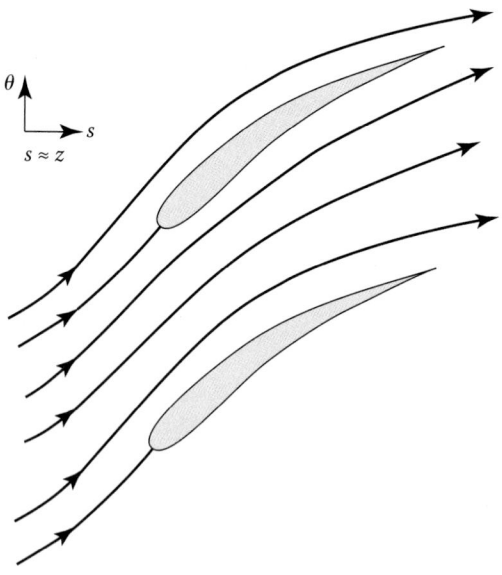

Fig. 9.6   Meridional projections of blade profiles.

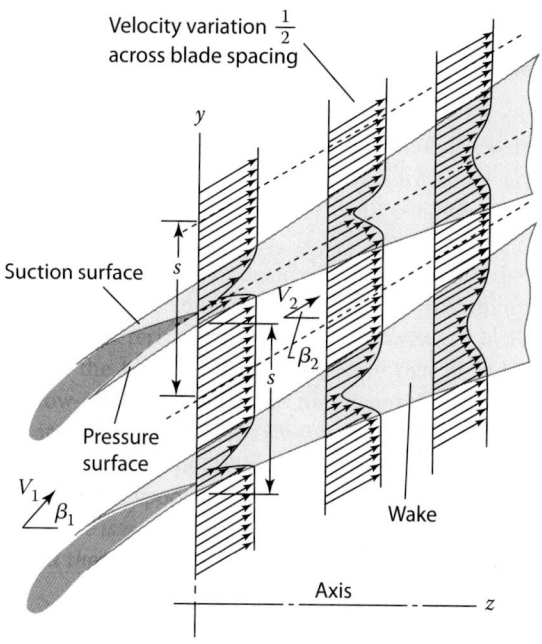

Fig. 9.7 Flowfield about cascade.

The most common method of obtaining performance data for different blade profiles is to run cascade tests. A set of airfoils of the desired blade shape is mounted in a conditioned flow stream, and the performance is experimentally measured. Figure 9.7 shows the complete flowfield about a typical cascade with the blades spaced a distance s apart.

The *secondary flowfield* exists because the fluid near the solid surfaces (in the boundary layer) of the blades and passage walls has a lower velocity than that in the *freestream* (external to the boundary layer). The pressure gradients imposed by the freestream will cause the fluid in the boundary layers to flow from regions of higher pressure to regions of lower pressure. Figure 9.8 shows the possible secondary flowfield within a stator row.

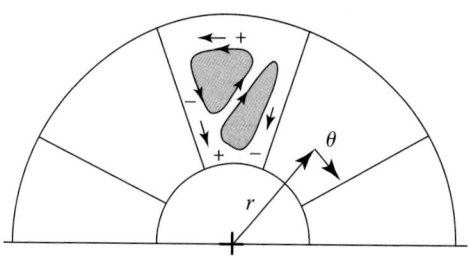

Fig. 9.8 Secondary flowfield within a stator row.

Designers try to minimize secondary flowfields because they generally represent losses and reduce component efficiency.

9.3.1 Two-Dimensional Flow Through Blade Rows

A cross section and a top view of a typical axial-flow compressor are shown in Fig. 9.9. Depending on the design, an inlet guide vane (IGV) may be used to deflect the incoming airflow to a predetermined angle toward the direction of rotation of the rotor. The rotor increases the angular velocity of the fluid, resulting in increases in total temperature, total pressure, and static pressure. The following stator decreases the angular velocity of the fluid, resulting in an increase in the static pressure, and sets the flow up for the following rotor. A compressor stage is made up of a rotor and a stator.

The basic building block of the aerodynamic design of axial-flow compressors is the *cascade*, an endless repeating array of airfoils (Fig. 9.10) that results from the "unwrapping" of the stationary (stators) and rotating (rotor) airfoils. *Each cascade passage, be it rotating or stationary, acts as a small diffuser*, and it is said to be well designed or "behaved" when it provides a large static pressure rise without incurring unacceptable total pressure losses and/or flow instabilities due to shock waves and/or boundary-layer separation.

The changes in fluid velocity induced by the blade and vane rows are related to changes in the fluid's thermodynamic properties in this section. The analysis is concerned with only the flow far upstream and far downstream of a cascade—the regions where the flowfields are uniform. In this manner, the details about the flowfield are not needed, and performance can be related to just the changes in fluid properties across each row.

In the analysis that follows, two different coordinate systems are used: one fixed to the compressor housing (absolute) and the other fixed to the rotating blades (relative). The static (thermodynamic) properties *do not*

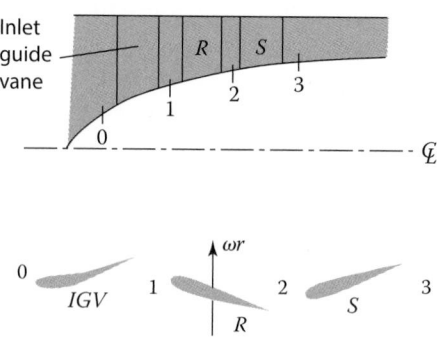

Fig. 9.9   Cross section and top view of a typical axial-flow compressor.

depend on the reference frame. However, the *total properties do depend on the reference frame.* The velocity of a fluid in one reference frame is easily converted to the other frame by the following simple, but very important, vector equation, which the reader would do well to memorize and understand:

$$V = V_R + U \qquad (9.6)$$

where

V = velocity in stationary coordinate system

V_R = velocity in moving coordinate system

U = velocity of moving coordinate system ($=\omega r$)

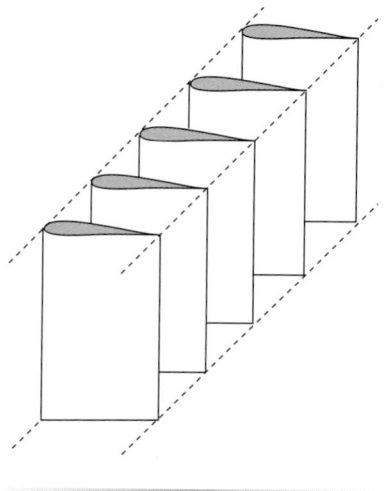

Fig. 9.10 Rectilinear cascade.

The graphical representation of vector Eq. (9.6) is velocity diagrams (or triangles), which are presented in the next section. Know them, understand them, and work with them until you are very comfortable with their use! In fact, if you draw velocity triangles to scale, you in essence define the airfoil shape at whatever section of the blade you are analyzing. Velocity diagrams are a great help in developing the relationships between turbomachinery thermodynamic property changes and the geometry of blade and vane rows. They are key to understanding blade and vane orientation relative to each other and relative to the airflow. Velocity diagrams provide the linkage between a rotating reference frame (rotors) and a stationary reference frame (stators).

9.3.2 Velocity Diagrams

The following development is performed at the midspan of the blades/vanes, appropriately called a mean-line analysis. Note that this analysis could be done anywhere along the blade/vane span, but a good starting point is midspan.

Consider the two compressor stages with inlet guide vanes (IGVs) shown in Figs. 9.11a and 9.11b. We use station numbers 0 and 1 for the IGVs and station numbers 1, 2, and 3 for each stage to indicate the entrance and exit from each row of rotor blades and stator blades. The flow enters the IGVs with velocity V_0 and is accelerated to velocity V_1, and tangential velocity is added to the air. (This is beneficial to the first stage—more about this in Sections 9.3.5 and 9.3.9.) The IGVs act as nozzles through which the static pressure decreases as the air velocity increases, and the fluid is

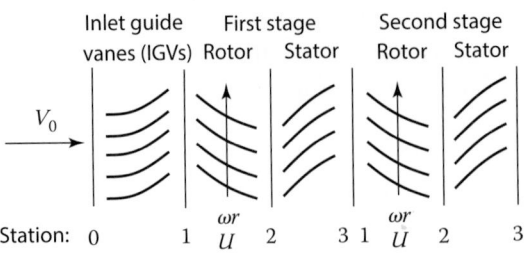

Fig. 9.11a   Two compressor stages with inlet guide vanes.

given a tangential (swirl) component in the direction of the rotor velocity. The flow enters the rotor with velocity V_1 (relative velocity V_{1R}) and leaves with velocity V_2 (relative velocity V_{2R}). The rotor is moving upward at velocity ωr (the symbol U is also used) and adds tangential velocity Δv. The flow enters the stator with velocity V_2 and leaves with velocity V_3. The stator is stationary and removes tandential velocity Δv.

We consider the flow through a typical stage, as shown in Fig. 9.12. Note that the velocity components of both the absolute and relative velocities are shown in Fig. 9.12 along with the flow angles, α and β. We use alpha (α) to represent the angle the absolute velocity vector makes with the axis of the compressor and beta (β) to represent the angle the relative velocity vector makes with the axis of the compressor. The velocity triangles at each station are composed of the axial and tangential velocity components u and v, respectively. Note that the absolute velocity vector V_1 is the vector sum of V_{1R} and U in accordance with Eq. (9.6).

Rather than keep the axial velocity constant, as is done in many textbooks, the following analysis permits variation in axial velocity from station to station. Application of the Euler equation

$$c_p(T_{t2} - T_{t1}) = \frac{\omega}{g_c}(r_2 v_2 - r_1 v_1)$$

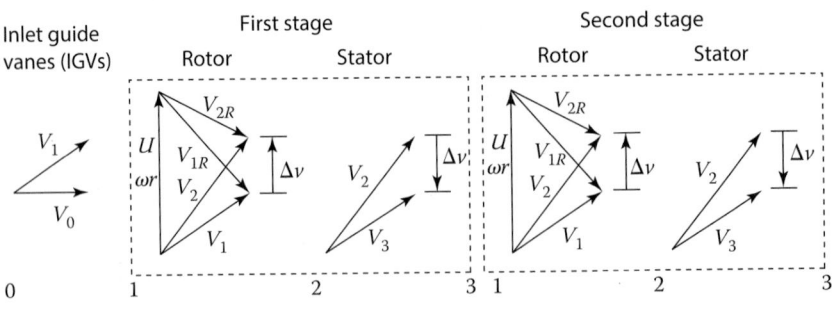

Fig. 9.11b   Velocity diagram for two compressor stages with inlet guide vanes.

Station: 1   2   3

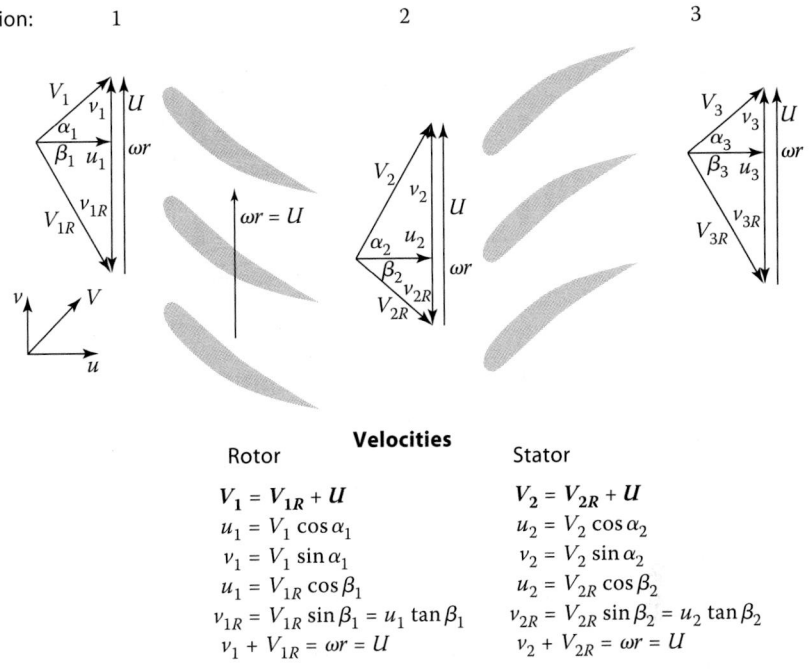

Velocities

Rotor   Stator

$V_1 = V_{1R} + U$ $V_2 = V_{2R} + U$

$u_1 = V_1 \cos \alpha_1$ $u_2 = V_2 \cos \alpha_2$

$v_1 = V_1 \sin \alpha_1$ $v_2 = V_2 \sin \alpha_2$

$u_1 = V_{1R} \cos \beta_1$ $u_2 = V_{2R} \cos \beta_2$

$v_{1R} = V_{1R} \sin \beta_1 = u_1 \tan \beta_1$ $v_{2R} = V_{2R} \sin \beta_2 = u_2 \tan \beta_2$

$v_1 + V_{1R} = \omega r = U$ $v_2 + V_{2R} = \omega r = U$

Fig. 9.12   Blade rows of a typical compressor.

to the rotor gives the following:

$$\text{For } r_2 \cong r_1, \quad c_p(T_{t2} - T_{t1}) = \frac{\omega r}{g_c}(v_2 - v_l) = \frac{U}{g_c}(v_2 - v_1) = \frac{U}{g_c}\Delta v \quad (i)$$

Using the velocity trigometric relationships of Fig. 9.12, we have

$$v_2 = \omega r - v_{2R} = \omega r - u_2 \tan \beta_2 = u_2 \tan \alpha_2$$

and

$$v_1 = \omega r - v_{1R} = \omega r - u_1 \tan \beta_1 = u_1 \tan \alpha_1$$

Equation (i) then becomes

$$c_p(T_{t2} - T_{t1}) = \frac{(\omega r)^2}{g_c}\frac{u_1}{\omega r}\left(\tan \beta_1 - \frac{u_2}{u_1}\tan \beta_2\right)$$

$$= \frac{U^2}{g_c}\frac{u_1}{U}\left(\tan \beta_1 - \frac{u_2}{u_1}\tan \beta_2\right) \quad (9.7a)$$

or

$$c_p(T_{t2} - T_{t1}) = \frac{(\omega r)^2}{g_c} \frac{u_1}{\omega r} \left(\frac{u_2}{u_1} \tan \alpha_2 - \tan \alpha_1 \right)$$

$$= \frac{U^2}{g_c} \frac{u_1}{U} \left(\frac{u_2}{u_1} \tan \alpha_2 - \tan \alpha_1 \right) \qquad (9.7b)$$

Hence the work done per unit mass flow can be determined from the rotor or blade speed ($U = \omega r$), the velocity ratios (u_1/U and u_2/u_1), and either the relative rotor flow angles (β_1 and β_2) or the absolute rotor flow angles (α_1 and α_2). Equations (9.7a) and (9.7b) are useful forms of the Euler equation for compressor stage design and show the dependence of the stage work on the rotor speed squared (U^2).

Turning to Fig. 9.12, the absolute velocity entering the rotor at station 1 is V_1. For a first-stage analysis with no IGVs ($\alpha = 0$), common for the first-stage fan of high-bypass turbofans, V_1 would equal u_1, the axial velocity component. Subtracting the rotor speed vr (U) from V_1 vectorially, we obtain the relative velocity V_{1R} entering the rotor. Note the use of Eq. (9.6), vector addition from the tip of V_{1R} to the tail of U. In the rotor blade row, the blade passages act as diffusers, reducing the relative velocity from V_{1R} to V_{2R} as the static pressure is increased from P_1 to P_2. The relative velocity leaving the rotor is V_{2R}. Combining V_{2R} vectorially with ωr, we get their sum V_2—the absolute velocity leaving the rotor.

The velocity of the air leaving the rotor and entering the stator at station 2 is V_2. The flow is further diffused in the stator vane passage, reducing the velocity to V_3 as the static pressure rises from P_2 to P_3. *Diffusion occurs in both the rotor and stator flow passages.* The change in absolute tangential velocity Δv best shown in Fig. 9.11b, is often called the "flow turning." It is directly related to the airfoil turning dictated in large part by the airfoil camber, which will be explored further in Section 9.3.4.3.

In general, the velocity V_3 entering the next rotor at station 3 does not have to be equal to V_1 entering the first-stage rotor. However, we often start a preliminary design with a repeating stage, repeating row (where V_1 would equal V_3) and then look at deviations from that design. The repeating stage, repeating row analysis is developed in Section 9.3.5.

The effects occurring in each compressor component are summarized in Table 9.1, where +, 0, and − mean increase, unchanged, and decrease, respectively. The table entries assume isentropic flow. In making entries in the table, it is important to distinguish between absolute and relative values. Because total pressure and total temperature depend on the speed of the gas, they have different values "traveling with the rotor" than for an observer not riding on the rotor. In particular, an observer on the rotor sees a force F (rotor on gas), but it is stationary; hence, in the rider's reference

Table 9.1 Property Changes[a] in an Isentropic Compressor

Property	Inlet Guide Vanes	Rotor	Stator
Absolute velocity	+	+	−
Relative velocity	n/a	−	n/a
Static pressure	−	+	+
Absolute total pressure	0	+	0
Relative total pressure	n/a	0	n/a
Static temperature	−	+	0
Absolute total temperature	0	+	0
Relative total temperature	n/a	0	n/a

[a] + = increase, − = decrease, 0 = unchanged, n/a = not applicable.

system, the force does no work on the gas. Consequently, the total temperature and total pressure do not change relative to an observer on the rotor as the gas passes through the rotor ($T_{t2R} = T_{t1R}$; $P_{t2R} = P_{t1R}$). An observer not on the rotor sees the force F (rotor on gas) moving at the rate ωr. Hence, to the stationary observer, work is done on the gas passing through the rotor, and the total temperature and total pressure increase.

9.3.3 Flow Annulus Area

For uniform total properties at a station i, the area of the flow annulus A_i can be obtained from the mass flow parameter (MFP) by using

$$A_i = \frac{\dot{m}\sqrt{T_{ti}}}{P_{ti}(\cos \alpha_i)\text{MFP}(M_i)} \tag{9.8}$$

where α_i is the angle that the velocity V_i makes with the centerline of the annulus. It is only the normal component of the gas that flows through the cross-sectional annulus area.

Example 9.1

Consider mean radius stage calculation—isentropic flow. Determine air flow properties, flow angles, and flowpath areas.
Given:

$$T_{t1} = 518.7°R, \quad P_{t1} = 14.70 \text{ psia}, \quad \omega = 1000 \text{ rad/s}, \quad r = 12 \text{ in.},$$
$$\alpha_1 = \alpha_3 = 40 \text{ deg}, \quad \dot{m} = 50 \text{ lbm/s}, \quad M_1 = M_3 = 0.7,$$
$$u_2/u_1 = 1.1, \quad P_{t3}/P_{t1} = 1.3$$

Gas is air.

(Continued)

Example 9.1 *(Continued)*

Note: For air, $\gamma = 1.4$, $c_p = 0.24\ \text{Btu}/(\text{lbm} \cdot {}^\circ\text{R})$, $Rg_c = 1716\ \text{ft}^2/(\text{s}^2 \cdot {}^\circ\text{R})$, $c_p g_c = 6006\ \text{ft}^2/(\text{s}^2 \cdot {}^\circ\text{R})$.

Solution:

$$T_1 = \frac{T_{t1}}{1 + [(\gamma - 1)/2]M_1^2} = -\frac{518.7{}^\circ\text{R}}{1 + 0.2 \times 0.7^2} = 472.4{}^\circ\text{R}$$

$$a_1 = \sqrt{\gamma Rg_c T_1} = \sqrt{1.4 \times 1716 \times 472.4} = 1065.3\ \text{ft/s}$$

$$V_1 = M_1 a_1 = 0.7 \times 1065.3\ \text{ft/s} = 745.71\ \text{ft/s}$$

$$u_1 = V_1 \cos \alpha_1 = 745.71\ \text{ft/s} \times 0.766 = 571.21\ \text{ft/s}$$

$$v_1 = V_1 \sin \alpha_1 = 745.71\ \text{ft/s} \times 0.6428 = 479.34\ \text{ft/s}$$

$$P_1 = \frac{P_{t1}}{\{1 + [(\gamma - 1)/2]M_1^2\}^{\gamma/(\gamma-1)}} = \frac{14.70\ \text{psia}}{(1 + 0.2 + 0.7^2)^{3.5}} = 10.60\ \text{psia}$$

$$\text{MFP}(M_1) = 0.4859$$

$$A_1 = \frac{\dot{m}\sqrt{T_{t1}}}{P_{t1}(\cos \alpha_1)\text{MFP}(M_1)} = \frac{50\sqrt{518.7}}{14.70 \times 0.7660 \times 0.4859} = 207.2\ \text{in.}^2$$

$$\omega r = 1000 \times \frac{12}{12} = 1000\ \text{ft/s}$$

$$v_{1R} = \omega r - v_1 = 1000 - 479.34 = 520.66\ \text{ft/s}$$

$$\beta_1 = \tan^{-1}\frac{v_{1R}}{u_1} = \tan^{-1} 0.9114 = 42.35\ \text{deg}$$

$$V_{1R} = \sqrt{u_1^2 + v_{1R}^2} = \sqrt{571.21^2 + 520.66^2} = 772.90\ \text{ft/s}$$

$$M_{1R} = \frac{V_{1R}}{a_1} = \frac{772.90}{1065.3} = 0.7255$$

$$T_{t1R} = T_1\left(1 + \frac{\gamma - 1}{2}M_{1R}^2\right) = 472.4(1 + 0.2 \times 0.7255^2) = 522.2{}^\circ\text{R}$$

$$P_{t1R} = P_1\left(\frac{T_{t1R}}{T_1}\right)^{\gamma/(\gamma-1)} = 10.60\left(\frac{522.1}{472.4}\right)^{3.5} = 15.04\ \text{psia}$$

$$P_{t2} = P_{t3} = 1.3 \times 14.70 = 19.11\ \text{psia}$$

$$T_{t3} = T_{t2} = T_{t1}\left(\frac{P_{t2}}{P_{t1}}\right)^{(\gamma-1)/\gamma} = 518.7(1.3)^{1/3.5} = 559.1{}^\circ\text{R}$$

(Continued)

Example 9.1 *(Continued)*

$$\tan \beta_2 = \frac{u_1}{u_2}\left[\tan \beta_1 - \frac{g_c c_p}{\omega r\, u_1}(T_{t2} - T_{t1})\right]$$

$$= \frac{1}{1.1}\left[\tan 42.35 - \frac{6006}{1000 \times 571.21}(559.1 - 518.7)\right] = 0.4425$$

$$\beta_2 = 23.87 \text{ deg}$$

$$u_2 = \frac{u_2}{u_1}u_1 = 1.1 \times 571.21 = 628.33 \text{ ft/s}$$

$$v_{2R} = u_2 \tan \beta_2 = 628.33 \times 0.4425 = 278.04 \text{ ft/s}$$

$$V_{2R} = \sqrt{u_2^2 + v_{2R}^2} = \sqrt{628.33^2 + 278.04^2} = 687.10 \text{ ft/s}$$

$$v_2 = \omega r - v_{2R} = 1000 - 278.04 = 721.96 \text{ ft/s}$$
$$\alpha_2 = \tan^{-1}\frac{v_2}{u_2} = \tan^{-1}\frac{721.96}{628.33} = 48.97 \text{ deg}$$

$$V_2 = \sqrt{u_2^2 + v_2^2} = \sqrt{628.33^2 + 721.96^2} = 957.10 \text{ ft/s}$$

$$T_{t2R} = T_{t1R} = 552.5 °R$$

$$P_{t2R} = P_{t1R} = 15.04 \text{ psia}$$

$$T_2 = T_{t2} - \frac{V_2^2}{2g_c c_p} = 559.1 - \frac{957.10^2}{2 \times 6006} = 482.8 °R$$

$$P_2 = P_{t2}\left(\frac{T_2}{T_{t2}}\right)^{\gamma/(\gamma-1)} = 19.11\left(\frac{482.8}{559.1}\right)^{3.5} = 11.43 \text{ psia}$$

$$a_2 = \sqrt{\gamma R g_c T_2} = \sqrt{1.4 \times 1746 \times 482.8} = 1077.0 \text{ ft/s}$$

$$M_2 = \frac{V_2}{a_2} = \frac{957.10}{1077.0} = 0.8887$$

$$M_{2R} = \frac{V_{2R}}{a_2} = \frac{687.10}{1077.0} = 0.6380$$

$$\text{MFP}(M_2) = 0.5260$$

$$A_2 = \frac{\dot{m}\sqrt{T_{t2}}}{P_{t2}(\cos \alpha_2)\text{MFP}(M_2)} = \frac{50\sqrt{559.1}}{19.11 \times 0.6566 \times 0.5260} = 179.1 \text{ in.}^2$$

$$T_3 = \frac{T_{t3}}{1 + [(\gamma-1)/2]M_3^2} = \frac{559.1 °R}{1 + 0.2 \times 0.7^2} = 509.2 °R$$

(Continued)

Example 9.1 *(Continued)*

$$P_3 = P_{t3}\left(\frac{T_3}{T_{t3}}\right)^{\gamma/(\gamma-1)} = 19.11\left(\frac{509.2}{559.1}\right)^{3.5} = 13.77 \text{ psia}$$

$$a_3 = \sqrt{\gamma R g_c T_3} = \sqrt{1.4 \times 1716 \times 509.2} = 1106.03 \text{ ft/s}$$

$$V_3 = M_3 a_3 = 0.7 \times 1106.03 = 774.22 \text{ ft/s}$$

$$u_3 = V_3 \cos \alpha_3 = 593.09 \text{ ft/s}$$

$$v_3 = V_3 \sin \alpha_3 = 497.66 \text{ ft/s}$$

$$A_3 = \frac{\dot{m}\sqrt{T_{t3}}}{P_{t3}(\cos \alpha_3)\text{MFP}(M_3)} = \frac{50\sqrt{559.1}}{19.11 \times 0.766 \times 0.4859} = 165.3 \text{ in.}^2$$

The results of this stage calculation for isentropic flow are summarized in Table 9.2, with the given data shown in bold type. (SI results are shown within parentheses.) Note that the flow through the rotor is adiabatic in the relative reference frame and, through the stator, is adiabatic in the absolute reference frame. Also note the decrease in flowpath area through the stage even though

Table 9.2   Results for Example 9.1 Axial-Flow Compressor Stage Calculation, Insentropic Flow

Property		Station				
		1	1R	2R	2	3
T_t	°R	**518.7**	522.1	522.1	559.1	559.1
	(K)	**(288.2)**	(290.1)	(290.1)	(310.6)	(310.6)
T	°R	472.4	472.4	482.8	482.8	509.2
	(K)	(262.4)	(262.4)	(268.2)	(268.2)	(282.9)
P_t	psia	**14.70**	15.04	15.04	19.11	**19.11**
	(kPa)	**(101.3)**	(103.7)	(103.7)	(131.7)	**(131.7)**
P	psia	10.60	10.60	11.43	11.43	13.77
	(kPa)	(73.05)	(73.05)	(78.85)	(78.85)	(94.96)
M		**0.700**	0.7255	0.6380	0.8887	0.700
V	ft/s	745.71	772.90	687.10	957.10	774.22
	(m/s)	(227.30)	(235.58)	(209.43)	(291.73)	(235.99)
u	ft/s	571.21	571.21	628.33	628.33	593.09
	(m/s)	(174.11)	(174.11)	(191.53)	(191.52)	(180.78)
v	ft/s	479.34	520.66	278.04	721.96	497.66
	(m/s)	(146.10)	(158.70)	(84.75)	(220.06)	(151.69)
α	deg	**40.00**	—	—	48.97	**40.00**
β	deg	—	42.35	23.87	—	—

Example 9.1 *(Continued)*

the axial velocity has increased. Can you explain this? This decrease in flow annulus area is characteristic of axial-flow compressors. Figure 9.13 is a representation of the flow properties for an isentropic stage on a *T-s* diagram.

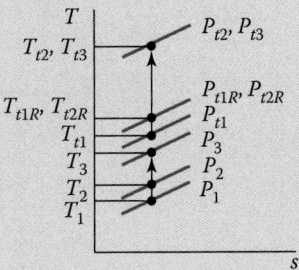

Fig. 9.13 Property changes of an isentropic compressor stage.

9.3.4 Stage Parameters

In this section, we present a number of compressor stage performance parameters. Some parameters provide similar information in a slightly different way, and not all of them are necessarily used within each engine company. Regardless, the reader will have at least been exposed to many of the more common stage parameters.

9.3.4.1 Efficiencies

Several efficiencies are used to compare the performance of compressor stage designs as developed in Chapter 6. The two efficiencies most commonly used are stage efficiency and polytropic efficiency. The *stage efficiency* of an adiabatic compressor is defined as the ratio of the ideal work per unit mass to the actual work per unit mass between the same total pressures, or

$$\eta_s = \frac{h_{t3s} - h_{t1}}{h_{t3} - h_{t1}} \tag{9.9}$$

For a calorically perfect gas, this simplifies to

$$\eta_s = \frac{T_{t3s} - T_{t1}}{T_{t3} - T_{t1}} = \frac{(P_{t3}/P_{t1})^{(\gamma-1)/\gamma} - 1}{T_{t3}/T_{t1} - 1} \tag{9.10}$$

where the states used in this equation for stage efficiency are plotted on the *T-s* diagram of Fig. 9.14a.

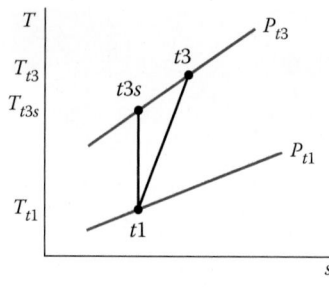

Fig. 9.14a Stages for definition of compressor stage efficiency.

The *polytropic efficiency* of an adiabatic compressor is defined as the ratio of the ideal work per unit mass to the actual work per unit mass for a differential pressure change, or

$$e_c = \frac{dh_{ti}}{dh_t} = \frac{dT_{ti}}{dT_t} = \frac{dT_{ti}/T_t}{dT_t/T_t} = \frac{\gamma - 1}{\gamma} \frac{dP_t/P_t}{dT_t/T_t}$$

For a constant polytropic efficiency e_c, integration between states $t1$ and $t3$ gives

$$e_c = \frac{\gamma - 1}{\gamma} \frac{\ln(P_{t3}/P_{t1})}{\ln(T_{t3}/T_{t1})} \tag{9.11}$$

A useful expression for stage efficiency, in terms of the pressure ratio and polytropic efficiency, can be easily obtained by using Eqs. (9.10) and (9.11). Solving Eq. (9.11) for the temperature ratio and substituting into Eq. (9.10) give

$$\eta_s = \frac{(P_{t3}/P_{t1})^{(\gamma-1)/\gamma} - 1}{(P_{t3}/P_{t1})^{(\gamma-1)/(\gamma e_c)} - 1} \tag{9.12}$$

In the limit, as the pressure ratio approaches 1, the stage efficiency approaches the polytropic efficiency. The variation in stage efficiency with stage pressure ratio is plotted in Fig. 9.14b for constant polytropic efficiency.

The compressor polytropic efficiency is useful in preliminary design of compressors for gas turbine engines to predict the compressor efficiency for a given level of technology. The value of the polytropic efficiency is mainly a function of the technology level (see Table 6.2). Axial-flow compressors designed in the 1980s have a polytropic efficiency of about 0.88, whereas the compressors being designed today have a polytropic efficiency of about 0.9.

9.3.4.2 Degree of Reaction

For compressible flow, the *degree of reaction* is defined as

$$°R_c = \frac{\text{rotor static enthalpy rise}}{\text{stage static enthalpy rise}} = \frac{h_2 - h_1}{h_3 - h_1} \tag{9.13a}$$

For a calorically perfect gas, the static enthalpy rises in the equation become static temperature rises, and the degree of reaction is

$$°R_c = \frac{T_2 - T_1}{T_3 - T_1} \tag{9.13b}$$

In the general case, it is desirable to have the degree of reaction in the vicinity of 0.5 because the rotor and stator rows then will "share the burden" of increasing the enthalpy of the flow. The degree of reaction for the stage data of Example 9.1 is approximately 0.28, which means the majority of the static temperature rise occurs in the stator.

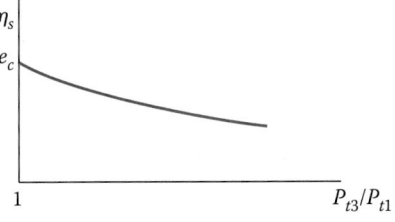

Fig. 9.14b Variation of stage efficiency with pressure ratio.

9.3.4.3 Cascade Airfoil Nomenclature and Loss Coefficient

The angles used in the analysis of Sections 9.3.2 and 9.3.3 are the flow angles relative to the rotating or stationary reference frames, betas and alphas, respectively. The flow angles do not necessarily coincide with the airfoil (metal) angles, although for most of the compressor operation they are close. Figure 9.15 shows the cascade airfoil nomenclature and the airfoil (metal) and flow angles. Subscripts i and e are used for the inlet and exit states, respectively. The ratio of the airfoil chord c to the airfoil spacing s is called the *solidity* σ, which traditionally has been around unity but is trending higher with modern designs. At design, the incidence angle, equivalent to the angle of attack for airplane airfoils, is nearly zero, goes positive (towards stall) at low flows, and can go negative at high flows—all of this for a constant blade speed U. The flow exit angle deviates from the airfoil exit metal angle primarily due to boundary layer growth and perhaps local flow separation at the trailing edge. The difference between the metal exit angle and flow angle is appropriately called the *deviation* δ_c and can be determined by using Carter's rule (Ref. 41):

$$\delta_c = \frac{\gamma_i - \gamma_e}{4\sqrt{\sigma}} \tag{9.14}$$

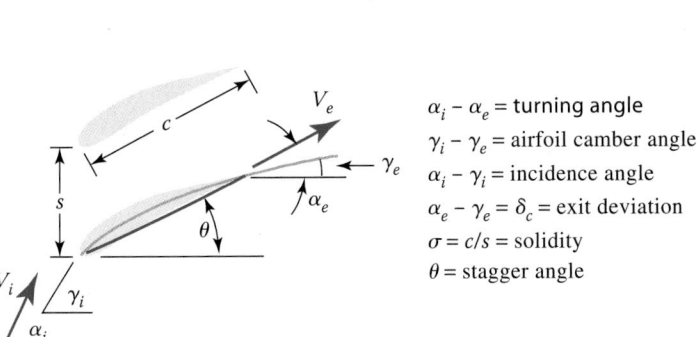

$\alpha_i - \alpha_e$ = turning angle
$\gamma_i - \gamma_e$ = airfoil camber angle
$\alpha_i - \gamma_i$ = incidence angle
$\alpha_e - \gamma_e = \delta_c$ = exit deviation
$\sigma = c/s$ = solidity
θ = stagger angle

Fig. 9.15 Cascade airfoil nomenclature.

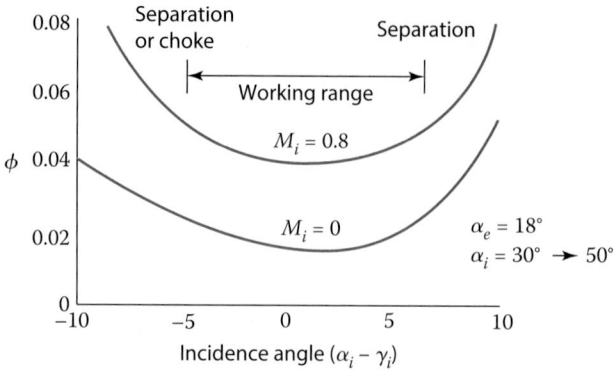

Fig. 9.16 Compressor cascade experimental behavior.

The airfoil angles of both the rotor and the stator can be calculated from the flow angles, given the incidence angle and solidity for each. To obtain the exit airfoil angle, Eq. (9.14) is rearranged to give

$$\gamma_e = \frac{4\alpha_e\sqrt{\sigma} - \gamma_i}{4\sqrt{\sigma} - 1}$$

For the data of Example 9.1, we obtain the following airfoil angles, assuming a solidity of 1 and a zero incidence angle for both rotor and stator:

$$\text{Rotor} \quad \gamma_i = 42.35 \text{ deg} \quad \gamma_e = 17.71 \text{ deg}$$
$$\text{Stator} \quad \gamma_i = 48.97 \text{ deg} \quad \gamma_e = 37.01 \text{ deg}$$

Losses in cascade airfoils are normally quantified in terms of the drop in total pressure divided by the dynamic pressure of the incoming flow. This ratio is called the *total pressure loss coefficient* and is defined as

$$\phi_c \equiv \frac{P_{ti} - P_{te}}{\rho V_i^2/(2g_c)} \tag{9.15}$$

Figure 9.16 shows the typical total pressure loss behavior of compressor airfoils obtained from cascade tests. Note that these losses increase when the Mach number and incidence angle vary from those that result in minimum loss (the design incidence). The loss coefficients shown in Fig. 9.16 include only the 2-D or "profile" losses and must be increased in compressor stage design to account for secondary and endwall losses (e.g., tip leakage, wall boundary layer, or cavity leakage).

From Fig. 9.16, the preceding discussion, and discussion in subsequent sections, one can easily understand the need and desire for variable stator vanes (VSVs) and variable inlet guide vanes (IGVs) initially presented in Section 1.4.2 (see Figs. 1.4b and 1.4c). In order to maintain high compressor efficiency (attempt to stay in the "sweet spot" of minimum total pressure loss) and stay away from compressor stall over a wide range of flow conditions, most core compressors of all types of gas turbine engines use IGVs and several rows of VSVs in the front stages of the compressor. Further, the variable vane hardware in the front stages helps considerably in low-speed and starting operation, as is discussed in Section 9.3.11.3. For this reason, the back stages do not incorporate variable vanes.

9.3.4.4 Diffusion Factor

The total pressure loss of a cascade depends on many factors. The pressure and velocity distribution about a typical cascade airfoil is shown in Fig. 9.17. The upper (suction) side of the compressor blade has a large static pressure rise (due to the deceleration from V_{max} to V_e), which can cause the viscous low-momentum boundary layer to separate. Boundary-layer separation is not desirable because of the associated higher losses in total pressure.

Cascade test results (see Fig. 9.18) show a direct correlation between total pressure loss and the deceleration (diffusion) on the upper (suction) side of blades. The amount of diffusion is measured by the diffusion factor D. The expression for the diffusion factor is based on the fact that the diffusion of the flow on the suction surface is approximated as

$$D \approx \frac{V_{max} - V_e}{V_{av}} \approx \frac{V_{max} - V_e}{V_i}$$

where

$$V_{max} \approx V_i + f\left(\frac{\Delta v}{\sigma}\right) \approx V_i + \frac{|\Delta v|}{2\sigma}$$

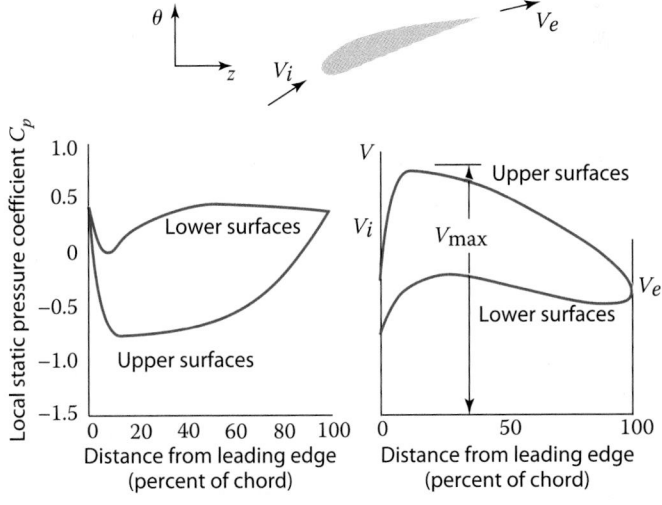

Fig. 9.17 Compressor airfoil pressure and velocity distribution (Ref. 41).

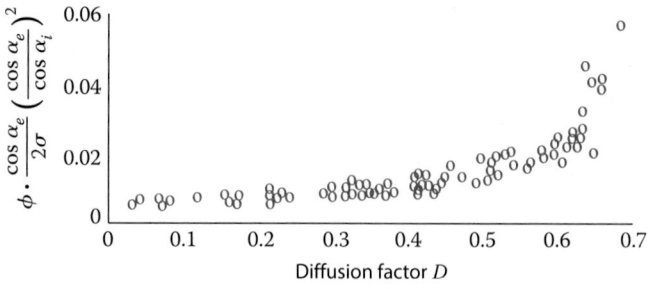

Fig. 9.18   Total pressure loss vs diffusion factor for typical compressor airfoil (Ref. 41).

Thus the *diffusion factor D* is defined by

$$D \equiv 1 - \frac{V_e}{V_i} + \frac{|v_i - v_e|}{2\sigma V_i} \tag{9.16}$$

where the subscripts i and e refer to the inlet and exit, respectively. The first term in Eq. (9.16) relates to the diffusion caused by the area increase in the flow passage, and the second term relates to the flow turning (the change in tangential velocity). Note that for a compressor stage, there is a diffusion factor for the rotor and another for the stator.

The diffusion factor for the rotor D_r is in terms of the relative velocities at stations 1 and 2, or

$$D_r = 1 - \frac{V_{2R}}{V_{1R}} + \frac{|v_{1R} - v_{2R}|}{2\sigma V_{1R}} \tag{9.17}$$

The diffusion factor for the stator D_s is in terms of the absolute velocities at stations 2 and 3, or

$$D_s = 1 - \frac{V_3}{V_2} + \frac{|v_2 - v_3|}{2\sigma V_2} \tag{9.18}$$

Figure 9.18 shows a loss parameter vs diffusion factor for typical compressor cascade airfoils. This figure is useful for estimating total pressure losses during preliminary design. Because the total pressure loss increases dramatically for diffusion factors above 0.6, *designs of axial compressors are limited to diffusion factors less than or equal to 0.6*. The reader will find it interesting to note that even though the data shown in Fig. 9.18 were reported in 1965, axial compressor designers today still use diffusion factors less than or equal to 0.6 as a general rule of thumb. The diffusion factors for the data of Example 9.1, with a solidity of 1, are

$$D_r = 1 - \frac{V_{2R}}{V_{1R}} + \frac{|v_{1R} - v_{2R}|}{2\sigma V_{1R}} = 1 - \frac{687.10}{772.90} + \frac{|520.66 - 278.04|}{2 \times 1 \times 772.90} = 0.2680$$

$$D_s = 1 - \frac{V_3}{V_2} + \frac{|v_2 - v_3|}{2\sigma V_2} = 1 - \frac{774.22}{957.10} + \frac{|721.96 - 497.66|}{2 \times 1 \times 957.10} = 0.3083$$

These values of diffusion factor show that both the rotor and stator of the example problem are lightly loaded. This stage could be designed to increase the total temperature more than it does (more work). One could try a higher pressure ratio and determine the resulting diffusion factor. In Section 9.3.5, a straightforward method of stage design is presented that starts with a speci-fied diffusion factor to determine the resultant pressure ratio.

9.3.4.5 Stage Loading and Flow Coefficients

One can think of the stage loading coefficient as an alternative or comp-lement to the rotor diffusion factor. Even though it is termed "stage," note from Eq. (9.19) that it's really referring to the rotor blade row because the stators do not contribute to the rise in total enthalpy. The ratio of the stage work to rotor speed squared is called the *stage loading coefficient* and is defined as

$$\psi \equiv \frac{g_c \Delta h_t}{(\omega r)^2} = \frac{g_c \Delta h_t}{U^2} \tag{9.19}$$

For a calorically perfect gas, the stage loading coefficient can be written as

$$\psi = \frac{g_c c_p \Delta T_t}{(\omega r)^2} = \frac{g_c c_p \Delta T_t}{U^2} \tag{9.20}$$

Modern axial-flow compressors used for aircraft gas turbine engines have stage loading coefficients in the range of 0.3–0.35 at the mean radius. Higher stage loadings can lead to blade stall (like $D = 0.6$). For Example 9.1, the stage loading coefficient is 0.24, a low value.

The ratio of the axial velocity to the rotor speed is called the *flow coeffi-cient* and is defined as

$$\Phi \equiv \frac{u_1}{\omega r} = \frac{u_1}{U} \tag{9.21}$$

The flow coefficients for modern axial-flow compressors of aircraft gas turbine engines are in the range of 0.45–0.55 at the mean radius. For Example 9.1, the flow coefficient is 0.57, a high value. Note that for a fixed blade speed U, lower flow coefficients tend towards blade stall. The motivated student would do well to review velocity triangles (Fig. 9.12) to verify this last statement.

9.3.4.6 Stage Pressure Ratio

Equation (9.10) can be rewritten as

$$\frac{P_{t3}}{P_{t1}} = \left(1 + \eta_s \frac{\Delta T_t}{T_{t1}}\right)^{\gamma/(\gamma-1)} \tag{9.22}$$

Likewise, Eq. (9.11) can be rewritten as

$$\frac{P_{t3}}{P_{t1}} = \left(\frac{T_{t2}}{T_{t1}}\right)^{\gamma e_c/(\gamma-1)} = \left(1 + \frac{\Delta T_t}{T_{t1}}\right)^{\gamma e_c/(\gamma-1)} \tag{9.23}$$

The stage pressure ratio can be calculated given the stage total temperature rise and the stage efficiency or polytropic efficiency from Eq. (9.22) or (9.23), respectively. We can see from these equations and Eq. (9.20) that a high stage pressure ratio requires high work loading (change in T_t). Equation (9.23) (plotted in Fig. 6.9) shows for the same total temperature change per stage that later stages will have a lower pressure ratio than early stages in a compressor due to the increase in inlet total temperature T_{t1}.

When the stage efficiency and polytropic efficiency are unknown, the stage pressure ratio can be determined by using loss coefficients based on cascade data and other losses. The total pressure for a compressor stage can be written in terms of loss coefficients of the rotor ϕ_{cr} and stator ϕ_{cs}. Noting that the total pressure loss of the rotor is based on the relative velocity, we write

$$\phi_{cr} \equiv \frac{P_{t1R} - P_{t2R}}{\rho_1 V_{1R}^2/(2g_c)} \tag{9.24}$$

Then

$$\frac{P_{t2R}}{P_{t1R}} = 1 - \phi_{cr}\frac{\rho_1 V_{1R}^2}{2g_c P_{t1R}} = 1 - \phi_{cr}\frac{\gamma P_1 M_{1R}^2}{2P_{t1R}}$$

or

$$\frac{P_{t2R}}{P_{t1R}} = 1 - \phi_{cr}\frac{\gamma M_{1R}^2/2}{\{1 + [(\gamma-1)/2]M_{1R}^2\}^{\gamma/(\gamma-1)}} \tag{9.25}$$

Likewise,

$$\phi_{cs} \equiv \frac{P_{t2} - P_{t3}}{\rho_2 V_2^2/(2g_c)} \tag{9.26}$$

then

$$\frac{P_{t3}}{P_{t2}} = 1 - \phi_{cs}\frac{\gamma M_2^2/2}{\{1 + [(\gamma-1)/2]M_2^2\}^{\gamma/(\gamma-1)}} \tag{9.27}$$

The total pressure ratio of a stage can be written in the following form:

$$\frac{P_{t3}}{P_{t1}} = \left(\frac{P_{t3}}{P_{t2}}\right)_{\phi_{cs}, M_2} \left(\frac{P_{t2}}{P_2}\right)_{M_2} \left(\frac{P_2}{P_{t2R}}\right)_{M_{2R}} \left(\frac{P_{t2R}}{P_{t1R}}\right)_{\phi_{cr}, M_{1R}}$$

$$\times \left(\frac{P_{t1R}}{P_1}\right)_{M_1R} \left(\frac{P_1}{P_{t1}}\right)_{M_1} \tag{9.28}$$

where the subscripts for the ratio inside the parentheses indicate the variables of the ratio. The loss coefficient of the rotor ϕ_{cr} required in Eq. (9.25) is obtained from cascade data with allowance for other losses (tip leakage, side wall boundary layer, etc.) and can range in value from 0.05 to 0.12. Likewise, the stator's loss coefficient ϕ_{cs} required in Eq. (9.26) is also obtained from cascade data with allowance for other losses and can range in value from 0.03 to 0.06.

Figure 9.19 is a T-s diagram of a compressor stage with losses. It shows the losses in total pressure in the rotor and the stator. Comparison with the T-s diagram for an ideal compressor stage (Fig. 9.13) can help one see the effects of losses on the process through a compressor stage.

9.3.4.7 Blade Mach Number M_b

Equation (9.7) can be written in terms of the flow angles (α_1 and β_2) and rearranged to give

$$\frac{\Delta T_t}{T_{t1}} = \frac{(\omega r)^2}{g_c c_p T_{t1}}\left[1 - \frac{u_2}{\omega r}\left(\tan\beta_2 + \frac{u_1}{u_2}\tan\alpha_1\right)\right]$$

or

$$\frac{\Delta T_t}{T_{t1}} = \frac{(\omega r)^2}{g_c c_p T_1}\frac{T_1}{T_{t1}}\left[1 - \frac{u_2}{\omega r}\left(\tan\beta_2 + \frac{u_1}{u_2}\tan\alpha_1\right)\right]$$

Note that

$$\frac{(\omega r)^2}{g_c c_p T_1} = \frac{\gamma R}{c_p}\frac{(\omega r)^2}{\gamma R g_c T_1} = (\gamma - 1)\frac{(\omega r)^2}{a_1^2}$$

$$= (\gamma - 1)M_b^2$$

where M_b is the blade tangential Mach number based on the upstream speed of sound a_1,

$$\frac{T_1}{T_{t1}} = \frac{1}{1 + [(\gamma - 1)/2]M_1^2} \quad \text{and}$$

$$\frac{u_2}{\omega r} = (\cos\alpha_1)\frac{M_1}{M_b}\frac{u_2}{u_1}$$

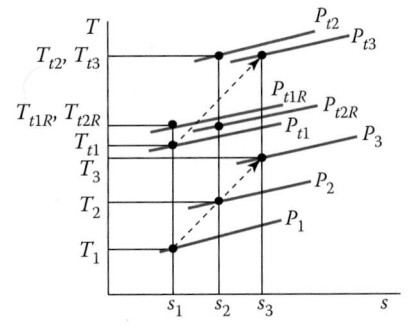

Fig. 9.19 The T-s diagram of a typical compressor stage with losses.

Then

$$\frac{\Delta T_t}{T_{t1}} = \frac{(\gamma - 1)M_b^2}{1 + [(\gamma - 1)/2]M_1^2} \left[1 - (\cos \alpha_1) \frac{M_1}{M_b} \left(\frac{u_2}{u_1} \tan \beta_2 + \tan \alpha_1\right)\right] \quad (9.29)$$

If we consider that the flow angles and the velocity ratio u_2/u_1 are functions of the geometry, then the stage temperature rise is mainly a function of M_b and the ratio M_1/M_b. From Eq. (9.29), we can see that high M_b is desirable to give higher stage temperature rise. The blade Mach number for Example 9.1 is about 0.94.

Example 9.2

Consider mean radius stage calculation—flow with losses.
Given:

$$T_{t1} = 288.16 \, \text{K}, \quad P_{t1} = 101.3 \, \text{kPa}, \quad \omega = 1000 \, \text{rad/s},$$
$$r = 0.3048 \, \text{m}, \quad \alpha_1 = \alpha_3 = 40 \, \text{deg}, \quad \sigma = 1,$$
$$\dot{m} = 22.68 \, \text{kg/s}, \quad M_1 = M_3 = 0.7, \quad u_2/u_1 = 1.1,$$
$$\Delta T_t = 22.43 \, \text{K}, \quad \phi_{cr} = 0.09, \quad \phi_{cs} = 0.03.$$

Gas is air. The input data are the same as in Example 9.1 with the same ΔT_t specified. Calculate performance and compare results to those of the ideal stage of Example 9.1.

Note: For air, $\gamma = 1.4$, $c_p = 1.004 \, \text{kJ/(kg} \cdot \text{K)}$, and $R = 0.287 \, \text{kJ/(kg} \cdot {}^\circ\text{R)}$.

Solution:

$$T_1 = \frac{T_{t1}}{1 + [(\gamma - 1)/2]M_1^2} = \frac{288.16 \, \text{K}}{1 + 0.2 \times 0.7^2} = 262.44 \, \text{K}$$

$$a_1 = \sqrt{\gamma R g_c T_1} = \sqrt{1.4 \times 287 \times 1 \times 262.44} = 324.73 \, \text{m/s}$$

$$V_1 = M_1 a_1 = 0.7 \times 324.73 = 227.31 \, \text{m/s}$$

$$u_1 = V_1 \cos \alpha_1 = 227.31 \times 0.7660 = 174.13 \, \text{m/s}$$

$$v_1 = V_1 \sin \alpha_1 = 227.31 \times 0.6428 = 146.11 \, \text{m/s}$$

$$P_1 = \frac{P_{t1}}{\{1 + [(\gamma - 1)/2]M_1^2\}^{\gamma/(\gamma-1)}}$$

$$= \frac{101.3 \, \text{kPa}}{(1 + 0.2 \times 0.7^2)^{3.5}} = 73.03 \, \text{kPa}$$

$$\text{MFP}(M_1) = 0.03700$$

(Continued)

Example 9.2 *(Continued)*

$$A_1 = \frac{\dot{m}\sqrt{T_{t1}}}{P_{t1}(\cos\alpha_1)\text{MFP}(M_1)} = \frac{22.68\sqrt{288.16}}{101{,}300 \times 0.7660 \times 0.0370}$$

$$= 0.134\,\text{m}^2$$

$$\omega r = 1000 \times 0.3048 = 304.8\,\text{m/s}$$

$$v_{1R} = \omega r - v_1 = 304.80 - 146.11 = 158.69\,\text{m/s}$$

$$\beta_1 = \tan^{-1}\frac{v_{1R}}{u_1} = \tan^{-1} 0.9113 = 42.34\,\text{deg}$$

$$V_{1R} = \sqrt{u_1^2 + v_{1R}^2} = \sqrt{174.13^2 + 158.69^2} = 235.59\,\text{m/s}$$

$$M_{1R} = \frac{V_{1R}}{a_1} = \frac{235.59}{324.73} = 0.7255$$

$$T_{t1R} = T_1\left(1 + \frac{\gamma-1}{2}M_{1R}^2\right) = 262.44(1 + 0.2 \times 0.7255^2) = 290.07\,\text{K}$$

$$P_{t1R} = P_1\left(\frac{T_{t1R}}{T_1}\right)^{\gamma/(\gamma-1)} = 73.03\left(\frac{290.07}{262.44}\right)^{3.5} = 103.67\,\text{kPa}$$

$$P_{t2R} = P_{t1R}\left(\frac{P_{t2R}}{P_{t1R}}\right)_{\phi_{cr},M_{1R}}$$

$$= P_{t1R}\left(1 - \phi_{cr}\frac{\gamma M_{1R}^2/2}{\{1 + [(\gamma-1)/2]M_{1R}^2\}^{\gamma/(\gamma-1)}}\right)$$

$$P_{t2R} = 103.67\left[1 - \frac{0.09 \times 0.7 \times 0.7255^2}{(1 + 0.2 \times 0.7255^2)^{3.5}}\right]$$

$$= 103.67 \times 0.9766 = 101.24\,\text{kPa}$$

$$T_{t2R} = T_{t1R} = 290.07\,\text{K}$$

$$T_{t2} = T_{t1} + \Delta T_t = 288.16 + 22.43 = 310.59\,\text{K}$$

$$\tan\beta_2 = \frac{u_1}{u_2}\left[\tan\beta_1 - \frac{g_c c_p}{\omega r u_1}(T_{t2} - T_{t1})\right]$$

$$\tan\beta_2 = \frac{1}{1.1}\left[\tan 42.34 - \frac{1004}{304.8 \times 174.13}(310.59 - 288.16)\right]$$

$$= 0.4426$$

$$\beta_2 = 23.87\,\text{deg}$$

(Continued)

Example 9.2 *(Continued)*

$$u_2 = \frac{u_2}{u_1} u_1 = 1.1 \times 174.13 = 191.54 \, \text{m/s}$$

$$v_{2R} = u_2 \tan \beta_2 = 191.54 \times 0.4426 = 84.78 \, \text{m/s}$$

$$V_{2R} = \sqrt{u_2^2 + v_{2R}^2} = \sqrt{191.54^2 + 84.78^2} = 209.46 \, \text{m/s}$$

$$v_2 = \omega r - v_{2R} = 304.80 - 84.78 = 220.02 \, \text{m/s}$$

$$\alpha_2 = \tan^{-1} \frac{v_2}{u_2} = \tan^{-1} \frac{220.02}{191.54} = 48.96 \, \text{deg}$$

$$V_2 = \sqrt{u_2^2 + v_2^2} = \sqrt{191.54^2 + 220.02^2} = 291.71 \, \text{m/s}$$

$$T_2 = T_{t2} - \frac{V_2^2}{2g_c c_p} = 310.59 - \frac{291.71^2}{2 \times 1 \times 1004} = 268.21 \, \text{K}$$

$$P_2 = P_{t2R} \left(\frac{T_2}{T_{t2R}} \right)^{\gamma/(\gamma-1)} = 101.24 \left(\frac{268.21}{290.07} \right)^{3.5} = 76.96 \, \text{kPa}$$

$$a_2 = \sqrt{\gamma R g_c T_2} = \sqrt{1.4 \times 287 \times 268.21} = 328.28 \, \text{m/s}$$

$$M_2 = \frac{V_2}{a_2} = \frac{291.71}{328.28} = 0.8886$$

$$M_{2R} = \frac{V_{2R}}{a_2} = \frac{209.46}{328.28} = 0.6381$$

$$P_{t2} = P_2 \left(\frac{T_{t2}}{T_2} \right)^{\gamma/(\gamma-1)} = 76.96 \left(\frac{310.59}{268.21} \right)^{3.5} = 128.61 \, \text{kPa}$$

$$\text{MFP}(M_2) = 0.04004$$

$$A_2 = \frac{\dot{m} \sqrt{T_{t2}}}{P_{t2}(\cos \alpha_2) \text{MFP}(M_2)} = \frac{22.68 \sqrt{310.59}}{128,610 \times 0.65659 \times 0.04004}$$
$$= 0.118 \, \text{m}^2$$

$$T_{t3} = T_{t2} = T_{t1} + \Delta T_t = 310.59 \, \text{K}$$

$$T_3 = \frac{T_{t3}}{1 + [(\gamma-1)/2]M_3^2} = \frac{310.59 \, \text{K}}{1 + 0.2 \times 0.7^2} = 282.87 \, \text{K}$$

(Continued)

Example 9.2 *(Continued)*

$$P_{t3} = P_{t2}\left(\frac{P_{t3}}{P_{t2}}\right)_{\phi_{cs},M_2}$$

$$= P_{t2}\left(1 - \phi_{cs}\frac{\gamma M_2^2/2}{\{1 + [(\gamma-1)/2]M_2^2\}^{\gamma/(\gamma-1)}}\right)$$

$$= 128.61\left[1 - \frac{0.03 \times 0.7 \times 0.8887^2}{(1 + 0.2 \times 0.8887^2)^{3.5}}\right]$$

$$= 128.61 \times 0.09900 = 127.32\,\text{kPa}$$

$$P_3 = P_{t3}\left(\frac{T_3}{T_{t3}}\right)^{\gamma/(\gamma-1)} = 127.32\left(\frac{282.87}{310.59}\right)^{3.5} = 91.79\,\text{kPa}$$

$$a_3 = \sqrt{\gamma R g_c T_3} = \sqrt{1.4 \times 287 \times 282.87} = 337.13\,\text{m/s}$$

$$V_3 = M_3 a_3 = 0.7 \times 337.13 = 235.99\,\text{m/s}$$

$$u_3 = V_3 \cos\alpha_3 = 180.78\,\text{m/s}$$

$$v_3 = V_3 \sin\alpha_3 = 151.69\,\text{m/s}$$

$$A_3 = \frac{\dot{m}\sqrt{T_{t3}}}{P_{t3}(\cos\alpha_3)\text{MFP}(M_3)} = \frac{22.68\sqrt{310.59}}{127,320 \times 0.7660 \times 0.0370} = 0.111\,\text{m}^2$$

$$^\circ R_c = \frac{T_2 - T_1}{T_3 - T_1} = \frac{268.21 - 262.44}{282.87 - 262.44} = 0.2824$$

$$D_r = 1 - \frac{V_{2R}}{V_{1R}} + \frac{|v_{1R} - v_{2R}|}{2\sigma V_{1R}} = 1 - \frac{209.46}{235.59} + \frac{|158.69 - 84.78|}{2 \times 1 \times 235.59} = 0.2678$$

$$D_s = 1 - \frac{V_3}{V_2} + \frac{|v_2 - v_3|}{2\sigma V_2} = 1 - \frac{235.99}{291.71} + \frac{|220.02 - 151.69|}{2 \times 1 \times 291.71} = 0.3081$$

$$\eta_s = \frac{(P_{t3}/P_{t1})^{(\gamma-1)/\gamma} - 1}{T_{t3}/T_{t1} - 1} = \frac{(127.32/101.30)^{1/3.5} - 1}{310.59/288.16} = 0.8672$$

$$e_c = \frac{\gamma-1}{\gamma}\frac{\ln(P_{t3}/P_{t1})}{\ln(T_{t3}/T_{t1})} = \frac{1}{3.5}\frac{\ln(127.32/101.30)}{\ln(310.59/288.16)} = 0.8714$$

(Continued)

Example 9.2 *(Continued)*

$$\psi = \frac{g_c c_p \Delta T_t}{(\omega r)^2} = \frac{1 \times 1004 \times 22.43}{304.8^2} = 0.2424$$

$$\Phi = \frac{u_1}{\omega r} = \frac{174.13}{304.8} = 0.5713$$

The results of this example stage calculation with losses are summarized in Table 9.3, with the given data listed in boldface type (results in English units are shown within parentheses). One sees, by comparison with the isentropic flow results of Example 9.1 (see Table 9.2), that all the properties are the same except the total and static pressures at stations 2R, 2, and 3. One might also want to compare the change in flow properties listed in Table 9.3 with those sketched in Fig. 9.19.

With losses, the resulting stage pressure ratio of Example 9.2 is 1.257. Without losses, the same amount of work per unit mass gives a pressure ratio of 1.300 for Example 9.1. Also note that the flow areas at stations 2 and 3 are larger in Example 9.2 than in Example 9.1 because of the lower total pressures resulting from losses [see Eq. (9.8)].

Table 9.3 Results for Example 9.2 Axial-Flow Compressor Stage Calculation, Flow with Loss

Property		Station				
		1	1R	2R	2	3
T_t	K	**288.2**	290.1	290.1	310.6	**310.6**
	(°R)	(518.7)	(522.1)	(522.1)	(559.1)	(559.1)
T	K	262.4	262.4	268.2	268.2	282.9
	(°R)	(472.4)	(472.4)	(482.8)	(482.8)	(509.2)
P_t	kPa	**101.3**	103.7	101.2	128.6	127.3
	(psia)	(14.69)	(15.04)	(14.68)	(18.65)	(18.46)
P	kPa	73.03	73.03	76.96	76.96	91.79
	(psia)	(10.59)	(10.59)	(11.16)	(11.16)	(13.31)
M		**0.700**	0.7255	0.6381	0.8886	**0.700**
V	m/s	227.31	235.59	209.46	291.71	235.99
	(ft/s)	(745.76)	(772.92)	(687.20)	(957.04)	(774.24)
u	m/s	174.13	174.13	191.54	191.54	180.78
	(ft/s)	(571.29)	(571.29)	(628.40)	(628.40)	(593.10)
v	m/s	146.11	158.69	84.78	220.02	151.69
	(ft/s)	(479.36)	(520.63)	(278.15)	(721.84)	(497.66)
α	deg	**40.00**	—	—	48.96	**40.00**
β	deg	—	42.34	23.87	—	—

9.3.5 Repeating-Stage, Repeating-Row, Mean-Line Design

The analysis and design of an axial-flow compressor are complex with many design choices. To simplify the design, we will consider a stage (Fig. 9.20) whose exit velocity and flow angle equal those at its inlet (repeating stage), made up of "repeating" (i.e., mirror-image) rows of airfoils. The analysis will be based on the behavior of the flow at the *average radius* (halfway between the hub radius and the tip radius), known henceforth as the *mean radius*. With this introduction in mind, the development of design tools for repeating-stage, repeating-row compressors follows (quite suitable for preliminary designs).

9.3.5.1 Assumptions

We make the following assumptions for this analysis:

1. Repeating-row, repeating-airfoil cascade geometry ($\alpha_1 = \beta_2 = \alpha_3$, $\beta_1 = \alpha_2 = \beta_3$, $u_1 = u_2 = u_3$)
2. Two-dimensional flow (i.e., no variation or component of velocity normal to the page)
3. Polytropic efficiency e_c representing stage losses
4. Constant mean radius

9.3.5.2 Analysis

We assume that the following data are given: D, M_1, γ, σ, and e_c. The analysis of the repeating-row, repeating-stage, mean-line design follows.

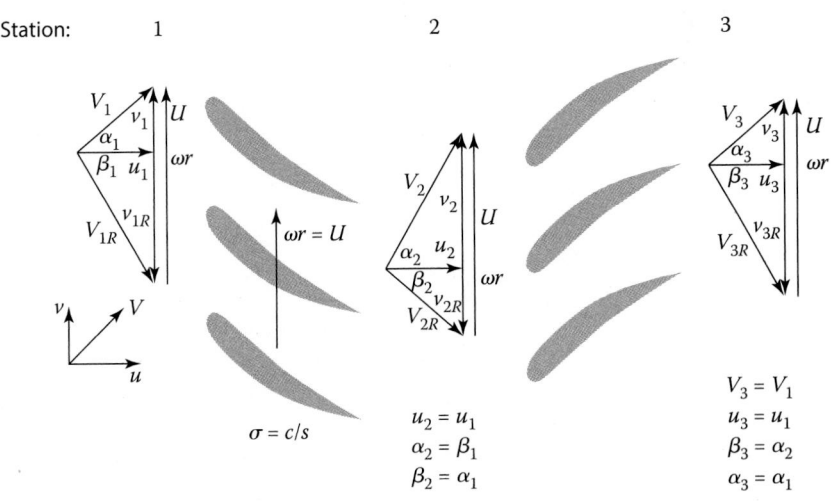

Station: 1 2 3

$$V_3 = V_1$$
$$u_3 = u_1$$
$$\beta_3 = \alpha_2$$
$$\alpha_3 = \alpha_1$$

$$u_2 = u_1$$
$$\alpha_2 = \beta_1$$
$$\beta_2 = \alpha_1$$

$$\sigma = c/s$$

Fig. 9.20   Repeating-row compressor stage nomenclature.

Conservation of mass: Application of this law for steady 1-D flow gives

$$\rho_1 u_1 A_1 = \rho_2 u_2 A_2 = \rho_3 u_3 A_3$$

or

$$\rho_1 A_1 = \rho_2 A_2 = \rho_3 A_3 \qquad (9.30)$$

Repeating-row constraint: Because $\beta_2 = \alpha_1$, then

$$v_{2R} = v_1 = \omega r - v_2$$

or

$$v_1 + v_2 = \omega r \qquad (9.31)$$

Also, because $\beta_3 = \alpha_2$, then $v_{3R} = v_2$ and $v_3 = v_1$; thus the stage exit conditions are indeed identical to those at the stage entrance.

Diffusion factor: Because both

$$D = 1 - \frac{V_{2R}}{V_{1R}} + \frac{v_{1R} - v_{2R}}{2\sigma V_{1R}} = 1 - \frac{V_3}{V_2} + \frac{v_2 - v_3}{2\sigma V_2}$$

and

$$D = 1 - \frac{\cos\alpha_2}{\cos\alpha_1} + \frac{\tan\alpha_2 - \tan\alpha_1}{2\sigma}\cos\alpha_2 \qquad (9.32)$$

are the same for the rotor and stator, D is evaluated only once for the stage. Rearranging Eq. (9.32) to solve for α_2, we find that

$$\cos\alpha_2 = \frac{2\sigma(1-D)\Gamma + \sqrt{\Gamma^2 + 1 - 4\sigma^2(1-D)^2}}{\Gamma^2 + 1} \qquad (9.33)$$

where

$$\Gamma \equiv \frac{2\sigma + \sin\alpha_1}{\cos\alpha_1} \qquad (9.34)$$

In other words, Eq. (9.33) shows that there is *only* one value of α_2 that corresponds to the chosen values of D and σ for each α_1. *Thus the entire flowfield geometry is indicated by those choices.*
Stage total temperature ratio: From the Euler equation, we have for constant radius

$$c_p(T_{t3} - T_{t1}) = \frac{\omega r}{g_c}(v_2 - v_1)$$

From Eq. (9.33)

$$\omega r = v_1 + v_2$$

then

$$c_p(T_{t3} - T_{t1}) = \frac{1}{g_c}(v_2 + v_1)(v_2 - v_1)$$

$$= \frac{v_2^2 - v_1^2}{g_c} = \frac{V_2^2 - V_1^2}{g_c}$$

Thus

$$\frac{\Delta T_t}{V_1^2/(g_c c_p)} = \frac{\cos^2 \alpha_1}{\cos^2 \alpha_2} - 1 \qquad (9.35a)$$

or

$$\tau_s \equiv \frac{T_{t3}}{T_{t1}} = \frac{(\gamma - 1)M_1^2}{1 + [(\gamma - 1)/2]M_1^2}\left(\frac{\cos^2 \alpha_1}{\cos^2 \alpha_2} - 1\right) + 1 \qquad (9.35b)$$

Stage pressure ratio: From Eq. (9.11), we can write

$$\pi_s \equiv \frac{P_{t3}}{P_{t1}} = \left(\frac{T_{t3}}{T_{t1}}\right)^{\gamma e_c/(\gamma-1)}$$

or

$$\pi_s = (\tau_s)^{\gamma e_c/(\gamma-1)} \qquad (9.36)$$

Degree of reaction: A special characteristic of repeating-stage, repeating-row compressor stages is that the degree of reaction must be exactly 0.5, as shown in the following:

$$°R_c = \frac{T_2 - T_1}{T_3 - T_1} = 1 - \frac{T_3 - T_2}{T_3 - T_1} = 1 - \frac{(V_2^2 - V_3^2)/(2g_c c_p)}{T_{t3} - T_{t1}}$$

$$= 1 - \frac{(V_2^2 - V_1^2)/(2g_c c_p)}{(V_2^2 - V_1^2)/(g_c c_p)} = \frac{1}{2}$$

Stage efficiency: From Eq. (9.10), we have

$$\eta_s = \frac{\pi_s^{(\gamma-1)/\gamma} - 1}{\tau_s - 1} \qquad (9.37)$$

Stage exit Mach number: Because

$$V_3 = V_1 \quad \text{and} \quad V^2 = M^2 \gamma R g_c T$$

then

$$\frac{M_3}{M_1} = \sqrt{\frac{T_1}{T_3}}$$

$$= \sqrt{\frac{1}{\tau_s\{1 + [(\gamma-1)/2]M_1^2\} - [(\gamma-1)/2]M_1^2}} \leq 1 \qquad (9.38)$$

Inlet velocity/wheel speed ratio $V_1/(\omega r)$: As will be shown in the next section, one of the most important trigonometric relationships for the stage is that between the total cascade inlet velocity V_1 and the mean wheel speed ωr. Because

$$V_1 = \frac{u_1}{\cos \alpha_1}$$

and

$$\omega r = v_1 + v_2 = u_1(\tan \alpha_1 + \tan \alpha_2)$$

then u_1 can be eliminated to yield

$$\frac{\omega r}{V_1} = (\cos \alpha_1)(\tan \alpha_1 + \tan \alpha_2) \qquad (9.39)$$

Stage loading and flow coefficients: The stage loading coefficient ψ for the repeating-stage, repeating-row, mean-line design can be expressed in terms of the flow angles α_1 and α_2 as

$$\psi = \frac{g_c c_p \Delta T_t}{(\omega r)^2} = \frac{\tan \alpha_2 - \tan \alpha_1}{\tan \alpha_1 + \tan \alpha_2} \qquad (9.40a)$$

Likewise, the flow coefficient can be expressed in terms of the flow angles α_1 and α_2 as

$$\Phi = \frac{u_1}{\omega r} = \frac{1}{\tan \alpha_1 + \tan \alpha_2} \qquad (9.40b)$$

9.3.5.3 General Solution

The behavior of all possible repeating-row compressor stages with given values of D, M_1, γ, σ, and e_c can now be computed. This is done by selecting any α_1 and using the following sequence of equations, expressed as functional relationships:

$$\alpha_2 = f(D, \sigma, \alpha_1) \qquad (9.33)$$

$$\Delta \alpha = \alpha_2 - \alpha_1$$

$$\frac{\Delta T_t}{V_1^2/(g_c c_p)} = f(\alpha_1, \alpha_2) \qquad (9.35a)$$

$$\tau_s = f(M_1, \gamma, \alpha_1, \alpha_2) \qquad (9.35b)$$

$$\pi_s = f(\tau_s, \gamma, e_c) \tag{9.36}$$

$$\eta_s = f(\tau_s, \pi_s, \gamma, e_c) \tag{9.37}$$

$$\frac{\omega r}{V_1} = f(\alpha_1, \alpha_2) \tag{9.39}$$

In most cases, T_{t1} is also known, so that selecting M_1 fixes V_1 and ωr is known.

We can now generate plots of α_2, $\Delta\alpha$ (flow turning), ΔT_t, M_{1R}/M_1, and $V_1/(\omega r)$ vs α_1. Note that M_{1R}/M_1 is included because it is important to keep the relative Mach numbers at the mean radius < 1.0 to avoid excessive losses in the blade tip regions due to shocks.

These calculations have been carried out for $D = 0.5$, $\gamma = 1.4$, $e_c = 0.9$, and 0 deg $< \alpha_1 < 70$ deg. The results are presented in Fig. 9.21a. The most notable characteristics of these data are that the most direct way to increase ΔT_t is to increase V_1 and that to operate at higher values of α_1, a large wheel speed is required. Also note that a stage without inlet swirl ($\alpha_1 = 0$) will have a temperature rise (ΔT_t) that is half that of a stage with $\alpha_1 = 50$ deg. Thus a stage without IGVs like the fan of a high-bypass turbofan engine is limited in the amount of temperature rise ΔT_t it can produce across the rotor blades. Figure 9.21a also shows that the amount of flow turning at the mean radius of a repeating-row stage is quite limited (less than 20 deg for inlet flow angle of 40 deg).

The stage pressure ratio π_s is dependent on the stage total temperature rise ΔT_t and the polytropic efficiency e_c. This relationship is shown in graphical form in Fig. 9.21b. Note that a 17.2% rise in total temperature

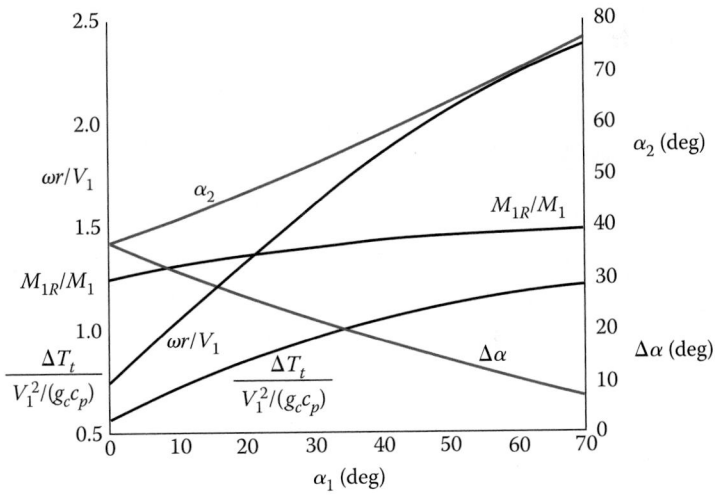

Fig. 9.21a Repeating compressor stage ($D = 0.5$, $\sigma = 1$, and $e_c = 0.9$).

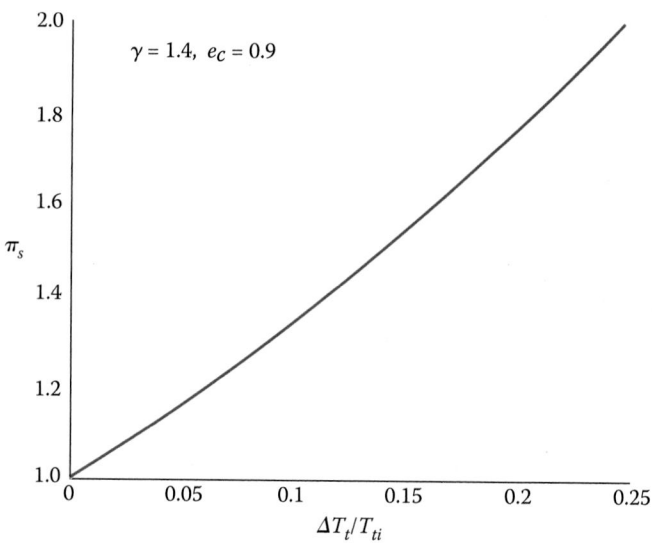

Fig. 9.21b   Compressor stage pressure ratio.

corresponds to a stage pressure ratio of 1.65. For air at 288.17 K (518.7°R), this corresponds to a total temperature rise of 50°C (90°F)—typical for the fan of a high-bypass turbofan engine.

Figure 9.21a shows that specific design values of α_1 and V_1 will give a stage ΔT_t. From Fig. 9.21b, we see that repeating this stage will decrease the pressure ratio of the following stages because the total temperature entering the stage is increasing and the stage ΔT_t is constant.

The following two simple examples illustrate the use of this method and are based on the parameters of Fig. 9.21a.

Example 9.3

Given:

$$M_1 = 0.6, \; a_1 = 360 \, \text{m/s, and} \; \omega r = 300 \, \text{m/s}$$

Then

$$\frac{\omega r}{V_1} = \frac{\omega r}{a_1 M_1} = 1.39$$

$$\frac{\Delta T_t}{V_1^2/(g_c c_p)} = 0.85$$

$$\alpha_1 = 22 \, \text{deg} \quad \Delta \alpha = 25 \, \text{deg}$$

$$\alpha_2 = 47 \, \text{deg} \quad \pi_s = 1.41$$

Example 9.4

Given:

$$a_1 = 320 \, \text{m/s and } M_1/\alpha_1 = 0.5/40 \, \text{deg and } 0.6/50 \, \text{deg}$$

Then

M_1	α_1	$\dfrac{\omega r}{V_1}$	$\dfrac{\omega r}{a_1}$	$\dfrac{\Delta T_t}{V_1^2/(g_c c_p)}$	τ_s	π_s
0.5	40 deg	1.852	0.926	1.053	1.1002	1.351
0.6	50 deg	2.075	1.245	1.1252	1.1511	1.558

Because a_1 is unaffected by the cascade and affected only slightly by M_1, the price of higher π_s is a greatly increased ωr (that is, $1.245/0.926 = 1.344$).

Figures 9.22 and 9.23 show how the performance of repeating compressor stages changes with diffusion factor and solidity, respectively. For a given inlet flow angle α_1 and inlet velocity V_1, we can see from Fig. 9.22 that increasing the diffusion factor D will increase the exit flow angle α_2, stage temperature rise ΔT_t, and wheel speed ωr. Likewise, for a given inlet flow angle α_1 and inlet velocity V_1, Fig. 9.23 shows that increasing the solidity σ will increase the exit flow angle α_2, stage temperature rise ΔT_t, and

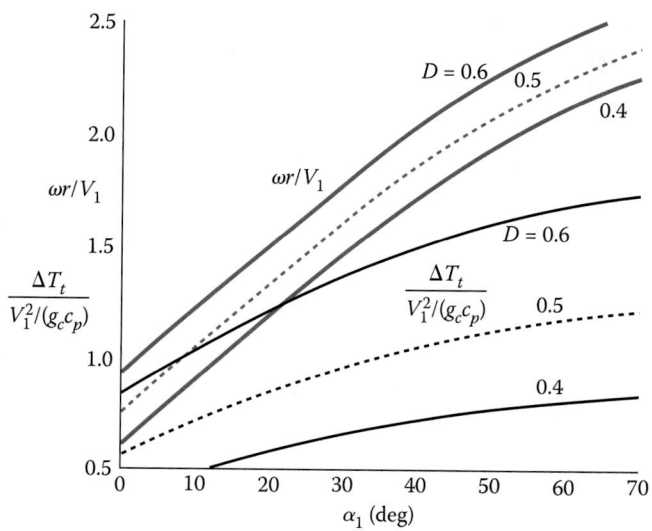

Fig. 9.22 Repeating compressor stage—variation with D ($\sigma = 1$, $e_c = 0.9$).

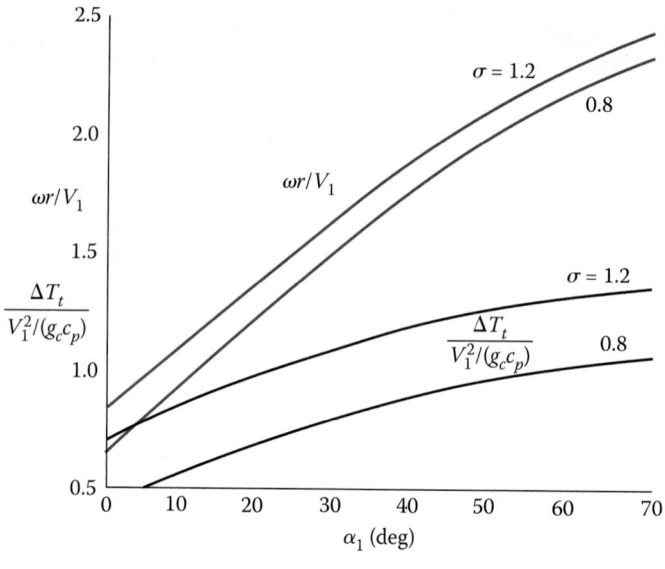

Fig. 9.23   Repeating compressor stage—variation with σ ($D = 0.5$, $e_c = 0.9$).

wheel speed ωr. Fan and compressor design trends in recent decades have been towards higher solidity (>1.0), most notably in wider chord blades and vanes.

For a multistage compressor composed of numerous stages designed by using the repeating-stage, repeating-row, mean-line design, the total temperature change across each of these stages will be the same. Also, the mean-line radius for each of these stages is the same. Because the ratio of M_3/M_1 is less than 1, the pressure ratio of any downstream repeating stages will be lower.

As previously mentioned, a repeating-stage, repeating-row, mean-line approach is often suitable for preliminary design. The COMPR program provided with this textbook allows for such a design, as well as others (i.e., constant tip radius).

Example 9.5

Given the following for a repeating-row, repeating-stage design:

$$\alpha_1 = 40 \deg, \quad D = 0.6, \quad \sigma = 1, \quad \gamma = 1.4, \quad M_1 = 0.5, \quad e_c = 0.9$$

Find the airflow velocities, flow angles, and performance parameters.

(Continued)

Example 9.5 *(Continued)*

Solution:

$$\Gamma = \frac{2\sigma + \sin\alpha_1}{\cos\alpha_1} = \frac{2 + \sin 40\,\text{deg}}{\cos 40\,\text{deg}} = 3.4499$$

$$\cos\alpha_2 = \frac{2 \times 0.4 \times 3.4499 + \sqrt{3.4499^2 + 1 - 4 \times 1^2 \times 0.4^2}}{3.4499^2 + 1} = 0.4853$$

$$\alpha_2 = 60.966\,\text{deg}$$

$$\frac{\Delta T_t}{V_1^2/(g_c c_p)} = \frac{\cos^2 40\,\text{deg}}{\cos^2 60.966\,\text{deg}} - 1 = 1.491$$

$$\tau_s = \frac{0.4 \times 0.5^2}{1 + 0.2 \times 0.5^2}\left(\frac{\cos^2 40\,\text{deg}}{\cos^2 60.966\,\text{deg}} - 1\right) + 1 = 1.142$$

$$\pi_s = 1.142^{3.5 \times 0.9} = 1.519$$

$$\eta_s = \frac{1.519^{1/3.5} - 1}{1.142 - 1} = 0.893$$

$$\frac{M_3}{M_1} = \sqrt{\frac{1}{1.142(1 + 0.2 \times 0.5^2) - 0.4 \times 0.5^2}} = 0.954$$

$$\frac{M_{1R}}{M_1} = \frac{\cos 40\,\text{deg}}{\cos 60.966\,\text{deg}} = 1.578$$

$$M_{1R} = 0.789$$

$$\frac{\omega r}{V_1} = (\cos 40\,\text{deg})(\tan 40\,\text{deg} + \tan 60.966\,\text{deg}) = 2.023$$

$$\psi = \frac{\tan 60.966\,\text{deg} - \tan 40\,\text{deg}}{\tan 40\,\text{deg} + \tan 60.966\,\text{deg}} = 0.364$$

$$\Phi = \frac{1}{\tan 40\,\text{deg} + \tan 60.966\,\text{deg}} = 0.379$$

The entire flowfield geometry is completely specified by the chosen values of D, σ, and α_1. The velocities and temperatures throughout this stage can now be determined by knowing either a velocity or a temperature at any station in the stage. For example, if $T_{t1} = 288.16$ K, then simple calculations give

$$a_1 = 332\,\text{m/s}, \quad V_1 = V_{2R} = V_3 = 166\,\text{m/s}, \quad \omega r = 336\,\text{m/s},$$
$$V_{1R} = V_2 = 261.9\,\text{m/s}, \quad \Delta T_t = 40.9\,\text{K}$$

9.3.6 Flow Path Dimensions

This section describes basic methods to estimate the physical dimensions, number of blades and vanes, and shapes of blades and vanes of the compressor preliminary design. The methods are simple enough to do hand calculations, but all are incorporated in the COMPR design program.

9.3.6.1 Annulus Area: Radial Dimensions

The preliminary design of multistage axial-flow compressors is typically based on determining each stage's flow properties along the compressor mean line (line along compressor axis that corresponds to the mean radius). The annulus area at any station is based on the flow properties (T_t, P_t, Mach number, and flow angle) at the mean radius and the total mass flow rate. Equation (9.8) is the easiest equation to calculate the flow area at any station i:

$$A_i = \frac{\dot{m}\sqrt{T_{ti}}}{P_{ti}(\cos \alpha_i)\text{MFP}(M_i)} \tag{9.8}$$

The *mean radius* of a flow annulus is defined as the average of the tip radius and hub radius. Consider Fig. 9.24, which shows a typical annulus area.

In many calculations, the flow area can be calculated and the mean radius is tied to the required rotor speed at the mean radius ωr_m. The designer can select ω and calculate the required mean radius. Then the root radius and tip radius are calculated directly from the mean radius and flow area. In some calculations, the designer may want to select the ratio of the hub radius to the tip radius r_h/r_t. The hub/tip ratio at the inlet to a multistage compressor normally is between 0.6 and 0.75,

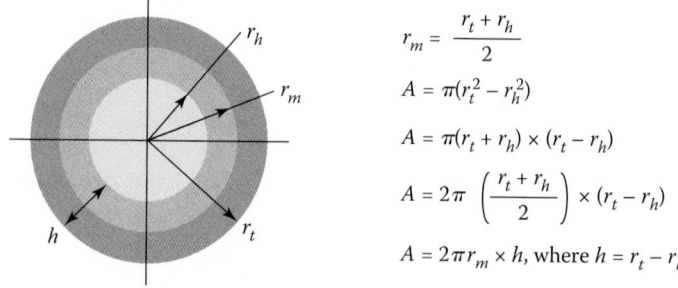

$$r_m = \frac{r_t + r_h}{2}$$

$$A = \pi(r_t^2 - r_h^2)$$

$$A = \pi(r_t + r_h) \times (r_t - r_h)$$

$$A = 2\pi \left(\frac{r_t + r_h}{2}\right) \times (r_t - r_h)$$

$$A = 2\pi r_m \times h, \text{ where } h = r_t - r_h$$

Fig. 9.24 Flow annulus dimensions.

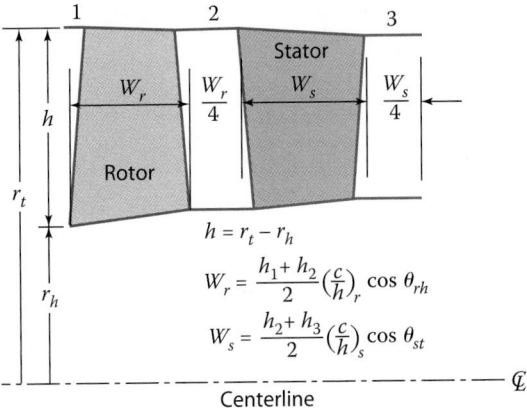

Fig. 9.25 Typical axial dimensions of a compressor stage.

whereas that at the compressor exit is in the range of 0.9 to 0.92. Then the tip, hub, and mean radii directly follow from the geometry:

$$A = \pi(r_t^2 - r_h^2) = \pi r_t^2 \left[1 - \left(\frac{r_h}{r_t} \right)^2 \right]$$

$$r_t = \sqrt{\frac{A}{\pi[1 - (r_h/r_t)^2]}} \qquad r_h = r_t \left(\frac{r_h}{r_t} \right) \qquad r_m = \frac{r_t + r_h}{2}$$

The variation of the flow area and the associated dimensions of the flow path from one station to the next can be calculated easily by using the preceding relationships and the results sketched. Figure 9.25 is a sketch of typical results. The calculation of the axial dimensions requires additional data.

9.3.6.2 Axial Dimensions

The axial dimensions of a typical stage are also shown in Fig. 9.25, and these can be used to estimate the axial length of a stage. Blade axial widths (W_r and W_s) of a stage are calculated in the COMPR program along with the blade spacings ($W_r/4$ and $W_s/4$) based on user-input chord-to-height ratios c/h for the rotor and stator blades and assuming a constant chord length for each blade. A minimum width of $1/4$ in. (0.006 m) and a minimum spacing of $1/8$ in. (0.003 m) are used in the plot of compressor cross section and calculation of axial length.

The value of the chord/height ratio c/h selected for a blade depends on such factors as the stage loading coefficient, diffusion factor, and so on. Typical values of c/h range from 0.2 to 0.8. Higher values of c/h lead to longer stages and fewer blades.

The chord of a blade c is obtained by multiplying its height h by the user-input chord/height ratio c/h. The blade width (W_r or W_s) is the maximum value of the axial chord c_x for that blade, which depends on the stagger angle θ as shown in Fig. 9.15. The maximum axial chord c_x normally occurs at the hub for a rotor blade and at the tip for a stator blade.

9.3.6.3 Number of Blades

The COMPR program calculates the number of blades n_b for the rotor and stator of a stage based on the user-input values of the solidity σ along the mean line. For a rotor or stator, the blade spacing s along the mean line is equal to the blade chord c divided by its solidity ($\sigma = c/s$). The number of blades is simply the circumference of the mean line divided by the blade spacing, rounded up to the next integer.

9.3.6.4 Blade Profile

The shapes of compressor rotor and stator blades are based on airfoil shapes developed specifically for compressor applications. One such airfoil shape is the symmetric NACA 65A010 compressor airfoil whose profile shape is shown in Fig. 9.26 and specified in Table 9.4. This airfoil has a thickness that is 10% of its chord c.

To obtain the desired change in fluid flow direction, the airfoil's camber line is curved and the symmetric shape of the airfoil profile is distributed about the camber line. The curved camber line of the airfoil and nomenclature used for flow through a compressor cascade are shown in Fig. 9.15. Normally, the curved camber line is that of a circular arc or a parabola. If a circular arc is used, then the stagger angle of the airfoil θ is the average of the camber line's inlet angle γ_i and exit angle γ_e. From Fig. 9.27 and basic trigonometry, the radius of the camber line is given by

$$r = \frac{c \sin \theta}{\cos \gamma_e - \cos \gamma_i} \tag{9.41}$$

The COMPR computer program uses the NACA 65A010 compressor profile with circular arc camber line to sketch the blade shapes for a stage. The user can change the blade thickness for the rotor or stator blades,

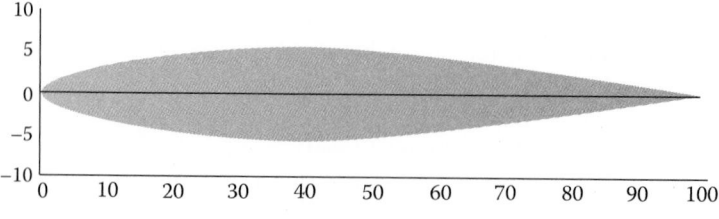

Fig. 9.26 NACA 65A010 blade shape.

Table 9.4 NACA 65A010 Compressor Airfoil[a,b]

x/c, %	y/c, %	x/c, %	y/c, %
0.0	0.0	40	4.995
0.5	0.765	45	4.983
0.75	0.928	50	4.863
1.25	1.183	55	4.632
2.5	1.623	60	4.304
5	2.182	65	3.809
7.5	2.65	70	3.432
10	3.04	75	2.912
15	3.658	80	2.352
20	4.127	85	1.771
25	4.483	90	1.188
30	4.742	95	0.604
35	4.912	100	0.0

[a]Leading-edge radius $= 0.00636c$. Trailing-edge radius $= 0.00023c$.
[b]Source: Ref. 50.

select the desired stage, specify the number of blades to be sketched, and specify the percentage of chord for stacking blade profiles. The user can view one or more blades in the rotor and stator at a time.

If the user selects to view just one blade, the program initially sketches the blade shape at the mean radius as a solid, and the shapes at the hub and tip are drawn in dashed outline stacked over the other profile at the user-selected stacking location. Figure 9.28 is a copy of the blade profile plot screen for one blade. The user can move the slider for percentage of blade height by using

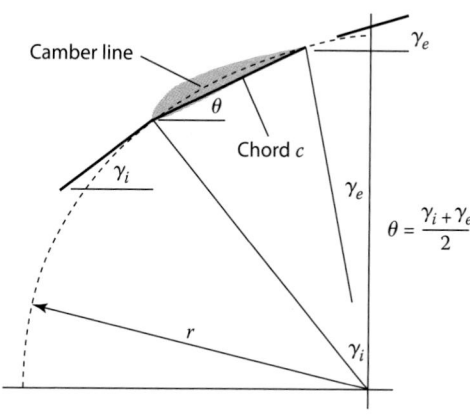

Fig. 9.27 Circular arc camber line.

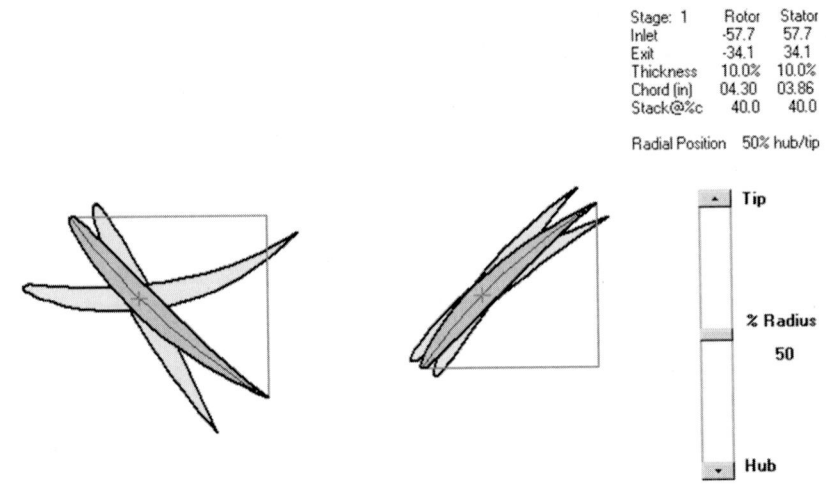

Stage: 1	Rotor	Stator
Inlet	-57.7	57.7
Exit	-34.1	34.1
Thickness	10.0%	10.0%
Chord (in)	04.30	03.86
Stack@%c	40.0	40.0

Radial Position   50% hub/tip

Tip

% Radius
50

Hub

Fig. 9.28   Single-blade profile sketch from COMPR.

the up and down arrows and replot the blade shape at a new height (radial location). If the user selects to view more than one blade, the program initially sketches the blade shapes at the mean radius.

9.3.7 Radial Variation

A practical goal of a design is to do a constant amount of work on the fluid passing through a stage that is independent of radius. For this to occur, the Euler pump equation [Eq. (9.1)] reveals that less "turning" of the fluid will be required as the radius increases. In addition, the static pressure must increase with the radius to maintain the radial equilibrium of the swirling flow. All airfoil and flow properties must, therefore, vary with radius. Up to this point, our analysis has focused on mean-line design.

In the following analysis, the relationships are developed for the radial variation of flow between the rows of airfoils (stage stations 1, 2, and 3). Initially, a general equation for radial equilibrium is developed, assuming no radial velocity. Then specific radial variations of the velocity are analyzed for the case of constant work on the fluid.

9.3.7.1 Radial Equilibrium Equation

The main features of the radial variation of flow *between* the rows of airfoils are accounted for in the following analysis. These equations are valid at stations 1, 2, and 3 at any radius. We assume the following:

1. Constant losses (entropy s = constant with respect to radius)
2. No circumferential variations
3. No radial velocity

The definition of total (stagnation) enthalpy with no radial velocity gives

$$dh_t = dh + \frac{d(u^2 + v^2)}{2g_c} \qquad \text{(i)}$$

Gibbs's equation can be written as

$$T\,ds = dh - \frac{dP}{\rho}$$

With $s = $ constant, Gibbs's equation becomes

$$dh = \frac{dP}{\rho} \qquad \text{(ii)}$$

Equations (i) and (ii) can be combined to give

$$dh_t = \frac{dP}{\rho} + \frac{u\,du}{g_c} + \frac{v\,dv}{g_c}$$

When h_t is a constant, this equation becomes the well-known Bernoulli equation. We rewrite the preceding equation as

$$\frac{dh_t}{dr} = \frac{1}{\rho}\frac{dP}{dr} + \frac{1}{g_c}\left(u\frac{du}{dr} + v\frac{dv}{dr}\right) \qquad \text{(iii)}$$

For radial equilibrium, the pressure gradient in the radial direction must balance the centrifugal acceleration, or

$$\frac{dP}{dr} = \frac{\rho v^2}{r g_c} \qquad \text{(iv)}$$

Equations (iii) and (iv) may be combined to yield

$$\frac{dh_t}{dr} = \frac{1}{g_c}\left(u\frac{du}{dr} + v\frac{dv}{dr} + \frac{v^2}{r}\right) \qquad (9.42)$$

This general form of the radial equilibrium equation prescribes the relationship between the radial variations of the three variables: h_t, u, and v. The designer can specify the radial variation of any two variables, and Eq. (9.42) specifies the radial variation of the third variable. This is done in the next two sections.

9.3.7.2 Free-Vortex Variation of Swirl

We consider the case where the axial velocity u and total enthalpy h_t do not vary with radius. For this case, Eq. (9.42) becomes

$$v\frac{dv}{dr} + \frac{v^2}{r} = 0 \quad \text{or} \quad \frac{dv}{v} + \frac{dr}{r} = 0$$

for which the integrated solution is

$$rv = \text{const}$$

or

$$v = v_m \frac{r_m}{r} \tag{9.43}$$

where the subscript m refers to values at the mean radius. Because v varies inversely with radius, this is known as *free-vortex* flow. Thus, if the flow at station 1 has $v_1 r = v_{m1} r_m$ and the rotor airfoils modify the flow to $v_2 r = v_{m2} r_m$, then the Euler equation confirms that this is a constant-work machine because

$$\omega r(v_2 - v_1) = \omega r \left(\frac{v_{m2} r_m}{r} - \frac{v_{m1} r_m}{r} \right)$$
$$= \omega r_m (v_{m2} - v_{m1})$$
$$= \text{const}$$

Equation (9.43) also shows that as long as r does not vary substantially from r_m (say, $\pm 10\%$), the airfoil and flow properties will not vary much from those of the original mean-line design.

If the compressor has constant axial velocity and repeating stages (not rows) for which $v_1 = v_3$, then the degree of reaction can be shown to be

$$^{\circ}R_c = 1 - \frac{v_1 + v_2}{2\omega r} \tag{9.44}$$

which, for a free-vortex machine, becomes

$$^{\circ}R_c = 1 - \frac{v_{m1} r_m + v_{m2} r_m}{2\omega r^2} = 1 - \frac{\text{const}}{(r/r_m)^2} \tag{9.45}$$

For a stage whose degree of reaction is 50% at the mean radius (where $(\omega r = v_1 + v_2)$, it becomes more difficult to design rotor airfoils at $r > r_m$ and stator airfoils at $r < r_m$. In fact, because $^{\circ}R_c = 0$ at $r = r_m/\sqrt{2}$, the rotor will actually experience accelerating flow for smaller radii, whereas this is never the case for the stator. Because of these problems, compressor designers have looked at other swirl distributions.

9.3.7.3 General Swirl Distributions

We now consider the case where the total enthalpy does not vary with the radius and where the swirl velocity at the inlet and exit to the rotor has the following general variation with radius:

$$v_1 = a \left(\frac{r}{r_m} \right)^n - b \frac{r_m}{r} \quad \text{and} \quad v_2 = a \left(\frac{r}{r_m} \right)^n + b \frac{r_m}{r} \tag{9.46}$$

where a, b, and r_m are constants. From the Euler equation, the work per unit mass flow is

$$\Delta h_t = \frac{\omega r(v_2 - v_1)}{g_c} = \frac{2b\omega r_m}{g_c} \qquad (9.47)$$

which is independent of the radius. Note from the preceding relationship that the constant b in Eq. (9.46) is determined by the total enthalpy rise across the rotor and the mean rotor speed. As will be shown next, constant a in Eq. (9.46) is related to the degree of reaction at the mean radius. Three cases of the swirl distribution of Eq. (9.46) are considered: $n = -1$, $n = 0$, and $n = 1$.

1. For $n = -1$:

$$v_1 = a\frac{r_m}{r} - b\frac{r_m}{r} \quad \text{and} \quad v_2 = a\frac{r_m}{r} + b\frac{r_m}{r} \qquad (9.48)$$

which is the *free-vortex* swirl distribution. For this case and constant h_t, Eq. (9.42) required that the axial velocity u not vary with the radius. Using Eq. (9.44) for a repeating stage ($v_3 = v_1$), we see that the degree of reaction is

$$°R_c = 1 - \frac{a}{\omega r_m}\left(\frac{r_m}{r}\right)^2 \qquad (9.49)$$

An equation for the constant a in Eq. (9.46) is obtained by evaluating the preceding equation at the mean radius. Thus

$$a = \omega r_m(1 - °R_{cm}) \qquad (9.50)$$

2. For $n = 0$:

$$v_1 = a - b\frac{r_m}{r} \quad \text{and} \quad v_2 = a + b\frac{r_m}{r} \qquad (9.51)$$

This is called the *exponential* swirl distribution. Before solving for the axial velocity distribution, we rewrite Eq. (9.42) for constant total enthalpy h_t in terms of the dimensionless radius r/r_m as

$$u\,du + v\,dV + v^2\frac{d(r/r_m)}{r/r_m} = 0$$

Integration of the preceding equation from the mean radius to any radius r gives

$$\frac{1}{2}(u^2 - u_m^2) = -\frac{1}{2}(v^2 - v_m^2) - \int_1^{r/r_m} v^2\frac{d(r/r_m)}{r/r_m} \qquad (9.52)$$

At the entrance to the rotor (station 1), substitution of the equation for v_1 into Eq. (9.52) and integration give the axial velocity profile

$$u_1^2 = u_{1m}^2 - 2\left(a^2 \ell n \frac{r}{r_m} + \frac{ab}{r/r_m} - ab\right) \qquad (9.53)$$

Likewise, at the exit from the rotor (station 2), we obtain

$$u_2^2 = u_{2m}^2 - 2\left(a^2 \ell n \frac{r}{r_m} - \frac{ab}{r/r_m} + ab\right) \qquad (9.54)$$

Because the axial velocity is not constant, we must start with Eq. (9.13b) to obtain an expression for the degree of reaction for the velocity distribution given by Eqs. (9.51), (9.53), and (9.54). We start by noting for repeating stages ($V_3 = V_1$) that

$$°R_c = \frac{T_2 - T_1}{T_3 - T_1} = \frac{T_{t2} - T_{t1}}{T_{t3} - T_{t1}} - \frac{V_2^2 - V_1^2}{2g_c c_p (T_{t3} - T_{t1})}$$

$$= 1 - \frac{V_2^2 - V_1^2}{2g_c c_p (T_{t3} - T_{t1})} = 1 - \frac{u_2^2 - u_1^2}{2g_c c_p (T_{t3} - T_{t1})} - \frac{v_2^2 - v_1^2}{2g_c c_p (T_{t3} - T_{t1})}$$

From the Euler equation, we write $g_c c_p (T_{t3} - T_{t1}) = \omega r (v_2 - v_1)$, and the preceding becomes

$$°R_c = 1 - \frac{u_2^2 - u_1^2}{2\omega r (v_2 - v_1)} - \frac{v_2^2 - v_1^2}{2\omega r (v_2 - v_1)}$$

$$°R_c = 1 + \frac{u_1^2 - u_2^2}{2\omega r (v_2 - v_1)} - \frac{v_2 + v_1}{2\omega r} \qquad (9.55)$$

For the case where $u_{1m} = u_{2m}$, Eqs. (9.53) and (9.54) give

$$u_1^2 - u_2^2 = 4ab\left(1 - \frac{r_m}{r}\right)$$

and from the swirl distribution of Eq. (9.51), we have

$$v_2 + v_1 = 2a \quad \text{and} \quad v_2 - v_1 = 2b\frac{r_m}{r}$$

Thus the degree of reaction for the exponential swirl distribution is given by

$$°R_c = 1 + \frac{a}{\omega r_m} - \frac{2a}{\omega r} \qquad (9.56)$$

3. For $n = 1$:

$$v_1 = a\frac{r}{r_m} - b\frac{r_m}{r} \quad \text{and} \quad v_2 = a\frac{r}{r_m} + b\frac{r_m}{r} \tag{9.57}$$

This is called *first-power* swirl distribution. At the entrance to the rotor (station 1), substitution of the equation for v_1 into Eq. (9.52) and integration give the axial velocity profile

$$u_1^2 = u_{1m}^2 - 2\left[a^2\left(\frac{r}{r_m}\right)^2 + 2ab\,\ell n\frac{r}{r_m} - a^2\right] \tag{9.58a}$$

Likewise, at the exit from the rotor (station 2), we obtain

$$u_2^2 = u_{2m}^2 - 2\left[a^2\left(\frac{r}{r_m}\right)^2 - 2ab\,\ell n\frac{r}{r_m} - a^2\right] \tag{9.58b}$$

An equation for the degree of reaction for the case of $u_{1m} = u_{2m}$ is obtained in the same manner. The resulting relationship is

$$^\circ R_c = 1 + \frac{2a\,\ell n(r/r_m)}{\omega r_m} - \frac{a}{\omega r_m} \tag{9.59}$$

9.3.7.4 Comparison

Equations (9.49), (9.56), and (9.59) give the radial variation of the degree of reaction for the free-vortex, exponential, and first-power swirl distributions, respectively. The value of the constant a is evaluated at the mean radius and is given by Eq. (9.50) for all three cases. Consider the case where the degree of reaction at the mean radius is 0.5.

Results from Eqs. (9.49), (9.56), and (9.59) are plotted in Fig. 9.29 for the range $0.5 < r/r_m < 1.5$. From Fig. 9.29, for r/r_m between about 0.85 and 1.15, there is very little difference in degree of reaction between the three choices for swirl distribution. The differences become more pronounced outside that range. Note that at any radius other than the mean, the free-vortex swirl distribution results in the lowest value for the degree of reaction, and the first-power swirl distribution gives the highest value.

Low, even negative values for the degree of reaction can occur in the hub region for stages having small hub/tip ratios r_h/r_t (large span blades and vanes). The variation with the radius ratio r/r_t for the degree of reaction

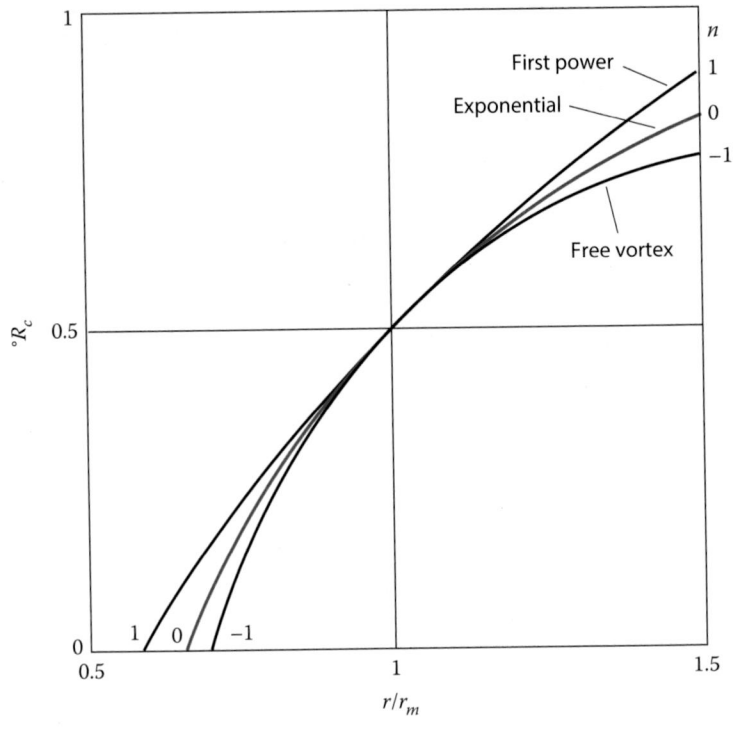

Fig. 9.29 Radial variation of degree of reaction (from Ref. 22).

can be obtained from Eqs. (9.49), (9.56), and (9.59) and the following relationship between r/r_t and r/r_m:

$$\frac{r}{r_m} = \frac{r}{r_m} \frac{1 + r_h/r_t}{2}$$

(9.60)

For the free-vortex swirl distribution with 0.5 deg of reaction at the mean radius, the degree of reaction is 0 at $r/r_m = 0.707$, which corresponds to the hub radius for a stage having a hub/tip ratio of 0.547. It is not uncommon for the initial stages of a fan or compressor to have a hub/tip ratio of 0.4. The degree of reaction at the hub can be increased by increasing the degree of reaction at the mean radius for stages with low hub/tip ratios. Another way to increase the degree of reaction at the hub is to change the swirl distribution.

For the first-power swirl distribution with 0.5 degree of reaction at the mean radius, the degree of reaction is 0 at $r/r_m = 0.6065$, which corresponds to the hub radius for a stage having a hub/tip ratio of 0.435. Even with this improvement over the free-vortex distribution, the degree of reaction at the mean radius will have to be greater than 0.5 to have a positive degree of reaction at the hub of a stage with a hub/tip ratio of 0.4.

Because of the problems encountered with the preceding swirl distribution, modern compressor designers have looked to nonconstant work variation with radius (h_t does vary with radius). However, these are absolutely dependent on advanced computational methods for their definition.

The free-vortex, exponential, and first-power swirl distributions are all included as design choices in the COMPR program. The program assumes that the solidity σ varies inversely with radius (constant chord length) for calculation of the radial variation in the diffusion factor. To have values for the total pressure and static pressures at stations 2 and 2R, the program requires a value of the stator loss coefficient ϕ_{cs} in addition to the polytropic efficiency for repeating-row, repeating-stage calculations.

Example 9.6

Consider the results using the COMPR program for a repeating-row, repeating-stage, mean-line stage design. The free-vortex swirl distribution is used for the radial variation of the tangential velocity v in this example.

Given:

$$\dot{m} = 22.68 \text{ kg/s } (50 \text{ lbm/s}), \quad P_t = 101.3 \text{ kPa } (14.70 \text{ psia}),$$

$$T_t = 288.16 \text{ K } (518.7°\text{R}), \quad \omega = 900 \text{ rad/s}, \quad e_c = 0.9, \quad \phi_{cs} = 0.03,$$

$$\text{on mean line: } \sigma = 1, \quad \alpha_1 = 45 \text{ deg}, \quad D = 0.5, \quad M_1 = 0.5$$

Solution:

Tables 9.5a and 9.5b summarize the stage results obtained from COMPR. The amount of information is remarkable, all obtained from the foundational material presented thus far in this chapter. One could take the information and construct velocity triangles at the hub, mean, and tip sections of stations 1, 2, and 3! Note that the Mach numbers at the hub of stations 1, 2, and 3 are higher than on the mean radius due to the swirl distribution. This affects both the static temperatures and the static pressures at the hub.

The cross section of this stage, as created by COMPR, is shown in Fig. 9.30. The stage length of 0.099 m is based on a chord/height ratio of 0.75 for both the rotor and stator blades.

The variation in absolute tangential (swirl) and axial velocities with radius across the rotor is shown in Fig. 9.31 for the free-vortex swirl distribution. To

(Continued)

Table 9.5a Flow Property Results for Example 9.6

Property		1h	1m	1t	1Rm	2Rm	2h	2m	2t	3h	3m	3t
T_t	K	288.2	288.2	288.2	303.1	303.1	318.1	318.1	318.1	318.1	318.1	318.1
	(°R)	(518.7)	(518.7)	(518.7)	(545.6)	(545.6)	(572.6)	(572.6)	(572.6)	(572.6)	(572.6)	(572.6)
T	K	272.7	274.4	275.7	274.4	289.4	284.5	289.4	293.1	303.0	304.4	305.4
	(°R)	(490.8)	(494.0)	(496.3)	(494.0)	(520.9)	(512.1)	(520.9)	(527.6)	(545.4)	(547.9)	(549.8)
P_t	kPa	101.3	101.3	101.3	120.9	117.7	139.3	139.3	139.3	138.3	138.3	138.3
	(psia)	(14.70)	(14.70)	(14.70)	(17.55)	(17.08)	(20.21)	(20.21)	(20.21)	(20.06)	(20.06)	(20.06)
P	kPa	83.5	85.4	86.8	85.4	100.1	94.3	100.1	104.6	116.7	118.5	119.9
	(psia)	(12.11)	(12.39)	(12.59)	(12.39)	(14.52)	(13.67)	(14.52)	(15.18)	(16.93)	(17.19)	(17.40)
M		0.533	0.500	0.475	0.723	0.487	0.769	0.704	0.653	0.499	0.475	0.456
V	m/s	176	166	158	240	166	260	240	224	174	166	160
	(ft/s)	(578)	(545)	(519)	(788)	(545)	(853)	(788)	(735)	(571)	(545)	(524)
u	m/s	117	117	117	117	117	117	117	117	117	117	117
	(ft/s)	(385)	(385)	(385)	(385)	(385)	(385)	(385)	(385)	(385)	(385)	(385)
v	m/s	132	117	106	209	117	232	209	191	128	117	108
	(ft/s)	(432)	(385)	(348)	(687)	(385)	(761)	(687)	(626)	(421)	(385)	(355)
α	deg	48.25	45.00	42.08	60.72	45.00	63.14	60.72	58.41	47.57	45.00	42.64
β	deg	—	—	—	60.72	45.00	—	—	—	—	—	—
radii	m	0.324	0.363	0.402	0.363	0.363	0.328	0.363	0.398	0.332	0.363	0.394
	(in.)	(12.76)	(14.30)	(15.83)	(14.30)	(14.30)	(12.91)	(14.30)	(15.68)	(13.07)	(14.30)	(15.52)

Example 9.6 *(Continued)*

Table 9.5b   Stage Results for Example 9.6

Hub	$°R_c = 0.3888$	$D_r = 0.5455$	$D_s = 0.5101$
Mean	$°R_c = 0.5000$	$D_r = 0.5000$	$D_s = 0.5000$
Tip	$°R_c = 0.5850$	$D_r = 0.4476$	$D_s = 0.4906$

$$\Delta T_t = 29.93\ \text{K} \quad \eta_s = 89.55\% \quad AN^2 = 1.184 \times 10^7\ \text{m}^2 \cdot \text{rpm}^2$$

$$\psi = \frac{g_c c_p \Delta T_t}{(\omega r)^2} = 0.281 \quad \Phi = \frac{u_1}{\omega r} = 0.359$$

obtain the 30.5°C (54.9°F) temperature rise across the rotor, the tangential velocity is increased 92 m/s (302 ft/s). The quantity AN^2, a common industry practice for representing blade centrifugal stress, is presented in Appendix I and discussed later in this chapter.

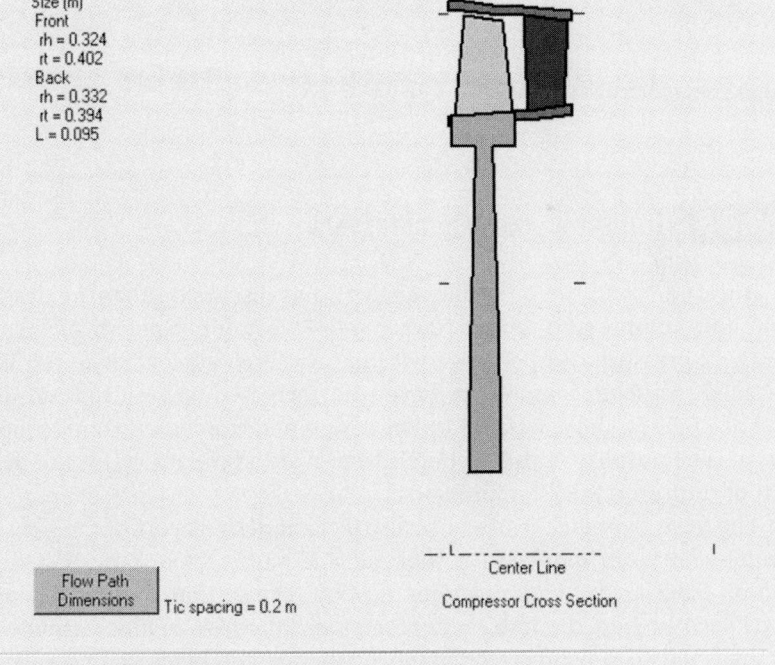

Size (m)
Front
rh = 0.324
rt = 0.402
Back
rh = 0.332
rt = 0.394
L = 0.095

Flow Path Dimensions   Tic spacing = 0.2 m

Center Line

Compressor Cross Section

Fig. 9.30   Cross section of a single stage from COMPR.

(Continued)

Example 9.6 (Continued)

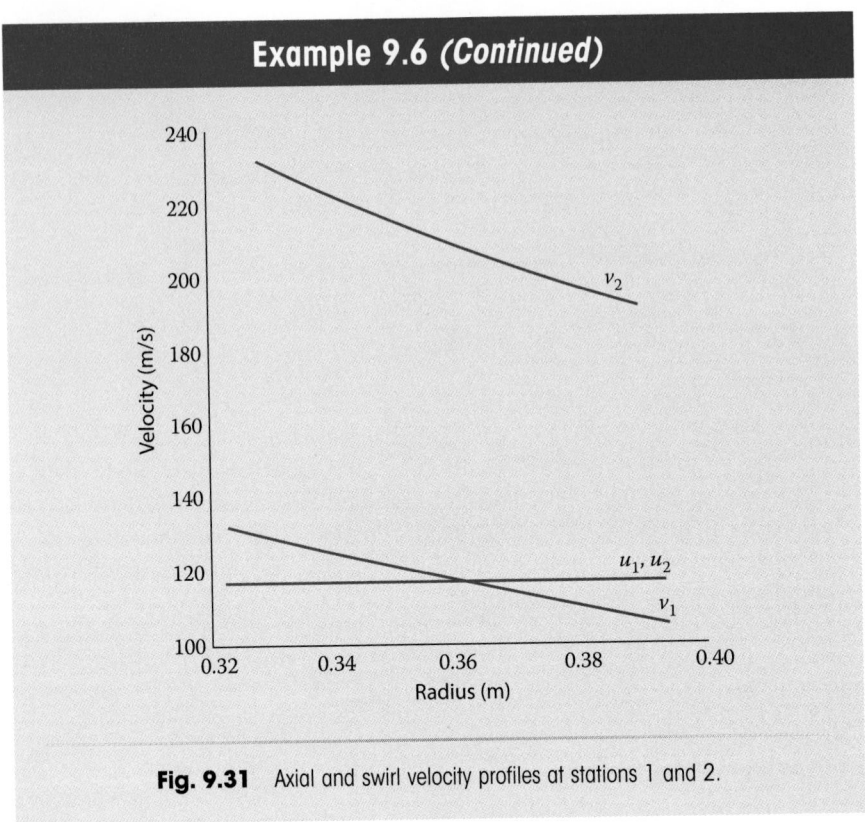

Fig. 9.31 Axial and swirl velocity profiles at stations 1 and 2.

9.3.8 Transonic Axial Compressors

It is clear from the preceding analysis and discussion that the design of axial-flow compressors trends toward more work input per stage, resulting in higher pressure rise per stage with improved efficiency. Since the 1960s, so-called *transonic compressors* have seen increased and widespread use in aircraft engines, most notably in fan design and, to some extent, propeller design. The purpose of this section is to provide a brief overview and discussion of transonic axial compressors.

The term *transonic* is used because in such designs part of the flow *relative to the rotating blade row* is subsonic and part is supersonic. A review of Section 9.3.2 on velocity diagrams quickly reveals that the portion of the blade span nearest the hub section or inner flow path engine casing would be subjected to subsonic relative flow, whereas the blade span nearest the tip or outer flow path engine casing would experience supersonic relative flow. It is common for the tip relative Mach number M_{1Rtip} of a modern fan design to be 1.3–1.4 at high (takeoff) rotational speed. A primary concern, then, for such designs is achieving the desired performance by tailoring the blade airfoil shapes so the resulting shock structure minimizes losses as well as other effects like aeroelastic and endwall and blade boundary layer interactions. This is one area where internal flows, which are generally more complex than external flows, have benefited more than external flows. The increased losses associated with shock formation on propeller blade tips limit the flight Mach number of the most modern turboprop aircraft to about 0.7. In contrast, one finds transonic axial compressors used for the fans of modern high-bypass turbofans, the multistage fans of modern low-bypass turbofans, and in some cases, the first few stages of the high-pressure compressor. These applications power a range of aircraft with top flight Mach numbers in the high subsonic to highly supersonic range.

As suggested by Fig. 9.32, the flow fields of transonic axial compressors are complex and highly 3-D. Secondary flows (any flows other than the streamwise flow) and their interactions are a major source of concern. Modern fan blades use blade twist, sweep (viewed from the axial cross section), and lean (observed looking axially through the fan), which also contribute to the 3-D flow effects (see Fig. 9.33b). These 3-D design features are even more pronounced when one examines the fan blade designs of modern high-bypass-ratio turbofans as in Figs. 1.8b–1.8f. In fact, a GE90-115B composite fan blade is literally a work of art, on display at the Museum of Modern Art's (MOMA's) Architecture and Design collection in New York City since 2004. Figure 9.33a shows an axial cross-section view of a first-generation fan blade and a newer one. Each blade has an average span of approximately 12 in.; the older blade has an average chord of around 3 in., the newer blade around 5.5 in. The transonic fan blade is the same in Figs. 9.33a and 9.33b as well as Fig. 9.34.

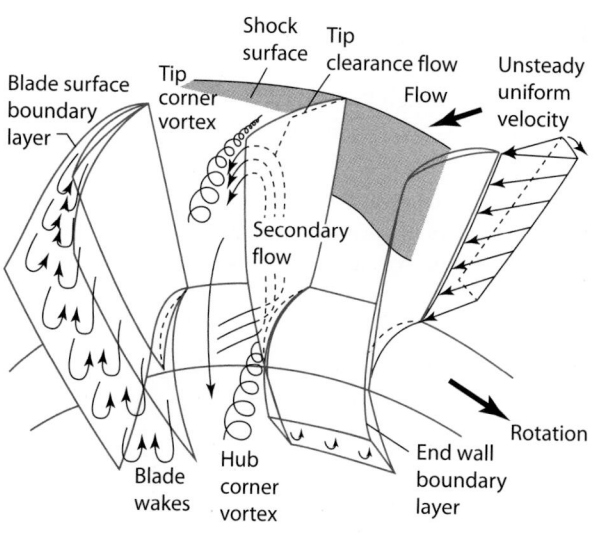

Fig. 9.32 Complex flow phenomena in transonic compressors (Ref. 51).

a) 1960s (left) and more modern fan blades

b) Transonic fan blade looking axially through

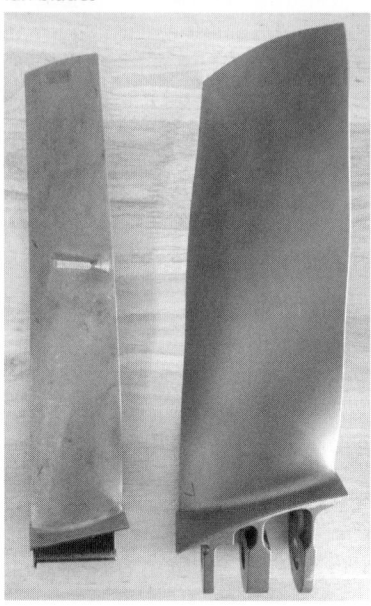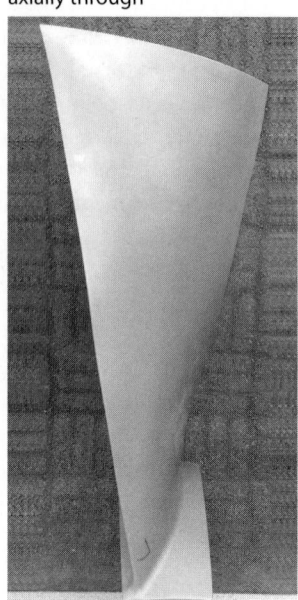

Fig. 9.33 Fan blades.

A significant consideration in the design of transonic blades is the control of shock location and strength to minimize aerodynamic losses without limiting flow. Not surprisingly, the airfoil contours used for transonic tip blades contain features we see on supersonic wings and include very sharp leading edges, very thin sections (low thickness/chord ratios), leading-edge sweep in some cases, and highly tailored suction surface shapes. The shape of the suction surface is key because it (1) influences the Mach number just ahead of the leading edge of the blade row, and (2) sets the maximum flow rate. One generally does not want choking of the flow in the fan or compressor. Figure 9.34 shows a tip airfoil of a modern transonic fan blade and provides a nice perspective of the blade twist from tip to hub.

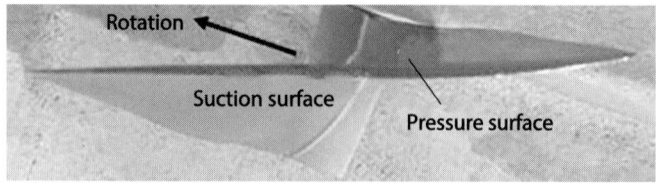

Fig. 9.34 Transonic fan blade: top view looking down.

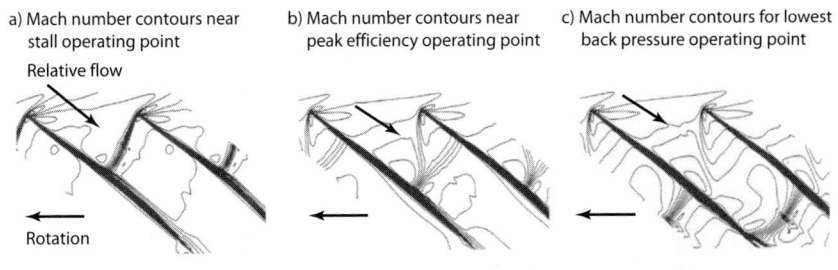

a) Mach number contours near stall operating point

b) Mach number contours near peak efficiency operating point

c) Mach number contours for lowest back pressure operating point

Fig. 9.35 Shock structure of transonic compressors (Ref. 52).

Figure 9.35 illustrates some typical features of the shock structure associated with transonic axial compressors at different operating conditions. The three operating conditions of near-stall, peak efficiency, and lowest back pressure in Fig. 9.35 can be associated with the compressor map operating conditions of stall, peak efficiency, and any low-pressure ratio point on the design 100% N_c speed line in Fig. 8.3. A grouping of Mach number contours in Fig. 9.35 indicates a shock wave. As shown in Fig. 9.35a, at the near-stall condition, we find a near-normal shock (normal to the relative flow) positioned at the leading edge of the blade row flow passage. At the peak efficiency condition, we observe an oblique leading-edge shock followed by a weak normal shock in the flow passage (Fig. 9.35b). When the blade is really unloaded, as is the case in Fig. 9.35c, the oblique shock appears quite weak whereas the normal shock in the passage has strengthened and moved downstream near the trailing-edge plane. For a first-order analysis, we have the tools from oblique and normal shock theory provided in Chapter 3 to analyze the flow passage shocks in transonic axial compressors (see Problem 3.23). Qualitatively, from our knowledge of oblique and normal shocks, the shock structure in Fig. 9.35b appears to correspond with minimal losses, thus correlating with peak efficiency.

9.3.9 Inlet Guide Vanes

Inlet guide vanes (IGVs) are required before the first stage of an axial-flow compressor when the flow entering the stage is required to be at a swirl angle α_1. As can be seen in Fig. 9.21a, higher values of the swirl angle α_1 give higher temperature rise ΔT_t per stage, which corresponds to a higher pressure ratio. Only high-bypass turbofan engines are currently produced without inlet guide vanes—the fan can produce the required pressure ratio with no entering swirl and the weight of the IGVs would be prohibitive.

Figure 9.36a shows a common inlet guide vane configuration—fixed strut/variable flap—for a modern fighter engine. The front part of the

Fig. 9.36a   Inlet guide vanes.

guide vane (fixed strut) is part of the engine structure that supports the front bearing for the low-pressure spool. The back portion of the guide vane is movable (variable flap) to improve off-design fan performance.

The inlet guide vanes change both the direction of the flow and its Mach number. If we model the flow through the inlet guide vane as adiabatic and with no losses (isentropic), the annulus flow area given by Eq. (9.8) can be used to give the following relationship between stations 0 and 1 (inlet and exit, respectively):

$$\cos \alpha_1 = \frac{A_0 \, \text{MFP}(M_0)}{A_1 \, \text{MFP}(M_1)} \tag{9.61}$$

This relationship is plotted for a given M_0 in Fig. 9.36b as α_1 versus M_1 for different values of A_1/A_0. We can see from this plot that for a specific value of inlet Mach number M_0 and exit swirl angle α_1, an increasing value of the exit annulus area A_1 reduces the exit Mach number M_1. If the area of the annulus is constant ($A_1/A_0 = 1$), increasing the swirl angle will increase the exit Mach number M_1.

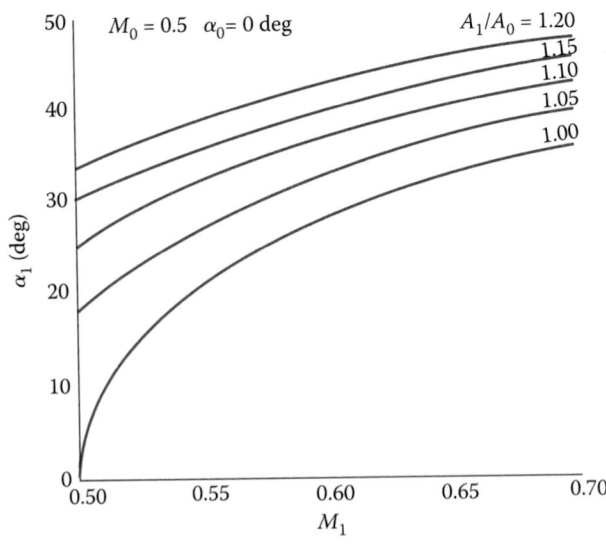

Fig. 9.36b   Mean exit conditions from inlet guide vanes.

9.3.10 Design Process

The theory and design tools presented in previous sections can now be applied to the design of a multistage axial-flow compressor. The design process requires both engineering judgment and knowledge of typical design values. We will consider the design of a compressor suitable for a simple turbojet gas turbine engine in the examples used throughout this section. It is assumed that such a compressor will require inlet guide vanes.

The complete design process for the compressor will include the following items:

1. Selection of rotational speed and annulus dimensions
2. Selection of the number of stages
3. Calculation of airflow angles for each stage at the mean radius
4. Calculation of airflow angle variations from the hub to tip for each stage
5. Selection of blading using experimental cascade data
6. Verification of compressor efficiency based on cascade loss data
7. Prediction of off-design performance
8. Rig testing of design

Items 1–4 will be covered in this section. Many of the remaining steps will be discussed in following sections. The design process is inherently iterative, often requiring a return to an earlier step when prior assumptions are found to be invalid.

Table 9.6 lists ranges for some design parameters of axial-flow compressors and can be used as a guide. The higher values of fan and low-pressure compressor pressure ratio are obtained with inlet guide vanes.

Many technical specialities are interwoven in a design (e.g., an axial-flow air compressor involves at least thermodynamics, aerodynamics, structures, materials, manufacturing processes, and controls). Design requires the active participation and disciplined communication of many technical specialists.

9.3.10.1 Selection of Rotational Speed and Annulus Dimensions

Selection of the rotational speed is not a simple matter. For a simple turbojet engine, it depends on a balance of the requirements of both components on the common shaft—the compressor and the turbine. Typically, the rotational speed can be found by assuming values for the rotor blade-tip speed, the hub/tip ratio, and axial velocity of the first stage. A check of the magnitude of AN^2 (related to blade stress) is also made (see Appendix I). For the first stage of typical compressors, the hub/tip ratio is between 0.6 and 0.75, the axial Mach number is between 0.48 and 0.6, and the rotor blade-tip speed is between 1150 and 1500 ft/s (350 to 460 m/s) (higher values for first-stage fans without inlet guide vanes).

Table 9.6 Range of Axial-Flow Compressor Design Parameters

Parameter	Design Range
Fan or low-pressure compressor	
Pressure ratio for one stage	1.5–2.0
Pressure ratio for two stages	2.0–3.5
Pressure ratio for three stages	3.5–5.0
Inlet corrected mass flow rate	40–44 lbm/(s · ft^2)
	[195–215 kg/(s · m^2)]
Tip speed	1400–1750 ft/s
	(427–533 m/s)
Hub/tip ratio at inlet	>0.3
Diffusion factor	0.50–0.55
Solidity	0.8–2.0
High-pressure compressor	
Pressure ratio for front stages	1.6–2.5
Pressure ratio for back stages	1.2–1.5
Inlet corrected mass flow rate	36–38 lbm/(s · ft^2)
	[175–185 kg/(s · m^2)]
Hub/tip ratio at inlet	0.60–0.75
Hub/tip ratio at exit	0.90–0.92
Maximum rim speed at exit	1300–1500 ft/s
	(396–457 m/s)
Flow coefficient	0.45–0.55
Stage loading coefficient	0.30–0.35
Diffusion factor	0.50–0.55
Solidity	0.8–1.5

Example 9.7

We will design a compressor for the simple turbojet gas turbine engine with a compressor pressure ratio of 10.2 and an air mass flow rate of 150 lbm/s (68.04 kg/s). The air enters the compressor with $P_t/T_t = 14.7$ psia/518.7°R (101.3 kPa/288.2 K).

Initially, we select a modest value for the axial Mach number M_0 of 0.5 entering the inlet guide vanes. Because the inlet guide vanes will accelerate and turn the flow to an angle of 30–45 deg, we will start with a compressor inlet Mach number M_1 of 0.6 and a high tip speed U_t of 1400 ft/s (426.7 m/s). From the definition of mass flow parameter, we can write

$$A = \frac{\dot{m}\sqrt{T_t}}{(\cos \alpha)(P_t)\mathrm{MFP}(M)}$$

(Continued)

Example 9.7 (Continued)

For the given mass flow rate and the inlet total pressure, total temperature, and Mach number, we can determine the flow annulus area at the entrance to the inlet guide vanes using Appendix D:

$$\text{MFP}(M_0) = 0.3969$$

$$A_0 = \frac{\dot{m}\sqrt{T_{t0}}}{P_{t0}\text{MFP}(M_0)} = \frac{150\sqrt{518.7}}{14.70 \times 0.3969}$$

$$= 585.7 \text{ in.}^2 \ (0.3780 \text{ m}^2)$$

Likewise, the area at the face of the first-stage rotor, assuming a flow angle of 40 deg, is

$$\text{MFP}(M_1) = 0.4476$$

$$A_1 = \frac{\dot{m}\sqrt{T_t}}{(\cos\alpha_1)(P_t)\text{MFP}(M)} = \frac{150\sqrt{518.7}}{0.7660 \times 14.70 \times 0.4476}$$

$$= 677.8 \text{ in.}^2 \ (0.4375 \text{ m}^2)$$

The tip radius r_t and mean radius r_m are directly related to the flow area A and hub/tip ratio r_h/r_t by

$$r_t = \sqrt{\frac{A}{\pi[1 - (r_h/r_t)^2]}}$$

$$r_m = r_t\frac{1 + r_h/r_t}{2}$$

The rotational speeds ω and N are related to the rotor tip speed by

$$\omega = \frac{U_t}{r_t} \quad \text{and} \quad N = \frac{30}{\pi}\omega$$

From the preceding relationships, we can calculate r_t, ω, N, and AN^2 over the hub/tip ratio range and obtain the data in Table 9.7a for a tip speed of 1400 ft/s (426.7 m/s).

With the information in Table 9.7a, it is most appropriate for the compressor designer to get together with the turbine designer to select a rotational speed. The first stages of turbines typically have rim speeds of about 900 ft/s (274.3 m/s), a hub/tip ratio of 0.85 to 0.9, and a Mach number entering the rotor of 1.1 at 65 deg to the centerline of the turbine. Assuming a fuel/air ratio of 0.04, the turbine mass flow rate is 156 lbm/s. For $\gamma = 1.3$ and a

(Continued)

Example 9.7 *(Continued)*

Table 9.7a Variation of First-Stage Size and Speed for a Tip Speed of 1400 ft/s

r_h/r_t	r_t, in.	r_m, in.	ω, rad/s	N, rpm	AN^2, in.$^2 \cdot$ rpm^2
0.60	18.36	14.69	914.8	8735.7	5.17×10^{10}
0.65	19.33	15.95	869.0	8298.2	4.67×10^{10}
0.70	20.57	17.49	816.6	7798.2	4.12×10^{10}
0.75	22.21	19.44	756.4	7222.7	3.54×10^{10}

5% total pressure loss through the combustor, the annulus area at the turbine rotor is

$$\text{MFP}(M) = 0.5140$$

$$A = \frac{\dot{m}\sqrt{T_t}}{(\cos \alpha)(P_t)\text{MFP}(M)} = \frac{156\sqrt{3200}}{\cos 65 \deg \times 142.40 \times 0.5140}$$

$$= 285.3 \text{ in.}^2 \ (0.1841 \text{ m}^2)$$

The rim speed U_r can be used to calculate the rotational speed of the shaft ω. We note from Fig. I.1 that the rim radius r_r is equal to the hub radius r_h minus the rim height h_r. Then the shaft speed ω can be written as

$$\omega = \frac{U_r}{r_r} = \frac{U_r}{r_h - h_r}$$

For a given value of hub/tip ratio r_h/r_t and flow area A, the shaft speed ω is

$$\omega = \frac{U_r}{r_r} = \frac{U_r}{\sqrt{\dfrac{A}{\pi[(r_t/r_h)^2] - 1]}} - h_r}$$

Table 9.7b gives the variation of the turbine first-stage shaft speed with hub/tip ratio for a rim speed U_r of 900 ft/s (274.3 m/s) for a rim height h_r of 2 in.

At this time, *we select a rotor speed of 800 rad/s for the compressor* to keep its AN^2 low while maintaining a turbine rim speed of 900 ft/s. This value will be used in the preliminary design of the compressor stages. The approximate mean radius of the compressor and turbine will be 18 and 16.7 in., respectively.

Table 9.7b Variation of Turbine Size and Speed for a Rim Speed U_r of 900 ft/s and 2-in. Rim Height h_r

r_h/r_t	r_t, in.	r_h, in.	r_m, in.	r_r, in.	ω, rad/s
0.85	18.09	15.38	16.73	13.38	807.4
0.875	19.68	17.22	18.45	15.22	709.4
0.9	21.86	19.68	20.77	17.68	611.0

9.3.10.2 Selection of the Number of Stages

The number of compressor stages depends on the overall pressure ratio and the change in total temperature of each stage. Normally the changes in total temperature of the first and last stages are less than those of the other stages. In the absence of inlet guide vanes, the inlet flow to the first stage is axial, and less work can be done than in a stage whose inlet flow has swirl. The exit flow from the last stage is normally axial, which reduces the work of this stage.

The preliminary compressor design is based on repeating-row, repeating-stage design. Thus the repeating-row, repeating-stage design equations and Figs. 9.21a through 9.23 can be used to estimate the number of compressor stages. We will base the design on a mean-line diffusion factor D of 0.5, a mean-line solidity σ of 1, and a polytropic efficiency e_c of 0.9.

Because the stage inlet Mach number decreases through the compressor, the stage pressure ratio of a repeating-row, repeating-stage design will decrease. However, *the change in total temperature across each stage will remain constant* for this type of design.

Example 9.8

For $M_1 = 0.6$ and α_1 between 30 and 40 deg, the data in Table 9.7b for the first stage can be obtained from the equations for repeating-row, repeating-stage design. Thus, for the assumed value of M_1 and range of α_1, we see from Table 9.7c that the total temperature change for a stage will be between 66.9 and 73.3°R. The total temperature rise for the whole compressor will be approximately

$$\Delta T_t = T_{ti}[(\pi_c)^{(\gamma-1)/(\gamma e_c)} - 1] = 518.7[(10.2)^{1/(3.5\times0.9)} - 1] = 565.5°R$$

The number of stages will then be either 9 (565.5/66.9 = 8.45) or 8 (565.5/73.3 = 7.71). *We select eight stages for this design.* Because the first stage has inlet guide vanes, we will use the same change in total temperature for this stage as for the other compressor stages. In addition, we will allow the last stage to have exit swirl (this can be removed by an additional set of stators) and will use the same change in total temperature for the last stage as for the other compressor stages. Thus each stage will have an equal temperature rise of 70.7°R (565.5/8). For the first stage, this corresponds to a temperature ratio of 1.136, which requires α_1 of about 36 deg and $V_1/(\omega r)$ of 0.57.

We select an inlet flow angle of 36 deg. The inlet air at $T_t = 518.7°R$ and $M = 0.6$ has a velocity of 647 ft/s. The resulting stage loading coefficient is 0.3302 and flow coefficient is 0.4609, both within the range of these coefficients for modern axial-flow compressors. For a rotor speed of 800 rad/s,

(Continued)

Example 9.8 *(Continued)*

Table 9.7c Variation in Repeating-Row, Repeating-Stage Designs with Inlet Flow Angle

α_1, deg	α_2, deg	τ_s	π_s	ΔT_t, R	$V_1/(\omega r)$	M_{1R}	ψ	Φ
30.0	51.79	1.1290	1.465	66.9	0.6250	0.840	0.375	0.541
32.0	52.94	1.1316	1.476	68.3	0.6051	0.844	0.359	0.513
34.0	54.10	1.1342	1.487	69.6	0.5867	0.848	0.344	0.486
36.0	55.28	1.1367	1.497	70.9	0.5697	0.852	0.330	0.461
38.0	56.47	1.1390	1.507	72.1	0.5541	0.856	0.318	0.437
40.0	57.67	1.1413	1.516	73.3	0.5396	0.859	0.306	0.413

the mean radius is about 17 in. The inlet annulus area of the first stage will be

$$\text{MFP}(M_1) = 0.44757$$

$$A_1 = \frac{\dot{m}\sqrt{T_{t1}}}{(\cos \alpha_1)(P_{t1})\text{MFP}(M_1)} = \frac{150\sqrt{518.7}}{\cos 36 \deg \times 14.70 \times 0.44757}$$

$$= 641.8 \text{ in.}^2 \ (0.4142 \text{ m}^2)$$

This area is 1.1 times the inlet area of the guide vanes. The corresponding tip radius and hub radius at the inlet to the first stage are about 20 and 14 in., respectively, giving a first-stage hub/tip ratio of 0.7.

Example 9.9

Consider the preliminary design of mean line. The COMPR program was used to perform the following analysis of a repeating-row, repeating-stage, mean-line design with a free-vortex swirl distribution. The following input data were entered:

Number of stages: 8	Mass flow rate: 150 lbm/s
Rotor angular speed ω: 800 rad/s	Inlet total pressure P_t: 14.70 psia
Inlet total temperature T_t: 518.7°R	Inlet flow angle α_1: 36 deg
Inlet Mach number M_1: 0.6	Diffusion factor D: 0.5
Solidity σ: 1.0	Polytropic efficiency e_c: 0.9
Rotor c/h: 0.6	Stator c/h: 0.6
Gas constant: 53.34 ft · lbf/(lbm · °R)	Ratio of specific heats γ: 1.4

The results of this extensive example that covers steps 3 and 4 of the design process are included in the online Chapter 9 Supporting Material.

9.3.11 Compressor Performance

We now turn our attention to a brief discussion on overall compressor performance (not just at the design point). As presented in Section 8.1.3, the pumping characteristics of axial-flow fans and compressors are best represented by a plot (based on experimental data) called the *compressor map*. The data are presented in terms of *corrected quantities* that are related to the performance of the compressor. A compressor is tested in a rig similar to the sketch of Fig. 9.37. The rotational speed is controlled by the electric motor and the mass flow rate by the valve. The inlet and exit conditions are measured (typically for each stage in a multistage design) along with the rotational speed and input power. These data are reduced to give the resulting fan or compressor map (see Figs. 9.38 and 9.39).

9.3.11.1 Corrected Quantities

The four corrected quantities shown in Table 9.8 are normally used to map a compressor's performance. The corrected rotational speed is directly proportional to the ratio of the axial to rotational velocity. The corrected mass flow rate is directly proportional to the Mach number of the entering flow. We note that

$$\dot{m}_i = \frac{P_{ti}}{\sqrt{T_{ti}}} A_i \, \mathrm{MFP}(M_i)$$

Then

$$\dot{m}_{ci} = \frac{\dot{m}_i \sqrt{T_{ti}}}{P_{ti}} \frac{P_{ref}}{\sqrt{T_{ref}}} = \frac{P_{ref}}{\sqrt{T_{ref}}} A_i \, \mathrm{MFP}(M_i)$$

Therefore,

$$\dot{m}_{ci} = f(M_i)$$

At engine station 2 (entrance to compressor or fan)

$$\dot{m}_{c2} = \frac{P_{ref}}{\sqrt{T_{ref}}} A_2 \, \mathrm{MFP}(M_2) \tag{9.62}$$

Thus the corrected mass flow rate is directly proportional to the Mach number at the entrance to the compressor.

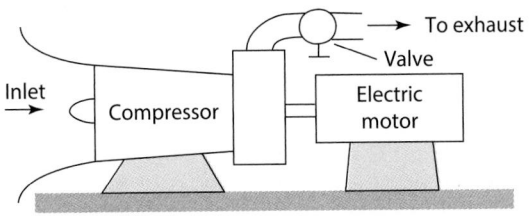

Fig. 9.37 Sketch of compressor test rig.

Table 9.8 Compressor Corrected Quantities

Item	Symbol	Corrected		
Pressure	P_{ti}	$\delta_i = \dfrac{P_{ti}}{P_{ref}}$ where $P_{ref} = 14.696$ psia		
Temperature	T_{ti}	$\theta_i = \dfrac{T_{ti}}{T_{ref}}$ where $T_{ref} = 518.7°R$		
Rotational speed	$N = $ RPM	$N_c = \dfrac{N}{\sqrt{\theta_i}}$		
Mass flow rate	$\dot{m}_i$	$\dot{m}_{ci} = \dfrac{\dot{m}_i \sqrt{\theta_i}}{\delta_i}$		

9.3.11.2 Compressor Map

The performance of two modern high-performance fan stages is shown in Fig. 9.38. They have no inlet guide vanes. One has a low tangential Mach number (0.96) to minimize noise. The other has supersonic tip speed and a considerably larger pressure ratio. Both have high axial Mach numbers.

Variations in the axial-flow velocity in response to changes in pressure cause the multistage compressor to have quite different mass flow vs pressure ratio characteristics from those of one of its stages. The performance map of a typical high-pressure-ratio compressor is shown in Fig. 9.39.

A limitation on fan and compressor performance of special concern is the stall or surge line. Steady operation above the line is impossible, and entering the region even momentarily is dangerous to the compressor and its application.

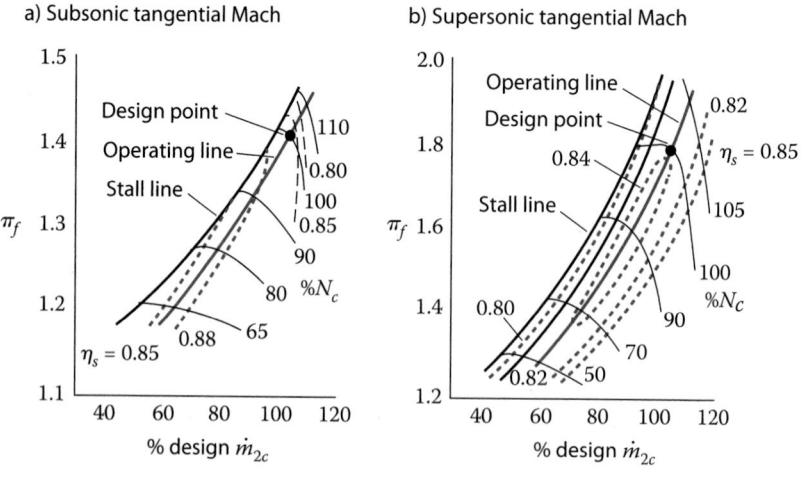

Fig. 9.38 Typical fan stage maps (Ref. 53).

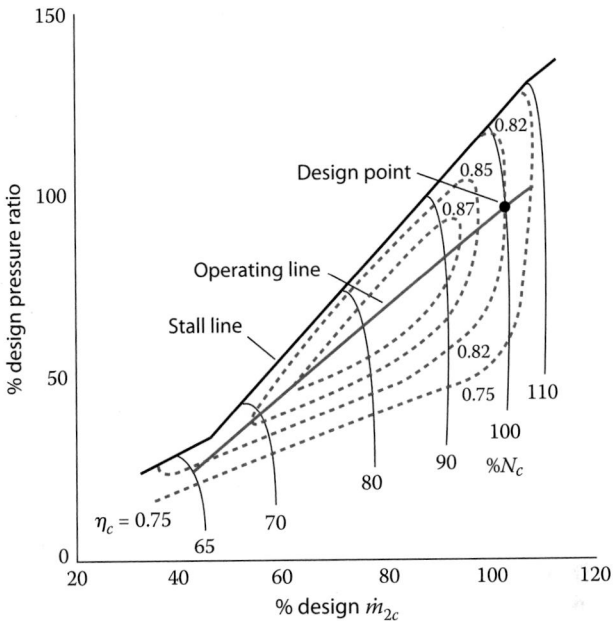

Fig. 9.39 Typical compressor map.

9.3.11.3 Compressor Starting Problems

The cross section of a multi-stage compressor is shown in Fig. 9.40. Also shown (in solid lines) at three locations in the machine are mean-line rotor inlet velocity diagrams for on-design operation. The design shown here has a constant mean radius, a constant axial velocity, and zero swirl at the rotor-inlet stations.

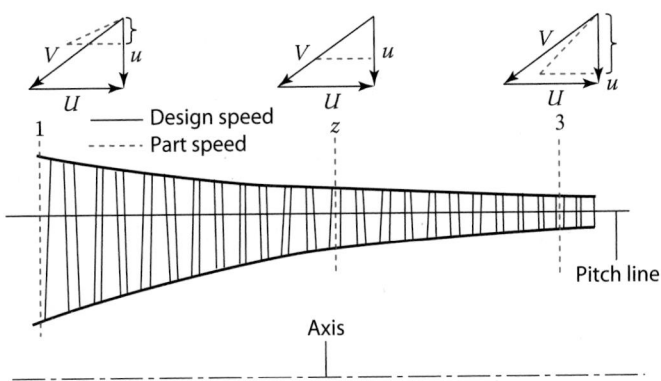

Fig. 9.40 Cross section of a multistage compressor. Rotor-inlet velocity diagrams are shown for inlet, middle, and exit stages (left to right).

Let us apply the continuity equation, which states that the mass flow rate is the same at all stations in the compressor. Then, at the inlets to the first and last stages,

$$\dot{m} = \rho_1 A_1 u_1 = \rho_3 A_3 u_3$$

or

$$\frac{u_1}{u_3} = \frac{\rho_3 A_3}{\rho_1 A_1} \qquad (9.63)$$

At the design point, because $u_3 = u_1$, then $\rho_3/\rho_1 = A_1/A_3$. When the compressor is operating at a lower speed, however, it is not able to produce as high a density ratio (or pressure ratio) as at design speed, and because the area ratio remains fixed, it follows from Eq. (9.63) that the inlet/outlet axial velocity ratio must have a value less than that at design.

In Fig. 9.40, the velocity diagrams shown (in dashed lines) indicate the rotor inlet conditions at partial speed. For each of the three stages represented, the blade speed is the same, but the inlet stage axial velocity is shown less than that for the outlet stage in accordance with Eq. (9.63). Because of the varying axial velocity, it is possible for only one stage near the middle of the machine (called the *pivot stage*) to operate with the same angle of attack at all speeds. At lower speeds, the blading forward of the pivot stage is pushed toward stall (highly loaded), and the blading aft of the pivot stage is windmilling with a tendency toward choking (highly unloaded). This is the basis for the so-called engine or compressor starting problem. At speeds higher than the design value, these trends are reversed.

Several techniques have been utilized to reduce the low-speed and starting effects:

1. Use bleed valves. Release air from the middle stages, reducing the tendency to windmill in later stages.
2. Use multispool compressors. Run different spools at their own suitable speeds.
3. Use variable stators in the early rows.
4. Use combinations of the preceding techniques.

An example of an early high-performance compressor is that used in the General Electric J79 turbojet engine. This compressor has 17 stages on a single spool that results in a pressure ratio of 13.5. Figure 9.41 shows the compressor case of this engine with the variable stators used on the first six stages and the manifold for the bleed valves on the ninth stage. Figure 1.4b shows a photo of the actual variable stator vane hardware, which also included a row of variable inlet guide vanes. Most modern engines have compressors with multiple spools, bleed valves, and variable vanes in the early rows of the core compressor.

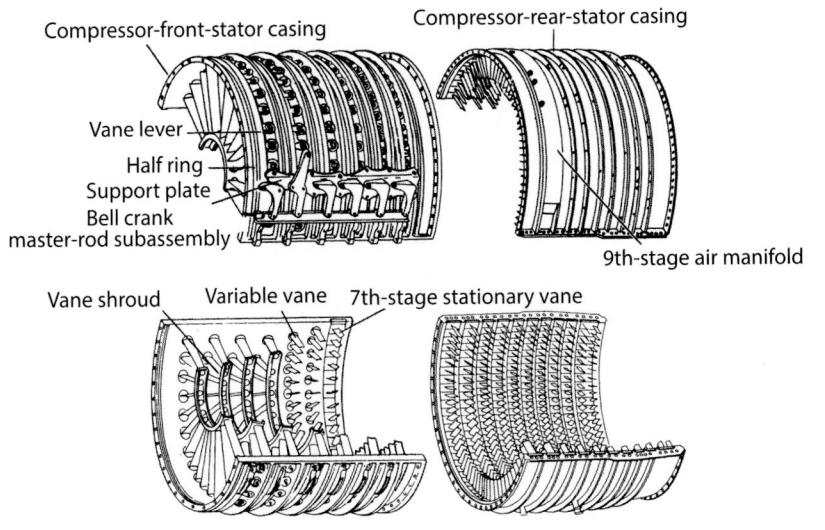

Fig. 9.41   Compressor case for the General Electric J79 turbojet engine.
(Courtesy of General Electric.)

9.4   Centrifugal-Flow Compressor Analysis

The centrifugal compressor has been around for many years. It was used in turbochargers before being used in the first turbojet engines of Whittle and von Ohain. Figures 9.42a and 9.42b show a sketch of a centrifugal-flow compressor with radial rotor (or impeller) vanes. In Fig. 9.42b, flow passes through the annulus between r_{1h} and r_{1t} at station 1 with no swirl ($v_1 = 0$) and enters the inducer section of the rotor (also called *rotating guide vanes*). Flow leaves the rotor at station 2 through the cylindrical area of radius r_2 and width b. The flow then passes through the diffuser, where it is slowed and then enters the collector scroll at station 3.

The velocity diagrams at the entrance and exit of the rotor (impeller) are shown in Fig. 9.43. The inlet flow is assumed to be axial of uniform velocity u_1. The relative flow angle of the flow entering the rotor increases from hub to tip and thus the twist of the inlet to the inducer section of the rotor. The flow leaves the rotor with a radial component of velocity w_2 that is approximately equal to the inlet axial velocity u_1 and a swirl (tangential) component of velocity v_2 that is about 90% of the rotor velocity U_t. The diffuser (which may be vaneless) slows the velocity of the flow v_3 to about 90 m/s (300 ft/s).

Application of the Euler equation to flow of a calorically perfect gas through a centrifugal-flow compressor with axial flow entering gives

$$T_{t3} - T_{t1} = \frac{v_2 U_t}{g_c c_p} \tag{9.64}$$

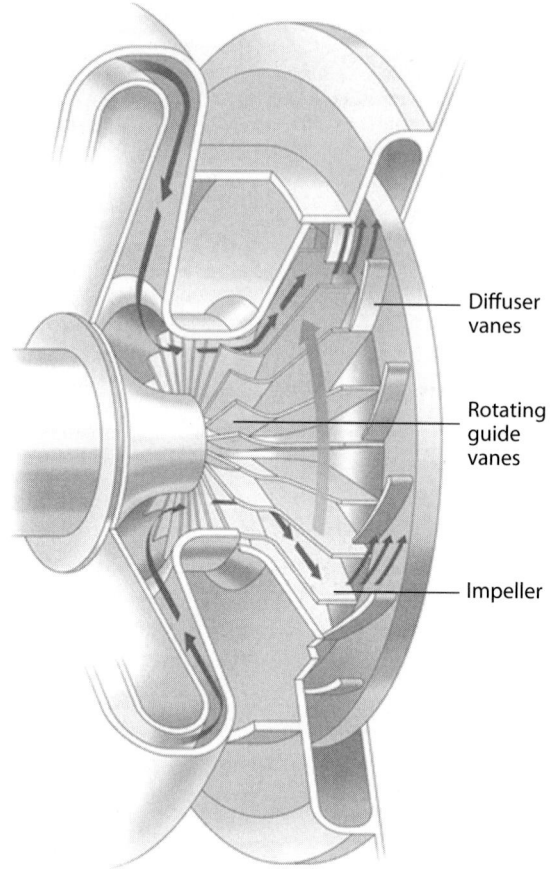

Fig. 9.42a   Centrifugal compressor. (Courtesy of Rolls Royce.)

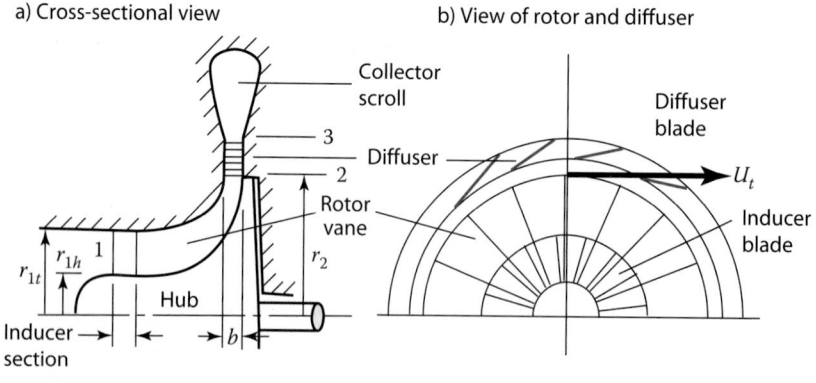

Fig. 9.42b   Centrifugal-flow compressor.

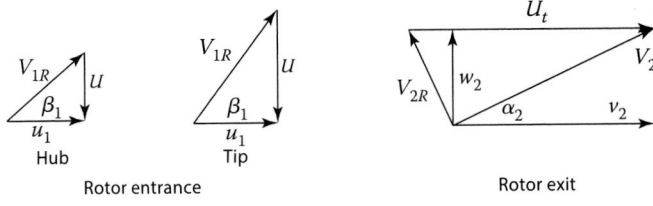

Fig. 9.43 Velocity diagrams for radial-vaned centrifugal compressor.

Ideally, the fluid leaving the rotor wheel has a swirl velocity v_2 equal to the rotor speed U_t. Because of slip between the rotor and fluid, the fluid leaving the rotor wheel attains only a fraction of the rotor speed. The ratio of the exit swirl velocity to the rotor speed is called the *slip factor ε*:

$$\varepsilon = \frac{v_2}{U_t} \tag{9.65}$$

The slip factor is related to the number of vanes on the rotor. As the number of vanes n is increased, the slip factor approaches 1 and the frictional losses of the rotor increase. Selection of the number of vanes is a balance between high slip factor and reasonable losses, which usually results in a slip factor of 0.9. A useful correlation between the slip factor ε and number of vanes n is

$$\varepsilon = 1 - \frac{2}{n} \tag{9.66}$$

Substitution of Eq. (9.65) into Eq. (9.64) gives a relationship for the compressor temperature rise in terms of the rotor speed U_t and slip factor ε:

$$T_{t3} - T_{t1} = \frac{\varepsilon U_t^2}{g_c c_p} \tag{9.67}$$

By using the polytropic compressor efficiency e_c and Eq. (9.67), the compressor pressure ratio can be expressed as

$$\pi_c = \frac{P_{t3}}{P_{t1}} = \left(1 + \frac{\varepsilon U_t^2}{g_c c_p T_{t1}}\right)^{\gamma e_c/(\gamma-1)} \tag{9.68}$$

From Eq. (9.68), compressor pressure ratio π_c is plotted vs rotor speed U_t in Fig. 9.44 for air [$\gamma = 1.4$, $c_p = 1.004$ kJ/(kg · K)] at standard conditions ($T_{t1} = 288.16$ K) with a slip factor ε of 0.9. For rotors with light alloys, U_t is limited to about 450 m/s (1500 ft/s) by the maximum centrifugal stresses of the rotor, which corresponds to compressor pressure ratios of about 4. More extensive materials permit higher speeds and pressure ratios up to about 8.

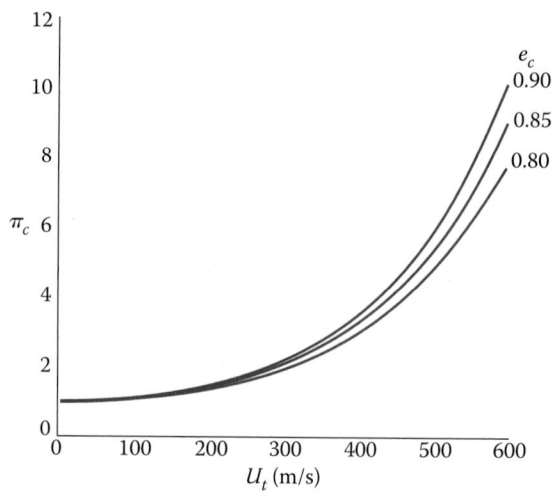

Fig. 9.44   Compressor pressure ratio vs rotor speed.

Equation (9.68) can also be written in terms of the compressor adiabatic efficiency η_c as

$$\pi_c = \frac{P_{t3}}{P_{t1}} = \left(1 + \frac{\eta_c \varepsilon U_t^2}{g_c c_p T_{t1}}\right)^{\gamma/(\gamma-1)} \tag{9.69}$$

The T-s diagram for the centrifugal compressor is shown in Fig. 9.45. This diagram is very useful during the analysis of centrifugal-flow compressors. Even though the swirl exit velocity v_2 may be supersonic, the flow leaving the rotor will choke only when the radial exit velocity w_2 is sonic.

For determining the total pressure at station 2, an estimate must be made of the split in total pressure loss between the rotor and diffuser. As an

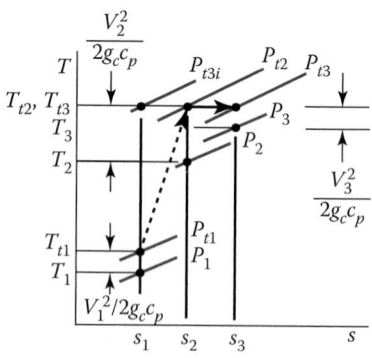

Fig. 9.45   The T-s diagram for centrifugal compressor.

estimate, we equate the total pressure ratio of the diffuser P_{t3}/P_{t2} to the actual/ideal total pressure ratio leaving the rotor P_{t2}/P_{t3s}, or

$$\frac{P_{t2}}{P_{t3s}} = \frac{P_{t3}}{P_{t2}} = \sqrt{P_{t3}/P_{t3s}} \qquad (9.70)$$

The radial component of velocity leaving the rotor w decreases with radius due to the increase in radial flow area. If frictional losses are neglected, the tangential momentum rv will remain constant, or

$$rv = \text{const} = r_2 v_2$$

As a result of the decreases in both the radial and tangential velocities, the velocity entering the diffuser is less than that leaving the rotor.

The diffuser further decreases the velocity and may turn the flow with blades. The flow leaving the diffuser is collected in the scroll (see Fig. 9.42b). The flow area of the scroll A_θ must increase in proportion to the mass flow being added to keep the flow properties constant. We can write

$$\frac{dA_\theta}{d\theta} = \frac{1}{\rho_3 V_3}\frac{d\dot{m}}{d\theta} = r_3 b_3 \qquad (9.71)$$

Centrifugal-flow compressors are used in gas turbine engines when the corrected mass flow rate is small (less than about 10 kg/s). For small corrected mass flow rates, the rotor blades of axial-flow compressors are very short, and the losses are high. The performance of a centrifugal compressor for low flow rates can be as good as or better than that of the axial-flow compressor.

The performance of a centrifugal-flow compressor is presented as a compressor map (similar to the map of an axial-flow compressor). The map of a typical centrifugal compressor is shown in Fig. 9.46. Note that the overall pressure ratio is limited to about 4 when constructed with modest strength-to-weight material alloy. Gas turbine engine cycles normally require a pressure ratio greater than 4, and additional stages of compression are needed. For engines with entering corrected mass flow rates greater than 10 kg/s, additional compression is obtained from axial-flow stages in front of the centrifugal compressor, the so-called "axi-centrif" compressor configuration quite common in turboshaft engines for helicopter application. When the corrected mass flow is less than 10 kg/s, higher cycle pressure ratios are obtained by using two or maybe three centrifugal compressors in series.

Centrifugal compressors find their aircraft engine application primarily in smaller mass flow rate machines (less than approximately 100 lbm/s) and as superchargers/turbochargers used in conjunction with reciprocating (piston-cylinder) aircraft engines. They are capable of much higher pressure rise per stage than an axial design, taking advantage of higher diffusion as the air gets "slung" radially outward from the centerline. Centrifugal compressors also tend to be more robust structurally, more tolerant of inlet flow distortion

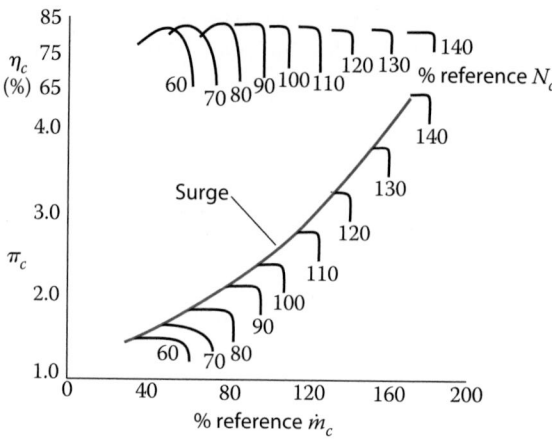

Fig. 9.46   Performance map of typical centrifugal compressor.

(more stall resistant), and less sensitive to tip clearances. However, they also tend to be less efficient than axial-flow designs (largely due to the flow turning from axial to radial back to axial) and are limited in practice to no more than about two stages back-to-back. Their radial dimension starts to get prohibitive for engines requiring much more than about 100 lbm/s air flow rates. Consequently, axial-flow compressors are the most common type found in aircraft applications primarily because of their high airflow per cross-sectional area and high overall pressure rise and efficiency achievable through multistaging.

Example 9.10

Consider a centrifugal compressor.
Given:

Mass flow rate $= 8$ kg/s	$P_{t1} = 101.3$ kPa
Pressure ratio $\pi_c = 4.0$	$T_{t1} = 288.16$ K
Polytropic efficiency $e_c = 0.85$	Slip factor $\varepsilon = 0.9$
Inlet root diameter $= 15$ cm	Inlet tip diameter $= 30$ cm
Outlet diameter of impeller $= 50$ cm	$w_2 = u_1$
$V_3 = 90$ m/s	

Find:

1. Rotational speed and rpm of the rotor
2. Rotor inlet Mach number, velocity, and relative flow angles at hub and tip

(Continued)

Example 9.10 *(Continued)*

3. Rotor exit velocity, Mach number, total temperature, total pressure, and direction
4. Depth of rotor exit s
5. Diffuser exit Mach number, area, total temperature, and total pressure

Solution:

$$U_t^2 = \frac{g_c c_p T_{t1}}{\varepsilon}(\pi_c^{(\gamma-1)/(\gamma e_c)} - 1)$$

$$= \frac{1004 \times 288.16}{0.9}(4^{1/(3.5 \times 0.85)} - 1)$$

$$U_t = 436.8 \text{ m/s}$$

$$N = \frac{60 U_t}{\pi d_2} = \frac{60(436.8)}{\pi(0.5)} = 16{,}685 \text{ rpm}$$

$$\text{MFP}(M_1) = \frac{\dot{m}\sqrt{T_{t1}}}{P_{t1}A_1} = \frac{8\sqrt{288.16}}{101{,}300 \times \pi(0.15^2 - 0.075^2)} = 0.025287$$

$$M_1 = 0.3966$$

$$u_1 = V_1 = \sqrt{2 g_c c_p T_{t1}\left\{1 - \frac{1}{1 + [(\gamma-1)/2]M_1^2}\right\}}$$

$$= \sqrt{2 \times 1004 \times 288.16\left(1 - \frac{1}{1 + 0.2 \times 0.3966^2}\right)} = 132.8 \text{ m/s}$$

$$v_{1Rh} = \frac{d_{1h}}{d_2}U_t = \frac{15}{50}(436.8) = 131.0 \text{ m/s}$$

$$v_{1Rt} = \frac{d_{1t}}{d_2}U_t = \frac{30}{50}(436.8) = 262.1 \text{ m/s}$$

$$\beta_{1h} = \tan^{-1}\frac{v_{1Rh}}{u_1} = 44.6 \text{ deg}$$

$$\beta_{1t} = \tan^{-1}\frac{v_{1Rt}}{u_1} = 63.1 \text{ deg}$$

$$T_{t3} = T_{t2} = T_{t1} + \frac{\varepsilon U_t^2}{g_c c_p} = 288.16 + \frac{0.9 \times 436.8^2}{1004} = 459.19 \text{ K}$$

$$\eta_c = \frac{(P_{t3}/P_{t1})^{\gamma/(\gamma-1)} - 1}{T_{t3}/T_{t1} - 1} = \frac{4^{1/3.5} - 1}{459.19/288.16 - 1} = 81.9\%$$

$$v_2 = \varepsilon U_t = 0.9 \times 436.8 = 393.1 \text{ m/s}$$

(Continued)

Example 9.10 (Continued)

$$w_2 = u_1 = 132.8 \, \text{m/s}$$

$$V_2 = \sqrt{w_2^2 + v_2^2} = \sqrt{132.8^2 + 393.1^2} = 414.9 \, \text{m/s}$$

$$\alpha_2 = \tan^{-1} \frac{w_2}{v_2} = 18.67 \, \text{deg}$$

$$M_2 = \sqrt{\frac{2}{\gamma - 1} \left[\frac{T_{t2}}{T_{t2} - V_2^2/(2g_c c_p)} - 1 \right]}$$

$$= \sqrt{5 \left(\frac{459.19}{459.19 - 85.73} - 1 \right)} = 1.071$$

$$\frac{P_{t3s}}{P_{t1}} = \left(\frac{T_{t3}}{T_{t1}} \right)^{\gamma/(\gamma-1)} = 5.108$$

$$\frac{P_{t2}}{P_{t3s}} = \frac{P_{t3}}{P_{t2}} = \sqrt{\frac{P_{t3}/P_{t1}}{P_{t3s}/P_{t1}}} = 0.8849$$

$$P_{t2} = 0.8849 \times 5.108 \times 101.3 = 457.9 \, \text{kPa}$$

$$P_{t3} = 4.0 \times 101.3 = 405.2 \, \text{kPa}$$

$$\text{MFP}(M_2) = 0.040326$$

$$A_2 = \frac{\dot{m}\sqrt{T_{t2}}}{P_{t2}\text{MFP}(M_2)(\cos \alpha_2)} = \frac{8\sqrt{459.19}}{457{,}900 \times 0.040326 \times 0.9474} = 0.0098 \, \text{m}^2$$

$$b = \frac{A_2}{\pi d_2} = \frac{0.0098}{\pi \times 0.5} = 0.624 \, \text{cm}$$

$$M_3 = \sqrt{\frac{2}{\gamma - 1} \left[\frac{T_{t3}}{T_{t3} - V_3^2/(2g_c c_p)} - 1 \right]}$$

$$= \sqrt{5 \left(\frac{459.19}{459.19 - 4.034} - 1 \right)} = 0.2105$$

$$\text{MFP}(M_3) = 0.014343$$

$$A_3 \cos \alpha_3 = \frac{\dot{m}\sqrt{T_{t3}}}{P_{t3}\text{MFP}(M_3)} = \frac{8\sqrt{459.196}}{405{,}200 \times 0.014343} = 0.0295 \, \text{m}^2$$

(Continued)

Example 9.10 *(Continued)*

Table 9.9   Results for Example 9.10 Centrifugal Compressor

Property		Station				
		1	1R	2R	2	3
T_t	K	**288.16**	296.70/322.36 (hub/tip)	383.23	459.19	459.19
T	K	279.37	279.37	272.50	373.50	455.16
P_t	kPa	**101.3**	112.2/150.0	243.1	457.9	**405.2**
P	kPa	90.9	90.9	222.2	222.2	392.9
M		0.3966	0.557/0.877	0.361	1.071	0.2105
V	m/s	132.8	186.54/293.82	139.8	414.9	90
u/w	m/s	132.8	132.8	132.8	132.8	
v	m/s	0	131.0/262.1	43.7	393.1	
r	cm		**15/30**	**50.0**	**50.0**	
α	deg	0	—	—	18.67	
β	deg	—	44.6/63.1	69.1	—	

The results of this example are summarized in Table 9.9. Note that different values of flow properties are listed for the hub and tip at station $1R$ and that M_2 is supersonic whereas M_{2R} is subsonic.

9.5 Axial-Flow Turbine Analysis

The mass flow of a gas turbine engine, which is limited by the maximum permissible Mach number entering the compressor, is generally large enough to require an axial turbine (even for centrifugal compressors). The axial turbine is essentially the reverse of the axial compressor including operating in a globally favorable pressure gradient. This permits greater angular changes, greater pressure changes, greater energy changes, and higher efficiency. However, there is more blade stress involved because of the higher work and temperatures. *It is this latter fact that generally dictates the blade shape.*

Conceptually, a turbine is a very simple device because it is fundamentally no different from a pinwheel that spins rapidly when air is blown against it. The pinwheel will turn in one direction or the other, depending on the direction of the impinging air, and a direction can be found for the air that causes no rotation at all. Thus it is important to properly direct the airstream if the desired motion and speed are to be obtained.

A modern turbine is merely an extension of these basic concepts. Considerable care is taken to establish a directed flow of fluid of high velocity by means of turbine nozzle vanes (see Fig. 8.5a), and then similar care is used in designing the blades on the rotating wheel so that the fluid applies the required force to these rotor blades most efficiently. Conceptually, the turbine is a distant cousin to the pinwheel, but such an analogy gives no appreciation for the source of the tremendous power outputs that can be obtained in a modern turbine. The appreciation comes when one witnesses a static test of the nozzle vanes that direct a gas stream to the rotor blades of a modern aircraft gas turbine. Such a stream exhausting into the quiescent air of a room will literally rip the paint off the wall at a point 6 ft away in line with the jet direction. When blades of the rotating element are pictured in place immediately at the exit of these directing blades, it then becomes difficult to imagine that the rotor could be constrained from turning and producing power.

Terminology can be a problem in the general field of turbomachinery. Compressor development grew out of aerodynamics and aircraft wing technology, whereas turbines have historically been associated with the mechanical engineers who developed the steam turbine. The symbols as well as the names used in these two fields differ, but for consistency and to minimize problems for the reader, the turbine will be presented using the nomenclature already established in the beginning of this chapter. Where a term or symbol is in such common use that to ignore it would mean an incomplete education, it will be indicated as an alternate. Thus turbine stator vanes are usually called *nozzles* and rotor blades are *buckets.*

In the gas turbine, the high-pressure, high-temperature gas from the combustion camber flows in an annular space to the stationary vanes (called stators or nozzles) and is directed tangentially against the rotating blade row (called rotor blades or buckets). A simple single-stage turbine configuration with nomenclature is shown in Fig. 9.47a. Modern aircraft gas turbine engines typically have a single-stage or two-stage core or high-pressure turbine (HPT). A picture of a core turbine rotor with shaft is shown in Fig. 9.47b.

Just like the axial compressor analysis, it is convenient to "cut" the blading on a cylindrical surface and look at the section of the stator and rotor blades

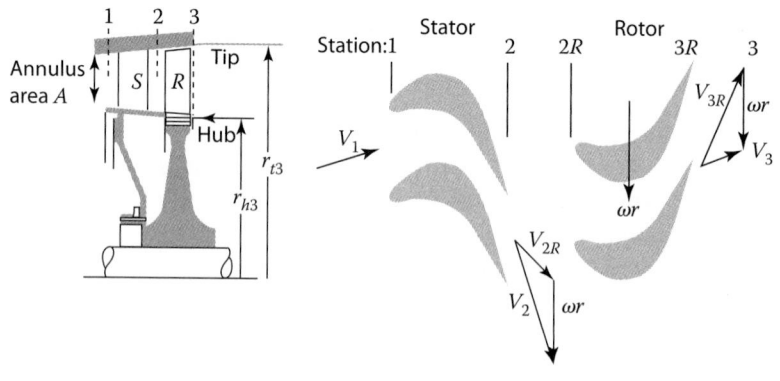

Fig. 9.47a Typical single-stage turbine and velocity diagram (after Ref. 22).

Fig. 9.47b   High-pressure turbine rotor and shaft.

in two dimensions. This leads to the construction of a vector or velocity diagram for the stage (see Fig. 9.47a) that shows the magnitude and direction of the gas velocities within the stage on the cylindrical surface.

In the stator or nozzle, the fluid is accelerated while the static pressure decreases and the tangential velocity of the fluid is increased in the direction of rotation. The rotor decreases the tangential velocity in the direction of rotation, tangential forces are exerted by the fluid on the rotor blades, and a resulting torque is produced on the output shaft. The absolute velocity of the fluid is reduced across the rotor. Relative to the moving blades, typically there is acceleration of the fluid with the associated decrease in static pressure and static temperature.

A low-pressure turbine (LPT) is always multistage, generally ranging from two to six stages. This is necessary because the highest energy gases from the combustor have already been used by the HPT and, unless speed reduction gearing is used, the large fan blades (see Figs. 1.8a–f) require a much reduced rotational speed. Compare the shafts of the HPT and LPT shown in Figs. 9.47b and 9.47c. Note the smaller diameter and longer length of the LPT shaft because it must fit within the HPT shaft and run the entire length of the engine core. Recall that these two assemblies are mechanically independent. Each has its own set of bearings that reside in static support structures called "frame" assemblies.

The following analysis of the axial-flow turbine stage is performed along the mean radius with radial variations being considered. In many axial-flow turbines, the hub and tip diameters vary little through a stage, and the hub/tip ratio approaches unity. There can be no large radial components of velocity between the annular walls in such stages, there is little variation in static pressure from root to tip, and the flow conditions are only slightly different at each radius.

For these stages of high hub/tip ratio, the 2-D analysis is sufficiently accurate. The flow velocity triangles are drawn for the mean-radius condition, but these triangles are assumed to be valid for the other radial sections. The mean-radius analysis presented in this section applies to the total flow for such 2-D stages—the flow velocity, blade speed, and pressures are assumed constant along the blade length.

From the Euler turbine equation [Eq. (9.4)], the energy per unit mass flow exchanged between fluid and rotor for $r_2 = r_3$ is

$$h_{t2} - h_{t3} = c_p(T_{t2} - T_{t3}) = \frac{\omega r}{g_c}(v_2 + v_3) \qquad (9.72)$$

Inspection of the velocity triangles (Fig. 9.48) shows that because of the large angle α_2 at the stator exit and the large turning possible in the rotor, the value of v_3 is often positive (positive α_3). As a result, the two swirl velocity terms on the right side of Eq. (9.72) add, giving large power output.

The large turning in stator and rotor is possible because, usually, the flow is accelerating through each blade row, that is, $V_2 > V_1$ and $V_{3R} > V_{2R}$. This means that the static pressure drops across both stator and rotor and, under such circumstances, a favorable pressure gradient exists for the boundary layers on the blade and wall surfaces, and separation can be avoided. Accelerating flow is an inherent feature of turbines and generally means that no flow breakdown similar to compressor stall will occur. Consequently, the diffusion factor D, so important in axial compressor design, is not used in turbine design and analysis.

Note that the vector diagram establishes the characteristics of a stage, and geometrically similar diagrams have the same characteristics. The angles of the vector diagram determine its shape and, therefore, become important design parameters. The velocity magnitudes are not important as performance parameters except in relation to the sonic velocity, which is quite high due to the high gas temperatures (i.e., except in terms of the

Fig. 9.47c   Low-pressure turbine rotor and shaft.

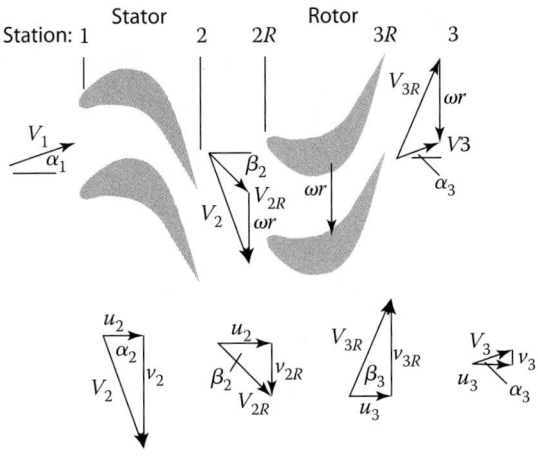

Fig. 9.48 Velocity triangles for a typical turbine stage.

Mach number). The angles may be used directly as design parameters or may be implied through the use of velocity ratios. Thus the magnitude of, say, $v_2 + v_3 = \Delta v$, although proportional to the absolute output, is not significant in defining the vector diagram, whereas the ratio $\Delta v/(\omega r)$ helps establish the vector diagram angles and, therefore, the stage characteristics.

The stage analysis in the following sections neglects the influence of cooling air. Thus the flow through the turbine nozzle is assumed to be adiabatic ($T_{t2} = T_{t1}$), as is the flow through the rotor in the relative reference frame ($T_{t3R} = T_{t2R}$). Turbine cooling is discussed at the end of this chapter.

Example 9.11

Consider mean-radius core turbine stage calculation—isentropic flow.
Given:

$$T_{t1} = 3200°R, \quad P_{t1} = 300\,\text{psia}, \quad \alpha_2 = 60\,\text{deg},$$
$$\alpha_3 = 0\,\text{deg}, \quad M_2 = 1.1, \quad \omega r = 1400\,\text{ft/s},$$
$$u_3 = u_2, \quad \gamma = 1.3, \quad R = 53.40\,\text{ft} \cdot \text{lbf}/(\text{lbm} \cdot °R)$$

Find the flow properties for this isentropic turbine stage and compare to those of the isentropic compressor stage of Example 9.1.

Solution:

$$T_{t2} = T_{t1} = 3200°R$$

$$T_2 = \frac{T_{t2}}{1 + [(\gamma - 1)/2]M_2^2} = \frac{3200}{1 + 0.15 \times 1.1^2} = 2708.4°R$$

(Continued)

Example 9.11 *(Continued)*

$$g_c c_p = g_c \left(\frac{\gamma}{\gamma - 1} \right) R = 32.174 \left(\frac{1.3}{0.3} \right) 53.40 = 7445 \ \text{ft}^2/(\text{s}^2 \cdot {}^\circ\text{R})$$

$$V_2 = \sqrt{2 g_c c_p (T_{t2} - T_2)} = \sqrt{2(7445)(3200 - 2708.4)} = 2705.5 \ \text{ft/s}$$

$$u_2 = V_2 \cos \alpha_2 = 2705.5 \cos 60 \ \text{deg} = 1352.8 \ \text{ft/s}$$

$$v_2 = V_2 \sin \alpha_2 = 2705.5 \sin 60 \ \text{deg} = 2343.0 \ \text{ft/s}$$

$$v_{2R} = v_2 - \omega r = 2343.0 - 1400 = 943.0 \ \text{ft/s}$$

$$V_{2R} = \sqrt{u_2^2 + v_{2R}^2} = \sqrt{1352.8^2 + 943.0^2} = 1649.0 \ \text{ft/s}$$

$$\beta_2 = \tan^{-1} \frac{v_{2R}}{u_2} = \tan^{-1} \frac{943.0}{1352.8} = 34.88 \ \text{deg}$$

$$M_{2R} = M_2 \frac{V_{2R}}{V_2} = 1.1 \left(\frac{1649.0}{2705.5} \right) = 0.670$$

$$T_{t2R} = T_2 + \frac{V_{2R}^2}{2 g_c c_p} = 2708.4 + \frac{1649.0^2}{2(7445)} = 2891.0 \ {}^\circ\text{R}$$

$$v_3 = 0$$

$$V_3 = u_3 = u_2 = 1352.8 \ \text{ft/s}$$

$$v_{3R} = v_3 + \omega r = 0 + 1400 = 1400 \ \text{ft/s}$$

$$V_{3R} = \sqrt{u_3^2 + v_{3R}^2} = \sqrt{1352.8^2 + 1400^2} = 1946.8 \ \text{ft/s}$$

$$\beta_3 = \tan^{-1} \frac{v_{3R}}{u_3} = \tan^{-1} \frac{1400}{1352.8} = 45.98 \ \text{deg}$$

$$T_{t3} = T_{t2} - \frac{\omega r}{g_c c_p} (v_2 + v_3) = 3200 - \frac{1400}{7445} (2343.0 + 0) = 2759.4 \ {}^\circ\text{R}$$

$$T_3 = T_{t3} - \frac{V_3^2}{2 g_c c_p} = 2759.4 - \frac{1352.8^2}{2(7445)} = 2636.5 \ {}^\circ\text{R}$$

$$M_3 = \sqrt{\frac{2}{\gamma - 1} \left(\frac{T_{t3}}{T_3} - 1 \right)} = \sqrt{\frac{2}{0.3} \left(\frac{2759.8}{2636.5} - 1 \right)} = 0.558$$

$$M_{3R} = M_3 \frac{V_{3R}}{V_3} = 0.558 \left(\frac{1946.8}{1352.8} \right) = 0.803$$

$$T_{t3R} = T_{t2R} = 2891.0 \ {}^\circ\text{R}$$

(Continued)

Example 9.11 *(Continued)*

$$P_{t2} = P_{t1} = 300 \, \text{psia}$$

$$P_2 = P_{t2}\left(\frac{T_2}{T_{t2}}\right)^{\gamma/(\gamma-1)} = 300\left(\frac{2708.4}{3200}\right)^{1.3/0.3} = 145.6 \, \text{psia}$$

$$P_{t2R} = P_2\left(\frac{T_{t2R}}{T_2}\right)^{\gamma/(\gamma-1)} = 145.6\left(\frac{2891.0}{2708.4}\right)^{1.3/0.3} = 193.2 \, \text{psia}$$

$$P_{t3R} = P_{t2R} = 193.2 \, \text{psia}$$

$$P_{t3} = P_{t2}\left(\frac{T_{t3}}{T_{t2}}\right)^{\gamma/(\gamma-1)} = 300\left(\frac{2759.4}{3200}\right)^{1.3/0.3} = 157.9 \, \text{psia}$$

$$P_3 = P_{t3}\left(\frac{T_3}{T_{t3}}\right)^{\gamma/(\gamma-1)} = 157.9\left(\frac{2636.5}{2759.4}\right)^{1.3/0.3} = 129.6 \, \text{psia}$$

Table 9.10 is a summary of the results for this axial-flow turbine stage with isentropic flow. The given data are listed in boldface type. With the exception

Table 9.10 Results for Example 9.11 Axial-Flow Turbine Stage Calculation, Isentropic Flow

Property		Station				
		1	2	2R	3R	3
T_t	°R	**3200**	3200.0	2891.0	2891.0	2759.4
	(K)	(1778)	(1777.8)	(1606.1)	(1606.1)	(1533.0)
T	°R		2708.4	2708.4	2636.5	2636.5
	(K)		(1504.7)	(1504.7)	(1464.7)	(1464.7)
P_t	psia	**300**	300	193.2	193.2	157.9
	(kPa)	(2068)	(2068)	(1332)	(1332)	(1089)
P_t	psia		145.6	145.6	129.6	129.6
	(kPa)		(1004)	(1004)	(893.6)	(893.6)
M			**1.10**	0.670	0.803	0.558
V	ft/s		2705.5	1649.0	1946.8	1352.8
	(m/s)		(824.6)	(502.6)	(593.4)	(412.3)
u	ft/s		1352.8	1352.8	1352.8	1352.8
	(m/s)		(412.3)	(412.3)	(412.3)	(412.3)
v	ft/s		2343.0	943.0	1400.0	0
	(m/s)		(714.2)	(287.4)	(426.7)	(0)
α	deg		**60.0**	—	—	0
β	deg			34.88	45.98	—

(Continued)

Example 9.11 *(Continued)*

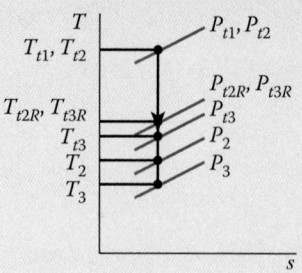

Fig. 9.49   Property changes of an isentropic turbine stage.

of the total pressure and total temperature, none of the properties are calculated at station 1. It is assumed that the flow exits the combustor with no swirl and low subsonic axial velocity. Even though the flow leaving the nozzle (station 2) is supersonic, the relative flow entering the rotor is subsonic. The flow through the rotor is turned through the axis 80.8 deg ($\beta_2 + \beta_3$). For comparison, the flow through the compressor rotor blades in Example 9.1 was turned 18.5 deg. Also note the total temperature changes: 441° across the turbine stage and 40° across the compressor stage of Example 9.1. Figure 9.49 shows the change in temperature and pressure for an isentropic turbine stage.

9.5.1   Stage Parameters

9.5.1.1 Adiabatic Efficiency

The adiabatic efficiency (the most common definition of efficiency for turbines) is the ratio of the actual energy output to the theoretical isentropic output (see Fig. 9.50a) for the same input total state and same exit total pressure:

$$\eta_t = \frac{\text{actual } \Delta h_t}{\text{ideal } \Delta h_t} = \frac{h_{t1} - h_{t3}}{h_{t1} - h_{t3s}} \tag{9.73}$$

For a calorically perfect gas, the efficiency can be written in terms of total temperatures and total pressures (see Fig. 9.50b) as follows:

$$\eta_t = \frac{h_{t1} - h_{t3}}{h_{t1} - h_{t3s}} = \frac{c_p(T_{t1} - T_{t3})}{c_p(T_{t1} - T_{t3s})}$$

$$\eta_t = \frac{1 - T_{t3}/T_{t1}}{1 - (P_{t3}/P_{t1})^{(\gamma-1)/\gamma}} \tag{9.74}$$

a) *h-s* diagram

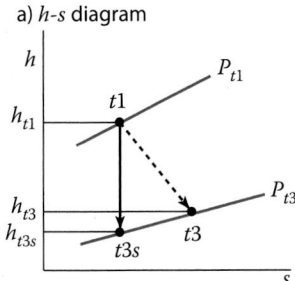

b) *T-s* diagram

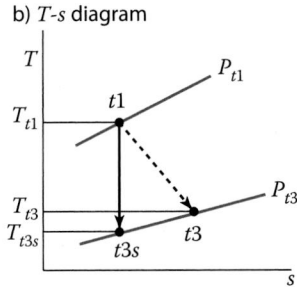

Fig. 9.50 Definition of turbine adiabatic efficiency.

The preceding definition is sometimes called the *total-to-total turbine efficiency* η_t, because the theoretical output is based on the leaving total pressure.

9.5.1.2 Exit Swirl Angle

The absolute angle of the leaving flow α_3 is called the *swirl angle* and, by convention, is positive when opposite to wheel speed direction (*backward-running*). The angle is important for two reasons. It is difficult in any fluid dynamic situation to efficiently convert kinetic energy to pressure or potential energy, and the kinetic energy in the flow leaving a turbine stage can be minimized by having $v_3 = u_3$, that is, by having zero swirl. Conversely, we see that the higher the swirl angle (if backward-running), the higher in magnitude v_3 will be, which generally means higher output from the stage [see Eq. (9.72) and Fig. 9.48]. Design is all about tradeoffs.

9.5.1.3 Stage Loading and Flow Coefficients

The *stage loading coefficient*, defined by Eq. (9.19), is the ratio of the stage work per unit mass to the rotor speed squared, or

$$\psi \equiv \frac{g_c \, \Delta h_t}{(\omega r)^2} = \frac{g_c \, \Delta h_t}{U^2} \qquad (9.75)$$

For a calorically perfect gas, we write [Eq. (9.20)]

$$\psi = \frac{g_c c_p \, \Delta T_t}{(\omega r)^2} = \frac{g_c c_p \, \Delta T_t}{U^2} \qquad (9.76)$$

The ratio of the axial velocity entering the rotor to the rotor speed is called the flow *coefficient* and is defined [similar to Eq. (9.21)] as

$$\Phi \equiv \frac{u_2}{\omega r} = \frac{u_2}{U} \qquad (9.77)$$

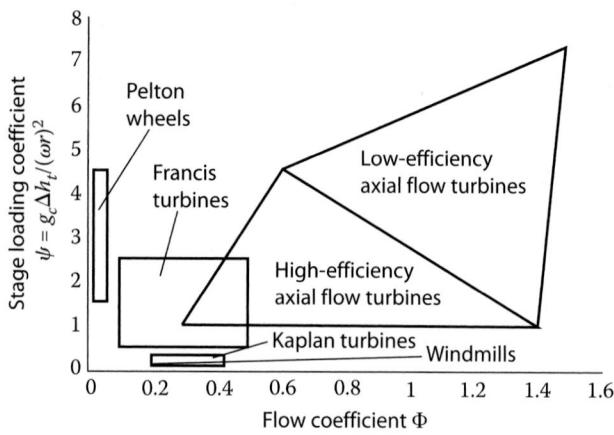

Fig. 9.51   Stage loading vs flow coefficient for different turbine types (Ref. 40).

The stage loading coefficient and flow coefficient for a turbine stage have a range of values much larger than those for a compressor stage. Figure 9.51 shows the range of these coefficients for several types of turbines. For Example 9.11 data, the stage loading coefficient is 1.67, and the flow coefficient is 0.962, which is well within the range for high-efficiency axial-flow

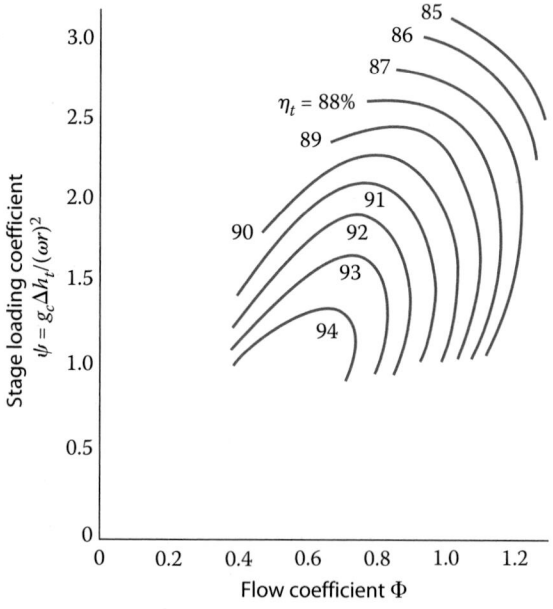

Fig. 9.52   Stage efficiency vs stage loading and flow coefficients, corrected for zero tip leakage (Ref. 40).

turbines. Both the stage loading and flow coefficients affect the turbine stage efficiency, as shown in Fig. 9.52. Modern high-pressure turbines used for aircraft gas turbine engines have stage loading coefficients in the range of 1.3–2.2 and flow coefficients in the range of 0.5–1.1.

From Fig. 9.48 and Eqs. (9.72), (9.20), and (9.77), we obtain the stage loading coefficient in terms of the flow coefficient, the axial velocity ratio u_3/u_2, and flow angles as

$$\psi = \frac{g_c c_p \Delta T}{(\omega r)^2} = \frac{v_2 + v_3}{\omega r} = \Phi\left(\tan \alpha_2 + \frac{u_3}{u_2}\tan \alpha_3\right) \qquad (9.78\text{a})$$

or

$$\psi = \frac{g_c c_p \Delta T}{(\omega r)^2} = \frac{v_2 + v_3}{\omega r} = \Phi\left(\tan \beta_2 + \frac{u_3}{u_2}\tan \beta_3\right) \qquad (9.78\text{b})$$

By using Fig. 9.48, the flow coefficient can be expressed in terms of the flow angles as

$$\Phi = (\tan \alpha_2 - \tan \beta_2)^{-1} \qquad (9.79)$$

9.5.1.4 Degree of Reaction

The *degree of reaction* is defined as

$$^\circ R_t = \frac{h_2 - h_3}{h_{t1} - h_{t3}} \qquad (9.80)$$

For a calorically perfect gas, we can write

$$^\circ R_t = \frac{T_2 - T_3}{T_{t1} - T_{t3}} \qquad (9.81)$$

That is, the degree of reaction is the ratio of the static enthalpy drop in the rotor to the drop in total enthalpy across both the rotor and stator. Figure 9.53 gives a complete picture for a general turbine stage. For Example 9.11, the degree of reaction is 0.166.

Three important basic designs of turbines are related to the choice of reaction—zero reaction, 50% reaction, and axial leaving velocity (variable reaction). It should be emphasized, however, that the designer is not limited to these three types and that in a 3-D turbine design, the reaction may vary continuously along the blades.

Zero reaction: From the definition of reaction, if the reaction is selected as zero for the case of constant axial velocity, then

$$h_3 = h_2 \quad \text{and} \quad \tan \beta_3 = \tan \beta_2$$

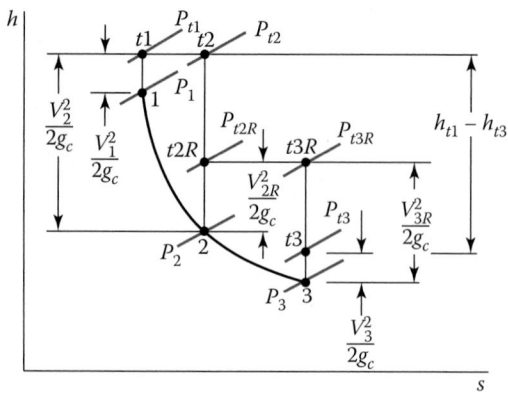

Fig. 9.53 The *h-s* diagram for general turbine stage.

Therefore,

$$\beta_3 = \beta_2 \quad \text{and} \quad V_{3R} = V_{2R}$$

Velocity triangles for this zero reaction stage are shown in Fig. 9.54, and the *h-s* diagram is drawn in Fig. 9.55.

If the flow is isentropic and the fluid is a perfect gas, then the condition of zero enthalpy drop ($h_3 - h_2 = 0$) implies no change in pressure across the rotor. The turbine is then *called* an impulse turbine, with the tangential load arising from impulsive forces only. *It is important to note that reaction is defined here not on the basis of pressure drops, but in terms of static enthalpy changes.* Note, however, that there is a pressure drop from P_2 to P_3 across the rotor (Fig. 9.55), and the stage is not, therefore, truly impulse. The *h-s* diagram for an "impulse" stage of zero pressure drop is shown in Fig. 9.56. There is an enthalpy increase from h_2 to h_3 across the rotor, and the relative velocity decreases. Thus, *the impulse stage is actually one of negative reaction!*

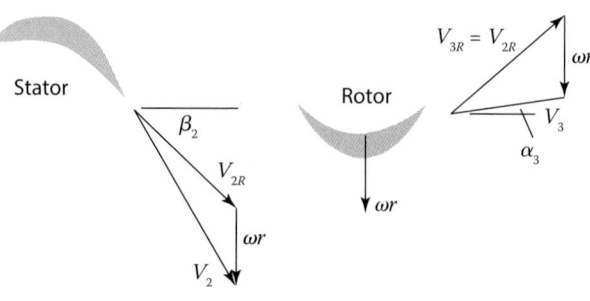

Fig. 9.54 Zero-reaction velocity diagram.

The stage loading coefficient for the impulse stage with constant axial velocity is

$$\psi = \frac{g_c c_p \, \Delta T_t}{(\omega r)^2} = \frac{\omega r \Delta v}{(\omega r)^2} = \frac{\Delta v}{\omega r} \quad (9.82)$$

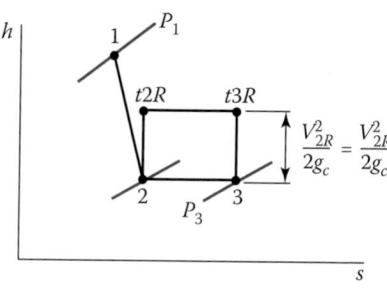

with

$$v_2 = u_2 \tan \alpha_2$$
$$v_3 = u_2 \tan \alpha_3 = u_2 \tan \alpha_2 - 2\omega r$$

Fig. 9.55 Zero-reaction h-s diagram.

then

$$\psi = 2(\Phi \tan \alpha_2 - 1) = 2\Phi \tan \beta_2 \quad (9.83)$$

We desire α_2 to be large. However, this leads to large v_2 and large v_{2R}, which lead to large losses. Thus α_2 is generally limited to less than 70 deg.

Also, if $\alpha_3 = 0$ (no exit swirl; see Fig. 9.54), then $\tan \alpha_2 = 2\omega r/u_2$ and

$$\psi = 2 \quad \text{no exit swirl} \quad (9.84)$$

Thus we see that the rotor speed ωr is proportional to $\sqrt{\Delta h_t}$. If the resultant blade speed is too high, we must go to a multistage turbine.

50% reaction: If there is equal enthalpy drop across rotor and stator, then $R_t = 0.5$ and the velocity triangles are symmetrical, as shown in Fig. 9.57. Then, $\alpha_2 = \beta_3$, $\alpha_3 = \beta_2$, and

$$\tan \beta_3 - \tan \beta_2 = \tan \alpha_2 - \tan \alpha_3 = \frac{\omega r}{u_2} = \frac{1}{\Phi}$$

The stage loading coefficient for this turbine with constant axial velocity is

$$\psi = \frac{\Delta v}{\omega r} = 2\Phi \tan \alpha_2 - 1 = 2\Phi \tan \beta_3 - 1 \quad (9.85)$$

Again, α_2 should be high but is limited to less than 70 deg. For *zero exit swirl*

$$\tan \beta_3 = \tan \alpha_2 = \frac{\omega r}{u}, \quad (9.86)$$
$$\beta_2 = 0, \quad \psi = 1$$

Thus, for the same ωr and $v_3 = 0$, the work per unit mass from a zero-reaction turbine is twice that from the 50% reaction turbine [compare Eqs. (9.84) and (9.86)].

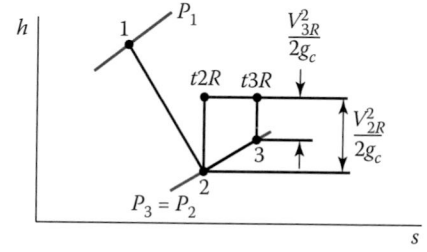

Fig. 9.56 Impulse turbine h-s diagram.

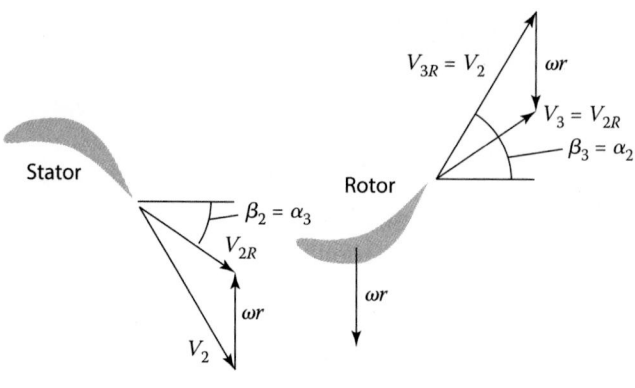

Fig. 9.57 Diagram of 50% reaction turbine velocity.

General zero-swirl case (constant axial velocity): If the exit swirl is to be zero (Fig. 9.58), then $\alpha_3 = 0$, $v_3 = 0$, and $\tan \beta_3 = \omega r/u$. From Eq. (9.82), the reaction is then

$$^{\circ}R_t = \frac{u}{2\omega r}\left(\frac{\omega r}{u} - \tan \beta_2\right) = \frac{1}{2} - \frac{u \tan \beta_2}{2\omega r} = 1 - \frac{v_2}{2\omega r}$$

$$^{\circ}R_t = 1 - \frac{v_2}{2\omega r} = 1 - \frac{\psi}{2} \tag{9.87}$$

This equation can be rewritten as

$$\psi = 2(1 - {}^{\circ}R_t) \tag{9.88}$$

Thus high stage loadings give a low degree of reaction. In aircraft gas turbine engines, engine weight and performance must be balanced. Weight can be reduced by increasing stage loading (reduces the number of turbine stages), but this normally leads to a loss in stage efficiency (see Fig. 9.52).

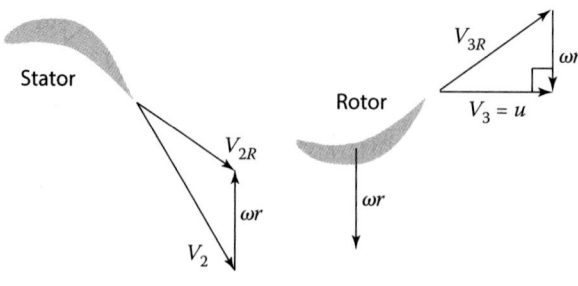

Fig. 9.58 Zero-exit-swirl turbine.

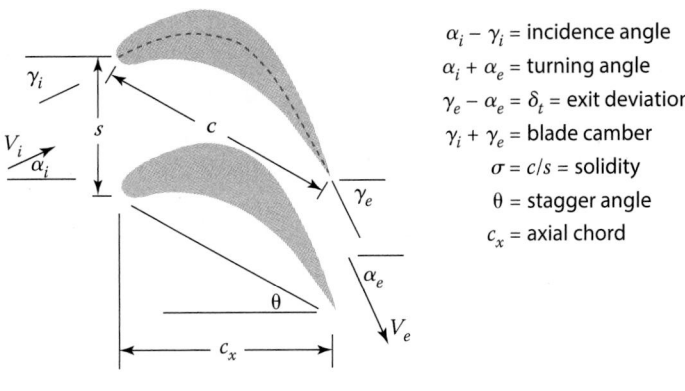

$\alpha_i - \gamma_i$ = incidence angle

$\alpha_i + \alpha_e$ = turning angle

$\gamma_e - \alpha_e = \delta_t$ = exit deviation

$\gamma_i + \gamma_e$ = blade camber

$\sigma = c/s$ = solidity

θ = stagger angle

c_x = axial chord

Fig. 9.59 Turbine airfoil nomenclature.

9.5.1.5 Turbine Airfoil Nomenclature and Design Metal Angles

The nomenclature for turbine airfoil cascades is presented in Fig. 9.59. The situation in unchoked turbines is similar to that in compressors except that the deviations are markedly smaller because of the thinner boundary layers. Hence

$$\delta_t = \frac{\gamma_i + \gamma_e}{8\sqrt{\sigma}} \tag{9.89}$$

is a good estimate of the turbine exit deviation. More importantly, however, when the turbine airfoil cascade exit Mach number is near unity, the deviation is usually negligible because the cascade passage is similar to a nozzle. In fact, the suction (or convex) surface of the airfoils often has a flat stretch before the trailing edge, which evokes the name *straight-backed*. Finally, the simple concept of deviation loses all meaning at large supersonic exit Mach numbers because expansion or compression waves emanating from the trailing edge can dramatically alter the final flow direction. This is a truly fascinating field of aerodynamics, but one that requires considerable study outside the realm of this textbook.

9.5.1.6 Stage Temperature Ratio τ_s

The stage temperature ratio ($\tau_s = T_{t1}/T_{t3}$) can be expressed as follows, by using the definition of the stage loading coefficient:

$$\tau_s = \frac{T_{t3}}{T_{t1}} = 1 - \psi \frac{(\omega r)^2}{g_c c_p T_{t1}} \tag{9.90}$$

Thus, for a given T_{t1} and ωr, the zero-reaction turbine stage will have a lower stage temperature ratio (greater work output per unit mass) than a 50% reaction turbine stage.

9.5.1.7 Stage Pressure Ratio π_s

Once the stage temperature ratio, flowfield, and airfoil characteristics are established, several avenues are open to predict the stage pressure ratio. The most simple and direct method is to employ the polytropic efficiency e_t. Recall from Section 6.5.8 that the polytropic efficiency is

$$e_t = \frac{dh_t}{dh_{t\,ideal}} = \frac{\gamma}{\gamma-1}\frac{dT_t/T_t}{dP_t/P_t}$$

Integration with constant γ and e_t yields the following equation for the stage pressure ratio

$$\pi_s = \frac{P_{t3}}{P_{t1}} = \left(\frac{T_{t3}}{T_{t1}}\right)^{\gamma/[(\gamma-1)e_t]} = \tau_s^{\gamma/[(\gamma-1)e_t]} \tag{9.91}$$

where the stage temperature ratio can be obtained from the total temperatures or an equation like Eq. (9.90).

Another approach involves the use of experimental or empirical cascade loss correlations, such as those shown in Figs. 9.60 and 9.61, to the stator and rotor in order to determine the total pressure loss. The *total pressure loss coefficient* for turbine cascade data is defined as

$$\phi_t \equiv \frac{P_{ti} - P_{te}}{P_{te} - P_e} \tag{9.92}$$

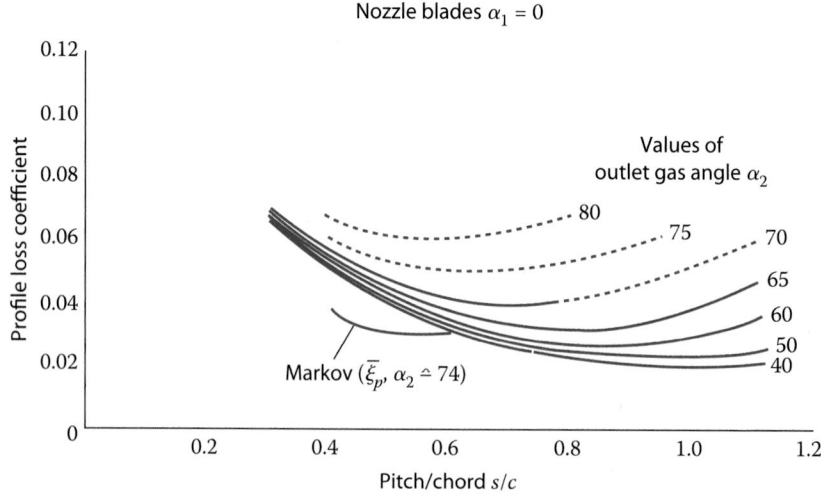

Fig. 9.60   Turbine stator cascade loss coefficient ($\alpha_1 = 0$) (Ref. 40).

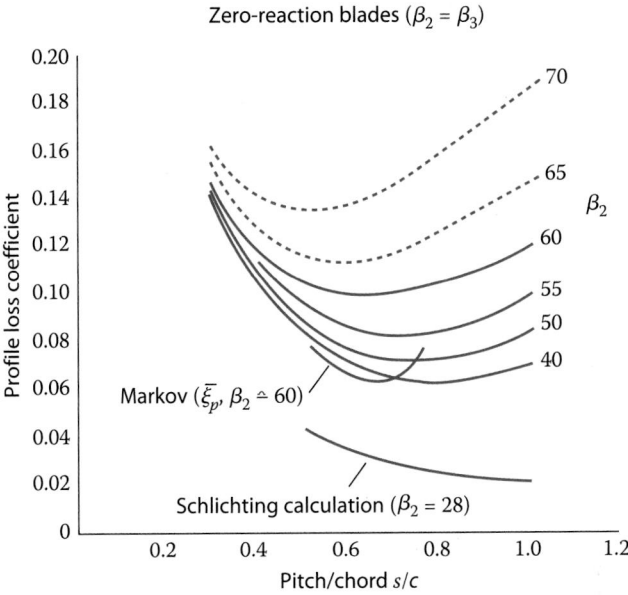

Fig. 9.61 Turbine rotor cascade loss coefficient ($\beta_2 = \beta_3$) (Ref. 40).

where subscripts i and e refer to the inlet and exit states, respectively. This equation can be rewritten for the cascade total pressure ratio as

$$\frac{P_{te}}{P_{ti}} = \frac{1}{1 + \phi_t(1 - P_e/P_{te})} \tag{9.93}$$

where P_e/P_{te} depends only on the usually known airfoil cascade exit Mach number M_e. Note that the total pressure loss coefficient for the rotor is based on the relative total states. The stage total pressure ratio can be written as

$$\frac{P_{t3}}{P_{t1}} = \left(\frac{P_{t2}}{P_{t1}}\right)_{\phi_{t\,\text{stator}}\,M_2} \frac{P_{t2R}}{P_{t2}}\left(\frac{P_{t3R}}{P_{t2R}}\right)_{\phi_{t\,\text{rotor}}\,M_{3R}} \frac{P_{t3}}{P_{t3R}} \tag{9.94}$$

where $\phi_{t\,\text{stator}}$ and $\phi_{t\,\text{rotor}}$ are the loss coefficients for the stator and rotor, respectively, and the subscripted total pressure ratios are obtained from Eq. (9.93) and cascade data. Additional losses are associated with tip leakage, annulus boundary layers, and secondary flows. Then, with all the stator, rotor, and stage properties computed, the stage efficiency can be computed from Eq. (9.74).

9.5.1.8 Blade Spacing

The momentum equation relates the tangential force of the blades on the fluid to the change in tangential momentum of the fluid. This force is equal and opposite to that which results from the difference in pressure between the pressure side and the suction side of the airfoil. Figure 9.62 shows the variation in pressure on both the suction and pressure surfaces of a typical turbine airfoil from cascade tests. On the pressure surface, the pressure is nearly equal to the inlet total pressure for 60% of the length before the fluid is accelerated to the exit pressure condition. However, on the suction surface, the fluid is accelerated in the first 60% of the length to a low pressure and then is decelerated to the exit pressure condition. The deceleration on the suction surface is limited and controlled because it can lead to boundary-layer separation.

Enough airfoils must be present in each row that the sum of the tangential force on each is equal to the change in tangential momentum of the fluid. Note that the same is true for the compressor, but the loading is low enough that blade spacing is dictated by the solidity, a key design choice for compressors. A simple expression for the relationship of the blade spacing to the fluid flow angles is developed in this section based on an incompressible fluid. This same expression correlates to the required blade spacing in a turbine stator or rotor row.

Referring to the cascade nomenclature in Fig. 9.59, we see that the tangential force per unit depth of blades spaced a distance s apart is

$$F_t = \frac{\rho u_i s (v_i + v_e)}{g_c} = \frac{\rho u_i^2 s}{g_c} \left(\tan \alpha_i + \frac{u_e}{u_i} \tan \alpha_e \right) \qquad (9.95)$$

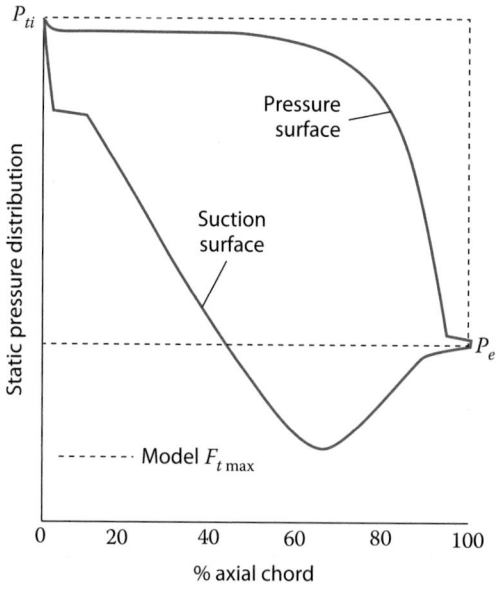

Fig. 9.62 Pressure distribution on a turbine cascade airfoil.

Zweifel [54] defines a tangential force coefficient that is the ratio of the force given by Eq. (9.95) to the maximum tangential force $F_{t\,max}$ that can be achieved efficiently, and $F_{t\,max}$ is obtained when

1. The pressure on the pressure surface is maintained at the inlet total pressure and drops to the exit static pressure at the trailing edge.
2. The pressure on the suction surface drops to the exit static pressure at the leading edge and remains at this value.

Thus the maximum tangential force is $F_{t\,max} = (P_{ti} - P_e)c_x$, where c_x is the axial chord of the blade (see Fig. 9.59). For reversible flow of an incompressible fluid, $F_{t\,max}$ can be written as

$$F_{t\,max} = \frac{\rho V_e^2 c_x}{2g_c} = \frac{\rho u_e^2 c_x}{2g_c \cos^2 \alpha_e} \tag{9.96}$$

The *Zweifel tangential force coefficient* Z is defined as

$$Z \equiv \frac{F_t}{F_{t\,max}} \tag{9.97}$$

From Eqs. (9.95) and (9.96), the equation of Z for a cascade airfoil becomes

$$Z = \frac{2s}{c_x}(\cos^2 \alpha_e)\left(\tan \alpha_t + \frac{u_e}{u_i}\tan \alpha_e\right)\left(\frac{u_i}{u_e}\right)^2$$

For the stator, we write

$$Z_s = \frac{2s}{c_x}(\cos^2 \alpha_2)\left(\tan \alpha_1 + \frac{u_2}{u_1}\tan \alpha_2\right)\left(\frac{u_1}{u_2}\right)^2 \tag{9.98a}$$

Likewise for the rotor, we write

$$Z_r = \frac{2s}{c_x}(\cos^2 \beta_3)\left(\tan \beta_2 + \frac{u_3}{u_2}\tan \beta_3\right)\left(\frac{u_2}{u_3}\right)^2 \tag{9.98b}$$

Because suction surface pressures can be less than the exit static pressure along the blade (see Fig. 9.62), *Z values near unity are attainable.* Care must be exercised in the design of the suction surface to prevent flow separation where the static pressure is increasing. It is through the specification of the Z value that rotor blade and nozzle vane spacing is determined in the turbine.

9.5.1.9 Radial Variations

Because the mass flow rate per unit area [that is, $\dot{m}/A = P_t/(MFP\sqrt{T_t})$] is higher in turbines than in compressors, turbine airfoils are correspondingly shorter. The result is little radial variation of aerodynamic properties from hub to tip except in the last few stages of the low-pressure turbine. Figure 9.63 is the rotor blade of a low-pressure turbine that shows radial

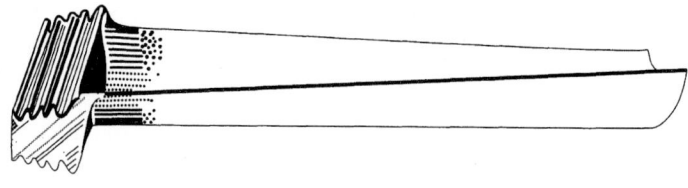

Fig. 9.63   Low-pressure turbine rotor blade. (Courtesy of Pratt & Whitney.)

variation from hub to tip. Typically, the degree of reaction varies from near zero at the hub (impulse) to about 40% at the tip.

If the aerodynamic design of these stages began as free vortex, the swirl distribution with radius is the same as for compressors, given by [Eq. (9.43)]

$$v = v_m \frac{r_m}{r}$$

For constant axial velocity ($u_2 = u_3$), the degree of reaction is

$$°R_t = \frac{T_2 - T_3}{T_{t1} - T_{t3}} = \frac{T_2 - T_3}{T_{t2} - T_{t3}} = 1 - \frac{V_2^2 - V_3^2}{2g_c c_p(T_{t1} - T_{t3})}$$

$$= 1 - \frac{v_2^2 - v_3^2}{2\omega r(v_2 + v_3)} = 1 - \frac{v_2 - v_3}{2\omega r}$$

Substituting Eq. (9.43), we write the degree of reaction at any radius in terms of the degree of reaction at the mean radius as

$$°R_t = 1 - \frac{v_{m2} - v_{m3}}{2\omega r}\frac{r_m}{r} = 1 - \frac{v_{m2} - v_{m3}}{2\omega r_m}\left(\frac{r_m}{r}\right)^2$$

$$°R_t = 1 - (1 - °R_{tm})\left(\frac{r_m}{r}\right)^2 \tag{9.99}$$

This is the same result as for compressors [Eqs. (9.49) and (9.50)]. Consequently, the most difficult airfoil contours to design would be at the hub of the rotating airfoils and at the tips of the stationary airfoils where the degree of reaction is low. It is, therefore, possible to find portions of some airfoils near the rear of highly loaded (i.e., high work per stage), low-pressure turbines where the static pressure actually rises across the cascade and boundary-layer separation is hard to avoid. In these cases, turbine designers have used advanced computational methods to develop nonfree or controlled vortex machines without these troublesome regions in order to maintain high efficiency at high loading.

Because of radial variations, the degree of reaction is lowest at the hub. Hence the Zweifel tangential force coefficient of the rotor Z_r times c_x/s will be maximum at the hub. Although the blade spacing varies directly with radius, $Z_r c_x/s$ is greatest at the hub and decreases faster than $1/r$ with increasing radius. Thus the value of $Z_r c_x/s$ at the rotor hub determines the spacing and number of rotor blades. For the stator, $Z_s c_x/s$ will be greatest at the tip, and its value determines the spacing and number of stator blades.

9.5.1.10 Velocity Ratio

The *velocity ratio* (VR) is defined as the ratio of the rotor speed ($U = \omega r$) to the velocity equivalent of the change in stage total enthalpy, or

$$\text{VR} \equiv \frac{U}{\sqrt{2g_c \, \Delta h_t}} = \frac{\omega r}{\sqrt{2g_c \, \Delta h_t}} \tag{9.100}$$

The velocity ratio is used by some turbine designers rather than the stage loading coefficient ψ, and one can show that

$$\text{VR} = \frac{1}{\sqrt{2\psi}} \tag{9.101}$$

The VR at the mean radius ranges between 0.5 and 0.6 for modern aircraft gas turbine engines. This range corresponds to stage loading coefficients ψ between 1.4 and 2.

9.5.2 Axial-Flow Turbine Stage

Consider the flow through a single-stage turbine as shown in Fig. 9.64. For generality, we will allow the axial velocity to change from station 2 to 3. The flows through the nozzle (stator) and rotor are assumed to be adiabatic. For solution, we assume that the following data are known: M_2, T_{t1}, T_{t3}, ωr, α_3, c_p, γ, and u_3/u_2. We will develop and write the equations for a general axial-flow turbine based on these known data. The TURBN program provided with this textbook allows for turbine design based on these known input data as well as other design choices.

To solve for the flow angle at station 2 (α_2), we first write the Euler turbine equation

$$c_p(T_{t2} - T_{t3}) = \frac{\omega r}{g_c}(v_2 + v_3) \tag{9.102}$$

Solving for v_2, we have

$$v_2 = \frac{g_c c_p \, \Delta T_t}{\omega r} - v_3$$

Then

$$\sin \alpha_2 = \frac{v_2}{V_2} = \frac{g_c c_p \, \Delta T_t}{\omega r V_2} - \frac{v_3}{V_2} \tag{i}$$

However,

$$\frac{v_3}{V_2} = \frac{u_3}{V_2}\tan \alpha_3 = \frac{u_3}{u_2}\frac{u_2}{V_2}\tan \alpha_3 = \frac{u_3}{u_2}\cos \alpha_2 \tan \alpha_3$$

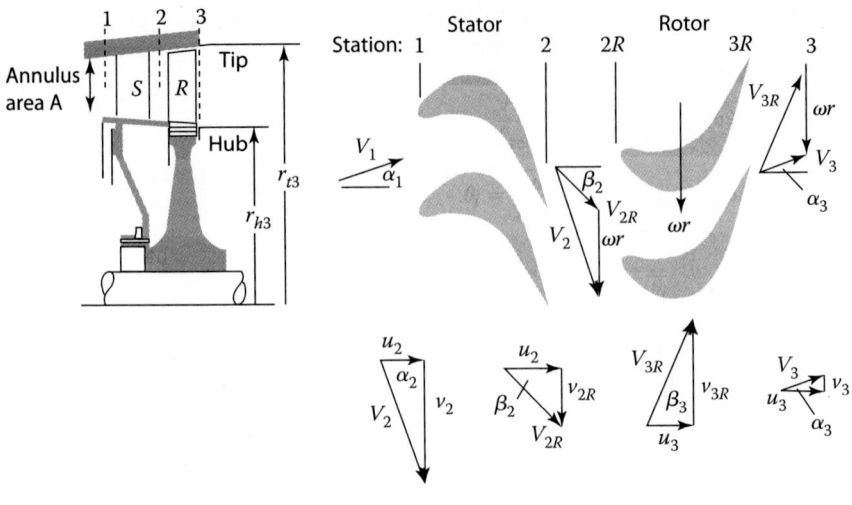

Fig. 9.64 Axial-flow turbine stage (after Ref. 22).

Thus, Eq. (i) becomes

$$\sin \alpha_2 = \frac{v_2}{V_2} = \frac{g_c c_p \Delta T_t}{\omega r V_2} - \frac{u_3}{u_2} \cos \alpha_2 \tan \alpha_3 \qquad \text{(ii)}$$

By using the stage loading parameter ψ, Eq. (ii) can be written as

$$\sin \alpha_2 = \psi \frac{\omega r}{V_2} - \frac{u_3}{u_2} \cos \alpha_2 \tan \alpha_3 \qquad \text{(9.103)}$$

The velocity at station 2 can be found from

$$V_2 = M_2 a_2 = \sqrt{\frac{2 g_c c_p T_{t2}}{1 + 2/[(\gamma - 1) M_2^2]}} \qquad \text{(9.104)}$$

If $\alpha_3 = 0$, then Eq. (9.103) simplifies to

$$\sin \alpha_2 = \psi \frac{\omega r}{V_2} \qquad \text{(9.105)}$$

If α_3 is not zero, Eq. (9.103) can be solved by substituting $\sqrt{1 - \sin^2 \alpha_2}$ for $\cos \alpha_2$, squaring both sides of the equation, and solving the resulting quadratic equation for $\sin \alpha_2$. The solution is

$$\sin \alpha_2 = \frac{\left(\psi \dfrac{\omega r}{V_2} \right) \pm \left(\dfrac{u_3}{u_2} \tan \alpha_3 \right) \sqrt{1 + \left(\dfrac{u_3}{u_2} \tan \alpha_3 \right)^2 - \left(\psi \dfrac{\omega r}{V_2} \right)^2}}{1 + \left(\dfrac{u_3}{u_2} \tan \alpha_3 \right)^2} \qquad \text{(9.106)}$$

When two solutions to the inlet flow angle α_2 exist, the Mach number at station $3R$ should be checked for each value of α_2.

The velocity at station 3 can be written in terms of that at station 2 and the two flow angles α_2 and α_3:

$$V_3 = \frac{u_3 \cos \alpha_2}{u_2 \cos \alpha_3} V_2 \tag{9.107}$$

The degree of reaction can be written in terms of the given data as follows:

$$°R_t = \frac{T_2 - T_3}{T_{t2} - T_{t3}} = \frac{T_{t2} - T_{t3} - (T_{t2} - T_2) + T_{t3} - T_3}{T_{t2} - T_{t3}}$$

$$= 1 - \frac{V_2^2 - V_3^2}{2 g_c c_p (T_{t2} - T_{t3})} = 1 - \frac{V_2^2 - V_3^2}{2 \psi (\omega r)^2}$$

$$= 1 - \frac{1}{2 \psi (\omega r)^2} \left(\frac{u_2^2}{\cos^2 \alpha_2} - \frac{u_3^2}{\cos^2 \alpha_3} \right)$$

$$°R_t = 1 - \frac{1}{2 \psi} \left(\frac{V_2}{\omega r} \right)^2 \left[1 - \left(\frac{u_3 \cos \alpha_2}{u_2 \cos \alpha_3} \right)^2 \right] \tag{9.108}$$

The Mach number at station 3 can be found from

$$M_3 = M_2 \frac{V_3}{V_2} \sqrt{\frac{T_2}{T_3}} \tag{9.109}$$

where

$$\frac{T_3}{T_2} = 1 - °R_t \frac{\Delta T_t}{T_{t2}} \left(1 + \frac{\gamma - 1}{2} M_2^2 \right) \tag{9.110}$$

An equation for the Mach number at station $2R$ can be developed as follows:

$$M_{2R} = M_2 \frac{V_{2R}}{V_2}$$

where

$$V_{2R} = \sqrt{u_2^2 + (v_2 - \omega r)^2} = V_2 \sqrt{\cos^2 \alpha_2 + \left(\sin \alpha_2 - \frac{\omega r}{V_2} \right)^2}$$

Thus

$$M_{2R} = M_2 \sqrt{\cos^2 \alpha_2 + \left(\sin \alpha_2 - \frac{\omega r}{V_2} \right)^2} \tag{9.111}$$

Likewise, an equation for the Mach number at station $3R$ is developed as follows:

$$M_{3R} = M_3 \frac{V_{3R}}{V_3}$$

where

$$V_{3R} = \sqrt{u_3^2 + (v_3 + \omega r)^2} = V_3 \sqrt{\cos^2 \alpha_3 + \left(\sin \alpha_3 + \frac{\omega r}{V_3}\right)^2}$$

Thus

$$M_{3R} = M_3 \sqrt{\cos^2 \alpha_3 + \left(\sin \alpha_3 + \frac{\omega r}{V_3}\right)^2} \qquad (9.112)$$

An equation for the rotor relative total temperature ($T_{t2R} = T_{t3R}$) can be developed by noting that

$$T_3 = T_{t3} - \frac{V_3^2}{2g_c c_p} = T_{t3R} - \frac{V_{3R}^2}{2g_c c_p}$$

Then

$$T_{t3R} = T_{t3} + \frac{V_{3R}^2 - V_3^2}{2g_c c_p}$$

or

$$T_{t3R} = T_{t3} + \frac{V_3^2}{2g_c c_p} \left[\cos^2 \alpha_3 + \left(\sin \alpha_3 + \frac{\omega r}{V_3}\right)^2 - 1 \right] \qquad (9.113)$$

Example 9.12

Consider mean-radius stage calculation—flow with losses.
Given:

$$T_{t1} = 1850 \text{ K}, \quad P_{t1} = 1700 \text{ kPa}, \quad M_1 = 0.4, \quad \alpha_1 = 0 \text{ deg},$$
$$T_{t3} = 1560 \text{ K}, \quad M_2 = 1.1, \quad \omega r = 450 \text{ m/s}, \quad \alpha_3 = 10 \text{ deg},$$
$$u_3/u_2 = 0.9, \quad \phi_{t \text{ stator}} = 0.06, \quad \phi_{r \text{ rotor}} = 0.15, \quad \gamma = 1.3,$$
$$R = 0.2873 \text{ kJ/(kg} \cdot \text{K)} \text{ [From Eq. (6.1), } c_p = 1.245 \text{ kJ/(kg} \cdot \text{K)]}$$

Find the flow velocities, properties, angles, and performance parameters.

(Continued)

Example 9.12 *(Continued)*

Solution:

$$T_1 = \frac{T_{t1}}{1 + [(\gamma - 1)/2]M_1^2} = \frac{1850\,\text{K}}{1 + 0.15 \times 0.4^2} = 1806.6\,\text{K}$$

$$V_1 = \sqrt{\frac{2g_c c_p T_{t1}}{1 + 2/[(\gamma - 1)M_1^2]}} = \sqrt{\frac{2 \times 1 \times 1245 \times 1850}{1 + 2/(0.3 \times 0.4^2)}} = 328.6\,\text{m/s}$$

$$u_1 = V_1 \cos \alpha_1 = 328.6\,\text{m/s}$$

$$v_1 = V_1 \sin \alpha_1 = 0$$

$$T_{t2} = T_{t1} = 1850\,\text{K}$$

$$T_2 = \frac{T_{t2}}{1 + [(\gamma - 1)/2]M_2^2} = \frac{1850\,\text{K}}{1 + 0.15 \times 1.1^2} = 1565.8\,\text{K}$$

$$V_2 = \sqrt{\frac{2g_c c_p T_{t2}}{1 + 2/[(\gamma - 1)M_2^2]}}$$

$$= \sqrt{\frac{2 \times 1 \times 1245 \times 18}{1 + 2/(0.3 \times 1.1^2)}} = 841.2\,\text{m/s}$$

$$\psi = \frac{g_c c_p (T_{t1} - T_{t3})}{(\omega r)^2} = \frac{1245(1850 - 1560)}{450^2} = 1.78296$$

$$\text{VR} = \frac{1}{\sqrt{2\psi}} = \frac{1}{\sqrt{2 \times 1.78296}} = 0.5296$$

$$\alpha_2 = \sin^{-1} \frac{\left(\psi\dfrac{\omega r}{V_2}\right) \pm \left(\dfrac{u_3}{u_2}\tan \alpha_3\right)\sqrt{1 + \left(\dfrac{u_3}{u_2}\tan \alpha_3\right)^2 - \left(\psi\dfrac{\omega r}{V_2}\right)^2}}{1 + \left(\dfrac{u_3}{u_2}\tan \alpha_3\right)^2}$$

$$\psi\frac{\omega r}{V_2} = 1.78296\left(\frac{450}{841.2}\right) = 0.95379$$

$$\frac{u_3}{u_2}\tan \alpha_3 = 0.9 \tan 10\,\text{deg} = 0.15869$$

$$\alpha_2 = \sin^{-1} \frac{0.95379 - 0.15869\sqrt{1 + 0.15869^2 - 0.95379^2}}{1.015869^2}$$

$$= \sin^{-1} 0.87776 = 61.37\,\text{deg}$$

(Continued)

Example 9.12 *(Continued)*

$$u_2 = V_2 \cos \alpha_2 = 841.2 \cos 61.37 \deg = 403.1 \, \text{m/s}$$

$$v_2 = V_2 \sin \alpha_2 = 841.2 \sin 61.37 \deg = 738.3 \, \text{m/s}$$

$$\Phi = \frac{u_2}{\omega r} = \frac{403.1}{450} = 0.8958$$

$$V_3 = \frac{u_3 \cos \alpha_2}{u_2 \cos \alpha_3} V_2 = 0.9 \left(\frac{\cos 61.37 \deg}{\cos 10 \deg} \right) (841.2) = 368.4 \, \text{m/s}$$

$$u_3 = V_3 \cos \alpha_3 = 368.4 \cos 10 \deg = 362.8 \, \text{m/s}$$

$$v_3 = V_3 \sin \alpha_3 = 368.4 \sin 10 \deg = 64.0 \, \text{m/s}$$

$$°R_t = 1 - \frac{1}{2\psi} \left(\frac{V_2}{\omega r} \right)^2 \left[1 - \left(\frac{u_3 \cos \alpha_2}{u_2 \cos \alpha_3} \right)^2 \right]$$

$$= 1 - \frac{1}{2 \times 1.78296} \left(\frac{841.2}{450} \right)^2 \left[1 - \left(0.9 \frac{\cos 61.37 \deg}{\cos 10 \deg} \right)^2 \right] = 0.2080$$

$$T_3 = T_2 - °R_t(T_{t1} - T_{t3}) = 1565.8 - 0.2080(1850 - 1560) = 1505.5 \, \text{K}$$

$$M_3 = M_2 \frac{V_3}{V_2} \sqrt{\frac{T_2}{T_3}} = 1.1 \left(\frac{368.4}{841.2} \right) \sqrt{\frac{1565.8}{1505.5}} = 0.4913$$

$$M_{2R} = M_2 \sqrt{\cos^2 \alpha_2 + \left(\sin \alpha_2 - \frac{\omega r}{V_2} \right)^2}$$

$$= 1.1 \sqrt{\cos^2 61.37 \deg + \left(\sin 61.37 \deg - \frac{450}{841.2} \right)^2} = 0.6481$$

$$M_{3R} = M_3 \sqrt{\cos^2 \alpha_3 + \left(\sin \alpha_3 + \frac{\omega r}{V_3} \right)^2}$$

$$= 0.4913 \sqrt{\cos^2 10 \deg + \left(\sin 10 \deg + \frac{450}{368.4} \right)^2} = 0.8390$$

$$T_{t3R} = T_{t3} + \frac{V_3^2}{2g_c c_p} \left[\cos^2 \alpha_3 + \left(\sin \alpha_3 + \frac{\omega r}{V_3} \right)^2 - 1 \right]$$

$$= 1560 + \frac{368.4^2}{2 \times 1 \times 1245} \left[\cos^2 10 \deg + \left(\sin 10 \deg + \frac{450}{368.4} \right)^2 - 1 \right]$$

$$= 1664.4 \, \text{K}$$

(Continued)

Example 9.12 *(Continued)*

$$T_{t2R} = T_{t3R} = 1664.4 \, \text{K}$$

$$\tau_s = \frac{T_{t3}}{T_{t1}} = \frac{1560}{1850} = 0.8432$$

$$\frac{Z_s c_x}{s} = (2\cos^2 \alpha_2)\left(\tan \alpha_1 + \frac{u_2}{u_1}\tan \alpha_2\right)\left(\frac{u_1}{u_2}\right)^2$$

$$= (2\cos^2 61.371\,\text{deg})\left(\tan 0\,\text{deg} + \frac{403.1}{328.6}\tan 61.37\,\text{deg}\right)\left(\frac{328.6}{403.1}\right)^2$$

$$= 0.6857$$

$$\beta_2 = \tan^{-1}\frac{v_2 - \omega r}{u_2} = \tan^{-1}\frac{738.3 - 450}{403.1} = 35.57\,\text{deg}$$

$$\beta_3 = \tan^{-1}\frac{v_3 + \omega r}{u_3} = \tan^{-1}\frac{64.0 + 450}{362.8} = 54.78\,\text{deg}$$

$$\frac{Z_r c_x}{s} = (2\cos^2 \beta_3)\left(\tan \beta_2 + \frac{u_3}{u_2}\tan \beta_3\right)\left(\frac{u_2}{u_3}\right)^2$$

$$= (2\cos^2 54.78\,\text{deg})(\tan 35.573\,\text{deg} + 0.9\tan 54.78\,\text{deg})\left(\frac{1}{0.9}\right)^2$$

$$= 1.6330$$

$$P_1 = P_{t1}\left(\frac{T_1}{T_{t1}}\right)^{\gamma/(\gamma-1)} = 1700\left(\frac{1806.6}{1850}\right)^{1.3/0.3} = 1533.8\,\text{kPa}$$

$$P_{t2} = \frac{P_{t1}}{1 + \phi_{t\,\text{stator}}[1 - (T_2/T_{t2})^{\gamma/(\gamma-1)}]}$$

$$= \frac{1700}{1 + 0.06[1 - (1565.8/1850)^{1.3/0.3}]} = 1649.1\,\text{kPa}$$

$$P_2 = P_{t2}\left(\frac{T_2}{T_{t2}}\right)^{\gamma/(\gamma-1)} = 1649.1\left(\frac{1565.8}{1850}\right)^{1.3/0.3} = 800.5\,\text{kPa}$$

$$P_{t2R} = P_2\left(\frac{T_{t2R}}{T_2}\right)^{\gamma/(\gamma-1)} = 800.5\left(\frac{1664.4}{1565.8}\right)^{1.3/0.3} = 1043.0\,\text{kPa}$$

$$P_{t3R} = \frac{P_{t2R}}{1 + \phi_{t\,\text{rotor}}[1 - (T_3/T_{t3R})^{\gamma/(\gamma-1)}]}$$

$$= \frac{1043.0}{1 + 0.15[1 - (1505.5/1664.4)^{1.3/0.3}]} = 990.6\,\text{kPa}$$

(Continued)

Example 9.12 (Continued)

$$P_3 = P_{t3R} \left(\frac{T_3}{T_{t3R}} \right)^{\gamma/(\gamma-1)} = 990.6 \left(\frac{1505.5}{1664.4} \right)^{1.3/0.3} = 641.3 \, \text{kPa}$$

$$P_{t3} = P_3 \left(\frac{T_{t3}}{T_3} \right)^{\gamma/(\gamma-1)} = 641.3 \left(\frac{1560}{1505.5} \right)^{1.3/0.3} = 748.1 \, \text{kPa}$$

$$\pi_s = \frac{P_{t3}}{P_{t1}} = \frac{748.1}{1700} = 0.441$$

$$\eta_s = \frac{1 - \tau_s}{1 - \pi_s^{(\gamma-1)/\gamma}} = \frac{1 - 0.8432}{1 - 0.4401^{0.3/1.3}} = 90.87\%$$

Although hand calculation of these results (and those in Example 9.13 to follow) is relatively straightforward, such calculations are tedious. The TURBN program is a convenient tool for facilitating these calculations. It, like any computational tool, is no substitute for sound engineering judgment and analysis based on solid understanding of the fundamentals. The student would do well to remember this always.

9.5.3 Flow Path Dimensions

In this section, we use the same basic methods as we did with the compressor (Section 9.3.6) to estimate the physical dimensions, number of blades and vanes, and shapes of blades and vanes of the turbine preliminary design. The methods are simple enough to do hand calculations, but all are incorporated in the TURBN preliminary design program.

9.5.3.1 Annulus Area

The annulus area (see Fig. 9.24) at any station of a turbine stage is based on the flow properties (T_t, P_t, Mach number, and flow angle) at the mean radius and the total mass flow rate. Equation (9.8) is the easiest equation to use to calculate the flow area at any station i:

$$A_i = \frac{\dot{m}\sqrt{T_{ti}}}{P_{ti}(\cos \alpha_i)\text{MFP}(M_i)}$$

By using the relationships of Fig. 9.24, the radii at any station i can be determined, given the flow annulus area [Eq. (9.8)] and either the mean radius r_m or the hub/tip ratio r_h/r_t.

9.5.3.2 Axial Dimensions and Number of Blades

We use the same method for determining radial and axial geometries for the turbine as we did for the compressor. Figure 9.65 shows the cross

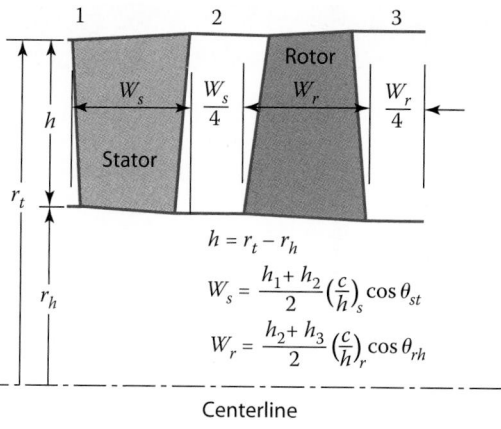

Fig. 9.65 Typical axial dimensions of a turbine stage.

section of a typical turbine stage that can be used to estimate its axial length. The chord/height ratio c/h of turbine blades varies from about 0.3 to 1.0. Assuming constant chord length and circular arc camber line, the program TURBN calculates the axial blade widths W_s and W_r of a stage, the blade spacings $W_s/4$ and $W_r/4$, and the number of blades based on user inputs of the tangential force coefficient Z and chord/height ratio c/h for both the stator and rotor blades. A minimum width of $\frac{1}{4}$ in. (0.6 cm) and spacing of $\frac{1}{8}$ in. (0.3 cm) are used in the plot of a turbine cross section and calculation of axial length.

The stagger angle θ of a blade depends on the shape of the camber line and blade angles γ_i and γ_e (see Fig. 9.59). For a circular arc camber line, the stagger angle θ is simply given by $\theta = (\gamma_e - \gamma_i)/2$. For constant-chord blades, the axial chord (and axial blade width) is greatest where the stagger angle is closest to zero. This normally occurs at the tip of the stator and hub of the rotor blades. For estimation purposes, a turbine blade's incidence angle is normally small and can be considered to be zero. Thus $\gamma_i = \alpha_i$. The blade's exit angle γ_e can be obtained using Eq. (9.89) for the exit deviation. However, Eq. (9.89) requires that the solidity ($\sigma = c/s$) be known. For known flow conditions (α_1, α_2, u_2/u_1, α_2, α_3, and u_3/u_2) and given tangential force coefficients (Z_s and Z_r), Eqs. (9.98a) and (9.98b) will give the required axial chord/spacing ratio c_x/s for the stator and rotor, respectively. An initial guess for the blade's solidity σ is needed to obtain the stagger angle θ from c_x/s.

After the solidities are determined at the hub, mean, and tip that give the desired tangential force coefficient Z, the number of required blades follows directly from the chord/height ratio c/h, the circumference, and the blade spacing at each radius. The following example shows the calculations needed to find the axial blade width and number of blades.

Example 9.13

Here we consider the turbine stator of Example 9.12 with a mass flow rate of 60 kg/s, a mean radius of 0.3 m, a tangential force coefficient Z_s of 0.9, and a chord/height ratio c/h of 1.0. The flow annulus areas and radii at stations 1 and 2 are as follows.

Station 1:

$$\text{MFP}(M_1) = 0.024569$$

$$A_1 = \frac{\dot{m}\sqrt{T_{t1}}}{P_{t1}\text{MFP}(M_1)(\cos \alpha_1)} = \frac{60\sqrt{1850}}{1{,}700{,}000 \times 0.024569 \times 1}$$

$$= 0.0617788 \text{ m}^2$$

$$h_1 = \frac{A_1}{2\pi r_m} = \frac{0.061788}{0.6\pi} = 0.03278 \text{ m}$$

$$r_{t1} = 0.3164 \text{ m} \quad r_{h1} = 0.2836 \text{ m}$$

$$v_{1h} = v_{1m} = v_{1t} = 0$$

Station 2:

$$\text{MFP}(M_2) = 0.039042$$

$$A_2 = \frac{\dot{m}\sqrt{T_{t2}}}{P_{t2}\text{MFP}(M_2)(\cos \alpha_2)}$$

$$= \frac{60\sqrt{1850}}{1{,}649{,}100 \times 0.039042 \times \cos 61.37 \text{ deg}} = 0.083654 \text{ m}^2$$

$$h_2 = \frac{A_2}{2\pi r_m} = \frac{0.083654}{0.6\pi} = 0.04438 \text{ m}$$

$$r_{t2} = 0.3222 \text{ m} \quad r_{h2} = 0.2778 \text{ m}$$

$$v_{2h} = v_{2m}\frac{r_m}{r_{2h}} = 738.3\left(\frac{0.3}{0.2778}\right) = 797.3 \text{ m/s}$$

$$\alpha_{2h} = \tan^{-1}\frac{v_{2h}}{u_2} = \tan^{-1}\frac{797.3}{403.1} = 63.18 \text{ deg}$$

$$v_{2t} = v_{2m}\frac{r_m}{r_{2t}} = 738.3\left(\frac{0.3}{0.3222}\right) = 687.4 \text{ m/s}$$

$$\alpha_{2t} = \tan^{-1}\frac{v_{2t}}{u_2} = \tan^{-1}\frac{687.4}{403.1} = 59.61 \text{ deg}$$

(Continued)

Example 9.13 *(Continued)*

The chord of the stator is

$$c = \frac{c}{h}\frac{h_1 + h_2}{2} = 1.0\frac{0.03278 + 0.04438}{2} = 0.03858 \text{ m}$$

For the specified tangential force coefficient Z_s, we calculate the stagger angle, solidity, and spacing of the stator at the mean line, hub, and tip.

Mean line:

$$Z_s\left(\frac{c_x}{s}\right)_m = (2\cos^2 \alpha_{2m})\left(\tan \alpha_{1m} + \frac{u_2}{u_1}\tan \alpha_{2m}\right)\left(\frac{u_1}{u_2}\right)^2$$

$$= (2\cos^2 61.37 \text{ deg})\left(\tan 0 \text{ deg} + \frac{403.1}{328.6}\tan 61.37 \text{ deg}\right)$$

$$\times \left(\frac{328.6}{403.1}\right)^2 = 0.6857$$

$$\left(\frac{c_x}{s}\right)_m = \frac{0.6857}{0.9} = 0.7619$$

$$\gamma_{1m} = \alpha_{1m} = 0$$

Initially, we assume a solidity σ of 1.0. Then,

$$\gamma_{2m} = \frac{\gamma_{1m} + 8\sqrt{\sigma_m}\, \alpha_{2m}}{8\sqrt{\sigma_m} - 1} = \frac{0 + 8\sqrt{1}\, 61.37}{8\sqrt{1} - 1} = 70.14 \text{ deg}$$

$$\theta_m = \frac{\gamma_{2m} - \gamma_{1m}}{2} = \frac{70.14}{2} = 35.07 \text{ deg}$$

$$\sigma_m = \frac{(c_x/s)_m}{\cos \theta_m} = \frac{0.7619}{\cos 35.07 \text{ deg}} = 0.9309$$

After several iterations, the results are $\gamma_{2m} = 70.49$ deg, $\theta_m = 35.25$ deg, and $\sigma_m = 0.9329$. The blade spacing s is 0.04135 m, and the axial chord is 0.03150 m.

Hub:

$$Z_s\left(\frac{c_x}{s}\right)_h = (2\cos^2 \alpha_{2h})\left(\tan \alpha_{1h} + \frac{u_2}{u_1}\tan \alpha_{2h}\right)\left(\frac{u_1}{u_2}\right)^2$$

$$= (2\cos^2 63.18 \text{ deg})\left(\tan 0 \text{ deg} + \frac{403.1}{328.6}\tan 63.18 \text{ deg}\right)$$

$$\times \left(\frac{328.6}{403.1}\right)^2 = 0.6565$$

(Continued)

Example 9.13 *(Continued)*

$$\left(\frac{c_x}{s}\right)_h = \frac{0.6565}{0.9} = 0.7294$$

$$\gamma_{1h} = \alpha_{1h} = 0$$

Initially, we assume a solidity σ of 1.0. Then,

$$\gamma_{2h} = \frac{\gamma_{1h} + 8\sqrt{\sigma_h}\ \alpha_{2h}}{8\sqrt{\sigma_h} - 1} = \frac{0 + 8\sqrt{1}\ 63.18}{8\sqrt{1} - 1} = 72.21\ \text{deg}$$

$$\theta_h = \frac{\gamma_{2h} - \gamma_{1h}}{2} = \frac{72.21}{2} = 36.10\ \text{deg}$$

$$\sigma_h = \frac{(c_x/s)_h}{\cos\theta_h} = \frac{0.7294}{\cos 361.0\ \text{deg}} = 0.9027$$

After several iterations, the results are $\gamma_{2h} = 72.73$ deg, $\theta_h = 36.37$ deg, and $\sigma_h = 0.9058$. The blade spacing s is 0.04259 m, and the axial chord is 0.03107 m.

Tip:

$$Z_x\left(\frac{c_x}{s}\right)_t = (2\cos^2\alpha_{2t})\left(\tan\alpha_{1t} + \frac{u_2}{u_1}\tan\alpha_{2t}\right)\left(\frac{u_1}{u_2}\right)^2$$

$$= (2\cos^2 59.61\ \text{deg})\left(\tan 0\ \text{deg} + \frac{403.1}{328.6}\tan 59.61\ \text{deg}\right)$$

$$\times\left(\frac{328.6}{403.1}\right)^2 = 0.7115$$

$$\left(\frac{c_x}{s}\right)_t = \frac{0.7115}{0.9} = 0.7905$$

$$\gamma_{1t} = \alpha_{1t} = 0$$

Initially, we assume a solidity σ of 1.0:

$$\gamma_{2t} = \frac{\gamma_{1t} + 8\sqrt{\sigma_t}\ \alpha_{2t}}{8\sqrt{\sigma_t} - 1} = \frac{0 + 8\sqrt{1}\ 59.61}{8\sqrt{1} - 1} = 68.13\ \text{deg}$$

$$\theta_t = \frac{\gamma_{2t} - \gamma_{1t}}{2} = \frac{68.13}{2} = 34.06\ \text{deg}$$

$$\sigma_t = \frac{(c_x/s)_t}{\cos\theta_t} = \frac{0.7905}{\cos 34.06\ \text{deg}} = 0.9542$$

After several iterations, the results are $\gamma_{2t} = 68.35$ deg, $\theta_t = 34.18$ deg, and $\sigma_t = 0.9555$. The blade spacing s is 0.04038 m, and the axial chord is 0.03192 m.

(Continued)

Example 9.13 *(Continued)*

Table 9.11   Summary of Example 9.13 Results

Location	Average Radius, m	Solidity	Spacing, m	Number of Blades	Axial Chord, m
Tip	0.3193	0.9555	0.04038	49.7	0.03192
Mean	0.3000	0.9329	0.04135	45.6	0.03150
Hub	0.2807	0.9058	0.04259	41.4	0.03107

For this stator blade with a chord of 0.03858 m, we require the information in Table 9.11. Thus the number of required stator blades is 50, which will have a mean-radius spacing s of 0.03770 m and solidity σ of 1.023 ($=0.03858/0.03770$) on the mean radius. The blade has an axial width W_s of 0.03192 m (see Fig. 9.65).

9.5.3.3 Blade Profile

The shapes of turbine stator and rotor blades are based on airfoil shapes developed specifically for turbine applications. Two airfoil shapes are included in the TURBN program to sketch the blade shapes for a stage: the C4 and T6 British profiles. The base profile of the C4 airfoil is listed in Table 9.12 and shown in Fig. 9.66 for a 10% thickness. Table 9.13 and Fig. 9.67 give the base profile of the T6 airfoil for a 10% thickness. The TURBN program assumes a circular arc mean line for sketching the blade shapes.

Example 9.14 uses the TURBN program to perform a preliminary design of a core turbine stage. Like the COMPR program, TURBN requires the specification of a stator or nozzle vane loss coefficient in addition to the turbine polytropic efficiency to be able to determine different values of total pressures at stations $2R$ and $3R$.

Table 9.12   C4 Airfoil Profile $(t/c = 0.10)^{a,b}$

x/c, %	y/c, %	x/c, %	y/c, %
0.0	0.0	40	4.89
1.25	1.65	50	4.57
2.5	2.27	60	4.05
5	3.08	70	3.37
7.5	3.62	80	2.54
10	4.02	90	1.60
15	4.55	95	1.06
20	4.83	100	0.0
30	5.00		

[a]Leading-edge radius $= 0.12t$.
[b]Trailing-edge radius $= 0.06t$.

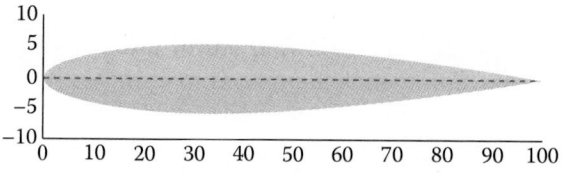

Fig. 9.66 The C4 turbine airfoil base profile.

Table 9.13 T6 Airfoil Profile $(t/c = 0.10)^{a,b}$

x/c, %	y/c, %	x/c, %	y/c, %
0.0	0.0	40	5.00
1.25	1.17	50	4.67
2.5	1.54	60	3.70
5	1.99	70	2.51
7.5	2.37	80	1.42
10	2.74	90	0.85
15	3.4	95	0.72
20	3.95	100	0.0
30	4.72		

[a]Leading-edge radius $= 0.12t$.
[b]Trailing-edge radius $= 0.06t$.

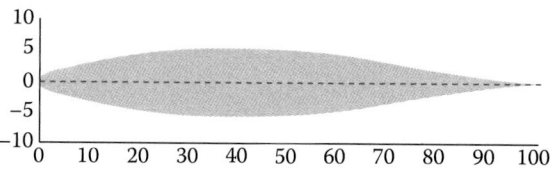

Fig. 9.67 The T6 turbine airfoil base profile.

Example 9.14

Consider a single-stage turbine using the TURBN computer program.
Given:

$$\dot{m} = 200 \text{ lbm/s}, \quad M_1 = 0.4, \quad T_{t1} = 3400°\text{R}, \quad T_{t3} = 2860°\text{R},$$
$$M_2 = 1.1, \quad \omega r = 1500 \text{ ft/s}, \quad r_m = 12 \text{ in.}, \quad P_{t1} = 250 \text{ psia},$$
$$\alpha_1 = 0 \text{ deg}, \quad \alpha_3 = 0 \text{ deg}, \quad \gamma = 1.3, \quad R = 53.40 \text{ ft} \cdot \text{lbf}/(\text{lbm} \cdot °\text{R}),$$
$$u_3/u_2 = 0.90, \quad \phi_{t \text{ stator}} = 0.06, \quad \phi_{t \text{ rotor}} = 0.15$$

(Continued)

Table 9.14 Results for Example 9.14 Axial-Flow Turbine Stage Calculation Using TURBN

Property		1h	1m	1t	2h	2m	2t	2Rm	3Rm	3h	3m	3t
T_t	°R	3400	3400	3400	3400	3400	3400	3052	3052	2860	2860	2860
T	°R	3320	3320	3320	2767	2878	2957	2878	2768	2767	2768	2769
P_t	psia	250.0	250.0	250.0	242.5	242.5	242.5	151.9	144.4	109.0	109.0	109.0
P	psia	225.6	225.6	225.6	99.2	117.7	132.4	117.7	94.6	94.4	94.6	94.7
M		0.400	0.400	0.400	1.236	1.100	1.100	0.636	0.827	0.474	0.471	0.469
V	ft/s	1089	1089	1089	3071	2789	2569	1611	2057	1178	1171	1167
u	ft/s	1089	1089	1089	1282	1282	1282	1282	1153	1153	1153	1153
v	ft/s	0	0	0	2791	2477	2226	977	1703	239	203	177
α	deg	0	0	0	65.34	62.64	60.07	—	—	11.68	10.00	8.74
β	deg	—	—	0	—	—	—	37.32	55.90	—	—	—
Radii	in.	11.04	12.00	12.96	10.65	12.00	13.35	12.00	12.00	10.20	12.00	13.80

Hub °R_t = −0.0005 A_1 = 144.31 in.²

Mean °R_t = 0.2034 A_2 = 203.72 in.²

Tip °R_t = 0.3487 A_3 = 271.04 in.²

$\tau_s = 0.8412$ $\pi_s = 0.4359$ RPM = 14,324 $\psi = 1.7868$ $\phi = 0.8544$

$\eta_s = 91.08\%$ and AN^2 at 2 = 4.18×10^{10} in.² · rpm²

Example 9.14 *(Continued)*

Solution:

The TURBN program is run with α_2 as the unknown. The stage results are given in Table 9.14 with the hub and tip tangential velocities based on free-vortex swirl distribution. As with the COMPR program, the amount of information is remarkable, all obtained from aerothermodynamic principles, trigonometric relationships (velocity triangles), and performance parameter definitions. One could take the information and construct velocity triangles at the hub, mean, and tip sections of stations 1, 2, and 3! The cross section of the stage sketched by the computer program is shown in Fig. 9.68 for stator $c/h = 0.8$ and rotor $c/h = 0.6$. As can be inferred from the material presented in Appendix I, the very high AN^2 of 4.18×10^{10} in.$^2 \cdot$ rpm^2 at a relative total temperature of 3052°R is not possible with current materials unless the blades are cooled. Also, the high rim speed of about 1206 ft/s would require existing materials to be cooled.

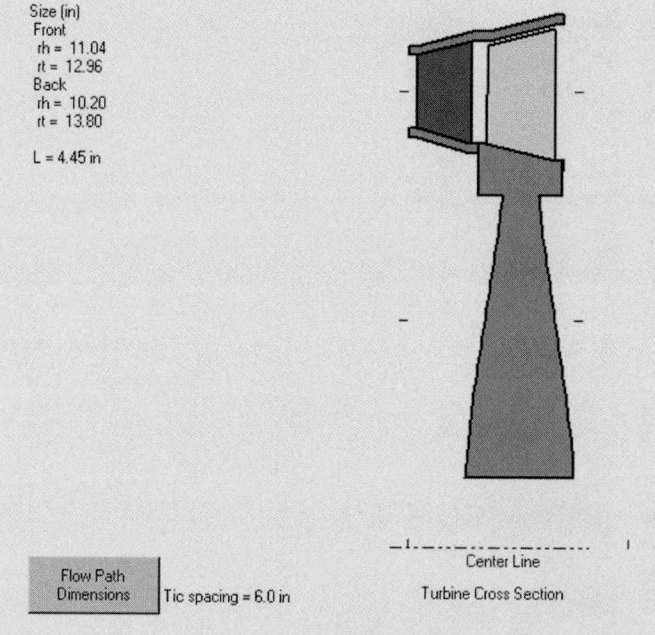

Size (in)
Front
rh = 11.04
rt = 12.96
Back
rh = 10.20
rt = 13.80

L = 4.45 in

Flow Path
Dimensions | Tic spacing = 6.0 in

Center Line

Turbine Cross Section

Fig. 9.68 Sketch of cross section for turbine stage of Example 9.14 from TURBN.

9.5.4 Axial-Flow Turbine Stage: α_2 Known

The preceding method of calculation assumed that α_2 was unknown. The TURBN program will handle the case when any one of the following four variables is unknown: α_2, T_{t3}, α_3, or M_2.

When T_{t3} is unknown, Eqs. (9.20) and (9.103) are solved for T_{t3}, giving

$$T_{t3} = T_{t1} - \frac{\omega r V_2}{g_c c_p}\left(\sin \alpha_2 + \frac{u_3}{u_2}\cos \alpha_2 \tan \alpha_3\right) \qquad (9.114)$$

When α_3 is unknown, Eqs. (9.20) and (9.103) are solved for α_3, giving

$$\tan \alpha_3 = \frac{1}{u_3/u_2}\left(\frac{\psi}{\cos \alpha_2}\frac{\omega \gamma}{V_2} - \tan \alpha_2\right) \qquad (9.115)$$

When M_2 is unknown, the velocity at station 2 (v_2) is obtained from Eq. (9.103), giving

$$V_2 = \frac{\psi \omega r}{\sin \alpha_2 + (u_3/u_2)\cos \alpha_2 \tan \alpha_3} \qquad (9.116)$$

Then M_2 is obtained from Eq. (9.104), rewritten as

$$M_2 = \frac{V_2}{\sqrt{(\gamma-1)g_c c_p T_{t2} - [(\gamma-1)/2]V_2^2}} \qquad (9.117)$$

9.5.5 Axial-Flow Turbine Stage Analysis: M_{3R} Known

A large work extraction ΔT_t per stage is desired to minimize the number of turbine stages in an engine. As can be seen in Eq. (9.102), this can be done with a combination of high rotor speed ωr and large change in tangential velocity across the rotor ($v_2 + v_3$).

In this section, we present an analysis that allows design choices of ωr, M_2, α_2, and M_{3R}. The equations for solution of a general axial-flow turbine stage based on these known data are developed along a constant mean radius using the model of Section 9.5.2. We assume adiabatic flow in the rotor and nozzle and constant axial velocity through the rotor ($u_2 = u_3 = u$). This analysis was first presented in Ref. 13 by Heiser and is repeated here. The results of the general turbine characteristics are shown in plots to help one choose appropriate values of M_2, α_2, and M_{3R}.

Analysis:

We assume the following data are known:

$$T_{t1}, \ \omega r, \ M_2, \ \alpha_2, \ M_{3R}, \ c_{pt}, \ \text{and} \ \gamma_t$$

We will make all velocities dimensionless by dividing by the velocity $\sqrt{g_c c_{pt} T_{t1}}$ denoted as V' (a quantity that is known at the entrance to the stage and directly related to the maximum or vacuum velocity that a fluid could reach when expanded into a vacuum, $V_{max} = \sqrt{2g_c c_{pt} T_{t1}}$). The symbol Ω is used for the dimensionless rotor speed ωr. Thus,

$$V' = \sqrt{g_c c_{pt} T_{t1}} \qquad (9.118a)$$

$$\Omega = \omega r/V' \qquad (9.118b)$$

1. Total velocity at station 2

$$\frac{V_2}{V'} = \sqrt{\frac{(\gamma_t - 1)M_2^2}{1 + (\gamma_t - 1)M_2^2/2}} \qquad (9.118c)$$

2. Stage axial velocity

$$\frac{u}{V'} = \frac{V_2}{V'}\cos\alpha_2 \qquad (9.118d)$$

3. Absolute tangential velocity at station 2

$$\frac{v_2}{V'} = \frac{V_2}{V'}\sin\alpha_2 \qquad (9.118e)$$

4. Rotor relative tangential velocity at station 2

$$\frac{v_{2R}}{V'} = \frac{v_2}{V'} - \Omega \qquad (9.118f)$$

5. Rotor relative flow angle at station 2

$$\tan\beta_2 = \frac{v_{2R}/V'}{u/V'} \qquad (9.118g)$$

6. Rotor relative total temperature

$$\frac{T_{t2R}}{T_{t1}} = 1 + \Omega^2\left(\frac{1}{2} - \frac{v_2/V'}{\Omega}\right) \qquad (9.118h)$$

7. Rotor relative flow angle at station 3

$$\tan\beta_3 = \sqrt{\frac{T_{t2R}/T_{t1}}{u^2/V'^2}\frac{(\gamma_t - 1)M_{3R}^2}{1 + (\gamma_t - 1)M_{3R}^2/2} - 1} \qquad (9.118i)$$

8. Rotor flow turning angle $= \beta_2 + \beta_3 \qquad (9.118j)$

9. Absolute tangential velocity at station 3

$$\frac{v_3}{V'} = \frac{u}{V'}\tan\beta_3 - \Omega \qquad (9.118k)$$

10. Stage exit flow angle

$$\tan\alpha_3 = \frac{v_3/V'}{u/V'} \qquad (9.118l)$$

11. Static temperature at station 3

$$\frac{T_3}{T_{t1}} = \frac{T_{t2R}/T_{t1}}{1 + (\gamma_t - 1)M_{3R}^2/2}$$

(9.118m)

12. Stage total temperature ratio

$$\tau_{ts} = \frac{T_{t3}}{T_{t1}} = \frac{T_3}{T_{t1}} + \frac{u^2}{V'^2}\frac{1 + \tan^2\alpha_3}{2}$$

(9.118n)

13. Rotor degree of reaction

$$°R_t = \frac{v_{3R}^2 - v_{2R}^2}{2(1 - \tau_{ts})V'^2}$$

(9.118o)

14. Rotor solidity based on axial chord c_x

$$\sigma_{xr} = \frac{c_x}{s} = \frac{2\cos^2\beta_3(\tan\beta_2 + \tan\beta_3)}{Z_r}$$

(9.118p)

9.5.5.1 General Solution

We may now use this analysis to explore the general behavior of turbine stages. The results will be more easily understood if, before proceeding, we describe the results by means of three useful generalizations.

First, we must consider two different types of stages, namely those having either choked or unchoked nozzle vanes or stators. The former are required as entrance stages for every turbine and will be represented in our computations by an M_2 of 1.1. The latter are required for all other turbine stages and will be represented in our computations by an M_2 of 0.9. Other supersonic or subsonic stator Mach numbers should be selected and evaluated by the reader, but these are quite representative.

Second, higher values of M_{3R} always improve stage performance, provided that some margin is allotted to avoid rotor choking. The reader should also independently confirm this assertion. Consequently, M_{3R} is set equal to the highest practical value of 0.9 in our computations.

Third, the analysis lends itself well to intuitively appealing examinations of the various turbine stage properties for the expected range of α_2 and Ω. The open literature strongly suggests that the best performance is obtained when 60 deg $< \alpha_2 < 75$ deg. The open literature also concludes that the larger values of Ω are better, the upper limit being presently in the range $0.2 < \Omega < 0.3$. The ensuing computations will justify these observations. The computations use $\gamma_t = 1.3$ and $g_c c_{pt} = 7386$ ft^2/(s$^2 \cdot$ °R) [=1.235 kJ/ (kg $\cdot$ K)], although the equations are formulated to allow you to choose any desired values.

General results. Figures 9.69, 9.70, and 9.71 present the general results. *Every measure of aerodynamic and thermodynamic performance is improved by increasing Ω.* This agrees with the open literature and the desire of the

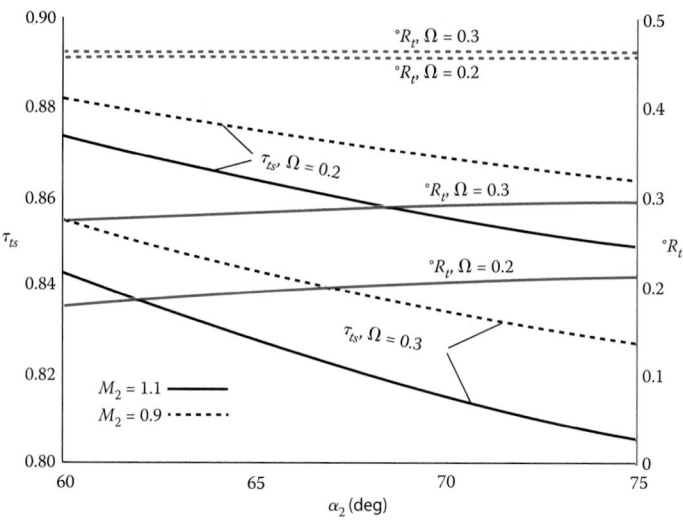

Fig. 9.69 Stage total temperature ratio and degree of reaction for $M_{3R} = 0.9$.

designer to increase the wheel speed ωr to the limit allowed by materials and structural technology. The remainder of the discussion therefore centers on the selection of α_2.

The goal of the designer is to accomplish the desired work extraction efficiently with the minimum number of stages, which is equivalent to finding the minimum practical value of τ_{ts}. The results presented in Fig. 9.69

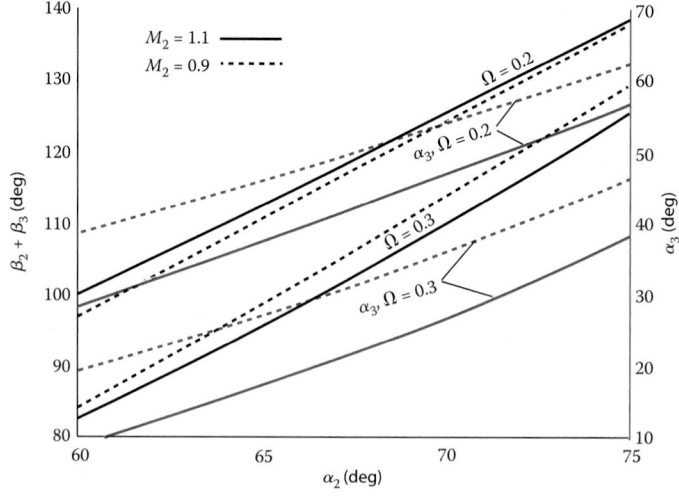

Fig. 9.70 Rotor flow turn angle $(\beta_2 + \beta_3)$ and stage exit flow angle (α_3) for $M_{3R} = 0.9$.

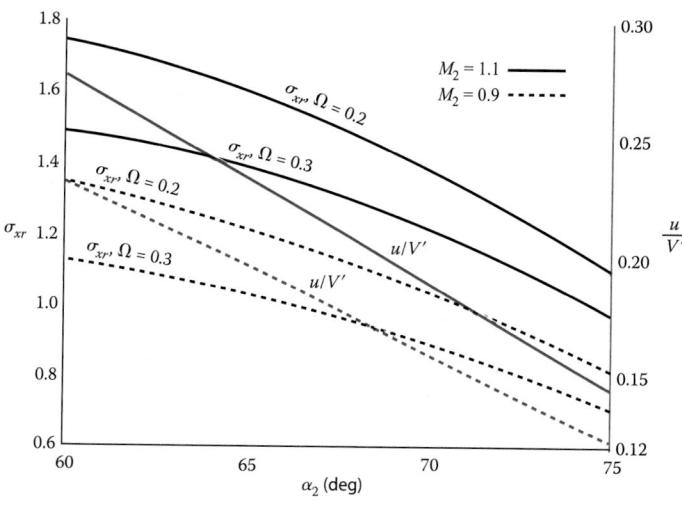

Fig. 9.71 Rotor solidity (based on $Z_r = 1$) and stage axial velocity for $M_{3R} = 0.9$.

demonstrate that τ_{ts} diminishes as α_2 increases over its expected range for both choked and unchoked stages. Furthermore, the results suggest that τ_{ts} values of less than 0.85 are attainable. (A turbine total temperature ratio of 0.85 is the approximate value associated with highly loaded stage with lots of exit swirl.) If higher values of τ_{ts} are required, they can easily be provided by reducing α_2 and/or Ω. However, they also reveal that it is difficult to reach τ_{ts} values much higher than about 0.88 unless the Mach number assumptions are relaxed. This often motivates designers to shift the work split between the low- and high-pressure compressors so that each turbine stage is working hard and the total number of low- and high-pressure turbine stages is minimized.

The rotor degree of reaction $^\circ R_t$ varies slightly with α_2 and thus places no limits on the α_2 value. The inlet Mach number M_2 has a major influence on $^\circ R_t$ for a fixed value of M_{3R}, and the inlet Mach number M_2 may be limited to have a positive degree of reaction.

As the results in Fig. 9.70 reveal, the tendency to increase stage work extraction by increasing α_2 is accompanied by a surprisingly rapid increase in the rotor flow turning angle. The streamline curvature effects within the passage between adjacent rotor airfoils caused by this flow turning are the principal contributor to the low suction surface static pressures and therefore to the potential for boundary-layer separation, especially when the average Mach number is near sonic and shock waves can arise. Carefully tailoring the pressure distribution around the complete airfoil profile by means of sophisticated computational fluid dynamics (CFD) in order to resist separation despite very large rotor flow turning angles is one of the major achievements of modern turbine technology (see Ref. 55). However, because the CFD must account for such exquisite effects as the state of the boundary layer (for example, laminar, turbulent, or transitional), freestream turbulence levels, airfoil roughness, airfoil heat transfer, and locally transonic flow, the application of these methods is beyond the scope of this textbook.

Instead, we will capitalize on our general observation that skillful designers are able to define rotor airfoil profiles that do not separate for flow turning angles $(\beta_2 + \beta_3)$ up to 120–130 deg, provided that the rotor degree of reaction is at least 0.20. That is, the average static temperature and pressure drop significantly across the rotor in order to partially counteract the adverse pressure gradient due to streamline curvature. Combining the results of Figs. 9.69 and 9.70, we see that these criteria are met as long as α_2 does not exceed approximately 70 deg, depending on the exact value of Ω. If some margin of safety is required, the selected value of α_2 should be further reduced.

The stage exit flow angle or swirl should be kept small whether or not another stage follows. In the former case, a small α_3 reduces the total flow turning for the succeeding inlet stator. In the latter case, the need for turbine exit guide vanes and their accompanying diffusion losses is reduced or avoided. The consensus of the turbine design community is that α_3 should not exceed about 40 deg. The results presented in Fig. 9.70 show that meeting this criterion depends strongly on both α_2 and Ω, as well as whether the stage is choked or unchoked. In some cases, this criterion could restrict α_2 to values less than 70 deg.

Figure 9.71 reveals that the minimum rotor solidity is of the order of one for $Z_r = 1$, and that it decreases rapidly as α_2 increases for both choked and unchoked stages. The designer is therefore encouraged to select higher values of α_2 in order to reduce the number of rotor airfoils. The dimensionless stage axial velocity is an indicator of the throughflow area that will be required by the stage and hence of the height of the airfoils and the rotor centrifugal stress (see Section I.2.1). The results presented in Fig. 9.71 show that u diminishes rapidly as α_2 increases for both choked and unchoked stages and is independent of Ω. The designer is therefore encouraged to choose values of α_2 less than 70 deg in order to increase u and thus obtain shorter, lighter airfoils, and rotor blades that have lower centrifugal stresses. This evidently creates another conflict between stage performance and airfoil life.

General conclusions: These results are clearly in agreement with the conventional wisdom of turbine stage design. In particular, they support the universal drive for increasing Ω to the limit allowed by materials and structures. Furthermore, they support the contention that the best choice of α_2 is in the range of 60–75 deg. Finally, they provide a sound basis for initial estimates for the detailed TURBN computations that must be carried out for your specific turbine stage design. Four sets of representative initial design choices that meet all the design criteria are presented in

Table 9.15 Summary of Representative Initial Turbine Stage Design Choices for $M_{3R} = 0.9$, $\gamma_t = 1.3$, $g_c c_p = 7378$ ft^2/(s$^2 \cdot$ °R) [$=1.235$ kJ/(kg $\cdot$ K)], and $Z = 1$

	M_2	1.1	0.9
$\Omega = 0.2$	α_2	67 deg	61 deg
	τ_{ts}	0.860	0.880
	$\beta_2 + \beta_3$	116 deg	103 deg
	°R_t	0.199	0.458
	α_3	41.2 deg	40.2 deg
	σ_{xr}	1.25	1.09
	u/V'	0.217	0.226
$\Omega = 0.3$	α_2	72 deg	72 deg
	τ_{ts}	0.811	0.831
	$\beta_2 + \beta_3$	120 deg	116 deg
	°R_t	0.294	0.463
	α_3	31.1 deg	40.0 deg
	σ_{xr}	0.92	0.67
	u/V'	0.171	0.144

Table 9.15 for your reference. The compilation indicates that, for a given Ω, increasing M_2 can decrease τ_{ts}, but the airfoil aerodynamics are nearer to the edge.

9.5.5.2 Multistage Turbine Design

Figure 9.69 shows that the temperature ratio ($\tau_{ts} = T_{t3}/T_{t1}$) for a stage is limited and an engine design may require a turbine whose temperature ratio is less than that given by the best stage. A multistage turbine will be required when this occurs. To obtain the choked flow in the first-stage stator (nozzle), the Mach number entering the rotor M_2 is slightly supersonic. The Mach number entering the rotor of the remaining stages is kept subsonic. The net result is that the first stage will have a greater temperature drop than will the following stage(s). The following example demonstrates the results for a two-stage turbine.

Example 9.15

We consider a two-stage axial-flow turbine with the following data: $\omega r = 1200$ ft/s, $\gamma_t = 1.3$, $c_{pt} = 0.295$ Btu/lbm $\cdot$ °R (1.235 kJ/kg $\cdot$ K), $\alpha_2 = 70$ deg, $Z_r = 1$, and $T_{t1} = 3200$°R entering the first stage. We desire the

(Continued)

Example 9.15 (Continued)

first stage choked and the second stage unchoked. Results are provided from TURBN at the mean radius.

Property	Stage 1	Stage 2
T_{t1}	3200°R	2701°R
α_2	65°	65°
M_2	1.1	0.9
M_{3R}	0.9	0.9
Ω	0.2468	0.2687
τ_s	0.8441	0.8510
ΔT_t	499°R	403°R
$°R_t$	0.2445	0.4622
$\beta_2 + \beta_3$	105.58°	101.81°
α_3	28.91°	34.04°
σ_{xr}	1.5070	1.0916

The result is a high-work (large ΔT_t) two-stage turbine with good thermo-dynamic and aerodynamic performance. Note that the temperature drop through the second stage is about 80% of that through the first stage.

9.5.6 Shaft Speed

The design rotational speed of a spool (shaft) having stages of compression driven by a turbine is initially determined by that component that limits the speed because of high stresses. For a low-pressure spool, the first stage of compression, because it has the greatest AN^2, normally dictates the rotational speed. The first stage of turbine on the high-pressure spool normally determines that spool's rotational speed because of its high AN^2 or high disk rim speed at elevated temperature. Refer to Appendix I for further analysis and discussion.

9.5.7 Design Process

The design process requires both engineering judgment and knowledge of typical design values. Table 9.16a gives the range of design parameters for axial-flow turbines that can be used for guidance. The comparison of turbines for Pratt & Whitney engines in Table 9.16b shows some interesting trends in turbofan engines and turbine technology that are relevant today as bypass ratios and turbine inlet temperatures continue to increase. Note the increases over the years in inlet temperature, mass flow rate, and output power.

From Table 9.16b, comparison of the JT3D and JT9D high-pressure tur-bines shows that the stage loading coefficient did not appreciably change

Table 9.16a Range of Axial-Flow Turbine Design Parameters

Parameter	Design Range
High-pressure turbine	
Maximum AN^2	4–5×10^{10} in.$^2 \cdot$ rpm^2
Stage loading coefficient	1.4–2.0
Exit Mach number	0.40–0.50
Exit swirl angle, deg	0–20
Low-pressure turbine	
Inlet corrected mass flow rate	40–44 lbm/(s $\cdot$ ft^2)
Hub/tip ratio at inlet	0.35–0.50
Maximum stage loading at hub	2.4
Exit Mach number	0.40–0.50
Exit swirl angle, deg	0–20

between the designs. However, the turbine inlet temperature increased to a value above the working temperature of available materials, requiring extensive use of cooling air. Current advances in cooling effectiveness, materials, and manufacturing aim to decrease the use of cooling air, ideally eliminating it. The stage loading coefficient for the low-pressure turbine increased dramatically, reducing the number of stages to four. If the stage loading coefficient of the low-pressure turbine of the JT3D were not increased significantly

Table 9.16b Comparison of Pratt & Whitney Engines

Parameter	JT3D	JT9D
Year of introduction	1961	1970
Engine bypass ratio	1.45	4.86
Engine overall pressure ratio	13.6	24.5
Core engine flow, lb/s	187.7	272.0
High-pressure turbine		
Inlet temperature, °F	1745	2500
Power output, hp	24,100	71,700
Number of stages	1	2
Average stage loading coefficient	1.72	1.76
Coolant plus leakage flow, %	2.5	16.1
Low-pressure turbine		
Inlet temperature, °F	1410	1600
Power output, hp	31,800	61,050
Number of stages	3	4
Average stage loading coefficient	1.44	2.47
Coolant plus leakage flow, %	0.7	1.4

in the design of the JT9D, about six or seven stages of low-pressure turbine would have been required—increasing both cost and weight.

9.5.7.1 Steps of Design

The material presented in previous sections can now be applied to the design of an axial-flow turbine. The complete design process for a turbine will include the following items:

1. Selection of rotational speed and annulus dimensions
2. Selection of the number of stages
3. Calculation of airflow angles for each stage at the mean radius
4. Calculation of airflow angle variations from the hub to the tip for each stage
5. Selection of blade material
6. Selection of blading using experimental cascade data
7. Selection of turbine cooling, if needed
8. Verification of turbine efficiency based on cascade loss data
9. Prediction of off-design performance
10. Rig testing of design

Items 1–5 will be covered in this section. The other steps are covered in Refs. 22, 37, 42, and 43. The design process is inherently iterative, often requiring the return to an earlier step when prior assumptions are found to be invalid. Many technical specialities are interwoven in a design (e.g., an axial-flow turbine involves at least thermodynamics, aerodynamics, structures, materials, heat transfer, and manufacturing processes). Design requires the active participation and disciplined communication by many technical specialists.

Example 9.16

We will consider the design of a turbine suitable to power the eight-stage, axial-flow compressor designed earlier in this chapter for a simple turbojet gas turbine engine (see Example 9.9). From engine cycle and compressor design calculations, a suitable design point for the turbine of such an engine at sea-level, standard-day conditions ($P = 14.696$ psia and $T = 518.7°R$) may emerge as follows:

Compressor pressure ratio: 10.41	Rotor speed ω: 800 rad/s
Compressor flow rate: 150 lbm/s	Turbine flow rate: 156 lbm/s
Compressor efficiency: 86.3%	T_t entering turbine: 3200°R
Compressor exit T_t: 1086°R	P_t entering turbine: 143.1 psia
Compressor γ: 1.4	Turbine γ: 1.3
Compressor R: 53.34 ft · lbf/(lbm · °R)	Turbine R: 53.40 ft · lbf/(lbm · °R)

(Continued)

Example 9.16 *(Continued)*

From these specified data, we can investigate the aerodynamic design of an axial-flow turbine. This detailed example is presented in the online Supporting Material for this chapter.

9.5.8 Turbine Cooling

The turbine components are subjected to much higher temperatures in the modern gas turbine engines being designed and built today than was possible 75 years ago. This is due mainly to improvements in metallurgy and cooling of turbine components. The cooling air to cool the turbine is bleed air from the compressor exit; it must be to provide the correct pressure differential to flow the air through the required turbine components. Please keep in mind that this cooling air is at temperatures as high as 1200–1300°F depending on the compressor pressure ratio. A schematic of a typical turbine cooling system is shown in Fig. 9.72. The nozzle vanes and the outer wall of

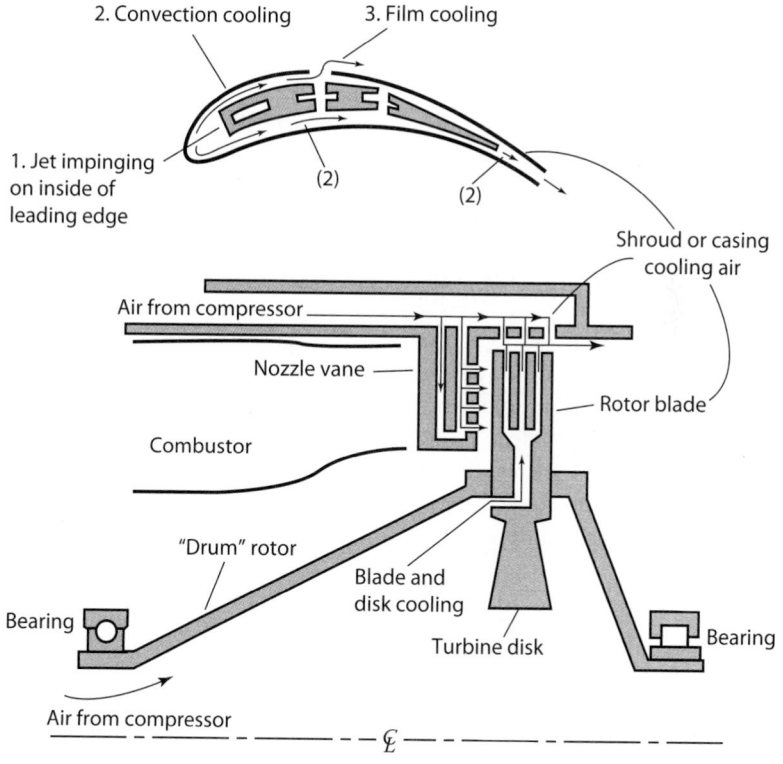

Fig. 9.72 Schematic of air-cooled turbine (from Ref. 21).

the turbine flow passage use cooling air that travels from the compressor between the combustor and outer engine case. The turbine rotor blades, disks, and inner wall of the turbine flow passage use cooling air that is routed through inner passageways.

The first-stage stator vanes (nozzles) are exposed to the highest turbine temperatures (including the hot spots from the combustor). The first-stage rotor blades are exposed to a somewhat lower temperature because of circumferential averaging, dilution of turbine gases with first-stage stator cooling air, and relative velocity effects. The second-stage stator vanes are exposed to an even lower temperature because of additional cooling air dilution and power extraction from the turbine gases. The turbine temperature decreases in a like manner through each blade row.

The cooling methods used in the turbine are illustrated in Fig. 9.73 and can be divided into the following categories:

1. Convection cooling
2. Impingement cooling
3. Film cooling
4. Full-coverage film cooling
5. Transpiration cooling

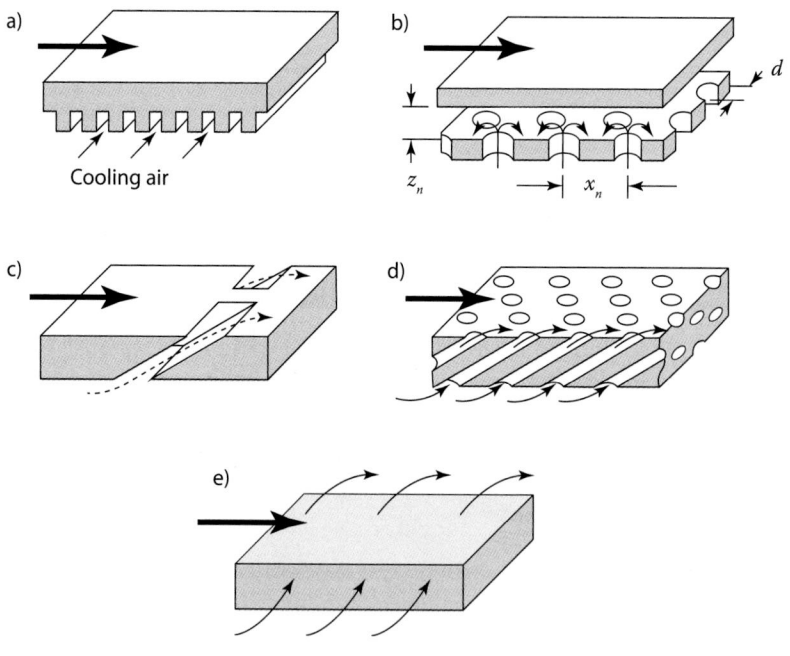

Fig. 9.73 Methods of turbine cooling (Ref. 42): a) convection cooling, b) impingement cooling, c) film cooling, d) full-coverage film cooling, e) transpiration cooling.

Applications of these five methods of cooling to turbine blades are shown in Fig. 9.74.

Figure 9.75 shows a typical application on a nozzle vane segment for a core turbine. This nozzle has cooling holes along its leading edge and

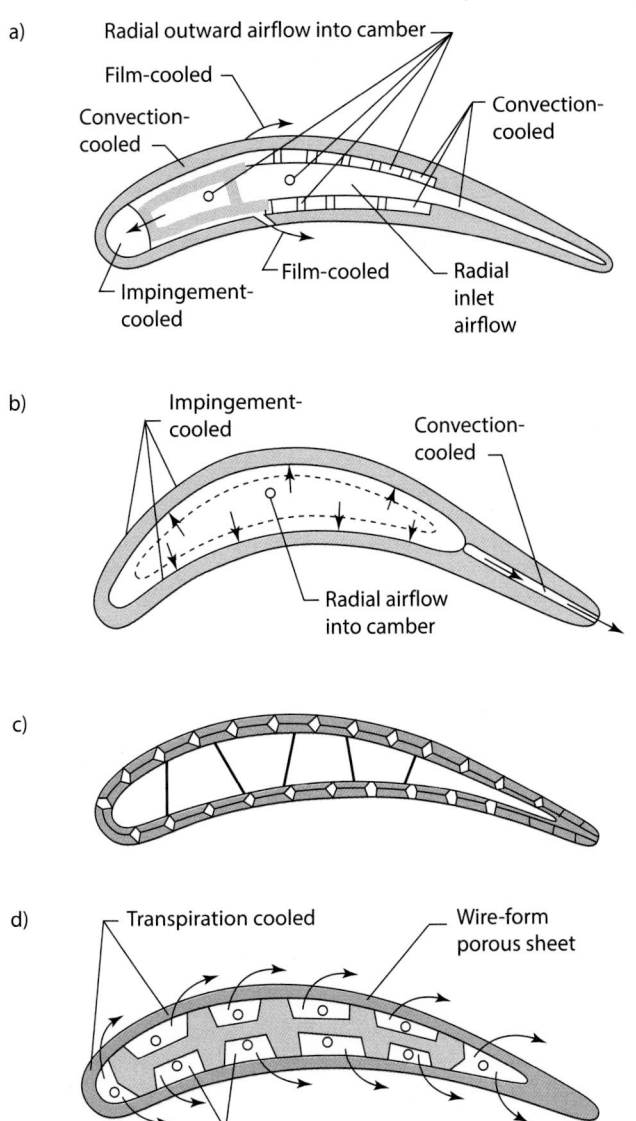

Fig. 9.74 Turbine blade cooling (Ref. 42): a) convection-, impingement-, and film-cooled blade configuration, b) convection- and impingement-cooled blade configuration, c) full-coverage film-cooled blade configuration, d) transpiration-cooled blade configuration.

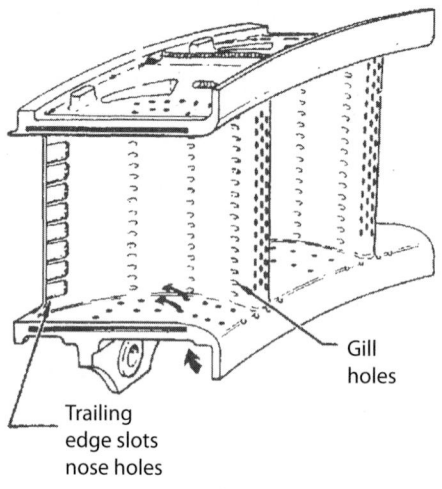

Fig. 9.75 Typical cooled turbine nozzle. (Courtesy of General Electric.)

pressure surface (gill holes) in addition to cooling flow exiting at its trailing edge. The cooling holes on the inner and outer annulus walls are also indicated.

A rotor blade of the General Electric CF6-50 engine is shown in Fig. 9.76. The cross section of the blade shows the elaborate internal cooling and flow exiting the blade tip (through its cap) and along the trailing edge. The blade isometric shows flow through the gill holes on the pressure surface in addition to the tip and trailing-edge cooling flows.

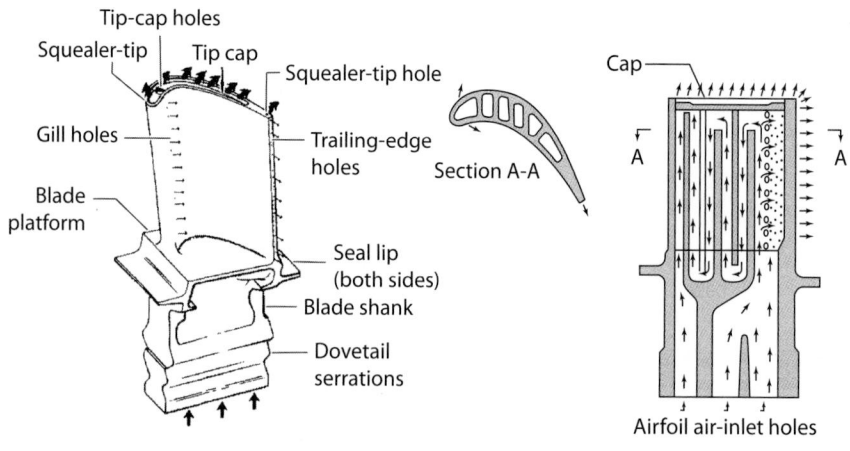

Fig. 9.76 Construction features of air-cooled turbine blade. (Courtesy of General Electric.)

9.5.9 Turbine Performance

We now turn our attention to a brief discussion on overall turbine performance (not just at the design point). To provide a quick review, the flow enters a turbine through stationary airfoils (typically called nozzle vanes) that turn and accelerate the fluid, increasing its tangential momentum. Then the flow passes through rotating airfoils (called rotor blades) that remove energy from the fluid as they change its tangential momentum. Successive pairs of stationary airfoils followed by rotating airfoils remove additional energy from the fluid. To obtain high output power to weight from a turbine, the flow entering the first-stage turbine rotor is normally supersonic, which requires the flow to pass through sonic conditions at the minimum passage area in the first-stage stators (nozzles). From Eq. (8.3), the corrected inlet mass flow rate based on this minimum passage area (throat) will be constant for fixed-inlet-area turbines. This flow characteristic is shown in the typical turbine flow map (Fig. 9.77) when the expansion ratio across the turbine (ratio of total pressure at entrance to total pressure at exit $P_{t4}/P_{t5} = 1/\pi_t$) is greater than about 2 and the flow at the throat is choked.

As presented in Section 8.1.3, the performance of a turbine is normally shown by using the total pressure ratio, corrected mass flow rate, corrected turbine speed, and component efficiency. This performance is presented in either two maps or one consolidated map. One map shows the interrelationship of the total pressure ratio, corrected mass flow rate, and corrected turbine speed, like that depicted in Fig. 9.77 for a single-stage turbine. Because the corrected-speed lines collapse into one line when the turbine is choked, the turbine efficiency must be shown in a separate map (see Fig. 8.5b). The constant corrected speed lines of Fig. 9.77 can be

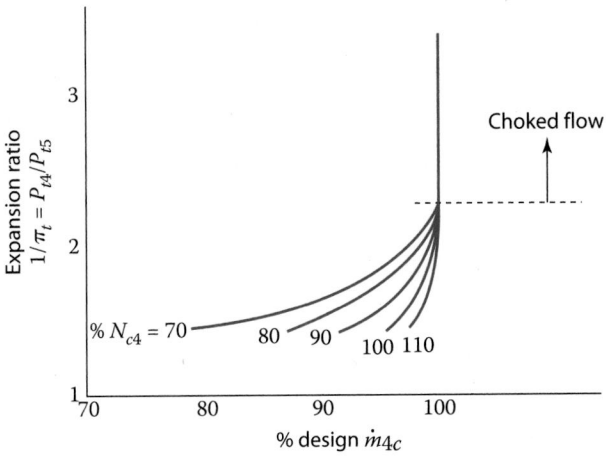

Fig. 9.77 Typical turbine flow map.

spread out horizontally by multiplying the corrected mass flow rate by the percentage of corrected speed. Now the turbine efficiency lines can be superimposed without cluttering the resulting performance map. Figure 9.78 shows the consolidated turbine performance map of a multistage turbine with all four performance parameters: total pressure ratio, corrected mass flow rate, corrected turbine speed, and efficiency.

For the majority of aircraft gas turbine engine operation, the turbine expansion ratio is constant and the turbine operating line can be drawn as a horizontal line in Fig. 9.78. (It would collapse to an operating point on the flow map of Fig. 9.77.) At off-design conditions, the corrected speed and efficiency of a turbine change very little from their design values. In the analysis of gas turbine engine performance of Chapter 8, we used this fact and considered that the turbine efficiency was constant.

To make these conclusions even more concrete, it is useful to look more closely at the turbine. According to the typical turbine performance map of Fig. 9.78, this machine can provide the same work (i.e., $1 - \tau_t$) for a wide range of $N_c = N_t$, while η_t varies only slightly. Please recall that as long as the flow in the turbine inlet guide vane and some downstream flow area remain choked and η_t does not vary significantly, then Eqs. (8.12a) and (8.12b) demonstrate that π_t and τ_t must be essentially constant. The question, then, is how the turbine flow conditions can adjust themselves in order to provide the *same* τ_t at *different* values of ωr_m. If the mechanics of adjustment are straightforward, then the entire process of component matching should be more easily comprehended.

Consider the single-stage, impulse, maximum work (i.e., no exit swirl), isentropic, constant-height turbine of Fig. 9.79. At its design point, this turbine has a choked inlet guide vane and an entirely subsonic flow relative

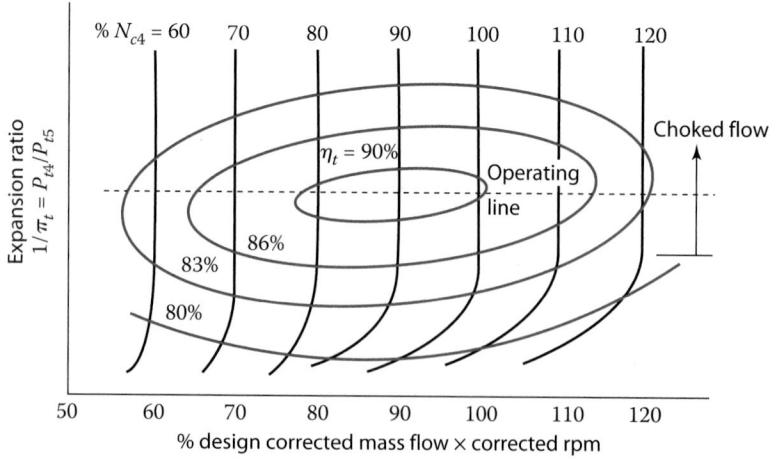

Fig. 9.78 Typical turbine consolidated performance map.

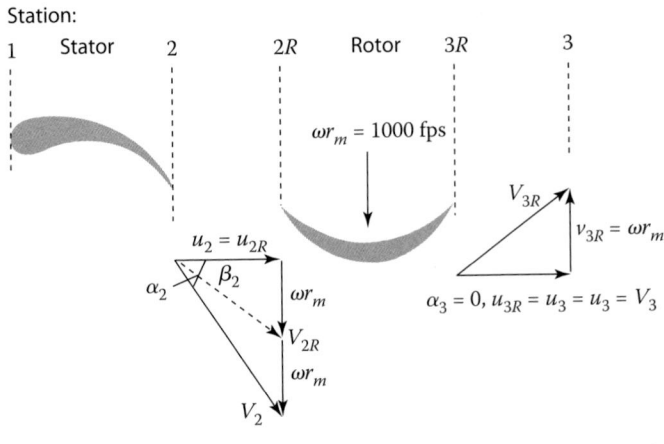

Station:

Fig. 9.79 Single-stage impulse turbine.

to the rotor. The flow angles are all representative of good practice. In short, this is a rather standard design.

Isentropic calculations have been performed at rotational speeds $\pm 10\%$ from design, which would encompass the entire operating range for most turbines. The necessary condition for a solution was that τ_t have a design point value of 0.896. The results are displayed in Table 9.17.

These results confirm the message of the Euler turbine equation. Because $(1 - \tau_t)$ is proportional to $\omega r_m(v_{2R} - v_{3R})$, then M_2, M_{2R}, and M_{3R} must increase in order to compensate for reductions in ωr_m, and vice versa. Nevertheless, even for such large differences in ωr_m, the aerodynamic results are far from disastrous. For one thing, the inlet guide vanes remain choked ($M_2 > 1$) and the rotor airfoils remain subsonic ($M_{2R} < 1$ and $M_{3R} < 1$) at all times. For another, the rotor airfoil relative inlet flow

Table 9.17 Turbine Off-Design Performance

Quantity	−10%	Design	+10%
ωr_m, ft/s	900	1000	1100
M_2	1.22	1.10	1.02
α_2, deg	50.6	52.0	52.4
β_2, deg	35.2	32.6	28.4
M_{2R}	0.947	0.804	0.708
M_{3R}	0.820	0.804	0.790
α_3, deg	−4.2	0	3.5
τ_t	0.899	0.896	0.898

$T_{t2} = 2800°R$; $\gamma = 1.33$; $g_c R = 1716 \text{ ft}^2/(\text{s}^2 - °R)$

angle β_2 and the inlet flow angle to the downstream stator airfoils α_3 are well within the low-loss operating range for typical turbine cascades. Finally, one might expect the frictional losses to increase and the efficiency to decrease as ωr_m decreases and the blade scrubbing velocity increases, but only gradually.

This turbine therefore performs gracefully as expected, providing the same τ_t with slight changes in η_t for a wide range of $N_c = N_t$, all the while remaining choked. As concluded in Section 8.2.8, the operating line of every component in the engine is a "free" byproduct of the off-design calculations, which are both correct and complete as presented in this textbook.

Problems

9.1 Relative velocities are an important concept in the analysis of turbomachines. To help understand relative velocity, consider that a baseball is thrown at a moving train as shown in Fig. P9.1. If the baseball has a velocity of 60 mph and the train is traveling at 80 mph, find the magnitude and direction of the baseball velocity relative to the train.

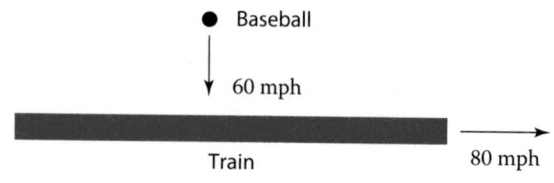

Fig. P9.1 Throwing baseball at train.

9.2 Perform Problem 9.1 for a baseball traveling at 100 km/h and a train traveling at 80 km/h.

9.3 A small fan used to circulate air in a room is shown in Fig. P9.2. The fan blades are twisted and change their angle relative to the centerline of the fan as the radius increases. If the relative velocity of the air makes the same angle to the centerline of the fan as the blades and if the blade at a radius of 0.2 m has an angle of 60 deg, determine the speed of the air for fan speeds of 1725 and 3450 rpm.

9.4 Air flows through an axial compressor stage with the following properties:

$$T_{t1} = 300 \text{ K}, \quad u_2/u_t = 1.0, \quad v_1 = 120 \text{ m/s}, \quad \alpha_1 = 0 \text{ deg},$$
$$\beta_2 = 45 \text{ deg}, \quad U = \omega r_m = 240 \text{ m/s}$$

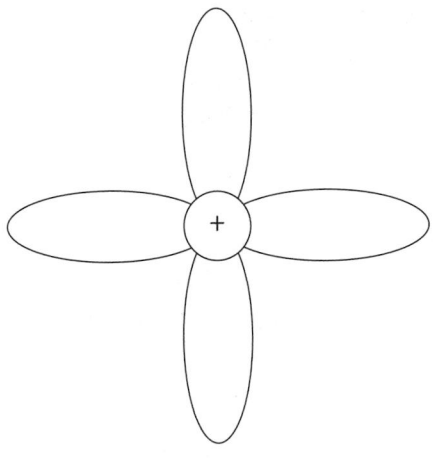

Fig. P9.2 Typical home fan.

Note: For air, use $\gamma = 1.4$ and $R = 0.286\ \text{kJ}/(\text{kg} \cdot \text{K})$. Determine the change in tangential velocity and the total pressure ratio of the stage for a stage efficiency of 0.88.

9.5   Perform Problem 9.4 with $\alpha_1 = 10$ deg and $\beta_2 = 40$ deg.

9.6   Air flows through an axial compressor stage with the following properties:

$$T_{t1} = 540\,°\text{R}, \quad u_2/u_1 = 1.0, \quad v_1 = 400\ \text{ft/s},$$
$$\alpha_1 = 0\ \text{deg}, \quad \beta_2 = 45\ \text{deg}, \quad U = \omega r_m = 800\ \text{ft/s}$$

Note: For air, use $\gamma = 1.4$ and $R = 53.34\ \text{ft} \cdot \text{lbf}/(\text{lbm} \cdot °\text{R})$. Determine the change in tangential velocity and the total pressure ratio of the stage for a stage efficiency of 0.9.

9.7   Perform Problem 9.6 with $\alpha_1 = 10$ deg and $\beta_2 = 40$ deg.

9.8   Air flows through an axial compressor stage with the following properties:

$$T_{t1} = 290\,\text{K}, \quad P_{t1} = 101.3\,\text{kPa}, \quad u_2/u_1 = 1.0, \quad M_1 = 0.6,$$
$$\alpha_1 = 40\ \text{deg}, \quad \alpha_2 = 57.67\ \text{deg}, \quad U = \omega r_m = 360\ \text{m/s}$$

Note: For air, use $\gamma = 1.4$ and $R = 0.286\ \text{kJ}/(\text{kg} \cdot \text{K})$. Determine the following:

a) V_1, u_1, and v_1

b) u_2, v_2, and V_2

c) ΔT_t and τ_s for the stage

d) π_s and P_{t3} for a polytropic efficiency of 0.88

9.9 Perform Problem 9.8 with $\alpha_2 = 60$ deg.

9.10 Air flows through an axial compressor stage with the following properties:

$$T_{t1} = 518.7°R, \quad P_{t1} = 14.696 \text{ psia}, \quad u_2/u_1 = 1.0, \quad M_1 = 0.6,$$
$$\alpha_1 = 40 \text{ deg}, \quad \alpha_2 = 57.67 \text{ deg}, \quad U = \omega r_m = 1200 \text{ ft/s}$$

Note: For air, use $\gamma = 1.4$ and $R = 53.34$ ft · lbf/(lbm · °R). Determine the following:

a) V_1, u_1, and v_1

b) u_2, v_2, and V_2

c) ΔT_t and τ_s for the stage

d) π_s and P_{t3} for a polytropic efficiency of 0.9

9.11 Perform Problem 9.10 with $\alpha_2 = 60$ deg.

9.12 Air enters a compressor stage that has the following properties:

$$\dot{m} = 50 \text{ kg/s}, \quad \omega = 800 \text{ rad/s}, \quad r = 0.5 \text{ m},$$
$$M_1 = M_3 = 0.5, \quad \alpha_1 = \alpha_3 = 45 \text{ deg}, \quad T_{t1} = 290 \text{ K},$$
$$P_{t1} = 101.3 \text{ kPa}, \quad u_2/u_1 = 1.0, \quad T_{t3} - T_{t1} = 45 \text{ K},$$
$$\phi_{cr} = 0.10, \quad \phi_{cs} = 0.03, \quad \sigma = 1$$

Note: For air, use $\gamma = 1.4$ and $R = 0.286$ kJ/(kg · K). Make and fill out a table of flow properties like Table 9.3, and determine the diffusion factors, degree of reaction, stage efficiency, polytropic efficiency, and flow areas and associated hub and tip radii at stations 1, 2, and 3.

9.13 Perform Problem 9.12 with $\alpha_1 = \alpha_3 = 40$ deg.

9.14 Air enters a compressor stage that has the following properties:

$$\dot{m} = 100 \text{ lbm/s}, \quad \omega = 1000 \text{ rad/s}, \quad r = 12 \text{ in.}, \quad M_1 = M_3 = 0.6,$$
$$\alpha_1 = \alpha_3 = 42 \text{ deg}, \quad T_{t1} = 518.7°R, \quad P_{t1} = 14.7 \text{ psia}, \quad u_2/u_1 = 1.0,$$
$$T_{t3} - T_{t1} = 50°R, \quad \phi_{cr} = 0.12, \quad \phi_{cs} = 0.03, \quad \sigma = 1$$

Note: For air, use $\gamma = 1.4$ and $R = 53.34$ ft · lbf/(lbm · °R). Make and fill out a table of flow properties like Table 9.3, and determine the diffusion factors, degree of reaction, stage efficiency, polytropic efficiency, and flow areas and associated hub and tip radii at stations 1, 2, and 3.

9.15 Perform Problem 9.14 with $\alpha_1 = \alpha_3 = 47$ deg.

9.16 Some axial-flow compressors have inlet guide vanes to add tangential velocity to the axial flow and thus set up the airflow for the first-stage rotor.

a) Assuming reversible, adiabatic flow through the inlet guide vanes, use the continuity equation and the mass flow parameter to show that

$$(\cos \alpha_1)(A_1)\mathrm{MFP}(M_1) = A_0\mathrm{MFP}(M_0)$$

where the subscripts 0 and 1 refer to the inlet and exit of the inlet guide vanes and the areas are those normal to the centerline of the axial-flow compressor.

b) For $A_0/A_1 = 1$ and 1.1, plot a curve of α_1 vs M_1 for $M_0 = 0.3, 0.4$, and 0.5 (see Fig. 9.36b).

9.17 In the analysis of turbomachinery, many authors use the dimensionless stage loading coefficient ψ, defined as

$$\psi = \frac{g_c c_p \Delta T_t}{(\omega r)^2}$$

For a repeating-stage, repeating-row compressor stage, show that the stage loading coefficient can be written as

$$\psi = \frac{g_c c_p \Delta T_t}{(\omega r)^2} = \frac{\tan \alpha_2 - \tan \alpha_1}{\tan \alpha_2 + \tan \alpha_1} = 1 - \frac{2 \tan \alpha_1}{\tan \alpha_2 + \tan \alpha_1}$$

9.18 Air enters a repeating-row, repeating-stage compressor that has the following properties:

$$\dot{m} = 40\,\mathrm{kg/s}, \quad T_{t1} = 290\,\mathrm{K}, \quad P_{t1} = 101.3\,\mathrm{kPa},$$
$$M_1 = 0.5, \quad \alpha_1 = 38\,\mathrm{deg}, \quad D = 0.5, \quad e_c = 0.9,$$
$$\omega = 1000\,\mathrm{rad/s}, \quad \phi_{cs} = 0.03, \quad \sigma = 1$$

Note: For air, use $\gamma = 1.4$ and $R = 0.286\,\mathrm{kJ/(kg \cdot K)}$. Make and fill out a table of flow properties like Table 9.3, and determine the temperature rise, pressure ratio, mean radius, and flow areas and associated hub and tip radii at stations 1, 2, and 3.

9.19 Perform Problem 9.18 with $\alpha_1 = 40$ deg and $D = 0.55$.

9.20   Air enters a repeating-row, repeating-stage compressor that has the following properties:

$$\dot{m} = 50\,\text{lbm/s}, \quad T_{t1} = 518.7^\circ\text{R}, \quad P_{t1} = 14.7\,\text{psia},$$
$$M_1 = 0.5, \quad \alpha_1 = 35\,\text{deg}, \quad D = 0.5, \quad e_c = 0.9,$$
$$\omega = 1200\,\text{rad/s}, \quad \phi_{cs} = 0.03, \quad \sigma = 1$$

Note: For air, use $\gamma = 1.4$ and $R = 53.34\,\text{ft} \cdot \text{lbf}/(\text{lbm} \cdot {}^\circ\text{R})$. Make and fill out a table of flow properties like Table 9.3, and determine the temperature rise, pressure ratio, mean radius, and flow areas and associated hub and tip radii at stations 1, 2, and 3.

9.21   Perform Problem 9.20 with $\alpha_1 = 42$ deg and $D = 0.52$.

9.22   For an exponential swirl distribution and the data of Problem 9.18, calculate and plot the variation of u_1, v_1, u_2, and v_2 vs radius from hub to tip (see Fig. 9.31).

9.23   For an exponential swirl distribution and the data of Problem 9.20, calculate and plot the variation of u_1, v_1, u_2, and v_2 vs radius from hub to tip (see Fig. 9.31).

9.24   For the data of Problem 9.18, determine the shape of both the rotor and the stator blades on the mean radius, using NACA 65-series airfoils with circular camber line. Assume $c/h = 0.3$ and 10% thickness.

9.25   For the data of Problem 9.20, determine the shape of both the rotor and the stator blades on the mean radius, using NACA 65-series airfoils with circular camber line. Assume $c/h = 0.3$ and 10% thickness.

9.26   a) For the data of Problem 9.18, determine AN^2 at station 1.

b) For the data of Problem 9.20, determine AN^2 at station 1.

9.27   At design speed and flow, the following mid-radius velocity triangle information is known for the second stage of a high-pressure-ratio axial compressor:

$$V_1 = 460\,\text{ft/s}, \quad V_2 = 787\,\text{ft/s}, \quad V_{1R} = 787\,\text{ft/s},$$
$$V_{2R} = 460\,\text{ft/s}, \quad \alpha_1 = \beta_2 = 30\,\text{deg}$$

a) On *one diagram and to scale*, sketch the inlet and exit velocity triangles. Be sure to clearly identify the blade mean radius tangential velocity and the change in air tangential velocity.

b) What is the total pressure ratio of the stage if the stage efficiency is 85% and the inlet total temperature is 520°R?

c) During starting, the axial velocity reaches only 50% of the design axial velocity after the rotor has reached design speed. To prevent stalling, variable stator vanes (VSVs) are to be used. How far and in what direction must the stator upstream of this stage be rotated to bring β_1 to the design value when u is only 50% of design?

d) On *one diagram and to scale*, sketch the stage inlet velocity triangles for the design and starting conditions.

9.28 A 0.4-m-diam rotor of a centrifugal compressor for air is needed to produce a pressure ratio of 3.8. Assuming a polytropic efficiency of 0.85, determine the angular speed ω, total temperature rise, and adiabatic efficiency. Determine the input power for a mass flow rate of 2 kg/s at 1 atm and 288.2 K. Assume a slip factor of 0.9.

9.29 A 12-in.-diam rotor of a centrifugal compressor for air is needed to produce a pressure ratio of 4. Assuming a polytropic efficiency of 0.86, determine the angular speed ω, total temperature rise, and adiabatic efficiency. Determine the input power for a mass flow rate of 10 lbm/s at 1 atm and 518.7°R. Assume a slip factor of 0.9.

9.30 Products of combustion flow through an axial turbine stage with the following properties:

$$T_{t1} = 1800\,\text{K}, \quad P_{t1} = 1000\,\text{kPa}, \quad u_3/u_2 = 1, \quad M_2 = 1.1,$$
$$U = \omega r_m = 360\,\text{m/s}, \quad \alpha_2 = 45\,\text{deg}, \quad \alpha_3 = 5\,\text{deg}$$

Note: For the gas, use $\gamma = 1.3$ and $R = 0.287\,\text{kJ}/(\text{kg}\cdot\text{K})$. Determine the following:

a) V_2, u_2, and v_2

b) u_3, v_3, and V_3

c) ΔT_t and τ_s for the stage

d) π_s and P_{t3} for a polytropic efficiency of 0.89

9.31 Perform Problem 9.30 with $M_2 = 1.05$.

9.32 Products of combustion flow through an axial turbine stage with the following properties:

$$T_{t1} = 3200°\text{R}, \quad P_{t1} = 200\,\text{psia}, \quad u_3/u_2 = 1,$$
$$M_2 = 1.05, \quad U = \omega r_m = 1200\,\text{ft/s}, \quad \alpha_2 = 40\,\text{deg}, \quad \alpha_3 = 25\,\text{deg}$$

Note: For the gas, use $\gamma = 1.3$ and $R = 53.4$ ft · lbf/(lbm · °R). Determine the following:

a) V_2, u_2, and v_2

b) u_3, v_3, and V_3

c) ΔT_t and τ_s for the stage

d) π_s and P_{t3} for a polytropic efficiency of 0.9

9.33   Perform Problem 9.32 with $M_2 = 1.1$.

9.34   In the preliminary design of turbines, many designers use the dimensionless work coefficient ψ, defined as

$$\psi \equiv \frac{g_c c_p \Delta T_t}{(\omega r)^2}$$

a) For a turbine stage with $\alpha_3 = 0$, show that the degree of reaction can be written as

$$°R_t = 1 - \frac{\psi}{2}$$

b) For a turbine stage with $\alpha_3 = 0$, $\gamma = 1.3$, $R = 0.287$ kJ/(kg · K), and $\psi = 1$ and 2, plot ΔT_t vs ωr for $250 < \omega r < 450$ m/s.

9.35   Products of combustion enter a turbine stage with the following properties:

$$\dot{m} = 40 \text{ kg/s}, \quad T_{t1} = 1780 \text{ K}, \quad P_{t1} = 1.40 \text{ MPa}, \quad M_1 = 0.3,$$
$$M_2 = 1.15, \quad \omega r = 400 \text{ m/s}, \quad T_{t3} = 1550 \text{ K}, \quad \alpha_1 = \alpha_3 = 0,$$
$$r_m = 0.4 \text{ m}, \quad u_3/u_2 = 1.0, \quad \phi_{t \text{ stator}} = 0.04, \quad \phi_{t \text{ rotor}} = 0.08$$

Note: For the gas, use $\gamma = 1.3$ and $R = 0.287$ kJ/(kg · K). Make and fill out a table of flow properties like Table 9.10 for the mean line, and determine the degree of reaction, total temperature change, stage efficiency, polytropic efficiency, and flow areas and associated hub and tip radii at stations 1, 2, and 3.

9.36   Perform Problem 9.35 with $M_1 = 0.7$ and $M_2 = 1.05$.

9.37   Perform Problem 9.35 with $T_{t1} = 1900$ K and $T_{t3} = 1630$ K.

9.38   Perform Problem 9.35 with $T_{t1} = 1850$ K and $T_{t3} = 1580$ K.

9.39 Products of combustion enter a turbine stage with the following properties:

$$\dot{m} = 105 \text{ lbm/s}, \quad T_{t1} = 3200°R, \quad P_{t1} = 280 \text{ psia}, \quad M_1 = 0.3,$$

$$M_2 = 1.2, \quad \omega r = 1400 \text{ ft/s}, \quad T_{t3} = 2700°R, \quad \alpha_1 = \alpha_3 = 0,$$

$$r_m = 14 \text{ in.}, \quad u_3/u_2 = 1.0, \quad \phi_{t\,\text{stator}} = 0.04, \quad \phi_{t\,\text{rotor}} = 0.08$$

Note: For the gas, use $\gamma = 1.3$ and $R = 53.34$ ft · lbf/(lbm · °R). Make and fill out a table of flow properties like Table 9.10 for the mean line, and determine the degree of reaction, total temperature change, stage efficiency, polytropic efficiency, and flow areas and associated hub and tip radii at stations 1, 2, and 3.

9.40 Perform Problem 9.39 with $M_1 = 0.7$ and $M_2 = 1.15$.

9.41 For axial flow turbine stage with $\alpha_2 = 65$ deg $M_2 = 1.1$, $M_{3R} = 0.9$, $T_{t2} = 1800$ K, and $U = 450$ m/s, determine T_{t3}, β_2, β_3, α_3, and $°R_t$. Assume $\gamma_t = 1.3$ and $g_c c_p = 7386$ ft²/(s² · °R)[=1.235 kJ/(kg · K)]. Check answer with Figs. 9.69 and 9.70.

9.42 For axial flow turbine stage with $\alpha_2 = 70$ deg $M_2 = 1.1$, $M_{3R} = 0.9$, $T_{t2} = 3200°R$, and $U = 1400$ ft/s, determine T_{t3}, β_2, β_3, α_3, and $°R_t$. Assume $\gamma_t = 1.3$ and $g_c c_p = 7386$ ft²/(s² · °R)[=1.235 kJ/(kg · K)]. Check answer with Figs. 9.69 and 9.70.

9.43 For the data of Problem 9.35, determine the shape of both the nozzle and the rotor blades on the mean radius, using a C4 profile with circular camber line for the nozzle blades and T6 profile with circular camber line for the rotor blades. Assume $c/h = 1$, $Z = 0.9$, and 10% thickness for both nozzle and rotor blades.

9.44 For the data of Problem 9.39, determine the shape of both the nozzle and the rotor blades on the mean radius, using a C4 profile with circular camber line for the nozzle blades and a T6 profile with circular camber line for the rotor blades. Assume $c/h = 1$, $Z = 0.9$, and 10% thickness for both nozzle and rotor blades.

9.45 For the data of Problem 9.35, determine AN^2 at station 2.

9.46 For the data of Problem 9.39, determine AN^2 at station 2.

9.47 Products of combustion [$\gamma = 1.3$, $R = 0.287$ kJ/(kg · K)] exit from a set of axial-flow turbine nozzles at a swirl angle of 60 deg, $M = 1.1$,

$P_t = 2.0$ MPa, and $T_t = 1670$ K. This gas then enters the turbine rotor.

a) For a mass flow rate of 150 kg/s, determine the area of the flow annulus upstream of the turbine rotor.

b) If the maximum centrifugal load on the turbine blade is 70 MPa, determine the maximum shaft speed (rpm, N). Assume a blade taper factor of 0.9 and material density of 8700 kg/m^3.

c) For the shaft speed determined in part b, the turbine disk designer has determined that the maximum radius of the hub is 0.6 m. Determine the tip radius, mean radius, and total temperature drop through the turbine rotor based on the mean rotor velocity and zero exit swirl.

9.48   Perform Problem 9.47 for $\alpha_2 = 70$ deg and $M_2 = 1.05$.

9.49   Products of combustion [$\gamma = 1.3$, $R = 53.34$ ft · lbf/(lbm · °R)] exit from a set of axial-flow turbine nozzles at a swirl angle of 65 deg, $M = 1.2$, $P_t = 300$ psia, and $T_t = 3000$°R. This gas then enters the turbine rotor.

a) For a mass flow rate of 325 lbm/s, determine the area of the flow annulus upstream of the turbine rotor.

b) If the maximum centrifugal load on the turbine blade is 10,000 psia, determine the maximum shaft speed (rpm, N). Assume a blade taper factor of 0.9 and material density of 0.25 lb/in.^{3}

c) For the shaft speed determined in part b, the turbine disk designer has determined that the maximum radius of the hub is 15 in. Determine the tip radius, mean radius, and total temperature drop through the turbine rotor based on the mean rotor velocity and zero exit swirl.

9.50   Perform Problem 9.49 for $\alpha_2 = 60$ deg and $M_2 = 1.1$.

9.51   A centrifugal turbine has products of combustion [$\gamma = 1.33$, $R = 0.287$ kJ/(kg · K)] entering at 1200 K and 1.5 MPa and leaving at 1000 K. For a rotor diameter of 0.3 m, flow angle α_2 of 60 deg, and polytropic efficiency of 0.9, determine the following:

a) Rotor tip speed and angular speed ω

b) Mach number at station 2

c) Total pressure at the exit

d) Adiabatic efficiency

9.52 Perform Problem 9.51 for $\alpha_2 = 65$ deg.

9.53 A centrifugal turbine has products of combustion $[\gamma = 1.33,$ $R = 53.34$ ft · lbf/(lbm · °R)] entering at 2100°R and 200 psia and leaving at 1800°R. For an 8-in. rotor diameter, flow angle α_2 of 55 deg, and polytropic efficiency of 0.85, determine the following:

a) Rotor tip speed and angular speed ω

b) Mach number at station 2

c) Total pressure at the exit

d) Adiabatic efficiency

9.54 Perform Problem 9.53 for $\alpha_2 = 60$ deg.

9.55 At a total pressure of 1 atm and total temperature of 288.2 K, 2.0 kg/s of air enters the centrifugal compressor of a turbocharger and leaves the compressor at a total pressure of 1.225 atm. This compressor is directly driven by a radial-flow turbine. Products of combustion with a mass flow rate of 2.06 kg/s enter the turbine at a total temperature of 900 K and leave the turbine at a total pressure of 1.02 atm. Assume adiabatic flow through both compressor and turbine, no loss of power to bearings from the drive shaft that connects the compressor and turbine, $e_c = 0.80$, $e_t = 0.82$, a slip factor of 0.9 for the compressor, and the following gas properties:

For air	$\gamma = 1.40$ and $c_p = 1.004$ kJ/(kg · K)
For products of combustion	$\gamma = 1.33$ and $c_p = 1.157$ kJ/(kg · K)

Determine the following:

a) Compressor rotor tip speed U_t in m/s and exit total temperature in Kelvins

b) Turbine rotor tip speed U_t in m/s and inlet total pressure in atmospheres

c) Turbine exit total temperatures in Kelvins

d) Tip radius of compressor rotor and tip radius of turbine rotor for $N = 20,000$ rpm

9.56 Perform Problem 9.55 for turbine inlet temperature of 850 K.

9.57 At a total pressure of 14.696 psia and total temperature of 518.7°R, 2.0 lbm/s of air enters the centrifugal compressor of a turbocharger and leaves the compressor at a total pressure of 18 psia. This compressor is directly driven by a radial-flow turbine. Products of combustion with a mass flow rate of 2.06 lbm/s enter the turbine at a total temperature of 1600°R and leave the turbine at a total pressure of 15.0 psia. Assume adiabatic flow through both compressor and turbine, no loss of power to bearings from the drive shaft that connects the compressor and turbine, $e_c = 0.80$, $e_t = 0.82$, a slip factor of 0.9 for the compressor, and the following gas properties:

For air	$\gamma = 1.40$ and $g_c c_p = 6000$ ft^2/(s$^2 \cdot$ °R)
For products of combustion	$\gamma = 1.33$ and $g_c c_p = 6860$ ft^2/(s$^2 \cdot$ °R)

Determine the following:

a) Compressor rotor tip speed U_t in ft/s and exit total temperature in degrees Rankine

b) Turbine rotor tip speed U_t in ft/s and inlet total pressure in psi absolute

c) Turbine exit total temperatures in degrees Rankine

d) Tip radius of compressor rotor and tip radius of turbine rotor for $N = 20{,}000$ rpm

9.58   Perform Problem 9.57 for turbine inlet temperature of 1400°R.

Axial Flow Turbomachinery Design Problems

9.D1   Perform the preliminary design of the turbomachinery for a turbojet engine having a thrust of 25,000 lbf at sea-level static conditions. Based on polytropic efficiencies of 0.9 for both compressor and turbine, the engineers in the engine cycle analysis group have determined the compressor and turbine inlet and exit conditions for a range of compressor pressure ratios that will give the required engine thrust. The results of their analysis are presented in Table P9.D1. Note that the mass flow rate through the compressor and turbine decreases with increasing compressor pressure ratio. To minimize engine weight, it is desirable to have the maximum compressor pressure that can be driven by a single-stage turbine with exit guide vanes. The number of compressor stages depends on both the compressor design and the turbine design. As you can see, there are many sets of compressor/turbine designs that meet the thrust need; however, some of the designs cannot be done with one turbine stage or may be too large (high mass flow rate). Data for other compressor pressure ratios between 7 and 13 not listed in the table can be obtained by interpolation.

Assume that a turbine disk material exists that permits a rim speed of 1200 ft/s and that the turbine rotor airfoils can withstand

Table P9.D1 Compressor and Turbine Design Data for a 25,000-lb Turbojet[a]

π_c	T_{t3}, °R	P_{t3}, psia	T_{t5}, °R	P_{t4}, psia	P_{t5}, psia	$\dot{m}_{compr}$, lbm/s	$\dot{m}_{turb}$, lbm/s	$\dot{W}_c$, kW	$\dot{W}_t$, kW
7.0	962.06	102.87	2853.21	98.76	56.85	200.28	208.66	22,298	22,523
7.5	983.36	110.22	2836.44	105.81	59.20	197.79	206.01	23,079	23,312
8.0	1003.72	117.57	2820.41	112.87	61.45	195.61	203.68	23,824	24,065
8.5	1023.22	124.92	2805.04	119.92	63.60	193.69	201.63	24,539	24,786
9.0	1041.96	132.26	2790.27	126.97	65.65	191.97	199.79	25,225	25,480
9.5	1060.00	139.61	2776.04	134.03	67.61	190.44	198.15	25,886	26,147
10.0	1077.40	146.96	2762.31	141.08	69.49	189.05	196.66	26,523	26,791
10.5	1094.22	154.31	2749.03	148.14	71.29	187.80	195.31	27,140	27,415
11.0	1110.50	161.66	2736.18	155.19	73.02	186.65	194.08	27,738	28,018
11.5	1126.28	169.00	2723.71	162.24	74.68	185.61	192.95	28,318	28,605
12.0	1141.60	176.35	2711.60	169.30	76.27	184.65	191.91	28,882	29,174
12.5	1156.49	183.70	2699.82	176.35	77.80	183.77	190.96	29,431	29,729
13.0	1170.98	191.05	2688.36	183.41	79.27	182.95	190.07	29,966	30,269

[a]Air entering compressor has $T_t = 518.7$°R and $P_t = 14.696$ psia. Gas enters turbine at 3200°R and P_{t4}. Polytropic efficiencies: $e_c = 0.90$ and $e_t = 0.90$. Also, $\dot{W}_t = \dfrac{\dot{W}_c}{0.99}$, $P_{t4} = 0.96 P_{t3}$.

an AN^2 of 5×10^{10} in.$^2 \cdot$ rpm^2. Use Fig. I.8 from Appendix I and computed AN^2 stresses to select material for each compressor stage.

For the compressor, use

$$\gamma = 1.4 \quad \text{and} \quad R = 53.34 \,\text{ft} \cdot \text{lbf}/(\text{lbm} \cdot {}^\circ\text{R})$$

For the turbine, use

$$\gamma = 1.3 \quad \text{and} \quad R = 53.39 \,\text{ft} \cdot \text{lbf}/(\text{lbm} \cdot {}^\circ\text{R})$$

Suggested Design Process

1. Select a set of compressor and turbine design data from Table P9.D1.

2. For the design turbine rim speed, estimate the mean turbine rotor speed ωr_m and determine the variation of turbine radii (hub, mean, and tip) and AN^2 vs rpm. Select a shaft speed (rpm).

3. Determine the turbine mean-line design using TURBN. Make sure the turbine meets the design criteria listed below.

4. For the selected shaft speed (rpm) and compressor pressure ratio, determine the number of compressor stages and the compressor mean-line design, using the repeating-row, repeating-stage design choice in COMPR. Make sure the compressor meets the design criteria listed below.

5. Check that 99% of power from the turbine equals the power required by the compressor.

6. Check the alignment of the compressor and turbine. If the mean radii are not more than 2 in. apart, go back to item 2 and select a new rpm.

7. Determine the inlet and exit flow conditions for the turbine exit guide vanes (include estimate of losses).

8. Determine the inlet and exit flow conditions for the inlet guide vanes (include estimate of losses).

9. Specify compressor material for each stage based on the rotor's relative total temperature and blade AN^2.

10. Make a combined scale drawing of the compressor and turbine flow path. Allow 12 in. between the compressor exit and turbine inlet for the combustor. Show the shaft centerline.

Compressor Design Criteria

1. Axial flow entering inlet guide vanes and leaving compressor
2. Reactions at all radii > 0
3. Diffusion at all radii < 0.6
4. M_{1R} at tip of first stage < 1.05
5. M_2 at hub of first stage < 1
6. Flow coefficient at mean radius between 0.45 and 0.55
7. Stage loading coefficient at mean radius between 0.3 and 0.35
8. Number of blades for rotor or stator of any stage < 85
9. Number of blades for inlet guide vanes < 85
10. AN^2 at entrance of rotor within material limits

Turbine Design Criteria

1. Axial flow entering turbine nozzle at $M = 0.3$
2. Axial flow leaving exit guide vanes of turbine
3. Reaction at all radii > -0.15
4. Number of blades for nozzle, rotor, or exit guide vanes < 85
5. $M_2 < 1.2$ at hub and > 1 at tip
6. M_{3R} at tip of rotor < 0.9
7. Velocity ratio at mean radius between 0.5 and 0.6
8. AN^2 at entrance of rotor within material limits
9. Tangential force coefficient Z for stator or rotor < 1.0

9.D2 Perform the preliminary design of the turbomachinery for a turbojet engine of Problem 9.D1 but with no inlet guide vanes for the compressor or exit guide vanes from the turbine. Thus the turbine must have zero exit swirl.

9.D3 Perform the preliminary design of the turbomachinery for a turbojet engine of Problem 9.D1 scaled for a thrust of 16,000 lbf at sea-level, static conditions. Thus the mass flow rates and powers will be 64% of the values listed in Table P9.D1.

9.D4 You are to perform the preliminary design of the turbomachinery for a turbojet engine of Problem 9.D3 but with no inlet guide vanes for the compressor or exit guide vanes from the turbine. Thus the turbine must have zero exit swirl.

Chapter 10 Rocket Propulsion

> How many more years I shall be able to work on the problem I do not know; I hope, as long as I live. There can be no thought of finishing, for "aiming at the stars," both literally and figuratively, is a problem to occupy generations, so that no matter how much progress one makes, there is always the thrill of just beginning.
>
> Robert H. Goddard, in letter to H. G. Wells, 1932

10.1 Introduction

The types of rocket propulsion and basic parameters used in performance evaluation of rocket propulsion are described in this chapter. First, the concepts of thrust, effective exhaust velocity, and specific impulse are presented for a gaseous continuum leaving a rocket engine. These concepts, initially presented in Section 1.6, are covered in more detail here. To determine the requirements of rocket vehicles, some introductory mechanics of orbital flight are investigated along with rocket mass ratios, rocket acceleration, and multistage rockets. This section on requirements and capabilities is followed by a general discussion of the various sources of rocket propulsion and a summary of their capabilities. Following this general introduction to rocket propulsion and rocket flight is a discussion of rocket nozzle types, a study of the detailed performance parameters for rocket engines expelling a continuum, detailed example problems, and discussion of liquid- and solid-propellant rockets. The reader would do well to review the material on converging-diverging (C-D) exhaust nozzles in Section 3.7.1.2, if needed, because the discussion in the chapter assumes a solid foundational understanding of C-D nozzle operation.

10.1.1 Rocket Thrust

The rocket thrust studied here will be applicable to rockets that eject a gaseous continuum material from a nozzle at a constant rate. Chemical and nuclear–H_2 rockets are examples of such rocket propulsion systems. The thrust of an ionic rocket that expels discrete ion particles at high speeds, on the other hand, is not related to properties of a continuum such as pressure and density. The thrust equation to be developed will not apply, therefore, to ionic rockets. The thrust of ionic rockets, however, still depends on the same basic principle of operation as chemical or nuclear–H_2 rockets (i.e., the reaction force resulting from imparting momentum to a mass).

10.1.2 Ideal Thrust Equation

The ideal rocket thrust is defined as the force required to hold a rocket at rest as the rocket ejects a propellant rearward under the following assumed conditions:

* Atmospheric pressure P_a acts everywhere on the rocket's external surface. (There is no interaction between the jet issuing from the rocket nozzle and the air in contact with the external surface of the rocket.)
* The propellant flows from the rocket at a constant rate in uniform 1-D flow at the nozzle exit section.
* The rate of change of the momentum within the rocket is negligibly small and may be neglected.

The rocket of Fig. 10.1a is held in place by a reaction strut and is ejecting a propellant to the rear under the preceding conditions. The shear force at section B-B of the reaction strut is equal in magnitude to the ideal rocket thrust. As shown by the freebody diagram in Fig. 10.1a, the thrust is produced by the internal force on the rocket shell. In a vacuum, the thrust would be

a) Forces on rocket shell

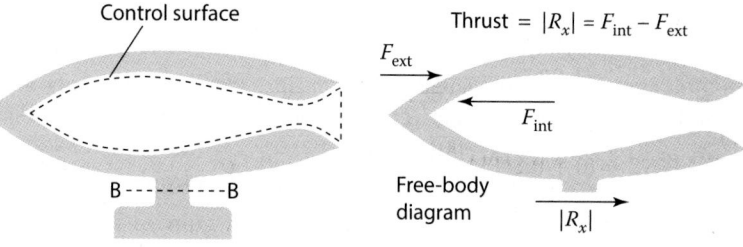

b) Evaluation of $|F_{int}|$ by momentum equation

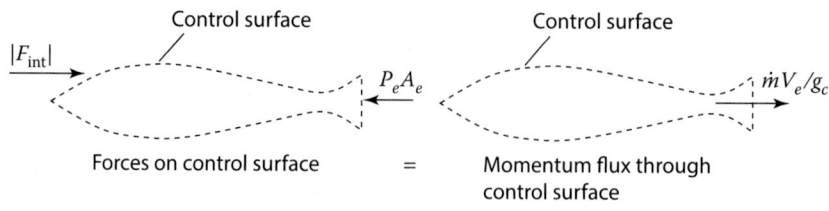

Fig. 10.1   Ideal rocket thrust.

identical to F_{int}, but is less than the internal force in the atmosphere because of the ambient pressure forces acting on the external surface of the rocket. The external force is due to atmospheric pressure forces only and is given by

$$F_{ext} = P_a A_e$$

If we assume internal isentropic flow, the interior force is due to pressure forces only. It is the sum of all of the pressure forces exerted by the propellant on the interior wall surfaces. The internal force is readily evaluated in terms of propellant flow properties by application of the momentum equation to the control surface indicated in Fig. 10.1b, and by using the fact that the force of the propellant on the internal walls equals, in magnitude, the force of the walls on the propellant. For the magnitude of the internal force, we obtain

$$|F_{int}| = P_e A_e + \frac{1}{g_c} \rho_e A_e V_e^2 = P_e A_e + \frac{\dot{m} V_e}{g_c}$$

The difference between F_{int} and F_{ext} is the ideal rocket thrust, F_i,

$$F_i = F_{int} - F_{ext}$$

$$F_i = \frac{\dot{m} V_e}{g_c} + (P_e - P_a) A_e \tag{10.1}$$

This equation, identical to that developed in Section 4.2.2 ($P_a = P_0$), gives the thrust in terms of the propellant flow properties, the exit area, and the ambient pressure. It is important to remember that only the term $P_a A_e$ on the right side of Eq. (10.1) represents *a real force acting on the rocket shell*.

10.1.3 Optimum Ideal Thrust

Let us consider a rocket ejecting a given propellant supersonically from a combustion chamber at a fixed temperature and pressure. Under these conditions, the mass flow through the rocket nozzle is constant at a value given [from Eqs. (3.8) and (3.11)] by

$$\dot{m} = \frac{P_t A}{\sqrt{T_t}} \sqrt{\frac{2 g_c}{R} \frac{\gamma}{\gamma - 1} \left\{ \left(\frac{P}{P_t}\right)^{2/\gamma} - \left(\frac{P}{P_t}\right)^{(\gamma+1)/\gamma} \right\}} \tag{10.2}$$

Let us further assume that the nozzle exit area is adjustable and that the ambient pressure P_a is constant. Now notice that as we vary A_e, the pressure P_e and the velocity V_e will vary, but mass flow rate $\dot{m}$ and P_a will remain constant. The following question arises: For what value of A_e will the corresponding values of V_e and P_e make the thrust of Eq. (10.1) a maximum? The simplest way to demonstrate the optimum thrust condition is to deal with the *real forces* acting on the rocket shell and to determine directly under what conditions these produce maximum thrust. First, we observe

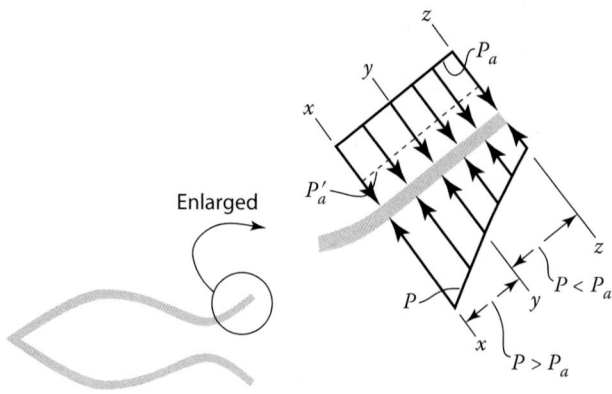

Fig. 10.2 Forces acting on rocket nozzle wall for optimum thrust consideration.

that as A_e is varied by adding a further divergent section on the nozzle exit, any change in thrust comes about by the net force acting on this added section. This follows from the fact that forces acting on the original surfaces of the rocket are not changed by the added section because atmospheric pressure still acts on the original external surfaces, and the original internal pressure distribution remains the same as long as supersonic flow exists at the nozzle exit. Because any thrust change, therefore, comes about by the action of the forces acting on the added surfaces introduced in changing A_e, we need only to examine the forces associated with the added surface to maximize thrust. Figure 10.2 shows an enlarged view of the pressure distribution on the external and internal surfaces near the exit of a rocket nozzle. The external pressure is constant at P_a. The internal pressure decreases from a value greater than P_a at x to P_a at y, and to a value lower than P_a at z, corresponding to an overexpanded nozzle. Now, as A_e is increased by adding surfaces up to station y, we find the thrust is increased because the internal pressure on these added surfaces is greater than P_a. If A_e is increased beyond that value corresponding to y, the thrust is decreased because there is a net drag force acting on the surfaces beyond y because the internal pressure is less than P_a between y and z. To eliminate the wall surface producing a net drag and to not eliminate any wall surface producing a net thrust, the nozzle should be terminated at y where $P_e = P_a$. The exit velocity corresponding to this condition will be denoted $(V_e)_{\text{opt}}$ so that

$$(F_i)_{\text{opt}} = \frac{\dot{m}(V_e)_{\text{opt}}}{g_c} \qquad (10.3a)$$

If the nozzle of Fig. 10.2 is at a higher altitude, the ambient pressure is reduced to a value of $P_{a'}$, indicated by the dashed line of Fig. 10.2. The pressure at this higher altitude is such that the internal pressure acting on the nozzle wall from y to z is greater than the new ambient pressure. At this higher altitude, then, the nozzle surface from y to z is a thrust-producing surface and should be retained. This means a higher nozzle area ratio at higher altitudes for optimum thrust.

10.1.4 Vacuum Ideal Thrust

In a vacuum (very high altitudes), $F_{ext} = 0$. Thus $(F_i)_{vac} = F_{int}$, where F_{int} can be evaluated by

$$(F_i)_{vac} = F_{int} = \frac{\dot{m}V_e}{g_c} + P_e A_e \qquad (10.3b)$$

Note that the thrust given in Appendix C for liquid-propellant rocket engines is the vacuum ideal thrust.

10.1.5 Thrust Variation with Altitude

To understand the variation of thrust with altitude, we consider two rockets, A and B, both of which have supersonic exhaust velocities at sea level. Rocket A has an infinitely adjustable area ratio, whereas rocket B's area ratio is fixed (constant) at a value that gives optimum expansion at sea level.

For each rocket, $F_i = F_{int} - F_{ext}$, and F_{int} remains the same for each rocket as altitude increases. For both rockets A and B, F_{ext} decreases the same as altitude increases. However, rocket A has thrust-producing surfaces (such as segment x-y of Fig. 10.2) added on to it as it climbs, allowing the exhaust gases to expand to the lower ambient pressures. Consequently, the thrust increase with altitude of rocket A exceeds that of rocket B, as sketched in Fig. 10.3.

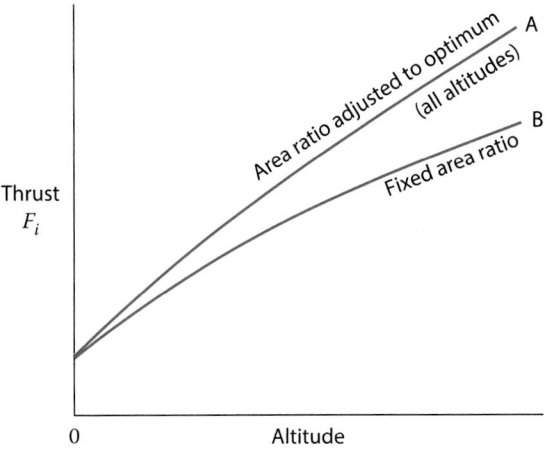

Fig. 10.3   Ideal thrust variation with altitude.

10.1.6 **Effective Exhaust Velocity** C **and Specific Impulse** (I_{sp})

The effective exhaust velocity C is defined in Chapter 1 by Eq. (1.53) as

$$C \equiv V_e + \frac{(P_e - P_a)A_e g_c}{\dot{m}_p} \qquad (10.4)$$

Using this definition of effective exhaust velocity in Eq. (10.1), the ideal thrust is simply

$$F_i = \frac{\dot{m}C}{g_c} \qquad (10.5)$$

The values of a rocket's exhaust velocity V_e and exhaust pressure P_e depend on the type of rocket engine and its design and operation. For launch from the surface of the Earth to Earth orbit, the atmospheric pressure P_a varies from a low of zero (vacuum) to a high of that at sea level. For a rocket engine where the mass flow rate and combustion chamber conditions are constant, the effective exhaust velocity C and ideal thrust F_i will vary from a low value for sea-level operation to a high value at vacuum conditions. Figure 1.42b shows the predicted variation of thrust with altitude for the space shuttle main engine (SSME) under these conditions. The variation of effective exhaust velocity C will be the same as the thrust times a constant. Even though the effective exhaust velocity C can vary, an average value of the effective exhaust velocity may be used for ease of analysis or calculation.

The specific impulse I_{sp} for a rocket is defined as the thrust per unit of propellant weight flow:

$$I_{sp} \equiv \frac{F_i}{\dot{w}} = \frac{F_i g_c}{\dot{m} g_0} \qquad (10.6)$$

where g_0 is the acceleration of gravity at sea level. Specific impulse has units of seconds. Using Eq. (10.5), the effective exhaust velocity is directly related to the specific impulse by

$$C = I_{sp} g_0 \qquad (10.7)$$

As just discussed, the effective exhaust velocity can vary during operation. From Eq. (10.7), the specific impulse I_{sp} for a rocket engine will vary in direct proportion to its effective exhaust velocity C. For ease of analysis or calculation, an average value of specific impulse may be used. Note that values of specific impulse at sea level and very high altitudes (vacuum) are provided for many of the liquid-propellant rocket engines in Appendix C.

Example 10.1

Appendix C lists the F-1 engine (see Fig. 3.47) used on the Saturn V rocket (mainly used to launch Apollo missions to the moon) as having a vacuum thrust of 1,726,000 lbf and vacuum I_{sp} of 305 s. Determine the mass flow rate at the vacuum conditions.

Solution

Solving Eq. (10.6) for the mass flow rate, we have

$$\dot{m} = \frac{F}{I_{sp}}\frac{g_c}{g_0} = \frac{1{,}726{,}000}{305}\frac{32.174}{32.174} = 5659 \, \text{lbm/s}$$

10.2 Rocket Propulsion Requirements and Capabilities

10.2.1 Requirements

The propulsion requirements for space flight missions are usually expressed in terms of speed increment that the propulsion system must supply to the space vehicle. When the appropriate increment in speed is supplied to a vehicle, the vehicle's guidance system must adjust the direction of the speed to attain the orbit desired. As a simple example, the ideal speed increment to place a vehicle in orbit about the Earth is that speed that will give a balance between the centrifugal and gravitational forces acting on the body in orbit. Considering the balance of these forces, we can determine

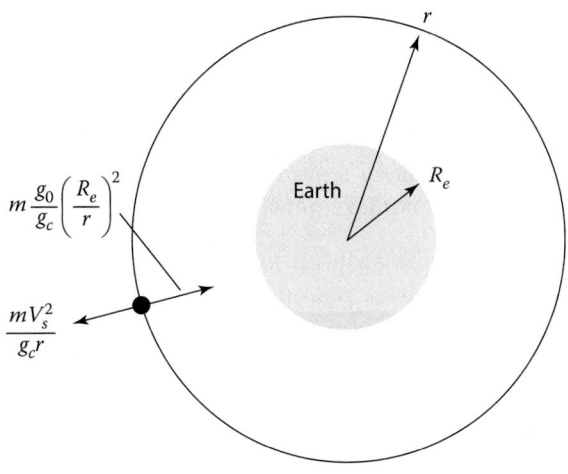

Fig. 10.4 Forces acting on near-Earth-orbiting mass.

the needed velocity increment. The force of gravity acting on a mass m at a distance r from the Earth's center is

$$m\frac{g_0}{g_c}\left(\frac{R_e}{r}\right)^2$$

where R_e is the Earth's radius (3959 miles, 6370 km), and g_0 is the acceleration of gravity on the surface of the Earth (32.174 ft/s², 9.807 m/s²). Referring to the force diagram of Fig. 10.4, we have

$$\frac{mV_s^2}{g_c r} = m\frac{g_0}{g_c}\left(\frac{R_e}{r}\right)^2$$

Solving for the satellite velocity V_s at a height h above the surface of the Earth ($r = R_e + h$) gives

$$V_s = \sqrt{g_0\frac{R_e^2}{R_e + h}} \qquad (10.8)$$

For a circular Earth orbit at 160 km (100 miles), Eq. (10.8) gives a satellite velocity of 7800 m/s (25,600 ft/s). Thus a rocket must increase the velocity from zero at launch to 7800 m/s at 160 km. The propulsion system's fuel is the original source of the kinetic energy and the higher potential energy possessed by the orbital mass of Fig. 10.4. The kinetic energy (KE) of the orbital mass is simply given by $\{mV_s^2/(2g_c)\}$, and the change in potential energy (ΔPE) is given by

$$\Delta \text{PE} = \int_{R_e}^{R_e+h} m\frac{g}{g_c}\,dr = mR_e^2\frac{g_0}{g_c}\int_{R_e}^{R_e+h}\frac{dr}{r^2} = m\frac{g_0}{g_c}\left(\frac{R_e}{R_e + h}\right) \qquad (10.9)$$

The sum of the changes in potential and kinetic energies per unit mass $\{\Delta(\text{pe} + \text{ke})\}$ of the orbiting mass relative to sea-level launch is given by

$$\Delta(\text{pe} + \text{ke}) = \frac{\Delta(\text{PE} + \text{KE})}{m} = \frac{g_0}{g_c}\left(\frac{R_e}{R_e + h}\right) + \frac{V_s^2}{2g_c} \qquad (10.10)$$

To place a mass in circular Earth orbit at 100 miles requires that the sum of the potential and kinetic energies be increased by 13,760 Btu/lbm (32 MJ/kg). This energy change is equal to the kinetic energy of a mass at a velocity of 26,250 ft/s (8000 m/s). In addition to this energy, the propulsion system must supply the energy that is delivered to the viscous atmosphere by the ascending vehicle due to aerodynamic drag forces. Thus the equivalent velocity increment requirement of a propulsion system to attain a near-Earth orbit is greater than 26,250 ft/s (8000 m/s) due to aerodynamic drag. Taking these items into account gives an equivalent velocity increment requirement

Table 10.1   Mission Equivalent Velocity Increment Requirements

Mission	ΔV_{equiv} (m/s)	ΔV_{equiv} (ft/s)
Low Earth orbit	9,140	30,000
Earth escape/lunar hit	12,800	42,000
High Earth orbit		
Lunar orbit	13,720	45,000
Mars/Venus probe		
Lunar soft landing	15,240	50,000
Lunar round trip	18,280	60,000
Escape from solar system		

of approximately 30,000 ft/s (9140 m/s) for a near-Earth orbit. Table 10.1 gives this velocity increment and those required for certain other space missions.

To escape the Earth's gravitational field ($h \rightarrow \infty$), solution of Eq. (10.9) gives the change in potential energy as Earth escape:

$$\Delta \text{pe} = \frac{g_0}{g_c} R_e \qquad (10.11)$$

This potential energy change is equal to the kinetic energy corresponding to the escape velocity V_{escape} given by

$$V_{\text{escape}} = \sqrt{2 g_0 R_e} \qquad (10.12)$$

Comparison of Eq. (10.12) for the escape velocity to Eq. (10.8) for the satellite velocity evaluated for low Earth orbit gives

$$V_{\text{escape}} = \sqrt{2} V_s \qquad (10.13)$$

This equation explains why the velocity increment of 42,000 ft/s (12,800 m/s) required for Earth escape (see Table 10.1) is 12,000 ft/s (3660 m/s) higher than that required for low Earth orbit.

The total mass of a rocket vehicle is the sum of its parts. For convenience, the vehicle mass is considered to be divided into three different masses: the payload mass m_{pl}, the propellant mass m_p, and the remaining mass, which will be called the dead weight mass m_{dw}. The dead weight mass includes the rocket engines, guidance and control system, tankage, rocket structure, and so on. The initial mass of a rocket vehicle m_o can be expressed as

$$m_o = m_{\text{pl}} + m_p + m_{\text{dw}} \qquad (10.14)$$

For convenience, we define the payload mass ratio λ and dead weight mass ratio δ as

$$\lambda \equiv \frac{m_{\text{pl}}}{m_o} \quad \text{and} \quad \delta \equiv \frac{m_{\text{dw}}}{m_o} \qquad (10.15)$$

If all of the propellant is consumed during firing, the resulting vehicle mass is defined as the burnout mass m_{bo}, or

$$m_{bo} = m_o - m_p = m_{pl} + m_{dw} \tag{10.16}$$

In addition to the mass ratios of Eq. (10.15), the vehicle mass ratio (MR) is defined as the initial mass m_o divided by the burnout mass m_{bo}, or

$$MR \equiv \frac{m_o}{m_{bo}} \tag{10.17}$$

Using Eq. (10.16) and the mass ratio defined in Eq. (10.15), the MR can be written as

$$MR = \frac{1}{\lambda + \delta} \tag{10.18}$$

10.2.2 Equation of Motion for an Accelerating Rocket

In this section, the velocity increment ΔV of a rocket is related to the propulsion system's burning time t_b, the ejected mass velocity V_e, and the space vehicle's MR. The equation of motion for an accelerating rocket (see Fig. 10.5) rising from the Earth under the influence of gravity and drag as it expels matter was developed in Chapter 1. The analysis resulted in the general expression for the differential velocity change given by Eq. (1.58), or

$$dV = -C\frac{dm}{m} - \left(\frac{D}{m/g_c} + g\cos\theta\right)dt \tag{10.19}$$

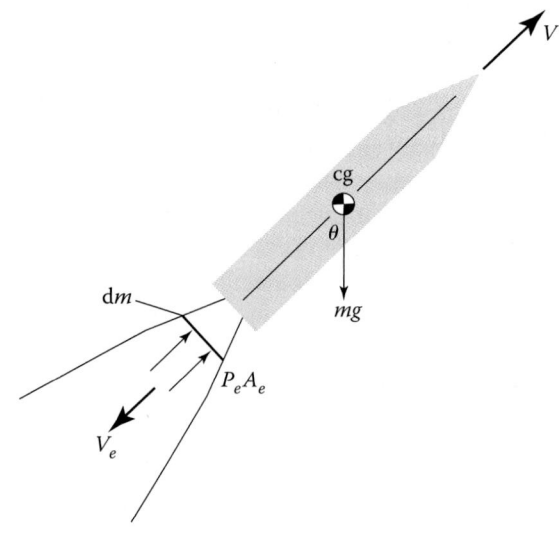

Fig. 10.5 Accelerating rocket.

where C is the effective exhaust velocity. For launch from the surface of the Earth, the acceleration of gravity g can be written in terms of the radius from the center of the Earth r, the acceleration of gravity at sea level g_0, and the radius of the Earth R_e, as follows:

$$g = g_0 \left(\frac{R_e}{r} \right)^2$$

Substitution of this expression into Eq. (10.19) yields

$$dV = -C \frac{dm}{m} - \left\{ \frac{D}{m/g_c} + g_0 \left(\frac{R_e}{r} \right)^2 \cos \theta \right\} dt \qquad (10.20)$$

In this equation, dV is the velocity increment supplied to the vehicle of mass m in time dt as mass dm is expelled rearward at effective velocity C relative to the vehicle. Notice that dm is a negative quantity for leaving mass, and hence $-C(dm/m)$ becomes a positive number exceeding the

$$\left\{ \frac{D}{m/g_c} + g_0 \left(\frac{R_e}{r} \right)^2 \cos \theta \right\} dt$$

when values of C are sufficiently high to produce a positive increment in V. When dm is zero (no burning), dV becomes a negative until the kinetic energy possessed by the vehicle at burnout is all converted into potential energy of the vehicle at zero velocity and maximum height. After this condition is reached, the vehicle begins to descend, and its potential energy is converted back into kinetic energy that, in turn, is dissipated into the atmosphere during reentry. To avoid this simple yo-yo motion, we assume that a guidance system has the rocket follow a specified trajectory. The trajectory is such that the vehicle's velocity vector is perpendicular to the radius r at the instant of burnout, and a near-Earth orbit is established corresponding to the velocity of 26,000 ft/s.

Integration of Eq. (10.20) from liftoff to burnout gives us the desired relation between the vehicle's velocity increment ΔV and the propulsion system's burning time t_b, the effective exhaust velocity C, the drag D, the trajectory, and the vehicle's MR. Thus

$$\Delta V = \int_0^{V_\infty} dV = -\int_{m_o}^{m_{bo}} C \frac{dm}{m} - g_0 \int_{t=0}^{t_{bo}} \left\{ \frac{D/m}{g_0/g_c} + \left(\frac{R_e}{r} \right)^2 \cos \theta \right\} dt \qquad (10.21)$$

where subscript bo denotes burnout conditions. Evaluation of this equation requires detailed knowledge of the trajectory and vehicle drag.

Using Eq. (10.21), the effective velocity increment $\Delta V_{effective}$, the drag velocity increment ΔV_{drag}, and the gravity velocity increment $\Delta V_{gravity}$ are defined by

$$\Delta V_{\text{effective}} \equiv - \int_{m_o}^{m_{\text{bo}}} C \frac{dm}{m}$$

$$\Delta V_{\text{drag}} \equiv g_c \int_{t=0}^{t_{\text{bo}}} \left\{ \frac{D}{m} \right\} dt \qquad (10.22)$$

$$\Delta V_{\text{gravity}} \equiv g_0 \int_{t=0}^{t_{\text{bo}}} \left\{ \left(\frac{R_e}{r} \right)^2 \cos \theta \right\} dt$$

Note that the effective velocity increment $\Delta V_{\text{effective}}$ is the specific vehicle impulse (engine thrust per unit mass of the vehicle integrated over time), or

$$\Delta V_{\text{effective}} \equiv - \int_{m_o}^{m_{\text{bo}}} C \frac{dm}{m} \equiv \int_{t=0}^{t_{\text{bo}}} \frac{\dot{m}C}{m} dt \equiv g_c \int_{t=0}^{t_{\text{bo}}} \frac{F}{m} dt \qquad (10.23)$$

Equation (10.21) for the vehicle's velocity increment can be rewritten as

$$\Delta V = \Delta V_{\text{effective}} - \Delta V_{\text{drag}} - \Delta V_{\text{gravity}} \qquad (10.24)$$

This equation shows the effect of drag and gravity on effective velocity increment that was discussed earlier in this section, namely, that the effective velocity increment $\Delta V_{\text{effective}}$ is greater than the vehicle's velocity increment ΔV due to vehicle drag losses and gravity.

Assuming constant effective exhaust velocity C, Eq. (10.22) gives the following expression for effective velocity increment:

$$\Delta V_{\text{effective}} = C \ell n \left(\frac{m_o}{m_{\text{bo}}} \right) = C \ell n(\text{MR}) = C \ell n \left(\frac{1}{\lambda + \delta} \right) \qquad (10.25)$$

Equation (10.25) is plotted in Fig. 10.6, which shows the exponential relationship between the mass ratio and the effective velocity ratio ($\Delta V_{\text{effective}}/C$). Both Fig. 10.6 and Eq. (10.25) show the extreme sensitivity of the mass ratio to the effective exhaust velocity for a given high-energy mission (large $\Delta V_{\text{effective}}$).

Many space missions require multiple $\Delta V_{\text{effective}}$. Consider a mission from Earth orbit to the surface of the moon and return. This mission will require at least six ΔV: one to exit Earth orbit and reach transit speed, a second to enter lunar orbit, a third to land on the moon, a fourth to reach lunar orbit again, a fifth to exit lunar orbit and reach transit speed, and a sixth to enter Earth orbit. Additional ΔV will be required for trajectory corrections, vehicle orientation, and the like. Each ΔV expends vehicle mass, and the mass ratio (m_o/m_{bo}) for the total mission of a single-stage vehicle can be

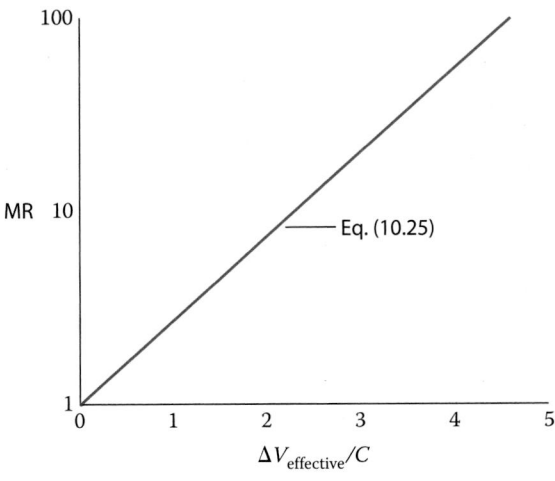

Fig. 10.6 Ideal rocket vehicle mass ratio.

obtained using Eq. (10.25), where $\Delta V_{\text{effective}}$ is the sum of all of the effective velocity changes ($\Delta V_{\text{effective}}$) in the mission.

For a near-Earth orbit, R_e/r is close to unity ($R_e = 3959$ miles, $r = 3959 + 100 = 4059$ miles, and $R_e/r = 0.975$), and $\Delta V_{\text{gravity}}$ can be approximated as $\Delta V_{\text{gravity}} \approx g_0 t_{\text{bo}}$. Assuming constant effective exhaust velocity C and negligible drag ($\Delta V_{\text{drag}} = 0$), Eqs. (10.24) and (10.25) give

$$\Delta V = V_{\text{bo}} = C\ell n\left(\frac{m_o}{m_{\text{bo}}}\right) - g_0 t_{\text{bo}} \qquad (10.26)$$

This simple equation provides a means of comparing the performance of different propulsion systems in terms of vehicle gross mass m_o required to put a given mass m_{bo} into a low Earth orbit requiring a vehicle velocity increment of V_{bo}.

Example 10.2

To demonstrate the importance of a high propellant exhaust velocity V_e, let us determine the orbital payload capabilities of various propulsion systems for a fixed initial vehicle gross weight. Table 10.2 gives approximate representative values of effective exhaust velocity C for four propulsion systems. The dead weight ratio δ listed consists of tankage, pumps, and structural members excluding the payload.

(Continued)

Example 10.2 *(Continued)*

Table 10.2 Performance of High-Thrust Rocket Propulsion Systems

Propulsion System Fuel and Propellant	Effective Exhaust Velocity, C		Assumed Dead Weight Ratio (δ)
	m/s	**ft/s**	
Solid	2440	8000	0.03
Liquid O_2–kerosene (RP)	3050	10,000	0.03
Liquid O_2 and H_2	4110	13,500	0.06
Nuclear fuel using H_2 propellant	8230	27,000	0.10

As a reference point, we determine first the gross weight at liftoff for a 2000-lbm payload placed in a near-Earth orbit by a liquid O_2–kerosene (LOX-RP) propulsion system. At burnout, the payload m_{pl} and the rocket dead weight m_{dw} have attained the velocity V_{bo} and both will orbit. Therefore,

$$m_{bo} = m_{pl} + m_{dw} = m_{pl} + \delta m_o = m_{pl} + 0.03\, m_o$$

We will assume for each propulsion system that the burning time is 100 s. We have, from Eq. (10.26),

$$MR = \frac{m_o}{m_{bo}} = \frac{m_o}{m_{pl} + 0.03\, m_o} = \exp\left\{\frac{V_{bo} + g_o t_{bo}}{C}\right\}$$

$$MR = \frac{m_o}{m_{bo}} = \frac{m_o}{2000 + 0.03\, m_o} = \exp\left\{\frac{30,000 + 3217}{10,000}\right\} = \exp\{3.32\} \approx 28$$

Thus

$$m_o = 28\{2000 + 0.03\, m_o\} = 56,000 + 0.84\, m_o$$
$$m_o = 56,000/0.16 = 350,000 \text{ lbm}$$

Thus to place a 2000-lbm payload in a near-Earth orbit requires a sea-level liftoff gross weight of 350,000 lbf. With this gross weight as a reference point, let us evaluate the orbiting payload capabilities of the other propulsion systems of Table 10.2.

For the propulsion system using liquid O_2 (LOX) and H_2 (LH2), we find a pad weight of 350,000 lbf is capable of placing 8890 lbm into a near-Earth orbit. The computation follows:

$$m_{pl} = m_{bo} - m_{dw} = m_o\left\{\exp\left(-\frac{V_{bo} + g_o t_{bo}}{C}\right) - 0.06\right\}$$

$$m_{pl} = 350,000\left\{\exp\left(-\frac{33,217}{13,500}\right) - 0.06\right\} = 350,000(0.0854 - 0.06)$$

$$m_o = 8890 \text{ lbm}$$

(Continued)

Example 10.2 *(Continued)*

Table 10.3   Near-Earth Orbital Payload for 350,000 lbm (158,760 kg) Launch Mass

	Solid Propellant	Liquid O_2–RP	Liquid H_2–O_2	Nuclear–H_2
C (ft/s)|(m/s)	8000|2440	10,000|3050	13,500|4115	27,000|8230
δ (dead weight ratio)	0.03	0.03	0.06	0.10
Payload (lbm)|(kg)	Nothing	2000|907	8890|4040	67,000|30,390
λ (payload ratio)	−0.0142	0.0057	0.0250	0.1914

Similar calculations give the results shown in Table 10.3. The solid-propellant propulsion system's performance is not sufficient to place any mass in orbit under the assumed conditions. The results in Table 10.3 indicate that a high value of effective exhaust velocity is extremely beneficial, if not mandatory. An increase in C by a factor of 2.7, in going from LOX–RP to nuclear–H_2, permits a 33-fold increase in the payload even though the dead weight of the nuclear–H_2 system is three times that of the LOX–RP propulsion unit. Obviously, it is advantageous to have a high exhaust velocity.

10.2.3 Rocket Vehicles in Free Space

When rockets are fired in "free space," there are no drag or gravitational penalties. Integration of Eq. (10.21) under these conditions, and assuming constant effective exhaust velocity C, yields

$$\Delta V / C = \ell n(\mathrm{MR}) = \ell n \left(\frac{1}{\lambda + \delta} \right) \tag{10.27a}$$

or

$$\mathrm{MR} = \frac{1}{\lambda + \delta} = \exp(\Delta V / C) \tag{10.27b}$$

For rocket vehicles in free space, the effective velocity increment $\Delta V_{\text{effective}}$ equals the vehicle's velocity increment ΔV, and thus Fig. 10.6 is also a plot of Eq. (10.27b).

10.2.4 Multiple-Stage Rocket Vehicles

The launch of a payload using single-stage rocket vehicles was investigated in the Section 10.2.2. When large energy changes are required (large ΔV) for a given payload, the initial vehicle mass of a single-stage vehicle and its associated cost become enormous. We consider a liquid H_2–O_2 chemical rocket ($C = 4115$ m/s, 13,500 ft/s) launching a 900-kg (2000-lbm) payload on a mission requiring a ΔV of 14,300 m/s (46,900 ft/s). From Fig. 10.6 or Eq. (10.27b), the required vehicle MR is 32.30. This high mass ratio for a payload of 900 kg and 3% dead weight results in an initial vehicle mass of 938,000 kg mass and a vehicle dead weight mass of about 29,070 kg. This large initial vehicle mass can be dramatically reduced by using a multistage vehicle.

For a multistage vehicle with N stages, the payload of a lower stage, $[m_{pl}]_i$, is the mass of all higher stages, $[m_o]_{i+1}$, or

$$[m_{pl}]_i = [m_o]_{i+1} \quad \text{for } i < N \tag{10.28}$$

From Eq. (10.27b), the initial mass of stage i and above (denoted by $[m_o]_i$) is given by

$$[m_o]_i = \frac{1}{\exp\{-(\Delta V/C)_i\} - \delta_i}[m_{pl}]_i \tag{10.29a}$$

and the payload ratio of stage i (denoted by λ_i) is given by

$$\lambda_i = \exp\{-(\Delta V/C)_i\} - \delta_i \tag{10.29b}$$

Using Eq. (10.27a), the total velocity change ΔV_{total} for an N-stage rocket can be written as

$$\Delta V_{total} = \sum_{i=1}^{N} C_i \ell n \left(\frac{1}{\lambda_i + \delta_i}\right) \tag{10.30}$$

The overall payload ratio λ_o for a multistage rocket is the product of all the λ_i and can be expressed as

$$\lambda_o = \prod_{i=1}^{N} \lambda_i \tag{10.31}$$

Equations (10.28) and (10.29) can be used to obtain the mass ratio required for each stage i of a multistage rocket.

Example 10.3

Consider both a two-stage vehicle and a three-stage vehicle for the launch of the 900-kg (2000-lbm) payload. Each stage uses a liquid H_2–O_2 chemical rocket ($C = 4115$ m/s, 13,500 ft/s), and the ΔV_{total} of 14,300 m/s (46,900 ft/s) is split evenly between the stages. Because each stage has the same ΔV and effective exhaust velocity C, the payload ratio λ_i of the stages of a multistage vehicle are the same. The payload ratio for a stage λ_i is 0.1460 for a two-stage vehicle and 0.2840 for a three-stage vehicle. Table 10.4 summarizes the results for one-, two-, and three-stage vehicles.

(Continued)

Example 10.3 *(Continued)*

Table 10.4   Multistage Vehicle Sizes for a 900-kg Payload at a ΔV of 14,300 m/s with Liquid H_2–O_2

	Single-Stage Vehicle	Two-Stage Vehicle	Three-Stage Vehicle
Stage ΔV	14,300 m/s	7150 m/s	4767 m/s
Stage δ_i	0.03	0.03	0.03
Stage λ_o	0.0009590	0.02130	0.02290
Vehicle m_o	938,480 kg	42,250 kg	39,300 kg
1st stage m_{dw}	28,170 kg	1270 kg	1180 kg
1st stage m_{pl}	900 kg	6170 kg	11,200 kg
2nd stage m_o		6170 kg	11,200 kg
2nd stage m_{dw}		185 kg	336 kg
2nd stage m_{pl}		900 kg	3180 kg
3rd stage m_o			3180 kg
3rd stage m_{dw}			95 kg
3rd stage m_{pl}			900 kg

Note that a very large reduction in initial vehicle mass is obtained when going from a single-stage vehicle to a two-stage vehicle (a two-stage vehicle is 4.5% of the mass of the single-stage vehicle), and only a slight reduction in mass is obtained going from a two-stage vehicle to a three-stage vehicle (a three-stage vehicle is 93% of the mass of the two-stage vehicle).

10.3   Rocket Propulsion Engines

Figure 10.7 shows the basic features of any propulsion system. The function of the propulsion system is to combine energy with matter to produce a directed stream of high-speed particles. That portion of the propulsion system that performs this function may be called the accelerator. In most systems, it is necessary to transport the propellant to the accelerator, and this implies pumps and a certain amount of plumbing. In some systems, it is necessary to supply the propellant in a special form. For example, the ion accelerator requires the propellant in the form of an ionized gas. We lump all of these functions under the propellant feed system in Fig. 10.7. In all propulsion systems, it is necessary to convert the energy stored onboard the vehicle into a form that is compatible with the particular mechanisms used to accelerate the exhaust particles. Thus there must be an energy conversion system as shown in Fig. 10.7. The energy conversion system may range from a single combustion chamber, wherein the energy of chemical

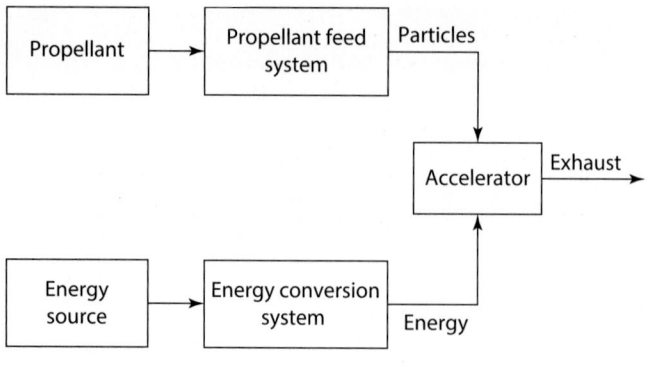

Fig. 10.7   Basic features of a rocket propulsion system.

bonds is released, to a nuclear–electric power plant. For the purposes of further discussion, we divide propulsion systems into two broad categories: 1) thermal systems and 2) electric systems.

10.3.1 Thermal Propulsion Systems

The basic principle of a thermal propulsion system is the paragon of simplicity. Energy from a chemical reaction, or from a nuclear reactor, or even from an electric arc discharge, is used to elevate the temperature of the propellant. The individual particles of the propellant thereby obtain considerable kinetic energy in the form of random thermal motion. The propellant is then expanded through a convergent-divergent nozzle whose purpose is to convert the random thermal energy of the propellant into a more or less unidirectional stream of high-speed particles. In this context, a thermal system is a hot gas generator with a nozzle for an accelerator. The nozzle-accelerator simply tries to make an ordered state of affairs out of the chaos of random thermal motion.

A key parameter in the analysis of propulsion systems in general is specific impulse I_{sp}. Thermodynamic analysis indicates, under some ideal assumptions, that the specific impulse of a thermal system is proportional to the quantity $T_c/\mathcal{M}$, where T_c is the propellant combustion chamber temperature and $\mathcal{M}$ is its molecular weight. To produce a high specific impulse in a thermal system, it is then desirable to have a high operating temperature in connection with a low molecular weight of exhaust products. The extent to which these two desirable objectives can be achieved is a rough basis for comparing thermal propulsion systems.

10.3.1.1 Chemical Propulsion Systems

Chemical propulsion systems are a subclass of thermal systems that use the energy released by exothermic chemical reaction in a combustion

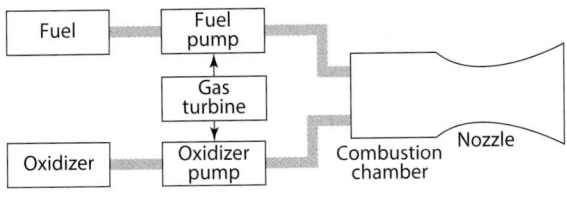

Fig. 10.8a Liquid-bipropellant rocket engine.

chamber. Chemical systems may be further divided into liquid-bipropellant engines, solid-propellant engines, and hybrid engines.

Figure 10.8a shows the essential features of a liquid rocket system. Two propellants (an oxidizer and a fuel) are pumped into the combustion chamber. The hybrid engine is shown in Fig. 10.8b, and the propellant is a solid fuel around the combustion chamber. We may summarize this system as follows:

1. *Energy conversion system*: combustion chamber
2. *Propellant feed system*: fuel pump, oxidizer pump, and gas turbine (not required in hybrid system)
3. *Accelerator*: nozzle

Figure 10.9 shows the essential features of a solid-propellant propulsion system. In this case, the fuel and oxidizer are mixed together and cast into a solid mass called the *grain*. The grain is usually formed with a hole down the middle called the perforation and is firmly cemented to the inside of the combustion chamber. After ignition, the grain burns radially outward, and the hot combustion gases pass down the perforation and are exhausted through the nozzle.

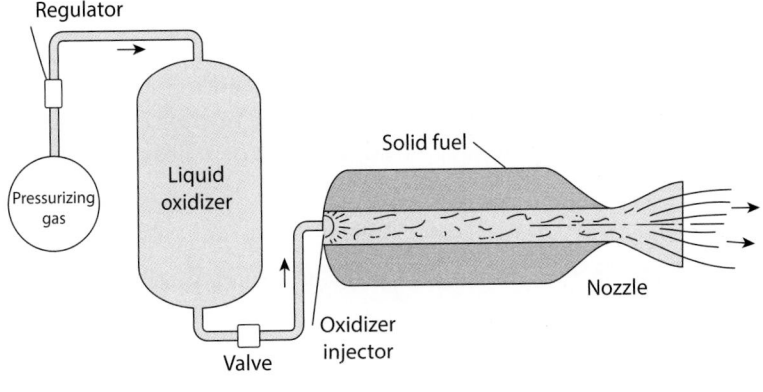

Fig. 10.8b Hybrid propulsion rocket engine.

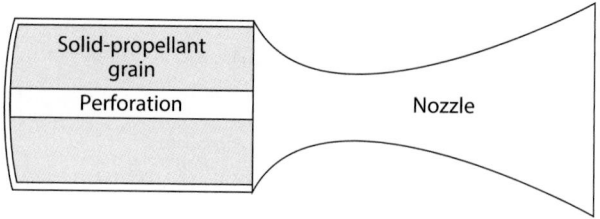

Fig. 10.9   Solid-propellant rocket engine.

We may summarize a solid-propellant system as follows:

1. *Energy conversion system*: combustion chamber
2. *Propellant feed system*: none
3. *Accelerator*: nozzle

The absence of a propellant feed system in the solid-propellant chemical rocket is one of its outstanding advantages.

Table 10.5 gives the pertinent characteristics of several different chemical propellants. It is important to note that the combustion temperature T_c and the molecular weight of exhaust products M are largely determined by the specific chemical reaction taking place in the combustion chamber. It is possible to select different combinations and thereby achieve an increase in specific impulse, but there are limitations because the products of chemical combustion inherently tend to have relatively high molecular weights.

This discussion serves to point out an important characteristic of chemical propulsion systems, namely, the oxidizer and fuel serve as both the source of energy and the source of particles for the propulsion system. Thus the process of supplying energy to the accelerator and the process of supplying particles to the accelerator are closely related. These fundamental facts place an upper bound of about 400 s on the specific impulse I_{sp} obtainable by chemical means.

10.3.1.2 Nuclear Heat Transfer Propulsion Systems

Figure 10.10 shows the essential features of the nuclear heat transfer propulsion system. As the name suggests, this system contains a nuclear reactor

Table 10.5   Characteristics of Chemical Propellants

Oxidizer and Fuel	T_c, K	M	I_{sp}, s
O_2–kerosene	3400	22	260
O_2–H_2	2800	9	360
F_2–H_2	3100	7.33	400
Nitrocellulose	2300–3100	22–28	200–230

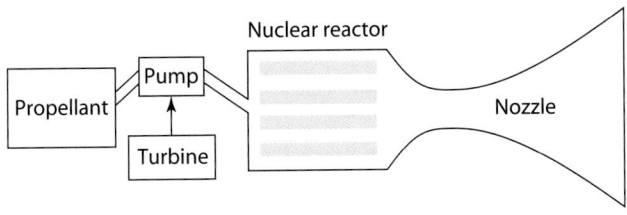

Fig. 10.10   Nuclear heat transfer propulsion system.

that serves two important purposes: 1) it produces the energy necessary to heat the propellant gas to a high temperature, and 2) it transfers this energy to the propellant gas. Thus the reactor serves as both an energy source and a heat exchanger. Once heated in the reactor, the hot propellant gas is expanded through a nozzle in the manner characteristic of all thermal systems. The major components of this system are:

1. *Energy conversion system*: solid core nuclear reactor
2. *Propellant feed system*: pump and turbine
3. *Accelerator*: nozzle

An important feature of the nuclear heat transfer system is that the source of exhaust particles and the source of energy are independent. The propellant supplies the particles and the reactor supplies the energy. Thus the propellant need not be selected on the basis of its energy content, but can be selected on the basis of its suitability as exhaust material for a thermal propulsion system. In other words, the primary consideration in the selection of a propellant is low molecular weight.

Table 10.6 lists some important characteristics of different propellants for nuclear propulsion. Note that hydrogen produces the highest specific impulse because it has the lowest molecular weight. This fact is the primary reason that hydrogen is most often mentioned as the propellant for nuclear systems. Note also that the temperature T_c is the same for all three propellants listed in the table. The temperature T_c is determined primarily by the structural limitations of the reactor–heat exchanger and not by the nature of the propellant.

Table 10.6   Characteristics of Propellants for Nuclear Propulsion

Propellant	T_c, K	M	Max I_{sp}, s
H_2	2800	2	1000
C_3H_8 (propane)	2800	5.8	530
NH_3 (ammonia)	2800	8.5	460

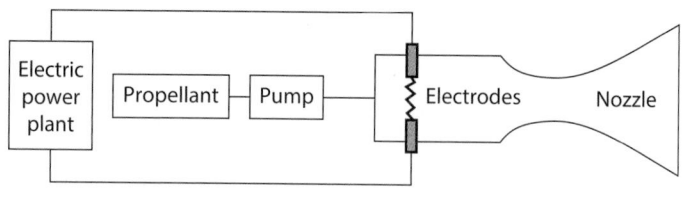

Fig. 10.11   Thermal arcjet propulsion system.

10.3.1.3 Electrothermal Propulsion Systems

Electrothermal propulsion systems use electric power to heat a propellant gas. One method of doing this is to pass the propellant around an electric arc as suggested by Fig. 10.11. The resulting propulsion system is sometimes called a thermal arcjet. A second possibility is to use tungsten heater elements (temperature increased by electrical energy) as a mechanism for imparting thermal energy to the propellant. In either case, the propellant can be heated to a rather high temperature, perhaps as high as 10,000 K. The temperature could be made higher, but beyond about 10,000 K, a significant portion of the input energy is used up in ionization and disassociation of the atoms of the propellant. This energy (called *frozen flow losses*) is, therefore, not available for the basic purpose of the device, which is to impart kinetic energy to a directed stream of particles. We should also add that at very high temperatures, the exhaust stream would contain significant pieces of the electrodes and nozzle. These two effects alone tend to place an upper bound on the temperature to which the propellant may be heated, and hence there is an upper bound ($\sim$2500 s) on the specific impulse of the electrothermal devices. The major components of the system are:

1. *Energy conversion system*: electric plant and arc chamber
2. *Propellant feed system*: propellant pump
3. *Accelerator*: nozzle

The electrothermal system is also characterized by the fact that the propellant plays no part in the process of supplying energy to the accelerator. The system also requires an electric power plant in the energy conversion system, and this feature is an important consideration in the design of vehicles propelled by electrothermal means. Because of the electric power plant, this system is often classified as an electric propulsion system. We have called it a thermal system because of the characteristic pattern of a high-temperature gas plus expansion through a nozzle.

10.3.2 Electric Propulsion Systems

Electric propulsion systems are characterized by an accelerator that makes use of the interaction of electromagnetic fields and charged particles.

In a rather crude sense, the electric charge on a particle provides a way for an electromagnetic field to regulate and direct the motion of the particle. In particular, it is possible to accelerate the particle to a high speed and eject it from a rocket vehicle. As should be clear by now, the ejection of high-speed particles is the very essence of rocket propulsion.

This simple picture of an electric propulsion system conceals some rather basic subtleties. First, the propellant feed system for such a system must do more than merely transport the propellant to the accelerator. It must also operate on the propellant to produce significant amounts of ionized particles at the entrance to the accelerator. Second, we note that electromagnetic fields are capable of accelerating charged particles to a very high specific impulse. However, it is necessary to supply the energy to the accelerator in electric form, and hence it is necessary to carry an electric power plant along on the mission. The electric power plant is the key to the problem because estimates of the power level for propulsion run into the megawatt range. In other words, the energy conversion system for an electric propulsion system is quite likely to be a very large and massive device. We will have more to say about this shortly.

10.3.2.1 Electromagnetic Propulsion Systems

Electromagnetic systems use a magnetic field to accelerate a collection of gas, called a plasma, to a speed on the order of 50,000 m/s (150,000 ft/s). A plasma is a fully or partially ionized gas containing essentially equal numbers of electrons and ions. Because the plasma is a rather good conductor of electricity, it will interact with electromagnetic fields. Because the plasma is also a gas, it displays the properties of a continuum or fluid. As a result, the basic description of the interaction of a magnetic field and a plasma combines all of the pleasantries of electromagnetic field theory and fluid mechanics into a whole new field of study called magnetohydrodynamics (MHD), or sometimes, magneto-fluid mechanics.

In this new field, it is quite possible to write down the governing equations, but it is quite another matter to solve them. The factors involved in the study of MHD include the electrical and fluid properties of the plasma, the time history of the magnetic field (e.g., pulsed or continuous), and the geometry of the accelerator. The upshot of all of this is that there are MHD forces acting on the plasma. These forces may act on the boundary of the plasma or may be distributed throughout the volume. In any case, the basic problem is to use these forces to drive the plasma out of the vehicle and thereby produce useful propulsive thrust.

Because of the variety of ways in which MHD forces can be made to act on a plasma, there is a bewildering array of plasma accelerators. The list of devices is constantly growing, and it is quite hopeless to try to classify them in any simple way that would do justice to the subject. In this case, we must be content with the schematic picture of the electromagnetic system shown in Fig. 10.12. We should recognize that the plasma generator

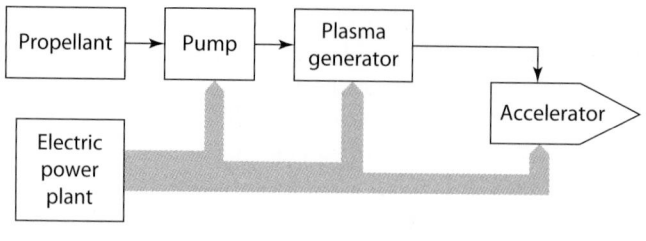

Fig. 10.12   Plasma propulsion system.

and plasma accelerator shown in the figure represent a variety of possible devices. However, they are all united by the joint feature of using the action of a magnetic field on a macroscopically neutral plasma. The major components of the system are:

1. *Energy conversion system*: electric power plant
2. *Propellant feed system*: pump and plasma generator
3. *Accelerator*: plasma accelerator

The terms MHD propulsion and plasma propulsion are often associated with this system. Although the specific impulse range for this system is not definitely established, the range of 2000–10,000 s is a reasonable consensus of the values given in open literature. Hydrogen, helium, and lithium are mentioned as propellants.

There is an important feature of the plasma system that is well worth mentioning at this point. The exhaust stream of this system is electrically neutral, and hence the neutralization problems of the ion engine, which we discuss next, are avoided in this case.

10.3.2.2 Electrostatic Propulsion Systems

The operating principle of this system, commonly referred to as an *ion engine*, is based on the static electric fields to accelerate and eject electrically charged particles. Figure 10.13 shows the main features of an ion engine. The propellant introduced into the engine is electrically neutral because charged particles cannot be stored in appreciable amounts. The ions are produced immediately before they are exposed to the accelerating field. This is accomplished by stripping off one or more electrons from the neutral propellant molecule, which is then a single- or multiple-charged positive ion. The positive ion is then accelerated by the electric field and ejected from the engine at a high speed. During this process of acceleration, it is necessary to focus and shape the ion beam to minimize the number of collisions between the ions and the structural members of the engine. Finally, both ions and electrons must be ejected from the engine to keep it electrically neutral. This is accomplished by gathering up the electrons available at the ion source and transporting them to the engine exit where they are injected into the ion stream, thus hopefully producing an electrically neutral propulsive beam. The three major parts of an ion engine are the ion source, accelerator section, and beam neutralizer (electron injector).

In the simplest terms, a charged particle acquires a kinetic energy equal to its loss of electric potential energy when it "falls" through a difference in electric potential. This basic idea leads to the following expression for the exhaust speed on an ion engine:

$$\frac{1}{2}m_i V_e^2 = q\Delta v \qquad (10.32)$$

where

Δv = net accelerating voltage, V
q = electric charge on the ion, C (Coulombs)
m_i = mass of the ion, kg
V_e = exhaust speed, m/s

The factors affecting the exhaust speed are the net accelerating voltage Δv and the charge to mass ratio q/m_i of the exhausted particles.

To have a compact, efficient engine, it is desirable to have a rather high accelerating voltage, something on the order of several tens of thousands of volts. With this constraint in mind, we can investigate the influence of charge-to-mass ratio on the selection of the propellant.

The value of specific impulse for the ion engine is several orders of magnitude higher than that obtainable from thermal systems (see Example 10.4). Thus it appears that the ion engine can produce incredibly high specific impulses.

However, several other considerations enter the picture to modify this apparently rosy situation. Performance analysis indicates there is an upper bound on the specific impulse so that the system has the capability of accomplishing useful missions. The reasons for this upper bound and

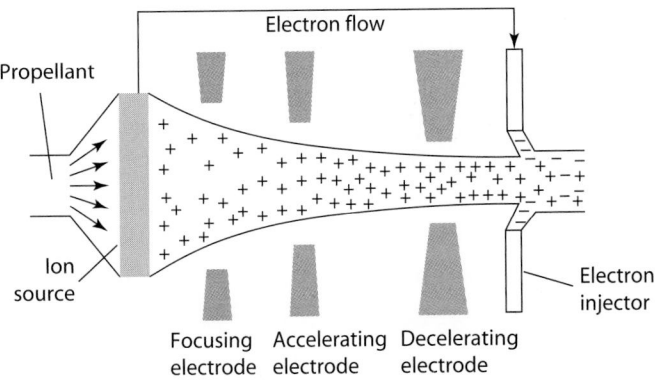

Fig. 10.13   Sketch of ion engine (electrostatic accelerator).

Example 10.4

If we assume 1) that the ions are protons (disassociated and ionized hydrogen, $q/m_i = 0.91 \times 10^8$ C/kg) and 2) a reasonable accelerating voltage of 10,000 V, we obtain an exhaust speed of 1.34×10^6 m/s. This corresponds to a specific impulse of 134,000 s.

the various factors involved will not emerge until one takes up the discussion of rocket performance. It suffices to say, at this point, that the ion engine is capable of producing specific impulses that are too high to be useful, and that a reasonable upper bound on specific impulse is something on the order of 100,000 s [31].

This situation points out one of the very profound differences between ion propulsion and thermal propulsion. In the latter, the propellant is often selected to make the specific impulse as high as possible. With ion propulsion, the problem is to select a propellant that will hold the specific impulse (at reasonable accelerator voltages) down to a useful value. For this reason, our interest centers on the heavier elements in the periodic table.

An additional consideration is the fact that the propellant must be supplied to the accelerator in ionized form, and this involves the expenditure of a certain amount of energy that is, therefore, not available for the purpose of propulsion. The amount of energy required for this purpose is measured by the ionization potential of an atom. Table 10.7 gives some important properties of the alkali metals. The alkali metals are the most promising propellants, because they combine the desirable features of low ionization potential together with reasonable accelerating voltages at useful specific impulses.

Figure 10.14 shows a schematic diagram of an electrostatic propulsion system. The major components may be identified as:

1. *Energy conversion system*: electric power plant
2. *Propellant feed system*: pump and ion source
3. *Accelerator*: ion engine

For obvious reasons, this system is also called an ion propulsion system. The preceding discussion points out there is an upper limit on the specific

Table 10.7 Characteristics of Propellants for Electrostatic Propulsion

Propellant	First Ionization Potential, ev	Atomic Weight	Accelerating Voltage at $I_{sp} = 10^5$ s
Sodium	5.14	23	12,000 V
Rubidium	4.18	85.4	45,000 V
Cesium	4.89	133.0	69,000 V

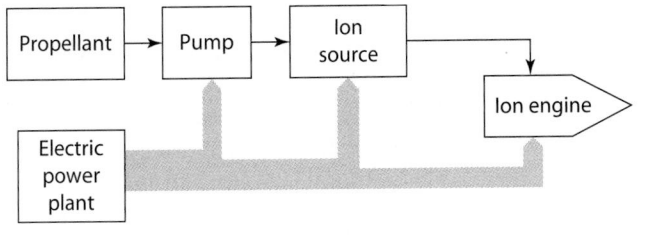

Fig. 10.14   Major features of an electrostatic (ion) propulsion system.

impulse for an ion system. It will also be difficult to obtain efficient operation at lower specific impulses. This lower limit arises due to a number of practical considerations such as electrode erosion at low accelerator voltages and radiation from the hot surfaces of the ion source. The lower limit appears to be about 7500 s.

10.3.2.3 Electric Power Plants for Space Propulsion

A major component in any electric propulsion system is the electric power plant, and so we examine the situation regarding the generation of power in space. Electrical power will be required for a variety of reasons, including communication, control, life support, and propulsion. There are a great many methods of power generation that could satisfy some of these needs. Our interest is the generation of electric power for the purposes of propulsion. The very stringent requirements of electric propulsion narrow the field of interest.

One important factor is the power levels of interest in electric propulsion. For Earth satellite missions, such as altitude control, the power levels are on the order of 50–100 kW. More ambitious missions, such as lunar flight, require 1–3 MW. Manned interplanetary flight will require 20–100 MW of electric power. As a matter of perspective, we note that the installed capacity of Hoover Dam is 2000 MW. In other words, we are talking about electric power plants that are comparable to Earth-bound station applications that must operate for long periods of time in the rather unfriendly environs of space.

Our discussion will center on two parameters of space power plants: efficiency and specific mass α_c. Specific mass is defined as the mass of the power plant per unit output (electrical) power, or

$$\alpha_c = \frac{m_c}{P_c} \tag{10.33}$$

where

α_c = the specific mass, kg/W
P_c = the electrical output power, W
m_c = the total mass of the power plant, kg

All space power plants consist of at least three basic elements: an energy source, a conversion device, and a waste heat radiator. The only two energy sources that seem to offer much hope of producing the power levels of interest are the sun and the nuclear fission reactor. Solar collectors for the megawatt range become extremely large and must be constructed in a very sophisticated (i.e., flimsy) manner. Such devices could stand only very small accelerations, and meteorite erosion of polished surfaces would be a problem. Nuclear power reactors are very powerful sources of energy. They have the important advantage that the power output is limited primarily by the capability of the reactor coolant to remove heat. They have a low specific mass and long life at high power levels. On the other hand, the reactor produces a radiation hazard for the crew and payload.

Both the solar collector and the nuclear reactor produce thermal energy. Among the possible methods of converting thermal energy into electrical energy are the following: thermionic, thermoelectric, and turboelectric. These three methods of conversion use vastly different mechanisms, but all three have one common feature—very low efficiency.

The inefficiency of power conversion schemes results in an increase in the specific mass of the power plant for two important reasons. First, the overall size and mass of the energy source must be increased to "supply" the losses. Second, the losses must somehow be dumped overboard by a waste heat rejection system. For the times of operation and power levels of interest in space propulsion, the only mechanism available for the rejection of heat waste is radiation into space. Because of the inherent low efficiency of the system, the waste heat radiator will be a very large and massive device. It will be susceptible to meteorite puncture and the attendant loss of working fluid from the system. For many designs, it appears that the waste heat radiator will be the dominant mass in the power plant for outputs exceeding a few hundred kilowatts.

Table 10.8 gives some projected characteristics of space power plants. As we have noted, all of the systems have low efficiency. The values of specific mass given in the table should be taken as tentative, particularly in the 1–10 MW range, because systems of this power level have not been built. Authorities differ widely in their estimate of the situation, and there is considerable uncertainty in the figures.

Table 10.8   Characteristics of Space Power Systems

| System | Efficiency | Specific Mass: α_c (kgm/W) | |
		10–100 kW	1–10 MW
Turboelectric	10–20%	0.02–0.04	0.003–0.004
Thermionic	5–10%	0.5–1.0	0.003–0.004
Thermoelectric	5–10%	0.5–1.0	—

In all of this, we should not lose sight of the fact that the purpose of the propulsion system is to impart kinetic energy to a directed stream of particles. In electric propulsion, this energy first appears as thermal energy and then must be converted into electrical energy before it is in a form suitable for the basic task at hand. As we have suggested, the process of converting thermal energy into electricity in space is not noted for its efficiency. For the foreseeable future, it appears that electric propulsion will continue to pay a severe penalty for the several conversion processes involved in the generation of electric power. It is no exaggeration to state that the use of electric propulsion will be paced by the development of lightweight, efficient power conversion systems.

10.3.3 Summary of Propulsion Systems

There are a number of ways in which propulsion systems can be classified to indicate certain fundamental features. In our discussion we have used the classifications of *thermal* and *electric*. This classification is based on the mechanism used by the accelerator to impart kinetic energy to the exhaust stream. Thus propulsion systems may be summarized as follows:

Thermal:
1. Chemical
2. Nuclear heat transfer
3. Electrothermal (thermal arc jet)

Electric:
1. Electrostatic (ion)
2. Electromagnetic (plasma)

There are other fundamental features of propulsion systems of no less importance, and consideration of these leads to other possible classifications. For example, in our discussion we noted that in certain systems, the propellant does not supply energy to the accelerator; that is, the source of energy and the source of particles are separate. Such propulsion systems are called separately powered. On the basis of this feature, we have the following classification:

Separately powered:
1. Nuclear heat transfer
2. Electrothermal (thermal arc jet)
3. Electromagnetic (plasma)
4. Electrostatic (ion)

Non-separately powered:
1. Chemical

Note that the chemical propulsion system is termed non-separately powered because the propellant supplies both energy and particles to the accelerator.

Propulsion systems can also be classified according to their performance capabilities as measured by the two key parameters, I_{sp} and thrust-to-weight (sea-level-weight) ratio. Table 10.9 presents typical values of these quantities for various propulsion systems. Note the tradeoff between specific impulse and thrust-to-weight for the rocket propulsion systems shown in Table 10.9.

The data in Table 10.9 offer several new ways to classify propulsion systems. For example, specific impulse values range over a fairly continuous spectrum from 200 (the lower end of the chemical range) up to 100,000 (the upper end of the ion range). Thus the order in which the systems are listed in Table 10.9 is a classification based on magnitude of specific impulse. There is, of course, some overlap, but in large measure, each of the systems fills a gap in the specific impulse spectrum.

A very important distinction between propulsion systems is found in Table 10.9. Note that there are several orders of magnitude difference between the thrust-to-weight ratio of electric propulsion systems and the chemical or nuclear propulsion systems. This leads to a classification of systems as either high thrust or low thrust.

High-thrust systems:
1. Chemical
2. Nuclear heat transfer

Low-thrust systems:
1. Electrothermal
2. Electromagnetic
3. Electrostatic

Note that all of the low-thrust systems require an electric power plant as the energy conversion system. This fact accounts for the low values of thrust-to-weight ratios for these systems. Significant increases in the thrust-to-weight capabilities of these systems can be achieved only by reducing the weight-per-unit power of space power plants.

Figure 10.15 is a summary of propulsion systems obtained by plotting the data in Table 10.9. Notice the very clear distinction between the high-thrust and low-thrust systems. Notice also that there is a fairly continuous range of specific impulse capability from 200 to 100,000 s with some significant overlapping among the electric propulsion systems.

Table 10.9 Rocket Performance Summary

Propulsion System	I_{sp}, s	Thrust-to-Weight Ratio
Chemical	300–400	up to 10^2
Nuclear heat transfer	400–1000	up to 30
Electrothermal	400–2500	up to 10^{-2}
Electromagnetic	2000–10^4	up to 10^{-3}
Electrostatic	7500–10^5	up to 10^{-4}

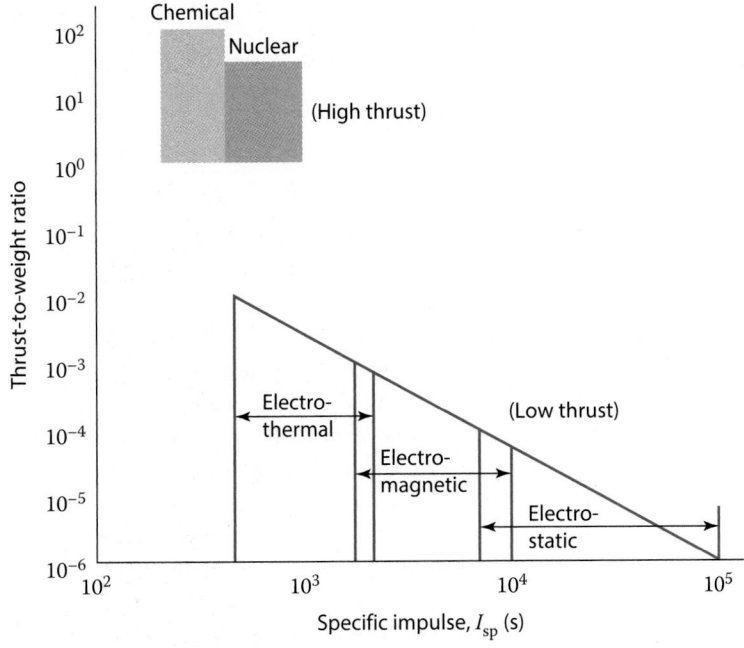

Fig. 10.15   Thrust-to-weight ratio vs specific impulse.

10.4 Types of Rocket Nozzles

The initial geometry of a circular-section rocket nozzle is fixed by the nozzle inlet, throat, and exit areas. The inlet area A_c is established by combustion chamber design considerations. The throat and exit areas are determined from a knowledge of the combustion chamber temperature and pressure, acceptable mean values of γ and R, the propellant mass flow rate, and the designed nozzle-pressure ratio $P_c/P_a = P_c/P_e$. (Review Section 3.7.1.2 as necessary.) With these data, the nozzle throat and exit areas can be obtained by application of the mass flow parameter (MFP). Having established A_c, A_t, and A_e (or the radii R_c, R_t, R_e), the problem of the nozzle contour design remains. Because of weight and space limitations, determination of the nozzle shape reduces to the problem of choosing a contour providing minimum length (a measure of volume and weight) and yielding maximum thrust. In addition, the selected contour must be sufficiently simple to manufacture.

10.4.1 Conical Nozzle

The conical nozzle represents a compromise of the length, thrust, and ease of manufacturing design criteria weighted somewhat in favor of the last factor. A conical nozzle consists of two truncated cones (Fig. 10.16),

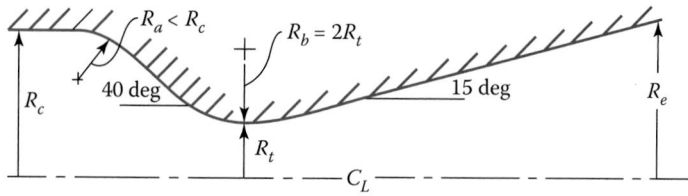

Fig. 10.16   Conical convergent-divergent nozzle.

joined top to top along their axis by a suitable radius to form the nozzle throat. The combustion chamber is similarly faired into the convergent nozzle section. The converging contour of the nozzle is not critical in regards to the flow, and a rather rapid change in cross section is permissible here with a conical apex half-angle on the order of 40 deg commonly used. The divergence angle of the supersonic portion of the nozzle, however, is limited by flow separation considerations and must not exceed a value of about 15 deg, discussed in the next paragraph. For divergence angles too much greater than 15 deg, the flow will separate from the nozzle walls short of the exit even when the nozzle is operating at design altitude $P_a = P_e$. Conversely, for a given divergence angle, the flow will separate if the nozzle back-pressure ratio P_a/P_e is too high ($P_a/P_e > 1$ when a nozzle is operating at an altitude lower than the design altitude). When separation occurs, an oblique shock wave is located inside the nozzle (Fig. 10.17), and the gas flow is turned and contained within the jet boundary or slip line as indicated. Figure 10.17 depicts the real flow situation that occurs (other than under ideal laboratory conditions) in place of normal shocks near the exit section as indicated in Fig. 3.46.

Summerfield has established a criterion for judging whether separation is likely to occur in the exit section of a conical nozzle [32]. The criterion is based on an accumulation of experimental separation data points from many sources. The experimental data give the pressure ratios P_a/P_{we} (P_{we} = wall pressure at exit = P_e for 15-deg conical nozzle) that produced

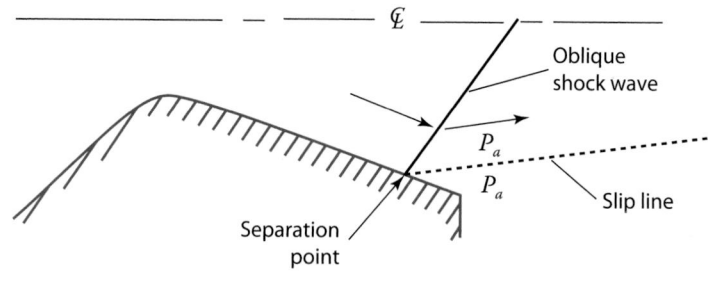

Fig. 10.17   Separation in overexpanded nozzle.

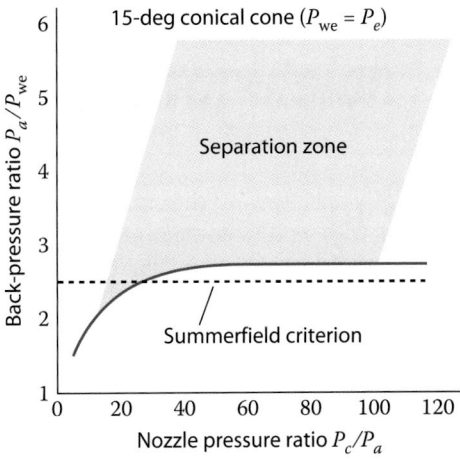

Fig. 10.18   Summerfield separation criterion.

flow separation in tests of 15-deg conical nozzles for various nozzle-pressure ratios. The data points are scattered throughout the cross-hatched region of Fig. 10.18. The Summerfield criterion states that for back-pressure ratios greater than 2.5, separation is likely to occur in the exit of a 15-deg conical nozzle for nozzle-pressure ratios in excess of 20. Current data on large rocket nozzles indicate that the critical back-pressure ratio for these large nozzles is 3.5. This increase in the permissible back-pressure ratio for no separation is to be expected as nozzle sizes increase because the boundary layer forms a smaller percentage of the total flow for larger nozzles. Using this modified Summerfield criterion value of 3.5, it should be noted that for large rocket nozzles, separation occurs when $P_a > 3.5 \, P_{we}$. For a 15-deg conical nozzle, $P_{we} = P_e$ because the stream properties at the exit section are essentially uniform at the values determined by 1-D flow analysis. In some nozzles, the streams are not uniform across the exit section and $P_{we} > P_e$ (Fig. 10.19). Separation will occur sooner (at lower values of P_a) in a conical nozzle than in those of the same area ratio for which $P_{we} > P_e$.

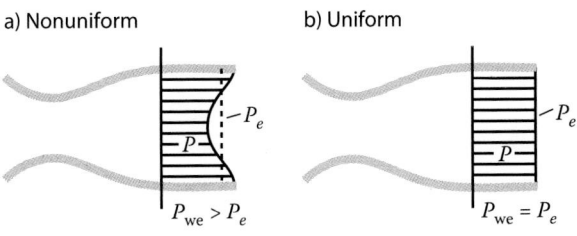

Fig. 10.19   Pressure profile at nozzle exit.

10.4.2 Bell Nozzle

The bell-shaped nozzle of the Atlas sustainer engine (shown in Fig. 10.20) is designed to reduce the thrust and length disadvantages of a conical nozzle. To reduce length, a bell-shaped nozzle employs a high divergence angle at the throat with a very rapid expansion of the gases from the throat. The flow is then turned rather abruptly back toward the axial direction. The comparative lengths of a bell-shaped (L_B) and a conical (L_C) nozzle having the same expansion ratio are indicated in Fig. 10.20. The bell-shaped contour is used on several current engines. The bell-shaped nozzle on the H-1 Saturn 1B engine has a length 20% less than the length that would be required for a 15-deg conical nozzle with the same 8:1 area ratio.

10.4.3 Free-Expansion (Plug) Nozzle

Figure 10.21 depicts the performance loss of conventional convergent-divergent (C-D) nozzles at off-design pressure ratios. To avoid the large penalties (Fig. 10.21) at liftoff pressure ratios, relatively low-area-ratio nozzles must be used. Note that high-area-ratio nozzles, however, improve performance at altitude. (An Atlas booster, used for low-altitude operation, has an area ratio of 8:1. An Atlas sustainer, used for high-altitude operation has an area ratio of 25:1.) Use of a high-altitude nozzle, such as the Atlas sustainer, at liftoff incurs a large performance degradation. To avoid such losses at off-design operation, a rocket nozzle with an adjustable area ratio that provides optimum expansion at each ascending altitude is needed (see

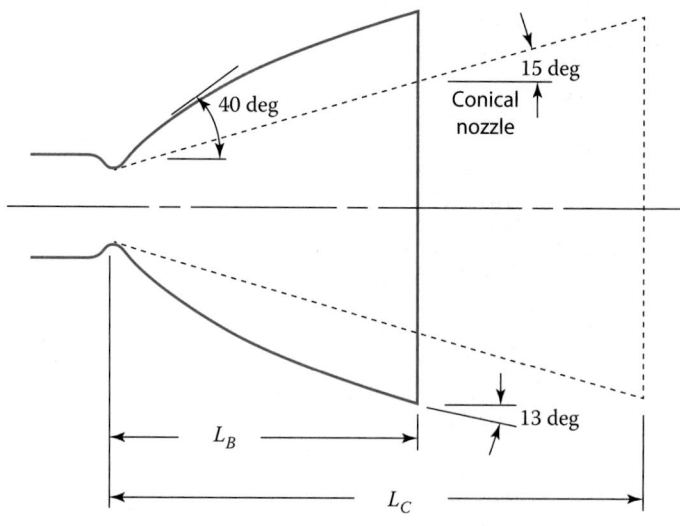

Fig. 10.20 Bell-shaped nozzle $A_e/A_t = 25$ (Atlas sustainer engine).

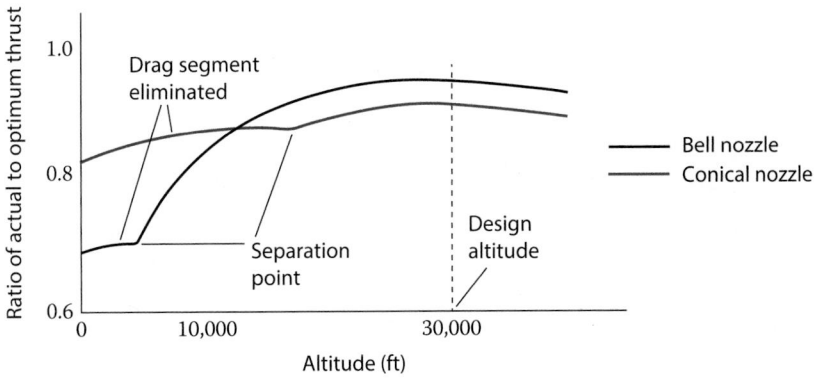

Fig. 10.21   Ratio of actual-to-optimum thrust for bell-shaped and conical nozzles vs altitude (pressure ratio).

Section 10.1.5). Obviously, a variable area C-D nozzle is not feasible because of hardware considerations. It is possible, however, to design a nozzle so that the expanding flow is not bound by immovable solid walls. In such free-expansion nozzles (Figs. 10.22 and 10.23), the expanding flow is bound by a solid surface and a free-to-move slip-line expansion surface. The adjustable slip-line boundary, in effect, produces a variable area ratio nozzle that accommodates itself to changing nozzle pressure ratios. Free-expansion nozzles, therefore, tend to operate at optimum expansion, with the ratio of thrust-to-optimum-thrust being quite insensitive to altitude variations

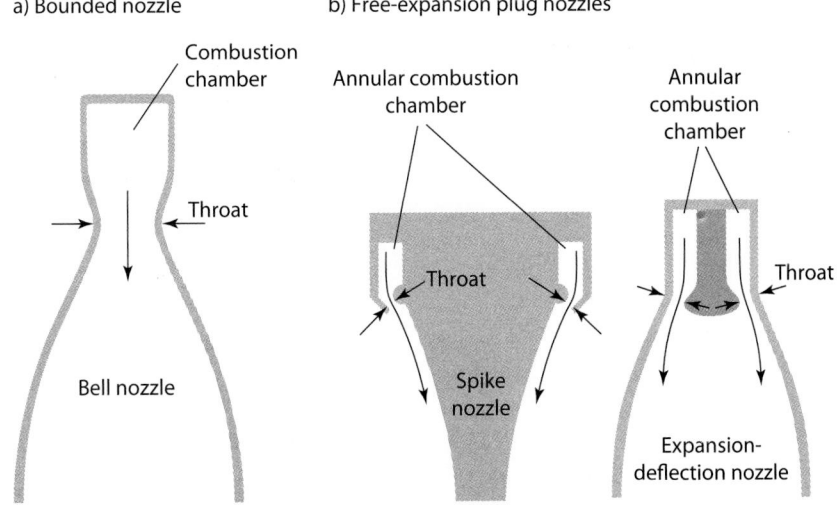

Fig. 10.22   Bounded and free-expansion nozzles.

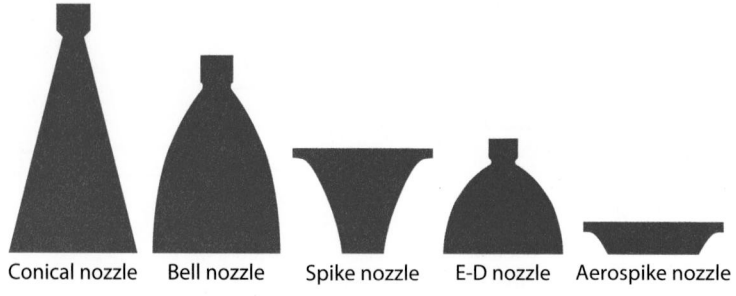

Conical nozzle   Bell nozzle   Spike nozzle   E-D nozzle   Aerospike nozzle

Fig. 10.23   Nozzle type and size comparison.

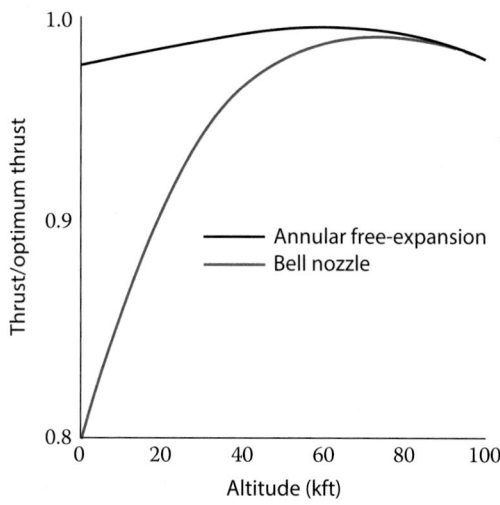

Fig. 10.24   Performance comparison of a bell nozzle and a free-expansion nozzle.

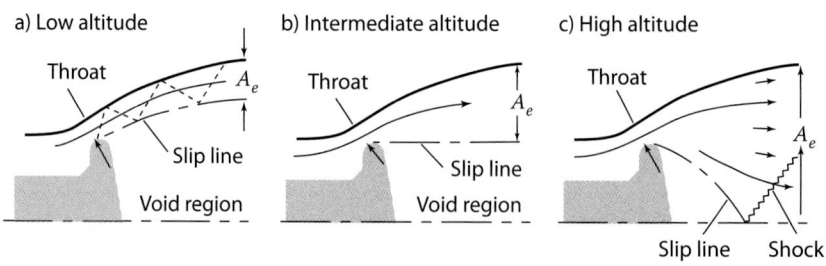

Fig. 10.25   Exhaust flow pattern from expansion-deflection (ED) plug nozzle.

a) Below design pressure ratio (low altitude)

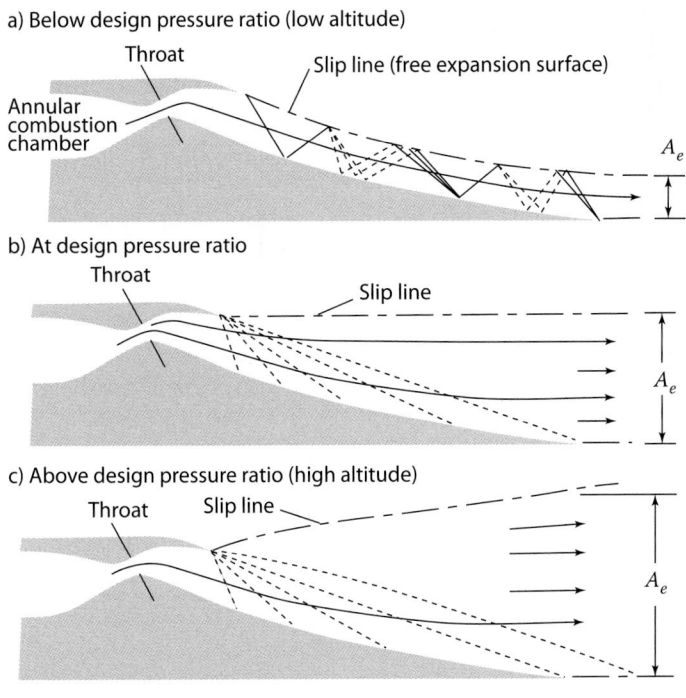

Fig. 10.26   Exhaust flow pattern from annular free-expansion nozzle.

(Fig. 10.24). The free-expansion nozzles shown in Fig. 10.22 have been dubbed "plug nozzles." The manner in which the flow adjusts to varying back-pressures for the two types of plug nozzles of Fig. 10.22 is shown in Figs. 10.25 and 10.26.

Note the performance gain of a free-expansion-type nozzle in the test data curves of Fig. 10.24. The improved performance of the free-expansion type of nozzle over the bell nozzle at lower off-design altitudes is evident. The absence of large thrust losses at lower altitudes suggests that a free-expansion-type nozzle would be particularly suited for use in both boosters or in single-stage ballistic missiles. However, this type of nozzle is more costly and complex to manufacture than the current conical nozzles.

10.5   Parameters for Chemical Rockets

The nozzle mass flow, ideal thrust equation, characteristic velocity, exit velocity, and area ratio for isentropic flow follow from the relations developed in Chapter 3. The purpose here is simply to write expressions for these quantities in forms that are particularly useful for substitution into the ideal chemical rocket-thrust coefficient equation to be developed later in this section. After deriving the mass flow and ideal thrust equations desired, the characteristic velocity C^* will be introduced. The expressions for mass flow rate, ideal thrust, C^*, V_e, and A_e/A_t will be written in terms of quantities relating to rocket nozzle hardware and gross parameters, such as chamber pressure and temperature. To identify the flow properties more closely with the physical hardware, the following notations will be adopted: the subscript c will denote combustion chamber; thus, P_c and T_c will designate the combustion chamber temperature and pressure, which we will assume corresponds to the total pressure and temperature of the ideal flow. The nozzle throat conditions will be represented by a subscript t and exit section quantities by a subscript e. Because the flow in the ideal rocket nozzle is isentropic, with sonic flow at the throat, we have $A_t = A^*$, $P_t = P^*$, and so on.

10.5.1 Rocket Nozzle Mass Flow

The mass flow through a nozzle is conveniently determined by applying the mass flow parameter equation at the nozzle throat where $M_t = 1.0$. We have, from Eq. (3.11), using the notation just introduced,

$$\frac{\dot{m}\sqrt{T_c}}{P_c A_t} = \frac{(1)\sqrt{\gamma g_c/R}}{\left\{1 + \dfrac{\gamma-1}{2}(1)^2\right\}^{\frac{\gamma+1}{2(\gamma-1)}}}$$

$$\frac{\dot{m}\sqrt{T_c}}{P_c A_t} = \sqrt{\frac{g_c}{R}\gamma\left\{\frac{\gamma+1}{2}\right\}^{-\frac{\gamma+1}{\gamma-1}}} = \frac{\Gamma}{\sqrt{R/g_c}} \tag{10.34}$$

where

$$\Gamma = \sqrt{\frac{\gamma}{\left\{\dfrac{\gamma+1}{2}\right\}^{\frac{\gamma+1}{\gamma-1}}}} \tag{10.35}$$

We introduce Γ because it is simpler to write than the square root of many exponential terms involving γ. Γ is a function of the combustion gases' specific heat ratio alone and is given in Table 10.10 for a range of γ values. The ratio Γ evaluated for air should give the value of Fig. 3.9 at $M = 1.0$ and $\gamma = 1.4$.

To illustrate the use of Table 10.10 and the important matter of units, let us evaluate $\Gamma/\sqrt{R/g_c}$ for $\gamma = 1.4$ and $R = 53.34$ ft · lbf/lbm · °R:

$$\frac{\dot{m}\sqrt{T_c}}{P_c A_t} = \frac{\Gamma}{\sqrt{R/g_c}} = \frac{0.6847}{\sqrt{53.34/32.174}} = 0.5318\frac{\text{lbm} \cdot \sqrt{°R}}{\text{lbf} \cdot \text{s}}$$

This value agrees with the GASTAB software.

Table 10.10 Γ Tabulated vs γ

γ	Γ	γ	Γ	γ	Γ
1.14	0.6366	1.21	0.6505	1.28	0.6636
1.15	0.6386	1.22	0.6524	1.29	0.6654
1.16	0.6406	1.23	0.6543	1.30	0.6673
1.17	0.6426	1.24	0.6562	1.31	0.6691
1.18	0.6446	1.25	0.6581	1.32	0.6726
1.19	0.6466	1.26	0.6599	1.33	0.6779
1.20	0.6485	1.27	0.6618	1.34	0.6847

Solving Eq. (10.34) for the mass flow rate, we obtain the following simple expression for rocket nozzle mass flow:

$$\dot{m} = P_c A_t g_c \frac{\Gamma}{\sqrt{R g_c T_c}} \tag{10.36}$$

The mass flow rate is seen to be determined by the nozzle throat area, the combustion chamber pressure, and the quantity $(\Gamma/\sqrt{R g_c T_c})$. The latter is determined by the combustion gas temperature T_c and the combustion gases' characteristics, γ and R.

Example 10.5

Determine the mass flow rate of a gas with a molecular weight of 18 g/mole and $\gamma = 1.25$ through a choked throat area of 2 ft^2 from a combustion chamber whose pressure is 2000 psia and temperature is 3000°R. From Eq. (10.35) or Table 10.10, $\Gamma = 0.6581$. For the gas, $R = \mathcal{R}/\mathcal{M} = 1545/18 = 85.83$ ft · lbf/(lbm · °R). Using Eq. (10.36), we have

$$\dot{m} = P_c A_t g_c \frac{\Gamma}{\sqrt{R g_c T_c}} = \frac{2000 \times 2 \times 144 \times 32.174 \times 0.6581}{\sqrt{85.83 \times 32.174 \times 3000}} = 4237 \, \text{lbm/s}$$

10.5.2 Ideal Thrust Equation

The ideal thrust is given by Eq. (10.1). Using the 1-D isentropic relationship of Chapter 3 for V_e, we get

$$V_e = \sqrt{\frac{2\gamma}{\gamma - 1} R g_c T_c \left\{ 1 - \left(\frac{P_e}{P_c} \right)^{\frac{\gamma - 1}{\gamma}} \right\}}$$

Combining this equation for V_e and Eq. (10.36) for the mass flow rate with Eq. (10.1), we get the following ideal thrust equation in terms of areas, pressures, and γ:

$$F_i = P_c A_t \left\{ \Gamma \sqrt{\frac{2\gamma}{\gamma-1} \left[1 - \left(\frac{P_e}{P_c}\right)^{\frac{\gamma-1}{\gamma}} \right]} + \left(\frac{P_e}{P_c} - \frac{P_a}{P_c}\right) \frac{A_e}{A_t} \right\} \qquad (10.37)$$

10.5.3 Characteristic Velocity, C^*

The quantity ($\sqrt{Rg_c T_c}/\Gamma$) of Eq. (10.36) has the dimensions of velocity, and as just noted, its value is determined by the combustion gas temperature T_c and the combustion gases' characteristics, γ and R. For a given propellant, therefore, the quantity ($\sqrt{Rg_c T_c}/\Gamma$) is a characteristic of the reaction process of the propellant because the propellant's combustion process will establish the flame temperature T_c and the composition of the gas mixture making up the products of combustion.

The composition of the combustion gases in turn establishes γ and the molecular weight $\mathcal{M}$ (hence R, which equals $\mathcal{R}/\mathcal{M}$, where $\mathcal{R}$ is the universal gas constant). This quantity, which has the dimensions of velocity and a value characterized by the propellant combustion process, is called the characteristic velocity, by convention denoted as C^*. (The superscript star bears no relation to the star notation used for sonic flow properties.) Thus, by definition

$$C^* = \frac{\sqrt{Rg_c T_c}}{\Gamma} \qquad (10.38)$$

The characteristic velocity can be determined theoretically by the methods of thermochemistry [16]. This theoretical value can be compared with an experimentally determined value of C^* to evaluate the actual combustion process against the "theoretical" combustion process. Experimental values of C^* can be obtained in a rocket motor test by measuring the propellant mass flow rate, the chamber pressure P_c, and the nozzle throat area A_t. These data and Eq. (10.36) give the experimental value C_x^*. Thus,

Experimental:

$$C_x^* = \frac{P_c A_t}{\dot{m}/g_c}$$

with P_c, A_t, and m obtained from measured data. Also, from Eq. (10.38) defining C_i^*

Theoretical:

$$C_i^* = \frac{\sqrt{Rg_c T_c}}{\Gamma}$$

with R, T_c, and Γ obtained from thermochemistry.

A well-designed combustion chamber should give a value of C_x^* in excess of 90% of the theoretical value computed by thermochemical means and

Eq. (10.38). An experimental value of C_x^* less than 0.9 of the theoretical value would be evidence of a poorly designed combustion chamber or injector. With present-day propellants, C^* ranges from 3000 to 7000 ft/s. The value of C^* for the design point operation of the Atlas sustainer rocket motor is 5440 ft/s.

10.5.4 Nozzle Exit Velocity, V_e

The expansion flow process through an ideal nozzle from a pressure P_c to P_e is shown on the temperature-entropy diagram of Fig. 10.27. The kinetic energy imparted to the expanding gas is proportional to the vertical distance indicated. The figure also depicts the maximum possible kinetic energy that could be delivered to the expanding gas. This maximum corresponds to an expansion to zero absolute temperature.

The diagram shows that $V_e^2/(2g_c\, c_p) = T_c - T_e$. Using this and the perfect gas relations accompanying the T-s diagram of Fig. 10.27, we obtain

$$V_e = \sqrt{\frac{2\gamma}{\gamma-1}\frac{\mathcal{R}}{\mathcal{M}}g_c T_c\left\{1 - \left(\frac{P_e}{P_c}\right)^{\frac{\gamma-1}{\gamma}}\right\}} \qquad (10.39)$$

The ratio of the exit velocity to the maximum possible exit velocity is

$$\frac{V_e}{V_{\max}} = \sqrt{1 - \left(\frac{P_e}{P_c}\right)^{\frac{\gamma-1}{\gamma}}} \qquad (10.40)$$

Because a high exit velocity means, for a given nozzle mass flow, a large momentum flux and, hence, thrust, let us examine the terms of Eq. (10.39)

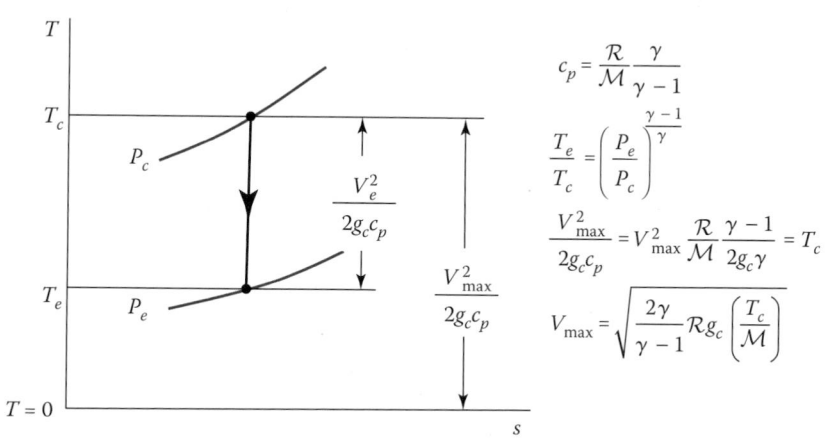

Fig. 10.27   Nozzle expansion process.

upon which V_e depends. These variables are 1) the pressure ratio P_c/P_e, 2) the chamber temperature T_c, and 3) the molecular weight $\mathcal{M}$.

Varying the specific heat ratio γ is of little practical use and will not be considered. Before discussing these factors, let us first note that for chemically propelled rockets, the original source of kinetic energy V_e^2 is the ejected material itself (i.e., the propellant's chemical energy). For chemical rockets, therefore, the choice of propellant is fixed by the selection of fuel and oxidizer. If, on the other hand, the rocket combustion chamber is replaced with a nuclear reactor, then the reactor's fuel becomes the original source of the propellant's kinetic energy. In this case, the energy is supplied by a source independent of the propellant, and one can choose a propellant without consideration of its chemical energy. This is pointed out here because T_c and $\mathcal{M}$ are somewhat interdependent for a chemical rocket but can be completely independent in other cases.

Allowance can be made for these different situations in considering the influence of T_c and $\mathcal{M}$ on V_e in the following:

1. *Pressure ratio P_c/P_e*: The pressure ratio determines what fraction of the maximum possible velocity is ideally attained. As P_c/P_e increases, V_e increases very rapidly (Fig. 10.28) up to a pressure ratio of 10. Above this value of P_c/P_e, the velocity increases less rapidly with pressure ratio. Increasing chamber pressure is thus seen to have a favorable influence on V_e. The favorable influence is limited, however, because a higher P_c means increased structural weight requirements and, for liquid chemical

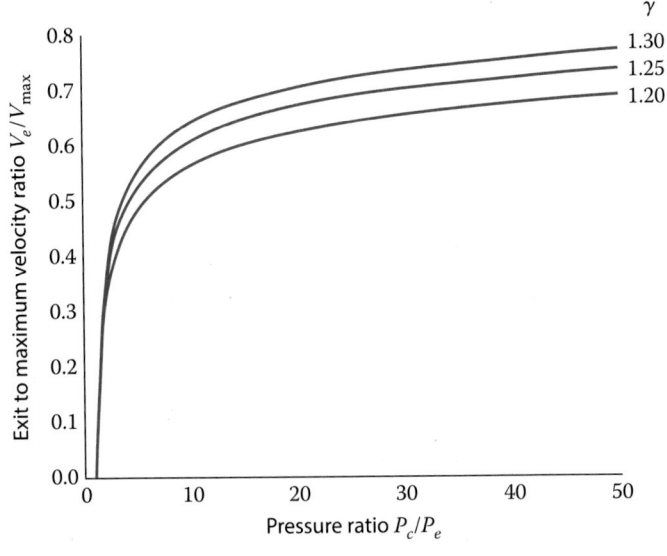

Fig. 10.28 Fraction of maximum velocity vs pressure ratio.

propellant rockets, increased propellant feed system weight. Note also that as P_c/P_e increases, the nozzle-area ratio must increase. The Atlas booster ($A_e/A_t = 8$) and sustainer ($A_e/A_t = 25$) operate with a pressure ratio (P_c/P_e) of 60 and 250, respectively.

2. *Chamber temperature* T_c: The exit velocity V_e increases as the square root of the chamber temperature T_c. The upper limit on T_c is fixed by either structural (in the case of a nonchemical nuclear rocket) or chemical considerations. Usually, in a liquid chemical rocket, cooling alleviates the structural limitation so that T_c is limited by chemical considerations. From a chemical point of view, it would appear that fuels with greater chemical energy will give an increasingly higher temperature T_c. This is true up to a temperature where dissociation of the combustion products—CO_2 dissociating into C and O_2, for example—occur and absorb energy that would otherwise be available for increasing T_c. Thus we find that for the best fuels, dissociation sets an upper temperature limit of about 6000°R.

3. *Molecular weight* $\mathcal{M}$: Because V_e varies inversely as the square root of $\mathcal{M}$, propellants with a low molecular weight are preferable. For chemical rockets, the molecular weight of the propellant is determined by the fuel and oxidizer and is on the order of 20. In a nuclear rocket, one has a freer choice of a propellant than in a chemical rocket, and hydrogen with a molecular weight of 2 can be used. A nuclear rocket, using H_2 for a propellant and having the same temperature T_c and pressure ratio P_c/P_e as a chemical rocket with $\mathcal{M} = 20$, would have a higher V_e factor $\sqrt{20/2} = \sqrt{10}$. This factor is approximately equal to the ratio of the exit velocities of the LOX–RP and the nuclear–H2 propulsion systems of Table 10.2. This is to be expected because the LOX–RP and the nuclear–H_2 systems each encounter the same structural limitations on T_c and P_c/P_e, so that the main difference in V_e between the two is due to the lower molecular weight of the nuclear system's propellant.

4. *Simultaneous effect of combustion temperature and molecular weight on V_e*: Up to this point, we have considered the combustion temperature and molecular weight effects on exit velocity independent of each other. For a given fuel–oxidizer combination in a chemical rocket, the combustion temperature T_c and the molecular weight $\mathcal{M}$ of the propellant are interdependent. They vary simultaneously with mixture ratio and chamber pressure. Consequently, the influence of these two variables on V_e must, in this case, be considered together as the mixture ratio is varied for a given chamber pressure. Figure 10.29 shows the variation of T_c and $\mathcal{M}$, along with V_e, for a chamber pressure of 300 psia and a pressure ratio of $P_c/P_e = 20.4$ for a LOX–gasoline fuel combination. As the mixture ratio is increased from 1.5, we observe that the combustion temperature and molecular weight each increase. The former increases more rapidly initially so that the net effect of the ratio $T_c/\mathcal{M}$ on the velocity V_e [Eq. (10.39)] is to increase V_e as shown. As the mixture ratio is increased

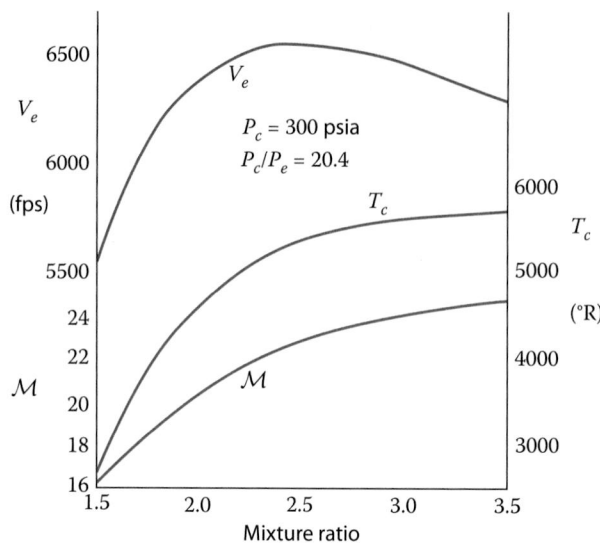

Fig. 10.29   LOX–gasoline performance vs mixture ratio.

beyond 2.5, the molecular weight increases more rapidly than the temperature, thus producing lower values of the ratio $T_c/\mathcal{M}$ and, hence, V_e. The optimum mixture ratio for the LOX–gasoline system is, therefore, a value of 2.5. This compares with a stoichiometric mixture ratio of 3.5 for LOX–gasoline. The mixture ratio of the LOX–RP-1 combination used in the Atlas booster engine (YLR89-NA-7 model) is 2.38.

The curves of Fig. 10.29 are for one chamber pressure. If a similar set of curves were drawn for a higher chamber pressure, the same general results would be obtained except that V_e would be higher due to the favorable effect of a higher pressure in reducing dissociation of the combustion products. Increasing the chamber pressure to obtain higher V_e, however, produces the unfavorable effect of increasing dead weight because of the higher stress levels introduced and the added fuel feed system pressure requirements. In any particular application, the operating chamber pressure selected must be a compromise of the several interrelated factors that chamber pressure affects.

10.5.5   Area Ratio and Velocity Ratio in Terms of the Pressure Ratio

The isentropic area ratio A/A^*, calculated in the GASTAB program and written as a function of the Mach number in Eq. (3.12), can be used to obtain the nozzle area ratio A_e/A_t when the exit Mach number is known. It is useful in rocket performance work to have the nozzle area expansion ratio A_e/A_t

expressed in terms of the ratio of the rocket chamber pressure to the exit section pressure P_c/P_e for isentropic flow. This relation can be obtained rather directly by use of the mass flow parameter given in terms of $\Gamma/\sqrt{R/g_c}$ [Eq. (10.34)] and pressure ratio [Eq. (10.2)]. Thus with $P_t = P_c$ and $T_t = T_c$, we have from Eq. (10.34)

$$\frac{\dot{m}\sqrt{T_c}}{P_c A_t} = \frac{\Gamma}{\sqrt{R/g_c}}$$

Writing Eq. (10.2) for the exit station $(P = P_e)$ and rearranging gives

$$\frac{\dot{m}\sqrt{T_c}}{P_c A_e} = \sqrt{\frac{2g_c}{R}\left(\frac{\gamma}{\gamma-1}\right)\left[\left(\frac{P_e}{P_c}\right)^{\frac{2}{\gamma}} - \left(\frac{P_e}{P_c}\right)^{\frac{\gamma+1}{\gamma}}\right]}$$

Dividing Eq. (10.34) by the preceding equation gives the desired expression for the area ratio A_e/A_t [Eq. (10.35)]:

$$\frac{A_e}{A_t} = \frac{\Gamma}{\sqrt{\frac{2\gamma}{\gamma-1}\left[\left(\frac{P_e}{P_c}\right)^{\frac{2}{\gamma}} - \left(\frac{P_e}{P_c}\right)^{\frac{\gamma+1}{\gamma}}\right]}} \tag{10.41}$$

Similarly, the ratio of the velocity at the exit V_e to the velocity at the throat V_t can be expressed as

$$\frac{V_e}{V_t} = \sqrt{\frac{\gamma+1}{\gamma-1}\left[1 - \left(\frac{P_e}{P_c}\right)^{\frac{\gamma-1}{\gamma}}\right]} \tag{10.42}$$

Equations (10.41) and (10.42) are plotted for several values of γ vs P_c/P_e on Fig. 10.30 as A_e/A_t and as V_e/V_t, respectively. Note that an area ratio (A_e/A_t) of 50 is required to obtain a value of 2 for V_e/V_t with $\gamma = 1.4$.

10.5.6 Thrust Coefficient

The thrust coefficient is defined as the thrust F divided by the product of the chamber pressure and throat area, $P_c A_t$, or

$$C_F \equiv \frac{F}{P_c A_t} \tag{10.43}$$

Why use thrust coefficient in place of thrust?

1. It is a dimensionless quantity.
2. Its numerical values have a convenient size that fall in the range 0.6–2.2.
3. It provides a figure of merit that is independent of the size of the rocket and forms a common basis with which to compare different-sized rockets.

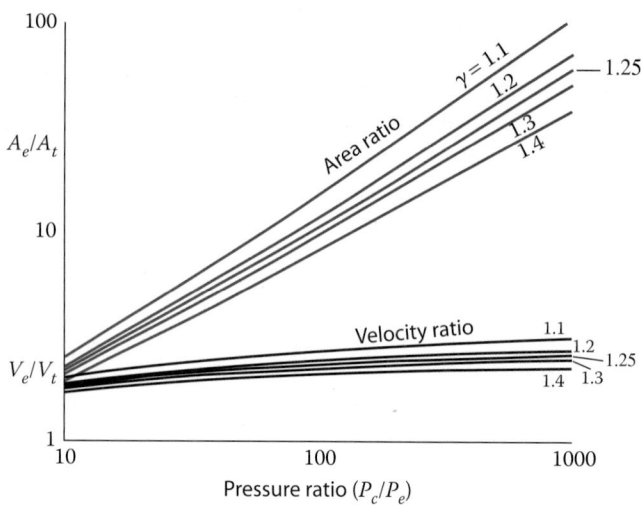

Fig. 10.30   Area and velocity ratios vs pressure ratio.

10.5.7 Ideal Thrust Coefficient, C_{Fi}

The proportionality factor between F_i and $P_c A_t$ is called ideal thrust coefficient. From Eqs. (10.37) and (10.43) we have

$$C_{Fi} = \Gamma \sqrt{\frac{2\gamma}{\gamma - 1} \left[1 - \left(\frac{P_e}{P_c} \right)^{\frac{\gamma-1}{\gamma}} \right]} + \left(\frac{P_e}{P_c} - \frac{P_a}{P_c} \right) \frac{A_e}{A_t} \qquad (10.44)$$

C_{Fi} is a function of A_e/A_t and P_c/P_a—the nozzle area and pressure ratios—for a given γ. Note that A_e/A_t determines P_e/P_c through Eq. (10.41). Using Eqs. (10.44) and (10.41), one can express the maximum value of the ideal thrust coefficient as

$$(C_{Fi})_{\max} = \Gamma \sqrt{\frac{2\gamma}{\gamma - 1}} \qquad (10.45)$$

10.5.8 Optimum Thrust Coefficient, $(C_{Fi})_{opt}$

For optimum ideal thrust, $P_e = P_a$ and Eq. (10.44) gives

$$(C_{Fi})_{opt} = \Gamma \sqrt{\frac{2\gamma}{\gamma - 1} \left[1 - \left(\frac{P_e}{P_c} \right)^{\frac{\gamma-1}{\gamma}} \right]} \qquad (10.46)$$

Note: It can be shown that

$$\frac{A_e}{A_t} = \frac{\gamma - 1}{2\gamma}\left(\frac{P_e}{P_c}\right)^{\frac{1}{\gamma}}\frac{(C_{Fi})_{\text{opt}}}{1 - \left(\dfrac{P_e}{P_c}\right)^{\frac{\gamma - 1}{\gamma}}} \tag{10.47}$$

10.5.9 Vacuum Thrust Coefficient, $(C_{Fi})_{\text{vac}}$

For vacuum, $P_a = 0$ and Eqs. (10.44) and (10.46) give

$$(C_{Fi})_{\text{vac}} = (C_{Fi})_{\text{opt}} + \frac{P_e A_e}{P_c A_t} \tag{10.48}$$

or

$$(C_{Fi})_{\text{vac}} = C_{Fi} + \frac{P_a A_e}{P_c A_t} \tag{10.49}$$

10.5.10 Graphs for Determining Ideal Thrust Coefficient

Equation (10.44) can be solved conveniently by use of graphs. In Fig. 10.31 $(C_{Fi})_{\text{vac}}$ is plotted vs γ and nozzle area ratio ε. Figure 10.30, which gives a plot of nozzle area ratio and velocity ratio vs P_c/P_e, is handy in computational work when the area ratio or P_c/P_e is known and the value of the other is required. Figures 10.32–10.35 present C_{Fi} for various

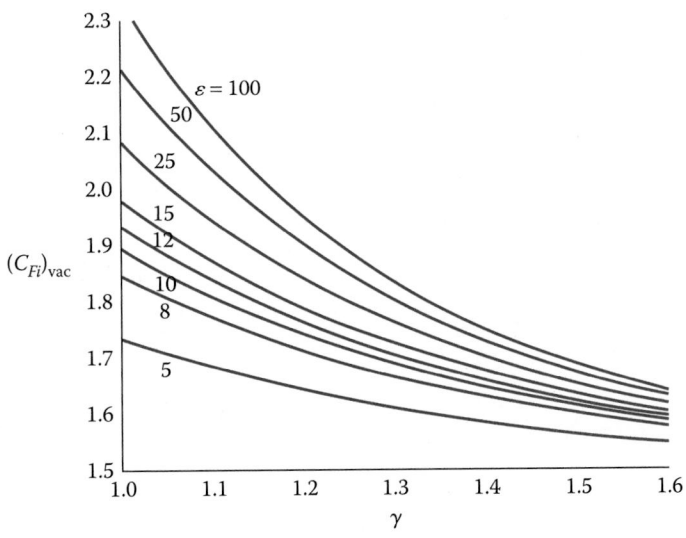

Fig. 10.31   Vacuum thrust coefficient, $(C_{Fi})_{\text{vac}}$.

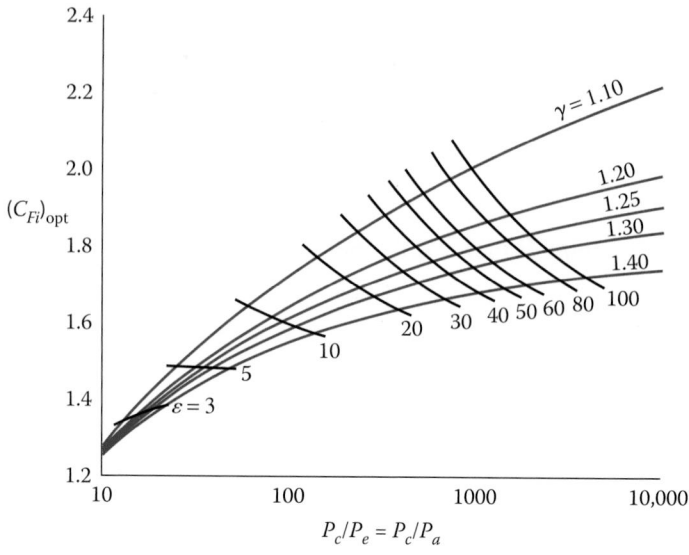

Fig. 10.32   Optimum thrust coefficient vs pressure ratio.

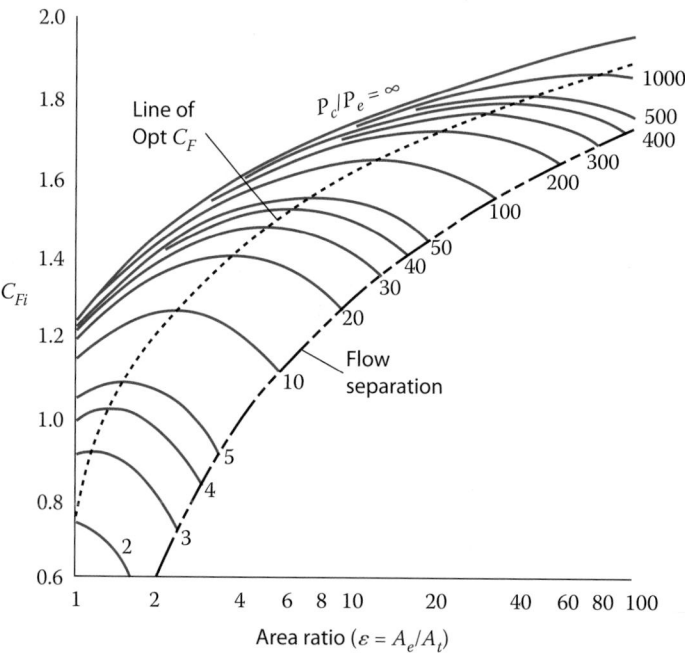

Fig. 10.33   Ideal thrust coefficient vs area ratio ($\gamma = 1.2$).

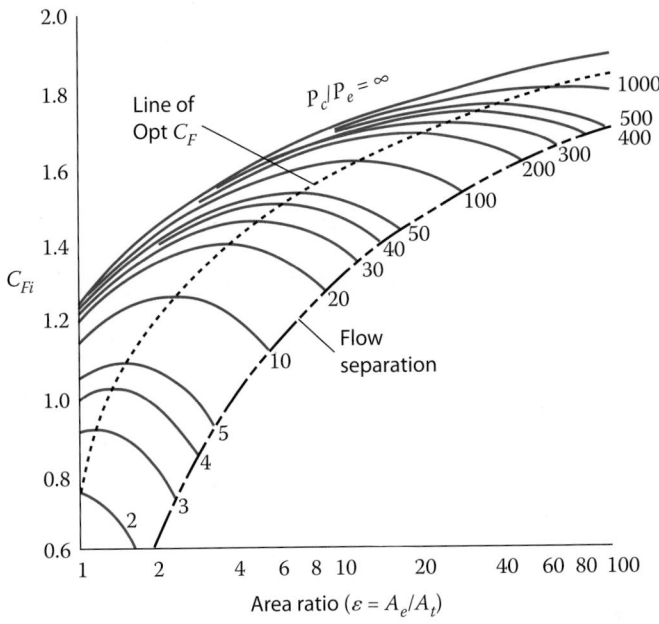

Fig. 10.34   Ideal thrust coefficient vs area ratio ($\gamma = 1.25$).

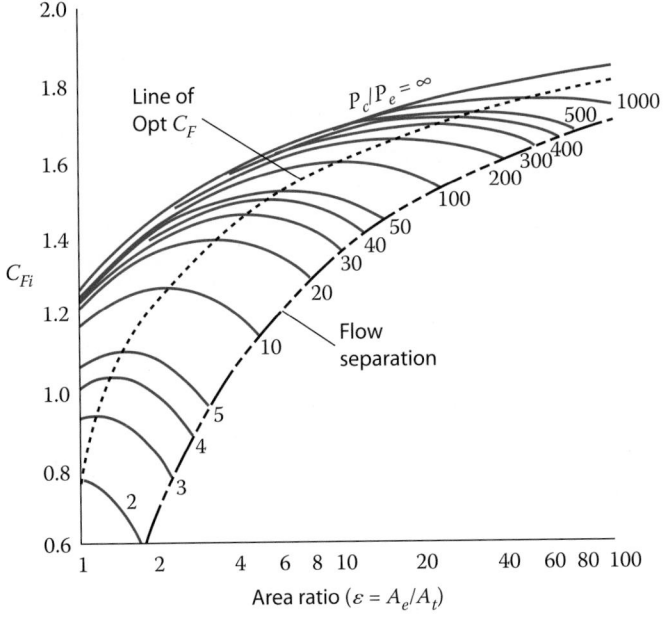

Fig. 10.35   Ideal thrust coefficient vs area ratio ($\gamma = 1.3$).

conditions. Figure 10.32 is a plot of $(C_{Fi})_{opt}$ vs nozzle pressure ratio with optimum expansion. The dashed curves in the lower right of Fig. 10.33 ($\gamma = 1.2$), Fig. 10.34 ($\gamma = 1.25$), and Fig. 10.35 ($\gamma = 1.3$) are plots of C_{Fi} with flow separation (Summerfield criterion for separation of $P_a > 3.5 \, P_{we}$ in Section 10.4.1). The remaining curves present Eq. (10.44) for various values of P_c/P_a.

Example 10.6

We can illustrate the use of the graph of Fig. 10.31 with an example. Let us find C_{Fi} for a rocket at 100,000-ft altitude with an area ratio of 25, $\gamma = 1.29$, and a chamber pressure $P_c = 500$ psia. Because the value of γ is not one of the specific values plotted in Figs. 10.30 and 10.32–10.35, use of Fig. 10.31 presents a convenient way of finding C_{Fi} without interpolation. First, we note that Eq. (10.45) can be written as

$$C_{Fi} = (C_{Fi})_{vac} - \frac{P_a A_e}{P_c A_t}$$

or

$$C_{Fi} = (C_{Fi})_{vac} - \Delta C_F \tag{10.50a}$$

where

$$\Delta C_F = \frac{P_a A_e}{P_c A_t} \tag{10.50b}$$

Thus C_{Fi} can easily be found by getting the value of $(C_{Fi})_{vac}$ from Fig. 10.31 and calculating the value of ΔC_F provided that the nozzle flow is not separated. From the standard atmosphere data in Appendix A, the atmospheric pressure P_a at 100,000 ft is 23.28 lbf/ft² (2116 × 0.0110) or 0.162 psia. Using the given value for P_c and this value of P_a gives a value of 3086 for the ratio P_c/P_a. A quick check of Figs. 10.34 and 10.35, with this value of P_c/P_a and the given value for nozzle area ratio ε, shows that the nozzle is operating underexpanded, and flow is not separated in the nozzle. Thus Fig. 10.31 and Eqs. (10.50a) and (10.50b) can be used to determine the ideal thrust coefficient. We obtain $(C_{Fi})_{vac} = 1.770$ from Fig. 10.31 corresponding to $\gamma = 1.29$ and $\varepsilon = 25$. Using the value of P_a determined in the preceding and the given data for P_c and ε in Eq. (10.50b) gives

$$\Delta C_F = \frac{P_a P_e}{P_c A_t} = \frac{0.162}{500} 25 = 0.008$$

Thus

$$C_{Fi} = (C_{Fi})_{vac} - \Delta C_F = 1.770 - 0.008 = 1.762$$

10.5.11 Relationship Among C, C^*, and C_{Fi}

Noting that from Eqs. (10.36) and (10.38), we can write

$$P_c A_t = \dot{m} C^* / g_c \qquad \text{(i)}$$

from Eqs. (10.5) and (10.43), we have

$$C_{Fi} = \frac{F_i}{P_c A_t} = \frac{\dot{m} C / g_c}{P_c A_t} \qquad \text{(ii)}$$

Substituting Eq. (i) into Eq. (ii) and solving for the effective exhaust velocity C gives

$$C = C_{Fi} C^* \qquad \text{(10.51)}$$

This equation shows that the effective exhaust velocity C can be expressed as the product of the ideal thrust coefficient (dependent primarily on the nozzle area and pressure ratios) times the characteristic velocity (dependent on thermochemistry).

Example 10.7

The space shuttle main engine (SSME) operates for up to 520 s in one mission at altitudes over 100 miles. The nozzle expansion ratio ε is 77:1, and the inside exit diameter is 7.54 ft. Assume a calorically perfect gas with the following properties: $\gamma = 1.25$, $P_c = 3000$ psia, $T_c = 6890°F$, and $R = 112$ ft · lbf/(lbm · °R).

We want to calculate the following:

1. Characteristic velocity C^*
2. Mass flow rate of gases through the nozzle
3. Mach number at exit M_e
4. Altitude that nozzle is "on-design"
5. Altitude that flow in nozzle is just separated (assume separation occurs when $P_a > 3.5 \, P_{we}$)
6. Ideal thrust coefficient C_{Fi} and the ideal thrust F_i in 5000-ft increments from sea level to 40,000 ft; in 10,000-ft increments from 40,000 to 100,000 ft; and at vacuum
7. $(I_{sp})_{vac}$ for the engine
8. Actual exhaust velocity V_e and effective exhaust velocity C at sea level

Solution

We assume isentropic flow of a perfect gas with constant specific heats.

a) *Characteristic velocity C^*.*

$$C^* = \frac{\sqrt{R g_c T_c}}{\Gamma}$$

(Continued)

Example 10.7 *(Continued)*

where $\Gamma = 0.6581$ for $\gamma = 1.25$ (Table 10.10) and $T_c = 7350°R$. Thus

$$C^* = \frac{\sqrt{112 \times 32.174 \times 7350}}{0.6581} = 7820\,\text{ft/s}$$

b) *Mass flow rate.*

$$\dot{m} = P_c A_t g_c / C^*$$

where

$$A_e = \frac{\pi D_e^2}{4} = \frac{\pi \times 7.54^2}{4} = 44.65\,\text{ft}^2$$

Then

$$A_t = \frac{A_e}{\varepsilon} = \frac{44.65}{77}\,144 = 83.50\,\text{in.}^2$$

Thus

$$\dot{m} = \frac{3000 \times 83.50 \times 32.174}{7820} = 1031\,\text{lbm/s}$$

c) *Mach number at exit M_e.* Using Eq. (3.12) for the flow between the throat and exit of the nozzle, we have

$$\frac{A_e}{A_t} = \frac{1}{M_e}\left\{\frac{2}{\gamma+1}\left(1 + \frac{\gamma-1}{2}M_e^2\right)\right\}^{\frac{\gamma+1}{2(\gamma-1)}} \tag{10.52}$$

This equation cannot be solved directly for the exit Mach number in terms of the area ratio A_e/A_t; however, it can be set into a form that permits fast iteration to the solution (which is used in the GASTAB program). Because there are two Mach numbers (a subsonic value and a supersonic value) that can give the same area ratio, we must be careful that we get the correct value. Solving Eq. (10.52) for the Mach number in the denominator gives an equation that will yield the subsonic Mach number corresponding to the area ratio A_e/A_t, or

Subsonic solution:

$$M_e = \frac{1}{A_e/A_t}\left\{\frac{2}{\gamma+1}\left(1 + \frac{\gamma-1}{2}M_e^2\right)\right\}^{\frac{\gamma+1}{2(\gamma-1)}} \tag{10.53}$$

Equation (10.53) can be used to iterate on a solution for the subsonic exit Mach number by starting with a guess (say 0.5) and substituting it into the right side of the equation and then solving for a new exit Mach number. The new value of the Mach number is substituted into the equation and another value of Mach number calculated. This iterative process is repeated until successive values do not change very much (say by about 0.001).

(Continued)

Example 10.7 *(Continued)*

The supersonic iteration equation for the exit Mach number is obtained from Eq. (10.52) by solving for the Mach number within the brace, or

Supersonic solution:

$$M_e = \sqrt{\frac{\gamma+1}{\gamma-1}\left\{M_e\frac{A_e}{A_t}\right\}^{\frac{2(\gamma-1)}{\gamma+1}} - \frac{2}{\gamma-1}} \qquad (10.54)$$

The nozzle pressure ratio P_c/P_e is great enough to give supersonic flow at the exit, and iterative use of Eq. (10.54) gives the results tabulated in Table 10.11 starting with a guess of 4.0 for the exit Mach.

d) *Nozzle design altitude.* The nozzle design altitude corresponds to that altitude where $P_c = P_a$. With both the exit Mach number and the chamber pressure known, the nozzle exit pressure can be determined using Eq. (3.8) for P_t/P written as

$$P_e = P_c\left(1 + \frac{\gamma-1}{2}M_e^2\right)^{-\gamma/\gamma-1} \qquad (10.55)$$

Thus $P_e = 2.187$ psia.

This exit pressure corresponds to a value for the atmospheric pressure ratio δ of 0.1488 (= 2.187/14.696). Using the standard atmospheric data of Appendix A, the nozzle design altitude is about 44,700 ft.

e) *Nozzle separation altitude.* The nozzle separation altitude corresponds to that altitude where $P_a = 3.5 \times P_{we} = 7.655$ psia. This pressure corresponds to a value for δ of 0.5209 (= 7.655/14.696). Using the standard atmospheric data of Appendix A, the nozzle separation altitude is about 17,000 ft.

Table 10.11 Solution to Exit Mach

Guess M_e	Eq. (10.54)
4.0	4.9148
4.9148	5.0657
5.0657	5.0881
5.0881	5.0913
5.0913	5.0918
5.0918	5.0919
5.0919	5.0919

Thus $M_e = 5.0919$. (This answer can be found quickly using the GASTAB program.)

(Continued)

Example 10.7 *(Continued)*

f) *Ideal thrust and thrust coefficient.* The ideal thrust coefficient C_{Fi} is given by Eq. (10.44) in terms of γ, P_c/P_e, A_e/A_t, and P_a/P_c. Equation (10.41) gives the nozzle expansion ratio A_e/A_t in terms of the nozzle pressure ratio P_c/P_e. When the flow in the nozzle is not separated, both the nozzle expansion ratio A_e/A_t and the nozzle pressure ratio P_c/P_e will remain constant at their design values of 77 and 1372 ($= 3000/2.187$), respectively. At altitudes below 17,000 ft, the flow in the nozzle will separate, and the effective nozzle expansion ratio A_e/A_t and nozzle pressure ratio P_c/P_e will be less than their design values.

The ideal thrust coefficient C_{Fi} and ideal thrust F_i will be obtained using Eqs. (10.44) and (10.43), respectively, and Eq. (10.41) will be used, when needed, to obtain the effective nozzle expansion ratio A_e/A_t. The required equations are Eq. (10.41):

$$\frac{A_e}{A_t} = \frac{\Gamma}{\sqrt{\dfrac{2\gamma}{\gamma-1}\left[\left(\dfrac{P_e}{P_c}\right)^{\frac{2}{\gamma}} - \left(\dfrac{P_e}{P_c}\right)^{\frac{\gamma+1}{\gamma}}\right]}}$$

Eq. (10.44):

$$C_{Fi} = \Gamma\sqrt{\frac{2\gamma}{\gamma-1}\left[1 - \left(\frac{P_e}{P_c}\right)^{\frac{\gamma-1}{\gamma}}\right]} + \left(\frac{P_e}{P_c} - \frac{P_a}{P_c}\right)\frac{A_e}{A_t}$$

and Eq. (10.43):

$$F_i = C_{Fi}P_c A_t$$

The values of C_{Fi} vs altitude are plotted in Fig. 10.36, and values of thrust are plotted in Fig. 1.42b. Note the change in slope of the plots at the separation altitude. As expected, the actual thrust is increased by the flow separation at altitudes below 17,000 ft. This is one of the few times when nature is kind. Table 10.12 shows the ideal thrust coefficient and ideal thrust vs altitude of the SSME.

g) *Calculate specific impulse in vacuum,* $(I_{sp})_{vac}$. Using the definition of specific impulse given by Eq. (10.6), we write

$$\left(I_{sp}\right)_{vac} = \frac{(F_i)_{vac}\,g_c}{\dot{m}g_o} = \frac{469{,}790 \times 32.174}{1030 \times 32.174} = 456.1\,\text{s}$$

h) *Actual exhaust velocity* V_e *and effective exhaust velocity* C *at sea level.* From Eq. (1.52), we can solve for the effective exhaust velocity C in terms of the thrust and mass flow rate to get

$$C = \frac{F_i g_c}{\dot{m}} = \frac{404{,}880 \times 32.174}{1030} = 12{,}647\,\text{ft/s}$$

(Continued)

Example 10.7 *(Continued)*

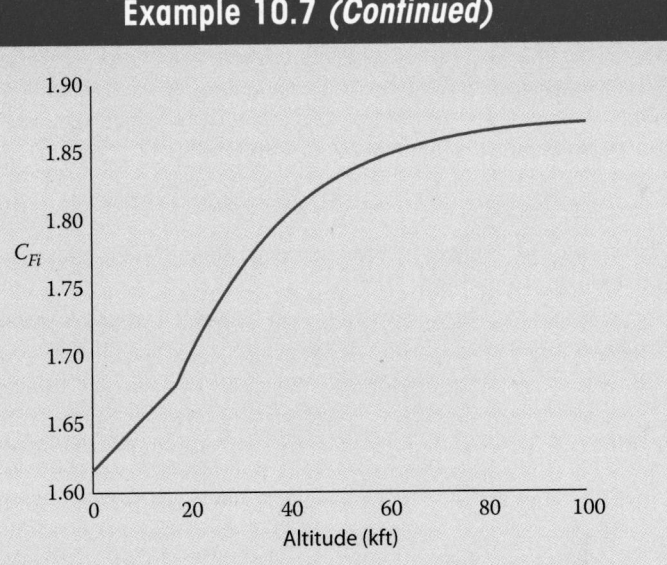

Fig. 10.36   Calculated ideal thrust coefficient of SSME.

Table 10.12   Ideal Thrust Coefficient and Ideal Thrust vs Altitude of SSME

Altitude, ft	P_a, psia	P_e, psia	A_e/A_t	C_{Fi}	F_i, lbf
0	14.696	4.199	46.7	1.61627	404,880
5000	12.228	3.494	53.8	1.63450	409,560
10,000	10.108	2.888	62.2	1.65345	414,190
15,000	8.297	2.371	72.4	1.67172	418,770
17,000	7.655	2.187	77.0	1.67892	420,570
20,000	6.759	2.187	77.0	1.70191	426,330
25,000	5.461	2.187	77.0	1.73523	434,670
30,000	4.373	2.187	77.0	1.76315	441,670
35,000	3.468	2.187	77.0	1.78638	447,490
40,000	2.730	2.187	77.0	1.80532	452,230
50,000	1.692	2.187	77.0	1.83197	458,910
60,000	1.049	2.187	77.0	1.84847	463,040
70,000	0.650	2.187	77.0	1.85871	465,610
80,000	0.404	2.187	77.0	1.86502	467,190
90,000	0.252	2.187	77.0	1.86893	468,170
100,000	0.160	2.187	77.0	1.87129	468,760
Vacuum	0	2.187	77.0	1.87540	469,790

(Continued)

Example 10.7 *(Continued)*

Because we have determined the value of C_{Fi} at sea level and C^*, Eq. (10.51) can also be used to determine the effective exhaust velocity C at sea level. Substituting the values into Eq. (10.51) gives

$$C = C_{Fi}C^* = 1.61627 \times 7820 = 12{,}639 \text{ ft/s}$$

This value of C is about 0.06% lower than the value determined that can be attributed to roundoff.

The actual exhaust velocity V_e can be written in terms of the effective exhaust velocity using Eq. (1.51) as

$$V_e = C - (P_e - P_a)A_e g_c / \dot{m}$$

Using the value of C calculated and the values of P_e, P_a, and A_e/A_t determined in part f for sea level, the exhaust velocity can be calculated:

$$V_e = 12{,}647 - \frac{(4.199 - 14.696)(46.7 \times 83.50)32.174}{1030}$$

$$V_e = 12{,}647 + 1279 = 13{,}926 \text{ ft/s}$$

Example 10.8

Determine the C_{Fi} of the H-l rocket engine at its design altitude of 80,000 ft using Fig. 10.31. The H-l rocket engine has a chamber pressure P_c of 633 psia and a nozzle expansion ratio ε of 8. Assume $\gamma = 1.33$.

Solution

Using the standard atmosphere data in Appendix A at 80,000 ft, we get $P_a = 0.406$ psia and by calculation obtain $P_a/P_c = 0.00064$. Because the nozzle is at its design altitude, the flow is not separated in the nozzle. For $\varepsilon = 8$ and $\gamma = 1.33$, we have $(C_{Fi})_{vac} = 1.660$ from Fig. 10.31. For $\varepsilon = 8$ and $P_a/P_c = 0.00064$, we get $\Delta C_F = 0.005$. Thus,

$$C_{Fi} = (C_{Fi})_{vac} - \Delta C_F = 1.655$$

10.5.12 Liquid-Propellant Rockets

Liquid-propellant rockets can be classified into two general propellant categories: bipropellant (fuel and oxidizer) and monopropellant. Ordinarily, bipropellants are used for applications requiring high thrust, such as launch vehicles, and monopropellants are used for small control rockets. The focus in this section is on bipropellant rockets.

A typical liquid-propellant rocket consists of four main parts: rocket engine, propellants, propellant feed system, and propellant tanks. Figure 10.37 shows a schematic diagram of a bipropellant rocket with a turbopump feed system. The turbopump feed system is usually used for high-thrust/long-duration rockets. In this system, about 1.4% of the main flow fuel and oxidizer are burned in the gas generator. The resultant high-temperature and high-pressure gas is expanded through the turbine before being exhausted. The

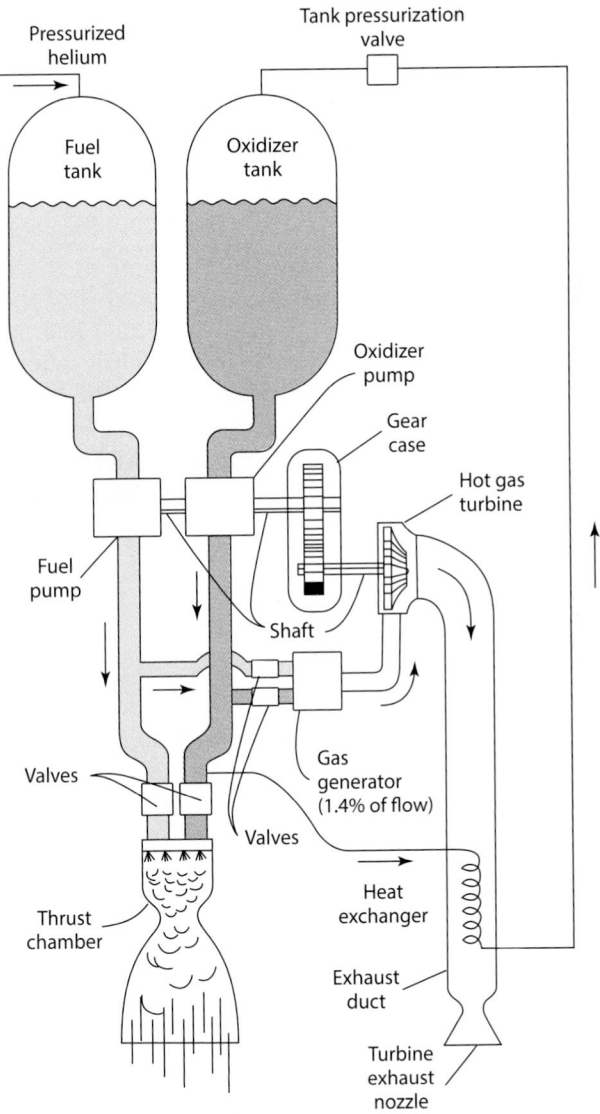

Fig. 10.37 Liquid-propellant rocket with turbopump feed system (from Ref. 31).

turbine powers both the fuel and oxidizer pumps. This turbopump-fed rocket engine cycle is referred to as an *open cycle* because the working fluid exhausted from the turbine is discharged overboard. In *closed-cycle* turbopump feed systems, all of the working fluid exhausted from the turbine is injected into the combustion chamber of the rocket engine.

For a liquid-propellant rocket, the steady-state mass flow rate of propellant and combustion chamber pressure P_c are determined by a balance between the mass flow rate entering the chamber from the fuel and oxidizer pumps and the mass flow rate leaving through the nozzle throat. The chamber pressure and mixture ratio of oxidizer to fuel directly affect the combustion process and its resultant products. For a given application, a high chamber pressure P_c is normally desirable because it permits higher nozzle thrust coefficient C_F and specific impulse I_{sp}. Most of the data presented in this section are for optimum expansion through a nozzle from a chamber pressure P_c of 1000 psia to an exhaust pressure P_e of 14.7 psia.

10.5.12.1 Propellants

Bipropellant liquid rockets use an oxidizer and a fuel. Typical oxidizers are liquid oxygen (O_2, also referred to as LOX), liquid fluorine (F), hydrogen peroxide (H_2O_2), chlorine trifluoride (CIF_3), nitric acid (HNO_3), and nitrogen tetroxide (N_2O_4). Common fuels include RP-1 (a hydrocarbon kerosene-like mixture), liquid hydrogen (H_2), hydrazine (N_2H_4), unsymmetrical dimethylhydrazine [$(CH_3)_2NNH_2$, also referred to as UDMH], monomethylhydrazine (CH_3NHNH_2), ethyl alcohol (C_2H_5OH), and liquid ammonia (NH_3). Table 10.13 shows some typical oxidizer/fuel combinations.

10.5.12.2 Performance Calculations and Results

It is beyond the scope of this chapter to cover the thermochemical theory and calculations required to determine the performance of a liquid-propellant rocket. (See Chapter 2 for fundamentals.) For the simple reaction between hydrogen (H_2) and oxygen (O_2), it is possible to form six products: water (H_2O), hydrogen (H_2), oxygen (O_2), hydroxyl (OH), atomic oxygen (O), and atomic hydrogen (H). The chemical equation for this reaction between a moles of hydrogen and b moles of oxygen can be written as

$$aH_2 + bO_2 \rightarrow n_{H_2O}H_2O + n_{H_2}H_2 + n_{O_2}O_2 + n_HH + n_OO + n_{OH}OH$$

Here n_{H_2O}, n_{H_2}, n_{O_2}, n_H, n_O, and n_{OH} are the molar quantities of the products.

An equilibrium solution of the preceding reaction in a combustion chamber can be obtained by use of the following principles:

1. Conservation of mass
2. Conservation of energy
3. Law of partial pressures

Table 10.13 Theoretical Performance[a] of Liquid Rocket Propellant Combinations[b] (from Ref. 31)

Oxidizer	Fuel	Mixture Ratio		Chamber Temperature, K	C^*, m/s	$\mathcal{M}$, kg/mol	I_{sp}, s	γ
		By Mass	By Volume					
Oxygen	75% ethyl alcohol	1.30	0.98	3177	1641	23.4	267	1.22
		1.43	1.08	3230	1670	24.1	279	1.25
	Hydrazine	0.74	0.66	3285	1871	18.3	301	1.26
		0.90	0.80	3404	1892	19.3	313	
	Hydrogen	3.40	0.21	2959	2428	8.9	387	1.24
		4.02	0.25	2999	2432	10.0	390	
	RP-1	2.24	1.59	3571	1774	21.9	286	1.25
		2.56	1.82	3677	1800	23.3	300	
	UDMH	1.39	0.96	3542	1835	19.8	295	1.33
		1.65	1.14	3594	1864	21.3	310	
Fluorine	Hydrazine	1.83	1.22	4553	2128	18.5	334	1.33
		2.30	1.54	4713	2208	19.4	363	
	Hydrogen	4.54	0.21	3080	2534	8.9	398	1.26
		7.60	0.35	3900	2549	11.8	410	
Nitrogen tetroxide	Hydrazine	1.08	0.75	3258	1765	19.5	283	1.24
		1.34	0.93	3152	1782	20.9	292	
	50% UDMH/50% hydrazine	1.62	1.01	3242	1652	21.0	278	1.22
		2.00	1.24	3372	1711	22.6	288	
Red fuming nitric acid (15% NO_2)	RP-1	4.10	2.12	3175	1594	24.6	258	1.22
		4.80	2.48	3230	1609	25.8	268	
	50% UDMH/50% hydrazine	1.73	1.00	2997	1682	20.6	272	
		2.20	1.26	3172	1701	22.4	279	

[a]Conditions: $P_c = 1000$ psia, $P_e = 14.7$ psia, optimum nozzle expansion ratio, and frozen flow through nozzle.
[b]For every propellant combination, there are two sets of values listed: the upper line refers to frozen flow through the nozzle, the lower line to equilibrium flow through the nozzle.

Figure 10.38 shows the results of these thermochemical calculations for liquid oxygen and the hydrocarbon fuel RP-1. Note that the concentration of carbon monoxide (CO) decreases and the oxidizer/fuel ratio is increased with a corresponding increase in carbon dioxide (CO_2). Also, note that the water (H_2O) reaches a maximum concentration for the given chamber pressure when the oxidizer/fuel ratio is about 2.7.

After reacting in the combustion chamber, the products then flow through the rocket nozzle. Two cases are considered for the flow of these products through the nozzle:

1. *Frozen flow.* The composition of the products does not change from those values determined for the combustion chamber. Here the mole fractions, molecular weight, and so on are the same at the nozzle exit as those at the exit of the combustion chamber.
2. *Equilibrium flow.* The composition of the products is maintained in chemical equilibrium as the pressure and temperature vary during the nozzle expansion process.

Figure 10.39 shows the equilibrium composition results leaving the nozzle for liquid oxygen and RP-1. Note that higher concentrations of carbon dioxide are shown in Fig. 10.39 leaving the nozzle than those shown leaving the combustion chamber in Fig. 10.38.

The actual flow through the nozzles falls between these easily calculated cases. The products are able to stay in chemical equilibrium to some point

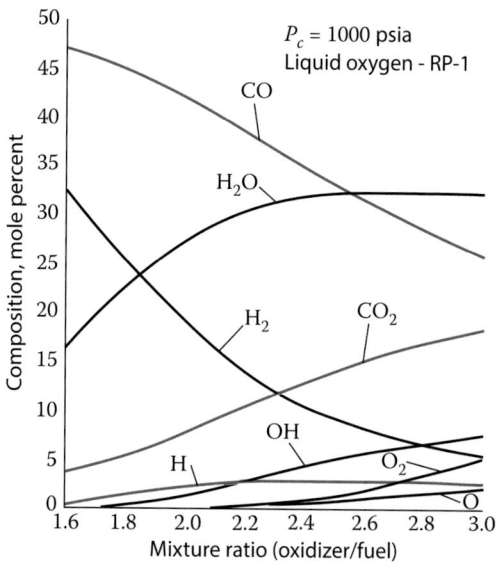

Fig. 10.38   Calculated chamber composition for liquid oxygen and RP-1 (from Ref. 31).

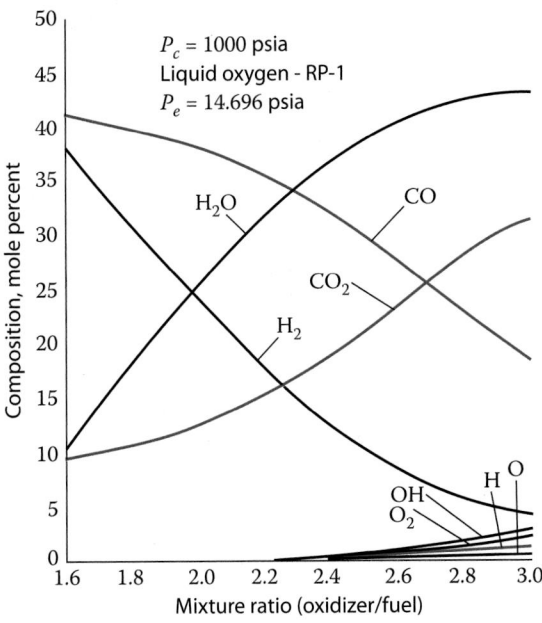

Fig. 10.39   Calculated nozzle exit equilibrium composition for liquid oxygen and RP-1 (from Ref. 31).

downstream of the nozzle where their composition freezes. The equilibrium flow calculations give lower exhaust nozzle temperatures because of the recombination of dissociated species and thus give higher performance. Both frozen and equilibrium flow results are presented in Table 10.13.

Example 10.9

Determine the I_{sp} of the oxygen–hydrogen rocket engine with a chamber temperature of 2960 K and an oxidizer-to-fuel mixture ratio by weight of 3.4:1 using the EQL software for both equilibrium and frozen flow through the nozzle. The rocket engine has a chamber pressure P_c of 1000 psia and an exit pressure P_e of 14.7 psia. Assume isentropic flow and perfect expansion.

Solution
The mixture ratio of 3.4:1 by weight equals a mole ratio of 0.214 (= 3.4 × 2.016/32). We use a pressure of 6893.2 kPa (= 1000 × 101.33/14.7) for the exit conditions from the combustion chamber. The nozzle inlet and exit

(Continued)

Example 10.9 *(Continued)*

states for both equilibrium and frozen flow are listed in Table 10.14. The exit velocity V_e follows from the definition of total enthalpy, Eq. (2.20).

Table 10.14 Oxygen–Hydrogen Rocket Engine Results Using EQL: Example 10.9

Property	Inlet	Exit, Equilibrium Flow	Exit, Frozen Flow
Pressure, kPa	6893.2	101.33	101.33
Temperature, K	2960.0	1308.3	1250.9
Molecular weight	8.80	8.87	8.80
Enthalpy, kJ/kg	371.3	−7835	−7651
V_e, m/s		4051	4006
I_{sp}, s		413	408

10.5.13 Solid-Propellant Rockets

As shown in Fig. 10.40, solid-propellant rockets are relatively simple in construction. They contain the propellant to be burned within the combustion chamber or case. The propellant charge contains all of the chemical elements required for complete burning and is formed into a solid mass called the *grain*. Once the propellant is ignited, it burns on the surfaces of the propellant that are not inhibited by the case. The rate of burning is proportional to the exposed surface area.

10.5.13.1 Propellants

There are three general types of propellants for this type of rocket: double-base (homogeneous), composite (or heterogeneous), and composite

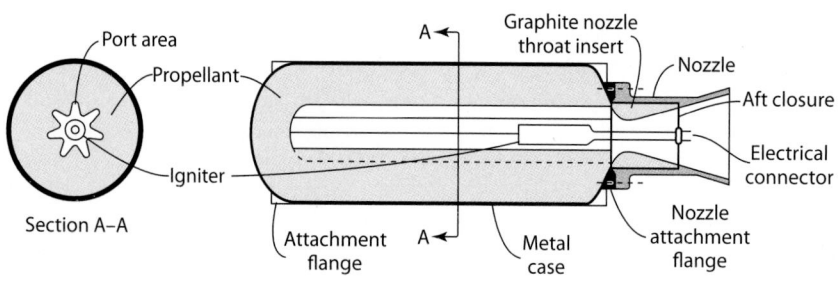

Fig. 10.40 Typical solid-propellant rocket motor (from Ref. 31).

Table 10.15 Representative Propellant Formulations in Percent Weight[a]

Propellant Ingredient	Double-Base (JPN)	Composite (PBAN)	Composite Modified Double-Base (CMDB)
Nitrocellulose	51.5		21.9
Nitroglycerine	43.0		29.0
Ammonium perchlorate		70.0	20.4
Aluminum powder		16.0	21.1
Polybutadiene–acrylic acid–acrylonitrile		11.78	
Triacetin			5.1
Diethyl phthalate	3.2		
Epoxy curative		2.22	
Ethyl centralite	1.0		
Potassium sulfate	1.2		
Additives	0.1		
Stabilizers			

[a]Source: U.S. Air Force Wright Laboratory, Rocket Branch, Edwards, CA.

modified double-base (CMDB). Table 10.15 gives representative formulations of these three types of propellants.

A homogeneous propellant has the fuel and oxidizer within the same molecule. The most common example of a homogeneous propellant is the double-base propellant of nitrocellulose and nitroglycerin with small amounts of additives. Figure 10.41 shows the variation in flame temperature and specific impulse with nitroglycerin (NG) concentration of the double-base propellant (JPN).

The composite propellant contains a heterogeneous mixture of oxidizer (as crystals) in a rubber-like binder. The binder is the fuel and must withstand a severe environment (high thermal and mechanical stresses). Hydroxyl terminated polybutadiene (HTPB) is a common binder. Common oxidizing crystals are ammonium perchlorate (AP), ammonium nitrate (AN), nitronium perchlorate (NP), potassium perchlorate (KP), potassium nitrate (KN), cyclotrimethylene trinitramine (RDX), and cyclotetramethylene tetranitramine (HMX).

Figure 10.42 shows the variation in flame temperature, molecular weight, and specific impulse with oxidizer concentration for HTPB-based composite propellants. Maximum flame temperature for RDX and HMX are associated with 100% concentration, and these are commonly referred to as stoichiometrically balanced fuel–oxidizer combinations.

The composite modified double-base (CMDB) propellant is a heterogeneous mixture of the other two types of propellants. Figure 10.43 shows

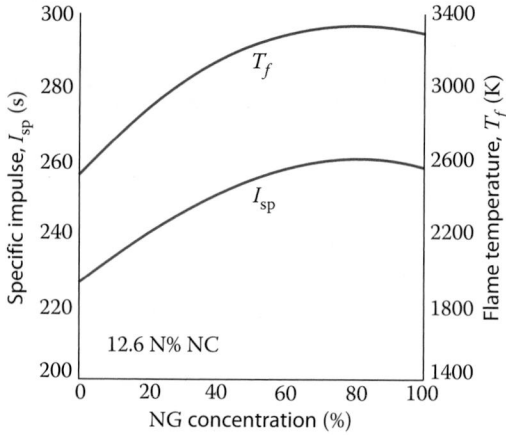

Fig. 10.41   Specific impulse and flame temperature vs nitroglycerin concentration of double-base propellants (from Ref. 33).

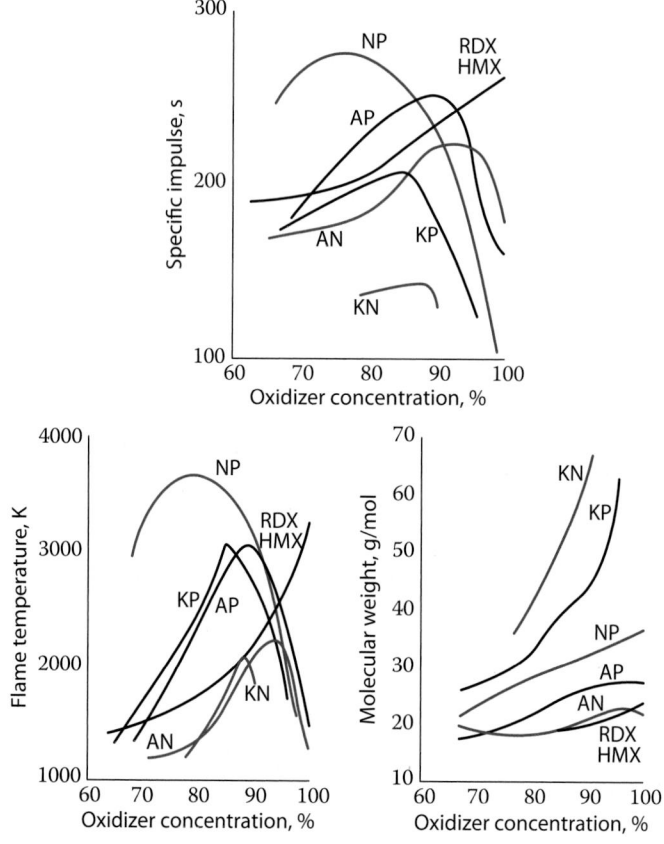

Fig. 10.42   HTPB-based composite propellants (from Ref. 33).

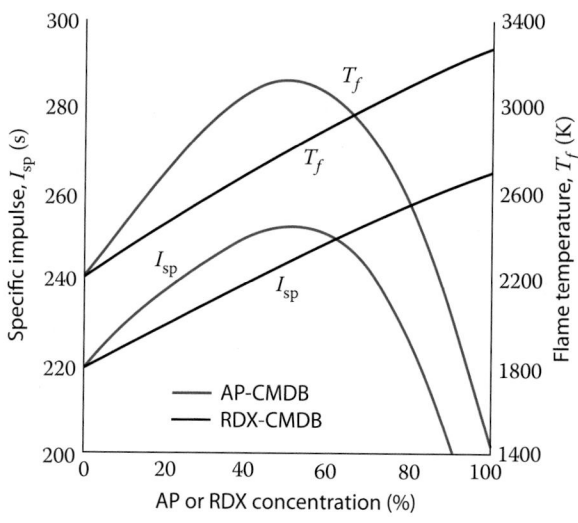

Fig. 10.43   Effects of percent concentration of AP or RDX on CMDB propellants (from Ref. 33).

the variation in flame temperature and specific impulse vs the concentration of AP and RDX in AP–CMDB and RDX–CMDB propellants.

Metal powders are commonly added to solid propellants to increase specific impulse and fuel density. Aluminum powders that constitute 12–20% of the propellant mass have provided the most successful experience.

The theoretical performances of three solid propellants are presented in Table 10.16. The results for specific impulse I_{sp} and effective exhaust velocity C are based on frozen flow through an optimum nozzle expansion ratio from a chamber pressure of 1000 psia to an exhaust pressure of 14.7 psia.

10.5.13.2 Burning Rate

The burning rate r for a solid propellant is defined as the recession of the propellant surface per unit time (see Fig. 10.44). The burning rates of typical solid propellants are given in Fig. 10.45. For a given solid propellant, this rate is mainly a function of the chamber pressure P_c and is commonly written as

$$r = aP_c^n \tag{10.56}$$

where a is an empirical constant influenced by the ambient temperature of the propellant prior to ignition, and n is known as the *burning rate pressure exponent*. When a single value of the burning rate is given, this value corresponds to a propellant ambient temperature of 70°F and a combustion pressure of 1000 psia unless otherwise stated. For the composite

Table 10.16   Theoretical Performance of Typical Solid-Rocket Propellant Combinations[a] (from Ref. 31)

Oxidizer	Fuel	Chamber Temperature, K	C^*, m/s	$\mathcal{M}$, g/mol	I_{sp}, s	γ
Ammonium nitrate	11% binder and 7% additives	1282	1209	20.1	192	1.26
Ammonium perchlorate 78–66%	18% organic polymer binder and 4–20% aluminum	2816	1590	25.0	262	1.21
Ammonium perchlorate 84–68%	12% polymer binder and 4–20% aluminum	3371	1577	29.3	266	1.17

[a]Conditions for I_{sp} and C: $P_c = 1000$ psia, $P_e = 14.7$ psia, optimum nozzle expansion ratio, and frozen flow through nozzle.

ammonium nitrate propellant curve in Fig. 10.45, corresponding to an ambient temperature of $-40°F$, the empirical constants a and n are 0.0034 in./s and 0.445, respectively, for the chamber pressure P_c expressed in units of psia.

Accounting for the density of both the solid propellant and the propellant gas, the net mass flow rate of propellant gas being generated can be expressed as

$$\dot{m} = (\rho_s - \rho_g)rA_b \tag{10.57}$$

where A_b is the burning area of the propellant, r the burning rate, ρ_s the solid propellant density at ambient temperature, and ρ_g the propellant gas density. Because A_b and r vary with time and chamber pressure P_c, the thrust of a solid-propellant rocket changes with time. The variation in burning area of the propellant A_b designed into the grain determines the variation in thrust over time. Figure 10.46 shows six grain designs, their

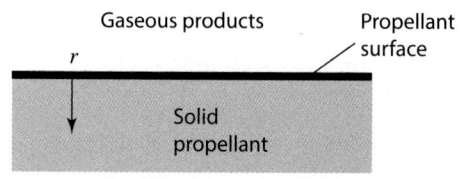

Fig. 10.44   Burning of a solid propellant.

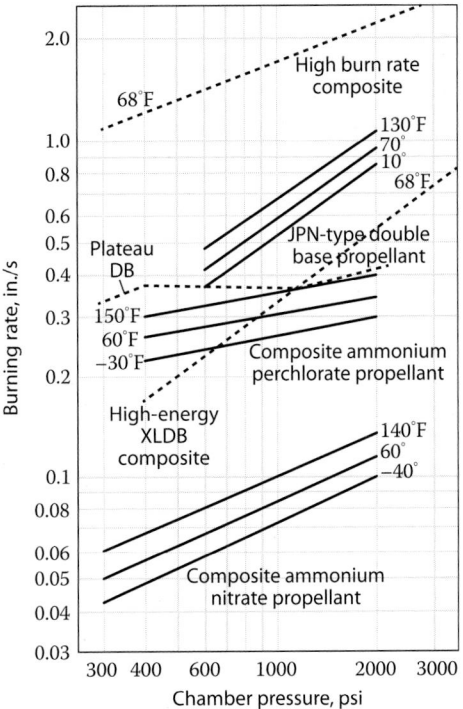

Fig. 10.45   Burning rates of typical solid propellants (from Ref. 31).

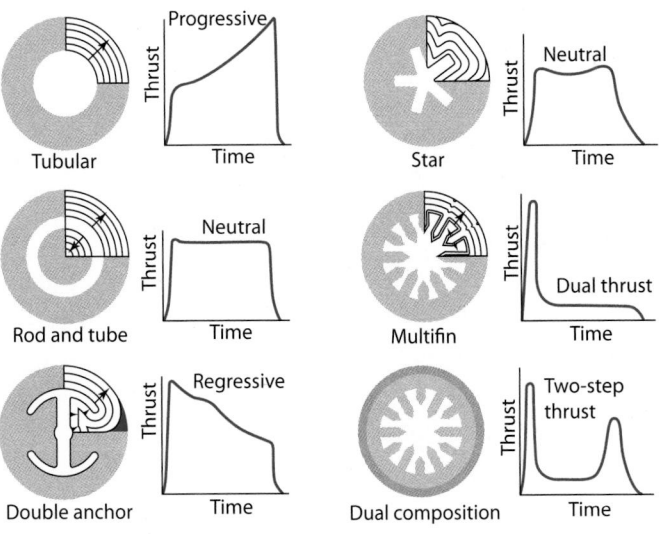

Fig. 10.46   Grain design and thrust vs time (from Ref. 37).

variations in thrust with time, and the corresponding general category of thrust variation.

10.5.13.3 Chamber Pressure and Stability

A steady-state chamber pressure P_c is reached when the net mass flow of propellant gas being generated equals the mass flow rate of propellant gas leaving the nozzle. Using Eqs. (10.36), (10.38), and (10.57), we write

$$\dot{m} = \frac{P_c A_t g_c}{C^*} = (\rho_s - \rho_g) r A_b$$

where A_t is the nozzle throat area and C^* is the characteristic velocity given by Eq. (10.38). For solid-propellant rockets, the characteristic velocity C^* depends mainly on the propellant. Substituting the empirical relationship of Eq. (10.56) for the burning rate r into the preceding equation gives

$$\dot{m} = \frac{P_c A_t g_c}{C^*} = (\rho_s - \rho_g) a P_c^n A_b \tag{i}$$

Plotting the nozzle mass flow rate and propellant gas generation rate vs the chamber pressure gives the general curves shown in Fig. 10.47. Two curves are drawn for the propellant gas generation rate (mass burn rate): one for $n < 1$ and another for $n > 1$. *Stability requires that burning rate pressure exponent n be less than unity.* If n is greater than unity, increases in chamber pressure above operating point will cause the propellant gas generation to increase faster than the nozzle flow rate. The result would lead to a rapid rise in chamber pressure with an accompanying explosion.

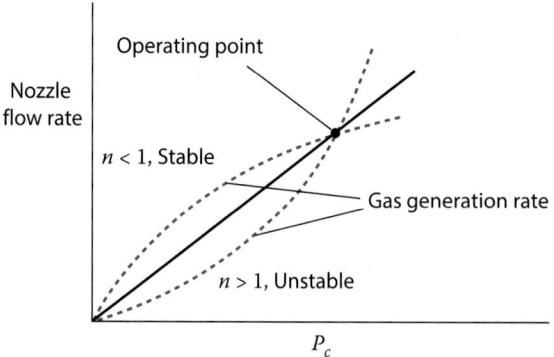

Fig. 10.47 Mass flow rate vs chamber pressure.

From Eq. (i), the chamber pressure P_c can be expressed as

$$P_c = \left\{ \frac{a(\rho_s - \rho_g)C^* }{g_c} \frac{A_b}{A_t} \right\}^{\frac{1}{1-n}} \tag{10.58}$$

This equation can be used to determine the chamber pressure when data are available for the terms on the right side of the equation. Because the density of the propellant gas is much smaller than that of the solid propellant, it is often neglected, and Eq. (10.58) is written as

$$P_c = \left\{ \frac{a\rho_s C^* }{g_c} \frac{A_b}{A_t} \right\}^{\frac{1}{1-n}} \tag{10.59}$$

The sensitivity of the chamber pressure P_c to changes in the burn-to-throat area ratio A_b/A_t can be investigated with this equation. As an example, for a burning rate pressure exponent n of 0.75, the change in chamber pressure is proportional to A_b/A_t raised to the fourth power, which can lead to large variations in P_c over the burn. It is for this reason that most propellants have burning rate pressure exponent n in the range of 0.2 to 0.3.

Example 10.10

As an example of the use of Eq. (10.59) to determine chamber pressure P_c, we consider a solid-propellant rocket with $A_b/A_t = 600$. The properties for the propellant are $\rho_s = 0.06$ lbm/in.^{3}, $n = 0.2$, $a = 0.05$ in./s, and $C^* = 5000$ ft/s. Using Eq. (10.59) gives

$$P_c = \left\{ \frac{0.05 \times 0.06 \times 5000}{32.174} 600 \right\}^{1.25} = 1144 \text{ psia}$$

Equation (10.59) can be used to determine the burn-to-throat area ratio A_b/A_t of a solid-propellant rocket when the other data are known. Solving Eq. (10.59) for this area ratio gives

$$\frac{A_b}{A_t} = \frac{g_c}{a\rho_s C^* } P_c^{1-n} \tag{10.60}$$

Example 10.11

Consider that a chamber pressure of 2000 psia is desired for the same propellant as used in the preceding example. The required burn-to-throat area ratio is

$$\frac{A_b}{A_t} = \frac{32.174}{0.05 \times 0.06 \times 5000} 2000^{0.8} = 938$$

10.5.13.4 Total Impulse and Times

The total impulse I_t of a rocket is the integral of the thrust over the burning time t_b, or

$$I_t = \int_0^{t_b} F \, dt = \bar{F} t_b \qquad (10.61)$$

where $\bar{F}$ is the average thrust over the burning time t_b, which is defined in Fig. 10.48, along with the action time t_a. Figure 10.48 shows representative plots of experimentally measured chamber pressure P_c vs time.

The specific impulse, effective exhaust velocity, and propellant mass flow rate cannot be measured directly in an experiment. The normal procedure is to determine the average specific impulse by finding the total impulse I_t and weight of propellant w used for a rocket test. The average specific impulse $(\bar{I}_{sp})$ is then given by

$$\bar{I}_{sp} = \frac{I_t}{w} \qquad (10.62)$$

The average effective exhaust velocity $\bar{C}$ and mass flow rate $\bar{m}$ then follow from Eq. (10.6) and the values for the weight of propellant w used and burning time t_b. Thus we have

$$\bar{C} = \bar{I}_{sp} g_o \qquad (10.63)$$

$$\bar{m} = \frac{w \, g_c}{t_b \, g_o} \qquad (10.64)$$

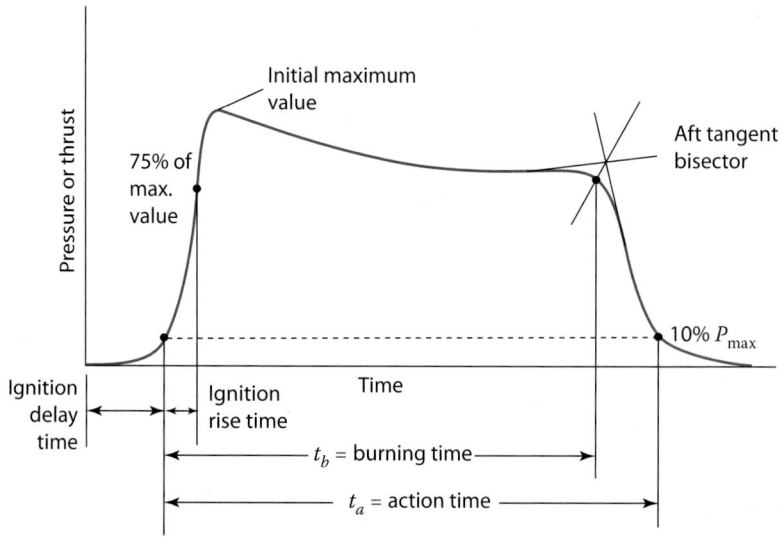

Fig. 10.48   Definition of burning time t_b and action time t_a (from Ref. 21).

Example 10.12

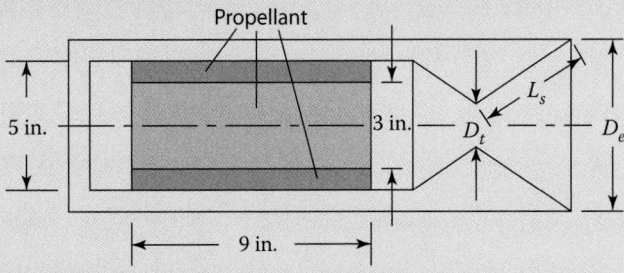

Fig. 10.49   Sketch of test rocket motor with convergent-divergent nozzle.

Two tests of a 5-in.-diam, center-perforated, solid-propellant rocket were performed at the U.S. Air Force Academy. The first test used a convergent-only nozzle, and the second test used a convergent-divergent nozzle. Figure 10.49 shows a sketch of the test rocket motors. The rocket propellant grain burns on both ends and in the center. The following data were provided by the manufacturer:

1. Ammonium perchlorate composite propellant with aluminum (70% NH_4ClO_4, 15% Al, 15% binder)
2. $C^* = 5118$ ft/s, $\mathcal{M} = 25.55$, $\gamma = 1.2$, $\rho_s = 0.064$ lbm/in.^{3}
3. Burning rate coefficients: $a = 0.0773$ in./s, $n = 0.21$, P_c in psia

The test data for the two tests are presented in Table 10.17. The results for Case I are calculated and discussed in the following. The calculations for Case II are left as a homework problem.

Case I results

$$\text{Average thrust} = \overline{F} = \frac{I_t}{t_b} = \frac{1257.6}{4.113} = 305.8 \text{ lbf}$$

$$\text{Average } P_c = \overline{P}_c = \left[\int_0^{t_b}(P_c - P_a)dt\right] \Big/ t_b + P_a$$

$$= 1161.6/4.113 + 23.16 \times 0.491 = 293.8 \text{ psia}$$

$$\text{Average } A_t = \overline{A}_t = \pi(1.1075)^2/4 = 0.963 \text{ in.}^2$$

$$\text{Burning rate} = r = a(\overline{P}_c)^n = 0.0773(293.8)^{0.21} = 0.255 \text{ in./s}$$

$$\text{Average burning rate} = \overline{r} = \frac{\text{Web thickness}}{t_b} = \frac{1.00}{4.113} = 0.243 \text{ in./s}$$

(Continued)

Example 10.12 (Continued)

$$\text{Initial burning surface} = A_b = 2\pi\left(r_i L + r_o^2 - r_i^2\right)$$
$$= 2\pi\left(1.5 \times 9 + 2.5^2 - 1.5^2\right) = 110\,\text{in.}^2$$

$$\text{Initial propellant flow rate} = \dot{m} = r\rho_s A_b$$
$$= 0.255 \times 0.064 \times 110 = 1.795\ \text{lbm/s}$$

$$\text{Average propellant flow rate} = \overline{\dot{m}} = \frac{wg_c}{t_b g_o} = \frac{7.32}{4.113}\frac{32.174}{32.174} = 1.780\,\text{lbm/s}$$

$$\text{Experimental } C^* = \frac{\overline{P_c}\,\overline{A_t}g_c}{\overline{\dot{m}}} = \frac{293.8 \times 0.963 \times 32.174}{1.780} = 5114\,\text{ft/s}$$

$$\text{Experimental } C_F = \frac{\overline{F}}{\overline{P_c}\,\overline{A_t}} = \frac{305.8}{293.8 \times 0.963} = 1.081$$

$$\text{Experimental } I_{\text{sp}} = \frac{I_t}{w} = \frac{1257.6}{7.32} = 171.8\,\text{s}$$

$$\text{Experimental } C = I_{\text{sp}}g_o = 171.8 \times 32.174 = 5527\,\text{ft/s}$$

Theoretical thrust coefficient: $\gamma = 1.2$, $\varepsilon = 1$

$$\frac{P_c}{P_a} = \frac{293.8}{23.16 \times 0.491} = 25.84, \quad \frac{P_c}{P_e} = \left(\frac{\gamma+1}{2}\right)^{\gamma/(\gamma-1)} = \left(\frac{2.2}{2}\right)^6 = 1.772$$

Table 10.17   Test Data of Two Small Solid-Propellant Rockets

Item	Case I	Case II
Barometric pressure P_a, Hg	23.16	23.33
Initial rocket case weight, lbf	17.42	17.35
Final rocket case weight, lbf	10.10	10.10
Nozzle type	Convergent	Convergent-divergent
Initial throat diameter d_t, in.	1.107	0.852
Final throat diameter d_t, in.	1.108	0.866
Exit diameter d_e, in.	N/A	2.225
Nozzle slant length L_s, in.	N/A	3.0
Burning time t_b, s	4.113	3.64
$\int_0^{t_b}(P_c - P_a)\,dt$, psig $\cdot$ s	1161.6	1921.6
$I_t = \int_0^{t_b} F\,dt$, lbf-s	1257.6	1570.8

(Continued)

Example 10.12 *(Continued)*

Using these data with Eq. (10.44) gives $C_F = 1.185$, and

$$\text{Ideal } C = C_F C^* = 1.185 \times 5118 = 6065 \text{ ft/s}$$

$$\text{Ideal } I_{sp} = \frac{C}{g_o} = \frac{6065}{32.174} = 188.5 \text{ s}$$

The experimental and manufacturer's values of C^* are the same. The average and initial burning rates are within 5% of each other. The average and initial mass flow rates are within 1% of each other. Closer agreement between experimental and ideal results could be obtained with a higher γ.

Problems

10.1   For propellants having a specific impulse of 360 s, determine the mass flow rate through a rocket nozzle to obtain a thrust of 10,000 N.

10.2   For propellants having a specific impulse of 265 s, determine the mass flow rate through a rocket nozzle to obtain a thrust of 1,500,000 lbf.

10.3   Determine the mass flow rate for the RS-68A rocket engine of Appendix C.

10.4   Determine the mass flow rate for the RD-181 rocket engine of Appendix C.

10.5   Exhaust gas with $R = 0.375$ kJ/kg $\cdot$ K and $\gamma = 1.26$ flows through an isentropic nozzle. If the gas enters the nozzle at 100 atm and 4000 K and exits to a pressure of 1 atm, determine the following for a choked nozzle throat:

a) The exhaust velocity V_e

b) The specific impulse I_{sp}

c) The mass flow rate and ideal thrust for a throat area A_t of 0.1 m^2

10.6   Determine the area ratio of the nozzle in Problem 10.5.

10.7   Determine the gross weight at liftoff for a single-stage rocket with a 2000-kg payload placed in a low Earth orbit using a liquid O_2–H_2 propulsion system. Assume a burn time of 100 s and an average specific impulse of 400 s.

10.8   Determine the gross weight at liftoff for a single-stage rocket with a 1000-kg payload placed in a low Earth orbit using a liquid

O_2–kerosene propulsion system. Assume a burn time of 120 s and an average specific impulse of 310 s.

10.9   Rework Example 10.3 with a liquid O_2–kerosene propulsion system having an average specific impulse of 310 s.

10.10   A calorically perfect gas with a molecular weight of 18 and $\gamma = 1.25$ has a combustion temperature T_c of 6000°R. Determine the characteristic velocity C^*.

10.11   Determine the characteristic velocity C^* for the data of Example 10.5.

10.12   Determine the area ratio (A_e/A_t) for a rocket nozzle operating at a pressure ratio (P_c/P_e) of 1000 with a calorically perfect gas where $\gamma = 1.2$.

10.13   A rocket nozzle has an ideal thrust coefficient C_{Fi} of 1.5, a chamber pressure P_c of 100 atm, and a throat area A_t of 0.15 m². Determine the ideal thrust F_i.

10.14   A rocket nozzle has an area ratio (A_e/A_t) of 25. If a calorically perfect gas with $\gamma = 1.25$ flows through the nozzle, determine its pressure ratio (P_c/P_e) and ideal thrust coefficient in a vacuum $(C_{Fi})_{vac}$.

10.15   Find the ideal thrust coefficient C_{Fi} for a rocket at 50-km altitude with an area ratio of 50, $\gamma = 1.28$, and a chamber pressure P_c of 40 atm.

10.16   Calculate the ideal thrust coefficient C_{Fi} and ideal thrust F_i for the space shuttle main engine (SSME) data of Example 10.7 at an altitude of 45,000 ft.

10.17   A chemical rocket motor has the following data during static testing at sea level:

$$P_c = 225 \text{ atm}, \quad T_c = 2800 \text{ K}, \quad M_e = 5.3,$$
$$\varepsilon = 62, \quad \text{exit diameter} = 2.15 \text{ m}$$

Assume a calorically perfect gas with $\gamma = 1.3$ and $R = 0.317$ kJ/kg · K. Determine the throat area A_t, exit pressure P_e, exit temperature T_e, exit velocity V_e, mass flow rate $\dot{m}$, and thrust at both sea-level and design altitude.

10.18   A rocket booster using H_2–O_2 requires an ideal thrust of 100,000 lbf at a design altitude of 80 kft. The booster will have a chamber temperature of 4840°R and a throat area of 0.5 ft². Assuming a specific impulse of 400 s and a calorically perfect gas with $\gamma = 1.26$ and $R = 173.6$ ft · lbf/lbm · °R, determine the following:

a) Mass flow rate

b) Characteristic velocity C^*

c) Chamber pressure P_c

d) Effective exhaust velocity C at 80-kft altitude

e) Ideal thrust coefficient C_{Fi} at 80-kft altitude

f) Nozzle expansion ratio ε and exit diameter

g) Altitude that flow in the nozzle is just separated (assume separation when $P_a > 3.5\, P_{we}$)

h) Ideal thrust at sea level

10.19   Determine the I_{sp} of the oxygen–hydrogen rocket engine with a chamber temperature of 3200 K and an oxidizer-to-fuel mixture ratio by weight of 3.0:1 using the EQL software for both equilibrium and frozen flow through the nozzle. The rocket engine has a chamber pressure P_c of 100 atm and an exit pressure P_e of 1 atm. Assume isentropic flow and perfect expansion.

10.20   Perform Problem 10.19 with a chamber temperature of 3600 K, a 3.2:1 ratio, a chamber pressure P_c of 80 atm, and an exit pressure P_e of 1 atm.

10.21   Determine the coefficients of Eq. (10.56) from Fig. 10.45 for high-energy XLDB composite propellant at chamber pressures of 500, 1000, 2000, and 3000 psi.

10.22   A solid-propellant rocket has $\rho_s = 0.06$ lbm/in.^{3}, $n = 0.2$, $a = 0.05$ in./s, and $C^* = 5000$ ft/s. Determine the burn-to-throat area ratio (A_b/A_t) for a chamber pressure of 1500 psia.

10.23   Determine the coefficients n and a of Problem 10.22 in SI units.

10.24   Determine and plot the variation in propellant burn area A_b, chamber pressure P_c, and burning rate r over time for the rocket motor of Example 10.12 (Case I). Assume the throat area A_t remains constant and time increments of 0.1 s. Use a chamber pressure of 40 psia for calculation of the burning rate during the first time increment. Neglect the time needed to fill the volume with combustion gas. Do the results make sense? Why?

10.25   Determine the performance of Case II in Table 10.17. Compare your results to Case I (Example 10.12).

10.26   From a search of the Internet, find a thrust vs time plot for a solid-propellant motor (e.g., Estes Rocket Motors) and determine the burning time, action time, impulse, average mass flow rate, and average thrust.

10.27   Test firing of a 2.75-in. solid rocket at the U.S. Air Force Academy on May 8, 1992, gave the data plotted in Fig. P10.1 [56]. The rocket motor used was the "Mighty Mouse" solid-propellant system used for the air-to-ground 2.75-in. folding fin aerial rocket. The motor uses a Class B propellant (MK43 MOD 1 propellant grain) with an internal burning eight-point star grain configuration. The motor has four straight, fixed, conical nozzles. The physical characteristics and published performance of the Mighty Mouse rocket motor based on 70°F sea-level operation are as follows:

Size and mass:

Overall length	39.4 in.
Nominal diameter	2.75 in.
Loaded mass	11.37 lbm
Expended mass	5.9 lbm
Propellant mass	6.40 lbm
Initial area of propellant burning surface	160.83 in.^{2}

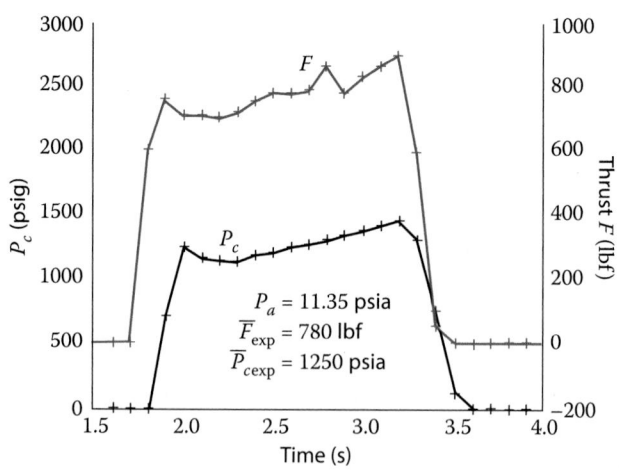

Fig. P10.1   Test firing of a 2.75-in. solid rocket at the U.S. Air Force Academy on May 8, 1992 (Ref. 56).

Manufacturer's performance characteristics (70°F sea-level operation):

Average chamber pressure P_c	1210 psia
Motor specific impulse	103 lbf · s/lbm
Action time impulse	1140 lbf · s
Total impulse	1170 lbf · s
Propellant	NC/NG/DEP/ other (50/36/11/4%)
Propellant specific impulse I_{sp}	183 s
Propellant density ρ_p	0.056 lbm/in.^{3}
Propellant burn rate (1000 psia) r	0.48 in./s
Ratio of specific heat for gas γ	1.17
Propellant mass fraction	0.563
Total nozzle throat area A^*	0.444 in.^{2}
Expansion cone half angle	10°15′
Nozzle expansion ratio ε	3.75
Characteristic exhaust velocity C^*_{manuf}	4159 ft/s
Thrust coefficient C_F	1.415

a) Calculate the exit Mach number M_e.

b) Using the experimental chamber pressure $P_{c\,exp}$ and calculated exit Mach number, calculate the exit pressure P_e.

c) Describe the exit flow as underexpanded or overexpanded (see Fig. 3.54).

d) Calculate the mass flow rate of the combustion products based on the initial area of propellant burning surface.

e) Calculate the characteristic exhaust velocity C^*_{exp} based on the mass flow rate of item c and the experimental chamber pressure $P_{c\,exp}$.

f) Calculate the thrust coefficient $C_{F\,calc}$ using the measured thrust and chamber pressure $P_{c\,exp}$.

g) Calculate the thrust produced by the rocket using the calculated thrust coefficient, measured chamber pressure, and nozzle throat area.

h) Calculate the experimental specific impulse $I_{sp\,exp}$.

i) Compare calculated results to manufacturer's performance characteristics.

Appendix A Altitude Tables

A.1 Units

Table A.1 British Engineering (BE) Units[a]

h, kft	δ, P/P_{std}	Standard Day θ, T/T_{std}	Cold Day θ, T/T_{std}	Hot Day θ, T/T_{std}	Tropical Day θ, T/T_{std}	h, kft
0	1.0000	1.0000	0.7708	1.0849	1.0594	0
1	0.9644	0.9931	0.7972	1.0774	1.0520	1
2	0.9298	0.9863	0.8237	1.0700	1.0446	2
3	0.8963	0.9794	0.8501	1.0626	1.0372	3
4	0.8637	0.9725	0.8575	1.0552	1.0298	4
5	0.8321	0.9656	0.8575	1.0478	1.0224	5
6	0.8014	0.9588	0.8575	1.0404	1.0150	6
7	0.7717	0.9519	0.8575	1.0330	1.0076	7
8	0.7429	0.9450	0.8575	1.0256	1.0002	8
9	0.7149	0.9381	0.8575	1.0182	0.9928	9
10	0.6878	0.9313	0.8565	1.0108	0.9854	10
11	0.6616	0.9244	0.8502	1.0034	0.9780	11
12	0.6362	0.9175	0.8438	0.9960	0.9706	12
13	0.6115	0.9107	0.8375	0.9886	0.9632	13
14	0.5877	0.9038	0.8312	0.9812	0.9558	14
15	0.5646	0.8969	0.8248	0.9738	0.9484	15
16	0.5422	0.8901	0.8185	0.9664	0.9410	16
17	0.5206	0.8832	0.8121	0.9590	0.9336	17
18	0.4997	0.8763	0.8058	0.9516	0.9262	18
19	0.4795	0.8695	0.7994	0.9442	0.9188	19
20	0.4599	0.8626	0.7931	0.9368	0.9114	20
21	0.4410	0.8558	0.7867	0.9294	0.9040	21
22	0.4227	0.8489	0.7804	0.9220	0.8965	22
23	0.4051	0.8420	0.7740	0.9145	0.8891	23
24	0.3880	0.8352	0.7677	0.9071	0.8817	24
25	0.3716	0.8283	0.7613	0.8997	0.8743	25

(Continued)

Table A.1   British Engineering (BE) Units[a] *(Continued)*

h, kft	δ, P/P_{std}	Standard Day θ, T/T_{std}	Cold Day θ, T/T_{std}	Hot Day θ, T/T_{std}	Tropical Day θ, T/T_{std}	h, kft
26	0.3557	0.8215	0.7550	0.8923	0.8669	26
27	0.3404	0.8146	0.7486	0.8849	0.8595	27
28	0.3256	0.8077	0.7423	0.8775	0.8521	28
29	0.3113	0.8009	0.7360	0.8701	0.8447	29
30	0.2975	0.7940	0.7296	0.8627	0.8373	30
31	0.2843	0.7872	0.7233	0.8553	0.8299	31
32	0.2715	0.7803	0.7222	0.8479	0.8225	32
33	0.2592	0.7735	0.7222	0.8405	0.8151	33
34	0.2474	0.7666	0.7222	0.8331	0.8077	34
35	0.2360	0.7598	0.7222	0.8257	0.8003	35
36	0.2250	0.7529	0.7222	0.8183	0.7929	36
37	0.2145	0.7519	0.7222	0.8109	0.7855	37
38	0.2044	0.7519	0.7222	0.8035	0.7781	38
39	0.1949	0.7519	0.7222	0.7961	0.7707	39
40	0.1858	0.7519	0.7222	0.7939	0.7633	40
42	0.1688	0.7519	0.7222	0.7956	0.7485	42
44	0.1534	0.7519	0.7095	0.7973	0.7337	44
46	0.1394	0.7519	0.6907	0.7989	0.7188	46
48	0.1267	0.7519	0.6719	0.8006	0.7040	48
50	0.1151	0.7519	0.6532	0.8023	0.6892	50
52	0.1046	0.7519	0.6452	0.8040	0.6744	52
54	0.09507	0.7519	0.6452	0.8057	0.6768	54
56	0.08640	0.7519	0.6452	0.8074	0.6849	56
58	0.07852	0.7519	0.6452	0.8091	0.6929	58
60	0.07137	0.7519	0.6452	0.8108	0.7009	60
62	0.06486	0.7519	0.6514	0.8125	0.7090	62
64	0.05895	0.7519	0.6609	0.8142	0.7170	64
66	0.05358	0.7521	0.6704	0.8159	0.7251	66
68	0.04871	0.7542	0.6799	0.8166	0.7331	68
70	0.04429	0.7563	0.6894	0.8196	0.7396	70
72	0.04028	0.7584	0.6990	0.8226	0.7448	72
74	0.03665	0.7605	0.7075	0.8255	0.7501	74
76	0.03336	0.7626	0.7058	0.8285	0.7553	76
78	0.03036	0.7647	0.7042	0.8315	0.7606	78
80	0.02765	0.7668	0.7026	0.8344	0.7658	80
82	0.02518	0.7689	0.7009	0.8374	0.7711	82

(Continued)

Table A.1 British Engineering (BE) Units[a] *(Continued)*

h, kft	δ, P/P_{std}	Standard Day θ, T/T_{std}	Cold Day θ, T/T_{std}	Hot Day θ, T/T_{std}	Tropical Day θ, T/T_{std}	h, kft
84	0.02294	0.7710	0.6993	0.8403	0.7763	84
86	0.02091	0.7731	0.6976	0.8433	0.7816	86
88	0.01906	0.7752	0.6960	0.8463	0.7868	88
90	0.01738	0.7772	0.6944	0.8492	0.7921	90
92	0.01585	0.7793	0.6927	0.8522	0.7973	92
94	0.01446	0.7814	0.6911	0.8552	0.8026	94
96	0.01320	0.7835	0.6894	0.8581	0.8078	96
98	0.01204	0.7856	0.6878	0.8611	0.8130	98
100	0.01100	0.7877	0.6862	0.8640	0.8183	100

[a]Density: $\rho = \rho_{std}$ $\sigma = \rho_{std}(\delta/\theta)$. Speed of sound: $a = a_{std}\sqrt{\theta}$.
Reference values: $P_{std} = 2116.2$ lbf/ft^2; $T_{std} = 518.69°$R; $\rho_{std} = 0.07647$ lbm/ft^3; $a_{std} = 1116$ ft/s.

Table A.2 Système International (SI) Units[a]

h, km	δ, P/P_{std}	Standard Day θ, T/T_{std}	Cold Day θ, T/T_{std}	Hot Day θ, T/T_{std}	Tropical Day θ, T/T_{std}	h, km
0	1.0000	1.0000	0.7708	1.0849	1.0594	0
0.25	0.9707	0.9944	0.7925	1.0788	1.0534	0.25
0.50	0.9421	0.9887	0.8142	1.0727	1.0473	0.50
0.75	0.9142	0.9831	0.8358	1.0666	1.0412	0.75
1.00	0.8870	0.9774	0.8575	1.0606	1.0352	1.00
1.25	0.8604	0.9718	0.8575	1.0545	1.0291	1.25
1.50	0.8345	0.9662	0.8575	1.0484	1.0230	1.50
1.75	0.8093	0.9605	0.8575	1.0423	1.0169	1.75
2.00	0.7846	0.9549	0.8575	1.0363	1.0109	2.00
2.25	0.7606	0.9493	0.8575	1.0302	1.0048	2.25
2.50	0.7372	0.9436	0.8575	1.0241	0.9987	2.50
2.75	0.7143	0.9380	0.8575	1.0180	0.9926	2.75
3.00	0.6920	0.9324	0.8575	1.0120	0.9866	3.00
3.25	0.6703	0.9267	0.8523	1.0059	0.9805	3.25
3.50	0.6492	0.9211	0.8471	0.9998	0.9744	3.50
3.75	0.6286	0.9155	0.8419	0.9938	0.9683	3.75
4.00	0.6085	0.9098	0.8367	0.9877	0.9623	4.00
4.25	0.5890	0.9042	0.8315	0.9816	0.9562	4.25
4.50	0.5700	0.8986	0.8263	0.9755	0.9501	4.50
4.75	0.5514	0.8929	0.8211	0.9695	0.9441	4.75

(Continued)

Table A.2 Système International (SI) Units[a] *(Continued)*

h, km	δ, P/P_{std}	Standard Day θ, T/T_{std}	Cold Day θ, T/T_{std}	Hot Day θ, T/T_{std}	Tropical Day θ, T/T_{std}	h, km
5.00	0.5334	0.8873	0.8159	0.9634	0.9380	5.00
5.25	0.5159	0.8817	0.8107	0.9573	0.9319	5.25
5.50	0.4988	0.8760	0.8055	0.9512	0.9258	5.50
5.75	0.4822	0.8704	0.8003	0.9452	0.9198	5.75
6.00	0.4660	0.8648	0.7951	0.9391	0.9137	6.00
6.25	0.4503	0.8592	0.7899	0.9330	0.9076	6.25
6.50	0.4350	0.8535	0.7847	0.9269	0.9015	6.50
6.75	0.4201	0.8479	0.7795	0.9209	0.8955	6.75
7.00	0.4057	0.8423	0.7742	0.9148	0.8894	7.00
7.25	0.3916	0.8366	0.7690	0.9087	0.8833	7.25
7.50	0.3780	0.8310	0.7638	0.9027	0.8773	7.50
7.75	0.3647	0.8254	0.7586	0.8966	0.8712	7.75
8.00	0.3519	0.8198	0.7534	0.8905	0.8651	8.00
8.25	0.3393	0.8141	0.7482	0.8844	0.8590	8.25
8.50	0.3272	0.8085	0.7430	0.8784	0.8530	8.50
8.75	0.3154	0.8029	0.7378	0.8723	0.8469	8.75
9.00	0.3040	0.7973	0.7326	0.8662	0.8408	9.00
9.25	0.2929	0.7916	0.7274	0.8601	0.8347	9.25
9.50	0.2821	0.7860	0.7222	0.8541	0.8287	9.50
9.75	0.2717	0.7804	0.7222	0.8480	0.8226	9.75
10.00	0.2615	0.7748	0.7222	0.8419	0.8165	10.00
10.25	0.2517	0.7692	0.7222	0.8358	0.8104	10.25
10.50	0.2422	0.7635	0.7222	0.8298	0.8044	10.50
10.75	0.2330	0.7579	0.7222	0.8237	0.7983	10.75
11.00	0.2240	0.7523	0.7222	0.8176	0.7922	11.00
11.25	0.2154	0.7519	0.7222	0.8116	0.7862	11.25
11.50	0.2071	0.7519	0.7222	0.8055	0.7801	11.50
11.75	0.1991	0.7519	0.7222	0.7994	0.7740	11.75
12.00	0.1915	0.7519	0.7222	0.7933	0.7679	12.00
12.25	0.1841	0.7519	0.7222	0.7940	0.7619	12.25
12.50	0.1770	0.7519	0.7222	0.7947	0.7558	12.50
12.75	0.1702	0.7519	0.7222	0.7954	0.7497	12.75
13.00	0.1636	0.7519	0.7222	0.7961	0.7436	13.00
13.25	0.1573	0.7519	0.7145	0.7968	0.7376	13.25
13.50	0.1513	0.7519	0.7068	0.7975	0.7315	13.50
13.75	0.1454	0.7519	0.6991	0.7982	0.7254	13.75

(Continued)

Table A.2 Système International (SI) Units[a] *(Continued)*

h, km	δ, P/P_{std}	Standard Day θ, T/T_{std}	Cold Day θ, T/T_{std}	Hot Day θ, T/T_{std}	Tropical Day θ, T/T_{std}	h, km
14.00	0.1399	0.7519	0.6914	0.7989	0.7193	14.00
14.25	0.1345	0.7519	0.6837	0.7996	0.7133	14.25
14.50	0.1293	0.7519	0.6760	0.8003	0.7072	14.50
14.75	0.1243	0.7519	0.6683	0.8010	0.7011	14.75
15.00	0.1195	0.7519	0.6606	0.8017	0.6951	15.00
15.25	0.1149	0.7519	0.6529	0.8024	0.6890	15.25
15.50	0.1105	0.7519	0.6452	0.8031	0.6829	15.50
15.75	0.1063	0.7519	0.6452	0.8037	0.6768	15.75
16.0	0.1022	0.7519	0.6452	0.8044	0.6708	16.0
16.5	0.09447	0.7519	0.6452	0.8058	0.6774	16.5
17.0	0.08734	0.7519	0.6452	0.8072	0.6839	17.0
17.5	0.08075	0.7519	0.6452	0.8086	0.6905	17.5
18.0	0.07466	0.7519	0.6452	0.8100	0.6971	18.0
18.5	0.06903	0.7519	0.6452	0.8114	0.7037	18.5
19.0	0.06383	0.7519	0.6531	0.8128	0.7103	19.0
19.5	0.05902	0.7519	0.6611	0.8142	0.7169	19.5
20.0	0.05457	0.7519	0.6691	0.8155	0.7235	20.0
20.5	0.05046	0.7534	0.6771	0.8169	0.7301	20.5
21.0	0.04667	0.7551	0.6851	0.8180	0.7367	21.0
21.5	0.04317	0.7568	0.6930	0.8204	0.7410	21.5
22.0	0.03995	0.7585	0.7010	0.8228	0.7453	22.0
23	0.03422	0.7620	0.7063	0.8277	0.7539	23
24	0.02933	0.7654	0.7036	0.8326	0.7625	24
25	0.02516	0.7689	0.7009	0.8374	0.7711	25
26	0.02160	0.7723	0.6982	0.8423	0.7797	26
27	0.01855	0.7758	0.6955	0.8471	0.7883	27
28	0.01595	0.7792	0.6928	0.8520	0.7969	28
29	0.01372	0.7826	0.6901	0.8568	0.8056	29
30	0.01181	0.7861	0.6874	0.8617	0.8142	30

[a]Density: $\rho = \rho_{std}$ $\sigma = \rho_{std}(\delta/\theta)$. Speed of sound: $a = a_{std}\sqrt{\theta}$.
Reference values: $P_{std} = 101{,}325$ N/m^2; $T_{std} = 288.15$ K; $\rho_{std} = 1.225$ kg/m^3; $a_{std} = 340.3$ m/s.

A.2 International Standard Atmosphere (ISA) 1975

A computer model (e.g., ATMOS program) of the standard day atmosphere can be written from the following material extracted from Ref. 2. All of the following is limited to geometric altitudes below 86 km—the original tables go higher, up to 1000 km. In addition, the correction for variation in mean molecular weight with altitude is very small, below 86 km, and so it is neglected.

The geometric or actual altitude h is related to the geo-potential altitude z, only used for internal calculations (a correction for variation of acceleration of gravity, used only for pressure and density calculations), by

$$z = r_0 h / (r_0 + h)$$

where $r_0 = 6356.577$ km is the Earth's radius.

The variation of temperature T with geo-potential altitude is represented by a continuous, piecewise linear relation,

$$T = T_i + L_i(z - z_i) \quad i = 0 \text{ through } 7$$

with fit coefficients as shown in Table A.3.

With $T_0 = 288.15$ K given, the corresponding values of temperature T_i can be readily generated from the given piecewise linear curve-fit. Note that z_7 corresponds exactly to $h = 86$ km.

The corresponding pressure P, also a piecewise continuous function, is given by

$$P = P_i \left(\frac{T_i}{T}\right)^{\left(\frac{g_0 \mathcal{M}}{\mathcal{R}^* L_i}\right)} \quad L_i \neq 0 \quad \text{or}$$

$$P = P_i \exp\left(\frac{-g_0 \mathcal{M}(z - z_i)}{\mathcal{R}^* T_i}\right) \quad L_i = 0$$

Table A.3   Standard Day Temperature Model

i	z_i, km	L_i, K/km
0	0	−6.5
1	11	0.0
2	20	+1.0
3	32	+2.8
4	47	0.0
5	51	−2.8
6	71	−2.0
7	84.852	–

Table A.4 Nonstandard Day Temperature Models

i	\multicolumn{2}{c}{Cold Day}		\multicolumn{2}{c}{Hot Day}		\multicolumn{2}{c}{Tropical Day}
	h_i, km	L_i, K/km	h_i, km	L_i, K/km	h_i, km	L_i, K/km
0	0	+25	0	−7.0	0	−7.0
1	1	0	12	+0.8	16	+3.8
2	3	−6.0	20.5	+1.4	21	+2.48
3	9.5	0				
4	13	−8.88				
5	15.5	0				
6	18.5	+4.6				
7	22.5	−0.775				

where $g_0 = 9.80665 \text{ m/s}^2$, $R^* = 8314.32 \text{ J/kmol} \cdot \text{K}$, $\mathcal{M} = 28.9644 \text{ Kg/kmol}$, and the pressure calculations start from $P_0 = 101{,}325 \text{ N/m}^2$.

The density ρ is given simply by the ideal gas law

$$\rho = \frac{P\mathcal{M}_0}{R^* T}$$

A.3 Cold, Hot, and Tropical Day Temperature Profiles

A computer model (e.g., ATMOS program) of the temperature profiles for cold, hot, and tropical days can be written from the following material extracted from linear curve-fits of the data in Refs. 58 and 59. The following is limited to pressure altitudes below 30.5 km. The variation of temperature T with pressure altitude is represented by a continuous, piecewise linear relation,

$$T = T_i + L_i(h - h_i), \quad i = 0 \text{ through } 7$$

with fit coefficients as shown in Table A.4.

With T_0 given in Table A.5 for the respective temperature profile, the corresponding values of temperature T_i can be readily generated from the given piecewise linear curve-fit.

Table A.5 Sea-Level Base Temperature

Day	Cold	Hot	Tropical
T_0, K	222.10	312.60	305.27

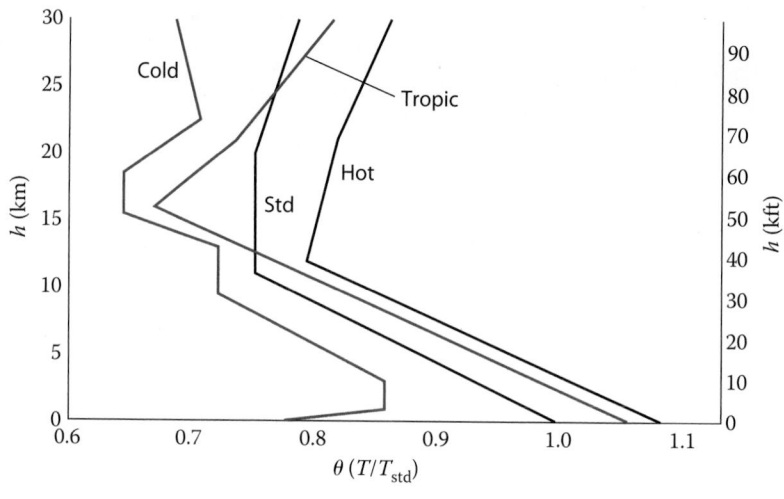

Fig. A.1   Four atmospheric temperature profiles vs pressure altitude h.

The pressure at the pressure altitude comes directly from the standard atmosphere calculation for the geometric or actual altitude h equal to that pressure altitude. Figure A.1 shows the three nonstandard day temperature profiles vs pressure altitude along with that of a standard day.

Appendix B

Gas Turbine Engine Data: Uninstalled Performance

Table B.1   Data for Some Military Gas Turbine Engines

Model No.	Type TJ,[e] TP[f], TS[g]	Max. Thrust or Power @ SLS klbf | kN eshp[h] | kW		SFC[a] at Max. lbm/(h·lbf) | g/(kN·s) lbm/(h·hp) | g/(MW·s)		Airflow lbm/s | kg/s		OPR[b] (Stages)[c]	TIT[d] °F | °C		Application
J57-P-23	TJ	16.0	71.2	2.10	59.5	165	74.8	11.5 (16)	1600	871	AB, F-102A, F-100D
J57-P-43WB	TJ	11.2	49.8	0.775	22.0	180	81.6	12 (16)	1600	871	Water-injected, KC-135
J58-P[i]	TJ	32.5	144.6			450	204.1	6 (9)	1600	871	AB, YF-12A, SR-71
J60-P-3	TJ	3.0	13.3	0.96	27.2	50	22.7	7 (9)	1600	871	T-39A, C-140A
J69-T-25	TJ	1.025	4.56	1.14	32.3	20.5	9.3	3.9 (0.1)	1525	830	T-37B
J75-P-17	TJ	24.5	109.0	2.15	60.9	252	114.3	12.0 (15)	1610	877	AB, F-106A|B
J79-GE-17	TJ	17.8	79.3	1.965	55.66	170	77.1	13.5 (17)	1210	655	AB, F-4E|G
J85-GE-5H	TJ	3.85	17.1	2.20	62.3	44	20.0	7 (8)	1640	894	AB, T-38A|B
J85-GE-17	TJ	2.85	12.7	0.99	28.04	44	20.0	7 (8)	1640	894	A-37B
J85-GE-21	TJ	5.0	22.2	2.13	60.3	51.9	23.5	8 (8)	1790	977	AB, F-5E|F
PT6A-42	TP	850	634	0.601	101.6	8.0	3.63	8 (3,1)			C-12E
PT6A-45R	TP	1197	893	0.553	93.4	8.6	3.9	8.7 (3,1)			C-23A
T400-CP-400	TS	1800	1342	0.606	102.4	6.51	2.95	7 (3,1)	1920	1049	Bell UH-1N
T406-AD-400 (AE1107C)	TS	6150	4585	0.424	71.6			16.7 (14)	1422	772	CV-22

Engine	Type		SFC		OPR (Stages)	TIT	Aircraft
T53-L-13	TS	1400 | 1044	0.58 | 98.0	12.2 | 5.53	7 (5,1)	1720 | 938	Bell UH-1H, AH-1G
T55-L-11	TS	3750 | 2796	0.52 | 87.9		8 (6,1)		Boeing CH-47C
T56-A-7	TP	3775 | 2815	0.528 | 89.2	32.5 | 14.7	9.45 (14)	1780 | 971	C-130B/E/F
T56-A-15	TP	4591 | 3423	0.54 | 91.2	32.5 | 14.7	9.55 (14)	1970 | 1077	C-130H/N/P
T58-GE-100	TS	1500 | 1119	0.606 | 102.4	14 | 6.35	8.4 (10)	1372 | 744	Sikorsky CH-3E, HH-3E, F
T64-GE-100	TS	4300 | 3207	0.487 | 82.3	29.3 | 13.3	14 (14)	1520 | 827	MH-53T
T700-GE-700	TS	1622 | 1210	0.46 | 77.7		15 (5,1)	1563 | 851	UH-60A
T76-G-10	TS	715 | 533	0.60 | 101.4	6.16 | 2.79	8.6 (2)	1818 | 992	OV-10A
TP400-D6	TP	11,000 | 8200		37.4 | 17.0	25		A400M
AE2100	TP	4152 | 3096 6100 | 4550			16.6 (14)		Saab 2000, C130J, C27J

[a] SFC = specific fuel consumption.
[b] OPR = overall pressure ratio (P_{t3}/P_{t2}).
[c] (Stages) = (axial, centrifugal) compressor stages.
[d] TIT = turbine inlet temperature.
[e] TJ = turbojet.
[f] TP = turboprop.
[g] TS = turboshaft.
[h] eshp = equivalent shaft horsepower.
[j] J-58 Reference: Miller, J., Lockheed SR-71 in *Aerofax Minigraph 1*, Aerofax, Inc., Arlington, TX, 1985.
Source: Manufacturers' literature.

Table B.2   Data for Some Military Turbofan Engines

Model No.	Thrust klbf | kN	TSFC[a] lbm/(h · lbf) | g/(kN · s)	Airflow lbm/s | kg/s	OPR[b]	TIT[c] °F | °C	FPR[d]	BR[e]	Application
F100-PW-229	29.0 | 129.0 17.8 | 79.2	2.05 | 58.1 0.74 | 21.0	248 | 112.5	23.0	2700 | 1537	3.8	0.4	F-15, F-16
F101-GE-102	30.78 | 136.9 17.39 | 77.4	2.46 | 69.7 0.562 | 15.9	356 | 161.5	26.8	2550 | 1400	2.31	1.91	B-1B
F103-GE-101	51.71 | 230.0	0.399 | 11.3	1476 | 670	30.2	2490 | 1366		4.31	KC-10A
F107-WR-101	0.635 | 2.82	0.685 | 19.4	13.6 | 6.17	13.8		2.1	1.0	Air Launch Cruise Missile
F108-CF-100	21.634 | 96.2	0.363 | 10.3	785 | 356	23.7	2228 | 1220	1.5	6.0	KC-135R
F110-GE-100	28.62 | 127.3 18.33 | 81.5	2.08 | 58.9 0.688 | 19.5	254 | 115.2	30.4		2.8	0.80	F-16
F117-PW-100	40.44 | 179.9	0.347 | 9.83	1264 | 573.3	29.6		1.71	5.44	(PW2040) C-17A
F118-GE-100	19.0 | 84.5	0.67 | 19.0	285 | 129.3	35.1		3.41	0.8	B-2
F119-PW-100	35.0 | 155.7							F-22 Raptor
F135-PW	43.0 | 191.3							F-35 JSF
F404-GE-F1D2	10.5 | 46.7	0.79 | 22.4	145.3 | 65.9	25		4.1	0.37	F-117A
F404-GE-400	16.0 | 71.2	1.85 | 52.4	142 | 64.4	26			0.34	F-18, F-5G
F414-GE-400	22.0 | 97.9			30				F/A-18E/F

Engine	Thrust	TSFC	Airflow	OPR	TIT	FPR	BR	Aircraft
JT3D-3B	18.0 | 80.1	0.535 | 15.2	458 | 207.7	13.6	1600 | 871	1.74	1.37	(TF33-102) EC/RC-135
JT8D-7B	14.5 | 64.5	0.585 | 16.6	318 | 144.2	16.9	1076 | 580		1.03	C-22, C-9, T-43A
TF30-P-111	25.1 | 111.6 14.56 | 64.8	2.450 | 69.4 0.686 | 19.4	260 | 117.9	21.8	2055 | 1124	2.43	0.73	F-111F
TF33-P-3	17.0 | 75.6	0.52 | 14.7	450 | 204.1	13	1600 | 871	1.7	1.43	B-52H
TF33-P-7	21.0 | 93.4	0.56 | 15.9	498 | 225.9	16	1750 | 954	1.9	1.21	C-141
TF34-GE-100	9.065 | 40.3	0.37 | 10.5	333 | 151.0	20	2234 | 1223	1.5	6.42	A-10
TF39-GE-1	40.805 | 181.5	0.315 | 8.92	1549 | 702.6	26	2350 | 1288	1.56	8.0	C-5A
TF41-A-1B	14.5 | 64.5	0.647 | 18.3	260 | 117.9	20	2165 | 1185	2.45	0.76	A-7D, K
TFE731-2	3.5 | 15.6	0.504 | 14.3	113 | 51.3	17.7		1.54	2.67	C-21A

[a]TSFC = thrust specific fuel consumption.
[b]OPR = overall pressure ratio (P_{t3}/P_{t2}).
[c]TIT = turbine inlet temperature.
[d]FPR = fan pressure ratio.
[e]BR = bypass ratio.
Source: Manufacturers' literature.

Table B.3   Data for Some Civil Gas Turbine Engines

Model No.	Takeoff Thrust klbf | kN	BR[a]	OPR[b]	Airflow lbm/s | kg/s	Cruise Alt kft | km	Mach	Thrust klbf | kN	TSFC[c] lbm/(h · lbf) | g/(s · kN)	Application
CF34-8	14.5 | 64.5	5	28					0.68 | 19.3	Bombardier CRJ700, Embraer 170/175
CF6-50-C2	52.5 | 233.5	4.31	30.4	1476 | 669.5	35 | 10.7	0.80	11.555 | 51.4	0.630 | 17.8	DC10-10, A300B, 747-200
CF6-80-C2	52.5 | 233.5	5.31	27.4	1650 | 748.4	35 | 10.7	0.80	12.0 | 53.4	0.576 | 16.3	787-200, -300, -200ER
GE90-B4	87.4 | 388.8	8.40	39.3	3037 | 1378	35 | 10.7	0.80	17.5 | 77.8		777
GE90-115B	115.3 | 512.9	9	42			0.85			777-200LR, -300ER, 777F
GEnx-1B70	69.8 | 310.5	9	44	2560 | 1161	30 | 9.1	0.80			787-8
JT8D-15A	15.5 | 68.9	1.04	16.6	327 | 148.3	35 | 10.7	0.80	4.92 | 21.9	0.779 | 22.1	727, 737, DC9
JT9D-59A	53.0 | 235.7	4.90	24.5	1639 | 743.4	35 | 10.7	0.85	11.95 | 53.1	0.646 | 18.3	DC10-40, A300B, 747-200
PW2037	38.25 | 170.1	6.0	27.6	1210 | 548.9	35 | 10.7	0.85	6.5 | 28.9	0.582 | 16.9	757-200
PW4052	52.0 | 231.3	5.0	27.5	1700 | 771.1					767, A310-300
PW4084	87.9 | 391.0	6.41	34.4	2550 | 1157		0.83			777
PW1124G	27.0 | 120.1	12.0							A320neo
CFM56-3C	23.5 | 104.5	6.0	22.6	655 | 297.1	35 | 10.7	0.80	5.54 | 24.6	0.648 | 18.4	737-300, -400, -500
CFM56-5C	31.2 | 138.8	6.6	31.5	1027 | 465.8	35 | 10.7	0.80	6.6 | 29.4	0.545 | 15.4	A340
CFM56-7B24	24.2 | 107.6	5.3	32.8	751 | 340.7					737-700, -800, -900

Engine		BR	OPR					TSFC	Application
LEAP-1A	24.5 | 109 32.9 | 146	11	40						A320neo
LEAP-1B	23 | 102 28 | 124	9	40						737 MAX
AE 3007	8.6 | 38.3	4.8	20						Embraer 37, Global Hawk UAV
RB211-524B	50 | 222	4.5	28.4	1513 | 686.3	0.85	35 | 10.7	11.00 | 48.9	0.643 | 18.21	L1011-200, 747-200
RB211-535E	40.1 | 178	4.3	25.8	1151 | 522.1	0.80	35 | 10.7	8.495 | 38.8	0.607 | 17.19	757-200
RB211-882	84.7 | 377	6.01	39	2640 | 1197	0.83	35 | 10.7	16.20 | 72.1	0.557 | 15.78	777
Trent 900	70 | 311 76.5 | 340	8.7–8.5		2655 | 1204 2745 | 1245					A380
Trent 1000	53 | 236 75 | 334	10–11		2400 | 1089 2670 | 1211					787
V2528-D5	28 | 125	4.7	30.5	825 | 374.2	0.80	35 | 10.7	5.773 | 25.7	0.574 | 16.26	MD-90
ALF502R-5	6.97 | 31	5.7	12.2		0.70	25 | 7.6	2.25 | 10.0	0.720 | 20.39	BAe 146-200
TFE731-5	4.5 | 20	3.34	14.4	140 | 63.5	0.80	40 | 12.2	0.986 | 4.39	0.771 | 21.84	BAe 125-800
PW300	4.75 | 21	4.5	23	180 | 81.6	0.80	40 | 12.2	1.113 | 4.95	0.675 | 19.12	BAe 1000
FJ44	1.9 | 8.45	3.28	12.8	63.3 | 28.7	0.70	30 | 9.1	0.60 | 2.67	0.750 | 21.24	
Olympus 593	38 | 169	0	11.3	410 | 186	2.0	53 | 16.2	10.03 | 44.6	1.190 | 33.71	Concorde
GP7270	70 | 311	8.7[d]	45.6[e]		0.85	35 | 10.7	12.633 | 56.2		A380

[a] BR = bypass ratio.
[b] OPR = overall pressure ratio (P_{t3}/P_{t2}).
[c] TSFC = thrust specific fuel consumption.
[d] At cruise.
[e] Max climb.
Source: Manufacturers' literature.

Table B.4   Temperature/Pressure Data for Some Engines

Temperature and Pressure	Pegas Turbofan — Separate Exhaust	J57 Turbojet — w/AB[a] Exhaust	JTD3 Turbofan — Separate Exhaust	JT8D Turbofan — Mixed Exhaust	JT9D Turbofan — Separate Exhaust	F100-PW-200 Turbofan — Mixed w/AB Exhaust
P_{t2}, psia | kPa	14.7 | 101.3	14.7 | 101.3	14.7 | 101.3	14.7 | 101.3	14.7 | 101.3	14.7 | 101.3
T_{t2}, °F | °C	59 | 15	59 | 15	59 | 15	59 | 15	59 | 15	59 | 15
$P_{t2.5}$, psia | kPa	36.1 | 248.9	54 | 372.3	63 | 434.4	60 | 413.7	32.1 | 221.3	47.5 | 327.5
$T_{t2.5}$, °F | °C	242 | 116.7	330 | 165.6	360 | 182.2	355 | 179.5	210 | 98.9	307 | 152.8
P_{t13}, psia | kPa	36.5 | 251.7		26 | 179.3	28 | 193.1	22.6 | 155.8	47.5 | 327.5
T_{t13}, °F | °C	257 | 125.0		170 | 76.7	190 | 87.8	130 | 54.5	307 | 152.8
P_{t3}, psia | kPa	216.9 | 1495	167 | 1151	200 | 1379	233 | 1606	316 | 2179	387.7 | 2673
T_{t3}, °F | °C	708 | 375.6	660 | 348.9	715 | 379	800 | 426.7	880 | 471.1	1000 | 537.8
P_{t4}, psia | kPa		158 | 1089	190 | 1310	220 | 1517	302 | 2.083	350.5 | 2417
T_{t4}, °F | °C	1028 | 553.3	1570 | 854.5	1600 | 871.1	1720 | 937.8	1970 | 1077	2629 | 1443
$P_{t5} \| P_{t6}$, psia | kPa	29.3 | 202.0	36 | 248.2			20.9 | 144.1	44.8 | 308.9
T_{t5} or T_{t6}, °F | °C	510 | 265.6	1013 | 545.0			850 | 454.5	1362 | 738.9
P_{t16}, psia | kPa						42.8 | 295.1

T_{t16}, °F | °C						331 | 166.1
P_{t6A}, psia | kPa				29 | 200	29 | 200	43.7 | 301.3
T_{t6A}, °F | °C				890 | 476.7	890 | 476.7	989 | 531.7
P_{t7}, psia | kPa		31.9 | 219.9	28 | 193.1	29 | 200	20.9 | 144.1	39.5 | 272.3
T_{t7}, °F | °C		2540 | 1393	890 | 476.7	890 | 476.7	850 | 454.5	3167 | 1742
P_{t17}, psia | kPa	36.5 | 251.7		26 | 179.3		22.4 | 154.4	
T_{t17}, °F | °C	257 | 125.0		170 | 76.7		130 | 54.5	
Bypass ratio α	1.4	0	1.36	1.1	5.0	0.63
Thrust, klbf | kN	21.5 | 95.6	16.0 | 71.2	18.0 | 80.1	14.0 | 62.3	43.5 | 193.5	23.77 | 105.7
Airflow, lbm/s | kg/s	444 | 201.4	167 | 75.75	460 | 208.7	315 | 142.9	1495 | 678.1	228 | 103.4

[a]AB = afterburner.

Appendix C

Data for Some Liquid-Propellant Rocket Engines

Table C.1 Data for Some Liquid-Propellant Rocket Engines

Engine[a]	Thrust (klbf | kN) in Vacuum	Expansion Ratio ($\varepsilon = A_e/A_t$)	Design Alt. (kft | km) ($P_e = P_a$)	I_{sp} (s) Vacuum	I_{sp} (s) SL	Application
LR-87-BC-13	16 | 71.2	45:1	60 | 18.3	293		Agena
LR-87-AJ-9	474 | 2108	8:1	20 | 6.1	283	257	Titan 2 – 1st stage
LR-87-AJ-11	548.0 | 2438	15:1		301		Titan 4 – 1st stage
LR-91-AJ-7	100.57 | 447.3	49:1	100 | 30.5	309		Titan 2 – 2nd stage
LR-91-AJ-11	105 | 467	49.2:1		316		Titan 4 – 2nd stage
LR-105*	83.7 | 372	24.9:1	100 | 30.5	304	212	Std. launch vehicle
F-1*	1726 | 7677	16:1	80 | 24.4	305	265	Saturn V – 1st stage
H-1*	219 | 974	8:1	80 | 24.4	287	255	Saturn 1B
J-2*	200 | 890	25.5:1	100 | 30.5	426	380	Saturn V – 2nd and 3rd stages
RS-27*	231.7 | 1031	8:1			263	Delta launch vehicle
RS-27A*	237 | 1054	12:1		302	255	Delta II
RS-68A*	800 | 3559	21.5:1		412	360	Delta IV
SSME*	470 | 2090	77:1	44.6 | 198.4	455	364	Space shuttle main engine
RL10A-4†	20.8 | 92.5			449		Atlas IIA upper stage
HM7B†	14.3 | 63.6	83:1		446		Ariane 5 upper stage
RD-0110‡	30.4 | 135.2	82:1		326		Soyuz stage 2
RD-180¶	860.2 | 3826 (SL)	36:1		338	311	Atlas V
RD-181¶	933.4 | 4152	37:1		338	311	Antares
NK-33¶	380 | 1690			331	297	Soviet N-1
AJ26-62	377 | 1677			331	291	Antares
Vulcain 2	286 | 1272	59:1		431		Ariane 5

[a]Manufacturers: BC = Bell Aerosystems Co.; AJ = Aerojet. * = Rocketdyne. ¶ = Pratt & Whitney. † = Snecma. ‡ = Russia.

Appendix D · *Compressible Flow Functions*

he tables in this appendix present the compressible flow functions for three values of the ratio of specific heats ($\gamma = 1.4$, 1.33, and 1.3). The mass flow parameter (MFP) is presented in dimensionless form where it has been multiplied by the square root of R/g_c—remember to convert back to MFP when using data. The Mach angle and Prandtl-Meyer function are listed for supersonic Mach numbers.

$$\frac{T}{T_t} = \left(1 + \frac{\gamma - 1}{2}M^2\right)^{-1}$$

$$\frac{P}{P_t} = \left(1 + \frac{\gamma - 1}{2}M^2\right)^{-\gamma/(\gamma-1)}$$

$$\frac{\rho}{\rho_t} = \left(1 + \frac{\gamma - 1}{2}M^2\right)^{-1/(\gamma-1)}$$

$$\frac{A}{A^*} = \frac{1}{M}\left[\frac{2}{\gamma + 1}\left(1 + \frac{\gamma - 1}{2}M^2\right)\right]^{(\gamma+1)/[2(\gamma-1)]}$$

$$MFP(M)\sqrt{\frac{R}{g_c}} = \frac{\dot{m}\sqrt{T_t}}{P_t A}\sqrt{\frac{R}{g_c}} = M\sqrt{\gamma}\left(1 + \frac{\gamma - 1}{2}M^2\right)^{-(\gamma+1)/[2(\gamma-1)]}$$

For air,

$$\sqrt{\frac{R}{g_c}} = 1.28758 \, \text{lbf} \cdot \text{sec}/(\text{lbm} \cdot \sqrt{°R}) \quad \text{or}$$

$$\sqrt{\frac{R}{g_c}} = 16.9115 \, \text{N} \cdot \text{sec}/(\text{kg} \cdot \sqrt{K})$$

For $M \geq 1$,
   Mach angle:

$$\mu = \sin^{-1}(1/M)$$

   Prandtl-Meyer function:

$$v(M) = \sqrt{\frac{\gamma + 1}{\gamma - 1}} \cdot \tan^{-1}\sqrt{\frac{\gamma - 1}{\gamma + 1}(M^2 - 1)} - \tan^{-1}\sqrt{M^2 - 1}$$

Table D.1   Compressible Flow Functions ($\gamma = 1.4$)

M	T/T_t	P/P_t	ρ/ρ_t	A/A^*	$\mathrm{MFP}\sqrt{R/g_c}$
0.00	1.000000	1.000000	1.000000	Indef	0.000000
0.01	0.999980	0.999930	0.999950	57.873843	0.011831
0.02	0.999920	0.999720	0.999800	28.942130	0.023659
0.03	0.999820	0.999370	0.999550	19.300542	0.035477
0.04	0.999680	0.998881	0.999200	14.481486	0.047283
0.05	0.999500	0.998252	0.998751	11.591444	0.059072
0.06	0.999281	0.997484	0.998202	9.665910	0.070840
0.07	0.999021	0.996578	0.997554	8.291525	0.082582
0.08	0.998722	0.995533	0.996807	7.261610	0.094295
0.09	0.998383	0.994351	0.995961	6.461342	0.105974
0.10	0.998004	0.993031	0.995017	5.821829	0.117614
0.11	0.997586	0.991576	0.993976	5.299230	0.129213
0.12	0.997128	0.989985	0.992836	4.864318	0.140766
0.13	0.996631	0.988259	0.991600	4.496859	0.152269
0.14	0.996095	0.986400	0.990267	4.182400	0.163717
0.15	0.995520	0.984408	0.988838	3.910343	0.175108
0.16	0.994906	0.982285	0.987314	3.672739	0.186436
0.17	0.994253	0.980030	0.985695	3.463509	0.197699
0.18	0.993562	0.977647	0.983982	3.277926	0.208892
0.19	0.992832	0.975135	0.982176	3.112259	0.220011
0.20	0.992063	0.972497	0.980277	2.963520	0.231053
0.21	0.991257	0.969733	0.978286	2.829294	0.242015
0.22	0.990413	0.966845	0.976204	2.707602	0.252892
0.23	0.989531	0.963835	0.974032	2.596812	0.263682
0.24	0.988611	0.960703	0.971771	2.495562	0.274380

0.25	0.987654	0.957453	0.969421	2.402710	0.284983
0.26	0.986660	0.954085	0.966984	2.317287	0.295488
0.27	0.985630	0.950600	0.964460	2.238471	0.305893
0.28	0.984562	0.947002	0.961851	2.165554	0.316192
0.29	0.983458	0.943291	0.959157	2.097928	0.326385
0.30	0.982318	0.939470	0.956380	2.035065	0.336467
0.31	0.981142	0.935540	0.953521	1.976507	0.346435
0.32	0.979931	0.931503	0.950580	1.921851	0.356287
0.33	0.978684	0.927362	0.947560	1.870745	0.366021
0.34	0.977402	0.923118	0.944460	1.822876	0.375633
0.35	0.976086	0.918773	0.941283	1.777969	0.385120
0.36	0.974735	0.914330	0.938029	1.735778	0.394481
0.37	0.973350	0.909790	0.934700	1.696086	0.403713
0.38	0.971931	0.905156	0.931297	1.658696	0.412813
0.39	0.970478	0.900430	0.927821	1.623433	0.421780
0.40	0.968992	0.895614	0.924274	1.590140	0.430611
0.41	0.967474	0.890711	0.920657	1.558673	0.439304
0.42	0.965922	0.885722	0.916971	1.528905	0.447857
0.43	0.964339	0.880651	0.913217	1.500718	0.456269
0.44	0.962723	0.875498	0.909398	1.474005	0.464538
0.45	0.961076	0.870267	0.905513	1.448672	0.472662
0.46	0.959398	0.864960	0.901566	1.424629	0.480639
0.47	0.957689	0.859580	0.897556	1.401795	0.488468
0.48	0.955950	0.854128	0.893486	1.380097	0.496147
0.49	0.954180	0.848607	0.889357	1.359468	0.503676
0.50	0.952381	0.843019	0.885170	1.339844	0.511053

(Continued)

Table D.1 Compressible Flow Functions ($\gamma = 1.4$) (Continued)

M	T/T_t	P/P_t	ρ/ρ_t	A/A^*	$MFP\sqrt{R/g_c}$
0.51	0.950552	0.837367	0.880927	1.321168	0.518277
0.52	0.948695	0.831654	0.876629	1.303388	0.525347
0.53	0.946808	0.825881	0.872278	1.286454	0.532263
0.54	0.944894	0.820050	0.867876	1.270321	0.539022
0.55	0.942951	0.814165	0.863422	1.254948	0.545626
0.56	0.940982	0.808228	0.858920	1.240294	0.552072
0.57	0.938985	0.802241	0.854371	1.226326	0.558360
0.58	0.936961	0.796206	0.849775	1.213007	0.564491
0.59	0.934911	0.790127	0.845135	1.200308	0.570463
0.60	0.932836	0.784004	0.840452	1.188200	0.576277
0.61	0.930735	0.777841	0.835728	1.176654	0.581931
0.62	0.928609	0.771639	0.830963	1.165646	0.587427
0.63	0.926458	0.765402	0.826160	1.155151	0.592764
0.64	0.924283	0.759131	0.821319	1.145148	0.597941
0.65	0.922084	0.752829	0.816443	1.135616	0.602960
0.66	0.919862	0.746498	0.811533	1.126535	0.607821
0.67	0.917616	0.740140	0.806590	1.117887	0.612523
0.68	0.915349	0.733758	0.801616	1.109655	0.617067
0.69	0.913059	0.727353	0.796612	1.101822	0.621454
0.70	0.910747	0.720928	0.791579	1.094373	0.625684
0.71	0.908414	0.714485	0.786519	1.087294	0.629758
0.72	0.906060	0.708025	0.781434	1.080571	0.633676
0.73	0.903685	0.701552	0.776324	1.074192	0.637439
0.74	0.901291	0.695068	0.771191	1.068144	0.641048
0.75	0.898876	0.688573	0.766037	1.062417	0.644503

M					
0.76	0.896443	0.682070	0.760863	1.056999	0.647807
0.77	0.893991	0.675562	0.755670	1.051881	0.650959
0.78	0.891520	0.669050	0.750460	1.047053	0.653961
0.79	0.889031	0.662536	0.745234	1.042505	0.656813
0.80	0.886525	0.656022	0.739992	1.038230	0.659518
0.81	0.884001	0.649509	0.734738	1.034219	0.662076
0.82	0.881461	0.643000	0.729471	1.030464	0.664488
0.83	0.878905	0.636496	0.724193	1.026959	0.666756
0.84	0.876332	0.630000	0.718905	1.023696	0.668882
0.85	0.873744	0.623512	0.713609	1.020669	0.670866
0.86	0.871141	0.617034	0.708306	1.017871	0.672710
0.87	0.868523	0.610569	0.702997	1.015297	0.674415
0.88	0.865891	0.604117	0.697683	1.012941	0.675984
0.89	0.863245	0.597680	0.692365	1.010798	0.677417
0.90	0.860585	0.591260	0.687044	1.008863	0.678716
0.91	0.857913	0.584858	0.681722	1.007131	0.679883
0.92	0.855227	0.578476	0.676400	1.005597	0.680920
0.93	0.852529	0.572114	0.671079	1.004258	0.681828
0.94	0.849820	0.565775	0.665759	1.003108	0.682610
0.95	0.847099	0.559460	0.660443	1.002145	0.683266
0.96	0.844366	0.553170	0.655130	1.001365	0.683798
0.97	0.841623	0.546905	0.649822	1.000763	0.684209
0.98	0.838870	0.540669	0.644520	1.000337	0.684501
0.99	0.836106	0.534460	0.639225	1.000084	0.684674
1.00	0.833333	0.528282	0.633938	1.000000	0.684731

(Continued)

Table D.1 Compressible Flow Functions ($\gamma = 1.4$) (*Continued*)

M	T/T_t	P/P_t	ρ/ρ_t	A/A^*	$MFP\sqrt{R/g_c}$	μ	v
1.01	0.830551	0.522134	0.628660	1.000083	0.684675	81.930699	0.044725
1.02	0.827760	0.516018	0.623391	1.000330	0.684506	78.635123	0.125688
1.03	0.824960	0.509935	0.618133	1.000738	0.684227	76.137568	0.229428
1.04	0.822152	0.503886	0.612887	1.001305	0.683839	74.057631	0.350983
1.05	0.819336	0.497872	0.607653	1.002029	0.683345	72.247210	0.487411
1.06	0.816513	0.491894	0.602432	1.002907	0.682746	70.629962	0.636687
1.07	0.813683	0.485952	0.597225	1.003938	0.682046	69.160306	0.797295
1.08	0.810846	0.480047	0.592033	1.005119	0.681244	67.808393	0.968040
1.09	0.808002	0.474181	0.586856	1.006449	0.680344	66.553383	1.147948
1.10	0.805153	0.468354	0.581696	1.007925	0.679347	65.380023	1.336201
1.11	0.802298	0.462567	0.576553	1.009547	0.678256	64.276740	1.532100
1.12	0.799437	0.456820	0.571427	1.011312	0.677072	63.234499	1.735039
1.13	0.796572	0.451114	0.566320	1.013219	0.675798	62.246077	1.944484
1.14	0.793701	0.445451	0.561232	1.015267	0.674435	61.305586	2.159960
1.15	0.790826	0.439829	0.556164	1.017454	0.672985	60.408154	2.381042
1.16	0.787948	0.434251	0.551116	1.019780	0.671450	59.549686	2.607346
1.17	0.785065	0.428716	0.546090	1.022242	0.669833	58.726702	2.838523
1.18	0.782179	0.423225	0.541085	1.024840	0.668135	57.936214	3.074255
1.19	0.779290	0.417778	0.536102	1.027573	0.666358	57.175632	3.314248
1.20	0.776398	0.412377	0.531142	1.030440	0.664504	56.442690	3.558233
1.21	0.773503	0.407021	0.526205	1.033440	0.662575	55.735397	3.805961
1.22	0.770606	0.401711	0.521292	1.036572	0.660573	55.051986	4.057198
1.23	0.767707	0.396446	0.516403	1.039835	0.658500	54.390884	4.311731
1.24	0.764807	0.391229	0.511539	1.043229	0.656358	53.750679	4.569357
1.25	0.761905	0.386058	0.506701	1.046753	0.654148	53.130102	4.829888

1.26	0.759002	0.380934	0.501888	1.050406	0.651873	52.528005	5.093147
1.27	0.756098	0.375857	0.497102	1.054189	0.649534	51.943343	5.358968
1.28	0.753194	0.370828	0.492342	1.058100	0.647133	51.375167	5.627196
1.29	0.750289	0.365847	0.487608	1.062138	0.644673	50.822606	5.897681
1.30	0.747384	0.360914	0.482903	1.066305	0.642154	50.284863	6.170286
1.31	0.744480	0.356029	0.478225	1.070598	0.639579	49.761202	6.444878
1.32	0.741576	0.351192	0.473575	1.075018	0.636949	49.250946	6.721334
1.33	0.738672	0.346403	0.468953	1.079565	0.634266	48.753467	6.999533
1.34	0.735770	0.341663	0.464361	1.084239	0.631532	48.268183	7.279366
1.35	0.732869	0.336971	0.459797	1.089038	0.628749	47.794554	7.560724
1.36	0.729970	0.332328	0.455263	1.093964	0.625918	47.332075	7.843507
1.37	0.727072	0.327733	0.450758	1.099015	0.623041	46.880275	8.127618
1.38	0.724176	0.323187	0.446283	1.104193	0.620119	46.438716	8.412965
1.39	0.721282	0.318690	0.441838	1.109496	0.617155	46.006983	8.699460
1.40	0.718391	0.314241	0.437423	1.114926	0.614150	45.584691	8.987020
1.41	0.715502	0.309840	0.433039	1.120481	0.611105	45.171476	9.275565
1.42	0.712616	0.305488	0.428686	1.126162	0.608022	44.766995	9.565018
1.43	0.709733	0.301185	0.424363	1.131969	0.604903	44.370926	9.855307
1.44	0.706854	0.296929	0.420072	1.137903	0.601749	43.982963	10.146361
1.45	0.703977	0.292722	0.415812	1.143963	0.598561	43.602819	10.438113
1.46	0.701105	0.288563	0.411583	1.150150	0.595341	43.230221	10.730501
1.47	0.698236	0.284452	0.407386	1.156463	0.592091	42.864910	11.023462
1.48	0.695372	0.280388	0.403220	1.162904	0.588812	42.506642	11.316937
1.49	0.692511	0.276372	0.399086	1.169471	0.585505	42.155184	11.610871
1.50	0.689655	0.272403	0.394984	1.176167	0.582172	41.810315	11.905209
1.51	0.686804	0.268481	0.390914	1.182991	0.578814	41.471824	12.199899

(Continued)

Table D.1 Compressible Flow Functions ($\gamma = 1.4$) *(Continued)*

M	T/T_t	P/P_t	ρ/ρ_t	A/A^*	$MFP\sqrt{R/g_c}$	μ	ν
1.52	0.683957	0.264607	0.386876	1.189943	0.575432	41.139510	12.494891
1.53	0.681115	0.260779	0.382870	1.197024	0.572028	40.813184	12.790139
1.54	0.678279	0.256997	0.378897	1.204234	0.568603	40.492661	13.085594
1.55	0.675447	0.253262	0.374955	1.211574	0.565159	40.177770	13.381214
1.56	0.672622	0.249573	0.371045	1.219044	0.561696	39.868342	13.676957
1.57	0.669801	0.245930	0.367168	1.226644	0.558215	39.564218	13.972780
1.58	0.666987	0.242332	0.363323	1.234376	0.554719	39.265248	14.268646
1.59	0.664178	0.238779	0.359510	1.242239	0.551207	38.971284	14.564516
1.60	0.661376	0.235271	0.355730	1.250235	0.547682	38.682187	14.860354
1.61	0.658579	0.231808	0.351982	1.258364	0.544144	38.397824	15.156125
1.62	0.655789	0.228389	0.348266	1.266626	0.540595	38.118064	15.451795
1.63	0.653006	0.225014	0.344582	1.275022	0.537035	37.842786	15.747333
1.64	0.650229	0.221683	0.340930	1.283553	0.533466	37.571869	16.042707
1.65	0.647459	0.218395	0.337311	1.292219	0.529888	37.305201	16.337887
1.66	0.644695	0.215150	0.333723	1.301021	0.526303	37.042671	16.632845
1.67	0.641939	0.211948	0.330168	1.309960	0.522712	36.784174	16.927552
1.68	0.639190	0.208788	0.326644	1.319037	0.519115	36.529607	17.221983
1.69	0.636448	0.205670	0.323152	1.328252	0.515513	36.278874	17.516110
1.70	0.633714	0.202593	0.319693	1.337606	0.511908	36.031879	17.809910
1.71	0.630986	0.199558	0.316264	1.347100	0.508300	35.788532	18.103358
1.72	0.628267	0.196564	0.312868	1.356735	0.504691	35.548744	18.396433
1.73	0.625555	0.193611	0.309502	1.366511	0.501080	35.312431	18.689110
1.74	0.622851	0.190697	0.306169	1.376430	0.497469	35.079511	18.981370
1.75	0.620155	0.187824	0.302866	1.386492	0.493859	34.849905	19.273192
1.76	0.617467	0.184990	0.299595	1.396698	0.490250	34.623535	19.564556

M							
1.77	0.614787	0.182195	0.296354	1.407049	0.486644	34.400330	19.855443
1.78	0.612115	0.179438	0.293145	1.417546	0.483040	34.180216	20.145835
1.79	0.609451	0.176720	0.289966	1.428190	0.479440	33.963124	20.435715
1.80	0.606796	0.174040	0.286818	1.438982	0.475844	33.748989	20.725064
1.81	0.604149	0.171398	0.283701	1.449923	0.472254	33.537744	21.013868
1.82	0.601511	0.168792	0.280614	1.461013	0.468669	33.329327	21.302110
1.83	0.598881	0.166224	0.277557	1.472255	0.465090	33.123677	21.589775
1.84	0.596260	0.163691	0.274530	1.483648	0.461519	32.920735	21.876849
1.85	0.593648	0.161195	0.271533	1.495194	0.457955	32.720443	22.163318
1.86	0.591044	0.158734	0.268566	1.506894	0.454399	32.522748	22.449169
1.87	0.588450	0.156309	0.265628	1.518750	0.450852	32.327593	22.734388
1.88	0.585864	0.153918	0.262720	1.530761	0.447314	32.134928	23.018963
1.89	0.583288	0.151562	0.259841	1.542930	0.443787	31.944701	23.302882
1.90	0.580720	0.149240	0.256991	1.555257	0.440269	31.756864	23.586135
1.91	0.578162	0.146951	0.254169	1.567743	0.436762	31.571368	23.868709
1.92	0.575612	0.144696	0.251377	1.580391	0.433267	31.388166	24.150595
1.93	0.573072	0.142473	0.248613	1.593200	0.429784	31.207215	24.431782
1.94	0.570542	0.140283	0.245877	1.606172	0.426313	31.028469	24.712260
1.95	0.568020	0.138126	0.243170	1.619309	0.422854	30.851886	24.992021
1.96	0.565509	0.135999	0.240490	1.632611	0.419409	30.677424	25.271055
1.97	0.563006	0.133905	0.237839	1.646080	0.415977	30.505044	25.549354
1.98	0.560513	0.131841	0.235215	1.659717	0.412559	30.334705	25.826910
1.99	0.558029	0.129808	0.232618	1.673523	0.409156	30.166370	26.103715
2.00	0.555556	0.127805	0.230048	1.687500	0.405767	30.000000	26.379761
2.02	0.550637	0.123888	0.224990	1.715971	0.399035	29.673015	26.929549
2.04	0.545756	0.120087	0.220037	1.745139	0.392365	29.353470	27.476222

(Continued)

Table D.1 Compressible Flow Functions ($\gamma = 1.4$) (Continued)

M	T/T_t	P/P_t	ρ/ρ_t	A/A^*	$MFP\sqrt{R/g_c}$	μ	ν
2.06	0.540915	0.116399	0.215190	1.775017	0.385761	29.041100	28.019730
2.08	0.536113	0.112823	0.210446	1.805614	0.379224	28.735654	28.560030
2.10	0.531350	0.109353	0.205803	1.836944	0.372756	28.436890	29.097081
2.12	0.526626	0.105988	0.201259	1.869016	0.366359	28.144581	29.630849
2.14	0.521942	0.102726	0.196814	1.901843	0.360036	27.858510	30.161301
2.16	0.517298	0.099562	0.192466	1.935437	0.353787	27.578468	30.688409
2.18	0.512694	0.096495	0.188212	1.969810	0.347613	27.304258	31.212148
2.20	0.508130	0.093522	0.184051	2.004975	0.341516	27.035692	31.732496
2.22	0.503606	0.090640	0.179981	2.040944	0.335497	26.772588	32.249433
2.24	0.499122	0.087846	0.176001	2.077731	0.329557	26.514775	32.762944
2.26	0.494677	0.085139	0.172110	2.115349	0.323697	26.262087	33.273015
2.28	0.490273	0.082515	0.168304	2.153811	0.317916	26.014366	33.779634
2.30	0.485909	0.079973	0.164584	2.193131	0.312216	25.771462	34.282794
2.32	0.481584	0.077509	0.160946	2.233323	0.306598	25.533229	34.782487
2.34	0.477300	0.075122	0.157390	2.274401	0.301060	25.299529	35.278709
2.36	0.473055	0.072810	0.153914	2.316381	0.295604	25.070228	35.771458
2.38	0.468850	0.070570	0.150516	2.359275	0.290230	24.845199	36.260731
2.40	0.464684	0.068399	0.147195	2.403100	0.284937	24.624318	36.746531
2.42	0.460558	0.066297	0.143950	2.447870	0.279725	24.407469	37.228860
2.44	0.456471	0.064261	0.140777	2.493602	0.274595	24.194537	37.707721
2.46	0.452423	0.062288	0.137677	2.540309	0.269546	23.985414	38.183121
2.48	0.448414	0.060378	0.134648	2.588010	0.264578	23.779995	38.655066
2.50	0.444444	0.058528	0.131687	2.636719	0.259691	23.578178	39.123564
2.52	0.440513	0.056736	0.128794	2.686453	0.254883	23.379868	39.588624
2.54	0.436620	0.055000	0.125968	2.737228	0.250155	23.184970	40.050257

2.56	0.432766	0.053319	0.123206	2.789063	0.245506	22.993394	40.508473
2.58	0.428949	0.051692	0.120507	2.841972	0.240935	22.805054	40.963286
2.60	0.425170	0.050115	0.117871	2.895975	0.236442	22.619865	41.414708
2.62	0.421429	0.048589	0.115295	2.951089	0.232027	22.437747	41.862753
2.64	0.417725	0.047110	0.112778	3.007331	0.227687	22.258621	42.307437
2.66	0.414058	0.045679	0.110320	3.064720	0.223424	22.082413	42.748775
2.68	0.410428	0.044292	0.107918	3.123274	0.219235	21.909050	43.186783
2.70	0.406835	0.042950	0.105571	3.183011	0.215121	21.738461	43.621478
2.72	0.403278	0.041650	0.103279	3.243951	0.211079	21.570578	44.052877
2.74	0.399757	0.040391	0.101039	3.306113	0.207111	21.405337	44.480999
2.76	0.396272	0.039172	0.098851	3.369515	0.203214	21.242672	44.905863
2.78	0.392822	0.037992	0.096714	3.434179	0.199387	21.082524	45.327487
2.80	0.389408	0.036848	0.094626	3.500123	0.195631	20.924832	45.745890
2.82	0.386029	0.035741	0.092587	3.567368	0.191943	20.769540	46.161093
2.84	0.382684	0.034669	0.090594	3.635934	0.188323	20.616590	46.573116
2.86	0.379374	0.033631	0.088648	3.705842	0.184771	20.465930	46.981980
2.88	0.376098	0.032625	0.086747	3.777113	0.181284	20.317508	47.387704
2.90	0.372856	0.031651	0.084889	3.849768	0.177863	20.171271	47.790312
2.92	0.369648	0.030708	0.083075	3.923829	0.174506	20.027172	48.189824
2.94	0.366472	0.029795	0.081302	3.999317	0.171212	19.885163	48.586261
2.96	0.363330	0.028910	0.079571	4.076255	0.167981	19.745198	48.979646
2.98	0.360220	0.028054	0.077879	4.154664	0.164810	19.607232	49.370000
3.00	0.357143	0.027224	0.076226	4.234568	0.161700	19.471221	49.757347
3.02	0.354098	0.026420	0.074612	4.315989	0.158650	19.337123	50.141707
3.04	0.351084	0.025641	0.073034	4.398950	0.155658	19.204897	50.523105
3.06	0.348102	0.024887	0.071494	4.483475	0.152723	19.074505	50.901562

(Continued)

Table D.1 Compressible Flow Functions ($\gamma = 1.4$) (Continued)

M	T/T_t	P/P_t	ρ/ρ_t	A/A^*	$MFP\sqrt{R/g_c}$	μ	v
3.08	0.345151	0.024156	0.069988	4.569587	0.149845	18.945906	51.277101
3.10	0.342231	0.023449	0.068517	4.657311	0.147023	18.819063	51.649744
3.12	0.339342	0.022763	0.067080	4.746670	0.144255	18.693941	52.019516
3.14	0.336483	0.022099	0.065676	4.837689	0.141541	18.570503	52.386438
3.16	0.333654	0.021455	0.064304	4.930393	0.138880	18.448715	52.750533
3.18	0.330854	0.020832	0.062964	5.024808	0.136270	18.328544	53.111824
3.20	0.328084	0.020228	0.061654	5.120958	0.133712	18.209957	53.470335
3.22	0.325343	0.019642	0.060374	5.218868	0.131203	18.092922	53.826087
3.24	0.322631	0.019075	0.059124	5.318566	0.128744	17.977409	54.179105
3.26	0.319947	0.018526	0.057902	5.420078	0.126332	17.863387	54.529410
3.28	0.317291	0.017993	0.056708	5.523429	0.123969	17.750827	54.877026
3.30	0.314663	0.017477	0.055541	5.628647	0.121651	17.639701	55.221976
3.32	0.312063	0.016977	0.054401	5.735759	0.119379	17.529982	55.564281
3.34	0.309490	0.016492	0.053287	5.844792	0.117152	17.421641	55.903965
3.36	0.306944	0.016022	0.052197	5.955774	0.114969	17.314654	56.241050
3.38	0.304425	0.015566	0.051133	6.068734	0.112829	17.208993	56.575559
3.40	0.301932	0.015125	0.050093	6.183699	0.110732	17.104635	56.907514
3.42	0.299466	0.014697	0.049076	6.300698	0.108675	17.001555	57.236938
3.44	0.297025	0.014282	0.048082	6.419760	0.106660	16.899729	57.563852
3.46	0.294610	0.013879	0.047111	6.540915	0.104684	16.799135	57.888279
3.48	0.292220	0.013489	0.046161	6.664192	0.102748	16.699749	58.210241
3.50	0.289855	0.013111	0.045233	6.789621	0.100850	16.601550	58.529759
3.52	0.287515	0.012744	0.044325	6.917231	0.098989	16.504516	58.846855
3.54	0.285199	0.012389	0.043438	7.047055	0.097166	16.408626	59.161552

3.56	0.282908	0.012044	0.042571	7.179122	0.095378	16.313860	59.473870
3.58	0.280640	0.011709	0.041723	7.313463	0.093626	16.220198	59.783831
3.60	0.278396	0.011385	0.040894	7.450111	0.091909	16.127620	60.091456
3.62	0.276176	0.011070	0.040083	7.589097	0.090226	16.036108	60.396766
3.64	0.273979	0.010765	0.039291	7.730452	0.088576	15.945643	60.699782
3.66	0.271804	0.010469	0.038516	7.874211	0.086959	15.856207	61.000525
3.68	0.269652	0.010182	0.037758	8.020404	0.085374	15.767781	61.299016
3.70	0.267523	0.009903	0.037017	8.169066	0.083820	15.680350	61.595275
3.72	0.265415	0.009633	0.036292	8.320230	0.082297	15.593896	61.889322
3.74	0.263330	0.009370	0.035584	8.473930	0.080804	15.508402	62.181178
3.76	0.261266	0.009116	0.034890	8.630199	0.079341	15.423852	62.470863
3.78	0.259223	0.008869	0.034212	8.789073	0.077907	15.340231	62.758397
3.80	0.257202	0.008629	0.033549	8.950585	0.076501	15.257523	63.043799
3.82	0.255201	0.008396	0.032901	9.114772	0.075123	15.175714	63.327089
3.84	0.253221	0.008171	0.032266	9.281668	0.073772	15.094787	63.608287
3.86	0.251261	0.007951	0.031646	9.451309	0.072448	15.014730	63.887411
3.88	0.249322	0.007739	0.031039	9.623732	0.071150	14.935527	64.164482
3.90	0.247402	0.007532	0.030445	9.798973	0.069878	14.857166	64.439517
3.92	0.245502	0.007332	0.029863	9.977069	0.068631	14.779632	64.712536
3.94	0.243622	0.007137	0.029295	10.158057	0.067408	14.702912	64.983558
3.96	0.241761	0.006948	0.028739	10.341975	0.066209	14.626994	65.252600
3.98	0.239919	0.006764	0.028194	10.528859	0.065034	14.551865	65.519681
4.00	0.238095	0.006586	0.027662	10.718750	0.063882	14.477512	65.784820

Table D.2 Compressible Flow Functions ($\gamma = 1.33$)

M	T/T_t	P/P_t	ρ/ρ_t	A/A^*	$\mathrm{MFP}\sqrt{R/g_c}$
0.00	1.000000	1.000000	1.000000	Indef	0.000000
0.01	0.999984	0.999934	0.999950	58.327677	0.011532
0.02	0.999934	0.999734	0.999800	29.168935	0.023060
0.03	0.999852	0.999402	0.999550	19.451621	0.034580
0.04	0.999736	0.998937	0.999200	14.594664	0.046087
0.05	0.999588	0.998339	0.998751	11.681852	0.057579
0.06	0.999406	0.997610	0.998202	9.741113	0.069050
0.07	0.999192	0.996748	0.997554	8.355846	0.080498
0.08	0.998945	0.995755	0.996807	7.317750	0.091917
0.09	0.998665	0.994632	0.995961	6.511104	0.103305
0.10	0.998353	0.993378	0.995017	5.866473	0.114656
0.11	0.998007	0.991994	0.993974	5.339673	0.125968
0.12	0.997630	0.990481	0.992834	4.901248	0.137236
0.13	0.997219	0.988840	0.991597	4.530805	0.148457
0.14	0.996776	0.987071	0.990264	4.213778	0.159626
0.15	0.996301	0.985176	0.988834	3.939486	0.170740
0.16	0.995794	0.983155	0.987308	3.699917	0.181796
0.17	0.995254	0.981010	0.985688	3.488945	0.192788
0.18	0.994682	0.978741	0.983973	3.301806	0.203715
0.19	0.994079	0.976349	0.982165	3.134737	0.214572
0.20	0.993443	0.973836	0.980263	2.984731	0.225356
0.21	0.992776	0.971202	0.978269	2.849351	0.236064
0.22	0.992077	0.968450	0.976184	2.726604	0.246691
0.23	0.991347	0.965580	0.974009	2.614845	0.257235
0.24	0.990585	0.962594	0.971743	2.512701	0.267691

0.25	0.989793	0.959494	0.969388	2.419020	0.278058
0.26	0.988969	0.956279	0.966946	2.332827	0.288332
0.27	0.988114	0.952953	0.964416	2.253293	0.298509
0.28	0.987229	0.949517	0.961800	2.179704	0.308587
0.29	0.986313	0.945972	0.959099	2.111448	0.318563
0.30	0.985367	0.942320	0.956314	2.047994	0.328433
0.31	0.984391	0.938563	0.953445	1.988877	0.338195
0.32	0.983385	0.934702	0.950495	1.933694	0.347846
0.33	0.982349	0.930740	0.947464	1.882088	0.357384
0.34	0.981283	0.926677	0.944353	1.833746	0.366806
0.35	0.980188	0.922516	0.941163	1.788389	0.376109
0.36	0.979064	0.918259	0.937895	1.745770	0.385290
0.37	0.977910	0.913908	0.934552	1.705669	0.394349
0.38	0.976728	0.909464	0.931133	1.667889	0.403281
0.39	0.975518	0.904930	0.927640	1.632254	0.412086
0.40	0.974279	0.900307	0.924075	1.598603	0.420760
0.41	0.973012	0.895597	0.920438	1.566794	0.429302
0.42	0.971717	0.890804	0.916731	1.536696	0.437711
0.43	0.970395	0.885928	0.912956	1.508193	0.445983
0.44	0.969045	0.880971	0.909113	1.481177	0.454117
0.45	0.967668	0.875936	0.905204	1.455551	0.462113
0.46	0.966264	0.870826	0.901230	1.431226	0.469967
0.47	0.964833	0.865641	0.897193	1.408120	0.477678
0.48	0.963376	0.860385	0.893093	1.386160	0.485246
0.49	0.961893	0.855059	0.888933	1.365277	0.492668
0.50	0.960384	0.849665	0.884714	1.345408	0.499944

(Continued)

Table D.2   Compressible Flow Functions ($\gamma = 1.33$) *(Continued)*

M	T/T_t	P/P_t	ρ/ρ_t	A/A^*	$\mathrm{MFP}\sqrt{R/g_c}$
0.51	0.958850	0.844207	0.880437	1.326495	0.507072
0.52	0.957290	0.838685	0.876104	1.308485	0.514051
0.53	0.955705	0.833102	0.871715	1.291329	0.520881
0.54	0.954095	0.827461	0.867273	1.274982	0.527559
0.55	0.952460	0.821763	0.862779	1.259400	0.534087
0.56	0.950802	0.816011	0.858234	1.244545	0.540461
0.57	0.949119	0.810206	0.853640	1.230380	0.546684
0.58	0.947413	0.804352	0.848999	1.216872	0.552752
0.59	0.945683	0.798450	0.844310	1.203989	0.558667
0.60	0.943931	0.792503	0.839577	1.191701	0.564427
0.61	0.942155	0.786512	0.834801	1.179982	0.570033
0.62	0.940357	0.780480	0.829982	1.168806	0.575483
0.63	0.938537	0.774409	0.825123	1.158149	0.580779
0.64	0.936694	0.768300	0.820225	1.147988	0.585919
0.65	0.934831	0.762158	0.815290	1.138302	0.590905
0.66	0.932945	0.755982	0.810318	1.129073	0.595735
0.67	0.931039	0.749776	0.805311	1.120281	0.600411
0.68	0.929112	0.743542	0.800271	1.111909	0.604931
0.69	0.927165	0.737281	0.795199	1.103941	0.609298
0.70	0.925198	0.730996	0.790097	1.096361	0.613510
0.71	0.923211	0.724689	0.784966	1.089155	0.617569
0.72	0.921204	0.718361	0.779807	1.082310	0.621475
0.73	0.919178	0.712016	0.774622	1.075813	0.625228
0.74	0.917133	0.705654	0.769412	1.069652	0.628829
0.75	0.915070	0.699277	0.764179	1.063815	0.632280

0.76	0.912989	0.692888	0.758923	1.058291	0.635580
0.77	0.910889	0.686489	0.753647	1.053071	0.638730
0.78	0.908772	0.680081	0.748352	1.048145	0.641732
0.79	0.906638	0.673667	0.743039	1.043504	0.644586
0.80	0.904486	0.667247	0.737708	1.039139	0.647294
0.81	0.902318	0.660824	0.732363	1.035042	0.649856
0.82	0.900134	0.654400	0.727004	1.031206	0.652273
0.83	0.897933	0.647977	0.721631	1.027623	0.654548
0.84	0.895717	0.641555	0.716248	1.024287	0.656680
0.85	0.893485	0.635137	0.710854	1.021190	0.658671
0.86	0.891239	0.628725	0.705451	1.018327	0.660523
0.87	0.888977	0.622319	0.700040	1.015692	0.662237
0.88	0.886701	0.615923	0.694623	1.013279	0.663813
0.89	0.884411	0.609536	0.689200	1.011084	0.665255
0.90	0.882106	0.603161	0.683773	1.009100	0.666563
0.91	0.879789	0.596799	0.678344	1.007324	0.667738
0.92	0.877458	0.590452	0.672912	1.005750	0.668783
0.93	0.875114	0.584121	0.667480	1.004376	0.669698
0.94	0.872757	0.577807	0.662048	1.003195	0.670486
0.95	0.870388	0.571512	0.656617	1.002206	0.671148
0.96	0.868007	0.565237	0.651189	1.001404	0.671686
0.97	0.865615	0.558984	0.645765	1.000785	0.672101
0.98	0.863210	0.552753	0.640345	1.000347	0.672395
0.99	0.860795	0.546546	0.634931	1.000086	0.672570
1.00	0.858369	0.540364	0.629524	1.000000	0.672628

(Continued)

Table D.2 Compressible Flow Functions ($\gamma = 1.33$) (Continued)

M	T/T_t	P/P_t	ρ/ρ_t	A/A^*	$MFP\sqrt{R/g_c}$	μ	ν
1.01	0.855932	0.534208	0.624124	1.000085	0.672571	81.930699	0.046082
1.02	0.853486	0.528080	0.618733	1.000340	0.672400	78.635123	0.129542
1.03	0.851029	0.521980	0.613352	1.000761	0.672117	76.137568	0.236534
1.04	0.848562	0.515910	0.607981	1.001346	0.671724	74.057631	0.361962
1.05	0.846086	0.509870	0.602621	1.002094	0.671223	72.247210	0.502809
1.06	0.843601	0.503861	0.597274	1.003001	0.670616	70.629962	0.656999
1.07	0.841108	0.497885	0.591940	1.004066	0.669904	69.160306	0.822978
1.08	0.838605	0.491942	0.586619	1.005288	0.669091	67.808393	0.999524
1.09	0.836095	0.486034	0.581314	1.006663	0.668176	66.553383	1.185639
1.10	0.833576	0.480160	0.576024	1.008192	0.667163	65.380023	1.380488
1.11	0.831050	0.474323	0.570751	1.009872	0.666053	64.276740	1.583356
1.12	0.828517	0.468522	0.565494	1.011701	0.664849	63.234499	1.793622
1.13	0.825976	0.462758	0.560256	1.013679	0.663552	62.246077	2.010742
1.14	0.823429	0.457033	0.555036	1.015804	0.662163	61.305586	2.234231
1.15	0.820875	0.451347	0.549836	1.018075	0.660686	60.408154	2.463653
1.16	0.818315	0.445700	0.544656	1.020491	0.659122	59.549686	2.698618
1.17	0.815748	0.440093	0.539496	1.023051	0.657473	58.726702	2.938768
1.18	0.813176	0.434527	0.534358	1.025754	0.655741	57.936214	3.183777
1.19	0.810598	0.429002	0.529241	1.028598	0.653927	57.175632	3.433347
1.20	0.808016	0.423519	0.524148	1.031584	0.652034	56.442690	3.687201
1.21	0.805428	0.418079	0.519077	1.034711	0.650064	55.735397	3.945085
1.22	0.802835	0.412682	0.514030	1.037977	0.648019	55.051986	4.206762
1.23	0.800238	0.407327	0.509008	1.041382	0.645900	54.390884	4.472010
1.24	0.797636	0.402017	0.504010	1.044927	0.643709	53.750679	4.740623
1.25	0.795031	0.396751	0.499038	1.048609	0.641448	53.130102	5.012410

1.26	0.792422	0.391529	0.494092	1.052430	0.639120	52.528005	5.287188
1.27	0.789809	0.386352	0.489172	1.056388	0.636725	51.943343	5.564788
1.28	0.787193	0.381221	0.484278	1.060483	0.634266	51.375167	5.845048
1.29	0.784574	0.376135	0.479413	1.064715	0.631745	50.822606	6.127819
1.30	0.781953	0.371095	0.474574	1.069083	0.629164	50.284863	6.412957
1.31	0.779328	0.366100	0.469764	1.073589	0.626523	49.761202	6.700326
1.32	0.776701	0.361153	0.464983	1.078231	0.623826	49.250946	6.989799
1.33	0.774073	0.356251	0.460230	1.083009	0.621074	48.753467	7.281254
1.34	0.771442	0.351397	0.455506	1.087924	0.618268	48.268183	7.574576
1.35	0.768809	0.346589	0.450813	1.092975	0.615411	47.794554	7.869653
1.36	0.766175	0.341828	0.446149	1.098163	0.612503	47.332075	8.166383
1.37	0.763540	0.337114	0.441515	1.103487	0.609548	46.880275	8.464665
1.38	0.760904	0.332448	0.436912	1.108949	0.606546	46.438716	8.764404
1.39	0.758267	0.327829	0.432339	1.114548	0.603499	46.006983	9.065510
1.40	0.755629	0.323257	0.427798	1.120283	0.600409	45.584691	9.367896
1.41	0.752991	0.318732	0.423288	1.126157	0.597278	45.171476	9.671478
1.42	0.750353	0.314255	0.418810	1.132169	0.594106	44.766995	9.976178
1.43	0.747715	0.309826	0.414364	1.138319	0.590896	44.370926	10.281921
1.44	0.745077	0.305443	0.409949	1.144607	0.587650	43.982963	10.588632
1.45	0.742439	0.301109	0.405567	1.151035	0.584368	43.602819	10.896243
1.46	0.739801	0.296821	0.401217	1.157603	0.581053	43.230221	11.204686
1.47	0.737165	0.292581	0.396900	1.164311	0.577705	42.864910	11.513899
1.48	0.734529	0.288387	0.392615	1.171159	0.574327	42.506642	11.823818
1.49	0.731895	0.284241	0.388364	1.178149	0.570920	42.155184	12.134386
1.50	0.729262	0.280142	0.384145	1.185281	0.567484	41.810315	12.445546
1.51	0.726630	0.276090	0.379959	1.192556	0.564023	41.471824	12.757242

(Continued)

Table D.2 Compressible Flow Functions ($\gamma = 1.33$) (Continued)

M	T/T_t	P/P_t	ρ/ρ_t	A/A^*	$MFP\sqrt{R/g_c}$	μ	v
1.52	0.724000	0.272084	0.375807	1.199974	0.560536	41.139510	13.069423
1.53	0.721371	0.268125	0.371688	1.207536	0.557026	40.813184	13.382037
1.54	0.718745	0.264212	0.367602	1.215242	0.553493	40.492661	13.695038
1.55	0.716121	0.260346	0.363550	1.223095	0.549940	40.177770	14.008377
1.56	0.713499	0.256525	0.359532	1.231093	0.546367	39.868342	14.322010
1.57	0.710879	0.252751	0.355547	1.239239	0.542775	39.564218	14.635894
1.58	0.708262	0.249022	0.351595	1.247533	0.539167	39.265248	14.949987
1.59	0.705648	0.245338	0.347678	1.255976	0.535542	38.971284	15.264248
1.60	0.703037	0.241700	0.343794	1.264569	0.531903	38.682187	15.578640
1.61	0.700429	0.238106	0.339943	1.273313	0.528251	38.397824	15.893124
1.62	0.697824	0.234557	0.336127	1.282208	0.524586	38.118064	16.207665
1.63	0.695222	0.231053	0.332344	1.291257	0.520910	37.842786	16.522227
1.64	0.692624	0.227592	0.328594	1.300459	0.517224	37.571869	16.836779
1.65	0.690030	0.224176	0.324879	1.309817	0.513529	37.305201	17.151286
1.66	0.687439	0.220803	0.321196	1.319330	0.509826	37.042671	17.465719
1.67	0.684852	0.217473	0.317548	1.329001	0.506116	36.784174	17.780046
1.68	0.682270	0.214187	0.313933	1.338830	0.502400	36.529607	18.094239
1.69	0.679691	0.210943	0.310351	1.348819	0.498680	36.278874	18.408271
1.70	0.677117	0.207741	0.306803	1.358968	0.494955	36.031879	18.722113
1.71	0.674547	0.204582	0.303288	1.369279	0.491228	35.788532	19.035740
1.72	0.671982	0.201464	0.299806	1.379754	0.487499	35.548744	19.349126
1.73	0.669421	0.198388	0.296358	1.390393	0.483768	35.312431	19.662248
1.74	0.666865	0.195353	0.292942	1.401199	0.480038	35.079511	19.975081
1.75	0.664314	0.192358	0.289559	1.412171	0.476308	34.849905	20.287604
1.76	0.661768	0.189404	0.286209	1.423312	0.472580	34.623535	20.599793

M							
1.77	0.659227	0.186490	0.282892	1.434623	0.468854	34.400330	20.911629
1.78	0.656691	0.183616	0.279608	1.446105	0.465131	34.180216	21.223090
1.79	0.654161	0.180781	0.276355	1.457761	0.461412	33.963124	21.534157
1.80	0.651636	0.177985	0.273136	1.469590	0.457698	33.748989	21.844811
1.81	0.649116	0.175228	0.269948	1.481596	0.453989	33.537744	22.155033
1.82	0.646602	0.172508	0.266792	1.493779	0.450287	33.329327	22.464806
1.83	0.644094	0.169827	0.263669	1.506140	0.446591	33.123677	22.774112
1.84	0.641592	0.167184	0.260577	1.518682	0.442903	32.920735	23.082935
1.85	0.639095	0.164577	0.257516	1.531407	0.439223	32.720443	23.391259
1.86	0.636605	0.162008	0.254487	1.544315	0.435551	32.522748	23.699068
1.87	0.634120	0.159474	0.251489	1.557408	0.431890	32.327593	24.006348
1.88	0.631642	0.156977	0.248523	1.570689	0.428238	32.134928	24.313084
1.89	0.629170	0.154516	0.245587	1.584158	0.424597	31.944701	24.619263
1.90	0.626704	0.152090	0.242682	1.597817	0.420967	31.756864	24.924870
1.91	0.624244	0.149698	0.239807	1.611669	0.417349	31.571368	25.229893
1.92	0.621792	0.147342	0.236963	1.625715	0.413743	31.388166	25.534320
1.93	0.619345	0.145019	0.234149	1.639957	0.410150	31.207215	25.838139
1.94	0.616905	0.142731	0.231366	1.654396	0.406570	31.028469	26.141337
1.95	0.614472	0.140475	0.228611	1.669035	0.403004	30.851886	26.443904
1.96	0.612046	0.138253	0.225887	1.683876	0.399452	30.677424	26.745829
1.97	0.609627	0.136064	0.223192	1.698920	0.395915	30.505044	27.047102
1.98	0.607214	0.133906	0.220526	1.714169	0.392393	30.334705	27.347712
1.99	0.604808	0.131781	0.217889	1.729625	0.388887	30.166370	27.647650
2.00	0.602410	0.129687	0.215281	1.745290	0.385396	30.000000	27.946907
2.02	0.597634	0.125593	0.210150	1.777257	0.378464	29.673015	28.543342
2.04	0.592886	0.121620	0.205133	1.810085	0.371601	29.353470	29.136948

(Continued)

Table D.2 Compressible Flow Functions ($\gamma = 1.33$) (Continued)

M	T/T_t	P/P_t	ρ/ρ_t	A/A^*	$MFP\sqrt{R/g_c}$	μ	ν
2.06	0.588168	0.117766	0.200226	1.843791	0.364807	29.041100	29.727664
2.08	0.583480	0.114028	0.195428	1.878392	0.358087	28.735654	30.315432
2.10	0.578821	0.110403	0.190738	1.913907	0.351443	28.436890	30.900199
2.12	0.574193	0.106888	0.186154	1.950353	0.344875	28.144581	31.481917
2.14	0.569595	0.103480	0.181673	1.987749	0.338387	27.858510	32.060539
2.16	0.565028	0.100177	0.177295	2.026114	0.331979	27.578468	32.636027
2.18	0.560492	0.096975	0.173017	2.065467	0.325654	27.304258	33.208341
2.20	0.555988	0.093872	0.168838	2.105829	0.319413	27.035692	33.777449
2.22	0.551515	0.090865	0.164756	2.147219	0.313256	26.772588	34.343319
2.24	0.547075	0.087953	0.160769	2.189657	0.307184	26.514775	34.905923
2.26	0.542666	0.085131	0.156875	2.233167	0.301199	26.262087	35.465238
2.28	0.538290	0.082397	0.153072	2.277768	0.295301	26.014366	36.021240
2.30	0.533946	0.079750	0.149360	2.323484	0.289491	25.771462	36.573910
2.32	0.529634	0.077186	0.145735	2.370337	0.283769	25.533229	37.123232
2.34	0.525355	0.074704	0.142196	2.418349	0.278135	25.299529	37.669189
2.36	0.521109	0.072300	0.138742	2.467545	0.272590	25.070228	38.211770
2.38	0.516896	0.069972	0.135371	2.517948	0.267134	24.845199	38.750965
2.40	0.512715	0.067719	0.132080	2.569583	0.261766	24.624318	39.286763
2.42	0.508568	0.065539	0.128869	2.622475	0.256486	24.407469	39.819159
2.44	0.504453	0.063428	0.125735	2.676650	0.251295	24.194537	40.348148
2.46	0.500372	0.061385	0.122678	2.732133	0.246192	23.985414	40.873725
2.48	0.496323	0.059407	0.119695	2.788951	0.241176	23.779995	41.395890
2.50	0.492308	0.057494	0.116784	2.847130	0.236248	23.578178	41.914642
2.52	0.488325	0.055642	0.113945	2.906700	0.231406	23.379868	42.429982
2.54	0.484375	0.053850	0.111175	2.967686	0.226651	23.184970	42.941913

2.56	0.480459	0.052117	0.108473	3.030119	0.221981	22.993394	43.450437
2.58	0.476575	0.050440	0.105838	3.094027	0.217396	22.805054	43.955560
2.60	0.472724	0.048817	0.103267	3.159440	0.212895	22.619865	44.457288
2.62	0.468905	0.047247	0.100760	3.226388	0.208477	22.437747	44.955627
2.64	0.465120	0.045728	0.098315	3.294902	0.204142	22.258621	45.450586
2.66	0.461367	0.044259	0.095931	3.365013	0.199889	22.082413	45.942173
2.68	0.457646	0.042838	0.093606	3.436754	0.195716	21.909050	46.430399
2.70	0.453957	0.041464	0.091338	3.510155	0.191624	21.738461	46.915273
2.72	0.450301	0.040134	0.089127	3.585252	0.187610	21.570578	47.396808
2.74	0.446677	0.038848	0.086971	3.662076	0.183674	21.405337	47.875015
2.76	0.443085	0.037604	0.084869	3.740663	0.179815	21.242672	48.349907
2.78	0.439525	0.036401	0.082819	3.821048	0.176032	21.082524	48.821498
2.80	0.435996	0.035237	0.080821	3.903265	0.172325	20.924832	49.289801
2.82	0.432499	0.034112	0.078872	3.987351	0.168691	20.769540	49.754832
2.84	0.429033	0.033024	0.076972	4.073342	0.165129	20.616590	50.216606
2.86	0.425598	0.031971	0.075120	4.161276	0.161640	20.465930	50.675137
2.88	0.422195	0.030953	0.073315	4.251191	0.158221	20.317508	51.130444
2.90	0.418822	0.029968	0.071554	4.343125	0.154872	20.171271	51.582541
2.92	0.415480	0.029016	0.069838	4.437118	0.151591	20.027172	52.031446
2.94	0.412168	0.028095	0.068165	4.533210	0.148378	19.885163	52.477176
2.96	0.408887	0.027205	0.066533	4.631440	0.145231	19.745198	52.919750
2.98	0.405636	0.026343	0.064943	4.731851	0.142149	19.607232	53.359185
3.00	0.402414	0.025510	0.063393	4.834484	0.139131	19.471221	53.795499
3.02	0.399223	0.024705	0.061882	4.939381	0.136177	19.337123	54.228712
3.04	0.396061	0.023925	0.060408	5.046587	0.133284	19.204897	54.658841
3.06	0.392928	0.023172	0.058972	5.156146	0.130452	19.074505	55.085908

(Continued)

Table D.2 Compressible Flow Functions ($\gamma = 1.33$) (*Continued*)

M	T/T_t	P/P_t	ρ/ρ_t	A/A^*	$MFP\sqrt{R/g_c}$	μ	ν
3.08	0.389825	0.022443	0.057572	5.268101	0.127679	18.945906	55.509929
3.10	0.386750	0.021738	0.056207	5.382498	0.124966	18.819063	55.930926
3.12	0.383704	0.021056	0.054876	5.499384	0.122310	18.693941	56.348918
3.14	0.380686	0.020397	0.053579	5.618806	0.119710	18.570503	56.763925
3.16	0.377697	0.019759	0.052314	5.740811	0.117166	18.448715	57.175966
3.18	0.374736	0.019142	0.051081	5.865447	0.114676	18.328544	57.585063
3.20	0.371802	0.018545	0.049879	5.992764	0.112240	18.209957	57.991235
3.22	0.368897	0.017968	0.048707	6.122812	0.109856	18.092922	58.394503
3.24	0.366018	0.017409	0.047564	6.255642	0.107523	17.977409	58.794887
3.26	0.363167	0.016869	0.046450	6.391304	0.105241	17.863387	59.192408
3.28	0.360343	0.016347	0.045364	6.529852	0.103008	17.750827	59.587086
3.30	0.357545	0.015841	0.044305	6.671339	0.100824	17.639701	59.978942
3.32	0.354774	0.015352	0.043273	6.815818	0.098686	17.529982	60.367997
3.34	0.352029	0.014879	0.042266	6.963344	0.096596	17.421641	60.754271
3.36	0.349310	0.014421	0.041285	7.113973	0.094550	17.314654	61.137785
3.38	0.346617	0.013978	0.040328	7.267762	0.092550	17.208993	61.518561
3.40	0.343950	0.013550	0.039395	7.424767	0.090593	17.104635	61.896617
3.42	0.341308	0.013135	0.038485	7.585047	0.088678	17.001555	62.271977
3.44	0.338691	0.012734	0.037598	7.748661	0.086806	16.899729	62.644659
3.46	0.336099	0.012346	0.036733	7.915668	0.084974	16.799135	63.014685
3.48	0.333532	0.011970	0.035889	8.086130	0.083183	16.699749	63.382075
3.50	0.330989	0.011607	0.035066	8.260109	0.081431	16.601550	63.746851
3.52	0.328470	0.011255	0.034264	8.437666	0.079717	16.504516	64.109031
3.54	0.325976	0.010914	0.033481	8.618866	0.078041	16.408626	64.468638

3.56	0.323505	0.010585	0.032718	8.803773	0.076402	16.313860	64.825692
3.58	0.321058	0.010265	0.031974	8.992451	0.074799	16.220198	65.180212
3.60	0.318634	0.009957	0.031248	9.184969	0.073231	16.127620	65.532220
3.62	0.316233	0.009658	0.030540	9.381392	0.071698	16.036108	65.881735
3.64	0.313855	0.009368	0.029849	9.581789	0.070199	15.945643	66.228777
3.66	0.311500	0.009088	0.029176	9.786230	0.068732	15.856207	66.573367
3.68	0.309167	0.008817	0.028519	9.994784	0.067298	15.767781	66.915525
3.70	0.306857	0.008555	0.027878	10.207523	0.065895	15.680350	67.255271
3.72	0.304568	0.008300	0.027253	10.424519	0.064524	15.593896	67.592623
3.74	0.302302	0.008054	0.026643	10.645845	0.063182	15.508402	67.927603
3.76	0.300057	0.007816	0.026048	10.871575	0.061870	15.423852	68.260229
3.78	0.297833	0.007585	0.025467	11.101785	0.060587	15.340231	68.590522
3.80	0.295631	0.007361	0.024901	11.336551	0.059333	15.257523	68.918500
3.82	0.293449	0.007145	0.024348	11.575950	0.058106	15.175714	69.244182
3.84	0.291288	0.006935	0.023809	11.820061	0.056906	15.094787	69.567589
3.86	0.289148	0.006732	0.023283	12.068964	0.055732	15.014730	69.888738
3.88	0.287028	0.006535	0.022769	12.322738	0.054584	14.935527	70.207650
3.90	0.284929	0.006345	0.022268	12.581466	0.053462	14.857166	70.524342
3.92	0.282849	0.006160	0.021779	12.845231	0.052364	14.779632	70.838833
3.94	0.280789	0.005981	0.021302	13.114115	0.051290	14.702912	71.151142
3.96	0.278748	0.005808	0.020837	13.388206	0.050240	14.626994	71.461288
3.98	0.276727	0.005640	0.020382	13.667587	0.049213	14.551865	71.769288
4.00	0.274725	0.005478	0.019939	13.952348	0.048209	14.477512	72.075161

Table D.3 Compressible Flow Functions ($\gamma = 1.3$)

M	T/T_t	P/P_t	ρ/ρ_t	A/A^*	$MFP\sqrt{R/g_c}$
0.00	1.000000	1.000000	1.000000	Indef	0.000000
0.01	0.999985	0.999935	0.999950	58.526145	0.011401
0.02	0.999940	0.999740	0.999800	29.268120	0.022798
0.03	0.999865	0.999415	0.999550	19.517690	0.034188
0.04	0.999760	0.998961	0.999200	14.644160	0.045565
0.05	0.999625	0.998377	0.998751	11.721390	0.056927
0.06	0.999460	0.997663	0.998202	9.774002	0.068269
0.07	0.999266	0.996821	0.997554	8.383977	0.079588
0.08	0.999041	0.995851	0.996807	7.342305	0.090879
0.09	0.998786	0.994752	0.995961	6.532869	0.102139
0.10	0.998502	0.993526	0.995016	5.886000	0.113364
0.11	0.998188	0.992173	0.993974	5.357364	0.124551
0.12	0.997845	0.990694	0.992834	4.917403	0.135694
0.13	0.997471	0.989089	0.991596	4.545655	0.146791
0.14	0.997069	0.987359	0.990262	4.227506	0.157838
0.15	0.996636	0.985506	0.988832	3.952236	0.168832
0.16	0.996175	0.983529	0.987306	3.711808	0.179767
0.17	0.995684	0.981430	0.985685	3.500075	0.190642
0.18	0.995164	0.979210	0.983969	3.312255	0.201453
0.19	0.994614	0.976870	0.982160	3.144575	0.212195
0.20	0.994036	0.974411	0.980257	2.994014	0.222865
0.21	0.993428	0.971834	0.978262	2.858130	0.233461
0.22	0.992792	0.969140	0.976176	2.734922	0.243979
0.23	0.992127	0.966331	0.973998	2.622739	0.254414
0.24	0.991434	0.963407	0.971731	2.520204	0.264765

0.25	0.990712	0.960371	0.969374	2.426161	0.275028
0.26	0.989962	0.957223	0.966929	2.339632	0.285200
0.27	0.989183	0.953965	0.964397	2.259784	0.295277
0.28	0.988377	0.950599	0.961778	2.185902	0.305257
0.29	0.987542	0.947126	0.959074	2.117371	0.315137
0.30	0.986680	0.943547	0.956285	2.053658	0.324914
0.31	0.985790	0.939865	0.953413	1.994298	0.334585
0.32	0.984872	0.936080	0.950459	1.938884	0.344148
0.33	0.983928	0.932195	0.947423	1.887060	0.353599
0.34	0.982956	0.928211	0.944306	1.838511	0.362936
0.35	0.981957	0.924130	0.941111	1.792957	0.372157
0.36	0.980931	0.919954	0.937838	1.750151	0.381260
0.37	0.979878	0.915684	0.934488	1.709872	0.390241
0.38	0.978799	0.911323	0.931062	1.671921	0.399099
0.39	0.977694	0.906872	0.927562	1.636123	0.407832
0.40	0.976563	0.902333	0.923989	1.602316	0.416436
0.41	0.975405	0.897708	0.920344	1.570357	0.424911
0.42	0.974222	0.892999	0.916628	1.540116	0.433255
0.43	0.973013	0.888209	0.912843	1.511475	0.441464
0.44	0.971780	0.883338	0.908990	1.484326	0.449539
0.45	0.970520	0.878389	0.905070	1.458572	0.457477
0.46	0.969236	0.873364	0.901085	1.434123	0.465275
0.47	0.967928	0.868266	0.897036	1.410899	0.472934
0.48	0.966594	0.863095	0.892924	1.388824	0.480451
0.49	0.965237	0.857855	0.888750	1.367830	0.487826
0.50	0.952381	0.843019	0.885170	1.339844	0.511053

(Continued)

Table D.3 Compressible Flow Functions ($\gamma = 1.3$) (Continued)

M	T/T_t	P/P_t	ρ/ρ_t	A/A^*	$MFP\sqrt{R/g_c}$
0.51	0.962450	0.847173	0.880225	1.328837	0.502140
0.52	0.961021	0.841736	0.875877	1.310727	0.509078
0.53	0.959569	0.836237	0.871472	1.293474	0.515868
0.54	0.958093	0.830679	0.867013	1.277032	0.522510
0.55	0.956595	0.825064	0.862501	1.261359	0.529003
0.56	0.955073	0.819393	0.857938	1.246415	0.535345
0.57	0.953530	0.813670	0.853324	1.232165	0.541537
0.58	0.951964	0.807896	0.848662	1.218574	0.547577
0.59	0.950376	0.802073	0.843953	1.205610	0.553465
0.60	0.948767	0.796203	0.839198	1.193244	0.559200
0.61	0.947136	0.790289	0.834399	1.181449	0.564783
0.62	0.945483	0.784333	0.829557	1.170199	0.570212
0.63	0.943810	0.778336	0.824674	1.159471	0.575489
0.64	0.942116	0.772301	0.819751	1.149240	0.580612
0.65	0.940402	0.766229	0.814789	1.139488	0.585581
0.66	0.938667	0.760124	0.809790	1.130193	0.590397
0.67	0.936913	0.753986	0.804756	1.121338	0.595059
0.68	0.935139	0.747819	0.799687	1.112904	0.599569
0.69	0.933345	0.741623	0.794586	1.104877	0.603925
0.70	0.931532	0.735401	0.789453	1.097239	0.608128
0.71	0.929701	0.729156	0.784291	1.089978	0.612179
0.72	0.927850	0.722888	0.779100	1.083080	0.616079
0.73	0.925982	0.716600	0.773882	1.076530	0.619827
0.74	0.924095	0.710294	0.768638	1.070319	0.623424
0.75	0.922190	0.703972	0.763370	1.064433	0.626871

0.76	0.920268	0.697636	0.758079	1.058863	0.630169
0.77	0.918328	0.691287	0.752766	1.053598	0.633318
0.78	0.916372	0.684927	0.747433	1.048629	0.636319
0.79	0.914399	0.678559	0.742082	1.043947	0.639173
0.80	0.912409	0.672183	0.736713	1.039542	0.641881
0.81	0.910403	0.665802	0.731327	1.035408	0.644444
0.82	0.908381	0.659418	0.725927	1.031535	0.646863
0.83	0.906343	0.653032	0.720513	1.027918	0.649140
0.84	0.904290	0.646646	0.715087	1.024549	0.651274
0.85	0.902222	0.640262	0.709650	1.021422	0.653268
0.86	0.900139	0.633880	0.704203	1.018530	0.655123
0.87	0.898041	0.627504	0.698748	1.015868	0.656840
0.88	0.895929	0.621134	0.693285	1.013430	0.658420
0.89	0.893803	0.614772	0.687816	1.011211	0.659865
0.90	0.891663	0.608420	0.682342	1.009206	0.661176
0.91	0.889510	0.602078	0.676865	1.007410	0.662354
0.92	0.887343	0.595749	0.671385	1.005819	0.663402
0.93	0.885163	0.589433	0.665904	1.004428	0.664321
0.94	0.882971	0.583133	0.660422	1.003234	0.665111
0.95	0.880766	0.576850	0.654941	1.002233	0.665776
0.96	0.878549	0.570585	0.649462	1.001421	0.666315
0.97	0.876321	0.564338	0.643986	1.000795	0.666732
0.98	0.874080	0.558113	0.638514	1.000351	0.667028
0.99	0.871828	0.551909	0.633048	1.000087	0.667204
1.00	0.869565	0.545728	0.627587	1.000000	0.667262

(Continued)

Table D.3 Compressible Flow Functions ($\gamma = 1.3$) (Continued)

M	T/T_t	P/P_t	ρ/ρ_t	A/A^*	$MFP\sqrt{R/g_c}$	μ	ν
1.01	0.867291	0.539571	0.622133	1.000087	0.667205	81.930699	0.046690
1.02	0.865007	0.533439	0.616688	1.000344	0.667033	78.635123	0.131267
1.03	0.862712	0.527334	0.611251	1.000771	0.666748	76.137568	0.239716
1.04	0.860407	0.521256	0.605825	1.001365	0.666353	74.057631	0.366881
1.05	0.858093	0.515207	0.600410	1.002123	0.665849	72.247210	0.509710
1.06	0.855769	0.509188	0.595006	1.003043	0.665238	70.629962	0.666106
1.07	0.853435	0.503198	0.589615	1.004124	0.664522	69.160306	0.834499
1.08	0.851093	0.497241	0.584238	1.005363	0.663703	67.808393	1.013653
1.09	0.848742	0.491315	0.578875	1.006760	0.662782	66.553383	1.202562
1.10	0.846382	0.485423	0.573528	1.008312	0.661762	65.380023	1.400381
1.11	0.844014	0.479565	0.568196	1.010018	0.660644	64.276740	1.606389
1.12	0.841637	0.473742	0.562881	1.011876	0.659431	63.234499	1.819961
1.13	0.839254	0.467955	0.557584	1.013886	0.658123	62.246077	2.040545
1.14	0.836862	0.462204	0.552306	1.016046	0.656724	61.305586	2.267653
1.15	0.834463	0.456490	0.547046	1.018355	0.655235	60.408154	2.500846
1.16	0.832058	0.450814	0.541807	1.020812	0.653658	59.549686	2.739729
1.17	0.829645	0.445177	0.536588	1.023416	0.651995	58.726702	2.983941
1.18	0.827226	0.439579	0.531390	1.026166	0.650248	57.936214	3.233154
1.19	0.824800	0.434021	0.526214	1.029061	0.648418	57.175632	3.487066
1.20	0.822368	0.428504	0.521060	1.032101	0.646508	56.442690	3.745399
1.21	0.819931	0.423027	0.515930	1.035286	0.644520	55.735397	4.007895
1.22	0.817488	0.417592	0.510823	1.038613	0.642455	55.051986	4.274315
1.23	0.815039	0.412199	0.505741	1.042083	0.640316	54.390884	4.544437
1.24	0.812585	0.406848	0.500683	1.045696	0.638103	53.750679	4.818052
1.25	0.810127	0.401540	0.495651	1.049451	0.635820	53.130102	5.094966

M							
1.26	0.807663	0.396276	0.490645	1.053348	0.633468	52.528005	5.374995
1.27	0.805195	0.391055	0.485665	1.057386	0.631049	51.943343	5.657968
1.28	0.802723	0.385879	0.480712	1.061565	0.628564	51.375167	5.943723
1.29	0.800246	0.380747	0.475787	1.065886	0.626017	50.822606	6.232107
1.30	0.797766	0.375660	0.470889	1.070348	0.623407	50.284863	6.522976
1.31	0.795282	0.370617	0.466020	1.074950	0.620738	49.761202	6.816193
1.32	0.792795	0.365621	0.461179	1.079694	0.618011	49.250946	7.111628
1.33	0.790305	0.360669	0.456368	1.084579	0.615227	48.753467	7.409158
1.34	0.787811	0.355764	0.451586	1.089605	0.612389	48.268183	7.708667
1.35	0.785315	0.350905	0.446833	1.094772	0.609499	47.794554	8.010043
1.36	0.782816	0.346092	0.442111	1.100081	0.606557	47.332075	8.313180
1.37	0.780314	0.341325	0.437420	1.105532	0.603567	46.880275	8.617978
1.38	0.777811	0.336605	0.432759	1.111124	0.600529	46.438716	8.924340
1.39	0.775305	0.331931	0.428130	1.116860	0.597445	46.006983	9.232173
1.40	0.772798	0.327304	0.423532	1.122738	0.594317	45.584691	9.541391
1.41	0.770288	0.322724	0.418965	1.128759	0.591147	45.171476	9.851909
1.42	0.767778	0.318191	0.414431	1.134924	0.587936	44.766995	10.163647
1.43	0.765266	0.313704	0.409929	1.141233	0.584685	44.370926	10.476528
1.44	0.762753	0.309265	0.405459	1.147687	0.581397	43.982963	10.790477
1.45	0.760239	0.304873	0.401022	1.154287	0.578073	43.602819	11.105425
1.46	0.757725	0.300527	0.396618	1.161032	0.574715	43.230221	11.421302
1.47	0.755210	0.296229	0.392247	1.167924	0.571323	42.864910	11.738045
1.48	0.752695	0.291977	0.387909	1.174964	0.567900	42.506642	12.055590
1.49	0.750179	0.287772	0.383604	1.182151	0.564448	42.155184	12.373878
1.50	0.747664	0.283614	0.379333	1.189488	0.560966	41.810315	12.692849
1.51	0.745148	0.279502	0.375096	1.196974	0.557458	41.471824	13.012450

(Continued)

Table D.3 Compressible Flow Functions ($\gamma = 1.3$) (*Continued*)

M	T/T_t	P/P_t	ρ/ρ_t	A/A^*	$MFP\sqrt{R/g_c}$	μ	ν
1.52	0.742633	0.275437	0.370892	1.204610	0.553924	41.139510	13.332626
1.53	0.740118	0.271418	0.366722	1.212398	0.550366	40.813184	13.653326
1.54	0.737605	0.267446	0.362587	1.220339	0.546784	40.492661	13.974500
1.55	0.735091	0.263519	0.358485	1.228433	0.543182	40.177770	14.296100
1.56	0.732579	0.259639	0.354418	1.236681	0.539559	39.866342	14.618081
1.57	0.730068	0.255805	0.350384	1.245084	0.535917	39.564218	14.940399
1.58	0.727558	0.252016	0.346385	1.253644	0.532258	39.265248	15.263009
1.59	0.725050	0.248272	0.342421	1.262361	0.528583	38.971284	15.585872
1.60	0.722543	0.244574	0.338490	1.271237	0.524892	38.682187	15.908947
1.61	0.720038	0.240921	0.334594	1.280273	0.521188	38.397824	16.232196
1.62	0.717535	0.237312	0.330733	1.289470	0.517470	38.118064	16.555582
1.63	0.715034	0.233749	0.326905	1.298828	0.513742	37.842786	16.879070
1.64	0.712535	0.230229	0.323113	1.308351	0.510003	37.571869	17.202624
1.65	0.710038	0.226754	0.319354	1.318038	0.506254	37.305201	17.526212
1.66	0.707544	0.223322	0.315630	1.327891	0.502498	37.042671	17.849801
1.67	0.705052	0.219934	0.311940	1.337911	0.498734	36.784174	18.173360
1.68	0.702563	0.216589	0.308284	1.348100	0.494965	36.529607	18.496859
1.69	0.700077	0.213287	0.304662	1.358460	0.491190	36.278874	18.820270
1.70	0.697593	0.210028	0.301075	1.368991	0.487412	36.031879	19.143564
1.71	0.695113	0.206811	0.297521	1.379695	0.483630	35.788532	19.466714
1.72	0.692636	0.203636	0.294002	1.390574	0.479847	35.548744	19.789694
1.73	0.690162	0.200503	0.290516	1.401629	0.476062	35.312431	20.112479
1.74	0.687692	0.197412	0.287064	1.412861	0.472277	35.079511	20.435044
1.75	0.685225	0.194361	0.283646	1.424273	0.468493	34.849905	20.757365
1.76	0.682762	0.191352	0.280262	1.435866	0.464711	34.623535	21.079420

1.77	0.680302	0.188383	0.276911	1.447642	0.460930	34.400330	21.401187
1.78	0.677847	0.185454	0.273593	1.459602	0.457154	34.180216	21.722643
1.79	0.675395	0.182565	0.270308	1.471748	0.453381	33.963124	22.043770
1.80	0.672948	0.179715	0.267057	1.484082	0.449613	33.748989	22.364545
1.81	0.670504	0.176905	0.263839	1.496605	0.445851	33.537744	22.684951
1.82	0.668065	0.174133	0.260653	1.509320	0.442095	33.329327	23.004969
1.83	0.665631	0.171400	0.257500	1.522229	0.438346	33.123677	23.324579
1.84	0.663200	0.168705	0.254380	1.535332	0.434605	32.920735	23.643766
1.85	0.660775	0.166047	0.251292	1.548632	0.430872	32.720443	23.962512
1.86	0.658354	0.163427	0.248236	1.562132	0.427149	32.522748	24.280800
1.87	0.655938	0.160844	0.245212	1.575832	0.423435	32.327593	24.598615
1.88	0.653526	0.158297	0.242220	1.589736	0.419732	32.134928	24.915942
1.89	0.651120	0.155787	0.239260	1.603845	0.416039	31.944701	25.232765
1.90	0.648719	0.153313	0.236331	1.618160	0.412359	31.756864	25.549071
1.91	0.646323	0.150874	0.233434	1.632685	0.408690	31.571368	25.864846
1.92	0.643932	0.148470	0.230568	1.647422	0.405034	31.388166	26.180076
1.93	0.641546	0.146101	0.227733	1.662371	0.401392	31.207215	26.494749
1.94	0.639166	0.143767	0.224928	1.677537	0.397763	31.028469	26.808852
1.95	0.636791	0.141466	0.222155	1.692921	0.394149	30.851886	27.122372
1.96	0.634421	0.139199	0.219411	1.708524	0.390549	30.677424	27.435299
1.97	0.632057	0.136966	0.216698	1.724351	0.386964	30.505044	27.747622
1.98	0.629699	0.134765	0.214015	1.740402	0.383396	30.334705	28.059328
1.99	0.627347	0.132597	0.211361	1.756680	0.379843	30.166370	28.370408
2.00	0.625000	0.130461	0.208737	1.773188	0.376307	30.000000	28.680852
2.02	0.620324	0.126284	0.203577	1.806903	0.369285	29.673015	29.299792
2.04	0.615673	0.122231	0.198533	1.841566	0.362334	29.353470	29.916075

(Continued)

Table D.3   Compressible Flow Functions ($\gamma = 1.3$) (Continued)

M	T/T_t	P/P_t	ρ/ρ_t	A/A^*	$MFP\sqrt{R/g_c}$	μ	ν
2.06	0.611045	0.118300	0.193603	1.877197	0.355457	29.041100	30.529630
2.08	0.606443	0.114487	0.188784	1.913817	0.348655	28.735654	31.140395
2.10	0.601866	0.110789	0.184077	1.951448	0.341932	28.436890	31.748310
2.12	0.597314	0.107205	0.179478	1.990112	0.335289	28.144581	32.353320
2.14	0.592789	0.103729	0.174985	2.029831	0.328728	27.858510	32.955374
2.16	0.588291	0.100361	0.170598	2.070628	0.322251	27.578468	33.554425
2.18	0.583819	0.097097	0.166313	2.112527	0.315860	27.304258	34.150430
2.20	0.579374	0.093934	0.162130	2.155553	0.309555	27.035692	34.743350
2.22	0.574957	0.090870	0.158047	2.199729	0.303338	26.772588	35.333146
2.24	0.570568	0.087902	0.154060	2.245083	0.297211	26.514775	35.919788
2.26	0.566207	0.085027	0.150170	2.291639	0.291173	26.262087	36.503243
2.28	0.561874	0.082243	0.146373	2.339424	0.285225	26.014366	37.083485
2.30	0.557569	0.079548	0.142669	2.388466	0.279369	25.771462	37.660488
2.32	0.553293	0.076938	0.139054	2.438792	0.273604	25.533229	38.234230
2.34	0.549046	0.074411	0.135528	2.490432	0.267930	25.299529	38.804692
2.36	0.544828	0.071966	0.132089	2.543414	0.262349	25.070228	39.371856
2.38	0.540640	0.069599	0.128734	2.597768	0.256860	24.845199	39.935705
2.40	0.536481	0.067308	0.125462	2.653524	0.251463	24.624318	40.496228
2.42	0.532351	0.065092	0.122272	2.710714	0.246157	24.407469	41.053411
2.44	0.528251	0.062947	0.119161	2.769370	0.240944	24.194537	41.607247
2.46	0.524180	0.060872	0.116128	2.829524	0.235821	23.985414	42.157725
2.48	0.520140	0.058865	0.113171	2.891209	0.230790	23.779995	42.704841
2.50	0.516129	0.056923	0.110288	2.954460	0.225849	23.578178	43.248590
2.52	0.512148	0.055045	0.107478	3.019311	0.220998	23.379868	43.788969
2.54	0.508197	0.053228	0.104739	3.085798	0.216237	23.184970	44.325975

2.56	0.504276	0.051471	0.102069	3.153956	0.211564	22.993394	44.859608
2.58	0.500385	0.049772	0.099468	3.223823	0.206979	22.805054	45.389869
2.60	0.496524	0.048129	0.096932	3.295436	0.202481	22.619865	45.916761
2.62	0.492693	0.046541	0.094462	3.368834	0.198069	22.437747	46.440286
2.64	0.488892	0.045005	0.092054	3.444056	0.193743	22.258621	46.960449
2.66	0.485121	0.043520	0.089709	3.521142	0.189502	22.082413	47.477254
2.68	0.481380	0.042084	0.087423	3.600133	0.185344	21.909050	47.990709
2.70	0.477669	0.040696	0.085197	3.681071	0.181269	21.738461	48.500820
2.72	0.473988	0.039354	0.083028	3.763998	0.177275	21.570578	49.007595
2.74	0.470336	0.038057	0.080915	3.848958	0.173362	21.405337	49.511044
2.76	0.466714	0.036803	0.078856	3.935995	0.169528	21.242672	50.011175
2.78	0.463122	0.035591	0.076851	4.025153	0.165773	21.082524	50.508000
2.80	0.459559	0.034420	0.074898	4.116480	0.162095	20.924832	51.001529
2.82	0.456025	0.033288	0.072996	4.210021	0.158494	20.769540	51.491774
2.84	0.452521	0.032194	0.071143	4.305826	0.154967	20.616590	51.978748
2.86	0.449047	0.031136	0.069338	4.403942	0.151515	20.465930	52.462463
2.88	0.445601	0.030114	0.067580	4.504419	0.148135	20.317508	52.942933
2.90	0.442184	0.029126	0.065869	4.607308	0.144827	20.171271	53.420171
2.92	0.438797	0.028171	0.064201	4.712660	0.141589	20.027172	53.894193
2.94	0.435438	0.027249	0.062578	4.820529	0.138421	19.885163	54.365013
2.96	0.432107	0.026357	0.060997	4.930967	0.135321	19.745198	54.832646
2.98	0.428805	0.025495	0.059457	5.044029	0.132288	19.607232	55.297109
3.00	0.425532	0.024663	0.057957	5.159772	0.129320	19.471221	55.758417
3.02	0.422287	0.023858	0.056497	5.278251	0.126417	19.337123	56.216587
3.04	0.419069	0.023080	0.055075	5.399524	0.123578	19.204897	56.671635
3.06	0.415880	0.022329	0.053690	5.523651	0.120801	19.074505	57.123578

(Continued)

Table D.3   Compressible Flow Functions ($\gamma = 1.3$) (Continued)

M	T/T_t	P/P_t	ρ/ρ_t	A/A^*	$MFP\sqrt{R/g_c}$	μ	ν
3.08	0.412718	0.021602	0.052341	5.650690	0.118085	18.945906	57.572435
3.10	0.409584	0.020900	0.051028	5.780704	0.115429	18.819063	58.018223
3.12	0.406478	0.020222	0.049750	5.913753	0.112832	18.693941	58.460958
3.14	0.403398	0.019567	0.048504	6.049902	0.110293	18.570503	58.900661
3.16	0.400346	0.018933	0.047292	6.189215	0.107811	18.448715	59.337348
3.18	0.397320	0.018321	0.046111	6.331757	0.105383	18.328544	59.771038
3.20	0.394322	0.017729	0.044961	6.477594	0.103011	18.209957	60.201750
3.22	0.391350	0.017157	0.043841	6.626796	0.100692	18.092922	60.629503
3.24	0.388404	0.016605	0.042751	6.779431	0.098425	17.977409	61.054315
3.26	0.385484	0.016070	0.041689	6.935569	0.096209	17.863387	61.476206
3.28	0.382591	0.015554	0.040655	7.095282	0.094043	17.750827	61.895195
3.30	0.379723	0.015055	0.039648	7.258643	0.091927	17.639701	62.311301
3.32	0.376881	0.014573	0.038667	7.425725	0.089858	17.529982	62.724543
3.34	0.374064	0.014107	0.037713	7.596605	0.087837	17.421641	63.134941
3.36	0.371272	0.013656	0.036783	7.771360	0.085862	17.314654	63.542515
3.38	0.368506	0.013221	0.035877	7.950066	0.083932	17.208993	63.947283
3.40	0.365764	0.012800	0.034995	8.132804	0.082046	17.104635	64.349266
3.42	0.363048	0.012393	0.034136	8.319655	0.080203	17.001555	64.748483
3.44	0.360355	0.012000	0.033299	8.510700	0.078403	16.899729	65.144953
3.46	0.357687	0.011619	0.032485	8.706024	0.076644	16.799135	65.538697
3.48	0.355043	0.011252	0.031691	8.905710	0.074925	16.699749	65.929734
3.50	0.352423	0.010896	0.030918	9.109847	0.073246	16.601550	66.318083
3.52	0.349826	0.010553	0.030165	9.318521	0.071606	16.504516	66.703765
3.54	0.347254	0.010220	0.029432	9.531822	0.070004	16.408626	67.086799

M							
3.56	0.344704	0.009899	0.028718	9.749841	0.068438	16.313860	67.467205
3.58	0.342177	0.009589	0.028022	9.972670	0.066909	16.220198	67.845002
3.60	0.339674	0.009288	0.027345	10.200404	0.065415	16.127620	68.220209
3.62	0.337193	0.008998	0.026685	10.433137	0.063956	16.036108	68.592847
3.64	0.334735	0.008717	0.026042	10.670967	0.062531	15.945643	68.962934
3.66	0.332299	0.008445	0.025415	10.913993	0.061138	15.856207	69.330491
3.68	0.329885	0.008183	0.024805	11.162315	0.059778	15.767781	69.695536
3.70	0.327493	0.007929	0.024211	11.416034	0.058450	15.680350	70.058089
3.72	0.325123	0.007683	0.023632	11.675255	0.057152	15.593896	70.418168
3.74	0.322774	0.007446	0.023067	11.940082	0.055884	15.508402	70.775795
3.76	0.320447	0.007216	0.022518	12.210623	0.054646	15.423852	71.130986
3.78	0.318141	0.006993	0.021982	12.486987	0.053437	15.340231	71.483762
3.80	0.315856	0.006778	0.021460	12.769283	0.052255	15.257523	71.834140
3.82	0.313592	0.006570	0.020952	13.057624	0.051101	15.175714	72.182141
3.84	0.311348	0.006369	0.020456	13.352124	0.049974	15.094787	72.527783
3.86	0.309125	0.006174	0.019973	13.652898	0.048873	15.014730	72.871084
3.88	0.306922	0.005986	0.019503	13.960065	0.047798	14.935527	73.212064
3.90	0.304739	0.005803	0.019044	14.273744	0.046748	14.857166	73.550739
3.92	0.302576	0.005627	0.018597	14.594056	0.045722	14.779632	73.887130
3.94	0.300432	0.005456	0.018162	14.921124	0.044719	14.702912	74.221254
3.96	0.298308	0.005291	0.017737	15.255074	0.043740	14.626994	74.553129
3.98	0.239919	0.006764	0.028194	10.528859	0.065034	14.551865	65.519681
4.00	0.238095	0.006586	0.027662	10.718750	0.063882	14.477512	65.784820

Appendix E Normal Shock Functions

For a 1-D normal shock, as shown in Fig. E.1, the following relationships are tabulated as functions of the upstream Mach number M_x for $\gamma = 1.4$:

$$M_y = \sqrt{\frac{M_x^2 + \dfrac{2}{\gamma - 1}}{\dfrac{2\gamma}{\gamma - 1}M_x^2 - 1}}$$

$$\frac{P_{ty}}{P_{tx}} = \left(\frac{\dfrac{\gamma + 1}{2}M_x^2}{1 + \dfrac{\gamma - 1}{2}M_x^2}\right)^{\gamma/(\gamma-1)} \qquad \left(\frac{2\gamma}{\gamma + 1}M_x^2 - \frac{\gamma - 1}{\gamma + 1}\right)^{-1/(\gamma-1)}$$

$$\frac{P_y}{P_x} = \frac{2\gamma}{\gamma + 1}M_x^2 - \frac{\gamma - 1}{\gamma + 1} \qquad \frac{\rho_y}{\rho_x} = \frac{P_y/P_x}{T_y/T_x}$$

$$\frac{T_y}{T_x} = \frac{\left(1 + \dfrac{\gamma - 1}{2}M_x^2\right)\left(\dfrac{2\gamma}{\gamma - 1}M_x^2 - 1\right)}{\dfrac{(\gamma + 1)^2}{2(\gamma - 1)}M_x^2}$$

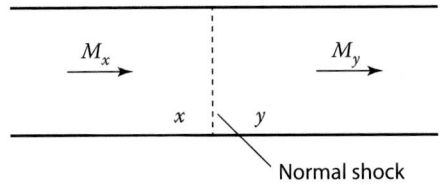

Fig. E.1   Normal shock wave.

Table E.1 Normal Shock Functions ($\gamma = 1.4$)

M_x	M_y	P_{ty}/P_{tx}	P_y/P_x	ρ_y/ρ_x	T_y/T_x
1.00	1.000000	1.000000	1.000000	1.000000	1.000000
1.01	0.990132	0.999999	1.023450	1.016694	1.006645
1.02	0.980519	0.999990	1.047133	1.033441	1.013249
1.03	0.971154	0.999967	1.071050	1.050240	1.019814
1.04	0.962025	0.999923	1.095200	1.067088	1.026345
1.05	0.953125	0.999853	1.119583	1.083982	1.032843
1.06	0.944445	0.999751	1.144200	1.100921	1.039312
1.07	0.935977	0.999611	1.169050	1.117903	1.045753
1.08	0.927713	0.999431	1.194133	1.134925	1.052170
1.09	0.919647	0.999204	1.219450	1.151985	1.058564
1.10	0.911770	0.998928	1.245000	1.169082	1.064938
1.11	0.904078	0.998599	1.270783	1.186213	1.071294
1.12	0.896563	0.998213	1.296800	1.203377	1.077634
1.13	0.889219	0.997768	1.323050	1.220571	1.083960
1.14	0.882042	0.997261	1.349533	1.237793	1.090274
1.15	0.875024	0.996690	1.376250	1.255042	1.096577
1.16	0.868162	0.996052	1.403200	1.272315	1.102872
1.17	0.861451	0.995345	1.430383	1.289610	1.109159
1.18	0.854884	0.994569	1.457800	1.306927	1.115441
1.19	0.848459	0.993720	1.485450	1.324262	1.121719
1.20	0.842170	0.992798	1.513333	1.341615	1.127994
1.21	0.836014	0.991802	1.541450	1.358983	1.134268
1.22	0.829986	0.990731	1.569800	1.376364	1.140541
1.23	0.824083	0.989583	1.598383	1.393757	1.146816
1.24	0.818301	0.988359	1.627200	1.411160	1.153094
1.25	0.812636	0.987057	1.656250	1.428571	1.159375
1.26	0.807085	0.985677	1.685533	1.445989	1.165661
1.27	0.801645	0.984219	1.715050	1.463412	1.171953
1.28	0.796312	0.982682	1.744800	1.480839	1.178251
1.29	0.791084	0.981067	1.774783	1.498267	1.184558
1.30	0.785957	0.979374	1.805000	1.515695	1.190873
1.31	0.780929	0.977602	1.835450	1.533122	1.197198
1.32	0.775997	0.975752	1.866133	1.550546	1.203533
1.33	0.771159	0.973824	1.897050	1.567965	1.209880
1.34	0.766412	0.971819	1.928200	1.585379	1.216239
1.35	0.761753	0.969737	1.959583	1.602785	1.222612

(Continued)

Table E.1 Normal Shock Functions ($\gamma = 1.4$) (*Continued*)

M_x	M_y	P_{ty}/P_{tx}	P_y/P_x	ρ_y/ρ_x	T_y/T_x
1.36	0.757181	0.967579	1.991200	1.620182	1.228998
1.37	0.752692	0.965344	2.023050	1.637569	1.235398
1.38	0.748286	0.963035	2.055133	1.654945	1.241814
1.39	0.743959	0.960652	2.087450	1.672307	1.248246
1.40	0.739709	0.958194	2.120000	1.689655	1.254694
1.41	0.735536	0.955665	2.152783	1.706988	1.261159
1.42	0.731436	0.953063	2.185800	1.724303	1.267643
1.43	0.727408	0.950390	2.219050	1.741600	1.274144
1.44	0.723451	0.947648	2.252533	1.758878	1.280665
1.45	0.719562	0.944837	2.286250	1.776135	1.287205
1.46	0.715740	0.941958	2.320200	1.793370	1.293765
1.47	0.711983	0.939012	2.354383	1.810582	1.300346
1.48	0.708290	0.936001	2.388800	1.827770	1.306948
1.49	0.704659	0.932925	2.423450	1.844933	1.313571
1.50	0.701089	0.929787	2.458333	1.862069	1.320216
1.51	0.697578	0.926586	2.493450	1.879177	1.326884
1.52	0.694125	0.923324	2.528800	1.896257	1.333574
1.53	0.690729	0.920003	2.564383	1.913308	1.340288
1.54	0.687388	0.916624	2.600200	1.930327	1.347026
1.55	0.684101	0.913188	2.636250	1.947315	1.353787
1.56	0.680867	0.909697	2.672533	1.964270	1.360573
1.57	0.677685	0.906151	2.709050	1.981192	1.367384
1.58	0.674553	0.902552	2.745800	1.998079	1.374220
1.59	0.671471	0.898901	2.782783	2.014931	1.381081
1.60	0.668437	0.895200	2.820000	2.031746	1.387969
1.61	0.665451	0.891450	2.857450	2.048524	1.394882
1.62	0.662511	0.887653	2.895133	2.065264	1.401822
1.63	0.659616	0.883809	2.933050	2.081965	1.408789
1.64	0.656765	0.879921	2.971200	2.098627	1.415783
1.65	0.653958	0.875988	3.009583	2.115248	1.422804
1.66	0.651194	0.872014	3.048200	2.131827	1.429853
1.67	0.648471	0.867999	3.087050	2.148365	1.436930
1.68	0.645789	0.863944	3.126133	2.164860	1.444035
1.69	0.643147	0.859851	3.165450	2.181311	1.451168
1.70	0.640544	0.855721	3.205000	2.197719	1.458330
1.71	0.637979	0.851556	3.244783	2.214081	1.465521

(*Continued*)

Table E.1 Normal Shock Functions ($\gamma = 1.4$) (*Continued*)

M_x	M_y	P_{ty}/P_{tx}	P_y/P_x	ρ_y/ρ_x	T_y/T_x
1.72	0.635452	0.847356	3.284800	2.230398	1.472742
1.73	0.632962	0.843124	3.325050	2.246669	1.479991
1.74	0.630508	0.838860	3.365533	2.262893	1.487270
1.75	0.628089	0.834565	3.406250	2.279070	1.494579
1.76	0.625705	0.830242	3.447200	2.295199	1.501918
1.77	0.623354	0.825891	3.488383	2.311279	1.509287
1.78	0.621037	0.821513	3.529800	2.327310	1.516687
1.79	0.618753	0.817111	3.571450	2.343292	1.524117
1.80	0.616501	0.812684	3.613333	2.359223	1.531578
1.81	0.614281	0.808234	3.655450	2.375104	1.539069
1.82	0.612091	0.803763	3.697800	2.390934	1.546592
1.83	0.609931	0.799271	3.740383	2.406712	1.554146
1.84	0.607802	0.794761	3.783200	2.422438	1.561732
1.85	0.605701	0.790232	3.826250	2.438112	1.569349
1.86	0.603629	0.785686	3.869533	2.453733	1.576999
1.87	0.601585	0.781125	3.913050	2.469301	1.584680
1.88	0.599569	0.776549	3.956800	2.484814	1.592393
1.89	0.597579	0.771959	4.000783	2.500274	1.600138
1.90	0.595616	0.767357	4.045000	2.515679	1.607916
1.91	0.593680	0.762743	4.089450	2.531030	1.615726
1.92	0.591769	0.758119	4.134133	2.546325	1.623568
1.93	0.589883	0.753486	4.179050	2.561565	1.631444
1.94	0.588022	0.748844	4.224200	2.576749	1.639352
1.95	0.586185	0.744195	4.269583	2.591877	1.647294
1.96	0.584372	0.739540	4.315200	2.606949	1.655268
1.97	0.582582	0.734879	4.361050	2.621964	1.663276
1.98	0.580816	0.730214	4.407133	2.636922	1.671317
1.99	0.579072	0.725545	4.453450	2.651823	1.679392
2.00	0.577350	0.720874	4.500000	2.666667	1.687500
2.02	0.573972	0.711527	4.593800	2.696181	1.703817
2.04	0.570679	0.702180	4.688533	2.725463	1.720271
2.06	0.567467	0.692839	4.784200	2.754511	1.736860
2.08	0.564334	0.683512	4.880800	2.783325	1.753586
2.10	0.561277	0.674203	4.978333	2.811902	1.770450
2.12	0.558294	0.664919	5.076800	2.840243	1.787453
2.14	0.555383	0.655666	5.176200	2.868345	1.804594

(*Continued*)

Table E.1   Normal Shock Functions ($\gamma = 1.4$) (Continued)

M_x	M_y	P_{ty}/P_{tx}	P_y/P_x	ρ_y/ρ_x	T_y/T_x
2.16	0.552541	0.646447	5.276533	2.896209	1.821876
2.18	0.549766	0.637269	5.377800	2.923834	1.839297
2.20	0.547056	0.628136	5.480000	2.951220	1.856860
2.22	0.544409	0.619053	5.583133	2.978365	1.874563
2.24	0.541822	0.610023	5.687200	3.005271	1.892409
2.26	0.539295	0.601051	5.792200	3.031936	1.910396
2.28	0.536825	0.592140	5.898133	3.058362	1.928527
2.30	0.534411	0.583295	6.005000	3.084548	1.946801
2.32	0.532051	0.574517	6.112800	3.110495	1.965218
2.34	0.529743	0.565810	6.221533	3.136202	1.983779
2.36	0.527486	0.557177	6.331200	3.161671	2.002485
2.38	0.525278	0.548621	6.441800	3.186902	2.021336
2.40	0.523118	0.540144	6.553333	3.211896	2.040332
2.42	0.521004	0.531748	6.665800	3.236653	2.059473
2.44	0.518936	0.523435	6.779200	3.261174	2.078760
2.46	0.516911	0.515208	6.893533	3.285461	2.098194
2.48	0.514929	0.507067	7.008800	3.309514	2.117773
2.50	0.512989	0.499015	7.125000	3.333333	2.137500
2.52	0.511089	0.491052	7.242133	3.356921	2.157374
2.54	0.509228	0.483181	7.360200	3.380279	2.177394
2.56	0.507406	0.475402	7.479200	3.403407	2.197563
2.58	0.505620	0.467715	7.599133	3.426307	2.217879
2.60	0.503871	0.460123	7.720000	3.448980	2.238343
2.62	0.502157	0.452625	7.841800	3.471427	2.258956
2.64	0.500477	0.445223	7.964533	3.493651	2.279717
2.66	0.498830	0.437916	8.088200	3.515651	2.300626
2.68	0.497216	0.430705	8.212800	3.537431	2.321685
2.70	0.495634	0.423590	8.338333	3.558991	2.342892
2.72	0.494082	0.416572	8.464800	3.580333	2.364249
2.74	0.492560	0.409650	8.592200	3.601458	2.385756
2.76	0.491068	0.402825	8.720533	3.622369	2.407412
2.78	0.489604	0.396096	8.849800	3.643066	2.429218
2.80	0.488167	0.389464	8.980000	3.663551	2.451173
2.82	0.486758	0.382927	9.111133	3.683827	2.473279
2.84	0.485376	0.376486	9.243200	3.703894	2.495536
2.86	0.484019	0.370141	9.376200	3.723755	2.517942

(Continued)

Table E.1 Normal Shock Functions ($\gamma = 1.4$) (*Continued*)

M_x	M_y	P_{ty}/P_{tx}	P_y/P_x	ρ_y/ρ_x	T_y/T_x
2.88	0.482687	0.363890	9.510133	3.743411	2.540500
2.90	0.481380	0.357733	9.645000	3.762864	2.563207
2.92	0.480096	0.351670	9.780800	3.782115	2.586066
2.94	0.478836	0.345701	9.917533	3.801167	2.609076
2.96	0.477599	0.339823	10.055200	3.820021	2.632237
2.98	0.476384	0.334038	10.193800	3.838679	2.655549
3.00	0.475191	0.328344	10.333333	3.857143	2.679012
3.02	0.474019	0.322740	10.473800	3.875414	2.702627
3.04	0.472868	0.317226	10.615200	3.893495	2.726394
3.06	0.471737	0.311800	10.757533	3.911387	2.750312
3.08	0.470625	0.306462	10.900800	3.929092	2.774381
3.10	0.469534	0.301211	11.045000	3.946612	2.798603
3.12	0.468460	0.296046	11.190133	3.963948	2.822977
3.14	0.467406	0.290967	11.336200	3.981103	2.847502
3.16	0.466369	0.285971	11.483200	3.998078	2.872180
3.18	0.465350	0.281059	11.631133	4.014875	2.897010
3.20	0.464349	0.276229	11.780000	4.031496	2.921992
3.22	0.463364	0.271480	11.929800	4.047943	2.947127
3.24	0.462395	0.266811	12.080533	4.064216	2.972414
3.26	0.461443	0.262221	12.232200	4.080319	2.997854
3.28	0.460507	0.257710	12.384800	4.096253	3.023446
3.30	0.459586	0.253276	12.538333	4.112020	3.049191
3.32	0.458680	0.248918	12.692800	4.127621	3.075088
3.34	0.457788	0.244635	12.848200	4.143059	3.101139
3.36	0.456912	0.240426	13.004533	4.158334	3.127342
3.38	0.456049	0.236290	13.161800	4.173449	3.153698
3.40	0.455200	0.232226	13.320000	4.188406	3.180208
3.42	0.454365	0.228232	13.479133	4.203205	3.206870
3.44	0.453543	0.224309	13.639200	4.217850	3.233685
3.46	0.452734	0.220454	13.800200	4.232341	3.260654
3.48	0.451938	0.216668	13.962133	4.246680	3.287776
3.50	0.451154	0.212948	14.125000	4.260870	3.315051
3.52	0.450382	0.209293	14.288800	4.274910	3.342479
3.54	0.449623	0.205704	14.453533	4.288804	3.370061
3.56	0.448875	0.202177	14.619200	4.302553	3.397797
3.58	0.448138	0.198714	14.785800	4.316158	3.425685

(*Continued*)

Table E.1   Normal Shock Functions ($\gamma = 1.4$) *(Continued)*

M_x	M_y	P_{ty}/P_{tx}	P_y/P_x	ρ_y/ρ_x	T_y/T_x
3.60	0.447413	0.195312	14.953333	4.329621	3.453728
3.62	0.446699	0.191971	15.121800	4.342944	3.481924
3.64	0.445995	0.188690	15.291200	4.356128	3.510273
3.66	0.445302	0.185467	15.461533	4.369175	3.538776
3.68	0.444620	0.182302	15.632800	4.382086	3.567433
3.70	0.443948	0.179194	15.805000	4.394864	3.596244
3.72	0.443285	0.176141	15.978133	4.407508	3.625208
3.74	0.442633	0.173143	16.152200	4.420021	3.654326
3.76	0.441990	0.170200	16.327200	4.432405	3.683598
3.78	0.441356	0.167309	16.503133	4.444661	3.713024
3.80	0.440732	0.164470	16.680000	4.456790	3.742604
3.82	0.440117	0.161683	16.857800	4.468794	3.772338
3.84	0.439510	0.158946	17.036533	4.480674	3.802225
3.86	0.438912	0.156258	17.216200	4.492432	3.832267
3.88	0.438323	0.153619	17.396800	4.504069	3.862463
3.90	0.437742	0.151027	17.578333	4.515586	3.892813
3.92	0.437170	0.148483	17.760800	4.526986	3.923317
3.94	0.436605	0.145984	17.944200	4.538268	3.953975
3.96	0.436049	0.143531	18.128533	4.549435	3.984788
3.98	0.435500	0.141122	18.313800	4.560488	4.015754
4.00	0.434959	0.138756	18.500000	4.571429	4.046875

Appendix F

Two-Dimensional Oblique Shock Functions

For a 2-D oblique shock, as shown in Fig. F.1, the following property ratios apply across the shock. Because a 2-D oblique shock acts as a normal shock perpendicular to the flow, the normal shock relations can be applied to oblique shocks. The normal relations of Appendix E apply, with M_x replaced by $M_1 \sin \beta$ and M_y replaced by $M_2 \sin (\beta - \theta)$. Thus, we can write

$$\frac{P_2}{P_1} = f_1(M_1 \sin \beta) = \frac{2\gamma}{\gamma + 1}(M_1 \sin \beta)^2 - \frac{\gamma - 1}{\gamma + 1} \tag{F.1}$$

$$\frac{T_2}{T_1} = f_2(M_1 \sin \beta) = \frac{\left[1 + \dfrac{\gamma - 1}{2}(M_1 \sin \beta)^2\right]\left[\dfrac{2\gamma}{\gamma - 1}(M_1 \sin \beta)^2 - 1\right]}{\dfrac{(\gamma + 1)^2}{2(\gamma - 1)}(M_1 \sin \beta)^2} \tag{F.2}$$

$$\frac{\rho_2}{\rho_1} = f_3(M_1 \sin \beta) = \frac{P_2/P_1}{T_2/T_1} \tag{F.3}$$

$$\frac{P_{t2}}{P_{t1}} = f_4(M_1 \sin \beta)$$

$$\frac{P_{t2}}{P_{t1}} = \left[\frac{\dfrac{\gamma + 1}{2}(M_1 \sin \beta)^2}{1 + \dfrac{\gamma - 1}{2}(M_1 \sin \beta)^2}\right]^{\gamma/(\gamma - 1)} \left[\frac{2\gamma}{\gamma + 1}(M_1 \sin \beta)^2 - \frac{\gamma - 1}{\gamma + 1}\right]^{-1/(\gamma - 1)} \tag{F.4}$$

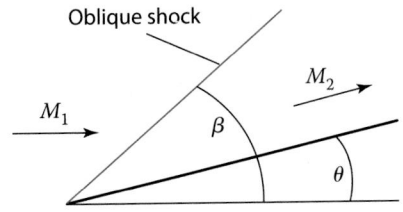

Fig. F.1 Oblique shock nomenclature.

Table F.1 Wave Angle β for Oblique Shock Waves ($\gamma = 1.4$)

M_1	\multicolumn{12}{c}{Flow Turning Angle θ, deg}		β	θ_{max}

M_1	0	2	4	6	8	10	12	14	16	18	20	22	β	θ_{max}
1.00														
1.01	81.93												0.05	85.36
1.02	78.63												0.14	83.49
1.03	76.14												0.26	82.08
1.04	74.06												0.40	80.93
1.05	72.25												0.56	79.94
1.06	70.63												0.73	79.06
1.07	69.16												0.91	78.27
1.08	67.81												1.10	77.56
1.09	66.55												1.30	76.90
1.10	65.38												1.52	76.30
1.11	64.28												1.73	75.73
1.12	63.23												1.96	75.21
1.13	62.25	70.93											2.19	74.72
1.14	61.31	68.71											2.43	74.26
1.15	60.41	67.00											2.67	73.82
1.16	59.55	65.56											2.92	73.41
1.17	58.73	64.28											3.17	73.02
1.18	57.94	63.11											3.42	72.66
1.19	57.18	62.05											3.68	72.31
1.20	56.44	61.05											3.94	71.98
1.21	55.74	60.12	68.09										4.21	71.66
1.22	55.05	59.24	66.03										4.47	71.36
1.23	54.39	58.40	64.47										4.74	71.07
1.24	53.75	57.60	63.15										5.01	70.80
1.25	53.13	56.84	61.99										5.29	70.54

1.26	70.29	5.56									60.93	56.12	52.53
1.27	70.05	5.83									59.97	55.42	51.94
1.28	69.82	6.11								67.38	59.06	54.75	51.38
1.29	69.60	6.39								65.05	58.22	54.10	50.82
1.30	69.40	6.66								63.46	57.42	53.47	50.28
1.31	69.19	6.94								62.16	56.67	52.87	49.76
1.32	69.00	7.22								61.03	55.95	52.28	49.25
1.33	68.82	7.49								60.02	55.26	51.72	48.75
1.34	68.64	7.77								59.09	54.60	51.17	48.27
1.35	68.47	8.05							66.91	58.23	53.97	50.63	47.79
1.36	68.31	8.33							64.29	57.43	53.36	50.11	47.33
1.37	68.15	8.60							62.69	56.68	52.77	49.61	46.88
1.38	68.00	8.88							61.43	55.96	52.20	49.12	46.44
1.39	67.85	9.15							60.34	55.28	51.65	48.64	46.01
1.40	67.72	9.43							59.37	54.63	51.12	48.17	45.58
1.41	67.58	9.70							58.48	54.01	50.60	47.72	45.17
1.42	67.45	9.97							57.67	53.42	50.10	47.27	44.77
1.43	67.33	10.25						63.95	56.91	52.84	49.61	46.84	44.37
1.44	67.21	10.52						62.31	56.19	52.29	49.14	46.42	43.98
1.45	67.10	10.79						61.05	55.52	51.76	48.68	46.00	43.60
1.46	66.99	11.05						59.98	54.87	51.24	48.23	45.60	43.23
1.47	66.88	11.32						59.04	54.26	50.74	47.79	45.20	42.86
1.48	66.78	11.59						58.19	53.68	50.25	47.37	44.82	42.51
1.49	66.68	11.85						57.41	53.11	49.78	46.95	44.44	42.16
1.50	66.59	12.11					64.36	56.68	52.57	49.33	46.54	44.06	41.81

(Continued)

Table F.1 Wave Angle β for Oblique Shock Waves ($\gamma = 1.4$) (Continued)

M_1	Flow Turning Angle θ, deg												β	θ_{max}
	0	2	4	6	8	10	12	14	16	18	20	22		
1.51	41.47	43.70	46.15	48.88	52.05	56.00	62.42						12.37	66.50
1.52	41.14	43.34	45.76	48.45	51.55	55.35	61.10						12.63	66.41
1.53	40.81	42.99	45.38	48.03	51.06	54.74	60.02						12.89	66.33
1.54	40.49	42.65	45.01	47.62	50.59	54.16	59.08						13.15	66.25
1.55	40.18	42.32	44.64	47.21	50.13	53.60	58.24						13.40	66.17
1.56	39.87	41.99	44.29	46.82	49.69	53.06	57.47						13.66	66.10
1.57	39.56	41.66	43.94	46.44	49.26	52.55	56.77						13.91	66.03
1.58	39.27	41.34	43.59	46.07	48.84	52.05	56.10	63.38					14.16	65.96
1.59	38.97	41.03	43.26	45.70	48.43	51.58	55.48	61.73					14.41	65.89
1.60	38.68	40.72	42.93	45.34	48.03	51.12	54.89	60.54					14.65	65.83
1.61	38.40	40.42	42.61	44.99	47.64	50.67	54.33	59.55					14.90	65.77
1.62	38.12	40.13	42.29	44.65	47.26	50.23	53.79	58.69					15.14	65.71
1.63	37.84	39.84	41.98	44.32	46.89	49.81	53.28	57.91					15.38	65.65
1.64	37.57	39.55	41.68	43.99	46.53	49.41	52.79	57.20					15.62	65.60
1.65	37.31	39.27	41.38	43.67	46.18	49.01	52.31	56.54					15.86	65.55
1.66	37.04	38.99	41.08	43.35	45.84	48.62	51.85	55.93	63.58				16.09	65.50
1.67	36.78	38.72	40.79	43.04	45.50	48.24	51.41	55.34	61.80				16.32	65.45
1.68	36.53	38.45	40.51	42.74	45.17	47.88	50.98	54.79	60.60				16.55	65.40
1.69	36.28	38.19	40.23	42.44	44.85	47.52	50.57	54.27	59.63				16.78	65.36
1.70	36.03	37.93	39.96	42.14	44.53	47.17	50.17	53.77	58.79				17.01	65.32
1.71	35.79	37.67	39.69	41.86	44.22	46.82	49.78	53.29	58.05				17.24	65.28
1.72	35.55	37.42	39.42	41.57	43.91	46.49	49.40	52.83	57.36				17.46	65.24
1.73	35.31	37.17	39.16	41.30	43.61	46.16	49.03	52.39	56.73				17.68	65.20
1.74	35.08	36.93	38.90	41.02	43.32	45.84	48.67	51.96	56.14				17.90	65.17
1.75	34.85	36.69	38.65	40.76	43.03	45.53	48.32	51.55	55.59	62.94			18.12	65.13

1.76	65.10	18.34			61.42	55.06	51.15	47.98	45.22	42.75	40.49	38.40	36.45	34.62
1.77	65.07	18.55			60.34	54.57	50.76	47.64	44.92	42.48	40.23	38.16	36.22	34.40
1.78	65.04	18.76			59.46	54.09	50.38	47.32	44.63	42.20	39.98	37.91	35.99	34.18
1.79	65.01	18.97			58.69	53.63	50.02	47.00	44.34	41.94	39.73	37.68	35.76	33.96
1.80	64.99	19.18			57.99	53.20	49.66	46.69	44.06	41.67	39.48	37.44	35.54	33.75
1.81	64.96	19.39			57.36	52.78	49.31	46.38	43.78	41.42	39.24	37.21	35.32	33.54
1.82	64.94	19.59			56.78	52.37	48.98	46.08	43.51	41.16	39.00	36.99	35.10	33.33
1.83	64.91	19.80			56.23	51.98	48.65	45.79	43.24	40.91	38.76	36.76	34.89	33.12
1.84	64.89	20.00			55.71	51.60	48.33	45.50	42.98	40.67	38.53	36.54	34.68	32.92
1.85	64.87	20.20		62.10	55.23	51.23	48.01	45.22	42.72	40.42	38.30	36.32	34.47	32.72
1.86	64.85	20.40		60.91	54.76	50.88	47.71	44.95	42.46	40.19	38.08	36.11	34.26	32.52
1.87	64.83	20.59		59.99	54.32	50.53	47.41	44.68	42.21	39.95	37.86	35.90	34.06	32.33
1.88	64.82	20.79		59.21	53.90	50.19	47.12	44.41	41.97	39.72	37.64	35.69	33.86	32.13
1.89	64.80	20.98		58.52	53.49	49.86	46.83	44.15	41.73	39.50	37.42	35.48	33.66	31.94
1.90	64.78	21.17		57.90	53.10	49.54	46.55	43.90	41.49	39.27	37.21	35.28	33.47	31.76
1.91	64.77	21.36		57.33	52.72	49.23	46.28	43.65	41.26	39.05	37.00	35.08	33.27	31.57
1.92	64.75	21.54		56.80	52.35	48.93	46.01	43.40	41.03	38.84	36.79	34.88	33.08	31.39
1.93	64.74	21.73		56.30	52.00	48.63	45.74	43.16	40.80	38.62	36.59	34.69	32.90	31.21
1.94	64.73	21.91		55.83	51.65	48.34	45.48	42.92	40.58	38.41	36.39	34.49	32.71	31.03
1.95	64.72	22.09	62.86	55.38	51.32	48.06	45.23	42.69	40.36	38.20	36.19	34.30	32.53	30.85
1.96	64.71	22.27	61.48	54.96	51.00	47.78	44.98	42.46	40.14	38.00	36.00	34.12	32.35	30.68
1.97	64.70	22.45	60.53	54.55	50.68	47.51	44.74	42.23	39.93	37.80	35.80	33.93	32.17	30.51
1.98	64.69	22.63	59.74	54.16	50.37	47.25	44.50	42.01	39.72	37.60	35.61	33.75	31.99	30.33
1.99	64.68	22.80	59.06	53.78	50.08	46.99	44.26	41.79	39.52	37.40	35.43	33.57	31.82	30.17
2.00	64.67	22.97	58.46	53.42	49.79	46.73	44.03	41.58	39.31	37.21	35.24	33.39	31.65	30.00

Table F.1 Wave Angle β for Oblique Shock Waves ($\gamma = 1.4$)

M_1	Flow Turning Angle, θ, deg										
	0	2	4	6	8	10	12	14	16	18	20
2.02	29.67	31.31	33.04	34.88	36.83	38.92	41.15	43.58	46.24	49.22	52.74
2.04	29.35	30.98	32.70	34.52	36.46	38.53	40.75	43.14	45.76	48.69	52.09
2.06	29.04	30.66	32.37	34.18	36.10	38.15	40.35	42.72	45.30	48.17	51.49
2.08	28.74	30.34	32.04	33.84	35.75	37.79	39.97	42.31	44.86	47.68	50.91
2.10	28.44	30.03	31.72	33.51	35.41	37.43	39.59	41.91	44.43	47.21	50.36
2.12	28.14	29.73	31.41	33.19	35.08	37.09	39.23	41.53	44.02	46.75	49.84
2.14	27.86	29.44	31.11	32.88	34.76	36.75	38.87	41.15	43.62	46.32	49.35
2.16	27.58	29.15	30.81	32.57	34.44	36.42	38.53	40.79	43.23	45.89	48.87
2.18	27.30	28.87	30.52	32.27	34.13	36.10	38.20	40.44	42.85	45.49	48.41
2.20	27.04	28.59	30.24	31.98	33.83	35.79	37.87	40.09	42.49	45.09	47.98
2.22	26.77	28.32	29.96	31.69	33.53	35.48	37.55	39.76	42.14	44.71	47.55
2.24	26.51	28.06	29.69	31.42	33.24	35.18	37.24	39.44	41.79	44.34	47.15
2.26	26.26	27.80	29.42	31.14	32.96	34.89	36.94	39.12	41.46	43.98	46.75
2.28	26.01	27.54	29.16	30.87	32.69	34.60	36.64	38.81	41.13	43.64	46.37
2.30	25.77	27.29	28.91	30.61	32.42	34.33	36.35	38.51	40.82	43.30	46.01
2.32	25.53	27.05	28.66	30.35	32.15	34.05	36.07	38.22	40.51	42.97	45.65
2.34	25.30	26.81	28.41	30.10	31.89	33.79	35.80	37.93	40.21	42.65	45.31
2.36	25.07	26.58	28.17	29.86	31.64	33.53	35.53	37.65	39.91	42.34	44.97
2.38	24.85	26.35	27.93	29.61	31.39	33.27	35.26	37.38	39.63	42.04	44.65
2.40	24.62	26.12	27.70	29.38	31.15	33.02	35.01	37.11	39.35	41.75	44.34
2.42	24.41	25.90	27.48	29.14	30.91	32.78	34.76	36.85	39.08	41.46	44.03
2.44	24.19	25.68	27.25	28.92	30.68	32.54	34.51	36.60	38.82	41.18	43.73
2.46	23.99	25.47	27.03	28.69	30.45	32.31	34.27	36.35	38.56	40.91	43.45
2.48	23.78	25.26	26.82	28.47	30.23	32.08	34.03	36.10	38.30	40.65	43.16
2.50	23.58	25.05	26.61	28.26	30.01	31.85	33.80	35.87	38.06	40.39	42.89
2.52	23.38	24.85	26.40	28.05	29.79	31.63	33.58	35.63	37.82	40.14	42.62
2.54	23.18	24.65	26.20	27.84	29.58	31.41	33.35	35.41	37.58	39.89	42.36
2.56	22.99	24.45	26.00	27.64	29.37	31.20	33.14	35.18	37.35	39.65	42.11
2.58	22.81	24.26	25.80	27.44	29.17	30.99	32.92	34.96	37.12	39.41	41.86
2.60	22.62	24.07	25.61	27.24	28.97	30.79	32.71	34.75	36.90	39.19	41.62
2.62	22.44	23.89	25.42	27.05	28.77	30.59	32.51	34.54	36.68	38.96	41.39
2.64	22.26	23.70	25.24	26.86	28.58	30.39	32.31	34.33	36.47	38.74	41.16
2.66	22.08	23.52	25.05	26.67	28.39	30.20	32.11	34.13	36.26	38.53	40.93
2.68	21.91	23.35	24.87	26.49	28.20	30.01	31.92	33.93	36.06	38.32	40.71
2.70	21.74	23.17	24.70	26.31	28.02	29.82	31.73	33.74	35.86	38.11	40.50
2.72	21.57	23.00	24.52	26.13	27.84	29.64	31.54	33.55	35.67	37.91	40.29
2.74	21.41	22.83	24.35	25.96	27.66	29.46	31.36	33.36	35.48	37.71	40.08
2.76	21.24	22.67	24.18	25.79	27.49	29.28	31.18	33.18	35.29	37.52	39.88
2.78	21.08	22.50	24.02	25.62	27.32	29.11	31.00	33.00	35.10	37.33	39.68
2.80	20.92	22.34	23.85	25.45	27.15	28.94	30.83	32.82	34.92	37.14	39.49
2.82	20.77	22.19	23.69	25.29	26.98	28.77	30.66	32.65	34.75	36.96	39.30
2.84	20.62	22.03	23.54	25.13	26.82	28.61	30.49	32.48	34.57	36.78	39.12
2.86	20.47	21.88	23.38	24.97	26.66	28.45	30.33	32.31	34.40	36.60	38.94
2.88	20.32	21.73	23.23	24.82	26.50	28.29	30.17	32.15	34.23	36.43	38.76
2.90	20.17	21.58	23.08	24.67	26.35	28.13	30.01	31.99	34.07	36.26	38.58
2.92	20.03	21.43	22.93	24.52	26.20	27.98	29.85	31.83	33.91	36.10	38.41
2.94	19.89	21.29	22.78	24.37	26.05	27.82	29.70	31.67	33.75	35.94	38.25
2.96	19.75	21.15	22.64	24.22	25.90	27.67	29.55	31.52	33.59	35.78	38.08
2.98	19.61	21.00	22.49	24.08	25.75	27.53	29.40	31.37	33.44	35.62	37.92
3.00	19.47	20.87	22.35	23.94	25.61	27.38	29.25	31.22	33.29	35.47	37.76

22	24	26	28	30	32	34	36	38	40	β	θ_{max}
57.39										23.31	64.65
56.46										23.65	64.64
55.63										23.98	64.63
54.87	61.28									24.30	64.63
54.17	59.77									24.61	64.62
53.52	58.62									24.92	64.62
52.91	57.65									25.23	64.62
52.34	56.81									25.52	64.62
51.79	56.05									25.82	64.62
51.28	55.36	62.70								26.10	64.62
50.79	54.72	60.85								26.38	64.62
50.32	54.12	59.63								26.66	64.63
49.87	53.56	58.65								26.93	64.64
49.44	53.03	57.82								27.19	64.64
49.03	52.54	57.08								27.45	64.65
48.63	52.06	56.41								27.71	64.66
48.25	51.61	55.79								27.96	64.67
47.88	51.18	55.22	61.97							28.20	64.68
47.52	50.77	54.69	60.65							28.45	64.70
47.17	50.37	54.18	59.66							28.68	64.71
46.84	49.99	53.71	58.83							28.91	64.72
46.52	49.62	53.26	58.11							29.14	64.74
46.20	49.27	52.83	57.47							29.36	64.75
45.90	48.93	52.43	56.88							29.58	64.77
45.60	48.60	52.04	56.33							29.80	64.78
45.31	48.28	51.66	55.83	64.27						30.01	64.80
45.04	47.97	51.30	55.36	62.05						30.22	64.81
44.76	47.67	50.96	54.91	60.94						30.42	64.83
44.50	47.38	50.63	54.49	60.08						30.62	64.85
44.24	47.10	50.31	54.09	59.35						30.81	64.87
43.99	46.83	50.00	53.71	58.72						31.01	64.88
43.75	46.56	49.70	53.34	58.14						31.19	64.90
43.51	46.31	49.41	52.99	57.62						31.38	64.92
43.28	46.05	49.12	52.66	57.14						31.56	64.94
43.05	45.81	48.85	52.33	56.69						31.74	64.96
42.83	45.57	48.59	52.02	56.26						31.92	64.97
42.61	45.34	48.33	51.72	55.87	63.25					32.09	64.99
42.40	45.11	48.08	51.44	55.49	62.00					32.26	65.01
42.19	44.89	47.84	51.16	55.13	61.14					32.42	65.03
41.99	44.68	47.60	50.89	54.79	60.43					32.59	65.05
41.79	44.47	47.38	50.62	54.46	59.83					32.75	65.07
41.60	44.26	47.15	50.37	54.14	59.29					32.91	65.09
41.41	44.06	46.93	50.13	53.84	58.80					33.06	65.11
41.23	43.86	46.72	49.89	53.55	58.35					33.21	65.13
41.04	43.67	46.51	49.65	53.27	57.93					33.36	65.15
40.87	43.48	46.31	49.43	53.00	57.54					33.51	65.16
40.69	43.30	46.12	49.21	52.74	57.17					33.65	65.18
40.52	43.12	45.92	49.00	52.49	56.83					33.80	65.20
40.36	42.95	45.73	48.79	52.25	56.50					33.94	65.22
40.19	42.78	45.55	48.59	52.01	56.18	63.67				34.07	65.24

(Continued)

M_1	Flow Turning Angle, θ, deg										
	0	2	4	6	8	10	12	14	16	18	20
3.05	19.14	20.53	22.01	23.59	25.26	27.03	28.90	30.86	32.92	35.10	37.38
3.10	18.82	20.20	21.68	23.26	24.93	26.69	28.55	30.51	32.57	34.74	37.02
3.15	18.51	19.89	21.37	22.94	24.60	26.37	28.23	30.18	32.24	34.40	36.67
3.20	18.21	19.59	21.06	22.63	24.29	26.05	27.91	29.86	31.92	34.07	36.34
3.25	17.92	19.29	20.76	22.33	23.99	25.75	27.60	29.56	31.61	33.76	36.02
3.30	17.64	19.01	20.48	22.04	23.70	25.46	27.31	29.26	31.31	33.46	35.71
3.35	17.37	18.73	20.20	21.76	23.42	25.17	27.03	28.98	31.02	33.17	35.42
3.40	17.10	18.47	19.93	21.49	23.15	24.90	26.75	28.70	30.75	32.89	35.13
3.45	16.85	18.21	19.67	21.23	22.88	24.64	26.49	28.44	30.48	32.62	34.86
3.50	16.60	17.96	19.42	20.97	22.63	24.38	26.24	28.18	30.22	32.36	34.60
3.55	16.36	17.71	19.17	20.73	22.38	24.14	25.99	27.94	29.98	32.12	34.35
3.60	16.13	17.48	18.93	20.49	22.14	23.90	25.75	27.70	29.74	31.88	34.11
3.65	15.90	17.25	18.70	20.26	21.91	23.67	25.52	27.47	29.51	31.65	33.88
3.70	15.68	17.03	18.48	20.03	21.69	23.44	25.30	27.25	29.29	31.42	33.65
3.75	15.47	16.81	18.26	19.81	21.47	23.23	25.08	27.03	29.07	31.21	33.44
3.80	15.26	16.60	18.05	19.60	21.26	23.02	24.87	26.82	28.86	31.00	33.23
3.85	15.05	16.39	17.84	19.40	21.05	22.81	24.67	26.62	28.66	30.80	33.03
3.90	14.86	16.20	17.64	19.20	20.85	22.61	24.47	26.42	28.47	30.61	32.83
3.95	14.66	16.00	17.45	19.00	20.66	22.42	24.28	26.23	28.28	30.42	32.65
4.00	14.48	15.81	17.26	18.81	20.47	22.23	24.09	26.05	28.10	30.24	32.46
4.05	14.29	15.63	17.07	18.63	20.29	22.05	23.91	25.87	27.92	30.06	32.29
4.10	14.12	15.45	16.89	18.45	20.11	21.88	23.74	25.70	27.75	29.89	32.12
4.15	13.94	15.27	16.72	18.27	19.94	21.70	23.57	25.53	27.58	29.72	31.95
4.20	13.77	15.10	16.55	18.10	19.77	21.54	23.41	25.37	27.42	29.56	31.79
4.25	13.61	14.94	16.38	17.94	19.60	21.37	23.24	25.21	27.27	29.41	31.64
4.30	13.45	14.77	16.22	17.78	19.44	21.22	23.09	25.06	27.11	29.26	31.49
4.35	13.29	14.62	16.06	17.62	19.29	21.06	22.94	24.91	26.97	29.11	31.35
4.40	13.14	14.46	15.90	17.46	19.13	20.91	22.79	24.76	26.82	28.97	31.20
4.45	12.99	14.31	15.75	17.31	18.99	20.77	22.65	24.62	26.68	28.83	31.07
4.50	12.84	14.16	15.61	17.17	18.84	20.62	22.50	24.48	26.55	28.70	30.94
4.55	12.70	14.02	15.46	17.02	18.70	20.48	22.37	24.35	26.42	28.57	30.81
4.60	12.56	13.88	15.32	16.88	18.56	20.35	22.24	24.22	26.29	28.44	30.68
4.65	12.42	13.74	15.18	16.75	18.43	20.22	22.11	24.09	26.16	28.32	30.56
4.70	12.28	13.60	15.05	16.61	18.30	20.09	21.98	23.97	26.04	28.20	30.44
4.75	12.15	13.47	14.92	16.48	18.17	19.96	21.86	23.85	25.92	28.09	30.33
4.80	12.02	13.34	14.79	16.36	18.04	19.84	21.74	23.73	25.81	27.97	30.22
4.85	11.90	13.22	14.66	16.23	17.92	19.72	21.62	23.61	25.70	27.86	30.11
4.90	11.78	13.09	14.54	16.11	17.80	19.60	21.50	23.50	25.59	27.76	30.00
4.95	11.66	12.97	14.42	15.99	17.68	19.49	21.39	23.39	25.48	27.65	29.90
5.00	11.54	12.85	14.30	15.88	17.57	19.38	21.28	23.29	25.38	27.55	29.80
5.25	10.98	12.29	13.75	15.33	17.03	18.85	20.78	22.79	24.90	27.08	29.34
5.50	10.48	11.79	13.24	14.84	16.55	18.39	20.32	22.35	24.47	26.66	28.93
5.75	10.02	11.33	12.79	14.39	16.12	17.97	19.92	21.96	24.09	26.30	28.57
6.00	9.59	10.91	12.37	13.98	15.73	17.59	19.55	21.61	23.75	25.97	28.26
6.25	9.21	10.52	11.99	13.61	15.37	17.24	19.22	21.29	23.45	25.67	27.97
6.50	8.85	10.16	11.64	13.27	15.04	16.93	18.92	21.01	23.17	25.41	27.71
6.75	8.52	9.83	11.32	12.96	14.74	16.64	18.65	20.75	22.93	25.17	27.48
7.00	8.21	9.53	11.02	12.67	14.46	16.38	18.40	20.51	22.70	24.96	27.28
7.25	7.93	9.24	10.74	12.40	14.21	16.14	18.18	20.30	22.50	24.76	27.09
7.50	7.66	8.98	10.48	12.16	13.98	15.92	17.97	20.10	22.31	24.58	26.92

22	24	26	28	30	32	34	36	38	40	β	θ_{max}
39.80	42.36	45.11	48.10	51.45	55.46	61.50				34.41	65.29
39.42	41.97	44.69	47.65	50.93	54.80	60.21				34.73	65.34
39.06	41.59	44.30	47.22	50.45	54.20	59.20				35.03	65.38
38.72	41.24	43.92	46.81	49.99	53.65	58.35				35.33	65.43
38.39	40.90	43.56	46.43	49.57	53.14	57.62				35.61	65.47
38.08	40.57	43.22	46.06	49.16	52.67	56.96				35.88	65.52
37.78	40.26	42.90	45.72	48.78	52.22	56.37	63.38			36.14	65.56
37.49	39.97	42.59	45.39	48.42	51.81	55.84	61.91			36.39	65.60
37.21	39.68	42.29	45.07	48.08	51.42	55.34	60.90			36.63	65.65
36.95	39.41	42.01	44.77	47.76	51.05	54.89	60.09			36.87	65.69
36.69	39.15	41.74	44.49	47.45	50.70	54.46	59.40			37.09	65.73
36.45	38.90	41.48	44.21	47.15	50.38	54.07	58.79			37.31	65.77
36.21	38.66	41.23	43.95	46.87	50.06	53.69	58.25			37.51	65.81
35.99	38.43	40.99	43.70	46.61	49.77	53.34	57.76			37.71	65.85
35.77	38.20	40.76	43.46	46.35	49.49	53.01	57.31			37.91	65.88
35.56	37.99	40.54	43.23	46.10	49.22	52.70	56.89	64.19		38.09	65.92
35.35	37.78	40.33	43.01	45.87	48.96	52.41	56.51	62.94		38.27	65.96
35.16	37.58	40.13	42.80	45.65	48.72	52.13	56.15	62.09		38.44	65.99
34.97	37.39	39.93	42.60	45.43	48.48	51.86	55.81	61.41		38.61	66.03
34.79	37.21	39.74	42.40	45.22	48.26	51.61	55.50	60.83		38.77	66.06
34.61	37.03	39.56	42.21	45.03	48.04	51.36	55.20	60.32		38.93	66.09
34.44	36.86	39.38	42.03	44.83	47.84	51.13	54.92	59.86		39.08	66.12
34.27	36.69	39.21	41.86	44.65	47.64	50.91	54.65	59.45		39.23	66.15
34.11	36.53	39.05	41.69	44.47	47.45	50.70	54.39	59.07		39.37	66.18
33.96	36.37	38.89	41.52	44.30	47.27	50.50	54.15	58.73		39.51	66.21
33.81	36.22	38.74	41.37	44.14	47.09	50.30	53.92	58.40		39.64	66.24
33.67	36.08	38.59	41.22	43.98	46.92	50.12	53.71	58.10		39.77	66.27
33.53	35.94	38.45	41.07	43.83	46.76	49.94	53.50	57.81		39.89	66.30
33.39	35.80	38.31	40.93	43.68	46.60	49.77	53.30	57.54	65.77	40.01	66.33
33.26	35.67	38.17	40.79	43.54	46.45	49.60	53.10	57.29	64.34	40.13	66.35
33.13	35.54	38.04	40.66	43.40	46.31	49.44	52.92	57.05	63.58	40.24	66.38
33.00	35.41	37.92	40.53	43.27	46.17	49.29	52.75	56.83	63.00	40.35	66.40
32.88	35.29	37.80	40.40	43.14	46.03	49.14	52.58	56.61	62.51	40.46	66.43
32.77	35.18	37.68	40.28	43.01	45.90	49.00	52.41	56.40	62.09	40.56	66.45
32.65	35.06	37.56	40.17	42.89	45.77	48.86	52.26	56.21	61.71	40.66	66.48
32.54	34.95	37.45	40.05	42.78	45.65	48.73	52.11	56.02	61.37	40.76	66.50
32.44	34.85	37.34	39.94	42.66	45.53	48.60	51.96	55.84	61.06	40.85	66.52
32.33	34.74	37.24	39.84	42.55	45.42	48.47	51.82	55.67	60.78	40.94	66.54
32.23	34.64	37.14	39.73	42.45	45.30	48.35	51.69	55.51	60.51	41.03	66.56
32.13	34.54	37.04	39.63	42.34	45.20	48.24	51.56	55.35	60.26	41.12	66.58
31.68	34.09	36.59	39.17	41.87	44.70	47.71	50.97	54.65	59.21	41.52	66.68
31.28	33.69	36.19	38.77	41.46	44.28	47.26	50.47	54.06	58.40	41.86	66.77
30.92	33.34	35.84	38.42	41.11	43.91	46.87	50.04	53.56	57.74	42.17	66.84
30.61	33.04	35.54	38.12	40.79	43.58	46.52	49.67	53.14	57.19	42.44	66.91
30.33	32.76	35.26	37.84	40.52	43.30	46.22	49.35	52.77	56.72	42.68	66.98
30.08	32.52	35.02	37.60	40.27	43.05	45.96	49.06	52.44	56.32	42.89	67.03
29.86	32.30	34.80	37.38	40.05	42.82	45.72	48.81	52.16	55.98	43.08	67.08
29.66	32.10	34.61	37.19	39.85	42.62	45.51	48.58	51.91	55.67	43.25	67.13
29.48	31.92	34.43	37.01	39.68	42.44	45.33	48.38	51.69	55.41	43.41	67.17
29.31	31.76	34.27	36.86	39.52	42.28	45.16	48.20	51.49	55.17	43.55	67.21

(Continued)

Table F.1 Wave Angle β for Oblique Shock Waves ($\gamma = 1.4$) *(Continued)*

M_1	Flow Turning Angle, θ, deg										
	0	2	4	6	8	10	12	14	16	18	20
7.75	7.41	8.73	10.24	11.93	13.76	15.72	17.78	19.92	22.14	24.42	26.76
8.00	7.18	8.50	10.02	11.71	13.56	15.53	17.60	19.76	21.98	24.27	26.62
8.25	6.96	8.28	9.81	11.51	13.37	15.35	17.44	19.60	21.84	24.14	26.49
8.50	6.76	8.08	9.61	11.33	13.20	15.19	17.29	19.46	21.71	24.01	26.37
8.75	6.56	7.89	9.43	11.15	13.04	15.04	17.15	19.33	21.58	23.89	26.26
9.00	6.38	7.71	9.25	10.99	12.88	14.90	17.02	19.21	21.47	23.79	26.16
9.25	6.21	7.54	9.09	10.84	12.74	14.77	16.90	19.10	21.36	23.69	26.06
9.50	6.04	7.38	8.94	10.69	12.61	14.65	16.78	18.99	21.27	23.60	25.97
9.75	5.89	7.22	8.79	10.56	12.48	14.53	16.68	18.90	21.18	23.51	25.89
10.00	5.74	7.08	8.65	10.43	12.37	14.43	16.58	18.81	21.09	23.43	25.82
11.00	5.22	6.56	8.17	9.99	11.96	14.06	16.24	18.50	20.81	23.16	25.56
12.00	4.78	6.14	7.77	9.63	11.64	13.77	15.98	18.26	20.58	22.96	25.37
13.00	4.41	5.78	7.44	9.33	11.38	13.54	15.77	18.07	20.41	22.79	25.22
14.00	4.10	5.48	7.17	9.09	11.16	13.35	15.60	17.91	20.27	22.66	25.09
15.00	3.82	5.21	6.93	8.88	10.99	13.19	15.46	17.79	20.15	22.56	24.99
16.00	3.58	4.98	6.73	8.71	10.84	13.06	15.35	17.68	20.06	22.47	24.91
17.00	3.37	4.78	6.55	8.56	10.71	12.95	15.25	17.60	19.98	22.40	24.84
18.00	3.18	4.61	6.40	8.43	10.60	12.86	15.17	17.52	19.91	22.33	24.79
19.00	3.02	4.45	6.27	8.32	10.51	12.78	15.10	17.46	19.86	22.28	24.74
20.00	2.87	4.31	6.15	8.22	10.43	12.71	15.04	17.41	19.81	22.24	24.70
21.00	2.73	4.19	6.04	8.14	10.36	12.65	14.99	17.36	19.77	22.20	24.66
22.00	2.61	4.07	5.95	8.06	10.29	12.60	14.94	17.32	19.73	22.17	24.63
23.00	2.49	3.97	5.87	8.00	10.24	12.55	14.90	17.29	19.70	22.14	24.60
24.00	2.39	3.88	5.79	7.94	10.19	12.51	14.87	17.26	19.67	22.11	24.58
25.00	2.29	3.79	5.73	7.88	10.15	12.47	14.84	17.23	19.65	22.09	24.56

22	24	26	28	30	32	34	36	38	40	β	θ_{max}
29.16	31.61	34.13	36.71	39.37	42.13	45.00	48.04	51.31	54.96	43.68	67.24
29.02	31.48	34.00	36.58	39.24	41.99	44.86	47.89	51.15	54.77	43.79	67.28
28.90	31.36	33.88	36.46	39.12	41.87	44.74	47.76	51.00	54.60	43.90	67.30
28.78	31.24	33.77	36.35	39.01	41.76	44.62	47.64	50.86	54.44	43.99	67.33
28.67	31.14	33.67	36.25	38.91	41.66	44.52	47.52	50.74	54.30	44.08	67.36
28.58	31.05	33.57	36.16	38.82	41.57	44.42	47.42	50.63	54.17	44.16	67.38
28.48	30.96	33.49	36.08	38.74	41.48	44.33	47.33	50.53	54.06	44.24	67.40
28.40	30.88	33.41	36.00	38.66	41.40	44.25	47.24	50.44	53.95	44.31	67.42
28.32	30.80	33.33	35.93	38.58	41.33	44.17	47.16	50.35	53.85	44.37	67.44
28.25	30.73	33.27	35.86	38.52	41.26	44.10	47.09	50.27	53.76	44.43	67.45
28.01	30.50	33.04	35.63	38.29	41.03	43.87	46.84	50.00	53.46	44.63	67.51
27.82	30.32	32.86	35.46	38.12	40.86	43.69	46.66	49.80	53.23	44.78	67.55
27.68	30.18	32.73	35.33	37.99	40.72	43.56	46.51	49.65	53.06	44.90	67.59
27.56	30.07	32.62	35.22	37.88	40.62	43.45	46.40	49.53	52.92	44.99	67.62
27.47	29.98	32.53	35.13	37.80	40.53	43.36	46.31	49.43	52.81	45.07	67.64
27.39	29.90	32.46	35.06	37.73	40.46	43.29	46.23	49.35	52.72	45.13	67.66
27.32	29.84	32.40	35.00	37.67	40.40	43.23	46.17	49.28	52.64	45.18	67.67
27.27	29.79	32.35	34.95	37.62	40.35	43.18	46.12	49.22	52.58	45.23	67.69
27.22	29.74	32.31	34.91	37.58	40.31	43.13	46.07	49.18	52.53	45.26	67.70
27.18	29.71	32.27	34.88	37.54	40.28	43.10	46.04	49.14	52.48	45.29	67.71
27.15	29.67	32.24	34.85	37.51	40.24	43.07	46.00	49.10	52.44	45.32	67.71
27.12	29.65	32.21	34.82	37.48	40.22	43.04	45.98	49.07	52.41	45.34	67.72
27.09	29.62	32.19	34.80	37.46	40.20	43.02	45.95	49.05	52.38	45.36	67.73
27.07	29.60	32.17	34.78	37.44	40.17	42.99	45.93	49.02	52.36	45.38	67.73
27.05	29.58	32.15	34.76	37.42	40.16	42.98	45.91	49.00	52.33	45.40	67.74

$$M_2 \sin(\beta - \theta) = f_5(M_1 \sin \beta) = \sqrt{\frac{(M_1 \sin \beta)^2 + \dfrac{2}{\gamma - 1}}{\dfrac{2\gamma}{\gamma - 1}(M_1 \sin \beta)^2 - 1}} \qquad \text{(F.5)}$$

The relationships for the static pressure ratio, static density ratio, total pressure ratio, and downstream Mach number M_2 are plotted, for $\gamma = 1.4$, vs upstream Mach number M_1 and flow deflection angle θ in Figs. 3.27, 3.28, 3.29, and 3.30, respectively.

The relationship among the upstream Mach number M_1, shock angle β, and flow deflection angle θ for an oblique shock is

$$\tan(\beta - \theta) = \frac{2 + (\gamma - 1)M_1^2 \sin^2 \beta}{(\gamma + 1)M_1^2 \sin \beta \cos \beta} \qquad \text{(F.6)}$$

which is given in chart form in Fig. 3.26. This relationship is also given as a table in the preceding pages. The maximum flow deflection angle θ_{max} and the corresponding shock angle β are also listed in this table for the upstream Mach number M_1.

Appendix G — Rayleigh Line Flow Functions

For simple flows with heat addition or heat rejection, the following relations apply:

$$\phi(M^2) = \frac{M^2\left(1 + \dfrac{\gamma - 1}{2}M^2\right)}{(1 + \gamma M^2)^2}$$

where

$$\frac{T_{t2}}{T_{t1}} = \frac{\phi(M_2^2)}{\phi(M_1^2)}$$

The following relationships apply for simple Rayleigh line flow (constant-area heating). They are tabulated on the following pages.

$$\frac{T_t}{T_t^*} = \frac{2(\gamma + 1)M^2}{(1 + \gamma M^2)^2}\left(1 + \frac{\gamma - 1}{2}M^2\right)$$

$$\frac{P_t}{P_t^*} = \frac{\gamma + 1}{1 + \gamma M^2}\left[\left(\frac{2}{\gamma + 1}\right)\left(1 + \frac{\gamma - 1}{2}M^2\right)\right]^{\gamma/(\gamma-1)}$$

$$\frac{T}{T^*} = \frac{(\gamma + 1)^2 M^2}{(1 + \gamma M^2)^2}$$

$$\frac{P}{P^*} = \frac{\gamma + 1}{1 + \gamma M^2}$$

Table G.1 Rayleigh Line Flow Functions ($\gamma = 1.4$)

M	$\phi(M^2)$	T_t/T_t^*	T/T^*	P_t/P_t^*	P/P^*
0	0	0	0	1.267876	2.400000
0.01	0.000100	0.000480	0.000576	1.267788	2.399664
0.02	0.000400	0.001918	0.002301	1.267522	2.398657
0.03	0.000898	0.004310	0.005171	1.267079	2.396980
0.04	0.001593	0.007648	0.009175	1.266460	2.394636
0.05	0.002484	0.011922	0.014300	1.265667	2.391629
0.06	0.003567	0.017119	0.020529	1.264700	2.387965
0.07	0.004838	0.023223	0.027841	1.263562	2.383648
0.08	0.006295	0.030215	0.036212	1.262256	2.378687
0.09	0.007932	0.038075	0.045616	1.260782	2.373089
0.10	0.009745	0.046777	0.056020	1.259146	2.366864
0.11	0.011729	0.056297	0.067393	1.257348	2.360021
0.12	0.013876	0.066606	0.079698	1.255394	2.352572
0.13	0.016182	0.077675	0.092896	1.253286	2.344528
0.14	0.018640	0.089471	0.106946	1.251029	2.335903
0.15	0.021242	0.101961	0.121805	1.248626	2.326709
0.16	0.023981	0.115110	0.137429	1.246083	2.316960
0.17	0.026850	0.128882	0.153769	1.243403	2.306672
0.18	0.029841	0.143238	0.170779	1.240592	2.295860
0.19	0.032946	0.158142	0.188410	1.237655	2.284539
0.20	0.036157	0.173554	0.206612	1.234596	2.272727
0.21	0.039465	0.189434	0.225333	1.231421	2.260440
0.22	0.042863	0.205742	0.244523	1.228136	2.247696
0.23	0.046341	0.222439	0.264132	1.224745	2.234512
0.24	0.049892	0.239484	0.284108	1.221255	2.220906
0.25	0.053508	0.256837	0.304400	1.217672	2.206897
0.26	0.057179	0.274459	0.324957	1.214000	2.192502
0.27	0.060898	0.292311	0.345732	1.210246	2.177740
0.28	0.064657	0.310353	0.366674	1.206416	2.162630
0.29	0.068448	0.328549	0.387737	1.202515	2.147190
0.30	0.072263	0.346860	0.408873	1.198549	2.131439
0.31	0.076094	0.365252	0.430037	1.194524	2.115395
0.32	0.079935	0.383689	0.451187	1.190446	2.099076
0.33	0.083779	0.402138	0.472279	1.186320	2.082502
0.34	0.087618	0.420565	0.493273	1.182153	2.065689
0.35	0.091446	0.43894	0.514131	1.177949	2.048656

(Continued)

Table G.1   Rayleigh Line Flow Functions ($\gamma = 1.4$) *(Continued)*

M	$\phi(M^2)$	T_t/T_t^*	T/T^*	P_t/P_t^*	P/P^*
0.36	0.095257	0.457232	0.534816	1.173714	2.031419
0.37	0.099044	0.475413	0.555292	1.169455	2.013997
0.38	0.102803	0.493456	0.575526	1.165175	1.996406
0.39	0.106528	0.511336	0.595488	1.160881	1.978663
0.40	0.110214	0.529027	0.615148	1.156577	1.960784
0.41	0.113856	0.546508	0.634479	1.152268	1.942785
0.42	0.117450	0.563758	0.653456	1.147960	1.924681
0.43	0.120991	0.580756	0.672055	1.143657	1.906487
0.44	0.124476	0.597485	0.690255	1.139364	1.888218
0.45	0.127901	0.613927	0.708037	1.135085	1.869887
0.46	0.131264	0.630068	0.725383	1.130825	1.851509
0.47	0.134561	0.645893	0.742278	1.126587	1.833097
0.48	0.137790	0.661390	0.758707	1.122377	1.814662
0.49	0.140948	0.676549	0.774659	1.118197	1.796219
0.50	0.144033	0.691358	0.790123	1.114053	1.777778
0.51	0.147044	0.705810	0.805091	1.109946	1.759350
0.52	0.149978	0.719897	0.819554	1.105882	1.740947
0.53	0.152836	0.733612	0.833508	1.101863	1.722579
0.54	0.155615	0.746952	0.846948	1.097892	1.704255
0.55	0.158315	0.759910	0.859870	1.093973	1.685985
0.56	0.160935	0.772486	0.872274	1.090109	1.667779
0.57	0.163474	0.784675	0.884158	1.086302	1.649643
0.58	0.165933	0.796478	0.895523	1.082556	1.631588
0.59	0.168311	0.807894	0.906371	1.078872	1.613619
0.60	0.170609	0.818923	0.916704	1.075253	1.595745
0.61	0.172826	0.829566	0.926527	1.071702	1.577972
0.62	0.174964	0.839825	0.935843	1.068221	1.560306
0.63	0.177022	0.849703	0.944657	1.064811	1.542754
0.64	0.179001	0.859203	0.952976	1.061475	1.525320
0.65	0.180902	0.868329	0.960806	1.058214	1.508011
0.66	0.182726	0.877084	0.968155	1.055031	1.490831
0.67	0.184474	0.885473	0.975030	1.051927	1.473785
0.68	0.186146	0.893502	0.981439	1.048904	1.456876
0.69	0.187745	0.901175	0.987391	1.045962	1.440109
0.70	0.189271	0.908499	0.992895	1.043104	1.423488
0.71	0.190725	0.915479	0.997961	1.040330	1.407014

(Continued)

Table G.1 Rayleign Line Flow Functions ($\gamma = 1.4$) (Continued)

M	$\phi(M^2)$	T_t/T_t^*	T/T^*	P_t/P_t^*	P/P^*
0.72	0.192109	0.922122	1.002598	1.037642	1.390692
0.73	0.193424	0.928435	1.006815	1.035041	1.374523
0.74	0.194672	0.934423	1.010624	1.032528	1.358511
0.75	0.195853	0.940095	1.014035	1.030104	1.342657
0.76	0.196970	0.945456	1.017057	1.027769	1.326964
0.77	0.198024	0.950515	1.019702	1.025525	1.311432
0.78	0.199016	0.955279	1.021980	1.023372	1.296064
0.79	0.199949	0.959754	1.023901	1.021311	1.280861
0.80	0.200823	0.963948	1.025477	1.019343	1.265823
0.81	0.201639	0.967869	1.026717	1.017468	1.250951
0.82	0.202401	0.971524	1.027633	1.015687	1.236247
0.83	0.203109	0.974921	1.028235	1.013999	1.221710
0.84	0.203764	0.978066	1.028533	1.012407	1.207341
0.85	0.204368	0.980968	1.028538	1.010909	1.193139
0.86	0.204924	0.983633	1.028260	1.009507	1.179106
0.87	0.205431	0.986069	1.027708	1.008200	1.165241
0.88	0.205892	0.988283	1.026894	1.006989	1.151543
0.89	0.206309	0.990282	1.025827	1.005875	1.138012
0.90	0.206682	0.992073	1.024516	1.004856	1.124649
0.91	0.207013	0.993663	1.022971	1.003934	1.111451
0.92	0.207304	0.995058	1.021201	1.003109	1.098418
0.93	0.207555	0.996266	1.019215	1.002381	1.085550
0.94	0.207769	0.997293	1.017023	1.001749	1.072846
0.95	0.207947	0.998145	1.014632	1.001215	1.060305
0.96	0.208089	0.998828	1.012052	1.000778	1.047925
0.97	0.208198	0.999350	1.009291	1.000437	1.035706
0.98	0.208274	0.999715	1.006357	1.000194	1.023646
0.99	0.208319	0.999930	1.003257	1.000049	1.011745
1.00	0.208333	1.000000	1.000000	1.000000	1.000000
1.01	0.208319	0.999931	0.996593	1.000049	0.988411
1.02	0.208277	0.999730	0.993043	1.000194	0.976976
1.03	0.208208	0.999400	0.989358	1.000437	0.965694
1.04	0.208114	0.998947	0.985543	1.000778	0.954563
1.05	0.207995	0.998376	0.981607	1.001215	0.943582
1.06	0.207853	0.997692	0.977555	1.001750	0.932749
1.07	0.207688	0.996901	0.973393	1.002381	0.922063

(Continued)

Table G.1   Rayleign Line Flow Functions ($\gamma = 1.4$) *(Continued)*

M	$\phi(M^2)$	T_t/T_t^*	T/T^*	P_t/P_t^*	P/P^*
1.08	0.207501	0.996006	0.969129	1.003110	0.911522
1.09	0.207294	0.995012	0.964767	1.003936	0.901124
1.10	0.207067	0.993924	0.960313	1.004858	0.890869
1.11	0.206822	0.992745	0.955773	1.005878	0.880753
1.12	0.206558	0.991480	0.951151	1.006995	0.870777
1.13	0.206278	0.990133	0.946455	1.008208	0.860937
1.14	0.205981	0.988708	0.941687	1.009519	0.851233
1.15	0.205668	0.987209	0.936853	1.010926	0.841662
1.16	0.205341	0.985638	0.931958	1.012430	0.832224
1.17	0.205000	0.984001	0.927005	1.014031	0.822915
1.18	0.204646	0.982299	0.922000	1.015729	0.813736
1.19	0.204278	0.980536	0.916946	1.017524	0.804683
1.20	0.203899	0.978717	0.911848	1.019415	0.795756
1.21	0.203509	0.976842	0.906708	1.021403	0.786952
1.22	0.203108	0.974916	0.901532	1.023488	0.778271
1.23	0.202696	0.972942	0.896321	1.025670	0.769709
1.24	0.202275	0.970922	0.891081	1.027949	0.761267
1.25	0.201845	0.968858	0.885813	1.030325	0.752941
1.26	0.201407	0.966754	0.880522	1.032798	0.744731
1.27	0.200961	0.964612	0.875209	1.035368	0.736635
1.28	0.200507	0.962433	0.869878	1.038035	0.728651
1.29	0.200046	0.960222	0.864532	1.040799	0.720777
1.30	0.199579	0.957979	0.859174	1.043660	0.713012
1.31	0.199106	0.955706	0.853805	1.046619	0.705355
1.32	0.198626	0.953407	0.848428	1.049675	0.697804
1.33	0.198142	0.951082	0.843046	1.052829	0.690357
1.34	0.197653	0.948734	0.837661	1.056081	0.683013
1.35	0.197159	0.946365	0.832274	1.059430	0.675771
1.36	0.196662	0.943976	0.826888	1.062878	0.668628
1.37	0.196160	0.941569	0.821505	1.066424	0.661584
1.38	0.195655	0.939145	0.816127	1.070068	0.654636
1.39	0.195147	0.936706	0.810755	1.073810	0.647784
1.40	0.194636	0.934254	0.805391	1.077652	0.641026
1.41	0.194123	0.931790	0.800037	1.081592	0.634360
1.42	0.193607	0.929315	0.794694	1.085631	0.627786
1.43	0.193090	0.926830	0.789363	1.089770	0.621301

(Continued)

Table G.1 Rayleign Line Flow Functions ($\gamma = 1.4$) *(Continued)*

M	$\phi(M^2)$	T_t/T_t^*	T/T^*	P_t/P_t^*	P/P^*
1.44	0.192570	0.924338	0.784046	1.094008	0.614905
1.45	0.192050	0.921838	0.778744	1.098346	0.608596
1.46	0.191528	0.919333	0.773459	1.102785	0.602373
1.47	0.191005	0.916823	0.768191	1.107324	0.596235
1.48	0.190481	0.914310	0.762942	1.111963	0.590179
1.49	0.189957	0.911794	0.757713	1.116703	0.584206
1.50	0.189432	0.909276	0.752504	1.121545	0.578313
1.51	0.188908	0.906757	0.747317	1.126489	0.572500
1.52	0.188383	0.904238	0.742152	1.131534	0.566765
1.53	0.187859	0.901721	0.737011	1.136682	0.561107
1.54	0.187334	0.899205	0.731894	1.141932	0.555525
1.55	0.186811	0.896692	0.726802	1.147285	0.550017
1.56	0.186288	0.894181	0.721735	1.152742	0.544583
1.57	0.185766	0.891675	0.716694	1.158302	0.539222
1.58	0.185244	0.889173	0.711680	1.163967	0.533931
1.59	0.184724	0.886677	0.706694	1.169736	0.528711
1.60	0.184205	0.884186	0.701735	1.175611	0.523560
1.61	0.183688	0.881702	0.696805	1.181591	0.518477
1.62	0.183172	0.879225	0.691903	1.187676	0.513461
1.63	0.182657	0.876754	0.687031	1.193869	0.508511
1.64	0.182144	0.874292	0.682188	1.200168	0.503626
1.65	0.181633	0.871839	0.677375	1.206574	0.498805
1.66	0.181124	0.869394	0.672593	1.213089	0.494047
1.67	0.180616	0.866958	0.667841	1.219711	0.489351
1.68	0.180111	0.864531	0.663120	1.226443	0.484715
1.69	0.179607	0.862115	0.658430	1.233284	0.480140
1.70	0.179106	0.859709	0.653771	1.240235	0.475624
1.71	0.178607	0.857314	0.649144	1.247297	0.471167
1.72	0.178110	0.854929	0.644549	1.254470	0.466766
1.73	0.177616	0.852556	0.639985	1.261754	0.462422
1.74	0.177124	0.850195	0.635454	1.269151	0.458134
1.75	0.176634	0.847845	0.630954	1.276661	0.453901
1.76	0.176147	0.845507	0.626487	1.284284	0.449721
1.77	0.175663	0.843181	0.622052	1.292021	0.445595
1.78	0.175181	0.840868	0.617649	1.299873	0.441521
1.79	0.174701	0.838567	0.613279	1.307841	0.437498

(Continued)

Table G.1   Rayleign Line Flow Functions ($\gamma = 1.4$) *(Continued)*

M	$\phi(M^2)$	T_t/T_t^*	T/T^*	P_t/P_t^*	P/P^*
1.80	0.174225	0.836279	0.608941	1.315925	0.433526
1.81	0.173751	0.834004	0.604636	1.324125	0.429604
1.82	0.173280	0.831743	0.600363	1.332443	0.425731
1.83	0.172811	0.829494	0.596122	1.340879	0.421907
1.84	0.172346	0.827259	0.591914	1.349434	0.418130
1.85	0.171883	0.825037	0.587738	1.358108	0.414400
1.86	0.171423	0.822829	0.583595	1.366903	0.410717
1.87	0.170966	0.820635	0.579483	1.375819	0.407079
1.88	0.170511	0.818455	0.575404	1.384856	0.403486
1.89	0.170060	0.816288	0.571357	1.394016	0.399937
1.90	0.169612	0.814136	0.567342	1.403300	0.396432
1.91	0.169166	0.811997	0.563359	1.412707	0.392970
1.92	0.168723	0.809873	0.559407	1.422240	0.389550
1.93	0.168284	0.807762	0.555488	1.431898	0.386171
1.94	0.167847	0.805666	0.551599	1.441683	0.382834
1.95	0.167413	0.803584	0.547743	1.451595	0.379537
1.96	0.166983	0.801517	0.543917	1.461635	0.376279
1.97	0.166555	0.799463	0.540123	1.471805	0.373061
1.98	0.166130	0.797424	0.536360	1.482104	0.369882
1.99	0.165708	0.795399	0.532627	1.492534	0.366740
2.00	0.165289	0.793388	0.528926	1.503096	0.363636
2.02	0.164460	0.789410	0.521614	1.524618	0.357539
2.04	0.163643	0.785488	0.514422	1.546678	0.351584
2.06	0.162838	0.781624	0.507350	1.569283	0.345770
2.08	0.162045	0.777816	0.500396	1.592441	0.340090
2.10	0.161263	0.774064	0.493558	1.616159	0.334541
2.12	0.160493	0.770368	0.486835	1.640446	0.329121
2.14	0.159735	0.766727	0.480225	1.665310	0.323824
2.16	0.158988	0.763142	0.473727	1.690759	0.318647
2.18	0.158252	0.759611	0.467338	1.716801	0.313588
2.20	0.157528	0.756135	0.461058	1.743446	0.308642
2.22	0.156815	0.752712	0.454884	1.770702	0.303807
2.24	0.156113	0.749342	0.448815	1.798578	0.299079
2.26	0.155422	0.746024	0.442849	1.827083	0.294455
2.28	0.154741	0.742758	0.436985	1.856227	0.289934
2.30	0.154071	0.739543	0.431220	1.886020	0.285510

(Continued)

Table G.1 Rayleign Line Flow Functions ($\gamma = 1.4$) *(Continued)*

M	$\phi(M^2)$	T_t/T_t^*	T/T^*	P_t/P_t^*	P/P^*
2.32	0.153412	0.736379	0.425554	1.916471	0.281183
2.34	0.152763	0.733264	0.419984	1.947589	0.276949
2.36	0.152125	0.730199	0.414509	1.979386	0.272807
2.38	0.151496	0.727182	0.409127	2.011871	0.268752
2.40	0.150878	0.724213	0.403836	2.045055	0.264784
2.42	0.150269	0.721291	0.398635	2.078948	0.260899
2.44	0.149670	0.718415	0.393523	2.113561	0.257096
2.46	0.149080	0.715585	0.388497	2.148905	0.253372
2.48	0.148500	0.712800	0.383556	2.184991	0.249725
2.50	0.147929	0.710059	0.378698	2.221831	0.246154
2.52	0.147367	0.707362	0.373923	2.259436	0.242656
2.54	0.146814	0.704708	0.369228	2.297818	0.239229
2.56	0.146270	0.702096	0.364611	2.336987	0.235871
2.58	0.145734	0.699525	0.360073	2.376958	0.232582
2.60	0.145207	0.696995	0.355610	2.417741	0.229358
2.62	0.144689	0.694506	0.351222	2.459349	0.226198
2.64	0.144178	0.692055	0.346907	2.501795	0.223101
2.66	0.143676	0.689644	0.342663	2.545091	0.220066
2.68	0.143181	0.687271	0.338490	2.589250	0.217089
2.70	0.142695	0.684935	0.334387	2.634285	0.214171
2.72	0.142216	0.682636	0.330350	2.680211	0.211309
2.74	0.141744	0.680374	0.326381	2.727039	0.208503
2.76	0.141281	0.678146	0.322476	2.774784	0.205750
2.78	0.140824	0.675954	0.318636	2.823459	0.203050
2.80	0.140374	0.673796	0.314858	2.873080	0.200401
2.82	0.139932	0.671672	0.311142	2.923659	0.197802
2.84	0.139496	0.669582	0.307486	2.975211	0.195251
2.86	0.139067	0.667523	0.303889	3.027751	0.192749
2.88	0.138645	0.665497	0.300351	3.081293	0.190293
2.90	0.138230	0.663502	0.296869	3.135853	0.187882
2.92	0.137820	0.661538	0.293443	3.191445	0.185515
2.94	0.137418	0.659604	0.290072	3.248086	0.183192
2.96	0.137021	0.657700	0.286754	3.305790	0.180910
2.98	0.136630	0.655825	0.283490	3.364573	0.178670
3.00	0.136246	0.653979	0.280277	3.424452	0.176471
3.02	0.135867	0.652161	0.277115	3.485442	0.174310

(Continued)

Table G.1   Rayleign Line Flow Functions ($\gamma = 1.4$) *(Continued)*

M	$\phi(M^2)$	T_t/T_t^*	T/T^*	P_t/P_t^*	P/P^*
3.04	0.135494	0.650371	0.274002	3.547559	0.172188
3.06	0.135127	0.648608	0.270938	3.610821	0.170104
3.08	0.134765	0.646872	0.267922	3.675243	0.168056
3.10	0.134409	0.645162	0.264954	3.740844	0.166044
3.12	0.134058	0.643478	0.262031	3.807639	0.164067
3.14	0.133712	0.641819	0.259153	3.875646	0.162124
3.16	0.133372	0.640185	0.256320	3.944883	0.160215
3.18	0.133036	0.638575	0.253530	4.015368	0.158339
3.20	0.132706	0.636989	0.250783	4.087118	0.156495
3.22	0.132381	0.635427	0.248078	4.160151	0.154681
3.24	0.132060	0.633888	0.245414	4.234486	0.152899
3.26	0.131744	0.632371	0.242790	4.310142	0.151146
3.28	0.131433	0.630877	0.240206	4.387137	0.149423
3.30	0.131126	0.629405	0.237661	4.465489	0.147729
3.32	0.130824	0.627954	0.235154	4.545219	0.146062
3.34	0.130526	0.626525	0.232684	4.626346	0.144423
3.36	0.130232	0.625116	0.230251	4.708889	0.142811
3.38	0.129943	0.623727	0.227854	4.792868	0.141225
3.40	0.129658	0.622359	0.225492	4.878303	0.139665
3.42	0.129377	0.621010	0.223166	4.965214	0.138130
3.44	0.129100	0.619681	0.220873	5.053622	0.136619
3.46	0.128827	0.618370	0.218614	5.143548	0.135133
3.48	0.128558	0.617078	0.216387	5.235012	0.133671
3.50	0.128293	0.615805	0.214193	5.328035	0.132231
3.52	0.128031	0.614549	0.212031	5.422639	0.130815
3.54	0.127773	0.613312	0.209899	5.518846	0.129420
3.56	0.127519	0.612091	0.207799	5.616676	0.128048
3.58	0.127268	0.610888	0.205728	5.716153	0.126696
3.60	0.127021	0.609701	0.203686	5.817298	0.125366
3.62	0.126777	0.608531	0.201674	5.920134	0.124056
3.64	0.126537	0.607378	0.199690	6.024684	0.122766
3.66	0.126300	0.606240	0.197734	6.130970	0.121495
3.68	0.126066	0.605117	0.195806	6.239015	0.120244
3.70	0.125836	0.604010	0.193904	6.348844	0.119012
3.72	0.125608	0.602919	0.192029	6.460479	0.117799
3.74	0.125384	0.601842	0.190179	6.573945	0.116603

(Continued)

Table G.1   Rayleigh Line Flow Functions ($\gamma = 1.4$) *(Continued)*

M	$\phi(M^2)$	T_t/T_t^*	T/T^*	P_t/P_t^*	P/P^*
3.76	0.125162	0.600780	0.188356	6.689265	0.115425
3.78	0.124944	0.599732	0.186557	6.806464	0.114265
3.80	0.124729	0.598698	0.184783	6.925566	0.113122
3.82	0.124516	0.597678	0.183034	7.046596	0.111996
3.84	0.124307	0.596672	0.181308	7.169580	0.110886
3.86	0.124100	0.595679	0.179605	7.294542	0.109792
3.88	0.123896	0.594699	0.177926	7.421508	0.108715
3.90	0.123694	0.593732	0.176269	7.550504	0.107652
3.92	0.123496	0.592778	0.174634	7.681556	0.106605
3.94	0.123299	0.591837	0.173021	7.814690	0.105573
3.96	0.123106	0.590908	0.171430	7.949932	0.104556
3.98	0.122915	0.589991	0.169860	8.087310	0.103553
4.00	0.122726	0.589086	0.168310	8.226849	0.102564

Fanno Line Flow Functions

F or simple flows with friction, the following relations apply:

$$\frac{4c_f L^*}{D} = \frac{1 - M^2}{\gamma M^2} + \frac{\gamma + 1}{2\gamma} \ln\left[\frac{M^2}{\left(\frac{2}{\gamma+1}\right)\left(1 + \frac{\gamma-1}{2}M^2\right)}\right]$$

$$\frac{I}{I^*} = \frac{1 + \gamma M^2}{M\sqrt{2(\gamma+1)\left(1 + \frac{\gamma-1}{2}M^2\right)}}$$

where $I = PA + \dot{m}V = PA(1 + \gamma M^2)$

$$\frac{T}{T^*} = \left[\left(\frac{2}{\gamma+1}\right)\left(1 + \frac{\gamma-1}{2}M^2\right)\right]^{-1}$$

$$\frac{P_t}{P_t^*} = \frac{1}{M}\left[\left(\frac{2}{\gamma+1}\right)\left(1 + \frac{\gamma-1}{2}M^2\right)\right]^{(\gamma+1)/[2(\gamma-1)]}$$

$$\frac{P}{P^*} = \frac{1}{M\sqrt{\left(\frac{2}{\gamma+1}\right)\left(1 + \frac{\gamma-1}{2}M^2\right)}}$$

Table H.1　Fanno Line Functions ($\gamma = 1.4$)

M	$4c_f L^*/D$	I/I^*	T/T^*	P_t/P_t^*	P/P^*
0.00	Indef	Indef	1.200000	Indef	Indef
0.01	7134.404538	45.649480	1.199976	57.873843	109.543416
0.02	1778.449882	22.833640	1.199904	28.942130	54.770065
0.03	787.081387	15.232315	1.199784	19.300542	36.511551
0.04	440.352214	11.434618	1.199616	14.481486	27.381747
0.05	280.020306	9.158370	1.199400	11.591444	21.903427
0.06	193.031082	7.642847	1.199137	9.665910	18.250849

(Continued)

Table H.1 Fanno Line Functions ($\gamma = 1.4$) (Continued)

M	$4c_f L^*/D$	I/I^*	T/T^*	P_t/P_t^*	P/P^*
0.07	140.655014	6.562023	1.198825	8.291525	15.641553
0.08	106.718216	5.752883	1.198466	7.261610	13.684309
0.09	83.496117	5.124867	1.198059	6.461342	12.161765
0.10	66.921560	4.623634	1.197605	5.821829	10.943513
0.11	54.687895	4.214608	1.197103	5.299230	9.946564
0.12	45.407962	3.874734	1.196554	4.864318	9.115592
0.13	38.207003	3.588055	1.195958	4.496859	8.412296
0.14	32.511306	3.343168	1.195314	4.182400	7.809317
0.15	27.931967	3.131716	1.194624	3.910343	7.286591
0.16	24.197830	2.947427	1.193887	3.672739	6.829072
0.17	21.115178	2.785508	1.193104	3.463509	6.425253
0.18	18.542654	2.642227	1.192274	3.277926	6.066183
0.19	16.375164	2.514642	1.191398	3.112259	5.744799
0.20	14.533266	2.400397	1.190476	2.963520	5.455447
0.21	12.956022	2.297584	1.189509	2.829294	5.193552
0.22	11.596054	2.204644	1.188495	2.707602	4.955370
0.23	10.416090	2.120287	1.187437	2.596812	4.737808
0.24	9.386481	2.043440	1.186333	2.495562	4.538289
0.25	8.483409	1.973200	1.185185	2.402710	4.354648
0.26	7.687567	1.908803	1.183992	2.317287	4.185054
0.27	6.983170	1.849600	1.182755	2.238471	4.027947
0.28	6.357214	1.795031	1.181474	2.165554	3.881988
0.29	5.798913	1.744617	1.180150	2.097928	3.746024
0.30	5.299253	1.697941	1.178782	2.035065	3.619057
0.31	4.850663	1.654640	1.177371	1.976507	3.500217
0.32	4.446743	1.614396	1.175917	1.921851	3.388741
0.33	4.082055	1.576931	1.174421	1.870745	3.283961
0.34	3.751953	1.541997	1.172883	1.822876	3.185286
0.35	3.452452	1.509377	1.171303	1.777969	3.092193
0.36	3.180118	1.478876	1.169682	1.735778	3.004218
0.37	2.931977	1.450322	1.168020	1.696086	2.920945
0.38	2.705448	1.423560	1.166317	1.658696	2.842004
0.39	2.498277	1.398450	1.164574	1.623433	2.767062
0.40	2.308493	1.374868	1.162791	1.590140	2.695819
0.41	2.134364	1.352700	1.160968	1.558673	2.628006
0.42	1.974366	1.331845	1.159107	1.528905	2.563377
0.43	1.827151	1.312209	1.157207	1.500718	2.501710

(Continued)

Table H.1   Fanno Line Functions ($\gamma = 1.4$) *(Continued)*

M	$4c_f L^*/D$	I/I^*	T/T^*	P_t/P_t^*	P/P^*
0.44	1.691525	1.293709	1.155268	1.474005	2.442804
0.45	1.566427	1.276267	1.153292	1.448672	2.386476
0.46	1.450911	1.259814	1.151278	1.424629	2.332557
0.47	1.344135	1.244285	1.149227	1.401795	2.280894
0.48	1.245341	1.229620	1.147140	1.380097	2.231346
0.49	1.153853	1.215767	1.145016	1.359468	2.183784
0.50	1.069060	1.202676	1.142857	1.339844	2.138090
0.51	0.990414	1.190299	1.140663	1.321168	2.094153
0.52	0.917418	1.178596	1.138434	1.303388	2.051873
0.53	0.849624	1.167526	1.136170	1.286454	2.011156
0.54	0.786625	1.157054	1.133873	1.270321	1.971916
0.55	0.728053	1.147146	1.131542	1.254948	1.934072
0.56	0.673571	1.137771	1.129178	1.240294	1.897550
0.57	0.622874	1.128899	1.126782	1.226326	1.862280
0.58	0.575683	1.120503	1.124353	1.213007	1.828199
0.59	0.531743	1.112559	1.121894	1.200308	1.795246
0.60	0.490822	1.105041	1.119403	1.188200	1.763364
0.61	0.452705	1.097930	1.116882	1.176654	1.732502
0.62	0.417197	1.091203	1.114330	1.165646	1.702610
0.63	0.384116	1.084842	1.111749	1.155151	1.673643
0.64	0.353299	1.078828	1.109139	1.145148	1.645558
0.65	0.324591	1.073144	1.106501	1.135616	1.618313
0.66	0.297853	1.067774	1.103834	1.126535	1.591871
0.67	0.272955	1.062704	1.101140	1.117887	1.566197
0.68	0.249775	1.057919	1.098418	1.109655	1.541257
0.69	0.228204	1.053405	1.095670	1.101822	1.517018
0.70	0.208139	1.049150	1.092896	1.094373	1.493452
0.71	0.189483	1.045143	1.090096	1.087294	1.470531
0.72	0.172149	1.041372	1.087272	1.080571	1.448227
0.73	0.156054	1.037825	1.084422	1.074192	1.426515
0.74	0.141122	1.034494	1.081549	1.068144	1.405372
0.75	0.127282	1.031369	1.078652	1.062417	1.384775
0.76	0.114468	1.028441	1.075731	1.056999	1.364704
0.77	0.102617	1.025700	1.072789	1.051881	1.345137
0.78	0.091672	1.023140	1.069824	1.047053	1.326055
0.79	0.081580	1.020752	1.066837	1.042505	1.307441
0.80	0.072290	1.018528	1.063830	1.038230	1.289277

(Continued)

Table H.1 Fanno Line Functions ($\gamma = 1.4$) *(Continued)*

M	$4c_f L^*/D$	I/I^*	T/T^*	P_t/P_t^*	P/P^*
0.81	0.063755	1.016463	1.060802	1.034219	1.271546
0.82	0.055932	1.014549	1.057753	1.030464	1.254233
0.83	0.048778	1.012781	1.054685	1.026959	1.237324
0.84	0.042256	1.011151	1.051598	1.023696	1.220803
0.85	0.036330	1.009654	1.048493	1.020669	1.204658
0.86	0.030965	1.008285	1.045369	1.017871	1.188875
0.87	0.026130	1.007039	1.042228	1.015297	1.173443
0.88	0.021795	1.005910	1.039069	1.012941	1.158349
0.89	0.017931	1.004895	1.035894	1.010798	1.143583
0.90	0.014512	1.003987	1.032702	1.008863	1.129133
0.91	0.011514	1.003184	1.029495	1.007131	1.114989
0.92	0.008913	1.002480	1.026273	1.005597	1.101143
0.93	0.006687	1.001872	1.023035	1.004258	1.087583
0.94	0.004815	1.001356	1.019784	1.003108	1.074302
0.95	0.003278	1.000929	1.016518	1.002145	1.061290
0.96	0.002057	1.000586	1.013240	1.001365	1.048540
0.97	0.001135	1.000325	1.009948	1.000763	1.036043
0.98	0.000495	1.000143	1.006644	1.000337	1.023792
0.99	0.000121	1.000035	1.003328	1.000084	1.011780
1.00	0.000000	1.000000	1.000000	1.000000	1.000000
1.01	0.000117	1.000034	0.996661	1.000083	0.988445
1.02	0.000459	1.000135	0.993312	1.000330	0.977108
1.03	0.001013	1.000300	0.989952	1.000738	0.965984
1.04	0.001769	1.000527	0.986582	1.001305	0.955066
1.05	0.002714	1.000813	0.983204	1.002029	0.944349
1.06	0.003838	1.001156	0.979816	1.002907	0.933827
1.07	0.005131	1.001553	0.976419	1.003938	0.923495
1.08	0.006585	1.002003	0.973015	1.005119	0.913347
1.09	0.008189	1.002503	0.969603	1.006449	0.903380
1.10	0.009935	1.003052	0.966184	1.007925	0.893588
1.11	0.011816	1.003647	0.962757	1.009547	0.883966
1.12	0.013823	1.004287	0.959325	1.011312	0.874510
1.13	0.015949	1.004970	0.955886	1.013219	0.865216
1.14	0.018188	1.005694	0.952441	1.015267	0.856080
1.15	0.020533	1.006458	0.948992	1.017454	0.847097
1.16	0.022977	1.007259	0.945537	1.019780	0.838265
1.17	0.025516	1.008097	0.942078	1.022242	0.829579

(Continued)

Table H.1   Fanno Line Functions ($\gamma = 1.4$) (Continued)

M	$4c_f L^*/D$	I/I^*	T/T^*	P_t/P_t^*	P/P^*
1.18	0.028142	1.008970	0.938615	1.024840	0.821035
1.19	0.030851	1.009876	0.935148	1.027573	0.812630
1.20	0.033638	1.010815	0.931677	1.030440	0.804362
1.21	0.036498	1.011784	0.928203	1.033440	0.796226
1.22	0.039426	1.012783	0.924727	1.036572	0.788219
1.23	0.042418	1.013810	0.921249	1.039835	0.780339
1.24	0.045471	1.014864	0.917768	1.043229	0.772582
1.25	0.048579	1.015944	0.914286	1.046753	0.764946
1.26	0.051739	1.017049	0.910802	1.050406	0.757428
1.27	0.054947	1.018178	0.907318	1.054189	0.750025
1.28	0.058201	1.019330	0.903832	1.058100	0.742735
1.29	0.061497	1.020503	0.900347	1.062138	0.735555
1.30	0.064832	1.021697	0.896861	1.066305	0.728483
1.31	0.068203	1.022911	0.893376	1.070598	0.721516
1.32	0.071607	1.024144	0.889891	1.075018	0.714652
1.33	0.075041	1.025394	0.886407	1.079565	0.707889
1.34	0.078504	1.026662	0.882924	1.084239	0.701224
1.35	0.081991	1.027947	0.879443	1.089038	0.694656
1.36	0.085503	1.029247	0.875964	1.093964	0.688183
1.37	0.089035	1.030562	0.872486	1.099015	0.681803
1.38	0.092586	1.031891	0.869011	1.104193	0.675513
1.39	0.096155	1.033233	0.865539	1.109496	0.669312
1.40	0.099738	1.034588	0.862069	1.114926	0.663198
1.41	0.103335	1.035955	0.858602	1.120481	0.657169
1.42	0.106943	1.037334	0.855139	1.126162	0.651224
1.43	0.110562	1.038723	0.851680	1.131969	0.645360
1.44	0.114189	1.040123	0.848224	1.137903	0.639577
1.45	0.117823	1.041532	0.844773	1.143963	0.633873
1.46	0.121462	1.042950	0.841326	1.150150	0.628245
1.47	0.125106	1.044377	0.837884	1.156463	0.622694
1.48	0.128753	1.045811	0.834446	1.162904	0.617216
1.49	0.132401	1.047253	0.831013	1.169471	0.611812
1.50	0.136050	1.048702	0.827586	1.176167	0.606478
1.51	0.139699	1.050158	0.824165	1.182991	0.601215
1.52	0.143346	1.051619	0.820749	1.189943	0.596021
1.53	0.146990	1.053086	0.817338	1.197024	0.590894
1.54	0.150631	1.054558	0.813935	1.204234	0.585833

(Continued)

Table H.1 Fanno Line Functions ($\gamma = 1.4$) *(Continued)*

M	$4c_f L^*/D$	I/I^*	T/T^*	P_t/P_t^*	P/P^*
1.55	0.154268	1.056035	0.810537	1.211574	0.580838
1.56	0.157899	1.057517	0.807146	1.219044	0.575906
1.57	0.161525	1.059002	0.803762	1.226644	0.571037
1.58	0.165143	1.060490	0.800384	1.234376	0.566229
1.59	0.168754	1.061982	0.797014	1.242239	0.561482
1.60	0.172357	1.063477	0.793651	1.250235	0.556794
1.61	0.175951	1.064974	0.790295	1.258364	0.552165
1.62	0.179535	1.066474	0.786947	1.266626	0.547593
1.63	0.183109	1.067975	0.783607	1.275022	0.543077
1.64	0.186673	1.069478	0.780275	1.283553	0.538617
1.65	0.190226	1.070982	0.776950	1.292219	0.534211
1.66	0.193766	1.072486	0.773635	1.301021	0.529858
1.67	0.197295	1.073992	0.770327	1.309960	0.525559
1.68	0.200811	1.075498	0.767028	1.319037	0.521310
1.69	0.204314	1.077004	0.763738	1.328252	0.517113
1.70	0.207803	1.078510	0.760456	1.337606	0.512966
1.71	0.211279	1.080016	0.757184	1.347100	0.508867
1.72	0.214740	1.081521	0.753920	1.356735	0.504817
1.73	0.218187	1.083025	0.750666	1.366511	0.500815
1.74	0.221620	1.084528	0.747421	1.376430	0.496859
1.75	0.225037	1.086030	0.744186	1.386492	0.492950
1.76	0.228438	1.087531	0.740960	1.396698	0.489086
1.77	0.231824	1.089029	0.737744	1.407049	0.485266
1.78	0.235195	1.090526	0.734538	1.417546	0.481490
1.79	0.238549	1.092021	0.731342	1.428190	0.477757
1.80	0.241886	1.093514	0.728155	1.438982	0.474067
1.81	0.245208	1.095004	0.724979	1.449923	0.470418
1.82	0.248512	1.096492	0.721813	1.461013	0.466811
1.83	0.251800	1.097977	0.718658	1.472255	0.463244
1.84	0.255070	1.099460	0.715512	1.483648	0.459717
1.85	0.258324	1.100939	0.712378	1.495194	0.456230
1.86	0.261560	1.102415	0.709253	1.506894	0.452781
1.87	0.264778	1.103888	0.706140	1.518750	0.449370
1.88	0.267979	1.105357	0.703037	1.530761	0.445996
1.89	0.271163	1.106823	0.699945	1.542930	0.442660
1.90	0.274328	1.108285	0.696864	1.555257	0.439360
1.91	0.277476	1.109744	0.693794	1.567743	0.436096

(Continued)

Table H.1   Fanno Line Functions ($\gamma = 1.4$) *(Continued)*

M	$4c_f L^*/D$	I/I^*	T/T^*	P_t/P_t^*	P/P^*
1.92	0.280606	1.111198	0.690735	1.580391	0.432867
1.93	0.283718	1.112649	0.687687	1.593200	0.429673
1.94	0.286812	1.114096	0.684650	1.606172	0.426513
1.95	0.289888	1.115538	0.681625	1.619309	0.423387
1.96	0.292946	1.116976	0.678610	1.632611	0.420295
1.97	0.295986	1.118409	0.675607	1.646080	0.417235
1.98	0.299008	1.119838	0.672616	1.659717	0.414208
1.99	0.302011	1.121263	0.669635	1.673523	0.411212
2.00	0.304997	1.122683	0.666667	1.687500	0.408248
2.02	0.310913	1.125508	0.660764	1.715971	0.402413
2.04	0.316756	1.128314	0.654907	1.745139	0.396698
2.06	0.322526	1.131100	0.649098	1.775017	0.391100
2.08	0.328225	1.133866	0.643335	1.805614	0.385616
2.10	0.333851	1.136610	0.637620	1.836944	0.380243
2.12	0.339405	1.139334	0.631951	1.869016	0.374978
2.14	0.344887	1.142035	0.626331	1.901843	0.369818
2.16	0.350299	1.144715	0.620758	1.935437	0.364760
2.18	0.355639	1.147372	0.615233	1.969810	0.359802
2.20	0.360910	1.150007	0.609756	2.004975	0.354940
2.22	0.366111	1.152619	0.604327	2.040944	0.350173
2.24	0.371243	1.155208	0.598946	2.077731	0.345498
2.26	0.376306	1.157774	0.593613	2.115349	0.340913
2.28	0.381302	1.160316	0.588328	2.153811	0.336415
2.30	0.386230	1.162835	0.583090	2.193131	0.332002
2.32	0.391092	1.165331	0.577901	2.233323	0.327672
2.34	0.395888	1.167803	0.572760	2.274401	0.323423
2.36	0.400619	1.170252	0.567666	2.316381	0.319253
2.38	0.405286	1.172677	0.562620	2.359275	0.315160
2.40	0.409889	1.175079	0.557621	2.403100	0.311142
2.42	0.414429	1.177456	0.552669	2.447870	0.307197
2.44	0.418907	1.179811	0.547765	2.493602	0.303324
2.46	0.423324	1.182141	0.542908	2.540309	0.299521
2.48	0.427680	1.184448	0.538097	2.588010	0.295787
2.50	0.431977	1.186732	0.533333	2.636719	0.292119
2.52	0.436214	1.188993	0.528616	2.686453	0.288516
2.54	0.440393	1.191230	0.523944	2.737228	0.284976
2.56	0.444514	1.193444	0.519319	2.789063	0.281499

(Continued)

Table H.1 Fanno Line Functions ($\gamma = 1.4$) *(Continued)*

M	$4c_f L^*/D$	I/I^*	T/T^*	P_t/P_t^*	P/P^*
2.58	0.448579	1.195634	0.514739	2.841972	0.278083
2.60	0.452588	1.197802	0.510204	2.895975	0.274725
2.62	0.456541	1.199947	0.505715	2.951089	0.271426
2.64	0.460441	1.202069	0.501270	3.007331	0.268183
2.66	0.464286	1.204169	0.496870	3.064720	0.264996
2.68	0.468079	1.206246	0.492514	3.123274	0.261863
2.70	0.471819	1.208301	0.488202	3.183011	0.258783
2.72	0.475508	1.210334	0.483933	3.243951	0.255755
2.74	0.479146	1.212345	0.479708	3.306113	0.252777
2.76	0.482735	1.214334	0.475526	3.369515	0.249849
2.78	0.486274	1.216302	0.471387	3.434179	0.246970
2.80	0.489765	1.218248	0.467290	3.500123	0.244138
2.82	0.493208	1.220173	0.463235	3.567368	0.241352
2.84	0.496604	1.222076	0.459221	3.635934	0.238612
2.86	0.499953	1.223959	0.455249	3.705842	0.235917
2.88	0.503257	1.225821	0.451318	3.777113	0.233265
2.90	0.506516	1.227662	0.447427	3.849768	0.230655
2.92	0.509731	1.229483	0.443577	3.923829	0.228088
2.94	0.512902	1.231284	0.439767	3.999317	0.225561
2.96	0.516030	1.233065	0.435996	4.076255	0.223074
2.98	0.519115	1.234826	0.432264	4.154664	0.220627
3.00	0.522159	1.236568	0.428571	4.234568	0.218218
3.02	0.525162	1.238290	0.424917	4.315989	0.215847
3.04	0.528125	1.239993	0.421301	4.398950	0.213512
3.06	0.531048	1.241677	0.417723	4.483475	0.211214
3.08	0.533932	1.243343	0.414182	4.569587	0.208951
3.10	0.536777	1.244989	0.410678	4.657311	0.206723
3.12	0.539584	1.246618	0.407210	4.746670	0.204529
3.14	0.542353	1.248228	0.403779	4.837689	0.202368
3.16	0.545086	1.249820	0.400384	4.930393	0.200240
3.18	0.547783	1.251394	0.397025	5.024808	0.198144
3.20	0.550444	1.252951	0.393701	5.120958	0.196080
3.22	0.553070	1.254490	0.390411	5.218868	0.194046
3.24	0.555661	1.256012	0.387157	5.318566	0.192043
3.26	0.558218	1.257517	0.383936	5.420078	0.190069
3.28	0.560741	1.259005	0.380749	5.523429	0.188125
3.30	0.563232	1.260477	0.377596	5.628647	0.186209

(Continued)

Table H.1 Fanno Line Functions ($\gamma = 1.4$) *(Continued)*

M	$4c_fL^*/D$	I/I^*	T/T^*	P_t/P_t^*	P/P^*
3.32	0.565690	1.261932	0.374476	5.735759	0.184321
3.34	0.568116	1.263371	0.371388	5.844792	0.182460
3.36	0.570510	1.264794	0.368333	5.955774	0.180626
3.38	0.572874	1.266201	0.365310	6.068734	0.178819
3.40	0.575207	1.267592	0.362319	6.183699	0.177038
3.42	0.577510	1.268968	0.359359	6.300698	0.175282
3.44	0.579783	1.270328	0.356430	6.419760	0.173552
3.46	0.582027	1.271674	0.353532	6.540915	0.171845
3.48	0.584242	1.273004	0.350664	6.664192	0.170163
3.50	0.586429	1.274320	0.347826	6.789621	0.168505
3.52	0.588588	1.275621	0.345018	6.917231	0.166870
3.54	0.590720	1.276907	0.342239	7.047055	0.165258
3.56	0.592825	1.278180	0.339489	7.179122	0.163668
3.58	0.594903	1.279438	0.336768	7.313463	0.162100
3.60	0.596955	1.280682	0.334076	7.450111	0.160554
3.62	0.598981	1.281913	0.331411	7.589097	0.159029
3.64	0.600982	1.283130	0.328774	7.730452	0.157524
3.66	0.602957	1.284334	0.326165	7.874211	0.156041
3.68	0.604909	1.285524	0.323583	8.020404	0.154577
3.70	0.606836	1.286701	0.321027	8.169066	0.153133
3.72	0.608739	1.287866	0.318498	8.320230	0.151709
3.74	0.610618	1.289018	0.315996	8.473930	0.150303
3.76	0.612474	1.290157	0.313519	8.630199	0.148917
3.78	0.614308	1.291283	0.311068	8.789073	0.147549
3.80	0.616119	1.292398	0.308642	8.950585	0.146199
3.82	0.617908	1.293500	0.306241	9.114772	0.144867
3.84	0.619675	1.294590	0.303865	9.281668	0.143552
3.86	0.621421	1.295669	0.301514	9.451309	0.142255
3.88	0.623145	1.296735	0.299186	9.623732	0.140974
3.90	0.624849	1.297791	0.296883	9.798973	0.139710
3.92	0.626532	1.298834	0.294603	9.977069	0.138463
3.94	0.628195	1.299867	0.292346	10.158057	0.137231
3.96	0.629838	1.300888	0.290113	10.341975	0.136015
3.98	0.631461	1.301899	0.287902	10.528859	0.134815
4.00	0.633065	1.302899	0.285714	10.718750	0.133631

Appendix I Turbomachinery Stresses and Materials

Introduction

E ven though the focus of this textbook is on the aerothermodynamics of the gas turbine engine, the engine structure is also very significant. Because of its importance, this appendix addresses the major stresses of the rotating components and the properties of materials used in these components. The rotating components of the compressor and turbine have very high momentum, and failure of one part can be catastrophic, with a resulting destruction of the engine and, in the extreme, the aircraft. This is especially true of the "critical" parts of the turbomachinery such as the long first-stage fan blades and the heavy airfoils and disks of the cooled high-pressure turbine.

Over their lives, the rotating parts must endure in a very harsh environment where the loads are very dependent on the engine use. For example, an engine developed for a commercial aircraft will not be subjected to as many throttle excursions as one developed for a fighter aircraft. As a result, the "hot section" [combustor and turbine(s)] of the fighter engine will be subjected to many more thermal cycles per hour of operation than that of the commercial aircraft.

The main focus of the analytical tools developed in this appendix is on the fundamental source of stresses in rotating components—the centrifugal force. To get some idea of the brutal climate in which these components must live, the centrifugal force experienced by an element of material rotating at 10,000 rpm with a radius of 1 ft (0.3 m) is equivalent to more than 34,000 g. Yet, there are many other forces at work that can consume the life of (or destroy) rotating parts, all of which must be considered in the design process. Some of the most important include:

- Stresses due to bending moments like those due to the lift on the airfoils or pressure differences across disks.
- Buffeting or vibratory stresses that occur as the airfoils pass through nonuniform incident flows such as the wakes of upstream blades. This can be most devastating when the *blade passing* frequency coincides with one of the lower natural frequencies of the airfoils.
- Airfoil or disk flutter, an aeroelastic phenomenon in which a natural frequency is spontaneously excited, the driving energy being extracted from the flowing gas. This is most often found in fan and compressor rows, and once it has begun, the life of the engine is measured in minutes.

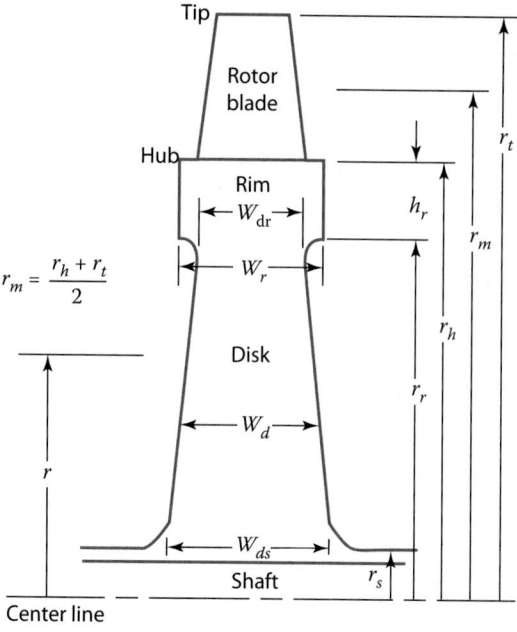

Fig. I.1   Turbomachinery rotor nomenclature.

* Torsional stresses that result from the transfer of power from the turbine to the compressor.
* Temperature gradients that can give rise to very high stresses. These can be very extreme during throttle transients when the engine is moving from one power setting to another. These cyclic thermal stresses extract life from components, especially in the hot section of fighter engines. This is commonly called *thermal* or *low-cycle fatigue (LCF)*.
* Vibratory stress due to high-frequency drivers from a number of sources to include blade/vane interactions, slight differences in parts due to manufacturing tolerances, slight differences in engine construction due to build tolerances, inlet distortion, and so on. These aeroelastic stresses, though low in amplitude (below the material's yield strength), also extract life from the components if subjected to enough cycles (millions, even billions of cycles). This is commonly called *high-cycle fatigue (HCF)*, and it affects static and rotating engine components.
* Local stress concentrations that result from holes, slots, corners, and cracks—the most feared of all.
* Foreign object damage (FOD) and domestic object damage (DOD) that result from external and internal objects, respectively. The need to withstand FOD or DOD can dramatically impact the design. The use of lightweight, nonmetallic materials for large fan blades has been prevented for several decades due to the requirement that they withstand bird strikes.

Each of these areas and many others are included in the design of engine components and material selection. After these are accounted for, an *allowable working stress* is developed and commonly used for the principal tensile stresses alone.

The next section analyzes dominant stresses in the rotating components. Figure I.1 depicts a turbine rotor with its airfoils, rim, and disk. A compressor or fan rotor is constructed similarly. The disk and rim are manufactured as one piece, as is implied in Fig. I.1. However, the portion of the airfoil stresses carried by the compressor or fan's rim is much greater than that of a turbine rotor; hence, its disk is either much smaller or nonexistent. We consider the principal stresses of each part, starting with the blades and working our way down through the structure shown in Fig. I.1.

I.2   Turbomachinery Stresses

I.2.1   Rotor Blade Centrifugal Stress σ_c

We start by considering the force in an airfoil at an arbitrary cross section A_b of a blade (see Fig. I.2). At any radius r, the force must restrain the centrifugal force on all the material beyond it. Thus the hub or base of the airfoil must experience the greatest force. The total centrifugal force acting on A_h is

$$F_c = \int_{r_h}^{r_t} \rho\omega^2 A_b r \, dr \tag{I.1}$$

so that the principal tensile stress is

$$\sigma_c = \frac{F_c}{A_h} = \rho\omega^2 \int_{r_h}^{r_t} \frac{A_b}{A_h} r \, dr \tag{I.2}$$

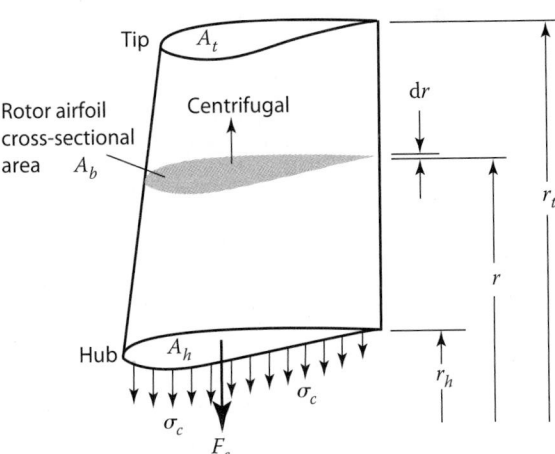

Fig. I.2   Rotor blade centrifugal stress nomenclature.

The airfoil cross-sectional area usually tapers down or reduces with increasing radius, which, according to Eq. (I.2), has the effect of reducing σ_c. If the taper is *linear*, we can write

$$\frac{A_b}{A_h} = 1 - \left(1 - \frac{A_t}{A_h}\right)\left(\frac{r - r_h}{r_t - r_h}\right) \tag{I.3}$$

and Eq. (I.2) becomes

$$\sigma_c = \rho\omega^2 \left[\frac{A}{2\pi} - \left(1 - \frac{A_t}{A_h}\right)\int_{r_h}^{r_t}\left(\frac{r - r_h}{r_t - r_h}\right) r\, dr\right] \tag{I.4}$$

where A is the flow path annulus area $\pi(r_t^2 - r_h^2)$. Integration of Eq. (I.4) gives

$$\sigma_c = \frac{\rho\omega^2 A}{4\pi}\left[2 - \frac{2}{3}\left(1 - \frac{A_t}{A_h}\right)\left(1 + \frac{1}{1 + r_h/r_t}\right)\right] \tag{I.5}$$

Equation (I.5) has an upper limit (corresponding to $r_h/r_t = 1$) of

$$\sigma_c = \frac{\rho\omega^2 A}{4\pi}\left(1 + \frac{A_t}{A_h}\right) \tag{I.6}$$

This equation reveals the basic characteristic that σ_c is proportional to $\rho\omega^2 A$. It also shows that tapering can, at most, reduce the stress of a *straight* airfoil (i.e., $A_t = A_h$) by half (i.e., $A_t = 0$).

Common practice in the industry is to refer to $\omega^2 A$ as the term AN^2 because it is easy to calculate and use. The COMPR computer program calculates and outputs AN^2 for each stage for proper material selection. For the TURBN computer program, the value of AN^2 is calculated and output for each stage.

We note that

$$AN^2 = A\omega^2\left(\frac{30}{\pi}\right)^2 \tag{I.7}$$

By using Eq. (I.7), we obtain for Eq. (I.6)

$$\frac{\sigma_c}{\rho} = AN^2\,\frac{\pi}{3600}\left(1 + \frac{A_t}{A_h}\right) \tag{I.8}$$

For a fixed value of AN^2, Eq. (I.8) shows the obvious reduction in centrifugal stress by using more lightweight materials [e.g., titanium with a density of about 9 slug/ft^3 (about 4600 kg/m^3)] rather than the heavier materials [densities of about 16 slug/ft^3 (about 8200 kg/m^3)]. This equation is given in graphical form in Fig. I.3 for two different airfoil taper ratios A_t/A_h for AN^2 values from 0 to 6×10^{10} in.$^2 \cdot$ rpm^2 (3.87×10^7 m$^2 \cdot$ rpm^2).

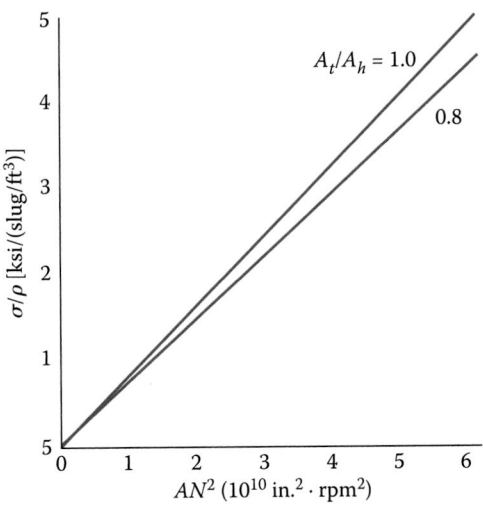

Fig. I.3 Variation of centrifugal stress-to-density ratio with AN^2.

Example I.1

In this example, we determine the centrifugal stresses at 10,000 rpm in typical airfoil materials of a compressor ($\rho = 9.0$ slug/ft^3) and turbine ($\rho = 15.0$ slug/ft^3) with flow annulus areas of 2.0 and 1.0 ft^2, respectively. Table I.1 summarizes the results, using Eq. (I.8) for an airfoil taper A_t/A_h of 0.8. For the compressor, $\sigma_c = (\pi/3.6)(9/1.44^2)(2.88)(1.8) = 19.63$ ksi. Even though the compressor centrifugal stress is higher than that in the turbine airfoil, the turbine airfoils are subjected to higher temperatures and will, most likely, have a lower *allowable working stress*.

Table I.1 Centrifugal Stress Results for Example I.1

Variable	Compressor	Turbine
ρ slug/ft^3 (kg/m^3)	9.0 4611	15.0 7686
N rpm	10,000	10,000
A ft^2 (m^2)	2.0 0.186	1.0 0.0929
AN^2 in.$^2 \cdot$ rpm^2 (m$^2 \cdot$ rpm^2)	2.88×10^{10} 1.86×10^7	1.44×10^{10} 9.29×10^6
A_t/A_h	0.8	0.8
σ_c psi (MPa)	19,630 135.3	16,360 112.8

Fig. I.4a Turbine disk section with blades inserted into rim.

I.2.2 Rim Web Thickness W_{dr}

The rotating airfoils are inserted into slots in an otherwise solid annulus of material known as the *rim* (see Fig. I.4a), which maintains their circular motion. The airfoil hub tensile stress σ_c is treated as though "smeared out" over the outer rim surface, so that

$$\bar{\sigma}_{blades} = \frac{\sigma_c n_b A_h}{2\pi r_h W_r} \qquad (I.9)$$

where n_b is the number of blades on the *wheel*.

Assuming uniform stress within the rim σ_r, we can use the force diagram of Fig. I.4b to determine the width of the disk where it attaches to the rim W_{dr} (see Fig. I.1) necessary to balance the blade and rim centrifugal forces. Please note that W_r and h_r are simply sensible initial choices, where W_r approximates the axial chord of the airfoil and h_r is similar in magnitude to W_r. It is important to realize that it will always be possible to design a rim large enough to "carry" the airfoils. The real question is whether the size of the rim is practical from the standpoint of space required, weight, installation, and manufacturing cost. Thus, there is no absolute solution, and the final choice must be based on experience and a sense of proportion.

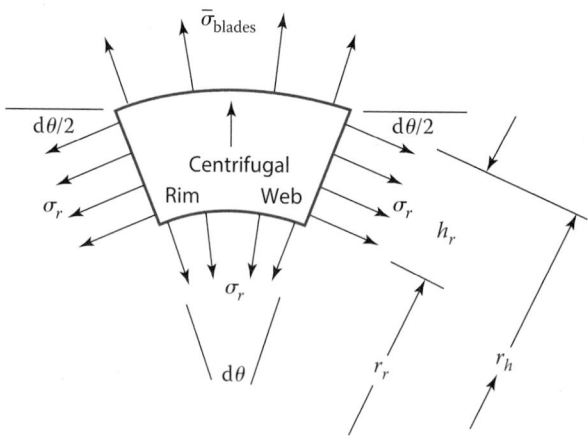

Fig. I.4b Rim segment radial equilibrium nomenclature.

A radial force balance of the differential rim element shown in Fig. I.4b gives

$$\overline{\sigma}_{\text{blades}} r_h W_r \, d\theta + \rho\omega^2 h_r W_r(r_r + h_r/2)^2 \, d\theta = \sigma_r r_r W_{\text{dr}} \, d\theta + 2\left(\sigma_r h_r W_r \frac{d\theta}{2}\right)$$

which, for any infinitesimal $d\theta$, becomes

$$\frac{W_{\text{dr}}}{W_r} = \left[\frac{\overline{\sigma}_{\text{blades}}}{\sigma_r}\left(1 + \frac{r_r}{h_r}\right) + \frac{\rho(\omega r_r)^2}{\sigma_r}\left(1 + \frac{h_r}{2r_r}\right)^2 - 1\right]\frac{h_r}{r_r} \qquad (\text{I.10})$$

If σ_r is sufficiently large, this equation clearly shows that W_{dr} can be zero or less, which means that the rim is *self-supporting*. Nevertheless, a token disk may still be required to transfer torque to the shaft and, of course, to keep the rim and airfoils in the right axial and radial position. Often, multiple disks are manufactured together as one piece, which helps facilitate self-supporting rims. These multiple-disk assemblies are called *drums*.

Example I.2

Using typical values of the terms in Eq. (I.10) for the compressor and turbine based on disk materials given in Table I.2, Eq. (I.10) yields $W_{\text{dr}}/W_r = 0.37$ for the compressor and $W_{\text{dr}}/W_r = 0.56$ for the turbine.

Table I.2 Typical Data for Example I.2

Variable	Compressor	Turbine
$\overline{\sigma}_{\text{blades}}/\sigma_r$	0.10	0.20
h_r/r_r	0.05	0.10
$\rho(\omega r_r)^2/\sigma_r$	6.0	4.0

I.2.3 Disk of Uniform Stress

The disk supports and positions the rim while connecting it to the shaft. Its thickness begins with a value of W_{dr} at the inside edge of the rim and must grow as the radius decreases because of the accumulating centrifugal force that must be resisted. Just as was found for the rim, a disk can always be found that will perform the required job, but the size, weight, and/or cost may be excessive. Thus the final choice usually involves trial and error and judgment.

It is impossible to overemphasize the importance of ensuring the structural integrity of the disks, particularly the large ones found in cooled high-pressure turbines. Because they are very difficult to inspect and because massive fragments that fly loose when they disintegrate cannot be contained, they must not be allowed to fail.

The most efficient way to use available disk materials is to design the disk for constant radial and circumferential stress. Because the rim and disk are one continuous piece of material, the design stress would be the same throughout ($\sigma_r = \sigma_d$).

Applying locally radial equilibrium to an infinitesimal element of the disk, as shown in Fig. I.5, leads to the equation

$$\rho(\omega r_r)^2 W_d \, dr \, d\theta = \sigma_d \left[\left(r - \frac{dr}{2} \right) \left(W_d - \frac{dW_d}{2} \right) d\theta \right.$$
$$\left. - \left(r + \frac{dr}{2} \right) \left(W_d + \frac{dW_d}{2} \right) d\theta + 2 \left(W_d \, dr \, \frac{d\theta}{2} \right) \right]$$

which becomes in the limit

$$\frac{dW_d}{W_d} + \frac{\rho\omega^2}{\sigma_d} d\left(\frac{r^2}{2} \right) = 0$$

This equation may be integrated, by starting from $r = r_r$ and $W_d = W_{dr}$, to yield the desired result:

$$\frac{W_d}{W_{dr}} = \exp\left\{ \frac{\rho(\omega r_r)^2}{2\sigma_d} \left[1 - \left(\frac{r}{r_r} \right)^2 \right] \right\} \tag{I.11}$$

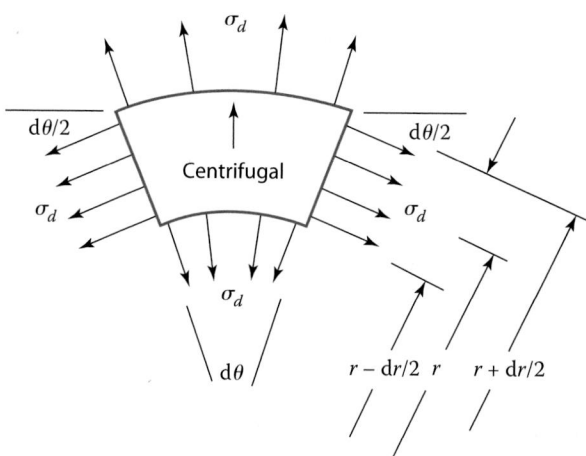

Fig. I.5   Disk element radial equilibrium nomenclature.

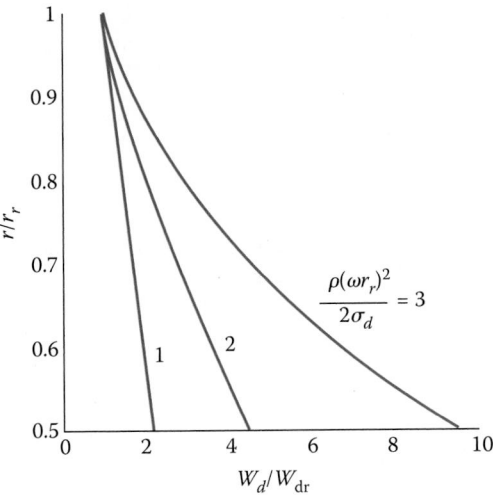

Fig. I.6 Disk thickness distributions.

The fundamental feature of Eq. (I.11) is that the disk thickness W_d grows exponentially in proportion to $(\omega r_r)^2$, which is the square of the maximum or rim velocity of the disk. Because this parameter has such a great influence on disk design, it is normal to hear structural engineers talk in terms of *allowable wheel speed*.

Figure I.6 shows the thickness distribution for typical values of the *disk shape factor* (DSF) $[= \rho(\omega r_r)^2/(2\sigma_d)]$ by using Eq. (I.11). From practical experience and judging by the looks of these thickness distributions in Fig. I.6, the maximum allowable value of the disk shape factor is not much more than 2, so that

$$\omega r_r \approx \sqrt{\frac{4\sigma_d}{\rho}} \tag{I.12}$$

Key point: It should be apparent that, on the low-pressure spool where the annulus flow area A is largest, the rotational speed (ω or N) is most likely to be limited by the allowable blade centrifugal stress [Eq. (I.8)]. Conversely, on the high-pressure spool, where the annulus flow area is considerably less, the rotational speed will probably be limited by the allowable wheel speed [Eq. (I.12)].

Both the COMPR and TURBN programs include a constraint analysis. The user can select maximum AN2 (and other constraints) as a convenient way to narrow down the design space for the compressor and turbine. The turbine constraint analysis includes the specification of a maximum wheel speed, and the compressor includes maximum tip speed (most relevant to transonic fan design).

Example I.3

Determine the allowable wheel speed for the typical compressor and turbine materials. The compressor material has $\sigma_d = 30$ ksi, $\rho = 9.0$ slug/ft^3, and thus $\sigma_d/\rho = 4.8 \times 10^5$ ft^2/s^2 (4.46 $\times$ 10^4 m^2/s^2). The turbine material has $\sigma_d = 20$ ksi, $\rho = 16.0$ slug/ft^3, and thus $\sigma_d/\rho = 1.8 \times 10^5$ ft^2/s^2 (1.67 $\times$ 10^4 m^2/s^2). Using Eq. (I.12), we see that the allowable wheel speed for the compressor is about 1400 ft/s (420 m/s) and for the turbine is about 850 ft/s (260 m/s). This indicates that the allowable wheel speed is more likely to limit the design of a turbine than the design of a compressor.

I.2.4 Disk Torsional Stress τ_d

The tangential disk shear stress required to transfer the shaft horsepower HP to the airfoils is easily calculated, because

$$\text{HP} = \text{shear stress} \times \text{area} \times \text{velocity}$$

Thus

$$\tau_d = \frac{\text{HP}}{2\pi r^2 W_d \omega} \tag{I.13}$$

Example I.4

Determine the disk torsional stress for the following typical values:

$$\text{HP} = 10,000 \, \text{hp} \ (7457 \, \text{kW})$$
$$r = 0.30 \, \text{ft} \ (0.0914 \, \text{m})$$
$$W_d = 0.10 \, \text{ft} \ (0.0305 \, \text{m})$$
$$\omega = 1000 \, \text{rad/s}$$

Equation (I.13) gives $\tau_d = 675$ psi (4.65 MPa), which makes a relatively small contribution to the overall stress.

I.2.5 Disk Thermal Stress σ_{tr} and $\sigma_{t\theta}$

For a disk of constant thickness with no center hole and a temperature distribution that depends only on radius [$T = T(r)$], it can be shown (Ref. 44) that the radial tensile stress is

$$\sigma_{tr} = \alpha E \left[\frac{1}{r_h^2} \int_0^{r_h} T(r)\,dr - \frac{1}{r^2} \int_0^r T(t)\,dt \right] \tag{I.14}$$

where α is the coefficient of linear thermal expansion and E is the modulus of elasticity, and the tangential tensile stress is

$$\sigma_{t\theta} = \alpha E \left[\frac{1}{r_h^2} \int_0^{r_h} T(r)\,dr + \frac{1}{r^2} \int_0^r T(r)\,dr - T \right] \tag{I.15}$$

both of which are zero if the temperature is constant. An interesting illustrative case is that of the linear temperature distribution $T = T_0 + \Delta T\,(r/r_h)$, for which Eq. (I.14) becomes

$$\sigma_{tr} = \frac{\alpha E\,\Delta T}{3}\left(1 - \frac{r}{r_h}\right) \tag{I.16}$$

and Eq. (I.15) becomes

$$\sigma_{t\theta} = \frac{\alpha E\,\Delta T}{3}\left(1 - 2\frac{r}{r_h}\right) \tag{I.17}$$

both of which have a maximum magnitude of $\alpha E\,\Delta T/3$ at $r = 0$.

Example I.5

For typical values of $\alpha = 1 \times 10^{-5}\,(^{\circ}\mathrm{F})^{-1}\,[1.8 \times 10^{-5}\,(^{\circ}\mathrm{C})^{-1}], E = 10 \times 10^6$ psi $(6.9 \times 10^4$ MPa), and $\Delta T = 100^{\circ}$F $(55.6^{\circ}$C), the maximum magnitude of σ_{tr} and $\sigma_{t\theta}$ is 6700 psi (46 MPa)! This simple case demonstrates forcefully that thermal stresses can be very large and, therefore, must be carefully accounted for and reduced as much as possible. This is especially true during transient operation.

With this perspective, it is possible to imagine that truly enormous stresses could be generated in the thin outer wall of the cooled turbine airfoils, if they are not very carefully designed. Because such stresses will be proportional to the temperature difference between the mainstream and the cooling air, there is also a limit to how cold a coolant may be used before it no longer truly *protects* the material.

I.3 Engine Materials

The harsh environment of the gas turbine engine requires special materials. Many of these materials are developed exclusively for this application by material experts, most of whom work for either an engine company or a materials company. Therefore, the properties of many materials used for critical components in gas turbine engines are corporate secrets, and published data are limited. However, some materials have been and continue to be developed in the public sector, and their properties can be found in references such as the *Aerospace Structural Metals Handbook* (Ref. 61)

Table I.3   Material Types

Material No.	Type	Density	
		Slug/ft³	kg/m³
1	Aluminum alloy	5.3	2716
2	Titanium alloy	9.1	4663
3	Wrought nickel alloy	16.0	8200
4	High-strength nickel alloy	17.0	8710
5	Single-crystal superalloy	17.0	8710

and *The Superalloys* (Ref. 57), textbooks (Refs. 63 and 64), and journals (Ref. 65).

To provide examples and general guidance, the data for several materials commonly found in critical parts of gas turbine engines are given in Table I.3 and Figs. I.7 and I.8. Figure I.7 gives the 1% *creep* curves of typical materials for 1000 h at temperatures between 0 and 2000°F. Creep is gradual and permanent elongation of material under high stress, especially challenging when that material is subjected to extended time at high temperature. Adequate creep strength is critical in turbine parts. Dividing the material's creep strength by its density gives the material's strength/weight ratio. The variations of strength/weight ratio with temperature are given in Fig. I.8.

The material used in gas turbine engines continues to change as better materials are developed. The fans of high-bypass turbofans are being

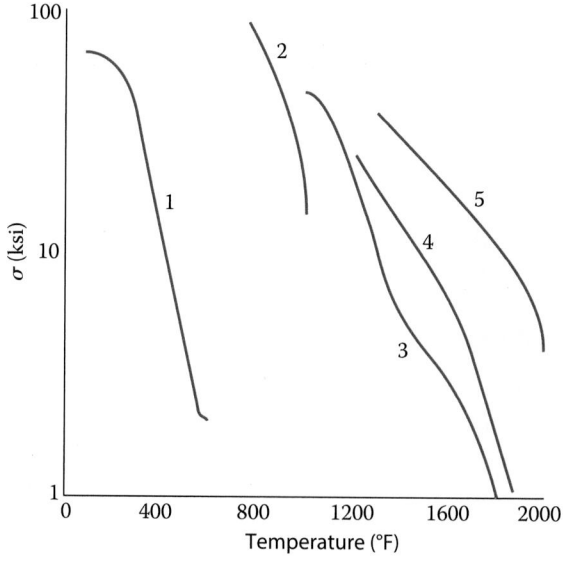

Fig. I.7   Strength at 1% creep for 1000 h.

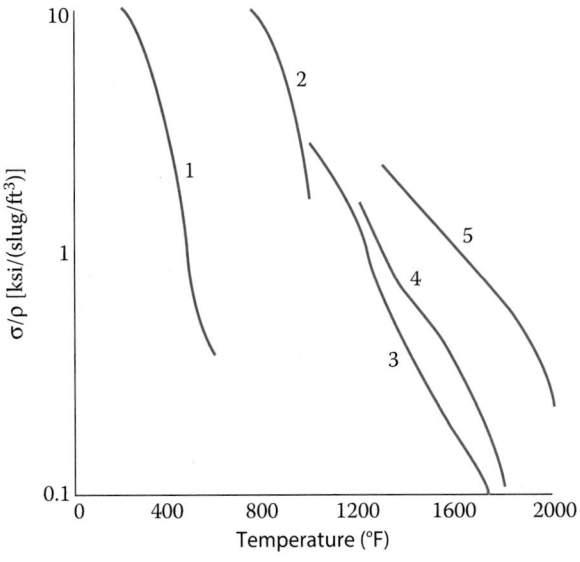

Fig. I.8   Strength-to-weight ratio at 1% creep for 1000 h.

made of composites with titanium spars. Ceramic coatings are used in the combustor and other parts in the high-temperature gas path. Figure I.9 shows the materials being used in a typical gas turbine engine.

I.3.1   Compressor and Fan Materials

Normal practice for compressor and fan blades is to base designs on creep and creep-rupture data. Common practice is to allow less than 1%

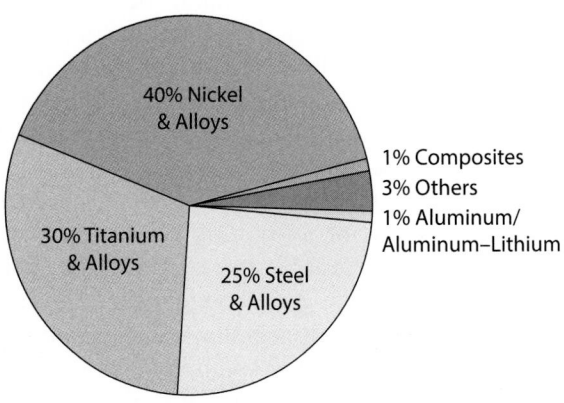

Fig. I.9   Typical engine materials. (Source: Rolls-Royce.)

creep during the life of the part. A reasonable estimate of the *allowable working stress* is 30–50% of the value obtained from Fig. I.7 for 1000 h. For example, Fig. I.7 gives a 1% creep stress for material 1 of about 32 ksi at 300°F (220 MPa at 150°C) and 1000 h. Thus the allowable working stress for this material at these conditions would be from 10 to 16 ksi (70–110 MPa).

During its life, a fan or compressor blade is subjected to billions of *high-cycle fatigue* cycles due to vibrations. Thus it is important that the *run-out* stress, which the material can apparently withstand forever, not be exceeded. Aluminum 2124 alloy has a low run-out stress of about 12 ksi at room temperature whereas titanium 6246 alloy has a high run-out stress of about 140 ksi (about 1 MPa) at room temperature [61]. Because of the poor fatigue characteristics of aluminum alloys, titanium alloys are typically used for fan and low-pressure compressor blades and disks.

Titanium's strength/weight ratio is severely reduced at temperatures above 900°F (480°C). Hence nickel-base alloys are commonly used for the back stages of high-pressure compressors.

I.3.2 Turbine Materials

The critical components of the turbine are exposed to very high temperatures. Many of these parts require a mixture of metals (alloy) specifically chosen in precise proportion for extended life at high temperatures, commonly called a *superalloy*. High-pressure turbine nozzles and blades nearly always require compressor air for cooling, as do some low-pressure turbine nozzles. Certain materials and environments also require protective coatings (usually ceramic), which generally add about 100–200°F (55–110°C) temperature capability. The superalloys are usually cobalt-base or nickel-base. Representative examples of these materials are Mar-M509, a high-chromium, carbide-strengthened, cobalt-base superalloy, and Rene 80, a cast, precipitation-hardenable, nickel-base superalloy [61]. It's common for these superalloys to contain 10–12 different metals that include elements like tantalum, tungsten, and rhenium—all metals with excellent high-temperature strength, but also very heavy (high density), and in some cases, challenging to mine. Consequently, the exact proportions of these metals are crucial and highly protected by engine manufacturers.

In addition to high-strength superalloys, carefully controlled materials processing is required when manufacturing turbine parts. Many of the newer superalloys for turbine rotor blades are cast and solidified in such a manner as to align the crystalline grain boundaries in the radial direction, called *directional solidification*, or even better, a single grain (instead of multiple radial grains) called *single crystal*. Grain boundaries are important for many of the mechanisms of creep, so turbine blades cast as single-crystal superalloys have been an enabling technology. The combined effects of single-crystal blades, improved cooling schemes, and protective coatings have been key to the nearly doubling of allowable turbine entry temperatures since the 1960s.

Finally, intermetallics, like titanium-aluminides (TiAl), and nonmetals, like ceramic-matrix composites (CMCs), seem poised for increased use in aircraft gas turbine engines [55, 73]. Both classes of materials have improved creep resistance and increased strength-to-weight over the materials just discussed, making them particularly attractive for even higher operating temperatures and reduced cooling requirements. The intermetallics and nonmetals have been studied for decades, but recent major investments by governments and engine companies have brought them to the forefront.

Appendix J · About the Software

A comprehensive set of software is included to assist you in learning the text material, solving some of the problems in the text, and completing your design problems. The software programs use pull-down menus and edit fields that support a mouse to make them very user-friendly. Like any computer tool, the programs facilitate calculations of results and in no way are a substitute for understanding the fundamental principles and underlying assumptions.

The AFPROP program defines the properties of air with hydrocarbon fuels and is provided in both executable and BASIC source code. The ATMOS program gives the properties of the atmosphere at any altitude for the user-selected temperature profile (standard, cold day, hot day, or tropic). The EQL program performs chemical equilibrium analysis for reactive mixtures of perfect gases. The GASTAB program gives complete 1-D gas dynamics tables for the user-selected ratio of specific heats.

PARA and PERF are a set of user-friendly computer programs written for the preliminary analyses of common airbreathing aircraft engine cycles. COMPR and TURBN are a set of computer programs that perform the preliminary analyses of axial-flow compressors and turbines, respectively. These programs allow you to solve a single problem, a number of "what if" solutions, or a design problem.

The current versions of PARA, PERF, COMPR, and TURBN are written in Microsoft Visual Basic 6 with plotting procedures from Olectra Chart 6. Comprehensive user guides provided with the software will help answer most of your questions about the software.

J.1 Getting Started

To get started, you need a PC-compatible system with the Microsoft Windows operating system (Vista or later). If you have an Apple Mac, you'll need a version of Microsoft Windows installed on a virtual machine. The software can be downloaded from the AIAA web site at http://arc.aiaa.org. Follow the instructions provided and enter the following password: **EoPGTnR2ed**.

Save the downloaded file and unzip into the files necessary for installation. Run the **Setup** program within the **EOP Software** folder, and then the opening installation window will display as shown in Fig. J.1.

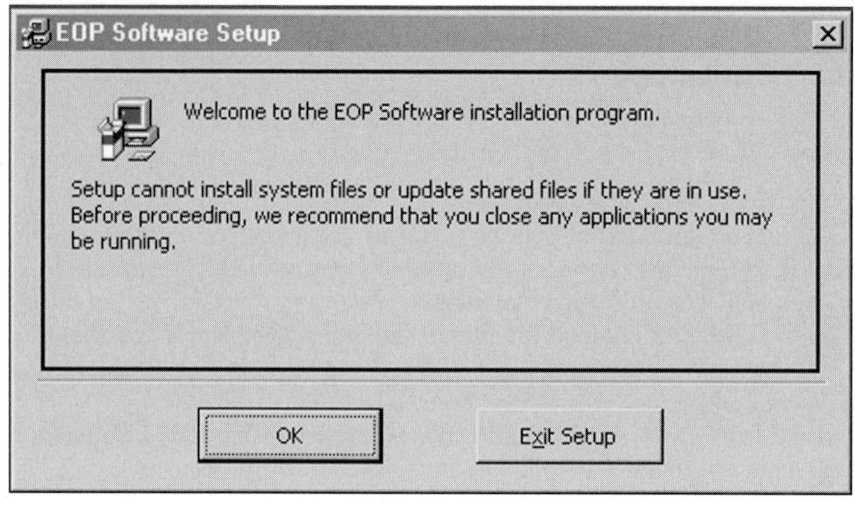

Fig. J.1 Opening installation window.

Click the OK button, and the software destination window is displayed as shown in Fig. J.2.

Select the highlighted button in the upper left to install the programs in the default location (C:/Programs/EOP). The program group window is displayed as shown in Fig. J.3. It is recommended that the user accept the **EOP Software** program group and press the **Continue** button. Software installation will then proceed.

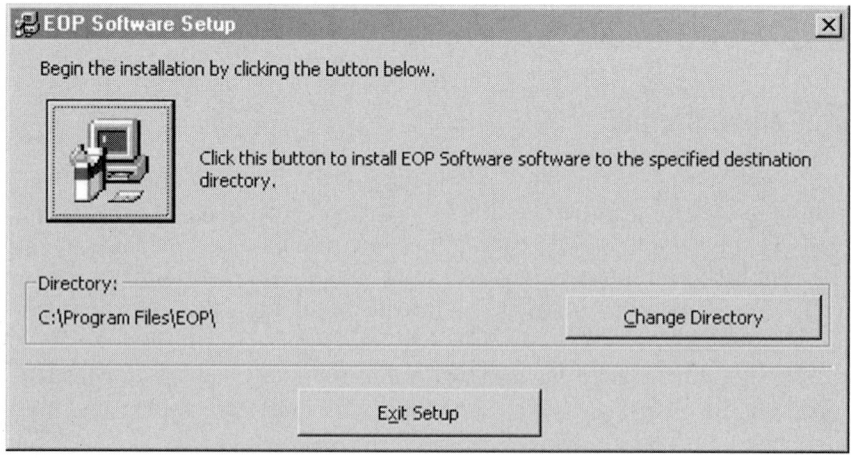

Fig. J.2 Installation destination window.

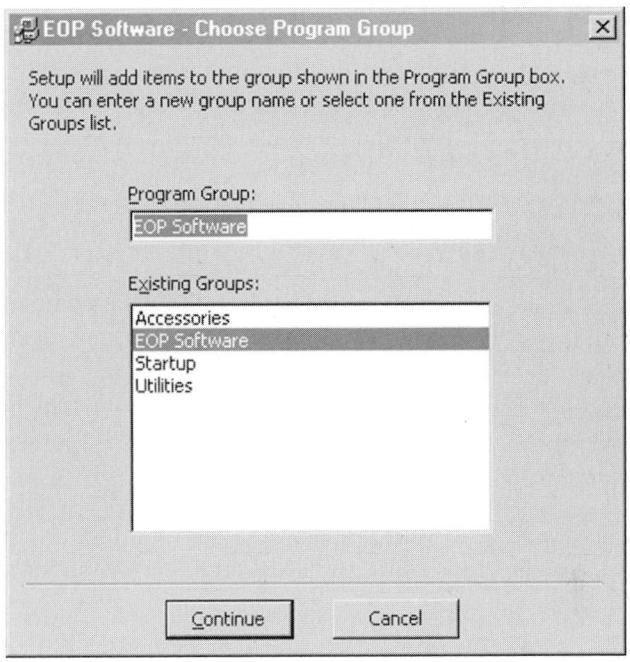

Fig. J.3 Program group window.

J.2 The Software

J.2.1 AFPROP.EXE (and AFPROP.BAS)

This program calculates the thermodynamic properties of air and fuel as a function of temperature and fuel/air ratio by using the equations in Chapter 2. Written in Microsoft Visual Basic 6, this program is supplied in both executable form (AFPROP.EXE) and source code (AFPROP.BAS). The source code can be used to write your own analysis programs with variable specific heats as discussed in the Supporting Materials for Chapters 7 and 8.

After selecting the unit system (English or SI), the user enters the fuel/air ratio and the temperature, or the enthalpy, or the reduced pressure. Input temperatures are limited to the range of 300°R (167 K) to 4000°R (2222 K), and the fuel/air ratio has a maximum value of 0.0676. The program will calculate the other properties in addition to the gas constant, specific heat at constant pressure, ratio of specified heats, and speed of sound.

J.2.2 ATMOS Program

This program calculates properties of the atmosphere for standard, cold, hot, and tropical days in either SI or English units.

J.2.3 EQL Program

The EQL program calculates equilibrium properties and process end states for reactive mixtures of perfect gases, for different problems involving hydrocarbon fuels. It supports the thermochemical analysis and problem in Chapters 2 and 10.

J.2.4 GASTAB.EXE

This program is equivalent to extensive, traditional, compressible flow appendices for the simple flows of calorically (constant specific heats) perfect gases. It includes isentropic flow; adiabatic, constant area frictional flow (Fanno flow); frictionless, constant area heating and cooling (Rayleigh flow); normal shock waves; oblique shock waves; multiple oblique shock waves; and Prandtl–Meyer flow. The user inputs the ratio of specific heats γ, the mean molecular weight $\mathcal{M}$, and the known property. Then the user clicks the check box beside this property and all other properties are calculated and displayed for that flow. The user can select the unit system (SI or English).

J.2.5 PARA.EXE

The PARA computer program determines the variation in gas turbine engine performance with cycle design variables such as the compressor pressure ratio. The program is based on the engine models contained in Chapters 5–7 of this book and is intended to be used with this textbook. The user can select the unit system (SI or English), the type of analysis model (ideal or real), the engine cycle (one of the eight engine cycles listed in Table J.1), and the iteration variable (one of the five variables listed in Table J.1) along with its applicable range and increment. Data are entered and/or changed on data screens through edit fields. Results can be plotted

Table J.1 PARA Engine Cycles and Iteration Variables

Engine Cycles	Variables
Turbojet—single spool	Single-point calculation
Turbojet—dual spool	Combustor exit temperature
Turbojet with afterburner	Compressor pressure ratio
Turbofan with two exhausts	Fan pressure ratio
Turbofan with mixed exhaust	Bypass ratio
Turbofan with afterburning	
Turboprop	
Ramjet	

on the screen by using defined ranges of plot data. In addition, the temperature vs entropy of the defined ideal engine cycle data can be plotted on the screen.

This program is very useful for examining the trends of a selected engine cycle's specific performance (specific thrust and thrust specific fuel consumption) with changes in applicable design variables. Up to 21 sets of results can be plotted at a time. When one variable is used for calculations, results can be plotted vs the iteration variable. If more than one variable is used for calculations, only the specific performance (thrust specific fuel consumption vs specific thrust) of the cycle can be plotted.

J.2.6   PERF.EXE

The PERF computer program determines the performance of a specific engine and its variation with flight condition and throttle. The program is based on the engine models contained in Chapter 8 of this book and is intended to be used with this textbook. Through pull-down menus, the user can select the specific engine cycle (one of the eight engine cycles listed in Table J.2) and its design, the iteration variable (one of the eight variables listed in Table J.2) along with its applicable range and increment, the control system limits (e.g., temperature leaving compressor), the unit system (SI or English), and the output device(s). Data are entered and/or changed on a data screen through edit fields. Results can be plotted on the screen by using defined ranges of plot data.

PERF is very useful for examining the variation of an engine's performance with changes in flight condition, throttle, and control system limits. Up to 21 sets of results can be plotted at a time. When one variable is used for calculations, results can be plotted vs the iteration variable. If more than one variable is used for calculations, only the thrust specific fuel consumption vs thrust of the engine can be plotted.

Table J.2   PERF Engine Cycles and Iteration Variables

Engine Cycles	Variables
Turbojet—single spool	Single point @ % thrust
Turbojet—dual spool	Mach number
Turbojet with afterburner	Altitude
Turbofan with separate exhausts	Ambient temperature
Turbofan with mixed exhaust	Ambient pressure
Turbofan with afterburning	T_{t4}—total temperature leaving combustor
Turboprop	T_{t7}—total temperature leaving afterburner
Ramjet	Exit nozzle pressure ratio (P_0/P_9)

Table J.3 COMPR Types of Design and
Swirl Distributions

Types of Design	Swirl Distributions
Constant tip radius	Free vortex
Constant mean radius	Exponential
Constant hub radius	First power
User-specified tip radius	
Repeating row/stage	

J.2.7 COMPR.EXE

The COMPR program calculates the change in properties along the mean line of a multistage, axial-flow compressor based on user input data. The program is based on the methods contained in Chapter 9 of this book and is intended to be used with this textbook. The user can select the type of design (one of the five types of designs listed in Table J.3), type of swirl distribution (one of the three swirl distributions listed in Table J.3), and unit system (SI or English). Data are entered and/or changed on a data screen through edit fields. The results are given for the hub, mean, and tip radius for both rotor and stator blades of each stage and include flow angles, diffusion factors, degree of reactions, pressures, Mach numbers, temperatures, velocities, number of blades, and blade centrifugal stress.

After stage calculations are completed, the user can have the computer sketch a cross-sectional drawing on the graphics screen (e.g., see Fig. 9.30). The user can also have the computer sketch the relative position and angle of rotor and stator blades, using the NACA 65A010 profile for a specified stage and number of blades. One can also look at the change in blade angles with changes in radius (e.g., see Fig. 9.28).

J.2.8 TURBN.EXE

The TURBN program calculates the change in properties along the mean line of a multistage, axial-flow turbine based on user input data. The program is based on the methods contained in Chapter 9 of this book and is intended to be used with this textbook. The user can select the unknown (one of the five cases listed in Table J.4), the loss model (one of the two listed in Table J.4), and the unit system (SI or English). Data are entered and/or changed on a data screen through edit fields. The results are given for the hub, mean, and tip radius for both stator and rotor blades of each stage and include flow angles, degree of reactions, pressures, Mach numbers, temperatures, velocities, number of blades, and solidities.

After stage calculations are completed, the user can have the computer sketch a cross-sectional drawing of the turbine on the graphics screen (e.g.,

Table J.4   TURBN Unknowns and Loss Models

Unknowns	Loss Models
Flow angle entering rotor	Polytropic efficiency
Flow angle leaving rotor—Case I	Loss coefficient
Flow angle leaving rotor—Case II	
Mach number entering rotor	
Total temperature leaving rotor	

see Fig. 9.68). The user can also have the computer sketch the relative position and angle of rotor and stator blades, using the British C4 or T6 profile for a specified stage and number of blades. One can also look at the change in spacing and blade angles with changes in radius, as is possible with the COMPR program.

Appendix K Answers to Selected Problems

1.1	50 kN
1.4	1973 ft/s
1.10	194,080 N, 33%, 10.23 (g/s)/kN
1.14	219.6 ft/s
1.15	78.5 m/s
1.18	(a) 7.906, 0.3162, 0.04; (b) 39.8 kft, 5690 lbf; (c) 45 kft, 4430 lbf; (d) 1140 nm
1.19	(a) 5.774, 0.3464, 0.06; (b) 6.65 km, 19,050 N; (c) 12.07 km, 19,050 N
1.21	(a) 1257 nm; (b) 1127 nm
1.23	(a) 10.2, 0.2449, 0.024, 0.7946; (c) 0.80; (d) 0.76
1.24	9970 s
1.27	45 kN, 91.84 s
1.28	(a) 14,700 m/s; (b) 568 kg/s
2.2	2970 lbf
2.6	28,950 lbf
2.8	53,800 N
2.9	135,900 lbf
2.13	772.9 K, 0.22 m^2, 0.19 m^2
2.15	18.34 MW (24,590 hp)
2.17	(a) 98.6 lbm/s; (b) 904°R, 66.5 psia; (c) 1.69 ft^2
2.20	(a) 81.3 kg/s; (b) 1009.4 K, 1.601; (c) 0.307 m^2, $-48,660$ N
2.22	20.43 MW, 0.0988 m^3/kg, 17.61 J/(kg · K)
2.23	2880.7°R, 29.30 MW, 0.0077 Btu/(lbm · °R)
3.10	1592 K, 752 kPa, 0.663 m^2, 0.0606 m^2
3.11	5.09
3.14	6.87 kg/s, 0.263, 295.9 K, 90.53 m/s
3.19	(a) 728 K, 735 kPa; (b) 241 kPa; (c) 697 K, 207 kPa
3.20	311.1°R, 12.78 psia, 1729 ft/s; 560°R, 72.09 psia; 5525°R, 57.51 psia, 648.5 ft/s
3.25	Case 1: 20 deg, 37.8 deg, 79.6 psia, 10.25 psia, 278.5
3.29	(a) 0.2117 m^2; (b) 0.03811 MPa, 125 K; (c) 0.4752, 0.4597 MPa, 0.3938 MPa, 350 K, 334.9 K; (d) 0.1797, 0.4495 MPa, 347.8 K
3.33	(a) 0.5007, 557.5 kPa, 998 K
4.3	108.9 kM, 1.64 kN
4.5	4400 lbf

4.7	7.7 kN
4.8	0.045, 0.060
4.11	268.2 N
5.9	3.916, 712.3 N/(kg/s), 35.8 (g/s)/kN
5.23	2.686
5.24	0.9425
6.3	0.9379
6.6	10+
6.8	(a) 0.8623, 0.8849, 0.8923, 0.9074; (b) 0.8932, 0.8739; (c) 0.9461, 1.8125, 0.8861, 2.037
6.9	0.990
6.10	0.686
6.12	(a) 0.9561, 0.9585; (b) 0.8594, 0.8737, 0.8996, 0.925; (c) 0.9225, 0.8924; (d) 30,035 hp (22.4 MW), 12,776 hp (9.53 MW), 57,702 hp (43.03 MW); (e) 105,160 hp (78.42 MW)
7.9	390 kPa, 540.6 K, 188.5 kPa, 1550.1 K, 1.724
7.11	(a) 5.50 ft^2; (b) 3.55 ft^2; (c) 8.165 lbm/s, 5.41 ft^2
7.14	Compressor pressure ratio range from 19 to 51
7.19	15.45 lbf/(lbm/s), 0.6075 (lbm/h)/lbf
7.28	111.9 lbf/(lbm/s), 1.852 (lbm/h)/lbf
7.31	Compressor pressure ratio range from 14.8 to 19.9 with a corresponding fan pressure ratio range of 2.85 to 2.87
7.36	$M_9 = 0.7656$, 1242 N/(kg/s), 15.8 (g/s)/kN
7.38	Compressor pressure ratio range from 10.8 to 17.7 with a corresponding turbine temperature ratio range of 0.60 to 0.55
8.2	(a) 0.0004508, 132.58; (b) 4.055, 16.28 kg/s; (c) 1620 K, 8, 25 kg/s, 1800 K, 6.92, 22.67 kg/s
8.12	2.2657, 14.316, 0.7750
8.13	$M_9 = 1.6276$, 18.622 kg/s, 10,530 N
8.15	(a) 97.0 lbm/s, 3.5124, 5.5672, 3223.6 ft/s, 10,129 lbf, 1.4445 (lbm/h)/lbf; (b) 4.7109, 86.01 lbm/s, 2.7578, 4.6393, 0.9129; (c) 5.3276, 92.74 lbm/s, 3.1188, 5.1032, 0.9574
8.22	570.8 lbm, 8170 lbf, 0.566 (lbm/h)/lbf
8.24	2100 lbm/s, 77,000 lbf, 0.3822 (lbm/h)/lbf
9.3	20.86 m/s, 41.72 m/s
9.4	120 m/s, 1.328
9.10	(a) 646.9 ft/s, 495.6 ft/s, 415.8 ft/s; (b) 495.6 ft/s, 783.0 ft/s, 926.7 ft/s; (c) 73.36°R, 1.1414; (d) 1.5168, 22.29 psia
9.18	$\beta_1 = 56.468$, Stage pressure ratio $= 1.3444$
9.26	(b) 3.129×10^{10} in.$^2 \cdot$ rpm^2

9.28	2133.5 rad/s, 163.2 K, 82.0%, 327.7 kW
9.30	(a) 827.82 m/s, 585.36 m/s, 585.36 m/s; (b) 585.36 m/s, 51.21 m/s, 587.6 m/s; (c) −184.26 K, 0.89763; (d) 0.59106, 591.06 kPa
9.45	$5.282 \times 10^6 m^2 \cdot rpm^2$
9.47	(a) 0.1569 m^2; (b) 5561 rpm; (c) 0.64027 m, 0.62014 m, 200.9 K
9.53	(a) 1440.5 ft/s, 2160.7 rad/s; (b) 0.850; (c) 96.3 psia; (d) 0.8614
9.55	(a) 155.46 m/s, 309.86 K; (b) 145.32 m/s, 1.128 atm; (c) 881.75 K; (d) 0.0742 m, 0.0694 m

10.2	5660 lbm/s
10.6	10.465
10.7	126,900 kg
10.10	6185 ft/s
10.13	227,980 N
10.15	1.820
10.18	(a) 250 lbm/s; (c) 758.9 psia; (e) 1.830; (g) 54 kft; (h) 78,830 lbf
10.27	(a) 2.529; (d) 4.323 lbm/s; (f) 1.405; (h) 180.4 s

Answers to Problems in Chapter 11 Online

11.2	1.636 ft
11.3	2.039 m
11.5	0.282 m
11.6	(a) 1.9228; (b) 1.697; (c) 0.7544; (d) 1.534
11.8	(a) 1.27924, 0.71204; (b) 0.52442
11.11	1.39, 0.6572
11.13	(a) 0.9122, 0.7209; (b) 0.79
11.22	$C_V = 0.9937$, $C_{fg} = 0.9764$, $F_{g\,actual} = 92{,}580$ N
11.25	0.4103 m
11.27	0.9930, 0.0909
11.29	0.99884, 0.025
11.31	10
11.33	98.5%
11.34	0.782 to 1.564 ft^3

REFERENCES

[1] *Random House Kernerman Webster's College Dictionary*, http://www.kdictionaries-online.com/DictionaryPage.aspx?ApplicationCode=18, 2016.

[2] International Organization for Standardization, "Standard Atmosphere," ISO 2533:1975, 1975.

[3] "Gas Turbine Engine Performance Station Identification and Nomenclature," Aerospace Recommended Practice (ARP) 755A, Society of Automotive Engineers, Warrendale, PA, 1974.

[4] Langston, L. S., "Not So Simple Machines," *Mechanical Engineering*, Jan. 2013, pp. 46–51.

[5] Norris, G., "Future Fans," *Aviation Week*, 25 Aug. 2014, pp. 19–20.

[6] "1989 AIAA/General Dynamics Corporation, Team Aircraft Design Competition, Engine Data Package," Director of Student Programs, AIAA, Washington, DC, 1988.

[7] Oates, G. C. (ed.), *Aerothermodynamics of Gas Turbine and Rocket Propulsion*, 3rd ed., AIAA Education Series, AIAA, Reston, VA, 1997.

[8] Mattingly, J. D., "Improved Methodology for Teaching Aircraft Gas Turbine Engine Analysis and Performance," *1992 ASEE Annual Conference Proceedings*, Vol. 1, ASEE, Washington, DC, 1992, pp. 240–247.

[9] Nicolai, L. M., *Fundamentals of Aircraft Designs*, METS, San Jose, CA, 1975.

[10] Raymer, D. P., *Aircraft Design: A Conceptual Approach*, 4th ed., AIAA Education Series, AIAA, Reston, VA, 2006.

[11] Hale, F. J., *Introduction to Aircraft Performance, Selection and Design*, Wiley, New York, 1984.

[12] Anderson, J. D., *Introduction to Flight*, 3rd ed., McGraw-Hill, New York, 1989.

[13] Mattingly, J. D., Heiser, W. H., and Pratt, D. T., *Aircraft Engine Design*, 2nd ed., AIAA Education Series, AIAA, Reston, VA, 2002.

[14] von Karman, T., "Supersonic Aerodynamics—Principles and Applications," The Tenth Wright Brothers Lecture, *Journal of the Aeronautical Sciences*, Vol. 14, July 1947, pp. 373–402.

[15] John, J., and Keith, T., *Gas Dynamics*, 3rd ed., Prentice Hall, Upper Saddle River, NJ, 2006.

[16] Penner, S. S., *Chemistry Problems in Jet Propulsion*, Pergamon Press, London, 1957.

[17] Keenan, J. H., and Kaye, J., *Gas Tables*, Wiley, New York, 1948.

[18] McKinney, J. S., "Simulation of Turbofan Engine (SMOTE)," AFAPL-TR-67-125, Air Force Aero Propulsion Laboratory, Wright-Patterson AFB, OH, Nov. 1967.

[19] Gordon, S., and McBride, B., "Computer Program for Calculation of Complex Chemical Equilibrium Compositions," NASA SP-273, 1971.

[20] Heiser, W. H., and Pratt, D. T., *Hypersonic Airbreathing Propulsion*, AIAA Education Series, AIAA, Washington, DC, 1994.

[21] Kerrebrock, J. L., *Aircraft Engines and Gas Turbines*, 2nd ed., MIT Press, Cambridge, MA, 1992.

[22] Cohen, H., Rogers, G. F. C., and Saravanamuttoo, H. I. H., *Gas Turbine Theory*, Wiley, New York, 1972.

[23] Milhouse, P. T., Kramer, S. C., King, P. I., and Mykytka, E. F., "Identifying Optimal Fan Compressor Pressure Ratios for the Mixed-Stream Turbofan Engine." *Journal of Propulsion and Power*, Vol. 16, No. 1, Jan.–Feb. 2000, pp. 79–86.

[24] Heiser, W. H., and Pratt, D. T., "Thermodynamic Cycle Analysis of Pulse Detonation Engines," *Journal of Propulsion and Power*, Vol. 18, No. 1, Jan.–Feb. 2002, pp. 68–76.

[25] Kailasanath, K., "Applications of Detonations to Propulsion: A Review," AIAA Paper 99-1067, 1999.

[26] Strehlow, R. A., *Combustion Fundamentals*, McGraw-Hill, New York, 1984.

[27] Shapiro, A. H., *The Dynamics and Thermodynamics of Compressible Fluid Flow*, Vol. 1, Ronald, New York, 1953.

[28] Pratt, D. T., Humphrey, J. W., and Glenn, D. E., "Morphology of Standing Oblique Detonation Waves," *Journal of Propulsion and Power*, Vol. 7, No. 5, 1991, pp. 837–845.

[29] Pratt, D. T., and Heiser, W. H., "Isolator–Combustor Interaction in a Dual-Mode Scramjet Engine," AIAA Paper 93-0358, 1993.

[30] Bleakney, W., Weimer, D. K., and Fletcher, C. H., "The Shock Tube: A Facility for Investigations in Fluid Dynamics," *Review of Scientific Instruments*, 1949, Vol. 20, pp. 807–815.

[31] Sutton, G. P., *Rocket Propulsion Elements*, 6th ed., Wiley, New York, 1992.

[32] Summerfield, M., Foster, C. R., and Swan, W. C., "Flow Separation in Overexpanded Supersonic Exhaust Nozzles," *Jet Propulsion*, Vol. 24, Sept.–Oct. 1954, pp. 319–321.

[33] Kubota, N. "Survey of Rocket Propellants and Their Combustion Characteristics," *Fundamentals of Solid-Propellant Combustion*, Vol. 90, Progress in Astronautics and Aeronautics, AIAA, New York, 1984.

[34] *Model Specification for Engines, Aircraft, Turbojet*, MIL-SPEC MIL-E-5008B, U.S. Dept. of Defense, Jan. 1959.

[35] Bathie, W. W., *Fundamentals of Gas Turbines*, Wiley, New York, 1972.

[36] Dixon, S. L., *Thermodynamics of Turbomachinery*, 3rd ed., Pergamon Press, Elmsford, NY, 1978.

[37] Hill, P. G., and Peterson, C. R., *Mechanics and Thermodynamics of Propulsion*, 2nd ed., Addison-Wesley, Reading, MA, 1992.

[38] Oates, G. C. (ed.), *Aerothermodynamics of Aircraft Engine Components*, AIAA Education Series, AIAA, Washington, DC, 1985.

[39] Horlock, J. H., *Axial Flow Compressors*, Krieger, Melbourne, FL, 1973.

[40] Horlock, J. H., *Axial Flow Turbines*, Krieger, Melbourne, FL, 1973.

[41] Johnsen, I. A., and Bullock, R. O. (eds.), *Aerodynamic Design of Axial-Flow Compressors*, NASA SP-36, 1965.

[42] Glassman, A. J. (ed.), *Turbine Design and Application*, Vols. 1–3, NASA SP-290, 1972.

[43] Wilson, D. G., *The Design of High-Efficiency Turbomachinery and Gas Turbines*, MIT Press, Cambridge, MA, 1984.

[44] Sorensen, H. A., *Gas Turbines*, Ronald, New York, 1951.

[45] Hess, W. J., and Mumford, N. V., *Jet Propulsion for Aerospace Applications*, 2nd ed., Pitman, New York, 1964.

[46] Treager, I. E., *Aircraft Gas Turbine Engine Technology*, 2nd ed., McGraw-Hill, New York, 1979.

[47] Glauert, H., *The Elements of Aerofoil and Airscrew Theory*, 3rd ed., Cambridge Univ. Press, Cambridge, UK, 1959.

[48] Theodorsen, T., *Theory of Propellers*, McGraw-Hill, New York, 1948.

[49] Theodorsen, T., "Theory of Static Propellers and Helicopter Rotors," Paper 326, 25th Annual Forum, American Helicopter Society, Alexandria, VA, May 1969.

[50] Abbott, I. H., and Von Doenhoff, A. E., *Theory of Wing Sections*, Dover, New York, 1959.

[51] Wisler, D. C., "Advanced Compressor and Fan Systems," lecture notes from UTSI short course Aeropropulsion, Dayton, OH, 7–11 Dec. 1987.

[52] Bloch, G. S., "Flow Losses in Supersonic Compressor Cascades," Ph.D. dissertation, Mechanical Engineering Dept., Virginia Polytechnic Institute and State Univ., Blacksburg, VA, July 1996.

[53] Nikkanen, J. P., and Brooky, J. D., "Single Stage Evaluation of Highly Loaded High Mach Number Compressor Stages V," NASA CR 120887 (PWA-4312), March 1972.

[54] Zweifel, O., "The Spacing of Turbomachinery Blading, Especially with Large Angular Deflection," *Brown Boveri Review*, Vol. 32, 1945, p. 12.

[55] *The Random House College Dictionary*, rev. ed., Random House, New York, 1975.

[56] Haven, B. A., and Wood, C. W., "The Rocket Laboratory in the USAF Aero-Propulsion Curriculum," AIAA Paper 93-2054, 1993.

[57] International Organization for Standardization, "Standard Atmosphere," ISO 2533:1975, 1975.

[58] *Climatic Information to Determine Design and Test Requirements for Military Equipment*, MIL-SPEC MIL-STD-210C, Rev. C, U.S. Dept. of Defense, Jan. 1997.

[59] *Climatic Information to Determine Design and Test Requirements for Military Equipment*, MIL-SPEC MIL-STD-210A, U.S. Dept. of Defense, Nov. 1958.

[60] *Aerospace Structural Metals Handbook*, Batelle Memorial Institute, Columbus Laboratories, Columbus, OH, 1984.

[61] Sims, C. T., and Hagel, W. C., *The Superalloys*, Wiley, New York, 1972.

[62] Smith, W. F., *Structure and Properties of Engineering Alloys*, 2nd ed., McGraw-Hill, New York, 1993.

[63] Brick, R. M., Pense, A. W., and Gordon, R. B., *Structure and Properties of Engineering Materials*, 4th ed., McGraw-Hill, New York, 1977.

[64] Imarigeon, J. P., "The Super Alloys: Materials for Gas Turbine Hot Section Components," *Canadian Aeronautics and Space Institute Journal*, Vol. 27, 1981, pp. 336–354.

[65] Zucker, R. D., and Biblarz, O., *Fundamentals of Gas Dynamics*, 2nd ed., Wiley, New York, 2002.

[66] Cotter, H. N., "Principles of Aircraft Gas Turbine Engines," University of Florida, Course Notes, West Palm Beach, FL, 2002.

[67] Bloch, G. S., Copenhaver, W. W., and O'Brien, W. F., "A Shock Loss Model for Supersonic Compressor Cascades," *Journal of Turbomachinery*, Vol. 121, No. 1, 1999, pp. 28–35.

[68] Norris, G., "Core Competency," *Aviation Week and Space Digest*, 8 June 2015, p. 94.

[69] Heiser, W. H., and Mattingly, J. D., "Supercruise Aircraft Range," *Journal of Aircraft*, Vol. 47, No. 3, May–June 2010, pp. 1066–1068.

[70] Airbus, "SNECMA Tackle Open-Rotor Integration," *Aviation Week and Space Digest*, 31 March 2014, p. 22.

[71] "Advanced Turboprop Project," NASA SP-495, 1988.

[72] Cumpsty, N., "Jet Propulsion: How, Why and Whither," Invited Lecture Session, ASME, IGTI Turbo Expo 2016, 16 June 2015, Montreal, Canada.

[73] Hunter, L., and Cawthon, J., "Improved Supersonic Performance Design for the F-16 Inlet Modified for the J79 Engine," AIAA Paper 84-1271, 1984.

[74] Conner, M., "Hans von Ohain: Elegance in Flight," AIAA.

ABOUT THE AUTHORS

JACK D. MATTINGLY has 50 years' experience analyzing and designing propulsion and thermodynamic systems. He has written three engineering textbooks, developed aerothermodynamic cycle analysis models, and created engineering software for preliminary design of air-breathing propulsion systems. Dr. Mattingly has more than 40 years of experience in Engineering Education. He served the majority of his Air Force career teaching at the Air Force Academy and Air Force Institute of Technology. After serving at the Aero Propulsion Lab-
oratory at Wright-Patterson AFB, he retired from active duty in 1989 and joined the faculty of Seattle University, retiring in 2000 as professor emeritus in Mechanical Engineering. He has since focused on writing, teaching professional short courses on preliminary engine design, and consulting. In addition to prior editions of this textbook, he is a co-author of *Aircraft Engine Design* (AIAA, 2002), winner of the AIAA Summerfield Book Award, and *Aircraft Engine Controls* (AIAA, 2009). He has been active with the AIAA Air Breathing Propulsion Technical Committee, being responsible for starting and directing the Team Aircraft Engine Design Competition. He received his BS and MS in mechanical engineering from the University of Notre Dame, and his PhD in aeronautics and astronautics from the University of Washington.

KEITH M. BOYER has been affiliated with Practical Aeronautics Inc. (PAI) since retiring as a Colonel from the U.S. Air Force in 2012 after 32-plus years of combined enlisted and officer active-duty service. As of the printing of this book, he is Vice President for Propulsion for PAI and works with Jack developing and teaching workforce development short courses on aeropropulsion systems. He has over 35 years' experience in aircraft-engine systems, the vast majority of it related to jet engines, to include hands-on flight line and

back shop maintenance, research and development, test and analysis, sustainment, systems engineering, logistics, supply chain, and multinational requirements management. He served on the faculty for 10 years in the Air Force Academy's Department of Aeronautics, designing and teaching aerothermodynamics and propulsion classes, was Associate Dean for Students at the Air Force Institute of Technology (AFIT), and was an adjunct faculty member at the Air Force Test Pilot School. From 1996 to 2009, Keith developed and taught a variety of jet engine short courses to over 1000 attendees in support of their professional development. He remains professionally active serving as Associate Editor for the ASME *Journal of Engineering for Gas Turbines and Power* and in numerous leadership roles for the ASME IGTI Aircraft Engine Committee, most recently as Committee Chair from 2014 to 2016. Keith received his PhD in Mechanical Engineering from Virginia Tech in 2001, master's in Aeronautical Engineering from the AFIT in 1992, and bachelor's in Aerospace Engineering from the University of Florida in 1987.

Index

Supporting Materials

To download supplemental material files, please go to AIAA's electronic library, Aerospace Research Central (ARC), and navigate to the desired book's landing page for a link to access the materials: arc.aiaa.org. Enter the password

EoPGTnR2ed

and follow the directions provided.

A complete listing of titles in the AIAA Education Series is available from ARC. Visit ARC frequently to stay abreast of product changes, corrections, special offers, and new publications.

AIAA is committed to devoting resources to the education of both practicing and future aerospace professionals. In 1996, the AIAA Foundation was founded. Its programs enhance scientific literacy and advance the arts and sciences of aerospace. For more information, please visit www.aiaafoundation.org.